JACK CARL KIEFER
COLLECTED PAPERS III

Jack Kiefer
1968

Jack Carl Kiefer
Collected Papers III
Design of Experiments

Published with the co-operation of the
Institute of Mathematical Statistics
and edited by

Lawrence D. Brown
Ingram Olkin
Jerome Sacks
Henry P. Wynn

Springer Science+Business Media, LLC

Lawrence D. Brown
Department of Mathematics
Cornell University
Ithaca, NY 14853
U.S.A.

Ingram Olkin
Department of Statistics
Stanford University
Stanford, CA 94305
U.S.A.

Jerome Sacks
Department of Mathematics
University of Illinois
Urbana, IL 61801
U.S.A.

Henry P. Wynn
Imperial College
London, England

AMS Subject Classification: 62-XX

Library of Congress Cataloging in Publication Data
Kiefer, Jack, 1924–1981
Jack Carl Kiefer collected papers.
"Published with the co-operation of the Institute of Mathematical Statistics."
Bibliography: p.
Contents: 1. Statistical inference and probability, 1951–1963—2. Statistical inference and probability, 1964–1984—3. Design of experiments.
1. Mathematical statistics—Collected works.
I. Brown, Lawrence D. II. Institute of Mathematical Statistics. III. Title.
QA276.A12K54 1984 519.5 84-10598

Originally published by Springer-Verlag New York Inc.
Softcover reprint of the hardcover 1st edition 1985

9 8 7 6 5 4 3 2 1

ISBN 978-1-4615-6662-5 ISBN 978-1-4615-6660-1 (eBook)
DOI10.1007/978-1-4615-6660-1

Preface

The theory of optimal design of experiments as we know it today is built on a solid foundation developed by Jack Kiefer, who formulated and resolved some of the major problems of data collection via experimentation. A principal ingredient in his formulation was statistical efficiency of a design.

Kiefer's theoretical contributions to optimal designs can be broadly classified into several categories:

He rigorously defined, developed, and interrelated statistical notions of optimality.

He developed powerful tools for verifying and searching for optimal designs; this includes the "averaging technique" [61] for approximate or exact theory, and "patchwork" [60] for exact theory. In [29] Kiefer and Wolfowitz provided a theorem now known as the *Equivalence Theorem*. This result has become a classical theorem in the field. One important feature of this theorem is that it provides a measure of how far a given design is from the optimal design.

He characterized and constructed families of optimal designs. Some of the celebrated ones are balanced block designs, generalized Youden designs, and weighing designs. He also developed combinatorial structures of these designs.

Kiefer's papers are sometimes difficult. In part this is due to the precision and care he exercised, which at times forced a consideration of pathologies and special cases. In contrast with the emphasis on the particular, he was a very generous scholar. There are numerous examples of expansive writing in his papers which give his wide perception and share his ideas and methodologies with the reader. A reading of his papers on design is replete with examples of his scholarship, his innovativeness, ingenuity, and strength as a researcher.

January, 1984 — Sam Hedayat

Contents

*Papers with asterisks have commentaries (see pages ix–xv).

Bibliography of Jack Kiefer

A. Papers (Published)

[1] Almost subminimax and biased minimax procedures, (with P. Frank). *Ann. Math. Statist.* **22** (1951), 465–468. [MR 13 (1952) 143, Zbl 43 (1952) 346]. (I)

[2] The inventory problem: I. Case of known distributions of demand, (with A. Dvoretzky and J. Wolfowitz). *Econometrica* **20** (1952), 187–222. [MR 13 (1952) 856, Zbl 46 (1953) 376]. (I)

[3] The inventory problem: II. Case of unknown distributions of demand, (with A. Dvoretzky and J. Wolfowitz). *Econometrica* **20** (1952), 450–466. [MR 14 (1953) 301, Zbl 48 (1953) 371]. (I)

[4] Stochastic estimation of the maximum of a regression function, (with J. Wolfowitz). *Ann. Math. Statist.* **23** (1952), 462–466. [MR 14 (1953) 299, Zbl 49 (1954) 366]. (I)

[5] Sequential minimax estimation for the rectangular distribution with unknown range. *Ann. Math. Statist.* **23** (1952), 586–593. [MR 14 (1953) 487, Zbl 48 (1953) 121]. (I)

[6] On minimum variance estimators. *Ann. Math. Statist.* **23** (1952), 627–629. [MR 15 (1954) 241, Zbl 48 (1953) 120]. (I)

[7] On Wald's complete class theorems. *Ann. Math. Statist.* **24** (1953), 70–75. [MR 14 (1953) 998, Zbl 50 (1954) 140]. (I)

[7a] Correction of a proof. *Ann. Math. Statist.* **24** (1953), 680. (I)

[8] Sequential minimax search for a maximum. *Proc. Amer. Math. Soc.* **4** (1953), 502–506. [MR 14 (1953) 1103, Zbl 50 (1954) 357]. (I)

[9] Sequential decision problems for processes with continuous time parameter. Testing hypotheses, (with A. Dvoretzky and J. Wolfowitz). *Ann. Math. Statist.* **24** (1953), 254–264. [MR 14 (1953) 997, 1279, Zbl 50 (1954) 148]. (I)

[9a] Corrections to "Sequential decision problems for processes with continuous time parameter testing hypotheses". *Ann. Math. Statist.* **30** (1959), 1265. (I)

[10] Sequential decision problems for processes with continuous time parameter. Problems of estimation, (with A. Dvoretzky and J. Wolfowitz). *Ann. Math. Statist.* **24** (1953), 403–415. [MR (1954) 242, Zbl 51 (1954) 366]. (I)

[11] On the optimal character of the (s, S) policy in inventory theory, (with A. Dvoretzky and J. Wolfowitz). *Econometrica* **21** (1953), 586–596. [MR 15 (1954) 333, Zbl 53 (1956) 279]. (I)

[12] On the theory of queues with many servers, (with J. Wolfowitz). *Trans. Amer. Math. Soc.* **78** (1955), 1–18. [MR 16 (1955) 601, Zbl 64 (1956) 133]. (I)

[13] On tests of normality and other tests of goodness of fit based on distance methods, (with M. Kac and J. Wolfowitz). *Ann. Math. Statist.* **26** (1955), 189–211. [MR 17 (1956) 55, Zbl 66 (1956–1957) 123]. (I)

[14] On the characteristics of the general queueing process, with applications to random walk, (with J. Wolfowitz). *Ann. Math. Statist.* **27** (1956), 147–161. [MR 17 (1956) 980, Zbl 70 (1957) 366]. (I)

[15] Asymptotic minimax character of the sample distribution function and of the classical multinomial estimator, (with A. Dvoretzky and J. Wolfowitz). *Ann. Math. Statist.* **27** (1956), 642–669. [MR 18 (1957) 772, Zbl 73 (1959–60) 146]. (I)

[16] Consistency of the maximum likelihood estimator in the presence of infinitely many incidental parameters, (with J. Wolfowitz). *Ann. Math. Statist.* **27** (1956), 887–906. [MR 19 (1958) 189, Zbl 73 (1959–60) 147]. (I)

[17] Sequential tests of hypotheses about the mean occurrence time of a continuous parameter Poisson process, (with J. Wolfowitz). *Naval Res. Logist. Quart.* **3** (1956), 205–219. [MR 18 (1957) 833]. (I)

[18] Some properties of generalized sequential probability ratio tests, (with L. Weiss). *Ann. Math. Statist.* **28** (1957), 57–74. [MR 19 (1958) 333, Zbl 79 (1959) 354]. (I)

[19] Invariance, minimax sequential estimation, and continuous time processes. *Ann. Math. Statist.* **28** (1957), 573–601. [MR 19 (1958) 1097, Zbl 80 (1959) 130]. (I)

[20] Optimum sequential search and approximation methods under minimum regularity assumptions. *J. Soc. Indust. Appl. Math.* **5** (1957), 105–136. [MR 19 (1958) 1097, Zbl 81 (1959) 385]. (I)

[21] On the deviations of the empiric distribution function of vector chance variables, (with J. Wolfowitz). *Trans. Amer. Math. Soc.* **87** (1958), 173–186. [MR 20 (1959) #5519, Zbl 88 (1961) 113]. (I)

[22] On the nonrandomized optimality and randomized nonoptimality of symmetrical designs. *Ann. Math. Statist.* **29** (1958), 675–699. [MR 20 (1959) #4910, Zbl 92 (1962) 361]. Corrections (and comments), unpublished. (III)

[23] Optimum designs in regression problems, (with J. Wolfowitz). *Ann. Math. Statist.* **30** (1959), 271–294. [MR 21 (1960) #3079, Zbl 90 (1961) 114]. (III)

[24] *K*-sample analogues of the Kolmogorov–Smirnov and Carmér-V. Mises tests. *Ann. Math. Statist.* **30** (1959), 420–447. [MR 21 (1960) #1668, Zbl 134 (1967) 367]. (I)

[25] Asymptotic minimax character of the sample distribution function for vector chance variables, (with J. Wolfowitz). *Ann. Math. Statist.* **30** (1959), 463–489. [MR 21 (1960) #6642, Zbl 93 (1962) 156]. (I)

[26] Optimum experimental designs. *J. Roy. Statist. Soc., Ser. B* **21** (1959), 272–319. [MR 22 (1961) #4101, Zbl 108 (1964) 153]. (III)

[27] A functional equation technique for obtaining Wiener process probabilities associated with theorems of Kolmogorov–Smirnov type. *Proc. Cambridge Philos. Soc.* **55** (1959), 328–332. [MR 22 (1961) #8557, Zbl 96 (1962) 334]. (I)

[28] Optimum experimental designs V, with applications to systematic and rotatable

designs. *Proc. 4th Berkeley Sympos. Math. Statist. and Prob.* **1** (1960), 381–405, Univ. California Press, Berkeley, Calif. [MR 24 (1962) #A3765, Zbl 134 (1967) 366]. (III)

[29] The equivalence of two extremum problems, (with J. Wolfowitz). *Canad. J. Math.* **12** (1960), 363–366. [MR 22 (1961) #8616, Zbl 93 (1962) 156]. (III)

[30] Distribution free tests of independence based on the sample distribution function, (with J. R. Blum and M. Rosenblatt). *Ann. Math. Statist.* **32** (1961), 485–498. [MR 23 (1962) #A2989, Zbl 139 (1968) 363]. (I)

[31] Optimum designs in regression problems, II. *Ann. Math. Statist.* **32** (1961), 298–325. [MR 23 (1962) #A735, Zbl 99 (1963) 135]. (III)

[32] On large deviations of the empiric d.f. of vector chance variables and a law of the iterated logarithm. *Pacific J. Math.* **11** (1961), 649–660. [MR 24 (1962) #A1732, Zbl 119 (1966) 349]. (I)

[33] Two more criteria equivalent to D-optimality of designs. *Ann. Math. Statist.* **33** (1962), 792–796. [MR 25 (1963) #701, Zbl 116 (1965) 113]. (III)

[34] An extremum result. *Canad. J. Math.* **14** (1962), 597–601. [MR 26 (1963) #1968, Zbl 134 (1967) 369]. (III)

[35] Channels with arbitrarily varying channel probability functions, (with J. Wolfowitz). *Information and Control* **5** (1962), 44–54. [MR 24 (1962) #B2506, Zbl 107 (1964) 345]. (I)

[36] Minimax character of Hotelling's T^2 test in the simplest case, (with N. Giri and C. Stein). *Ann. Math. Statist.* **34** (1963), 1524–1535. [MR 27 (1964) #6331, Zbl 202 (1971) 495]. (I)

[37] Asymptotically optimum sequential inference and design, (with J. Sacks). *Ann. Math. Statist.* **34** (1963), 705–750. [MR 27 (1964) #893, Zbl 255 (1973) 62063]. (I)

[38] Local and asymptotic minimax properties of multivariate tests, (with N. Giri). *Ann. Math. Statist.* **35** (1964), 21–35. [MR 28 (1964) #2605, Zbl 133 (1967) 418]. (II)

[39] Minimax character of the R^2-test in the simplest case, (with N. Giri). *Ann. Math. Statist.* **35** (1964), 1475–1490. [MR 29 (1965) #6579, Zbl 137 (1967) 368]. (II)

[40] Optimum extrapolation and interpolation designs, I, (with J. Wolfowitz). *Ann. Inst. Statist. Math.* **16** (1964), 79–108. [MR 31 (1966) #2806, Zbl 137 (1967) 131]. (III)

[41] Optimum extrapolation and interpolation designs, II, (with J. Wolfowitz). *Ann. Inst. Statist. Math.* **16** (1964), 295–303. [MR 31 (1966) #2806, Zbl 137 (1967) 131]. (III)

[42] Admissible Bayes character of T^2-, R^2-, and other fully invariant tests for classical multivariate normal problems, (with R. Schwartz). *Ann. Math. Statist.* **36** (1965), 747–770. [MR 30 (1965) #5430; 50 (1975) #11567, Zbl 137 (1967) 36; 249 (1973) 62058]. (II)

[42a] Correction to "Admissible Bayes character of T^2-, R^2-, and other fully invariant tests for classical multivariate normal problems". *Ann. Math. Statist.* **43** (1972), 1742. (II)

[43] On a problem connected with the Vandermonde determinant, (with J. Wolfowitz). *Proc. Amer. Math. Soc.* **16** (1965), 1092–1095. [MR 32 (1966) #115, Zbl 142 (1968) 269]. (III)

[44] On a theorem of Hoel and Levine on extrapolation designs, (with J. Wolfowitz). *Ann. Math. Statist.* **36** (1965), 1627–1655. [MR 32 (1966) #3230, Zbl 138 (1967) 140]. (III)

[45] Multivariate optimality results. *Multivariate Analysis. Proceedings of an International Symposium* (ed. by P. R. Krishnaiah), (1966), 255–274, Academic Press, New York. [MR 37 (1969) #2372, Zbl 218 (1972) 448]. (II)

[46] Optimum multivariate designs, (with R. H. Farrell and A. Walbran). *Proc. Fifth Berkeley Sympos. Math. Statist. and Probability* **1** (1967), 113–138. [MR 35 (1968) #5099, Zbl 193 (1970) 171]. (III)

[47] On Bahadur's representation of sample quantiles. *Ann. Math. Statist.* **38** (1967), 1323–1342. [MR 36 (1968) #933, Zbl 158 (1960) 370]. (II)

[48] Statistical inference, (panel discussion with G. A. Barnard, L. M. LeCam, and L. J. Salvage). *The Future of Statistics* (Proceedings of a Conference on the Future of Statistics held at the University of Wisconsin, Madison, Wisconsin, June 1967) (ed. by D. G. Watts), (1968), 139–160, Academic Press, New York. (II)

[49] Statistical inference. *The Mathematical Sciences*, (1969), 60–71, The M.I.T. Press, Cambridge. [Reprinted in *Math. Spectrum.* **3** (1970–71), 1–11]. (II)

[50] On the deviations in the Skorokhod–Strassen approximation scheme. *Z. Wahrsch. Verw. Gebiete.* **13** (1969), 321–332. [MR 41 (1971) #1117, Erratum 41, p. 1965, Zbl 176 (1969) 482]. (II)

[51] Old and new methods for studying order statistics and sample quantiles. *Nonparametric Techniques in Statistical Inference* (ed. by M. L. Puri), (1970), 349–357, Cambridge University Press, London. [MR 44 (1972) #3442]. (II)

[52] Deviations between the sample quantile process and the sample df. *Nonparametric Techniques in Statistical Inference* (ed. by M. L. Puri), (1970), 299–319, Cambridge University Press, London. [MR 43 (1972) #2808]. (II)

[53] Optimum experimental designs. *Actes du Congrès International des Mathématiciens, Nice*, **3** (1970), 249–254. [MR 54 (1977) #8993, Zbl 237 (1972) 62050]. (III)

[54] Iterated logarithm analogues for sample quantiles when $p_n \downarrow 0$. *Proc. Sixth Berkeley Symp. on Mathematical Statistics and Probability* **1** (1970), 227–244, Univ. California Press, Berkeley. [MR 53 (1977) #6696, Zbl 264 (1974) 62815]. (II)

[55] The role of symmetry and approximation in exact design optimality. *Statistical Decision Theory and Related Topics.* (*Proc. Symp.*) (ed. by S. S. Gupta and J. Yackel), (1971), 109–118, Academic Press, New York. [MR 50 (1975) #3447, Zbl 274 (1974) 62050]. (III)

[56] Skorohod embedding of multivariate rv's, and the sample df. *Z. Wahrsch. Verw. Gebiete.* **24** (1972), 1–35. [MR 49 (1975) #6382, Zbl 267 (1974) 60034]. (II)

[57] Optimum designs for fitting biased multiresponse surfaces. *Multivariate Analysis —III. Proceedings of the Third International Symposium on Multivariate Analysis* (ed. by P. R. Krishnaiah), (1973), 287–297, Academic Press, New York. [MR 51 (1976) #2188, Zbl 291 (1975) 62093]. (III)

[58] General equivalence theory for optimum designs (approximate theory). *Ann. Statist.* **2** (1974), 849–879. [MR 50 (1975) #8856, Zbl (1975) 62092]. (III)

[59] Discussion on the paper: Planning experiments for discriminating between models, by A. C. Atkinson and D. R. Cox. *J. Roy. Statist. Soc. Ser. B* **36** (1974), 345–346. (III)

[60] Balanced block designs and generalized Youden designs, I. Construction (patchwork). *Ann. Statist.* **3** (1975), 109–118. [MR 51 (1976) #4578, Zbl 305 (1976) 62052]. (III)

[61] Construction and optimality of generalized Youden designs. *A Survey of Statistical Design and Linear Models* (ed. by J. N. Srivastava), (1975), 333–353, North-Holland Pub. Co, Amsterdam. [MR 52 (1976) #15877, Zbl 313 (1975) 62057]. (III)

[62] Review of the paper: Bayesian analysis of generic relations in Agaricales, by R. E. Machol and R. Singer. *Mycologia* **67** (1975), 203–205. (II)

[63] Optimal design: Variation in structure and performance under change of criterion. *Biometrika* **62** (1975), 277–288. [MR 52 (1976) #2064, Zbl 321 (1976) 62086]. (III)

[64] Optimal designs for large degree polynomial regression, (with W. J. Studden). *Ann. Statist.* **4** (1976), 1113–1123. [MR 54 (1977) #11676, Zbl 357 (1978) 62051]. (III)

[65] Asymptotically minimax estimation of concave and convex distribution functions, (with J. Wolfowitz). *Z. Wahrsch. Verw. Gebiete.* **34** (1976), 73–85. [MR 53 (1977) #1829, Zbl 354 (1978) 62035]. (II)

[66] Large sample comparison of tests and empirical Bayes Procedures, (with D. S. Moore). *On the History of Statistics and Probability: Proc. of a Symp. on the American Mathematical Heritage* (ed. by D. B. Owen), (1976), 347–365, M. Dekker, New York. (II)

[67] Admissibility of conditional confidence procedures. *Ann. Statist.* **4** (1976), 836–865. [MR 55 (1978) #11454, Zbl 353 (1975) 62008]. (II)

[68] Asymptotically minimax estimation of concave and convex distribution functions. II, (with J. Wolfowitz). *Statistical Decision Theory and Related Topics, II* (ed. by S. S. Gupta and D. S. Moore), (1977), 193–211, Academic Press, New York. [MR 56 (1978) #1572, Zbl 418 (1980) 62031]. (II)

[69] Conditional confidence statements and confidence estimators. *J. Amer. Statist. Assoc.* **72** (1977), 789–827. [MR 58 (1979) #24638, Zbl 375 (1978) 62023]. (II)

[70] The ideas of conditional confidence in the simplest setting, (with C. Brownie). *Comm. Statist.—Theor. Methods.* **A6**(8) (1977), 691–751. [MR 56 (1978) #3993, Zbl 392 (1979) 62002]. (II)

[71] Conditional confidence and estimated confidence in multidecision problems (with applications to selection and ranking). *Multivariate Analysis—IV. Proceedings of the Fourth International Symposium on Multivariate Analysis* (ed. by P. R. Krishnaiah), (1977), 143–158, North-Holland, Amsterdam. [MR 58 (1979) #18810, Zbl 381 (1979) 62009]. (II)

[72] Comparison of rotatable designs for regression on balls, I (Quadratic), (with Z. Galil). *J. Statist. Plann. Inference* **1** (1977), 27–40. [MR 58 (1979) #24769, Zbl 394 (1979) 62058]. (III)

[73] The foundations of statistics—are there any? *Synthese* **36** (1977), 161–176. [MR 58 (1979) #31488, Zbl 375 (1978) 60005]. (II)

[74] Comparison of design for quadratic regression on cubes, (with Z. Galil). *J. Statist. Plann. Inference* **1** (1977), 121–132. [MR 58 (1979) #24770, Zbl 381 (1979) 62062]. (III)

[75] Comparison of simplex designs for quadratic mixture models, (with Z. Galil). *Technometrics* **19** (1977), 445–453. [MR 57 (1978) #17972, Zbl 372 (1978) 62058]. (III)

[76] Comparison of Box–Draper and D-optimum designs for experiments with mixtures, (with Z. Galil). *Technometrics* **19** (1977), 441–444. [MR 58 (1979) #3233, Zbl 369–389 (1978) 62087]. (III)

[77] Asymptotic approach to familes of design problems. *Comm. Statist.—Theory Methods* **A7** (1978), 1347–1362. [MR 82g (1982) 62106, Zbl 389 (1979) 62058]. (III)

[78] A Diophantine problem in optimum design theory. *Utilitas Math.* **14** (1978), 81–98. [MR 80b (1980) 62091, Zbl 391 (1979) 62055]. (III)

[79] Comment on paper: Pseudorandom number assignment in statistically designed simulation and distribution sampling experiments, by L. W. Schruben and B. H. Margolin. *J. Amer. Statist. Assoc.* **73** (1978), 523–524. (III)

[80] Extrapolation designs and Φ_p-optimum designs for cubic regression on the q-ball, (with Z. Galil). *J. Statist. Plann. Inference* **3** (1979), 27–38. [MR 81a (1981) 62074, Zbl 412 (1980) 62055]. (III)

[81] Sequential statistical methods. *Studies in Probability Theory* (ed. by M. Rosenblatt), (1978), 1–23, Studies in Math., 18, Math. Assoc. Amer., Washington, D.C. [MR 80m (1980) 62077, Zbl 412 (1980) 62056]. (II)

[82] Comments on taxonomy, independence, and mathematical models (with reference to a methodology of Machol and Singer). *Mycologia* **71** (1979), 343–378. (II)

[83] Optimal design theory in relation to combinatorial design. *Ann. Discrete Math.* **6** (1980), 225–241. [MR 82a (1982) 62107, Zbl 463 (1982) 62066]. (III)

[84] Designs for extrapolation when bias is present. *Multivariate Analysis — V. Proc. Fifth International Symp.* (ed. by P. R. Krishnaiah), (1980), 79–93, North-Holland Pub. Co., Amsterdam. [MR 81i (1981) 62130, Zbl 458 (1982) 62063]. (III)

[85] *D*-optimum weighing designs, (with Z. Galil). *Ann. Statist.* **8** (1980), 1293–1306. [MR 82g (1982) 62104, Zbl 466 (1982) 62066]. (III)

[86] Time- and space-saving computer methods, related to Mitchell's DETMAX, for finding D-optimum designs, (with Z. Galil). *Technometrics* **22** (1980), 301–313. [MR 81j (1981) 62147, Zbl 459 (1982) 62060]. (III)

[87] Optimum weighing designs, (with Z. Galil). *Recent Developments in Statistical Inference and Data Analysis. Proceedings of the International Conference in Statistics in Tokyo* (ed. by K. Matusita), (1980), 183–189, North-Holland Pub. Co. Amsterdam. [MR 82a (1982) 62108, Zbl 462 (1982) 62059]. (III)

[88] Optimum balanced block and Latin square designs for correlated observations, (with H. P. Wynn). *Ann. Statist.* **9** (1981), 737–757. [MR 82h (1982) 62122]. (III)

[89] The interplay of optimality and combinatorics in experimental design. *Canad. J. Statist.* **9** (1981), 1–10. [MR 82m (1982) 62167]. (III)

[90] Relationships of optimality for individual factors of a design, (with J. Eccleston). *J. Statist. Plann. Inference* **5** (1981), 213–219. [MR 83c (1983) 62115, Zbl 481 (1982) 62059]. (III)

[91] Optimum rates for non-parametric density and regression estimates, under order restrictions. *Statistics and Probability, Essays in Honor of C. R. Rao* (ed. by G. Kallianpur, P. R. Krishnaiah, J. K. Ghosh), (1982), 419–428, North-Holland Pub. Co., Amsterdam. (II)

[92] On the characterization of D-optimum weighing designs for $n \equiv 3 \pmod 4$, (with Z. Galil). *Statistical Decision Theory and Related Topics III* (ed. By S. S. Gupta and J. O. Berger), **1** (1982), 1–35, Academic Press, New York. (III)

[93] Eight lectures on mathematical statistics. (Chinese) *Advances in Mathematics (Beijing)* (Shuxue Jin Zhan) **10** (1981), 94–130. (II)

[94] Conditional inference. *Encyclopedia of Statistical Sciences*, Volume 2 (ed. By S. Kotz, N. L. Johnson, and C. B. Read), (1982), 103–109, Wiley-Interscience. (II)

[95] Construction methods for *D*-optimum weighing designs when $n \equiv 3 \pmod 4$, (with Z. Galil). *Ann. Statist.* **10** (1982), 502–510. [MR 83i (1983) 62139]. (III)

[96] Autocorrelation-robust design of experiments, (with H. P. Wynn). *Scientific Inference, Data Analysis, and Robustness* (ed. by T. Leonard and C. -F. Wu), (1983), 279–299, Academic Press, New York. (III)

[97] Comparison of designs equivalent under one or two criteria, (with Z. Galil). *J.*

Statist. Plann. Inference **8** (1983), 103–116. (III)

[98] Optimum and minimax exact treatment designs for one-dimensional autoregressive error processes, (with H. P. Wynn). *Ann. Statist.* **12** (1984), 414–450. (III)

B. Book Reviews

[99] Review of *The Advanced Theory of Statistics*, Volume 2, "Inference and Relationship," M. G. Kendall and A. Stuart. *Ann. Math. Statist.* **35** (1964), 1371–1380. (II)

[100] Review of *The Savory Wild Mushroom*, M. McKenney. (Second edition, revised and enlarged by D. E. Stuntz). *Quarterly Review Biology* **47** (1972), 342–343. (II)

[101] Review of *A Field Guide to Western Mushrooms*, A. H. Smith. *Quarterly Review Biology* **52** (1977), 91. (II)

C. Books, Edited Volumes, Lecture Notes

[102] *Sequential identification and ranking procedures, with special reference to Koopman–Darmois populations*, (with R. E. Bechhofer and M. Sobel), Statistical Research Monographs, Vol. 3, 1968, The University of Chicago Press, Chicago, Illinois. [MR 39 (1970) #6445, Zbl 208 (1971) 446].

[103] Jacob Wolfowitz, selected papers (ed. by J. Kiefer, with the assistance of U. Augustin and L. Weiss). 1980, Springer-Verlag, New York. [MR 83d (1983) 01080, Zbl 447 (1981) 62001].

[104] Lectures on design theory, (1974). Mimeograph Series #397, Department of Statistics, Purdue University.

[105] Contributions to the theory of games and statistical decision functions. Doctoral dissertation, 1952, Columbia University.

[106] Notes on decision theory (1953), (notes recorded by J. Sacks).

D. Papers (Unpublished)

[107] Note on asymptotic efficiency of M.L. estimators in nonparametric problems, (with J. Wolfowitz), (circa 1960). (II)

[108] Mathematics 371 Final Examination (with D. Kiefer), (1972). (II)

[109] Lecture notes on statistical inference, (1973), to be published by Springer-Verlag, 1984.

[110] *D*-optimality of the GYD for $v \geq 6$, (1974). (III)

[111] Optimality criteria for designs, (1975). (III)

Jack Kiefer's Contributions to Experimental Design

H. P. WYNN

Imperial College, London

1. History. Careful experimentation is part and parcel of the scientific method developed in the eighteenth and nineteenth century. John Stuart Mill was probably the first to give clear prescriptions on how to carry out experiments. He separated experiments into "spontaneous" experiments, what we would now call observational studies and "artificial" experiments, namely controlled experiments. Mill and others were firmly of the belief that controlled experimentation was better, if the subject matter allowed it. This was carried through into this century with the "crucial experiment" becoming the cornerstone of the falsification ideas of Karl Popper and his followers. The details of experimental strategy, however, were neglected by the philosophers except that it was recognized that careful variation in the levels of "agents" $A, B, C, \ldots$ would yield an analysis of their effects $a, b, c, \ldots$.

The breakthrough into a more versatile approach to experimental design came with the work of Ronald A. Fisher and his followers, notably Frank Yates, at the Rothamstead Experimental Station in England. A number of useful concepts were introduced such as balance, orthogonality, blocking and aliasing. This led to an explosion of work on combinatorial design which took seed in the U.S.A. through the work of Raj Chandra Bose and collaborators.

Here and there in the combinatorial literature the idea of efficiency—usually relative to some standard design—had been discussed. However, at the end of the second World War the theory of optimum design was almost nonexistent except for a remarkable early paper by Smith (1918) and the important paper of Wald (1943). It is no accident that the modern theory of optimum design has its roots in the decision theory school of U.S. statistics founded by Abraham Wald. The idea of "risk," developed formally by Wald and arising out of the earlier work of Neyman and Pearson, was the most important innovation of that school. There were parallel developments in utility theory, mathematical programming, and mathematical economics so that the early history of the subjects are interwoven. Together with Wald,

Jacob Wolfowitz and Jack Kiefer were leading members of this school. They started the second great advance in the science of experimentation this century by applying decision-theoretic ideas, and over the subsequent 20 years Jack Kiefer himself nurtured this science to maturity.

2. Continuous Theory. Since Wald's paper a number of papers had appeared by Elfving (1952), Hoel (1958), Guest (1958), and important works by Box and Draper and co-workers at the University of Wisconsin on response surface design. Thus the literature leading up to the first Kiefer–Wolfowitz paper [23] had gradually liberated the allowable region of experimentation to grow from one-at-a-time methods through combinatorial design to multifactor experimentation and response surface design. It had reached the point where Kiefer and Wolfowitz could allow an almost arbitrary design region $\mathscr{X}$ in the same way that decision theory had earlier allowed an almost arbitrary action space. The other brave step, technical rather than conceptual, was to abolish in a stroke computational difficulties involved in changing the sample size. Thus a design became a probability measure ξ over the design space $\mathscr{X}$, and a "continuous theory" was born.

It is worth restating briefly the basic Kiefer–Wolfowitz set-up. At each point in $\mathscr{X}$ there is a potential observation Y_x whose expectation is

$$E(Y_x) = \sum_{i=1}^{k} \theta_i f_i(x) = \theta^T f(x),$$

where $\theta = (\theta_1, \ldots, \theta_k)$ are unknown parameters and the f_i's are continuous functions on $\mathscr{X}$ which is taken to be compact. For an (exact) experiment, observations $Y_{x_1}, \ldots, Y_{x_N}$ are taken and assumed to be uncorrelated with equal variance σ^2. The $n \times n$ information matrix is $X^T X$, where $X^T = [f(x_1), \ldots, f(x_N)]$ and $\mathrm{cov}(\hat{\theta}) = \sigma^2 (X^T X)^{-1}$ is the covariance matrix of the least squares estimate of θ based on the observations.

The normalized version of $X^T X$, namely $(1/N) X^T X$, generalizes to the moment matrix

$$M(\xi) = \int_{\mathscr{X}} f(x) f(x)^T \xi(dx).$$

A key mathematical benefit of this approach is that the set of all $M(\xi)$, the moment space, is closed and convex. In addition, many of the optimality criteria which had been introduced in earlier work, when extended to $M(\xi)$, involved minimization of a convex functional. The most important of these was D-optimality (introduced by Wald):

$$\min_{\xi} \{ -\log\det M(\xi) \}.$$

This is one of a wider class of Φ_p criteria introduced by Kiefer in later work:

$$\min_{\xi} \left[\mathrm{trace}\left(M(\xi)^{-p} \right) \right]^{1/p},$$

$-1 \leq p \leq \infty$. The case $p = 0$ gives D-optimality, $p = \infty$ the so-called E-optimality (minimizing the maximum eigenvalue of $M(\xi)^{-1}$), and $p = 1$, A-optimality which had been studied mostly in block design settings. The power of the extension to measures is demonstrated by the proof of the beautiful General Equivalence Theorem (GET) [K29]. This showed that D-optimality was equivalent to G-optimality which achieves

$$\min_{\xi} \max_{x \in \chi} f(x)^T M(\xi)^{-1} f(x).$$

The quantity $f(x)^T M(\xi)^{-1} f(x)$ is the generalization of the (normalized) variance function, the variance of the estimated response $\hat{Y}_x$.

Casting the problem as a convex program brought it into the arena with Lagrangian theory, game theory, and minimax/saddle point theory. It was clear then that more general results than the GET could be established. Jack Kiefer tied up much of this in [K58] for a general smooth class of optimality criteria. Earlier, Kariln and Studden had given a game theoretical proof of the GET. More recently, Pukelsheim (1980) has used the Fenchel duality theorem to achieve duality theorems for Φ_p-optimality following an emphasis on the duality approach by Silvey and Titterington (1973). Thus the core of continuous optimum design can now be seen to have an extra existence as a fascinating subculture of optimization theory.

The ability of the equivalence theorem to throw up rich examples led to pioneering work by Jack Kiefer and his co-workers. The connection with orthogonal polynomials proved vital to this analysis. This arose because of the simple observation that the (generalized) variance function satisfied, for polynomial models,

$$d(x, \xi) = \sum_{i=1}^{k} \phi_i^2(x),$$

where $\phi_1, \ldots, \phi_k$ are orthonormal polynomials with respect to the design. The technique was to guess at a nice class of symmetric-looking designs possibly using the invariance of D-optimality under linear transformations of the parameters. The class would be defined up to unknown weights $\alpha, \beta, \gamma, \ldots$ at certain support points. A general expression for the orthonormal polynomials would be found and the optimal α, β, γ calculated using two additional features of the GET: (i) the minmax value of $d(x, \xi)$ is k (the number of parameters), and (ii) $\max_x d(x, \xi)$ is achieved at a support point for optimal ξ. This technique proved very successful for quadratic regression on a simplex and hypercube ([31], [46]). It transpired that the classical orthogonal fractions were not necessarily optimum but in a mysterious way provided the support for the optimum designs. A beautiful paper by Kiefer and Studden [64] exploited the classical theory of orthogonal polynomials to find limiting designs for large k. After such analytic methods became difficult (as for example when $\mathscr{X}$ was a sphere), Kiefer and Galil used a mixture of analysis search and direct computation to find solutions, as in [80].

One particularly difficult area computationally is the so-called singular case which arises in the extension of D-optimality (and ϕ_p-optimality) to subsets of parameters. The difficulty arises basically from the fact that although a chosen vector of parameters $\alpha = B\theta$ may be estimable under a particular design (and indeed at the optimum), so that $BM(\xi)B^T$ is nonsingular, $M(\xi)$ itself may be singular. This leads to technical problems in the specification of a solution. A discussion of this case with some history appears in the recent book by Silvey (1980).

3. Exact Theory. Despite the success of the continuous theory Jack Kiefer always had in mind that it was important for the development of the subject to solve outstanding problems in the exact theory, that is, to find optimum designs for fixed sample size or under more rigid combinatorial restrictions. Indeed his very first paper on design [22] established under mild conditions the optimality in a wide sense of incomplete block designs and Latin squares. He made use of a delightfully simple but very useful lemma (Proposition 1, [61]) for symmetric matrices.

In block design settings, or more general m-way layouts, one often works with the so-called C-matrix written C_d, where d refers to the design. This is the information matrix for estimating treatment contrasts. If there are v treatments (v treatment parameters) the c_d will be a $v \times v$ matrix but have maximal rank $v-1$. The lemma says that C_d, and hence d, is optimal if, within the class of all C_d, (a) it is completely symmetric (that is, all diagonal elements equal and all off diagonal elements equal), (b) it has maximum trace. Optimal here means "universally optimal," that is, it maximizes any convex, nonincreasing, permutation invariant function on C_d. It therefore includes A-, E-, and D-optimality. This was the starting point for two main lines of research.

The first of the problems on which Kiefer spent a considerable effort was on what happens in the Latin square type situations when the combinatorial restrictions are relaxed to allow v treatments in a row and column design with b_1 rows and b_2 columns ($b_1 = b_2 = v$ is the Latin square case). He defined a generalized Youden square (GYS). This is based on a balanced block design (BBD), that is a block design with v treatments in b blocks of size k. Treatment i appears n_{ij} times in block j and (1) all the $r_i = \Sigma_j n_{ij}$ are equal, (2) all the $\lambda_{ih} = \Sigma_j n_{ij} n_{hj}$ are equal, and (3) $|n_{ij} - k/v| < 1$ for all i and j. A GYD is a design in which the treatment/row design and treatment/column design are each separately BBD's. He had proved [22] the universal optimality of a GYD in the so-called regular case when v divides b_1 (or b_2). It was the failure of D-optimality in some non-regular settings that led Jack Kiefer on a long and difficult quest. First A- and E-optimality could be proved. He then produced a remarkable theorem ([61], [83]) which says that D-optimality holds when $v \neq 4$ and clarified the case when $v = 4$. In this work we see him at his most ingenious, using completely original analytic arguments to solve an extremely hard combinatorial problem. There was an immediate spin-off in the study of GYD's both by himself [60] and others (Ash (1981), Ruiz and Seiden (1974), Seiden and Wu (1978)). C. -S. Cheng (1981) extended some of the work to Youden hyper-rectangles.

The other problem was what happens when no BIBD exists so that universal optimality is not so apparent. The natural place to look for optimal designs is among partially balanced block designs with two associate classes or, more particularly, group divisible designs. In a series of papers C. -S. Cheng (for example, Cheng (1978)) proved optimality for certain members of a class of designs called regular graph designs. He exploited important links between optimum combinatorial design and certain problems in graph theory concerned with finding graphs with a maximal number of spanning trees. It is a tribute to the work of Kiefer and Cheng that upon translation into the language of graph theory their theorems gave new results.

The blending of optimality and combinatorics continues to be very live research area. For example, there has been a burst of recent work using Schur-convexity, a slightly weaker notion than universal optimality; see Giovagnoli and Wynn (1981), the unpublished work by Gregory Constantine, and the paper of Cheng (1979).

Jack Kiefer had a most fruitful collaboration on experimental design in the last ten years with Zvi Galil. The second part of this consisted of an in-depth study of the maximization of $\det(X^T X)$ when X only has elements with values ± 1. This is a linear regression problem in which each variable is allowed to take values ± 1. It is usually called a weighing design problem because it derives from an experiment to estimate weights when objects are weighed, possibly together, on a chemical balance. When X is square then the solution when 4 divides k is to take $X = H$ a Hadamard matrix if such a matrix exists—a well-known but unsolved conjecture (now established for all $k \leq 200$). They address a far less studied case than the Hadamard conjecture when $k = 1, 2, 3 \bmod 4$ concentrating mostly on the most interesting case $k = 3 \pmod 4$. Ehlich (1964a, 1964b) had made some original contributions to this problem, which Kiefer and Galil extended to an almost complete solution (modulo the truth of the Hadamard conjecture). Taking X to be an $m \times k$ matrix ($m = 4l - 1$), they showed that, for $m \geq 2k - 5$, the following "easy" solution X_e is D-optimum. Take an $(m+1) \times (m+1)$ Hadamard matrix, delete a row, and select any k columns. They also filled in many cases between the bound and the saturated case $m = k$ [85], [92], [95].

Jack Kiefer had a barely disguised love for combinatorics although his work was never really in the main stream of combinatorial theory. His profound originality proves that he could have turned his hand to algebra on number theory as easily as to mathematical statistics and probability theory.

4. Algorithms. An outgrowth from the continuous theory was the introduction and development of algorithms of the steepest ascent type to generate optimum designs computationally. They consisted of augmenting design measures, thus

$$\xi_{n+1} = (1 - \alpha_n)\xi_n + \alpha_n \xi_n' (0 \leq \alpha_n \leq 1)$$

where ξ_n was the previous design measured and ξ_n' a measure giving the "direction" of movement, often a single point mass. Wynn (1970) dealt with the natural case $\alpha_n = 1/(n+1)$, work which was simultaneous with that of V. V. Fedorov (1972). The work took firm hold in the United Kingdom notably with papers by Silvey and Titterington (1973), and in the U.S.S.R. and Eastern Europe (Fedorov (1972), Pazman (1974)). The main problems were with the unboundedness of the functionals ($-\log\det M(\xi)$ is infinite when $M(\xi)$ is singular) and with the "singular case" referred to above. These were essentially infinite dimensional algorithms since they dealt with measures (see Wu and Wynn (1978)). C. -F. Wu (1978) did important work bringing to bear the extensive literature in optimization algorithms from the optimization literature. By keeping the support of the design finite, finite dimensional algorithms could be developed.

Mitchell (1974) had developed special algorithms involving "excursions" in which new points would be added and old points thrown out (the original Wynn algorithm could be considered as a single infinite excursion). Kiefer and Galil improved on the Mitchell algorithm [86] using a variety of sophisticated computing techniques. They were, as they claimed, "15 to 50 times faster." The algorithm was used to back up and explore the more theoretical work on weighing designs. It was an excellent example of the computer being the slave of mathematicians and a fine blend of sophisticated mathematics and computation. There remain many interesting problems concerned with the speed of these "optimum subset" algorithms. More recently Welch (1982) has developed a branch and bound integer program leading to exact optima.

5. Nonstandard Models. A criticism of the classical optimum design theory was that whereas the mathematics was nice, problems of investigating different models had been ignored. Naturally Kiefer was aware of this. From his files and research it is clear that he was an authority on the variance-bias approach introduced by Box and Draper (1959) in their pioneering paper following earlier work on response surfaces. He was particularly familiar with the extensions introduced by Karson, Manson, and Hader (1969). Roughly, the idea is to guard against the possible presence of a more complex model, for example, a quadratic model when a linear model is fitted. In [57] Kiefer gave an illuminating and thorough discussion of the Box–Draper and Karson–Manson–Hader approaches pointing out some defects and giving methodology to get around some of them. He explored the issues further in [63] and [75] and showed in particular that, for dimensions greater than two, the D-optimum design for quadratic regression on a simplex is more protective against a cubic departure than the Box–Draper approach. Results of Draper and Herzberg based on the same approach are discussed in [84]. The problem still seems largely unresolved and it is possible that D-optimum designs will continue to be robust in higher dimensions. Other authors have taken different approaches, for example Marcus and Sacks (1977).

Models in which errors are uncorrelated have a small but long history, going back to a paper on analysis of variance by Papadakis (1937). Jack Kiefer in his Berkeley Symposium paper, [28], tackled head-on a problem posed by Williams to give optimum one-dimensional exact designs to estimate treatment differences in the presence of autocorrelated errors in discrete time. At each point in time one observes

$$Y_t = \alpha_{[t]} + \varepsilon_t,$$

where ε_t is the process and $\alpha_{[t]}$ one of k treatment parameters allocated at time t. He showed for an autoregressive process of order two that, asymptotically, the best designs lay among one of the following patterns (letters are treatments): (1) $AA \ldots BB \ldots$, (2) $ABAB \ldots$, (3) $AABBAABB \ldots$, or (with three or more treatments) (4) $ABCABC \ldots$. Over the two years up to 1981 he and Wynn [98] have given a fairly complete theoretical solution to the problem for a pth order process and a complete combinatorial solution in the case $p = 3$. They were also able to push forward the combinatorial theory into higher dimensions and develop a close connection with the theory of stationary discrete state processes [96]. As in Jack Kiefer's previous work on exact design the door was opened on a new class of combinatorial objects, and it was gratifying to find connections in other fields, in this case communication theory. They had also investigated the robustness of classical designs, Latin squares, and BIBD's, for simple auto-correlated models [88].

There has been parallel work in continuous time, and the seminal work of Jerome Sacks and Donald Ylvisaker ((1969) and other papers) should be mentioned. Like them, Kiefer had seen the importance of studying time-dependent and spatial processes and the possibilities this gives for extending the scope of optimum data collection into less controlled environments.

6. Personal Note. To work with Jack Kiefer was a privelege and a joy. His untimely death was a dreadful loss to his family, friends, and colleagues, and his work remains a monument to a great scholar and a delightful human being.

REFERENCES

Ash, A. S. (1981). Generalized Youden designs: construction and tables. *J. Statist. Plan. Inference*, **5**, 1–25.

Box, G. E. P. and Draper, N. R. (1959). A basis for the selection of a response surface design. *J. Amer. Statist. Assoc.* **54**, 622–654.

Cheng, C. -S. (1978). Optimality of certain asymmetrical experimental designs. *Ann. Statist.* **6**, 1239–1261.

Cheng, C. -S. (1979). Optimal incomplete block designs with four varieties. *Sankhya*, *B* **41**, 1–14.

Cheng, C. -S. (1981). Optimality and construction of pseudo-Youden designs. *Ann. Statist.* **9**, 201–205.

Ehlich, H. (1964a). Determinantenabschatzungen fur binare Matrizen. *Z. Wahrsch. Verw. Geliete* **83**, 123–132.

Ehlich, H. (1964b). Determinantenabschatzungen fur binare Matrizen mit $n \equiv 3 \bmod 4$. *Z. Wahrich. Verw. Geliete* **84**, 438–447.

Elfving, G. (1952). Optimum allocation in linear regression theory. *Ann. Math. Statist.* **23**, 255–262.

Fedorov, V. V. (1972). *Theory of Optimal Experiments*. Academic Press, New York.

Hoel, P. G. (1958). Efficiency problems in polynomial estimation. *Ann. Math. Statist.* **29**, 1134–1146.

Guest, P. G. (1958). The spacing of observations in polynomial regression. *Ann. Math. Statist.* **29**, 214–299.

Karlin, S. and Studden, W. J. (1966). Optimal experimental designs. *Ann. Math. Statist.* **37**, 783–815.

Karson, M. J., Manson, A. R. and Hader, R. J. (1969). Minimum bias estimation and experimental designs for response surfaces. *Technometrics* **11**, 461–476.

Giovagnoli, A. and Wynn, H. P. (1981). Optimum continuous block designs. *Proc. Roy. Soc. London Ser. A* **377**, 405–416.

Hoel, P. G. (1958). Efficiency problems in polynomial estimation. *Ann. Math. Statist.* **29**, 1134–1146.

Papadakis, J. S. (1937). Methode statistique pour des experiences sur champ. *Bull. Inst. Amél. Plantes a Salonique* **23**.

Pazman, A. (1974). A convergence theorem in the theory of *D*-optimum experimental designs. *Ann. Math. Statist.* **2**, 216–218.

Pukelsheim, F. (1980). On linear regression designs. *J. Statist. Plan. Inference* **4**, 339–364.

Marcus, M. B. and Sacks, J. (1977). "Robust designs for regression problems" Statistical Decision Theory and Related Topics II. (Ed. S. S. Gupta and D. S. Moore), Academic Press, New York.

Mitchell, T. J. (1974). An algorithm for the construction of "*D*-optimal" experimental designs. *Technometrics* **16**, 203–210.

Ruiz, F. and Seiden, E. (1974). On construction of some families of generalized Youden designs. *Ann. Statist.* **2**, 503–519.

Sacks, J. and Ylvisaker, D. (1969). Designs for regression problems with correlated errors. III. *Ann. Math. Statist.* **41**, 2057–2074.

Silvey, S. D. and Titterington, D. M. (1973). A geometric approach to optimal design theory. *Biometrika* **60**, 21–32.

Silvey, D. M. (1980). *Optimum Design*. Chapman Hall, London.

Seiden, E. and Wu, C. -Z. (1978). A geometric construction of some families of generalized Youden designs for ra prime power. *Ann. Statist.* **6**, 451–460.

Smith, K. (1918). On the standard deviations of adjusted and interpolated values of an observed polynomial function and its constants and the guidance they give towards a proper choice of the distribution of observations. *Biometrika* **12**, 1–85.

Wald, A. (1943). On the efficient design of statistical investigations. *Ann. Math. Statist.* **14**, 134–140.

Welch, W. J. (1982). Branch-and-bound search for experimental designs based on *D*-optimality and other criteria. *Technometrics* **24**, 41–8.

Wu, C. -F. (1978). Some algorithmic aspects of the theory of optimal design. *Ann. Statist.* **6**, 1286–1301.

Wu, C. -F. and Wynn, H. P. (1973). The convergence of general step length algorithms for regular optimum design criteria. *Ann. Statist.* **6**, 1273–1285.

Wynn, H. P. (1970). The sequential generation of *D*-optimal experimental designs. *Ann. Math. Statist.* **41**, 1055–1064.

JACK KIEFER
China, 1980

Reprinted from THE ANNALS OF MATHEMATICAL STATISTICS
Vol. 29, No. 3, September, 1958
Printed in U.S.A.

ON THE NONRANDOMIZED OPTIMALITY AND RANDOMIZED NONOPTIMALITY OF SYMMETRICAL DESIGNS[1]

BY J. KIEFER[2]

Cornell University

0. Summary. Many commonly employed symmetrical designs such as Balanced Incomplete Block Designs (BIBD's), Latin Squares (LS's), Youden Squares (YS's), etc., are shown to have optimum properties among the class of *non-randomized*[1] designs (Section 3). This represents an extension of a property first proved by Wald for LS's in [1]; a similar property demonstrated by Ehrenfeld for LS's in [2] (as well as a third optimum property considered here) is shown to be an immediate consequence of the Wald property, and the Wald property is shown to be the more relevant when one considers optimality rigorously (Section 2). Surprisingly, all of these optimum properties fail to hold if *randomized*[1] designs are considered (Section 4); the results of Sections 2 and 3, as well as those appearing previously in the literature (as in [1], [2], [3]) must be interpreted in this sense. Generalizations of the BIBD's and YS's, for which analogous results hold, are introduced.

1. Introduction. Wald [1] stated an optimality criterion (called E-optimality in Section 2) for designs used in testing hypotheses in the setting of two-way soil heterogeneity where LS's are commonly employed, and succeeded in proving that a slightly different criterion (called D-optimality in Section 2) is satisfied by the LS design. Wald also stated that an analogous result holds for Graeco-Latin Squares and higher Latin Squares. This statement gives rise to speculation when one considers that, in a 3×3 Graeco-Latin Square (or, more generally, in an $n \times n$ square of order $n - 1$), there are no degrees of freedom for error: this implies that any test (e.g., of the hypothesis H_0 that there are no treatment effects) whose size (= supremum of the power function under H_0) is α, has a power function whose infimum over any of the contours usually considered ($\psi(\mu)/\sigma^2$ = constant, as discussed in the sequel) is $\leqq \alpha$. It is easy to construct a better design, i.e., one for which the infimum of the power function of some test over such a contour is $>$ the size of the test; for example, for each of the two

Received July 8, 1957; revised January 22, 1958.

[1] One of the referees of this paper felt that the following remark on nomenclature should be included: Throughout this paper, the term *randomized design* is used in describing a statistical procedure which chooses according to a prescribed probability mechanism a member of a given class of ordinary designs, the chosen design being the one actually used; a precise definition is given in the text. The properties of such a procedure take into account the probabilities of the various possible choices. A *nonrandomized* design chooses one member of the given class with probability one. The customary usage of the phrase *randomized design* in the design of experiments can be viewed as a special case of the decision-theoretic usage employed here, but the reader is warned not to interpret the phrase in that narrower sense.

[2] Research sponsored by the Office of Naval Research.

factors, with probability $\frac{1}{2}$ use an ordinary LS design on the three levels of that factor holding the level of the other factor fixed.[3]

The phenomenon just described makes one wonder whether the optimality result for ordinary LS's also fails to hold if one permits comparison with randomized designs.[1] At the same time, the question arises whether an analogue of the limited optimality property of the LS (or Graeco-LS) design holds in a wide class of design settings for designs with suitable symmetry properties, and whether these designs fail to be optimum when compared with randomized designs.[1] This paper answers these questions affirmatively.

In Section 2A we define four optimality criteria (D-, E-, M-, and L-optimality) for designs (especially, for the normal case); Wald [1] and Ehrenfeld [2] proved D- and E-optimality, respectively, for the LS design. It is indicated why M-optimality, the strongest and least artificial of the four, seems very difficult to verify in most problems (although L-optimality, which is a local version of M-optimality, can sometimes be verified). At the same time, we list briefly for later reference the known results on the Analysis of Variance Test which are used in optimality considerations, and point out the incorrectness of tacitly assuming (as previous work in this area has done) that one should use that test, whatever design is chosen. In Section 2B we indicate by example why E-optimality seems, at least in the present state of knowledge indicated in 2A, the least satisfactory of the criteria considered; the connection of D-optimality with Isaacson's notion of type D tests [11] is examined. In Section 2C it is shown in a general setting where there is suitable symmetry that D-optimality implies E-optimality and L-optimality.

In Section 3A it is indicated why the treatment of LS's is much simpler than that of YS's, BIBD's, etc., and the general treatment of incomplete block designs

[3] It should be evident that the example of the 3×3 Graeco-Latin square, as well as the example discussed in the fourth paragraph below wherein two observations are taken, are of no *practical* importance; these simple examples are given to illustrate the general principles of Section 4. Those principles show that a precise study of certain optimality criteria for designs associated with familiar problems of testing hypotheses, can lead to the unexpected conclusion that certain intuitively unappealing randomized designs are superior to certain intuitively appealing nonrandomized symmetrical designs. The principles are less transparent (although applicable) in the context of applicationally meaningful problems such as those of Section 4, than in the simple examples; hence, the latter examples are discussed first. The present comments are included because two referees apparently read these simple examples as practical suggestions. In the same light, it is clear that the design δ in the fourth paragraph below, as well as its analogues in Section 4, is not suggested to the practical worker who wants *estimates* of all treatment effects; for these designs illustrate a non-optimality property of classical nonrandomized symmetrical designs in *hypothesis testing*, and a local property at that (see Section 5.4). In fact, the results of Section 4 are not even relevant for most estimation problems (see Section 5.2). To the practical worker who objects (as at least one has) to the conclusions of Section 4 on the grounds that one should not use a design which does not estimate all treatment effects, it should be pointed out that (1) the classical nonrandomized symmetrical design may still possibly possess certain *global* optimality properties (see Section 5.4), and (2) perhaps his problem is not really one of testing hypotheses.

of Bose [4] is briefly recalled; this treatment proves more useful in Section 3C than the more direct least squares approach used in [1] and [2] would be. In Section 3B several algebraic propositions (emphasizing the role of symmetry) are verified, which can be used to prove D- and E-optimality in important examples. Several such examples are considered in Section 3C, including generalizations of the BIBD's and the YS's.

Section 4 contains two theorems the consequences of which are that non-randomized symmetrical designs are not optimum if randomization is permitted. In Section 4B it is shown that, whether or not the variance is known, for α sufficiently small there is a randomized design whose power function is uniformly larger than that of the symmetrical design in some neighborhood of the hypotheses H_0 that all treatment effects are the same. This is slightly less transparent than the result of Section 4A, which gives an analogous result for *all* α when the above H_0 is replaced by the hypothesis that all treatment effects are equal to some specified value. The latter result can best be understood by considering the simplest example[3]: Suppose X_{ij} normal with unit variance and mean μ_i and that all X_{ij} are independent $(i, j = 1, 2)$. Our problem is to select (before observation) exactly two of the X_{ij} and use them to test $\mu_1 = \mu_2 = 0$ against some class of alternatives. The symmetrical design d(say) selects X_{11} and X_{21} and uses the usual χ^2 test, and obviously has constant power $> \alpha$ on the contour $\mu_1^2 + \mu_2^2 = c > 0$, while either of the designs d_i $(i = 1, 2)$, where d_i uses X_{i1} and X_{i2}, has α for the infimum of the power function on this contour. Let δ be the randomized design[1] obtained by using d_1 or d_2 with probability $\frac{1}{2}$ each. It is easily seen that, for μ_1 and μ_2 near 0, the power function of δ is $\alpha + c_1(\mu_1^2 + \mu_2^2) +$ terms of higher order, where $c_1 > 0$. Thus, on the contour $\mu_1^2 + \mu_2^2 = c > 0$ with c small, the power function of δ is almost constant and hence approximately equal to the value at $\mu_1 = \mu_2 = (c/2)^{\frac{1}{2}}$. Thus, in comparing d and δ near H_0, we may to a first approximation assume $\mu_1 = \mu_2$. But δ is clearly optimum for testing $\mu_1 = \mu_2 = 0$ *assuming* $\mu_1 = \mu_2$, while d (whose test is based on $X_1^2 + X_2^2$) is not. This explains why, for c small, δ has a power function greater than that of d.

Many of the results of this paper have counterparts for problems of point and interval estimation, for other distributions, etc. Such extensions and generalizations, as well as various other remarks, are stated in Section 5.

In design settings where no suitably symmetric design exists, it is often tedious algebraically to show that a design which is "closest to symmetrical" is optimum (if it *is* optimum: see the example of Section 2B), and we omit such considerations here. On the other hand, the conclusions of Section 4 have little to do with whether or not symmetrical designs are being considered.

Throughout this paper, except where explicitly stated to the contrary, Y will denote an N element column vector whose components Y_i are independent normal random variables with common variance σ^2 (it will be explicitly stated whenever σ^2 is assumed known; whether or not σ^2 is known has very little effect on our results); μ is an unknown m-vector, X_d is a known $N \times m$ matrix depending on an index d (the "design") and which will be described further below, and the expected value of Y when μ and σ^2 are the parameter values and when the design d is used is

$$E_{\mu,\sigma;d}Y = X_d\mu. \tag{1.1}$$

X_d is, within limits, subject to choice by the experimenter. (In many applications it is a matrix of zeros and ones.) We denote by Δ the set of choices of the index d which are available to the experimenter. A randomized design[1] δ is a probability measure on Δ (the latter will usually be finite in this paper, and measurability considerations will be trivial otherwise) which is used by selecting a d from Δ according to this measure and then using the selected d. We denote the class of available δ by Δ_R.

In many problems, one imposes an additional assumption of the form $\Gamma\mu = \gamma$, where Γ and γ are known $g \times m$ and $g \times 1$ matrices. Such an assumption can be absorbed into (1.1) and we suppose this to have been done, with no loss of generality.

A hypothesis H_0 will in this paper be of the form $R\mu = 0$, where R is a specified $r \times m$ matrix ($r \leqq m$) which we can take to be of rank r with no loss of generality. For simplicity, we can think of the class H_1 of alternatives as being all μ for which $R\mu \neq 0$. (For simplicity, we assume that σ^2 is either known exactly or else is known only to be positive, under both H_0 and H_1.) A hypothesis of the form $R\mu = \rho$ is easily reduced to the above form by letting p satisfy $Rp = \rho$ and replacing Y by $Y^* = Y - X_d p$ and μ by $\mu^* = \mu - p$ in (1.1).

We introduce some notation to be used in Section 2. We denote the $k \times k$ identity matrix by I_k. The transpose of a matrix A is written A'. It may or may not be that all r elements of $R\mu$ are estimable when a given design d is used. Suppose that there are s_d linearly independent linear combinations of the elements of $R\mu$ which have unbiased estimators when d is used, but not $s_d + 1$ such combinations. Then there is an $s_d \times r$ matrix Q_d such that there exist linear unbiased estimators of all components of $Q_d R\mu$ when design d is used; let t_d be the s_d-vector of such estimators with minimum variance ("best linear estimators" or b.l.e.'s), and let $\sigma^2 V_d$ be the convariance matrix of the components of t_d. When $s_d = r$, we may take Q_d to be the identity; for this choice of Q_d, we shall denote V_d by $\bar{V}_d$. Let b_d be the rank of X_d. Then there are b_d linearly independent combinations of the components of μ which are estimable when d is used. Of these, s_d of them can be taken to be the elements of $Q_d R\mu$; thus, there exists a $(b_d - s_d) \times m$ matrix J_d of rank $b_d - s_d$ whose rows are orthogonal to those of $Q_d R$ (i.e., $J_d' Q_d R = 0$) and such that all components of $J_d\mu$ have unbiased estimates when d is used. Let L_d be the $b_d \times m$ matrix whose first $b_d - s_d$ rows are J_d and whose last s_d rows are $Q_d R$. Let $\bar{S}_d$ be the usual best unbiased estimator of σ^2 (if it is unknown), so that $(N - b_d)\,\bar{S}_d/\sigma^2$ has the χ^2-distribution with $h_d = N - b_d$ degrees of freedom (it may be that $h_d = 0$ and there is no $\bar{S}_d$). For any test ϕ_d associated with d, let $\beta_{\phi_d}(\mu, \sigma^2)$ be the power function of ϕ_d (of course, β_{ϕ_d} actually depends on μ only through $L_d\mu$). For $0 < \alpha < 1$ we denote by

$H_d(\alpha)$ the class of all ϕ_d of size α, i.e., all ϕ_d for which

$$\beta_{\phi_d}(\mu, \sigma^2) \leqq \alpha \text{ whenever } R\mu = 0; \tag{1.2}$$

and by $H_d^*(\alpha)$, the class of similar tests of size α, i.e., those for which (1.2) holds with the inequality sign replaced by equality. Finally, let $F_{d,\alpha}$ denote the usual F-test of H_0 of size α with s_d and h_d degrees of freedom, based on $t_d' V_d^{-1} t_d / s_d \bar{S}_d$ (if σ^2 is known, this is replaced by the appropriate χ^2-test).

The symbol $g_{i,j}(\alpha)$ is used to denote the derivative at H_0 of the power function of the F-test of size α and i, j degrees of freedom, with respect to (a common choice of) the parameter on which it depends; specifically, if $r = m = i$, $N - r = j$, the matrices R, Q_d, and V_d are the identity, and the true values of μ and σ^2 are such that $\mu'\mu/\sigma^2 = \lambda$, then, as $\lambda \to 0$, the power function of $F_{d,\alpha}$ is

$$\alpha + g_{i,j}(\alpha)\lambda + O(\lambda^2). \tag{1.3}$$

The results of this paper can be stated in a very general setting involving invariance of Δ, of the restriction $R\mu = 0$, and of a generalization of the function ψ considered below, as well as of certain designs, under an appropriate group of permutations of the components of μ. However, in order to make our proofs (and, in particular, the role of symmetry) as transparent as possible, we will carry them out in two cases; the reader will not find it difficult to state our results more generally by making appropriate linear transformations, etc. The two cases (Δ and X_d being further specified in particular examples; the role of the function ψ which distinguishes contours on which the power function is examined, will be seen in Section 2A) are:

CASE I: $$\psi(\mu) = \sum_1^u \mu_i^2 \text{ and } R = R_{\text{I}};$$

CASE II: $$\psi(\mu) = \sum_1^u (\mu_i - \bar{\mu})^2 \text{ and } R = R_{\text{II}};$$

here we have written $\mu' = (\mu_1, \cdots, \mu_m)$, and $\bar{\mu} = \sum_1^u \mu_i/u$, while R_{I} is the $u \times u$ identity followed by $m - u$ columns of zeros (so $R_{\text{I}}\mu = 0$ means $\mu_1 = \cdots = \mu_u = 0$), and R_{II} is a $(u - 1) \times u$ matrix P followed by $m - u$ columns of zeros, where P consists of the last $u - 1$ rows of a $u \times u$ orthogonal matrix $\bar{O}$ whose first row elements are all $1/\sqrt{u}$ (so $R_{\text{II}}\mu = 0$ means $\mu_1 = \cdots = \mu_u$). The optimality results which hold in Case I are usually much more trivial to obtain than those of Case II, and Section 3B will therefore be mainly devoted to results applicable to the latter case, it being clear how to obtain the corresponding results in the former case.

2. Optimality criteria.

2A. *Preliminaries.* For a fixed design d, the test $F_{d,\alpha}$ is known to have several optimum properties, which we now list (there are obvious analogues when σ^2 is known):

(a) If $s_d = 1$ (and only then), among tests in $H_d(\alpha)$ which are unbiased (this implies that the tests are in $H_d^*(\alpha)$), $F_{d,\alpha}$ is uniformly most powerful (UMP). See [5] (a trivial completeness argument characterizing similar tests is all that is required to allow the $J_d\mu$ which is not present in [5] to be introduced, carrying through the argument there for each fixed value of the b.l.e. of $J_d\mu$).

(b) Among tests in $H_d(\alpha)$, $F_{d,\alpha}$ is UMP invariant (under the usual group of transformations when the problem is reduced to canonical form). See [5].

(c) (Wald's theorem) Among tests in $H_d^*(\alpha)$, for each $c > 0$, $\sigma^2 > 0$, and value of $J_d\mu$, the test $F_{d,\alpha}$ maximizes the Lebesgue integral of $\gamma_{\phi_d}(\nu, J_d\mu, \sigma^2)$ on the sphere $\nu'\nu = c$, where $\nu = G_dQ_dR\mu$ with G_d nonsingular $s_d \times s_d$ is such that the b.l.e.'s of the components of ν have σ^2 times the identity for their covariance matrix (i.e., ν is the vector of parameters about which H_0 is concerned in the canonical form of the problem), and where $\gamma_{\phi_d}(G_dQ_dR\mu, J_d\mu, \sigma^2) = \beta_{\phi_d}(\mu, \sigma^2)$. See [6] or [7] (the parenthetical remark at the end of (a) is relevant to [7] here).

(d) (Hsu's theorem, a consequence of (c)) Among tests in $H_d(\alpha)$ whose power function depends only on $\lambda_d = \mu'R'Q_d'V_d^{-1}Q_dR\mu/\sigma^2$ (this implies that the tests are in $H_d^*(\alpha)$), $F_{d,\alpha}$ is UMP. See [8].

(e) Among tests in $H_d(\alpha)$, $F_{d,\alpha}$ is minimax (over H_1) for a variety of weight functions, e.g., any nonnegative function of the λ_d of (d); in particular, $F_{d,\alpha}$ maximizes the minimum power on the contour $\lambda_d = c$ for each $c > 0$. See [9] or [10] (the result follows from (c) if we restrict consideration to $H_d^*(\alpha)$).

(f) (A special case of (e)) $F_{d,\alpha}$ is most stringent in $H_d(\alpha)$. See [9] or [10].

(g) (A consequence of (c)) $F_{d,\alpha}$ is of type D in $H_d(\alpha)$. (See [11] or Section 2B below for definition of type D, and Section 2B for a proof.)

It is to be noted that all the above criteria of optimality of the test $F_{d,\alpha}$ are relative to the design d. Thus, it is an error to assume (as has been done in previous papers on optimum designs) in a logical approach to optimum design problems that one should automatically use the test $F_{d,\alpha}$, whatever the chosen d, *when a reasonable criterion for optimality of a design, or of a test for a given design, may dictate the use of a test other than* $F_{d,\alpha}$. In fact, the example of Section 2B really illustrates that the use of $F_{d,\alpha}$ need not lead to an optimum design or test for many reasonable definitions of optimality; and the fact that it seems difficult (for many reasonable optimality criteria such as M-optimality, and for many common design problems) to characterize the appropriate test, is what makes it much harder than it has been thought to give a rigorous demonstration of the optimality of various common designs. We now list four optimality criteria for designs (there are many other obvious similar ones); the discussion of their meaning immediately follows the fourth definition.

M-optimality: For $c > 0$ and $0 < \alpha < 1$, a design d^* is said to be *$M_{\alpha,c}$-optimum in* Δ if, for some $\phi_{d^*}^*$ in $H_{d^*}(\alpha)$,

$$\inf_{\Gamma_c} \beta_{\phi^*_{d^*}}(\mu, \sigma^2) = \max_{d\varepsilon\Delta} \sup_{\phi\varepsilon H_d(\alpha)} \inf_{\Gamma_c} \beta_\phi(\mu, \sigma^2), \tag{2.1}$$

where Γ_c is the set of all μ, σ^2 for which $\psi(\mu)/\sigma^2 = c$.

L-optimality: A design is said to be *L_α-optimum in* Δ if, for some $\phi^*_{d^*}$ in $H_{d^*}(\alpha)$,

$$\lim_{c\to 0} [a_{\phi^*_{d^*}}(c) - \alpha]/[b(c) - \alpha] = 1, \tag{2.2}$$

where $a_{\phi^*_{d^*}}(c)$ and $b(c)$ are the expressions on the left and right sides of (2.1), respectively. A design is said to be *L-optimum in* Δ if it is L_α-optimum in Δ for $0 < \alpha < 1$.

D-optimality: A design d^* is said to be *D-optimum in* Δ if

$$\det \bar{V}_{d^*} = \min_{d\varepsilon\Delta'} \det \bar{V}_d, \tag{2.3}$$

where Δ' is the set of d in Δ for which $s_d = r$, and if $d^* \varepsilon \Delta'$.

E-optimality: A design d^* is said to be *E-optimum in* Δ if

$$\pi(\bar{V}_{d^*}) = \min_{d\varepsilon\Delta'} \pi(\bar{V}_d) \tag{2.4}$$

and if d^* is a member of Δ', where $\pi(\bar{V}_d)$ is the maximum eigenvalue of $\bar{V}_d$.

The above definitions will also be used with Δ replaced by Δ_R. In that case, for any δ, $\bar{V}_\delta^{-1}$ is defined to be the expected value under δ of $\bar{V}_d^{-1}$, the latter being replaced by the inverse of the covariance matrix of the b.l.e. of the estimable components of $R\mu$ (with zeros adjoined to this inverse in appropriate places to make it $r \times r$) if $s_d < r$; Δ_R' is then the set of δ for which $\bar{V}_\delta^{-1}$ is nonsingular. (This $\bar{V}_\delta^{-1}$ appears in computing certain β_{ϕ_δ} near H_0.)

D-optimality and E-optimality have been discussed in [1] and [2] and will also be discussed in Section 2B, where it will be seen that they have to do with local properties (near H_0) or optimum properties *assuming the use of* $F_{d,\alpha}$. Unfortunately, $M_{\alpha,c}$-optimality in Δ (or, better, $M_{\alpha,c}$-optimality in Δ simultaneously for all c) seems very difficult to verify, even in many simple problems, although it does not require much temerity to conjecture that it holds in such cases as those discussed in Section 2C. A similar remark applies to L-optimality (see, however, Lemma 2.2), a local version (near H_0) of M-optimality. The source of this difficulty in verifying M-optimality is illustrated by the example of Section 2B; it is simply that for fixed d the test which achieves the supremum over ϕ on the right side of (2.1) need not be $F_{d,\alpha}$ and is generally hard to compute (as is therefore the right side of (2.1)).

2B. *D- and E- optimality.* We begin by describing the meaning of E-optimality (which criterion is stated in [1] and is verified for the LS design in [2]). Suppose for fixed α, that we agreed to restrict ourselves to using $F_{d,\alpha}$, whatever d is chosen. The power function of $F_{d,\alpha}$ is then a strictly increasing function of λ_d (defined in Section 2A(d)). Now, in either Case I or II, for any $c > 0$, if we want a design d for which $F_{d,\alpha}$ maximizes the minimum power on the contour $\psi(\mu)/\sigma^2 = c$ (i.e., *which is $M_{\alpha,c}$-optimum in Δ under the additional restriction that we use $F_{d,\alpha}$*), we may restrict our attention to Δ' (since, for $s_d < r$, the infimum of $\beta_{F_{d,\alpha}}$ on the contour $\psi(\mu)/\sigma^2 = c$ is α; if Δ' is empty, there is no problem). $F_{d,\alpha}$ has the same number of numerator degrees of freedom for all d in Δ'; if also

b_d is the same for each d in Δ' (this is often the case in important examples such as those of Section 3C) so that the denominator degrees of freedom are the same for all $F_{d,\alpha}$, then a design which maximizes the minimum power on $\psi(\mu)/\sigma^2 = c$ *simultaneously for all* c is precisely one which maximizes the minimum of λ_d subject to $\psi(\mu)/\sigma^2 = c$. Since $\psi(\mu) = (R\mu)'(R\mu)$ in both Cases I and II, this means maximizing $\min_{\xi'\xi=1} \xi'\bar{V}_d^{-1}\xi = 1/\pi(\bar{V}_d)$. This is precisely the criterion of E-optimality.

One can cite many practical examples to illustrate that the restriction to using $F_{d,\alpha}$, which is imposed in order to make E-optimality meaningful, can have serious detrimental consequences. The simplest possible situation will suffice as an example: Suppose $N > 2$, $r = m = 2$, $R = R_{\mathrm{I}}$, and Δ' to consist of two designs with

$$\bar{V}_{d_1} = \begin{pmatrix} 1 & 0 \\ 0 & 1 \end{pmatrix}, \qquad \bar{V}_{d_2} = \begin{pmatrix} 1+\epsilon & 0 \\ 0 & \epsilon \end{pmatrix},$$

where $\epsilon > 0$. Clearly, d_1 is E-optimum. Moreover, if d_1 is used, optimum property (e) above states that, for every c, $F_{d_1,\alpha}$ maximizes the minimum power on the contour $(\mu_1^2 + \mu_2^2)/\sigma^2 = c$ among all tests in $H_{d_1}(\alpha)$. However, if d_2 is used, $F_{d_2,\alpha}$ does not have this property. For example, if d_2 is used, let ϕ' be the test which with probability $(1+\epsilon)/(1+2\epsilon)$ uses the F-test (with 1 and N-2 degrees of freedom) of size α of the hypothesis $\mu_1 = 0$, and which with probability $\epsilon/(1+2\epsilon)$ uses the F-test of size α of the hypothesis $\mu_2 = 0$. The power function of ϕ' near $(\mu_1^2 + \mu_2^2)/\sigma^2 = 0$ is then

$$\alpha + g_{1,N-2}(\alpha)\,(\mu_1^2 + \mu_2^2)/(1+2\epsilon)\sigma^2 + o([\mu_1^2 + \mu_2^2]/\sigma^2),$$

while that of $F_{d_2,\alpha}$ is

$$\alpha + g_{2,N-2}(\alpha)\left(\frac{\mu_1^2}{1+\epsilon} + \frac{\mu_2^2}{\epsilon}\right)/\sigma^2 + o([\mu_1^2 + \mu_2^2]/\sigma^2).$$

The infimum of the expression multiplying $g_{2,N-2}(\alpha)$, taken on the contour $(\mu_1^2 + \mu_2^2)/\sigma^2 = c$, is $c/(1+\epsilon)$, compared with $c/(1+2\epsilon)$ for the coefficient of $g_{1,N-2}(\alpha)$; since $g_{1,N-2}(\alpha)/g_{2,N-2}(\alpha) \to 2$ as $\alpha \to 0$ (see Lemma 4.3 below) the assertion three sentences above regarding $F_{d_2,\alpha}$ is verified. Moreover, since the power function of $F_{d_1,\alpha}$

$$\alpha + g_{2,N-2}(\alpha)(\mu_1^2 + \mu_2^2)/\sigma^2 + o([\mu_1^2 + \mu_2^2]/\sigma^2),$$

we see similarly that, at least for α, ϵ, and c sufficiently small, d_1 is *not* $M_{\alpha,c}$-optimum or L_α-optimum, ϕ' being locally uniformly more powerful than $F_{d_1,\alpha}$; thus, the assertion of the first sentence of this paragraph regarding E-optimality is verified.

Of course, for any fixed α, ϵ, and c we have not asserted that the test ϕ' (considered above only for illustrative purposes) is $M_{\alpha,c}$-optimum. If one uses d_2, the power functions of ϕ', $F_{d_2,\alpha}$, etc., are not constant on $(\mu_1^2 + \mu_2^2)/\sigma^2 = c$ (the same is true of the test which minimizes the integral of the power function on that contour), and the computation of the supremum over ϕ on the right side of (2.1) does not seem easy (this will be discussed further in Section 5). Thus, the above example also illustrates why M-optimality (or L-optimality) seems so difficult to verify in many problems.

In order to see the meaning of D-optimality, we turn to the notion of a type D test as defined in [11] (we discuss the case where σ^2 is unknown, the other case being similar): For fixed d, let the function $\bar{\beta}_\phi(\eta, \tau, \sigma^2)$ be defined by $\bar{\beta}_\phi(Q_d R\mu, L_d\mu, \sigma^2) = \beta_\phi(\mu, \sigma^2)$ and let $\beta_\phi^i(\tau, \sigma^2)$ (resp., $\beta_\phi^{ij}(\tau, \sigma^2)$) be the derivative of $\bar{\beta}_\phi(\eta, \tau, \sigma^2)$ with respect to the ith (resp., ith and jth) component of η, evaluated at $\eta = 0$ (these derivatives always exist). A test ϕ in $H_d(\alpha)$ is said to be locally (near H_0) strictly unbiased if

(a) $\phi \; \varepsilon \; H_d^*(\alpha)$,

(b) $\beta_\phi^i(\tau, \sigma^2) = 0$ for all i, τ, and σ^2,

(c) the matrix $B_\phi(\tau, \sigma^2) = \| \beta_\phi^{ij}(\tau, \sigma^2) \|$ is positive definite for all τ and σ^2.

Clearly, (c) can be satisfied only if $d \; \varepsilon \; \Delta'$. Suppose then that $d \; \varepsilon \; \Delta'$ and that Q_d = identity (we have mentioned the fact that we can make this choice of Q_d when $d \; \varepsilon \; \Delta'$). For any ϕ satisfying (a), (b), (c) just above, det $B_\phi(\tau, \sigma^2)$ is the Gaussian curvature of the surface given by the graph of $\bar{\beta}_\phi(\eta, \tau, \sigma^2)$ as a function of η for fixed τ, σ^2, at $\eta = 0$. A test ϕ is defined in [11] to be of type D if it maximizes this curvature for all τ and σ^2, among all locally strictly unbiased tests. This criterion of optimality, although a local one, has certain appealing features; for example, it is invariant under all one-to-one transformations of the parameter space which leave $\eta = 0$ fixed and which at $\eta = 0$ are twice differentiable with non-vanishing Jacobian [11]. Now, since without loss of generality we are taking Q_d = identity, we can compare the behavior of the type D tests for various designs in Δ', assuming b_d to be the same for all d in Δ'. A design for which the Gaussian curvature at $\eta = 0$ of the test of maximum Gaussian curvature (for a given design) is a maximum (over all designs) is thus, if it exists, that d which maximizes $\max_{\phi_d}$ det $B_{\phi_d}(\tau, \sigma^2)$ simultaneously for all τ, σ^2. That such a design is precisely one which is D-optimum follows immediately from the following lemma[4] (there is an obvious analogue when σ^2 is known):

LEMMA 2.1. *For d in Δ' and $0 < \alpha < 1$, the test $F_{d,\alpha}$ is of type D.*

PROOF. $F_{d,\alpha}$ is clearly locally strictly unbiased. We again put Q_d = identity, and a nonsingular linear transformation reduces the proof to the case where G_d = identity (see Section 2A(c)), so that $\nu = \eta$. Wald's theorem can then be stated as

$$(2.5) \qquad \int_{\eta'\eta=c} [\bar{\beta}_{F_{d,\alpha}}(\eta, \tau, \sigma^2) - \alpha]A(d\eta) \geqq \int_{\eta'\eta=c} [\bar{\beta}_\phi(\eta, \tau, \sigma^2) - \alpha]A(d\eta)$$

for every $c > 0$, $\sigma^2 > 0$, and ϕ in $H_d^*(\alpha)$, where $A(d\eta)$ is Lebesgue measure on the sphere $\eta'\eta = c$. Noting that

$$(2.6) \qquad \int_{\eta'\eta=c} \eta_i \eta_j A(d\eta) = \begin{cases} K(c, r) & \text{if } i = j, \\ 0 & \text{if } i \neq j, \end{cases}$$

where η_i is the ith component of η and $K(c, r)$ is positive and depends only on c

[4] The author understands that Isaacson gave a longer, unpublished proof, earlier.

and r, we obtain from (2.5) by normalizing properly and letting $c \to 0$, for any ϕ satisfying conditions (a) and (b) above,

$$\sum_{i=1}^{r} \beta_{F_{d,\alpha}}^{ii}(\tau, \sigma^2) \geqq \sum_{i=1}^{r} \beta_{\phi}^{ii}(\tau, \sigma^2). \tag{2.7}$$

Since $B_{F_{d,\alpha}}(\tau, \sigma^2)$ is a constant times the identity in our reduction, using the inequality of the geometric and arithmetic means and the fact that the determinant of a positive-definite matrix is no greater than the product of its diagonal elements, we obtain (omitting some appearances of τ, σ^2),

$$\begin{aligned} \det B_{\phi}(\tau, \sigma^2) &\leqq \prod_{i=1}^{r} \beta_{\phi}^{ii}(\tau, \sigma^2) \leqq \left[\sum_{i=1}^{r} \beta_{\phi}^{ii}/r\right]^r \\ &\leqq \left[\sum_{i=1}^{r} \beta_{F_{d,\alpha}}^{ii}/r\right]^r = \prod_{i=1}^{r} \beta_{F_{d,\alpha}}^{ii} = \det B_{F_{d,\alpha}}(\tau, \sigma^2), \end{aligned} \tag{2.8}$$

which completes the proof.

To summarize, D-optimality and L-optimality, although local properties, seem more reasonable criteria than E-optimality, which is tied to the ad hoc assumption that $F_{d,\alpha}$ should always be used; M-optimality (and to a lesser extent L-optimality) seems difficult to verify in many examples.

2C. *Relationship among optimality criteria in symmetric cases.* For future reference we state the following simple result (which was alluded to in Section 0 in reference to the relation between [1] and [2]):

LEMMA 2.2. *Suppose b_d is constant for d in Δ'. If d^* is D-optimum and $\bar{V}_{d^*}$ is a multiple of the identity, then d^* is E-optimum and L-optimum.*

PROOF. E-optimality is obvious from the nature of $\bar{V}_{d^*}$. If d^* were not L-optimum, since $F_{d^*,\alpha}$ has property 2A(c), for some other design d' there would by (2.2) be an associated test $\phi_{d'}$ in $H_{d'}^*(\alpha)$ with

$$\inf_{\tau,\sigma^2} \det B_{\phi_{d'}}(\tau, \sigma^2) > \det B_{F_{d^*,\alpha}}(\tau, \sigma^2) \tag{2.9}$$

(the right side of (2.9) is constant); by Lemma 2.1, equation (2.9) is a fortiori true if $\phi_{d'}$ is replaced by $F_{d',\alpha}$; this yields the contradiction that $\det \bar{V}_{d'} <$ $\det \bar{V}_{d^*}$.

In many examples of Case I where symmetrical designs exist, the condition on $\bar{V}_{d^*}$ in the hypothesis of Lemma 2.2 will be obvious. In Case II, as discussed in Section 3A, it is often convenient to write the normal equations in the form $C_d t_d^* = Z_d$, where C_d is a $u \times u$ matrix of rank $u - 1$, Z_d is a u-vector of linear forms in Y with covariance matrix C_d, and for any solution t_d^* of these equations one obtains the best linear estimator of any contrast $\sum_1^u c_i\mu_i$ with $\sum c_i = 0$ by forming $\sum c_i t_{di}^*$ where the t_{di}^* are the components of t_d^*. Clearly, Pt_d^* is the b.l.e. t_d of $R_{\text{II}}\,\mu$. Hence, if every diagonal element of C_d has the same value and if all off-diagonal elements have the same value, the fact that the first row of the orthogonal matrix $\bar{O}$ defined in Case II of Section 1 is constant immediately yields the fact (see Section 3A) that $\bar{V}_d^{-1} = PC_dP'$ is a multiple of the identity, so that

Lemma 2.2 may be applicable in such cases. For future reference, we state this simple computation (put $a + (u - 1)c = 0$) in

LEMMA 2.3. *If U is a $u \times u$ matrix with diagonal elements a and off-diagonal elements c, then*

$$\bar{O}U\bar{O}' = \begin{pmatrix} a + (u - 1)c & 0 \\ 0 & (a - c)I_{u-1} \end{pmatrix}. \tag{2.10}$$

We remark that the form of R_{II} (associated with $\bar{O}$) used here makes computations and proofs simpler and emphasizes more the role of symmetry (e.g., as it appears in the form of $\bar{V}_d^{-1}$ just noted, when C_d has appropriate symmetry), than would be the case if R_{II} were replaced by a matrix obtained by adjoining a column of 1's and $m - u$ columns of 0's to I_{u-1}, as in [1] and [2].

3. Optimality of symmetrical designs.

3A. *Preliminaries.* The results of this section will be proved for the case where σ^2 is unknown, the other case being handled similarly. The setting of two-way heterogeneity where the LS design is employed is much easier to analyze (and thereby obtain an optimality proof) than other settings considered in Section 3B such as those where the YS and BIBD are used (and the remarks at the end of Section 2 indicate how this analysis can be made even simpler than in [1] and [2]). The reason for this is that in this setting where the LS is used, whether μ is considered to have $3u$ components (u each for row, column, and treatment effects in the $u \times u$ case) or $3u - 2$ components (to make $X_d'X_d$ nonsingular when $s_d = b_d = u - 1$), $X_d'X_d$ becomes particularly simple, having large blocks of 1's (each row and column occur together once, etc.) or multiples of an identity (rows by rows, etc.) in the former case, and large blocks of 0's (especially if $\bar{O}$ is used in reducing X_d) and multiples of an identity, in the latter. Other design situations yield more complicated forms of $X_d'X_d$. Therefore, although the examples of Section 3C could be analyzed in a manner analogous to that used for the LS in [1] and [2], it appears algebraically simpler to use the incomplete block design analysis of Bose [4], to which end we now briefly outline the notation. Of course, we are concerned here with the more difficult Case II, which includes most of the important examples.

The form of the Z_d and C_d mentioned in Section 2C depends on the design setting and, in particular, in this section, on whether we are in a setting of one-way or two-way heterogeneity of (for example) soil (since all block sizes will be the same in our example of the former, it could be considered as a special case of the latter under further restrictions on μ). We shall first state the pertinent results which apply in both of these settings, and then specify the particular forms (see [4] for details). The $u \times u$ symmetric matrix C_d has row (or column) sums equal to zero, and the sum of the components of the u-vector Z_d is zero. The covariance matrix of Z_d is $\sigma^2 C_d$ and the expected value of Z_d is $C_d\mu^{(u)}$, where $\mu^{(u)}$ is the vector of the first u components of μ. We may assume $d \varepsilon \Delta'$, which means the design d is connected and that C_d has rank $u - 1$. If t_d^* satisfies $C_dt_d^* =$

Z_d and P is the $(u-1) \times u$ matrix defined in Case II in Section 1, then $t_d = Pt_d^*$ is the vector if b.l.e.'s of $R_{II}\mu$; the last $u-1$ rows of the equation $\bar{O}C_d\bar{O}'\bar{O}t_d^* = \bar{O}Z_d$ are thus $PC_dP't_d = PZ_d$ (the first row and column of $\bar{O}C_d\bar{O}'$ are zero), so that $t_d = (PC_dP')^{-1}PZ_d$ (the inverse may be taken for d in Δ') and thus the components of t_d have covariance matrix $(PC_dP')^{-1}$.

In the one-way heterogeneity setting we have u treatments, to be planted in b blocks; in our example, each block will contain the same number k of plots, one "planting" to be allowed per plot. The component of Y corresponding to an appearance of treatment i in block j has expected value $\mu_i + b_j$; thus, $m = u + b$, with $\mu_{u+j} = b_j$. Let n_{dij} be the number of appearances of treatment i in block j. We do *not* restrict n_{dij} to be 0 or 1, as is often done. Thus, D consists of those d for which X_d is any matrix of 0's and 1's for which each row contains exactly one 1 among the first m elements and one 1 among the last b elements and for which the last b columns each contain k one's; of course, $N = bk$. Let $r_{di} = \sum_j n_{dij}$ = number of replications of treatment i, let T_{di} = sum of all components of Y corresponding to treatment i, and let B_{dj} = sum of all components of Y arising from block j. The ith component Z_{di} of Z_d ("adjusted yield of treatment i") is $Z_{di} = T_i - \sum_j n_{ij} B_j/k$, and the (i, j)th component c_{dij} of C_d is

$$c_{dij} = \delta_{ij} r_{di} - \lambda_{dij}/k, \tag{3.1}$$

where δ_{ij} is the Kronecker delta and $\lambda_{dij} = \sum_s n_{dis} n_{djs}$.

In the setting of two-way heterogeneity, we have u treatments and a $k_1 \times k_2$ array of plots, and the expected value of a component of Y corresponding to treatment i in row j and column h is $\mu_i + b_j^{(1)} + b_h^{(2)}$; thus, $m = u + k_1 + k_2$ with $b_j^{(1)} = \mu_{m+j}$ and $b_h^{(2)} = \mu_{m+k_1+h}$. Let $n_{dij}^{(1)}$ (resp., $n_{dih}^{(2)}$) be the number of times treatment i appears in row j (resp., column h), and let T_{di} be as before and $B_{dj}^{(1)}$ (resp., $B_{dh}^{(2)}$) be the sum corresponding to the jth row (resp., hth column). r_{di} is as above, while $\lambda_{dij}^{(q)} = \sum_s n_{dis}^{(q)} n_{djs}^{(q)}$ for $q = 1, 2$. In this case $Z_{di} = T_{di} - \sum_j n_{dij}^{(1)} B_{dj}^{(1)}/k_2 - \sum_h n_{dih}^{(2)} B_{dh}^{(2)}/k_1 + r_{di} \sum_s T_{ds}/k_1k_2$ and

$$c_{dij} = \delta_{ij} r_{di} - \frac{\lambda_{dij}^{(1)}}{k_2} - \frac{\lambda_{dij}^{(2)}}{k_1} + \frac{r_{di} r_{dj}}{k_1 k_2}. \tag{3.2}$$

Many other design settings can be treated similarly; the above two will be used in the examples of Section 3C to illustrate our methods of proving optimality.

3B. *Algebraic results*. We now demonstrate the algebraic results used in proving optimality in the examples of Section 3C and which will be useful in other examples of Case II. The results proved here are meant to apply elsewhere than in the settings of Section 3A. We suppose in the present Section 3B that we are given a class $\{K_d, d \in \Delta'\}$ of $u \times u$ symmetric nonnegative definite matrices of rank $u - 1$ with row and column sums zero and define $W_d = PK_dP'$ (in our applications, $W_d = \bar{V}_d^{-1}$). The elements of $\bar{O}$, K_d, and W_d will be denoted by $\bar{o}_{ij}$,

k_{dij}, and w_{dij}, respectively. In Lemma 3.2 we consider an orthogonal matrix $\tilde{O} = \| o_{ij} \|$, not necessarily $\bar{O}$, and a diagonal matrix $D = \| d_{ij} \|$.

Our first lemma merely translates into terms of K_d the obvious fact that, if W_{d^*} has equal eigenvalues and if the sum of the eigenvalues (= trace) of W_d is a maximum for $d = d^*$, then the product of eigenvalues (= determinant) of W_d is a maximum for $d = d^*$.

LEMMA 3.1. *If all diagonal elements of K_{d^*} are equal and all off-diagonal elements of K_{d^*} are equal and $\sum_i k_{dii}$ is a maximum for $d = d^*$, then* det W_d *is a maximum for $d = d^*$.*

PROOF. Since $\bar{o}_{ij} = 1/\sqrt{u}$ and $\sum_{i,j} k_{dij} = 0$, the upper left-hand element of $\bar{O}K_d\bar{O}'$ is zero. Since the traces of $\bar{O}K_d\bar{O}'$ and K_d are equal, we conclude that the traces of K_d and W_d are equal, so that the trace of W_d is a maximum for $d = d^*$. The result now follows from Lemma 2.3 (follow the steps of (2.8) with W_d for B_ϕ and W_{d^*} for $B_{F_{d,\alpha}}$).

We shall actually prove in Theorems 3.1 and 3.2 that the trace of the matrix PC_dP' is a maximum and that all eigenvalues are equal when d is a BBD or GYS, so that Lemma 3.1 is relevant. However, there are settings in which the next three lemmas are more useful for proving D- or E-optimality directly when the hypothesis of Lemma 3.1 is difficult to verify or is false.

LEMMA 3.2. *For $u > 1$ if $\tilde{O}$ is orthogonal $u \times u$, D is diagonal $u \times u$, K is symmetric nonnegative definite $u \times u$ with row and column sums zero, and $\tilde{O}D\tilde{O}' = K$, then*

$$\left(\frac{u-1}{u}\right)^u \left(\prod_{i=1}^{u-1} d_{ii}\right)^{u/(u-1)} \leqq \prod_{i=1}^{u} k_{ii}. \tag{3.3}$$

PROOF. We assume $d_{uu} = 0 < d_{ii}$ for $i < u$, or the result is trivial. Since, then,

$$0 = \sum_{i=1}^{u}\sum_{j=1}^{u} k_{ij} = \sum_{i=1}^{u}\sum_{j=1}^{u}\sum_{s=1}^{u-1} o_{is}\, o_{js}\, d_{ss} = \sum_{s=1}^{u-1} d_{ss}\left(\sum_{i=1}^{u} o_{is}\right)^2, \tag{3.4}$$

we conclude that the first $u - 1$ columns of $\tilde{O}$ are orthogonal to the vector of ones. Hence, $o_{ju} = 1/\sqrt{u}$ (or its negative, which is treated in the same way).

Let the coordinates of a point ϵ in $u(u-1)$-dimensional Euclidean space be denoted by ϵ_{ij} $(i = 1, \cdots, u; j = 1, \cdots, u-1)$, and let B be the set of points ϵ in this space for which all $\epsilon_{ij} \geqq 0$, for which $\sum_j \epsilon_{ij} = (u-1)/u$ for all i, and for which $\sum_i \epsilon_{ij} = 1$ for all j. We shall prove below that ϵ in B implies

$$\prod_{i=1}^{u}\left(\sum_{j=1}^{u-1} \epsilon_{ij} d_{jj}\right) \geqq \left(\frac{u-1}{u}\right)^u \left(\prod_{i=1}^{u-1} d_{ii}\right)^{u/u-1}; \tag{3.5}$$

since the left side of (3.5) with $\epsilon_{ij} = o_{ij}^2$ gives the right side of (3.3) and since the restrictions on the ϵ_{ij} in B must be satisfied by the o_{ij}^2 (the orthogonality restrictions on the o_{ij} are omitted in defining B), (3.5) implies (3.3).

Call the left side of (3.5) $f(\epsilon)$. It is easy to verify that $-\log f(\epsilon)$ is convex in ϵ on $u(u-1)$-space, and hence on B. Moreover, B is a convex body in $u(u-1)$

space, and any extreme point of B is either

$$\begin{pmatrix} \epsilon_{11} & \cdots & \epsilon_{1,u-1} \\ \cdot & \cdots & \cdot \\ \cdot & \cdots & \cdot \\ \cdot & \cdots & \cdot \\ \epsilon_{u1} & \cdots & \epsilon_{u,u-1} \end{pmatrix} = \begin{pmatrix} & \frac{u-1}{u} I_{u-1} & \\ \frac{1}{u} & \cdots\cdots\cdots & \frac{1}{u} \end{pmatrix} \tag{3.6}$$

or is obtained by permuting the rows of the matrix on the right side of (3.6). Since a convex function on a convex set attains its maximum at an extreme point, we conclude that the minimum of f is attained at one of these extreme points. But f has the same value at any of these extreme points, namely,

$$\min_B f(\epsilon) = \left(\sum_{i=1}^{u-1} d_{ii}/u\right) \prod_{i=1}^{u-1} \left(\frac{u-1}{u} d_{ii}\right). \tag{3.7}$$

Thus, it remains only to prove that the right side of (3.7) is no less than the right side of (3.5), i.e., that

$$\prod_{i=1}^{u-1} d_{ii}^{1/(u-1)} \leqq \sum_{i=1}^{u-1} d_{ii}/(u-1); \tag{3.8}$$

but (3.8) is merely the well-known inequality between the geometric and arithmetic means.

The form of Lemma 3.2 which is useful in many applications is the following:

LEMMA 3.3. *If* $\prod_1^u k_{dii}$ *is a maximum for* $d = d^*$ *and if* K_{d^*} *has all diagonal elements equal and all off-diagonal elements equal, then* det W_d *is a maximum for* $d = d^*$.

PROOF: We use Lemma 3.2 with the product on the left side of (3.3) going from 2 to u, in order to conform to previous notation. In this form, with $\tilde{O} = \bar{O}$, it follows from Lemma 2.3 that the left and right sides of (3.3) are equal for $K = K_{d^*}$. Hence, from Lemma 3.2, $\prod_i w_{dii}$ is a maximum for $d = d^*$. Since $\prod_i w_{dii} \geqq$ det W_d with equality for the diagonal matrix W_{d^*}, the proof is complete.

The following lemma could be used in the case of the YS, and in more complicated problems where D-optimality is hard to prove or false, to prove E-optimality directly (i.e., without the use of Lemma 2.2):

LEMMA 3.4. *For* $u > 1$, *if* $m(W_d)$ *is the minimum eigenvalue of* W_d, *then*

$$m(W_d) \leqq \frac{u}{u-1} \min_i k_{dii}; \tag{3.9}$$

if all diagonal elements of K_d *are equal and all off-diagonal elements are equal, equality holds in* (3.9).

PROOF. Let δ_i be a u-vector with ith element one and all other elements zero. Let $\xi_i = P\delta_i$. Clearly, $\sqrt{u/(u-1)}\, \xi_i$ has unit length. Hence,

$$\begin{aligned} k_{dii} &= \delta_i' K_d \delta_i = (\bar{O}\delta_i)'(\bar{O}K_d\bar{O}')(\bar{O}\delta_i) \\ &= \xi_i' W_d \xi_i \geqq \frac{u-1}{u} \min_{a'a=1} a'W_d a = \frac{u-1}{u} m(W_d), \end{aligned} \tag{3.10}$$

which proves (3.9); the result on equality follows from Lemma 2.3.

The results for Case I analogous to those proved for Case II in this subsection are trivial (since in Case I the analogue of K_d will be nonsingular and K_{d^*} will be a multiple of the identity), and will be omitted.

3C. *Examples.*[5] (1). *Optimality of* BIBD's. In the setting of one-way heterogeneity described in Section 3A (with $u > 1$), suppose b, u, and k to be such that there exists a design d^* for which all n_{d^*ij} are k/u if k/u is an integer and are either of the two integers closest to k/u otherwise, for which all r_{d^*i} are equal, and for which all λ_{d^*ij} are equal for $i \neq j$. Such a design is called a BIBD if $k < u$, but we do not impose this last restriction here, and therefore call such a design a Balanced Block Design (BBD). (For example, if $b = 2$, $u = 2$, $k = 3$, such a d^* is that for which $n_{d^*11} = n_{d^*22} = 1$ and $n_{d^*12} = n_{d^*21} = 2$.) Our result is:

THEOREM 3.1. *If a* BBD d^* *exists, it is D-optimum, E-optimum, and L-optimum.*

PROOF. From (3.1) we have

$$\sum_{i=1}^{u} c_{dii} = N - \textstyle\sum_i \sum_s n_{dis}^2/k; \tag{3.11}$$

since $\sum_i \sum_s n_{dis} = N$, it is clear that (3.11) is a maximum for $d = d^*$. The result now follows from Lemma 3.1 and Lemma 2.2.

(2). *Optimality of* YS's. In the setting of two-way heterogeneity described in Section 3A (with $u > 1$), suppose k_1, k_2, and u to be such that there exists a design d^* for which all r_{d^*i} are equal, for which all $\lambda^{(1)}_{d^*ij}$ are equal for $i \neq j$, for which all $\lambda^{(2)}_{d^*ij}$ are equal for $i \neq j$, and for which all $n^{(q)}_{d^*ij}$ are equal to k_q/u if k_q/u is an integer and are either of the two integers closest to k_q/u otherwise ($q = 1, 2$). Thus, d^* is a BBD when either the rows or the columns are considered to be the blocks. Such a design d^* is usually called a YS if $k_1 < u$ (and k_2/u is an integer); we do not impose this condition, and shall hence call such a design d^* a Generalized Youden Square (GYS). (For example, if $u = 2$, $k_1 = 4$, $k_2 = 3$, such a design d^* is easily constructed.) If $k_1 = k_2 = u$, such a d^* is of course a LS. Our result is:

THEOREM 3.2. *If k_1/u or k_2/u is an integer and if a* GYS d^* *exists, then d^* is D-optimum, E-optimum, and L-optimum.*

PROOF. We shall show that $\sum_i c_{dii}$ is a maximum for $d = d^*$; Lemma 3.1 then yields the desired result. In this proof only we write $[x]$ = greatest integer $\leqq x$. Let r be an integer. Subject to the restrictions that $\sum_1^k m_j = r$ and that all m_j are integers, the expression $\sum_1^k m_j^2$ is minimized by taking $k - r + k[r/k]$ of the m_i to be $[r/k]$ and $r - k[r/k]$ of them to be $[r/k] + 1$, the corresponding minimum of $\sum m_j^2$ being $r + (2r - k)[r/k] - k[r/k]^2 = h(r, k)$ (say). We may assume

[5] The Editor has informed the author that E-optimality of the BIBD's (as a subclass of the BBD's) has been proved independently by V. L. Mote, and that the minimization of the average variance (see numbered paragraph 2 of Section 5) and of the generalized variance (i.e., the attainment of D-optimality) achieved by the BIBD's and YS's (a subclass of the GYS's) has been proved independently by A. M. Kshirsagar; both of these authors prove their results under the restriction and that the n_{dij} and $n^{(q)}_{dij}$ are all 0 or 1. Under this restriction, these special cases of the results of this paper are a consequence of the following line of argument: the trace of C_d is the same for all d, and the results follow at once from the symmetry of the BIBD and YS.

k_1/u is an integer. From (3.2) we have, for any d,

$$(3.12)\quad k_1 k_2(k_1 k_2 - \sum_i c_{dii}) \geqq \sum_i \{k_2 h(r_{di}, k_2) - r_{di}^2\} + \sum_i k_1 h(r_{di}, k_1),$$

with equality in the case of a GYS. The theorem will be proved if we show that each of the two sums on the right side of (3.12) attains its minimum for $d = d^*$. Now, $h(r, k) \geqq r^2/k$, since the latter is the minimum of $\sum m_j^2$ subject to $\sum m_j = r$ without the restriction that the m_j be integers. Hence, the first sum on the right side of (3.12) is at least zero. Moreover, this lower bound is achieved by the first sum on the right in (3.12) when $d = d^*$, since $r_{d^*i}/k_2 = k_1/u$ is an integer. It remains to consider the last sum of (3.12). We shall show that, subject to $\sum_1^m z_i = c$, the expression

$$(3.13)\quad q(z_1, \cdots, z_m) = \sum_{i=1}^{m} \{(2z_1 - 1)[z_i] - [z_i]^2\}$$

is a minimum when all z_i are equal; putting $z_i = [r_{di}/k_1]$, we see that this will yield the desired conclusion regarding the last sum of (3.12). The proof regarding (3.13) is by induction: assuming the conclusion to be true of $m = M$, in proving the case $m = M + 1$ we may put $z_1 = \cdots = z_M = s$ and $z_{M+1} = c - Ms$ in (3.13). The resulting expression is continuous in s and, except on a discrete set, has a derivative with respect to s which is equal to $2M([s] - [c - Ms])$. The latter is $\leqq 0$ if $s < c/(M + 1)$ and is $\geqq 0$ if $s > c/(M + 1)$, so that $s = c/(M + 1)$ yields a minimum. This completes the proof of Theorem 3.2.

We remark that, without the assumption that k_1/u or k_2/u is an integer, the above proof fails and Lemma 3.3 also fails to be applicable generally. To see this, consider the case $k_1 = k_2 = 6$, $u = 4$. A GYS d^* exists here, e.g., that one whose successive rows are (134324), (412233), (241342), (124123), (313412), (321441). We obtain $c_{d^*ii} = 25/4$ for all i. Let d' be the design whose rows are (133442), (213344), (421334), (442133), (344213), and (334421). Then $c_{d'11} = c_{d'22} = 5$, $c_{d'33} = c_{d'44} = 8$, $c_{d'12} = -1$, $c_{d'34} = -4$, and all other $c_{d'ij} = -2$. Thus, we obtain $\sum_i c_{d'ii} = 26 > 25 = \sum_i c_{d^*ii}$ and even $\prod_i c_{d'ii} = 1600 > (25/4)^4 = \prod_i c_{d^*ii}$. However, det $\bar{V}_{d'}^{-1} = 576 < (25/3)^3 =$ det $\bar{V}_{d^*}^{-1}$. Thus, between the designs d^* and d', the former is D-optimum, although Lemmas 3.1 and 3.3 cannot be used to prove it. Lemma 3.4 could still have been used to prove the E-optimality of d^* directly.

(3) *Other examples.* Many other design settings can be analyzed in a manner differing only slightly from the above examples and we mention but a few. One can treat similarly problems where the test concerns the b_j and $b_j^{(q)}$ of Section 3A. Problems involving Graeco-Latin Squares or higher Latin Squares, with or without replications, admit similar treatments. Higher-dimensional analogues (more than two directions of heterogeneity) can also be considered in a like fashion, as can complete or partial factorial arrangements. Many of the Case I analogues, such as the analogue of the BIBD treatment which assumes the b_j to be known, are trivial.

Other problems such as those for which E-optimality is considered in [2] (e.g., Hotelling's weighing problem and certain problems in the analysis of covariance) could be considered regarding D- and L-optimality by similar methods.

The treatment of some problems is in part parallel, but entails other considerations in addition to symmetry; such a problem is to test whether a regression function $\sum_{j=1}^{m} \mu_j f_j(x)$ is actually such that $\mu_1 = \cdots = \mu_r = 0$, where the f_j are given and N x's must be chosen from a given region of some space. (Many problems in the analysis of covariance involve similar considerations.) D- and E-optimality are also relevant in estimation problems (see Section 5.2).

The consideration of some of these other examples will appear elsewhere, in a paper by J. Wolfowitz and the author.

4. Nonoptimality of symmetrical nonrandomized designs among randomized designs.[1]

4A. Case I. We consider here the simplest general setting of Case I, namely, the extension of the example of Section 1 to more observations N and more treatments u. Other examples, such as the Case I analogues of the examples of Section 4B, have parallel analyses, and we omit them. We shall carry out the treatment when σ^2 is unknown, the treatment when σ^2 is known being similar. The underlying probabilistic property (of the normal distribution) which is relevant here will now be stated in a lemma. Let U/σ^2 have a non-central χ^2 distribution with N_1 degrees of freedom (d.f) and non-central parameter $\lambda = EU/\sigma^2 - N_1$, and let V/σ^2 have the central χ^2 distribution with N_2 d.f., with U and V independent. Let $P_{N_1,N_2}(\lambda; \alpha)$ denote the power function of the F-test of size α for testing $\lambda = 0$ based on N_2U/N_1V, and, as in (1.3), let $g_{N_1,N_2}(\alpha)$ denote the derivative of this power function with respect to λ at $\lambda = 0$.

Lemma 4.1. *If* $N_1 \leqq N_1'$ *and* $N_1 + N_2 \geqq N_1' + N_2'$ *with at least one of these a strict inequality, then* $P_{N_1,N_2}(\lambda; \alpha) > P_{N_1',N_2'}(\lambda; \alpha)$ *for* $\lambda > 0$ *and* $0 < \alpha < 1$, *and* $g_{N_1,N_2}(\alpha) > g_{N_1',N_2'}(\alpha)$ *for* $0 < \alpha < 1$.

Proof. Let U/σ^2 have a χ^2 distribution with parameter λ and N_1 d.f., and let V_1/σ^2, V_2/σ^2, and V_3/σ^2 have central χ^2 distributions with N_2', $N_1' - N_1$, and $N_1 + N_2 - N_1' - N_2'$ d.f., respectively (if any of the d.f.'s is 0, so is the corresponding V_i). U, V_1, V_2, V_3 are independent. For testing the hypothesis $\lambda = 0$ against alternatives $\lambda > 0$ based on U, V_1, V_2, V_3, it is easy to prove that the F-test based on $N_2U/N_1(V_1 + V_2 + V_3)$ is UMP unbiased of size α and is of type A, and is the unique (up to sets of measure zero) test with each of these properties; in particular, this is true in comparison with the F-test based on $N_2'(U + V_2)/N_1'V_1$, which proves the lemma.

The above lemma indicates both that the numerator d.f. should be as small as possible without affecting λ, which is also true when σ^2 is known, and also that for fixed $N_1 + N_2$, decreasing N_1 helps even more if σ^2 is *unknown*, since N_2 is increased (compare (4.5) and (4.7) below).

We now consider the following problem: Y_{ij} are independent and normally distributed random variables with unknown mean μ_i ($j = 1, \cdots, n_i$; $i = 1$,

$\cdots, u$) and variance σ^2 (we use a convenient notation for the example, rather than that introduced in Section 1). The problem is to test $H_0: \mu_1 = \mu_2 = \cdots = \mu_u = 0$, and a design d in Δ is a specification of nonnegative integers n_i whose sum is N. For any such d, we denote by $M(d)$ the set of i for which $n_i > 0$; by $k(d)$, the number of integers in $M(d)$; by τd, the design associated with the values $n_i = n^*_{\tau(i)}$ when d is associated with the values $n_i = n^*_i$, where τ is any element of the symmetric group S_u on u symbols; by δ_d, the design in Δ_R which assigns probability $1/u!$ to each τd for τ in S_u; by $f_{d,\alpha}$ the test associated with δ_d which is obtained by using the appropriate F-test of size α with whatever τd is chosen by δ_d. We shall also use the symbol $a_\phi(c)$ of (2.2), with $\psi(\mu) = \sum_1^u \mu_i^2$, and shall denote by a'_ϕ its derivative with respect to c at $c = 0$. We shall also use the symbols $g_{ij}(\alpha)$ introduced in (1.3). Our result, which implies that the "symmetrical" design associated with $k(d) = u$ and all n_i equal (or as nearly so as possible) is not L_α-optimum in Δ_R, and that the δ_d associated with the d for which $n_1 = N$ (this δ_d chooses each i with probability $1/u$ and takes all Y_{ij} with the chosen i) is locally best among the δ_d, is the following:

THEOREM 4.1. *For every d, α, and c,*

$$a_{F_{d,\alpha}}(c) \leqq a_{f_{d,\alpha}}(c); \tag{4.1}$$

$a'_{f_{d,\alpha}}$ is strictly decreasing in $k(d)$, and the same is true of $a_{f_{d,\alpha}}(c)$ for all c in some neighborhood of $c = 0$.

PROOF. (4.1) is trivial, and we proceed to the rest of the proof. The numerator $t'_d V_d^{-1} t_d$ of $F_{d,\alpha}$ is of course

$$U_d = \sum_{i \varepsilon M(d)} n_i \left(\sum_{j=1}^{n_i} Y_{ij}/n_i \right)^2,$$

and U_d/σ^2 has a χ^2 distribution with $k(d)$ d.f. and non-central parameter

$$\sum_{i \varepsilon M(d)} n_i \mu_i^2/\sigma^2.$$

The denominator of $F_{d,\alpha}$ has $N - k(d)$ d.f. Write $\lambda = \sum_1^u \mu_i^2/\sigma^2$. From (1.3) we have, as $\lambda \to 0$,

$$\begin{aligned} \beta_{f_{d,\alpha}}(\mu, \sigma^2) &= \sum_{\tau \varepsilon S_u} \beta_{F_{\tau d,\alpha}}(\mu, \sigma^2)/u! \\ &= \sum_{\tau \varepsilon S_u} [\alpha + g_{k(d),N-k(d)}(\alpha) \sum_{\tau \varepsilon S_u} n_{\tau(i)}\mu_i^2/\sigma^2 + 0(\lambda^2)]/u! \\ &= \alpha + g_{k(d),N-k(d)}(\alpha) \sum_i (\sum_\tau n_{\tau(i)}/u!)\mu_i^2/\sigma^2 + 0(\lambda^2) \\ &= \alpha + \frac{N}{u} g_{k(d),N-k(d)}(\alpha)\lambda + 0(\lambda^2). \end{aligned} \tag{4.2}$$

The desired conclusion now follows from Lemma 4.1.

Existing tables and charts of the power functions of the F-test and χ^2-test are presented in such forms (in terms of $\sqrt{\lambda/(k(d) + 1)}$, usually in inverted form and with wide spacing of arguments) as to make accurate comparisons of the

$\beta_{f_{d,\alpha}}$ difficult. This difficulty is made the worse by the fact that $\beta_{f_{d,\alpha}}$ is not (with an obvious exception) constant on the contour λ = constant, making it somewhat of a task to obtain $a_{f_{d,\alpha}}(c)$. It is not true, as might be supposed, that this minimum power on the contour λ = constant is always attained for a μ with all components equal, or else is always attained for a μ with all components except one equal to zero. To see this, consider the problem of Section 1 ($N = u = 2$, σ^2 known). Let C_α be the value such that, if Y is a normal random variable with 0 mean and unit variance, then $P\{ | Y | > C_\alpha\} = \alpha$. A direct computation of the power function of δ near $\lambda \equiv \mu_1^2 + \mu_2^2 = 0$ yields

$$\beta_\delta(\mu) = \alpha + \frac{C_\alpha \exp(-C_\alpha^2/2)}{2\sqrt{2\pi}} \cdot \{2(\mu_1^2 + \mu_2^2) + (C_\alpha^2 - 3)(\mu_1^4 + \mu_2^4)/3 + O(\lambda^3)\}. \tag{4.3}$$

Hence, when c is sufficiently small, the minimum of $\beta_\delta(\mu)$ on the contour $\lambda = c$, neglecting the term $O(\lambda^3)$, is located at $\mu_1 = \sqrt{c}$, $\mu_2 = 0$ (or $\mu_2 = \sqrt{c}$, $\mu_1 = 0$) if $C_\alpha \leqq \sqrt{3}$ and at $\mu_1 = \mu_2 = \sqrt{c/2}$ if $C_\alpha \geqq \sqrt{3}$. When we include terms of higher order in μ, it is no longer even evident that the minimum must be attained at one of these two values of μ.

We see from (4.3) that $g_{1,\infty}(\alpha) = (2\pi)^{-\frac{1}{2}} C_\alpha \exp(-C_\alpha^2/2)$ and it is not hard to show that $g_{2,\infty}(\alpha) = -\alpha (\log \alpha)/2$ (see [12], equation (6.27), where λ is our $\lambda/2$). Thus, a comparison of $a'_{f_{d,\alpha}}$ for $k(d) = 1$ and 2 is given in this example by the following table:

α	$g_{1,\infty}(\alpha)$	$g_{2,\infty}(\alpha)$
.01	.037	.023
.05	.114	.075
.10	.175	.115
.20	.225	.161
.30	.242	.181
.50	.214	.173
.90	.050	.047

The following lemma shows that, as $\alpha \to 0$, the ratio of the second to third column above goes to 2 and, more generally, that $g_{i,\infty}(\alpha)/g_{j,\infty}(\alpha) \to j/i$ (this gives a comparison of the various δ_d for general N and u and for various $k(d)$ when σ^2 is known, as $\alpha \to 0$; see Lemma 4.3 for the case when σ^2 is unknown):

LEMMA 4.2. *As* $\alpha \to 0$,

$$g_{j,\infty}(\alpha) = -[1 + o(1)]\alpha(\log \alpha)/j. \tag{4.5}$$

PROOF. Fix j. Let k_α be such that if Y is a random variable with central χ^2 distribution with j d.f., then $P\{Y > k_\alpha\} = \alpha$. Let f_λ be the χ^2 density function with j d.f. and non-central parameter λ. A simple calculation shows that $df_\lambda(u)/d\lambda$ at $\lambda = 0$ is $f_0(u)[(u/2j) - 1/2]$. Hence, as $k_\alpha \to \infty$,

$$g_{j,\infty}(\alpha) = \int_{k_\alpha}^{\infty} f_0(u)[(u/2j) - 1/2]\, du = 1 + o(1))f_0(k_\alpha)k_\alpha/j, \tag{4.6}$$

by partial integration. On the other hand, an integration by parts shows that

$\alpha = 2f_0(k_\alpha)[1 + o(1)]$ as $k_\alpha \to \infty$, and hence that $k_\alpha = -2[1 + o(1)] \log \alpha$. This completes the proof.

4B. CASE II. We again treat the case where σ^2 is unknown, the other case being handled similarly (mainly, use Lemma 4.2 for Lemma 4.3). We first prove two simple lemmas.

LEMMA 4.3. *As* $\alpha \to 0$,

$$g_{ji}(\alpha) = i\alpha/2j + o(\alpha). \tag{4.7}$$

(This does not contradict (4.5), since j is fixed in (4.7).)

PROOF: Fix j and i. Let h_α be such that if Y has a central F-distribution with j and i d.f., then $P\{Y > h_\alpha\} = \alpha$. Let G_λ be the F density function with j and i d.f. and non-central parameter λ. From [12], equation (6.29) (with λ there replaced by our $\lambda/2$), it is easy to compute that $dG_\lambda(u)/d\lambda$ at $\lambda = 0$ is $G_0(u)$ $[(j + i)\, u/j(1 + u) - 1]/2$. Hence, as $k_\alpha \to \infty$,

$$g_{ji}(\alpha) = \frac{1}{2}\int_{k_\alpha}^{\infty} G_0(u)\left[\frac{i}{j} - \frac{j+i}{j}\cdot\frac{1}{1+u}\right] du = i\alpha/2j + o(\alpha). \tag{4.8}$$

In the next lemma, we use the following notation: n_i ($i = 1, \cdots, u$) are again nonnegative integers with sum N. S_u is the symmetric group on u symbols and, for τ in S_u, $\bar{\mu}(\tau) = N^{-1} \sum_i n_{\tau(i)}\mu_i$; finally, $\bar{\mu} = u^{-1} \sum_i \mu_i$.

LEMMA 4.4. *For all* $u > 1$, μ, *and* N,

$$\sum_{\tau \varepsilon S_u} \sum_i n_{\tau(i)}(\mu_i - \bar{\mu}(\tau))^2 = u(u-2)![N - N^{-1}\sum_i n_i^2] \sum_i (\mu_i - \bar{\mu})^2. \tag{4.9}$$

PROOF. Since

$$\sum_{\tau \varepsilon S_u} n_{\tau(i)}^2 = (u-1)! \sum_i n_i^2 \tag{4.10}$$

and, for $i \neq j$,

$$\sum_{\tau \varepsilon S_u} n_{\tau(i)} n_{\tau(j)} = (u-2)! \sum_{i \neq j} n_i n_j = (u-2)![N^2 - \sum_i n_i^2], \tag{4.11}$$

we have

$$\begin{aligned} N^2 \sum_{\tau \varepsilon S_u} \bar{\mu}(\tau)^2 &= \sum_{i,j} \mu_i \mu_j \sum_{\tau \varepsilon S_u} n_{\tau(i)} n_{\tau(j)} \\ &= (u-1)! \sum_i n_i^2 \sum_j \mu_j^2 \\ &\quad + (u-2)![N^2 - \sum_i n_i^2][u^2\bar{\mu}^2 - \sum_j \mu_j^2]. \end{aligned} \tag{4.12}$$

Also,

$$\sum_{\tau \varepsilon S_u} \sum_i n_{\tau(i)} \mu_i^2 = \sum_i \mu_i^2 \sum_\tau n_{\tau(i)} = (u-1)! N \sum_i \mu_i^2. \tag{4.13}$$

Equations (4.12) and (4.13), together with

$$\sum_\tau \sum_i n_{\tau(i)}(\mu_i - \bar{\mu}(\tau))^2 = \sum_\tau \sum_i n_{\tau(i)} \mu_i^2 - N \sum_\tau \bar{\mu}(\tau)^2, \tag{4.14}$$

give (4.9).

The maximum for fixed $k(d)$ of the factor in square brackets on the right side of (4.9) will of course be nondecreasing in $k(d)$. It is the factor $g_{k(d)-1,h_d(\alpha)}$ which will increase rapidly enough as $k(d)$ is decreased to more than make up for the decrease in this term in brackets.

We are now ready to give our nonoptimality result in several illustrative examples of Case II, including those of Section 3C(1) and 3C(2). In all of these examples we ignore the divisibility properties; considerations when the design does not "divide up" properly (e.g., when $k(d)$ does not divide N in Example (1) below) are messier and their consideration does not help in the understanding of the phenomenon we are illustrating; thus, we shall assume whatever divisibility properties of N are needed to make our examples simple.

(1). *One-way analysis of variance.* In our first and simplest example, the setup is that of Section 4A, except that we now are testing $\mu_1 = \cdots = \mu_u$, and the appropriate F-tests are changed accordingly. Our result has the same implication as that stated just above Theorem 4.1, except that it now holds only when α is sufficiently small, and the optimum δ chooses each pair (i, j) $(i \neq j)$ with equal probability and sets $n_i = n_j = N/2$.

THEOREM 4.21. *For every d, α, and c, (4.1) holds; for fixed $k(d)$, $a'_{f_{d,\alpha}}$ is strictly decreasing in $\sum_i n_i^2$, attaining its maximum for $n_1 = \cdots = n_{k(d)} = N/k(d)$; for this choice of the n_i and for all α in some neighborhood of 0, $a'_{f_{d,\alpha}}$ is strictly decreasing in $k(d)$ for $k(d) > 1$; the results just stated for $a'_{f_{d,\alpha}}$ hold also for $a_{f_{d,\alpha}}(c)$ for all c in some neighborhood of 0.*

PROOF. From Lemma 4.4 and an argument like that of (4.2), we have, setting $\lambda = \sum_i(\mu_i - \bar{\mu})^2/\sigma^2$,

$$\beta_{f_{d,\alpha}}(\mu, \sigma^2) = \alpha + g_{k(d)-1,N-k(d)}(\alpha)(u-1)^{-1}(N - N^{-1}\sum_i n_i^2)\lambda + O(\lambda^2). \tag{4.15}$$

When $n_1 = \cdots = n_{k(d)} = N/k(d)$, the ratio of values of $a'_{f_{d,\alpha}}$ corresponding to two values k and k' of $k(d)$ with $1 < k < k'$ is thus

$$\frac{g_{k-1,N-k}(\alpha)(1 - 1/k)}{g_{k'-1,N-k'}(\alpha)(1 - 1/k')}; \tag{4.16}$$

as $\alpha \to 0$, by Lemma 4.3, this ratio approaches

$$\frac{(N-k)/k}{(N-k')/k'} > 1, \tag{4.17}$$

completing the proof.

For a numerical example, suppose $N = 6$, $u = 3$, with σ^2 known. Comparing the δ_d's for which $k = 2$ and $k' = 3$, we see that $(1 - 1/k)/(1 - 1/k') = \frac{3}{4}$; thus, the ratio of the two $a'_{f_{d,\alpha}}$ in this example is $\frac{3}{4}$ times the ratio of second to third column in the table above Lemma 4.2. For $\alpha < .3$, then, the design with $k(d) = 2$ is locally better than that with $k(d) = 3$, in this example.

(2). *Several-way analysis of variance.* With or without interactions, the considerations are very similar to those of Example (1), and we omit them.

(3). *One-way heterogeneity.* In the setting described in Section 3A, suppose for fixed b, k, and u that BBD's exist for two possible choices u_1 and u_2 of the "number of treatments" to be tested, say for u_1 and u_2 with $1 < u_1 < u_2 \leqq u$. Let $d_i(i = 1, 2)$ be the design which uses the BBD with parameters b, k, and u_i to test the hypothesis $\mu_1 = \cdots = \mu_{u_i}$, and let δ_{d_i} be the corresponding randomized design which replaces the subscripts $1, \cdots, u_i$ here by $\tau(1), \cdots, \tau(u_i)$ with probability $1/u!$ for each τ (or, which is the same thing, which chooses each of the possible subsets of u_i treatments with equal probability). Otherwise, we use the same notation as in Example (1) of this section.

For any design setting, the parameter of the non-central χ^2 variable $t_d' V_d^{-1} t_d/\sigma^2$ is $(Q_d R\mu)' V_d^{-1}(Q_d R\mu)$, and by Lemma 2.3 and equation (3.1) this reduces in the case of a BBD d^* with parameters b, k, and u to

$$[r_{d^*1} - (\lambda_{d^*11} - \lambda_{d^*12})/k] \sum_i (\mu_i - \bar{\mu})^2/\sigma^2. \tag{4.18}$$

For the sake of arithmetical simplicity only, suppose that k/u_i is either an integer or is < 1 (the phenomenon to be studied persists without this assumption). Then, for $d^* = d_i$, the term in square brackets in (4.18) is easily computed to be

$$f(u_i) = \begin{cases} b(k-1)/(u_i - 1) & \text{if } k/u_i \leqq 1, \\ bk/u_i & \text{if } k/u_i \geqq 1. \end{cases} \tag{4.19}$$

Using now the counterpart of (4.18) for the designs d_i and the fact that, for $n_1 = \cdots = n_{u_q} = 1$ and all other $n_j = 0$, (4.9) becomes

$$\sum_{\tau \varepsilon S_u} \sum_i n_{\tau(i)}(\mu_i - \bar{\mu}(\tau))^2/u! = (u-1)^{-1}(u_q - 1) \sum_{i=1} (\mu_i - \bar{\mu})^2, \tag{4.20}$$

we obtain, corresponding to (4.16),

$$\frac{a'_{f_{d_1,\alpha}}}{a'_{f_{d_2,\alpha}}} = \frac{g_{u_1-1, bk-u_1-b+1}(\alpha)(u_1 - 1)f(u_1)}{g_{u_2-1, bk-u_2-b+1}(\alpha)(u_2 - 1)f(u_2)}. \tag{4.21}$$

By Lemma 4.3, as $\alpha \to 0$ this ratio approaches

$$\frac{(bk - u_1 - b + 1)f(u_1)}{(bk - u_2 - b + 1)f(u_2)}. \tag{4.22}$$

It is trivial to verify that $(bk - u - b + 1)f(u)$ is strictly decreasing in u for $u > 1$, so that the expression of (4.22) is >1. Thus, we have proved

THEOREM 4.22. *For fixed b, k, u and all α in some neighborhood of* 0, $a'_{f_{d_i,\alpha}}$ *is strictly decreasing in u_i for $i > 1$; the same is true for $a_{f_{d_i,\alpha}}(c)$ for all c in some neighborhood of* 0.

This result implies that, if k is even, the locally best δ_{d_i} is that which chooses each pair of treatments with equal probability and assigns each of the two chosen treatments to $k/2$ of the plots in *every* block.

(4). *Two-way heterogeneity.* Using (3.2) in place of (3.1), the analogue of Theorem 4.22 can be proved for the YS design by an argument very similar to that of Example (3) just above, and which we therefore omit. One can even give an example of the lack of optimality of the YS in Δ_R without resorting to this analysis: for the case $k_1 = 2$, $k_2 = 3$, $u = 3$, the usual YS gives no d.f. to error, while the design which chooses two treatments at random and assigns each treatment to three plots, at least once in each row and column (full symmetry is impossible here) is uniformly more powerful for all α and all alternatives.

(5) *Other examples.* Examples like those mentioned in Section 3C (3) can be considered similarly, with analogous results. In particular, a trivial example in the case of a higher LS has already been mentioned in the first paragraph of Section 1.

5. Remarks and extensions. We list a few of the variants of the examples considered in this paper for which similar results hold, and make a few comments on questions which arise in connection with the paper, some of which present unanswered research problems.

1. A few of the other problems to which modifications of our method apply have been mentioned in Section 3C, and some of these will be considered elsewhere. Some such results hold under various non-normal probability laws (the point of the results of Section 4 is not merely that they hold for *many* models, but that they hold for the simplest, classical, normal model). Of course, a design which is optimum for one model may fail to be optimum for another, and vice versa; in particular, the results are obviously sensitive to change in the function ψ (even to changes to other quadratic forms and for a fixed d, as indicated in Section 2). Optimality criteria can be altered in other ways; e.g., one can consider $M_{\alpha,c,\sigma}$-optimality, in imitation of 2A(c). The extent of completeness of nonoptimality results like those on the higher LS design (first paragraph of Section 1) and YS design (Section 4B(4)) obviously depends on whether or not σ^2 is known. The results for Model II and certain mixed models of the analysis of variance differ considerably from those for the model considered herein, since the dependence of the power function on the design (and on the test, for a fixed design) is so different; however, similar methods can be used there.

2. Besides changing the model, one can also change the decision space. From the examples cited just above regarding higher LS and YS designs, it is clear that *nonoptimality* results for some classical symmetrical designs hold for many decision problems. For normal and certain nonparametric point estimation problems, the discussion of [2] and [3] indicates why Section 3 yields *optimality* results (these actually hold for many weight functions other than squared error). Another typical estimation result is contained in the fact that the designs d^* of Theorems 3.1 and 3.2 maximize the trace of V_d^{-1} and that V_{d^*} is a multiple of the identity; from these it follows at once that *average variance of* t_d (= trace of $\sigma^2 V_d/(u-1)$) *is a minimum for* d^*. However, the results of Section 4 are meaningless for many common weight functions, since V_δ is not the covariance matrix of b.l.e.'s. Similar results hold for some interval estimation problems; for estimating $\psi(\mu)/\sigma^2$ (e.g., in "multiple comparison" problems), Section 4 is now sometimes relevant. Multiple classification and ranking problems can be treated in like

manner. Of course, a D-optimum design minimizes the approxiate *generalized variance* in point estimation problems.

3. As we have mentioned, nonoptimality results like those of Section 4 do not depend on the nonrandomized design being symmetrical. Much more difficult is the problem of characterizing optimum designs in the sense of Section 3 when there is no appropriate symmetry. (Even the considerations of Sections 3B(2) and 4B(3 and 4) become messier without the restrictions on k_i/u and k/u; it would be nice if neat proofs could be given in such cases.) It seems often to be true that a design which is "closest to being symmetrical" in an appropriate sense (e.g., note the dependence on $\sum n_i^2$ in Theorem 4.21) is optimum, but the algebra involved in proving this can be tedious. Problems like that cited in the next to last paragraph of Section 3C(3) can be similarly unwieldy under heteroscedasticity. In connection with a general symmetry-invariance approach like that mentioned below (1.3), we note that appropriate symmetry of X_d is useful as a partial sufficient condition for some optimality results, but that appropriate symmetry of $X_d'X_d$ is what is really relevant (for the functions ψ we have considered).

4. We have mentioned in Section 2 some of the difficulties present in verifying M- (or sometimes L-) optimality. If b_d is not a constant for d in Δ', or if randomized designs are considered, this difficulty is increased by the nonconstancy of the d.f. for $\bar{S}_d$, etc. (We have not considered here a thorough investigation of the optimality properties of the procedures δ_d of Section 4). The difficulty encountered in connection with M-optimality in the nonconstancy of the power functions of competing tests on appropriate contours also manifests itself when one tries to find a *most stringent* design (the "envelope power function" being obtained by taking the supremum of β_ϕ over all ϕ in $H_d(\alpha)$ and all d in Δ or Δ_R). The method of invariance used to prove 2A(f) cannot even supply a start here, and the method of [6] or [7] used to prove 2A(c) yields no analogue here where d is not fixed. Thus, even in such a simple example as that of Section 2B, the stringency problem seems extremely difficult.

It is interesting to note that the δ_d of Section 4 lack a "consistency" property if $k(d) < r$, in that $a_{f_{d,\alpha}}(c)$ does not approach 1 as $c \to \infty$ (in fact, it is easy to see that the μ for which one component of $R\mu$ is $\sigma\sqrt{c}$ and all others are 0 is asymptotically worst on the contour $\psi(u)/\sigma^2 = c$ as $c \to \infty$, giving power approaching $[k(d) + (r - k(d))\alpha]/r$). Nevertheless, the question remains open as to whether any of these δ_d, or some other design and associated test which lacks this consistency property, is nevertheless most stringent.

The reader will not find it difficult in considerations like those of Section 3B to supply the details which show, in some problems, that the D-optimum (or L- or E-optimum) design is unique. When uniqueness is not present (e.g., for some α and ϵ, both designs in Section 2B will be L-optimum), questions of global admissibility arise. A related problem is to look not at a fixed contour or family of contours in the manner of Section 2, but rather to characterize complete classes of designs in the manner of [3]; in such considerations, especially for problems of testing hypotheses, Section 4 shows that results like those of [3] must be altered if Δ_R is considered rather than Δ.

Finally, we may remark that, for a fixed d, the problem of characterizing an L_α-optimum test is unsolved; the generalized Neyman-Pearson Lemma does not seem to yield explicit results easily, although it is not difficult to show that an L_α-optimum test is obtained by replacing the numerator of the F-test by some other quadratic form.

REFERENCES

[1] A. WALD, "On the efficient design of statistical investigations," *Ann. Math. Stat*, Vol. 14 (1943), pp. 134–140.

[2] S. EHRENFELD, "On the efficiency of experimental designs," *Ann. Math. Stat.*, Vol. 26 (1953), pp. 247–255.

[3] S. EHRENFELD, "Complete class theorems in experimental design," *Third Berkeley Symposium on Probability and Statistics*, Vol. 1, University of California Press, 1946.

[4] R. C. BOSE, "Least Squares Aspects of Analysis of Variance" (mimeographed notes), Institute of Statistics, North Carolina.

[5] E. L. LEHMANN, "Notes on Testing Hypotheses" (mimeographed), Associated Student Store, University of California, Berkeley.

[6] A. WALD, "On the power function of the analysis of variance test," *Ann. Math. Stat.*, Vol. 13 (1942), pp. 434–439, and Vol. 15 (1944), pp. 330–333.

[7] J. WOLFOWITZ, "The power of the classical tests associated with the normal distribution," *Ann. Math. Stat.*, Vol. 20 (1949), pp. 540–551.

[8] P. L. HSU, "Analysis of variance from the power function standpoint," *Biometrika*, Vol. 32 (1941), p. 62.

[9] G. A. HUNT AND C. STEIN, "Invariant tests," unpublished.

[10] J. KIEFER, "Invariance, sequential minimax estimation, and continuous time processes," *Ann. Math. Stat.*, Vol. 28 (1957).

[11] S. ISAACSON, "On the theory of unbiased tests of simple statistical hypotheses specifying the values of two or more parameters," *Ann. Math. Stat.*, Vol. 22 (1951), pp. 217–234.

[12] H. B. MANN, *Analysis and Design of Experiments*, Dover, New York, 1949.

Corrections (and Comments) to "On the Nonran.. Opt'y..."

by J. Kiefer, Ann. Math. Stat. 29, 1958, pp. 675-699

p. 694, line below (4.7): i for j.

p. 681, above "E-optimality" insert

"A-optimality: $\operatorname{tr} \bar{V}_{d^*} = \min_{d \in \Delta'} \operatorname{tr} \bar{V}_d$."

p. 684, Lemma 2.2. This is correct, but a more unified approach is given on p. 112 of the 1971 Gupta volume paper, according to which

$$\left\{ \begin{array}{ll} \text{(i)} & C_{d^*} \text{ of the form } a\mathbf{I}+b\mathbf{J} \\ & (\mathbf{J} \text{ is a matrix of 1's}) \\ \text{(ii)} & \max_{d \in \Delta'} \operatorname{tr} C_d = \operatorname{tr} C_{d^*} \end{array} \right\} \Rightarrow d^* \text{ optimum (minimizing) for}$$

functions $\Phi(V_d^{-1})$ which are convex, symmetric, and for which $\Phi(bC) \leq \Phi(C)$ for $b > 1$ and C pos. def. This yields $\{(i),(ii)\} \Rightarrow$ A-, D-, E- optimality (and, if also $N\text{-}b_d$ is max'd by d*, L-opt'y). In a forthcoming paper (ii) will be altered (as will the conclusion) to treat cases where a sym. design may not be optimum for all such Φ.

Lemma 2.2 itself can be generalized; e.g., $\{$(i), d*D-opt.$\}$ $\Rightarrow$ $\{$(i), d*A-opt.$\}$ $\Rightarrow$ $\{$(i), d*E-opt.$\}$.

p. 687, Lemma 3.2. The lemma is correct but the proof as given is wrong. The error is that the extreme points of B are not necessarily of the form (3.6); for example, when $u = 4$,

$$\begin{bmatrix} 0 & 1/4 & 1/2 \\ 1/4 & 0 & 1/2 \\ 0 & 3/4 & 0 \\ 3/4 & 0 & 0 \end{bmatrix}$$

is extreme. However, a correct proof is much more elementary:

$$\Pi_{j=1}^{u-1} d_{jj}^{1/(u-1)} \leq \Sigma_{j=1}^{u-1} \frac{e_{ij} d_{jj}}{(\frac{u-1}{u})} = \frac{u}{u-1} k_{ii} \ \forall i$$

by the geom.-arithm-mean ineq.; and taking $\Pi_{i=1}^{u}$ of both sides $\Rightarrow$ (3.3).

<u>MOREOVER</u>, the use of Lemmas 3.1-3.3 is better replaced by the treatment indicated just above (commenting on Lemma 2.2). (Lemma 3.4 is of separate use, for cases where (ii) is false but d^* is E-optimum.)

p. 689, <u>Theorems 3.1-3.2</u>: Insert "A-optimum" in statements and shorten treatment by replacing ref. to Lemmas 2.2 and 3.1 by the unified approach given above.

p. 690 contains some unnecessary repetitions. Replace the last Σ of (3.12) by $\Sigma_i k_1 (\lambda_{d\ ii}^{(1)})^2$, and replace line, "We shall..." through line 19 (end of ∯) by "This is minimized by d^*, as we found in the discussion of h on the previous page."

p. 687, line 10, $\bar{o}_{ij}$ should be $\bar{o}_{1j}$.

Reprinted from THE ANNALS OF MATHEMATICAL STATISTICS
Vol. 30, No. 2, June, 1959
Printed in U.S.A.

OPTIMUM DESIGNS IN REGRESSION PROBLEMS

BY J. KIEFER[1] AND J. WOLFOWITZ[2]

Cornell University

1. Introduction and Summary. Although regression problems have been considered by workers in all sciences for many years, until recently relatively little attention has been paid to the optimum design of experiments in such problems. At what values of the independent variable should one take observations, and in what proportions? The purpose of this paper is to develop useful computational procedures for finding optimum designs in regression problems of estimation, testing hypotheses, etc. In Section 2 we shall develop the theory for the case where the desired inference concerns just one of the regression coefficients, and illustrative examples will be given in Section 3. In Section 4 the theory for the case of inference on several coefficients is developed; here there is a choice of several possible optimality criteria, as discussed in [1]. In Section 5 we treat the problem of global estimation of the regression *function*, rather than of the individual coefficients.

We shall now indicate briefly some of the computational aspects of the search for optimum designs by considering the problem of Section 2 wherein the inference concerns one of k regression coefficients. For the sake of concreteness, we shall occasionally refer here to the example of polynomial regression on the real interval $[-1, 1]$, where all observations are independent and have the same variance. The quadratic case is rather trivial to treat by our methods, so we shall sometimes refer here to the case of cubic regression. In the latter case we suppose all four regression coefficients to be unknown, and we want to estimate or test a hypothesis about the coefficient a_3 of x^3. If a fixed number N of observations is to be taken, we can think of representing the proportion of observations taken at any point x by $\xi(x)$, where ξ is a probability measure on $[-1, 1]$. To a first approximation (which is discussed in Section 2), we can ignore the fact that in what follows $N\xi$ can take only integer values. We consider three methods of attacking the problem of finding an optimum ξ:

A. The direct approach is to compute the variance of the best linear estimator of a_3 as a function of the values of the independent variable at which observations are taken or, equivalently, as a function of the moments of ξ. Denoting by μ_i the ith moment of ξ, and assuming ξ to be concentrated entirely on more than three points (so that a_3 is estimable), we find easily that the *reciprocal* of

Received April 21, 1958; revised November 25, 1958.

[1] Research under contract with the Office of Naval Research.

[2] The research of this author was supported by the U. S. Air Force under Contract No. AF 18(600)-685, monitored by the Office of Scientific Research.

this variance is proportional to

$$\frac{\begin{array}{r}\mu_5^2(\mu_1^2 - \mu_2) + 2\mu_5(\mu_2^2\mu_3 + \mu_3\mu_4 - \mu_1\mu_3^2 - \mu_1\mu_2\mu_4) \\ - \mu_4^3 + \mu_4^2(\mu_2^2 + 2\mu_1\mu_3) - 3\mu_4\mu_2\mu_3^2 + \mu_3^4\end{array}}{\mu_4(\mu_2 - \mu_1^2) - \mu_3^2 - \mu_2^3 + 2\mu_1\mu_2\mu_3} + \mu_6$$

in the case of cubic regression.

The problem is to find a ξ on $[-1, 1]$ which maximizes this expression. Thus, this direct approach leads to a calculation which appears quite formidable. This is true even if one uses the remark on symmetry of the next paragraph and restricts attention to symmetrical ξ, so that $\mu_i = 0$ for i odd. For polynomials of higher degree or for regression functions which are not polynomials, the difficulties are greater.

B. The results of Section 2 yield the following approach to the problem: Let $c_0 + c_1x + c_2x^2$ be a best Chebyshev approximation to x^3 on $[-1, 1]$, i.e., such that the maximum over $[-1, 1]$ of $|x^3 - (c_0 + c_1x + c_2x^2)|$ is a minimum over all choices of the c_i, and suppose B is the subset of $[-1, 1]$ where the maximum of this absolute value is taken on. Then ξ must give measure one to B, and the weights assigned by ξ to the various points of B (there are four in this case) can be found either by solving the *linear* equations (2.10) or by computing these weights so as to make ξ a maximin strategy for the game discussed in Section 2. Two points should be mentioned:

(1) In the general polynomial case, where there are k parameters ($k = 4$ here), the results described in [10], p. 42, or in Section 2 below imply that there is an optimum ξ concentrated on at most k points. Thus, even if we use this result with the approach of the previous paragraph, we obtain the following comparison in a k-parameter problem in Section 2:

Method A: minimize a nonlinear function of $2k - 1$ real variables.

Method B: solve the Chebyshev problem and then solve $k - 1$ simultaneous *linear* equations.

The fact that the solution of the Chebyshev problem can often be found in the literature (e.g., [2]) makes the comparison of the second method with the first all the more favorable.

(2) Although the computational difficulty cannot in general be reduced further, in the case of polynomial regression on $[-1, 1]$ there is present a kind of symmetry (discussed in Section 2) which implies that there is an optimum ξ which is symmetrical about 0 and which is concentrated on four points; thus, in the case of cubic regression, this fact reduces the computation under Method A to a minimization in 3 variables, but Method B involves only the solution of a single linear equation.

C. A third method, which rests on the game-theoretic results of Section 2, and which is especially useful when one has a reasonable guess of what an optimum ξ is, involves the following steps: first guess a ξ, say ξ^*, and compute the minimum on the left side of (2.8); second, if this minimum is achieved for $c = c^*$, compute the square of the maximum on the right side of (2.9); then, if these two computations yield the same number, ξ^* is optimum. If one has a guess of a class of ξ's depending on one or several parameters, among which it is thought that there is an optimum ξ, then one can maximize over that class at the end of the first step and, the maximum being at ξ^*, go through the same analysis as above. This method is illustrated in Example 3.5 and Example 4. Of course, the remarks (1) and (2) of the previous paragraph can be used in applying Method C, as in these examples.

In the example of cubic regression just cited, the optimum procedure turns out to be $\xi(-1) = \xi(1) = \frac{1}{6}$, $\xi(\frac{1}{2}) = \xi(-\frac{1}{2}) = \frac{1}{3}$. It is striking that any of the commonly used procedures which take equal numbers of observations at equally spaced points on $[-1, 1]$ requires over 38% more observations than this optimum procedure in order to yield the same variance for the best linear estimator of a_3 (see Example 3.1); the comparison is even more striking for higher degree regression. The unique optimum procedure in the case of degree h is given by (3.3).

The comparison of a direct computational attack, analogous to that of A above, with the methods developed in Sections 4 and 5 for the problems considered there, indicates even more the inferiority of the direct attack. In particular cases, e.g., Example 5.1, special methods may prove useful.

Among recent work in the design of experiments we may mention the papers of Elfving [3], [4], Chernoff [5], Williams [11], Ehrenfeld [12], Guest [13], and Hoel [15]. Only Guest and Hoel explicitly consider computational problems of the kind discussed below. Our methods of employing Chebyshev and game theoretic results seem to be completely new. The results obtained in the examples below are also new, except for some slight overlap with results of [13] and [15], which is explicitly described below.

We shall consider elsewhere some further problems of the type considered in this paper.

2. The optimum design relative to 1 out of k regression coefficients. Let $f_1, \cdots, f_k$ be k real-valued functions on a given space $\mathfrak{X}$. Throughout this section we assume a topology is given on $\mathfrak{X}$ in which

$$\mathfrak{X} \text{ is compact}; f_1, \cdots, f_k \text{ are continuous.} \tag{2.1}$$

We also assume

$$f_1, \cdots, f_k \text{ are linearly independent on } \mathfrak{X}. \tag{2.2}$$

Since we will be considering a regression problem in which the f_i are known functions and $\sum_i a_i f_i$ is the regression function, (2.2) is really only an assumption of identifiability of the a_i which will avoid trivial circumlocutions. Without some assumption like the first part of (2.1), there may trivially exist procedures which estimate some of the regression coefficients with arbitrarily small variance, as can be seen in the example of estimation of the slope of a straight line on $\mathfrak{X}$ = real line. The assumption of continuity of the f_i can be somewhat weakened, as will be clear from our proofs.

We consider the following regression setup: For any point x (value of the independent variable) in $\mathfrak{X}$, one can observe a random variable Y_x for which

$$EY_x = \sum_1^k a_i f_i(x), \qquad \text{Var}(Y_x) = \sigma^2, \tag{2.3}$$

where $a = (a_1, \cdots, a_k)$ is the vector of regression coefficients, an unknown element of $\mathfrak{A}$. The value of σ^2 will usually be unknown. (The case where σ^2 can depend on x in a way which is known except for a proportionality constant will be discussed in the last paragraph of this section.) An integer n is given (usually $n > k$), and the experimenter must select a collection $\mathbf{X} = (x_1, \cdots, x_n)$ of n points in $\mathfrak{X}$ at which the independent random variables $Y_{x_1}, \cdots, Y_{x_n}$ are to be observed. The x_i need not be distinct, but if $i \neq j$ and $x_i = x_j$ we shall still, without confusion, write Y_{x_i} and Y_{x_j} for two *independent* random variables.

Any X can be viewed as a measure η on $\mathfrak{X}$ which assigns to each point x a mass equal to the number of x_i in X which are equal to x. Dividing this measure by n, we obtain a discrete probability measure ξ on $\mathfrak{X}$ which assigns to each point of $\mathfrak{X}$ a measure equal to an integral multiple of $1/n$. In the present section (a similar discussion applying in Sections 4 and 5), we shall be concerned with choosing a ξ (hence, an X) to maximize a quantity of the form

$$\min_c \int_{\mathfrak{X}} H_c(x)\eta(dx) = \min_c \int_{\mathfrak{X}} nH_c(x)\xi(dx), \tag{2.4}$$

where the form of H_c is determined by the problem at hand. The fact that ξ can only take on multiples of $1/n$ as its values makes this problem of maximization quite unwieldy in general. We shall treat, instead, a problem whose solution will sometimes give a solution to the original problem and which will usually give a good approximation to the latter: *Find a probability measure ξ^* on $\mathfrak{X}$ for which the right side of* (2.4) *is a maximum*: i.e., we maximize (2.4) with no restriction on ξ. Of course, the maximum does not depend on n. Thus, if n is such that $n\xi^*$ takes on only integral values, this yields an exact solution η to the original problem. We shall see in Sections 3 and 5 that, in two typical examples, ξ^* takes on only values which are multiples of $1/(2k - 2)$ (Example 3.1) or $1/k$ (Example 5.1), so that this situation is not vacuous. Moreover, there will typically be a ξ^* which is concentrated on approximately k points; thus, when $n\xi^*$ does not take on only integral values, obvious integral approximations η' to $n\xi^*$ will yield values of (2.4) whose ratio to the maximum tends to 1 as $n \to \infty$ (it is easy to give a bound on the difference of this ratio from unity). Thus, the characterization of a single ξ^* which yields an *almost* optimum design for all large n, in distinction to finding the best ξ which may depend in a complicated fashion on n, seems to be of practical value.

We therefore define Ξ to be the space of all discrete probability measures ξ on $\mathfrak{X}$. We could, more generally, specify a Borel field $\mathfrak{B}$ on $\mathfrak{X}$ and let Ξ be the class of all measures ($\mathcal{B}$) on $\mathfrak{X}$; however, in all of our applications (see Theorem 2) it will suffice to let $\mathcal{B}$ consist of countable sets and their complements.

In the present section we are concerned with statistical inference about the single parameter a_k, where all a_i are assumed unknown. We shall give a precise definition of optimality in the next paragraph. What this definition means is that we restrict ourselves to designs for which a_k is estimable (i.e., for which there exist linear unbiased estimators of a_k; in practice, of course, n will have to be suitably large for there to exist such designs), and seek a design for which the linear unbiased estimator of a_k with minimum variance (best linear estimator, or b.l.e.) has a variance which is a minimum over all designs, within the approximation noted two paragraphs above. It is well known that such a design is optimum for problems of point estimation of a_k if the Y_x are assumed to be normal, in the sense that (for example) it yields a minimax procedure for any of a wide variety of weight functions; when the distributions of Y_x are assumed to belong to any larger class, the same result holds for the squared error loss function. For problems of interval estimation and hypothesis testing or m decisions, similar optimality results hold under normality if σ^2 is known. If σ^2 is unknown, such results hold provided every design for which a_k is estimable yields as many degrees of freedom to error as does the design we obtain; see Example 3.4 in Section 3 for further discussion.

We now define precisely the term "optimum" as used in this section. There are a few preliminaries. In the original description of a design, let X be a design for which a_k is estimable. Let $h_1, \cdots, h_{k-1}$ be numbers such that the function $f_k^* = f_k - \sum_1^{k-1} h_j f_j$ is orthogonal to f_i for $i < k$ in the sense that

$$\sum_{r=1}^{n} f_i(x_r) f_k^*(x_r) = 0, \qquad i < k. \tag{2.5}$$

Let $a^* = (a_1^*, \cdots, a_k^*)$ be such that $\sum_1^k a_i f_i = \sum_1^{k-1} a_i^* f_i + a_k^* f_k^*$; thus, $a_k^* = a_k$. For the least squares setup in terms of a^*, the orthogonality of f_k^* to the f_i for $i < k$ makes the last of the normal equations

$$\sum_{r=1}^{n} [f_k^*(x_r)]^2 a_k^* = \sum_{r=1}^{n} f_k^*(x_r) Y_{x_r}, \tag{2.6}$$

so that σ^2 times the *reciprocal* of the variance of the b.l.e. of $a_k^* = a_k$ is $\sum_r [f_k^*(x_r)]^2$. Since f_k^* is orthogonal to $f_1, \cdots, f_{k-1}$, this last sum is just the square of the distance of the n-vector $(f_k(x_1), \cdots, f_k(x_n))$ from the linear space spanned by the vectors $(f_i(x_1), \cdots, f_i(x_n))$ for $i < k$, namely,

$$\min_c \sum_r [f_k(x_r) - \sum_{j=1}^{k-1} c_j f_j(x_r)]^2, \tag{2.7}$$

where we have written c for $(c_1, \cdots, c_{k-1})$. Since (2.7) is σ^2 times the inverse of the variance of the b.l.e. of a_k, a design X will minimize that variance if it maximizes (2.7). Thus, finally, in terms of the probability measures ξ we have introduced above, we make the following

DEFINITION. *A measure ξ^* in Ξ is said to be an optimum design (for the pa-*

rameter a_k) if

$$\min_c \int [f_k(x) - \textstyle\sum_1^{k-1} c_j f_j(x)]^2 \xi^*(dx) \tag{2.8}$$
$$= \max_{\xi \epsilon \Xi} \min_c \int [f_k(x) - \textstyle\sum_1^{k-1} c_j f_j(x)]^2 \xi(dx).$$

For any ξ in Ξ, the ratio of the left side of (2.8) (*with ξ for ξ^**) *to the right will be called the efficiency $e(\xi)$ of ξ.*

Of course, the practical meaning of efficiency is that, if one design has r times the efficiency of the second design, then the latter requires r times as many observations as the former in order to obtain the same value for the left side of (2.4). We note that it is a consequence of this definition that an optimum design is optimum for all values of σ^2.

The form of (2.8) is very suggestive of a game, and we shall exploit that fact presently. However, the main aspect of our technique for computing an optimum ξ^* has nothing to do with the game formulation, so we treat that aspect first. Our technique is to throw the main computational difficulties into a Chebyshev approximation problem, which can often be solved by standard methods and which, for many important $\{f_i\}$, even has a solution which can be found in the literature. We shall call $c^* = (c_1^*, \cdots, c_{k-1}^*)$ a *Chebyshev coefficient vector* if $\sum_1^{k-1} c_j^* f_j$ is a best approximation to f_k on $\mathfrak{X}$ in the sense of Chebyshev, i.e., in the uniform norm:

$$\min_c \max_{x \epsilon \mathfrak{X}} |f_k(x) - \sum_1^{k-1} c_j f_j(x)| = \max_{x \epsilon \mathfrak{X}} |f_k(x) - \sum_1^{k-1} c_j^* f_j(x)|. \tag{2.9}$$

Let $m(c^*)$ denote the right side of (2.9), and let $B(c^*)$ be the set of points x for which $| f_k(x) - \sum_1^{k-1} c_j^* f_j(x)| = m(c^*)$. Our first result gives a simple geometric sufficient condition for a ξ to be optimum; this is valid even without the conditions that yield the game-theoretic results of Theorem 2.

THEOREM 1. *If c^* is Chebyshev and $\xi(B(c^*)) = 1$ and*

$$\int [f_k(x) - \textstyle\sum_1^{k-1} c_j^* f_j(x)] f_i(x) \xi(dx) = 0 \tag{2.10}$$

for $i < k$, then ξ is optimum.

PROOF: According to (2.10), $\sum_1^{k-1} c_j^* f_j$ is the projection relative to ξ of f_k on the linear space spanned by $f_1, \cdots, f_{k-1}$. Hence, for any element ξ' of Ξ,

$$\min_c \int [f_k(x) - \textstyle\sum_1^{k-1} c_j f_j(x)]^2 \xi(dx)$$
$$= \int [f_k(x) - \textstyle\sum_1^{k-1} c_j^* f_j(x)]^2 \xi(dx) \tag{2.11}$$
$$= [m(c^*)]^2 \geqq \int [f_k(x) - \textstyle\sum_1^{k-1} c_j^* f_j(x)]^2 \xi'(dx)$$
$$\geqq \min_c \int [f_k(x) - \textstyle\sum_1^{k-1} c_j f_j(x)]^2 \xi'(dx),$$

which proves the desired result.

The question arises as to whether there always exists a ξ which satisfies the hypotheses of Theorem 1 and whether, in fact, the conditions of the theorem are also *necessary* for a ξ to be optimum. There also arises the question of whether we can find a useful bound such that there is an optimum ξ which assigns positive probability to at most the number of points given by this bound. These questions can be answered directly algebraically, but since the results we require already appear in the literature in connection with the analysis of certain games, we shall therefore consider the following *zero-sum two-person game associated with the design problem*: player 1 (resp., 2) has $\mathfrak{X}$ (resp., C = Euclidean $(k-1)$-space) as his space of pure strategies; the payoff function is $K(x, c) = [f_k(x) - \sum_1^{k-1} c_j f_j(x)]^2$; the space of mixed strategies of player 1 is Ξ, while that of player 2 is immaterial, since the convexity of K in C implies, according to Jensen's inequality, that for any randomized strategy of player 2 there is a nonrandomized strategy which is at least as good for all x. Of course the important thing is that an optimum (maximin) strategy for player 1 represents an optimum design. We now state the simple modifications of certain results of [6] which we require.

LEMMA. *The game of Ξ vs. C is determined, player 2 has a nonrandomized minimax strategy c^*, and player 1 has a maximin strategy ξ^* which is concentrated on at most $k - p$ points, where p is the dimensionality of the convex set of nonrandomized minimax strategies of player 2.*

PROOF: Let C_N be the set of all c for which $c'c = \sum_1^{k-1} c_i^2 \leqq N^2$, and let $\bar{C}_N$ be the complement of C_N. Since the f_i are linearly independent, there is a finite subset H of $\mathfrak{X}$ such that, for every c with $c'c = 1$, $\sum_1^{k-1} c_i f_i(x)$ is nonzero for at least one x in H. Hence, if ξ' assigns positive probability to each x in H, we clearly have $|\sum_1^{k-1} c_i \sum_{x \epsilon H} f_i(x)\xi'(x)| > \epsilon > 0$ for all c such that $c'c = 1$, and thus this absolute value is $> N\epsilon$ for $c'c = N^2$. Since f_k is bounded, we conclude that $\inf_{c \epsilon \bar{C}_N} K(\xi', c) \to \infty$ as $N \to \infty$. Hence, there is an N' such that for any c in $\bar{C}_{N'}$ there is a c' in $C_{N'}$ with $\sup_\xi K(\xi, c') < \sup_\xi K(\xi, c)$. Thus c^* is minimax if and only if $c^* \epsilon C_{N'}$ and c^* is minimax when the space of player 2 is restricted to $C_{N'}$. Since C_N is compact and K is continuous, the game of Ξ vs. C_N is determined, and there exists, for all $N > N'$, a minimax strategy c^* which we can take to be a fixed member of $C_{N'}$. Let p be the dimension of the (convex) set of such minimax strategies in $C_{N'}$. There also exists a maximin strategy ξ_N^* for the game of Ξ vs. C_N, and by [6] we can for $N > N'$ take ξ_N^* to be concentrated on at most $k - p$ points. Let $\xi_j = [(j-1)\xi_j^* + \xi']/j$. Clearly, for each j there is an N_j such that $K(\xi_j, c^*) < K(\xi_j, c)$ for all c in $\bar{C}_{N_j}$. Thus, since ξ_j^* is maximal with respect to c^*, we have, for $N_j > N'$,

$$\begin{aligned} \sup_{\xi \epsilon \Xi} \inf_{c \epsilon C} K(\xi, c) &\geqq \inf_{c \epsilon C} K(\xi_j, c) = \inf_{c \epsilon C_{N_j}} K(\xi_j, c) \\ &\geqq \left(1 - \frac{1}{j}\right) \inf_{c \epsilon C_{N_j}} K(\xi_j^*, c) = \left(1 - \frac{1}{j}\right) K(\xi_j^*, c^*) \qquad (2.12) \\ &= \left(1 - \frac{1}{j}\right) \sup_\xi K(\xi, c^*) \geqq \left(1 - \frac{1}{j}\right) \inf_{c \epsilon C} \sup_\xi K(\xi, c). \end{aligned}$$

Letting $j \to \infty$, we see that the game of Ξ vs. C is determined, that c^* is minimax, and that if $\{\xi^*_{j_i}\}$ is a subsequence of the $\{\xi^*_j\}$ which converges to a limit ξ^* which is concentrated on no more than $k - p$ points (such a subsequence and limit exist, by the compactness of $\mathfrak{X}$) and c'' minimizes $K(\xi^*, c)$, we have

$$(2.13)\qquad \begin{aligned} \sup_c \inf_{c \epsilon C} K(\xi, c) &= \lim_{i\to\infty} \inf_{c \epsilon C} K(\xi_{j_i}, c) \leq \lim_{i\to\infty} K(\xi_{j_i}, c'') \\ &= K(\xi^*, c'') = \inf_{c \epsilon C} K(\xi^*, c), \end{aligned}$$

so that ξ^* is maximin. Thus, the lemma is proved.

We mention in passing several other related points: The bound $k - p$ is indicated in [6] not to be the best possible and is reduced under conditions (c^* in the boundary of a compact C) for which it is difficult to find general counterparts here. Also, it is evident that c^* is unique ($p = 0$) if $K(x, c)$ is strictly convex in c, but strict convexity is clearly not a useful condition in our problem. If $\mathfrak{X}$ is not compact or the f_j are not continuous, suitable assumptions will still imply determinateness, but the other results will have to be stated in terms of ϵ-optimum strategies.

The above lemma indicates one method for trying to compute a ξ^*: For simplicity, assume $p = 0$ or that we have no knowledge of p. The ξ's on $\mathfrak{X}$ which are concentrated on at most k points form a $(2k - 1)$-parameter family. One can thus, in principle, maximize $\min_c K(\xi, c)$ with respect to these $2k - 1$ parameters and obtain an optimum ξ^*. As we have indicated in the introduction, this is usually an unrewarding task, and the method indicated in Theorem 1 seems far superior in practical examples. The consequences of the lemma for the method of Theorem 1 may be summarized as follows:

THEOREM 2. *If ξ is maximal with respect to c^* while c^* is minimal with respect to ξ, then ξ is optimum and c^* is Chebyshev. Every optimum ξ satisfies the conditions of Theorem* 1 *for every Chebyshev c^*. There exists an optimum ξ concentrated on at most $k - p$ points, where p is the dimensionality of the Chebyshev vectors.*

PROOF: The Chebyshev vectors clearly coincide with the minimax strategies. If ξ is maximin and c^* is minimax, then determinateness implies that c^* is minimal with respect to ξ, i.e., $\min_c K(\xi, c) = K(\xi, c^*)$. Thus, $\sum_1^{k-1} c_j^* f_j$ is the projection, relative to ξ, of f_k on the linear space spanned by $f_1, \cdots, f_{k-1}$, so that (2.10) clearly holds. Since, by (2.11), max min $K(\xi, c) = [m(c^*)]^2$ is the value of the game, $\xi(B_{c_*}) = 1$. The last assertion of the theorem is taken directly from the lemma, while the first is a general result in the theory of games. We note that any optimum ξ must give measure one to the intersection of all $B(c^*)$ for c^* Chebyshev.

We have mentioned, in Theorem 1 and in the second paragraph below the proof of the lemma, two computational approaches. The first sentence of Theorem 2 indicates a useful approach if one can make a good guess of ξ: guess a ξ' and compute $\min_c K(\xi', c) = K(\xi', c')$ (say); compute $\max_x K(x, c')$; if these two are equal, then ξ' is optimum. This is an approach which is standard in game theory and which has proved useful in many examples; it sometimes helps to let ξ' depend on a few parameters, with respect to which one maximizes

$\min_c K(\xi', c)$. A comparison of the various methods for obtaining an optimum ξ was given in an example in Section 1.

In the next section we shall give several examples of the computation of optimum ξ's. We shall not bother to list in detail all of the standard results in approximation theory which are useful in such computations. We mention here for future reference only the classical generalized Chebyshev theorem [2, p. 74], which states that if $\mathfrak{X}$ is a compact real interval and if no nontrivial linear combination of $f_1, \cdots, f_{k-1}$ has more than $k - 2$ zeros (in this case, these f_i are called a *Chebyshev system*), then the Chebyshev vector c^* is unique and is characterized by the fact that there are at least k points at which $f_k - \sum_1^{k-1} c_j^* f_j$ attains its maximum absolute deviation from zero, the maximum being taken on with successive alternations in sign. (The literature contains generalizations of this result to other spaces.)

Before proceeding further it is relevant here to point out the following connections with earlier results:

1) Elfving [3] considered the special case where $\mathfrak{X}$ contains a finite number of discrete points. It follows from his elegant geometrical argument that the optimum ξ is concentrated on at most k points and satisfies (2.10).

2) Consider the case of polynomial regression ($\mathfrak{X}$ a closed interval of the real line, $f_i(x) = x^{i-1}$). Then $p = 0$ by the Chebyshev theorem cited above. Theorem 2 then says, inter alia, that there exists an optimum ξ concentrated on at most k points. This result (for this important particular case) is already well known in the theory of moment problems ([10], p. 42). It holds *identically in* σ^2. If it did not hold for all σ^2 it would be useless in our problem when σ^2 is unknown. This result holds even when (2.18) below is true, with fixed v.)

We now give a simple result on the uniqueness of the optimum ξ^*.

THEOREM 3. *If $\mathfrak{X}$ is a compact real interval, $f_1, \cdots, f_{k-1}$ is a Chebyshev system, and $B(c^*)$ contains exactly k points, then the optimum ξ^* is unique.*

PROOF: Let $x_1, \cdots, x_k$ be the ordered members of $B(c^*)$, and let Q be the $(k - 1) \times k$ matrix whose (i, j)th element is $(-1)^j f_i(x_j)$. Let ξ denote a k-vector whose jth component is the number $\xi(x_j)$. According to (2.10), which, by Theorem 2, is necessary, and the Chebyshev theorem cited above, any optimum ξ must satisfy

$$Q\xi = 0. \tag{2.14}$$

(Of course, it must also satisfy $\xi(B(c^*)) = 1$.) Now, Q has rank $k - 1$, since, if it had smaller rank, a nontrivial weighted sum of rows of Q would be 0 and the f_i could not be a Chebyshev system. The linear equations (2.14) thus have a one-dimensional set of solutions ξ, and clearly at most one of these can be a probability measure. This completes the proof.

If $B(c^*)$ consists of more than k points, an analysis like that above will give information on how large the class of optimum ξ's can be.

Remark on symmetry (invariance): As we have indicated in Section 1, it will sometimes be easy, as in the case of polynomial regression, to infer that there is an optimum ξ with some symmetry property. Formally, suppose that there is a group G of transformations on $\mathfrak{X}$ such that for each g in G there is a transformation g' on $\mathfrak{A}$ such that, writing $(g'a)_i$ for the ith coordinate of $g'a$, we have $(g'a)_k = a_k$ for g in G and

$$\sum_i a_i f_i(x) = \sum_i (g'a)_i f_i(gx) \tag{2.15}$$

for all x and all $(a_1, \cdots, a_k)$. (One may let g' act on the vector of functions f_i instead of on a.) Then the problem in terms of the parameters $(g'a)_i$ and the independent variable gx coincides with the original problem. Hence, if ξ is optimum for the original problem, it is also optimum for the above problem in terms of gx and hence the measure ξ_g defined by

$$\xi_g(A) = \xi(g^{-1}A) \tag{2.16}$$

is optimum for the original problem in terms of x. Suppose for the moment that G contains a finite number, say L, of elements. Write

$$\bar{\xi} = \sum_{g \varepsilon G} \xi_g/L. \tag{2.17}$$

It is easy to prove that, if ξ is optimum, then so is $\bar{\xi}$; in fact, this is obvious statistically, since the variance of the average of L b.l.e.'s from the L independent experiments ξ_g with N/L observations each cannot be less than that of the b.l.e. from $\bar{\xi}$ with N observations (since $\bar{\xi}$ can be broken up into such experiments), but is clearly equal to the variance of the b.l.e. from ξ based on N observations. Thus, we have:

There exists an optimum design which is symmetric with respect to (invariant under) G.

The analogous result can be proved for G compact or satisfying conditions which yield the usual minimax invariance theorem in statistics; see, e.g., [7].

The fact that there exists an optimum symmetric design and an optimum design concentrated on (e.g.) k points does not imply the existence of an optimum design with both of these properties. For example, if $\mathfrak{X} = [-1, 1]$, $k = 2$, $f_1(x) = 1$, and $f_2(x) = x^2$, there is an optimum design concentrated on the two points 0 and 1, but the only symmetric design requires the three points 0, -1, and 1. However, in the event that g' does not act (as it does in the example just cited) as the identity for every g, we may be able to obtain some simplification. For example, without discussing the most general possibility, let us suppose that Q is a set of integers containing k and such that $(g'a)_j = a_j$ for all g if $j \varepsilon Q$, while $\sum_g (g'a)_j = 0$ for j not in Q. Consider the problem of finding an optimum design ξ on the space of equivalence classes of $\mathfrak{X}$ under the equivalence $x \sim x'$ if $x' = gx$ for some g, where the regression function is $\sum_{j \varepsilon Q} a_j f_j(x)$ (at the equivalence class of x). If there are q integers in Q, there is by Theorem 2 an optimum τ^* concentrated on at most q points. This τ^* corresponds to a unique symmetric (with respect to G) measure ξ^* on $\mathfrak{X}$, and it is easy to see that (2.10) is satisfied for *all* $i < k$. Thus, if there are L elements in G, this ξ^* is concentrated on at most qL points. For example, in the case of polynomial regression of even degree h ($= k - 1$) on $[-1, 1]$, G contains two elements and the set Q corresponds to the $q = 1 + h/2$ even powers, and we obtain that there is a symmetric optimum ξ concentrated on at most $h + 2$ points. The actual case (see Ex. 3.1) is that there is a symmetric optimum ξ concentrated on $k = h + 1$ points; the previous argument did not give the best result because τ^* gave positive probability to the equivalence class of 0, which corresponds to only one point of $\mathfrak{X}$. The best result could, however, be obtained using another argument: since, according to Theorem 3, the optimum ξ is *unique*, our discussion of two paragraphs above implies that it must be symmetric, and it is thus concentrated on $h + 1$ points. Similarly, one could conclude that there is a symmetric optimum design concentrated on $h + 1$ points when h is odd, either by using Theorem 3, or else by invoking an obvious modification of the previous argument for the case when $(g'a)_k = \pm a_k$. A similar result holds in the setup of Ex. 3.5.

Remark on heteroscedasticity and variable cost: Suppose the second line of (2.3) is replaced by

$$\text{Var}(Y_x) = [v(x)\sigma]^2, \tag{2.18}$$

where v is a known positive continuous function on $\mathfrak{X}$. To avoid trivialities, assume $v(x)$ bounded away from 0. Then, replacing Y_x by $Y_x^* = Y_x/v(x)$ and $f_i(x)$ by $f_i^*(x) = f_i(x)/v(x)$, it is clear that the entire discussion of this section goes through exactly as before (i.e., assuming (2.3)) since the a_i for which $EY_x = \sum a_i f_i(x)$ are the same a_i as those for which $EY_x^* = \sum a_i f_i^*(x)$, and the latter setup satisfies the original condition (2.3) of this section.

If there is a cost $c(x)$ of taking an observation at the point x, and the total cost rather than the total number of observations is to be kept constant, it is easily seen that an optimum design is obtained by going through the analysis of this section with $v(x)$ above replaced by $v(x)[c(x)]^{1/2}$.

Similar remarks will apply to the problems considered in Sections 4 and 5.

3. Examples of optimum designs in the case of Section 2.

Example 3.1. *Polynomials on* $[\alpha, \beta]$. One of the most important practical examples is that where $\mathfrak{X}$ is the closed finite nondegenerate interval $[\alpha, \beta]$ of reals, $k = h + 1$ for some $h > 0$, and $f_j(x) = x^{j-1}$ for $1 \leqq j \leqq h + 1$; we hereafter write $b_{j-1} = a_j$, $b = (b_0, \cdots, b_h)$, $d_{j-1} = c_j$, and $d = (d_0, \cdots, d_{h-1})$. Thus, assuming that the regression function is a polynomial of degree $\leqq h$, we may want to test the hypothesis that it is actually of degree $\leqq h - 1$, i.e., that $b_h = 0$. (In Section 4 we consider the possibility of testing that the degree is $\leqq h - m$ where m is specified). We first note that we can write

$$\textstyle\sum_{j=0}^h b_j x^j = \sum_{j=0}^h b_j'[(2x - \alpha - \beta)/(\beta - \alpha)]^j,$$

where $b_h' = [(\beta - \alpha)/2]^h b_h$; since $(2x - \alpha - \beta)/(\beta - \alpha)$ takes on values in $[-1, 1]$, an optimum strategy for arbitrary $[\alpha, \beta]$ is immediately obtained by an obvious change in location and scale from an optimum strategy in the case $[-1, 1]$, and we may hereafter limit our attention to the latter. Next, we note that b_h is obviously not estimable unless ξ gives positive probability to at least $h + 1$ points (of course, in practice we need $n > h + 1$ if σ^2 is unknown and $n \geq h + 1$ if σ^2 is known). Hence, by Theorem 2 (or by the result of [10] cited in Section 2) there exists an optimum ξ concentrated on exactly $(h + 1)$ points. We shall actually find a unique ξ^* which satisfies (2.8) and gives positive probability to exactly $h + 1$ points.[3] Thus, the phenomenon concerning degrees of freedom in the estimate of σ^2 which was discussed in the sixth paragraph of Section 2, and which is illustrated in Example 3.4 below, cannot occur in the present example.

The unique Chebyshev d^* (i.e., c^*) is well known in this example: $x^h - \sum_0^{h-1} d_j^* x^j$ is simply the hth Chebyshev polynomial (see, e.g., [2]),

$$(3.1)\qquad \begin{aligned} x^h - \sum\nolimits_0^{h-1} d_j^* x^j &= 2^{1-h} \cos(h \cos^{-1} x) \\ &= 2^{-h}\{[x + (x^2 - 1)^{1/2}]^h + [x - (x^2 - 1)^{1/2}]^h\}. \end{aligned}$$

Moreover, $m(d^*) = 2^{1-h}$, and this extreme value is attained in magnitude (with successive alterations in sign) by $x^h - \sum_0^{h-1} d_j^* x_j$ at the $h + 1$ points

$$(3.2)\qquad x_j = -\cos \frac{j\pi}{h}, \qquad 0 \leq j \leq h.$$

Thus, $B(d^*)$ consists of these $h + 1$ points. Moreover, the above d^* is the unique Chebyshev vector, since $x^0, x^1, \cdots, x^{h-1}$ form a Chebyshev system.

According to Theorem 3, the optimum ξ^* is unique. We now show that the unique optimum ξ^* is

$$(3.3)\qquad \begin{aligned} \xi^*(-1) &= \xi^*(1) = \tfrac{1}{2}h, \\ \xi^*\left(\cos \frac{j\pi}{h}\right) &= 1/h, \qquad 1 \leq j \leq h - 1. \end{aligned}$$

To prove this, we shall verify (2.14) for $\xi = \xi^*$, since this is just (2.10), which by Theorems 1 and 2 is necessary and sufficient for an optimum ξ. Since the d_j^*'s of (3.1) are zero if $j + h$ is odd, the polynomial of (3.1) is clearly orthogonal (with respect to ξ^*) to x^t when $t + h$ is odd. When $t + h$ is even, we can combine the weights $\xi^*(-1)$ and $\xi^*(1)$ and rewrite (2.14) as

$$(3.4)\qquad \sum_{j=0}^{h-1} (-1)^j \left(\cos \frac{\pi j}{h}\right)^t = 0.$$

Since $\cos^t\theta$ can be written as a linear combination of $\cos t\theta$, $\cos(t - 2)\theta, \cdots$, it suffices to prove (3.4) with $\cos^t(\pi j/h)$ replaced by $\cos(rj\pi/h)$, where $h + r$

[3] For $h = 1$ and 2, the solution is given in [14]. The general solution (3.3) of the problem of Example 3.1 for a design optimum in the sense of Section 2, is also given in the abstract [11] of the apparently contemporaneous work of E. J. Williams. The methods of this author are probably different from ours because he does not seem to use probability measures ξ. The authors are indebted to H. L. Lucas for calling their attention to [11] which appeared after submission of the present manuscript.

is even and $0 \leqq r \leqq h$. But for such r we have

$$
(3.5) \qquad \begin{aligned} \sum_{j=0}^{h-1} (-1)^j \cos(rj\pi/h) &= \operatorname{Re}\left\{\sum_{j=0}^{h-1} \exp[ji\pi(1 + r/h)]\right\} \\ &= \operatorname{Re}\left\{\frac{1 - \exp[i\pi(h + r)]}{1 - \exp[i\pi(1 + r/h)]}\right\} = 0. \end{aligned}
$$

It is interesting to compare the design ξ^* of (3.3) with the often used design $\xi^{h,M}$ (say) which assigns measure $1/M$ to each of the values $(2i - M - 1)/(M - 1)$, $i = 1, 2, \cdots, M$; thus $\xi^{h,M}$ takes an equal number of observations at each of M equally spaced points ranging from -1 to 1. Of course, $M > h$. For such a design with M observations on the interval $[0, M - 1]$, Fisher [8, p. 153] has calculated the left side of (2.4) to be $(h!)^4 M(M^2 - 1)(M^2 - 4) \cdots (M^2 - h^2)/(2h)!(2h + 1)!$ To obtain the corresponding quantity for the interval $[-1, 1]$, we must divide by $[(M - 1)/2]^{2h}$, and we must divide also by M in order to obtain the left side of (2.4) with η replaced by $\xi^{h,M}$. Since $[m(d^*)]^2 = 2^{2-2h}$, we obtain for the efficiency (see the definition following (2.8)) of $\xi^{h,M}$

$$
(3.6) \qquad e(\xi^{h,M}) = \frac{2^{4h-2}(h!)^4}{(2h)!(2h + 1)!} \prod_{i=1}^{h} \frac{M^2 - i^2}{(M - 1)^2}.
$$

The best choice of M varies: it is $h + 1$ if $h = 1$ or 2, $h + 2$ if $h = 3$, etc. For the often used procedure $\xi^{h,h+1}$, we have

$$
(3.7) \qquad e(\xi^{h,h+1}) = \frac{2^{4h-2}(h!)^4}{(2h)!\,h^{2h}(h + 1)}.
$$

Of course, (3.7) becomes 1 for $h = 1$, since $\xi^{1,2} = \xi^*$ for $h = 1$; for $h = 2$, (3.7) becomes 8/9, for $h = 3$ it is 256/405 (the best procedure, $\xi^{3,5}$, has efficiency .72), etc.; for large h, by Stirling's approximation, it is approximately $\pi^{3/2}h^{1/2}2^{2h-1}e^{-2h}$, which goes to zero very rapidly. For $\xi^{h,M}$ with $M \to \infty$, the efficiency (3.6) approaches $2^{4h-2}(h!)^4/(2h)!(2h + 1)!$, which as $h \to \infty$ is approximately $\pi/8$.

To the experimenter who protests at the above comparison that the design $\xi^{h,M}$ for some $M > h$ is more to his liking than is the ξ^* of (3.3) because the former will permit him to estimate regression coefficients a_j up to a_{M-1} (instead of up to a_h), we can only answer that his problem is not the one of the present example, that he is probably using a method of inference (to "choose the polynomial of correct degree") whose properties are questionable, and that a precise statement of his decision problem would probably lead to a procedure far superior to $\xi^{h,M}$. In Sections 4 and 5 we shall consider some other related problems which may be what the experimenter is faced with, rather than the problem of the present example. The problem of "fitting the polynomial of best degree" is more unwieldy, depending strongly on the somewhat arbitrary choice of losses which are to be assigned to errors in estimation as compared with the penalty for using a polynomial of large degree.

Example 3.2. *An example where $p > 0$.* It is easy to construct examples where the p of Theorem 2 is not 0 as it is in the case of a Chebyshev system. We illustrate the situation with a very simple example. Suppose $\mathfrak{X} = [-1, 1]$, $k = 3$, $f_1(x) = 1$, $f_2(x) = x^2$, $f_3(x) = x + 1$. The expression $x + 1 - c_1 - c_2x^2$ has, within $[-1, 1]$, derivative equal to 0 at $x = \frac{1}{2}c_2$ if $| c_2 | \geqq \frac{1}{2}$ and is monotone on $\mathfrak{X}$ if $| c_2 | < \frac{1}{2}$. Thus, a routine computation of $\max_x | x + 1 - c_1 - c_2x^2 |$ leads to the conclusion that any c with $c_1 + c_2 = 1$ and $| c_2 | \leqq \frac{1}{2}$ is Chebyshev; i.e., $p = 1$. Hence, $k - p = 2$, and indeed the design ξ^* for which $\xi^*(-1) = \xi^*(1) = \frac{1}{2}$ is optimum. The heart of the matter is that $(1, x^2)$ is not a Chebyshev system and that is is possible to estimate a_3 optimally without estimating a_2 at all.

Example 3.3. *An example where the optimum ξ^* is not unique.* There are many obvious examples of this kind, as we have indicated in the paragraph following the proof of Theorem 3. For example, one simple example is given by $\mathfrak{X} = [-1, 1]$, $k = 2$, $f_1(x) = 1$, $f_2(x) = 1 + \sin 10\, x$ (any ξ which assigns measure $\frac{1}{2}$ to each of the sets where $\sin 10\, x = 1$ or -1 satisfies (2.10)); an even more trivial one is $k = 1$, $f_1(x) = 1$, where every strategy is optimum.

Example 3.4. *An example where a nonoptimum ξ may be preferable.* This example illustrates the phenomenon alluded to in the text, wherein a design ξ which is not optimum in the sense defined in Section 2 may be preferable to an optimum design ξ^* for use (e.g., in testing a hypothesis about a_2) because the latter yields one less degree of freedom for the estimate of σ^2. Let ϵ be a fixed small positive number, and suppose that $\mathfrak{X}$ consists of the three integers 0, 1, and 2, that $k = 2$, and that $f_1(x) = x^2$ and $f_2(x) = 1 + (1 + \epsilon)x$. It is easily computed that the Chebyshev c^* is $1 + 3\epsilon/5$, that $B(c^*)$ consists of the points 1 and 2, that $m(c^*) = 1 + 2\epsilon/5$, and that the optimum ξ^* is given by $\xi^*(1) = 1 - \xi^*(2) = 4/5$. Thus, the efficiency of the design which takes all observations at $x = 0(\xi(0) = 1)$ and estimates a_2 in the obvious way, is $(1 + 2\epsilon/5)^{-2}$; when ϵ is small, this is more than offset by the extra degree of freedom for estimating σ^2 (e.g., 4 for the latter design against 3 for ξ^*, when 5 observations are taken), for the problem of testing a hypothesis about a_2 or giving a confidence interval on a_2.

Example 3.5. *A multidimensional example.* Let $\mathfrak{X}$ be the set of all points (x_1, x_2) in the Euclidean plane for which $| x_1 | \leqq 1$ and $| x_2 | \leqq 1$. Let $k = 6$ and suppose that the functions f_i are, in order, 1, x_1, x_2, x_1^2, x_2^2, and x_1x_2; thus, for example, we may be testing the hypothesis that a quadratic function of two variables has no interaction term $a_6x_1x_2$, i.e., that $a_6 = 0$. An easy approach to obtaining an optimum ξ is the third method mentioned in Section 2: An obvious guess of a ξ which might be optimum is that measure ξ' (say) which assigns probability $\frac{1}{4}$ to each corner of the square $\mathfrak{X}$. Thus, writing $c_1 + c_4 + c_5 = \bar{c}$, we see that $K(\xi', c)$ is symmetric in each of the variables c_2, c_3, and $\bar{c}$ (which are the only quantities on which it depends), so that $\min_c K(\xi', c) = K(\xi', c') = 1$ is attained for any c' for which $c_2 = c_3 = \bar{c} = 0$. Let c'' have all five of its components equal to zero. Then, clearly, $\max_x K(x, c'') = 1$. Thus, by the discussion following Theorem 2, we have proved that ξ' is optimum. Another way of verifying the optimality of

ξ' is to note that, in the terminology of the remark on symmetry of Section 2, G is the group of symmetries of the square, and an analogue of the last argument mentioned there for the case of polynomial regression with h odd obtains ξ' from the optimum design τ^* which assigns mass 1 to (1, 1) for the problem of estimating a_6 on $0 \leq x \leq y \leq 1$ when the regression function is $a_6 xy$. We note that only a_2, a_3, a_6, and $a_1 + a_4 + a_5$ are estimable for this design. The fact that only four linearly independent estimable linear parametric functions exist here is reflected in the fact that, in the notation of Section 2, $p = 2$. This can be seen by noting that, if $c' = (\epsilon + \delta, 0, 0, -\epsilon, -\delta)$, where ϵ and δ are sufficiently small, then $\max_x K(x, c')$ is still equal to unity, so c' is Chebyshev.

Other examples. Many other examples of optimum designs can be obtained from the extensive literature on Chebyshev approximation problems. For example, Section 37 of [2] can be used to obtain such a design for the setup of Example 3.1 wherein f_k is altered to $f_k(x) = 1/(x - c)$ with $c > b$.

4. The case of several regression coefficients. We consider now the setup of (2.1)–(2.3) (see also (2.18)) in the case where we are interested in inference about more than one of the a_i. In some estimation problems, a treatment like that of Section 5, wherein the behavior of the function $\sum a_i f_i$ rather than that of the a_i themselves is considered, will seem appropriate. However, in most problems of testing hypotheses, as well as in many problems of estimation (especially where the inference is not about all of the a_i), the treatment of the present section may seem appropriate.

We must first choose a criterion of optimality of a design for a problem of estimation or testing hypotheses about s of the a_i, say $a_{k-s+1}, \cdots, a_k$. Of course, it is easy to specify a loss function and a criterion (minimax, etc.) for choosing a design and associated decision procedure; but, as shown in [1], such a simple criterion as that of maximizing the minimum power of a test on an appropriate contour (M-optimality) will usually lead to most unwieldy computations. Even the corresponding local criterion on the power near the null hypothesis (L-optimality) will lead to difficult computations. Two other criteria considered in [1] are D-optimality and E-optimality. In the present setting, n being fixed, a design d^* is said to be D-optimum if $a_{k-s+1}, \cdots, a_k$ are all estimable under d^* and if, among all designs for which these parameters are estimable, denoting by $\sigma^2 V_d$ the covariance matrix of the b.l.e.'s of these parameters when design d is used, det V_d is a minimum for $d = d^*$. A design is said to be E-optimum in the above setting if the maximum eigenvalue of V_d is a minimum for $d = d^*$. The relevance of these criteria for problems of testing hypotheses and of estimation was indicated in [1] and the reference cited there. It was shown that D-optimality is generally more meaningful. There is an additional reason why this is so in problems of the type considered here: Consider the polynomial setup of Example 3.1 for any value $k > 2(h > 1)$ and $s > 1$. It is clear that the change of scale $x' = hx$ does *not* leave invariant the criterion of E-optimality: a change in the scale of measurement can change the E-optimum design. This is unsatisfactory from both an intuitive point of view (the optimum design depends on the choice of a unit of scale) and from a practical one; one would have to table optimum designs in such problems, as a function of α, β. (A similar remark, of course, applies to L-optimum and M-optimum designs.) On the other hand, D-optimality is invariant under such transformations. The same result is true under a change of origin (or a change of both scale and origin) in this polynomial example: D-optimality is invariant, but E-optimality is not.

Thus, although D-optimality is not an appropriate criterion in all problems, for the reasons given in the previous paragraph it seems reasonable to investigate this criterion as a first attack on the problem of finding optimum designs. We shall thus develop a method for obtaining D-optimum designs in the remainder of this section, except that we shall indicate briefly at the end of this section how various other criteria can be treated similarly.

Proceeding as in Section 2, let h_{tj} be numbers such that, for $i \leqq k - s < t$, the functions f_i are orthogonal to the functions $f_t^* = f_t - \sum_{j=1}^{k-s} h_{tj} f_j$ in the sense of (2.5), i.e.,

$$\sum_{r=1}^{n} f_i(x_n) f_t^*(x_r) = 0, \qquad i \leqq k - s < t. \tag{4.1}$$

Then, as in the discussion of (2.6), we see that σ^2 times the inverse of the covariance matrix $\sigma^2 V_d$ of best linear estimators of $a_{k-s+1}, \cdots, a_k$ has elements $\sum_r f_i^*(x_r) f_j^*(x_r)$, $k - s < i, j \leqq k$. For $t > k - s$, let $f_t^{**} = f_t^* - \sum_{j<t} g_{tj} f_j^*$ be orthogonal to f_j^* for $k - s < j < t$. Since the linear transformation which takes the f_t^* into the f_t^{**}, $k - s < t \leqq k$, has determinant 1, and since $\sum_r f_i^{**}(x_r)$ $f_j^{**}(x_r) = 0$ if $k - s < i < j$, we obtain

$$\det V_d^{-1} = \prod_{i>k-s} \sum_r [f_i^{**}(x_r)]^2. \tag{4.2}$$

Now, f_i^{**} is clearly f_i minus the projection of f_i on the linear space spanned by $f_1, f_2, \cdots, f_{i-1}$. Thus, the ith term in the product of (4.2) is just the expression of (2.7) with k replaced by i. Finally, then, making the same approximation as in Section 2 regarding the representation of the class of all designs by the class of all probability measures ξ on $\mathfrak{X}$, we have demonstrated, to within this approximation, the validity of the following definition, wherein $c^{(j)}$ denotes a vector $(c_1^{(j)}, \cdots, c_{j-1}^{(j)})$ of $j - 1$ components:

Definition. *A measure ξ^* in Ξ is said to be D-optimum (for the parameters $a_{k-s+1}, \cdots, a_k$) if*

$$\begin{aligned} &\prod_{j>k-s} \min_{c^{(j)}} \int [f_j(x) - \textstyle\sum_1^{j-1} c_i^{(j)} f_i(x)]^2 \xi^*(dx) \\ &\qquad = \max_{\xi \in \Xi} \prod_{j>k-s} \min_{c^{(j)}} \int [f_j(x) - \sum_1^{j-1} c_i^{(j)} f_i(x)]^2 \xi(dx) \end{aligned} \tag{4.3}$$

Of course, (4.3) reduces to (2.8) in the case $s = 1$. When f_1 is a constant, a ξ which is optimum for $s = k - 1$ is also optimum for $s = k$.

We note that it is a consequence of this definition that an optimum design is optimum for all values of σ^2.

For the special case where $s = k$ and $\mathfrak{X}$ consists of k points, it is easy to prove that the unique optimum ξ puts mass $1/k$ on each point. For if A is the matrix whose (i, j) element is $f_i(x_j)$ and B is the diagonal matrix with $\xi(x_j)$ the diagonal element in the jth row, an optimum design maximizes $\det(ABA') = (\det A)^2 \det B$. This argument has been employed by Hoel in the problem considered by him; see Example 4 below.

The methods of Section 2 do not directly yield anything here for the general problem. The analogue of Theorem 1 is essentially empty, since the various $B(c^{(j)})$'s for $c^{(j)}$ Chebyshev will not in general coincide. The game-theoretic approach is inapplicable because the product on the left side of (4.3) is not linear in ξ^*; moreover, the product of the integrals (before minimizing over the $c^{(j)}$) is not convex in the $c^{(j)}$'s, since u^2v^2 is not a convex function of u and v. The following analysis will, however, yield a method for obtaining an optimum ξ.

For $j > k - s$, let

$$F_j(\xi) = \min_{c^{(j)}} \int [f_j(x) - \sum_1^{j-1} c_i^{(j)} f_i(x)]^2 \xi(dx). \tag{4.4}$$

In s-dimensional Euclidean space R^s, let S be the set of all points $F(\xi) = (F_{k-s+1}(\xi), \cdots, F_k(\xi))$ for ξ in Ξ. Although S may not be convex, it possesses the following "upper convexity" property, which is all we require: For any ξ_1 and ξ_2 in Ξ and any λ with $0 < \lambda < 1$,

$$F_j(\lambda\xi_1 + (1 - \lambda)\xi_2) \geqq \lambda F_j(\xi_1) + (1 - \lambda)F_j(\xi_2) \tag{4.5}$$

for all $j > k - s$. In fact, (4.5) is an immediate consequence of the linearity in ξ of the integral of (4.4).

Let $u_{k-s+1}, \cdots, u_k$ be the coordinate functions of R^s. For $\delta > 0$, let G_δ be the set of all points in R^s with all coordinates positive and $\prod_j u_j \geqq \delta$. Let G_δ' be the subset of G_δ where $\prod_j u_j = \delta$. We note that G_δ is convex. Suppose that S is closed (this is easily proved from (4.4) if $\mathfrak{B}$ is large enough so that Ξ is compact; the modification which is needed if Ξ is not closed is trivial, anyway), and let δ_0 be the largest value of δ such that G_δ and S have a nonempty intersection. (Such a δ_0 exists since S has points with all coordinates positive.) If T is the convex hull of S, property (4.5) implies that δ_0 is also the largest value of δ such that G_δ and T have a nonempty intersection. Hence, applying the separation theorem for G_{δ_0} and T, we conclude that there is a hyperplane L with positive direction cosines such that L separates G_{δ_0} and S. Thus, any point $F(\xi^*)$ in $G_{\delta_0} \cap S$ clearly maximizes $\prod_j F_j(\xi)$ (i.e., that ξ^* satisfies (4.3)); and, for positive numbers λ_j' for which L is given by $\sum \lambda_j' U_j =$ constant, that point maximizes $\sum_j \lambda_j' F_j(\xi)$. Finally, since all points of G_σ' are extreme, L intersects G_{δ_0} in exactly one point, as does therefore S.

Before summarizing the above results, we note that, for $\lambda = (\lambda_{k-s+1}, \cdots, \lambda_k)$

with all $\lambda_i > 0$, the payoff function

(4.6) $$K_\lambda(x, c) = \sum_{j>k-s} \lambda_j[f_j(x) - \sum_{i<j} c_i^{(j)} f_i(x)]^2,$$

where $c = (c^{(k-s+1)}, \cdots, c^{(k)})$, satisfies all of those conditions satisfied by the function K of Section 2 which were used in the proof of the game-theoretic results of the lemma there. Thus, that lemma is valid when K is replaced by K_λ.[4] The function K_λ is of course no longer in a form suitable to make use of Chebyshev approximation results. However, for any λ, if c_λ^* is minimax for the payoff function K_λ, we can still characterize maximin ξ_λ's in terms of the set $\bar{B}_\lambda(c_\lambda^*)$ (say), defined to be the set of x for which $K_\lambda(x, c_\lambda^*)$ achieves its maximum. With this interpretation of symbols, the analogue of (2.10) is proved here exactly as in (2.11).

We have thus proved that following,[5] where C now stands for the set of vectors $c = (c^{(k-s+1)}, \cdots, c^{(k)})$ and $c_\lambda^* = \{c_i^{(j)*}\}$ stands for a vector of this type:

THEOREM 4. *The game of Ξ vs. C with payoff function K_λ is determined. If ξ_λ is maximal with respect to c_λ^* while c_λ^* is minimal with respect to ξ_λ, then ξ_λ is maximin. Thus, if c_λ^* is minimax and*

(4.8) $$\xi_\lambda(\bar{B}_\lambda(c_\lambda^*)) = 1$$

and

(4.9) $$\int [f_j(x) - \sum_{i<j} c_i^{(j)*} f_i(x)] f_i(x) \xi_\lambda(dx) = 0$$

for $i < j$ and $k - s < j \leq k$, then ξ_λ is maximin; moreover, every maximin ξ_λ satisfies (4.8) and (4.9) for every minimax c_λ^*. There is, to within a multiplicative constant, a unique value λ^* of λ such that $\prod_i F_i(\xi_\lambda)$ is a maximum for $\lambda = \lambda^*$ and some ξ_{λ^*}. Those ξ_{λ^*} which maximize $\prod_i F_i(\xi_{\lambda^*})$, and no other ξ's, are optimum. $F(\xi_{\lambda^*})$ is the same for any optimum ξ_{λ^*}.

We now consider an example.

Example 4. Consider the setup of Example 3.1, where (see the end of the second paragraph of the present section) we may suppose $\alpha = -1$, $\beta = 1$. Suppose $k = 3$ ($h = 2$), and $s = 2$; as we have remarked earlier, the optimum design obtained below will also obviously be D-optimum for the case $s = 3$. An

[4] That part ot the lemma which concerns the number $k - p$ is valid when k is replaced by $1 + s(2k - s - 1)/2$ ($= 1 +$ number of components of c) in the statement of the lemma. However, this is of no use to us since it may be that no maximin strategy on the specified number of points is optimum. For example, in the set-up of Example 5.2 below with $s = k = 2$, one can verify that the λ^* of Theorem 4 is (15/4, 1), and that any ξ_λ^* with first and second components equal is maximin, but only (4/15, 4/15, 7/15) is optimum.

It is trivial that the optimum strategy need be concentrated on no more than $1 + k(k + 1)/2$ points. For the criterion of optimality (4.3) involves ξ only through the elements (5.2) below of the matrix $M(\xi)$. These matrices form a convex body of dimensionality at most $k(k + 1)/2$, spanned by matrices of ξ's concentrated on a single point Hence any $M(\xi)$ is a linear convex combination of at most $1 + k(k + 1)/2$ extreme elements.

[5] See also footnote 6.

elegant solution to this problem for general k and $s = k$, has been given by P. G. Hoel [15] (see also Example 5.1 below). The case $s < k - 1$ does not seem to yield to his attack. The present problem is discussed here as an illustration of our methods. We may take 1 and γ for the components of λ, and write $K'_\gamma(x, d) = (x - d'_0)^2 + \gamma(x^2 - d''_1 x - d''_0)^2$ in place of K_λ. For fixed γ, one may *guess* that there will be a maximin strategy ξ'_γ of the form $\xi'_\gamma(-1) = \xi'_\gamma(1) = \alpha_\gamma$, $\xi'_\gamma(0) = 1 - 2\alpha_\gamma$, for some α_γ. With respect to such a ξ_γ, the minimal strategy (which must merely satisfy the orthogonality relation (4.9)) is obviously $d'_0 = d''_1 = 0$, $d''_0 = 2\alpha_\gamma$. For this choice d_γ (say) of d, we obtain $K'_\gamma(\xi'_\gamma, d_\gamma) = \gamma[2\alpha_\gamma - 4\alpha_\gamma^2] + 2\alpha_\gamma$. This is maximized by $a_\gamma = \min(\frac{1}{2}, (\gamma+1)/4\gamma)$, and for the strategy ξ_γ. corresponding to this value of α_γ we obtain

$$\text{(4.10)} \qquad \min_d K'_\gamma(\xi^*_\gamma, d) = \begin{cases} (\gamma + 1)^2/4\gamma & \text{if } \gamma > 1, \\ 1, & \text{if } \gamma \leqq 1. \end{cases}$$

On the other hand,

$$\text{(4.11)} \qquad \min_d \max_x K'_\gamma(x, d) \leqq \max_x K'_\gamma(x, d_\gamma).$$

Since $K'_\gamma(x, d_\gamma)$ is convex in x^2, its maximum is attained at either $x^2 = 0$ or $x^2 = 1$, and an easy computation shows that the right side of (4.11) is in fact equal to the right side of (4.10). Thus, we have proved that ξ^*_γ is maximin. Finally, $F_2(\xi^*_\gamma)F_3(\xi^*_\gamma) = 4\alpha_\gamma^2(1 - 2\alpha_\gamma)$, which is maximized by $\alpha_\gamma = \frac{1}{3}$. Thus, an optimum design for this problem is $\xi(-1) = \xi(0) = \xi(1) = \frac{1}{3}$. Of course the optimum designs for a given set of f_i will depend on s, as exemplified by the different results obtained in Example 3.1 and Example 4.

We shall now mention briefly methods for obtaining designs which are optimum in two other senses. Although it is not difficult to characterize E-optimum procedures in *simple* examples, they often seem much harder to calculate than D-optimum ones. Somewhat easier is the characterization of that design which minimizes the maximum eigenvalue of the covariance matrix of best linear estimators of the regression coefficients of the f_t^{**} (the regression function being expressed in terms of the f_t for $t \leqq k - s$ and of the f_t^{**} for $t > k - s$); i.e., of $L_d V_d L'_d$ where L_d is a square matrix with ones on the main diagonal and zeros above it. (The f_t^{**} depend on the design, which indicates the intuitive weakness of this criterion; however, as pointed out in [1], the criterion of E-optimality, which has often been considered in the literature, suffers from a similar shortcoming.) Again making the approximation that we do not restrict $n\xi$ to be integer-valued, this criterion amounts to finding that ξ which maximizes $\min_{j>k-s} F_j(\xi)$, i.e., if δ' is the largest value of δ for which the orthant $H_\delta = \{\min_j u_j \geqq \delta\}$ intersects S nonvacuously, those ξ for which $F(\xi)$ is in $H_{\delta'} \cap S$ are the optimum procedures with respect to this criterion. Another criterion which has been considered in the literature, especially in estimation problems, is that of minimizing the "average variance", $\sigma^2 s^{-1}$ trace (V_d). Defining $F_j^*(\xi')$ to be the expression of (4.4) with the sum in the integrand taken only from 1 to $k - s$, this criterion amounts to minimizing $\sum_{j>k-s} F_j^*(\xi)$. Replacing S by the set of points $F^*(\xi) = (F_{k-s+1}^*(\xi), \cdots, F_k^*(\xi))$, and restricting the sum over i in (4.6) to values $\leq k - s$, this amounts to finding the maximin ξ's for a λ with all components equal. These maximin ξ's for the *original* S and K_λ (with all λ_j equal) would of course minimize the average variance of the b.l.e.'s of regression coefficients of the f_i^{**}; i.e., would minimize the trace of $L_d V_d L_d'$. Criteria like that of minimizing the average variance are subject to the same criticisms as E-optimality.

Remarks. As in the problem of Section 2, one can prove that the symmetry condition (2.15) and the obvious analogue of the condition of the line above (2.15) imply the existence of a symmetrical optimum ξ for any of the criteria considered in the present section. For example, from (4.5) it follows at once that, if ξ is D-optimum, then the symmetrical $\bar{\xi}$ defined by (2.17) is also D-optimum. Remarks analogous to those of Section 2 on the number of points at which a symmetrical optimum ξ will be concentrated, clearly hold in the problems of this section. We note that the choice of the form of ξ_γ' in Example 4 is motivated by symmetry considerations, although the optimum weights must be computed in any approach.

The remark concerning the modification of (2.18) applies also to the problems of this section.

5. Estimation of the whole regression function. In the setup described by (2.1–(2.3), suppose the problem is one of estimation concerning all the a_i. One approach has been indicated in Section 4. Another approach is to think of the problem not as one of estimating the parameters a_i, but rather as one of estimating the entire function $\sum a_i f_i$. Thus, if g is the estimate of $\sum a_i f_i$, it is desired to make some measure of the average deviation of g from $\sum a_i f_i$ small in some sense, by choosing an appropriate design. The most obvious possibilities of such measures are perhaps (1) $\sup_a EW(\sup_x | g(x) - \sum a_i f_i(x) |)$, where W is nondecreasing; (2) the integral with respect to some measure μ on $\mathfrak{X}$ of $\sup_a EW(| g(x) - \sum a_i f_i(x) |)$; (3) the supremum on $\mathfrak{X}$ and $\mathfrak{A}$ of $EW(| g(x) - \sum a_i f_i(x) |)$. Of these three possibilities, the first is perhaps the most meaningful for most applications (with perhaps the inclusion of a weight function $h(x)$ multiplying $| g(x) - \sum a_i f_i(x) |$) but is computationally much more difficult to treat than the others; the second possibility is by far the easiest computationally, but is least satisfactory from a practical point of view because of the necessity of choosing μ—for example, if $\mathfrak{X}$ is a line segment, the optimum design will not be invariant under homeomorphisms of $\mathfrak{X}$, if μ is always chosen to be Lebesgue measure; the third possibility is a compromise between the first two and, as a first attack on the problem, is what we consider in this section, with $W(t) = t^2$. We note that a remark of [9, p. 215] indicates that Box and Hunter are considering the second approach for certain polynomial multiple regression problems when $W(t) = t^2$ and μ is Lebesgue measure on a Euclidean set. We note that it is a consequence of all three definitions of optimality discussed in this paragraph that an optimum design is optimum for all values of σ^2.

We shall not have to concern ourselves here with the choice of the function g: for example, the remarks of Section 2 extend here to show that, for a given design, if $\hat{a}$ is the b.l.e. of a, then $\sup_{a,x} E_a[\sum a_i f_i(x) - g(x)]^2$ is a minimum for $g(x) = \sum \hat{a}_i f_i(x)$. We therefore assume this choice of g in what follows. Thus, we are led to consider the minimization with respect to the design d of the expression

$$\max_x E[\textstyle\sum_i(\hat{a}_i - a_i)f_i(x)]^2 = \sigma^2 \max_x f(x)'V_d f(x), \tag{5.1}$$

where we have written $f(x)$ for the vector of $f_i(x)$'s. Using again the representation of a design as a measure ξ, the analogue of V_d is the inverse of the matrix $M(\xi)$ whose (i, j)th element is

$$m_{ij}(\xi) = \int f_i(x)f_j(x)\xi(dx). \tag{5.2}$$

Thus, making an approximation like that of Section 2 in not requiring $n\xi$ to be integral, we define a design ξ^* to be optimum for the problem of this section if $M(\xi^*)$ is nonsingular and

$$\max_x f(x)'M(\xi^*)^{-1}f(x) = \min_{\xi\epsilon\Xi} \max_x f(x)'M(\xi)^{-1}f(x). \tag{5.3}$$

It seems more difficult here than in Section 2 to give a useful general computing algorithm. We now describe one device which seems useful in many examples. Let D_ξ be a non-singular matrix such that the vector $g_\xi = D_\xi f$ consists of functions $g_{\xi,i}$ which are orthonormal with respect to ξ; it is clear that such a D_ξ exists for any ξ in Ξ for which $M(\xi)$ is non-singular. Since the (i, j)th element of $D_\xi M(\xi)D_\xi'$ is the integral with respect to ξ of $g_{\xi,i}g_{\xi,j}$, we obtain

$$f(x)'M(\xi)^{-1}f(x) = g_\xi'(x)(D_\xi M(\xi)D_\xi')^{-1}g_\xi(x) = \textstyle\sum_i [g_{\xi,i}(x)]^2 \tag{5.4}$$

(Since the left side of (5.4) does not depend on D_ξ, neither does the right side; thus, in searching for a ξ which minimizes the maximum with respect to x of (5.4), it suffices to consider for each ξ only that D_ξ and g_ξ which are computationally most convenient.) Since the $g_{\xi,i}$'s are orthonormal with respect to ξ, the integral with respect to ξ of the last expression of (5.4) is k, and this cannot be greater than the maximum with respect to x of (5.4). Thus, a sufficient condition for a given ξ to be an optimum design is

$$\max_x \sum_{i=1}^{k} [g_{\xi,i}(x)]^2 = k. \tag{5.5}$$

Of course, a necessary condition for (5.5) to be satisfied is that ξ give measure one to the set of x where $\sum[g_{\xi,i}(x)]^2 = k$, and it is useful to keep this in mind in examples.

Suppose (5.5) is satisfied for a ξ' concentrated on k points, say $x_1, \cdots, x_k$. Then the $k \times k$ matrix whose (i, j)th element is $g_{\xi',i}(x_j)[\xi'(x_j)]^{\frac{1}{2}}$ has orthonormal rows and, hence, orthonormal columns: $\sum_i[g_{\xi',i}(x_j)]^2\xi_j = 1$ for $1 \leq j \leq k$.

Hence, $\xi_j' > 0$, and by (5.5) $\xi_j' \geqq 1/k$. Hence, each ξ_j is $1/k$. We summarize our results.

THEOREM 5. *If* (5.5) *holds, then* ξ *is optimum.*[6] *If* (5.5) *holds for a* ξ *concentrated on* k *points, then* ξ *gives measure* $1/k$ *to each of these points.*

If the setup is that of Example 3.1 it follows from the results of [10] cited in Section 2 that there exists an optimum ξ concentrated on exactly k points. This will not be true in general (see Example 5.2 below).

In the special case where $\mathfrak{X}$ consists of k points, the argument of the paragraph preceding the present theorem, applied to the distribution $\xi_j = 1/k, j = 1, \cdots, k$, shows that (5.5) is satisfied for this distribution, and hence the latter is optimum. Combining this with a remark which follows (4.3) we conclude that, when $\mathfrak{X}$ consists of k points, the design which puts mass $1/k$ on each point is the unique optimum design according to both the definition (4.3) for $s = k$ (the problem of Section 4) and the definition (5.3) (the problem of the present section).

Example 5.1. *The setup of Example* 3.1. It is possible to solve this problem by our methods and such a solution was given in the original draft of this paper. In the meantime, however, a solution has been published by Guest [13], so that there is no point to repeating the details of our solution. An earlier discussion by Smith [14] gave details of designs up to $k = 7$. The optimum design assigns mass $1/k$ to the points $+1$, -1, and the roots of $L_h'(x) = 0$, where L_h' is the derivative of the Legendre polynomial. (5.5) is satisfied ([13], equation (10)). It therefore follows from Theorem 6 below that this design is also optimum in the sense of definition (4.3) for $s = k$ (problem of Section 4) for this setup; i.e., a special case of Theorem 6 asserts that Hoel's design [15] is the same as that of Guest [13].[7] This last fact was noted by Hoel through an examination of the explicit results in the polynomial case.

Example 5.2. This example illustrates the use of Theorem 5 where the optimum ξ is concentrated on more than k points and does not give equal measure to all of them. Let $k = 2$ and let $\mathfrak{X}$ consist of three points. Thus, we hereafter write the f_i and ξ and S as triples, where $S(x) = \sum_i [g_{\xi,i}(x)]^2$. Suppose $f_1 = (1, 1, 0)$ and $f_2 = (0, 1, 2)$. For $\xi = (\xi_1, \xi_2, \xi_3)$, we obtain easily

$$S = (\xi_1\xi_2 + 4\xi_2\xi_3 + 4\xi_1\xi_3)^{-1}(\xi_2 + 4\xi_3, \xi_1 + 4\xi_3, 4\xi_1 + 4\xi_2)$$

We have $\sum \xi_i S_i = 2$, identically in ξ. Suppose $\xi_1 = 0$, $\xi_2 > 0$, $\xi_3 > 0$. Then either 1) $S_2 = S_3$, in which case $\xi_2 = \xi_3 = \frac{1}{2}$ and $S_1 > S_2 = 2$, or 2) $\max(S_2, S_3) > 2$. Thus, in either case $\max_i S_i > 2$. A similar argument applies if either of the other ξ_i's is 0, and two ξ_i's can obviously not be 0. Thus, $\max_i S_i$ can be 2 only if all ξ_i are positive and all S_i are equal to 2. The unique optimum ξ is thus easily seen to be (4/15, 4/15, 7/15).

---

[6] The converse of this statement is true. In fact, it will be proved in a subsequent paper (the results were obtained too late for inclusion in the present paper) that the following three statements are equivalent: (a) the design ξ is optimum in the sense of Section 4 with $s = k$; (b) the design ξ is optimum in the sense of Section 5; (c) the design ξ satisfies (5.5).

[7] This is a special case (for polynomial regression) of the result described in footnote 6.

It is obvious how to give examples like those of Section 3 where the optimum ξ is not unique, etc.

The argument just after (2.17) is easily modified to apply to the expression on the left side of (5.1), so that we can again conclude that there exists an optimum symmetrical ξ if (2.15) and the obvious analogue of the condition of the line above (2.15) hold. The question of the number of points at which an optimum symmetrical ξ will be concentrated is difficult, as is the corresponding question for general optimum ξ.

The modification of (2.18) can be made in the problem of this section, exactly as in Section 2.

We shall conclude this section with a result which sheds some light on the connection between the problem of Section 4 for $s = k$ and the problem of the present section. This result has already been cited in Example 5.1.

THEOREM 6.[8] *If the design which puts mass* $1/k$ *on each of* k *points satisfies* (5.5) *and is optimum in the sense of* (5.3) (*problem of Section* 5), *then this design is also optimum in the sense of* (4.3) (*problem of Section* 4) *with* $s = k$.

PROOF: Let ξ_0 be a design, optimum in the sense of (5.3), such that ξ_0 assigns mass $1/k$ to each of the points $x_1, \cdots, x_k$ in $\mathfrak{X}$, and such that (5.5) is satisfied for $\xi = \xi_0$. Since a design optimum for the problem of Section 4 with $s = k$ is invariant under a linear transformation on the f_i, it will suffice to prove that ξ_0 is optimum for this problem assuming $f_i = g_{\xi_0, i}$; henceforth we make this assumption. Thus

$$\max_x \sum_i f_i^2(x) = k \tag{5.6}$$

and

$$m_{ij}(\xi_0) = \delta_{ij},$$

and we have to prove that ξ_0 maximizes $\det M(\xi)$. Now from (5.6) for any ξ we have

$$\sum_i m_{ii}(\xi) \leqq k$$

and hence

$$\det M(\xi) \leqq \prod_i m_{ii}(\xi) \leqq 1 = \det M(\xi_0)$$

This proves the theorem.

The authors are obliged to Professor G. Elfving for helpful comments.

REFERENCES

[1] J. KIEFER, "On the nonrandomized optimality and randomized nonoptimality of symmetrical designs", *Ann. Math. Stat.*, Vol. 29 (1958), pp. 675–699.

[2] N. I. ACHIESER, *Theory of Approximation*, Ungar Pub. Co., New York, 1956.

[8] Theorem 6 is a very special case of the results announced in footnote 6.

[3] G. Elfving, "Optimum allocation in linear regression theory", *Ann. Math. Stat.*, Vol. 23 (1952), pp. 255–262.

[4] G. Elfving, "Geometric allocation theory", *Skand. Akt.* 1955, pp. 170–190.

[5] H. Chernoff, "Locally optimum designs for estimating parameter" *Ann. Math. Stat.*, Vol. 24 (1953), pp. 586–602.

[6] H. F. Bohnenblust, S. Karlin, and L. Shapley, "Games with continuous, convex payoff", *Ann. Math. Studies*, No. 24, pp. 181–192.

[7] J. Kiefer, "Invariance, minimax sequential estimation, and continuous time processes", *Ann. Math. Stat.*, Vol. 28 (1957), pp. 573–601.

[8] R. A. Fisher, *Statistical Methods for Research Workers*, tenth edition, Oliver and Boyd, Edinburg, 1946.

[9] G. E. P. Box and J. S. Hunter, "Multi-factor experimental designs for exploring response surfaces", *Ann. Math. Stat.* Vol. 28 (1957), pp. 195–241.

[10] J. A. Shohart and J. D. Tamarkin, *The Problems of Moments*, *Math. Surveys*, No. 1, Amer. Math. Soc., New York, 1943.

[11] E. J. Williams, "Optimum allocation for estimation of polynomial regression," (abstract) *Biometrics*, Vol. 14 (1958), p. 573.

[12] S. Ehrenfeld, "Complete class theorems in experimental design", *Proceedings of the Third Berkeley Symposium on Mathematical Statistics and Probability*, University of California Press, 1955.

[13] P. G. Guest, "The spacing of observations in polynomial regression", *Ann. Math. Stat.*, Vol. 29 (1958), pp. 294–299.

[14] K. Smith, "On the standard deviations of adjusted and interpolated values of an observed polynomial function and its constants and the guidance they give towards a proper choice of the distribution of observations", *Biometrika*, Vol. 12 (1918), pp. 1–85.

[15] P. G. Hoel, "Efficiency problems in polynomial estimation", *Ann. Math. Stat.*, Vol. 29 (1958), pp. 1134–46.

Optimum Experimental Designs

By J. Kiefer*

Cornell and Oxford Universities

[Read before a Research Methods meeting of the Royal Statistical Society, May 27th, 1959, Professor G. A. Barnard in the Chair]

SUMMARY

After some introductory remarks, we discuss certain basic considerations such as the nonoptimality of the classical symmetric (balanced) designs for hypothesis testing, the optimality of designs invariant under an appropriate group of transformations, etc. In section 3 we discuss complete classes of designs, while in section 4 we consider methods for verifying that designs satisfy certain specific optimality criteria, or for computing designs which satisfy such criteria. Some of the results are new, while part of the paper reviews pertinent results of the author and others.

1. Introductory Remarks

This paper will survey various recent developments in the theory of the determination of optimum experimental designs, in the course of which a few new results will be given. Our purpose will not be to state things in the most general possible abstract setting, nor to try to apply the discussion to the more complicated practical problems; it is not only the computational difficulties which dictate this—rather, it is that our primary aim will be to stress certain ideas and principles which are most easily set forth in simple situations. These ideas and principles have just as much content in the complicated settings which one often encounters in practice, but their transparency is often obscured there by the arithmetic.

We shall be concerned with methods for verifying whether or not given designs satisfy certain optimality criteria, and for computing designs which satisfy such criteria. Our approach is in the spirit of Wald's decision theory. Thus, problems of constructing designs which satisfy certain *algebraic* conditions (resolvability, revolvability, etc.), as distinguished from problems of constructing designs which satisfy certain optimality criteria, will not primarily concern us; such algebraic conditions will be of interest to us only in so far as they may be proved to entail some property of optimality. Perhaps the traditional development of the subject of experimental design has been too much concerned with the former problems and too little with the latter, although recent developments show that progress in finding useful and efficient designs cannot and should not rest on this traditional approach of being satisfied to find designs which merely satisfy some intuitively appealing algebraic property

Our discussion will not include all of the recent papers which fit into the above framework, but some of these excluded topics will now be mentioned, since they are important

*Research sponsored by the U.S. Office of Naval Research.

for future research. For example, we shall not treat any problems in the very important area of sequential design. In a sense, the first result in this area was really the proof of the optimum character of Wald's sequential probability ratio test (Wald and Wolfowitz, 1948), and other sequential decision theory can be viewed in the same way. However, in the case in which we are really interested in the present paper, wherein several possible experiments are available at each stage, there are very few explicit results, although the general setup was formulated some time ago (Wald, 1950). The earliest result of this kind was obtained by Stein (1948), but explicit results are so difficult to obtain that recent papers have obtained them in only a few simple settings such as those considered by Bradt and Karlin (1957), Bradt, Johnson and Karlin (1957), and Sobel and Groll (1959). There are also a few asymptotic optimality results for sequential designs, such as those obtained by Chung (1954), Hodges and Lehmann (1955), and Sacks (1958) for the stochastic approximation (sequential search) designs of Robbins and Monro (1951), Kiefer and Wolfowitz (1952), Blum (1954), and Sacks (1958), and the result of Robbins (1952). The much more difficult nonasymptotic problem of finding optimum designs with stopping rules (or even nonoptimum but reasonably efficient stopping rules with prescribed confidence properties) in these cases remains unanswered, although corresponding search problems where errors are negligible have been treated (Kiefer, 1953, 1957*b*). Although practical people often employ techniques like those of Box and Wilson (1951) in problems of this kind (such methods were first suggested by Friedman and Savage (1947) and were further developed by the author (1948)), such methods cannot in their *present* state have any role in satisfactorily solving these problems, since (as pointed out in the discussion to the paper by Box and Wilson) they have no guaranteed probability (confidence) properties (not even asymptotic ones such as those enjoyed by the stochastic approximation methods cited above), and in fact are often not even well-defined rules of operation; the main value of such intuitive considerations to the approach of the present paper is that they might suggest well-defined procedures which can be analysed precisely (under suitable restrictions on the regression function, Hotelling (1941) has obtained an optimum nonsequential solution to such a problem). These search problems usually fall into the area of *nonparametric* design problems, an important field which deserves more attention but which we shall not treat here.

We now discuss the assumptions we shall make in this paper, on the class of possible distributions which occur; this class can be nonparametric, but only trivially so: in the theory of point estimation, it is well known (see, e.g., Hodges and Lehmann, 1950) that best linear estimators possess certain more meaningful optimality properties than merely being best among *linear* estimators (e.g., they are minimax for certain weight functions) if the class of possible distributions contains certain normal distributions (this can also be shown for randomized designs, defined below). This fact can be used with any of the estimation criteria of the section 4 to yield a minimax result. We shall usually assume normality, so our results do have this nonparametric extension in point estimation problems. (Although best linear estimators and certain related confidence intervals are minimax in the normal case (see Stein and Wald, 1947; Wolfowitz, 1950; Kiefer, 1957*a*), if we add the risk functions of several estimators (rather than to consider the vector risk), they may sometimes have to be modified slightly in order to be admissible (Stein, 1955)). This justification for choosing a design on the basis of the obvious comparison between the covariance matrices of best linear estimators does not apply to such problems as those of testing hypotheses, as we shall see in Section 2A. It should be obvious that the considerations of that section apply with appropriate modifications under various other parametric assumptions (for example, the results for the classical randomized block setting are also mentioned in Section 2A); the striking thing is that these phenomena exist in the common, unpathological, normal case. A similar remark applies to other portions of this paper. For example, optimality results like those of Sections 4A–B can also be proved for appropriate components of variance (or mixed) models.

Throughout this paper, then, except where explicitly stated to the contrary, Y_d will denote an N element column vector whose components Y_{di} are independent (often normal) random variables with common variance σ^2 (unknown unless stated to the contrary, although this matters little); θ is an unknown m-vector with components θ_i, Θ is the m-space of possible θ's, X_d is a known $N \times m$ matrix depending on an index d (the "design") and which is, within limits, subject to choice by the experimenter; and the expected value of Y_d when θ and σ^2 are the parameter values and the design d is used is

$$E_{\theta,\sigma,d}\, Y_d = X_d\theta. \tag{1.1}$$

Additional restrictions of the form $A\theta = B$ can be assumed already to have been absorbed into (1.1). The consideration of various designs with different numbers of observations is easily accomplished; for example, if we are only interested in point estimation, or if σ^2 is known, we need only append a suitable number of rows of zeros to various X_d's. See Section 2D for further discussion of our assumptions.

We denote by Δ the set of all choices of the index d which are available to the experimenter, and by Δ_R a class of probability measures on Δ (usually these measures will have finite support, and measurability considerations will be trivial otherwise), the class of *randomized designs* (the classical use of this term is a special case of the present usage, and the reason for the classical use of such designs, under different probability models, is entirely different from ours) available to the experimenter. A randomized design δ is used by choosing a d from Δ according to this measure and then using the selected d. We think of Δ_R as including Δ.

Write $A_d = X_d'X_d$ (primes on vectors and matrices denote transposes) for the "information matrix" of the design d. If X_d (hence A_d) is of rank b_d, then there is a $b_d \times m$ matrix L_d, of rank b_d, such that the distribution of Y_d depends on θ only through the vector $L_d\theta$, whose best linear estimators (b.l.e.'s) are the components t_{di} of $t_d = L_d t$ where t is any solution of the normal equations $A_d t = X_d' Y_d$. If $b_d = m$, we can take L_d to be the identity matrix, and will always do so. If $\bar{S}_d$ is the usual best unbiased estimator of σ^2 (so that $(N - b_d)\bar{S}_d/\sigma^2$ has the χ^2 distribution with $N - b_d$ degrees of freedom), then $(t_d, \bar{S}_d)$ is a minimal sufficient statistic.

The covariance matrix of t_d will be denoted by $\sigma^2 V_d$; of course, $V_d = A_d^{-1}$ if $b_d = m$.

If a randomized design δ assigns probability one to a set of d's all of which have the same L_d, then the covariance matrix of b.l.e.'s under δ is the expectation with respect to δ of $\sigma^2 V_d$, although the distribution is not generally normal. The role of such designs for problems of point estimation is considered in Section 3. If δ gives positive probability to a d for which a linear parametric function $q'\theta$ is not estimable, then the expected squared error for any estimator of $q'\theta$ is infinite when δ is used, although δ may still be useful for interval estimation, hypothesis testing, ranking, etc., involving $q'\theta$. In fact, as we shall see in Section 2A, the appropriate distribution theory—e.g., computation of δ's and associated tests with good power functions—is quite complicated, and the classical approach of resting considerations on the F-test and hence a comparison of V_d's (or of δ's which give probability one to a set of d's all of which have the same L_d and V_d) is untenable. Thus, for example, in the case of testing a hypothesis of equality of "treatment effects" in the setup of block designs, if we are interested in alternatives which are close to the hypothesis of equality, we shall see that it is not the average according to δ of V_d's, but rather of V_d^{-1}'s, which is relevant. More precisely, we shall have occasion to consider the average (with respect to δ) A_δ of a set of A_d's with common value b_d (this last condition is irrelevant if σ^2 is known). Of course, this A_δ is meaningless as far as variance is concerned; e.g., if A_δ is nonsingular, A_δ^{-1} is not a covariance matrix of b.l.e.'s unless δ gives probability one to a set of d's with identical A_d's. However, such an average will be meaningful as the inverse of a covariance matrix in another setting: if a new experiment is formed by replicating each of a set of d's a number of time $\eta(d)$, then the integral $A(\eta)$ of A_d with respect to the measure η is meaningful in the same way that A_d is. This circumstance will most often arise in regression problems, where we are given a set of row vectors x_d, each with m elements, the expected value of an observation corresponding to experiment d is $x_d\theta$, and we must choose an experiment d' consisting of N observations; if $X_{d'}$ has $\eta(d)$ rows equal to x_d, then $A_{d'}$ is just $A(\eta)$, but it will be more useful to think of the experiment in the latter form. See also Sections 2C and 3 in this connection. It will also be convenient in such regression problems to think of being given an m-vector f of real functions $f_1, \ldots, f_m$ on a space $\mathscr{X}$, an observation at the point x' of $\mathscr{X}$ yielding the experiment $x_d = (f_1(x') \ldots f_m(x'))$. In such settings we shall always assume $\mathscr{X}$ compact in a topology for which the f_i are continuous. This assumption merely insures that suprema of certain functions are attained, so that optimum designs exist; it is easily weakened, and is completely unnecessary in such places as our invariance considerations (Sections 2E, 3, and 4A). We also assume the f_i to be linearly independent; this is an assumption of identifiability which is easily dispensed with, as is often done in settings like those of Section 4B.

The reader is cautioned to keep clear the distinction between the measures δ and η; the first symbol will always refer to a randomized design; the second, to a non-randomized design (η is replaced by a probability measure $\xi = \eta/N$ in the approximate development described in Section 2C).

Many of the topics covered in this paper were treated also in a paper by the author (1958) and in a paper by the author and Wolfowitz (1959), which we will hereafter refer to as I and II, respectively.

2. Basic Considerations

Before taking up our complete class and optimality results, it is convenient to discuss a few topics which are relevant to considerations of this kind.

A. *Randomization and degrees of freedom.*—It was indicated in the previous section that new considerations arise when the problem at hand is not one of point estimation. There are two phenomena here:

(i) It can happen that a design which is inefficient for point estimation is preferred for problems of testing hypotheses, etc., when σ^2 is *unknown*, due to the fact that the loss of accuracy in the relevant b.l.e.'s is more than offset by an increase in the number of degrees of freedom associated with $\bar{S}_d$. A trivial example of this occurs in the degenerate case of the "one-way analysis of variance" where $m = N$ and Δ consists of all matrices for which each row has $m-1$ zeros and a single one; here the only design in Δ_R which estimates all components of θ with finite variance takes one observation from each population (X_d is a permutation matrix), but this design is useless for testing hypotheses since $N - b_d = 0$; on the other hand, if r is a positive integer less than m and δ chooses at random among those X_d which have r columns of zeros (i.e., the m observations are taken from $m - r$ populations), we obtain an example of a randomized design for which, for example, we can obtain on the contour $\sigma^{-2}\theta'\theta = c > 0$ a power function for testing the hypothesis $\theta = 0$, whose infimum is greater than the size of the test. A slightly less trivial type of example, which has nothing to do with randomized designs, is given by the regression problem where in $m = 2$, we are interested in inference about θ_2, and the three possible x_d's are $x_1 = (0, 1)$, $x_2 = (1, 1)$, $x_3 = (1, 3\cdot2)$. It is easy to show by the method of Section 4C that, when N is even, the unique best design d^* for minimizing the variance of the b.l.e. of θ_2 is given by $\eta(1) = 0$, $\eta(2) = \eta(3) = N/2$, the variance of the b.l.e. being $1/(1\cdot21N)$. The design d^{**} for which $\eta(1) = N$, $\eta(2) = \eta(3) = 0$, yields the larger variance $1/N$, but also gives one more degree of freedom to error. If, e.g., we are testing the hypothesis that $\theta_2 = 0$ by using the two-tailed Student's test of size α, then for fixed N and small α the power at the alternative $|\theta_2|/\sigma = c$ is approximately $\alpha[1 + 1\cdot21N(N-2)c^2/2]$ under d^* and $\alpha[1 + N(N-1)c^2/2]$ under d^{**} when $|c|$ is small (see I). Thus, if we are mainly interested in alternatives which are close to the null hypothesis, and if α is not too large, then the design d^{**} is superior to d^* when $N = 2$, 4, or 6, although d^* is better for point estimation of θ_2. (Throughout our discussion, "α sufficiently small" will include commonly employed values like $0\cdot01$ and $0\cdot05$.)

(If the reader dismisses these and subsequent examples on the grounds that they are not practical problems, let him be reminded that our examples are being chosen to be simple, and that similar examples exist in more complicated practical settings. Moreover, it seems to the author that the "burden of proof" in such practical settings rests on the proponents of various designs, who should seriously take up the question of whether or not the "intuitively appealing" designs being suggested by them are actually reasonably efficient or not.)

(ii) Even when σ^2 is known, it is generally true in settings where incomplete block designs, orthogonal arrays, Latin squares, etc., are used, that the symmetrical designs (like those just named) which are generally used are ***not*** optimum for testing the usual hypotheses against alternatives which are near the null hypothesis. This phenomenon can be seen in the following simple example: Suppose $N = 6$, $m = 3$, and again we have the "one-way analysis of variance" setup wherein each row of X_d must have a single one and two zeros, and an experiment d can thus be specified by the numbers of observations n_1, n_2, n_3 taken from each of the three populations. σ^2 is known, so we can assume $\sigma^2 = 1$. We consider three problems: (a) point estimation of θ, (b) testing the hypothesis $\theta_1 = \theta_2 = \theta_3 = 0$, (c) testing the hypothesis $\theta_1 = \theta_2 = \theta_3$; if in (b) and (c) we are mainly interested in alternatives close to the null hypothesis, and the size α in (c) is $<0\cdot3$, then we shall see that *each of these three problems dictates the use of a different design.* For problem (a), with any of the relevant definitions of optimality of Section 4, the nonrandomized design d^* for which $n_1 = n_2 = n_3 = 2$ is obviously optimum. For problem (b), let us compare d^* with the randomized design δ_1 which assigns probability 1/3 to each of the three possible designs where one n_i equals 6; i.e., δ_1 chooses a population at random and takes all 6 observations from the single chosen population. If d^* is used, we use the usual test based on the χ^2 distribution with three degrees of freedom, and the power function is of the form $\alpha + c_3\lambda + O(\lambda^2)$ when λ is small, where λ, the parameter on which the power function depends, is just the excess of the expected value of the χ^2-distributed statistic over what it is under the null hypothesis; thus, $\lambda = 2(\theta_1^2 + \theta_2^2 + \theta_3^2)$. When δ_1 is employed, if population i is chosen we use the two-tailed normal test of the hypothesis $\theta_i = 0$, or, in other words, the test based on the χ^2-distribution with one degree of freedom, and with a notation parallel to that above we have for the conditional power (given that population i is chosen) $\beta_i = \alpha + c_1\lambda_i + O(\lambda_i^2)$, where $\lambda_i = 6\theta_i^2$; thus, the *un*conditional power associated with the design δ_1 is $(\beta_1 + \beta_2 + \beta_3)/3 = \alpha + c_1\lambda + O(\lambda^2)$. Thus, we are led to compare c_1 and c_3, and it is easily shown that $c_1 > c_3$ for all α (and $c_1/c_3 \to 3$ as $\alpha \to 0$). In fact, a simple way to prove this without computation is to consider the problem of testing whether or not a normal random variable Z, with unit variance, has mean $\mu = 0$, on the basis of Z, W_1, and W_2, where the W_i are normal, independent of each other and of Z, and have unit variances and means known to be zero: the test based on large values of Z^2 is well known to be superior to that based on $Z^2 + W_1^2 + W_2^2$, and c_1 and c_3 are just the derivatives of the two power functions with respect to μ^2, at the origin. One can compare other designs with δ_1 similarly, and conclude that δ_1 is the unique best design for maximizing the minimum power on the contour $\Sigma_i\theta_i^2 = b^2$ for all b in a neighbourhood of the origin. We must still consider problem (c). Here the behavior of the power functions is slightly more delicate. Let δ_2 be the design which chooses each possible pair of populations with probability 1/3 and then takes 3 observations from each of the two chosen populations. Writing $\bar{\theta} = \Sigma\theta_i/3$ and $\lambda' = 2\,\Sigma(\theta_i - \bar{\theta})^2$, an argument similar to that used in discussing problem (b) yields $\alpha + c_2\lambda' + O(\lambda'^2)$ for the power function associated with d^* (where c_2 has the corresponding meaning when there are two degrees of freedom) and

$$\alpha + (3/4)\, c_1\lambda' + O(\lambda'^2)$$

for that associated with δ_2 and which arises from the usual test of $\theta_i = \theta_j$ for whichever pair (i, j) of populations is chosen. One computes easily that $(3/4)\, c_1 > c_2$ if $\alpha < 0{\cdot}3$, and a similar comparison of δ_2 with other designs thus shows that δ_2 maximizes the minimum power on the contour $\Sigma(\theta_i - \bar{\theta})^2 = b^2$ for all b in a neighbourhood of the origin, whenever $\alpha < 0{\cdot}3$.

When σ^2 is unknown, the phenomena of (i) and (ii) reinforce each other. As was remarked earlier, these phenomena are present in all of the standard block and array design situations (see I); the intuitively appealing symmetrical designs which are good for point estimation are relatively poor for testing hypotheses against alternatives near the null hypothesis, while randomized designs which choose as few different "treatments" as possible (one in problems like (ii)(b), two in problems like (ii)(c)) for actual observation, are optimum against such alternatives. For various other problems, e.g., ranking problems and certain problems of interval estimation, similar results hold.

Randomized procedures of the type illustrated by δ_1 and δ_2 above have a certain unappealing property, despite their "local" optimality: in problem (ii)(b), for example, the minimum power on the contour $\Sigma\theta_i^2 = b^2$ when δ_1 is used approaches $(1 + 2\alpha)/3$ as $b \to \infty$. This lack of a desirable consistency property (which requires that this minimum power should approach one as $b \to \infty$) for δ_1 makes one wonder if d^* has some desirable *global* optimality properties, after all. For example, we can define the property of *stringency* of a design, analogous to that of a most stringent test of a hypothesis, as follows: under the restriction to procedures of size α for testing a given hypothesis, let the "envelope power function" $e(\theta')$ be the maximum possible power at the *particular* alternative $\theta = \theta'$ (with obvious modifications if σ^2 is unknown); if β_γ denotes the power function of the procedure γ (this is a design δ together with a test for each possible choice of d if δ is randomized), we define a *most stringent design of size* α to be one for which $L(\gamma) = \sup_\theta[e(\theta) - \beta_\gamma(\theta)]$ is a minimum.

The calculation of a most stringent design is in general quite a formidable task, and is usually much more difficult than the calculation of a most stringent test for a fixed design. However, the calculation can be performed easily in a few simple cases. For example, let us consider a slightly simpler problem than that of (ii)(b), where now $N = m = 2$. Here we shall let d' be the design for which $n_1 = n_2 = 1$, d_i will be the design for which $n_i = 2$, and δ' will be the design for which d_1 and d_2 are each chosen with probability 1/2; d' and δ' are thus the analogues of d^* and δ_1 in problem (ii)(b). The hypothesis is again $\theta_1 = \theta_2 = 0$. Here $e(\theta)$ is easy to compute: for example, if $\theta' = (\theta_1', \theta_2')$ with $\theta_1' > \theta_2' > 0$, then the best procedure against the particular alternative $\theta = \theta'$ is obviously to use design d_1 and reject the null hypothesis if the sum of the observations is large. Thus, we obtain $e(\theta) = \Phi[-k_\alpha + \sqrt{2}\max(|\theta_1|, |\theta_2|)]$, where Φ is the standard normal distribution function and $\Phi(-k_\alpha) = \alpha$. For the power function associated with δ' we obtain

$$\Sigma_i[\Phi(-k_{\alpha/2} + \sqrt{2}\,\theta_i) + \Phi(-k_{\alpha/2} - \sqrt{2}\,\theta_i)]/2,$$

while for d' it is the appropriate non-central χ^2 probability. Noting that $e(\theta)$ depends only on $\max(|\theta_1|, |\theta_2|)$ and that both d' and δ' attain their minimum power on the contour $\max(|\theta_1|, |\theta_2|) = c$ when $\theta_1 = c$ and $\theta_2 = 0$, one can compute directly that $e(\theta) - \beta_\gamma(\theta)$ has a maximum slightly greater than $0\cdot5$ when d' is used, and slightly less than $0\cdot5$ when δ' is used, when $\alpha = 0\cdot01$. A similar calculation for other designs shows that δ' *is indeed the most stringent design in this case.* (How general this phenomenon is, is not known.) One can certainly conceive of situations in which consistency (in the sense used above) is not the most important factor in choosing a design.

Designs such as δ' have also been criticized on the ground that "they do not provide estimates of all θ_i". If the problem confronting the experimenter is really one of hypothesis testing, this intuitive objection must be rejected as being irrelevant.

In the setting where classical "randomization" is employed, e.g., to eliminate intrablock bias in randomized block designs, an argument like that of I shows that the locally best design for hypothesis testing must again choose (at random) as few treatments as possible to be used, and should then choose their positions at random in the blocks.

What are the consequences of the facts discussed thus far in this section? Evidently, that if one is really interested in problems such as ones of hypothesis testing, and in criteria such as the local behaviour of the power function (which is a natural extension of various local Neyman-Pearson criteria for a fixed design) or stringency, then one must alter the widely held view that the design to use is necessarily the same one that would be used for point estimation. Moreover, we shall see in Section 3 that, even in problems of point estimation, randomized designs may be needed. (These phenomena also occur in regression problems like those discussed in Section 4C, although we shall see there that there are certain problems in which these phenomena do not occur.) Thus, essentially all of the known "optimality" results in the literature which refer to such designs must be understood to be incorrect for testing hypotheses unless we add the restriction that we limit our consideration to nonrandomized designs, or somewhat alter our optimality criteria. For example, if we use criteria of local optimality or stringency of the type considered above, but add the proviso that the design must yield a power function which is consistent in the sense described above, then the results of Section 4B imply that the classical designs have certain optimum properties. But one must be careful with this kind of hedging; for, if examples like those we have considered may sometimes help us to reconsider our goals more carefully and to realize, for example, that we may not really have a problem of hypothesis testing, nevertheless the rational approach is still to state the problem and the optimality criterion and then to find the appropriate design, and not to alter the statements of the problem and criterion just to justify the use of a design to which we are wedded by our prejudiced intuition.

B. *The use and misuse of the F-test.*—It is generally assumed in the normal case that, whatever design is employed, the appropriate F-test (or the corresponding χ^2-test if σ^2 is known) should be used in testing a hypothesis. Although this assumption is warranted if the design is appropriately symmetrical and we are equally concerned with alternatives in all directions from the null hypothesis, it can be drastically wrong in other cases. For an arithmetically simple example with important practical implications, suppose that with $m = 2$, $N = 4$, $\sigma^2 = 1$, and the problem of testing the hypothesis $\theta_1 = \theta_2 = 0$ in the one-way analysis of variance setup, we use the nonrandomized design d with 2 observations from each population. We are equally interested in alternatives in all directions, and might specify this fact by saying that we want a test of size α which maximizes the minimum power on the contour $\theta_1^2 + \theta_2^2 = c^2$. (Such considerations are discussed in detail by Neyman and Pearson, 1938.) It is well known that, for every $c > 0$, the usual test based on the χ^2-distribution with two degrees of freedom does the job (this is originally an unpublished result of Hunt and Stein; for a proof, see Kiefer, 1957*a*). Now suppose that we start with the design d, but are plagued by a missing observation and thus end up with the design d' for which

$$A_{d'} = \begin{pmatrix} 10 \\ 02 \end{pmatrix}.$$

Now, presumably our equal interest in alternatives in all directions should not be "conveniently" changed in the way appropriate to justify the use of the χ^2-test (based on $t_{d'1}^2 + 2t_{d'2}^2$), by the same accident which reduced us from d to d'. But if we still want a test of size α which maximizes the minimum power on the contour $\theta_1^2 + \theta_2^2 = c^2$, then this test no longer satisfies the criterion, and hence should not be used! Of course, similar considerations apply in examples such as that of A(ii)(c) of this section. We note also that if we had instead specified our interest in various alternatives above by saying that we were equally interested in all θ for which max $(|\theta_1|, |\theta_2|) = c$ (one can certainly envisage this possibility), and hence wanted to maximize the minimum power on such a contour, then, even for d, the χ^2-test is not appropriate.

In this age of high-speed machinery, it should be easy to determine and table approximately optimum test procedures in situations like the above; it seems to be mainly the inertia of tradition which delays such computations.

Computational difficulties are somewhat reduced if we limit our considerations to the local theory (i.e., to the behavior of the power function near the null hypothesis). Thus, for either of the contours mentioned just above, if we let c approach 0 and rephrase our criterion in terms of derivatives of the power function, the problem reduces to that of finding a regular Neyman-Pearson test of type C (Neyman and Pearson, 1938; the author regrets that this reference was omitted from I through an oversight). In the case of design d, the appropriate test is then the usual one based on the χ^2 distribution, while in the case of design d' the critical region is of the form $at_{d'1}^2 + bt_{d'2}^2 > 1$, where the positive constants a and b can be obtained numerically. This was shown by Neyman and Pearson; a simple demonstration can be obtained by the usual minimax technique of considering Bayes procedures with respect to the *a priori* distribution which assigns probability p_1 to each of $\theta = (0, \epsilon)$ and $(0, -\epsilon)$, probability p_2 to $(\epsilon, 0)$ and $(-\epsilon, 0)$, and probability $1 - 2p_1 - 2p_2$ to $(0, 0)$, and then letting ϵ approach 0. It is useful to note that, when α is small, the local behaviour of the regular type C region is well approximated by that of a test with rectangular acceptance region, and the latter is easy to compute.

Thus, if we consider only nonrandomized designs and want to find that design and associated test which approximately maximize the minimum power on small spheres (with a similar treatment for a given family of ellipsoids) about the null hypothesis (we shall call these "designs of type L" in Section 4), we can conclude that a regular type C test is to be used with whatever design is chosen. For randomized designs, the situation is slightly more complicated.

Although it seems reasonable that the statistician should often be able to specify the relative importance of alternatives in each direction, the type C test has often been criticized on the grounds that it *does* depend on such a specification (although perhaps the subconscious motivation for such criticism was often the fact that this criterion did not always lead to the use of the classical test!). Thus, if we make a one-to-one transformation on the parameter space which is twice differentiable and has nonvanishing Jacobian at the null hypothesis, which hypothesis the transformation leaves fixed, then the regular type C tests for the new and old parametrizations will differ unless the transformation is locally a homogeneous stretching times an orthogonal transformation. Motivated by this, Isaacson (1951) defined a type D test of a simple null hypothesis to be a locally strictly unbiased test which maximizes the Gaussian curvature of the power function at the null hypothesis, and noted that a type D test remains invariant under all reparametrizations like the above. Thus, type D tests will appeal to statisticians who want an optimality criterion which does not require any specification of the relative importance of various alternatives, although obviously one must be warned against the use of such a criterion as a panacea for ignorance or laziness. It is shown in I that, in the normal case, the usual F-test is a type D test (or, more precisely, what Isaacson calls a type E test, which is a type D test for each fixed value of the nuisance parameters), with the analogous result for the test based on the χ^2-distribution if σ^2 is known.

It should be noted that this last criterion is the *only* simple optimality criterion which is satisfied by these classical tests no matter what V_d may be. Criteria like those associated with the theorems of Hsu (1941) and Wald (1942) (see also Wolfowitz, 1949), and which look at the power function on just those contours where the F-test has constant power, must be excluded as being unnatural in themselves, except in some cases where these contours are spheres or where, because of the relative importance of various errors, the design was chosen precisely to achieve the particular elliptical contours at hand.

C. *Discreteness.*—A computational nuisance in the determination of optimum designs is the fact that such designs are often easy to characterize for many values of N, but that slight irregularities occur for other values of N. For example, to minimize the generalized variance in the case of cubic regression on the interval $[-1, 1]$, one should take 1/4 of the observations at each of the values ± 1, $\pm 5^{-1/2}$, if N is divisible by 4. If $N = 5$, the appropriate design takes one observation at each of the values 0, $\pm 0{\cdot}511$, ± 1, and there are similar irregularities for the other values of N which are not divisible by 4. While it is of interest, especially for small values of N, to have a table of (exactly) optimum designs, the first task would seem to be the computation of a design pattern like that on the four points above, which will yield approximately optimum designs for *all* large N. Thus, in the above example, if N is not divisible by 4 and we take all observations at these four points with as equal a division as possible (many other similar designs will clearly be suitable), then we obviously obtain a design for which the generalized variance is at most $[1 + O(1/N)]$ times the minimum possible. The departure from exact optimality is far less than that of the commonly employed "equal spacing" designs in this situation.

Not only are approximately optimum designs of the type just illustrated convenient from the point of view of tabling designs; in addition, many useful algorithms will be seen below to exist for obtaining such designs, whereas the *exact* optimum designs are in general very difficult to compute in even quite simple situations. This is a familiar obstacle in extremum problems involving discrete variables.

We shall also see, in the discussion following equation (3.3), that randomized designs are unnecessary in point estimation, to within such approximations.

Thus, although we shall sometimes give examples in Sections 3 and 4 which illustrate the differences between the discrete (exact) theory and the continuous (sometimes approximate) theory, we shall most often consider the latter in problems of regression. In the notation introduced at the end of Section 1, these approximate considerations are accomplished by not requiring η to be integer valued. We shall usually write $\eta = N\xi$ and

$$M(\xi) = A(\eta)/N$$

(the information matrix of ξ, per observation), and thus shall be considering probability measures ξ on a space $\mathscr{X}$. These measures will always have finite support, this being the only type which it is practically meaningful or necessary to consider, although other measures can be included without theoretical difficulty (the support of ξ is the smallest closet set of unit ξ measure). The reader is cautioned again not to confuse such a probability measure ξ, which represents a nonrandomized design (more precisely, $N\xi$ does, approximately) with the probability measures δ on Δ, which are randomized designs.

In the setups where balanced block designs, orthogonal arrays, Latin square designs, etc., are customarily employed, we shall see in Section 4B that these classical symmetrical designs possess various optimum properties. What if the restrictions on the problem, e.g., the block sizes and number of blocks, are such that no such design exists? It is tempting to conjecture that a design which is in some sense as "near as possible" to being symmetrical (balanced) is optimum, but this is difficult to make precise in general. Thus, in simple cases like the one-way analysis of variance setup, it is obvious that a design which splits the N observations as equally as possible among the m populations minimizes both the average variance and the generalized variance, but in several more complex situations a corresponding result is not so easy to obtain, and optimum designs are often difficult to characterize when appropriately symmetrical ones do not exist. We shall not be concerned with the tedious optimality calculations in these cases, although for small N a (machine) enumeration of appropriate designs would be of practical use.

When N is large in a regression experiment, or the block size is large in a block experiment, discrete (exact) approximations to the ξ's in an approximate complete class will yield an (exact) ϵ-complete class (Wolfowitz, 1951) with ϵ small, with an analogous result for specific optimality criteria, as illustrated in the first paragraph of this subsection.

D. *Heteroscedasticity and variable cost.*—If our assumption of constant variance is replaced by the assumption that the covariance matrix of Y_d under design d is $\sigma^2 R_d$, which we assume for convenience to be positive definite, and if B_d is the positive definite symmetric square root of R_d, then replacing Y_d by $B_d^{-1} Y_d$ and X_d by $B_d^{-1} X_d$ puts us in the previous framework. If the experiment d costs c_d and the total cost rather than total number of observations is to be kept constant, then replacing B_d by $c_d^{\frac{1}{2}} B_d$ in the above again returns us to a problem in the previous framework, whose solution yields the desired minimum cost for the problem at hand. (An obvious modification handles the case where the cost of *analyzing* each d is known in advance; such considerations are more difficult for randomized designs.) In the regression setup, analogous remarks apply with c_d and the scalar R_d referring to the observation with expectation $x_d \theta$.

Some further remarks on this subject appear just above the example of Section 3.

E. *Invariance (symmetry).*—In many regression problems the functions f_i will be appropriately symmetric with respect to a group of transformations on $\mathscr{X}$, to enable us to conclude that there is a symmetric ξ which is optimum in a given sense. For example, in the case of polynomial regression on the interval $[-1, 1](f_i(x) = x^{i-1})$, the generalized variance is minimized (i.e., det $M(\xi)$ is maximized) by a ξ which is symmetric about 0 in the sense that $\xi(x) = \xi(-x)$ for each x in $\mathscr{X}$.

To give an example of one such general symmetry result, suppose G is a group of transformations on $\mathscr{X}$ such that, for each g in G, there is a linear transformation $\bar{g}$ on Θ which can be represented as an $m \times m$ matrix of determinant one for which

$$\Sigma_i \theta_i f_i(x) = \Sigma_i (\bar{g}\theta)_i f_i(gx)$$

for all x and θ, where we have written $(\bar{g}\theta)_i$ for the ith component of $\bar{g}\theta$. Then, for fixed g, the problem in terms of the parameters $\bar{\theta} = \bar{g}\theta$ and variable $\bar{x} = gx$, coincides with the original problem. Hence, if ξ minimizes the generalized variance (maximizes det $M(\xi)$) for the original problem in terms of x and θ, it is also optimum in this sense for the problem in terms of $\bar{x}$ and $\bar{\theta}$. But then the measure ξ_g defined by

$$\xi_g(L) = \xi(g^{-1}L)$$

is optimum for the problem in terms of $x = g^{-1}\bar{x}$ and $\bar{\theta}$. Thus, since $\bar{g}$ has determinant one, the inverse $\sigma^{-2}M(\xi_g)$ of the covariance matrix of b.l.e.'s of $\theta = \bar{g}^{-1}\bar{\theta}$ when ξ_g is used has the same determinant as the corresponding matrix $\sigma^{-2}M(\xi)$ which is relevant when ξ is used. Suppose for the moment that G contains a finite number, say p, of elements. Write

$$\bar{\xi} = \sum_{g \epsilon G} \xi_g / p.$$

Then $\bar{\xi}$ is a probability measure (design) on $\mathscr{X}$ which is symmetric with respect to (invariant under) G in the sense that $\bar{\xi}(L) = \bar{\xi}(gL)$ for every subset L of $\mathscr{X}$ and every g in G. We want to show that $\bar{\xi}$ is optimum.

Now, the experiment $\bar{\xi}$ can be "broken up" into the experiments ξ_g/p. Thus, for estimating θ under $\bar{\xi}$, we clearly have

$$M(\bar{\xi}) = p^{-1} \Sigma_g M(\xi_g), \tag{2.1}$$

and since the $M(\xi_g)$ all have the same determinant, the fact that $\bar{\xi}$ minimizes the generalized variance (maximizes det $M(\xi)$) will follow from the proposition that if det $M(\xi)$ is a maximum for $\xi = \xi'$ and for $\xi = \xi''$, then it is also a maximum for $\xi = \lambda\xi' + (1 - \lambda)\xi''$, where $0 < \lambda < 1$. To prove this last, let C be a nonsingular $m \times m$ matrix such that $CM(\xi')C'$ is the identity and $CM(\xi'')C'$ is diagonal with diagonal elements g_i and (since det $M(\xi')$ = det $M(\xi'')$) $\prod_i g_i = 1$. Since

$$M(\lambda\xi' + (1 - \lambda)\xi'') = \lambda M(\xi') + (1 - \lambda) M(\xi''),$$

our proposition comes down to proving that

$$\prod_i (1 - \lambda + \lambda g_i) \geqslant 1, \tag{2.2}$$

since strict inequality here is impossible by the maximality of det $M(\xi')$. Now,

$$- \Sigma_i \log (1 - \lambda + \lambda g_i)$$

is convex in λ, and hence is $\leqslant (1 - \lambda)[- \Sigma_i \log 1] + \lambda[- \Sigma_i \log g_i] = 0$. Equation (2.2) follows at once. In fact, this last argument is valid for any optimum ξ' and ξ''; thus, we have proved:

There is a D-optimum ξ which is invariant under G; the set of all D-optimum ξ is a convex set which is invariant under G.

(*D*-optimality, which is discussed in Section 4, refers to minimizing the generalized variance, and invariance of a set of ξ's means that ξ_g is in the set whenever ξ is.) Of course, it may still require considerable computation to determine which symmetric ξ's are optimum.

Similar results hold for other optimality criteria. For example, from the orthogonality of the transformations $\bar{g}$ and the invariance under orthogonal transformations of the trace of the covariance matrix of best linear estimators we obtain that, for extimating θ, the average variance is the same under all ξ_g. Now, assuming G is finite, an unbiased estimator of θ_i under $\bar{\xi}$ can be obtained from the average of the b.l.e.'s of that θ_i under the subexperiments ξ_g/p which go to make up $\bar{\xi}$, and the b.l.e. under $\bar{\xi}$ must do at least as well. We conclude that $\bar{\xi}$ also minimizes the average variance, and again the set of all such minimizers is a convex invariant set.

Similar considerations for the case where we are interested in a subset of the components of θ are given in II. An improvement on these results will be given in Theorem 4.3, where the best possible result of this nature is proved.

Analogous results hold when G is compact (one of the conditions which yield the usual minimax invariance theorem in statistics and games; see, e.g., Kiefer (1957*a*)).

Throughout this paper, we shall limit our invariance considerations to nonrandomized designs; the reader will find it easy to obtain analogues for randomized designs. (For examples, note the designs δ_1 and δ_2 in Section 2A(ii).)

Of course, the validity of our results on invariant designs depends crucially on the invariance of the optimality criterion under the group of transformations. Thus, for example, in the trivial case of linear regression on $[-1, 1]$, where $f_i(x) = x^{i-1}$ for $i = 1, 2$, we can conclude that there is a symmetrical design which is optimum for estimating θ_1, one which is optimum for estimating θ_2, or one that minimizes the average (or the generalized) variance for estimating θ_1, θ_2; but for estimating $\theta_1 + \theta_2$, the unique optimum design is $\xi(1) = 1$. Thus, our symmetry considerations do not generally yield anything useful for the complete class results of Section 3 in the way that they do for particular optimality criteria; in fact, the polynomial example of Section 3 shows that ξ can be admissible without the corresponding $\bar{\xi}$ being admissible. However, we shall see in Theorem 3.3 that a subset of the designs which are invariant under a group of transformations each of which leaves fixed the same s parameters, say $\theta_1, \ldots \theta_s$, is an essentially complete class for any problem concerned only with $\theta_1, \ldots, \theta_s$.

It is to be noted that the optimum properties of balanced block designs, Latin square designs, orthogonal arrays, etc. (see Section 4B), are really symmetry results in the same sense as those we have considered here, but in a much more complex setting wherein we are interested only in certain linear functions of the θ_i; the proof that an appropriately symmetric design (if it exists) is optimum in the exact theory, is therefore more difficult.

The reader should keep in mind that our use of the word "symmetry" has nothing to do with its usage in reference to balanced incomplete block designs where the parameters b and u are equal. This is also true in Section 4B.

3. Admissibility and Complete Classes

We begin with some perfectly general considerations, and shall use the example of polynomial regression on a real interval for illustration at the end of the section.

By a *statistical problem* π we mean a specification of the possible states of nature, decisions, and losses (including costs of various possible experiments and of their analyses), with perhaps some restriction on the class of procedures we consider. We shall say of two designs δ and δ' in Δ_R that δ' is *at least as good* (π) as δ if for each risk function for problem π which is attainable by some procedure under δ, there is a risk function attainable by some procedure under δ' which is nowhere larger. If δ' is at least as good as δ for each of a collection P of problems π, we shall say that δ' is at least as good (P) as δ. If δ' is at least as good (P) as δ but not vice versa, we shall say that δ' is *better* (P) than δ, or that δ' *dominates* (P) δ. If δ' is such that no δ'' dominates (P) δ', then we say that δ' is *admissible* (P). A class **C** of designs is *complete* (P) (resp., *essentially complete* (P)) if for any δ not in **C** there is a δ' in **C** which is better (P) than (resp., as good (P) as) δ. If no proper subset of **C** has this property, **C** is said to be minimal complete (P) (resp., minimal essentially complete (P)).

The reader should be aware that the complete class considerations of Ehrenfeld (1955*b*) and of the present section, in the nonparametric case, have more restricted meaning than is usual in decision theory: we consider only best linear (minimax) estimators, and a complete class of experiments supplies us with a good experiment (and estimators) for any such problem. Broader complete class results which also compare nonlinear estimators depend on the extent of the nonparametric class.

Let $\mathscr{F} = \{F\}$ be the space of possible states of nature. For d in Δ, let $\mathscr{Y}_d$ be the space of possible complexes of observed values under d (thus, if Δ consists of experiments with N real observations, we could take each $\mathscr{Y}_d$ to be Euclidean N-space). An appropriate Borel field is given on each $\mathscr{Y}_d$. Let $H_F(A, d)$ be the probability that the outcome of the experiment will be in the measurable subset A of $\mathscr{Y}_d$, when experiment d is performed and F is the true state of nature. An experiment d' is said to be *sufficient* for the experiment d'' if there is a function $q(A, y)$ which for each y is a probability measure on $\mathscr{Y}_{d''}$ and for each A is a measurable function on $\mathscr{Y}_{d'}$, and such that

$$H_F(A, d'') = \int_{\mathscr{Y}_{d'}} q(A, y)\, H_F(dy, d') \tag{3.1}$$

for every measurable subset A of $\mathscr{Y}_{d''}$ and every F in $\mathscr{F}$. This is really the same notion of sufficiency that one is used to for comparing two random vectors for a fixed experiment; it says that d' is sufficient for d'' if we can imitate d'' probabilistically from d' in the way indicated here: observe the outcome y (say) of d', use the measure $q(A, y)$ to obtain a random variable Y'' with values in $\mathscr{Y}_{d''}$, and on integrating over possible values of y note that Y'' has the same distribution according to each F as would the outcome of experiment d''. Of course, q depends on d' and d'', but not on F.

Let $\mathscr{Z}$ be the space of all pairs (d, y_d) for y_d in $\mathscr{Y}_d$ and d in Δ. If all $\mathscr{Y}_d$ can be taken to be the same space $\mathscr{Y}$, then $\mathscr{Z}$ is just the cartesian product of Δ and $\mathscr{Y}$. In any event, we assume a Borel field to be given on $\mathscr{Z}$ which includes the measurable subsets of $\mathscr{Y}_d$ as sections at d and which also includes each set which is the union over d in Δ' of all $\mathscr{Y}_d$, whenever Δ' is any subset of Δ which is a member of the Borel field with respect to which the δ's in Δ_R are measures. Since we are usually to be concerned with δ's which give measure one to a finite set, these considerations will be trivial. We now think of δ as being a probability measure on Δ. When δ is used and F is the true state of nature, the probability that the chosen design r (say) and subsequent outcome y_r of the experiment will satisfy $(r, y_r) \in B$, where B is a measurable subset of $\mathscr{Z}$, is just

$$G_F(B;\ \delta) = \int_{\Delta} H_F(B_r, r)\, \delta(dr),$$

where $B_r = \{y_r\colon\ (r, y_r) \in B\}$. For randomized experiments, then, we say that δ' is *sufficient for* δ'' if there is a function $Q(B, z)$, which is a measure on $\mathscr{Z}$ for each z and a measurable function on $\mathscr{Z}$ for each B, such that

$$G_F(B;\ \delta'') = \int_{\mathscr{Z}} Q(B, z)\, G_F(dz;\ \delta') \tag{3.2}$$

for each measurable B and each F in $\mathscr{F}$. This is the appropriate generalization of the previous definition, it now being necessary to think of δ as yielding an "observed value" in $\mathscr{Z}$.

In the case of nonrandomized designs, sufficiency and its equivalence to certain other criteria have been treated extensively by D. Blackwell, S. Sherman, C. Stein, and others (see, e.g., Blackwell, 1950). The most important result for our considerations is that, under standard measurability assumptions, if P^* is the class of all possible problems (usually a much smaller subclass will suffice), then d' is at least as good (P^*) as d'' if and only if d' is sufficient for d''. Of course, the "if" part of this is an immediate consequence of the definition of sufficiency.

From the definition of sufficiency for randomized designs (in particular, the correspondence of (3.2) to (3.1) when $\mathscr{X}$ is made to correspond with $\mathscr{Y}_{d'}$ and $\mathscr{Y}_{d''}$, which enables us to think of a randomized design as a nonrandomized design with sample space $\mathscr{X}$), we see at once that all of the equivalence results just mentioned have natural analogues in the case of randomized designs; in particular, δ' is at least as good (P^*) as δ'' if and only if δ' is sufficient for δ''.

We shall now consider the setup of Section 1, where we shall discuss some results of Ehrenfeld (1955*b*). (Ehrenfeld's definition of completeness and "at least as good" differ from ours in that θ is fixed throughout. This does not matter in the linear theory now to be discussed, but yields different results in the nonlinear asymptotic case to be discussed a little later, where it yields a smaller "'essentially complete class" which perhaps conforms less to the usual decision-theoretic meaning of such classes.)

If A and B are $m \times m$ matrices, we shall write $A \geqslant B$ if, for every m-vector w, we have $w'Aw \geqslant w'Bw$; or, equivalently, if $A - B$ is nonnegative definite. If $A \geqslant B$ and $A \neq B$, we write $A > B$. Ehrenfeld (1955*b*, pp. 59–61) shows by an algebraic argument that, if $A_{d''} \geqslant A_{d'}$, then any linear parametric function $\Sigma_i c_i \theta_i$ which is estimable under d' is estimable under d'', and its b.l.e. under d'' has a variance no greater than the variance of the b.l.e. under d'. We now give a transparent proof of this result: Assume normality and that σ^2 is known to be 1, since the linear theory (consideration of b.l.e.'s and their variances) does not depend on this. If $A_{d''} - A_{d'} = B$, which is nonnegative definite, let d^* be a "fictitious experiment" such that $X_{d^*}' X_{d^*} = B$ (such a matrix X_{d^*} clearly exists), with Y_{d^*} independent of $Y_{d'}$. Let $T_d = X_d' Y_d$; then T_d is a sufficient statistic under d, and $T_{d'} + T_{d^*}$ is sufficient under the combination of experiments d' and d^*. But $T_{d''}$ has the same distribution as $T_{d^*} + T_{d'}$, from which it is immediately clear that d'' is sufficient for d' and hence that Ehrenfeld's result holds. In fact, denoting by P_N^* the class of *all* problems under the assumption of normality and known variance, and by P_L the corresponding class of all nonparametric problems of *point estimation* of any linear parametric functions, where we restrict ourselves to linear estimators (as mentioned earlier in this section, the consequent limited meaning of complete (P_L) class results should be kept in mind), we have the following slightly more precise version of Ehrenfeld's result:

Theorem 3.1. *The following are equivalent*:

(a) $A_{d''} \geqslant A_{d'}$,
(b) d'' *is at least as good* (P_N^*) *as* d',
(c) d'' *is at least as good* (P_L) *as* d'.

These equivalences hold with "$\geqslant$" *replaced by* "$>$" *and* "*at least as good as*" *by* "*better than*".

For randomized designs, (c) is equivalent to (a) with A_d replaced by the inverse of the average of V_d with respect to δ, with an obvious modification in the singular case.

It is clear that we could also replace P_L in (c) by problems of vector estimation of all vectors of linear parametric functions; for example, we could treat one vector estimator as being at least as good as another if the variances of components for the former are all no greater than the corresponding variances of the latter (vector risk), or if the covariance matrix of the latter minus that of the former is nonnegative definite, or if the average (or generalized) variance is no greater for the former than the latter; no generality is gained in any case. The essential fact is that, if B and C are positive definite symmetric matrices, then $C \geqslant B$ if and only if $B^{-1} \geqslant C^{-1}$.

If we restrict ourselves to using nonrandomized designs, we see at once that d is admissible (P_N^* or P_L) if and only if there is no d' for which $A_{d'} > A_d$. Thus, if the class of possible A_d for d in Δ is compact as a set of points in m^2-space (this can be weakened), and if we restrict ourselves to designs in Δ, we have an immediate characterization of the minimal complete class (P_N^* or P_L).

If σ^2 is unknown and if we do not restrict ourselves to the point estimation problems in P_N^*, we must also consider the varying number of degrees of freedom associated with different designs, as discussed in Section 2A. This should also be kept in mind in evaluating the meaning of some further complete class results which will be discussed below.

What happens to the above results if we consider Δ_R rather than Δ? As noted in Section 2A(ii), further considerations are necessary in even the simplest problems of normal hypothesis testing; if we consider only the *local* theory there and σ^2 is known, and if we compare δ's which always yield the same number of degrees of freedom and use the F-test (see, however, Section 2B), then the matrices A_δ take the place of A_d in all of the above, although such considerations do not suffice outside of the local theory. For problems of linear (or normal) estimation with quadratic loss, it is the average of V_d with respect to δ that is relevant; more precisely, if c_d is the variance of the b.l.e. of a given linear parametric function when d is used, then the variance under δ is the average with respect to δ of c_d, with a corresponding statement for joint estimation of several linear parametric functions. If δ gives positive probability to a set of d's for which a given linear parametric function is not estimable, then the corresponding variance (more precisely, risk) under δ is infinite.

We have seen in Section 2A that randomized designs cannot in general be ruled out of consideration in hypothesis testing, and that is also true in estimation theory. For example, if $m = 2$ and $N = 3$ in the setup of Section 2A(ii), the generalized variance is a minimum for that δ which assigns probability $\frac{1}{2}$ to each of the two designs for which one n_i is 1 and the other n_i is 2. However, if Δ is sufficiently rich, such uses of randomized designs in estimation problems will be unnecessary. For suppose μ is a probability measure on Δ. Then μ may make sense as a *nonrandomized* experiment ξ (see Sections 1 and 2C), with corresponding information matrix $M(\mu)$, but we can also think of a *randomized* experiment δ as being represented by μ. We assert that *the former is always at least as good* (P_L) *as the latter.* It will be enough to consider the case where μ gives measure one to a set of d's for which A_d is nonsingular, other cases being handled by an appropriate passage to the limit, or by considering appropriate linear transformations and submatrices. The assertion is then a consequence of the following trivial lemma, which in fact tells us that, if μ makes sense as a nonrandomized experiment, then as a randomized experiment it is either *inadmissible* (P_L) or else gives measure one to a set of d's *all of which have the same* A_d:

Lemma. If A and B are symmetric, positive definite $m \times m$ matrices and $0 < \lambda < 1$, *then*

$$\lambda A + (1 - \lambda) B \geqslant (\lambda A^{-1} + (1 - \lambda) B^{-1})^{-1}, \tag{3.3}$$

with equality if and only if $A = B$.

This follows at once on simultaneously diagonalizing A and B. A generalization will be proved in Lemma 3.2 below.

Thus, to within the approximation adopted in Section 2C, if σ^2 is known it is unnecessary to use randomized designs for P_L; thus, *the usefulness of randomized designs for P_L will be greatest in the small sample-size or small block-size case*, where the discrete effects discussed in Section 2C are the largest.

Of course $M(\mu) = A_\mu$ in the above discussion, but this does not mean that randomized designs are unnecessary in normal hypothesis testing, even if σ^2 is known; we have discussed this in Section 2A.

It can sometimes happen that we are interested only in a subclass P_L' of problems in P_L which are concerned with a fixed, proper subset of the m parameters (or of a set of less than m linear parametric functions), say $\theta_1, \ldots, \theta_s$. Let

$$A_d = \begin{pmatrix} B_d & C_d \\ C_d' & D_d \end{pmatrix},$$

where B_d is $s \times s$. Then, if D_d and $D_{d'}$ are nonsingular (with obvious modifications if they are not), we have that d is at least as good (P_L') as d' if

$$B_d - C_d D_d^{-1} C_d' \geqslant B_{d'} - C_{d'} D_{d'}^{-1} C_{d'}',$$

the latter being the inverse (if it exists) of the covariance matrix of b.l.e.'s of $\theta_1, \ldots, \theta_s$ under d'. One can now imitate here our previous discussion in the case $s = m$. Instead of doing all this in detail, we shall only give a needed generalization of the previous lemma, and shall then apply this to obtain a generalization to the present case of the invariance considerations of Section 2E.

Lemma 3.2. *For $j = 1, \ldots, r$, let C_j be $s \times (m - s)$, let D_j be nonsingular positive definite and symmetric, and suppose $\lambda_j > 0$, $\Sigma\lambda_j = 1$.* Then

$$[\Sigma\, \lambda_j C_j][\Sigma\, \lambda_j D_j]^{-1}\, [\Sigma\, \lambda_j C_j]' \leqslant \Sigma\, \lambda_j C_j D_j^{-1} C_j', \tag{3.4}$$

with equality if and only if all $C_j D_j^{-1}$ are equal.

Once the result is proved for the case $r = 2$, it follows easily for general r by induction. In the case $r = 2$, let F be such that FD_1F' is the $(m - s) \times (m - s)$ identity I and FD_2F' is a diagonal matrix Q, and let $E_i = C_iF'$. Write $\lambda_1 = \lambda$. Then (3.4) can be written

$$[\lambda E_1 + (1 - \lambda) E_2][\lambda I + (1 - \lambda) Q]^{-1} [\lambda E_1 + (1 - \lambda) E_2]' \leqslant \lambda E_1 E_1' + (1 - \lambda) E_2 Q^{-1} E_2',$$

which, on writing out inverses and collecting terms, becomes

$$-\lambda(1 - \lambda)[E_1 - E_2Q^{-1}][\lambda Q^{-1} + (1 - \lambda) I]^{-1} [E_1 - E_2Q^{-1}]' \leqslant 0;$$

but this inequality is always true, with equality if and only if $E_1 = E_2Q^{-1}$. This proves the lemma.

Now suppose that G is a group satisfying the conditions of Section 2E, and in addition that each $\bar{g}$ acts trivially on $\theta_1, \ldots, \theta_s$; i.e. that $(\bar{g}\theta)_i = \theta_i$ for $1 \leqslant i \leqslant s$, and hence

$f_i(gx) = f_i(x)$ for $1 \leqslant i \leqslant s$. Suppose ξ is any design, with $M(\xi)$ partitioned as

$$\begin{pmatrix} B(\xi) & C(\xi) \\ C(\xi)' & D(\xi) \end{pmatrix},$$

as above. For any g we have

$$M(\xi_g) = \begin{pmatrix} B(\xi) & C(\xi)\, F_g \\ F_g' C(\xi)' & F_g' D(\xi)\, F_g \end{pmatrix},$$

where F_g is a nonsingular $(m-s) \times (m-s)$ matrix. Hence, if $D(\xi)$ is nonsingular (an obvious modification again sufficing otherwise), we have

$$B(\xi) - C(\xi)\, D(\xi)^{-1}\, C(\xi)' = B(\xi_g) - C(\xi_g)\, D(\xi_g)^{-1}\, C(\xi_g)'.$$

Suppose now that G is finite (other groups are handled as in Kiefer (1957*a*), the determinant of $\bar{g}$ no longer being required to be 1 as it was in Section 2E). An application of Lemma 3.2 and the fact that $B(\xi_g) = B(\xi)$ yield

$$B(\bar{\xi}) - C(\bar{\xi})\, D(\bar{\xi})^{-1}\, C(\bar{\xi})' \geqslant B(\xi) - C(\xi)\, D(\xi)^{-1}\, C(\xi)'.$$

Thus, recalling the remark which follows the proof of (3.3), we have the following result in the approximate theory:

Theorem 3.3. *If G is as specified above and every $\bar{g}$ leaves $\theta_1, \ldots, \theta_s$ fixed, then the designs which are invariant under G form an essentially complete (P_L') class.*

As noted in Section 2E we could extend this result to Δ_R and thus obtain an essentially complete class relative to problems in P_N^* concerned with $\theta_1, \ldots, \theta_s$.

In Section 4A we shall give a corollary of Theorem 3.3 which refers to specific optimality criteria.

We now turn to the question of characterizing admissibility and completeness (P_L) in regression experiments in terms of the x_d defined in Section 1. As noted by Ehrenfeld, if $x_{d'} = a x_d$ with $|a| > 1$ (Ehrenfeld requires $a > 1$, but that is unnecessary), then $x_{d'}' x_{d'} > x_d x_d'$; hence, defining x_d to be maximal if there is no $x_{d'} = a x_d$ with $|a| > 1$, if $\{x_d: d \epsilon \Delta\}$ is a compact set in R^m we conclude at once that, in either the approximate or exact theory, the class of all experiments η which use only maximal x_d is complete (P_L).

If the set of all A_d is convex (in particular, if we consider the approximate theory) and compact, there is a revealing game-theoretic approach: consider the zero-sum, two-person game with payoff function $K(d, t) = t' A_d t$, where the space of player 1's pure strategies is Δ and that of player 2 is $\{t : t't = 1\}$. It is well known (Wald, 1950) that a complete class for player 1 consists of all A_d which are maximal with respect to some mixed strategy of player 2, and this is clearly the same as the class of all A_d which maximize $\mathrm{tr}(TA_d)$ for some symmetric $T \geq 0$. Thus (note also the remark preceding (3.3)), we have

Theorem 3.4.—*If $\{A_d : d \epsilon \Delta\}$ is convex, closed, and bounded, then*

$$\bigcup_{T \geq 0} \{d' : \mathrm{tr}\, TA_{d'} = \max_d \mathrm{tr}\, TA_d\}$$

is complete (P_L).

The characterization of a minimal complete class can be given in the manner of Wald and Wolfowitz (1950).

In the setting of Theorem 3.4, it is also evident that the set of all X_d for which $t'X'_d$ is maximal for some t (i.e., for which there is a t such that for no a with $|a| > 1$ is there a d' with $t'X_{d'} = at'X_d$) is complete. It is obvious how to improve this result.

In the exact theory of P_L, Ehrenfeld considers further the case where the convex closure of $\{x_d : d\epsilon\Delta\}$, or of the set of maximal x_d, has r extreme points (from the discussion of maximal x_d above, it is evident that it will suffice to consider a set which is often somewhat smaller than the set of all x_d; for example, each x_d can be replaced by $\pm x_d$ where the sign is chosen to make the first nonzero coordinate of x_d positive). In this case in the exact theory any N-observation experiment can be replaced by a better $(N+r)$-observation experiment using only the r extreme points (of course, $N+r$ would be replaced by N in the approximate theory). These developments are really all direct consequences of the simple inequality,

$$[\alpha X_1 + (1-\alpha) X_2][\alpha X_1 + (1-\alpha) X_2]' \leqslant \alpha X_1 X_1' + (1-\alpha) X_2 X_2' \qquad \text{for } 0 \leqslant \alpha \leqslant 1.$$

In such a simple problem as that of quadratic regression on a finite interval (see below), there are infinitely many extreme points and the above criterion is useless in the exact theory. A more useful device in such cases would be a result like that claimed by de la Garza (1954), who stated in the polynomial case ($\mathscr{X} = [-1, 1], f_i(x) = x^{i-1}$ for $1 \leqslant i \leqslant m$) that for any N-observation experiment there is another N-observation experiment with the same information matrix, and which takes observations at no more than m different values in $\mathscr{X}$; unfortunately, this result appears to be incorrect (e.g., when $m = 3$, $N = 4$, and observations are taken at $-1, -\frac{1}{2}, \frac{1}{2}, 1$), and the correct results seems complicated. The corresponding result in the approximate theory is trivial: it is an elementary fact in considerations of the moment problem (see, e.g., Shohat and Tamarkin, 1943) that for any probability measure ξ on $[-1, 1]$ there is a probability measure ξ' with the same moments of orders, $1, 2, \ldots, 2m-2$ (these are elements of $M(\xi)$), and such that ξ' has support on at most m points. More generally, if all f_i are continuous functions on a compact $\mathscr{X}$, and if there are H different functions of the form $f_i f_j (i \leqslant j)$, then the set of all matrices $M(\xi)$ can be viewed as a closed subset S of m^2-dimensional space which is convex (since $M(\xi)$ is linear in ξ) and of dimension $J \leqslant H$ (here J is equal to the number of nonconstant, linearly independent functions $f_i f_j$, $i \leqslant j$). Any extreme point of S is attainable as the $M(\xi)$ corresponding to a ξ with support at a single point. Thus (see the remark following the proof of (3.3)), *the set of all ξ with support on at most $J+1$ points is essentially complete (P_L) in the approximate theory.*

We shall investigate questions of admissibility (minimality of such classes) in an example, a little later. Of course, all these results are generally meaningful only as long as we do not consider problems such as ones of hypothesis testing.

Ehrenfeld has also considered *asymptotic essential completeness* in the case where the expected value of an observation is a *nonlinear* function of θ, say $F_d(\theta)$ corresponding to x_d (for convenience, we still assume normality). Under appropriate regularity conditions (and a precise statement of the results which takes account of Hodges' superefficiency phenomenon), one can characterize an asymptotically essentially complete class by replacing x_d in our previous (linear) considerations by the vector $x_d(\theta) = \{\partial F_d(\theta)/\partial\theta_i, 1 \leqslant i \leqslant m\}$ (note that, under suitable assumptions, $x_d(\theta)\, x_d(\theta)'$ is a normalized limiting covariance matrix for the corresponding maximum likelihood estimator). However, the class of all experiments on the set $\{d : x_d(\theta)$ is maximal for some $\theta\}$, which is what Ehrenfeld considered, is not asymptotically essentially complete in the usual decision-theoretic sense, but only in the sense that for each *fixed* θ it yields a set of $M(\theta, \xi)$ (corresponding to our earlier $M(\xi)$) which dominate all other $M(\theta, \xi)$. Thus, for example, if $m = 1$, $\mathscr{X} = [0, 1]$, $0 \leqslant \theta \leqslant 1$ (a slightly smaller interval will achieve the same result while keeping $\partial F_x(\theta)/\partial\theta$ bounded), and $F_x(\theta) = \sqrt{(\theta x)} + \sqrt{[(1 - \theta)(1 - x)]}$, we obtain easily that $\partial F_x(\theta)/\partial\theta$ is maximal only for $x = 0$ (if $\theta \geqslant \frac{1}{2}$) or $x = 1$ (if $\theta \leqslant \frac{1}{2}$); but no allocation of N observations at the two values $x = 0$, $x = 1$ can achieve an asymptotic variance function which is at least as good *for all* θ as the experiment which puts all observations at $x = \frac{1}{2}$. What is needed here, in the same role as the Bayes procedures which enter into complete class theorems in the usual decision theory, is the class of all experiments ξ for which $\int M(\theta, \xi)\mu\,(d\theta)$ cannot be dominated for at least one probability measure μ on the parameter space, or perhaps the closure of this class in a suitable sense (see, e.g. Wald, 1950 or LeCam, 1955; it will usually suffice, for example, to take the closure of the set of ξ's corresponding to μ's with finite support, both here and also in an exact or nonasymptotic development). Thus, as in Theorem 3.4, we can obtain an asymptotically essentially complete class as *the class of all* ξ *which, for some* T *and* μ, *maximize* $\int \operatorname{tr}[M(\theta, \xi)\, T]\, \mu(d\theta)$ (or as a closure of this class, as indicated above). In these asymptotic considerations, there is essentially no difference between the exact and approximate theories.

If the linear problems P_L are modified by allowing the matrix R_d of Section 2D to depend on θ, then considerations of the above type are relevant.

We now turn to an important example, to illustrate some of the concepts of this section.

The polynomial case.—Suppose $\mathscr{X}$ is a compact interval, which we can take to be $[-1, 1]$ without affecting the results below, and that $f_i(x) = x^{i-1}$, $1 \leqslant i \leqslant m$; we hereafter write $m = k + 1$ for convenience. For any design ξ, the $(i, j)^{\text{th}}$ element of $M(\xi)$ is the moment of order $i + j - 2$ of ξ:

$$m_{ij}(\xi) = \mu_{i+j-2}(\xi) = \int_{-1}^{1} x^{i+j-2}\, \xi(dx), \qquad 1 \leqslant i, j \leqslant k + 1.$$

Suppose now that $M(\xi') \geqslant M(\xi'')$. Of course, $m_{11}(\xi') = m_{11}(\xi'') = 1$. If t is an m-vector with first component $t_1 = 1$, $(r + 1)st$ component $t_{r+1} = u$, and all other components $t_i = 0$, we have

$$0 \leqslant t'[M(\xi') - M(\xi'')]\, t = 2u[\mu_r(\xi') - \mu_r(\xi'')] + u^2[\mu_{2r}(\xi') - \mu_{2r}(\xi'')]$$

for all u. Hence, $\mu_r(\xi') = \mu_r(\xi'')$ for $0 \leqslant r \leqslant k$. Repeating this argument with $t_{q+1} = 1$, $t_{s+1} = u$ (with $s > q$), and all other $t_i = 0$, first for $q = k$ and then for successively larger q, we obtain $\mu_r(\xi') = \mu_r(\xi'')$ for $0 \leqslant r \leqslant 2k - 1$. Thus, finally, sufficiency being obvious,

Lemma 3.5.—*$M(\xi') > M(\xi'')$ if and only if $\mu_r(\xi') = \mu_r(\xi'')$ for $0 \leqslant r \leqslant 2k - 1$ and $\mu_{2k}(\xi') > \mu_{2k}(\xi'')$.*

Of course, this criterion can be used in both the exact and approximate developments, but we shall now see that we obtain more elegant results in the latter. (If we compare two procedures with different values of N, the two values of m_{11} differ and Lemma 3.5 does not apply.)

Consider in $2k$-dimensional space the set Q of all points $\mu(\xi) = (\mu_1(\xi), \ldots, \mu_{2k}(\xi))$ for ξ a probability measure on $\mathscr{X}$. Write Q^* for the corresponding set of points

$$\mu^*(\xi) = (\mu_1(\xi), \ \ldots \ , \mu_{2k-1}(\xi)).$$

From the work of Karlin and Shapley (1953) (especially Theorems 11.1 and 20.2 and the discussion on pp. 60 and 64, where the results given for the interval [0, 1] are easily transformed to the present setting) we have that Q is a convex body whose extreme points correspond to ξ's giving measure one to a single point; that μ in Q corresponds to a unique ξ if and only if μ is in the boundary of Q; that for μ^* in the interior of Q^* there is a nondegenerate line of points $\mu(\xi)$ for which $\mu^*(\xi) = \mu^*$, and among these $\mu_{2k}(\xi)$ is maximized by a ξ whose support is $k + 1$ points including 1 and -1; and that for μ^* in the boundary of Q^* there is a unique ξ for which $\mu^*(\xi) = \mu^*$, any such ξ being a limit of ξ's for which $\mu^*(\xi)$ is in the interior of Q^* and $\mu_{2k}(\xi)$ a maximum for that μ^*. Using these facts the well known result (Shohat and Tamarkin (1943), p. 42) that there is at most one ξ with given moments $\mu_1, \ldots, \mu_{2k}$ whose support is a set of at most $k + 1$ points containing a given point, the remark following the proof of (3.3), and Lemma 3.5, we have the following result in the continuous development:

Theorem 3.6. *The minimal complete and minimal essentially complete class of admissible ξ in the polynomial case consists of all ξ whose support consists of at most $k + 1$ points, at most $k - 1$ of which are in the interior of $\mathscr{X}$.*

Before turning to the exact theory, let us look briefly at the changes which occur in the above results if we replace the space $[-1, 1]$ in this example by a finite set of points, say $r_1, \ldots, r_L$. We can no longer limit our consideration to ξ's supported by $k + 1$ points. Writing $\xi = (\xi_1, \ldots, \xi_L)$, to determine whether or not there is a ξ'' better than a given ξ' we write $\xi'' - \xi' = \gamma = (\gamma_1, \ldots, \gamma_L)$ and find, as in Lemma 3.5, that such a ξ'' exists if and only if, for some $c > 0$, the linear equations

$$\sum_{i=1}^{L} \gamma_i r_i^j = 0, \qquad 0 \leqslant j < 2k,$$

$$\sum_{i=1}^{L} \gamma_i r_i^{2k} = c$$

have a solution $\gamma(c)$ such that all components of $\xi' + \gamma(c)$ are nonnegative. Let $c_{\xi'}$ be the largest value of c for which such a solution exists. The set of all ξ of the form $\xi' + \gamma(c_{\xi'})$ (for all ξ') is then the minimal complete (P_L) class. Thus, the determination of admissible ξ in this case becomes a linear programming problem.

Now let us consider the exact theory with $\mathscr{X} = [-1, 1]$. To avoid too much arithmetic, we shall limit our considerations to procedures in Δ. In the linear case ($k = 1$) it is easy to characterize the admissible procedures. For, applying Lemma 3.5, and noting that for a given value of μ_1 we maximize μ_2 if and only if at most one observation is in the interior of $\mathscr{X}$, we conclude that the class of designs with this property is the minimal essentially complete and minimal complete (P_L) class for any given N. For $k > 1$, the situation is more complicated, as we shall now indicate by mentioning a phenomenon which occurs when $k = 2$. When $N = 3$, it is easy to see that a given set of moments μ_1, μ_2, μ_3 is achievable by at most one design, so all designs are admissible. For larger N, however, a design supported at three points may or may not be admissible. For a given N, we can think of the set of possible $\mu(\xi)$ as the subset Q' of Q for which ξ takes on only multiples of $1/N$ as values. For a given μ^* in Q^*, there may be zero or some positive number of points $\mu(\xi)$ in Q' for which $\mu^*(\xi) \epsilon Q^*$, and for each μ^* for which such a ξ exist, we must again select that ξ for which $\mu^*(\xi) = \mu^*$ and $\mu_{2k}(\xi)$ is a maximum. Roughly speaking, the larger N, the more points will there be in Q' corresponding to a given μ^*. For example, consider the symmetrical ξ's on at most three points, say $\xi(b) = \xi(-b) = J/N$ and $\xi(0) = 1 - 2J/N$ (where J is an integer), which we represent as the pair (b, J). For a given value $\mu_2 > 0$, the designs

$$([N\mu_2/2J]^{\frac{1}{2}}, J), \qquad N\mu_2/2 \leqslant J \leqslant N/2,$$

all have $\mu^* = (0, \mu_2, 0)$. Thus, a design (b, J) is admissible among the symmetric distributions on at most three points if and only if no design $(b', J - 1)$ of this class exists with the same μ^*; i.e., if and only if $b > [(J-1)/J]^{\frac{1}{2}}$. We see here the way in which the continuous theory approximates the exact for large N. The general considerations are arithmetically rather messy.

4. SPECIFIC OPTIMALITY CRITERIA

We now turn from the characterization of complete classes to the determination of designs which satisfy particular optimality criteria. We shall limit our considerations to designs in Δ; see the previous sections and I for discussions of necessary modifications if Δ_R is considered.

A. *Various criteria and their relationship.*—Suppose we are interested in inference concerning s given linearly independent parametric functions $\psi_j = \Sigma_i c_{ji}\theta_i$, $1 \leqslant j \leqslant s$. Let Δ' be the class of designs in Δ for which all ψ_j are estimable, and let $\sigma^2 v_d$ be the covariance matrix of b.l.e.'s of the ψ_j for a d in Δ'. For testing the hypothesis that all $\psi_j = 0$ under the assumption of normality, when a given design d is used, let $\bar{\beta}_\varphi(c)$ be the infimum of the power function of the test ϕ over all alternatives for which $\Sigma_j\psi_j^2/\sigma^2 = c$, and let $\bar{\beta}(d, c, \alpha)$ be the supremum of $\bar{\beta}_\phi(c)$ over all ϕ of size α. As in I, we now consider five optimality criteria.

For $c > 0$ and $0 < \alpha < 1$, a design d^* is said to be $M_{\alpha,c}$*-optimum in* Δ if

$$\bar{\beta}(d^*, c, \alpha) = \max_{d \epsilon \Delta} \bar{\beta}(d, c, \alpha).$$

A design d^* is said to be L_α*-optimum in* Δ if

$$\lim_{c \to 0} [\bar{\beta}(d^*, c, \alpha) - \alpha]/[\sup_{d \epsilon \Delta} \bar{\beta}(d, c, \alpha) - \alpha] = 1.$$

A design d^* is said to be *D-optimum in* Δ if $d^* \epsilon \Delta'$ and

$$\det v_{d^*} = \min_{d \epsilon \Delta'} \det v_d.$$

A design d^* is said to be *E-optimum in* Δ if $d^* \epsilon \Delta'$ and

$$\pi(v_{d^*}) = \min_{d \epsilon \Delta'} \pi(v_d),$$

where $\pi(v_d)$ is the maximum eigenvalue of v_d,

A design d^* is said to be *A-optimum in* Δ if $d^* \epsilon \Delta'$ and

$$\text{trace } v_{d^*} = \min_{d \epsilon \Delta'} \text{trace } v_d.$$

These definitions are meaningful whether or not σ^2 is known, and the last three do not require the assumption of normality. (Actually, normality has nothing to do with any of these definitions, but only with some of the interpretations below.) By replacing A_d by $NM(\xi)$, we obtain corresponding definitions for the approximate theory.

These criteria are discussed extensively in I and II, and we shall merely summarize some of the important points here. M-optimality is generally extremely difficult to characterize, even in very simple situations. L-optimality, which is a local version of M-optimality, involves the use of type C regions, as discussed in Section 2B. E-optimality was first considered in hypothesis testing (Wald (1943), Ehrenfeld (1955*a*)) because, if σ^2 is known or all b_d are equal for d in Δ', it is the design for which the associated F-test of size α maximizes the minimum power on the contour $\Sigma_j \psi_j^2 = c$, for every α and c; this throws serious doubt on the acceptability of this criterion for hypothesis testing, since (see Section 2B) the F-test may not be the one which, for a given design, maximizes this minimum power. For point estimation, an E-optimum design minimizes the maximum over all $(a_1, \ldots a_s)$ with $\Sigma a_i^2 = 1$ of the variance of the b.l.e. of $\Sigma a_i \psi_i$. An A-optimum design minimizes the average variance of the b.l.e.'s of $\psi_1, \ldots, \psi_s$, and thus of any s linear parametric functions obtained from the ψ_i by an orthogonal transformation. A D-optimum design minimizes the generalized variance of the b.l.e.'s of the ψ_i, and thus, under normality with σ^2 known or else all $N - b_d$ the same for d in Δ', minimizes the volume (or expected volume, if σ^2 is unknown) of the smallest invariant confidence region on $\psi_1, \ldots, \psi_s$ of any given confidence coefficient. For testing hypotheses under these same conditions of normality, it follows from the result on regions of type D discussed in Section 2B that, for each σ (and each set of values of the parameters other than $\psi_1, \ldots, \psi_s$), a D-optimum design achieves a test whose power function has maximum Gaussian curvature at the null hypothesis, among all locally unbiased tests of a given size.

Other criteria can be considered similarly. For example, the contour considered in hypothesis testing can be altered from $\Sigma \psi_i^2 = c$, or one can consider maximizing trace v_d^{-1} in place of A-optimality (some examples in Section 4B throw a bad light on the latter possibility). The often-considered criterion of restricting oneself to designs for which the b.l.e.'s of $\psi_1, \ldots, \psi_s$ all have equal variances, and of minimizing this variance, will in unsymmetrical settings often produce a design inferior to that which minimizes the maximum diagonal element of v_d without restriction.

D-optimality has an appealing invariant property which is not possessed by the other criteria we have mentioned. Let $\psi_1', \ldots, \psi_s'$ be related to $\psi_1, \ldots, \psi_s$ by a non-singular linear transformation. Then, if d^* is D-optimum for the functions $\psi_1, \ldots, \psi_s$, it is also D-optimum for $\psi_1', \ldots, \psi_s'$. The analogue for other criteria is false in even the simplest settings. For hypothesis testing, D-optimality is also invariant under non-linear transformations, as discussed in Section 2B.

The invariance of D-optimality is well illustrated by the problem of polynomial regression (see Section 3) with $\psi_i = \theta_i$. For polynomial regression on the interval $[a, b]$, a

D-optimum design is obtained from that on $[-1, 1]$ by simply transforming the loci of observations according to the linear transformation which takes $[-1, 1]$ onto $[a, b]$. For any other of the above criteria, even a simple change in units (consideration of $[-h, h]$ instead of $[-1, 1]$) will change the optimum design if $k > 1$. This is both intuitively unappealing (having the choice of design depend on whether measurements are recorded in inches or in feet), and also has the disadvantage of requiring us to give a table of designs which depend on a and b.

A precise statement of losses will obviously entail the use of any of a large number of designs, not always the D-optimum design. However, the discussion of the previous paragraphs should give some workers a good justification for using D-optimum designs in many settings. A further appealing property of D-optimum designs will now be described.

In the regression setup, suppose we are interested in estimating the whole regression function, $\Sigma\theta_i f_i$. As indicated in II, various criteria of optimality can be suggested. A design which minimizes the expected maximum deviation over $\mathscr{X}$ between estimated and true regression function (or the square of this deviation) will be different under normality from what it is under other assumptions, and will generally be very difficult to calculate. Another criterion which has been suggested is to minimize the integral of var $[\Sigma_i(t_{\alpha i} - \theta_i) f_i(x)]$ with respect to some measure on $\mathscr{X}$; the arbitrariness present in choosing the measure, the lack of invariance of seemingly "natural" measures on $\mathscr{X}$ under certain transformations, and the fact that the variance may be very large at some points while the average is small, are some of the shortcomings here. A criterion which has been considered by several authors is the minimization of $\sup_x$ var $[\Sigma_i(t_{\alpha i} - \theta_i) f_i(x)]$ (see II for a discussion of the optimality of using $\Sigma t_{\alpha i} f_i$ rather than some other estimator of the function $\Sigma\theta_i f_i$ in this problem). Let us call a design which satisfied this last global criterion *G-optimum in* Δ.

The discussion of the present paragraph will refer to the approximate theory. In a remarkable paper written in 1918, perhaps the first systematic computation of optimum regression designs, K. Smith (1918) determined the G-optimum designs for the polynomial case with $k \leqslant 6$. Guest (1958) characterized the G-optimum designs in this case for arbitrary k in terms of the zeros of the derivative of a Legendre polynomial. Hoel (1958) computed the D-optimum design in this case, and noted that his D-optimum design coincided with Guest's G-optimum design. It was proved in II that this phenomenon holds more generally, and finally the following result in the approximate theory, which was announced in II and will appear elsewhere, was proved:

Theorem 4.1.—*If $\mathscr{X}$ is a compact space on which the f_i are continuous and linearly independent, then ξ is D-optimum for $\theta_1, \ldots, \theta_m$ if and only if it is G-optimum.*

It is not possible to prove an analogue of Theorem 4.1 in the discrete case. For example, in the case of quadratic regression on $[-1, 1]$ with $N = 4$, a D-optimum design takes observations at ± 1, 0, while a G-optimum design takes observations at ± 1, $\pm(5^{\frac{1}{2}} - 2)$. This is another illustration of the usefulness of considering the continuous theory, where many results are valid which are false in the discrete theory, but which are of practical value for large N.

Theorem 4.1 will also be helpful in the computation of D-optimum designs in problems like those of Section 4C, since it will permit us to exploit the interplay between different criteria for D- and G-optimality.

As Hoel also noted, if $x_1, \ldots, x_m$ are any m points of $\mathscr{X}$, writing $F = ||f_i(x_j)||$ we see that the generalized variance of the b.l.e.'s of the regression function at the m points

$x_1, \ldots, x_m$, which is merely the determinant of $\sigma^2 F v_d F'$, is minimized by a D-optimum design. (No analogous property for s points, $1 < s < m$, is generally valid.)

Having stated the relationship between G- and D-optimality in Theorem 4.1, we now turn to the question of the relationships among the other criteria. In general, D-, E-, A-, and L-optimality are unrelated, in either the exact or approximate development. For example, if $m = r = 2$ and there are two possible x_d's, (1, 0) and (1, 1), to which the design ξ assigns measures ξ_1 and $\xi_2 = 1 - \xi_1$, it is easy to compute that $\xi_2 = \frac{1}{2}$ for D-optimality, $2^{\frac{1}{2}} - 1$ for A-optimality, and 2/5 for E-optimality; an L_α-optimum design is not explicitly known. However, in certain situations such as those where balanced block designs, orthogonal arrays, Latin squares, etc., are customarily employed, it happens that these criteria are related, due to the symmetric way in which the ψ_i enter into the problem (see the last paragraph of Section 2E). This is expressed in the following simple lemma, which was employed in I (it is not useful in most regression problems of the type treated in II):

Lemma 4.2.—*If d^* maximizes the trace of v_d^{-1} and also $N - b_d$, over all d for which v_d is nonsingular, and if v_{d^*} is a multiple of the idenity, then d^* is A, D-, E-, and L_α-optimum (for all α) in Δ.*

Of course, the maximization of $N - b_d$ is unnecessary if σ^2 is known or if we are only interested in point estimation.

Under the stated hypothesis on the form of v_{d^*}, other similar results can be stated (e.g., D-optimality implies A-, E-, and L-optimality); but other relationships generally need additional conditions for their validity. For example, it is easy to find situations where an E-optimum design d for which v_d is the identity is neither A-, D-, nor L_α-optimum (see I).

We shall make further remarks on these relationships in Section 4B.

We now turn to the question of invariance. We shall extend the result of Section 2E to the case of s parameters ($1 \leqslant s \leqslant m$). Suppose in the setting of Section 2E that G is the direct product of groups G_1 and G_2, where G_1 is as in Section 2E, G_2 is as in Theorem 3.3, $\bar{g}$ leaves $\theta_1, \ldots, \theta_s$ fixed if $g \epsilon G_2$, and $\bar{g}$ leaves $\theta_{s+1}, \ldots, \theta_m$ fixed if $g \epsilon G_1$. By Theorem 3.3, we can restrict our consideration to ξ's which are invariant under G_2, and we hereafter do so. We shall use the decomposition of $M(\xi)$ employed in proving Theorem 3.3. Suppose that G_1 has p elements (which leave $\theta_{s+1}, \ldots, \theta_m$ fixed), and let ξ be a G_2-invariant D-optimum design. Then, for g in G_1 and $D(\xi)$ nonsingular (the singular case being treated as before), we clearly have $B(\xi_g) = H_g B(\xi) H_g'$, $C(\xi_g) = H_g C(\xi)$, and $D(\xi_g) = D(\xi)$, where H_g has determinant one. Hence,

$$B(\bar{\xi}) - C(\bar{\xi})\, D(\bar{\xi})^{-1}\, C(\bar{\xi})' = p^{-1}\, \Sigma_g H_g[B(\xi) - C(\xi)\, D(\xi)^{-1}\, C(\xi)']\, H_g'. \qquad (4.1)$$

Since all H_g have determinant one, we can argue from (4.1) exactly as we did from (2.1) in Section 2E. Thus, we have (extending the result to nonfinite G as before),

Theorem 4.3.—*If $G = G_1 \times G_2$ is as specified above, with $\bar{g}$ leaving $\theta_1, \ldots, \theta_s$ fixed if $g \epsilon G_2$ and $\bar{g}$ leaving $\theta_{s+1}, \ldots, \theta_m$ fixed if $g \epsilon G_1$, then there is a G-invariant ξ which is D-optimum for $\theta_1, \ldots, \theta_s$.*

Similarly, from the orthogonality of $\bar{g}$ for g in G_1, we obtain the same result for A-optimality.

B. *Block designs, arrays, etc.*—The settings where balanced block designs, Latin squares, orthogonal arrays, etc., are customarily employed are characterized by the fact that X_d is a matrix of 1's and 0's (and $-$ 1's, in the case of certain weighing experiments) satisfying certain restrictions. One of the first optimality results for such designs was proved by Wald (1943), who showed that, in the setting of two-way heterogeneity where $k \times k$ Latin square and higher Latin square designs are usually employed, these designs are actually D-optimum in Δ for inference on any full set of $k - 1$ linearly independent contrasts of "treatment effects". Shortly afterwards, Hotelling (1944) began the careful study of weighing problems, some ideas on this subject originating in earlier work of Yates (1935), Kishen, and Banerjee. A comprehensive treatment of weighing problems was given by Mood (1946), who considered problems of N weighings on m objects ($N \geqslant m$) on both spring and chemical balances, proved in the latter case the D-optimality of 2-level orthogonal arrays when they exist, as well as the minimization by them of the variances of b.l.e.'s of weights among all d for which the diagonal elements of v_d are all the same, and obtained optimum designs in the case of spring balances and also in certain cases where no orthogonal arrays exist (for small N), where the two optimality criteria just mentioned were noted not always to agree. At the same time, more general orthogonal arrays were considered independently by Plackett and Burman (1946) (see also Plackett, 1946), who proved their optimality in the multifactorial setup, in the second sense mentioned above in connection with Mood's results. An extensive study by Tocher (1952) considered also the settings where incomplete block designs and Youden squares are customarily employed, and proved that these designs are optimum in the sense of minimizing the variances of b.l.e.'s of treatment differences $\theta_i - \theta_j$, among all designs for which these differences are all estimated with the same variance. These last three papers also considered various methods of construction, which are not the subject of the present paper.

It should be mentioned at this point that although criteria like those mentioned in connection with the last two references happen to lead to designs which are optimum in other senses in situations where sufficiently symmetrical designs exist, these criteria are not intuitively acceptable in themselves; for there are many problems where the restriction to "equal precisions" is attainable only by relatively poor designs, and where there exist better designs which give unequal but better precisions to all estimates.

Ehrenfeld (1955*a*) proved the E-optimality of the Latin square design (for an appropriate set of treatment contrasts) and of orthogonal arrays in weighing problems. Under the restriction to designs for which each variety appears at most once in each block (or in each row and column), Mote (1958) proved the E-optimality of the balanced incomplete block design and Kshirsagar (1958) proved the A- and D-optimality of this design and of the Youden square. At the same time, without this restriction, more general results were obtained in I, although the approach described below, which leads to trivial proofs in the standard cases (e.g., Youden squares), can entail somewhat more arithmetic in the general settings (e.g., generalized Youden squares, defined below). We shall summarize some of these results.

We begin by indicating a simple approach to optimality proofs in all such symmetrical situations. In problems such as the usual k-way analysis of variance setup, weighing problems, and multifactorial problems, where the "treatment effects" $\theta_1, \ldots, \theta_u$ (say) in which we are customarily interested can all be estimated, the results are easily obtained using Lemma 4.2, since (with the partition of A_d used in Section 3) it is not difficult to find a bound on the trace of $B_d - C_d D_d^{-1} C_d'$, and to show that this bound is attained by the appropriate symmetrical design, for which this matrix becomes a multiple of the identity. In settings like those where we are only interested in (or can only estimate) *contrasts* of $\theta_1, \ldots, \theta_u$, such as those where balanced incomplete block designs and Youden squares are employed, the above $u \times u$ matrix is singular. If the $u-1$ linearly independent contrasts $\theta_1, \ldots, \theta_{u-1}$ which can be estimated are not chosen in a suitable way, the computation of v_d^{-1} may be quite messy. The most expeditious choice in many settings is to let $\psi_i = \Sigma_{j=1}^{u} o_{ij}\theta_j$, $1 \leqslant i \leqslant u-1$, where $\|o_{ij}\| = \bar{O}$ is a $u \times u$ orthogonal matrix with $o_{uj} = u^{-\frac{1}{2}}$, $1 \leqslant j \leqslant u$. (This implies that our optimality criteria which refer to hypothesis testing are concerned with the power function on the contours $\Sigma_1^u (\theta_i - \bar{\theta})^2 = c\sigma^2$, where $\bar{\theta} = \Sigma_1^u \theta_i/u$.) The development can be carried out through direct consideration of the possible A_d's, but somewhat less arithmetic is needed if we use Bose's $\mathscr{C}$-matrices corresponding to the incomplete block (or h-way heterogeneity) setting at hand. In this approach, letting the first u θ_i's represent the treatment effects (the other θ_i's representing block effects or row and column effects, etc.), for any design there is a matrix $\mathscr{C}_d$ of rank at most $u-1$ and a u-vector Z_d of linear functions of the Y_{di} such that the b.l.e. of any contrast $\Sigma_1^u c_i\theta_i$ (where $\Sigma c_i = 0$) which is estimable under d is given by $\Sigma c_i t_{di}^*$ where t_d^* is any solution of the reduced normal equations $\mathscr{C}_d t_d^* = Z_d$. Also, $\mathscr{C}_d$ has row and column sums equal to zero, and Z_d has covariance matrix $\sigma^2\mathscr{C}_d$. Now, it is often easy to give a bound on the trace of all $\mathscr{C}_d$ in terms of N, etc. Moreover, appropriately symmetrical designs such as balanced incomplete block designs, Youden squares, etc., will generally have the property that all diagonal elements of $\mathscr{C}_d$ are equal (and $\mathscr{C}_d$ is not zero, and hence has rank $u-1$). Suppose $\mathscr{C}_{d*}$ is of this form and has maximum possible trace. Then $\bar{O}\mathscr{C}_{d*}\bar{O}'$ is easily verified to have the same positive constant for each of its first $u-1$ diagonal elements, and is zero elsewhere. Since the upper left hand $(u-1) \times (u-1)$ submatrix of $\bar{O}\mathscr{C}_d\bar{O}'$ is just the v_d^{-1} for the b.l.e.'s of the $\psi_1, \ldots, \psi_{u-1}$ defined just above, we have the following elementary lemma:

Lemma 4.4. *If $\mathscr{C}_{d*}$ has maximum possible trace, all diagonal elements equal, and all off-diagonal elements equal, then v_{d*}^{-1} has maximum possible trace and is a multiple of the identity.*

Thus, although any choice of $u-1$ linearly independent contrasts $\psi_i, \ldots, \psi_{u-1}$ will lead to the same D-optimum design(s), the choice given above makes the arithmetic by far the least cumbersome in most applications. For if $\mathscr{C}_{d*}$ has the form hypothesized in Lemma 4.4, we can combine the conclusion of this lemma with Lemma 4.2 and thus obtain the desired optimum properties.

This method can be employed to prove the D-, E-, L-, and A-optimality of balanced block designs (a generalization of balanced incomplete block designs, to be defined below), regular generalized Youden squares (defined below), higher Latin squares, orthogonal arrays, and other appropriately symmetrical designs in all of the customarily encountered settings where we are interested in contrasts of treatment effects; see I. As remarked earlier, the corresponding results are even simpler in the case where we are interested in all u effects, all of which are estimable.

In many optimality proofs, authors have not really proved the appropriate results, since they have restricted considerations to a subclass or Δ (for example, to incomplete block designs where each treatment occurs at most once per block). Such restrictions are quite unnecessary.

To illustrate these ideas suppose we have b blocks of size k and u varieties in the usual incomplete block design setting, except that we do *not* assume $k \leqslant u$. Such situations arise often. Generalizing the notion of a balanced incomplete block design, we define a a design d in the above setting to be a *balanced block design* (BBD) if (a) the number of times n_{dij} that variety i appears in block j is k/u if this is an integer, and is one of the two closest integers otherwise; (b) the number $\Sigma_j n_{dij}$ of replications of variety i is the same for all i; and (c) for each pair i_1, i_2 with $i_1 \neq i_2$, $\Sigma_j n_{di_1j} n_{di_2j}$ is the same. Certain designs of this type with $k > u$ have been considered by Tocher (1952) and others. Appropriate modifications of some of the constructive methods which are used when $k < u$ will sometimes work here. In the setting described here, it is easy to verify that, for any design d (not necessarily a BBD), the trace of $\mathscr{C}_d$ is

$$bk - \Sigma_{i,j}\, n_{dij}^2/k. \tag{4.2}$$

Since $\Sigma_{i,j}\, n_{dij} = bk$, expression (4.2) is clearly maximized by a BBD (if one exists), and the $\mathscr{C}_d$ of such a design is of the form hypothesized in Lemma 4.2. Thus, we have proved that BBD's are A-, D-, E-, and L-optimum.

It is interesting to note that many designs maximize (4.2) (all that is required is that the n_{dij} be as nearly equal as possible); however, of these designs, only a BBD will have the form required by Lemma 4.4. However, expression (4.2) and Lemmas 4.2 and 4.4 again suggest the idea, mentioned in Section 2C, that if no appropriately symmetrical (balanced) design (a BBD in this case) exists, then a design which is as close as possible to such symmetry in some sense will be optimum. Our next example shows, however, that considerable delicacy will be needed to make these notions precise, since optimality can be difficult to prove even for a design of maximum symmetry (balance).

In the setting of two-way heterogeneity (expected value of an observation = row effect + column effect + variety effect) with k_1 rows, k_2 columns, and u varieties, we say that d is a *generalized Youden square* (GYS) if it is a BBD when rows are considered to be blocks and also when columns are considered to be blocks. A GYS is said to be *regular* if at least one of k_1/u and k_2/u is an integer. Using Lemmas 4.2 and 4.4, it was shown in I that a regular GYS is A-, D-, E-, and L-optimum; the argument for proving that the trace of $\mathscr{C}_d$ is a maximum when d is a regular GYS is somewhat more complicated here than in the case of a BBD. In fact, if the GYS is not regular, its $\mathscr{C}_d$ may not have maximum trace, as was illustrated by an example in I in the case $k_1 = k_2 = 6$, $u = 4$. It still seems likely that such nonregular GYS's are optimum, but a different argument from that based on Lemmas 4.2 and 4.4 needs to be developed in this case and in certain other settings which represent extensions of the classical situations.

We mention one constructional aspect of GYS's: the method of Hartley and Smith (1948) can be extended to prove the following:

Lemma 4.5. *If* $L = k_2/u$ *is an integer and there is a BBD with parameters* u, $b = k_2$, $k = k_1$, *then there is a (regular) GYS with parameter values* u, k_1, k_2.

In fact, thinking of the BBD as a k_1 (rows) $\times$ k_2 (columns) array, we let $m_{ij} = \max\,(0, [-L +$ number of times variety i appears in row j]) and $\Sigma_{i,j} m_{ij} = M$. Following Hartley and Smith, we can give a method for reducing M by at least one if $M > 0$, and then use induction: one has only to go through the demonstration of these authors, replacing the occurrence of a variety 0, 1, or more than 1 time in a row by its occurrence less than L, L, or more than L times, respectively. This method of construction cannot be modified to work for nonregular GYS's merely by trying to make the n_{dij} in rows as nearly equal as possible; for example, in the case $k_1 = 6$, $k_2 = 6$, $u = 4$, one can construct a design whose columns are a BBD and whose rows have all n_{dij} as nearly equal as possible, but which is not a GYS.

Similar optimality results for symmetrical designs can also be obtained for appropriate components of variance models and mixed models.

C. *Methods for computing optimum regression designs.*—Pioneering work was done by Elfving in several papers on this subject (e.g., Elfving, 1952, 1955*a*), and some of the earliest work is also due to Chernoff (1953), whose asymptotic results apply also to the nonlinear case. These authors present geometric descriptions of optimum combinations of a given class of information matrices (using A-optimality in the case of more than one parameter). Box (1952) noted the optimality in the sense of Plackett and Burman (1946) (see Section 4B above) of orthogonal arrays obtained by rotations from theirs, in the problem of estimation of a linear function of several variables; presumably the observations are limited to x_d's in a sphere about the origin, if all of these designs are to be meaningful and optimum as asserted. Scheffé (1958) considered that allocation of observations in certain simplex experiments which satisfies a criterion specified by him. Elfving (1955*b*) treated also the subject of nonrepeatable observations. Entirely different approaches are given in II, and we shall now summarize briefly some of the methods of that paper (Lemma 4.2 is usually useless here). We shall consider the approximate theory, throughout, and omit the appearance of N in variances, etc.

Suppose, first, that we are interested only in one of the m parameters, say θ_m, and want to select a ξ which minimizes the variance $\sigma^2 v(\xi)$ (say) of the b.l.e. of θ_m. A direct approach would be to write $v(\xi)$ as a function of the elements of $M(\xi)$ which one then tries to minimize. This is usually much too difficult; for example, in the case of polynomial regression (where we might be interested in testing whether or not a polynomial of degree $\leqslant m-1$ is actually of degree $\leqslant m-2$), we obtain a rational function of the first $2m-2$ moments of ξ, and it seems extremely difficult to find the maximum of this expression directly. The approach suggested in II is this: Let ξ be a design for which θ_m is estimable, and let

$$f_m^* = f_m - \Sigma_{j<m} a_j f_j$$

be orthogonal to f_i, $1 \leqslant i \leqslant m-1$, with respect to ξ. We can rewrite the regression function as

$$\Sigma_1^m \theta_i f_i = \Sigma_1^{m-1} \theta_i^* f_i + \theta_m^* f_m^*,$$

so that $\theta_m^* = \theta_m$. For the least squares setup in terms of the θ_i^*, the last normal equation contains only θ_m^*, with coefficient equal to the integral with respect to ξ of f_m^{*2}, and this coefficient is also $1/v(\xi)$. But, by definition,

$$\int f_m^{*2}(x)\, \xi(dx) = \min_c \int [f_m(x) - \Sigma_{j<m} c_j f_j(x)]^2\, \xi(dx), \tag{4.3}$$

where $c = (c_1, \ldots, c_{m-1})$. Hence, writing $K(\xi, c)$ for the integral on the right side of (4.3), a design ξ^* is optimum (minimizes $v(\xi)$) if

$$\min_c K(\xi^*, c) = \sup_\xi \min_c K(\xi, c). \tag{4.4}$$

This is very suggestive of a zero-sum two-person game, in which player 1 chooses a measure

ξ on $\mathscr{X}$ and player 2 chooses an $(m-1)$-vector c (since $K(\xi, c)$ is convex in c, player 2 need only consider pure strategies). It is shown in II that this game is determined. The maximin "strategies" ξ of player 1 are precisely the optimum designs. Most helpful of all is the consideration of minimax strategies. If $c^* = (c_1^*, \ldots, c_{m-1}^*)$ is minimax, we have

$$\max_x \left| f_m(x) - \Sigma_{j<m} c_j^* f_j(x) \right| = \min_c \max_x \left| f_m(x) - \Sigma_{j<m} c_j f_j(x) \right|; \qquad (4.5)$$

in other words, the coefficient vector c^* is a solution of the well known Chebyshev approximation problem, of finding the best uniform approximation to f_m of the form $\Sigma_{j<m} c_j f_j$. Now, there is a vast literature on such problems (see, e.g., Achiezer, 1956), from which we can thus often find a minimax c^*; it is well known that this knowledge simplifies the computation of a maximin (optimum) ξ^* considerably. Because of the form of $K(\xi, c)$, this computation can be reduced to the following simple problem of solving linear equations:

Let c^* satisfy (4.5) *and let $B(c^*)$ be the subset of $\mathscr{X}$ where $\left| f_m - \Sigma_{j<m} c_j^* f_j \right|$ attains its maximum. Then ξ^* is optimum if and only if $\xi^*(B(c^*)) = 1$ and*

$$\int [f_k(x) - \Sigma_{j<k} c_j^* f_j(x)] f_i(x)\, \xi^*(dx) = 0, \qquad i < k. \qquad (4.6)$$

Since $B(c^*)$ is often a finite set, (4.6) is just a set of linear equations in the unknowns $\xi^*(x)$, $x \epsilon B(c^*)$. It can also be shown that there is an optimum ξ^* on a set consisting of at most k, and sometimes fewer, points. Another method for determining an optimum ξ^*, which is especially useful if we know no c^* but have what we *guess* to be an optimum ξ', is given by a standard game-theoretic method:

Guess a ξ' and compute $\min_c K(\xi', c) = K(\xi', c')$ (say); if $K(\xi', c') = \max_x K(x, c')$, then ξ' is optimum.

Further results and several examples are given in II, and we shall mention only one example here: in the polynomial case defined in Section 3, the unique optimum ξ^* is given by

$$\xi^*(-1) = \xi^*(1) = 1/2k,$$

$$\xi^*\left(\cos\frac{j\pi}{k}\right) = 1/k, \qquad 1 \leqslant j \leqslant k-1.$$

It is shown in II that, for $k > 1$, a considerable loss in efficiency results from using any of the common "equally-spaced observations" designs instead of this ξ^*. We note also that this design is optimum for hypothesis testing even when σ^2 is unknown: θ_m is estimable only for designs on at least m different points, so the phenomenon of the second example of Section 2A(i) is impossible.

In the case where we are interested in s out of the m parameters, say $\theta_1, \ldots, \theta_s$, the development used above is not available to us. One possible technique developed in II is to consider, for $1 \leqslant j \leqslant s$, the functions

$$F_j(\xi) = \min_{c^{(j)}} \int [f_j(x) - \Sigma_{i>j} c_i^{(j)} f_i(x)]^2\, \xi(dx), \qquad (4.7)$$

with $c^{(j)} = (c_{j+1}^{(j)}, \ldots, c_m^{(j)})$. If we write $c = (c^{(1)}, \ldots, c^{(s)})$ and $\lambda = (\lambda_1, \ldots, \lambda)_s$

where all $\lambda_i > 0$, and consider for each λ the game with payoff function

$$K_\lambda(x, c) - \sum_{j=1}^{s} \lambda_j[f_j(x) - \Sigma_{i>j} c_i^{(j)} f_i(x)]^2, \tag{4.8}$$

we again have determinateness. Thus, for example, if c_λ^* is minimax, if ξ_λ gives measure 1 to the set where $K_\lambda(x, c_\lambda^*)$ attains its maximum, and if

$$\int [f_j(x) - \Sigma_{i>j} c_i^{(j)*} f_i(x)] f_i(x)\, \xi_\lambda(dx) = 0 \tag{4.9}$$

for $i > j$ and $1 \leqslant j \leqslant s$, then we can show that ξ_λ is maximin, and conversely. It can then be shown that, if ξ_λ denotes a maximin strategy for K_λ, then there is a unique value λ^* of λ such that $\prod_{i \leqslant s} F_i(\xi_\lambda)$ is a maximum for $\lambda = \lambda^*$ and some ξ_{λ^*}. *Such a* ξ_{λ^*} *is then D-optimum.* Similarly, if we define K_λ^* by (4.8) with the inner sum only over $i > s$, then we find that a maximin ξ for $\lambda = (1, 1, \ldots, 1)$ is A-optimum. Further details, and methods for satisfying other optimality criteria, are given in II.

In the case $s = m$, we have at our disposal for obtaining a D-optimum design the method outlined above, the method of directly maximizing det $M(\xi)$, and also, by Theorem 4.1, any methods which suggest themselves for the calculation of G-optimum designs, because of the definition of the latter. We mention here the following result (see II for further discussion):

Lemma 4.6. *A design* ξ *is G-optimum (hence, D-optimum) if and only if* $M(\xi)$ *is non-singular and*

$$\max_x f(x)' M(\xi)^{-1} f(x) = m, \tag{4.10}$$

where $f(x)$ *is the m-vector with components* $f_1(x), \ldots, f_m(x)$.

It is very useful in applications to use these different methods simultaneously. For example, if one suspects that an optimum ξ lies within a given parametric family W, it may be quite messy to compute the minimum of the lefte side of (4.10) over all ξ in W, but it may be quite easy to maximize det $M(\xi)$ over W (although this maximization over *all* ξ may be quite difficult to perform directly). If ξ' maximizes det $M(\xi)$ over W, we can then see whether or not (4.10) is satisfied for $\xi = \xi'$; if it is, our guess has been good, and ξ' is D- and G-optimum. Several examples of this approach will appear in a forthcoming paper.

For problems of testing hypotheses, the considerations of Section 2A are again relevant: in such a simple problem as that of linear regression on $[-1, 1]$, the locally best design for testing the hypothesis that both parameters equal zero (or any other specified values) is the randomized design which takes all observations at 1 or -1, with equal probabilities.

We have already mentioned, in Section 4A, the results of Smith, Guest, and Hoel in the polynomial case. We mention here that Guest presents an interesting comparison of his designs with certain "equal spacing" designs. It should also be mentioned that Hoel compares various designs in the polynomial case for various types of dependence (see Cox (1952) for a discussion of related work in other settings); but the determination of optimum designs in these cases is still an open problem.

REFERENCES

ACHIEZER, M. I. (1956), *Theory of Approximation*. New York: Ungar.

BLACKWELL, D. (1950), "Comparison of experiments", *Proc. Second Berkeley Symposium*, 93–102, Univ. of California Press.

BLUM, J. (1954), "Multidimensional stochastic approximation methods", *Ann. Math. Statist.*, **25**, 737–744.

BOX, G. E. P. & WILSON, K. B. (1951), "On the experimental attainment of optimum conditions", *J. R. Statist. Soc.* B, **13**, 1–45.

BOX, G. E. P. (1952), "Multi-factor designs of first order", *Biometrika*, **39**, 49–57.

BRADT, R. N., JOHNSON, S. M. & KARLIN, S. (1957), "On sequential designs for maximizing the sum of *n* observations", *Ann. Math. Statist.*, **28**, 1,060–1,074.

—— & KARLIN, S. (1957), "On the design and comparison of certain dichotomous experiments", *Ann. Math. Statist.*, **28**, 390–409.

CHERNOFF, H. (1953), "Locally optimum designs for estimating parameters", *Ann. Math. Statist.*, **24**, 586–602.

COX, D. R. (1952), "Some recent work on systematic experimental designs", *J. R. Statist. Soc.* B, **14**, 211–219.

CHUNG, K. L. (1954), "On a stochastic approximation method", *Ann. Math. Statist.*, **25**, 463–483.

DE LA GARZA, A. (1954), "Spacing of information in polynomial regression", *Ann. Math. Statist.*, **25**, 123–130.

EHRENFELD, S. (1955*a*), "On the efficiency of experimental designs", *Ann. Math. Statist.*, **26**, 247–255.

—— (1955*b*), "Complete class theorems in experimental design", *Proc. Third Berkeley Symposium*, vol. 1, 57–67. Univ. of California Press.

ELFVING, G. (1952), "Optimum allocation in linear regression theory", *Ann. Math. Statist.*, **23**, 255–262.

—— (1955*a*), "Geometric allocation theory", *Skand. Akt.*, **37**, 170–190.

—— (1955*b*), "Selection of nonrepeatable observations for estimation", *Proc. Third Berkeley Symposium*, vol. 1, 69–75. Univ. of California Press.

FRIEDMAN, M. & SAVAGE, L. J. (1947), "Experimental determination of the maximum of a function", *Selected Techniques of Statistical Analysis*, 363–372. New York: McGraw-Hill,

GUEST, P. G. (1958), "The spacing of observations in polynomial regression", *Ann. Math. Statist.*, **29**, 294–299.

HARTLEY, H. O. & SMITH, C. A. B. (1948), "The construction of Youden squares", *J. R. Statist. Soc.* B, **10**, 262–263.

HODGES, J. L., Jr. & LEHMANN, E. L. (1950), "Some problems in minimax point estimation", *Ann. Math. Statist.*, **21**, 182–197.

—— —— (1955), "Two approximations to the Robbins-Monro process", *Proc. Third Berkeley Symposium*, vol. 1, 95–104. Univ. of California Press.

HOEL, P. G. (1958), "Efficiency problems in polynomial estimation", *Ann. Math. Statist.*, **29**, 1,134–1,145.

HOTELLING, H. (1941), "Experimental determination of the maximum of a function", *Ann. Math. Statist.*, **12**, 20–46.

—— (1944), "Some improvements in weighing and other experimental techniques", *Ann. Math. Statist.*, **15**, 297–306.

HSU, P. L. (1941), "Analysis of variance from the power function standpoint", *Biometrika*, **32**, 62.

ISAACSON, S. (1951), "On the theory of unbiased tests of simple statistical hypotheses specifying the values of two or more parameters", *Ann. Math. Statist.*, **22**, 217–234.

KARLIN, S. & SHAPLEY, L. S. (1953), *Geometry of Moment Spaces*, vol. 12 of *Amer. Math. Soc. Memoirs*.

KIEFER, J. (1948), "Sequential determination of the maximum of a function", M.I.T. (thesis).

—— (1953), "Sequential minimax search for a maximum", *Proc. Amer. Math. Soc.*, **4**, 502–506.

—— (1957*a*), "Invariance, sequential minimax estimation, and continuous time processes", *Ann. Math. Statist.*, **28**, 573–601.

—— (1957*b*), "Optimum sequential search and approximation methods under minimum regularity conditions", *J. Soc. Ind. Appl. Math.*, **5**, 105–136.

—— (1958), "On the nonrandomized optimality and randomized nonoptimality of symmetrical designs", *Ann. Math. Statist.*, **29**, 675–699. (Referred to in the present paper as I.)

—— & WOLFOWITZ, J. (1952), "Stochastic estimation of the maximum of a regression function", *Ann. Math. Statist.*, **23**, 462–466.

—— —— (1959), "Optimum designs in regression problems", *Ann. Math. Statist.*, **30**. (Referred to in the present paper as II.)

KSHIRSAGAR, A. M. (1958), "A note on incomplete block designs", *Ann. Math. Statist.*, **29**, 907–910.

LE CAM, L. (1955), "An extension of Wald's theory of statistical decision functions", *Ann. Math. Statist.*, **26**, 69–781.

MOOD, A. (1946), "On Hotelling's weighing problem", *Ann. Math. Statist.*, **17**, 432–446.

MOTE, V. L. (1958), "On a minimax property of a balanced incomplete block design", *Ann. Math. Statist.*, **29**, 910–913.

NEYMAN, J. & PEARSON, E. S. (1938), "Contributions to the theory of testing statistical hypotheses, III", *Stat. Res. Memoirs*, **2**, 25–27.

PLACKETT, R. L. (1946), "Some generalizations in the multifactorial design", *Biometrika*, **33**, 328–332.

—— & BURMAN, J. P. (1946), "The design of optimum multifactorial experiments", *Biometrika*, **33**, 296–325.

ROBBINS, H. (1952), "Some aspects of the sequential design of experiments", *Bull. Amer. Math. Soc.*, **58**, 527–535.
—— & MONRO, S. (1951), "A stochastic approximation method", *Ann. Math. Statist.*, **22**, 400–407.
SACKS, J. (1958), "Asymptotic distribution of stochastic approximation procedures", *Ann. Math. Statist.*, **29**, 373–405.
SCHEFFÉ, H. (1958), "Experiments with mixtures", *J. R. Statist. Soc.* B, **20**, 344–360.
SHOHAT, J. A. & TAMARKIN, J. D. (1943), *The Problem of Moments*, *Amer. Math. Soc. Surveys*, No. 1.
SMITH, K. (1918), "On the standard deviations of adjusted and interpolated values of an observed polynomial function and its constants and the guidance they give towards a proper choice of the distribution of observations", *Biometrika*, **12**, 1–85.
SOBEL, M. & GROLL, P. A. (1959), "On group testing with a finite population". (To be published.)
STEIN, C. (1948), "On sequences of experiments" (abstract), *Ann. Math. Stat.*, **19**, 117–118.
—— (1955), "Inadmissibility of the usual estimator for the mean of a multivariate normal distribution", *Proc. Third Berkeley Symposium*, vol. 1, 197–206. Univ. of California Press.
—— & WALD, A. (1947), "Sequential confidence intervals for the mean of a normal distribution with known variance", *Ann. Math. Statist.*, **18**, 427–433.
TOCHER, K. D. (1952), "The design and analysis of block experiments", *J. R. Statist. Soc.* B, **14**, 45–100.
WALD, A. (1942), "On the power function of the analysis of variance test", *Ann. Math. Statist.*, **13**, 434–439.
—— (1943), "On the efficient design of statistical investigations", *Ann. Math. Statist.*, **14**, 134–140.
—— (1950), *Statistical Decision Functions*, New York: John Wiley.
—— & WOLFOWITZ, J. (1948), "Optimum character of the sequential probability ratio test", *Ann. Math. Statist.*, **19**, 326–339.
—— —— (1950), "Characterization of the minimal complete class of decision functions when the number of distributions and decisions is finite", *Proc. Second Berkeley Symposium*, 149–157. Univ. of California Press.
WOLFOWITZ, J. (1949), "The power of the classical tests associated with the normal distribution", *Ann. Math. Statist.*, **20**, 540–551.
—— (1950), "Minimax estimates of the mean of a normal distribution with known variance", *Ann. Math. Statist.*, **21**, 218–230.
—— (1951), "On ε-complete classes of decision functions", *Ann. Math. Statist.*, **22**, 461–465.
YATES, F. (1935), "Complex experiments", *J. R. Statist. Soc. Sup.*, **2**, 181–247.

DISCUSSION ON DR. KIEFER'S PAPER

Dr. K. D. TOCHER: It is almost exactly seven years ago since I read a paper to this Section of the Society which I also hoped would set the design of experiments on a sound mathematical basis. It is interesting to look back on the results that have flowed from that attempt. The paper managed to prove some results which had been known for 20 years in three lines instead of three pages and produced a number of designs which I now realize to be extraordinarily silly.

Perhaps we can spare a few moments to see what led to this rather fruitless effort. I started with the assumption that design of an experiment consisted of arranging that the variance matrix of some estimates had some required optimum property. It is a sad fact that almost all attempts to do this led to a requirement for a diagonal variance matrix and to an orthogonal design. When the combinatorial restraints do not allow such a design, we must then turn our attention from the absolute values of the variance matrix to the relative values of the elements, i.e. to the structure of the matrix. It is then easy to impose conditions for the design to have any exactly specified structure, but in many cases there will exist other designs which approximately meet the structure requirements and which have much lower absolute values.

In his contribution to the discussion on my paper, Dr. Yates commented that it showed both the power and the weakness of a mathematical approach. The power lies in that once the real problem has been formalized, it is possible repeatedly to manipulate the symbols, giving no further thought to their meaning until a result has been synthesized. The ease of manipulation prevents a critical examination of the meaning of the intermediate answers. If this were done, it would often show some glaring error which could be traced back not to the mathematics, but to the assumptions in the model being analysed.

Now Dr. Kiefer, in his attempt to set the theory of design on a firm foundation, has taken a much more sweeping standpoint than I did seven years ago. He starts from the obviously acceptable thesis that in order to give a precise design for an experiment, one must have a precise objective. Whilst this cannot be denied, the implication that the experimenters must have a precise objective so that mathematicians can help experimenters seems quite unwarranted. An experiment is performed to find something out about the real world, and the number of cases where we are only asking one question are very rare indeed. We want our experiment to be robust in the sense that it will give useful, if not the optimum amount of information, about a number of different questions we may wish to ask.

Consider Dr. Kiefer's example on page 276 where he sets up three alternative objectives:

(i) to estimate the values of 3 parameters θ_1, θ_2, θ_3,
(ii) to test the hypothesis that they are all equal, and
(iii) to test the hypothesis that they are all zero,

and shows by quite faultless arguments that with these three different objectives we require three different designs. A more likely situation than any of these is that the experimenter or one of his colleagues has, hidden up his sleeve, a complete mathematical theory of the subject under discussion if the θ's are zero. However, even before the experiment, it is quite clear that it is possible that this will disprove this assumption, and then the theoretician has considered what he will do if it turns out they cannot be considered zero. He then sees how to generalize his analysis so that he can, at the expense of a little complication, cover the case when the θ's are all equal. If this turns out to be false, a theoretical analytic approach is reluctantly abandoned and one substitutes a numerical analysis. This will require the best estimates of the θ's. The objective of the experiment is to provide the best information to carry out this programme of activity and to completely specify it needs several parameters. The appropriate values are strongly determined by the prior probabilities of the different contingencies and I am sure Dr. Kiefer would feel this is leading too far from his own standpoint. This example, I think, is typical of the sort of robustness that an experimenter wants from his experiment. Thus the natural reaction at being told to perform an experiment that will not give any information about a quantity in which one is interested is perfectly justified, and the proposition only arises because the experiment is considered to have a single limited objective.

Most of the really startling results of this paper flow from objectives of testing hypotheses, and I think Dr. Kiefer's comment that experimenters are under the misconception that experiments designed for estimation are suitable for hypothesis testing, is itself a misconception. The whole school of British experimental designers, who are a very practical race of people, are certain that the purpose of experiments is to estimate parameters. There may be, as in the example above, an interim flirtation with hypothesis testing.

Dr. Kiefer has modelled his theory of design on the theory of decision functions which has grown out of the work of Neyman and Pearson over 20 years ago. Their original basic idea was, in its time, a brilliant new way of clearly expressing the factors involved in a simple balance of judgments, but attempts to build a general theory on this idea have led to more and more arbitrary elements being introduced. I think it is now recognized amongst practical statisticians that statistical decision function theory cannot be a complete theory of inference, and because of the enormous complexities that arise in all but the simplest problem, a new foundation is probably needed. This attempt to found design theory on similar lines reinforces that viewpoint.

The tilts made in the paper against some of the more exotic block designs are, I think, quite justified, but then it is fairly widely recognized that these designs are not used at all extensively in practice, and there is a distinct impression that they have become the playthings of mathematicians who have a flair for combinatorial analysis.

I am sorry to see that factorial designs are not mentioned in this paper and I would be most interested to hear Dr. Kiefer's comments on the implications of this approach to the factorial designs.

One of the practical considerations underlying an experiment is that accidents in the conduct of the experiment are bound to occur and the experiment must still be capable of giving some results. I would like to ask Dr. Kiefer how his experiments stand up to this criteria.

I found the section on optimum experiment for polynomial regression most interesting, but not as startling as I think Dr. Kiefer expected me to. I would argue, in my naive, way roughly as follows:

The practical way of fitting a regression to equally spaced data is by the use of orthogonal polynomials which, of course, lead to a diagonal variance matrix for the coefficients of those polynomials. Since an optimum experiment is associated with a diagonal matrix, if we want minimum variance estimates of the coefficients in the polynomial, we have to use a set of points so that the orthogonal polynomials associated with that set of points turn out to be exactly the powers of x and this, of course, brings us immediately to the problems described by Dr. Kiefer.

I felt a little disappointed, after the great play made of present day design methods being based on arbitrary criteria, to find towards the end of the paper another set of arbitrary criteria, some of which I suspect have only been put up as Aunt Sallys to be knocked down.

It may be that my comments have been verging on the critical and, therefore, I feel that some constructive note perhaps is rather necessary. This, however, is much more difficult, and in fact, any advice on how to set up a mathematical theory is likely to fall into the trap that has already been sprung in this paper. I am sure that any future for theory of experimental design will lie in a sequential theory in which it is a reasonable approximation to have for each stage of the experiment a precise objective, providing that the theory gives a freedom of choice of new objectives after each stage of the experiment. It is just this sort of freedom of choice given by the evolutionary operational technique developed by Box which makes it so attractive to practical experimenters.

In conclusion I should like to say how much I have enjoyed the mathematics of this paper which uses some very elegant and powerful modes of argument which, I am sure, we will all remember and attempt to use ourselves in other connections.

In this paper, I think Dr. Kiefer has shown conclusively just how useful mathematics is in the theory of design and, if for no other reason, this gives me great pleasure to propose a vote of thanks.

Mr. QUENOUILLE: In this paper, the author has presented us with an exhaustive and, at first reading exhausting, account of considerations affecting choice of experimental designs. In reading it, I must admit to having had considerable difficulty in interpreting some of Dr. Kiefer's results. Much further digestion (if this is the correct word in my case) would be necessary before I felt able to discuss these. My remarks are therefore of a general nature and are to some extent foreshadowed by Dr. Tocher's remarks.

The main message that I have received in reading Dr. Kiefer's paper may be summarized thus: "Given an experimental problem, given also a specific experimental set-up and given finally a method of measuring the effectiveness of its solution, there exists an optimum design. The design depends upon the problem, the set-up and the method of measurement". Trivial though this statement may sound, it is one which is often overlooked and if this paper does no more than to cause the experimenter to re-examine these assumptions and objectives, it will have served a useful purpose.

But, of course, this paper does much more than this. The author gives methods of deriving designs and examines various objectives, and criteria of optimality. It is in this examination that the experimenter will need to consider whether Dr. Kiefer is dealing with situations approximating to reality. For example, in significance testing it might be asked whether the power function in the heighbourhood of the null hypothesis is the appropriate thing to consider. Many experimenters, I suggest, are not much interested in small effects. More important is whether the objectives are as clear-cut as Dr. Kiefer supposes. Often neither the future applications of the experimental results, nor (as is particularly true for the regression problem) can their likely form be foretold. What is needed here seems to be an extension of a further idea from significance testing, namely robustness: a "robust" design being one which answers a range of questions that may turn out to be important as well as those originally considered. We might add to this property, "robustness" in the sense of continuing to work well under the usual experimental hazards such as lost and interchanged observations. I suspect that most of the classical experimental designs in addition to having the properties listed in this paper are favoured because of these properties. A final point which will probably bear little weight with Dr. Kiefer is that most fiducialists will be very wary of any procedure which concentrates solely upon maximizing power.

I have much pleasure in seconding the vote of thanks to Dr. Kiefer.

The vote of thanks was put to the meeting and carried unanimously.

Dr. C. L. MALLOWS: I would like to put in a plea for intuition, which I feel Dr. Kiefer has maligned too strongly in his otherwise excellent paper. Dr. Kiefer gives some interesting examples of situations in which what he calls randomized designs (another name for them is random balance designs) are optimum (in various senses), in contrast to the classical "rigid balance" designs. A major part of Dr. Kiefer's contribution is that he is forcing us to consider very carefully what it is we want from a design. But it seems to me to be no more reprehensible to start with an intuitively attractive design and then to search for optimality criteria which it satisfies, than to follow the approach of the present paper, starting from (if I may call it so) an intuitively attractive criterion, and then to search for designs which satisfy it. Either way one is liable to be surprised by what comes out; but the two methods are complementary. Certainly we should consider what it is about the classical designs that makes them so attractive; and the present paper demonstrates that superficially attractive criteria (like the "local" optimality used in section 2A(ii)) can lead to designs that are hopelessly non-optimum in other senses. Incidentally I would like to ask Dr. Kiefer whether he has considered criteria based on the use of conditional power functions, i.e. conditional on the design actually used; these often seem more relevant than his unconditional power functions. Indeed, Dr. Kiefer does not seem to be consistent in this; in section 2B he has a design with a missing observation, and he considers modifying the χ^2 test to preserve an optimality property conditional on this design; to be consistent with his earlier approach, he should have enlarged his set of possible designs, assigning positive probability to these possibilities, and then have continued to use the average power.

I would like also to ask Dr. Kiefer whether he would agree that where in his introduction he says that techniques such as those of Box and Wilson cannot have any role in satisfactorily solving certain problems, he is referring to strictly mathematical problems; and that while it might be possible to define a procedure which could be proved to have desirable properties in certain idealized conditions, this would not necessarily help the practical man, who will never know whether the assumptions are true or not. The precise results may not be relevant in practice, without a discussion of the effect of departures from the assumptions.

In this last connection, it is perhaps relevant to mention some work carried out recently by Box and Draper into the effect of the presence of higher order terms when fitting a polynomial regression. Suppose we are interested in a regression function over a finite region in k-dimensional factor space; we suppose this region to be given *a priori.* Then Box and Draper consider the problem of choosing a design to minimize the integral over the region of the mean square error; this can be split into two components, the integrated variance, arising from the experimental error, and the integrated squared bias, due to the higher order terms. They found that the variance component was likely to be swamped by the bias component; so that it seems more important to choose the design points to minimize bias, than to minimize variance. One gets quite different designs from the two approaches. The rule for obtaining a design to minimize the integrated squared bias is quite simple; one has merely to match up, as far as possible, the moments of the design with the moments of the given region. This is correct even if we have a weight function in this region, and are using weighted least squares. It is a very general result, and is equivalent to choosing an integration formula which will be accurate for polynomial functions of as high degree as possible.

It is interesting to note that if we try to choose a design to minimize the maximum expected bias, due to the presence of a higher term, we have a problem of Tchebycheff approximation similar to that mentioned by Dr. Kiefer in his section 4C, but with the higher function to be approximated. Thus in this case the solution is just to use a design appropriate for estimating the bias term in addition to those already considered.

Mr. R. N. CURNOW: I agree with Dr. Kiefer that there are some situations in which the aims of the experimenter can best be met by not observing the effects of all the treatments. Where I differ from him is in thinking that these aims are concerned not with hypothesis testing but rather with the minimization of loss functions. There is an interesting resemblance, for example, between Dr. Kiefer's random selection of treatments for actual observation and the random selection of varieties for inclusion in a varietal selection trial. As a very simple example, suppose that a large number of varieties are available for inclusion in a field trial, at the end of which the variety with the highest mean yield is to be selected: the aim of the selection procedure is to maximize the expected yield of the selected variety. If all the available varieties are included in the trial the errors may be so large that the expected yield of the selected variety is nowhere near the maximum possible expected yield. If this is so, then the number of varieties actually tested should be reduced by randomly selecting from those available only a sufficient number to achieve the stated aim of maximizing expected yield.

There are two fundamental differences between Dr. Kiefer's random choice of treatments for actual observation and the random selection of varieties for inclusion in a field trial. Firstly, the parameters in the hypothesis testing situation are fixed constants whereas the yields of the varieties are thought of as random variables following some prior distribution. Secondly, the aims are so different. In one the aim is to maximize the expected power of a test and in the other the aim is to maximize the expected yield of the selected variety. The author has shown that designs with two treatments are often optimal. In contrast, if, in the very simple selection situation I have just described, the prior distribution is Normal, then the optimal number of varieties to include in the trial is never less than 5 and may be very much larger. Furthermore, selection trials do not necessarily have to provide estimates of error variance and single replicate designs are often optimal.

Dr. F. YATES: With reference to the remarks of the last speaker, I may mention that some time ago I discussed the question of the number of lines to be included for testing in a plant selection programme and showed how this could be related to the genetic variance and the experimental error variance so as to maximize the genetical advance with a given amount of experimental resources (Yates, *Emp. J. Exp. Agric.*, **8** (1940), 223–230). I concluded that for this purpose it paid to include a larger number of lines than would be the case if the objective were to test whether the lines were in fact different.

I may say I am fully in agreement with Dr. Tocher's remarks, and would like to emphasize that practically all experiments, except those which are undertaken solely for purposes of estimating treatment differences, are potentially multi-purpose. If, for example, the θ's of Dr. Kiefer's example of three treatments are found to differ from zero, we shall wish to know whether the θ's differ amongst themselves, and if this is established, we are likely to require estimates of the magnitudes of the different θ. Dr. Kiefer has ignored this requirement, but even if we accept his limited objectives, he has reached some rather remarkable conclusions. In particular, in his second case, when he is testing the composite hypothesis $\theta_1 = \theta_2 = \theta_3 = 0$, I think that any practical man would be somewhat surprised if he were told that an experiment had established this, and then learned that only one of the three θ's had been tested. He might point out that if one of the θ's was in fact large and the other two zero, there would be a $\frac{2}{3}$ chance of accepting the hypothesis, whereas if all three θ's had been tested the hypothesis would almost certainly have been refuted.

TABLE

Probabilities of Obtaining a Significant Result ($P = 0\cdot05$) *for Three Procedures*

$2(\theta_1^2+\theta_2^2+\theta_3^2)$	Procedure (*a*)	(b_1)	(b_2)
0	0·05	0·05	0·05
1	0·11	0·17	0·17
2	0·19	0·26	0·29
3	0·27	0·32	0·41
4	0·35	0·34	0·52
6	0·52	0·36	0·69
8	0·65	0·37	0·81
10	0·76	0·37	0·88
12	0·84	0·37	0·93

Dr. Kiefer has, of course, not overlooked this point entirely, but by concentrating on "alternatives close to the null hypothesis" he has obscured it. The above table gives the probabilities of obtaining a significant result (a) when $n_1 = n_2 = n_3 = 2$, and (b) when one treatment i is selected at random for test and $n_i = 6$.* In the latter case the probabilities have been calculated for the case (b_1) where $\theta_2 = \theta_3 = 0$, and the case (b_2) where $\theta_1 = \theta_2 = \theta_3$. It will be seen that, although the probabilities for (a) are always less than those for (b_2), they are only less than (b_1) when the probabilitity of obtaining a significant result by any type of experiment is small. When the probability for (a) is large there is no great difference between (a) and (b_2), whereas (b_1) is only slightly greater than $\frac{1}{3}$.

In practice if the probability of detecting a substantial deviation from the hypothesis is small the experiment must be regarded as of inadequate accuracy. The accuracy of the experiment must, in fact, be related to the magnitude of the deviations from the hypothesis which are considered to be of practical importance. Nothing is easier than to "prove" that a hypothesis is true by testing it by an experiment which is sufficiently inaccurate. Consequently the experimenter is only really interested in cases where the probabilities of obtaining a significant result are large when there are deviations from the hypothesis which are of practical importance. This means he is concerned with the bottom part of the table where (b_1) is very markedly inferior.

It may also be noted that if we are concerned with the two tests, (1) that the θ's are the same (or similar) but differ from zero, and (2) that the θ differ amongst themselves, test (1) will be the same whether design (a) or design (b) is used (since observations on all three θ's will be pooled), and will give the probabilities tabulated in (b_2) above when the θ's are in fact the same. Consequently the advantage of design (a) is wholly illusory.

There are, of course, cases where a random selection of items for test is necessary, though I think Dr. Kiefer's term of "randomized design" for this process is a confusing one, which should not be used. In addition to the plant selection problem mentioned above, one that frequently arises in agriculture is the estimation of responses to treatments such as fertilizers, where the responses vary from centre to centre, often in association with known conditions such as soil characteristics. Here we have to decide the numbers of centres at which trials should be conducted and the accuracy required at each centre.

Dr. Kiefer appears to be under the misapprehension that the standard types of experimental design in current use have been adopted because they were "intuitively appealing". Of course we required designs which did not need excessive labour for their analysis, but fortunately simple algebraic solutions frequently go hand in hand with maximum efficiency. Dr. Kiefer can rest assured that logical reasoning has been applied to questions of efficiency.

* Since the meeting the values for (a) have been computed more accurately on the electronic computer.

Mr. E. M. L. BEALE: I would like to add my thanks to Dr. Kiefer for a most stimulating paper, and to say a few words about sequential design. In the introduction Dr. Kiefer mentions the problem of finding optimum sequential designs, and says that techniques like those of Box and Wilson cannot in their present state have any role in satisfactorily solving these problems. Speaking theoretically, and assuming that computation is infinitely cheap compared with experimentation, I agree that the Box-Wilson techniques are not optimum. For one thing, they do not always use all the relevant information in the analysis. In Princeton last year I spent some time thinking about an alternative approach to this problem. I came up with a rather complicated procedure which seemed to me intuitively reasonable as a method of saving some experimentation at the expense of having an untidy experimental design and much more computation. I wrote a technical report* about this, but the work is still in a rather preliminary state.

I did not try to work from, or prove, any optimality properties. One difficulty about such a fundamental approach to this problem seems to me to be that a good practical procedure must take advantage of the prior knowledge, or feeling, that the response surface is probably well behaved. It seems difficult to put this feeling into precise mathematical terms, and perhaps it means different things in different circumstances. A typical meaning might be that the partial second derivatives are probably not too large and do not vary too rapidly, except perhaps over certain critical hypersurfaces. This information might be incorporated in a prior probability distribution for the response surface, but this would be awkward to manipulate, and might be found to imply more than one really meant.

Dr. D. R. COX: In the problems of experimental design and of the allocation of observations that Dr. Kiefer has dealt with so impressively and elegantly, there seem often to be at least three things for consideration:

(a) the control of systematic errors;
(b) the minimization of the effect of random errors;
(c) the examination of interactions.

Under (a) are included the effects of sources of variation that it has not been possible to randomize or balance, the effects of using a wrong form of response curve, etc. As for (c), even in the simplest type of experiment, to compare a treatment with a control, it is very often desirable to use experimental units in such a way that we can examine what limits the validity of any treatment effect found. Dr. Kiefer has concentrated on (b).

I should like to say something about (c). To a certain extent the problems it raises are similar to those that Dr. Kiefer has discussed. Suppose that two treatments are under comparison with a given number, $2N$, of experimental units, and that the units can be arranged in groups. If there are r groups we suppose there to be $2N/r$ units per group. For example, the groups may be sites in an agricultural experiment. There are numerous ways of setting about choosing r.

(i) We may regard the groups as "random" and maximize the precision of the treatment main effect. This is equivalent to the usual allocation problem in two-stage sampling and leads to a value of r determined by the ratio of the interaction variance σ_{tg}^2 to the within-group error variance, and by a cost factor.

(ii) In the situation of (i) we may require a sensitive significance test of the interaction σ_{tg}^2. This is exactly the problem of §2A (ii). It is difficult to conceive of this being the primary requirement.

(iii) Often it will be more profitable to regard groups as systematically classified. If the interaction is appreciable and complicated we shall regard each group as having its own treatment effect to be estimated separately. Then we may require there to be sufficient observations in each group for the standard error to be reasonably small for each estimated effect.

(iv) Let there be R groups that could possibly be included in the experiment. Suppose

* Beale, E. M. L. (1958), "On an iterative method for finding a local minimum of a function of one variable", *Statistical Techniques Research Group Report* No. 25.

that it is required to maximize the probability that a particular group having a treatment effect Δ leads to a difference significant at the α per cent. level; this probability is of course the probability r/R that the group is included in the experiment multiplied by the probability of a significant difference given that it is included. This leads us in the direction of large r.

(v) The groups may have a factorial structure, possibly fractionally replicated. If the true treatment effect is constant, the precision is unaffected by including the group variations. If there are important interactions between treatment effect and the group factors, it will depend on the order of these interactions how many treatment effect parameters are to be estimated. For example we might include group factors up to the point at which a treatment effect of satisfactory precision can be obtained corresponding to any combination of two group factors.

Of course, the restriction to a single experiment of preassigned size makes this discussion rather artificial.

Now in principle it is undoubtedly possible to include (a), (b) and (c), and any other relevant considerations into Dr. Kiefer's approach and to derive an optimum design. Nuisance parameters will usually have to be removed by averaging or by a minimax criterion and a compromise between conflicting requirements reached by appropriate weighting. I find it however difficult to believe that this concentration on optimum designs is, in complex cases, really the most useful way of presenting the results of this type of theoretical work. Will not comparative numerical figures for the efficiency of designs according to various criteria be more helpful?

A final point concerns Dr. Kiefer's criticism of combinatorial conditions such as resolvability. The statistical advantages of resolvability were established by Dr. Yates in his papers on the recovery of inter-block information, and I would have imagined that it is for these statistical reasons alone that the condition has received attention.

Note added after the meeting: Dr. Kiefer's very interesting example, where to test $\theta_1 = \theta_2 = \theta_3$ it appears better not to make observations about one of the θ's, seems to illustrate the difficulties of not adopting conditional inferences where possible. From the conditional viewpoint, if say we have observations relevant to θ_1 and θ_2, the questions that can be answered can concern only θ_1 and θ_2. The general conclusion then seems very acceptable that where we start with many parameters and limited resources, it may, from some points of view, be desirable to investigate only a few of the parameters. If we do this we must then recognize, when we analyse the results, that we have changed the question being asked. I realize that this is not the precise import of the example, but would welcome Dr. Kiefer's comments on whether the interpretation of results should in such cases be made conditionally.

Professor G. A. BARNARD: I would like first to say how grateful we are to Dr. Kiefer for taking so much trouble to set out his approach to the problem of experimental design, with such clarity and definiteness. Dr. Kiefer understands very well that one of our aims was to have a sharp discussion of the issues involved, concentrating more on the methodological approach than on detailed results, and I imagine that some of his formulations have been made with the object of promoting discussion rather than with the object of stating all the qualifications which Dr. Kiefer himself would wish to impose.

I would first like to mention two small matters of detail. It is stated that the methods of Box and Wilson were first suggested by Friedman and Savage and were further dedeloped by the author. I would like to ask whether Dr. Kiefer's own work, like that of Friedman and Savage, was concerned with changing a single variable at a time, since there are very important differences between this and Box and Wilson's approach.

The second point of detail concerns the statement that it is generally believed that in the normal case an F test should be used. As far as statisticians in this country are concerned I think it would be fair to say that this illusion had some currency before the war (as instanced, for example, by Newman's first tabulation of the studentized range) but it is clear, for example, from *Biometrika Tables for Statisticians*, page 52, that this is no longer the case.

Now to come to the general points. Some indication of my major philosophical differences with Dr. Kiefer is given on page 273 where the procedures of Box and Wilson are said to be "often not even well-defined rules of operation". There is a suggestion here that this is a defect; but it should be pointed out that in the field of practical human activity rules of operation which are not well-defined may be preferable to the rules which are. For we are dealing with thinking beings and a procedure which forces the exercise of the human capacity to think is often better than one which can be carried through entirely mechanically.

The essential point of divergence is concisely stated at the end of section A "The rational approach is to state the problem and the optimality criterion and then to find the appropriate design, and not to alter the statements of the problem and criterion just to justify the use of the design to which we are wedded by our prejudiced intuition". I would agree with this statement if the word "deductive" were inserted in place of the word "rational". As a rationalist I feel that the word "rational" is one which indicates a high element of desirability, and I think it is much broader in its meaning than "deductive". In fact what appears to me to be the rational approach is to take designs which are in use already, to see what is achieved by these designs, by consideration of the general aims to evaluate such designs, in a provisional way, and then to seek to find designs which improve on existing practice. Having found such designs the cycle should be repeated again. The important thing about this approach is that we are always able to adjust our optimality criteria to designs as well as adjusting our designs to our optimality criteria.

To take the argument to a somewhat more technical level I feel it to be a serious weakness of Dr. Kiefer's approach that he adopts wholeheartedly the decision-theoretic formulation of statistics, without accepting what seems to me to be an inevitable consequence of such acceptance, namely, the abandonment of the search for uniquely optimum procedures. I have long been satisfied that the decision-theoretic approach has its place in statistics, and I have equally long been satisfied that the only way to apply this approach is via a full-blooded use of Bayes's theorem, associated with some sort of plausible prior distribution. The continued search for "optimality", in any sense independent of any prior distribution, is therefore, to my mind, misguided. If in a special case there really is a "uniformly best" procedure, it will be found by the Bayesian approach, since, *a fortiori*, it will be best for any particular prior distribution.

Finally, however, I must repeat my conviction that the decision-theoretic approach is inadequate for a full account of problems of experimental design, as it is for most other problems of practical statistics. When, as in most of the cases dealt with here, we are concerned to select "preferred" hypotheses from a specified set, we shall in fact look, *a posteriori*, at the likelihood function yielded by our experiment. A "good" experiment would then be one which gave a "good" likelihood function; and a "good" likelihood function would be one which factorized in such a way as to allow the parameters to be considered separately, which was "highly concentrated", and which had such multiplicity of other features as would defy any attempt at quantification.

Mr. R. C. Curnow: Speaking purely on criteria which must have validity in the practical world, and as a non-member of the Research Section I would like to comment on the assumption that "long-run" desirability is useful in a criterion for experimental design.

Discussing briefly the problem of selection raised by my namesake, and formally posed by Dr. Yates in *J. Emp. Exp. Agr.*, 1948, it would seem that long-run and short-run approaches may conflict, whereas it is often clear that the latter is more desirable. The problem is one of allocating a fixed amount of experimentation between the number of varieties chosen and the amount of testing or evaluation carried out on each in order to select the best or best small proportion.

In the long-run sense, for most practical cases encountered, the weighting is heavily on the side of choosing a large number of varieties. But in the short run, what is really wanted is to maximize the minimum advance gained by selection and this is very dependent on the ratio of test to genetic variance. Generally speaking, the short-run approach is much preferred practically.

It would seem too that criticism can be made from the practical angle if an experimenter were to concentrate on hypothesis-testing, as for example in Dr. Kiefer's case of 3 catalysts A, B, C. It would seem not useful telling the experimenter that he has done the best possible thing if no effects are present. What is wanted is an experiment which is most powerful over all the possible outcomes. I suggest that since this involves specification of all the prior probabilities, this boils down in practice to the experimenter weighting his investigations by the degree of plausibility, which seems desirable. I suspect, moreover, that one other criticism would emerge in practice should a random balance design be accepted. The possibility of a long run of failures even to have considered the "right" catalyst would, in a competitive world, ensure that a short run view of experimentation would be adopted!

Dr. M. Stone: In §4 Dr. Kiefer stated that D, E and A-optimalities are in general unrelated. I would like to describe how, when we have normality, these criteria (with an additional one closely related to trace v_d^{-1}) are special cases of a criterion based on the Shannon-information about ψ when σ^2 is known. The specializations which have to be made to produce the criteria may throw some light on the nature of the differences between them. The design-oriented use of the Shannon-information measure was introduced by Lindley (*Ann. Math. Statist.*, **27** (1956), 986).

(i) *D-optimality*.—Suppose $\varphi_1, \varphi_2, \ldots, \varphi_{m-s}$ are additional linearly independent syzygies of θ such that ψ and φ are also linearly independent. Denote the covariance matrix of the b.l.e.'s of (ψ, φ) by

$$\begin{pmatrix} v_d & u_d \\ u_d' & w_d \end{pmatrix} \sigma^2.$$

Taking a multivariate-normal prior distribution for (ψ, φ) with covariance matrix

$$P = \begin{pmatrix} B & E \\ E' & C \end{pmatrix},$$

it can be deduced from §6 of my paper (*Ann. Math. Statist.*, **30** (1952), 55–70) that the information in Y_d about ψ, say I_ψ, is a decreasing function of $|v_d|$ when *either* the matrices B and C are large compared with v_d and w_d and P is not near degeneracy (indicating that we should use $|v_d|$ when the experiment is "powerful" for both ψ and φ and the prior correlations between ψ and φ are not near unity) *or* the matrix B is large and E and u_d are null (indicating $|v_d|$ when the experiment is "powerful" for ψ; ψ and φ have prior independence and the b.l.e.'s of ψ are uncorrelated with those of φ).

(ii) *R-optimality*.—From the last reference, we find that I_ψ is a decreasing function of $|L_d|$ where L_d is the covariance matrix of ψ in its posterior distribution, i.e. the leading $s \times s$ submatrix of

$$\left[\begin{pmatrix} v_d & u_d \\ u_d' & w_d \end{pmatrix}^{-1} \sigma^{-2} + P^{-1}\right]^{-1}$$

$$\sim P - P \begin{pmatrix} v_d & u_d \\ u_d' & w_d \end{pmatrix}^{-1} P \sigma^{-2},$$

if

$$\begin{pmatrix} v_d & u_d \\ u_d' & w_d \end{pmatrix}^{-1} P \sigma^{-2}$$

is small (a condition equivalent to saying that the experiment is "weak" for all parameters).

When E is null,

$$|L_d| = -|B| \text{ trace } (B\, a_d)\, \sigma^{-2},$$

where a_d is the $s \times s$ leading submatrix of the "information" matrix of (ψ, φ). If *R-optimality* is defined on trace $(B\, a_d)$, then its use leads to designs maximizing I_ψ when the experiment is "weak" for all parameters and there is no prior correlation between ψ and φ.

When $B \propto I$ and $s = m$, the criterion becomes trace a_d or trace v_d^{-1}.

R-optimality has the incidental superiority (assisting sequential design) that if d_1 is R-better than d_2 then $d_1 + d_3$ is R-better than $d_2 + d_3$ for any other (weak) design d_3; for the information-matrices of the independent experiments based on d_1 (or d_2) and d_3 add to give the information-matrix of the combined (non-randomized) design. This property does not hold for trace v_d^{-1} when $s < m$.

(iii) *A-optimality*.—Consider a hypothetical stochastic transformation from (ψ, φ) to μ whereby μ is distributed as $N[(\psi, \varphi), G]$. The prior distribution for μ will have covariance matrix $P + G$. After some calculation, we find that the covariance matrix of the b.l.e.'s of μ is

$$H = (P + G)\, P^{-1}\, V_d\, \sigma^2\{I + [P^{-1} + V_d^{-1}\, \sigma^{-2}]\, G\}.$$

When P is large, the information about μ is a decreasing function of $|H|$ (see (i)), i.e. of $|V_d \sigma^2 + G|$, or of trace $(G^{-1} V_d)$ when G is large (representing a "strong" stochastic transformation). We can say that this criterion determines A_G-optimality. With

$$G = \operatorname{diag}(\underset{(1,}{K}, \underset{2,}{K}, \underset{\ldots}{\ldots}, \underset{s)}{K}, \infty, \infty, \ldots, \infty)$$

where K is large, we obtain trace v_d or *A-optimality*. Thus the design minimizing trace v_d maximizes the information about $\mu_1, \ldots, \mu_s$. This is a sort of information about ψ; we have, by the use of this special G, effectively transformed φ out of consideration.

(iv) *E-optimality*.—This can be regarded as a special case of A_G-optimality. For consider all G's of the form

$$G(\alpha) = \begin{pmatrix} (\alpha\alpha' + \epsilon I)^{-1} & O \\ O & \operatorname{diag}(\infty, \ldots, \infty) \end{pmatrix}$$

where ϵ is a small constant and α is an s-vector with $\alpha'\alpha = 1$. Then

$$\begin{aligned} \text{trace } (G(\alpha)^{-1} V_d) &= \text{trace } [(\alpha\alpha' + \epsilon I)\, v_d] \\ &= \alpha'\, v_d\, \alpha + \epsilon \text{ trace } v_d \\ &\sim \alpha'\, v_d\, \alpha, \text{ since } \epsilon \text{ is small.} \end{aligned}$$

Now $\pi(v_d) = \max_\alpha \alpha' v_d \alpha = \max$ trace $(G(\alpha)^{-1} V_d)$. To interpret this, consider $\beta'\mu$ where $\beta'\beta = 1$. Under the transformation determined by $G(\alpha)$,

$$\text{var } (\beta'\mu/\psi, \varphi) = \infty \text{ unless } \beta_{s+1} = \ldots = \beta_m = 0.$$

When the latter condition holds and we write $\beta(s) = (\beta_1, \ldots, \beta_s)$, we have

$$\text{var } (\beta'\mu/\psi, \varphi) = \beta(s)'\, (\alpha\alpha' + \epsilon I)^{-1}\, \beta(s).$$

Using the identity

$$\begin{aligned} (I + \alpha\alpha'/\epsilon)^{-1} &= I - \alpha(1 + \alpha'\alpha/\epsilon)^{-1}\, \alpha'/\epsilon \\ &= I - \alpha\alpha'/(1 + \epsilon), \end{aligned}$$

we find var $(\beta'\mu/\psi, \varphi) = \epsilon^{-1} - [\beta(s)'\, \alpha]^2/\epsilon(1 + \epsilon)$ which is $O(1/\epsilon)$, unless $\beta(s) = \alpha$, when it is unity. Hence trace $(G(\alpha)^{-1} V_d)$ effectively determines (as a decreasing function) the information about the stochastic transform of $\alpha'\psi$. Therefore an *E-optimal* design maximizes the minimum "stochastic information" among standardized syzygies of ψ, under certain conditions.

Dr. KIEFER replied briefly at the meeting and more fully in writing, as follows: I must admit being somewhat disappointed to see that such a large proportion of the comments have been engendered by a careless reading of my paper and by what seems to be a lack of understanding of the spirit of decision theory; while the first of these causes is undoubtedly largely the fault of my writing, I cannot accept the blame for the latter. From reading the practical literature, I would never have guessed that I would be accused by a group of British statisticians of being too strong an advocate of hypothesis testing. As a matter of fact, I carefully stated in conjunction with the examples of Section 2A(ii) my scepticism at the idea that one really wanted to test hypotheses, and emphasized that these examples were chosen for the sake of arithmetical simplicity, supposing (incorrectly, I am sorry to say) that readers would know that mathematical counter-examples (to certain optimum properties of symmetrical designs) do not constitute sweeping recommendations for certain other procedures. Moreover, for many years now the mathematical statisticians who have been influenced by Wald's work have repeatedly criticized the classical dichotomy, exemplified so often in the practical literature, wherein most problems are classified either as ones of estimation or as ones of hypothesis testing, complex problems being treated as combinations of these two basic types. Thus, for example, in recent years a variety of useful ranking procedures have replaced the classical usage of a sequence of F-tests in many important practical applications which were formerly inadequately treated as analysis of variance problems. Perhaps the greatest relevance of decision theory to the practical man is that it tells him to set out, in advance of the experiment, the list of possible decisions (or statements) he may want to make, and then (taking into consideration the seriousness of various possible wrong decisions) to compute an appropriate procedure.

It is in this choice of a procedure that we see the difference between the classical approach and that of decision theory. Many examples of the former can be found in the literature; typically, one tests various hypotheses, each suggested by previous tests based on the same data, and at the end throws in a list of certain point or interval estimates of "significant differences". The probability or confidence meaning of the intersection of the various stated conclusions is not considered. (There are a few recent papers by Paul, Bechhofer, and others, in which the effect of preliminary tests in certain settings is computed.) Why not instead, from the outset, construct a procedure which, in the setting of Example 2A(ii) (c), but with a different decision space, can yield any of various conclusions such as "all θ_i are about equal," or "θ_1 and θ_2 are about equal, and θ_3 is 2 units larger," etc.? The approach of decision theory is to list in advance these possible statements, and then to design a good procedure for obtaining one of these conclusions, and to know its probabilistic properties. As far as I know, the only rigorous discussion of whether or not the classical method of superimposing various tests and estimation procedures in such problems can lead to an admissible procedure, is contained in two recent papers of E. L. Lehmann.

I am sorry to have found it necessary to use the space to make the above comments, which should have been unnecessary if the spirit of modern statistical theory were properly understood. The extensive discussion of Dr. Tocher, Dr. Yates, and several others, which scolds me for testing a hypothesis, is irrelevant and misleading. It is not, as Dr. Yates says, that I have ignored his requirement of giving estimates in certain cases; I have merely restricted my attention in Section 2A to simple examples to illustrate certain ideas, as has been stated repeatedly in the paper. There is no argument at all but that a list of decisions of the type given in the previous paragraph is more meaningful than that of hypothesis testing, in most practical problems. The computation of appropriate procedures for such decision spaces is difficult. With modern computing machines, though, it is feasible, and I believe that it is only our arithmetical ineptitude and laziness which has prevented decision theory from producing more useful new procedures and thus from refuting Dr. Tocher's thesis regarding its failure. In fact, one has only to look at the advances made in recent years in a broad area of complex production, allocation, and inventory problems, to see what improvements the rational approach and big computing machines can bring about. Statisticians are far behind their colleagues in these other areas in introducing such improvements.

It seems to me that the discussion of Dr. Tocher and Mr. Quenouille regarding what they call "robustness" of a design stems largely from thinking of procedures in terms of tests and estimates, instead of in terms of a complete set of possible decisions, set out in advance. To say that a design must be useful for estimation as well as for testing because the result of the test may lead one to decide to give estimates, is to say that one is going to use a sequence of classical tests and estimates in the manner I have already criticized. Why not think in terms of the right space of decisions from the outset? I might add that the possibilities of accidents to the experiment, which Dr. Tocher mentions, should be included in the space of possible states of nature, in choosing a procedure in the manner of decision theory.

I regret that Dr. Tocher seems also to have misunderstood my mathematics, although I am happy that his incorrect argument about polynomial regression led him not to be startled, since I did not (as he expected) expect or plan to startle him. There is no question of a diagonal matrix being associated with optimality, or of making the powers of x be orthogonal with respect to the optimum ξ; for example, it should be obvious to everyone that x and x^3 cannot be orthogonal unless $\xi(0) = 1$. Thus, Dr. Tocher's explanation is meaningless.

I am not guilty of the dishonesty Dr. Tocher accuses me of, in setting up criteria just to knock them down; these criteria have all been used previously, and their rationale has been discussed extensively.

Regarding the factorial designs that Dr. Tocher asks about, the results of Sections 2 and 4 apply also to them, as was stated also in my 1958 paper.

Professor Barnard and Dr. Tocher have both alluded to the freedom of action at each stage in procedures like those of Box and Wilson, as a desirable property. It seems to me that the complexity of these problems, which makes the definition of any reasonable procedure more complicated than it would be in simpler settings, should not be taken as an excuse for not defining procedures precisely. It is a confusion to think that a "freedom of action" which evades such a precise definition by listing many possible courses of action without always telling the practical man when to use which one, will accomplish more for him than a carefully stated rule of action which tells him what to do in each possible circumstance. I think the same confusion is in part responsible for the notion that there are problems of inference which fall outside the realm of decision theory; the immense complexity of the spaces of possible decisions, experiments, and states of nature in some problems makes it require less effort to think of such problems in a lazy, unprecise fashion, than to list all of the possibilities. I do not say that writing down the problem carefully is much of a theoretical accomplishment (just as I agree with Mr. Quenouille that the main message he received from reading my paper, which is neither mine or new, is trivial); but it seems to me that a careful statement of the problem is essential and is a first step toward a careful description of a procedure.

Thus, I think "freedom of action" is a deception. A precisely stated rule of action for each stage of a sequential experiment, determined by the previous observations so as to achieve certain goals which are stated in advance, will also often appear to show great freedom of action because of the many situations it must deal with. But the end is *not* freedom of action; rather, it is the achievement of certain goals, e.g., the estimation of the maximum of a response surface with a certain precision. We run into extreme computational difficulties when we try to compute the properties of various procedures for such problems, and at this stage of development of the subject we must probably be satisfied with finding procedures which, while not optimum, are good, *and which have known probabilistic properties.* That is why I view the contributions of Blum and of Sacks, which were mentioned in Section 1 and which at least give known asymptotic properties, as the foremost advances to date on this problem of finding a maximum, and it is why I view the development of a stopping rule with appropriate properties as the next step. Let me say that I view my own work on this subject (1948), about which Professor Barnard inquired, as being no more satisfactory than that of Box and Wilson, whose empirical studies of the subject undoubtedly greatly exceeded mine; my methods also allowed several variables to be varied at once, but I cannot agree with Professor Barnard's implication that this in itself is much of an accomplishment: if one can assert nothing about the precision of such a method (and my work was unsatisfactory because I could assert very little), I find it difficult to give the procedure much credit for its intuitive appeal.

I think an anecdote is relevant here. A British statistician told me that, before he visited America, his reading of theoretical papers in statistics made him wonder what in the world people on the other side of the Atlantic advised when consulting with practical people; he was relieved to discover that they usually advised exactly what he would have recommended. The lesson, I think, is not so much that theoretical statisticians do not practice what they preach, but that their arithmetic is too poor for them to be able to derive the procedures they would like to use. Obviously, if a scientist asks my advice about a complex problem for which I cannot compute a good procedure in the near future, I am not going to tell him to cease his work until such a procedure is found. But when I give him the best that my intuition and reason can now produce, I am not going to be satisfied with it, no matter how clever a procedure it may appear on the surface. The aim of the subject is not the construction of nice looking procedures with intuitive appeal, but the determination of procedures which are proved to be good. I have alluded previously to the advances made in recent years in other fields of optimization, with the aid of modern computing machinery, and it seems evident to me that a replica of this kind of effort is what is now wanted in the complex problems of statistics.

I look forward very much to reading about Mr. Beale's improvements on the Box-Wilson techniques.

In reply to Dr. Mallows, I would feel less confidence than he would in choosing my criterion to justify the procedure, that being the temptation into which I fear his approach (a similar approach being intimated by Professor Barnard) would lead us. Regarding his question about conditional power functions, in the present context I think this is mainly a denial of the *raison d'être* of randomized procedures in statistics, where (unconditional) risk functions are being compared. The example of Section 2B which he cites is one in which *a priori* probabilities would usually not be known; if they were and if the integration he suggests were performed, we would still reach the conclusion which the example sets out to demonstrate: the lack of optimality of the χ^2 or F-test.

I do not understand the reason for Dr. Mallows's discussion of the conditions which the practical man may not know to be satisfied, since we all constantly use procedures whose behaviour we know to depend in some way on the underlying probability mechanism. The whole subject of what is now sometimes called robustness in statistics can be viewed as a discussion of (nonparametric) problems in which we want the risk function to be very small for a certain class of mechanisms and not too large for the others.

I was not acquainted with the Box-Draper work which Dr. Mallows mentions, and look forward to seeing it; it should make the users of equally spaced observations happy. However, it seems to me that Dr. Mallows rushes by an important aspect of the problem. For these considerations entail a specification of which bias terms are of interest—they do not let one escape from stating a precise model, since, for example, the question of whether the bias term or other error term is larger depends on what bias terms are being considered, i.e., on which other functions are known to be possible contributors to the regression. Moreover, it seems to me that if, for example, the problem is one of curve fitting where we think we have a quadratic but might have anything up to a quintic, then the problem might best be formulated as one of curve fitting where there are various losses to be attributed to misestimation as well as to using many terms in the estimated regression function, and a procedure must, *inter alia*, prescribe the degree of the estimated regression curve. Even for a fixed design, the determination of a good estimation procedure in problems like this, as distinct from the classical methods which are based on a sequence of F-tests, is difficult, and perhaps requires the aid of computing machines, as has been mentioned.

These integral-minimizing criteria have been discussed in Section 4 and in reference II. It still seems to me that I would be happier knowing that the *maximum* expected error of my regression functon is small, than knowing that some *average* is small. Moreover, the lack of invariance of these integral-minimizers is unappealing; for example, if instead of estimating $a + be^x$ we estimate $a + by$ where $y = e^x$, it seems reasonable that an optimum allocation of observations in terms of x, say at $x_1, \ldots, x_n$, should be transformable into an optimum allocation in terms of y, say at $y_1, \ldots, y_n$, by putting $y_i = e^{x_i}$. Integral-minimizing designs do not have this property.

I look forward to seeing the details of Mr. R. N. Curnow's example, although I am sceptical about the assumption of a normal *a priori* distribution.

Regarding Dr. Yates's comments, I have already mentioned that many of his remarks reflect a misreading of my paper. I will only add that a more careful reading will show him that I am indeed concerned with the behaviour of the power function at alternatives other than those which are close to the null hypothesis, and in the fact that the local behaviour is not necessarily of prime concern; one of my mathematical counter-examples has been taken by him as a sweeping recommendation by me of these randomized designs. I am grateful to Dr. Yates for computing the table, since I had only enough stamina to compute the derivative of the power function in my earlier paper (1958). I must sadly remain unconvinced by Dr. Yates's reassurance that logical reasoning has been generally applied in choosing experimental designs.

The kind of dissection of a problem that Dr. Cox has given (as well as the more general use of the intuition which I have been accused of maligning) will sometimes help one to guess at optimum procedures, but I still see no other course for the eventual verification of optimality except by the methods of which he is sceptical. To give comparative numerical figures of efficiency according to various criteria, as he suggests, does not yield an escape route; for one is still going to pick a design to be used, by some (perhaps subconscious) combination of these criteria, and that is no different from using from the outset the approach about which he expresses doubts, with this final criterion. Regarding conditional power, which has also come up before, the fact remains that the unconditional power has perfectly valid meaning from the point of view of the outset of experimentation. There is an emptiness in trying to apply this conditional point of view (which is, of course, useful in constructing certain statistical procedures) in every possible place; carried to an extreme, we find ourselves trying to argue conditionally given the entire outcome of the experiment, which gives us nothing at all.

I have already answered some of Professor Barnard's comments. Regarding his mention of page 52 of the *Biometrika* tables, I hope it is clear that I am not talking about statistics like the studentized range, but rather about *improvements* on the F-test in certain settings. Regarding "uniformly best procedures", I think we are all agreed that they rarely exist. As for the use of Bayes's theorem, it is a wonderful tool; it yields an optimum procedure with much less work than that required, for example, to compute a minimax procedure (this last, I suspect, being a big reason for its popularity); the only difficulty is, *where do you dig up the* a priori *distribution*? Now, we have a result like Savage's *tour de force*, which, under certain assumptions about rational behaviour, yields a weight function and *a priori* probabilities for each individual. Ignoring the difficulty of deciding which of the sets of probabilities of two co-workers we are to use, let me insert another practical comment (having been bombarded by "practical people" all afternoon)—is it really more useful to find an individual's *a priori* probabilities and weight functions by asking him an infinite (or even, for a good approximation, a large finite) number of questions, than to present him with a complete class of operating characteristics and to ask him to choose among them? The questions are easier (and more numerous) in the former system, but is this nice theoretical result really going to give us useful *a priori* distributions in practical problems?

Regarding Professor Barnard's final paragraph, it is easy to give examples which show that the criterion of factorizability of the likelihood function may lead to the choice of the uniformly worst of two experiments. By now it is too clear to profit from further comment, that I cannot agree with his feeling that the desirable features of an experiment must defy any attempt at quantification.

As for Mr. R. C. Curnow's comments, I invite him to find "an experiment which is most powerful over all possible outcomes"! The *a priori* probabilities he desires have already been discussed. His final comment on short and long run points of view must involve some misunderstanding of the meaning of the risk function of a design, which will reflect expected losses relative to the appropriate loss function for the problem.

I am very happy to have received Dr. Stone's written comments after the meeting, having been somewhat disappointed by the lack of mathematical comments at the meeting. I would like to take a few lines to give my views on this "information theory" approach. In Lindley's paper on this subject (*Ann. Math. Statist.*, **27** (1956), 986) he gave an example to show that this approach could lead to inadmissible experiments: experiment A could be sufficient for experiment B, but experiment B could yield a larger measure of information. It is thus a little puzzling that one should generally devote much effort to this approach. In particular cases, the method may be useful, and Dr. Stone shows in his 1959 paper that some of Ehrenfeld's results which are discussed in Section 3 of my present paper can be obtained from his results. For example, in terms of his measure of information $\Delta I_F(\mathbf{A})$ associated with information matrix F and *a priori* normal distribution on the parameters with covariance matrix $\mathbf{A}$, he shows that $\Delta I_F(\mathbf{A}) \geqslant \Delta I_G(\mathbf{A})$ for all $\mathbf{A}$ if and only if $F \geqslant G$. In his Theorem 5.1 he considers the integral of the power function near the origin and obtains a result which can also be obtained from Wald's theorem or from my 1958 result on type D tests (stated in Section 2B of the present paper). He demonstrates the D-optimality of certain designs in a few symmetrical settings which fall within the general symmetrical framework considered in my 1958 paper (and which is mentioned in Section 4B of the present paper). He obtains certain results on the number of points needed to support an optimum ξ, which are related to Chernoff's results and some of the results of reference II. I need not reiterate my view on the meaningfulness of the *a priori* distribution in question; thus, the most useful aspect of Dr. Stone's approach seems to me to be that one in which the matrix $\mathbf{A}$ is viewed as a mathematical tool, just as *a priori* distributions were viewed by Wald in his complete class and minimax results. The consideration of $\Delta I_F(\mathbf{A})$ thus yields, among other things, some of the results just mentioned and which can also be obtained without this information theory approach. In his present contribution, Dr. Stone has extended to various optimality criteria the kind of discussion which appeared in his 1959 paper, where the criterion of D-optimality was obtained indirectly by considering $\Delta I_F(\mathbf{A})$ when $\mathbf{A}$ is large. It is a bit misleading to say that these criteria are special cases of Shannon information; it is as meaningful to say that they are special cases of the consideration of $\mathbf{A}_d$. I must confess that I remain unsure as to what advantage is obtained by expressing these things in terms of information-theoretic concepts. I will only add that Dr. Stone's R-optimality has been mentioned in Section 4A as one to which little recommendation can be given, and that in the paragraph following equation (4.2) we have seen just how meaningless this criterion is.

It is tempting to conclude this reply with a parallel to Dr. Tocher's proposal of the vote of thanks, by expressing my gratitude at being shown just how useful anti- and non-mathematical discussions are, but I would rather end on a note of hope that more people will see fit to work on the development of optimum designs in complex situations.

Reprinted from
J. Roy. Statist. Soc. **B21** (2), 272–319 (1959)

OPTIMUM EXPERIMENTAL DESIGNS V, WITH APPLICATIONS TO SYSTEMATIC AND ROTATABLE DESIGNS

J. KIEFER
CORNELL UNIVERSITY

1. Introduction and summary

In this paper we continue to develop the theory of construction of optimum experimental designs along the lines of [15], [19], [16], and [17]. Section 2 of the paper considers further general developments in both the exact and approximate theories, while in section 3 we apply the theory to construct optimum designs in the settings where systematic designs (subsection 3.1) and rotatable designs (subsection 3.2) are often employed, and in the setting of linear regression on an arbitrary Euclidean subset (subsection 3.3). Open problems are mentioned throughout the paper.

2. Generalities

2.1. *Notation and preliminaries.* Throughout this paper we shall achieve brevity by considering mainly a linear model. Corresponding asymptotic results in nonlinear problems hold and are obtainable without serious difficulty. One example of such a problem will be found in subsection 3.1 (ρ unknown), and further examples of explicit computations in certain nonlinear problems will be found in Chernoff [6] and Box and Lucas [4], while complete classes of designs for such problems were treated by the author (see [16], pp. 290–291). We shall also be primarily concerned with nonsequential designs, although one sequential problem is treated below theorem 3.1.2. The main idea in the construction of many such asymptotic sequential designs goes back to Wald [24], while recent work can be found in the papers of Chernoff [7] and his students.

We assume, then that $f_1, f_2, \cdots, f_k$ are k given real functions on a space $\mathfrak{X}$. Write f for the column vector of functions f_i. Let θ denote a real unknown column k-vector. Corresponding to each x in $\mathfrak{X}$, there is a random variable Y_x for which

$$E_\theta Y_x = \theta' f(x). \tag{2.1.1}$$

(Throughout this paper transposes are denoted by primes and subscripts on E

Research sponsored by the Office of Naval Research.

or P refer to the distribution under which an expectation or probability is computed.)

In various applications there will be specified the possible distributions of Y_x, the dependence among various Y_x, and so forth.

An *exact* or *discrete* design will now be defined. An integer N, the total number of observations to be taken, is specified. An exact or discrete design d is a choice of N points $x_1, x_2, \cdots, x_N$ in $\mathfrak{X}$. Sometimes it will be permitted that several x_i are equal, in which case the Y_{x_i} corresponding to two equal x_i will usually not be the same random variable. It may also be the case that certain restrictions are imposed on the allowable choices of $x_1, \cdots, x_N$. For example, in the setting of two-way heterogeneity where $v \times v$ Latin square designs are customarily employed ($N = v^2$), we can take $\mathfrak{X}$ to be the space of v^3 triples (i, j, k), $1 \leqq i, j, k \leqq v$, a design being restricted to a choice of $x_1, \cdots, x_N$ for which no two x_i agree in both of their first two coordinates.

In many applications the Y_{x_i} are assumed to be uncorrelated and to have common (perhaps unknown) variance σ^2. In this case, for any design $d = (x_1, \cdots, x_N)$, the matrix

$$A_d = \sum_{i=1}^{N} f(x_i)f(x_i)' \tag{2.1.2}$$

is called the *information matrix* of the design d. If all components of θ are estimable under d, then $\sigma^2 A_d^{-1}$ is the covariance matrix of best linear estimators (b.l.e.). See [15] and [19] for a discussion of why it suffices to consider linear estimators.

The computation of optimum designs in the above setting will be discussed briefly in subsection 2.2. This is the *exact* or *discrete* theory.

Suppose that there is no restriction on the choice of the x_i in the above setting. Let $\xi_d(x)$ be the proportion of x_i for $1 \leqq i \leqq N$ which are equal to x when design d is used. Then ξ_d can be thought of as a probability measure on $\mathfrak{X}$. If ξ is any probability measure on $\mathfrak{X}$ (there will never be any measure-theoretic difficulties), we write

$$m_{ij}(\xi) = \int f_i(x)f_j(x)\xi(dx) \tag{2.1.3}$$

and $M(\xi) = ||m_{ij}(\xi)||$. Thus, for an exact design d, we have $A_d = NM(\xi_d)$. We shall call $M(\xi)$ the *information matrix of* ξ. A typical optimality criterion in the design of experiments is to choose d to minimize some simple real functional Q of A_d, as we shall discuss in the next two subsections. An exact design d is a probability measure ξ_d taking on only values which are integral multiples of $1/N$. Suppose we find a probability measure ξ^* which minimizes $Q[M(\xi)]$ over *all* probability measures ξ on $\mathfrak{X}$. Clearly, it can happen that ξ^* takes on values other than multiples of $1/N$, and thus does not correspond to an exact design. Nevertheless, we shall consider this problem of minimizing $Q[M(\xi)]$ over all ξ, calling this the *approximate* or *continuous* theory and calling any probability measure ξ on $\mathfrak{X}$ an *approximate* design.

There are three reasons for considering the approximate theory: (1) the exact theory will often exhibit a fine structure dependence on N, necessitating a lengthy table of optimum designs for a given problem, whereas one optimum approximate design is relevant for all N; (2) the optimum approximate design immediately yields an exact design for each N which is optimum to within order N^{-1}, and often will turn out to be an exact design for many N; (3) the exact theory often presents a difficult combinatorial problem admitting no simple method of solution, whereas the approximate theory admits simple computational algorithms such as those mentioned in subsection 2.3.

In the present paper we shall consider the computation of designs which are optimum with respect to certain specific criteria. The reader is referred to Kiefer [16] and to Elfving [12] for proofs and listings of results on such related topics in optimum design theory as admissibility and complete classes of designs, the role of randomized designs, the computations associated with other optimality criteria, and so forth.

2.2. *The exact theory.* Results in the exact theory have been obtained mainly in the settings where incomplete block designs, factorial designs, et cetera, are often employed, rather than in regression experiments where $\mathfrak{X}$ is a continuum. Many scattered results for various settings and optimality criteria were obtained by various authors (see [16] and [12] for listings), all of these results being obtainable from an elementary approach of the author [15], [16]. Let Q be a $v \times v$ orthogonal matrix whose first row Q_1 is constant, and let Q_2 denote its last $v - 1$ rows. If C_d is Bose's $v \times v$ information matrix of design d for the varieties in a block design setting where there are v varieties (of course, $v < k$, where k was defined in subsection 2.1), then $Q_2C_dQ_2'$ is proportional to the inverse of the covariance matrix V_d of best linear estimators of contrasts $\psi = Q_2\bar{\theta}$ of the variety effects $\bar{\theta}' = (\theta_1, \cdots, \theta_v)$. One proves easily that, if d^* *maximizes the trace of* C_d *and* C_{d^*} *has all diagonal elements equal and all off-diagonal elements equal,* then tr V_d is maximized by d^*, and V_{d^*} is a multiple of the identity. From this we conclude that any design d^* with the above italicized properties is optimum according to any of a wide variety of optimality criteria which were considered separately by various authors. These criteria include:

(a) D-optimality: minimizing the generalized variance, or det V_d;

(b) A-optimality: minimizing the average variance, or tr V_d;

(c) E-optimality, called minimaxity with respect to all standard parametric forms in [12]: minimizing the largest eigenvalue of V_d;

(d) called minimaxity with respect to single parameters in [12]: minimizing the maximum diagonal element of V_d;

(e) maximizing the average efficiency, that is, minimizing the average of the variances of best linear estimators of $\theta_i - \theta_i$, this average being easily proved to be proportional to tr V_d;

(f) L-optimality: maximizing, in the Gaussian case, the minimum power on spheres $\psi'\psi = c^2$ as $c \to 0$, for testing the hypothesis $\psi = 0$. If σ^2 is unknown, d^* must also maximize the number of degrees of freedom for error among designs for which ψ is estimable, to insure L-optimality.

As discussed in [15] and [16], properties such as (a) through (e) also have interpretations in terms of power properties for hypothesis testing problems in the Gaussian case; remarkably enough, these optimality properties for testing problems are no longer possessed by the standard symmetrical designs if they are compared with certain intuitively less appealing randomized designs.

The interpretation of these criteria with regard to confidence region problems in the Gaussian case, such as those considered by Scheffé [22], is also well known.

In a given design setting, for example, that where balanced incomplete block designs are customarily used, one need only verify that the design d^* has the property italicized above in order to conclude the optimality of d^* in all of these senses. The corresponding results in settings where $\bar{\theta}$ and not merely ψ is to be estimated, are usually even simpler to obtain. This often involves only elementary arithmetic. A comparatively difficult example is that of the generalized Youden square, for which optimality has not yet been proved if neither the number of rows nor the number of columns is divisible by v; in fact, the above method fails in this case (see [15]).

Of course, there are many design settings, especially where $\mathfrak{X}$ is a continuum, wherein the various criteria (a) through (f) above need not lead to the choice of the same design. In such cases, if, as is often the case, one does not have a well-specified loss function, one may want to choose one of these criteria. The criterion (a) seems to the author to have several appealing properties in such circumstances. For example, if the problem is that of polynomial regression of degree $k-1$ on a given interval, this criterion alone among those listed yields a design which does not depend on the scale of measurement or on which k linear functions of coefficients are chosen as parameters. Another optimality criterion one might consider is

(g) G-optimality: minimize the maximum (over $\mathfrak{X}$) variance $N^{-1}\sigma^2\bar{d}(\xi)$ of the estimated regression function.

A second appealing property of (a) is that, in the approximate theory, it is equivalent to (g) [20]. These two criteria had both been considered often in the past, but as different criteria; see, for example, [4], p. 89. Further properties of the various optimality criteria are discussed in [15] and [16].

G-optimality is not generally equivalent to D-optimality in the exact theory. For example, in the problem of linear regression on the interval $[-1, 1]$, with $N = 3$, let ξ_1 be the exact design for which $\xi_1(-1) = \xi_1(0) = \xi_1(1) = 1/3$, and let ξ_2 be the exact design for which $\xi_1(-1) = 1/3$, $\xi_2(1) = 2/3$. It is easy to verify that $\det M(\xi_1) = 2/3$, $\bar{d}(\xi_1) = 5/2$, $\det M(\xi_2) = 8/9$, $\bar{d}(\xi_2) = 3$, and that ξ_1 is G-optimum among all exact designs, while ξ_2 is D-optimum among all exact designs. The same result holds if $\mathfrak{X}$ is replaced by the set consisting only of the three points $-1, 0, 1$. Nevertheless, it is often true, especially in settings where each f_i can only take on two values, that G-optimality and D-optimality are still equivalent, although this does not follow from theorem 2.3.1 below; it would be enlightening to investigate this relationship further (bounds like those of subsection 2.3 might prove useful). A related problem is that of making precise the idea that, in settings where an appropriate symmetrical design (for example, a *BIBD*) does not exist, that design which is "closest" to it in some sense will often have optimum properties. The exact design which is closest to an approximate design which is optimum, seems to possess similar properties.

In some examples of the discrete theory, one can also obtain the desired result by applying the methods of the next subsection and noting that an approximately optimum design turns out to be exactly optimum. For example, we shall describe two methods for proving the *D*-optimality of the Latin square design d^* in the setting described in subsection 2.1, and it is easy to verify that *D*-optimality implies the other types of optimality because of the structure of V_{d^*} in this case.

Let $\theta' = (\alpha', \beta', \gamma', \mu)$, where α, β, and γ are the v-vectors of variety, row, and column effects, respectively, and μ is the "grand mean." We may assume $\sum \alpha_i = \sum \beta_j = \sum \gamma_k = 0$. Thus, $EY_{ijk} = \alpha_i + \beta_j + \gamma_k + \mu$. Let Q be a $v \times v$ orthogonal matrix of the type described earlier. Let $\bar{\alpha} = Q_2\alpha$, $\bar{\beta} = Q_2\beta$, $\bar{\gamma} = Q_2\gamma$, and let $\phi' = (\bar{\alpha}', \bar{\beta}', \bar{\gamma}', \mu)$, so that ϕ has maximal dimension among estimable vectors. Since A_{d^*}, in terms of θ, consists of four diagonal blocks each of which is a multiple of an identity and six pairs of off-diagonal blocks each of which is a constant matrix, we obtain easily that A_{d^*}, in terms of ϕ, consists of four diagonal matrices each of which is a multiple of the identity, and is zero elsewhere. It follows at once that, if ξ^* is the measure corresponding to d^*, then the function $D_{\bar{\alpha}}(x, \xi^*)$ of the next subsection is constant on the v^3 points of $\mathfrak{X}$, so that d^* is *D*-optimum for estimating contrasts of the α_i. The *D*-optimality of d^* for estimating *all* parameters ϕ is a consequence of the even more obvious fact that the variance $N^{-1}\sigma^2 d(x, \xi^*)$ of the estimated regression is the same at each of the v^3 points x. We have purposely refrained from explicit computation of A_{d^*} here; it was unnecessary, only the form of A_{d^*} being important, in view of theorem 2.3.2!

A second method of proof, which uses the invariance results referred to in the next section, is even shorter. These results imply that the invariant design $\bar{\xi}$ which assigns measure $1/v^3$ to each point of $\mathfrak{X}$, is (in the approximate theory) *D*-optimum for $\bar{\alpha}$ and also for ϕ. Since $M(\xi^*) = M(\bar{\xi})$, we conclude that the Latin square design d^* is (exactly) *D*-optimum.

Unfortunately, these approaches do not work in all settings; for example, this is the case for a Youden square which is not a Latin square, and we would have to fall back on the earlier method of proof in that problem.

We end this subsection by mentioning a common misconception which has occurred repeatedly in the literature. Authors have often restricted their attention to designs for which the b.l.e. are orthogonal, or are orthogonal in or between intuitively appealing sets (for example, blocks and varieties), apparently assuming that optimum designs are to be found among such designs. While it is well known that some optimum designs possess such orthogonality properties, it is only orthogonality *in combination with some other property such as the trace maximization we have mentioned* which yields optimality. However, in experiments such as those concerned with polynomial regression, we rarely have such orthogonality, although the misconception persists even in that familiar setting (see, for example, the discussion of [16], p. 306, lines 9–13 and p. 316, lines 17–23). For another example, we note that it is common in designing experiments where a time trend is to be removed, to restrict attention to designs for which the b.l.e. of time trend is orthogonal to the b.l.e. of effects in which one is interested. In subsection 3.1, we shall give an example to illustrate the bad consequences of such an orthogonality restriction in this setting.

2.3. *The approximate theory; invariance; bounds; extensions.* The earliest general development of algorithms for computing optimum designs is due to Elfving [11] in the case of the average variance, further developments being due to Chernoff [6]. Kiefer and Wolfowitz [16] developed algorithms for various optimality criteria, of which we shall be concerned in the present paper with the generalized variance. Further results for that criterion were obtained by the author[17].

If ξ is an approximate design for which θ is estimable, that is, for which $M(\xi)$ is nonsingular, then the variance of the b.l.e. of the estimated regression at the point x is $\sigma^2N^{-1}d(x, \xi)$, where

$$d(x, \xi) = f(x)'M^{-1}(\xi)f(x). \tag{2.3.1}$$

Assume for simplicity that f is continuous in a topology for which $\mathfrak{X}$ is compact, and write $\bar{d}(\xi) = \max_x d(x, \xi)$. The equivalence of D- and G-optimality alluded to in subsection 2.2 is contained in the following theorem [20]:

THEOREM 2.3.1. *ξ is D-optimum if and only if it is G-optimum, and if and only if $\bar{d}(\xi) = k$. For all D-optimum ξ, $M(\xi)$ is the same.*

This theorem has been the basis for computing optimum designs in problems of polynomial regression on a simplex or hypercube [17]. A generalization of theorem 2.3.1, of use when we are interested in a subset $\theta^{(1)} = (\theta_1, \cdots, \theta_s)'$ of the parameters, has been proved in [17]. Partition θ into $\theta^{(1)}$ and $\theta^{(2)}$[a $(k - s)$-vector], f into $f^{(1)}$ and $f^{(2)}$, and $M(\xi)$ into

$$M(\xi) = \begin{Vmatrix} M_1(\xi) & M_2(\xi) \\ M_2(\xi)' & M_3(\xi) \end{Vmatrix}, \tag{2.3.2}$$

where M_1 is $s \times s$. The information matrix for estimating $\theta^{(1)}$ can be written as $M^* = M_1 - M_2M_3^{-1}M_2'$ if M is nonsingular. Thus, ξ is D-optimum for $\theta^{(1)}$ if it maximimizes $\det M^*(\xi)$. Write $D_{\theta^{(1)}}(x, \xi) = d(x, \xi) - f^{(2)}(x)'M_3^{-1}(\xi)f^{(2)}(x)$ and $D(\xi) = \max_x D_{\theta^{(1)}}(x, \xi)$. One of the results of [17] is

THEOREM 2.3.2. *Suppose $M(\xi^*)$ is nonsingular. Then ξ^* is D-optimum for $\theta^{(1)}$ if and only if ξ^* minimizes $D(\xi)$, and if and only if $D(\xi^*) = s$.*

The reader is referred to [17] for a determination of the structure of the class of matrices $M(\xi)$ for which ξ is D-optimum for $\theta^{(1)}$, for the modifications necessary when $\theta^{(1)}$ is estimable but all of θ is not, et cetera. When $s = k$ we obtain theorem 2.3.1, and when $s = 1$ we obtain a different proof of the algorithm of [19] for this case. When $1 < s < k$, the algorithm of [19] differs from that of theorem 2.3.2, the latter so far seeming to have yielded simpler arithmetic in examples (see [17]).

Invariance. The subject of invariant optimum designs, invariant complete classes for $\theta^{(1)}$, et cetera, has been dealt with extensively in [16]. Here we will only comment briefly on a slightly different approach to invariance where D- (hence, G-) optimality is involved. Instead of working with $\det M(\xi)$ or $\det M^*(\xi)$ as in [16], we can work with $d(x, \xi)$ or $D(x, \xi)$. For example, let $\mathcal{G}$ be a compact group of transformations on $\mathfrak{X}$ with Haar measure μ, $\mu(\mathcal{G}) = 1$. Suppose that, for g in $\mathcal{G}$, there is an associated transformation $\bar{g}$ on the space of ξ, such that

$$d(gx, \xi) = d(x, \bar{g}\xi). \tag{2.3.3}$$

Writing

$$\bar{\xi} = \int (\bar{g}\xi)\mu(dg), \tag{2.3.4}$$

the trivial fact that $\lambda A^{-1} + (1 - \lambda)B^{-1} - [\lambda A + (1 - \lambda)B]^{-1}$ is nonnegative definite for $0 \leqq \lambda \leqq 1$ if A and B are, yields

$$\begin{aligned} \sup_x d(x, \xi) &\geqq \sup_x \int d(gx, \xi)\mu(dg) = \sup_x \int d(x, \bar{g}\xi)\mu(dg) \\ &= \sup_x f(x)' \left\{\int M^{-1}(\bar{g}\xi)\mu(dg)\right\} f(x) \\ &\geqq \sup_x f(x)' \left\{\int M(\bar{g}\xi)\mu(dg)\right\}^{-1} f(x) = \sup_x d(x, \bar{\xi}). \end{aligned} \tag{2.3.5}$$

Since $\bar{\xi}$ is an invariant design, that is, $\bar{\xi}(gA) = \bar{\xi}(A)$ for all g and A, we conclude:

THEOREM 2.3.3. *Under the above conditions, there is a $\mathcal{G}$-invariant D- (and G-) optimum design for estimating θ.*

This result can be extended without serious difficulty to noncompact groups (just as with the usual invariance theory in statistics, for example, as in [18]), as well as to the estimation of $\theta^{(1)}$, in the same way that such extensions were obtained in [17].

Bounds. For computational purposes in obtaining designs which are almost optimum in complex settings, it is useful to know when one has obtained a design which is sufficiently close to optimality for practical purposes. For the criterion of G-optimality of θ, we have such a method: $[\bar{d}(\xi) - k]/k$ is, by theorem 2.3.1, the relative excess of $\bar{d}(\xi)$ over the minimum attainable. No such simple expression is available in terms of $\det M(\xi)$, but we can use this expression in terms of $\bar{d}(\xi)$ to obtain a bound on $\det M(\xi)/\max_{\xi'} \det M(\xi')$. Thus, one can in practice compute $\bar{d}(\xi)$ and, if it is close to k, conclude that $\det M(\xi)$ is within a bound we shall derive, from the maximum attainable. For completeness we shall give such bounds in both directions. For brevity we shall consider only the case $s = k$; the case $s < k$ can be treated similarly.

Write $\Delta = \max_\xi \det M(\xi)$. Suppose η is D-optimum. Fix ξ. An elementary computation of the first two derivatives of $q(\alpha) = \log \det M[\alpha\eta + (1-\alpha)\xi]$ for $0 < \alpha < 1$ (see [20] or [17] for details) yields

$$
\begin{aligned}
q'(0) &= \frac{d}{d\alpha} q(\alpha)\Big|_{\alpha=0} \leqq \bar{d}(\xi) - k, \\
\frac{d^2}{d\alpha^2} q(\alpha) &\leqq 0,
\end{aligned}
\tag{2.3.6}
$$

and hence

$$
\frac{d}{d\alpha} q(\alpha) \leqq \bar{d}(\xi) - k \qquad \text{for} \quad 0 \leqq \alpha \leqq 1. \tag{2.3.7}
$$

Hence, $q(1) - q(0) \leqq \bar{d}(\xi) - k$, and thus

$$
\frac{\det M(\xi)}{\Delta} \geqq \exp [k - \bar{d}(\xi)]. \tag{2.3.8}
$$

Inequality (2.3.8) is easily seen to be strict unless ξ is D-optimum.

In the other (less useful) direction, suppose $\bar{d}(\xi) = k + \epsilon$. For simplicity, assume $\epsilon \leqq 1$. Again following the derivation of [20] or [17], we conclude that there is an η (not necessarily optimum) for which $q'(0) = \epsilon$. Let A be nonsingular and such that $AM(\xi)A'$ is the identity and $AM(\eta)A'$ is diagonal with diagonal entries d_i. Then, for $0 \leqq \alpha \leqq 1$,

$$
\frac{d^2}{d\alpha^2} q(\alpha) = - \sum_{i=1}^{k} \frac{(d_i - 1)^2}{(1 - \alpha + \alpha d_i)^2}. \tag{2.3.9}
$$

Also, the fourth derivative of $q(\alpha)$ is nonpositive, so that $d^2q(\alpha)/d\alpha^2$ is concave and thus attains its minimum L on $0 \leqq \alpha \leqq r \leqq 1$ at 0 or r. Thus,

$$
L = -\max \left[\sum_i (d_i - 1), \sum_i \frac{(d_i - 1)^2}{(1 - r + rd_i)^2} \right]. \tag{2.3.10}
$$

Now, $\sum (d_i - 1)^2$ is convex in the d_i on the set $B = \{(d_1, \cdots, d_k) |$ all $d_i \geqq 0$, $\sum d_i = k + \epsilon\}$, which must contain the actual $(d_1, \cdots, d_k)$ corresponding to η, since $q'(0) = \sum (d_i - 1)$. The maximum of $\sum (d_i - 1)^2$ on B, taken on at an extreme point, is $k - 1 + (k + \epsilon - 1)^2$.

Consider now the second term in the expression (2.3.10) for L. Let $h(u) = (u-1)^2/(1 - r + ru)^2$. Then $d^2h(u)/du^2 \geqq 0$ if $0 \leqq u \leqq 1 + 1/2r$. Since $d_i \leqq k + \epsilon \leqq k + 1$ on B, we conclude that, if $r = 1/2k$, this second expression of (2.3.10) has its maximum over B at an extreme point, where this maximum is easily estimated to be less than k^2.

We conclude that, if $r = 1/2k$, we have $L \geqq -k(k+1)$. Hence, $q(\alpha) \geqq \epsilon\alpha - k(k+1)\alpha^2/2$ for $0 \leqq \alpha \leqq 1/2k$. Putting $\alpha = \epsilon^2/2k(k+1)$ and $\xi' = \alpha\eta + (1-\alpha)\xi$ for this α, we obtain

$$
\log [\det M(\xi')/\det M(\xi)] \geqq \epsilon^2/2k(k+1), \tag{2.3.11}
$$

where again the inequality is strict unless $\epsilon = 0$. Summarizing our results, we have

THEOREM 2.3.4. *For any* ξ,

$$\frac{\det M(\xi)}{\max_{\xi'} \det M(\xi')} \geqq \exp\,[k - \bar{d}(\xi)] \tag{2.3.12}$$

and, if $\bar{d}(\xi) - k \leqq 1$,

$$\frac{\det M(\xi)}{\max_{\xi'} \det M(\xi')} \leqq \exp\left\{-\frac{[\bar{d}(\xi) - k]^2}{2k(k+1)}\right\}, \tag{2.3.13}$$

with strict inequality unless ξ *is D-optimum.*

In the unlikely situation that one has computed a lower bound on $\det M(\xi)/\Delta$ and wants to obtain an upper bound on $\bar{d}(\xi)$, one would invert (2.3.13). Thus, if $\Delta/\det M(\xi) \leqq \exp\,(1/2k[k+1])$, we obtain, from (2.3.11),

$$\bar{d}(\xi) \leqq k + \left[2k(k+1)\log\frac{\Delta}{\det M(\xi)}\right]^{1/2}. \tag{2.3.14}$$

Extension to the vector case. Various extensions of our theory are possible, to models where Y_x is a vector. For example, if Y_x is an h-vector of components Y_{xi} and θ and f are hr-vectors ($k = hr$) of components θ_{ij} and f_{ij} with $E_\theta Y_{xi} = \sum_j \theta_{ij} f_{ij}(x)$, the Y_x being uncorrelated with $E_\theta Y_x Y_x' = \sigma^2 I$ (other cases are easily reduced to this form), then the information matrix of any ξ breaks up into r diagonal $h \times h$ blocks $M_j(\xi)$, and the previously discussed theory goes over into this setting with $d(x, \xi)$ becoming $\sum_j f^{(j)}(x)' M_j^{-1}(\xi) f^{(j)}(x)$ where $f^{(j)}$ is the h-vector of f_{ij}. Other vector models can also be treated.

The number of points supporting a design. In general, if H is the dimension of the linear space spanned by the functions $f_i f_j$, $i \leqq j$, then for any ξ there is a ξ' supported by no more than $H + 1$ points of $\mathfrak{X}$ and for which $M(\xi') = M(\xi)$. Often more can be said. For example, if H is the maximum possible $[k(k+1)/2]$, then there is a D-optimum design on H points, since the optimum design yields a boundary point in the convex body of $M(\xi)$. Various results for the number of points needed when $s < k$ can be found in [11], [6], [19], [16], [23]. In certain cases, improvements are possible because of the nature of the f_i. For example in the case of polynomial regression of degree $m = k - 1$ on a real interval, we have $H = 2m$, but for any ξ there is a ξ' on at most $m + 1$ points with $M(\xi') = M(\xi)$. The admissible ξ were characterized in this example by the author in [16]. An outstanding problem is the characterization, in terms of $\mathfrak{X}$ and the f_i, of the smallest number q such that, for any ξ, there is a ξ' supported by at most q points, and with $M(\xi') = M(\xi)$.

We cite an example to indicate the theoretical importance of the above considerations (the practical importance being obvious). The author [17] has computed D-optimum ξ for the problem of quadratic regression on an m-cube when $m \leqq 5$, these optimum designs being supported by the vertices, midpoints of edges, and midpoints of two-dimensional faces. When $m = 6$, a design with such support can not be optimum. Already for $m = 4$ and $m = 5$ the optimum designs of this structure are supported by more than $H + 1$ points, which, incidentally, exemplifies the fact that the optimum ξ is no longer unique as it is in the case $m = 1$. It is reasonable to suppose that, if one could obtain D-optimum designs supported by fewer points when $m = 4$ or 5, this might suggest a structure which would also work when $m \geqq 6$.

Note added in proof. Such designs for all m have been obtained recently by Dr. R. H. Farrell and the author. Details will appear elsewhere.

Another question is this: Suppose we restrict our attention, as has often been done in the literature (see, for example, [4]) to designs supported by k points. The best [for example, in terms of det $M(\xi)$] design of this structure need not be optimum among all designs, but by how much does it miss? Using the concavity of log det $M(\xi)$, one can develop bounds of the sort developed earlier in this section, but it would be useful if these could be improved by using the form of the f_i.

The results we have cited on the number of points needed to support designs do not apply to the exact theory as has sometimes, especially in the polynomial case, been assumed. However, the error introduced by assuming these results to apply is only of order $1/N$, as indicated in subsection 2.1.

3. Applications

3.1. *Systematic Designs.* Systematic designs arise in contexts where observations are taken over time. Thus, if at time $t = 1, 2, \cdots, T$, one or more observations can be taken at points in a space S, we can view $\mathfrak{X}$ as S^T; however, there will usually be restrictions on the number of observations which can be taken at each time point t. Thus, although it can happen that an approximate design, obtained without imposing any restrictions, either turns out to satisfy these practical restrictions, or is close to a design which satisfies them, this will usually not be the case, and the theory of subsection 2.3 is thus not always useful. What would often be useful here is a theory which is approximate with respect to S but exact with respect to T.

There are many recent papers in this area, some of which are those of Williams [25], Cox [8], [9], [10], Box [1], Box and Hay [2], and Patterson [21]; further references can be found in these papers.

In the simplest of these problems, the model is such that the dependence on the time variable can be treated as the block effect in a *BIBD* or as the row or column effect in a Youden square. The theory of subsection 2.2 can then be used to give optimum designs.

In more complex models, this device will not suffice. As mentioned in subsection 2.2, it is often erroneously assumed in such problems that an optimum design is to be found among those designs for which the b.l.e. of treatment effects are orthogonal to the b.l.e. of time effects (see, for example, [1], [2]). A simple counterexample is the following: Suppose that S is the interval $[-1, 1]$, that $T = 4$, and that the expected value of an observation at the point z in S at time t is $\alpha z + \beta_0 + \beta_1 t + \beta_2 t^2$, where α and the β_i are unknown. All observations have the same variance σ^2 and are independent. We wish to estimate α.

For simplicity of computation in this counterexample, we suppose that we are allowed one observation at each of the four times t; thus, a design is a quadruple (z_1, z_2, z_3, z_4), where z_t is the point at which an observation is to be taken at time t. One sees easily that the restriction to designs for which the b.l.e. of α is orthogonal to those of the β_i means that the design is of the form $(z_1, -3z_1, 3z_1, -z_1)$, and the variance of the b.l.e. of α is obviously a minimum among designs of this form, namely $9\sigma^2/20$, when $z_1 = \pm 1/3$. On the other hand, the design $(-1, 1, -1, 1)$, for which we do not have this orthogonality, yields a b.l.e. of α with variance $5\sigma^2/16$. The variance for the best orthogonal design is thus 44% larger!

For settings where S is higher-dimensional, the computation of optimum designs will usually be very tedious. The neat approach of Box [1], which yields designs with the above computationally useful orthogonality property, will not yield optimum designs, both because of the phenomenon illustrated in the previous paragraph, and also because the method of construction used in [1] is not geared to the specification of a fixed S with respect to which all designs are to be compared. For example, if S is replaced by the unit disc in the example of the previous paragraph with $T \geqq 5$, the method of [1] yields as a design a T-tuple $(z_1, z_2, \cdots, z_T)$, where $z_t = (z_{t1}, z_{t2})$ is a two-vector and the T-vectors $(z_{11}, z_{21}, \cdots, z_{T1})$ and $(z_{12}, z_{22}, \cdots, z_{T2})$ are chosen so as to yield the orthogonality property. But the method does not insure that $z_{t1}^2 + z_{t2}^2 \leqq 1$ for all t; thus, to use such a design for the S we have specified, it would be necessary to change the original design of [1] by dividing all z_{tj} by the factor $\max_t (z_{t1}^2 + z_{t2}^2)^{1/2}$. Hence, the various choices of $(z_1, \cdots, z_T)$ in [1] which yield the same information matrix there do not yield the same matrix when we scale them down to our specified S. We shall again encounter this need for a careful scaling of the designs, which are usually scaled in a different manner in the literature, in our consideration of rotatable designs in subsection 3.2.

Williams' model. We shall now consider in detail the model studied by Williams [25], wherein the effect of time appears entirely through a correlation among observations. There are k treatments, any one of which can be tested at time t for $t = 1, 2, \cdots, T$. Thus, a design d_T is a T-tuple $(v_1, v_2, \cdots, v_T)$, where v_t is the label number of the treatment tested at time t, $(1 \leqq v_t \leqq k)$. If treatment i is tested at time t, the expected value of the observation Y_t, say, is θ_i.

The first order model. We shall first consider Williams' first order model, wherein $Y_t = \rho Y_{t+1} + \epsilon_t$, the ϵ_t being uncorrelated with common variance σ^2 for $t < 1$. The variance σ_1^2 of Y_1 is assumed positive, but will be shown not otherwise to concern us, since we will be concerned with an asymptotic theory as $T \to \infty$. It is often customary to assume var (Y_1) to be such as to make $Y_t - EY_t$ stationary in the wide sense, but this is unnecessary. For the present, we assume ρ known, $-1 < \rho < 1$; actually, we shall see that it is only necessary to know that $-1 < \rho \leqq 0$ or $0 \leqq \rho < 1$ in order to choose an asymptotically optimum design. Later we shall consider a minimax approach when nothing is known about ρ.

The problem is to estimate all contrasts of the θ_i, and we shall find designs which are asymptotically optimum as $T \to \infty$, in any of the senses (a) through (f) of subsection 2.2. Our considerations are thus asymptotic, but for exact designs.

Williams discussed in detail two types of designs for this problem: A design is of type II(a) if (i) each unordered pair of distinct integers between 1 and k, inclusive, appears the same number of times among the pairs (v_1, v_2), (v_2, v_3), $\cdots$, (v_{T-1}, v_T), and (ii) $v_{t-1} \neq v_t$ for all t. A design is of type II(b) if (i) is satisfied and if each unordered pair of different integers appears twice as often as each pair of equal integers. Examples of these two types of designs when $k = 4$ are (2123423143142) and (11234413224133421). In constructing such designs, it is often convenient, as here, to have $v_{rk+2}, v_{rk+3}, \cdots, v_{(r+1)k+1}$ constitute a permutation of $1, 2, \cdots, k$ for each integer $r \geq 0$. Williams showed that, between these two types of designs, as $T \to \infty$, II(a) is the better in the sense of criterion (e) of subsection 2.2 if $\rho > 0$, while II(b) is the better if $\rho < 0$. It has often been conjectured (see, for example, Cox [9]) that these designs are optimum among *all* designs in the two respective cases. This conjecture turns out to be true for II(a) but false for II(b) [although II(b) will turn out to have a different optimum property], as we shall now see.

Let $W_t = Y_t - \rho Y_{t-1}$ for $t > 1$. The W_j are uncorrelated with common variance σ^2. Since $\sigma_1^2 > 0$ and $|\rho| < 1$, it is easy to see that, for any design, T^{-1} times the information matrix associated with $Q_2\theta$ (in the notation of subsection 2.2) is asymptotically the same as $T \to \infty$, whether based on $Y_1, Y_2, \cdots, Y_T$ (or, which is equivalent, on $Y_1, W_2, \cdots, W_T$) or on $W_2, \cdots, W_T$, and approaches a positive definite limit for the asymptotically optimum designs obtained below. It follows that we can ignore Y_1 and base our considerations on $W_2, \cdots, W_T$ in the asymptotic considerations which follow.

Writing $W^{(T)\prime} = (W_2, \cdots, W_T)$ and $EW^{(T)} = J_T\theta$, we thus must consider the information matrix $J_T'J_T = A_{d_T}$, say, of the $W^{(T)}$ associated with any design $d_T = (v_1, \cdots, v_T)$ based on T observations. Write $A_d = ||a_{dij}||$. We see at once that

$$
\begin{aligned}
(3.1.1) \qquad a_{dii} = {} & (\text{number of } r,\ 2 \leq r \leq T, \text{ for which } v_r = i) \\
& + \rho^2(\text{number of } r,\ 1 \leq r \leq T-1, \text{ for which } v_r = i) \\
& - 2\rho(\text{number of } r,\ 1 \leq r \leq T-1, \text{ for which } v_r = v_{r+1} = i)
\end{aligned}
$$

and, for $i \neq j$,

$$
\begin{aligned}
(3.1.2) \qquad a_{dij} = -\rho[&\text{number of } r,\ 1 \leq r \leq T-1, \text{ for which } (v_r, v_{r+1}) \\
& = (i, j) \text{ or } = (j, i)].
\end{aligned}
$$

For any d based on T observations, write

$$
(3.1.3) \qquad \frac{1}{T} QA_dQ' = \begin{Vmatrix} b_d & h_d' \\ h_d & Z_d \end{Vmatrix},
$$

where Z_d is $(k-1) \times (k-1)$. The covariance matrix of the b.l.e. of $Q_2\theta$ based on $W^{(T)}$ is $\sigma^2T^{-1}(Z_d - h_db_dh_d')^{-1}$. Since $h_db_dh_d'$ is nonnegative-definite, the discussion of subsection 2.2 shows that d_T^* is optimum in all of the senses (a) through (f) of subsection 2.2, provided that (1) $Z_{d_T^*}$ is a multiple of the identity and $h_{d_T^*} = 0$, and (2) tr Z_{d_T} is maximized by d_T^*. We shall exhibit a design d_T^* for which $Z_{d_T^*}$ approaches a positive definite limit as $T \to \infty$ and for which (1) and (2) are satisfied to within order T^{-1}; these designs are thus asymptotically optimum as $T \to \infty$.

Since Q_2 has row sums equal to zero, (1) is satisfied to within order T^{-1} by any sequence, with T, of designs $\{d_T^*\}$ for which (3) for all T and some finite constant c, independent of T, $|a_{d_Tii} - a_{d_Tjj}| < c$ and $|a_{d_Tij} - a_{d_Tef}| < c$ for all $i \neq j$ and $e \neq f$. Moreover, if we define

(3.1.4) $$\pi_{d_T} = T^{-1}(\text{number of } r,\ 1 \leqq r \leqq T-1, \text{ for which } v_r = v_{r+1}),$$

we obtain easily, since $b_{d_T} = k^{-1}(1 - T^{-1})(1 - \rho)^2$ for all d_T,

(3.1.5) $$\begin{aligned} \operatorname{tr} Z_{d_T} &= T^{-1} \operatorname{tr} A_{d_T} - b_{d_T} = (1 - T^{-1})(1 + \rho^2) - 2\rho\pi_{d_T} - b_{d_T} \\ &= (1 + \rho^2)(1 - T^{-1})(1 - k^{-1}) - 2\rho[\pi_{d_T} - k^{-1}(1 - T^{-1})]. \end{aligned}$$

Thus, condition (2) is satisfied to within order T^{-1} by any sequence of designs d_T^* for which (4a) $T\pi_{d_T^*} < c$ for all T, if $\rho \geqq 0$, or (4b) $T(1 - \pi_{d_T^*}) < c$ for all T, if $\rho \leqq 0$. Noting that our omission of Y_1 changed Z_{d_T} by order T^{-1}, we conclude:

THEOREM 3.1.1. *A sequence $\{d_T^*\}$ of designs is asymptotically optimum to within relative error T^{-1} as $T \to \infty$ in Williams' first order model provided $\{d_T^*\}$ satisfies condition* (3) *of the previous paragraph and also condition* 4(a) *(respectively* 4(b)) *if $\rho \geqq 0$ (respectively, $\rho \leqq 0$).*

The statement of necessary and sufficient conditions for asymptotic optimality without having the error term of order T^{-1}, is obvious. Theorem 3.1.3, for the second order case, will be stated in that form.

Thus, when $\rho > 0$, Williams' designs of type II(a), or ones of approximately this structure (which exist for *any* T), are asymptotically optimum. *However, when $\rho < 0$, Williams' designs of type II(b), although better than those of type II(a), are not asymptotically optimum.* Rather, an approximately optimum design is now one which observes approximately T/k treatments of type 1, then T/k of type 2, et cetera (or which is almost of this structure): $d_T^* = (1, 1, \cdots, 1, 2, 2, \cdots, 2, \cdots, k, k, \cdots, k)$. For a design of type II(b), $\pi_{d_T} \to 1/k$ as $T \to \infty$; thus, by (3.1.5), the "relative efficiency" of such designs for $\rho < 0$, compared with optimum ones (in terms of the variance of the b.l.e. of any contrast) as $T \to \infty$, is $1 - 2(-\rho)/(1 - \rho)^2$.

It is clear that, for fixed T with the appropriate divisibility property, designs of the above forms will be *exactly* optimum if σ_1^2/σ^2 is sufficiently large. On the other hand, if σ_1^2/σ^2 is sufficiently small, an exactly optimum design will have $v_j \neq v_1$ for all $j > 1$. It would be interesting to delimit the sets of values of σ_1^2/σ^2 wherein these extreme designs and others between them are exactly optimum.

Of course, all designs for which the treatments appear with (approximately) equal frequency are asymptotically optimum when $\rho = 0$.

Suppose ρ were not known exactly, but that we only knew in advance of the experiment that $\rho \geqq 0$ (or, similarly, that $\rho \leqq 0$). Then ρ can be estimated consistently from the data (see [24] for details) and one finds that the same limiting formulas hold as before (of course, estimators are no longer linear). Thus, theorem 3.1.1 specifies asymptotically optimum designs for this situation, too.

Next, suppose we know in advance of the experiment only that $-1 < \rho < 1$. We seek a design which is minimax over ρ as $T \to \infty$, with respect to any of the criteria (a) through (f) of subsection 2.2. Since ρ can be estimated as indicated in the previous paragraph, the problem thus reduces to choosing d_T so as to maximize $\min_\rho \operatorname{tr} Z_{d_T}$ as $T \to \infty$. The result is clear:

THEOREM 3.1.2. *If ρ is unknown, $\{d_T^*\}$ is asymptotically minimax over ρ for criteria (a) through (f) if and only if, for $i \neq j$ and $e \neq f$,*

$$\lim_{T\to\infty} T^{-1}(a_{d_T^* ii} - a_{d_T^* jj}) = 0,$$

$$\lim_{T\to\infty} T^{-1}(a_{d_T^* ij} - a_{d_T^* ef}) = 0, \tag{3.1.6}$$

$$\lim_{T\to\infty} \pi_{d_T^*} = 1/k.$$

Williams' type II(b) designs satisfy these conditions.

Thus, the type II(b) designs do possess an optimum property, but not the one they are usually thought to possess.

It is easy to give sequential designs which improve upon those specified in theorem 3.1.2: It is only necessary, in standard fashion, to decide after $n(T)$ observations, where $n(T) \to \infty$ and $T^{-1}n(T) \to 0$ as $T \to \infty$, whether $\rho \geqq 0$ or $\rho < 0$, and to use the appropriate design of theorem 3.1.1 for the remaining $T - n(T)$ observations, as dictated by this decision. As $T \to \infty$, the resulting designs yield the same limiting information matrix as would have been obtained if ρ had been known.

The second order model. Williams' second order model differs from the first order model only in that we now assume $Y_t + \rho_1 Y_{t-1} + \rho_2 Y_{t-2} = \epsilon_t$, $t > 2$, where the ϵ_t are again uncorrelated with common variance σ^2. The joint distribution of Y_1 and Y_2 is assumed to be nonsingular, and it is assumed that $-1 < \rho_2 < 1$ and $-(1 + \rho_2) < \rho_1 < (1 + \rho_2)$. The model for EY_t, and the notation for a design, are as before.

Williams considers in this setting designs of type III, which satisfy the conditions of a design of type II(a) and also the condition that $v_{t-2} \neq v_t$ for all t and that each unordered pair of integers between 1 and k, inclusive, appears the same number of times among the pairs (v_{t-2}, v_t), $3 \leqq t \leqq T$. Thus, $k \geqq 3$ for such a design to exist. An example of such a design for $k = 4$ is $d = (2, 4, 1, 2, 3, 4, 2, 1, 3, 4, 1, 3, 2, 4)$. There is no corresponding analogue of type II(b) designs here.

We now define π_{d_T} as before, and also

$$\gamma_{d_T} = T^{-1}(\text{number of } r,\ 1 \leqq r \leqq T-2, \text{ for which } v_r = v_{r+2}). \tag{3.1.7}$$

We now replace our analysis in terms of the W_t by one in terms of $V_t = Y_t + \rho_1 Y_{t-1} + \rho_2 Y_{t-2}$ for $3 \leqq t \leqq T$. A simple computation now shows that

$$\text{(3.1.8)} \qquad \operatorname{tr} Z_{d_T} = T^{-1} \operatorname{tr} A_{d_T} - b_{d_T} = 1 + \rho_1^2 + \rho_2^2 + 2\{\rho_1(1 + \rho_2)\pi_{d_T} + \rho_2 \gamma_{d_T}\} - k^{-1}[1 + \rho_1 + \rho_2]^2 + O(T^{-1}).$$

As before, we will obtain designs which are approximately symmetrical with respect to treatments; that is, d_T will be such as to make the $a_{d_T ii}$ almost equal and the $a_{d_T ij}$ for $i \neq j$ almost equal, while maximizing $\operatorname{tr} Z_{d_T}$.

Since $\rho_2 + 1 > 0$, the maximization of (3.1.8) breaks up into four cases according to the signs of ρ_1 and ρ_2. If ρ_1 and ρ_2 are both positive, optimality necessitates $\pi_{d_T} \to 1$ and $\gamma_{d_T} \to 1$ as $T \to \infty$; for example, one can take approximately T/k consecutive observations on treatment 1, then the same number on treatment 2, et cetera. If $\rho_1 < 0 < \rho_2$, we want $\gamma_{d_T} \to 1$ and $\pi_{d_T} \to 0$; this can be achieved while maintaining the approximate symmetry with respect to treatments by taking observations in consecutive blocks of approximately $2T/k(k-1)$ observations each, each block being of the form $ijij \cdots ij$ for a different pair $i < j$.

If $\rho_1 < 0$ and $\rho_2 < 0$, there are two cases to consider, according to whether $k = 2$ or $k > 2$. If $k > 2$, we can obviously achieve optimality by using an approximately symmetric design with $\pi_{d_T} \to 0$ and $\gamma_{d_T} \to 0$; for example, we can construct a design consisting of $k(k-1)(k-2)$ blocks of approximately equal numbers of observations, each of the form $hijhij \cdots hij$ for a different triple of unequal integers h, i, j. On the other hand, when $k = 2$, we cannot achieve both $\pi_{d_T} \to 0$ and $\gamma_{d_T} \to 0$. The set of achievable points $(\pi_{d_T}, \gamma_{d_T})$ has as its limiting set (topologically, not set-theoretically) a set B in the plane which is clearly convex. We must determine the lower left hand boundary B' of B. It is easy to see that, in any approximately symmetric design containing the same number of 1's and 2's (which is all we need consider, since any d_T can be replaced by a d_{2T} with approximately the same π and γ and with the two treatments appearing approximately symmetrically), any appearance of blocks of three or more consecutive 1's can be broken up by exchanging some 1's with 2's from a corresponding block of 2's, without increasing π or γ. Thus, in determining B', it suffices to consider designs which never contain more than two consecutive 1's or 2's. It is now easy to see that B' consists of the line segment $\{2\pi + \gamma = 1, 0 \leqq \gamma \leqq 1\}$. It follows that an asymptotically optimum sequence $\{d_T\}$ satisfies (i) $\pi_{d_T} \to 0$, $\gamma_{d_T} \to 1$ if $\rho_1(1 + \rho_2) < 2\rho_2$, (ii) $\pi_{d_T} \to 1/2$, $\gamma_{d_T} \to 0$ if $\rho_1(1 + \rho_2) > 2\rho_2$, and (iii) $2\pi_{d_T} + \gamma_{d_T} \to 1$ if $\rho_1(1 + \rho_2) = 2\rho_2$. Examples of such designs in the respective cases are (i) $d = (1212 \cdots 12)$, (ii) $d = (11221122 \cdots 1122)$, and (iii) many designs, including the previous two and $d = (112112 \cdots 112; 221221 \cdots 221)$.

When $\rho_1 > 0 > \rho_2$, we must determine the lower right hand boundary B'' of B. This time if there are many occurrences in d of triples of the form 1a1, 1b1, 1c1, et cetera, it is easy to see how to combine them (for example, into 111abc111 here) so as not to increase γ or decrease π. Thus, we need only consider designs which

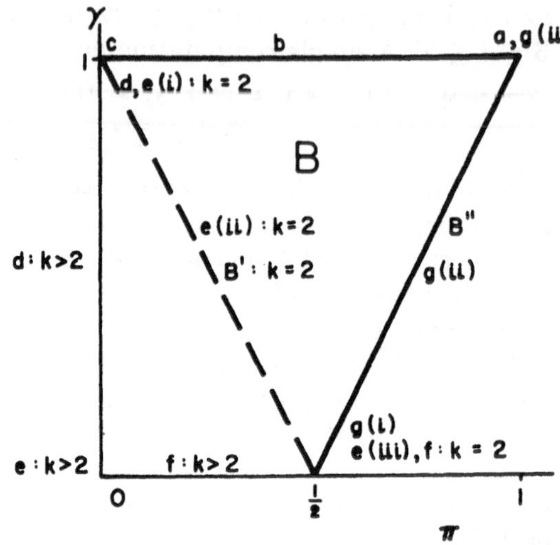

FIGURE 1

contain blocks of at least two i's in a row, every time any i occurs. Symmetrically arranged blocks of exactly $m \geqq 2$ equal integers yield, approximately, $\pi = (m - 1)/m$, $\gamma = (m - 2)/m$, and it is easy to verify that convex combinations of these points generate B''; thus, $B'' = \{\gamma + 1 = 2\pi, 0 \leqq \gamma \leqq 1\}$. We conclude that an optimum $\{d_T\}$ satisfies (i) $\pi_{d_T} \to 1/2$, $\gamma_{d_T} \to 0$ if $\rho_1(1 + \rho_2) < -2\rho_2$, (ii) $\pi_{d_T} \to 1$, $\gamma_{d_T} \to 1$ if $\rho_1(1 + \rho_2) > -2\rho_2$, and (iii) $2\pi_{d_T} - \gamma_{d_T} \to 1$ if $\rho_1(1 + \rho_2)$

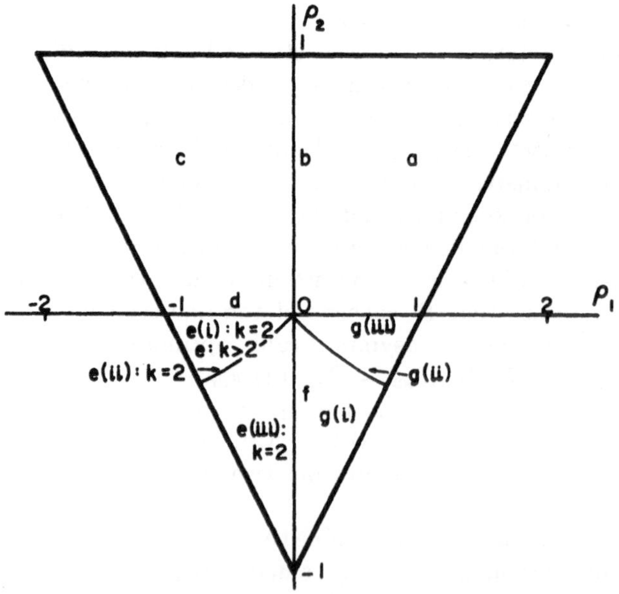

FIGURE 2

$= -2\rho_2$. Examples are (i) the design described above with $m = 2$; for example, $d = (112233113322 \cdots)$ for $k = 3$; (ii) the above form with $m \to \infty$ as $T \to \infty$; and (iii) the above form for any $m \geqq 2$.

We summarize our results in figures 1 and 2 and in the following statement, which traces counterclockwise around the boundary of B, starting with the point $(1, 1)$, and around the (ρ_1, ρ_2) domain starting with the region $\rho_1 > 0$, $\rho_2 \geqq 0$:

THEOREM 3.1.3. *The sequence* $\{d_T\}$ *is asymptotically optimum in senses* (a) *through* (f) *for Williams' second order model, if and only if* (1) $T^{-1}A_{d_T}$ *approaches a limit with diagonal elements equal and off-diagonal elements equal, and* (2) $\pi_{d_T} \to \pi$ *and* $\gamma_{d_T} \to \gamma$, *where*

(a) $\pi = \gamma = 1$	*if* $\rho_1 > 0, \rho_2 \geqq 0$;
(b) $\gamma = 1$	*if* $\rho_1 = 0, \rho_2 > 0$;
(c) $\gamma = 1, \pi = 0$	*if* $\rho_1 < 0, \rho_2 > 0$;
(d) $\pi = 0$ (*hence, if* $k = 2, \gamma = 1$)	*if* $\rho_1 < 0, \rho_2 = 0$;
(e) $k > 2$: $\pi = 0, \gamma = 0$	*if* $\rho_1 < 0, \rho_2 < 0$;
$k = 2$: (i) $\pi = 0, \gamma = 1$	*if* $\rho_1(1 + \rho_2) < 2\rho_2 < 0$;
(ii) $2\pi + \gamma = 1$	*if* $\rho_1(1 + \rho_2) = 2\rho_2 < 0$;
(iii) $\pi = 1/2, \gamma = 0$	*if* $2\rho_2 < \rho_1(1 + \rho_2) < 0$;
(f) $\gamma = 0$ (*hence, if* $k = 2, \pi = 1/2$)	*if* $\rho_1 = 0, \rho_2 < 0$;
(g) (i) $\pi = 1/2, \gamma = 0$	*if* $0 < \rho_1(1 + \rho_2) < -2\rho_2$;
(ii) $2\pi - \gamma = 1$	*if* $0 < \rho_1(1 + \rho_2) = -2\rho_2$;
(iii) $\pi = \gamma = 1$	*if* $0 < -2\rho_2 < \rho_1(1 + \rho_2)$;

and where, in cases (b), (d) *for* $k > 2$, (e) (ii) *for* $k = 2$, (f) *for* $k > 2$, *and* (g) (ii), *the vector* $(\pi_{d_T}, \gamma_{d_T})$ *need not approach a limit, but only the designated line.*

The version of theorem 3.1.3 with error term T^{-1} (as in theorem 3.1.1) is obvious.

We note that Williams' type III designs are thus optimum only in cases (d), (e), (f) for $k > 2$.

When $\rho = 0$, the same designs are asymptotically optimum as in the case of the linear model.

The case where ρ_1 and ρ_2 are not known exactly, but are only known to fall in a specified one of the regions listed in theorem 3.1.3, is treated as in the linear case. So is the case where ρ_1 and ρ_2 are completely unknown, but where a sequential design can be used.

Finally, a nonsequential minimax (over ρ_1 and ρ_2) design is obtained by making the coefficients of terms other than 1, ρ_1^2, and ρ_2^2 in the expression (3.1.8), equal zero. We obtain:

THEOREM 3.1.4. *If* ρ_1 *and* ρ_2 *are unknown,* $\{d_T\}$ *is asymptotically minimax over* ρ_1 *and* ρ_2 *for criteria* (a) *through* (f) *if and only if it satisfies the first two lines of equation* (3.1.6) *and also*

$$\lim_{T\to\infty} \pi_{d_T} = \lim_{T\to\infty} \gamma_{d_T} = \frac{1}{k} \quad (3.1.9)$$

Since the point $(1/k, 1/k)$ is an interior point of B, there are many ways of obtaining asymptotically minimax designs. Two examples when $k = 3$ are $d = (112113112113 \cdots ; 221223221 \cdots ; 331332331 \cdots 332)$ and $d = [11 \cdots 122 \cdots 233 \cdots 3$ of length $T/3$ together with $123123 \cdots 123$ of length $2T/3]$.

Higher order dependence. Models with higher order dependence can be investigated in the same fashion.

3.2. *Polynomial regression on an m-ball; rotatable designs.* Let m and d be positive integers. We now treat the problem where $\mathfrak{X}$ is the unit m-dimensional ball, consisting of those points $x = (x^{(1)}, \cdots, x^{(m)})$ in Euclidean m-space for which $\sum_j (x^{(j)})^2 \leqq 1$, and where the f_i are all the functions $\Pi_j (x^{(j)})^{r_j}$ for which the r_j are nonnegative integers satisfying $\sum_j r_j \leqq d$; thus, $k = \binom{d+m}{m}$. We refer to this as the problem of dth degree regression on the m-ball. We shall consider the approximate theory, and shall treat the problem of D-optimal estimation of all regression coefficients, which by theorem 2.3.1 is the same as G-optimal estimation of the entire regression function.

The optimum orthogonal-invariant approximate design characterized in theorem 3.2.1 below is not a discrete measure. In theorem 3.2.1 we also give an upper bound on the number of points needed to support a discrete measure which is D-optimum (see subsection 2.3). We shall not be concerned here with the actual construction of these discrete ξ, except for brief mention of the cases $d = 1$ and $d = 2$; these considerations when $d > 2$ and $m > 2$ are considerably more difficult, since it is no longer generally possible to replace uniformly distributed measure on an $(m - 1)$-sphere by a uniform discrete distribution on an appropriate finite set of "uniformly spaced" points as in the case $m = 2$. The construction of exact rotatable designs, which is the subject of much recent literature, is also not our concern here. In fact, such designs may be far from being optimum, as we shall exemplify below in the case $d = 2$, $m = 2$. In fact, our results do *not* imply that there exists an exact rotatable design which is within order N^{-1} of being optimum, but only that there is an exact design which is within order N^{-1} of being rotatable [in the value of its $M(\xi)$] and which is within order N^{-1} of being optimum. The term "rotatable" will be used by us in reference not only to exact, but also to approximate, designs.

The case $m = 1$, where $\mathfrak{X}$ is an interval, was treated in full by Guest [13] and Hoel [14]. Unfortunately neither of their elegant methods is applicable when $m > 1$, and we shall not obtain such explicit results for general d as those of Guest and Hoel.

Rotatable designs were invented by Box and Hunter [3], and their intuitive appeal (which has never been justified until the present paper) has attracted considerable interest and usage. As mentioned in the first part of subsection 3.1, the usual treatment in the literature of design problems where such designs are employed does not make a precise specification of $\mathfrak{X}$. Thus, in [3] and in subsequent papers on the subject, it is standard to employ the normalization $\sum_{i=1}^{N} x_i^{(j)}$

$= 0$, that is, $\int (x^{(j)}) d\xi = 0$, and the normalization $\sum_{i=1}^{N} (x_i^{(j)})^2 = N$, that is, $\int (x^{(j)})^2 d\xi = 1$, and then to compare the variance of the estimated regression for various designs at each distance ρ from the origin. Now, the use of two different designs with the same such normalization will usually entail taking observations in balls of quite different radii, that is, $\max_i \sum_j (x_i^{(j)})^2$ will usually not be the same for two such designs. Thus, the comparison of various rotatable designs in this manner is of questionable practical value! If the set of points x at which one is interested in estimating the regression function and at which observations can be taken is actually a ball of radius R, it is meaningless to insist on the above normalization, since it can exclude many good admissible rotatable designs and can include many poor ones, namely, those taking all observations within a ball of radius $< R$, when $mN < R^2$. In fact, a trivial consequence of the behavior of $\mathfrak{X}$ (and the f_i) under multiplication by R/R' is that any design which takes all observations in a ball of radius R' with $R' < R$ yields a larger generalized variance and a larger $\bar{d}(\xi)$ than the corresponding design in the ball of radius R. We shall return to this point in the final paragraph of this subsection.

The reader is warned that certain formulas which are relevant to the material of this section will differ from corresponding formulas of [3] and other papers on the subject, because of the above-mentioned normalization used in these papers (see, for example, the second paragraph above theorem 3.2.2, below).

We shall use, in this section only, a notation which is more convenient in this particular setting than is the general notation employed in the rest of the paper. Instead of using a single subscript, we label the elements of θ and of the vector f, as well as the rows and columns of $M(\xi)$, by m-tuples $(r_1, r_2, \cdots, r_m)$, corresponding to the function $\Pi_j (x^{(j)})^{r_j}$.

Let $\mathcal{G}$ be the orthogonal group on m-space. For g in $\mathcal{G}$, the design $g\xi$ is as usual defined by $(g\xi)(A) = \xi(g^{-1}A)$. Since the problem (set of possible regression functions, et cetera) looks the same with respect to $g\xi$ as with respect to ξ, we clearly have $d(gx, g\xi) = d(x, \xi)$, or $d(gx, \xi) = d(x, g^{-1}\xi)$. Hence, theorem 2.3.3 is applicable and we conclude that there is an optimum ξ^* which is invariant under $\mathcal{G}$; that is, for which $\xi^*(A) = \xi^*(gA)$ for every orthogonal transformation g and Borel set A. Such a ξ^* can be factored into the form $\xi_1^* \times \xi_2^*$:

$$\xi^*(A) = \int_0^1 \xi_1^*(\rho^{-1}[A \cap S_\rho]) \xi_2^*(d\rho), \tag{3.2.1}$$

where S_ρ is the $(m - 1)$-sphere of radius ρ, ξ_1^* is the uniform probability measure on S_1, the integrand is taken to be 1 or 0 when $\rho = 0$ according to whether or not $0 \in A$, and ξ_2^* is a probability measure on the interval [0, 1]. Write

$$\mu_s = \int_0^1 \rho^s \xi_2^*(d\rho) \tag{3.2.2}$$

and

$$F(s_1, s_2, \cdots, s_m) = \int_{S_1} \Pi_j (x^{(j)})^{s_j} \xi_2^*(d\rho). \tag{3.2.3}$$

Then the elements of $M(\xi^*)$ for $\xi^* = \xi_1^* \times \xi_2^*$ can be computed as follows:

$$m_{(\alpha_1,\cdots,\alpha_m),(\beta_1,\cdots,\beta_m)}(\xi^*) = \mu_{\Sigma(\alpha_i+\beta_i)}F(\alpha_1+\beta_1, \cdots, \alpha_m+\beta_m). \tag{3.2.4}$$

In particular, this is zero if $\alpha_i + \beta_i$ is odd for any i, and is positive otherwise, unless $\xi_2^*(0) = 1$.

Now, suppose ξ^* is D-optimum and invariant, as above. Then $d(x, \xi^*)$ is easily seen to be a polynomial in ρ^2, of degree $\leq d$ in ρ^2, say $d^*(\rho^2, \xi^*) = \sum_{j=0}^{d} q\, \rho^{2j}$. From the behavior as $x^{(1)} \to \infty$, we see that $q_d > 0$. Now, $\int q(x, \xi)\xi(dx) = k$ for any ξ (see subsection 2.3), so that, by theorem 2.3.1, $d^*(\rho^2, \xi^*) = k$ on a set of ξ^*-measure one if ξ^* is optimum. It is easy to prove for any invariant ξ^* that $M(\xi^*)$ is singular if ξ_2^* gives measure one to a set consisting of fewer than $(d+1)/2$ points of the interval $0 \leq \rho \leq 1$, where the origin is counted as $1/2$ point (the converse of this for general invariant ξ^* is also simple to prove, but will not be needed here, since it will follow from other considerations that ξ_2^* has exactly $(d+1)/2$ points of support for a D-optimum invariant ξ^*). Hence, for an optimum ξ^*, the function $d^*(\tau, \xi^*)$ is a polynomial of degree d with $q_d > 0$ and such that $d^*(\tau, \xi^*) = k$ at least $(d+1)/2$ points.

If d is even, say $d = 2b$, it follows from elementary algebraic considerations that $d^*(\tau, \xi^*)$ is equal to its maximum k on $0 \leq \tau \leq 1$ at $\tau = 0$ and at b other points, of which one is $\tau = 1$, since $d^*(\tau, \xi)$ must be increasing at $\tau = 1$. Similarly, if d is odd, say $d = 2b + 1$, then $d^*(\tau, \xi^*)$ is equal to its maximum k at $b + 1$ points, all different from zero, one of them being 1. In either case, write $\tau_1, \cdots, \tau_{b+1}$ for these values, so that ξ^* is supported by $\{\tau_1^{1/2}, \cdots, \tau_{b+1}^{1/2}\}$.

Thus, an invariant design can be optimum only if ξ_2^* has exactly $(d+1)/2$ points of support, one of which is at 1.

According to theorem 2.3.1, $M(\xi)$ and $d(x, \xi)$ are the same for all D-optimum ξ, whether or not ξ is invariant. Thus, all D-optimum approximate designs ξ are rotatable, in that $d(x, \xi)$ is a function only of ρ for all of them. Since the function $d(x, \xi)$ is the same for all D-optimum designs, and since $d(x, \xi)$ attains its maximum on a set of ξ-measure one if ξ is D-optimum, we conclude that every D-optimum ξ (invariant or not) gives measure one to the union of the same $b + 1$ $(m - 1)$-spheres (one of which may degenerate to the origin) of radii $\tau_1^{1/2}, \cdots, \tau_{b+1}^{1/2}$.

Any D-optimum ξ can be integrated over $\mathcal{G}$ to yield an invariant D-optimum ξ^* with $\xi_2^*(\rho) = \xi(S_\rho)$ for all ρ. For fixed values of $\tau_1, \cdots, \tau_{b+1}$, the function log det $M(\xi)$ is strictly concave in the positive weights $\gamma_i = \xi_2(\tau_i^{1/2})$ among invariant designs, so that the optimum weights, as well as the τ_i, are unique. For any invariant designs ξ' for which ξ_2' is supported by $(d+1)/2$ points $\lambda_1^{1/2}, \cdots, \lambda_{b+1}^{1/2}$ with $\lambda_{b+1} = 1$ and with $\lambda_1 = 0$ if d is even, write $d^*(\tau, \xi') = d(\tau; \gamma_1, \cdots, \gamma_{b+1}, \lambda_1, \cdots, \lambda_{b+1}) = d(\tau; \gamma, \lambda)$ to exhibit the support points and weights. Consider the equations

$$d(\lambda_j; \gamma, \lambda) = k;$$

$$(3.2.5)\qquad \frac{\partial}{\partial\tau} d(\tau;\gamma,\lambda)|_{\tau=\lambda_j} = 0, \qquad \begin{cases} 1 \leqq j \leqq b \text{ if } d \text{ is odd,} \\ 2 \leqq j \leqq b \text{ if } d \text{ is even;} \end{cases}$$

$$\lambda_{b+1} = 1;$$

$$\lambda_1 = 0 \qquad \text{if } d \text{ is even.}$$

The first line contains only b independent equations. Clearly, the unique optimum choices of γ and λ satisfy these equations. Conversely, any positive probability vector γ and any λ, with $0 \leqq \lambda_i \leqq 1$, satisfying these equations clearly define an invariant ξ^* for which $\bar{d}(\xi^*) = k$, so that this ξ^* is D-optimum.

The integer H of Section 4 is the number of functions of the form $\Pi_j(x^{(j)})^{s_j}$ with $s_j \geqq 0$ and $1 \leqq \sum_j s_j \leqq 2d$, which is just $\binom{2d+m}{m} - 1$.

We summarize our results.

THEOREM 3.2.1. *For the problem of dth degree regression on the unit m-ball, there are numbers* $\lambda_1 < \lambda_2 < \cdots < \lambda_{b+1} = 1$ *with* $b =$ *(greatest integer* $\leqq d/2$*) and* $\lambda_1 = 0$ *(respectively,* > 0*) if d is even (respectively, odd), and positive numbers* $\gamma_1, \cdots, \gamma_{b+1}$ *whose sum is unity, such that* ξ *is D-optimum if and only if it satisfies*

(a) $\xi(S_{\lambda_j^{1/2}}) = \gamma_j, \qquad 1 \leqq j \leqq b+1,$

(b) ξ *is rotatable, that is,* $d(x, \xi)$ *depends only on* $\rho^2 = \sum_{j=1}^m (x^{(j)})^2$.

In particular, there is a unique orthogonal-invariant design satisfying these conditions, and there are designs supported by at most $\binom{2d+m}{m}$ *points which satisfy these conditions.*

The γ_j *and* λ_j *can be obtained as the unique solution of* (3.2.5) *satisfying* $\gamma_j > 0$, $\sum\gamma_j = 1, 0 \leqq \lambda_j \leqq 1$.

It should be noted that the common practice of combining nonrotatable designs on spheres in such a way as to yield a rotatable design will often lead to a design on more than $(d + 1)/2$ spheres, which thus cannot be optimum.

We now consider some examples of optimum designs.

When $d = 1$, we have the trivial result that $\xi(S_1) = 1$ if ξ is optimum. Examples of discrete ξ's which are D-optimum are the uniform distribution on the $m + 1$ vertices of an inscribed regular simplex or on the vertices of any other inscribed regular polygon.

When $d = 2$, we need only consider rotatable designs which assign measure δ, say, to S_1 and measure $1 - \delta$ to the origin. Using formula (3.2.4) (or, renormalizing formula (49) of [3] in the manner discussed earlier, using $V(\delta^{-1/2}\rho)$ where V is given by that formula), one can without difficulty write out the function $d^*(\tau, \xi)$ for such a ξ and use its convexity in τ (see the next paragraph) to compute the desired result. However, it is even unnecessary to do this in order to obtain the result, since a simpler argument suffices to compute $d^*(0, \xi)$ and $d^*(1, \xi)$ in this example, as we shall now see.

The average U of "observations" at the origin is an unbiased estimator of

$\theta_{(0,0,\cdots,0)}$ and yields no information about any other parameters. Moreover, U is the b.l.e. of $\theta_{(0,\cdots,0)}$, since otherwise it is easy to see that $M(\xi')$ would be nonsingular, where ξ' is the uniform measure on S_1, and this last is false by the argument leading up to theorem 3.2.1. We conclude that U is the b.l.e. of the regression function at the origin. Hence, $d(0, \xi) = (1 - \delta)^{-1}$. Now, if $d(0, \xi) = k$, we automatically obtain $d(1, \xi) = k$, since $\int d(x, \xi)\xi(dx) = k$ for all ξ; by the convexity of the second degree polynomial $d(\tau, \xi)$ with positive coefficient of τ^2, we will then have $\bar{d}(\xi) = k$, so that ξ will be D-optimum. Thus, we obtain

THEOREM 3.2.2. *When $d = 1$, ξ is D- and G-optimum if and only if it is rotatable and $\xi(S_1) = 1$. When $d = 2$, ξ is optimum if and only if it is rotatable and $\xi(0) = 1 - \xi(S_1) = 2/(m + 1)(m + 2)$.*

The authors of [3] say on page 215 that they do not claim any optimum properties for their designs, but we now see that some, but not all, of their designs are indeed D-optimum. Thus, when $m = 2$ and $d = 2$, their design which takes one observation at the origin and one at each of five equally spaced points on the unit circle (or any multiple of this design) is D-optimum. The reader is invited to consult table 1 of [3] to see how the considerations with the normalization used there are, as we have mentioned earlier, misleading for the problem we have considered unless one is careful about translating their meaning. For example, "$\rho = 1$" there does not mean "$\rho = 1$" for us, but rather "$\rho = \delta^{1/2}$" for designs of the structure considered above; thus, values of V for a given value of ρ there actually correspond to different points of $\mathfrak{X}$ for different designs, and one must look at the behavior of various designs on different domains of ρ there, in order to obtain their behavior on the same domain $\mathfrak{X}$ in our problem.

3.3. *Linear regression on a Euclidean subset.* Let $\mathfrak{X}$ be a compact subset of Euclidean m-space, which we assume not to lie in an $(m - 1)$-dimensional hyperplane (if it did, θ would not be estimable for any ξ). Writing $x = (x^{(1)}, \cdots, x^{(m)})$, we consider the problem of estimating all of θ, or the whole regression function, when $k = m + 1$ and $f_i(x) = x^{(i)}$ for $1 \leq i \leq m$ and $f_{m+1}(x) = 1$.

For any ξ for which θ is estimable, $d(x, \xi)$ is quadratic in x and is in fact strictly convex. Hence, if $\mathfrak{X}'$ is the convex closure of $\mathfrak{X}$, the function $d(x, \xi)$ achieves its maximum on $\mathfrak{X}'$ only on a set B of extreme points of $\mathfrak{X}'$, which are clearly points of $\mathfrak{X}$. Thus, $d(x, \xi)$ attains its maximum on $\mathfrak{X}$ precisely on the set B, and thus, as in previous examples, $\xi^*(B) = 1$ if ξ^* is D-optimum. Since $d(x, \xi^*)$ is quadratic and is equal to k on B if ξ^* is optimum, there must therefore exist an ellipsoid T (by which we mean the hypersurface, not the solid) which can be circumscribed about $\mathfrak{X}$ in such a way that $T \cap \mathfrak{X} = B$ and $d(x, \xi) = m + 1$ for x in T. Thus, $d(x, \xi^*) = (x - c)'C(x - c) + c_0$ for a suitable vector c, positive real c_0, and positive definite symmetric matrix C, T being the set where $(x - c)'C(x - c) = m + 1 - c_0$,

If A is an $m \times m$ matrix such that $A'A = (m + 1 - c_0)C$, the affine transformation $y = A(x - c)$ takes T onto the unit sphere $S = \{y|y'y = 1\}$. Under this transformation, ξ^* is transformed into a measure ξ' on S. Since $(x^{(1)}, \cdots,$

$x^{(m)}, 1)M^{-1}(\xi^*)(x^{(1)}, \cdots, x^{(m)}, 1)' = (x - c)'C(x - c) + c_0$, we conclude that

$$M^{-1}(\xi^*) = \begin{Vmatrix} C & -Cc \\ -c'C & c'Cc + c_0 \end{Vmatrix}, \tag{3.3.1}$$

and hence that

$$M(\xi^*) = \begin{Vmatrix} C^{-1} + c_0^{-1}cc' & c_0^{-1}c \\ c_0^{-1}c' & c_0^{-1} \end{Vmatrix}. \tag{3.3.2}$$

Hence,

$$\begin{aligned} C^{-1} + c_0^{-1}cc' &= \int xx'\xi^*(dx) = \int (A^{-1}y + c)(A^{-1}y + c)'\xi'(dy), \\ c_0^{-1}c &= \int x\xi^*(dx) = \int (A^{-1}y + c)\xi'(dy), \\ c_0^{-1} &= \int 1\xi^*(dx) = 1, \end{aligned} \tag{3.3.3}$$

from which we conclude at once that

$$\begin{aligned} \int y\xi'(dy) &= 0, \\ \int yy'\xi'(dy) &= m^{-1}I. \end{aligned} \tag{3.3.4}$$

Since $\sum_i \int x_i^2\xi(dx) = 1$ for any ξ on S, we easily compute that $H = (m^2 + 3m - 2)/2$.

We summarize our results:

THEOREM 3.3.1. *For the problem of linear regression on a compact subset $\mathfrak{X}$ of Euclidean m-space which does not lie in an $(m - 1)$-dimensional hyperplane, there is an affine transformation t of $\mathfrak{X}$ into the unit m-ball such that the intersection of $t\mathfrak{X}$ with the unit sphere supports a probability measure ξ' for which*

$$\begin{aligned} \int y_i\xi'(dy) &= 0, & 1 \leqq i \leqq k, \\ \int y_iy_j\xi'(dy) &= m^{-1}\delta_{ij}, & 1 \leqq i, j \leqq k; \end{aligned} \tag{3.3.5}$$

a D- and G-optimum ξ^* is then defined by

$$\xi^*(W) = \xi'(tW). \tag{3.3.6}$$

Conversely, each optimum ξ^* can be obtained in this way from such a ξ', for some t. Whenever such a ξ' exists for a given t, there exists such a ξ' supported by $m(m + 3)/2$ or fewer points.

Of course, the optimum ξ^* need not be unique.

The ellipsoid T which was circumscribed about $\mathfrak{X}$ (or $\mathfrak{X}'$) above, and which is "closest" to it in the sense appropriate for our considerations, does not seem to have been considered in the literature on circumscribing figures about convex bodies. The explicit determination of T or t for a given $\mathfrak{X}$ or $\mathfrak{X}'$ poses an interesting geometric problem. In the case $m = 2$, using complex notation, ξ' must be such that

$$\int z\xi'(dz) = 0, \qquad \int z^2\xi'(dz) = 0. \tag{3.3.7}$$

If the subset $\mathfrak{X}$ of the unit circle supports such a measure, then a subset of $t\mathfrak{X}$ consisting of 5 or fewer points will also support such a measure. Fewer points may suffice, as when $\mathfrak{X}$ is a triangle, square, or circle; if $\mathfrak{X}$ is a regular pentagon, five points are needed.

REFERENCES

[1] G. E. P. Box, "Multi-factor designs of first order," *Biometrika*, Vol. 39 (1952), pp. 49–57.

[2] G. E. P. Box and W. A. Hay, "A statistical design for the efficient removal of trends occurring in a comparative experiment with an application in biological assay," *Biometrics*, Vol. 9 (1953), pp. 304–319.

[3] G. E. P. Box and J. S. Hunter, "Multi-factor experimental designs for exploring response surfaces," *Ann. Math. Statist.*, Vol. 28 (1957), pp. 195–241.

[4] G. E. P. Box and H. L. Lucas, "Design of experiments in nonlinear situations," *Biometrika*, Vol. 46 (1959), pp. 77–90.

[5] G. E. P. Box and K. B. Wilson, "On the experimental attainment of optimum conditions," *J. Roy. Statist. Soc., Ser. B*, Vol. 13 (1951), pp. 1–45.

[6] H. Chernoff, "Locally optimum designs for estimating parameters," *Ann. Math. Statist.*, Vol. 24 (1953), pp. 586–602.

[7] ———, "Sequential design of experiments," *Ann. Math. Statist.*, Vol. 30 (1959), pp. 755–770.

[8] D. R. Cox, "Some systematic experimental designs," *Biometrika*, Vol. 38 (1951), pp. 312–323.

[9] ———, "Some recent work on systematic experimental designs," *J. Roy. Statist. Soc., Ser. B*, Vol. 14 (1952), pp. 211–219.

[10] ———, "The design of an experiment in which certain treatment arrangements are inadmissible," *Biometrika*, Vol. 41 (1954), pp. 287–295.

[11] G. Elfving, "Optimum allocation in linear regression theory," *Ann. Math. Statist.*, Vol. 23 (1952), pp. 225–262.

[12] ———, "Design of linear experiments," *Probability and Statistics* (Cramér volume), New York, Wiley, 1959, pp. 58–74.

[13] P. G. Guest, "The spacing of observations in polynomial regression," *Ann. Math. Statist.*, Vol. 29 (1958), pp. 294–299.

[14] P. G. Hoel, "Efficiency problems in polynomial regression," *Ann. Math. Statist.*, Vol. 29 (1958), pp. 1134–1145.

[15] J. Kiefer, "On the nonrandomized optimality and randomized nonoptimality of symmetrical designs," *Ann. Math. Statist.*, Vol. 29 (1958), pp. 675–699.

[16] ———, "Optimum experimental designs," *J. Roy. Statist. Soc., Ser. B*, Vol. 21 (1959), pp. 273–319.

[17] ———, "Optimum designs in regression problems, II," to appear in *Ann. Math. Statist.*

[18] ———, "Invariance, sequential estimation, and continuous time processes," *Ann. Math. Statist.*, Vol. 28 (1957), pp. 573–601.

[19] J. Kiefer and J. Wolfowitz, "Optimum designs in regression problems," *Ann. Math. Statist.*, Vol. 30 (1959), pp. 271–294.

[20] ———, "The equivalence of two extremum problems," *Canad. J. Math.*, Vol. 12 (1960), pp. 363–366.

[21] H. D. PATTERSON, "The construction of balanced designs for experiments involving sequences of treatments," *Biometrika*, Vol. 39 (1952), pp. 32–48.
[22] H. SCHEFFÉ, "A method for judging all contrasts in the analysis of variance," *Biometrika*, Vol. 40 (1953), pp. 87–104.
[23] M. STONE, "Application of a measure of information to the design and comparison of regression experiments," *Ann. Math. Statist.*, Vol. 30 (1959), pp. 55–70.
[24] A. WALD, "Asymptotic minimax solutions of sequential point estimation problems," *Proceedings of the Second Berkeley Symposium on Mathematical Statistics and Probability*, Berkeley and Los Angeles, University of California Press, 1951, pp. 1–12.
[25] R. M. WILLIAMS, "Experimental designs for serially correlated observations," *Biometrika*, Vol. 39 (1952), pp. 151–167.

Reprinted from
Proc. Fourth Berkeley Sympos. Math. Statist. and Probability
1, 381–405 (1960)

THE EQUIVALENCE OF TWO EXTREMUM PROBLEMS

J. KIEFER AND J. WOLFOWITZ

1. Introduction. Let $f_1, \ldots, f_k$ be linearly independent real functions on a space X, such that the range R of $(f_1, \ldots, f_k)$ is a compact set in k-dimensional Euclidean space. (This will happen, for example, if the f_i are continuous and X is a compact topological space.) Let S be any Borel field of subsets of X which includes X and all sets which consist of a finite number of points, and let $C = \{\xi\}$ be any class of probability measures on S which includes all probability measures with finite support (that is, which assign probability one to a set consisting of a finite number of points), and which are such that

$$m_{ij}(\xi) = \int_X f_i(x) f_j(x) \xi(dx) \qquad i, j = 1, \ldots, k$$

is defined. In all that follows we consider only probability measures ξ which are in C. Write $M(\xi)$ for the $k \times k$ matrix $||m_{ij}(\xi)||$. When $M(\xi)$ is non-singular, write $[M(\xi)]^{-1} = ||m^{ij}||$. (We shall not always exhibit dependence on ξ.) Letting $f(x)$ denote the column vector with components $f_i(x)$, and letting primes denote transposes, we define

$$d(x; \xi) = f(x)'[M(\xi)]^{-1} f(x)$$

whenever $M(\xi)$ is non-singular.

We consider two extremum problems. The first is to choose ξ so that

(1) $$\xi \text{ maximizes } \det M(\xi).$$

The second is to choose ξ so that

(2) $$\xi \text{ minimizes } \max_x d(x; \xi).$$

We also note that the integral with respect to ξ of $d(x; \xi)$ is k; hence, $\max_x d(x; \xi) \geqslant k$, and thus a sufficient condition for ξ to satisfy (2) is

(3) $$\max_x d(x; \xi) = k.$$

The result of this note is that (1), (2), and (3) are equivalent. This result, which seems to have interest *per se*, also strengthens and extends results of the authors **(1)** on the optimum design of regression experiments. A brief description of the connection with the design of such experiments is given below. The proof of the theorem is elementary and brief.

Received March 30, 1959 Research of J. Kiefer was sponsored by the Office of Naval Research. Research of J. Wolfowitz was supported by the United States Air Force under Contract no. AF 18(600)–685 monitored by the Office of Scientific Research.

2. The theorem. For every ξ consider $M(\xi)$ as a point in Euclidean k^2-space, let T be the totality of such points for all ξ in C, and let $\bar{T}$ be the convex closure of T. It is clear that every extreme point of $\bar{T}$ can be achieved by a ξ which assigns probability one to a single point. Since C contains every ξ with finite support, it follows that $T = \bar{T}$. The class C need not, of course, be convex. However, since our argument will be concerned only with the $M(\xi)$, we may argue below as if C were convex. Thus, if ξ_1 and ξ_2 are in C and

$$\frac{\xi_1 + \xi_2}{2}$$

is not, we may still discuss

$$M\left(\frac{\xi_1 + \xi_2}{2}\right),$$

because there exists a ξ in C with finite support, say ξ_3, such that

$$M(\xi_3) = M\left(\frac{\xi_1 + \xi_2}{2}\right).$$

Moreover, if $H - 1$ is the dimension of the linear space spanned by the functions f_if_j, $i \leqslant j$, any $M(\xi)$ is equal to an $M(\xi')$ where the support of ξ' consists of at most H points. This can often be impoved, as in the case where X is the unit interval and $f_i(x) = x^{i-1}$.

Call a subset D of C *linear* if the following condition holds: For every α, $0 \leqslant \alpha \leqslant 1$, and every pair ξ_1, ξ_2 in D, $\alpha\xi_1 + (1 - \alpha)\xi_2$ is in D whenever it is in C. Thus, if C is convex, D is also convex.

We shall prove the following:

THEOREM. *Conditions* (1), (2), *and* (3) *are equivalent. The set B of all ξ satisfying these conditions is linear, and $M(\xi)$ is the same for all ξ in B.*

This result has a function space corollary which may be of interest. Suppose ξ satisfies (3) and that Q is a real $k \times k$ matrix such that $QM(\xi)Q'$ is the identity. Then $g = Qf$ is a vector of orthonormal functions with respect to ξ, and $g(x)'g(x) = d(x; \xi)$. Thus we have

COROLLARY. *If $f_1, \ldots, f_k$ are linearly independent, continuous, real functions on a compact space X, then there is a probability measure ξ on X and a linear transformation $g_i = \sum_j a_{ij}f_j$ such that $g_1, \ldots, g_k$ are orthonormal with respect to ξ and*

$$\max_x \sum_{i=1}^{k} g_i^2(x) = k.$$

The set of all such ξ is the set B of the theorem.

Proof of the theorem. We shall say that ξ is a *local solution* of (1) if $\det M(\xi) > 0$ and if, for every ξ',

(4) $$\frac{\partial}{\partial\alpha}\log\det M([1-\alpha]\xi+\alpha\xi')|_{\alpha=0+}\leqslant 0.$$

Now, if $\det M(\xi)>0$, A is such that $AM(\xi)A'$ is the identity, and $AM(\xi')A'$ is diagonal with diagonal elements b_i, then $\det M([1-\alpha]\xi+\alpha\xi')=\det A^{-2}\,\Pi_i[1-\alpha+\alpha b_i]$, from which we easily compute that $-\log\det M([1-\alpha]\xi+\alpha\xi')$ is convex in $\alpha(0<\alpha<1)$ and is strictly convex unless all $b_i=1$ (that is, unless $M(\xi)=M(\xi')$). Hence, if $\det M(\xi')>\det M(\xi)$, equation (4) cannot hold for that ξ'. We conlcude that local solutions of (1) are actual solutions of (1), and of course the converse is true. Moreover, if $\det M(\xi)=\det M(\xi')=h>0$, we have $\det M(\xi/2+\xi'/2)>h$ unless $M(\xi)=M(\xi')$, so that ξ and ξ' cannot both satisfy (1) unless $M(\xi)=M(\xi')$. It follows from this and the linearity in ξ of $M(\xi)$ that, if ξ and ξ' both satisfy (1), then so does $\alpha\zeta+(1-\alpha)\zeta'$, whenever it is in C.

It now suffices to prove that $\det M(\xi)>0$ and ξ satisfies (4) for all ξ', if and only if ξ satisfies (2), and only if it satisfies (3). First suppose ξ satisfies (4) and that $\det M(\xi)>0$. Performing the differentiation in (4), and denoting by M_{ij} the cofactor of m_{ij}, we have

(5) $$\begin{aligned} 0 &\geqslant [\det M(\xi)]^{-1}\sum_{i,j}\frac{\partial\det M}{\partial m_{ij}}\frac{\partial m_{ij}([1-\alpha]\xi+\alpha\xi')}{\partial\alpha}\Bigg|_{\alpha=0}\\ &=[\det M(\xi)]^{-1}\sum_{i,j}\left(\frac{\partial}{\partial m_{ij}}\sum_q m_{iq}M_{iq}\right)[m_{ij}(\xi')-m_{ij}(\xi)]\\ &=[\det M(\xi)]^{-1}\sum_{i,j}M_{ij}(\xi)[m_{ij}(\xi')-m_{ij}(\xi)]=\sum_{i,j}m^{ij}(\xi)m_{ij}(\xi')-k.\end{aligned}$$

Letting ξ' give measure one to the point x, we obtain

(6) $$[f(x)]'M(\xi)^{-1}f(x)\leqslant k$$

for all x. Thus, (3) is satisfied and, as we have remarked, this implies (2).

Finally, if (2) is satisfied, we must have (6) for all x, since we have just seen that there always exist ξ's satisfying (3). Hence, for any ξ' with finite support, we obtain $\sum_{i,j}m^{ij}(\xi)m_{ij}(\xi')\leqslant k$. Hence this inequality is valid for all ξ', and (5) is satisfied. This completes the proof of the theorem.

3. Extensions and applications. We remark that it is easy to see that, if R is bounded but not compact, and if $\{\xi_i\}$ is a sequence of measures on S, then $\lim_i\det M(\xi_i)$ is a maximum if and only if $\lim_i\sup_x d(x;\xi_i)$ is a minimum, and if and only if $\lim_i\sup_x d(x;\xi_i)=k$. Similarly, the first part of the corollary holds with the replacement $\sup_x\sum_i g_i^2(x)<k+\epsilon$, for any $\epsilon>0$.

We now describe briefly the statistical applications of the results. An integer N is given, and the statistician must choose N points $x_1,\ldots,x_N$ (not necessarily distinct) corresponding to which he obtains observations on uncorrelated random variables Y_i $(1\leqslant i\leqslant N)$ with common variance σ^2 (perhaps unknown) and with expectation $\sum_{j=1}^k\theta_j f_j(x_i)$, where the θ_j are unknown real parameters. If $\xi(x)$ denotes the proportion of x_i's which are equal to x, we find that the covariance matrix of best linear estimators of $\theta_1, \ldots, \theta_k$ is $N^{-1}\sigma^2[M(\xi)]^{-1}$. The function ξ is called the experiment or the experimental design. A criterion often adopted for choosing a design is to minimize the determinant of the above covariance matrix (the "generalized variance"). Another possible criterion is to minimize the maximum over x of the variance $N^{-1}\sigma^2 d(x; \xi)$ of the "best linear estimator," given ξ, of the "regression function" $\sum_j \theta_j f_j(x)$. If we consider not merely the class C_N of probability measures ξ which take on only integral multiples of N^{-1} as values, but rather *all* probability measures ξ in C, then our result is that the two optimality criteria are equivalent. Moreover, for any ξ with support on H points which satisfies (1), (2), and (3), there is clearly a ξ' in C_N which achieves (1), (2), and (3) to within a multiplicative factor $1 + 0(N^{-1})$, and is easy to write down from ξ. Since the *exactly* optimum designs are often difficult to obtain, depend on N, and differ for the two criteria, we see the practical importance of our considerations.

It is very helpful to use the interplay of the two criteria (1) and (2) in obtaining a solution. For example, one can sometimes guess that a solution exists which is a member of a class of ξ which depend on several parameters. One may use (1) as the more convenient initial approach, maximize det $M(\xi)$ over the parametric class, and then verify whether the maximum just obtained is indeed a maximum over *all* ξ (which may be difficult in terms of (1)) by verifying (3). It is useful to note that, if ξ has a set consisting of k points as its support, then it gives equal measure to each of these points. (This is part of Theorem 5 of **(1)**.) Examples which make use of such methods will appear elsewhere, as will generalizations such as one concerned with the minimization of the determinant of a principal minor of $M(\xi)^{-1}$.

REFERENCE

1. J. Kiefer and J. Wolfowitz, ***Optimum designs in regression problems***, Ann. Math. Stat., *30* (1959).

Cornell University

Reprinted from
Canad. J. Math. **12**, 363–366 (1960)

Reprinted from THE ANNALS OF MATHEMATICAL STATISTICS
Vol. 32, No. 1, March, 1961
Printed in U.S.A.

OPTIMUM DESIGNS IN REGRESSION PROBLEMS, II

BY J. KIEFER[1]

Cornell University

0. Summary. Extending the results of Kiefer and Wolfowitz [10], [11], methods are obtained for characterizing and computing optimum regression designs in various settings, and examples are given where D-optimum designs are computed.

In Section 1 we introduce the main definitions and notation which will be used in the paper, and discuss briefly the roles of invariance, randomization, number of points at which observations are taken, and nonlinearity of the model, in our results.

In Section 2 we prove the main theoretical results. We are concerned with the estimation of s out of the k parameters, extending an approach developed in [10] and [11] in the case $s = k$. There is no direct way of ascertaining whether or not a given design ξ^* is D-optimum for (minimizes the generalized variance of the best linear estimators of) the s chosen parameters, and Theorems 1 and 2 provide algorithms for determining whether or not a given ξ^* is D-optimum. If all k parameters are estimable under ξ^*, we can use (2.7) to decide whether ξ^* is D-optimum, while if not all k parameters are estimable we must use the somewhat more complicated condition (2.17) (of which part (a) or (b) is necessary for optimality, while (a), (c), or (d) is sufficient). An addition to Theorem 2 near the end of Section 3 provides assistance in using (2.17) (b). Theorem 3 of Section 2 characterizes the set of information matrices of the D-optimum designs.

In Section 3 we give a geometric interpretation of the results of Section 2, and compare the present approach with that of [10]. In the case $s = k$, the present approach reduces to that of Section 5 of [10] and of [11]. When $1 < s < k$, we obtain an algorithm which differs from that of Section 4 of [10] and which appears to be computationally easier to use. When $s = 1$, the results of the present paper are shown to reduce to those of Section 2 of [10]; in particular, we obtain the game-theoretic results without using the game-theoretic machinery of [10].

In Section 4 we determine D-optimum designs for the problems of quadratic regression on a q-cube and polynomial regression on a real interval with $1 < s < k$.

Part II of the paper is devoted entirely to the determination of D-optimum designs for various problems in the setting of simplex designs considered by Scheffé [12].

Received December 28, 1959; revised October 12, 1960.
[1] Research sponsored by the Office of Naval Research and the Army Signal Corps.

Various unsolved problems are mentioned throughout the paper. Further examples will be published elsewhere.[2]

PART I. GENERAL CONSIDERATIONS

1. Introduction. The design of optimum experiments in regression settings can be a tedious computational problem. In the present paper we are concerned with the development and application of algorithms which make this task easier. Early work in this area was done by Elfving [4], Chernoff [1], and Ehrenfeld [2]. The characterization of optimum designs in general symmetrical settings where block designs are customarily employed was given by the present author in [8]. As described in Section 0, the present paper continues the work of Kiefer and Wolfowitz [10], [11].

We now introduce our terminology and notation, which is essentially that of [9] and [11]. Let $f_1, f_2, \cdots, f_k$ be given functions on a space $\mathfrak{X}$. To avoid trivial circumlocutions (see [11]), we assume $\mathfrak{X}$ compact and the f_i linearly independent and continuous. By $f(x)$ we denote the column k-vector with components $f_i(x)$. For any discrete probability measure ξ on $\mathfrak{X}$, write

$$m_{ij}(\xi) = \int_{\mathfrak{X}} f_i(x)f_j(x)\xi(dx). \tag{1.1}$$

(As discussed in [10] and [11], other probability measures can be considered, but are not needed.) Write $M(\xi)$ for the $k \times k$ matrix $\{m_{ij}(\xi)\}$. Any such ξ is called an *experiment*, and the set of all ξ will be denoted by Ξ.

The practical meaning of these notions is this: We are concerned with inference regarding an unknown k-vector θ, an element of a k-dimensional Euclidean space Ω. A single observation at the point x (value of the independent variable) in $\mathfrak{X}$ yields a random variable Y_x for which

$$\begin{aligned} EY_x &= \theta' f(x) = \sum_1^k \theta_i f_i(x), \\ \operatorname{Var}(Y_x) &= \sigma^2. \end{aligned} \tag{1.2}$$

Thus, $\theta' f(x)$ is the regression function. If $\eta(x)$ observations out of the total of n uncorrelated observations available in a given experiment are taken at the point x and we let $\xi = n^{-1}\eta$, we obtain $n\sigma^{-2}M(\xi)$ for the "information matrix" of the experiment; thus, for example, if all components of θ are estimable, the covariance matrix of the best linear estimators (b.l.e.'s) of components of θ is $n^{-1}\sigma^2 M^{-1}(\xi)$. The justification for considering only b.l.e.'s in the sequel (whether or not σ^2 is known), and the trivial modifications which are needed in our developments if the Y_x are not uncorrelated with equal variances or do not cost the same amounts, are discussed in [9] and [10]. Also discussed there is the relevance of considering designs ξ which take on values other than multiples of $1/n$. Briefly, such considerations allow us to develop useful computational

[2] Optimum designs for certain problems in the settings where systematic designs and rotatable designs are employed, will appear in the *Proceedings of the Fourth Berkeley Symposium on Probability and Statistics*.

techniques which are completely absent if we restrict the values of ξ, and they allow us to find one optimum ξ (rather than a different one for each n) which can immediately be altered into an *actual* design (i.e., a ξ with restricted values) which is within $O(n^{-1})$ of being optimum. Thus, we shall always consider, without restriction, the whole space Ξ of designs.

Suppose we are interested in inference regarding s independent linear parametric functions of θ, which we can take to be $\theta_1, \theta_2, \cdots, \theta_s$ without loss of generality. We partition $M(\xi)$ and $M^{-1}(\xi)$ as

$$\begin{Vmatrix} M_1(\xi) & M_2(\xi) \\ M_2'(\xi) & M_3(\xi) \end{Vmatrix} \quad \text{and} \quad \begin{Vmatrix} M^{(1)}(\xi) & M^{(2)}(\xi) \\ M^{(2)'}(\xi) & M^{(3)}(\xi) \end{Vmatrix},$$

respectively, where if $M(\xi)$ is singular we take $M^{-1}(\xi)$ to be a pseudo-inverse (see, e.g., [1] and the next section for details). Here $M_1(\xi)$ and $M^{(1)}(\xi)$ are $s \times s$, and $n^{-1}\sigma^2 M^{(1)}(\xi)$ is the nonsingular covariance matrix of b.l.e.'s of $\theta_1, \cdots, \theta_s$ if all of these parameters are estimable when the design ξ is used. We shall say that ξ^* *is D-optimum for* $\theta_1, \cdots, \theta_s$ if

$$\det M^{(1)}(\xi^*) = \min_{\xi \in \Xi} \det M^{(1)}(\xi), \tag{1.3}$$

i.e., if ξ^* minimizes the generalized variance of the b.l.e.'s of $\theta_1, \cdots, \theta_s$. Extensive discussions in [8], [9], and [10] are concerned with the meaningfulness of this criterion, with certain intuitively appealing properties of the criterion (e.g., invariance under certain transformations), and with a comparison of this criterion with certain other optimality criteria. In the present paper we shall be entirely concerned with D-optimality and an equivalent criterion which is discussed just after (1.4) below; results concerning other optimality criteria are contained in [1], [2], [4], [8], [9], [10], and in the references cited there.

We shall also partition θ (resp., $f(x)$) into $\theta^{(1)}$ and $\theta^{(2)}$ (resp., $f^{(1)}(x)$ and $f^{(2)}(x)$), where $\theta^{(1)}$ (resp., $f^{(1)}(x)$) is an s-vector.

When $s = k$, it will sometimes be convenient to state a problem in terms of the k-dimensional vector space F spanned by the f_i; clearly, D-optimality depends only on F and not on the choice of the f_i used to represent it (analogous remarks apply when $s < k$).

The variance of the b.l.e. of $\theta' f(x)$, the true regression function evaluated at the point x, when the design ξ (for which all components of θ are estimable) is used, is $\sigma^2 n^{-1} d(x, \xi)$, where

$$d(x, \xi) = f(x)' M^{-1}(\xi) f(x). \tag{1.4}$$

Another possible optimality criterion when $s = k$ is for ξ to minimize $\max_x d(x, \xi)$. One of the results of [11] is that this criterion and D-optimality are equivalent, and this fact proves very useful in constructing optimum designs. Thus, the direct maximization of det $M(\xi)$ over *all* ξ will usually be very difficult. However, as in the first example of Section 4, one can often guess a simple (finite-dimensional) subclass of Ξ, compute the ξ^* which maximizes det $M(\xi)$ over this subclass, and then compute $\max_x d(x, \xi^*)$. If this last quantity equals k (and only then), ξ^* is indeed optimum among *all* members of Ξ, as was proved in [10] and [11]. Thus, it would appear useful to find a generalization of $d(x, \xi)$ and the criterion in terms of it, when $s < k$, and to prove the equivalence of this criterion to D-optimality, so as to yield a computational technique in the case $s < k$ which is analogous to that just discussed for the case $s = k$. (Since the matrix $M^*(\xi)$ of (1.5) is not linear in ξ if $s < k$, and since $M(\xi^*)$ can be singular, there will be complications which did not arise in the case $s = k$.) This is the task of Section 2. The method of constructing D-optimum designs developed in this paper is thus to guess a ξ^* and to compute $\max_x d(x, \xi^*)$ (with the definition of (2.3) in place of (1.4)) or, if not all components of θ are estimable, $\max_\xi D(\xi, \xi^*)$. Paralleling the procedure outlined above in the case $s = k$, we then use (2.7) or (2.17) (a) to test the optimality of ξ^*.

Before turning to Section 2, we shall mention a few results which are relevant for the remainder of the paper.

Invariance. If A and B are $s \times s$ nonnegative definite symmetric matrices, write $A \geqq B$ if $A - B$ is nonnegative definite. Without recourse to a specific optimality criterion such as D-optimality, one can study complete classes of designs, admissible designs, etc., and this was first done in the regression setting for $s = k$ by Ehrenfeld [3], who exploited the usefulness of $M(\xi_1) \geqq M(\xi_2)$ as a criterion equivalent to "ξ_1 is at least as good as ξ_2 for all linear estimation problems". As discussed by the present author [9], the idea can also be explained as the sufficiency (in the sense of Blackwell) in the normal case of ξ_1 for ξ_2, criteria other than Ehrenfeld's for completeness can be given, and the whole theory can be extended to the case where we are only interested in s out of the k parameters. In this case, ξ_1 is at least as good as ξ_2 if and only if $M^{(1)}(\xi_1) \leqq M^{(1)}(\xi_2)$ or, equivalently, if $M^*(\xi_1) \geqq M^*(\xi_2)$, where

$$(1.5) \qquad M^*(\xi) = [M^{(1)}(\xi)]^{-1} = M_1(\xi) - M_2(\xi)M_3^{-1}(\xi)M_2'(\xi).$$

(The modification in the singular case is obvious.) If A is a nonsingular $k \times k$ matrix of the form

$$A = \begin{Vmatrix} A_1 & A_2 \\ 0 & A_3 \end{Vmatrix},$$

where A_1 is $s \times s$, and if $J(\xi) = AM(\xi)A'$, then, with an obvious notation, we have

$$(1.7) \qquad J^*(\xi) = A_1 M^*(\xi) A_1'.$$

Suppose $\bar{G}$ is a group of linear transformations on Ω of the form (1.6) with $A_1 = I$ and $A_2 = 0$ (so that $\bar{G}$ leaves $\theta^{(1)}$ fixed), where for each $\bar{g}$ in $\bar{G}$ there is a transformation g on $\mathfrak{X}$ for which

$$(1.8) \qquad \theta' f(x) = (\bar{g}\theta)' f(gx)$$

for all x and θ. Suppose also that $G = \{g\}$ satisfies the usual conditions of the transformation group in the invariance theory in statistics (see, e.g., Kiefer [7]). Then, as proved by the author in Theorem 3.3 of [9], we have

Complete class invariance theorem. Under the above conditions, the class of designs ξ which are G-invariant, i.e., for which

$$\xi(gB) = \xi(B) \quad \text{for all } g \text{ and } B, \tag{1.9}$$

is essentially complete for linear estimation of $\theta^{(1)}$. This is a convex set of measures, invariant under G.

Fundamental in the proof of this result is the following lemma (Lemma 3.2 of [9]), which we shall also need in the next section:

LEMMA 1. *For $j = 1, 2, \cdots, r$, let C_j be $s \times (k - s)$, let D_j be positive definite symmetric $s \times s$, and suppose $\lambda_j > 0$, $\sum \lambda_j = 1$. Then*

$$[\sum \lambda_j C_j][\sum \lambda_j D_j]^{-1}[\sum \lambda_j C_j'] \leqq \sum \lambda_j C_j D_j^{-1} C_j', \tag{1.10}$$

with equality if and only if the matrix $C_j D_j^{-1}$ is the same for all j. (The extension of Lemma 1 to the case of singular D_j's is discussed shortly before the statement of Theorem 3.)

In the case where we are interested in a specific optimality criterion, the group G in the above theorem can be more general; for example, in the case of D-optimality the A_1's do not have to be the identity, but only a group of matrices of determinant one for which the usual invariance theorem in statistics holds. We obtain (Theorem 4.3 of [9])

Invariance theorem for D-optimality. Under the above conditions, there is a G-invariant ξ which is D-optimum for $\theta^{(1)}$.

This is proved by using the previous invariance theorem and also the following trivial lemma, which we shall also have occasion to use in the next section:

LEMMA 2. *If A and B are nonnegative definite symmetric $s \times s$ matrices, then $-\log \det (\lambda A + (1 - \lambda)B)$ is convex in λ for $0 \leqq \lambda \leqq 1$, and is strictly convex unless $A = B$ or $A + B$ is singular.*

Invariance theorems for other optimality criteria are considered in [9]. The invariance theorem for D-optimality is extremely useful in problems like those considered in Section 4 and Part II of the present paper, where it enables us to limit our search to suitably symmetrical ξ rather than to all of Ξ.

Randomization. As pointed out by the present author [8] and [9], in the exact small sample theory it can happen that, in terms of certain criteria (especially in problems of hypothesis testing), some randomized design may be better than any of the nonrandomized designs which are customarily employed in block experiments (this does not merely refer to the classical reason for employing randomization in such experiments). In the considerations of the present paper, we need not be concerned with randomized designs, since they are not needed. The reason for this is that a special case of Lemma 1 says that

$$\lambda M(\xi_1) + (1 - \lambda)M(\xi_2) \geqq [\lambda M^{-1}(\xi_1) + (1 - \lambda)M^{-1}(\xi_2)]^{-1} \tag{1.11}$$

for $0 < \lambda < 1$; if $M(\xi_1)$ and $M(\xi_2)$ are nonsingular (the singular case requiring only obvious modifications), equality can hold if and only if $M(\xi_1) = M(\xi_2)$. The right side of (1.11) is proportional to the inverse of the covariance matrix when ξ_1 is used with probability λ and ξ_2 is used with probability $(1 - \lambda)$; the left side is proportional to the information matrix of the nonrandomized design $\lambda\xi_1 + (1 - \lambda)\xi_2$; thus, the latter design is always at least as good as, and usually better than, the former (randomized) design, for linear estimation.

The number of points needed in the support of an optimum design. Elfving [4] and Chernoff [1] gave upper bounds on the number of points needed to support a design which minimizes the *average* variance of the b.l.e.'s (i.e., the trace of $M^{(1)}(\xi)$); Chernoff's result gives $s(2k - s + 1)/2$ as an upper bound in the general case. Chernoff's geometrical argument can be duplicated in the case of the generalized variance; for example, when $s = k$ this amounts to noting that the $M(\xi)$ can be considered as a closed convex set in Euclidean $k(k + 1)/2$-space with extreme points obtainable from ξ's with single points for support, and that $\det(aM)$ is increasing in $a > 0$ if M is positive definite, so that any D-optimum $M(\xi)$ must be a boundary point of the set. The bound in the case $s = 1$ is obtained by a different method in [10]. An alternative but less direct approach to obtain the bound $s(2k - s + 1)/2$ is to use Stone's characterization [13] of this as the number of points needed to support a design which maximizes $\det L_\lambda^*(\xi)$ where $L_\lambda(\xi) = I + \lambda M(\xi)$; letting λ go to infinity gives the desired result.

The bound obtained in this fashion can be very poor, as can be seen in the examples of Part II. A method of sharpening the bound slightly when $s = 1$ is given in [10]. As discussed (with slight inaccuracy) in [9] and [11], a method for sharpening the results in many cases is to count the dimension $H - 1$ (say) of the range of the set of convex linear combinations of the functions $f_i f_j$, $i \leq j$, and then to note that any $M(\xi)$ can be achieved by a ξ whose support is H or fewer points.

Even this result gives a poor bound in many cases, as can be seen in the examples of Part II where, as is often the case, there exists a D-optimum design for θ with support on k points (obviously the minimum number possible). In the case of polynomial regression of degree $k - 1$ or less on a real interval ($f_i(x) = x^{i-1}$), it is an old and well known fact that any $M(\xi)$ (i.e., any set of values for the moments up to the $2(k - 1)$st) can be achieved by a ξ with support on at most k points, and this has been sharpened by the present author [9] to state that the minimal (essentially) complete class of admissible designs consists of those ξ whose support consists of at most k points, at most $k - 2$ of which are in the interior of the interval. A difficult and important mathematical problem is to extend such results as the well known ones just cited on the moment problem to the case of other f_i, so as to characterize for any given F the maximum number of points needed for the support of ξ's in an essentially complete class. The results for polynomials in one variable, which depend on certain properties of orthogonal polynomials, are not directly extendable.

A final problem in this area is to combine the invariance results with those just discussed. Thus, it can happen in some settings that there is an optimum design (in some sense) on P points without there being a G-invariant optimum design on P points. How many points are needed for the latter?

The nonlinear case. If the regression function is not linear in θ as in (1.2), it is still possible to obtain relevant asymptotic results. Thus, the results of the next section can be applied in exactly the way Chernoff's [1] are for the average variance, in such cases. When $s = 1$, the average and generalized variances of course coincide, and the results of Section 2 of [10] and of the present paper thus yield a computational algorithm for the problems treated by Chernoff; such an algorithm for minimizing the average variance when $s > 1$ can be found in Section 4 of [10].

The author thanks Professor J. Wolfowitz for helpful discussions.

2. The main results. Since the case where a D-optimum ξ for $\theta^{(1)}$ has *nonsingular* $M(\xi)$ is slightly easier to handle than is the singular case, and since the non-singular case yields sharper results, we shall treat this case first, although we now make some definitions which apply generally. We shall say that ξ^* yields a global minimum of det $M^{(1)}(\xi)$, or is D-optimum for $\theta^{(1)}$, if (1.3) is satisfied; of course, $M^{(1)}(\xi)$ is well defined and finite if $\theta^{(1)}$ is estimable under ξ, whether or not $M(\xi)$ is nonsingular; we define $\det M^{(1)}(\xi) = \infty$ if $\theta^{(1)}$ is not estimable under ξ. Thus, in this formula as well as in those which follow, determinants and inverses can always be computed in the case where $M(\xi)$ is singular by computing them with $M(\xi)$ replaced by $M(\xi) + \lambda I$ and then letting λ approach zero. Equation (1.3) can of course be written (using (1.5)) as

$$(2.1) \quad \det M^*(\xi^*) = \det M(\xi^*)/\det M_3(\xi^*) = \max_\xi \det M^*(\xi);$$

so that (1.3) can be rephrased to state that ξ^* yields a *global maximum* of det $M^*(\xi)$. We shall say that ξ^* yields a *local maximum* of det $M^*(\xi)$ if det $M^*(\xi^*) > 0$ and

$$(2.2) \qquad \frac{\partial}{\partial\alpha} \log \det M^*([1 - \alpha]\xi^* + \alpha\xi) \,|_{\alpha=0+} \leqq 0 \quad \text{for all } \xi.$$

We generalize the definition of (1.4) in the case $s < k$ to define

$$(2.3) \qquad d(x, \xi) = f(x)'M^{-1}(\xi)f(x) - f^{(2)}(x)'M_3^{-1}(\xi)f^{(2)}(x);$$

the form of this in the case of singular $M(\xi)$ will be seen more explicitly, later (in the functions D and $\bar{D}$ introduced below; $D(x, \xi)$ is the direct analogue of $d(x, \xi)$ in the singular case, but is not the appropriate function to yield an analogue of Theorem 1, as we shall see). If $M(\xi)$ is non-singular, the integrals with respect to ξ of the two terms on the right side of (2.3) are easily seen to be k and $k - s$. Hence,

$$(2.4) \qquad \int d(x, \xi)\xi(dx) = s$$

(this is similarly true whenever $\theta^{(1)}$ is estimable, whether or not $M(\xi)$ is nonsingular, as we shall see just above (2.18)), and thus

$$\max_x d(x, \xi) \geqq s. \tag{2.5}$$

Consider now the problem of determining ξ^* so that

$$\max_x d(x, \xi^*) = \min_\xi \max_x d(x, \xi). \tag{2.6}$$

It follows from (2.5) that a *sufficient* condition for ξ^* to satisfy (2.6) is

$$\max_x d(x, \xi^*) = s. \tag{2.7}$$

(Obviously, a necessary but not sufficient condition for ξ^* to satisfy (2.7) is that ξ^* give unit measure to the set of x for which $d(x, \xi^*) = s$.)

We now prove

THEOREM 1. *If $M(\xi^*)$ is nonsingular, equations* (2.1) (*D-optimality of ξ^* for $\theta^{(1)}$*), (2.2), (2.6), *and* (2.7) *are equivalent.*

PROOF. Clearly, (2.1) implies (2.2), and we have already seen that (2.7) implies (2.6). We first show that (2.2) implies (2.7). Denoting by $M_{ij}(\xi)$ the cofactor in $M(\xi)$ of $m_{ij}(\xi)$ and by $M^{ij}(\xi)$ the (i, j)th element of $M^{-1}(\xi)$, we have, as in [11], that, for $M(\xi^*)$ nonsingular (sometimes omitting the argument ξ^* for typographical ease),

$$\begin{aligned}
&\frac{\partial}{\partial\alpha} \log \det M([1 - \alpha]\xi^* + \alpha\xi) \mid_{\alpha=0+} \\
&\quad = \det M^{-1}(\xi^*) \sum_{i,j} \frac{\partial \det M}{\partial m_{ij}} \frac{\partial m_{ij}([1 - \alpha]\xi^* + \alpha\xi)}{\partial\alpha} \Bigg|_{\alpha=0+} \\
&\quad = \det M^{-1}(\xi^*) \sum_{i,j} \left(\frac{\partial}{\partial m_{ij}} \textstyle\sum_q m_{iq} M_{iq}\right) [m_{ij}(\xi) - m_{ij}(\xi^*)] \\
&\quad = \det M^{-1}(\xi^*) \sum_{i,j} M_{ij}(\xi^*)[m_{ij}(\xi) - m_{ij}(\xi^*)] \\
&\quad = \sum_{i,j} m^{ij}(\xi^*) m_{ij}(\xi) - k.
\end{aligned} \tag{2.8}$$

(A somewhat neater derivation proceeds by letting $BM(\xi^*)B' = I$ and $BM(\xi)B' = D$, a diagonal matrix; the left side of (2.2) is immediately seen to be $\mathrm{tr}D - k$, and we have $\mathrm{tr}D = \mathrm{tr}[BM(\xi)B'] = \mathrm{tr}[(B'B)M(\xi)] = \mathrm{tr}[M^{-1}(\xi^*)M(\xi)]$.) Similarly, writing $M_3^{-1} = \{\bar{m}^{ij}; i, j > s\}$, we have

$$\frac{\partial}{\partial\alpha} \log \det M_3([1 - \alpha]\xi^* + \alpha\xi) \mid_{\alpha=0+} = \sum_{i,j>s} \bar{m}^{ij}(\xi^*) m_{ij}(\xi) - (k - s). \tag{2.9}$$

Hence, (2.2) can be written as

$$\begin{aligned}
&\mathrm{tr}\, [M^{-1}(\xi^*)M(\xi)] - \mathrm{tr}[M_3^{-1}(\xi^*)M_3(\xi)] \\
&\quad = \sum_{i,j} m^{ij}(\xi^*) m_{ij}(\xi) - \sum_{i,j>s} \bar{m}^{ij}(\xi^*) m_{ij}(\xi) \leqq s \text{ for all } \xi.
\end{aligned} \tag{2.10}$$

In particular, if ξ gives measure 1 to the point x, the left side of the inequality (2.10) becomes $d(x, \xi^*)$. Thus, (2.2) implies (2.7).

Conversely, if (2.7) holds, we have (2.10) for every ξ which gives measure one to a single point. Since $m_{ij}(\xi)$, and thus the left side of (2.10), is linear in ξ, we obtain (2.10) for all ξ. Thus, (2.7) implies (2.2).

Now, a design ξ^* satisfying (2.7) always exists (since a ξ^* satisfying (2.1) exists by compactness, and we have seen that (2.1) implies (2.7)). We conclude from (2.5) that (2.6) implies (2.7).

It remains to prove that (2.2) implies (2.1). From Lemma 1 of the previous section, we have, for $0 < \alpha < 1$,

$$M^*([1-\alpha]\xi^* + \alpha\xi) \geqq (1-\alpha)M^*(\xi^*) + \alpha M^*(\xi). \tag{2.11}$$

From (2.11) and Lemma 2, we obtain

$$\begin{aligned} -\log \det M^*([1-\alpha]\xi^* + \alpha\xi) & \\ &\leqq -\log\det\,[(1-\alpha)M^*(\xi^*) + \alpha M^*(\xi)] \\ &\leqq -(1-\alpha)\log\det M^*(\xi^*) - \alpha\log\det M^*(\xi). \end{aligned} \tag{2.12}$$

(We shall later, but not now, use the fact that the first inequality is strict unless $M_2(\xi^*)M_3^{-1}(\xi^*) = M_2(\xi)M_3^{-1}(\xi)$, and that the second one is strict unless $M^*(\xi^*) = M^*(\xi)$.) Thus, log det $M^*([1-\alpha]\xi^* + \alpha\xi)$ is concave in α, and thus this function has positive derivative at $\alpha = 0$ if det $M^*(\xi) >$ det $M^*(\xi^*)$; this last inequality holds for some ξ if (2.1) does not hold, and hence (2.2) is also violated in this case. This completes the proof of Theorem 1.

We now turn to the case where $\theta^{(1)}$ is estimable but $M(\xi^*)$ is singular. If we try to duplicate the proof of Theorem 1, we find, exactly as before, that (2.1) and (2.2) are equivalent. What happens to the rest of the proof is most clearly seen by considering a linear transformation of θ into $(A')^{-1}\theta$, where A is of the form (1.6); this transforms $M^*(\xi)$ into the form (1.7), and thus leaves unchanged the various criteria considered in Theorem 1. We shall use such a transformation in order better to display what occurs. In all that follows, ξ^* is fixed and $\theta^{(1)}$ is estimable under ξ^*. We can choose A of the form (1.6) so that $J(\xi^*) = AM(\xi^*)A'$ is a diagonal matrix with its first $s + r$ diagonal elements unity and the rest zero. For a fixed ξ, we can at the same time choose A so that

$$J(\xi) = \left\|\begin{matrix} J_1 & J_2 & J_4 & 0 \\ J_2' & J_3 & J_6 & 0 \\ J_4' & J_6' & J_5 & 0 \\ 0 & 0 & 0 & 0 \end{matrix}\right\| \tag{2.13}$$

where J_1 is $s \times s$, J_3 is $r \times r$, J_5 is $p \times p$, J_5 is nonsingular, $J_6 = 0$, and J_1, J_3, and J_5 are diagonal.

If we try to go through the proof of Theorem 1, duplicating the arguments in the present case, we obtain

$$J^*([1-\alpha]\xi^* + \alpha\xi) = (1-\alpha)I + \alpha D - \alpha^2 J_2([1-\alpha]I + \alpha J_3)^{-1}J_2' - \alpha J_4 J_5^{-1} J_4', \tag{2.14}$$

and thus, in place of (2.10),

$$\sum_{j \le s} \mu_{jj}(\xi) - \rho(\xi, \xi^*) \le s, \tag{2.15}$$

where we denote the elements of $J(\xi)$ by $\mu_{ij}(\xi)$ and where $\rho(\xi, \xi^*)$ is the trace of $J_4 J_5^{-1} J_4'$. To put (2.15) in a form which more closely resembles (2.10), write $\mu^{ij}(\xi^*)$ for the elements of the inverse of the upper left $(r+s) \times (r+s)$ submatrix $\bar{J}(\xi^*)$ of $J(\xi^*)$, and $\bar{\mu}^{ij}(\xi^*)$, $s < i, j \le s + r$, for the elements of the inverse of $\{\mu_{ij}(\xi^*), s < i, j \le s + r\}$. Then (2.15) becomes

$$\begin{aligned} &\operatorname{tr}[\bar{J}^{-1}(\xi^*)\hat{J}(\xi)] - \operatorname{tr}[J_3^{-1}(\xi^*)\hat{J}_3(\xi)] \\ &= \operatorname{tr}[\bar{J}^{-1}(\xi^*)\bar{J}(\xi)] - \operatorname{tr}[J_3^{-1}(\xi^*)J_3(\xi)] - \operatorname{tr}[J_1^{-1}(\xi^*)J_4(\xi)J_5^{-1}(\xi)J_4'(\xi)] \\ &= \sum_{i,j \le s+r} \mu^{ij}(\xi^*)\mu_{ij}(\xi) - \sum_{\substack{s < i,j \\ \le s+r}} \bar{\mu}^{ij}(\xi^*)\mu_{ij}(\xi) - \rho(\xi, \xi^*) \le s, \end{aligned} \tag{2.16}$$

where $J_7' = (J_4' J_6')$ and $\hat{J}(\xi) = \bar{J}(\xi) - J_7(\xi)J_5^{-1}(\xi)J_7'(\xi)$ is proportional to the information matrix of ξ for the first $s + r$ components of $(A')^{-1}\theta$ (analogous to J^* for the first s components) and $\hat{J}_3 = J_3 - J_6 J_5^{-1} J_6'$. In fact, without requiring $J(\xi^*)$ and the $J_i(\xi)$ to be of these special forms which facilitated the computation of (2.15), we clearly have (2.16) whenever $AM(\xi^*)A'$, of rank $r + s$, has zeros outside of the upper left hand $(r+s) \times (r+s)$ matrix, where J_5 is no longer necessarily of full rank (so that we can take it to be $(k - s - r) \times (k - s - r)$) and $\rho(\xi, \xi^*)$ is again the trace of the product of $J_1^{-1}(\xi^*)$ by the matrix $\lim_{\lambda \to 0} J_4(\xi)[J_5(\xi) + \lambda I]^{-1}J_4'(\xi)$; the matrix $\hat{J}$ has the same meaning as above. Thus, for a given ξ^* and A, the same formula (2.16) holds for all ξ. In fact, it is easy to give an invariant, geometric definition of the left side of (2.16), as we shall see in Section 3, and we note here that the first form on the left side of (2.16) could have been obtained by using (2.2) and (2.10) on $\alpha\hat{J}(\xi) + (1-\alpha)\bar{J}(\xi^*)$ with k replaced by $r + s$. If we denote by $^{\#}$ the operation * of (1.5) when $k = r + s$, the essence of the matter is that, in the singular or nonsingular case,

$$(\hat{M})^{\#} = M^*.$$

This is easy to prove directly or in terms of the J's. We hereafter denote the expression on the left side of inequality (2.16) by $D(\xi, \xi^*)$; we also write $\bar{D}(\xi, \xi^*) = D(\xi, \xi^*) + \rho(\xi, \xi^*)$; this is the left side of (2.16) ignoring the $\rho(\xi, \xi^*)$ term. We also denote by $D(x, \xi^*)$ and $\bar{D}(x, \xi^*)$ and $\rho(x, \xi^*)$ the corresponding expressions when ξ gives measure one to the point x.

If $M(\xi^*)$ is singular, the functions $\bar{D}$ and ρ depend on the choice of A, but the function D does not.

It is thus clear from (2.16) that, in analogy to the implication of (2.2) by (2.7) in Theorem 1, (2.2) is now implied by the *first* or the *third* of the following four statements, and implies the *first* and *second*:

$$
\begin{aligned}
&\text{(a)}\ \max_\xi D(\xi, \xi^*) = s,\\
&\text{(b)}\ \max_x D(x, \xi^*) = s,\\
&\text{(c)}\ \max_\xi \bar{D}(\xi, \xi^*) = s,\\
&\text{(d)}\ \max_x \bar{D}(x, \xi^*) = s.
\end{aligned}
\tag{2.17}
$$

Moreover, (2.17) (c) and (d) are clearly equivalent.

By using the transformation A which makes $AM(\xi^*)A'$ the identity of order $(r + s)$ together with zeros (as above) and writing $g(x) = Af(x)$, we see that the first $r + s$ components of g are orthonormal functions with respect to ξ^*, while the other components vanish on a set of unit ξ^* measure. Hence, $D(x, \xi^*) = \sum_1^s g_i^2(x) - \rho(x, \xi^*)$, where $\rho(x, \xi^*) = 0$ on the support of ξ^*. Hence, $D(\xi^*, \xi^*) = \bar{D}(\xi^*, \xi^*) = s$, and thus, analogous to (2.5), we have

$$
\begin{aligned}
&\max_\xi \bar{D}(\xi, \xi^*) = \max_x \bar{D}(x, \xi^*)\\
&\geqq \max_\xi D(\xi, \xi^*) \geqq \max_x D(x, \xi^*) \geqq s,
\end{aligned}
\tag{2.18}
$$

for *every* ξ^* (optimum or not). Thus, if in analogy to (2.6) we set ourselves the problem of determining ξ^* so that

$$
\begin{aligned}
&\text{(a)}\ \max_\xi D(\xi, \xi^*) = \min_{\xi'} \max_\xi D(\xi, \xi'),\\
&\text{or}\\
&\text{(b)}\ \max_x D(x, \xi^*) = \min_{\xi'} \max_x D(x, \xi'),\\
&\text{or}\\
&\text{(c)}\ \max_\xi \bar{D}(\xi, \xi^*) = \min_{\xi'} \max_\xi \bar{D}(\xi, \xi'),\\
&\text{or}\\
&\text{(d)}\ \max_x \bar{D}(x, \xi^*) = \min_{\xi'} \max_x \bar{D}(x, \xi'),
\end{aligned}
\tag{2.19}
$$

it follows from (2.18) that (2.17) (a) (resp., (b), (c), (d)) is a sufficient condition for (2.19) (a) (resp., (b), (c), (d)) to be satisfied. (Of course, (2.19) (c) and (d) are equivalent.) Moreover, from (2.16) we see that there exists at least one ξ^* (namely, any satisfying (2.1)) which satisfies (2.17) (a) and (b), so that these two conditions are in fact equivalent to (2.19) (a) and (b), respectively.

We summarize our results:

THEOREM 2. *If $\theta^{(1)}$ is estimable under ξ^*, equations* (2.1), (2.2), (2.17) (a), *and* (2.19) (a) *are equivalent. Moreover,* (2.1) (*and thus any of the above*) *implies* (2.17) (b), *which is equivalent to* (2.19) (b), *while* (2.1) *is implied by* (2.17) (c) (*or, equivalently,* (d)).

An addition to this theorem, which simplifies the use of (2.17) (b), will be found in Section 3. The fact that (2.17) (c) (or equivalently, (d)) implies (2.19) (c) (or, equivalently, (d)) has not been stated as part of Theorem 2 since the latter two do not have the same intrinsic interest that (2.6) does when $s = k$. The second sentence of the theorem is not of primary interest, but (2.17) (b) is useful in *eliminating* various ξ^*'s from optimality considerations; for example, it can be used in certain problems to show that a D-optimum design cannot have such a simple structure as any of those encountered in the examples of Part II or the first example of Section 4. Of course, (2.17) (d) is a useful *sufficient* condition for D-optimality. Of primary interest is the question of whether or not (2.17) (b), (c), or (d) is equivalent to (2.17) (a), since (2.17) (b) or (d) would seem on the surface to be a more natural analogue of (2.7) than is (2.17) (a). Unfortunately (from the viewpoint of computations as well as of esthetics!), the answer in general is "no". This is easy to see by examples in the case of either of the three criteria, and we shall content ourselves here with seeing why (2.17) (b) need not entail (2.17) (a).

To this end, suppose $k = 2$, $s = 1$, and that $\mathfrak{X}$ consists of three points, with $f(x_1)' = (1,0)$, $f(x_2)' = (0,1)$, and $f(x_3)' = (b, 1)$ with $b^2 > 4$. Let ξ^* give measure 1 to x_1. Then (2.17) (b) is easily seen to be satisfied. However, if $\xi(x_2) = \xi(x_3) = \frac{1}{2}$, we have $D(\xi, \xi^*) = b^2/4 > 1$. The difficulty is really that we have lost the linearity which permitted us to go from (2.7) to (2.10), the "convexity" of Lemma 1 working in the wrong direction here.

Needless to say, there is no general equivalence of (2.17) (c) and (d) to (2.19) (c) and (d).

We end this section with a description of the set of D-optimum ξ's. It is no longer the case when $s < k$, as it was when $s = k$ (treated in [11]), that $M(\xi)$ is the same for all D-optimum ξ. From the concavity of log det $M^*([1 - \alpha]\xi^* + \alpha\xi)$ proved in (2.12) (which is valid whether or not $M(\xi)$ and $M(\xi^*)$ are nonsingular), it is clear that, if ξ^* and ξ both maximize det $M^*(\xi)$, then so does $[1 - \alpha]\xi^* + \alpha\xi$ for $0 \leqq \alpha \leqq 1$; i.e., the set of D-optimum ξ's is convex. Suppose now that $M(\xi^*)$ is nonsingular and ξ^* is D-optimum, and write $M^*(\xi^*) = R$, $M_2(\xi^*)M_3^{-1}(\xi^*) = E$, and

$$B = \begin{Vmatrix} I & -E \\ 0 & I \end{Vmatrix}. \tag{2.20}$$

If ξ is also D-optimum and $M(\xi)$ is nonsingular, we must have equality in (2.12) (otherwise the parenthetical remark following (2.12) implies that $(\xi + \xi^*)/2$ would be better). But then $M^*(\xi) = R$, $M_2(\xi)M_3^{-1}(\xi) = E$, and hence

$$BM(\xi)B' = \begin{Vmatrix} R & 0 \\ 0 & M_3(\xi) \end{Vmatrix}. \tag{2.21}$$

Conversely, if for some T we have

$$M(\xi) = B^{-1}\begin{pmatrix} R & 0 \\ 0 & T \end{pmatrix}(B^{-1})', \tag{2.22}$$

then $M^*(\xi) = R$, $M_2(\xi)M_3^{-1}(\xi) = E$, and hence ξ is D-optimum.

If $M(\xi^*)$ is singular, the characterization of (2.22) must be modified slightly. In Lemma 1 with $r = 2$, suppose the C_j and D_j are of the form

$$D_1 = \begin{Vmatrix} Q_1 & 0 & 0 \\ 0 & Q_3 & 0 \\ 0 & 0 & 0 \end{Vmatrix}, \quad C_1 = \|L_1\ L_3\ 0\|, \quad D_2 = \begin{Vmatrix} Q_2 & 0 & 0 \\ 0 & 0 & 0 \\ 0 & 0 & Q_4 \end{Vmatrix}, \quad C_2 = \|L_2\ 0\ L_4\|,$$

where the Q_i are nonsingular. Then the conclusion is modified to state that equality holds if and only if $L_1Q_1^{-1} = L_2Q_2^{-1}$. If the M_3's and M_2's are reduced to the form of the D's and C's above through a simultaneous diagonalization, we see easily that the conclusions of the previous paragraph are still valid if $M(\xi^*)$ is nonsingular. If $M(\xi^*)$ is singular, the modified conclusions can be stated in several ways, perhaps the simplest being that, if $J(\xi^*)$ is of the form prescribed above (2.13) (nothing special being assumed about the form of $J(\xi)$), then $J_2(\xi) = 0$.

Writing out the form of (2.22), we obtain

THEOREM 3. *The set of D-optimum ξ's is convex. If ξ^* is D-optimum and $M(\xi^*)$ is nonsingular, then the set of all D-optimum ξ's consists of those ξ's for which $M(\xi)$ is of the form*

$$M(\xi) = \begin{Vmatrix} R + ETE' & ET \\ TE' & T \end{Vmatrix}, \tag{2.23}$$

where $R = M^*(\xi^*)$, $E = M_2(\xi^*)M_3^{-1}(\xi^*)$, and T is arbitrary. If ξ^* is optimum and if $J(\xi^*)$ is as prescribed above (2.13), then ξ is optimum if and only if $J_2(\xi) = 0$ and $J^*(\xi) = I$.

In any problem where at least one optimum ξ^* exists for which $M(\xi^*)$ is nonsingular, the characterization of (2.23) can be used. The characterization of the final sentence of the theorem in the case of singular $M(\xi^*)$ can easily be given a geometric formulation in the manner of Section 3.

Of course, $M^*(\xi)$ is the same for all D-optimum ξ, which thus all perform identically for problems of linear estimation of $\theta^{(1)}$. All D-optimum (for $\theta^{(1)}$) designs are thus admissible for linear estimation of $\theta^{(1)}$, but clearly a design can be D-optimum for $\theta^{(1)}$ and inadmissible for linear estimation of the full vector θ. The designs of the form (2.23) which are also admissible for linear estimation of θ are easily characterized (see [9]) as those for which T is maximal in the sense that if a design of the form (2.23) exists with T replaced by $\bar{T}$ and with $\bar{T} \geq T$, then $\bar{T} = T$.

3. Other forms and relationship to previous results. First suppose $M(\xi^*)$ is nonsingular and that $AM(\xi^*)A' = I$, where A is of the form (1.6). A trivial computation yields

$$M^{-1}(\xi^*) - \begin{pmatrix} 0 & 0 \\ 0 & M_3^{-1}(\xi^*) \end{pmatrix} = \begin{pmatrix} A_1' \\ A_2' \end{pmatrix} (A_1\ A_2). \tag{3.1}$$

Now, $(A_1\ A_2)\ M(\xi^*)(0\ A_4)' = 0$, and since A_4 is nonsingular this means that $(A_1 A_2)(M_2'\ (\xi^*)M_3\ (\xi^*))' = 0$. Thus, the expression (3.1) is of the form $\beta'(\beta M(\xi^*)\beta')^{-1}\beta$, where β is any $s \times k$ matrix of rank s whose rows are orthogonal to those of $(M_2'(\xi^*)M_3(\xi^*))$ (so that $(A_1 A_2) = L^{-1}\beta$ where L is $s \times s$ and $LL' = \beta M(\xi^*)\beta')$. Writing $\beta = (B_1\ B_2)$ where B_1 is $s \times s$, we can write

$$\beta M(\xi^*)\beta' = B_1 M_1(\xi^*)B_1' - B_2 M_3(\xi^*)B_2'.$$

We can now give a geometric description of $d(x, \xi)$. Let $g_i(x) = \sum_j \beta_{ij}f_j(x)$, $1 \leqq i \leqq s$, be linearly independent with respect to (i.e. on, the support of) ξ^*, and also orthogonal (ξ^*) to all $k - s$ functions of $f^{(2)}$. Writing

$$g(x)' = (g_1(x), \cdots, g_s(x))$$

and

$$(g_i, g_j)_\xi = \int g_i(x)g_j(x)\xi(dx), \tag{3.2}$$

and denoting by $G(\xi)$ the matrix $\{(g_i, g_j)_\xi\}$, we have

$$d(x, \xi^*) = g(x)'G^{-1}(\xi^*)g(x). \tag{3.3}$$

In particular, if the g_i are chosen to be mutually orthogonal (ξ^*), we obtain

$$d(x, \xi^*) = \textstyle\sum_{i=1}^{s} g_i^2(x)/(g_i, g_i)_{\xi^*}. \tag{3.4}$$

Thus, for example, we obtain (3.4) if we let $\beta_{ii} = 1$ and choose the other β_{ij} so as to minimize (for each i)

$$\int [f_i(x) + \sum_{j \neq i} \beta_{ij}f_j(x)]^2\xi^*(dx) \tag{3.5}$$

or, with $\beta_{ij} = 0$ for $j < i$, so as to minimize (for each i)

$$\int [f_i(x) + \sum_{j > i} \beta_{ij}f_j(x)]^2\xi^*(dx). \tag{3.6}$$

In the case of (3.5) (resp., (3.6)), g_i is the part of f_i orthogonal (ξ^*) to the linear space spanned by the f_j with $j \neq i$ (resp., $j > i$); i.e., g_i is f_i minus the projection (ξ^*) of f_i on that linear space.

In many examples, (3.5) will be the more convenient form to use; for, if the components of $f^{(1)}$ enter the problem symmetrically, it will only be necessary to carry out the computation of the β_{ij} for a single value of i.

We shall now indicate the relationship of (3.4) with the choice (3.6) to the results of Section 4 of [10]. In the notation of the present paper, the approach of Section 4 of [10] is to consider, for $\lambda = (\lambda_1, \cdots, \lambda_s)$ with all $\lambda_i > 0$, the zero-sum two-person game with payoff function

$$K_\lambda(x, \beta) = \sum_{i=1}^{s} \lambda_i[f_i(x) + \sum_{j>i} \beta_{ij}f_j(x)]^2, \tag{3.7}$$

where $\beta = \{\beta_{ij};\ 1 \leqq i \leqq s, i < j \leqq k\}$. It is shown there that if ξ_λ is a maximin strategy for this determined game, then the D-optimum ξ's are those ξ_λ's which maximize $\prod_{i=1}^{s} F_i(\xi_\lambda)$, where

$$(3.8)\qquad F_i(\xi) = \min_{\{\beta_{ij}\}} \int [f_i(x) + \sum_{j>i} \beta_{ij} f_j(x)]^2 \xi(dx),$$

and that there is, to within a multiplicative constant, a unique value λ^* of λ at which the maximum is attained. The results of the present paper give us additional information. Write $\lambda_i(\xi) = 1/F_i(\xi)$ and $\lambda(\xi) = \{\lambda_i(\xi)\}$. Then, clearly, $d(x, \xi) = K_{\lambda(\xi)}(x, \beta(\xi))$ where $\beta(\xi)$ is minimal with respect to ξ for the payoff function $K_{\lambda(\xi)}$; hence,

$$(3.9)\qquad \max_x d(x, \xi) = \max_x K_{\lambda(\xi)}(x, \beta(\xi)) \geqq K_{\lambda(\xi)}(\xi, \beta(\xi)) = s,$$

and a D-optimum ξ^* will be one *which is maximin when* $\lambda = \lambda(\xi^*)$ (assuming still that $M(\xi^*)$ is nonsingular for that ξ^*), and we will have equality in (3.9) for ξ equal to such a ξ^*. Hence, $\lambda^* = \lambda(\xi^*)$ for such a ξ^*, and the value of the game with $\lambda = \lambda(\xi^*) = \lambda^*$ is s. Thus, the essence of the matter is a fixed point theorem which we have proved. Hereafter calling the case where there exists a D-optimum ξ^* with nonsingular $M(\xi^*)$ the *regular case* (of $f^{(1)}$ relative to f), we have:

In the regular case there is a ξ' *such that* $\xi' = \xi_{\lambda(\xi')}$, *and any such* ξ' *is D-optimum.*

Of course, not all D-optimum ξ's need have $M(\xi)$ non-singular in the regular case. It will be easy to see how the above results must be modified in the singular case, but we note here that there are many examples where $\theta^{(1)}$ is estimable if and only if θ is estimable, and the above results of the regular case apply to such examples.

We remark that the considerations of Section 4 of [10] require only obvious modifications to apply to the case where (3.6) is replaced by (3.5), or where (3.4) is replaced by the general form (3.3).

To compare the method of Section 4 of [10] with the method which uses the results of Section 2 of the present paper (and which is described in the paragraph containing (1.4)), we consider the trivial problem treated in Example 4 of [10]. In the present notation, $k = 3$, $s = 2$, $\mathfrak{X} = [-1, 1]$, and $f_i(x) = x^{3-i}$. As in [10], we might begin by guessing that there is an optimum ξ of the form $\xi(-1) = \xi(1) = a$, $\xi(0) = 1-2a$. For such a ξ, we have $g_1(x) = x^2 - 2a$, $g_2(x) = x$, and

$$(3.10)\qquad d(x, \xi) = \frac{(x^2 - 2a)^2}{2a(1-a)} + \frac{x^2}{2a}.$$

This is a convex function of x^2 for $0 \leqq x^2 \leqq 1$, and thus has its maximum at $x^2 = 0$ or 1:

$$(3.11)\qquad \max_x d(x, \xi) = \max\left(\frac{2a}{1-2a}, \frac{2-2a}{2a}\right).$$

This last expression equals 2 if and only if $a = 1/3$; this choice thus yields a D-optimum design. It can be seen that the computations here were exceedingly simple. A less trivial example will be found in Example 4.2. In such more complicated examples, it appears that the present method may often involve considerably less computation than that of [10].

In the case $s = 1$, we can take $\lambda = \lambda_1 = 1$ and write K for K_λ, in conformity with the notation of Section 2 of [10]. Our criterion $d(x, \xi^*) \leqq 1$ for the D-optimality of ξ^* becomes

$$\max_x K(x, \beta(\xi^*)) = K(\xi^*, \beta(\xi^*)), \tag{3.12}$$

where again $\beta(\xi^*)$ is minimal with respect to ξ^*. It follows at once from (3.12) that the game is determined and that ξ^* is maximin (and maximal with respect to $\beta(\xi^*)$) while $\beta(\xi^*)$ is minimax (and minimal with respect to ξ^*). Moreover, these assertions regarding $\beta(\xi^*)$ and ξ^* imply that $\max_x d(x, \xi^*) = 1$. Thus, the game-theoretic results of Section 2 of [10], which were proved by entirely different methods there, have been obtained at once from Theorem 1 of the present paper in the case where there exists a D-optimum ξ^* for which $M(\xi^*)$ is nonsingular. The result in the singular case can be derived with only slightly more manipulation.

In the application of our method when $s = 1$, we shall use the following notation: Suppose we are interested only in estimating θ_j, where now j need not be 1. In order to investigate the D-optimality of a ξ^* for which $M(\xi^*)$ is nonsingular, we find a column vector c of k components which is orthogonal to all columns of $M(\xi^*)$ other than the jth, and for which $c'M(\xi^*)c = 1$. Writing $\delta_j(x, \xi^*) = c'f(x)$, we then have that ξ^* is D-optimum for θ_j if and only if $\max_x |\delta_j(x, \xi^*)| = 1$.

In [11] a function space corollary of our results was stated in the case $s = k$: *There exists a* ξ^* *and a nonsingular* $k \times k$ *matrix* C *such that the functions* $h_i(x)$, *where* $h(x) = Cf(x)$, *are orthonormal with respect to* ξ^* *and* $\max_x \sum_1^k h_i^2(x) = k$. Of course, $M(\xi^*)$ is nonsingular for a D-optimum ξ^* when $s = k$, so the result there is not complicated by the possibility of singularity. One obtains a close analogue of the above result in the case $s < k$ in the regular case; in fact, given $f_1, \cdots, f_M$, the result we shall state is stronger in the case $s = N < k = M$ than is the result stated above in the case $s = k = N$. The analogue is:

Corollary to Theorem 1. *Given* $\mathfrak{X}$ *and* $f = \{f_i, 1 \leqq i \leqq k\}$ *in the regular case for* $f^{(1)}$ *(the first* s *components of* f*) relative to* f*, there exists a probability measure* ξ^* *on* $\mathfrak{X}$ *and an* $s \times k$ *matrix* C *of rank* s *such that the functions* $h_i(x)$*, where* $h(x) = Cf(x)$*, are orthonormal* (ξ^*)*, are orthogonal* (ξ^*) *to the* f_j *with* $j > s$*, and satisfy* $\max_x \sum_1^s h_i^2(x) = s$.

We now turn to the case where $M^*(\xi^*)$ is nonsingular but $M(\xi^*)$ is singular (of course, the discussion which follows reduces to the preceding discussion if $M(\xi^*)$ is nonsingular). Let $g_i(x) = \sum_j \beta_{ij} f_j(x)$, $1 \leqq i \leqq s$, be linearly independent (ξ^*) and orthogonal (ξ^*) to all f_j, $j > s$. Let $g_i(x) = \sum_{j>s} \beta_{ij} f_j(x)$, $s < i \leqq s + r$, be a maximal set of linearly independent (ξ^*) functions of this form. Of course, each g_i is orthogonal (ξ^*) to each g_j, $1 \leqq i \leqq s < j \leqq s + r$. As in the development of (3.4), we can and do choose the g_i, $1 \leqq i \leqq s + r$, to be mutually orthogonal (ξ^*); the reader will have no difficulty in supplying the modifications needed in what follows if these functions are not so chosen.

Finally, let β_{ij}, $i > s+r$, $j > s$, be chosen so that the matrix $\{\beta_{ij}, 1 \leq i, j \leq k\}$ is nonsingular (where $\beta_{ij} = 0$ for $1 \leq j \leq s < i \leq k$), and write $q_i(x) = \sum_{j>s} \beta_{ij} f_j(x)$, $s + r < i \leq k$. Then the q_i are all zero on the support of ξ^*.

As in (3.4), we have

$$\bar{D}(x, \xi^*) = \sum_{i=1}^{s} g_i^2(x)/(g_i, g_i)_{\xi^*}, \tag{3.13}$$

and we have linearity in obtaining $\bar{D}(\xi, \xi^*)$ from $\bar{D}(x, \xi^*)$:

$$\bar{D}(\xi, \xi^*) = \int \bar{D}(x, \xi^*)\xi(dx) = \sum_{i=1}^{s} (g_i, g_i)_\xi/(g_i, g_i)_{\xi^*}. \tag{3.14}$$

We must still exhibit $\rho(x, \xi^*)$ and $\rho(\xi, \xi^*)$. Let $e_i(x) = \sum_{j>s+r} \gamma_{ij} q_j(x)$ be a maximal set of linearly independent (ξ) functions of this form, $1 \leq i \leq t$, where again for simplicity we choose the e_i to be orthogonal (ξ). Then it is easy to see that

$$\rho(\xi, \xi^*) = \sum_{\substack{j \leq s \\ m \leq t}} (g_j, e_m)_\xi^2/[(g_j, g_j)_{\xi^*}(e_m, e_m)_\xi]. \tag{3.15}$$

The functions $\bar{D}$ and ρ depend on the β_{ij}, but D does not. If ξ gives measure one to the point x, we obtain $\rho(x, \xi^*)$, and this takes on a particularly simple form since t must then be 0 or 1. If $t = 0$, we have $\rho(x, \xi^*) = 0$. If $t = 1$, consider the special diagonalization of $M(\xi)$ (now of rank 1) of Section 2, wherein J_1, J_3, and J_5 are diagonal and $J_6 = 0$. J_5 is a positive scalar, which we can choose to be unity. Since $J_6 = 0$, we must have $J_3 = 0$, since otherwise $J(\xi)$ would have rank >1. Similarly, at most one element, say the first, of J_1 can be other than zero, and if this element is h^2 the first element of J_4 is $\pm h$ (possibly $h = 0$), and all other elements of $J(\xi)$ except for J_5 are zero. A trivial computation yields $\bar{D}(x, \xi) = h^2$ and $D(x, \xi) = 0$ in this case (compare the example of Section 2). Thus, we see how easy it is for ξ^* to be D-optimum without (2.17)(d) being satisfied. More important, from our two results in the cases $t = 0$ and $t = 1$ we have the following sharpening of a part of the results of Section 2, which obviously shortens the computation needed to use (2.17) (b) to eliminate nonoptimum designs:

ADDITION TO THEOREM 2: *Let* $Z(\xi^*) = \{x : q_i(x) = 0, i > s + r\}$. *Then*

$$\max_x D(x, \xi^*) = \max_{x \in Z(\xi^*)} D(x, \xi^*) = \max_{x \in Z(\xi^*)} \bar{D}(x, \xi^*). \tag{3.16}$$

Equation (3.15) makes clear the lack of linearity in ξ of $\rho(\xi, \xi^*)$, which causes the complications in the singular case.

The reader will not find it difficult to write down analogues in the singular case of (3.9), (3.12), the Corollary to Theorem 1, etc.

4. Some examples.

Example 4.1. *Quadratic regression on a q-cube.* Let $\mathfrak{X}$ be the q-dimensional cube consisting of all points $x = (x_1, \cdots x_q)$ for which $-1 \leq x_i \leq 1$, $1 \leq i \leq q$. The problem of linear regression on $\mathfrak{X}$ is trivial: a D-optimum ξ is that measure which assigns measure 2^{-q} to each corner of the cube (at least for $q > 3$, there are other optimum designs,[3] since the bound H of Section 1 here becomes $H = q(q+3)/2 + 1 < 2^q$). We therefore turn to the problem of quadratic regression. The unique D-optimum ξ is well known when $q = 1$ to put equal weights on the points $x_1 = -1, 0, 1$. In what follows we restrict our attention to the case $q \geqq 2$.

It will be convenient, for the purpose of partitioning $M(\xi)$, to write the f_i in the following order: $f_1(x) = 1; f_{1+j}(x) = x_j^2, 1 \leqq j \leqq q; f_{q+1+j}(x) = x_j, 1 \leqq j \leqq q$; $f_i(x)$ for $2q + 2 \leqq i \leqq (q+1)(q+2)/2$ are the functions $x_p x_r$, $p < r$, in any order. Thus, $k = (q+1)(q+2)/2$, and it is easy to compute that $H = [(q+1)(q+2)(q+3)(q+4)]/24$. We shall seek an optimum ξ with support on $r = 2^{q-3}[8 + 4q + q(q-1)]$ points, of the following form: ξ assigns positive measure α to each of the 2^q corners of the cube, positive measure β to the midpoint of each of the $q2^{q-1}$ edges, and positive measure γ to the center of each of the $q(q-1)2^{q-3}$ two-dimensional (square) faces. We shall obtain such a design, and will verify its optimality, for $q = 2, 3, 4, 5$. We note that $r < H$ when $q = 2$ or 3, but that $r > H$ when $q > 3$, so that other optimum ξ exist in at least these latter cases.[4]

Although the set of points supporting the optimum ξ just described is of the same form for $q = 2, 3, 4, 5$ (the design for the case $q = 1$ is also of this form), the ratios among α, β and γ change with q. It is interesting to contrast this with the optimum ξ mentioned above for linear regression on a q-cube, or those optimum ξ's of the example of Section 6 for linear or quadratic regression on a q-simplex, where equal weights suffice in all cases. In fact, in the present example, a ξ with support of the form we are considering can no longer be optimum when $q \geqq 6$, as we shall discuss below.

For ξ of the above form, write

$$\begin{aligned} u &= \int x_1^2 \xi\,(dx) = 2^{q-3}[8\alpha + 4(q-1)\beta + (q-1)(q-2)\gamma], \\ v &= \int x_1^2 x_2^2\, \xi\,(dx) = 2^{q-3}[8\alpha + 4(q-2)\beta + (q-2)(q-3)\gamma]. \end{aligned} \tag{4.1}$$

then

$$M(\xi) = \begin{Vmatrix} 1 & F & 0 & 0 \\ F' & G & 0 & 0 \\ 0 & 0 & uI_q & 0 \\ 0 & 0 & 0 & vI_{q(q-1)/2} \end{Vmatrix}, \tag{4.2}$$

where I_q is the $q \times q$ identity, F is a row-vector of q u's, G is a $q \times q$ matrix with

[3] In fact, for $q \geqq 3$, an optimum design assigning measure $1/h$ to each of h points of a proper subset of the 2^q corners can be obtained from an Hadamard matrix or orthogonal array of strength 2 which describes a design for the corresponding factorial problem with q factors at 2 levels. Here h can be taken to be $\leqq 2q$ (an easily improvable bound), so that we see again how poor the bound H can be. These results on linear regression are much simpler than the corresponding results on quadratic regression which are mentioned in footnote 5.

[4] See footnote 5 in this connection.

diagonal elements u and off-diagonal elements v, and the symbol 0 denotes any matrix of zeros. From this we obtain easily

$$M(\xi)^{-1} = \begin{Vmatrix} a & B & 0 & 0 \\ B' & C & 0 & 0 \\ 0 & 0 & u^{-1}I_q & 0 \\ 0 & 0 & 0 & v^{-1}I_{q(q-1)/2} \end{Vmatrix} \tag{4.3}$$

where $a = [(q-1)v + u]/[(q-1)v + u - qu^2]$, each of the q elements of B is $b = -u/[(q-1)v + u - qu^2]$, and C has diagonal elements

$$c = [(q-2)v + u - (q-1)u^2]/(u-v)[(q-1)v + u - qu^2]$$

and off-diagonal elements $d = [u^2 - v]/(u-v)[(q-1)v + u - qu^2]$. Also, from (4.2), we have

$$\det M(\xi) = u^q v^{q(q-1)/2}(u-v)^{q-1}[u + (q-1)v - qu^2]. \tag{4.4}$$

Since the problem at hand is illustrative of many similar examples, we now indicate two methods for "guessing" values α, β, γ for which one can verify that $\max_x d(x, \xi) = (q+1)(q+2)/2$. Firstly, as mentioned in the introduction, we can try to maximize $\det M(\xi)$ among ξ of this form, by solving the equations $\partial \log \det M(\xi)/\partial u = \partial \log \det M(\xi)/\partial v = 0$ in the region where α, β, and γ are all positive. Secondly, we can used (4.3) to write out $d(x, \xi)$, say $d(x, \xi) = P + Q\sum_i x_i^2 + R\sum_i x_i^4 + S\sum_{i \neq j} x_i^2 x_j^2$, where P, Q, R, S are functions of u and v, and then try to determine u and v so as to make $d(x, \xi)$ have some simple form for which it is obvious that $\max_x d(x, \xi) = (q+1)(q+2)/2$; for example, we can try to find u and v such that $P = (q+1)(q+2)/2$, $R = -Q \geq 0$, $S = 0$. Either of these approaches leads to the same formal solution in the present cases, neglecting for the moment the question of positivity of α, β, γ:

$$\begin{aligned} u &= \frac{(q+3)}{4(q+1)(q+2)^2}\{(2q^2 + 3q + 7) + (q-1)[4q^2 + 12q + 17]^{\frac{1}{2}}\}, \\ v &= \frac{(q+3)}{8(q+2)^3(q+1)} \\ &\quad \cdot \{(4q^3 + 8q^2 + 11q - 5) + (2q^2 + q + 3)[4q^2 + 12q + 17]^{\frac{1}{2}}\}. \end{aligned} \tag{4.5}$$

For this choice of u and v we obtain after some reduction

$$d(x, \xi) = (q+1)(q+2)/2 - c\sum_i (x_i^2 - x_i^4), \tag{4.6}$$

whose maximum over $\mathfrak{X}$ is clearly the desired value $(q+1)(q+2)/2 = k$, since c, defined just below (4.3), is easily seen to be positive. The corresponding values of α, β, and γ which are obtained from the equations (4.1) and the equation $2^{q-3}[8\alpha + 4q\beta + q(q-1)\gamma] = 1$, are

$$\alpha = 2^{-q-1}[(q-1)(q-2) - 2q(q-2)u + q(q-1)v],$$

$$\beta = 2^{-q+1}[(2q-3)u - (q-1)v - (q-2)],$$
$$\gamma = 2^{2-q}[1 + v - 2u];$$

more explicitly,

$$\begin{aligned}\alpha = [2^{q+4}(q+2)^3(q+1)]^{-1}\{(4q^6 + 12q^5 \\ - 25q^4 - 107q^3 + 85q^2 + 479q + 128) \\ - (2q^2 - q - 19)q\,(q-1)(q+3)[4q^2 + 12q + 17]^{\frac{1}{2}}\},\end{aligned}$$

$$\begin{aligned}\beta = [2^{q+2}(q+2)^3(q+1)]^{-1}\{-(4q^5 + 16q^4 - 11q^3 - 143q^2 \\ -149q + 139) + (q+3)(q-1)(2q^2 + q - 15)[4q^2 + 12q + 17]^{\frac{1}{2}}\}\end{aligned} \tag{4.7}$$

$$\begin{aligned}\gamma = [2^{q+1}(q+2)^3(q+1)]^{-1}\{(4q^4 + 24q^3 + 43q^2 - 24q - 119 \\ -(q+3)(2q^2 + 3q - 11)[4q^2 + 12q + 17]^{\frac{1}{2}}\}.\end{aligned}$$

Thus, (4.7) provides an optimum ξ, provided that the α, β, γ given here are all nonnegative. This is the case for $q \leqq 5$, and the following is a table of numerical values:

q	α	β	γ
1	.250	.500	.000
2	.1458	.08015	.0962
3	.071975	.01895	.03280
4	.03705	.0038375	.01185
5	.01928	.0003125	.004475

For comparison, we note that, when $q = 2$, the ξ which assigns measure $\frac{1}{9}$ to each of the nine points supporting the optimum ξ, yields a value of det $M(\xi)$ which is about 15 per cent lower and a value of $\max_x d(x, \xi)$ which is about 21 per cent higher, than does the optimum design. For larger q, the comparison is even more striking.

To see what happens to the above solution when $q > 5$, it will suffice to consider the case $q = 6$. Equation (4.7) no longer gives a solution, since $\beta < 0$ (i.e., the solution can no longer be obtained by solving $\partial \log \det M(\xi)/\partial u = \partial \log \det M(\xi)/\partial v = 0$). This suggests that we look for a D-optimum ξ of the form we have been considering, but with $\beta = 0$. If, in fact, we investigate the behavior of the expression (4.4) on the region $\{\alpha \geqq 0, \beta \geqq 0\ \gamma \geqq 0\} = \{u \leqq (v+1)/2, u \leqq (10 + 15v)/24, u \geqq (4 + 5v)/9\}$, we find that the maximum is attained at $(u, v) = ([5v' + 4]/9, v')$, where $v = v'$ is the solution between .7 and .8 of the equation $350v^3 - 190v^2 - 139v + 60 = 0$ (this last equation is obtained by solving $\partial \log \det M(\xi)/\partial v = 0$ on the line $9u = 4 + 5v$, and it is not hard to prove that this gives the desired solution). For the corresponding ξ (for which $\beta = 0$) we obtain, at $x = 0$, $d(0, \xi) = 3(25v' + 2)/5(1 - v')(5v' - 2) > 28 = k$. Hence, we have proved that *the best ξ of the form we have considered* (i.e., *over all choices of* α, β, γ) *is not D-optimum when* $q = 6$. The corresponding result also holds when $m > 6$, and a D-optimum design for the case $q \geq 6$ is still unknown.[5]

Example 4.2. *The case of polynomial regression on a real interval when* $1 < s < k$. The problem of polynomial regression on a real interval was solved by Guest [5] and Hoel [6] in the case $s = k$ and by Kiefer and Wolfowitz [10, Section 3] in the case $s = 1$. The other cases are more difficult to handle. A trivial example (quadratic regression, $k = 3$, $s = 2$) was treated in the previous section and in [10], and we now illustrate the more complicated problems which can arise by considering two computationally more difficult examples for the case $s = 2 < k$. In both examples it is obvious from the outset that we are in the regular case of Section 3.

First consider the problem of estimating the quadratic and cubic regression coefficients in the case of cubic regression; i.e. $s = 2$, $k = 4$, $\mathfrak{X} = [-1, 1]$, and $f_i(x) = x^{4-i}$, $i = 1, 2, 3, 4$; we want a D optimum design for estimating θ_1 and θ_2 (the coefficients of x^3 and x^2), and the comments of Section 1 suggest that we seek one of the form $\xi(a) = \xi(-a) = \alpha/2$, $\xi(1) = \xi(-1) = (1 - \alpha)/2$, where $0 < a < 1$. We easily compute that the $g_i(x)$ of Section 3 can be taken to be $x^3 - cx$ and $x^2 - b$, where $c = (1 - \alpha + \alpha a^4)/(1 - \alpha + \alpha a^2)$ and $b = 1 - \alpha + \alpha a^2$. Writing $x^2 = u$ and $a^2 = A$, we obtain

$$(4.8)\quad d(x, \xi) = \frac{(u - b)^2}{\alpha(1 - \alpha)(1 - A)^2} + \frac{(1 - \alpha + \alpha A)}{(1 - \alpha)\alpha A(1 - A)^2} u(u - c)^2.$$

If ξ is D-optimum, we must have $d(1, \xi) = 2$; i.e.,

$$(4.9)\quad 2z^2 + (A - 1)z - 2A = 0,$$

where we have written $z = (1 - \alpha)/\alpha$. If $d(1, \xi) = 2$, we must also have $d(a, \xi) = 2$, and since the expression (4.8) is a cubic in u we will clearly have $d(x, \xi) \leq 2$ for all x if

$$(4.10)\quad \begin{aligned} \partial d(x, \xi)/\partial u \,|_{u=A} &= 0, \\ \partial^2 d(x, \xi)/\partial u^2 \,|_{u=A} &< 0. \end{aligned}$$

The first half of (4.10) yields $z = 3A^2/(1 - 4A)$; substituting this into (4.9), we obtain, finally,

$$(4.11)\quad \begin{aligned} a &= A^{\frac{1}{2}} = [(11 - 73^{\frac{1}{2}})/12]^{\frac{1}{2}}, \\ \alpha &= (z + 1)^{-1} = (73^{\frac{1}{2}} - 5)/6, \end{aligned}$$

---

[5] Recently Dr. R. H. Farrell and the author have obtained optimum designs for all values of q. For $q > 5$, the support of such designs must contain points of the 3^q array which are midpoints of faces of dimension >2. The invariant designs of this form (which are not unique for $q > 2$) can always be obtained by choosing weights analogous to α, β, and γ above in such a way as to make the moments defined by the lefthand equations of (4.1) equal to the quantities defined by (4.5). The designs obtained in this way will be supported by more than H points if $q > 5$. Designs on fewer than H points (in fact, on $O(q^3)$ points) can be obtained by combining certain orthogonal arrays of strength 4. Results of the type discussed here and in footnote 3 will appear elsewhere.

and it is easy to check that (4.9) and (4.10) are satisfied by these values. Thus, (4.11) gives a D-optimum design.

Next, suppose with the same cubic setup that we only have $k = 3$ (i.e., the constant term is missing). Surprisingly, the arithmetic is now more complicated. One obtains $z^3 - (1 - 2A^2)z^2 - (2A - A^3)z - A^3 = 0$ in place of (4.9) and $(3A - 1)z^2 + (5A^3 - A^2)z + (4A^4 - 2A^3) = 0$ for the first half of (4.10), and more effort is required to solve these than in the previous case. We obtain, finally, that a D-optimum design ξ of the same structure as above is now given by

$$a = [(5 \cdot 33^{1/2} - 21)/24]^{1/2}, \tag{4.12}$$
$$\alpha = (3 + 33^{1/2})/20.$$

PART II. SIMPLEX EXPERIMENTS

5. Preliminaries. Scheffé [12] has given an interesting account of experiments in which $\mathfrak{X}$ is the *q-simplex* S_q consisting of all $(q + 1)$-vectors $(x_1, x_2, \cdots, x_{q+1})$ for which all x_i are nonnegative and $\sum_i x_i = 1$. (Scheffé uses $q - 1$ to denote the dimensionality of the simplex, but we shall find the present notation more convenient, and will adhere to it throughout.) The reader is referred to the fundamental paper [12] for discussions of the construction and use of such experiments, including modifications in the case where $\mathfrak{X}$ is only a part of S_q. We shall be concerned here with optimum properties which are possessed (or not possessed) by certain of Scheffé's designs, namely, those designs in which ξ gives measure 1 to the *(q, m)-lattice* $S_{q,m}$ consisting of those $\binom{m+q}{m}$ points of S_q all of whose coordinates are integral multiples of $1/m$, and, in particular, the design $\xi_{q,m}$ whch assigns equal measure to each of these points.

In the footnote on page 353 of [12], Scheffé mentions the desirability of investigating the optimality of his designs (in the case $s = k$ of our Section 1) in precisely the sense discussed in Part I of the present paper, i.e., in the sense of minimizing $\max_x d(x, \xi)$. We shall investigate the optimality of $\xi_{q,m}$ or certain simple modifications of it in various cases of polynomial regression on $\mathfrak{X}$. Thus, in the case where all polynomials on S_q of degree m or less are possible regression functions, the set $\{f_i\}$ of Section 1 can be chosen in various ways (see [12]) as a set of $\binom{m+q}{m}$ linearly independent polynomials of degree $\leq m$. We shall also discuss certain other cases considered by Scheffé, in which only a proper subset of the polynomials of degree m are possible.

Before proceeding to these investigations, it is necessary to verify a conjecture of Scheffé regarding designs on $S_{q,m}$:

Orthogonal polynomials and identifiability for designs on a (q, m) lattice. Scheffé makes a conjecture on page 346 of [12] which is equivalent to the statement that, for mth degree regression, any design which gives positive measure to all $\binom{m+q}{m}$

points of the (q, m) lattice $S_{q,m}$ enables all $\binom{m+q}{m}$ regression coefficients to be estimated. (He verifies this for $m = 1, 2, 3$.) We now verify this conjecture by proving the existence of a system of $\binom{m+q}{m}$ polynomials of degree $\leq m$ such that, for any point of the lattice, there is a polynomial in the system which is not zero at that point, but which vanishes at all other points of the lattice. (This system is thus orthogonal for any design whose support is the (q, m) lattice.) Since there are exactly as many points of the lattice as there are regression coefficients, this will imply the validity of Scheffé's conjecture.[6]

Fix q. Such a system of polynomials obviously exists when $m = 1$. Suppose such a system exists when $m = M - 1$, where $M > 1$. Let p be a point of $S_{q,M}$. Since $M > 1$, there is a bounding hyperplane L of the simplex S_q on which $S_{q,M}$ is the lattice, such that $p \not\in L$. Since $T = S_{q,M} - L$ is essentially a $(q, M - 1)$ lattice, there is a polynomial Φ of degree at most $M - 1$ which vanishes everywhere on T except at p. But then, if f is a linear function which vanishes on L but not on T, the function Φf is a polynomial of degree at most M which vanishes everywhere on $S_{q,M}$ except at p. This completes the proof.

6. Quadratic regression on the q-simplex.

A D-optimum design for all coefficients in quadratic regression on the q-simplex. We shall now show that, when $\mathfrak{X}$ is the q-simplex S_q and F consists of all polynomials of degree ≤ 2, the design $\xi_{2,q}$ which assigns measure $2/(q+1)(q+2)$ to each of the points of the $(q, 2)$ lattice $S_{q,2}$ on $\mathfrak{X}$, is D-optimum. To this end, we compute $d(x, \xi_{2,q})$. This can be done directly by computing $M(\xi_{2,q})^{-1}$ and thus $f(x)'M(\xi_{2,q})^{-1}f(x)$ for the usual choice $\{f_i(x)\} = \{x_r, 1 \leq r \leq q+1$ and $x_r x_s, 1 \leq r < s \leq q+1\}$, but a somewhat quicker method is to note that a system of $(q+1)(q+2)/2$ quadratic orthonormal polynomials with respect to $\xi_{2,q}$, each of which vanishes except at one point of the lattice, consists of the functions $[2(q+1)(q+2)]^{\frac{1}{2}} x_i(x_i - \frac{1}{2})$, $1 \leq i \leq q+1$, and the functions $[8(q+1)(q+2)]^{\frac{1}{2}} x_i x_j$, $1 \leq i < j \leq q+1$. Hence, $d(x, \xi_{2,q})$ is just the sum of squares of these functions (see Section 3), and we obtain, denoting by $\sum'$ the summation over all j not equal to i (for fixed i),

$$\begin{aligned}\frac{2}{(q+1)(q+2)} d(x, \xi_{2,q}) &= 4 \sum_i x_i^2 (x_i - 1/2)^2 + 16 \sum_{i<j} x_i^2 x_j^2 \\ &= \sum_i (1 - \textstyle\sum' x_j)(4x_i^3 - 4x_i^2 + x_i) + 8 \sum_{i \neq j} x_i^2 x_j^2\end{aligned}$$

---

[6] It will be seen that it is unnecessary to exhibit these polynomials explicitly in carrying out the inductive proof which follows, although that induction can be used to obtain them explicitly. Professor Scheffé has informed the author that Professor L. J. Savage had independently constructed and communicated to him the formula for a polynomial of degree m on S_q which vanishes at all points of the (q, m) lattice except for the point $(z_1, z_2, \cdots, z_{q+1})$, where it is unity. Savage's expression is

$$\prod_{i=1}^{q+1} \left\{ [(mz_i)!]^{-1} \prod_{j=0}^{mz_i - 1} (mx_i - j) \right\}.$$

$$= 1 - \sum_i 4x_i^2(1 - x_i) - \sum_i (2x_i - 1)^2 x_i {\sum}' x_j + 8\sum_{i \neq j} x_i^2 x_j^2$$

$$= 1 - \sum_{i \neq j} x_i x_j \{4x_i + (2x_i - 1)^2\} + 8\sum_{i \neq j} x_i^2 x_j^2$$

$$= 1 - \sum_{i \neq j} x_i x_j \{2x_i + 2x_j + (2x_i - 1)^2/2 + (2x_j - 1)^2/2\} + 8\sum_{i \neq j} x_i^2 x_j^2$$

$$= 1 - \sum_{i \neq j} x_i x_j \{2(x_i - x_j)^2 + (1 - 4x_i x_j)\}.$$

The last expression in braces is always nonnegative. Hence, $d(x, \xi_{2,q}) \leqq (q + 1)(q + 2)/2$ for all x, and $\xi_{2,q}$ is indeed optimum.

It is striking to note how much simpler the treatment of the present example is, than is that of quadratic regression on the q-cube in Section 4. Unfortunately, the cases where $m \geqq 3$ are not so simple.

An optimum design for estimating only the coefficients of the quadratic terms of a quadratic on S_2. This example will illustrate the use of our theory when $1 < s < k$, and contains a good example of the type of geometric argument which is often useful. Write the f_i's in order as x_2x_3, x_1x_3, x_1x_2, x_1, x_2, x_3. We seek a design which minimizes the generalized variance of the three b.l.e.'s of coefficients of f_1, f_2, f_3. It is to be noted that any D-optimum design for this problem is also D-optimum for the problem where f_1, f_2, f_3 are replaced by x_1^2, x_2^2, x_3^2, since the transformation which takes one problem into the other is of the form (1.6).

We shall search for an optimum design among those designs $\xi^{(\alpha)}$ which, for some α, assign measure $\alpha/3$ to each vertex of S_2 and measure $(1 - \alpha)/3$ to the midpoint of each edge of S_2. Denoting by $[a, b]$ a 3×3 matrix with diagonal elements a and off-diagonal elements b, we obtain for such a design $\xi^{(\alpha)}$,

$$3M(\xi^{(\alpha)}) = \begin{Vmatrix} [\alpha/16, 0] & [0, \alpha/8] \\ [0, \alpha/8] & [1 - \alpha/2, \alpha/4] \end{Vmatrix}$$

and thus

$$\frac{1}{3}M^{-1}(\xi^{(\alpha)}) - \frac{1}{3}M_2^{-1}(\xi^{(\alpha)}) = \begin{Vmatrix} \left[\frac{8(2-\alpha)}{\alpha(1-\alpha)}, \frac{4}{1-\alpha}\right] & \left[0, \frac{-2}{1-\alpha}\right] \\ \left[0, \frac{-2}{1-\alpha}\right] & \left[\frac{2\alpha - \alpha^2}{(4-3\alpha)(1-\alpha)}, \frac{\alpha}{4-3\alpha}\right] \end{Vmatrix}$$

and (using the fact that $\sum x_i = 1$)

$$\frac{1}{3}d(x, \xi^{(\alpha)}) = \frac{(2-\alpha)\alpha}{(4-3\alpha)(1-\alpha)} \sum_i x_i^2 + \frac{2\alpha}{4-3\alpha}\sum_{i<j} x_i x_j$$

$$- \frac{4}{1-\alpha}\sum_{i \neq j} x_i^2 x_j + \frac{8(2-\alpha)}{\alpha(1-\alpha)}\sum_{i<j} x_i^2 x_j^2 + \frac{8}{1-\alpha} x_1 x_2 x_3.$$

Of course, a necessary condition for optimality is that $d(x, \xi^{(\alpha)}) = 3$ on a set of unit $\xi^{(\alpha)}$-measure. It is only necessary to check this condition at the point

(1, 0, 0), since it then follows for other relevant points from symmetry and the fact that the integral of $d(x, \xi^{(\alpha)})$ with respect to $\xi^{(\alpha)}$ is automatically 3. We obtain $\alpha = \bar{\alpha}$, where

$$\bar{\alpha} = \frac{9 - 17^{\frac{1}{2}}}{8} = .6530.$$

In order to prove that $\xi^{(\bar{\alpha})}$ is optimum, we must show that $d(x, \xi^{(\bar{\alpha})}) \leqq 3$ on S_2. First we note that if we consider the function $d(x, \xi^{(\bar{\alpha})})$ not merely on S_2 but on the whole plane $P = \{\sum x_i = 1\}$, it is obviously a quartic which is nonnegative (see (3.4)) and which, on the line $x_3 = 0$, is symmetric about $(\frac{1}{2}, \frac{1}{2}, 0)$ and equal to 3 on this line at $x_1 = 0, \frac{1}{2}$, and 1. We conclude without any computation that $d(x, \xi^{(\bar{\alpha})}) \leqq 3$ on that part of the line $x_3 = 0$ which is part of S_2, and thus on the whole boundary of S_2.

Next, we compute easily that $d(x', \xi^{(\bar{\alpha})}) < 3$, where $x' = (\frac{1}{3}, \frac{1}{3}, \frac{1}{3})$. Furthermore, it is not hard to compute that x' is a local strict maximum of $d(x, \xi^{(\alpha)})$. From this and the fact that d is positive quartic on the plane P which is $\leqq 3$ on the boundary of S_2, we conclude easily that $d(x, \xi^{(\bar{\alpha})}) \leqq 3$ on that part of any line of P through x' which is contained in S_2. Hence, $d(x, \xi^{(\bar{\alpha})}) \leqq 3$ throughout S_2, and thus $\xi^{(\bar{\alpha})}$ is D-optimum.

7. Cubic and higher regression on the q-simplex. The cases where $m \geqq 3$ are computationally much more difficult. We already know from the results of Guest [5] and Hoel [6] that, even in the case $q = 1$, any design on the (q, m) lattice (regardless of whether or not ξ assigns equal measure to the points) is *not* D-optimum when $m \geqq 3$.

For the sake of brevity, we will limit our discussion to the case $q = 2$, $m = 3$. We shall briefly discuss three different models. In the general cubic case (in Scheffé's terminology) we can take the f_i to be the ten functions x_i, $x_i x_j$, $x_i x_j(x_i - x_j)$, and $x_1 x_2 x_3$ (here $1 \leqq i < j \leqq 3$). Scheffé's "special cubic" omits the functions $x_i x_j(x_i - x_j)$. In the "cubic without 3-way effect" we shall consider the nine functions other than $x_1 x_2 x_3$; it is clear in what sense the meaning of this name is to be taken, and the physical significance of each of the three models is clear (see also [12]).

In the case of the cubic without 3-way effect, for $0 < b < \frac{1}{2}$ consider the design ξ_b which puts measure $\frac{1}{9}$ on each of the three points $x_i = 1$, $x_j = x_k = 0$ and each of the six points $x_i = 1 - x_j = b$, $x_k = 0$. It is not too difficult to compute that

$$\det M(\xi_b) = \text{const}\ V^{12}(1 - 4V)^3,$$

where $V = b(1 - b)$. Hence, $V = \frac{1}{5}$, or $b = (1 - 5^{-\frac{1}{2}})/2$, gives the optimum design *among designs of this structure*. It is interesting to note that *this value of b also gives the D-optimum design in the case $q = 1$, $m = 3$*, with equal weights at each of the points $x_1 = 0, b, 1 - b$, and 1.

For the general cubic, if we consider designs which assign measure $\frac{1}{10}$ to each of the nine points supporting ξ_b in the previous paragraph, and also to the point

$x_1 = x_2 = x_3 = \frac{1}{3}$, the best choice of b changes to $(1 - 3^{-\frac{1}{2}})/2$. In fact, it is far from clear that we should expect the D-optimum ξ to be of this form or to be supported by only 10 points; the situation appears to be more complex than that of quadratic regression on a square (discussed in Section 4).

A D-optimum design for the special cubic on S_2. We turn now to the case of the special cubic, where we shall show that Scheffé's design $\bar{\xi}$ which assigns measure $\frac{1}{7}$ to each of the six points of the (2, 2) lattice $S_{2,2}$ and also to the point $x_1 = x_2 = x_3 = \frac{1}{3}$, is indeed D-optimum. We cannot, in imitation of our development in Section 6, take $d(x, \bar{\xi})$ to be the sum of squares of the seven orthonormal cubic functions each of which vanishes on all but one of these seven points; for these cubics will not all be linear combinations of *only* the seven functions we began with. Rather than to compute appropriate orthogonal functions, we shall in this example compute M^{-1} directly. Writing the seven functions in the order $x_1, x_2, x_3, x_2x_3, x_3x_1, x_1x_2, x_1x_2x_3$, and denoting by $[a, b]$ a 3×3 matrix with diagonal elements a and off-diagonal elements b, and by $[c]$ a 3×1 matrix of elements c, we obtain

$$7M(\bar{\xi}) = \begin{Vmatrix} \left[\frac{29}{18}, \frac{13}{36}\right] & \left[\frac{1}{27}, \frac{35}{216}\right] & \left[\frac{1}{81}\right] \\ \left[\frac{1}{27}, \frac{35}{216}\right] & \left[\frac{97}{1296}, \frac{1}{81}\right] & \left[\frac{1}{243}\right] \\ \left[\frac{1}{81}\right]' & \left[\frac{1}{243}\right]' & \left[\frac{1}{729}\right] \end{Vmatrix}$$

and thus

$$\frac{1}{7} M^{-1}(\bar{\xi}) = \begin{Vmatrix} [1, 0] & [0, -2] & [3] \\ [0, -2] & [24, 4] & [-60] \\ [3]' & [-60]' & 1188 \end{Vmatrix}.$$

Hence, we obtain

$$\frac{1}{7}d(x, \bar{\xi}) = \sum_i x_i^2 - 4\sum_{i \neq j} x_i^2 x_j + 24\sum_{i<j} x_i^2 x_j^2 + 8 \sum_i x_i x_1 x_2 x_3 + 6x_1x_2x_3 - 120x_1x_2x_3\sum_{i<j} x_i x_j + 1188x_1^2x_2^2x_3^2. \tag{7.1}$$

The fourth term on the right is of course just $8x_1x_2x_3$. The first term on the right can be written as

$$\sum_i x_i^2 = 1 - 2\sum_{i<j} x_i x_j = 1 - 2 \sum_{\substack{i<j \\ i \neq k \neq j}} x_i x_j(x_i + x_j + x_k) = 1 - 2\sum_{i \neq j} x_i^2 x_j - 6x_1x_2x_3. \tag{7.2}$$

We substitute this last expression in (7.1) and, in the resulting form, substitute for the expression $-6\sum_{i \neq j} x_i^2 x_j$ the last of the following expressions:

$$
\begin{aligned}
-6\sum_{i\neq j} x_i^2 x_j &= -6\sum_{i\neq j} x_i^3 x_j - 6\sum_{i\neq j} x_i^2 x_j(1 - x_i) \\
&= -6\sum_{i\neq j} x_i^3 x_j - 6 \sum_{\substack{i\neq j \\ i\neq k\neq j}} x_i^2 x_j(x_j + x_k) \qquad (7.3)\\
&= -6\sum_{i\neq j} x_i^3 x_j - 12\sum_{i<j} x_i^2 x_j^2 - 12x_1x_2x_3 .
\end{aligned}
$$

We obtain, finally,

$$
\begin{aligned}
\tfrac{1}{7}d(x,\bar\xi) &= 1 - 6\sum_{i\neq j} x_i^3 x_j + 12\sum_{i<j} x_i^2 x_j^2 - 4x_1x_2x_3 \\
&\qquad - 120x_1x_2x_3\sum_{i<j} x_i x_j + 1188x_1^2x_2^2x_3^2 \qquad (7.4)\\
&= 1 - \{6\sum_{i<j} x_i x_j(x_i - x_j)^2 + 4x_1x_2x_3(1 - 27x_1x_2x_3) \\
&\qquad + 120x_1x_2x_3(\sum_{i<j} x_i x_j - 9x_1x_2x_3)\}.
\end{aligned}
$$

Each of the three terms inside the curly braces is easily seen to be nonnegative on the simplex. Hence, $d(x, \bar\xi) \leqq 7$ for all x in the simplex, and thus $\bar\xi$ is indeed D-optimum.

An optimum design for estimating only the coefficient of the cubic term of a special cubic on S_2. Scheffé showed that, among the designs which assign measure one to the set of seven points which supports the $\bar\xi$ of the previous example, the one which minimizes the variance of the b.l.e. of the coefficient of $x_1x_2x_3$ is the measure ξ' which assigns measure $\frac{1}{24}$ to each vertex of S_2, $\frac{4}{24}$ to the midpoint of each side of S_2, and $\frac{9}{24}$ to the centroid of S_2. We now show that, in fact, ξ' is optimum among *all* designs.

The proof is quite simple. Using the notation of the previous example, we obtain

$$
24M(\xi') = \left\| \begin{matrix} [4, 2] & [1/3, 5/6] & [1/9] \\ [1/3, 5/6] & [13/36, 1/9] & [1/27] \\ [1/9]' & [1/27]' & 1/81 \end{matrix} \right\| \qquad (7.5)
$$

A column vector c which is orthogonal to the first six columns of $M(\xi')$ and for which $c'M(\xi')c = 1$ is given by $c' = (1, 1, 1, -8, -8, -8, 72)$. Thus, in the notation of Section 3,

$$
\begin{aligned}
\delta_7(x, \xi') &= \sum_i x_i - 8\sum_{i<j} x_i x_j + 72x_1x_2x_3 \qquad (7.6)\\
&= 1- 8\{x_1x_2 + x_1x_3 + x_2x_3 - 9\, x_1x_2x_3\}.
\end{aligned}
$$

The term in braces is easily seen to have a maximum of $\frac{1}{4}$ and a minimum of 0 on S_2. Hence, $\max_x |\delta_7(x, \xi')| = 1$, and thus (see Section 3) ξ' is optimum.

REFERENCES

[1] H. Chernoff, "Locally optimum designs for estimating parameter," *Ann. Math. Stat.*, Vol. 24 (1953), pp. 586–602.

[2] S. EHRENFELD, "On the efficiency of experimental designs," *Ann. Math. Stat.*, Vol. 26 (1955), pp. 247–255.

[3] S. EHRENFELD, "Complete class theorems in experimental design," *Proceedings Third Berkeley Symposium on Mathematical Statistics and Probability*, University California Press, Berkeley 1955.

[4] G. ELFVING, "Optimum allocation in linear regression theory," *Ann. Math. Stat.*, Vol. 23 (1952), pp. 255–262.

[5] P. G. GUEST, "The spacing of observations in polynomial regression," *Ann. Math. Stat.*, Vol. 29 (1958), pp. 294–299.

[6] P. G. HOEL, "Efficiency problems in polynomial estimation," *Ann. Math. Stat.*, Vol. 19 (1958), pp. 1134–46.

[7] J. KIEFER, "Invariance, minimax sequential estimation, and continuous time processes," *Ann. Math. Stat.*, Vol. 28 (1957), pp. 573–601.

[8] J. KIEFER, "On the nonrandomized optimality and randomized nonoptimality of symmetrical designs," *Ann. Math. Stat.*, Vol. 29 (1958), pp. 675–699.

[9] J. KIEFER, "Optimum experimental designs," *J.R.S.S.* (Ser. B), Vol. 21 (1959), pp. 272–319.

[10] J. KIEFER AND J. WOLFOWITZ, "Optimum designs in regression problems," *Ann. Math. Stat.*, Vol. 30 (1959), pp. 271–294.

[11] J. KIEFER AND J. WOLFOWITZ, "The equivalence of two extremum problems," *Can. Jnl. Math.*, Vol. 12 (1960), pp. 363–366.

[12] H. SCHEFFÉ, "Experiments with mixtures," *J.R.S.S.* (Ser. B), Vol. 20 (1958), pp. 344–360.

[13] M. STONE, "Application of a measure of information to the design and comparison of regression experiments," *Ann. Math. Stat.*, Vol. 30 (1959), pp. 55–70.

Reprinted from THE ANNALS OF MATHEMATICAL STATISTICS
Vol. 33, No. 2, June, 1962
Printed in U.S.A.

TWO MORE CRITERIA EQUIVALENT TO D-OPTIMALITY OF DESIGNS[1]

BY J. KIEFER

Cornell University

0. Summary. Two minimax "regret" criteria for global optimality of designs in terms of estimation of the entire regression function are shown to be equivalent to minimizing the generalized variance. (The equivalence for the minimax criterion *without* modification was proved by Kiefer and Wolfowitz [9].) A consequent algorithm which is helpful in computing optimum designs is given.

1. Introduction. Let f be a column k-vector of continuous functions f_i on a compact space X. The expected value of an observation Y_x corresponding to the value x of the independent variable is $\theta' f(x) = \sum_1^k \theta_i f_i(x)$, where the vector θ of regression coefficients is unknown. A design is a (discrete) probability measure ξ on X, where for each x the value $\xi(x)$ denotes the proportion of observations taken at x. The observations are assumed uncorrelated and have common variance σ^2. Let $M(\xi)$ denote the $k \times k$ matrix

$$M(\xi) = \int_X f(x)f(x)'\xi(dx).$$

If N observations are taken according to ξ, then $N\sigma^{-2}M(\xi)$ is the information matrix of this design; if nonsingular, this is the inverse of the covariance matrix of best linear estimators of θ. In the approximate theory wherein Elfving [3], Chernoff [1], Kiefer and Wolfowitz [8], and others have successfully characterized designs which are optimum in various senses for such regression problems, we do not restrict $N\xi$ to be integral-valued, but instead allow ξ to be an arbitrary (discrete) probability measure. This permits useful results to be obtained, and a single optimum ξ can be used to yield, for each N, an actual design which is optimum to within order $1/N$. We shall be working in the approximate theory in this note.

A design ξ^* is said to be D-*optimum* (minimizes the generalized variance of best linear estimators of θ) if

$$\det M(\xi^*) = \max_\xi \det M(\xi). \tag{1}$$

If $M(\xi)$ is nonsingular, write $d(x, \xi) = f(x)'M^{-1}(\xi)f(x)$. Then $N^{-1}\sigma^2 d(x, \xi)$ is the variance of the (best linear) estimated regression at x. We also define $d(x, \xi)$ to conform with this if x is such that $\theta' f(x)$ is estimable but $M(\xi)$ is

Received June 27, 1961; revised December 4, 1961.
[1] Research sponsored by the Office of Naval Research.

singular, and put $d(x, \xi) = \infty$ for those x for which $\theta' f(x)$ is not estimable under ξ. A design ξ^* is said to be G-*optimum* (globally optimum) if

$$\max_x d(x, \xi^*) = \min_\xi \max_x d(x, \xi). \tag{2}$$

It is not hard to show that $\max_x d(x, \xi) \geqq k$ for every ξ. It was proved by Kiefer and Wolfowitz [9] that ξ^* is D-optimum if and only if it is G-optimum, and if and only if

$$\max_x d(x, \xi^*) = k. \tag{3}$$

This equivalence was extended by the author to the case where we are interested in s out of the k parameters, and it has proved useful in the computation of optimum designs in various settings [6], [7].

Let

$$d(x) = \min_\xi d(x, \xi). \tag{4}$$

This quantity is proportional to the variance of the best linear estimator of $\theta' f(x)$ for the design which is best for that x. An optimality criterion which is analogous to the minimax regret criterion of decision theory was suggested to the author by Professor Herman Rubin: ξ^* will be said to be MR-*optimum* (optimum in the sense of multiplicative regret) if

$$\max_x [d(x, \xi^*)/d(x)] = \min_\xi \max_x [d(x, \xi)/d(x)]. \tag{5}$$

A similar criterion in terms of additive regret is to call ξ^* AR-*optimum* if

$$\max_x [d(x, \xi^*) - d(x)] = \min_\xi \max_x [d(x, \xi) - d(x)]. \tag{6}$$

The purpose of this note is to give a very simple proof of the following.

THEOREM. ξ^* *is* D- *(and* G-*) optimum if and only if it is* MR-*optimum, and if and only if it is* AR-*optimum.*

This result incidentally yields a computational algorithm which is discussed in Section 3.

We make three brief remarks before proceeding to the proof. Firstly, the analogous results in the *exact* theory in the symmetric settings where symmetric block designs (Latin squares, BIBD's, etc.) are customarily used (the f_i taking on values 0 and 1) follow at once from the constancy of $d(x)$ and the results of [4]; this is not of much practical interest, since estimated regression is of less concern than estimated contrasts of treatment or variety effects in such settings, where block effects, row effects, etc., are of no interest. Secondly, the function-space interpretation of the equivalence of (1), (2) and (3) which was mentioned in [9] has an addition in terms of (5) and (6) which the reader will find it no difficulty to state. Thirdly, since the analogue of $d(x, \xi)$ in the case $s < k$ (see [6]) has an interpretation which depends on ξ (it is not merely proportional to the variance of the best linear estimator of $\sum_1^s \theta_i f_i(x)$), there is no obvious meaningful extension of the theorem of this note to that case.

2. Proof of the theorem. We first prove that (3) implies (5). Let ξ_x assign measure one to the point x. The regression $\theta' f(x)$ is clearly estimable at x if the design ξ_x is used, with N[variance of estimated regression at x under ξ_x]$/\sigma^2 = 1$. Hence,

$$d(x) \leqq 1. \tag{7}$$

Suppose ξ^* is D-optimum. It follows from (7) and the sentence following (2) that, for every design ξ,

$$\max_x \, [d(x, \xi)/d(x)] \geqq k. \tag{8}$$

We shall show that

$$\max_x \, [d(x, \xi^*)/d(x)] = k, \tag{9}$$

which will thus prove that ξ^* is MR-optimum.

If (9) is not satisfied, then there is a design ξ' and a value x' such that

$$d(x', \xi^*)/d(x', \xi') > k. \tag{10}$$

By replacing ξ' by $\epsilon\xi^* + (1 - \epsilon)\xi'$ with ϵ small, we can assume (10) is satisfied by a nonsingular ξ'. We can assume $M(\xi^*)$ to be the identity and $M(\xi')$ to be diagonal with diagonal elements d_i, since otherwise these matrices can be so diagonalized by a linear transformation without affecting the proof. According to (10),

$$1/k > \sum_{i=1}^{k} \left\{ f_i^2(x') \Big/ \sum_{j=1}^{k} f_j^2(x') \right\} d_i^{-1}. \tag{11}$$

Hence, at least one d_i^{-1} is $<k^{-1}$, say $d_1 > k$. But then

$$\begin{aligned} &\frac{\partial}{\partial \alpha} \log \det M(\alpha\xi' + (1 - \alpha)\xi^*) \Big|_{\alpha=0+} \\ &\quad = \frac{\partial}{\partial \alpha} \sum_i \log (1 - \alpha + \alpha d_i) \Big|_{\alpha=0} \\ &\quad = \sum_i (d_i - 1) > 0, \end{aligned} \tag{12}$$

contradicting the fact (1) that $\det M(\xi)$ is a maximum for $\xi = \xi^*$.

The converse is trivial: if ξ^* is MR-optimum, it must satisfy (9), since any D-optimum design satisfies (9); but then, by (7), we have $\max_x d(x, \xi^*) \leqq k$, so that ξ^* is D-optimum.

To prove the equivalence of D-optimality to AR-optimality, it is only necessary, in the previous two paragraphs, to replace every appearance of (8) by

$$\max_x \, [d(x, \xi) - d(x)] \geqq k - 1, \tag{8'}$$

of (9) by

$$\max_x \, [d(x, \xi^*) - d(x)] = k - 1, \tag{9'}$$

of (10) by

(10′) $$d(x', \xi^*) - d(x', \xi') > k - 1,$$

and of (11) by

(11′) $$k^{-1} \geqq 1 - \frac{k-1}{\sum_j f_j^2(x')} > \sum_i \{f_i^2(x')/\sum_j f_j^2(x')\} d_i^{-1},$$

the first half of (11′) being a consequence of (3) and the second half following from (10′). The rest of the proof then reads as before.

3. A computational algorithm. It follows from (7), (9), and the fact [9] that any D-optimum ξ^* assigns measure one to a set of values x for which $d(x, \xi^*) = k$, that any such ξ^* assigns measure one to the set

$$B = \{x : d(x) = 1\}.$$

Thus, if B is much smaller than X, the characterization of B can be of aid in computing a D-optimum ξ^*. In the present section we describe a characterization of B in terms of a family of related Chebyshev approximation problems, bringing out a relationship between the problems of estimating one and k parameters.

Before obtaining this characterization we remark that, if ξ^* is admissible in the sense of Ehrenfeld [2] (see also [5]), then, since multiples of $\theta' f(x)$ are the only estimable linear parametric functions when ξ_x is used, we must have $d(x) = 1$ and thus $x \in B$. This indicates that B need not be much of a reduction from X; for example, in the case of polynomial regression in one variable, the admissible ξ's were characterized in [5] and include all ξ_x's, so that $B = X$ in that case. However, it is easy to give other examples where B is smaller than X. Moreover, one can sometimes find a proper subset B' of B which must support an optimum ξ^*: if x in $B - B'$ implies that there is a ξ' on B' which is at least as good as ξ_x (that is, such that $M(\xi') - M(\xi_x)$ is nonnegative-definite), then this is the case.

We now describe an algorithm for computing $d(x_0)$. Suppose, without loss of generality, that $f_1(x_0) \neq 0$. (Any x for which all $f_i(x) = 0$ can be deleted from X.) Let $\phi_1 = \theta' f(x_0)$ and $\phi_i = \theta_i$ for $i \geqq 2$, and let $g_1(x) = f_1(x)/f_1(x_0)$ and $g_i(x) = f_i(x) - f_i(x_0) g_1(x)$ for $i \geqq 2$. Then

$$\sum \phi_i g_i(x) = \sum \theta_i f_i(x),$$

and the problem of estimating $\theta' f(x_0)$ when the regression is $\theta' f(x)$ is the same as that of estimating ϕ_1 when the regression is $\phi' g(x)$. The latter problem was first attacked in [3], and the following algorithm was given in [8]: Let $c^* = (c_2^*, \cdots, c_k^*)$ yield a best linear Chebyshev approximation of g_1 by $g_2, \cdots, g_k$; that is, the quantity

$$m(c) = \max_{x \in X} |g_1(x) - \sum_2^k c_i g_i(x)|$$

is minimized by the choice $c = c^*$. A design which minizes $d(x_0)$ can then be obtained easily from c^* in a manner described in [8]; for our present considerations, we need only mention that $d(x_0) = [m(c^*)]^{-2}$, which can be used to tell us whether or not $x_0 \varepsilon B$.

Finally, we remark that the Chebyshev approximation problem just described in terms of the g_i's can be rewritten as a "modified Chebyshev problem" in terms of the original f_i's, namely, to minimize

$$\max_x |[1 + \sum_2^k c_i f_i(x_0)]f_1(x)/f_1(x_0) - \sum_2^k c_i f_i(x)|.$$

For computational purposes, it is often convenient to solve this problem by first solving the restricted Chebyshev problem of minimizing

$$\max_x |f_1(x)/f_1(x_0) - r^{-1} \sum_2^k c_i f_i(x)|$$

subject to $\sum_2^k c_i f_i(x_0) = r - 1$, then multiplying the resulting minimum by r and minimizing with respect to r.

REFERENCES

[1] CHERNOFF, H. (1953). Locally optimum designs for estimating parameters. *Ann. Math. Statist.* **24** 586–602.

[2] EHRENFELD, S. (1955). Complete class theorems in experimental designs. *Proc. Third Berkeley Symp. Math. Stat. Prob.* **1** 57–67. Univ. of California Press.

[3] ELFVING, G. (1952). Optimum allocation in linear regression theory. *Ann. Math. Statist.* **23** 225–262.

[4] KIEFER, J. (1958). On the nonrandomized optimality and randomized nonoptimality of symmetrical designs. *Ann. Math. Statist.* **29** 675–699.

[5] KIEFER, J. (1959). Optimum experimental designs. *J. Roy. Statist. Soc., Ser. B* **21** 272–319.

[6] KIEFER, J. (1961). Optimum designs in regression problems, II. *Ann. Math. Statist.* **32** 298–325.

[7] KIEFER, J. (1961). Optimum experimental designs, *V*, with applications to rotatable and systematic designs. *Proc. Fourth Berkeley Symp. Math. Stat. Prob.* **1** 381–405. Univ. of California Press.

[8] KIEFER J. and WOLFOWITZ, J. (1959). Optimum designs in regression problems. *Ann. Math. Statist.* **30** 271–294.

[9] KIEFER J. and WOLFOWITZ, J. (1960). The equivalence of two extremum problems. *Can. J. Math.* **12** 363–366.

AN EXTREMUM RESULT

J. KIEFER

1. Introduction. The main object of this paper is to prove the following:

THEOREM.† *Let* $f_1, \ldots, f_k$ *be linearly independent continuous functions on a compact space* $\mathfrak{X}$. *Then for* $1 \leqslant s \leqslant k$ *there exist real numbers* a_{ij}, $1 \leqslant i \leqslant s$, $1 \leqslant j \leqslant k$, *with* $\{a_{ij}, 1 \leqslant i, j \leqslant s\}$ *non-singular, and a discrete probability measure* ξ^* *on* $\mathfrak{X}$, *such that*

(a) *the functions* $g_i = \sum_{j=1}^{k} a_{ij} f_j$ $1 \leqslant i \leqslant s$, *are orthonormal* ($\xi^*$) *and are orthogonal* ($\xi^*$) *to the* f_j *for* $s < j \leqslant k$;

(b)
$$\max_{x \in \mathfrak{X}} \sum_{1}^{s} f_i^2(x) = \int_{\mathfrak{X}} \sum_{1}^{s} f_i^2(x) \xi^*(dx) = s.$$

The result in the case $s = k$ was first proved in **(2)**. The result when $s < k$, which because of the orthogonality condition of (a) is more general than that when $s = k$, was proved in **(1)** under a restriction which will be discussed in § 3. The present proof does not require this *ad hoc* restriction, and is more direct in approach than the method of **(2)** (although involving as much technical detail as the latter in the case when the latter applies). The latter proof involved showing the equivalence of two extremum problems encountered in the theory of optimum statistical designs, under the restriction mentioned. The analogous equivalence result when the restriction is not satisfied is more difficult to state and to prove, and cannot be used to obtain the result of the present paper without proving an additional fact, discussed in § 3, which the methods of **(1)** do not seem to yield. On the other hand, the present result implies that additional fact in the design setting, and yields a neater proof of a major part of the equivalence result when the mentioned restriction is not necessarily satisfied. With or without the restriction, the remainder of the equivalence theorem can then be given a short proof.

The idea of the present proof is to reduce the result to one of the ontoness of a certain natural mapping induced by the problem. This ontoness, which seems intuitively plausible, is not so easy to prove, and the author is indebted

Received July 31, 1961. Research for this paper was sponsored by the Office of Naval Research of the United States.

†The same conclusion obviously holds if $\dim\{(f_1(x), \ldots, f_k(x)); x \in \mathfrak{X}\} = s + \dim\{(f_{s+1}(x), \ldots, f_k(x)); x \in \mathfrak{X}\}$, and the necessary modification in the statement of the theorem in other cases is obvious. There is a choice of ξ^* whose support consists of no more than $s(2k-s+1)/2$ points **(1)**. The a_{ij}, $i > j$, can be taken to be zero.

to his colleague Professor Namioka for supplying a proof of a much more general theorem **(3)** which yields this ontoness (as well as other interesting results), and for many helpful discussions.

2. Proof of the theorem. The proof will be divided into three parts for convenience.

I. Let $\mathfrak{C}$ be the set of vectors $c = (c_{ij}, 1 \leqslant i \leqslant s, i < j \leqslant k)$. Let $\Lambda = \{(\lambda_1, \ldots, \lambda_s) : \lambda_i \geqslant 0, \sum \lambda_i = 1\}$. For $\lambda \in \Lambda$, write

$$K_\lambda(x, c) = \sum_{i=1}^{s} \lambda_i \Big[f_i(x) - \sum_{j>i} c_{ij} f_j(x) \Big]^2 .$$

We consider the game $\{K_\lambda, \mathfrak{X}, \mathfrak{C}\}$ with $\mathfrak{X}$ and $\mathfrak{C}$ the spaces of pure strategies and K_λ the payoff function. Writing $K_\lambda(\xi, c) = \int K_\lambda(x, c)\xi(dx)$, we as usual define c^* to be minimax if $\max_x K_\lambda(x, c^*) = \min_c \max_x K_\lambda(x, c)$, and ξ^* to be maximin if $\min_c K_\lambda(\xi^*, c) = \max_\xi \min_c K_\lambda(\xi, c)$. Since K_λ is convex in c, it suffices to consider pure strategies c for player 2, but we must allow mixed strategies ξ for player 1. The range of $\{f_j, 1 \leqslant j \leqslant k\}$ is a compact Euclidean set which could actually be regarded as $\mathfrak{X}$, the associated class Ξ of Borel measures being weakly compact, which will be used below. There are never any measure-theoretic difficulties, and in fact any ξ can be replaced by a ξ' with finite support and such that $K_\lambda(\xi, c) = K_\lambda(\xi', c)$ for all λ and c.

Let $\mathfrak{C}_N$ be the subset of vectors c satisfying $\sum_{j=i+1}^{k} c_{ij}^2 \leqslant N$, $1 \leqslant i \leqslant s$, and let $\mathfrak{C}_i' = \{(c_{i,i+1}, \ldots, c_{i,k}) : \sum_{j=i+1}^{k} c_{ij}^2 = 1\}$. For each i, $1 \leqslant i \leqslant s$, the quantity

$$b_i(\xi) = \min_{\{c_{ij}\} \in \mathfrak{C}_i'} \int \Big[\sum_{j>i} c_{ij} f_j(x) \Big]^2 \xi(dx)$$

is positive for some $\xi \in \Xi$ because the f_i are linearly independent. Hence (averaging the ξ's for different i's) there is a $\xi' \in \Xi$ and an $\epsilon > 0$ such that $b_i(\xi') > \epsilon$ for $1 \leqslant i \leqslant s$. Thus, there is a constant d such that

$$\min_{c \notin \mathfrak{C}_N} K_\lambda(\xi', c) > N\epsilon - d$$

for all $\lambda \in \Lambda$. It follows that there is a value N' of N such that, for any $c \notin \mathfrak{C}_{N'}$,

$$\max_\xi K_\lambda(\xi, c) > \min_{c'} \max_\xi K_\lambda(\xi, c'),$$

for all λ.

Hence, c^*_λ is minimax for the game $\{K_\lambda, \mathfrak{X}, \mathfrak{C}\}$ if and only if it is minimax for $\{K_\lambda, \mathfrak{X}, \mathfrak{C}_{N'}\}$. The latter game, having compact strategy spaces and continuous payoff, is determined. For $N > N'$, if ξ^*_N is maximin for the game $\{K_\lambda, \mathfrak{X}, \mathfrak{C}_N\}$ and c^*_λ is minimax (for all $N > N'$), and if $\xi_j = [(j-1)\xi^*_j + \xi']/j$ and $N_j > N'$ is such that $K_\lambda(\xi_j, c^*_\lambda) < K_\lambda(\xi_j, c)$ for all $c \notin \mathfrak{C}_{N_j}$, we have (since $K_\lambda(\xi^*_j, c^*_\lambda) = \max_\xi K(\xi, c^*_\lambda)$ by the determinateness),

$$\sup_\xi \inf_{c \in \mathfrak{C}} K_\lambda(\xi, c) \geqslant \inf_{c \in \mathfrak{C}} K_\lambda(\xi_j, c)$$

$$= \inf_{c \in \mathfrak{C}_{N_j}} K_\lambda(\xi_j, c) \geqslant \left(1 - \frac{1}{j}\right) \inf_{c \in \mathfrak{C}_{N_j}} K_\lambda(\xi_j^*, c)$$

$$= \left(1 - \frac{1}{j}\right) K_\lambda(\xi_j^*, c_\lambda^*) = \left(1 - \frac{1}{j}\right) \sup_\xi K_\lambda(\xi, c_\lambda^*)$$

$$\geqslant \left(1 - \frac{1}{j}\right) \inf_{c \in \mathfrak{C}} \sup_\xi K_\lambda(\xi, c).$$

Letting $j \to \infty$, we see that the game $\{K_\lambda, \mathfrak{X}, \mathfrak{C}\}$ is also determined.

Moreover, the set C_λ (say) of (pure) minimax strategies for this last game is a subset of $\mathfrak{C}_{N'}$, for all λ. Since $\max_x K_\lambda(x, c)$ is convex in c (being a maximum of convex functions), we have that C_λ is convex. Let Ξ_λ be the (mixed) maximin strategies for this game. Since $K_\lambda(\xi, c)$ is linear in ξ, we similarly have that Ξ_λ is convex.

Let $\{\lambda_n\}$ be a sequence in Λ, converging to λ_0. If $c^*_{\lambda_n}$ and $\xi^*_{\lambda_n}$ are any elements of C_{λ_n} and Ξ_{λ_n}, we can, by the compactness of $\mathfrak{C}_{N'}$ and Ξ, select a subsequence $\{n_m\}$ of n such that $\{c^*_{\lambda_{n_m}}\}$ and $\{\xi^*_{\lambda_{r_m}}\}$ converge to limits c_0 and ξ_0 in $\mathfrak{C}_{N'}$ and Ξ. Since $\min_c K_\lambda(\xi, c)$ and $\max_x K_\lambda(x, c)$ are clearly continuous in (λ, ξ) and (λ, c), respectively (because the f_i are continuous), we see at once that $c_0 \in C_{\lambda_0}$ and $\xi_0 \in \Xi_{\lambda_0}$. We conclude that, if in the space $\Lambda \times \mathfrak{C}_{N'} \times \Xi$ we let $V_\lambda = (\{\lambda\} \times C_\lambda \times \Xi_\lambda)$ and

$$G = \cup_{\lambda \in \Lambda} V_\lambda,$$

then G is closed and each section V_λ is a compact, convex subset of the compact space $\mathfrak{C}_{N'} \times \Xi$.

II. Let k_λ denote the value $\min_c \max_x K_\lambda(x, c) = \max_\xi \min_c K_\lambda(\xi, c)$ of the above game. Since

$$\min_c K_\lambda(\xi, c) \geqslant \min_i \min_{\{c_{ij}\}} \int \left[f_i(x) - \sum_{j>i} c_{ij} f_j(x)\right]^2 d\xi$$

and the inner minimum is the square of the $L_2^{(\xi)}$ norm of $(f_i - \text{proj}_{\{f_{i+1}, \ldots, f_k\}} f_i)$, and since for each i there is (by linear independence) a ξ_i for which this squared norm is $> \epsilon > 0$, we clearly have

$$k_\lambda \geqslant \min_c K_\lambda\left(\sum_1^s \xi_i/s, c\right) > \epsilon/s > 0$$

for all λ. We hereafter write

$$K^{(i)}(x, c) = \left[f_i(x) - \sum_{j>i} c_{ij} f_j(x)\right]^2.$$

Let F be the following mapping from G into the $(s-1)$-dimensional simplex $E = \{(e_1, \ldots, e_s) : e_i \geqslant 0,\ 1 \leqslant i \leqslant s,\ \sum e_i = 1\}$:

$$F(\lambda, c_\lambda, \xi_\lambda) = \left(\frac{\lambda_1 K^{(1)}(\xi_\lambda, c_\lambda)}{k_\lambda}, \ldots, \frac{\lambda_s K^{(s)}(\xi_\lambda, c_\lambda)}{k_\lambda}\right).$$

Since $k_\lambda = \sum \lambda_j K^{(j)}(\xi_\lambda, c_\lambda)$ when ξ_λ is maximin and c_λ is minimax, the range of F is indeed in E. It is easy to see that F is continuous.

We want to show that F is onto. Obviously, for λ restricted to a vertex of Λ, F maps V_λ onto the corresponding vertex of E, and similarly F maps the part $\cup_{\lambda \in T} V_\lambda$ of G above any subsimplex (edge, face, etc.) T of Λ into the corresponding subsimplex of E. It follows from the theorem of Namioka **(3)** cited above that F is onto. (In general, $F(V_\lambda)$ need not be a point, which is why we could not directly define a mapping $F : \Lambda \to E$; there need not exist a cross-section from Λ into G, so we need Namioka's result.)

III. Since F is onto, there is a point $(\bar\lambda, c_{\bar\lambda}', \xi_{\bar\lambda}')$ of G which F maps into $(s^{-1}, s^{-1}, \ldots, s^{-1})$. Let ξ^* (of the statement of the theorem) $= \xi_{\bar\lambda}'$ and

$$g_i(x) = \Big[f_i(x) - \sum_{j>i} c'_{\bar\lambda ij} f_j(x)\Big] \Big/ [K^{(i)}(\xi'_{\bar\lambda}, c'_{\bar\lambda})]^{1/2}.$$

(The denominator, being $[k_{\bar\lambda}/\bar\lambda_i s]^{1/2}$ where $\bar\lambda \in \operatorname{Int}\Lambda$, is finite and not zero.) Clearly, $\int g_i^2 d\xi_{\bar\lambda}' = 1$; and since the $c_{\bar\lambda ij}'$ for each i are chosen to minimize $K^{(i)}(\xi_{\bar\lambda}', c)$, we see that the g_i are orthonormal $(\xi_{\bar\lambda}')$, $1 \leqslant i \leqslant s$, and are orthogonal $(\xi_{\bar\lambda}')$ to the f_j, $j > s$. Finally,

$$\sum_{i=1}^{s} g_i^2(x) = \sum_{i=1}^{s} \frac{K^{(i)}(x, c'_{\bar\lambda})}{K^{(i)}(\xi'_{\bar\lambda}, c'_{\bar\lambda})} = \sum_{i=1}^{s} \frac{s\bar\lambda_i K^{(i)}(x, c'_{\bar\lambda})}{k_{\bar\lambda}} = s\,\frac{K_{\bar\lambda}(x, c'_{\bar\lambda})}{K_{\bar\lambda}(\xi'_{\bar\lambda}, c'_{\bar\lambda})}.$$

By the game-theoretic results,

$$\max_x K_{\bar\lambda}(x, c'_{\bar\lambda}) = K_{\bar\lambda}(\xi'_{\bar\lambda}, c'_{\bar\lambda}),$$

proving the desired result.

3. Relationship to previous method and results. Let f denote the column vector of f_j's and $M(\xi)$ the matrix $\int f(x) f(x)' \xi(dx)$. It was shown in **(2)** that ξ^* maximizes $\det M(\xi)$ if and only if ξ^* minimizes $\max_x f(x)' M^{-1}(\xi) f(x)$ among ξ for which $M(\xi)$ is non-singular, and if and only if $\max_x f(x)' M^{-1}(\xi) f(x) = k$. This yields the theorem of the present paper when $s = k$. If $s < k$, let $\bar f$ be the last $k - s$ functions of f and $\bar M(\xi) = \int \bar f(x) \bar f(x)' \xi(dx)$. It was shown in **(1)** that, if $M(\xi^*)$ is non-singular, then ξ^* minimizes the determinant of the upper left-hand $s \times s$ submatrix of $M^{-1}(\xi)$ (which can be defined in an obvious way even if $M(\xi)$ is singular) if and only if ξ^* minimizes

$$d(\xi) = \max_x [f(x)' M^{-1}(\xi) f(x) - \bar f(x)' \bar M^{-1}(\xi) f(x)],$$

and if and only if $d(\xi^*) = s$. When $s = k$, non-singularity is no restriction, but when $s < k$ the result of the present paper is now obtained in **(1)** only under the artificial *ad hoc* assumption that there exists a minimizing ξ^* for which $M(\xi^*)$ is non-singular. The result of **(1)** in the general case is obtained only by replacing $d(\xi)$ above by the quantity $\max_{\xi'} D(\xi', \xi)$, where

$$D(\xi', \xi) = \bar{D}(\xi', \xi) - \rho(\xi', \xi)$$
$$= \mathrm{tr}[\bar{J}^{-1}(\xi)\bar{J}(\xi')] - \mathrm{tr}[J_3^{-1}(\xi)J_3(\xi')] - \rho(\xi', \xi),$$

where $J(\xi) = AM(\xi)A'$ is of rank $r + s$ with zero elements outside the first principal $(r + s) \times (r + s)$ submatrix $\bar{J}(\xi)$ and with A triangular (zeros below the main diagonal), $J_3(\xi)$ is the lower right-hand $r \times r$ submatrix of $\bar{J}(\xi)$, $\bar{J}(\xi')$ and $J_3(\xi')$ have the same meanings for the same A (for which $J(\xi')$ need not have the same properties as $J(\xi)$), and

$$\rho(\xi', \xi) = \mathrm{tr}[J_1^{-1}(\xi) \lim_{\lambda \to 0} J_4(\xi')[\dot{J}_5(\xi') + \lambda I]^{-1} J_4(\xi')']$$

where J_1, J_5, and J_4 are, respectively, the first principal $s \times s$ submatrix of J, the last principal $(k - s - r) \times (k - s - r)$ submatrix of J, and the corresponding $s \times (k - s - r)$ submatrix of J.

Although D is invariantly defined (does not depend on the choice of A), $\bar{D}$ and ρ are not. Clearly, the condition $\max_{\xi'} \bar{D}(\xi', \xi) = s$ is sufficient for the result $\max_{\xi'} D(\xi', \xi) = s$, but it is not necessary (except under the *ad hoc* assumption, which implies that $\rho = 0$ and $\max_{\xi'} D(\xi', \xi^*) = d(\xi^*)$). The result of the present paper is precisely that, for ξ^* satisfying the above-mentioned extremum criteria of **(1)**, there is a choice of A for which $\max_{\xi'} \bar{D}(\xi', \xi^*) = s$. Conversely, this last result, combined with that of **(1)**, would yield the theorem of the present paper, but the methods of **(1)** do not yield this result on the choice of A. Thus, the present theorem states a somewhat stronger result than that of **(1)** in the general case. One can also try to obtain the present result by a passage to the limit from the case satisfying the extra assumption, but considerable delicacy is involved.

References

1. J. Kiefer, *Optimum designs in regression problems, II*, Ann. Math. Stat., *32* (1961), 298–325.
2. J. Kiefer and J. Wolfowitz, *The equivalence of two extremum problems*, Can. J. Math., *12* (1960), 363–366.
3. I. Namioka, *On certain onto maps*, Can. J. Math., *14* (1962), 461–466.

Cornell University

Reprinted from
Canad. J. Math. **14**, 597–601 (1962)

OPTIMUM EXTRAPOLATION AND INTERPOLATION DESIGNS, I

J. KIEFER[1] AND J. WOLFOWITZ[2]

Summary

For regression problems where observations may be taken at points in a set X which does not coincide with the set Y on which the regression function is of interest, we consider the problem of finding a design (allocation of observations) which minimizes the maximum over Y of the variance function (of estimated regression). Specific examples are calculated for one-dimensional polynomial regression when Y is much smaller than or much larger than X. A related problem of optimum estimation of two regression coefficients is studied. This paper contains proofs of results first announced at the 1962 Minneapolis Meeting of the Institute of Mathematical Statistics. No prior knowledge of design theory is needed to read this paper.

1. Introduction

Let $f=(f_0, f_1, \cdots, f_m)'$ be a column vector (the accent on a vector or matrix denotes a transpose) of $m+1$ continuous real-valued functions on $X \cup Y$ where X and Y are two (not necessarily disjoint) compact sets. The f_i are assumed to be linearly independent on X. A design is a probability measure ξ (which can always be assumed discrete) on X. (For a discussion of this see Kiefer and Wolfowitz [10].)

Write

$$(1.1) \qquad m_{ij}(\xi)=\int f_i f_j d\xi, \quad M(\xi)=\{m_{ij}(\xi),\ 0\leq i,\ j\leq m\}\ ,$$

and if $M(\xi)$ is nonsingular,

$$(1.2) \qquad M^{-1}(\xi)=V(\xi)=\{v_{ij}(\xi),\ 0\leq i,\ j\leq m\}\ ,$$

[1] John Simon Guggenheim Memorial Foundation Fellow. Research supported in part by the Office of Naval Research under Contract No. Nonr 266(04)(NR 047-005).

[2] The research of this author was supported in part by the U.S. Air Force under Contract No. AF 18(600)-685.

and

(1.3) $$d(y, \xi)=f(y)'V(\xi)f(y) \text{ for } y \text{ in } Y, \quad D_Y(\xi)=\max_{y \in Y} d(y, \xi).$$

If N uncorrelated observations with equal variance σ^2 (known or unknown) are made, taking $N\xi(x)$ observations at x for each x, and if the expected value of an observation at x is $\theta' f(x)=\sum_0^m \theta_i f_i(x)$ where $\theta=(\theta_0, \cdots, \theta_m)'$, then $\sigma^2 N^{-1}V(\xi)$ is the covariance matrix of best linear estimators of the vector θ, and $\sigma^2 N^{-1}d(y, \xi)$ is the variance of the best linear estimator of the regression function $\theta' f(y)$ at the point y. Thus, $d(y, \xi)=\infty$ if $\theta' f(y)$ is not estimatable under ξ, while $d(y, \xi)$ has an obvious (finite) definition if $\theta' f(y)$ is estimatable, even if $M(\xi)$ is singular. X is the set of points at which we are permitted to take observations, while Y is the set on which the regression function is of interest to us.

As has been discussed in other papers, we do not restrict ξ to take on values which are integral multiples of N^{-1}. This allows us to obtain optimum design algorithms which cannot be obtained under the restriction, and at the same time yields designs which can be implemented in practice through the use of closely related ξ's which do take on only values which are integral multiples of N^{-1} (see [10]).

Most previous developments in optimum design theory have been concerned with the case $X=Y$. However, there clearly are problems of practical importance where $X \neq Y$. Thus, when $X \subset Y$, we have, in addition to consideration of the regression function on X, the problem of *extrapolation* of the estimated regression to the set $Y-X$. On the other hand, when $Y \subset X$ we have what we will refer to as an *interpolation* problem, and are especially interested in the regression function on a subset of the set of points at which observations can be taken. The importance of this last problem has been discussed by Behnken and Box [2], whose discussion really applies to the more general situation where Y is not necessarily a subset of X, but is in some reasonable sense "small" compared with X.

A natural possible criterion is to choose ξ so as to minimize $D_Y(\xi)$. Unfortunately, we are unable to obtain a general algorithm analogous to that which holds when $X=Y$ (see [11]). Even for such simple problems as that where $X=[-1, 1]$, $Y=Y_a=[-a, a]$, and $f_i(x)\equiv x^i$, the actual evaluation of a ξ optimum in this sense is formidable, the dependence on a being complicated. In the present paper we make only a beginning on this problem. We consider only this simple polynomial problem, and study in detail only the interpolation problem as $a \to 0$ and the extrapolation problem as $a \to \infty$. The problems are formulated precisely in sections 3, 5, and 8. In the rest of this section we give a brief discussion of these problems which is intended to make clear why these problems cannot be solved (in the way certain other related extrapolation problems can) by direct application of previous results, and what some of the difficulties are. This discussion will be immediately intelligible only to one who has some knowledge of the previous results in this subject. Other readers may wish to return to this discussion after reading the rest of the paper.

Briefly and very roughly, our attack on these problems is to show that, for a very small or very large, if ξ^* minimizes $D_{Y_a}(\xi)$, then $d(x, \xi^*)$ attains its maximum on $[-a, a]$ at the points $\pm a$. We are thus led to the problem of finding ξ^* to minimize $\max[d(a, \xi), d(-a, \xi)]$. Because of the symmetry of the problem, this is the same as minimizing $[d(a, \xi)+d(-a, \xi)]/2$, or of minimizing $d(a, \xi)$ over symmetric designs ξ. When a is small (resp., large), we approximate $d(a, \xi)$ by $v_{00}(\xi)+a^2v_{11}(\xi)$ (resp., $a^{2m}(v_{mm}(\xi)+a^{-2}v_{m-1,\,m-1}(\xi))$ and minimize this quantity. Whether we think of the problem in terms of $d(a, \xi)$ and $d(-a, \xi)$, or in terms of $v_{00}(\xi)$ and $v_{11}(\xi)$ (or $v_{mm}(\xi)$ and $v_{m-1,\,m-1}(\xi)$), our problem is thus concerned with estimation of *two* linear functions of the parameters θ_i. Such problems have been treated in [10], [7], and [8], and one tool for solving them is given in Theorem 2.3 of section 2.

In typical examples, these problems lead to more difficult calculations and less elegant results than do problems where a single linear function of the parameters is estimated. As is proved in [10] (see Theorem 2.2 below), problems of the latter type can be solved by studying an associated Chebyshev approximation problem. In the notation of the first paragraph of this section, the best design for estimating θ_m is obtained easily (in fact, by solving a set of linear equations for the weights of ξ) once the support of ξ is determined, and the support is obtained from the best (uniform norm) approximation on X to f_m of the form $\sum_0^{m-1} q_i f_i$ where the q_i are real. One application of this device is the following: Suppose, instead of being concerned with $d(x, \xi)$ at two points (a and $-a$), we are interested in minimizing $d(b, \xi)$ for a single b. Now, $d(b, \xi)$ is proportional to the variance of the best linear estimator of $\theta' f(b)$. As shown in [9], the Chebyshev theory can be applied after the following simple reduction: Suppose $f_m(b)\neq 0$. (This is no loss of generality, since $f_i(b)\neq 0$ for some i or else $\theta' f(b)=0$ for all θ.) Write

$$\phi_m=\sum_0^m \theta_i f_i(b)\,,$$

$$\phi_i=\theta_i,\quad 0\leq i<m\,,$$

(1.4)

$$g_m(x)=f_m(x)/f_m(b)\,,$$

$$g_i(x)=f_i(x)-f_i(b)g_m(x),\quad 0\leq i<m\,.$$

Then

$$\sum_0^m \phi_i g_i(x) = \sum_0^m \theta_i f_i(x) ,$$

and the problem of estimating $\theta' f(b)$ when the regression is $\theta' f(x)$ is the same as that of estimating ϕ_m when the regression is $\phi' g(x)$. Applying the theory of [10] to the problem in the latter form, we have (after multiplying through by $f_m(b)$ and writing $q_i' = q_i f_m(b)$) the Chebyshev problem of choosing $q_0', \cdots, q_{m-1}'$ to minimize

$$\max_{x \in X} | f_m(x) - \sum_0^{m-1} q_i'(f_i(x) - h_i f_m(x)) | , \tag{1.5}$$

where $h_i = f_i(b)/f_m(b)$.

Such a result could be helpful in the corresponding one-sided extrapolation problem where $X = [-1, 1]$ and $Y = [-1, b]$ with $b > 1$; for, if ξ^* minimizes $d(\xi, b)$ and if $D_Y(\xi^*) = d(\xi^*, b)$, then ξ^* clearly minimizes $D_Y(\xi)$. One would expect this to be the case, for example, in polynomial regression with b large.

In the polynomial case $f_i(x) \equiv x^i$ with $X = [-1, 1]$, $d(b, \xi)$ is well known (see [9]) to be minimized when $-1 \leq b \leq 1$ by the design which assigns measure one to the point b. When $b > 1$, an elegant solution to this problem has been announced recently by P. G. Hoel and A. Levine [12], namely, that the support of the optimum ξ is the same (while the weights vary) for all $b > 1$, and is the same as the support for the optimum design for estimating θ_m, obtained in [10] (see Theorem 2.6 below).

We remark, finally, that the approach of the present paper can be used with appropriate modifications in other problems where (a) f_i is even or odd according to whether i is even or odd, and (b) for the case $a \to 0$, we have $b_0(0) \neq 0$, $f_1(0) = 0$, and $f_i(x) = o(f_1(x))$ for $i > 1$ and $|x| \to 0$, or (b′) for the case $a \to \infty$, we have $f_i(x) = o(f_m(x))$ for $i < m$ and $|f_m(x)| \nearrow \infty$ as $|x| \to \infty$. However, the computational analogues of the results of sections 4, 7, and 8 will be less likely to be available in the literature of approximation theory.

2. Auxiliary results

In this section we list those results of optimum design theory and related game and Chebyshev approximation theory which will be needed in subsequent sections. Here and in the remainder of the paper we use the notation introduced in section 1. In addition, throughout the paper, we use c_i' and c_i'' to denote positive constants, and b_i, d_i to denote real constants. All integrations in sections 3 through 8 are over the interval

$[-1, 1]$. The symbols $\sum'$ and $\sum''$ denote, respectively, summation over odd indices j and even indices j, $0 \leq j \leq m$; similarly, $\sum''_{j>2}$ denotes summation over even j, $2 < j \leq m$. A probability measure ν on $[-1, 1]$ is said to be symmetric if $\nu(A) = \nu(-A)$ for every Borel set A. The symbols $O(a)$, $o(1)$, etc., refer to the case $a \to 0$ in sections 3 through 7, and to the case $a \to \infty$ in section 8.

The considerations of [10] include the study of the game with payoff function

$$K(\xi, q) = \int_X [f_0(x) - \sum_1^m q_j f_j(x)]^2 \xi(dx) \tag{2.1}$$

where the f's are again continuous on a compact X; the minimizing player may (by convexity of K in q) be restricted to pure strategies which are vectors $q = (q_1, \cdots, q_m)$ of real components, while the maximizing player has probability measures ξ (which can be taken to be discrete) on X for his mixed strategies. The importance of this game is given in the following simple result of section 2 of [10]:

THEOREM 2.1. *The variance $\sigma^2 N^{-1} v_{00}(\xi)$ for estimating θ_0 when design ξ is used satisfies*

$$v_{00}(\xi) = 1/\min_q K(\xi, q)\,. \tag{2.2}$$

Hence, for the problem of estimating θ_0, ξ^* is optimum in the sense of minimizing $v_{00}(\xi)$ if and only if it is maximin for the game with payoff K; i.e., if and only if

$$\min_q K(\xi^*, q) = \max_\xi \min_q K(\xi, q)\,. \tag{2.3}$$

If $q^* = (q_1^*, \cdots, q_m^*)$ is any solution to the associated Chebyshev approximation problem (which will be discussed further below) of choosing q to minimize $\max_{x \in X} | f_0(x) - \sum_1^m q_j f_j(x) |$, we shall call q^* a *Chebyshev vector*. Let $B(q^*)$ be the set where $| f_0(x) - \sum_1^m q_j^* f_j(x) |$ achieves its maximum. The following results are proved in the Lemma and Theorems 1 and 2 of [10]:

THEOREM 2.2 *The game with payoff K is determined, the minimizing player has a pure minimax strategy q^*, and the maximizing player has a maximin strategy ξ^* on at most $m+1$ points. The minimax strategies coincide with the Chebyshev vectors. The design ξ^* is maximin if, and only if, for any Chebyshev vector q^*, $\xi^*(B(q^*)) = 1$ and ξ^* satisfies the " orthogonality relations "*

$$\int_X [f_0(x) - \sum_1^m q_j^* f_j(x)] f_i(x) \xi^*(dx) = 0\,, \quad 1 \leq i \leq m\,. \tag{2.4}$$

It follows that ξ^* assigns probability one to the intersection of all $B(q^*)$

for q^* Chebyshev.

The above approach can be extended to other problems. For example, write

$$(2.5) \qquad 1/K'(\xi, (b, d)) = \bar{K}((b, d), \xi)$$
$$= \frac{1}{\int_X [f_0(x) - \sum_{j \neq 0} d_j f_j(x)]^2 \xi(dx)} + \frac{A^2}{\int_X [f_1(x) - \sum_{j \neq 1} b_j f_j(x)]^2 \xi(dx)}$$

and

$$(2.6) \qquad h(A, \xi) = v_{00}(\xi) + A^2 v_{11}(\xi) ,$$

$$(2.7) \qquad H(A) = \min_{\xi} h(A, \xi) .$$

A consequence of Theorem 2.1 is

THEOREM 2.3.

$$(2.8) \qquad h(A, \xi) = \frac{1}{\min_{\{d_j\}} \int_X [f_0(x) - \sum_{j \neq 0} d_j f_j(x)]^2 \xi(dx)} + \frac{A^2}{\min_{\{b_j\}} \int_X [f_1(x) - \sum_{j \neq 1} b_j f_j(x)]^2 \xi(dx)}$$
$$= \max_{b, d} \bar{K}((b, d), \xi) ,$$

so that ξ^* minimizes $h(A, \xi)$ if and only if it is maximin for the game with payoff K'.

Thus if one were interested in estimating both θ_0 and θ_1, one might try, for specified A, to find a ξ^* for which $h(A, \xi^*) = H(A)$. If a design which minimizes the generalized variance $D = v_{00}v_{11} - v_{01}^2$ for estimating θ_0 and θ_1 must have $v_{01} = 0$ (this being the case in polynomial regression with $f_0(x) \equiv x^r$, $f_1(x) \equiv x^s$, $r - s$ odd, for reasons of symmetry discussed below), then this minimizing design can be found from the design which minimizes (2.8), by varying A. This is similar to the method given in section 4 of [10] for solving the generalized variance problem for a subset of the parameters, another method being given by Kiefer ([7], [8]). The problem of minimizing (2.8) in our polynomial setting is considered in sections 4, 6, and 7, after its relation to the interpolation problem is developed in sections 3 and 5; this problem of minimizing (2.8) is also of interest for the reasons just given. (The extrapolation problem of section 7 involves a similar but more complex expression.) The interaction of the two integrals of (2.8) makes this problem more complex than that of (2.1), and we are able to make the limited progress we do make only because of the special polynomial structure.

In section 6 we consider, for odd m and for fixed real s, a game with payoff function

$$K_s(\mu, \bar{q}) = \int (1 - sx^2 - \sum_{j>2}'' q_{2j} x^{2j})^2 \mu(dx) , \tag{2.9}$$

where $\bar{q} = (q_4, q_6, \cdots, q_{m-1})$ and μ is restricted to be symmetric on $[-1, 1]$ and to satisfy $\mu(0) = 0$. The restriction to even indices and symmetric μ results from symmetry considerations discussed below and in section 6. By applying Theorem 2.2 to the symmetrized problem with μ on $[-1, 1]$ replaced by 2μ on $(0, 1]$, we obtain

THEOREM 2.4. *The game with payoff (2.9) is determined, the minimizing player has a pure minimax strategy, and the maximizing player has a symmetric maximin strategy on at most $m-1$ points of $[-1, 1]$.*

We shall use the following fact about Chebyshev approximation (see Achieser [1], especially pages 51–55):

THEOREM 2.5 *If F and G are continuous real functions on $[-1, 1]$, not identically zero, then the quantity*

$$L(q) = \max_{-1 \leq x \leq 1} | F(x) - G(x) \sum_0^n q_j x^j | \tag{2.10}$$

is uniquely minimized with respect to $q = (q_0, q_1, \cdots, q_n)$ by that (unique) choice for which $F(x) - G(x) \sum_0^n q_j x^j$ attains its maximum in absolute value, with alternating signs, at no fewer than $n+2$ points of $[-1, 1]$.

The best-known example of Chebyshev approximation is the case $F(x) = x^h$, $G(x) \equiv 1$, $n = h-1$. In that case (see [1]) it is well known that, for the Chebyshev vector q^*,

$$x^h - \sum_0^{h-1} q_j^* x^j = 2^{1-h} \cos(h \cos^{-1} x) , \tag{2.11}$$

$\cos(h \cos^{-1} x)$ being the hth Chebyshev polynomial of the first kind. The function (2.11) attains its maximum magnitude 2^{1-h} with alternating signs at the $h+1$ points (of $B(q^*)$)

$$x_r = \cos(\pi r/h) , \qquad 0 \leq r \leq h . \tag{2.12}$$

Using these facts and Theorem 2.2, the following result was proved in [10] (for clarity we replace the m of the previous paragraphs by h, since the result will be used in section 4 with $h = m-1$):

THEOREM 2.6. *The unique optimum design for minimizing the variance $v_{00}(\xi)$ of the best linear estimator of the coefficient of x^h in the* *problem of h-th degree regression on $[-1, 1]$ is given by*

$$\xi^*(x_r)=\begin{cases} 1/h & \text{if } 0<r<h \\ 1/2h & \text{if } r=0 \text{ or } h . \end{cases} \tag{2.13}$$

In particular, (2.4) becomes

$$\sum_{r=0}^{h} \xi^*(x_r)x_r^j \cos(h \cos^{-1}x_r)=0 , \qquad 0\leqq j\leqq h-1 . \tag{2.14}$$

The value of the associated game with payoff given by (2.1) is

$$1/v_{00}(\xi^*)=2^{2-2h} . \tag{2.15}$$

Theorem 2.6 is used in sections 4 and 8. In section 7 we use the more difficult Chebyshev approximation result for Solotarev's first problem ([1], p. 280), where (in the notation of Theorem 2.5) $F(x)=x^p-p\sigma x^{p-1}$ with σ arbitrary and fixed, $G(x)\equiv 1$, and $n=p-2$. The required details of the solution to this problem are stated in section 7.

The problems treated in the remainder of this paper are all ones for which the symmetry and admissibility results of Kiefer ([5], [6]), and the fact that $D_Y(\xi)=\infty$ if ξ is supported by fewer than $m+1$ points, can be applied to yield

THEOREM 2.7. *If $X=[-1, 1]$, $Y=[-a, a]$ and $f_i(x)\equiv x^i$, $0\leqq i\leqq m$, then $D_Y(\xi)$ is minimized by a symmetric probability measure ξ on $m+1$ points, two of which are ± 1, and this minimizing measure is unique if $m>0$, $a>0$.*

We list the results of [5] and [6] which are used to prove this, restricting attention to the nontrivial case $m>0$, $a>0$: (1) ([6], Theorem 2.3.3) For any ξ and the symmetric $\bar{\xi}$ defined by $\bar{\xi}(A)=[\xi(A)+\xi(-A)]/2$, we have $D_Y(\bar{\xi})\leqq D_Y(\xi)$. We now take ξ to minimize $D_Y(\xi)$, and consider the corresponding $\bar{\xi}$. (2) ([5], p. 284 and Theorem 3.6) Define ξ_1 to be at least as good as ξ_2 if $M(\xi_1)-M(\xi_2)$ is nonnegative definite. This implies that ξ_1 is at least as good as ξ_2 for every linear estimation problem, and, in particular, $d(x, \xi_1)\leqq d(x, \xi_2)$ and $D_Y(\xi_1)\leqq D_Y(\xi_2)$. The ξ' for which $D_Y(\xi')<\infty$ and which are admissible according to this definition are shown to have $m+1$ points of support including ± 1, and we now consider such a ξ' which is at least as good as $\bar{\xi}$. (3) ([5], Lemma 3.5) This lemma states that ξ' is at least as good as $\bar{\xi}$ if and only if $\mu_i(\xi')=\mu_i(\bar{\xi})$ for $0\leqq i<2m$ and $\mu_{2m}(\xi')\geqq\mu_{2m}(\bar{\xi})$. Since $\mu_i(\bar{\xi})=0$ for i odd and $<2m$, it must therefore also be that $\mu_i(\xi')=0$ for i odd and $<2m$. (4) Since ξ' is admissible, is supported by $m+1$ points including ± 1, and has $\mu_i(\xi')=0$

for i odd and $<2m$, it must be symmetric; for the symmetrization $\bar{\xi}'$ of ξ' has the same moments as ξ', so that either (a) the support of ξ' is a symmetric set and thus $\bar{\xi}'$ is identical with ξ' (since the first $2m$ moments uniquely determine a measure on $\leqq m+1$ points, this also yields the uniqueness part of Theorem 2.7), or else (b) the support of ξ' is asymmetric and that of $\bar{\xi}'$ consists of $>m+1$ points, so that $\bar{\xi}'$ is inadmissible, contradicting the admissibility of ξ' since $M(\xi')=M(\bar{\xi}')$.

The conclusion of Theorem 2.7 is clearly valid if minimization of $D_Y(\xi)$ is replaced by any other criterion for which parts (1) and (2) of the proof hold.

In sections 3-7 we shall be concerned with $X=[-1, 1]$, $Y=Y_a=[-a, a]$ with $a\to 0$, $f_i(x)\equiv x^i$, $0\leqq i\leqq m$. In that setting we shall consider families $\{\xi_a\}$ of designs, the designs in a family depending on the parameter a. We have then

THEOREM 2.8. *If* $\{\xi_a\}$ *is a family of designs for which*

$$\lim_{a\to 0} D_{Y_a}(\xi_a)\leqq 1\,, \tag{2.16}$$

then equality holds in (2.16) and $\xi_a\to\xi_0$ *weakly, where* $\xi_0(1)=1$. *There exists a family satisfying (2.16) with equality.*

PROOF: We have $D_{Y_a}(\xi_a)\geqq d(0, \xi_a)$, and it is well known that $\min_\xi d(0, \xi)=d(0, \xi_0)=1$. (For example, from Theorem 3.6 of [5], ξ_0 is admissible; since only θ_0 is estimatable under ξ_0, ξ_0 must minimize $d(0, \xi)$.) This proves $D_{Y_a}(\xi_a)\geqq 1$ for all a, and hence the first part of the theorem. On the other hand, if ξ' is any design supported by at least m points other than 0 and $0<\epsilon<1$, then $M((1-\epsilon)\xi_0+\epsilon\xi')$ is nonsingular, and by the continuity of D_{Y_a} in a we have

$$\lim_{a\to 0} D_{Y_a}((1-\epsilon)\xi_0+\epsilon\xi')=d(0, (1-\epsilon)\xi_0+\epsilon\xi')\leqq(1-\epsilon)^{-1}\,, \tag{2.17}$$

the last inequality following from consideration of the estimator of θ_0 which is obtained by averaging the observations taken at $x=0$. The existence of a family satisfying (2.16) with equality follows from (2.17).

Of course, (2.16) is a very crude estimate, and it will be improved upon in Theorems 3.1 and 5.1.

3. Polynomial interpolation, m even. Reduction of the problem

We now consider the case $X=[-1, 1]$, $Y=Y_a=[-a, a]$, $a\to 0$, $f_i(x)\equiv x^i$, $i=0, 1, \cdots, m$, m even. The case $m=0$ is trivial, so we assume $m>0$. We would like to find a design (probability measure) on X, more precisely, a family of designs which depend on the positive parameter a, such that, as $a \to 0$, the maximum, in Y_a, of the variance of the best linear estimator of the regression function $\theta' f(y)$, will be approximately a minimum. We now put this last condition precisely. We seek a family of designs on X, say $\{\xi_a^*\}$, with the following property: Let $\{\xi_a\}$ be any family of designs on X. Then the inequality

$$\underline{\lim}_{a \to 0} \frac{D_{Y_a}(\xi_a) - 1}{D_{Y_a}(\xi_a^*) - 1} \geq 1 \tag{3.1}$$

holds. The reason for subtracting one in both numerator and denominator is this: Recall that Theorem 2.8 implies $D_{Y_a}(\xi_a^*) = 1 + o(1)$ for an optimum family. We shall see below that there are many designs $\{\xi_a\}$ for which

$$D_{Y_a}(\xi_a) = 1 + O(a) .$$

If we used the criterion based on

$$\frac{D_{Y_a}(\xi_a)}{D_{Y_a}(\xi_a^*)} \tag{3.2}$$

instead of (3.1) we would not be able to distinguish among these designs. On the other hand, we shall see below that (3.1) does distinguish satisfactorily among all designs.

The reason we consider only the limiting form (3.1) rather than exact optimality for each value a, is that the latter computation admits no simple solution.

Let ϵ_a be a positive function of a such that $\lim_{a \to 0} \epsilon_a = 0$. It follows from Theorems 2.7 and 2.8 that, for the solution of our problem, it is sufficient to consider only designs of the form

$$\xi_a = (1 - \epsilon_a)\xi_0 + \epsilon_a \mu_a , \tag{3.3}$$

where $\xi_0(0) = 1$, ϵ_a is a properly chosen function of a, and μ_a (which depends on a) is a properly chosen symmetric design on m points of $[-1$ $1]$, among which are the points ± 1. We are unable to solve the prob lem in this generality. Instead we will seek a solution only among the class of designs ξ_a of the form

$$\nu_a \equiv \nu(\epsilon_a, \mu) = (1 - \epsilon_a)\xi_0 + \epsilon_a \mu , \tag{3.4}$$

where μ (which now does *not* depend on a) is some (fixed) symmetric design on m points of $[-1, 1]$, two of which are ± 1. Our solution will therefore consist of a design μ^* (i.e., $\mu = \mu^*$) and a function ϵ_a^*. Moreover, the competitive designs $\{\xi_a\}$ with which we will compare $\nu_a^* = \nu(\epsilon_a^*, \mu^*)$ in (3.1) will also be of the form (3.4), with, of course, possibly different ϵ_a and μ. Thus, we shall prove

THEOREM 3.1.[3] *When m is even the family of designs $\{\nu(\epsilon_a^*, \mu^*)\}$ defined by (3.4) and (4.11) below, satisfies (3.1) among all families $\{\nu(\epsilon_a, \mu)\}$ which satisfy (3.4).*

Actually, our proof will also show that $\{\nu(\epsilon_a^*, \mu^*)\}$ satisfies (3.1) among all families $\{\xi_a\}$ for which

(3.5) $$d(a, \xi_a) \geqq v_{00}(\xi_a) + a^2 v_{11}(\xi_a) + o(a)$$

((3.5) is satisfied by $\nu(\epsilon_a^*, \mu^*)$). This is a less natural and less practical condition than (3.4).

Consider the family of designs (3.4) for some μ and ϵ_a. The determinant of the ith principal minor of $M(\nu)$ is clearly $c_i'\epsilon_a^{-1+m}(1+o(1))$, for $i>0$, while a computation by minors of the first row of M yields $\det M = c_0'\epsilon_a^m(1+o(1))$. Hence

(3.6) $$v_{ii}(\nu_a) = c_i \epsilon_a^{-1}(1+o(1)), \qquad i>0 .$$

Obviously

(3.7) $$1 = 1/m_{00}(\nu_a) \leqq v_{00}(\nu_a) = 1 + O(\epsilon_a) .$$

Hence, by Schwarz's inequality,

(3.8) $$v_{0i}(\nu_a) = O(\epsilon_a^{-1/2}), \qquad i \geqq 1 .$$

For odd i we have the stronger result

(3.9) $$v_{0i}(\nu_a) = 0, \qquad i \text{ odd},$$

by symmetry. Also

(3.10) $$v_{ij}(\nu_a) = O(\epsilon_a^{-1}) \qquad i, j \geqq 1 .$$

Hence

(3.11) $$d(a, \nu_a) = v_{00}(\nu_a) + a^2 v_{11}(\nu_a)(1+o(1)) .$$

We will obtain in section 4 a μ and ϵ_a, say μ', ϵ_a', which satisfy (3.12), (3.13), and (3.14), namely: Let μ^0 and ϵ_a^0 be any pair μ, ϵ_a, and write $\nu_a' = \nu(\epsilon_a', \mu')$ and $\nu_a^0 = \nu(\epsilon_a^0, \mu^0)$. Then

(3.12) $$\varliminf_{a \to 0} \frac{v_{00}(\nu_a^0) - 1 + a^2 v_{11}(\nu_a^0)}{v_{00}(\nu_a') - 1 + a^2 v_{11}(\nu_a')} \geqq 1 ,$$

(3.13) $$D_{Y_a}(\nu_a') - d(a, \nu_a') = o(a) ,$$

3) The authors have recently proved that the conclusion of Theorem 3.1 holds without any restriction whatsoever (such as (3.4)) on the families considered. The additional argument required will appear in a later paper.

(3.14) $$d(a, \nu_a') = v_{00}(\nu_a') + a^2 v_{11}(\nu_a') + o(a) = 1 + O(a)\,.$$

We will also show that

(3.15) $$\inf_{\{\epsilon_a\},\,\mu} \varliminf_{a\to 0} \frac{v_{00}(\nu(\epsilon_a, \mu)) - 1 + a^2 v_{11}(\nu(\epsilon_a, \mu))}{a} > 0\,.$$

It follows from (3.11), the inequality of (3.7), and (3.14), that any family $\{\nu_a^0\}$, for which

(3.16) $$\lim\,[d(a, \nu_a^0) - 1]/[d(a, \nu_a') - 1] < 1$$

on a sequence of a's approaching zero, must satisfy on that sequence

$$a^2 v_{11}(\nu_a^0) = O(a)$$

and hence, from (3.11),

(3.17) $$d(a, \nu_a^0) = v_{00}(\nu_a^0) + a^2 v_{11}(\nu_a^0) + o(a)$$

holds on that sequence. But then (3.12) and (3.16) contradict each other. We conclude that $\xi_a^* = \xi_a'$, $\mu^* = \mu'$, is a solution of our problem (where we limit ourselves to families of the form (3.4)). Moreover, it will appear that μ^* is unique and ξ_a^* can be varied only in terms of higher order, so that the solution to this problem is not trivial as would have been that based on (3.2).

We conclude that Theorem 3.1 will follow with $\epsilon_a^* = \epsilon_a'$, $\mu^* = \mu'$, if we prove (3.15) and find an ϵ_a', μ' which satisfy (3.12) - (3.14).

4. Polynomial interpolation, m even. Conclusion of the proof of Theorem 3.1

It follows from Theorem 2.1 that

$$(1-\epsilon) + \epsilon \cdot \min_{\{a_j\}} \int (1 - \sum_{j=1}^{m} d_j x^j)^2 d\mu = [v_{00}((1-\epsilon)\xi_0 + \epsilon\mu)]^{-1}$$

and

$$\epsilon \cdot \min_{\{b_j\}} \int (x - \sum_{j\neq 1} b_j x^j)^2 d\mu = [v_{11}((1-\epsilon)\xi_0 + \epsilon\mu)]^{-1}\,.$$

Define

(4.1) $$H(a) = \min_{\epsilon,\mu} \left\{ \frac{1}{(1-\epsilon) + \epsilon \min_{\{a_j\}} \int (1 - \sum_{j=1}^{m} d_j x^j)^2 d\mu} \right.$$

$$+\frac{a^2}{\epsilon \min_{\{b_j\}} \int (x-\sum_{j\neq 1} b_j x^j)^2 d\mu}\Big\} .$$

We shall now seek to find $H(a)$. Whatever μ may be, the d_j ($j=1, \cdots, m$) can clearly be chosen so as to make the first integrand in (4.1) zero at the m points supporting μ. (This is merely the fact that the best linear estimator of θ_0 is the average of the observations at $x=0$.) Also we can set $b_j=0$ for j even, by the symmetry of μ. Hence

$$H(a)=\min_{\epsilon}\left\{\frac{1}{1-\epsilon}+\frac{a^2}{\epsilon \max_{\mu} \min_{\{b_j\}} \int (x-\sum_{j\neq 1}' b_j x^j)^2 d\mu}\right\} . \tag{4.2}$$

We shall now try to solve the game with payoff function

$$\int (x-\sum_{j\neq 1}' b_j x^j)^2 d\mu , \tag{4.3}$$

where the maximizing player chooses μ and the minimizing player chooses $\{b_j\}$. As described in Theorem 2.2, this means solving the problem of Chebyshev approximation of x by $\{x^3, x^5, \cdots, x^{m-1}\}$. The function

$$g(x)=(-1)^{(m+2)/2}(m-1)^{-1}\cos[(m-1)\cos^{-1}x] , \tag{4.4}$$

a multiple of the Chebyshev polynomial of the first kind, is an odd polynomial in x of degree $m-1$ with coefficient of x equal to one. (See Erdelyi [3], page 185, equation (2.2).) Since it has maximum absolute value $(m-1)^{-1}$, attained with alternating signs at the m points $x_r=\cos[\pi r/(m-1)]$, $r=0, \cdots, m-1$, it follows from Theorem 2.5 (with $F(x)=x$, $G(x)=x^3$, $n=m-4$), that $x-g(x)$ is the unique best approximating polynomial in the Chebyshev sense. From (2.14) we have, putting $\alpha_r=1/2$ for $r=0$ and $r=m-1$, and $\alpha_r=1$ for $1\leqq r\leqq m-2$, that

$$\sum_{r=0}^{m-1} \alpha_r g(x_r)x_r^j=0 , \qquad 0\leqq j\leqq m-2 . \tag{4.5}$$

Now, using the fact (Jolley [4], page 82, formula 445) that

$$\sum_{r=0}^{m-2} \sec^2[\pi r/(m-1)]=(m-1)^2, \tag{4.6}$$

we can define a probability measure λ on the x_r by

$$\lambda(x_r)=(m-1)^{-2}\alpha_r/x_r^2 , \qquad 0\leqq r\leqq m-1 . \tag{4.7}$$

It then follows from (4.5) that

$$\sum_{r=0}^{m-1} \lambda(x_r)g(x_r)x_r^j=0 , \qquad 2\leqq j\leqq m , \tag{4.8}$$

and of course (4.8) holds for $j=0$ because g is odd and λ is symmetric. Thus, we have proved the orthogonality relations (2.4) for λ and the $\{b_j\}$ corresponding to g (these require (4.8) only for odd $j>1$), so that λ is maximin. Since $g^2(x_r)=(m-1)^{-2}$, this is the value of the game.

Returning now to (4.2), we obtain

$$H(a)=\min_{\epsilon}\left\{\frac{1}{1-\epsilon}+\frac{a^2(m-1)^2}{\epsilon}\right\}=[1+a(m-1)]^2, \tag{4.9}$$

achieved for

$$\epsilon=a(m-1)/[1+a(m-1)]\,. \tag{4.10}$$

When a is small the minimizing value of ϵ is approximately $(m-1)a$.

We will now show that

$$\epsilon_a'=(m-1)a,\quad \mu'=\lambda \tag{4.11}$$

is a solution of our problem. It is trivial that (3.14) holds. That (3.12) and (3.15) hold follows from the manner in which we obtained the value of $H(a)$. It remains only to verify (3.13). From (4.11) it follows easily that (3.6) and (3.7) can be strengthened, respectively, to

$$v_{ii}(\nu((m-1)a,\,\lambda))=c_i''a^{-1}(1+o(1))\,,\qquad i>0\,, \tag{4.12}$$

$$v_{00}(\nu((m-1)a,\,\lambda))=1+c_0''a+o(a)\,. \tag{4.13}$$

Hence, if $|y|\leqq a$, we have

$$d(y,\,\nu((m-1)a,\,\lambda))=1+c_0''a+c_1''y^2/a+o(a)\,, \tag{4.14}$$

from which (3.13) follows easily. This completes the proof of Theorem 3.1.

5. Polynomial interpolation, m odd. Reduction of the problem

We now consider the case $X=[-1,\,1]$, $Y=Y_a=[-a,\,a]$, $a\to 0$, $f_i(x)\equiv x^i$, $i=0,\,\cdots,\,m$, m odd. We shall assume in sections 5-7 that $m>1$; when $m=1$, an optimum design for all a is given by $\xi^*(-1)=\xi^*(1)=1/2$. Because we are unable to solve the general problem we restrict ourselves to families of designs of the form

$$\gamma_a\equiv\gamma(\epsilon_a,\,\delta_a,\,\mu)=(1-\epsilon_a)\xi_{0a}+\epsilon_a\mu\,, \tag{5.1}$$

where, as always, ϵ_a is a positive function of a which approaches zero as $a\to 0$, μ is now any fixed (as a varies) symmetric probability measure on $(m-1)$ points (none of which can therefore be zero) including ± 1, and ξ_{0a} assigns probability 1/2 to each of the two points $\pm\delta_a$, where δ_a

is a positive function of a, $|\delta_a|<1$, and $\delta_a \to 0$ as $a \to 0$. The general problem would be to allow μ to depend on a, similarly to the m even case discussed in section 3. In that form it follows from Theorem 2.7 that it is no restriction to assume that the measures ξ_{0a} and μ are symmetric and are supported respectively by 2 points and by $m-1$ points including ± 1. The mathematical difficulties which lead us to restrict the form to (5.1) are like those of section 3. By Theorem 2.8, the conditions $\epsilon_a \to 0$, $\delta_a \to 0$ are not really restrictions. The reason for using the criterion (3.1) is discussed in section 3. Our solution will therefore consist of an ϵ_a^*, δ_a^*, and μ^*. Also the competitive designs will consist of an ϵ_a, δ_a, and μ. We shall prove

THEOREM 5.1.[4] *When m is odd the family of designs defined by (5.1) and (6.34) below, satisfies (3.1) among all families $\{\gamma(\epsilon_a, \delta_a, \mu)\}$ which satisfy (5.1).*

As in section 3.1, our optimum μ^* is unique and our optimum ϵ_a^*, δ_a^* can only be changed by terms of higher order.

Before proceeding to the proof we shall devote this paragraph to an outline of it. Define

$$h(\gamma_a, a) = v_{00}(\gamma_a) + a^2 v_{11}(\gamma_a) . \tag{5.2}$$

Suppose first that, in addition to (5.1), which will always be required, we impose the requirement (to be removed later) that

$$\delta_a = o(\epsilon_a^{1/4}) . \tag{5.3}$$

We will then show that

$$d(a, \gamma_a) \geqq v_{00}(\gamma_a) + a^2 v_{11}(\gamma_a)(1+o(1)) . \tag{5.4}$$

We will obtain in section 6 a design $\gamma_a^* = \gamma(\epsilon_a^*, \delta_a^*, \mu^*)$ such that

$$D\gamma_a(\gamma_a^*) = h(\gamma_a^*, a) + o(a) \tag{5.5}$$

and

$$h(\gamma_a^*, a) = \min_{\xi} h(\xi, a) + o(a) , \tag{5.6}$$

where the minimum is over all designs ξ (or equivalently, since a is fixed, over all designs which satisfy (5.3)). It will be apparent from the argument that

$$\inf_{\gamma_a} \underline{\lim}_{a \to 0} \frac{v_{00}(\gamma_a) - 1 + a^2 v_{11}(\gamma_a)}{a} > 0 \tag{5.6a}$$

4) The authors have recently proved that the conclusion of Theorem 5.1 holds without any restriction whatsoever (such as (5.1)) on the families considered. The additional argument required will appear in a later paper.

and

$$h(\gamma_a^*,\ a)=1+O(a)\ . \tag{5.7}$$

Then, using an argument like that of (3.16) - (3.17), the conclusion of Theorem 5.1 follows from (5.4) - (5.7) in the same way that Theorem 3.1 followed from (3.12) - (3.15), provided we restrict consideration to families which satisfy the additional condition (5.3). However, it will be shown in the second paragraph below that, if (5.3) does not hold on a sequence of a's approaching zero, then, on that sequence of a's,

$$D_{\Gamma_a}(\gamma_a)-1\neq O(a)\ . \tag{5.8}$$

From this we conclude that Theorem 5.1 is a consequence of (5.4) - (5.7). The above outline is similar to that given in section 3 for the proof of Theorem 3.1, modifications being necessary here because of the presence of ξ_{0a}, which makes it more difficult to compute the orders of the v_{ij}.

Define

$$A(\gamma_a)=\min_{\{d_j\}}\int(1-\sum_{j\neq 0}d_jx^j)^2\gamma_a(dx) \tag{5.9}$$

$$=\min_{\{d_j\}}\int(1-\sum_{j\neq 0}{}''\ d_jx^j)^2\gamma_a(dx)$$

and

$$B(\gamma_a)=\min_{\{b_j\}}\int(x-\sum_{j\neq 1}b_jx^j)^2\gamma_a(dx) \tag{5.10}$$

$$=\min_{\{b_j\}}\int(x-\sum_{j\neq 1}{}'b_jx^j)^2\gamma_a(dx)\ .$$

(The last equalities in (5.9) and (5.10) are a consequence of symmetry, just as was (4.2) in section 4.) We have

$$h(\gamma_a,\ a)=(A(\gamma_a))^{-1}+a^2(B(\gamma_a))^{-1}\ . \tag{5.11}$$

We now prove (5.8). We have

$$v_{00}^{-1}(\gamma_a)=A(\gamma_a)\ . \tag{5.12}$$

Regarding $A(\gamma_a)$ as the square of the length of a projection, we have

$$A(\gamma_a)\leq 1-\frac{\left(\int x^2\gamma_a(dx)\right)^2}{\int x^4\gamma_a(dx)}\leq 1-\frac{(1-\epsilon_a)^2\delta_a^4}{\epsilon_a+(1-\epsilon_a)\delta_a^4} \tag{5.13}$$

$$=\frac{\epsilon_a[1+(1-\epsilon_a)\delta_a^4]}{\epsilon_a[1+(1-\epsilon_a)\delta_a^4]+(1-\epsilon_a)^2\delta_a^4}\ .$$

Hence

(5.14)
$$d(0, \gamma_a)=1/A(\gamma_a)\geqq 1+c_0\frac{\delta_a^4}{r_a} .$$

If (5.3) does not hold then (5.14) implies (5.8). This proves that (5.3) must hold.

For any design γ_a of the form (5.1), write $M(\gamma_a)$ by first writing the even $(0, 2, 4, \cdots, m-1)$ rows and columns and then writing the odd $(1, 3, 5, \cdots, m)$ rows and columns. Since the odd moments of γ_a are zero we obtain

(5.15)
$$M(\gamma_a)=\begin{pmatrix} M_E(\gamma_a) & 0 \\ 0 & M_0(\gamma_a) \end{pmatrix} ,$$

where

(5.16)
$$M_E(\gamma_a)=\begin{pmatrix} z_0, & z_2, & \cdots, & z_{m-1} \\ z_2, & z_4, & \cdots, & z_{m+1} \\ \cdot & \cdot & \cdot & \cdot \\ z_{m-1}, & z_{m+1}, & \cdots, & z_{2m-2} \end{pmatrix}$$

$$M_0(\gamma_a)=\begin{pmatrix} z_2, & z_4, & \cdots, & z_{m+1} \\ z_4, & z_6, & \cdots, & z_{m+3} \\ \cdot & \cdot & \cdot & \cdot \\ z_{m+1}, & z_{m+3} & \cdots, & z_{2m} \end{pmatrix} .$$

Here

(5.17)
$$z_i=\epsilon_a h_i+(1-\epsilon_a)\delta_a^i ,$$

and h_i is the ith moment of μ. From (5.3) we have

(5.18)
$$z_{2i}=\epsilon_a h_{2i}(1+o(1)) , \qquad i\geqq 2 .$$

Write

$$G_1=\begin{pmatrix} 1, & h_2, & \cdots, & h_{m-1} \\ h_2, & h_4, & \cdots, & h_{m+1} \\ \cdot & \cdot & \cdot & \cdot \\ h_{m-1}, & h_{m+1}, & \cdots, & h_{2m-2} \end{pmatrix}$$

$$G_2=\begin{pmatrix} h_2, & h_4, & \cdots, & h_{m+1} \\ h_4, & h_6, & \cdots, & h_{m+3} \\ \cdot & \cdot & \cdot & \cdot \\ h_{m+1}, & h_{m+3}, & \cdots, & h_{2m} \end{pmatrix} .$$

Since μ has $(m-1)$ symmetric points of support, none of them zero, it follows that the $(m+1)/2\times(m+1)/2$ matrices G_1 and G_2 have rank $(m-1)/2$ and are singular, while all their minors are nonsingular. (This is easily verified by considering the elements h_{2i} of G_1 and G_2 as the ith moments of the measure τ on $(m-1)/2$ points of $(0,1]$, defined by $\tau(x^2)=2\mu(x)$.) Let $G_k(i, j)$ and $G_k(i, r; j, s)$ be, respectively, the minor, in G_k, of the (i, j)th element, and of the i, r rows and j, s columns. We will number the rows and columns of G_1 by $0, 2, \cdots, m-1$, and those of G_2 by $1, 3, \cdots, m$.

We shall presently show that

$$v_{0i}(\gamma_a)=o(v_{11}(\gamma_a)), \qquad i=2, 4, \cdots, (m-1) \tag{5.19}$$

and

$$v_{1i}(\gamma_a)=O(v_{11}(\gamma_a)), \qquad i=3, 5, \cdots, m\ . \tag{5.20}$$

It follows from (5.19) that

$$\sum_{j=1}^{(m-1)/2} v_{0,2j}(\gamma_a)a^{2j}=o(a^2v_{11}(\gamma_a))\,, \tag{5.21}$$

and from (5.20) that

$$\sum_{j=1}^{(m-1)/2} v_{1,2j+1}(\gamma_a)a^{2j+2}=o(a^2v_{11}(\gamma_a))\ . \tag{5.22}$$

Since $M^{-1}(\gamma_a)$ is positive definite,

$$\sum_{i,j\geq 2} v_{ij}(\gamma_a)a^{i+j}\geq 0\ . \tag{5.23}$$

From (5.21), (5.22), and (5.23) we obtain the desired result (5.4). It remains to prove (5.19) and (5.20).

From (5.18) and the fact that the minors of G_2 are nonsingular we have

$$\frac{v_{1i}(\gamma_a)}{v_{11}(\gamma_a)}=\frac{\det G_2(1, i)}{\det G_2(1, 1)}(1+o(1))\,, \qquad i=3, 5, \cdots, m\,, \tag{5.24}$$

which proves (5.20).

We now prove (5.19). We shall make use of the trivial estimate

$$v_{11}(\gamma_a)\geq 1/m_{11}(\gamma_a)>1/[\epsilon_a h_2+\delta_a^2]\ . \tag{5.25}$$

Expanding $\det M_E(\gamma_a)$ in minors of its first row and using (5.3), (5.17), and (5.18) (and, in particular, that $z_0z_4=h_4\epsilon(1+o(1))$ while $z_2^2=o(\epsilon)$), we obtain

$$\det M_E(\gamma_a)=\epsilon_a^{(m-1)/2}\det G_1(0, 0)(1+o(1))\ . \tag{5.26}$$

Similarly, the minor of the $(0, i)$th element of $M_E(\gamma_a)$ is

(5.27) $$v_{0i}(\gamma_a) \det M_E(\gamma_a) = \det G_1(0, i) O(\epsilon_a^{(m-3)/2}[\epsilon_a + \delta_a^2]) , \quad i=2, 4, \cdots, m-1 .$$

Hence, from (5.25) - (5.27), for $i=2, 4, \cdots, m-1$,

(5.28) $$\frac{v_{0i}(\gamma_a)}{v_{11}(\gamma_a)} \leq [\epsilon_a h_2 + \delta_a^2] O\left(1 + \frac{\delta_a^2}{\epsilon_a}\right) = O(\epsilon_a) + O(\delta_a^2) + O\left(\frac{\delta_a^4}{\epsilon_a}\right) .$$

The desired result (5.19) now follows from (5.28) and (5.3).

We next prove (5.5). Since γ_a^* satisfies (5.3), it satisfies (5.19) and (5.20). We shall see in section 6 that

(5.29) $$v_{11}(\gamma_a^*) = O(a^{-1})$$

and

(5.30) $$(\delta_a^{*2})(\epsilon_a^*)^{-1} = O(1) , \qquad \frac{a}{\epsilon_a^*} = O(1) .$$

Assuming (5.29) and (5.30), we will prove in the next paragraph that

(5.31) $$v_{ii}(\gamma_a^*) = O(a^{-1}) , \qquad i \geqq 2 .$$

From (5.31) by Schwarz's inequality we obtain

(5.32) $$v_{ij}(\gamma_a^*) = O(a^{-1}) , \qquad i, j \geqq 2 .$$

From (5.19), (5.20), (5.31), and (5.32), an argument like that used to prove (5.21) and (5.22) yields

(5.33) $$\sup_{|x| \leq a} |d(x, \gamma_a^*) - h(\gamma_a^*, x)| = o(a) ,$$

from which (5.5) follows at once. It remains to prove (5.31); the reader will verify (5.29) and (5.30) in section 6.

We have, for i odd and $\neq 1$,

(5.34) $$\frac{v_{ii}(\gamma_a^*)}{v_{11}(\gamma_a^*)} = \frac{\det G_2(i, i) + (\delta_a^{*2}/\epsilon_a^*) \det G_2(1, i; 1, i)}{\det G_2(1, 1)} (1 + o(1)) ,$$

by (5.3) and the same argument which led to (5.24). From (5.29), (5.30), and (5.34) we obtain (5.31) for i odd and $\neq 1$. When i is even and $\neq 0$ we have, by (5.26) and an application, of the argument which led to (5.26), to the computation of the principal minors of $M_E(\gamma_a)$, that

(5.35) $$v_{ii}(\gamma_a^*) = \frac{(\epsilon_a^*)^{(m-3)/2} \det G_1(0, i; 0, i)}{\det M_E(\gamma_a^*)} (1 + o(1)) = O(\epsilon_a^{*-1}) .$$

Application of (5.30) to (5.35) completes the proof of (5.31) for i even and $\neq 0$, and hence the proof of (5.31).

We now recapitulate the present status of the proof of Theorem 5.1. We have already proved (5.3), (5.4), and (5.5). It remains to verify (5.6), (5.6a), (5.7), (5.29), and (5.30). This will be done in section 6.

6. Polynomial interpolation, m odd. Conclusion of the proof of Theorem 5.1

We now proceed to the task of obtaining ϵ_a^*, δ_a^*, and μ^* so as to satisfy (5.6). Consider $B(\gamma_a)$ defined in (5.10). If the b_j are chosen to be those values b'_j (say) for which $(x-\sum_{j\neq 1}' b_j x^j)$ vanishes on the support $\{\pm x_i,\ i=1, \cdots, (m-1)/2\}$ (say) of μ, then

(6.1) $$\delta_a^{-2}\int (x-\sum_{j\neq 1}' b'_j x^j)^2 \gamma_a(dx) = 1+o(1)\,.$$

Hence,

(6.2) $$(\delta_a)^{-2} B(\gamma_a) \leqq 1+o(1)$$

for every γ_a. We shall now show that ϵ_a^* and δ_a^* (the existence of which will follow from subsequent developments) must be such that

(6.3) $$\lim_{a\to 0} (\delta_a^*)^{-2} B(\gamma_a^*) = 1\,.$$

Suppose to the contrary that on a sequence $\{a_n\}$ tending to 0 we have

(6.4) $$\lim_{n\to\infty} (\delta_{a_n}^*)^{-2} B(\gamma_{a_n}^*) = s < 1\,.$$

(By (6.2), it cannot be that $s>1$.) Let $b_{3a}, b_{5a}, \cdots, b_{ma}$ be, respectively, the (as it will turn out, unique) minimizing values of $b_3, \cdots, b_m$ in the last expression of (5.10) with $\gamma_a = \gamma_a^*$. Obviously

(6.5) $$B(\gamma_{a_n}^*) \geqq (1-\epsilon_{a_n}^*)\delta_{a_n}^{*2}(1-\sum_{j\neq 0}'' b_{(j+1)a_n} \delta_{a_n}^{*j})^2.$$

Hence for a subsequence $\{e_n\}$ of $\{a_n\}$ we must have, for some $s' \leqq s < 1$,

(6.6) $$\lim_{n\to\infty} (1-\sum_{j\neq 0}'' b_{(j+1)e_n} \delta_{e_n}^{*j})^2 = s' < 1\,.$$

Now (6.6) could hold only if, for a subsequence $\{g_n\}$ of $\{e_n\}$ and some odd i, $3 \leqq i \leqq m$,

(6.7) $$b_{ig_n}(\delta_{g_n}^*)^{i-1} > c_1\,, \qquad n=1, 2, \cdots$$

so that

(6.8) $$b_{ig_n} > c_1(\delta^*_{g_n})^{-2}, \qquad n=1, 2, \cdots.$$

Consider the non-homogeneous linear equations in the variables $z_3, \cdots, z_m$:

(6.9) $$x_k - \sum_{j \neq 1}^{m} x_k^j z_j = 0, \qquad k=1, 2, \cdots, (m-1)/2.$$

Their determinant is a non-zero multiple of a Vandermonde determinant, so that their solution is unique. Let $z'_3, \cdots, z'_m$ be this solution. It follows that

(6.10) $$\int (x - \sum_{j \neq 1}{}' b_j x^j)^2 \mu(dx)$$

is a positive definite quadratic form in $(b_3 - z'_3), \cdots, (b_m - z'_m)$. Hence, when (6.8) holds (so that $|b_{ig_n}| \to \infty$), for n sufficiently large,

(6.11) $$B(\gamma^*_{g_n}) \geqq c_2 \epsilon^*_{g_n} (\delta^*_{g_n})^{-4}.$$

However, (6.11) and (5.3) contradict (6.4). This proves (6.3).

Let $d_{2a}, \cdots, d_{(m-1)a}$ be, respectively, the (as it will turn out, unique) $d_2, \cdots, d_{(m-1)}$ which minimize the last expression of (5.9) with $\gamma_a = \gamma^*_a$. We will now prove

(6.12) $$d_{ka}(\delta^*_a)^k = o(a), \qquad \text{all even } k \geqq 4.$$

This is an immediate consequence of

(6.13) $$d_{ka}(\delta^*_a)^2 = O(a^{1/2}), \qquad \text{all even } k \geqq 2,$$

and

(6.14) $$(\delta^*_a)^2 = o(a^{1/2}),$$

which we will now demonstrate. To prove (6.14), we note that (5.5) and (5.7), which will be demonstrated later for the γ^*_a defined by (6.34), and (5.14) imply a sharpening of (5.3), namely,

(6.15) $$(\delta^*)^4 / \epsilon^*_a = O(a).$$

Thus, (6.14) follows from (6.15) and the fact that $\epsilon^*_a = o(1)$. Turning to (6.13), if it does not hold then, for a sequence $\{a_n\}$ tending to zero and for some even $k \geqq 2$,

(6.16) $$\lim_{n \to \infty} a_n^{-1/2} |d_{ka_n}| (\delta^*_{a_n})^2 = \infty.$$

From (6.16) and (6.14), we have $\lim|d_{ka_n}|=\infty$. Hence, from (6.16) and an argument like that which led from (6.8) to (6.11) we obtain that, for large n,

$$A(\gamma^*_{a_n})\geqq a_n\epsilon^*_{a_n}(\delta^*_{a_n})^{-4}K_n \tag{6.17}$$

where $\lim K_n=\infty$. On the other hand, by setting all $d_j=0$ we see that

$$A(\gamma^*_{a_n})\leqq 1\,. \tag{6.18}$$

From (6.17) and (6.18) we have

$$\lim_{n\to\infty}(\delta^*_{a_n})^4/a_n\epsilon^*_{a_n}=\infty\,. \tag{6.19}$$

This contradicts (6.15), completing the proof of (6.13) and thus of (6.12).

In the next paragraph we shall use the estimates (6.12) and (6.13) in (6.24), but we shall otherwise let the d_{2j}'s vary without restriction. This is justified by the fact that this development yields d_{2j}'s which do not depend on a and hence satisfy (6.12) - (6.13), and yields μ^*, $\{\delta^*_a\}$, $\{\epsilon^*_a\}$ which satisfy the restrictions previously obtained and minimize $h(\gamma_a, a)$ to within $o(a)$.

Now, for fixed d_2, define

$$F(d_2,\mu)=\min_{\{d_j,\,j>2\}}\int(1-\sum_{j\neq 0}{}''d_jx^j)^2\mu(dx) \tag{6.20}$$

and

$$[F(d_2)]^{1/2}=\min_{\{d_j,\,j>2\}}\max_{|x|\leq 1}|1-\sum_{j\neq 0}{}''d_jx^j|\,. \tag{6.21}$$

By Theorem 2.4 we have

$$F(d_2)=\max_{\mu}F(d_2,\mu)\,. \tag{6.22}$$

Now, according to (6.2) and (6.3), in our problem of minimizing $h(\gamma_a, a)$ to within $o(a)$ we can take the second term on the right side of (5.11) to be $(a/\delta^*_a)^2(1+o(1))$ and simply consider the problem of minimizing $(A(\gamma_a))^{-1}$ with respect to μ and without restrictions. That is, we can take $\mu^*_{d_2}$ to satisfy

$$F(d_2)=F(d_2,\mu^*_{d_2}) \tag{6.23}$$

and, for an appropriate choice d^*_2 of d_2, we will then obtain $\mu^*=\mu^*_{d_2^*}$. Write $\theta_a=-\epsilon^*_a-d_2(\delta^*_a)^2/2$. By (6.13) and (5.1), $\theta_a=o(1)$. Hence, from (5.11), (6.12), (6.23), and (6.3), we have

$$h(\gamma_a^*, a)=[\min_{d_2}\{\epsilon_a^* F(d_2)+(1-\epsilon_a^*)[1-d_2(\delta_a^*)^2+o(a)]^2\}]^{-1}$$

(6.24)
$$+a^2(\delta_a^*)^{-2}(1+o(1))$$
$$=[\min_{d_2}\{1-\epsilon_a^*(1-F(d_2)+o(1))-2d_2(\delta_a^*)^2(1+\theta_a)+o(a)\}]^{-1}$$
$$+a^2(\delta_a^*)^{-2}(1+o(1))\,.$$

We now set

$$\delta_a^*=h_a a^{1/2}, \quad \epsilon_a^*=e_a a \tag{6.25}$$

in the right member of (6.24), which becomes

$$[\min_{d_2}\{1-a[e_a(1-F(d_2))+2d_2h_a^2(1+\theta_a)]\}]^{-1} \tag{6.26}$$
$$+ah_a^{-2}+o(a)\,.$$

For small a (6.26) is not less than

$$1+a\max_{d_2}[e_a(1-F(d_2))+2d_2h_a^2(1+\theta_a)]+ah_a^{-2}+o(a)\,. \tag{6.27}$$

We shall now consider the problem of minimizing to within $o(a)$, with respect to e and h, the expression

$$1+a\max_{d_2}[e(1-F(d_2))+2d_2h^2]+ah^{-2}+o(a)\,, \tag{6.28}$$

by minimizing

$$h^{-2}+\max_{d_2}[e(1-F(d_2))+2d_2h^2]\,. \tag{6.29}$$

We shall see that the minimum is achieved for unique finite non-zero e and h, say $e=e^*$, $h=h^*$. Moreover, it will be clear from continuity considerations and the fact that $\theta_a=o(1)$ that the choice $e_a=e^*$, $h_a=h^*$ minimizes (6.27) to within $o(a)$. Since also (6.26) and (6.27) are clearly equal for this choice, we conclude that these values minimize $h(\gamma_a, a)$ to within $o(a)$. Moreover, if $[e^*(1-F(d_2))+2d_2(h^*)^2]$ is maximized by $d_2=d_2^*$, it will become clear (see especially (7.2)) that $\mu^*=\mu^*_{d_2^*}$ is the optimum choice of μ.

Define

(6.30)
$$\lambda_i=\int x^{2i}\mu^*_{d_2}(dx), \qquad \Lambda=\{\lambda_{i+j+2},\ 1\leq i,\ j\leq(m-3)/2\}\,,$$
$$\rho=(\lambda_2, \lambda_3, \cdots, \lambda_{(m-1)/2})',\ \eta=(\lambda_3, \lambda_4, \cdots, \lambda_{(m+1)/2})',$$
$$\bar d=(d_4, d_6, \cdots, d_{m-1})',\ \ U=(x^4, x^6, \cdots, x^{m-1})'.$$

We note that Λ is nonsingular. From (6.20) and (6.23) we have

$$F(d_2)=\min_{\bar{d}}\int(1-d_2x^2-\bar{d}'U)^2\mu^*_{d_2}(dx)=\int(1-d_2x^2)^2\mu^*_{d_2}(dx)$$

$$(6.31)\qquad -\left[\int(1-d_2x^2)U\mu^*_{d_2}(dx)\right]\left[\int UU'\mu^*_{d_2}(dx)\right]^{-1}\left[\int(1-d_2x^2)U\mu^*_{d_2}(dx)\right]$$

$$=1-2d_2\lambda_1+d_2^2\lambda_2-(\rho-d_2\eta)'\Lambda^{-1}(\rho-d_2\eta),$$

where the second equality follows from consideration of the regression of the vector $(1-d_2x^2)$ (on the support of $\mu^*_{d_2}$) on the vectors x^4, x^6, $\cdots$, x^{m-1}. From (6.31) we obtain that

$$(6.32)\qquad e(1-F(d_2))+2d_2h^2+h^{-2}=e(2d_2\lambda_1-d_2^2\lambda_2+\rho'\Lambda^{-1}\rho-2d_2\eta'\Lambda^{-1}\rho$$
$$+d_2^2\eta'\Lambda^{-1}\eta)+2d_2h^2+h^{-2}.$$

Now $\lambda_2-\eta'\Lambda^{-1}\eta$ is the reciprocal of the element in the first row and first column of the positive definite matrix

$$\{\lambda_{i+j},\ 1\leqq i,\ j\leqq(m-1)/2\}^{-1},$$

and is therefore positive. Hence the coefficient of d_2^2 in the right member of (6.32) is negative, and the maximum with respect to d_2 of the expression (6.32) is attained where the derivative (with respect to d_2) of the right member of (6.32) is zero. Therefore this maximum of (6.32) with respect to d_2 is

$$(6.33)\qquad h^{-2}+e\rho'\Lambda^{-1}\rho+\frac{(e\lambda_1+h^2-e\eta'\Lambda^{-1}\rho)^2}{e[\lambda_2-\eta'\Lambda^{-1}\eta]}.$$

Since $\rho'\Lambda^{-1}\rho>0$ it follows that, as $e\to\infty$, (6.33) approaches ∞. If $h\to 0$ then also (6.33) approaches ∞. If e remains bounded and $h\to\infty$ then also (6.33) approaches ∞. It follows that the minimum, with respect to e and h, of (6.33), is attained where e and h are both finite. Also (6.33) is the sum of constant multiples of e and h^2, and of positive multiples of h^{-2} and h^4/e. All of these summands are convex in (positive) e and h^2 jointly, and $c_1h^{-2}+c_2h^4e^{-1}$ is strictly convex in e and h^2. Hence so is (6.33), and the latter therefore has a unique minimum with respect to e and h.

In the next section we shall make some comments on the problem of computing e^*, h^*, and μ^*. The reader will notice that the unfinished business of section 5, the verification of (5.6), (5.6a), (5.7), (5.29) and (5.30), is now trivial.

We conclude that

$$\mu^*=\mu^*_{d_2^*}$$
$$(6.34)\qquad \delta^*_a=h^*a^{1/2}$$
$$\epsilon^*_a=e^*a$$

where $e=e^*$, $h=h^*$ minimize (6.33) and d_2^* is defined just above (6.30).

7. Some computational aspects of the e^*, h^*, and μ^* of Theorem 5.1

We shall devote most of this section to consideration of the game with payoff function (2.9) and value (6.22) (with $d_2=s$). Before discussing the relationship of this game to a problem of Solotarev, we shall show how to obtain μ^*, e^*, and h^* for small odd m by direct computation. We first note that we can conveniently change the order of the various minimizing and maximizing operations, as follows: For fixed e and h, the game of $(d_2, \bar{d})$ against μ with payoff $K_{e,h}((d_2, \bar{d}), \mu) = -2d_2h^2 + e\int(1-d_2x^2-\bar{d}'U)^2 d\mu$ being determined, we have the last of the following expressions equal to the first:

$$(7.1)\qquad \min_{d_2}\min_{\mu}\max_{\bar{d}} K_{e,h}((d_2, \bar{d}), \mu) \geq \min_{d_2}\max_{\bar{d}}\min_{\mu} K_{e,h}((d_2, \bar{d}), \mu)$$

$$\geq \max_{\mu}\min_{d_2}\min_{\bar{d}} K_{e,h}((d_2, \bar{d}), \mu)\,.$$

Hence, all three expressions are equal, and therefore

$$\min_{h,e}\max_{d_2}\{e-eF(d_2)+2d_2h^2+h^{-2}\}$$

$$(7.2)\qquad =\min_{h,e}\max_{d_2}\{e+h^{-2}-\max_{\mu}\min_{\bar{d}} K_{e,h}((d_2, \bar{d}), \mu)\}$$

$$=\min_{h,e,\mu}\{e+h^{-2}+\max_{d_2}[2d_2h^2-eF(d_2, \mu)]\}\,.$$

Thus, our problem is reduced to minimizing, with respect to e, h^2, and μ, the expression obtained from (6.33) by putting μ for $\mu^*_{d_2}$ (and thus $F(d_2, \mu)$ for $F(d_2)=F(d_2, \mu^*_{d_2})$) in (6.30) - (6.32). Write $g=h^2$ and, with μ replacing $\mu^*_{d_2}$ in (6.30), let

$$A' \equiv A'(\mu) = \{\rho'\Lambda^{-1}\rho\lambda_2 - \rho'\Lambda^{-1}\rho\eta'\Lambda^{-1}\eta + \lambda_1^2 + (\rho'\Lambda^{-1}\eta)^2 - 2\eta'\Lambda^{-1}\rho\lambda_1\}^{1/2}\,,$$

$$(7.3)\qquad B' \equiv B'(\mu) = \lambda_2 - \eta'\Lambda^{-1}\eta\,,$$

$$C' \equiv C'(\mu) = \lambda_1 - \eta'\Lambda^{-1}\rho\,.$$

The expression (6.33), with μ replacing $\mu^*_{d_2}$, then becomes

$$(7.4)\qquad g^{-1}+[(A')^2e+2gC'+g^2e^{-1}]/B'\,.$$

The minimum of this with respect to e, achieved at $e=g/A'$, is $g^{-1}+2g(A'+C')/B'$. The minimum of this with respect to g, achieved at $g=[B'/2(A'+C')]^{1/2}$, is $2[2(A'+C')/B']^{1/2}$. Thus, finally, we must choose μ to minimize $(A'+C')/B'$.

The case $m=3$ is trivial, since μ^* is immediately seen to assign measure 1/2 to each of the two points ± 1, and one obtains easily $e^*=1/2$, $h^*=2^{-1/2}$, and $H(a)=1+4a+o(a)$. In the case $m=5$, let μ assign measure $(1-\varphi)/2$ (resp., $\varphi/2$) to each of the values ± 1 (resp., $\pm b$). Write $\beta=b^2$. We then have $\lambda_i=(1-\varphi)+\varphi\beta^i$, so that

$$\lambda_4 A'=\{\lambda_4\lambda_2^3+\lambda_1^2\lambda_4^2-2\lambda_1\lambda_2\lambda_3\lambda_4\}^{1/2}=\beta(1-\beta)\varphi^{1/2}(1-\varphi)^{1/2}[(1-\varphi)+\varphi\beta^4]^{1/2}, \tag{7.5}$$

$$\lambda_4 B'=\lambda_2\lambda_4-\lambda_3^2=\beta^2(1-\beta)^2\varphi(1-\varphi)\,, \tag{7.6}$$

$$\lambda_4 C'=\lambda_1\lambda_4-\lambda_2\lambda_3=\beta(1-\beta)^2(1+\beta)\varphi(1-\varphi)\,. \tag{7.7}$$

Hence,

$$(A'+C')/B'=\frac{1}{\beta(1-\beta)}\left[\frac{1}{\varphi}+\frac{\beta^4}{1-\varphi}\right]^{1/2}+\frac{(1+\beta)}{\beta}\,. \tag{7.8}$$

The minimum of this with respect to φ, at $\varphi=(1+\beta^2)^{-1}$, is $2/\beta(1-\beta)$, which has its minimum at $\beta=1/2$. We conclude that μ^* assigns measure 1/10 (resp., 2/5) to ± 1 (resp., $\pm 2^{-1/2}$), and that $h^*=[B'/2(A'+C')]^{1/4}=1/2$, $e^*=g^*/A'=5/4$, and $H(a)=1+2[2(A'+C')/B']^{1/2}a+o(a)=1+8a+o(a)$. The order of computation of section 6, which first solves the game with payoff (2.9) and then minimizes with respect to e and g, appears to be longer.

The family of solutions to the game considered in (2.9), as $s=d_2$ is varied, essentially coincides with the family of solutions, as σ is varied, of Solotarev's first problem (see Achieser [1], p. 280), that of finding the best Chebyshev approximation on $[-1, 1]$ to $t^p-p\sigma t^{p-1}$ (for fixed σ) among polynomials of degree at most $p-2$, where $p=(m-1)/2\geq 2$. (The fixed coefficient of t^{p-1} is written as $p\sigma$ in order to conform to the notation in [1].) To see this, we first note that the transformation $y=(t+1)/2$, $\psi=p(\sigma+1)/2$ takes Solotarev's problem into that of Chebyshev approximation of $y^p-\psi y^{p-1}$ by a polynomial of degree at most $p-2$ on $[0, 1]$. If $r_\psi(y)=\sum_0^{p-2}q_i(\psi)y^i$ is Solotarev's (unique) solution, then $R_\psi(y)=y^p-\psi y^{p-1}-r_\psi(y)$ attains its maximum in absolute value, with alternating signs, on a set B_ψ consisting of $p+1$ or p points (depending on whether or not $\sigma=0$) including 0 if $\sigma\geq 0$ or $\sigma\leq -\tan^2(\pi/2p)$. (The figures of [1] for $\sigma>0$ are reflected through 0 to yield the corresponding figures when the value σ is replaced by $-\sigma$.) Now, $q_0(\psi)=R_\psi(0)\neq 0$ except for one value ψ_0 of ψ which we shall temporarily exclude. (This value ψ_0 corresponds to the value $-\sigma_0$ of σ where $\sigma_0=\tan^2(\pi/4p)$ is the value for which, in the language of p. 281 of [1], $0<\sigma<\tan^2(\pi/2p)$ and $y_1(1)=0$, which is equivalent to $R_\psi(1)=0$ in our notation. This situation occurs between those of Fig. 6 and Fig. 7 of [1].) Define $k_i(\psi)=-q_i(\psi)/q_0(\psi)$ for $i=1, 2, \cdots,$

$p-2$, and $k_p(\psi)=1/q_0(\psi)$ and $k_{p-1}(\psi)=-\psi/q_0(\psi)$. Then, for $\psi\neq\psi_0$,

(7.9) $$-R_\psi(y)/q_0(\psi)=1-k_1(\psi)y-\sum_2^p k_i(\psi)y^i .$$

Because of the p extrema of alternating signs and equal magnitudes, we conclude from Theorem 2.5 (with $G(y)=y^2$ and $n=p-2$) that $\sum_2^p k_i(\psi)y^i$ is the best Chebyshev approximation of $1-k_1(\psi)y$ among functions of the form $\sum_2^p c_i y^i$. Putting $x^2=y$ in (6.20) - (6.21) or (2.9), we see that this is precisely the problem we had to solve, with $d_2=s=k_1(\psi)$. Thus, for each ψ (or σ) Solotarev's problem corresponds to ours with $d_2=k_1(\psi)$. Conversely, if, for fixed d_2, the quantity

(7.10) $$\max_{0\leq y\leq 1}\left|1-d_2y-\sum_2^p d_jy^j\right|$$

is minimized by $d_j=d_j^*(d_2)$, and if $d_p^*(d_2)\neq 0$, we can similarly divide all members of (7.10) by $d_p^*(d_2)$ and obtain a solution to Solotarev's problem with $\psi=\psi(d_2)=-d_{p-1}^*(d_2)/d_p^*(d_2)$. If $d_p^*(d_2)=0$, it cannot be that $d_{p-1}^*(d_2)$ is also zero, since $1-d_2y-\sum_2^{p-2}d_j^*(d_2)y^j$ would then have too few alternations of sign to correspond to a Chebyshev solution; hence, if $d_p^*(d_2)=0$, we can divide through by $d_{p-1}^*(d_2)$ and conclude that our problem corresponds to the (simpler) Chebyshev problem of approximating y^{p-1} by a polynomial of the form $\sum_0^{p-2}c_jy_j$; this can be thought of as the case $\sigma=\infty$ (more precisely, $\sigma\to\infty$) of Solotarev's problem. With this convention, we conclude that the solutions to the family of Solotarev problems as σ varies correspond to the solutions to our problem associated with (6.21) or (2.9) as d_2 varies (with $\sigma=-\sigma_0$ corresponding to $d_2=\infty$).

Unfortunately, this correspondence is not easy to write down explicitly. However, we can exhibit the relationship between the maximin strategies for the two games (in analogy to (4.7)), as follows: For y in $B_{\psi(d_2)}$, let $\nu_{d_2}^*(y)$ be the probability assigned to the value y by a maximin strategy for the game of (2.9) or (6.20) - (6.21) (in terms of $y=x^2$). The orthogonality relations (2.4) become

(7.11) $$\sum_{y\in B_{\psi(d_2)}}\left(1-d_2y-\sum_2^p d_j^*(d_2)y^j\right)y^i\nu_{d_2}^*(y)=0\,,\qquad 2\leqq i\leqq p\,.$$

In discussing (7.11) we need only be concerned with values d_2 for which $\nu_{d_2}^*(0)<1$, since in fact the optimum μ^* is known to have $\mu^*(0)=0$. Hence, for $\psi=\psi(d_2)$ and such a d_2, we can define a probability measure ξ_ψ on B_ψ by writing

(7.12) $$\xi_\psi(y)=\nu_{d_2(\psi)}^*(y)y^2\Big/\sum_{y\neq 0}\nu_{d_2(\psi)}^*(y)y^2 .$$

This yields

$$\sum_y (y^p - \psi y^{p-1} - \sum_0^{p-2} q_i(\psi)) y^i \xi_\psi(y) = 0\,, \qquad 0 \leq i \leq p-2\,. \tag{7.13}$$

Since the support of ξ_ψ is a set on which the maximum of (7.10) is attained, we conclude from (7.13) by Theorem 2.2 that ξ_ψ is maximin. Conversely, a maximin ξ_ψ with $\xi_\psi(0)=0$ yields a corresponding $\nu^*_{d_2(\psi)}$.

Designate the values y in the support of ξ_ψ by y_j, $1 \leq j \leq W$, with $0 < y_1 < y_2 < \cdots < y_W$. Then (7.13) is equivalent to

$$\sum_{j=1}^{W} (-1)^j y_j^i \xi_\psi(y_j) = 0\,, \qquad 0 \leq i \leq p-2\,. \tag{7.14}$$

Since the matrix $\{(-1)^j y_j^i;\ 0 \leq i \leq p-2,\ 1 \leq j \leq W\}$ has rank W if $W \leq p-1$, we conclude from (7.14) that $W \geq p$. Recalling that $y=0$ corresponds to $t=-1$, we conclude from knowledge of Solotarev's solution (Achieser [1], p. 281, the reflection of Fig. 6 or Fig. 7 but not Fig. 8) that we must have $0 \geq \sigma > -\tan^2(\pi/2p)$. The solution obtained above in the case $m=5$, $p=2$ actually corresponds to $\sigma=0$, when Solotarev's problem degenerates into the classical one of Chebyshev approximation of t^p by $\sum_0^{p-2} q_j t^j$ on $[-1, 1]$. We have not been able to verify whether or not this is also the case for $p>2$.

8. Polynomial extrapolation

This is the problem described in section 1, where $Y_a=[-a, a]$, $a \to \infty$, and m can be even or odd. If we now ask for a design for which $\lim_{a\to\infty} a^{-2m} D_{Y_a}(\xi)$ is minimized over ξ, the preponderance of $a^{2m} v_{mm}(\xi)$ in $d(a, \xi)$ makes it clear that the unique solution is the design ξ^* which minimizes $v_{mm}(\xi)$, namely, that of (2.12) - (2.13) with $h=m$. For this choice, we obtain, from (2.15), $\lim_{a\to\infty} a^{-2m} D_{Y_a}(\xi^*) = v_{mm}(\xi^*) = 2^{2m-2}$. We are thus motivated (similarly to the first paragraph of section 3) to consider families $F=\{\xi_a\}$ and to seek a family F^* which minimizes

$$\overline{\lim_{a\to 0}}\, a^2[a^{-2m} D_{Y_a}(\xi_a) - 2^{2m-2}]\,. \tag{8.1}$$

By an analogue of Theorem 2.8, it is easy to show that $\{\xi_a\}$ cannot minimize (8.1) unless $\xi_a \to \xi^*$ weakly. (This is also a simple consequence of the result of [12].) Hence, it is no restriction to assume $\{\xi_a\}$ is of this form. Moreover, by Theorem 2.7, we may again restrict consideration to families of symmetric designs ξ_a on $m+1$ points (varying with a), including ± 1. We then treat the problem by approximately minimizing $d(a, \xi_a)$ among such designs, since it is easily verified that $D_{Y_a}(\xi_a) = d(a, \xi_a)$ for an optimum family when a is large. The quantity to be minimized to within a term of order $o(1)$ is then

$$a^2(v_{mm}(\xi_a)-2^{2m-2})+v_{m-1,\,m-1}(\xi_a)+2v_{m,\,m-2}(\xi_a)\,. \tag{8.2}$$

This yields a more complex problem than that associated with (4.1) or (5.11), since we now have a sum of three terms instead of two. (It should be noted, however, that we need not assume an analogue of (3.4) or (5.1) here, since in the present problem we do not have the complication of $M(\xi_a^*)$ approaching singularity for an optimum family $\{\xi_a^*\}$.) We have not yet obtained a solution to this problem in terms of classical solutions to Chebyshev problems in the manner of sections 4, 6 and 7. For small m, it is not difficult to compute $M(\xi_a)$ for ξ_a obtained by slight perturbation from the ξ^* of (2.12) - (2.13) with $h=m$, and thus to perform the minimization directly. Thus, for $m=2$, an optimum family is given by

$$\begin{aligned} &\xi_a(0)=1/2-3/8a^2\,,\\ &\xi_a(\pm 1)=1/4+3/16a^2\,, \end{aligned} \tag{8.3}$$

for which $d(a,\,\xi_a)=4a^4-17a^2/4+O(1)$, a relative improvement of approximately $9/16a^2$ over $d(a,\,\xi^*)=4a^4-2a^2+O(1)$. For $m=3$, an optimum family is given by

$$\begin{aligned} &\xi_a(\pm 1/2\mp 1/16a^2)=1/3-1/9a^2\,,\\ &\xi_a(\pm 1)=1/6+1/9a^2\,, \end{aligned} \tag{8.4}$$

with $d(a,\,\xi_a)=16a^6-21a^4+O(a^2)$, a relative improvement of approximately $5/16a^2$ over $d(a,\,\xi^*)=16a^6-16a^4+O(a^2)$.

CORNELL UNIVERSITY

REFERENCES

[1] N. I. Achieser, *Theory of Approximation*, Ungar Pub. Co., New York, 1956.

[2] W. Behnken and G. E. P. Box, "Simplex sum designs," *Ann. Math. Stat.*, 31 (1960), 838-864.

[3] A. Erdelyi (editor), *Higher Transcendental Functions*, 2, McGraw-Hill, New York, 1953.

[4] L. B. W. Jolley, *Summation of Series*, Dover, New York, 1961.

[5] J. Kiefer, "Optimum experimental designs," *J. Roy. Stat. Soc. (B)*, 21 (1959), 273-319.

[6] J. Kiefer, "Optimum experimental designs V," *Proc. 4th Berkeley Symp.*, I (1960), 381-405.

[7] J. Kiefer, "Optimum designs in regression problems II," *Ann. Math. Stat.*, 32 (1961), 298-325.

[8] J. Kiefer, "An extremum result," *Can. J. Math.*, 14 (1962), 597-601.

[9] J. Kiefer, "Two more criteria equivalent to D-optimality of designs," *Ann. Math. Stat.*, 33 (1962), 792-796.

[10] J. Kiefer and J. Wolfowitz, "Optimum designs in regression problems," *Ann. Math. Stat.*, 30 (1959), 271-294.

[11] J. Kiefer and J. Wolfowitz, "The equivalence of two extremum problems," *Can. J. Math.*, 12 (1960), 363-366.
[12] P. G. Hoel and A. Levine, "Optimal spacing and weighting in polynomial prediction," to appear in *Ann. Math. Stat.*

Reprinted from
Ann. Inst. Statist. Math. **16**, 79–108 (1964)

OPTIMUM EXTRAPOLATION AND INTERPOLATION DESIGNS II

J. KIEFER[1] AND J. WOLFOWITZ[2]

9. Introduction

In our paper [1], some familiarity with which will be assumed and whose notation we henceforth adopt, we proved Theorem 3.1 not for the most general family of designs [1], (3.3), but for the restricted family [1], (3.4), and we proved Theorem 5.1 for the restricted family [1], (5.1), and not for the general family

$$\gamma_a \equiv \gamma(\epsilon_a, \delta_a, \mu_a) = (1-\epsilon_a)\xi_{0a} + \epsilon_a \mu_a \,. \tag{9.1}$$

We imposed the restrictions [1], (3.4) and (5.1) because we were unable to solve the general problems. The purpose of the present paper is to show that Theorems 3.1 and 5.1 hold under the most general conditions, i.e., under [1], (3.3), and (9.1) respectively.

10. Polynomial interpolation, *m* odd

Let $m_0=(m+1)/2$. Let the positive points of support of γ_a of (9.1) be

$$0<\delta_a=z_{1a}<z_{2a}<\cdots<z_{m_0 a}=1 \,.$$

Define $L_{ia}(x)$, $-m_0 \leqq i \leqq m_0$, $i\neq 0$, by

$$L_{ia}(x)=\prod_{j\neq i}(x-z_{ja})/\prod_{j\neq i}(z_{ia}-z_{ja}) \,,$$

where the products are over non-zero $j(\neq i)$ from $-m_0$ to $+m_0$. It is well known (e.g., [2]) that

$$d(x, \gamma_a)=\sum_{i=-m_0}^{m_0} L_{ia}^2(x)(\gamma_a(z_{ia}))^{-1} \,. \tag{10.1}$$

---

1) Research supported by the U.S. Office of Naval Research under contract No. Nonr 266 (04) (NR 047-005).

2) Research supported by the U.S. Air Force under contract No. AR 18(600)-685, monitored by the Office of Scientific Research.

In [1] we constructed a family $\{\gamma_a\}$ of designs for which

(10.2) $$D_{\gamma_a}(\gamma_a)=1+O(a).$$

Since

(10.3) $$z_{ia}^2/(z_{ia}^2-\delta_a^2)>1, \quad i\geq 2,$$

we have from (10.1) that

(10.4) $$|L_{ia}(0)|>\frac{1}{2}, \quad i=\pm 1.$$

Hence from (10.1) (or from [1], (5.14)) we obtain that

(10.5) $$d(0, \gamma_a)>4L_{1a}^2(0)/(1-\epsilon_a)>1/(1-\epsilon_a)>1+\epsilon_a.$$

If the family $\{\gamma_a^*\}$ which we constructed in [1], section 5, does not satisfy the optimality condition [1], (3.1), against *all* competitors, then there exists a competing family which is better on a sequence of a's approaching zero. This better competing family must therefore satisfy, on this sequence of a's, (10.2) and

(10.6) $$\epsilon_a=O(a).$$

From these two conditions we will deduce the desired result. In the remainder of this section $\{\gamma_a\}$ will represent any family which satisfies (10.2) and (10.6) on a sequence of a's approaching zero.

In order to avoid tedious repetition we will use the symbols $\lim_{a\to 0}$, $\overline{\lim}_{a\to 0}$, $O(\ )$, $o(\ )$, as if $a\to 0$, but actually they will refer to a sequence of a's approaching zero. If then we select a subsequence of a's this will be a subsequence of the first sequence.

Consider $L_{2a}(0)$ and $L_{-2a}(0)$. Since

(10.7) $$z_{ia}^2/(z_{ia}^2-z_{2a}^2)>1, \quad i\geq 3,$$

and

$$(z_{2a}^2-z_{1a}^2)<1,$$

we have that

(10.8) $$|L_{ia}(0)|>\delta_a^2/2, \quad i=\pm 2.$$

The inequality between the second and last terms of (10.5), together with (10.1) and (10.6), implies that $2L_{2a}^2(0)/\gamma_a(z_{2a})=O(a)$. Hence, from (10.8),

(10.9) $$\delta_a^4/\gamma_a(z_{2a})=O(a),$$

and hence, from (10.6),

(10.10) $$\delta_a = O(a^{1/2}) .$$

Suppose $\delta_a \leqq a$. Then, since

$$L_{1a}(\delta_a) = 1 ,$$

we have

$$L_{1a}^2(\delta_a)/[(1-\epsilon)/2] > 2 ,$$

in violation of (10.2). Hence

(10.11) $$\delta_a > a .$$

Suppose now that, on a sequence of a's tending to zero,

(10.12) $$a^{-1/2}\delta_a \to 0 .$$

From (10.11) we have that

(10.13) $$\frac{z_{ia}^2 - a^2}{z_{ia}^2 - \delta_a^2} > 1, \quad i \geqq 2 .$$

Hence

(10.14) $$| L_{1a}(a) | > (a+\delta_a)/2\delta_a$$

and

(10.15) $$| L_{-1a}(a) | > (-a+\delta_a)/2\delta_a .$$

Consequently

(10.16) $$d(a, \gamma_a) > [L_{-1a}^2(a) + L_{1a}^2(a)]/\gamma_a(\delta_a) = \frac{2}{1-\epsilon_a}\left[\left(\frac{a+\delta_a}{2\delta_a}\right)^2 + \left(\frac{-a+\delta_a}{2\delta_a}\right)^2\right]$$

$$> 1 + a^2\delta_a^{-2} = 1 + (\delta_a^2/a)^{-1}a .$$

Then (10.12) and (10.16) contradict (10.2). We conclude that, for some c_2' satisfying $0 < c_2' < \infty$,

(10.17) $$\varliminf_{a \to 0} a^{-1/2}\delta_a = c_2' .$$

We shall use below the fact that the inequality between the second and last expressions of (10.16), together with (10.1) and (10.2), imply that

(10.18) $$L_{2a}^2(a)/\gamma_a(z_{2a}) = O(a) .$$

In the next few paragraphs we shall use the fact that, for some $c_2'' > 0$,

(10.19) $$|L_{2a}(a)| > c_2''\delta_a^2/(z_{2a}^2-\delta_a^2)\,,$$

which is a consequence of (10.13) and the definition of $L_{2a}(x)$.

We shall now prove that

(10.20) $$\varliminf_{a\to 0} z_{2a} > 0\,.$$

In the relationship between z_{2a} and z_{1a} there are two possibilities:

(10.21) $$z_{1a}=o(z_{2a})\,,$$

(10.22) $$\varliminf_{a\to 0}(z_{2a}-z_{1a})/z_{1a}<\infty\,.$$

If (10.21) holds then, from (10.17) and (10.19),

$$|L_{2a}(a)| > c_4'a/z_{2a}^2$$

and hence from (10.6) and (10.18),

$$a^2/[z_{2a}^4 O(a)]=O(a)\,,$$

from which the desired (10.20) follows.

It remains to show that (10.22) cannot hold. If (10.22) holds (10.19) implies that, on some subsequence for which $a\to 0$,

$$|L_{2a}(a)| > c_5'\delta_a\,,$$

which, with (10.17) and (10.18), implies

$$a/\gamma(z_{2a})=O(a)\,.$$

This violates (10.6), proves that (10.22) cannot hold, and completes the proof of (10.20).

From (10.17), (10.19), (10.20), and the fact that $\delta_a\to 0$, we obtain

(10.23) $$\varliminf_{a\to 0}|L_{2a}(a)|/a>0\,.$$

From (10.18) and (10.23) we obtain

(10.24) $$\varliminf_{a\to 0}\gamma_{2a}(z_{2a})/a>0\,.$$

Proceeding in this manner and considering the $L_{ia}(a)$ successively, we easily obtain

(10.25) $$\varliminf_{a\to 0}(z_{ia}-z_{(i-1)a})>0,\quad i\geqq 3$$

and

(10.26) $$\underline{\lim}_{a\to 0} \gamma_{ia}(z_{ia})/a > 0, \quad i \geq 3 .$$

From (10.17), (10.20), (10.24), (10.25), and (10.26) we will now deduce the conclusion that the family of designs which was shown in [1] to satisfy the conclusion of Theorem 5.1 of [1] (i.e., the optimality condition [1], (3.1)) when the competing families of designs are forced to satisfy restriction [1], (5.1), also satisfies the conclusion of Theorem 5.1 when the competing families of designs are the most general possible (i.e., satisfy only condition (9.1)). This is the desired result of the present paper for odd m.

Suppose then that [1], (3.1) does not hold for $\{\xi_a^*\}$ equal to the family $\{\gamma_a^*\}$ obtained in [1]. Then there is a family $\{\gamma_a^{00}\}$ (satisfying only (9.1) and not necessarily [1], (5.1)) such that, for some sequence $\{a_i'\}$ approaching zero and some K,

(10.27) $$\lim_{n\to\infty} \frac{D_{\gamma_{a_n'}}(\gamma_{a_n'}^{00}) - 1}{D_{\gamma_{a_n'}}(\gamma_{a_n'}^{*}) - 1} = \frac{K}{\lim_{a\to 0} a^{-1}[D_{\gamma_a}(\gamma_a^*) - 1]} < 1 .$$

(The limit in the denominator of the second expression of (10.27) was shown to exist in [1], section 3.) By the above, $\{\gamma_{a_n'}^{00}\}$ satisfies (10.17), (10.20), and (10.24) - (10.26). Let $a_1, a_2, \cdots$ be a subsequence of $a_1', a_2', \cdots$, such that

(10.28) $$\lim_{n\to\infty} a_n^{-1/2}\delta_{c_n} = g_1 > 0$$

(10.29) $$\lim_{n\to\infty} z_{ic_n} = g_i > 0, \quad i \geq 2$$

(10.30) $$\lim_{n\to\infty} \gamma_{ia_n}^{00}(z_{ia_n})/a = h_i > 0, \quad i \geq 2$$

and

(10.31) $$g_2 < g_3 < \cdots < g_{m_0} \equiv 1 .$$

The existence of such a sequence is assured by (10.17), (10.20), and (10.24) - (10.26). Let μ_0 be the symmetric probability measure which assigns probability h_i/e, $e = 2(h_2 + \cdots + h_{m_0})$, to the point g_i, $i \geq 2$. Let $\{\gamma_a^0\}$ be the family of designs defined by

(10.32) $$\gamma_a^0 = (1 - ea)\xi_{0a} + ea\mu_0 .$$

It follows easily from the construction of the sequence $\{a_n\}$ that

(10.33) $$\lim_{n\to\infty} (D_{\gamma_{a_n}}(\gamma_{a_n}^{00}) - 1)/(D_{\gamma_{a_n}}(\gamma_{a_n}^{0}) - 1) = 1 .$$

By (10.32) and [1], (6.34), $\lim_{a\to 0} a^{-1}[D_{Y_a}(\gamma_a^0)-1]$ exists. By (10.33) and the inequality of (10.27) we have, for the family $\{\gamma_a^*\}$ of [1],

$$\lim_{a\to 0}(D_{Y_a}(\gamma_a^0)-1)/(D_{Y_a}(\gamma_a^*)-1)<1 . \tag{10.34}$$

According to (10.32), the family $\{\gamma_a^0\}$ satisfies [1], (5.1). Hence, (10.34) contradicts Theorem 5.1 (as proved in [1] under the restriction (5.1)). This proves the desired result.

11. Polynomial interpolation, *m* even

The general family of designs for this problem is given in [1], (3.3), as

$$\xi_a=(1-\epsilon_a)\xi_0+\epsilon_a\mu_a .$$

Let $m'=m/2$, m even, $z_{0a}=0$, and let

$$z_{1a}\equiv T_a<z_{2a}<z_{3a}<\cdots<z_{m'a}\equiv 1 \tag{11.1}$$

be the positive points of support of μ_a. Thus the ith ($i=0, \cdots, m'$) point of support of μ_a is z_{ia}, and the ith ($i=-1, \cdots, -m'$) point of support of μ_a is $-z_{ia}$. Define the ith Lagrange polynomial ($-m'\leqq i\leqq m'$) by

$$L_{ia}(x)=\prod_{j\neq i}(x-z_{ja})/\prod_{j\neq i}(z_{ia}-z_{ja}) , \tag{11.2}$$

where in (11.2) the product is over j from $-m'$ to $+m'$, including $j=0$. As in (10.1) we have

$$d(x, \nu_a)=\sum_{i=-m'}^{m'} L_{ia}^2(x)(\nu_a(z_{ia}))^{-1} . \tag{11.3}$$

In [1] we constructed a family $\{\nu_a\}$ of designs for which

$$D_{Y_a}(\nu_a)=1+O(a) . \tag{11.4}$$

On the other hand, it follows from (11.3) or [1], (2.2), that

$$d(0, \nu_a)>(1-\epsilon_a)^{-1}>1+\epsilon_a . \tag{11.5}$$

Exactly as in section 10, if the conclusion of Theorem 3.1 does not hold when general families (which satisfy [1], (3.3), but not necessarily [1], (3.4)) are admitted in competition, then there exists a family which is better than the family $\{\nu_a^*\}$ of [1] (in the sense of (3.1)) on some sequence of a's approaching zero. On this sequence this family satisfies (11.4) and

(11.6) $$\epsilon_a = O(a) .$$

Throughout the remainder of this section $\{\nu_a\}$ will represent any family which is better than $\{\nu_a^*\}$ in the sense of [1], (3.1), on a sequence of a's approaching zero, and which therefore satisfies (11.4) and (11.6) on that sequence. Henceforth we adopt the notational convention of the paragraph below (10.6).

Now suppose that $T_a \leqq a$. Since $L_{1a}(T_a)=1$ we have, from (11.3) and (11.6), that

(11.7) $$d(T_a, \nu_a) \geqq 1/\epsilon_a .$$

Because of (11.6) this violates (11.4). Hence $T_a > a$.

Now consider $L_{1a}(a)$. For $j \geqq 2$ we have

(11.8) $$(z_{ja}^2 - a^2)/(z_{ja}^2 - T_a^2) > 1 .$$

Hence from (11.2), (11.8), and the fact that $T_a > a$,

(11.9) $$| L_{1a}(a) | > a/2T_a .$$

From (11.9), (11.3), (11.4), and (11.6), we obtain, for some $c' > 0$,

(11.10) $$T_a^2 > a^2/4L_{1a}^2(a) \geqq a^2/4d(a, \nu_a)\nu_a(T_a) > c'a .$$

From (11.3) and (11.9) we have

(11.11) $$d(a, \nu_a) > a^2/4T_a^2\nu_a(T_a) + L_{0a}^2(a)/(1-\epsilon_a) .$$

Using (11.10) and (11.2) the second term in the right member of (11.11) is easily seen to be $1+O(a)$. Hence, by (11.4),

(11.12) $$a^2/T_a^2\nu_a(T_a) = O(a) .$$

From (11.6) and (11.12) we conclude that

(11.13) $$\varliminf_{a \to 0} T_a > 0$$

and

(11.14) $$\varliminf_{a \to 0} \nu_a(T_a)/a > 0 .$$

We now consider $L_{2a}(a)$. If T_a is replaced by z_{2j} in (11.8), the resulting inequality holds for $j \geqq 3$. Using (11.3) and the fact that $z_{2a} > z_{1a}$, one then obtains, in analogy with (11.9),

$$| L_{2a}(a) | > c''a/(z_{2a} - T_a) \quad \text{for some } c'' > 0 .$$

From an obvious analogue of (11.11) we then obtain, in analogy with

(11.12),

$$a^2/(z_{2a}-T_a)^2\nu_a(z_{2a})=O(a)\,. \tag{11.15}$$

From (11.6) and (11.15) we conclude that

$$\underline{\lim}_{a\to 0}\ \nu_a(z_{2a})/a>0 \tag{11.16}$$

and

$$\underline{\lim}_{a\to 0}\ (z_{2a}-T_a)>0\,. \tag{11.17}$$

Proceeding in this manner for $i=3, \cdots, m'$, we obtain that, for $1\leqq i\leqq m'$,

$$\underline{\lim}_{a\to 0}\ \nu_a(z_{ia})/a>0 \tag{11.18}$$

and

$$\underline{\lim}_{a\to 0}\ (z_{ia}-z_{(i-1)a})>0\,. \tag{11.19}$$

The desired result (for m even) is now obtained from (11.13), (11.18), and (11.19) in the same manner that the result for m odd was obtained from (10.17), (10.20), and (10.24) - (10.26). This completes the proof.

CORNELL UNIVERSITY

REFERENCES

[1] J. Kiefer and J. Wolfowitz, "Optimum extrapolation and interpolation designs I" *Ann. Inst. Stat. Math.*, 16 (1964), 79-108.

[2] P. G. Guest, "The spacing of observations in polynomial regression," *Ann. Math. Statist.*, 29 (1958), 294-299.

Remark about our paper [1].

There is an omission in that part of the proof of Theorem 2.8 of [1] which is devoted to proving that $\xi_a\to\xi_0$ weakly. This conclusion is an immediate consequence of a compactness argument and the fact that ξ_0 is the *unique* optimum design for estimating θ_0. Uniqueness can be proved as follows: Clearly, for no design supported at one point, other than the design ξ_0, is θ_0 estimable. Hence, if ξ' is optimum for estimating θ_0 and $\xi'\neq\xi_0$, it follows that ξ' is supported by at least two points. Consequently, some linear parametric function other than θ_0 is estimable under ξ', while (as mentioned in the proof of Theorem 2.8) only θ_0

is estimable under ξ_0. It follows that ξ' is better than ξ_0, contradicting the admissibility of ξ_0 (mentioned in the proof).

In a future paper we shall publish generalizations of the results of Hoel and Levine which were mentioned in [1].

Reprinted from
Ann. Inst. Statist. Math. **16**, 295–303 (1964)

ON A PROBLEM CONNECTED WITH THE VANDERMONDE DETERMINANT

J. KIEFER[1] AND J. WOLFOWITZ[2]

Let m be a positive integer. We denote by $a=(a_0, a_1, \cdots, a_m)$, $x=(x_0, \cdots, x_{m-1})$ vectors with real components. Define

$$U = \{x \mid \text{all } |x_i| \leqq 1,\ x_i \neq x_j,\ i \neq j\},$$

$$(1) \qquad P^*(x, a) = \begin{vmatrix} 1 & 1 & \cdot & 1 & a_0 \\ x_0 & x_1 & \cdot & x_{m-1} & a_1 \\ x_0^2 & x_1^2 & \cdot & x_{m-1}^2 & a_2 \\ \cdot & \cdot & \cdot & \cdot & \cdot \\ \cdot & \cdot & \cdot & \cdot & \cdot \\ \cdot & \cdot & \cdot & \cdot & \cdot \\ x_0^m & x_1^m & \cdot & x_{m-1}^m & a_m \end{vmatrix},$$

$$A^* = \{a \mid P^*(x, a) \neq 0 \text{ for every } x \in U\}.$$

This paper is devoted to the problem of characterizing A^*, a problem which is of interest in the theory of the design of statistical experiments (see [1]) and is perhaps of interest per se.

When $m=1$, it is trivial that $A^* = \{(a_0, a_1) \mid |a_0| < |a_1|\}$. We hereafter assume $m>1$. Define

$$A(c) = A^* \cap \{a \mid a_m = c\}.$$

We shall prove the following:

(I) $A(1)$ is a set A described precisely below.

(II) For $c \neq 0$, we have $(a_0, a_1, \cdots, a_{m-1}, c) \in A(c)$ if and only if $(a_0/c, \cdots, a_{m-1}/c, 1) \in A$.

(III) $A(0)$ is empty.

Of these, (II) is obvious, and (III), which we shall prove in the paragraph after next, is only slightly less so. The problem is to describe A.

Let $S_0(x) \equiv 1$ and

Received by the editors August 24, 1964.

[1] Research supported by the U. S. Office of Naval Research under contract No. Nonr 266(04) (NR 047–005).

[2] Research supported by the U. S. Air Force under contract No. AR 18(600)-685, monitored by the Office of Scientific Research.

$$S_j(x) \equiv (-1)^j \sum x_{i_1}x_{i_2} \cdots x_{i_j}, \tag{2}$$

where the summation is with respect to $i_1, \cdots, i_j$, $0 \leqq i_1 < i_2 < \cdots < i_j < m$. Let

$$P(x, a) \equiv \sum_{j=0}^{m} a_{m-j}S_j(x). \tag{3}$$

Expanding the determinant P^* of (1) in minors of the last column shows that

$$P^*(x, a) = P(x, a)K(x),$$

where the polynomial K is the Vandermonde determinant which is the minor of a_m in the determinant P^* of (1); thus, K is never zero on U. Hence in the definition of A^* we may replace $P^*(x, a)$ by $P(x, a)$.

We shall now prove that $A(0)$ is empty. Let a be any $(m+1)$-vector with $a_m = 0$. If all $a_j = 0$, clearly $a \notin A^*$. Suppose then that j_0 is the smallest integer for which $a_{m-j_0} \neq 0$, $0 < j_0 \leqq m$. We shall find two points x' and x'' in the convex subset T of U where $-1 \leqq x_0 < x_1 < \cdots < x_{m-1} \leqq 1$ such that

$$(-1)^{j_0} a_{j_0} P(x', a) < 0 < (-1)^{j_0} a_{j_0} P(x'', a), \tag{4}$$

from which it follows that $P(x, a) = 0$ for some x in T; it follows that $A(0)$ is empty. Let $x'' = (\epsilon, 2\epsilon, \cdots, m\epsilon)$ with $0 < \epsilon < 1$. Each term in the sum in (2) is then positive, and thus $(-1)^{j_0} S_{j_0}(x'') > \epsilon^{j_0}$, while $S_j(x'') = o(\epsilon^{j_0})$ as $\epsilon \to 0$ if $j > j_0$. Hence, for ϵ sufficiently small, the second half of (4) follows from (3). The first half of (4) follows similarly if j_0 is odd upon taking $x' = (-m\epsilon, -(m-1)\epsilon, \cdots, -\epsilon)$, and if j_0 is even upon taking $x' = (-(m-1)\epsilon, -(m-2)\epsilon, \cdots, -\epsilon, 1)$.

Let

$$Q_h(x) \equiv \sum (1 - x_{i_1})(1 - x_{i_2}) \cdots (1 - x_{i_h})(1 + x_{i_{h+1}}) \cdots \cdot(1 + x_{i_m}) \Big/ \binom{m}{h}, \quad 0 \leqq h \leqq m, \tag{5}$$

where the summation is over the $\binom{m}{h}$ choices of the disjoint sets $(i_1, i_2, \cdots, i_h)$ and $(i_{h+1}, \cdots, i_m)$ of distinct integers between 0 and $m-1$, inclusive. Define the points

$$a^{(h)} = (a_0^{(h)}, \cdots, a_m^{(h)}), \quad 0 \leqq h \leqq m, \tag{6}$$

by

$$Q_h(x) = \sum_{j=0}^{m} a_{m-j}^{(h)} S_j(x). \tag{7}$$

We note that $a_m^{(h)} = 1$ for all h, and that

$$a^{(0)} = ((-1)^m, (-1)^{m-1}, \cdots, (-1)^0),$$
$$a^{(m)} = (1, 1, \cdots, 1).$$

We will now prove:

THEOREM. *For $m > 1$, A is the closed m-simplex with extreme points $a^{(0)}, \cdots, a^{(m)}$, minus the (closed) edge which connects $a^{(0)}$ and $a^{(m)}$.*

Let $\overline{U}$ be the closure of U. If x is a vertex of $\overline{U}$ with r coordinates -1 and $(m-r)$ coordinates $+1$, then

$$Q_h(x) = 2^m \Big/ \binom{m}{h} \quad \text{if } h = r,$$
$$= 0 \quad \text{if } h \neq r.$$

Hence the Q_h are linearly independent functions. Moreover, each $Q_h(x) \geqq 0$ on U. If $h \neq 0$ or m, $Q_h(x) > 0$ on U, since at most one x_i can take the value $+1$ and at most one x_i can take the value -1 on U.

According to (3), $P(x, a)$ is linear in a. According to (7), $P(x, a^{(h)}) = Q_h(x)$. Combining these facts, if $\alpha_0, \cdots, \alpha_m$ are real numbers,

$$P\left(x, \sum_0^m \alpha_h a^{(h)}\right) = \sum_0^m \alpha_h Q_h(x). \tag{8}$$

Since the Q_h are linearly independent, it follows from (7) that the vectors $a^{(i)}$, $0 \leqq i \leqq m$, are linearly independent. Hence, $a^{(0)}, \cdots, a^{(m)}$ lie in the hyperplane $a_m = 1$, but in no other hyperplane. Hence, any point a in the hyperplane $a_m = 1$ can be written in a unique way as

$$a = \sum_0^m \alpha_h a^{(h)},$$

where $\sum \alpha_h = 1$. Suppose now that $\alpha_h \geqq 0$ for all h, and $\alpha_h > 0$ for some $h \neq 0$ or m. Then, from (8) and the paragraph below the statement of the theorem, we have $P(x, a) > 0$ for all x in U. From this it follows that A is at least as large as stated in the theorem. It remains to prove that it is no larger.

If $\alpha_0 + \alpha_m = 1$ and all other $\alpha_h = 0$, we have $P(x, a) = 0$ for any x in U such that $x_0 = 1$, $x_1 = -1$. Hence the closed segment $[a^{(0)}, a^{(m)}]$ is not in A.

Suppose now that $\alpha_{i_0}<0$. If 0 is the origin then of course $P(0, a)=1$. It follows from the paragraph below the statement of the theorem that there is an x^* in U such that (i) the half-open segment $(0, x^*]$ is in U, (ii) $Q_{i_0}(x^*)>\frac{1}{2}$, and (iii) for $j\neq i_0$, $Q_j(x^*)<|\alpha_{i_0}|/2\sum|\alpha_h|$. From (8) we have $P(x^*, a)<0$. Hence, for some point $y\in(0, x^*)$ (so that $y\in U$) we must have $P(y, a)=0$. Hence a cannot be in A. The proof of the theorem is complete.

We remark that the points a of the form $(b^m, b^{m-1}, \cdots, b^1, 1)$ with $|b|<1$ are easily verified to be in A; this is of interest in applications.

Reference

1. J. Kiefer and J. Wolfowitz, *On a theorem of Hoel and Levine on extrapolation designs*, Ann. Math. Statist. (to appear).

Cornell University

Reprinted from
Proc. Amer. Math. Soc. **16**, 1092–1095 (1965)

Reprinted from THE ANNALS OF MATHEMATICAL STATISTICS
Vol. 36, No. 6, December, 1965
Printed in U.S.A.

ON A THEOREM OF HOEL AND LEVINE ON EXTRAPOLATION DESIGNS

BY J. KIEFER[1] AND J. WOLFOWITZ[2]

Cornell University

0. Summary. Recent results [5] of Hoel and Levine (1964), which assert that designs on $[-1, 1]$ which are optimum for certain polynomial regression extrapolation problems are supported by the "Chebyshev points," are extended to cover other nonpolynomial regression problems involving Chebyshev systems. In addition, the large class of linear parametric functions which are optimally estimated by designs supported by these Chebyshev points is characterized.

1. Introduction. Let $f = (f_0, f_1, \cdots, f_m)$ be a row vector of $m + 1$ continuous real-valued functions on a compact set X, where $m > 0$. (Unprimed vectors will ordinarily denote row vectors, and the transpose of a vector or matrix will be denoted by a prime.) The f_i are assumed to be linearly independent on X. A design is a probability measure ξ (which can always be assumed discrete) on X. (For a discussion of this see [8].)

Write

$$m_{ij}(\xi) = \int f_i f_j \, d\xi, \qquad M(\xi) = \{m_{ij}(\xi), 0 \leqq i, j \leqq m\}. \tag{1.1}$$

If N uncorrelated observations with equal variance σ^2 (known or unknown) are made, taking $N\xi(x)$ observations at x for each x in X, and if the expected value of an observation at x is $\theta f(x)' = \sum_0^m \theta_i f_i(x)$ where $\theta = (\theta_0, \cdots, \theta_m)$ with the θ_i unknown real parameters, then, if $M(\xi)$ is nonsingular, $\sigma^2 N^{-1} M^{-1}(\xi)$ is the covariance matrix of best linear estimators of the vector θ. Moreover, setting

$$V(a, \xi) = aM^{-1}(\xi)a' \tag{1.2}$$

where $a = (a_0, a_1, \cdots, a_m)$ with the a_i real, $\sigma^2 N^{-1} V(a, \xi)$ is the variance of the best linear estimator of the linear parametric function $\theta a'$. The function V of (1.2) is defined to have the same meaning even if M is singular; in particular, $V(a, \xi) = \infty$ if $\theta a'$ is not estimatable under ξ.

As has been discussed in other papers, we do not restrict ξ to take on values which are integral multiples of N^{-1}. This allows us to obtain optimum design characterizations which cannot be obtained under that restriction, and at the same time yields designs which can be implemented in practice through the use of closely related ξ's which do take on only values which are integral multiples of N^{-1}.

Received 21 April 1965.

[1] Research supported by the Office of Naval Research under Contract No. Nonr 266(04) (NR 047-005).

[2] Research supported by the U. S. Air Force under Contract No. AF 18(600)-685, monitored by the Office of Scientific Research.

One problem of interest is to characterize, for each a, a design ξ, termed *a-optimum*, which minimizes $V(a, \xi)$; work on this problem has been done in [4] Elfving (1952), [3] Chernoff (1953), and [8] Kiefer and Wolfowitz (1959). A particular way in which this problem can arise, and which is of considerable practical importance, is that f is extended continuously so as to be defined on the set $Y \cup X$, and that for some point e in Y it is required to choose a design which estimates optimally the regression $\theta f(e)'$ at the point e, the design ξ still being restricted to be a probability measure on the set X of points at which we are permitted to take observations. For e in $Y - X$ (resp., $Y \cap X$) this may be called the problem of *extrapolation* (resp., *interpolation*) of the estimated regression to the point e of the set Y on which the regression function is of interest to us. This is clearly the problem of finding an a-optimum design when $a = f(e)$. (Other extrapolation and interpolation problems, such as that of minimizing $\max_{y \epsilon Y} V(f(y), \xi)$ for certain sets Y other than those of the type mentioned in the next two paragraphs, are considered in [9] Kiefer and Wolfowitz (1964a, b); when $Y = X$, this problem has been considered extensively in other papers.)

Hoel and Levine (1964) [5] have considered this extrapolation problem in the important case of univariate polynomial regression, where $X = [-1, 1]$, $Y = (-\infty, \infty)$, $f_i(y) = y^i$ for y in Y. Their elegant main result is that, for $|e| > 1$, there is an $f(e)$-optimum design which, for every e, is supported by the same set of $m + 1$ points (with weights which depend on e), namely, the "Chebyshev points" which were shown in [8] to be the support of the a-optimum design when $a = (0, 0, \cdots, 0, 1)$. (It will often be helpful to think of this a as $\lim_{e \to \pm\infty} [f_m(e)]^{-1} f(e)$.) It is well known (see Example 2 of Section 5) that this conclusion cannot be extended to the $f(x)$-optimum design for $|x| \leqq 1$.

(We are indebted to Dr. T. J. Rivlin for pointing out that part of the above-mentioned development on p. 1556 of [5] which shows that $\sum_0^m |L_i(e)|$ is maximized by the Chebyshev points, where the L_i are Lagrange interpolation polynomials, rediscovers a result proved by S. Bernstein on p. 186 of [2].)

A second result of Hoel and Levine is that, for e sufficiently large, their $f(e)$-optimum design also minimizes $\max_{-1 \leqq y \leqq e} V(f(y), \xi)$, which is proportional to the maximum over the interval $[-1, e]$ of the variance of the estimated regression. (In most of the literature to which we refer, $V(f(y), \xi)$ is denoted by $d(y, \xi)$.)

It is natural to ask then, whether in this polynomial case there are vectors a other than constant multiples of those in the one-dimensional set $\{f(e), |e| > 1\}$, such that there is an a-optimum design supported by this same set of Chebyshev points. One of the results of the present paper (Theorems 1 and 2 of Section 4 and Example 2 of Section 5) is that there are many such a, in fact, an $(m + 1)$-dimensional set of them in the $(m + 1)$-dimensional space of all a. This apparent anomaly (in view of the low dimensionality of these designs in the space of all admissible designs) is discussed after the introduction of necessary nomenclature, in the first paragraph of Section 4, and is illustrated in Example 2(b) of Section 5.

It is also natural to ask whether the two Hoel-Levine results and the result discussed in the previous paragraph can be extended to examples other than that of polynomial regression. Theorems 4, 5, 1 and 2 give such extensions under assumptions stated in the next section; the main assumptions (1 and 2) are related to the behavior of the f_i for a related Chebyshev approximation problem. (The proof of Theorem 4 reduces, in the polynomial case, to one which differs from that of [5].) It seems interesting to determine, under these assumptions, the set of vectors a for which there is an optimum design supported by (that is, which assigns positive probability to, and only to) the Chebyshev points, and Theorems 1 and 2 show that this is the set T^* defined in the next section. Also of interest is the set $\bar{T}^*$ of vectors for which there is an optimum design supported by some subset of the Chebyshev points. $\bar{T}^*$, which is not generally merely the closure of T^* (as will be seen in Example 2(b)), does not permit as simple an analysis as T^*. For the most part we are concerned with a subset R^* of T^*, where a related Chebyshev approximation problem has a solution of a particular form, and where the optimum design is unique (as it need not be in $T^* - R^*$). The sets R^* and $T^* - R^*$ are not difficult to characterize explicitly (see (2.20), Theorem 2, and (2.27)), but Theorem 3 describes the inclusion in R^* of a set A^* of vectors which is sometimes also easy to compute, namely, the set of vectors a for which $a'\theta$ is not estimatable for any design on fewer than $m + 1$ points. Theorem 3 is used in establishing Theorem 4.

Section 2 contains nomenclature, definitions, assumptions, and statements of results from previous papers which we shall use. Section 3 contains proofs of auxilliary lemmas which are used in the proofs of our main results (Theorems 1–5) in Section 4; Remark 5 in the latter section describes some further extensions of our results. Finally, Section 5 contains some examples which illustrate the relationship among T^*, R^*, A^*, and other sets we shall consider.

The preliminary propositions, examples, and theorems of Section 2 will be numbered decimally (2.x). All other theorems and examples, and all lemmas and remarks, will be numbered consecutively without indication of section.

2. Preliminaries. Our basic model of f, X, and Y will be as described in Section 1. Throughout this paper, except when explicitly stated to the contrary (as, in particular, in the extensions of Remark 5), X will be $[-1, 1]$ and Y will be $(-\infty, \infty)$. We shall usually denote points of X, Y and $Y - X$ by x, y, and e, respectively. As described in Section 1, we always assume, without further statement in the sequel,

ASSUMPTION 0. $f_0, f_1, \cdots, f_m$ are linearly independent on $[-1, 1]$ and continuous on $(-\infty, \infty)$.

The functions $f_0, \cdots, f_k$ are called a *Chebyshev system* on the set U if every linear combination $\sum_0^k c_i f_i$, with not all of the real constants c_i zero, has k or fewer zeros on U. This condition can be rephrased in any of several equivalent forms which will be useful in what follows: For distinct $x_0, x_1, \cdots, x_k$ in U,

(i) the vector $\sum_0^k c_i(f_i(x_0), \cdots, f_i(x_k))$ is not the zero vector unless all c_i are 0;

(ii) the matrix $\{f_i(x_j), 0 \leq i, j \leq k\}$ has rank $k + 1$;

(iii) the vector $(f_0(x_k), \cdots, f_k(x_k))$ cannot be represented as $\sum_0^{k-1} c_i(f_0(x_i), \cdots, f_k(x_i))$.

In the lemmas and theorems of Sections 3 and 4, we shall make use of the following five assumptions, stating explicitly where any of them is made. The first two assumptions are used in all five of the theorems.

ASSUMPTION 1. The functions $f_0, f_1, \cdots, f_{m-1}$ constitute a Chebyshev system on $[-1, 1]$.

If F is a continuous real-valued function on $[-1, 1]$, we shall say that F *changes direction* at x_0 if $-1 < x_0 < 1$ and if F has a local maximum or minimum at x_0. In particular, if F is constant on any open subinterval of $[-1, 1]$, it will be said to have infinitely many changes of direction.

ASSUMPTION 2. For each q in E^{m+1} (Euclidean $(m + 1)$-space), the function fq' on $[-1, 1]$ either has fewer than m changes of direction, or else is constant on $[-1, 1]$.

In proving our generalizations of the Hoel-Levine results we shall also use

ASSUMPTION 3. The continuous functions $f_0, f_1, \cdots, f_m$ are a Chebyshev system on $(-\infty, \infty)$;

ASSUMPTION 4. For $0 \leq i < m$, we have $\lim_{e \to \pm\infty} f_i(e)/f_m(e) = 0$;

ASSUMPTION 5. $|f_m(e)|$ is strictly increasing (resp., decreasing) when e(resp., $-e$) is sufficiently large; $\lim_{e \to \pm\infty} |f_m(e)| = +\infty$; and, for $0 \leq i < m$, the quantity

$$\sup_{0 < |\Delta| \leq 1} |f_i(e + \Delta) - f_i(e)|/|f_m(e + \Delta) - f_m(e)|$$

remains bounded as $e \to \pm\infty$.

Assumption 5 can of course be phrased in terms of derivatives, if they exist. We next remark briefly on Assumptions 1 and 2.

REMARK 1(a). We shall prove in Lemma 2 that Assumption 2 implies that $f_0, f_1, \cdots, f_m$ constitute a Chebyshev system on $[-1, 1]$ with $\sum_0^m d_j f_j(x) \equiv 1$ for some $d_0, \cdots, d_m$. In the other direction we have the obvious

PROPOSITION 2.1. *If* $f_0 \equiv 1$, *the* f_i *are differentiable, and* $\{df_i(x)/dx, 1 \leq i \leq m\}$ *constitute a Chebyshev system on* $[-1, 1]$, *then Assumption 2 holds.*

However, simple examples show that Assumption 2 is stronger than $\{f_i, 0 \leq i \leq m\}$ being a Chebyshev system. An illustration is

EXAMPLE 2.1. If $\{f_i, 0 \leq i \leq m\}$ satisfies Assumption 2 and h is positive and continuous on $[-1, 1]$, then $\{hf_i, 0 \leq i \leq m\}$ is Chebyshev but need not satisfy Assumption 2 (e.g., let $f_i(x) = x^i$ and $h(x) = 2 + \sin 10x$).

REMARK 1(b). The form of Assumptions 1 and 2 appears to single out f_m for special treatment. This asymmetric form of the first two assumptions can be replaced by a more symmetric form. For example, Assumptions 1 and 2 can be replaced by the hypotheses of the following:

PROPOSITION 2.2 *Assumption 2 and the assumption that* $f_0, f_1, \cdots, f_m$ *is Chebyshev on a set* $[-1, 1] \cup \{v\}$ *for some point* $v \not\in [-1, 1]$ *imply that Assumptions 1 and 2 are satisfied for* $\{\bar{f}_i, 0 \leq i \leq m\}$ *where* $\bar{f}' = Qf'$ *for some nonsingular* Q. *In particular, this conclusion is implied by Assumptions 2 and 3.*

(The second assumption of Proposition 2.2, without the v, is the conclusion of Lemma 2 obtained under Assumptions 1 and 2. Example 2.1 above shows that Assumption 3 alone does not imply Assumption 2.)

PROOF OF PROPOSITION 2.2. We clearly need only prove that Assumption 1 is satisfied for $\bar{f}_0, \cdots, \bar{f}_{m-1}$ with some nonsingular Q. By linear independence there is a nonsingular Q such that $\bar{f}(v)' \equiv Qf(v)' = (0, 0, \cdots, 0, 1)'$. But then, writing $x_m = v$, if $0 \leqq x_0 < x_1 < \cdots < x_{m-1} \leqq 1$ we have

$$\det \{\bar{f}_i(x_j), 0 \leqq i, j < m\} = \det \{\bar{f}_i(x_j), 0 \leqq i, j \leqq m\}$$
$$= \det Q \det \{f_i(x_j), 0 \leqq i, j \leqq m\} \neq 0,$$

the last by the Chebyshev assumption for $[-1, 1] \cup \{v\}$. This proves the desired result.

The following example shows that the Chebyshev nature of $\{f_0, f_1, \cdots, f_m\}$ on $[-1, 1]$ does not by itself imply that $\{\bar{f}_0, \cdots, \bar{f}_{m-1}\}$ is Chebyshev for some Q, even if $f_0(x) \equiv 1$:

EXAMPLE 2.2. Suppose $f_0(x) \equiv 1$ on $[-1, 1]$, and let $f_1(x) = \cos \pi x$ and

$$f_2(x) = \sin \pi x \qquad \text{if} \quad -1 \leqq x \leqq \tfrac{1}{2},$$
$$= (1 + \sin \pi x)/2 \qquad \text{if} \quad \tfrac{1}{2} \leqq x \leqq 1.$$

This system $\{f_0, f_1, f_2\}$ was given in [11] Volkov (1958) as an example of a Chebyshev system on $[-1, 1]$ which cannot be extended to be Chebyshev on a larger interval (our example must of course be of this nature by Proposition 2.2). To see that this system is indeed Chebyshev, it is only necessary to graph the function $c_1f_1 + c_2f_2$ (with c_1 and c_2 not both zero) in each of the four cases $0 \leqq \pm c_1 \leqq c_2$ and $0 \leqq c_2 < \pm c_1$ and to note that this function assumes each value at most twice. If now $\{\bar{f}_0, \bar{f}_1\}$ were Chebyshev on $[-1, 1]$ with $\bar{f}_i = \sum_j q_{ij}f_j$, some linear combination of $\bar{f}_0$ and $\bar{f}_1$ would be of the form $g_1 = c_1f_1 + c_2f_2$ and (by the Chebyshev property) would have at most one zero on $[-1, 1]$. An examination of the four cases described above shows that any function of this form has two zeros on $[-1, 1]$, except for multiples of $f_1 + cf_2$ with $c > 2$. For g_1 of this last form we have $g_1'(r) = 0$ for a single r satisfying $-1 < r < -\frac{1}{2}$, where we use primes to denote derivatives in this paragraph. Let g_0 be a linear combination of $\bar{f}_0$ and $\bar{f}_1$ with g_0 positive throughout $[-1, 1]$; the existence of such a g_0 is shown in the first paragraph of Example 1 of Section 5. The development of that paragraph also shows (since $g_1(-1) < 0 < g_1(1)$) that g_1/g_0 is nondecreasing on $[-1, 1]$. Since $(g_1/g_0)' = g_1'/g_0 - g_1g_0'/g_0^2$, we see that $(g_1/g_0)'(r) \geqq 0$ if and only if $g_0'(r) \geqq 0$. If $g_0'(r) > 0$, then, since $g_0'(-1) < 0$ for g_1/g_0 to be increasing (because $g_1(-1) < 0, g_1'(-1) < 0$), there is an s with $-1 < s < r$ and $h'(s) = 0$; but then $(g_1/g_0)'(s) < 0$, a contradiction. Hence $g_0'(r) = 0$. But then, since $\{1, g_1, f_1\}$ span the same vector space as $\{f_0, f_1, f_2\}$ and $f_1'(r) \neq 0$, we conclude that $g_0 = k_0 + k_1g_1$ for some constants k_i with $k_0 \neq 0$. But then $g_1 + k_3(g_0 - k_1g_1)$ is a linear combination of $\bar{f}_0$ and $\bar{f}_1$ which, for a suitable choice of the constant k_3, has two zeros in a small neighborhood of r (since $g_1''(r) \neq 0$).

For reference in the proofs of Sections 3 and 4, we now summarize in Theorems 2.1–2.3 the known results of Chebyshev approximation theory and optimum design theory which will be used. Suppose $h_0, h_1, \cdots, h_m$ are continuous real-valued functions on $[-1, 1]$. Then $\sum_0^{m-1} c_i^* h_i$ is called a *best Chebyshev (uniform) approximation* to h_m on $[-1, 1]$ if $\max_{-1 \leqq x \leqq 1} |h_m(x) - \sum_0^{m-1} c_i h_i(x)|$ is minimized, over all choices of the real constants c_i, by the choice $c_i = c_i^*$ $(0 \leqq i < m)$. The vector $c^* = (c_0^*, \cdots, c_{m-1}^*)$ is then called a *Chebyshev vector*. A classical result of approximation theory ([1] Achieser (1956), p. 74) is

THEOREM 2.1. *If $h_0, \cdots, h_{m-1}$ form a Chebyshev system, then there is a unique Chebyshev vector c^*, and it is characterized by the fact that there are at least $m + 1$ points at which the residual $h_m - \sum_0^{m-1} c_i^* h_i$ attains its maximum in absolute value, this maximum being assumed with at least $m + 1$ successive alternations in sign.*

If c^* is a Chebyshev vector, we shall denote by $B(c^*)$ the set where $|h_m - \sum_0^{m-1} c_j^* h_j|$ attains its maximum on $[-1, 1]$.

We shall be concerned both with cases where $h_0, \cdots, h_{m-1}$ are a Chebyshev system, and with cases where they are not (see Lemma 8, etc.). It will be shown in Lemma 3 that, if Assumptions 1 and 2 hold, which in particular imply that Theorem 2.1 holds with $h_i = f_i$ $(0 \leqq i \leqq m)$, the set $B(c^*)$ consists of exactly $m + 1$ points $-1 = x_0^* < x_1^* < \cdots < x_m^* = 1$. We shall denote by X_m^* the set of these *Chebyshev points*.

If $h_0, h_1, \cdots, h_m$ are continuous real-valued linearly independent functions on $[-1, 1]$, and if $\sum_0^m \psi_i h_i(x)$ is the expected value of an observation at x, where the ψ_i are unknown real parameters and the observations are uncorrelated with common variance σ^2, we are led to consider the game with payoff function

$$K(\xi, q) = \int_{-1}^{1} [h_m(x) - \textstyle\sum_0^{m-1} q_j h_j(x)]^2 \xi(dx), \tag{2.1}$$

where the minimizing player may (by convexity of K in q) be restricted to pure strategies which are vectors $q = (q_0, \cdots, q_{m-1})$ of real components, while the maximizing player has probability measures ξ (which can be taken to be discrete) for his mixed strategies. From Section 2 of [8], we have the following results:

THEOREM 2.2. *The variance $\sigma^2 N^{-1} v_m(\xi)$ of the best linear estimator of ψ_m when the design ξ is used satisfies*

$$v_m(\xi) = 1/\min_q K(\xi, q). \tag{2.2}$$

Hence, ξ^* is optimum for estimating ψ_m if and only if it is maximin for the game with payoff K; i.e., *if and only if*

$$\min_q K(\xi^*, q) = \max_\xi \min_q K(\xi, q). \tag{2.3}$$

THEOREM 2.3. *The game with payoff K is determined, the minimizing player has a pure minimax strategy q^*, and the maximizing player has a maximin strategy*

ξ^* on $m + 1 - p$ points, where p is the dimensionality of the convex set of Chebyshev vectors for approximating h_m by $h_0, \cdots, h_{m-1}$ on $[-1, 1]$. The minimax strategies coincide with the Chebyshev vectors. (Hence, from standard game theory, ξ is maximal relative to q and q is minimal with respect to ξ, if and only if ξ is optimum and q is Chebyshev.) The design ξ^* is maximin if and only if, for any Chebyshev vector q^*, $\xi^*(B(q^*)) = 1$ and ξ^* satisfies the "orthogonality relations"

$$\int_{-1}^{1} [h_m(x) - \textstyle\sum_0^{m-1} q_j^* h_j(x)] h_i(x) \xi^*(dx) = 0, \qquad 0 \leqq i < m. \tag{2.4}$$

In the setting of Section 1 and the first paragraph of the present section, we considered linear parametric functions $\theta a'$ for a in $B^* = E^{m+1} - \{0\}$ where E^k denotes Euclidean k-space and 0 denotes the origin in whatever Euclidean space is under consideration (or, where appropriate, a matrix of zeros). Clearly, for every a and every real $\lambda \neq 0$, ξ is a-optimum if and only if it is (λa)-optimum. Thus, from the point of view of characterizing a-optimum designs, one could replace B^* by the real projective m-space P^m of its equivalence classes under the equivalence $a \sim \lambda a$ for all $\lambda \neq 0$. Throughout most of the developments of Sections 3 and 4, it is not profitable to do this, and therefore *all starred sets, such as A^*, R^*, and T^* will be regarded as subsets of B^**. However, because of the special role of the last coördinate θ_m in Assumption 1 (where f_m, of the $m + 1$ functions f_i, is absent), it is sometimes more convenient, especially in explicit representations of the optimum designs for various parametric functions, to consider the a's in terms of two disjoint sets $B = E^m$ and $B_0 = P^{m-1}$, as follows: To each a with $a_m = 1$ we make correspond the vector $b = (b_0, \cdots, b_{m-1})$ in B defined by $b_i = a_i$ $(0 \leqq i < m)$, and conversely; we shall thus think of the m-vector b as corresponding to the linear parametric function $\theta_m + \sum_0^{m-1} b_i\theta_i$ and shall call a design which is optimum for this linear parametric function *b-optimum*. Of course, for any a with $a_m \neq 0$ (whether or not $a_m = 1$), the a-optimum designs will then coincide with the b-optimum designs with

$$b = (a_0/a_m, \cdots, a_{m-1}/a_m).$$

Similarly, the a's with $a_m = 0$ but not all $a_i = 0$ correspond to points in B_0. We let Γ denote the mapping of B^* onto $B \cup B_0$ under this identification. Throughout this paper we shall use a to mean an element of B^* and b to mean an element of B. The disadvantages of working in terms of B and B_0 rather than B^* will be seen in Examples 1 and 2(c) of Section 5.

It will sometimes be useful in the developments which follow to work not with the space Ξ_m of probability $(m + 1)$-vectors with positive components (to be thought of as being on X_m^*), but rather with the set $\Xi_m^* = \{\eta: \eta = \lambda\xi$ for some ξ in Ξ_m and some real $\lambda \neq 0\}$, which can be regarded as the union of two congruent convex cones in E^{m+1}.

We shall apply Theorems 2.2–2.3 to the problem of determining a-optimum designs in the setting of the first paragraph of this section by using the following simple reduction of [7] Kiefer (1962), p. 795: For fixed a and i_0 with $a_{i_0} \neq 0$, write

$$\begin{aligned} \varphi_{i_0} &\equiv \varphi_{i_0,a} = \textstyle\sum_0^m a_i\theta_i/a_{i_0}, \\ \varphi_i &= \theta_i, & i \neq i_0, \\ g_{i_0} &= f_{i_0}, \\ g_i &\equiv g_{i,a} = f_i - (a_i/a_{i_0})f_{i_0}, & i \neq i_0. \end{aligned} \tag{2.5}$$

Then $\sum_0^m \varphi_i g_i = \sum_0^m \theta_i f_i$, and the problem of estimating $a\theta'$ when the regression is $\theta f(x)'$ for $x \,\varepsilon\, [-1, 1]$ is the same as that of estimating φ_{i_0} when the regression is $\varphi g(x)'$ for $x \,\varepsilon\, [-1, 1]$. Thus, we can apply Theorems 2.2–2.3 to the problem of finding a-optimum designs by setting

$$\begin{aligned} \psi_m &= \varphi_{i_0}, & \{\psi_i, 0 \leqq i < m\} &= \{\varphi_i, i \neq i_0\}, \\ h_m &= g_{i_0}, & \{h_i, 0 \leqq i < m\} &= \{g_i, i \neq i_0\}. \end{aligned} \tag{2.6}$$

In particular, the payoff function (2.1) becomes

$$K(\xi, q) = \int_{-1}^1 [f_{i_0}(x) - \textstyle\sum_{i \neq i_0} q_i(f_i(x) - (a_i/a_{i_0})f_{i_0}(x))]^2 \xi(dx), \tag{2.7}$$

the orthogonality relations (2.4) become

$$\begin{aligned} \int_{-1}^1 [a_{i_0} f_{i_0}(x) - \textstyle\sum_{j \neq i_0} q_j^*(a_{i_0} f_j(x) - a_j f_{i_0}(x))] \\ \cdot [a_{i_0} f_i(x) - a_i f_{i_0}(x)] \xi^*(dx) = 0, \quad i \neq i_0, \end{aligned} \tag{2.8}$$

and the related Chebyshev problem is that of approximating f_{i_0} on $[-1, 1]$ by $\{f_i - (a_i/a_{i_0})f_{i_0}, i \neq i_0\}$. All of these depend on the a and i_0 under consideration (although the latter dependence will be seen to be irrelevant).

The functions g_i of course satisfy Assumption 0 if the f_i do. However, the same is not true of Assumption 1 (as will be seen in Lemma 8). Thus, although Theorem 2.1 can be applied under Assumption 1 to help characterize a 0-optimum design (for estimating θ_m, where $b = 0$), and to find X_m^*, we cannot apply Theorem 2.1 in the same way to find other a-optimum designs.

We now define the sets whose study will be the chief concern of this paper. A parallel notation will be used throughout: A starred symbol D^* (say) will always be defined as a subset of B^* which is invariant under multiplication by a nonzero scalar, and we will then always write

$$D = \Gamma(D^*) \cap B, \qquad D_0 = \Gamma(D^*) \cap B_0; \tag{2.9}$$

the starred sets D^* will be our main objects of interest in Sections 3 and 4, but it will be useful to consider the unstarred sets in the examples of Section 5. Under Assumptions 1 and 2, as we have mentioned, there is, corresponding to $\{f_0, f_1, \cdots, f_m\}$, the set of $m + 1$ Chebyshev points $\{x_0^*, x_1^*, \cdots, x_m^*\} = X_m^*$. We define, as used in Section 1,

$$T^* = \{a: a \,\varepsilon\, B^* \text{ and there is an } a\text{-optimum design supported by the entire set } X_m^*\}. \tag{2.10}$$

One could instead study the set $\bar{T}^*$ (say) where there is an a-optimum design supported by a subset of X_m^*, but the set $\bar{T}^* - T^*$ is less susceptible to study by our methods. Let U be the subset of E^m defined by

$$(2.11) \qquad U = \{(x_0, x_1, \cdots, x_{m-1}): \text{ all } |x_i| \leq 1 \text{ and all } x_i \text{ different}\},$$

and write, for $\tilde{x} = (x_0, \cdots, x_{m-1})$ in E^m and a in B^*,

$$(2.12) \qquad P^*(\tilde{x}, a) = \det \begin{pmatrix} f_0(x_0) & \cdots & f_0(x_{m-1}) & a_0 \\ \vdots & \vdots\vdots\vdots & \vdots & \vdots \\ f_m(x_0) & \cdots & f_m(x_{m-1}) & a_m \end{pmatrix}.$$

We define

$$(2.13) \qquad A^* = \{a: a \in B^* \text{ and } P^*(\tilde{x}, a) \neq 0 \text{ for all } \tilde{x} \text{ in } U\}.$$

We also define

$$(2.14) \qquad N^* = \{a: a \in B^* \text{ and } a\theta' \text{ is only estimatable for designs supported by at least } m+1 \text{ points of } [-1, 1]\}.$$

It is easy to see (Lemma 1 below) that $N^* = A^*$. The usefulness of this set is of course that, for a in A^*, any a-optimum design has at least $m + 1$ points of support, so that X_m^* is at least a possible candidate for this support.

For fixed a in B^*, suppose $a_{i_0} \neq 0$, and consider the system of m linear equations in the $m + 1$ unknowns η_j,

$$(2.15) \qquad \sum_{j=0}^{m} (-1)^j \eta_j [a_{i_0} f_i(x_j^*) - a_i f_{i_0}(x_j^*)] = 0, \qquad 0 \leq i \leq m, \quad i \neq i_0.$$

We shall also consider the related system

$$(2.16) \qquad \sum_{j=0}^{m} (-1)^j \xi_j [a_{i_0} f_i(x_j^*) - a_i f_{i_0}(x_j^*)] = 0, \qquad i \neq i_0;$$
$$\sum_{j=0}^{m} \xi_j = 1.$$

For $i_0 = m$, putting $b = \Gamma(a)$, (2.16) becomes

$$(2.17) \qquad \sum_{j=0}^{m} (-1)^j \xi_j [f_i(x_j^*) - b_i f_m(x_j^*)] = 0, \qquad 0 \leq i < m;$$
$$\sum_{j=0}^{m} \xi_j = 1.$$

(It will be clear from the derivation of (2.20) and (2.27) that for fixed a, an η in Ξ_m^* is a solution of (2.15) or (2.23) for some i_0 for which $a_{i_0} \neq 0$, if and only if it is a solution for every such i_0.) The form (2.15) will be of chief concern in Sections 3 and 4. We shall use (2.17) extensively in Example 2(b) of Section 5. We now define

$$(2.18) \qquad \begin{aligned} R^* &= \{a: a \in B^* \text{ and for some } i_0 \text{ with } a_{i_0} \neq 0 \text{ (2.15) has a solution } \eta \text{ in } \Xi_m^*\} \\ &= \{a: a \in B^* \text{ and for some } i_0 \text{ with } a_{i_0} \neq 0 \text{ (2.16) has a solution } \xi \text{ in } \Xi_m\}. \end{aligned}$$

Write

$$F_{R^*} = \{(-1)^j f_i(x_j^*), \quad 0 \leqq i, \ j \leqq m\}, \tag{2.19}$$
$$F_{S^*} = \{f_i(x_j^*), \quad 0 \leqq i, \ j \leqq m\}.$$

By Lemma 2, Assumption 2 implies that these matrices are nonsingular, a fact which we shall use repeatedly. Suppose $a = \eta F'_{R^*}$ for some η in Ξ_m^*. Since F_{R^*} is nonsingular by Assumption 2, $a_{i_0} \neq 0$ for some i_0, and a clearly satisfies (2.15), since the latter can be written as $a_{i_0}(F_{R^*}\eta')_i = a_i(F_{R^*}\eta')_{i_0}$. Also, this form of (2.15) shows that every a in R^* can be obtained in this way. Thus,

$$R^* = \{a: a = \eta F'_{R^*} \quad \text{for some} \quad \eta \quad \text{in} \quad \Xi_m^*\} \tag{2.20}$$
$$= (\Xi_m^*)F'_{R^*}.$$

We thus have an explicit representation of R^* as a pair of congruent open convex cones obtained from the linear mapping F_{R^*} acting on Ξ_m^*. One of these cones is spanned by the $m + 1$ half-lines consisting of the positive multiples of column vectors of F_{R^*}; the other, of the negative multiples.

From (2.18) we also have

$$R = \{b: b \,\varepsilon\, B \quad \text{and (2.17) has a solution in} \quad \Xi_m\}$$
$$= \{b: b_i = \textstyle\sum_{j=0}^m (-1)^j \xi_j f_i(x_j^*)/\sum_{j=0}^m (-1)^j \xi_j f_m(x_j^*), \tag{2.21}$$
$$0 \leqq i < m, \quad \text{for some} \quad \xi \quad \text{in} \quad \Xi_m\}.$$

The importance of the set R^* is given in Theorem 1. For a in R^*, it will turn out that the residual of the best Chebyshev approximation of f_{i_0} by $\{f_i - (a_i/a_{i_0})f_{i_0}, i \neq i_0\}$, mentioned below (2.8), is *oscillatory* (that is, satisfies the condition of Theorem 2.1) even though (Lemma 8) $\{f_i - (a_i/a_{i_0})f_{i_0}, i \neq i_0\}$ is not a Chebyshev system for $a \,\varepsilon\, R^* - A^*$. It will also turn out that this residual attains its maximum in absolute value at the x_j^*. Hence, the residual at x_j^* (first factor of the integrand of (2.8)) is a multiple of $(-1)^j$ and thus, writing

$$\xi_j = \xi^*(x_j^*), \tag{2.22}$$

the orthogonality relations (2.8) will be seen to reduce to (2.16). The application of the game theory of Theorem 2.3 will be used, in the proof of Theorem 1, to show that $R^* \subset T^*$.

We have mentioned in the previous paragraph that $a \,\varepsilon\, R^*$ implies the oscillatory nature of a certain Chebyshev approximation problem. One could also study the designs in $T^* - R^*$, which are also supported by X_m^*, but for which (by Lemma 5 and Theorem 1) the solution to this Chebyshev approximation problem has *constant* nonzero residual. Paralleling the development indicated in the previous paragraph, we now consider, in place of (2.15), (2.16), (2.17), the systems

$$\textstyle\sum_{j=0}^m \eta_j[a_{i_0} f_i(x_j^*) - a_i f_{i_0}(x_j^*)] = 0, \qquad 0 \leqq i \leqq m, \quad i \neq i_0; \tag{2.23}$$

$$(2.24)\quad \textstyle\sum_{j=0}^{m} \xi_j[a_{i_0}f_i(x_j^*) - a_i f_{i_0}(x_j^*)] = 0, \qquad i \neq i_0, \\ \textstyle\sum_{j=0}^{m} \xi_j = 1;$$

$$(2.25)\quad \textstyle\sum_{j=0}^{m} \xi_j[f_i(x_j^*) - b_i f_m(x_j^*)] = 0, \qquad 0 \leqq i < m, \\ \textstyle\sum_{j=0}^{m} \xi_j = 1.$$

In place of (2.18) we now define

$$(2.26)\quad \begin{aligned} S^* &= \{a: a \in B^* \text{ and for some } i_0 \text{ with } a_{i_0} \neq 0 \text{ (2.23) has a solution } \eta \text{ in } \Xi_m^*\} \\ &= \{a: a \in B^* \text{ and for some } i_0 \text{ with } a_{i_0} \neq 0 \text{ (2.24) has a solution } \xi \text{ in } \Xi_m\}. \end{aligned}$$

Using the second half of (2.19) we obtain in place of (2.20),

$$(2.27)\quad \begin{aligned} S^* &= \{a: a = \eta F'_{S^*} \text{ for some } \eta \text{ in } \Xi_m^*\} \\ &= (\Xi_m^*)F'_{S^*}. \end{aligned}$$

In place of (2.21) we now have

$$(2.28)\quad \begin{aligned} S &= \{b: b \in B \text{ and (2.25) has a solution in } \Xi_m\} \\ &= \{b: b_i = \textstyle\sum_{j=0}^{m} \xi_j f_i(x_j^*)/\sum_{j=0}^{m} \xi_j f_m(x_j^*), \quad 0 \leqq i < m, \\ &\qquad \text{for some } \xi \text{ in } \Xi_m\}. \end{aligned}$$

Using the results indicated at the outset of the present paragraph, we shall show in Theorem 2 that $T^* - R^* = S^*$. A major difference between R^* and S^*, which will be illustrated in Example 2(b) of Section 5, is that the a-optimum design is unique for $a \in R^*$, while for $a \in S^*$ we can have other a-optimum designs whose supports are not X_m^*.

While the set $\bar{T}^*$ defined just below (2.10) will be illustrated in Example 2(b), we shall not analyze $\bar{T}^*$ in general. Such an analysis would be more complicated than that of T^* because of the variety of forms the residual can now have and the necessity of determining when the orthogonality relations parallel to (2.15) or (2.23) do indeed correspond to an a for which the residual attains its maximum absolute value on the relevant subset of X_m^*. In particular, for $m > 1$ the set $\bar{T}^*$ is not generally the closure of T^*.

The closure of R^* or S^* (and, hence, of T^*) in $E^{m+1} - \{0\}$ is obviously obtained by replacing Ξ_m^* in (2.20) or (2.27) by the closure of Ξ_m^* in $E^{m+1} - \{0\}$; that is, by the set of all $(m + 1)$-vectors not all of whose components are zero, but whose nonzero components all have the same sign.

The definition of the generalization of the set considered by Hoel and Levine is

$$(2.29)\quad H^* = \{a: a \in B^* \text{ and } a = \lambda f(e)' \text{ for some real } \lambda \neq 0 \text{ and some real } e \text{ with } |e| > 1\}.$$

(If $f(e) = 0$, $\theta f(e)'$ can of course be estimated without error; such e are excluded from H^*, and under Assumption 3 they obviously can not exist.) In particular, if $f_m(e) \neq 0$ for $|e| > 1$, H_0 is empty and

$$\Gamma(H^*) = H = \{b: b_i = f_i(e)/f_m(e), \quad 0 \leqq i < m, \quad \text{for some} \quad e \quad \text{with} \quad |e| > 1\}. \tag{2.30}$$

The examples of Section 5 illustrate the sets defined in this section.

We add the definition of a concept which arises in the first paragraph of Section 4 and in Example 2(b) of Section 5, that of an *admissible design* ξ, which is a design such that for no ξ' is $M(\xi') - M(\xi)$ nonnegative definite and not the zero matrix. The meaning of this concept is discussed in [6] Kiefer (1959).

In Example 2(a) we shall introduce some additional material from the literature, which is used only there.

3. Auxiliary lemmas. The lemmas of this section will be used in proving our main results in the next section.

LEMMA 1. $N^* = A^*$.

PROOF. Since $\theta f(x_j)'$ is the expected value of an observation at x_j, any linear parametric function which is estimatable under a design supported at $\{x_0, x_1, \cdots, x_{m-1}\}$ or a subset thereof must be of the form $\sum_0^{m-1} \gamma_j \theta f(x_j)'$ for some real $\gamma_0, \cdots, \gamma_{m-1}$. Hence, $\theta a'$ is estimatable under a design on $\{x_0, x_1, \cdots, x_{m-1}\}$ if and only if there exist $\gamma_0, \cdots, \gamma_{m-1}$ such that $\sum_0^{m-1} \gamma_j \theta f(x_j)' = \theta a'$ for all θ; that is, such that $\sum_0^{m-1} \gamma_j f(x_j) = a$. This last is equivalent to $P^*(\bar{x}, a) = 0$. This completes the proof.

LEMMA 2. *Under Assumption 2, $f_0, f_1, \cdots, f_m$ form a Chebyshev system on $[-1, 1]$ and there are numbers d_j such that $\sum_0^m d_j f_j(x) \equiv 1$ for x in $[-1, 1]$.*

PROOF. If $f_0, f_1, \cdots, f_m$ are not a Chebyshev system, there is a vector q other than the zero vector such that $q'f$ has at least $m + 1$ zeros on $[-1, 1]$. Since $q'f$ clearly has at least one change of direction at some point strictly between any two successive zeros, it follows that $q'f$ has at least m changes of direction.

To prove the second assertion, write $h(x) = f(x) - f(0)$. Since $ch(0)' = 0$ for all c, Assumption 2 implies that, for each c, ch' either has fewer than m changes of direction or else is identically zero on $[-1, 1]$. If the latter holds for some c which is not the zero vector, we must have $cf(0)' \neq 0$ (since otherwise $cf(x)' \equiv 0$, contradicting Assumption 0), and $d_j = c_j/cf(0)'$ then yields the desired result. If $ch(x)'$ is not identically zero for all nonzero c, the h_i are linearly independent so that, by continuity, there are $m + 1$ points x_j $(0 \leqq j \leqq m)$ in $[-1, 1]$, none of them zero, such that $H = \{h_i(x_j), 0 \leqq i, j \leqq m\}$ is nonsingular. The $m + 1$ linear equations $cH = (0, \cdots, 0, 1)$ then have a solution $c = \bar{c}$ (say), and $\bar{c}h'$ then vanishes at the $m + 1$ points $0, x_0, x_1, \cdots, x_{m-1}$ and thus has at least m changes of direction, which is a contradiction.

LEMMA 3. *Under Assumptions 1 and 2 the residual $f_m - \sum_0^{m-1} c_j^* f_j$ of the unique best Chebyshev approximation to f_m on $[-1, 1]$ of the form $\sum_0^{m-1} c_j f_j$, attains its maximum in absolute value at exactly $m + 1$ points $-1 = x_0^* < x_1^* \cdots < x_m^* = 1$, the residual alternating in sign at successive x_j^*.*

PROOF. The alternating nature of the residual on $m + 1$ points follows from Assumption 1 and Theorem 2.1. It then follows from Assumption 2 that the residual cannot take on its maximum in absolute value at more than $m + 1$ points, and that $x_0^* = -1, x_m^* = 1$.

The reader is reminded of the definition which follows Theorem 2.1, according to which $\{x_0^*, x_1^*, \cdots, x_m^*\}$ will be called the *Chebyshev points of* $\{f_0, f_1, \cdots, f_m\}$.

LEMMA 4. *Assumption* 1 *implies that* $0 \varepsilon A$.

PROOF. The proof of Lemma 1, with $a = (0, 0, \cdots, 0, 1) = a^*$ (say), shows that θ_m is estimatable for a design on $\{x_0, x_1, \cdots, x_{m-1}\}$ if and only if $0 = P^*(\bar{x}, a^*) = \det \{f_i(x_j), 0 \leq i, j < m\}$. The latter is not zero if the x_i are distinct, by Assumption 1.

LEMMA 5. *Under Assumptions* 1 *and* 2, *if an a-optimum design is supported by at least* $m + 1$ *points, then either* $a \varepsilon R^*$ *or else, for each* i_0 *for which* $a_{i_0} \neq 0$, *every best Chebyshev approximation of* f_{i_0} *by* $\{f_i - (a_i/a_{i_0})f_{i_0}, i \neq i_0\}$ *on* $[-1, 1]$ *has constant nonzero residual.*

PROOF. Suppose there is a best approximation $\sum_{i \neq i_0} c_i'[f_i - (a_i/a_{i_0})f_{i_0}]$ such that the residual $r(x) = f_{i_0}(x) - \sum_{i \neq i_0} c_i'[f_i(x) - (a_i/a_{i_0})f_{i_0}(x)]$ is not constant. By Theorem 2.3 and the hypothesis of the lemma, there are $m + 1$ points $x_0 < x_1 < \cdots < x_m$ in the support of ξ at which $|r(x)|$ attains its maximum on $[-1, 1]$. It follows easily from Assumption 2 that, if $r(x)$ is not constant, $r(x)$ alternates in sign at $x_0, x_1, \cdots, x_m$ and thus has m zeros on $[-1, 1]$. In that case the coefficient of f_m in r is not zero, since, if it were, r would be a linear combination of $f_0, \cdots, f_{m-1}$ which has m zeros but which is not identically zero, contradicting Assumption 1.

Writing $r(x) = q[f_m(x) - \sum_0^{m-1} h_i f_i(x)]$, it follows from the oscillation property of $q^{-1}r(x)$ at $x_0, x_1, \cdots, x_m$ and Theorem 2.1 that $\sum_0^{m-1} h_i f_i$ is the best Chebyshev approximation of f_m by $f_0, \cdots, f_{m-1}$. Hence $x_0, x_1, \cdots, x_m$ are the Chebyshev points and by Lemma 3 they in fact constitute the entire support of ξ. Thus, $a \varepsilon R^*$.

Finally, if $r(x) \equiv 0$, then by Theorem 2.2 the variance of the best linear estimator of $\theta a'$ is infinite, contradicting the fact that any design on $m + 1$ points yields an estimator with finite variance.

LEMMA 6. *For fixed a with* $a_{i_0} \neq 0$, *there are real constants* c_i *and* K *such that*

$$(3.1) \quad f_{i_0}(x) - \textstyle\sum_{i \neq i_0} c_i[f_i(x) - (a_i/a_{i_0})f_{i_0}(x)] \equiv K \quad \textit{for} \quad x \varepsilon [-1, 1],$$

if and only if $K \neq 0$ *and there are unique numbers* d_i *such that*

$$(3.2) \qquad \textstyle\sum_0^m d_i f_i(x) \equiv 1 \quad \textit{for} \quad x \varepsilon [-1, 1],$$

and

$$(3.3) \qquad K = a_{i_0}[\textstyle\sum_{i=0}^m a_i d_i]^{-1}, \qquad c_i = -K d_i .$$

For fixed a with $a_{i_0} \neq 0$ *and fixed reals* c_i', *there are real constants* c_i *and* K *such that*

$$(3.4) \quad f_{i_0}(x) - \textstyle\sum_{i \neq i_0} c_i[f_i(x) - (a_i/a_{i_0})f_{i_0}(x)]$$
$$\equiv K \textstyle\sum_{i=0}^m c_i' f_i(x) \quad \textit{for} \quad x \varepsilon [-1, 1]$$

if and only if $\sum_0^m a_i c_i' \neq 0$ *and*

$$K = a_{i_0}[\sum_0^m a_i c_i']^{-1}, \qquad c_i = -Kc_i'. \tag{3.5}$$

Proof. First suppose (3.1) holds. Since it is impossible for all c_i to be 0 while $1 + \sum_{i \neq i_0} a_i c_i / a_{i_0}$ is also 0, the left side of (3.1) cannot be identically 0, by the linear independence of the f_i. Hence $K \neq 0$. The existence of numbers d_i satisfying (3.2) now follows, and their uniqueness is a consequence of the linear independence of the f_i. Substituting $K \sum_0^m d_i f_i(x)$ for K in (3.1), for each i ($0 \leq i \leq m$) the coefficients of f_i on both sides must be the same, again by linear independence. This yields (3.3). The converse is obvious.

Finally, assuming (3.4), equality of the coefficients of f_i on both sides yields (3.5). The converse is again clear.

Lemma 7. *Under Assumptions* 1 *and* 2, *suppose that* $\sum_0^{m-1} c_j^* f_j$ *is the best Chebyshev approximation of* f_m *by* $\{f_0, f_1, \cdots, f_{m-1}\}$ *and write* $c_m^* = -1$; *furthermore, let* $d_0, \cdots, d_m$ *be the numbers whose existence is guaranteed by Lemma* 2. *Then* $a \, \varepsilon \, R^*$ *implies* $\sum_0^m a_i c_i^* \neq 0$, *and* $a \, \varepsilon \, S^*$ *implies* $\sum_0^m a_i d_i \neq 0$.

Proof. Suppose $a \, \varepsilon \, R^*$ but that $\sum_0^m a_i c_i^* = 0$. Multiply the ith orthogonality relation (2.15) by c_i^* and sum over $i \neq i_0$. We obtain

$$\sum_{j=0}^m (-1)^j \eta_j [\sum_{i=0}^m c_i^* f_i(x_j^*)] = 0. \tag{3.6}$$

Since the term in square brackets in (3.6) is some nonzero constant times $(-1)^j$, this leads to a contradiction. Similarly, if $a \, \varepsilon \, S^*$ but $\sum_0^m a_i d_i = 0$, multiplying the ith relation of (2.23) by d_i and summing over $i \neq i_0$ yields

$$\sum_{j=0}^m \eta_j [\sum_{i=0}^m d_i f_i(x_j^*)] = 0, \tag{3.7}$$

which yields a contradiction since $\sum_0^m d_i f_i \equiv 1$.

Lemma 8. *Suppose* $a_{i_0} \neq 0$. *Then* $\{a_{i_0} f_i - a_i f_{i_0}, i \neq i_0\}$ *is a Chebyshev system on* $[-1, 1]$ *if and only if* $a \, \varepsilon \, A^*$.

Proof. For $0 \leq i \leq m$ and $i \neq i_0$, subtract a_i/a_{i_0} times the i_0th row of the matrix of (2.12) from the ith row. We obtain

$$P^*(\tilde{x}, a) = \pm a_{i_0} \det \{f_i(x_j) - (a_i/a_{i_0}) f_{i_0}(x_j), i \neq i_0, 0 \leq j < m\},$$

which at once yields the conclusion.

Lemma 9. *Under Assumption* 3, $H^* \subset A^*$.

The proof is immediate.

Remark 2. Lemma 9 really uses something weaker than Assumption 3, namely, the nonvanishing of $\det \{f_i(x_j), 0 \leq i, j \leq m\}$ when m different x_i's are in $[-1, 1]$ and one x_i is outside $[-1, 1]$.

4. Principal results. Our first result is that $a \, \varepsilon \, R^*$ implies that the unique a-optimum design is supported by the Chebyshev points, and that R is m-dimensional (and hence R^* is $(m + 1)$-dimensional, which we already knew from (2.20)). This last is perhaps surprising in view of the fact that, in such a simple example as that of polynomial regression, the designs on the Chebyshev points are only an m-parameter family out of the $(2m - 1)$-parameter family of designs on $m + 1$ points including ± 1, all of which are admissible (see [6]), and the Hoel-Levine set H is one-dimensional. We shall see why this is possible in Example 2(b), where it will be seen that infinitely many different supporting sets may yield a-optimum designs for the same a.

Theorem 1. *Under Assumptions 1 and 2, if $a \varepsilon R^*$ the orthogonality relations (2.16) have a unique solution $(\xi_0, \cdots, \xi_m)$, and the corresponding design (2.22) (which is supported by the Chebyshev points) is the unique a-optimum design. (Thus, $R^* \subset T^*$.) Furthermore, R contains a neighborhood of the origin in E^m (that is, R^* contains a neighborhood of $(0, 0, \cdots, 0, 1)$ in E^{m+1}).*

Proof. Suppose $a \varepsilon R^*$. Let $c_0^*, \cdots, c_m^*$ be as in Lemma 7. By Lemma 7, $\sum_0^m a_i c_i^* \neq 0$. Hence, by Lemma 6 with $c_i' = -c_i^*$ and i_0 such that $a_{i_0} \neq 0$, there are values $c_i = \bar{c}_i$ (say) and $K \neq 0$ (by (3.5)) such that, on $[-1, 1]$,

$$f_{i_0} - \sum_{i \neq i_0} \bar{c}_i [f_i - (a_i/a_{i_0}) f_{i_0}] = K[f_m - \sum_0^{m-1} c_i^* f_i]. \tag{4.1}$$

By Lemma 3 the right member of (4.1) attains its maximum in absolute value on $[-1, 1]$ at, and only at, the Chebyshev points $x_0^*, \cdots, x_m^*$, and this is therefore true of the left member. Hence, ξ^* is maximal with respect to $\bar{c} = \{\bar{c}_i, i \neq i_0\}$ for the game with payoff (2.7) if and only if the support of ξ^* is a subset of X_m^*. On the other hand, since $f_{i_0}(x_j^*) - \sum_{i \neq i_0} \bar{c}_i [f_i(x_j^*) - (a_i/a_{i_0}) f_{i_0}(x_j^*)]$ is some nonzero constant times $(-1)^j$, by (2.22) the orthogonality relations (2.8) reduce to (2.16), and thus $\bar{c}$ is minimal relative to any nonnegative solution to (2.16). One such strictly positive solution is of course guaranteed by the definition of R^*. Since ξ^* is maximal with respect to $\{\bar{c}_i, i \neq i_0\}$, and the latter is minimal with respect to ξ^*, standard game theory results, mentioned in Theorem 2.3, assert that ξ^* is maximin for the game with payoff (2.7). Hence ξ^* is a-optimum. Moreover, these same results assert that the left member of (4.1) is the residual of the Chebyshev approximation of f_{i_0} on $[-1, 1]$ by a linear combination of the functions $\{(f_i - (a_i/a_{i_0}) f_{i_0}), i \neq i_0\}$, a fact of which we shall make use in a moment.

We now turn to the proof of uniqueness. According to Theorem 2.3, every a-optimum design ξ is maximal relative to the $\bar{c}$ of the previous paragraph, and hence, as deduced there, is supported by some subset of X_m^* and (by Theorem 2.3) satisfies (2.16) (with the interpretation (2.22)), with all ξ_i nonnegative (perhaps some zero). If there were more than one such solution, the $(m + 1) \times (m + 1)$ matrix L_a whose (i, j)th element is

$$\begin{aligned} (L_a)_{i,j} &= f_i(x_j^*) - (a_i/a_{i_0}) f_{i_0}(x_j^*), \quad 0 \leqq i \leqq m, \quad i \neq i_0, \quad 0 \leqq j \leqq m, \\ &= (-1)^j, \qquad\qquad\qquad\qquad\qquad\qquad i = i_0, \quad 0 \leqq j \leqq m, \end{aligned} \tag{4.2}$$

would be singular, since (2.16) can be written as $L_a(\xi_0, -\xi_1, \xi_2, \cdots, (-1)^m \xi_m)' = (0, \cdots, 0, 1, 0, \cdots, 0)'$, with the 1 in i_0th place in the last vector. Now $(-1)^j$ can be written as $K_0 \sum_0^m c_i^* f_i(x_j^*)$ for some nonzero constant K_0. Hence (4.2) yields

$$L_a = \begin{pmatrix} I_{i_0} & \rho_1' & 0 \\ & K_0 c^* & \\ 0 & \rho_2' & I_{m-i_0} \end{pmatrix} F_{S^*} \tag{4.3}$$

where I_r is the $r \times r$ identity matrix, $c^* = (c_0^*, c_1^*, \cdots, c_m^*)$, and $\rho_1 = (-a_0/a_{i_0}, \cdots, -a_{i_0-1}/a_{i_0})$, $\rho_2 = (-a_{i_0+1}/a_{i_0}, \cdots, -a_m/a_{i_0})$. The second factor on the right side of (4.3) is nonsingular by Lemma 2. The determinant of the first factor, which can be computed by adding $-K_0c_i^*$ times the ith row to the i_0th row for each $i \neq i_0$, is $K_0a_{i_0}^{-1} \sum_0^m a_ic_i^*$, which is nonzero by Lemma 7. Hence, L_a is nonsingular and there is a unique a-optimum design.

It remains to show that R contains an open neighborhood of the origin. When $b = 0$, there is an optimum design ξ^* on the Chebyshev points by Lemma 3 and Theorem 2.3 with $h_i = f_i$ and $\psi_i = \theta_i$, and all ξ_i^* are then positive, since otherwise θ_m would be estimatable on fewer than $m + 1$ points, in violation of Lemmas 1 and 4. This shows that $0 \varepsilon R$. Moreover, if $a = (b, 1)$ in the first factor on the right side of (4.3) is varied by varying b in a small enough neighborhood of 0, $L_{(b,1)}$ remains nonsingular (since $\sum_0^{m-1} b_jc_j^* \neq -c_m^*$ for b near 0) and the coordinates ξ_j of the solution to (2.16), which will vary continuously with b, will remain positive as they were when $b = 0$. (Alternatively, this last sentence may be replaced by (2.20).) This completes the proof of Theorem 1.

THEOREM 2. *Under Assumptions* 1 *and* 2, $T^* - R^* = S^*$; *and, if* $a \varepsilon S^*$, *the orthogonality relations* (2.24) *have a unique solution which corresponds to the design* (2.22) *on the entire set* X_m^*. *There is no other a-optimum design supported by* X_m^* *or a subset thereof.* (*There may be other a-optimum designs.*)

PROOF. The proof parallels that of Theorem 1, so we merely outline the differences. Suppose $a \varepsilon S^*$. By Lemma 7, $\sum_0^m a_i d_i \neq 0$. By Lemmas 2 and 6 with $a_{i_0} \neq 0$, there are constants c_i^0 and $K \neq 0$ such that, on $[-1, 1]$,

$$f_{i_0}(x) - \sum_{i \neq i_0} c_i^0[f_i(x) - (a_i/a_{i_0})f_{i_0}(x)] \equiv K. \tag{4.4}$$

Hence, every ξ^* is maximal with respect to $c^0 = \{c_i^0, i \neq i_0\}$ for the game with payoff (2.7). By (2.22) and (4.4), the orthogonality relations (2.8) become (2.24). Therefore c^0 is minimal relative to any nonnegative solution of (2.24), while, as we have already seen, the latter is maximal relative to c^0. Hence, by the standard game theory results cited in the proof of Theorem 1, any nonnegative solution of (2.24) is a-optimum. One strictly positive solution of (2.24) is guaranteed by the definition of S^*, and this is surely a-optimum.

If there were two a-optimum designs with subsets of X_m^* for support, there would be more than one solution to (2.24), which can be written as $M_a(\xi_0, \xi_1, \cdots, \xi_m)' = (0, \cdots, 0, 1, 0, \cdots, 0)$, where M_a is obtained from the L_a of (4.2) by replacing $(-1)^j$ by 1 in the i_0th row. Since $\sum_0^m d_if_i(x) \equiv 1$, the equation for M_a corresponding to (4.3) is obtained by replacing the i_0th row of the first factor on the right side by $(d_0, d_1, \cdots, d_m)$. Adding $-d_i$ times the ith row of this factor to the i_0th row for $i \neq i_0$, we obtain $a_{i_0}^{-1}\sum_0^m a_i d_i$ for the determinant of this factor, which is thus nonzero by Lemma 7. Hence there is only one a-optimum design supported by a subset of X_m^*.

By Lemma 5, $T^* = R^* \cup S^*$, so that it remains to show that R^* and S^* are disjoint. If, to the contrary, there were an a in $R^* \cap S^*$, then for $a_{i_0} \neq 0$ there would, by our previous development, be two different Chebyshev approximations to f_{i_0} by $\{f_i - (a_i/a_{i_0})f_{i_0}, i \neq i_0\}$, one with constant residual and one with oscillatory residual. The Chebyshev vectors are thus at least one-dimensional. Applying Theorem 2.3 with $p \geqq 1$, we conclude that there is an a-optimum design supported by m or fewer points. Since $a \varepsilon R^*$, this contradicts the conclusion of Theorem 1. The proof of Theorem 2 is now complete.

REMARK 3. Example 2(b) of Section 5 will illustrate the lack of uniqueness of a-optimum designs for $a \varepsilon S^*$, as well as the fact mentioned in Section 2 that the set $\bar{T}^*$, defined just below (2.10) and discussed above (2.29), has a more complicated structure than T^* (in particular, that $\bar{T}^*$ is not merely the closure of T^*).

THEOREM 3. *Under Assumptions* 1 *and* 2, $A^* \subset R^*$.

PROOF. Suppose $a \varepsilon A^*$. Let i_0 be such that $a_{i_0} \neq 0$. By Lemma 8, $\{(a_{i_0}f_i - a_if_{i_0}), i \neq i_0\}$ is Chebyshev, and hence by Theorem 2.1 the best Chebyshev approximation of $a_{i_0}f_{i_0}$ by $\{(a_{i_0}f_i - a_if_{i_0}), i \neq i_0\}$ has an oscillatory residual. Since $a \varepsilon A^*$, any a-optimum design is, by Lemma 1, supported by at least $m + 1$ points. Lemma 5 now yields $a \varepsilon R^*$.

The next (and last) two theorems of this section are direct generalizations of the Hoel-Levine results discussed in Section 1, since their example of polynomial regression satisfies the assumptions of these theorems.

THEOREM 4. *Under Assumptions* 2 *and* 3, $H^* \subset R^*$. *If also* $f_m(e) \neq 0$ *for* $|e| > 1$, *then* $\Gamma(H^*) = H \subset R$.

PROOF. By Proposition 2.2 of Remark 1(b), Assumptions 2 and 3 imply that Assumptions 1 and 2 are satisfied for $\{\bar{f}_i, 0 \leqq i \leqq m\}$ where $\bar{f}' = Qf'$, for some nonsingular Q. Let A_1^* and R_1^* be the sets defined by (2.13) and (2.18) if f is replaced by $\bar{f}$ and θ is replaced by $\bar{\theta} = \theta Q^{-1}$ (so that $\theta f' = \bar{\theta}\bar{f}'$). Then $a\bar{\theta}' = aQ'^{-1}\theta'$, so that the vector a in A_1^* must be multiplied by Q'^{-1} to give the corresponding vector in A^*; that is, $A^* = A_1^*Q'^{-1}$, and similarly $R^* = R_1^*Q'^{-1}$. Since $A_1^* \subset R_1^*$ by Theorem 3, we thus obtain $A^* \subset R^*$. Lemma 9 now completes the proof that $H^* \subset R^*$. The remainder of the theorem follows from (2.30).

REMARK 4. Assumptions 2 and 3 may be replaced in Theorem 4 by Assumptions 1 and 2 and the assumption indicated in Remark 2.

A consequence of our conclusion that $H^* \subset R^* \subset T^*$ under Assumptions 1–3 is the result of Hoel and Levine [5] mentioned in Section 1, that $H^* \subset T^*$ if $f_i(x) \equiv x^i$.

The last theorem of this section concerns $V(f(y), \xi) = f(y)M^{-1}(\xi)f(y)'$ which, we recall, is proportional to the variance of the best linear estimator of the regression function $\theta f(y)'$ at the point y of Y when the design ξ on $[-1, 1]$ is used.

THEOREM 5. *Under Assumptions* 1, 2, 4, *and* 5, *if* k *is a real function of* e, $1 < e < \infty$, *such that always* $k(e) \leqq e$ *and* $\liminf_{e \to +\infty} k(e) > -\infty$, *then for* e *(resp.,* $-e$*) sufficiently large, the unique* $f(e)$*-optimum design* $\xi^{(e)}$ *minimizes* $\max_{k(e) \leqq y \leqq e} V(f(y), e)$ *(resp.,* $\max_{e \leqq y \leqq -k(-e)} V(f(y), e)$*)*.

PROOF. We shall prove only the first conclusion (as $e \to +\infty$), the case $e \to -\infty$ being treated similarly.

By Assumption 4, $f(e) = f_m(e)(o(1), \cdots, o(1), 1)$ as $e \to +\infty$, so that for e

sufficiently large the second part of Theorem 1 shows that the $f(e)$-optimum design $\xi^{(e)}$ is unique and is supported by X_m^*. Moreover, the proof of Theorem 1 shows that

$$\lim_{e\to+\infty} \xi^{(e)}(x_i^*) = \xi^*(x_i^*), \qquad 0 \leqq i \leqq m, \tag{4.5}$$

where ξ^* is the unique optimum design for estimating θ_m.

Since $\xi^{(e)}$ minimizes $V(f(e), \xi)$, the theorem will follow if we prove that, for some real e_0,

$$\max_{k(e) \leqq y \leqq e} V(f(y), \xi^{(e)}) = V(f(e), \xi^{(e)}) \qquad \text{for } e > e_0. \tag{4.6}$$

It was shown in the proof of Theorem 1 that the unique optimum design ξ^* for estimating θ_m has $\xi^*(x_i^*) > 0$ for $0 \leqq i \leqq m$. It follows from Lemma 2 that $M(\xi^*)$ is nonsingular. Hence, from (4.5), $M(\xi^{(e)})$ is nonsingular for sufficiently large, and we can write

$$M^{-1}(\xi^{(e)}) = M^{-1}(\xi^*) + \{o(1)\} \qquad \text{as } e \to +\infty \tag{4.7}$$

where $\{o(1)\}$ is a matrix whose elements approach 0 as $e \to +\infty$. Since f is continuous and $V(f(y), \xi) = f(y)M^{-1}(\xi)f(y)'$, it follows from the nonsingularity of $M(\xi^*)$ and (4.7) that, for every compact set K,

$$\max_{y \epsilon K} |V(f(y), \xi^{(e)}) - V(f(y), \xi^*)| \to 0 \qquad \text{as } e \to +\infty. \tag{4.8}$$

We shall show below that there is an $\epsilon > 0$ and real k_0 and e_1, $k_0 < e_1$, such that $e > e_1$ implies

(4.9)
(a) $V(f(y), \xi^{(e)})$ is strictly increasing in y for $y \geqq e_1$;
(b) $V(f(e_1 + 1), \xi^{(e)}) - V(f(e_1), \xi^{(e)}) > \epsilon$;
(c) $k(e) \geqq k_0$;
(d) $V(f(e_1), \xi^*) = \max_{k_0 \leqq y \leqq e_1} V(f(y), \xi^*)$.

If we let K be the interval $[k_0, e_1]$, we can by (4.8) find an e_2 such that the left side of (4.8) is $<\epsilon/2$ for $e > e_2$. Then (4.9) implies that $V(f(e_1 + 1), \xi^{(e)}) = \max_{k_0 \leqq y \leqq e_1+1} V(f(y), \xi^{(e)})$ if $e > \max(e_1 + 1, e_2) = e_3$ (say); consequently, from (4.9) (a), we obtain (4.6) for $e_0 = e_3$.

We now prove (4.9). The hypothesis of the theorem on k implies (c) if e_1 is sufficiently large. (4.9) (d) follows from the validity of (4.9) (a) with $\xi^{(e)}$ replaced by ξ^*, which will be proved below, and from the fact that

$$\lim_{e\to+\infty} V(f(e), \xi^*) = \infty;$$

the latter follows from Assumption 4, according to which $V(f(e), \xi^*) = f_m^2(e)v_m(\xi^*)(1 + o(1))$ as $e \to +\infty$, where $v_m(\xi^*)$ is the lower right element of $M^{-1}(\xi^*)$, and from the fact (Assumption 5) that $f_m^2(e)$ approaches $+\infty$ with e. Next, we note that

$$\begin{aligned} V(f(y + \Delta), \xi) &- V(f(y), \xi) \\ &= [f(y + \Delta) + f(y)]M^{-1}(\xi)[f(y + \Delta) - f(y)]'. \end{aligned} \tag{4.10}$$

By Assumption 4, $f(y + \Delta) + f(y) = (f_m(y + \Delta) + f_m(y))(o(1), o(1), \cdots, o(1), 1)$ as $y \to +\infty$, with the $o(1)$ terms uniform for positive Δ. Similarly, by Assumption 5, $f(y + \Delta) - f(y) = (f_m(y + \Delta) - f_m(y))(O(1), O(1), \cdots, O(1), 1)$ as $y \to +\infty$, with the $O(1)$ terms uniform for $0 < \Delta \leqq 1$. We also note, from Assumption 5, that

$$f_m^2(y + \Delta) - f_m^2(y) > 0 \tag{4.11}$$

for y sufficiently large and all $\Delta > 0$. From these and (4.7), we have

$$V(f(y + \Delta), \xi^{(e)}) - V(f(y), \xi^{(e)}) = [f_m^2(y + \Delta) - f_m^2(y)]v_m(\xi^*)(1 + o(1)) \tag{4.12}$$

as min $(y, e) \to +\infty$, uniformly for $0 < \Delta \leqq 1$. (4.11) and (4.12) yield (4.9) (a) and (b) for e_1 sufficiently large. (4.9) (a) with $\xi^{(e)}$ replaced by ξ^* is proved in the same way. This completes the proof of Theorem 5.

REMARK 5. *Extensions.* The results of this section can be extended by altering the nature of X and Y. For example, it is well known that much of the Chebyshev approximation theory, in particular Assumption 1 and Theorem 2.1, apply if X is a subset of the 1-sphere (boundary of the unit circle). Without going into further detail, we note that a case of practical importance which can be treated by our methods is that where there are open intervals in $[-1, 1]$ where observations are prohibited for technological reasons; $[-1, 1]$ is then replaced by a union of closed intervals. Similarly, Y can be altered from $(-\infty, \infty) - [-1, 1]$; for example, it may be that it only makes sense to define f on $[-1, \infty)$ because $x + 1$ is inherently nonnegative; for another example, if X is a union of disjoint intervals as mentioned just above, Y might be $(-\infty, \infty) - X$. For the required approximation theory results, see, e.g., [10a], section 2.3.3. These results apply, in particular, to the polynomial csae where X is two intervals, studied independently by [4a] Hoel (1965), some of whose arguments can be simplified by use of this theory.

5. Examples.

EXAMPLE 1. *The case* $m = 1$. For the sake of completeness (and for use in Example 2.2) we first determine the possible Chebyshev systems when $m = 1$, and then, in the next paragraph, show that under the stronger Assumptions 1 and 2 there is essentially only one example. We first show, then, that *if* $\{f_0, f_1\}$ *is a continuous Chebyshev system on* $[-1, 1]$, *then, for some nonsingular* D *and* $g = fD$, *we have* $g_0(x) > 0$ *and* $h(x) = g_1(x)/g_0(x)$ *strictly increasing for all* x. This result is probably known (although we did not succeed in finding a reference), and the proof is quite simple: We first show there is a linear combination $g_0 = fa'$ which is positive throughout $[-1, 1]$. By the Chebyshev assumption, there are linear combinations $G_i = fa^{(i)\prime}$, $i = \pm 1$, with $G_i(i) = 0$, $G_i(-i) = 1$. If $G_0 = G_1 + G_2$, then either (i) G_0 is such a g_0; or else (ii) G_0 has at least two zeros (contradicting the Chebyshev assumption); or else (iii) G_0 has a single zero at q (say) with $-1 < q < 1$, in which case $G_0 -$ (sgn $G_1(q))G_1/2$ has at least two zeros. With the existence of a g_0 thus established, we need only observe that the Chebyshev nature of $\{1, h\}$ follows from that of $\{f_0, f_1\}$, and that h is hence strictly monotone and can be taken as increasing by making a change of sign if necessary.

If now we also impose Assumptions 1 and 2, since $G_0(\pm 1) = 1$ we see that $G_0(x) \equiv 1$, because otherwise G_0 would be a nonconstant function with at least one change of direction. Hence, in this case we can find D such that $g_0(x) \equiv 1$ on $[-1, 1]$ and $g_1(\pm 1) = \pm 1$, with g_1 strictly increasing on $[-1, 1]$. Write $f\theta' = g\psi'$. Now map X onto another copy Z of $[-1, 1]$ using the mapping g_1. The regression problem on Z with regression $\psi_0 + \psi_1 z$ then corresponds to that on X with regression $\psi_0 + \psi_1 g_1(x)$ in such a manner that if ξ' is a-optimum on Z, then an a-optimum design ξ on X is defined by $\xi(x) = \xi'(g_1(x))$.

For the linear regression problem on Z just described, it is easily verified that $A = \{b_0 : |b_0| < 1\}$ and that A_0 is empty (since ψ_0 can be estimated by an observation at $z = 0$). The Chebyshev points are $x_0^* = -1$, $x_0^* = 1$, so that

$$F_{R^*} = \begin{pmatrix} 1 & -1 \\ -1 & -1 \end{pmatrix}.$$

Hence, writing $\alpha = \eta_0 - \eta_1$ and $\sigma = -\eta_0 - \eta_1$, we have from (2.20) that a general element a of R^* has the form (α, σ). Since $\sigma \neq 0$ for $\eta \in \Xi_1^*$, we see that R_0 is empty. Since the range of α/σ for η in Ξ_1^* is the interval $(-1, 1)$, we conclude that $R = A$. The points $b_0 = \pm 1$ of B correspond to optimum designs on one point: $\xi(\pm 1) = 1$. All admissible designs in this problem are well known to be supported on X_1^* or a subset thereof, and from this or (2.27) we see that S^* consists of all points of B^* except R^* and $\Gamma^{-1}(b_0)$ for each of the two additional points $b_0 = \pm 1$ of B. (As Example 2 (b) and (c) shows, no such simple result holds when $m > 1$.) The point $S_0 = T_0$ corresponds to estimating ψ_0, which can be done optimally both by the design ξ' (say) for which $\xi'(1) = \xi'(-1) = \frac{1}{2}$ and also by any of an infinite number of inadmissible designs, the simplest of which is the design ξ'' (say) for which $\xi''(0) = 1$; it is easily verified that $M(\xi') - M(\xi'')$ is nonnegative definite, of rank 1.

The above characterizations also hold for the regression $g\psi'$ on X. The linear transformation which took f into g can then be used to characterize the corresponding sets for the original problem with regression $f\theta'$ on X. For example, as in the proof of Theorem 4, if $g' = Qf'$, then $(A^* \text{ for } g)Q'^{-1} = A^* \text{ for } f$. However, R need no longer be connected. For example, if $f_0(x) = 1 + x/2$, $f_1(x) \equiv 1$, so that

$$Q = \begin{pmatrix} 0 & 1 \\ 2 & -2 \end{pmatrix},$$

one obtains $(a_0, a_1) = (\bar{a}_0 + \bar{a}_1/2, \bar{a}_0)$, where (a_0, a_1) refers to f and $(\bar{a}_0, \bar{a}_1)$ refers to g (treated in the previous paragraph). Thus, for f we obtain

$$R = A = (-\infty, \tfrac{1}{2}) \cup (\tfrac{3}{2}, \infty),$$

with $R_0 = A_0 =$ "point at infinity." This unnecessary complication points up the advantage of working in terms of R^* (as described by (2.20)), whose geometric characteristics are unchanged by the linear transformation Q.

As for H, suppose we extend the map of $X \to Z$ to $(-\infty, \infty) \to (-\infty, \infty)$ by the identity map on $(-\infty, \infty) - [-1, 1]$. Write $\psi h(z)'$ for the regression function on $(-\infty, \infty)$ as extended from Z, so that $h_i(z) = z^i$ for $z \in Z$. Under Assumption 3 it is easy to see that the graph of h_1 crosses (and is not merely tangent to) that of h_0 at 1 and at no other point of $(-\infty, \infty)$, and that $h_1(z) = 0$ only at $z = 0$. Under Assumption 4 (for example, if $h_i(z) = z^i$ on $(-\infty, \infty)$) we obtain $H = (-1, 0) \cup (0, 1) = A - \{0\}$ (another result which does not hold if $m > 1$); if Assumption 4 does not hold (for example, if $h_0(z) = 1$ and $h_1(z) = 2z/[1 + |z|]$ for $|z| > 1$), H is a proper subset of $A - \{0\}$.

EXAMPLE 2. *Polynomial regression* $(f_i(x) = x^i)$, $m > 1$.

(a) *General results.* The Chebyshev points in the polynomial case are well known (for example, see [1]) to be $x_j^* = -\cos(j\pi/m)$, $0 \leqq j \leqq m$. Thus, R^* and S^* can be described explicitly from (2.20) and (2.27), as we shall do in detail below for $m = 2, 3$.

The set A^* has been characterized in [10] as follows: Define real-valued functions S_j and Q_h on E^m (whose points we write as $x = (x_0, \cdots, x_{m-1})$) by

$$S_j(x) = (-1)^j \textstyle\sum^{(j)} x_{i_1} x_{i_2} \cdots x_{i_j}, \qquad 1 \leqq j \leqq m, \tag{5.1}$$

where $\sum^{(j)}$ denotes summation over the set $0 \leqq i_1 < i_2 < \cdots < i_j < m$, and

$$Q_h(x) = \textstyle\sum^{(h)} (1 - x_{i_1})(1 - x_{i_2}) \cdots (1 - x_{i_h})(1 + x_{i_{h+1}}) \cdots (1 + x_{i_m})/\binom{m}{h}, \quad 0 \leqq h \leqq m, \tag{5.2}$$

where the m subscripts in the summand are distinct ($\sum^{(0)}$ consists of one term). Define the points $b^{(h)} = (b_0^{(h)}, \cdots, b_{m-1}^{(h)})$, $0 \leqq h \leqq m$, by

$$Q_h(x) = 1 + \textstyle\sum_{j=1}^m b_{m-j}^{(h)} S_j(x). \tag{5.3}$$

In particular, $b^{(m)} = (1, 1, \cdots, 1)$ and $b^{(0)} = ((-1)^m, (-1)^{m-1}, \cdots, (-1)^1)$. The points $b^{(0)}, \cdots, b^{(m)}$ can be shown not to lie in any hyperplane of E^m, so that they span an m-dimensional simplex. Let Δ_m denote this simplex minus the closed edge containing $b^{(0)}$ and $b^{(m)}$. The main result of [10] is

THEOREM 6. *For polynomial regression with* $m > 1$, $A = \Delta_m$ *and* A_0 *is empty.*

As we shall see in Example 2(c), R_0 is not generally empty.

We note that H is the twisted curve $\{(t^m, t^{m-1}, \cdots, t^1) : 0 < |t| < 1\}$, whose two open components have end-points $b^{(0)}, b^{(m)}$, and (in common) $\mathbf{0}$.

(b) *The case* $m = 2$. As in the case $m = 1$, a complete analysis of the a-optimum designs, for a in B^*, is possible here, but would be much more complicated as m increases, as will be seen in (c). We begin by describing the structure of b-optimum designs for all b in the (b_0, b_1)-plane B. From Theorem 6, we have

$A =$ triangle with vertices $(-1, 0)$, $(1, 1)$, $(1, -1)$,

minus closed segment joining the latter two.

Recalling the first paragraph of Example 2(a), we have $X_2^* = \{-1, 0, 1\}$ and

$$F_{R^*} = \begin{pmatrix} 1 & -1 & 1 \\ -1 & 0 & 1 \\ 1 & 0 & 1 \end{pmatrix}, \qquad F_{S^*} = \begin{pmatrix} 1 & 1 & 1 \\ -1 & 0 & 1 \\ 1 & 0 & 1 \end{pmatrix}.$$

Setting $\alpha = \eta_0 + \eta_2$ and $\beta = \eta_2 - \eta_0$, we have from (2.20) that a general element a of R^*, obtained as a linear combination of the columns of F_{R^*}, is of the form $(\alpha - \eta_1, \beta, \alpha)$. Since $\alpha \neq 0$ for $\eta \varepsilon \Xi_m^*$, we obtain that R_0 is empty. Moreover, R is the set of points of the form $(b_0, b_1) = (1 - \eta_1/\alpha, \beta/\alpha)$. For $\eta \varepsilon \Xi_m^*$, the variables η_1/α and β/α can vary independently of each other over domains $(0, \infty)$ and $(-1, 1)$, respectively. Hence, $R = \{(b_0, b_1): b_0 < 1, |b_1| < 1\}$. Similarly, by (2.27), a point of S^* is of the form $(\alpha + \eta_1, \beta, \alpha)$, so that S_0 is empty, $S = \{(b_0, b_1): b_0 > 1, |b_1| < 1\}$. Thus, T_0 is empty and $\Gamma(T^*) = T = \{(b_0, b_1): b_0 \neq 1, |b_1| < 1\}$.

In subdividing the plane B into regions where the b-optimum designs are of various forms, we shall encounter repeatedly the parabola $b_0 = b_1^2$, which consists of $\mathbf{0}$, the set $H = \{(t^2, t): 0 < |t| < 1\}$, and the set J (say) where $b_0 = b_1^2 \geqq 1$. The point (t^2, t) of J with $|t| \geqq 1$ corresponds to the linear parametric function $f(t^{-1})\theta'$. Since $|t^{-1}| \leqq 1$, this linear parametric function can be estimated by the design $\xi^{(t)}$ (say) for which $\xi^{(t)}(t^{-1}) = 1$. It was shown in [6] that $\xi^{(t)}$ is admissible for $|t| \leqq 1$. Since $f(t^{-1})\theta'$ and its multiples are the only linear parametric functions estimatable under $\xi^{(t)}$, it follows that $\xi^{(t)}$ must be $f(t^{-1})$-optimum for $0 < |t| \leqq 1$. ($\xi^{(0)}$ will be discussed with B_0.) No two of these $\xi^{(t)}$'s allow estimation of the same linear parametric function. Hence, if there were a design other than $\xi^{(t)}$ which was also $f(t^{-1})$-optimum, it would have to be supported by at least two points, and thus it would allow estimation of some linear parametric function not estimatable under $\xi^{(t)}$, from which it follows easily that $\xi^{(t)}$ would be inadmissible. Thus, we have shown that

$$J \equiv \{(t^2, t): |t| \geqq 1\} = \{\text{points of } B \text{ where there is an optimum design supported by one point}\}.$$

(The analogue of this holds for general m, with $J = \{(t^m, t^{m-1}, \cdots, t): |t| \geqq 1\}$.)

In analyzing B further we shall use the fact that, for a design supported by *more than one point*, the residual to the best Chebyshev approximation of x^2 on $[-1, 1]$ by $\{1 - b_0x^2, x - b_1x^2\}$, being quadratic and attaining its maximum in absolute value at the points of support, must be of one of the following forms:

(i) a multiple of $x^2 - \frac{1}{2}$, with support a subset of X_2^*;

(ii) a constant;

(iii) a quadratic with derivative 0 at q, $0 < |q| \leqq 1$,

and with values of equal magnitude and opposite sign, at -1 and q if $q > 0$, and at 1 and q if $q < 0$, these two values being the support in the respective cases (the case $q = 0$ is covered by (i) above);

(iv) a multiple of $x^2 - L$ with $L < \frac{1}{2}$ and support $\{-1, 1\}$;

(v) a quadratic or linear function with nonzero derivatives of the same sign at ± 1, and with values of equal magnitude and opposite sign at the two points of support $-1, 1$.

Corresponding to these, there are three forms of the orthogonality relation (2.4) which we shall consider:

(I) the Equations (2.17), where we no longer demand that all ξ_j be positive, but only that two be positive and one nonnegative; this corresponds to (i) and, with $\xi(0) = 0$, to (iv) and the case of (ii) where the support is $\{-1, 1\}$;

(II) corresponding to (iii) and (v), the equations (a) and (b) for $q > 0$ and $q < 0$, respectively:

$$(5.4)\quad \begin{array}{ll} \text{(a)} & \begin{aligned} \xi(-1)(1 - b_0) \quad &- \xi(q)(1 - q^2 b_0) = 0, \\ \xi(-1)(-1 - b_1) &- \xi(q)(q - q^2 b_1) = 0, \\ \xi(-1) \quad &+ \xi(q) \qquad\qquad = 1; \end{aligned} \\ \text{(b)} & \begin{aligned} \xi(1)(1 - b_0) \quad &- \xi(q)(1 - q^2 b_0) = 0, \\ \xi(1)(1 - b_1) \quad &- \xi(q)(q - q^2 b_1) = 0, \\ \xi(1) \quad &+ \xi(q) \qquad\qquad = 1; \end{aligned} \end{array}$$

(III) corresponding to the part of case (ii) not covered in (I), equations which will be discussed below, and which lead to (5.8).

It is trivial that for each fixed b there exists a vector $c = (c_0, c_1)$ which yields a residual with each of the possible sets of extrema and oscillations of sign represented by (I), (II), and (III). Hence, in each of these three cases, any ξ with the given support is maximal relative to any c yielding such a residual, and if the orthogonality relations are satisfied then c is minimal with respect to ξ. From Theorem 2.3 it then follows that ξ is b-optimum; it is unnecessary to go back to (i)–(v) and compare residuals to find which approximation is best, where the best approximation is not unique, etc.

The regions where these three forms hold can be described as follows: partition B into disjoint sets B_{I}, B_{II}, B_{III}, J, defined by

$$(5.5)\quad \begin{aligned} B_{\text{I}} &= \{(b_0, b_1): b_0 \leqq 1, |b_1| \leqq 1\} - \{(1, 1)\} - \{(1, -1)\}, \\ B_{\text{II}} &= \{(b_0, b_1): |b_1| > 1, b_0 < b_1^2\}, \\ B_{\text{III}} &= \{(b_0, b_1): b_0 > \max(b_1^2, 1)\}. \end{aligned}$$

We shall show that, for $L = \text{I, II, III}$, the orthogonality relations of case L have a solution on two or more points if and only if $b \in B_L$.

Case I is treated by the same computation which yields R; in fact, the Equations (2.17) have a nonnegative solution on the closure of R,

$$(5.6)\quad \begin{aligned} cl(R) &= B_{\text{I}} \cup \{(1, 1) \cup \{(1, -1)\} \\ &= R \cup \{(b_0, 1): b_0 < 1\} \cup \{(b_0, -1): b_0 < 1\} \cup \{(1, b_1): |b_1| < 1\} \\ &\quad \cup \{(1, 1)\} \cup \{(1, -1)\}. \end{aligned}$$

In this last partition of (5.6) we have, respectively, none of the three $\xi(x_j^*)$'s zero, only $\xi(-1) = 0$, only $\xi(1) = 0$, only $\xi(0) = 0$, $\xi(1) = 1$, and $\xi(-1) = 1$, the last two being points of J. (The point corresponding to $\xi(0) = 1$, which does not arise from (2.17), will be seen later to be in B_0.)

In describing Case II, we shall use the partition of B_{II} into disjoint sets L_s, $-\infty < s < \infty$, defined as follows:

$$L_s = \{(b_0, b_1): b_0 - sb_1 = 1 + s,\ b_0 < b_1^2\} \quad \text{if } s \geqq 0,$$
$$\qquad = \{(b_0, b_1): b_0 - sb_1 = 1 - s,\ b_0 < b_1^2\} \quad \text{if } s \leqq 0. \tag{5.7}$$

Thus, L_s is that portion not in $cl(B_{III})$ of a line passing through $(1, -1)$ if $s \geqq 0$ and through $(1, 1)$ if $s \leqq 0$; in particular, $L_0 = \{(1, b_1): |b_1| > 1\}$. Consider now the orthogonality relations (5.4) (a) in the case $0 < q < 1$. Equating the ratios $\xi(-1)/\xi(q)$ in the first two equations, one obtains $b_0 - sb_1 = s + 1$ where $s = (1 - q)/q > 0$; from the positivity condition $0 < \xi(-1)/\xi(q) < \infty$ one obtains $0 < (1 - b_0q^2)/(1 - b_0) < \infty$ or $\{b_0 > q^{-2}\} \cup \{b_0 < 1\}$, which with $b_0 - sb_1 = s + 1$ yields L_s as the subset of B for which the b-optimum design is supported by the two points -1, q and the residual has opposite signs at these two points. (The support $\{-1, q\}$ arises in case III with constant residual.) The case $-1 < q < 0$ of (5.4) (b) similarly yields L_s with $s = (1 + q)/q < 0$. Finally, the case $q = 1$ of (5.4) (a) coincides with $q = -1$ in (5.4) (b) and yields L_0 as the subset of B for which the b-optimum design is supported by the two points $1, -1$ with residual of opposite sign at the two points. (The set $\{(1, b_1): |b_1| < 1\}$ encountered in case I also has support $\{1, -1\}$, but with residual of the same sign at the two points.)

For any $b \varepsilon S$ every best Chebyshev approximation has constant residual (Theorem 2 and Lemma 5). Since for b in $B_I \cup B_{II}$ the residual is not constant, as we have seen, it follows that $S \subset B_{III}$. On $B_I \cup B_{II}$ the optimum design is unique because the orthogonality relations have a unique solution. For b in B_{III} there is no uniqueness of the b-optimum design. In fact, while the b-optimum design for each b in $B - B_{III}$ is unique and hence admissible, for each b in B_{III} there are infinitely many different supporting sets of admissible b-optimum designs, and also infinitely many different supporting sets (including supporting sets with an arbitrarily large finite number, or an infinite number, of points) of inadmissible b-optimum designs. Since the admissible b-optimum designs are of greater theoretical and practical interest, we shall exhibit only the totality of these, for each b in B_{III}. We shall then indicate by an example the existence of inadmissible b-optimum designs.

It was shown in [6] that the supports of admissible designs are of the form $\{-1, q, 1\}$ with $-1 < q < 1$, or subsets thereof, and conversely. The orthogonality relations (2.4) for the set $\{-1, q, 1\}$ in case III are

$$\begin{aligned}
&\xi(-1)(1 - b_0) + \xi(q)(1 - b_0q^2) + \xi(1)(1 - b_0) = 0,\\
&\xi(-1)(-1 - b_1) + \xi(q)(q - b_1q^2) + \xi(1)(1 - b_1) = 0,\\
&\xi(-1) + \xi(q) + \xi(1) = 1.
\end{aligned} \tag{5.8}$$

We seek a nonnegative solution to these for which $0 < \xi(q) < 1$; this condition is equivalent to that of finding a solution for which at least two of the components are positive (to eliminate J) and for which $\xi(q) > 0$ (to eliminate the part of (ii) included in case I, namely, the interval $\{(1, b_1): |b_1| < 1\}$ denoted by L_0' below). In describing such solutions, it is convenient to write

$$\begin{aligned} L_s' &= \{(b_0, b_1): b_0 - sb_1 = 1 + s, b_0 > b_1^2\} && \text{if } s \geqq 0, \\ &= \{(b_0, b_1): b_0 - sb_1 = 1 - s, b_0 > b_1^2\} && \text{if } s \leqq 0; \\ M_r' &= \{(b_0, b_1): b_0 - rb_1 = 1 - r, b_0 > b_1^2\} && \text{if } r > 1, \\ &= \{(b_0, b_1): b_0 - rb_1 = 1 + r, b_0 > b_1^2\} && \text{if } r < -1, \\ &= \{(b_0, b_1): |b_1| = 1, b_0 > 1\} && \text{if } r = \infty. \end{aligned} \tag{5.9}$$

Thus, L_s', L_s and the two points (q^{-2}, q^{-1}) and $(1, -\text{sign } q)$ of J (or $(1, 1)$ and $(1, -1)$ if $s = 0$) constitute a partition of the line encountered in conjunction with (5.7). For $r \neq \infty$, M_r' is the intersection with B_{III} of a line of slope $1/r$ through $(1, 1)$ if $r > 0$ and through $(1, -1)$ if $r < 0$, while M_∞' consists of two half-lines in B_{III}.

The Equations (5.8) have the formal solution

$$\begin{aligned} \xi(q) &= (b_0 - 1)/b_0(1 - q^2), \\ \xi(1) &= [1 + b_1(1 - q) - b_0 q]/2b_0(1 - q), \\ \xi(-1) &= [1 - b_1(1 + q) + b_0 q]/2b_0(1 + q). \end{aligned} \tag{5.10}$$

The condition $0 < \xi(q) < 1$ is equivalent to $b_0 > 1$. We also require the nonnegativity of the numerators of $\xi(1)$ and $\xi(-1)$ in (5.10), with at least one being positive. It is easy to verify that $\xi(1) = 0$ on the line through $(1, -1)$ of slope $q/(1 - q)$, and that $\xi(-1) = 0$ on the line through $(1, 1)$ of slope $q/(1 + q)$. We conclude that, for $-1 < q < 1$, (5.8) has a solution for which $0 < \xi(q) < 1$ for b in the set V_q defined by

$$\begin{aligned} V_q &= \{\text{triangle with vertices } (1, 1)(1, -1), (q^{-2}, q^{-1})\} \cap B_{\text{III}} \\ & \qquad\qquad \text{if } 0 < |q| < 1, \\ V_0 &= \{(b_0, b_1): b_0 > 1, |b_1| \leqq 1\}. \end{aligned} \tag{5.11}$$

In each case, all three components of ξ are positive if b is in the interior of V_q, while two components are positive on that part of the boundary which is in V_q. The latter is M_∞' if $q = 0$ and, if $q \neq 0$, consists of the two open line segments L_s' and M_r', where $s = (1 - q)/q$ and $r = (1 + q)/q$ if $q > 0$, and $s = (1 + q)/q$ and $r = (1 - q)/q$ if $q < 0$. The rest of the boundary of V_q of course consists of the interval L_0' of B_{I} and the points $(1, 1)$, $(1, -1)$, and (if $q \neq 0$)(q^{-2}, q^{-1}) of J. Thus, for any point $b = (b_0, b_1)$ in B_{III} there is an admissible design supported by q and one or both of the points $1, -1$, provided that $b \in V_q$. From the condition of nonnegativity of $\xi(1)$ and $\xi(-1)$ in (5.10), this interval of q-values, always of positive length for b in B_{III}, is

$$\{q: (b_1 - 1)/(b_0 - b_1) \leqq q \leqq (b_1 + 1)/(b_1 + b_0)\},$$

the endpoints corresponding to designs for which ξ has only two nonzero components. Hence, for each b in B_{III}, there are infinitely many different supporting sets of admissible b-optimum designs.

As an illustration of inadmissible a-optimum designs for $\Gamma(a)$ in B_{III}, consider (2, 0, 1)-optimality, that is, optimality for estimating $2\theta_0 + \theta_2$. Among the admissible designs for this problem, obtained above, two examples are $q = 0$, $\xi(-1) = \xi(1) = \frac{1}{4}$, $\xi(0) = \frac{1}{2}$, for which

$$(5.12) \qquad M(\xi) = \begin{pmatrix} 1 & 0 & \frac{1}{2} \\ 0 & \frac{1}{2} & 0 \\ \frac{1}{2} & 0 & \frac{1}{2} \end{pmatrix}, \qquad M^{-1}(\xi) = \begin{pmatrix} 2 & 0 & -2 \\ 0 & 2 & 0 \\ -2 & 0 & 4 \end{pmatrix},$$

for which $V(a, \xi) = (2, 0, 1)M^{-1}(\xi)(2, 0, 1)' = 4$, and the design with $q = \frac{1}{2}$, $\xi(1) = 0$, $\xi(-1) = \frac{1}{3}$, $\xi(\frac{1}{2}) = \frac{2}{3}$, for which $M(\xi)$ is singular, but for which $V(a, \xi)$ is again 4. Among the many inadmissible designs are symmetric designs supported by $\{(1 - \epsilon), -(1 - \epsilon), 0\}$ with $0 < \epsilon < 1 - 2^{-\frac{1}{2}}$ and with $\xi'(\pm(1 - \epsilon)) = \frac{1}{4}(1 - \epsilon)^2$. For such a design

$$M(\xi') = \begin{pmatrix} 1 & 0 & \frac{1}{2} \\ 0 & \frac{1}{2} & 0 \\ \frac{1}{2} & 0 & (1 - \epsilon)^2/2 \end{pmatrix},$$

$$M^{-1}(\xi') = [4/(1 - 4\epsilon + 2\epsilon^2)] \begin{pmatrix} (1 - \epsilon)^2/2 & 0 & -\frac{1}{2} \\ 0 & (1 - 4\epsilon + 2\epsilon^2)/2 & 0 \\ -\frac{1}{2} & 0 & 1 \end{pmatrix},$$

so that again $V(a, \xi) = 4$. The inadmissibility of such designs is exhibited in the fact that, for ξ given by (5.12), $M(\xi) - M(\xi')$ (or $M^{-1}(\xi') - M^{-1}(\xi)$) is nonnegative definite of rank one. It is not difficult to obtain inadmissible b-optimum ξ's here supported by any number of points $\geqq 2$ (a 2-point design being given by ξ' just above when $\epsilon = 1 - 2^{-\frac{1}{2}}$), or even with ξ absolutely continuous with positive Lebesgue density on $[-1, 1]$.

It remains to consider B_0. The unique optimum design for estimating θ_0 is, by the same argument used in discussing J, that for which $\xi(0) = 1$. For $-\infty < s < \infty$, the unique optimum design for estimating $s\theta_0 + \theta_1$, to which corresponds the problem of approximating x on $[-1, 1]$ by $\{x^2, 1 - sx\}$, is easily found by calculations parallel to those for B (solutions to the orthogonality relations now existing only in the case corresponding to II above). We obtain an optimum design supported by $\{-1, q\}$ if $q > 0$ and $s = (1 - q)/q$, and by $\{1, q\}$ if $q < 0$ and $s = (1 + q)/q$. In particular, for $s = 0$ we obtain the unique optimum design for estimating θ_1, for which $\xi(-1) = \xi(1) = \frac{1}{2}$. We note, then, that if we think of $B_0 = P^1 = \{b_{(s)}, -\infty < 1/s \leqq \infty\}$ in the usual manner, $b_{(s)}$ being the "point at infinity" of all lines in B of slope $1/s$, $-\infty < 1/s \leqq \infty$, then the optimum designs for these points $b_{(s)}$ of B_0 can be obtained, for

$-\infty < 1/s < \infty$, as limits of the corresponding designs for the family L_s; for $s = \infty$ we have the optimum design for estimating θ_0 considered at the outset of this paragraph, which can be thought of conveniently as the limit of designs as $b \to \infty$ in R or in B_{I} or in B_{III} or in J.

We note also that, in the notation of and the sentence following (2.10),

$$\bar{T} = \{(b_0, b_1): |b_1| \leqq 1\} \cup \{(1, b_1): -\infty < b_1 < \infty\},$$
$$\bar{T}^* = \Gamma^{-1}(\bar{T}) \cup \{b_{(0)}\} \cup \{b_{(\infty)}\}.$$

As an example of the explicit computation of how large the "sufficiently large" of Theorem 5 is, we consider the case $k(e) \equiv -1$, $e > 1$. For $b = \Gamma(f(e)) = (e^{-2}, e^{-1})$, writing $e^{-1} = t$, (2.17) yields

$$\xi^{(e)} = (\xi_0^{(e)}, \xi_1^{(e)}, \xi_2^{(e)}) = [2(2-t^2)]^{-1}(1-t, 2(1-t^2), 1+t),$$

so that

$$M(\xi^{(e)}) = [1/(2-t^2)]\begin{pmatrix} 2-t^2 & t & 1 \\ t & 1 & t \\ 1 & t & 1 \end{pmatrix},$$

$$M^{-1}(\xi^{(e)}) = [(2-t^2)/(1-t^2)]\begin{pmatrix} 1 & 0 & -1 \\ 0 & 1 & -t \\ -1 & -t & 2 \end{pmatrix},$$

and thus

$$V(f(y), \xi^{(e)}) \cdot (1-t^2)/(1-2t^2) = 1 - y^2 - 2ty^3 + 2y^4 = p_t(y) \quad \text{(say)}.$$

The function p_t is easily seen to have local minima at $y = [3t \pm (9t^2 + 16)^{\frac{1}{2}}]/8$ and a local maximum at $y = 0$, all three of these points being in $[-1, 1]$. Thus, $\max_{-1 \leqq y \leqq e} p_t(y) = \max (p_t(-1), p_t(0), p_t(t^{-1}))$. Since $p_t(-1) > p_t(0)$, we seek t such that $p_t(-1) - p_t(t^{-1}) \leqq 0$, that is, such that $2t^4 - t^3 + t^2 + 2t - 2 \leqq 0$. This last is satisfied for $t \leqq .694$, or $t^{-1} \geqq 1.44$. Thus, for $e \geqq 1.44$ the design $\xi^{(e)}$ minimizes $\max_{-1 \leqq y \leqq e} V(f(y), \xi)$. We remark that it is even easier to conclude that, since $p_t(y)$ is increasing for $y \geqq 1$, the design $\xi^{(e)}$ minimizes $\max_{1 \leqq y \leqq e} V(f(y), \xi)$ for $e \geqq 1$.

(c) *The case* $m = 3$. The set $A = \Delta_3$ of Theorem 6 is determined by the points

$$b^{(0)} = (-1, 1, -1),$$
$$b^{(1)} = (1, -\tfrac{1}{3}, -\tfrac{1}{3}),$$
$$b^{(2)} = (-1, -\tfrac{1}{3}, \tfrac{1}{3}),$$
$$b^{(3)} = (1, 1, 1).$$

The set X_3^* is $\{-1, -\frac{1}{2}, \frac{1}{2}, 1\}$, and

$$F_{R^*} = \begin{pmatrix} 1 & -1 & 1 & -1 \\ -1 & \frac{1}{2} & \frac{1}{2} & -1 \\ 1 & -\frac{1}{4} & \frac{1}{4} & -1 \\ -1 & \frac{1}{8} & \frac{1}{8} & -1 \end{pmatrix}.$$

Setting $\alpha = \eta_0 + \eta_3$, $\beta = \eta_0 - \eta_3$, $\gamma = \eta_1 - \eta_2$, $\sigma = \eta_0 + \eta_1 + \eta_2 + \eta_3$, we obtain from (2.20) that a general element of R^* has the form

$$a = (\beta - \gamma, (\sigma - 3\alpha)/2, (4\beta - \gamma)/4, (\sigma - 9\alpha)/8). \tag{5.13}$$

Thus, R_0 is no longer empty as it was when $m = 2$. To obtain R, we consider η to be in Ξ_m and thus $\sigma = 1$, and find

$$\begin{aligned} R = \{(8[\beta - \gamma]/[1 - 9\alpha], 4[1 - 3\alpha]/[1 - 9\alpha],& \\ 2[4\beta - \gamma]/[1 - 9\alpha]) : (\alpha, \beta, \gamma) \,\varepsilon\, \{0 < \alpha < 1,& \\ \alpha \neq \tfrac{1}{9}\} \cap \{-1 < \beta/\alpha < 1\} \cap \{-1 < \gamma/(1 - \alpha) < 1\}\};& \end{aligned} \tag{5.14}$$

the value $\alpha = \frac{1}{9}$ yields points in R_0, discussed below. The variables α, β/α $\gamma/(1 - \alpha)$ vary independently in (5.14), so that for b in R the range of b_1 is $(-\infty, 1) \cup (4, \infty)$. For each fixed value k of b_1 (that is, for each fixed value of α) the range of (b_0, b_2) in (5.14) is an open parallelogram $R(k)$ (say) in the plane $b_1 = k$, symmetric about $(0, k, 0)$, but whose dimensions and angles depend on k. Thus,

$$R = \bigcup_{b_1 < 1 \text{ or } > 4} R(b_1)$$

is no longer connected as it was when $m = 2$.

The set R_0 can be obtained as the set of elements of (5.13) with $\sigma = 9\alpha = 1$ and $b_1 = (\sigma - 3\alpha)/2 = \frac{1}{3}$; this is, by an analysis similar to that of (5.14), the set of ratios $(b_0/b_1, b_2/b_1) = (3(\beta - \gamma), 3(\beta - \gamma/4))$ in the region $|\beta| < \frac{1}{9}$, $|\gamma| < \frac{8}{9}$. This can be thought of as a "parallelogram at infinity" corresponding to the ratios $(b_0/b_1, b_2/b_1)$ of (5.14) as $\alpha \to \frac{1}{9}$.

The set S^* can be analyzed similarly. As was the case with R, the sets S and T no longer have the simple structure of the case $m = 2$. As in the next to last paragraph of Example 1, this again points up the greater simplicity of working with R^*, S^*, and T^*. The convexity of the cones which constitute half of $R^*$and S^* can of course be carried over to R and S in a different parametrization, one in which R_0 and S_0 are empty so that $R = \Gamma(R^*)$ can be thought of as a base (section) of a cone which constitutes half of R^*, and similarly for S. Thus, in place of x^3 we seek a function $\tilde{f}_3(x) = \sum_0^3 \lambda_i x^i$ such that, if we work in terms of $\tilde{f} = (\tilde{f}_0, \tilde{f}_1, \tilde{f}_2, \tilde{f}_3)$ instead of f, the quantity $\sum_0^3 (-1)^j \eta_j \tilde{f}_3(x_j^*)$, which corresponds to the last element of (5.13), and the quantity $\sum_0^3 \eta_j \tilde{f}_3(x_j^*)$ for the corresponding development for S^*, are never 0 for $\eta \,\varepsilon\, \Xi_3^*$. Writing $\lambda = (\lambda_0, \lambda_1, \lambda_2, \lambda_3)$, this says that all non-zero elements (there is at least one such) of $(\zeta_0, \zeta_1, \zeta_2, \zeta_3) = \lambda F_{R^*}$ must be of the same sign, and all non-zero elements of $(\zeta_0, -\zeta_1, \zeta_2, -\zeta_3) = \lambda F_{S^*}$ must be of the same sign. Hence, either (i) ζ_0 or $\zeta_2 \neq 0$, $\zeta_0\zeta_2 \geqq 0$, $\zeta_1 = \zeta_3 = 0$, or else (ii) ζ_1 or $\zeta_3 \neq 0$, $\zeta_1\zeta_3 \geqq 0$, $\zeta_2 = \zeta_4 = 0$. Since

$$\begin{pmatrix} -1 & 1 & 4 & -4 \\ -1 & 2 & 1 & -2 \\ 1 & 2 & -1 & -2 \\ 1 & 1 & -4 & -4 \end{pmatrix} F_{R^*} = \begin{pmatrix} 6 & 0 & 0 & 0 \\ 0 & \frac{3}{2} & 0 & 0 \\ 0 & 0 & \frac{3}{2} & 0 \\ 0 & 0 & 0 & 6 \end{pmatrix},$$

the solutions in case (i) are easily seen to be $\lambda = \pm[k_0(-1, 1, 4, -4) + k_2(1, 2, -1, -2)]$ with $k_0 \geqq 0$, $k_2 \geqq 0$, $k_0 + k_2 > 0$, and in case (ii) they are $\lambda = \pm[k_1(-1, 2, 1, -2) + k_3(1, 1, -4, -4)]$ with $k_1 \geqq 0$, $k_3 \geqq 0$, $k_1 + k_3 > 0$. For any such λ and $\tilde{f}_3$ we can, for example, take $\tilde{f}_i = x^i$ for $0 \leqq i \leqq 2$ and the transformation from f to $\tilde{f}$ will be nonsingular.

A development analogous to that of the previous paragraph can be carried out for general m.

We shall not analyze B^* further in the manner of Example 2(b). The number of cases to be treated and the complexity of the resulting regions increase with m, as is evident even from the above characterization of R.

EXAMPLE 3. Other Chebyshev systems are discussed in the literature of approximation theory. As illustrated in Example 2.1 of Remark 1, Assumption 2 is somewhat stronger than the assumption that $\{f_i, 0 \leqq i \leqq m\}$ is Chebyshev. The sufficient condition for Assumption 2 which is given in Proposition 2.1 of Remark 1(a) is useful in applications, as is the condition of Proposition 2.2.

REFERENCES

[1] ACHIESER, N. I. (1956). *Theory of Approximation.* Ungar, New York.

[2] BERNSTEIN, S. (1926). *Lecons sur les Propriétés Extrémales et la Meilleure Approximation des Fonctions Analytiques d'une Variable Réele.* Gauthier-Villars, Paris.

[3] CHERNOFF, H. (1953). Locally optimum designs for estimating parameters. *Ann. Math. Statist.* **24** 586–602.

[4] ELFVING, G. (1952). Optimum allocation in linear regression theory. *Ann. Math. Statist.* **23** 255–262.

[4a] HOEL, P. G. (1965). Optimum designs for polynomial extrapolation. *Ann. Math. Statist.* **36** 1483–1493.

[5] HOEL, P. G. and LEVINE, A. (1964). Optimal spacing and weighting in polynomial prediction. *Ann. Math. Statist.* **35** 1553–1560.

[6] KIEFER, J. (1959) Optimum experimental designs. *J. Roy. Statist. Soc. Ser. B* **21** 272–319.

[7] KIEFER, J. (1962). Two more criteria equivalent to D-optimality of designs. *Ann. Math. Statist.* **33** 792–796.

[8] KIEFER, J. and WOLFOWITZ, J. (1959). Optimum designs in regression problems. *Ann. Math. Statist.* **30** 271–294.

[9] KIEFER, J. and WOLFOWITZ, J. (1964 a, b). Optimum extrapolation and interpolation designs, I and II. *Ann. Inst. Statist. Math.* **16** 79–108 and 295–303.

[10] KIEFER, J. and WOLFOWITZ, J. (1965). On a problem connected with the Vandermonde determinant. *Proc. Amer. Math. Soc.* **16**.

[10a] TIMAN, A. F. (1963). *Theory of Approximation of Functions of a Real Variable,* Macmillan, N.Y.

[11] VOLKOV, V. I. (1958). Some properties of Chebyshev systems. *Kalinin Gos. Ped. Inst. Uč. Zap.* **26** 41–48.

OPTIMUM MULTIVARIATE DESIGNS

R. H. FARRELL,[1] J. KIEFER,[1] and A. WALBRAN
CORNELL UNIVERSITY

1. Introduction

1.1. *Notation and preliminaries.* This paper is concerned with the computation of optimum designs in certain multivariate polynomial regression settings.

Let $f = (f_1, \cdots, f_k)$ be a vector of k real-valued continuous linearly independent functions on a compact set X. We shall work in the realm of the approximate theory discussed in many of the references, wherein a design is a probability measure ξ (which can be taken to be discrete) on X. The information matrix $M(\xi)$ of the design ξ for problems where the regression function is $\sum_1^k \theta_i f_i(x)$ (with $\theta = (\theta_1, \cdots, \theta_k)$ unknown and with uncorrelated homoscedastic observations and quadratic loss considerations of best linear unbiased estimators) has elements $m_{ij}(\xi) = \int f_i f_j \, d\xi$. Thus, $\det M^{-1}(\xi)$ is proportional to the generalized variance of the best linear estimators of all θ_i. We denote by Γ the space of all $M(\xi)$. We shall have occasion to consider the set of all *distinct* functions of the form $f_i f_j$, $i \geq j$, and shall write them as $\{\phi_t, 1 \leq t \leq p\}$. We then write $\mu_t(\xi) = \int \phi_t \, d\xi$. Whether or not some ϕ_t is a nonzero constant (as it is in our polynomial examples), we define $\phi_0(x) \equiv 1$ and $\mu_0 = 1$.

The main results of this paper characterize, for certain X and f, some designs ξ^* which are D-optimum; that is, for which

$$\det M(\xi^*) = \max_{\xi} \det M(\xi). \tag{1.1}$$

Define, for $M(\xi)$ nonsingular,

$$\begin{aligned} d(x, \xi) &= f(x) M^{-1}(\xi) f(x)', \\ \bar{d}(\xi) &= \max_{x \in X} d(x, \xi). \end{aligned} \tag{1.2}$$

The quantity $d(x, \xi)$ is proportional to the variance of the best linear estimator of the regression $f(x)\theta'$ at x. A result of Kiefer and Wolfowitz [8] is that ξ^* satisfies (1.1) if and only if it satisfies the G-(global-) optimality criterion

$$\bar{d}(\xi^*) = \min_{\xi} \bar{d}(\xi), \tag{1.3}$$

and that (1.1) and (1.3) are satisfied if and only if

$$\bar{d}(\xi^*) = k. \tag{1.4}$$

If the support of an optimum design is exactly k points, then ξ is uniform on those points. Our main way of finding D- and G-optimum (hereafter simply

[1] Research supported by ONR Contract Nonr-401(03).

113

called "optimum") designs and of verifying their optimality is thus to guess a ξ^* (perhaps by minimizing det $M(\xi)$ over some subset of designs depending on only a few parameters) and then to verify (1.4). We also record here the fact that all optimum ξ^* have the same $M(\xi^*)$, and that they all satisfy

$$\xi^*(\{x:d(x, \xi^*) = k\}) = 1. \tag{1.5}$$

It is often the case that there is a compact group $G = \{g\}$ of transformations on X, with associated transformations $\{g\}$ on $\{\xi\}$, and such that $d(gx, \xi) = d(x, g\xi)$. In such a case (see Kiefer [6]), there is G-invariant optimum design ξ^* (that is, such that $\xi^*(gA) = \xi^*(A)$ for all g and A), and the function $d(\cdot, \xi^*)$ and set of (1.5) are G-invariant.

Whereas some of our discussion refers to general X and f, our detailed examples of sections 2, 3, and 4 all treat problems of *polynomial regression in q variables, of degree $\leq m$*. Here X is a compact q-dimensional Euclidean set whose points we usually denote by $x = (x_1, \cdots, x_q)$, and the $f_i(x)$ are of the form $\Pi_{j=1}^q x_j^{m_j}$ where the m_j are nonnegative integers with sum $\leq m$. It is well known in this case that

$$k = \binom{m + q}{q}. \tag{1.6}$$

Moreover, since the f_if_j are all the monomials of degree $\leq 2m$, we have

$$p = \binom{2m + q}{q}. \tag{1.7}$$

The three examples we shall treat in detail are (in section 4) the unit q-ball

$$\left\{x: \sum_1^q x_i^2 \leq 1\right\}; \tag{1.8}$$

(in section 3) the q-cube

$$\{x: \max_{1 \leq i \leq q} |x_i| \leq 1\}; \tag{1.9}$$

and (in section 2) the unit simplex, which it is more convenient to represent in barycentric coordinates $x = (x_0, x_1, \cdots, x_q)$ as

$$\left\{x: \min_{0 \leq i \leq q} x_i \geq 0, \sum_0^q x_i = 1\right\}. \tag{1.10}$$

These are perhaps the three generalizations which are simplest, most natural, and of greatest practical importance, of the unit interval ($q = 1$), which is discussed in section 2. Unfortunately, the simple structure which is present when $q = 1$ and which is reflected in the elegant results of Guest [3] and Hoel [4] does not carry over to $q > 1$, and the results depend strongly on the shape of X; even in the case of the simplex where at least some analogous results seem to hold, they cannot be obtained by the same methods. We now indicate how this is reflected in the geometry of Γ.

1.2. *The geometry of* Γ. The set Γ can clearly be regarded as a convex body in p-dimensional Euclidean space with coordinates μ_t, $1 \leq t \leq p$; of course,

$p \leq k(k+1)/2$. Write $a = (a_0, a_1, \cdots, a_p)$. Let $\sum_1^p a_t\mu_t + a_0 = 0$ be a supporting hyperplane of Γ with $\sum_1^p a_t\mu_t + a_0 \geq 0$ in Γ. Clearly, the supporting polynomial $T(x; a) = \sum_1^p a_t\phi_t(x) + a_0$ is nonnegative on X.

(For future reference, the reader should note in connection with the previous and next paragraphs that, if ξ^* is optimum and $\gamma^* = M(\xi^*)$, then ξ^* is admissible and hence γ^* is a boundary point, and $k - d(x, \xi^*)$ supports Γ at γ^*.)

Let $\gamma_0 = M(\xi_0)$ be a boundary point of Γ. A supporting polynomial $T(\cdot\,; a^0)$ which supports Γ at γ_0 is then ≥ 0 on X and is 0 on the support of ξ_0. Thus, an analysis of what the set of zeros of a T of the above form can be can yield information about the boundary points of Γ (while the extreme points are clearly a subset of the points corresponding to ξ's with one-point support). For example, in the well-known univariate polynomial case $X = [0, 1]$, $k = m + 1$, $f_i(x) = x^{i-1}$, any such T is a nonnegative polynomial on X of degree $\leq 2m$, which (if not identically zero) therefore has at most $m + 1$ zeros, at most m of which are in the interior of X. In this example, moreover, if $\gamma = M(\xi)$ is an arbitrary point of Γ and $\xi^{(0)}(0) = 1$, the line from the boundary point $M(\xi^{(0)})$ through γ passes through another boundary point $\gamma' = M(\xi')$, so that $\gamma = M(\lambda\xi^{(0)} + (1 - \lambda)\xi')$ with $0 \leq \lambda \leq 1$; thus one concludes that any point of Γ can be represented as $M(\xi'')$ for a ξ'' supported by at most $m + 1$ points. One can also characterize the admissible ξ easily in this example as the boundary points with at most $m - 1$ points of support in the interior of X (Kiefer [5]).

Unfortunately, the examples studied in the present paper (as well as non-polynomial, and especially non-Chebyshev systems in one dimension) do not yield such simple analyses. This is clear when one considers the more complex sets on which a T can now vanish. For example, in the case of linear regression ($m = 1$) on the square (1.9) with $q = 2$, any supporting T which is not identically zero, being quadratic, vanishes either on a subset of the corners of X, or at a single point of X, or on a line of X. In the latter case we invoke the one-dimensional result to conclude that at most two points are needed to support a ξ yielding this $M(\xi)$; thus, every boundary point of Γ is obtainable from a ξ supported either by a subset of the corners or else by at most two other points. Replacing $\xi^{(0)}$ in the argument of the previous paragraph by the measure which assigns all probability to the point $(-1, -1)$, we conclude that every point of Γ can be obtained from a ξ which is supported either by subset of the corners or else by at most three points of X. If we replace (1.9) by (1.10) with $q = 2$, we obtain that at most 3 points rather than 4 are needed. The admissible points can be characterized similarly, but it is clear that the difficulty of obtaining such characterizations will be much greater for larger q and m. As for the optimum design, it is the uniform distribution on the 3 corners in the case (1.10), on the 4 corners in the case (1.9), and, for another example, on the 5 corners if X is a symmetric pentagon. The uniqueness in all three cases can be proved by the method given in the next paragraph, but in other cases, such as (1.8), there is no uniqueness. Section 3.3 of [6] characterizes optimum designs for linear regression on general compact X in q dimensions.

The increased complexity in higher dimensions is also present in the uniqueness question: given $\gamma = M(\xi^*)$, when is there no other ξ with $M(\xi) = \gamma$? This can sometimes be answered as follows. Suppose γ is a boundary point and that a supporting polynomial T at γ has exactly L zeros $x^{(1)}, x^{(2)}, \cdots, x^{(L)}$ on X. Any ξ with $M(\xi) = \gamma$ must be supported by a subset of $\{x^{(1)}, \cdots, x^{(L)}\}$, and must satisfy $\sum_j \phi_t(x^{(j)})[\xi^*(x^{(j)}) - \xi(x^{(j)})] = 0$ for $0 \leq t \leq p$. Hence, if rank $\{\phi_t(x^{(j)}), 0 \leq t \leq p, 1 \leq j \leq L\} = L$, then ξ^* is the unique design yielding γ. In the univariate polynomial example of the second paragraph above, each boundary point γ can be proved by this device to be yielded by a unique ξ^*.

The prescription outlined just below (1.4) for verifying optimality, and which has worked well when $q = 1$ or $m \leq 2$, is difficult to apply in other cases. This is because $k - d(x, \xi^*)$ can no longer be written as a sum of a small number of obviously nonnegative simple polynomials, but may instead require a large number of rational functions for such a representation. The decision procedures (Tarski, Henkin, and others) for representing or verifying nonnegativity of such polynomials are unwieldy to implement in these problems. The example of the simplex (1.10) with $q = 2$, $m = 3$, treated in section 2 by direct analysis, illustrates the increased complexity. In other cases we have been unable to obtain analytical verifications of optimality and have used machine methods to obtain results which are satisfactory from a practical point of view but which, theoretically, only yield statements of results which hold to within a certain accuracy, rather than complete proofs of the exact results.

We end this subsection with a simple observation which is often useful in optimum design theory for polynomial regression on a q-dimensional set X. If B is a subset of X such that for some $q \times q$ orthogonal matrix A and some scalar b with $|b| > 1$, the set $bAB = \{x: b^{-1}A^{-1}x \in B\}$ is also a subset of X, then no design ξ supported by B can be optimum. This follows at once upon defining ξ' by $\xi'(C) = \xi(b^{-1}A^{-1}C)$ for $C \subset X$ and computing $\det M(\xi') = b^k \det M(\xi)$. In particular, if X is such that $x \in X \Rightarrow ax \in X$ for $0 \leq a < 1$, then the support of any optimum design must contain at least one point of the boundary of X. The considerations of this paragraph can be modified in an obvious way for admissibility questions.

1.3. *Number of points needed for an optimum design.* An aspect of the geometry of Γ which is of particular practical importance is the minimum number N of points such that there is an optimum design supported by N points. (It will be clear how to modify much of the discussion which follows to treat this question for points of Γ other than those corresponding to D-optimum designs, but for brevity we will treat only the latter.) An optimum design will be called *minimal* if no proper subset of its support is the support of an optimum design. We shall see that this property is broader than that of being an optimum design on N points; the latter will be called *absolutely minimal.*

Clearly $N \geq k$. On the other hand, if there is a matrix B of rank b such that $\sum_{j=1}^{p} b_{ij}\phi_j(x)$ is a constant function of x for each i, then Γ has dimension $\leq p - b$.

Since the extreme points of Γ can be obtained from ξ's with one-point support, we obtain the trivial bounds

$$k \leq N \leq \min(p - b + 1, k(k+1)/2), \tag{1.11}$$

where the well-known bound $k(k+1)/2$, which is relevant only when $p = k(k+1)/2$ and $b = 0$, is a consequence of the fact that optimum designs correspond to certain boundary points of Γ. In the polynomial case we have $b = 1$ (since 1 is a ϕ_t) and thus, from (1.6) and (1.7),

$$\binom{m+q}{q} \leq N \leq \binom{2m+q}{q}. \tag{1.12}$$

Of greater use is the upper bound one can obtain once one knows some optimum design ξ^*. Let $V = \{x: d(x, \xi^*) = k\}$ (see (1.5)), and let W be the support of ξ^*. We denote the number of points in these sets by v and w. (When v or w is infinite, as in the example of the ball (1.8) for $q \geq 2$ in section 4, it is easy to see that (1.14) below still holds, but we shall usually treat the finite case.) Let $U = V$ or W (and $u = v$ or w). The $p + 1$ linear equations

$$\sum_{x \in U} \phi_t(x)\xi(x) = \mu_t(\xi^*), \qquad 0 \leq t \leq p \tag{1.13}$$

in the unknowns $\xi(x)$ are consistent (since $\{\xi^*(x)\}$ is a solution), so that the dimensionality of the linear set H (say) of solutions of (1.13) is $u - h$ where $h = \text{rank}\{\phi_t(x), 0 \leq t \leq p, x \in U\}$.

Considering H as a set in the u-dimensional space with coordinates $\xi(x)$, $x \in U$, we know that ξ^*, with all coordinates nonnegative, is in H, and conclude easily that H contains a point with all coordinates nonnegative and with at least h zero coordinates. Hence,

$$N \leq \text{rank}\{\phi_t(x), 0 \leq t \leq p, x \in U\}. \tag{1.14}$$

(Of course, if U is replaced by X, this becomes the $p - b + 1$ of (1.11).) We will illustrate the use of this in the polynomial case (where $1 \in \{\phi_t\}$ so that the domain of t in (1.14) can be taken to be $1 \leq t \leq p$) in section 3.3, in the case of quadratic regression on the q-cube (1.9). We can think of such polynomial applications in terms of finding a matrix C of rank c such that $\sum_{j=1}^{p} c_{ij}\phi_j(x) \equiv 0$ on W. We can then conclude that, if N_W is the minimum number of points in W supporting an optimum design, then (paralleling (1.11))

$$N_W \leq p - c. \tag{1.15}$$

On the other hand, it is obvious from (1.13) that, for $U = V$ or W,

$$N_U = u \quad \text{if rank} \quad \{\phi_t(x), 0 \leq t \leq p, x \in U\} = u. \tag{1.16}$$

It is easy to give examples which illustrate the fact that we can have $N_W > N$; that is, that minimality and absolute minimality do not coincide. For example, in the case $m = 1$, $q = 2$ of linear regression on the unit disc (1.8), the discussion of the fourth paragraph of section 1.2 shows that the set of j equally spaced points on the boundary is minimal if $j = 3$, 4, or 5, but is absolutely minimal only when $j = 3$.

Also, the designs whose optimality is easiest to verify are often ones which are symmetric, that is, invariant in the sense described below (1.5), and there is no reason why minimal designs should be of this form. For example, in the case of the q-cube (1.9) with $m = 1$, the uniform distribution on the 2^q corners is the only optimum design invariant under the symmetries of the q-cube, but if q is such that there exists a $(q + 1) \times (q + 1)$ Hadamard matrix (for example, if $q + 1$ is a power of 2), then it is well known that there is an optimum design on $k = q + 1$ corners (namely, the corners of an inscribed regular q-simplex). This example illustrates another technique for reducing an upper bound on N or N_W; in sections 3 and 4 we shall see how the use of various known results on orthogonal arrays and rotatable configurations can be used similarly.

The search for absolutely minimal designs can be described as a programming problem, of finding a nonnegative solution of (1.13) with $U = V$, which has a minimum number of nonzero elements. Analytical or machine methods for solving this problem would seem important.

2. The simplex

We have mentioned in section 1 that this case (1.10) evidences the most regular mathematical behavior among q-dimensional sets X. In the linear and quadratic cases it has been known for some time that the simplex exhibits a behavior (described precisely, below) very much like that present when $q = 1$. This phenomenon appears to carry over to cubic and perhaps higher degree regression, although we have as yet proved only one small fragment of the conjectured general result, and have machine computations in only two other cases. To describe these results, let E_m be the set of $m + 1$ points supporting the Guest-Hoel design when $q = 1$. (Thus, $E_1 = \{x_0 = 0, 1\}$; $E_2 = \{x_0 = 0, \frac{1}{2}, 1\}$; $E_3 = \{x_0 = 0, (1 \pm 5^{-1/2})/2, 1\}$; $E_4 = \{x_0 = 0, \frac{1}{2}, 1, (1 \pm (\frac{3}{7})^{1/2})/2\}$; and so on.)

The results in the linear and quadratic cases for general dimension q (Kiefer [7]) can be summarized by stating that, for degrees $m = 1$ and 2, the unique optimum design assigns equal measure to each of the points which is in the E_m of some edge of the q-simplex (when that edge is considered as a 1-simplex). We cannot hope for this pattern for $m > 2$, since the E_m points on all edges will be fewer in number than the k of (1.6). However, one can still conjecture that one or all of the following are true: (1) there is an optimum design whose support includes the E_m points on all edges (and no other points on edges); (2) there is an optimum design which assigns equal measure to each of k points; (3) the optimum design is unique; (4) generalizing the vertex- and edge-stationarity of (1), for fixed m there are optimum designs for dimension q which have the same support on the r-dimensional faces of X for $q \geq r$; (5) the design of (4) has points of support only on faces of dimension $\leq \min(m - 1, q)$.

What we have succeeded in treating analytically is the case $m = 3$, $q = 2$, and the details will be found at the end of this section. We have also observed, by machine search, that (a) the optimum design in the case $m = 3$, $q = 3$ appears to give equal weights to the E_3-points on edges (including vertices) and the midpoints of 2-dimensional faces, just as in the cases $m = 3$, $q = 1$ or 2; (b) the optimum design in the case $m = 4$, $q = 2$ appears to give equal weights to the E_4-points on edges and to the three points of the form $\{x_h = 0.567, x_i = x_j\}$. One can also prove analytically for $m = 3$ and general k that, among all designs which assign equal weights to the vertices, midpoints of 2-dimensional faces, and points on edges satisfying $x_i = b = 1 - x_j$, the choice $b = (1 \pm 5^{-1/2})/2$ minimizes the generalized variance for each k. (The generalized variance for such designs for $q \geq 2$ is proportional to $[v(1 - 2b)^{1/2}]^{-2q(q+1)}$, where $v = b(1 - b)$.) All of the above results conform with the conjectures of the previous paragraph.

If any of the general conjectures are true, they would constitute a deep new result in the area of multidimensional moment and approximation theory. Evidently a new approach is needed, perhaps even to verify analytically (b) and (c) of the previous paragraph. The technique employed for low dimensions and/or degrees, for example, by Kiefer [7] and Uranisi [11], has been that described at the end of section 1.2, and the difficulties encountered for larger q or m are as described there. Even in the case $m = 3$, $q = 2$ which we now consider, a much more brutal approach is used, and it does not suffice when $q = 3$.

THEOREM 2.1. *For $m = 3$, $q = 2$, the unique optimum design ξ^* assigns measure $\frac{1}{10}$ to each of the three vertices, the point $x_0 = x_1 = x_2 = \frac{1}{3}$, and the six points $\{x_h = 0, x_i = 1 - x_j = (1 + 5^{-1/2})/2.\}$.*

PROOF. We shall show that $10 - d(x, \xi^*) \geq 0$ on X, with equality only at the ten points supporting ξ^*. Since the function d is the same for all optimum designs, any optimum design must have this same support, and the weights are unique since there are 10 points and 10 functions. This yields uniqueness.

It is convenient to consider, in place of the coordinates x_i, the coordinates $\beta, t(-\frac{1}{2} \leq \beta \leq 1, 0 \leq t \leq 1)$ satisfying $3x_0 = 1 - t, 3x_1 = 1 - \beta t, 3x_2 = 1 + (\beta + 1)t$ on the portion $0 \leq x_0 \leq x_1 \leq x_2$ of X which, because of the symmetry of ξ^*, is all we need consider. For fixed β, variation of t from 0 to 1 yields a segment from center to edge of X. Write $L = \beta^2 + \beta + 1$ (so that $\frac{3}{4} \leq L \leq 3$). A simple computation yields

$$
\begin{aligned}
&9 \sum_{i<j} x_i x_j = 3 - Lt^2, \\
&81 \sum_{i<j} x_i^2 x_j^2 = 3 - 6(L - 1)t^3 + L^2 t^4, \\
&729 \sum_{i<j} x_i^3 x_j^3 = 3 + 3Lt^2 - 21(L - 1)t^3 + 3L^2 t^4 \\
&\qquad + 3L(L - 1)t^5 + (-L^3 + 3L^2 - 6L + 3)t^6, \\
&27 \prod_i x_i = 1 - Lt^2 + (L - 1)t^3.
\end{aligned}
\tag{2.1}
$$

A straightforward computation of $M(\xi^*)$ and $d(x, \xi^*)$ (for example, in terms of the functions x_i, $x_i x_j$, $x_i x_j(x_i - x_j)$, and $\Pi_i x_i$, with $i < j$) yields

$$(2.2)\qquad 1 - d(x, \xi^*)/10 = 12 \sum_{i<j} x_i x_j - 120 \sum_{i<j} x_i^2 x_j^2 + 300 \sum_{i<j} x_i^3 x_j^3 + \prod_i x_i[-102 + 410 \sum_{i<j} x_i x_j - 1512 \prod_i x_i].$$

From (2.1) and (2.2) we have, writing $g(t, L) = 729[1 - d(x, \xi^*)/10]/6t^2$,

$$(2.3)\qquad g(t, L) = 131L - 318(L - 1)t - 77L^2t^2 + 449L(L - 1)t^3 - [50L^3 + 102(L - 1)^2]t^4.$$

We must show $g(t, L) \geq 0$ for $0 < t \leq 1$, $\frac{3}{4} \leq L \leq 3$. We note that $g(1, L) = 2(3 - L)(5L - 6)^2$, so that the zeros of g on the boundary of X are precisely the vertex ($L = 3$) and E_3-point $L = \frac{6}{5}$.

Writing $f(t, L) = [g(t, L) - g(1, L)]/(1 - t)$, we obtain

$$(2.4)\qquad \begin{aligned} f(t, L) &= [50L^3 - 270L^2 + 563L - 216] \\ &\quad + [50L^3 - 270L^2 + 245L + 102]t \\ &\quad + [50L^3 - 347L^2 + 245L + 102]t^2 + [50L^3 + 102(L - 1)^2]t^3 \\ &= D(L) + C(L)t + B(L)t^2 + A(L)t^3 \quad \text{(say)}. \end{aligned}$$

We shall show that $f > 0$ for $0 \leq t \leq 1$, $\frac{3}{4} \leq L \leq 3$, and this will complete the proof.

We have $D(\frac{3}{4}) > 0$ and $D'(L) = 150L^2 - 540L + 563 > 0$. Thus,

$$(2.5)\qquad A(L) > 0, \qquad D(L) > 0, \qquad \tfrac{3}{4} \leq L \leq 3.$$

Also, one sees easily that, for $\frac{3}{4} \leq L \leq 1$, we have $B(L) \geq 50L^3$ and $C(L) \geq 0$. We conclude that $f(t, L) > 0$ for $0 \leq t \leq 1$, $\frac{3}{4} \leq L \leq 1$. We divide the region $1 \leq L \leq 3$ into two parts, the division λ being the zero of $C(L)$ in $1 \leq L \leq 3$ ($1.6 < \lambda < 1.7$).

For $\lambda \leq L \leq 3$, we have $C(L) \leq 0$. We shall show the positivity of something $\leq f$, namely,

$$(2.6)\qquad \begin{aligned} f(t, L) + C(L)(1 - t)^2(1 + t) &= [100L^3 - 540L^2 + 808L - 114] \\ &\quad - 77L^2t^2 + [100L^3 - 168L^2 + 41L + 204]t^3 \\ &= E(L) - 77L^2t^2 + F(L)t^3 \text{ (say)}. \end{aligned}$$

We first note that $-23L^2 + 132L - 114$ is positive at $L = 1.6$ and $L = 3$ and, hence, for $\lambda \leq L \leq 3$. Therefore,

$$(2.7)\qquad \begin{aligned} 0 &< 100L(L - 2.6)^2 - 23L^2 + 132L - 114 = E(L) - 3L^2 \\ &< E(L) - \frac{4(77)^3}{27(84)^2} L^2 = E(L) + L^2 \min_{0 \leq t \leq 1} t^2(84t - 77) \\ &\leq E(L) - 77L^2t^2 + 84L^2t^3. \end{aligned}$$

Thus, the expression (2.6) will be proved positive if we show that $0 < F(L) - 84L^2 = h(L)$ (say). But an easy computation shows that $h(1.6) > 0$, $h'(1.6) > 0$, and $h''(L) > 0$ for $L \geq 1.6$.

In the region $1 \leq L \leq \lambda$, we have $C(L) \geq 0$, and thus need only show that

$D(L) + B(L)t^2 + A(L)t^3 > 0$. Because of (2.5), this is immediate if $B(L) \geq 0$, so we need only consider the possibility $B(L) < 0$, in which case

$$\begin{aligned} (2.8) \qquad & D(L) + B(L)t^2 + A(L)t^3 \\ & \geq [D(L) + B(L) \max_{0 \leq t \leq 1} (t^2 - t^3/2)] + t^3[A(L) + B(L)/2] \\ & = [D(L) + B(L)/2] + t^3[A(L) + B(L)/2] = R(L) + S(L)t^3 \text{ (say).} \end{aligned}$$

One sees easily that $R(L) > 0$ for $1 \leq L \leq 2$ and $S(L) \geq 0$ for $L \geq 1$, completing the proof.

3. Symmetric regions; the cube

The case of the q-cube (1.9) exhibits less regularity than either the simplex or ball. This is seen even in the linear case described in section 1.3, where more than k points of support may be required (as when $q = 2$ and the unique optimum design is uniform on the 4 corners); and, even more, in the quadratic case, where optimum designs can be written down explicitly almost immediately for the ball and simplex, but require at least some consideration for the cube, regarding weights assigned to the points of the 3^q array J of points with coordinates 0, 1, -1. We shall now study this quadratic case in considerable detail. We begin by characterizing some properties of optimum quadratic designs for more general symmetric regions. (For general linear regression see Kiefer [6].)

3.1. *Quadratic regression on symmetric regions.* We introduce some of the ideas by considering, in the present paragraph, the q-cube. The fact that, when $m = 2$, the support of every optimum ξ^* is a subset of the 3^q array J, is easily seen as follows: $d(x, \xi^*)$ for any optimum ξ^* is symmetric under the group of symmetries of the cube (see discussion just below (1.5)), goes to $+\infty$ with $|x|$, and is a positive quartic on Euclidean q-space. Writing B for the subset of X where $d(x, \xi^*) = k$ (so that the support of ξ^* is contained in B), we will show that the existence of points in $B - J$ leads to a contradiction. Calling vertices, edges, and so on, the 0-, 1-, $\cdots$, skeleton of X, suppose that x' in $B - J$ lies in the r-skeleton, and hence in the relative interior of some r-cube G of that skeleton. By symmetry of B, there is another point x'' of $B - J$ which is also in the relative interior of G. The function d attains its maximum on G at x' and x'', and hence cannot be a positive quartic on q-space unless it is a constant, in which case it does not go to $+\infty$ with $|x|$.

We turn now to more general symmetric regions X to which we can apply some similar arguments.

We consider quadratic regression in q variables $x_1, \cdots, x_q$ on a symmetric region X of Euclidean q-space. The meaning of saying X is symmetric is that X is invariant under permutations $(x_1, \cdots, x_q) \rightarrow (x_{\sigma_1}, \cdots, x_{\sigma_q})$ and is invariant under sign changes $(x_1, \cdots, x_q) \rightarrow (\epsilon_1 x_1, \cdots, \epsilon_q x_q)$, $\epsilon_1 = \pm 1, \cdots, \epsilon_q = \pm 1$. The discussion just below (1.5) states that there are optimum designs which are symmetric; also, it implies that the function $d(\cdot, \xi^*)$ for any optimum design ξ^*

is a symmetric polynomial in the variables $x_1^2, \cdots, x_q^2$, of degree 2 in these variables. If we write $s = x_1^2 + \cdots + x_q^2$ and $t = x_1^4 + \cdots + x_q^4$, then the general polynomial of this type is

$$P(s, t) = as^2 + bs + c + dt. \tag{3.1}$$

The map $h: (x_1, \cdots, x_q) \to (x_1^2 + \cdots + x_q^2, x_1^4 + \cdots + x_q^4)$ maps the region X to a region X^* in the s, t plane. Clearly, $d(h^{-1}(s, t), \xi) = d^*((s, t), \xi)$ (say) is well-defined for any symmetric ξ and any (s, t) in X^*. We will be concerned with an examination of the values of $d^*(\cdot, \xi^*)$ at points of X^*, for an optimum ξ^*. We know from (1.4) and (1.6) that, throughout X,

$$d(x, \xi^*) - (q + 1)(q + 2)/2 \leq 0, \tag{3.2}$$

the equality holding at points including the support of ξ. We now show that there are two possibilities:

(i) the zeros of $d^*(\cdot, \xi) - (q + 1)(q + 2)/2$ lie entirely on the boundary of X^*;

(ii) the coefficient $d = 0$ in (3.1), so that the polynomial has the form $as^2 + bs + c$. (In this case the design is a rotatable design which can be shown to be optimum for the problem wherein X is replaced by the smallest ball centered at the origin and containing X, minus the largest open ball contained in its complement (which subtraction is vacuous if X contains the origin).

To see the validity of this assertion, suppose (s_0, t_0) is an interior point of X^* at which $P(s_0, t_0) - (q + 1)(q + 2)/2 = 0$. In view of (3.2) and continuity of the map h, (s_0, t_0) is a local maximum of the polynomial P. Therefore, the first partial derivatives vanish at (s_0, t_0), so that $d = 0$ follows. That proves the assertion.

An analysis of which of (i) and (ii) holds requires more precise knowledge of X, as we see by contrasting the cases (1.8) and (1.9). Although we already know from the first paragraph of this subsection that (i) holds for the q-cube (1.9), we shall continue our analysis for that example along the present lines, both to illustrate this method which can be applied to other symmetric regions similarly, and also because we will then use the method for cubic regression on the q-cube.

Thus, we now suppose X is given by (1.9). We will see that X^* is a closed bounded set which may be described in terms of an upper and lower boundary curve. The upper curve consists of q pieces:

$$\{t = (s - i)^2 + i, \quad i \leq s \leq i + 1\}, \qquad i = 0, 1, \cdots, q - 1. \tag{3.3}$$

The lower curve may be described by the single equation

$$\{qt = s^2, 0 \leq s \leq q\}. \tag{3.4}$$

This last assertion follows at once by the Cauchy-Schwarz inequality since

$$s^2 = (x_1^2 + \cdots + x_q^2)^2 \leq q(x_1^4 + \cdots + x_q^4) = qt. \tag{3.5}$$

We observe that equality holds in (3.5) if and only if $x_1^2 = x_2^2 = \cdots = x_q^2$.

To obtain the upper boundary we suppose the value of $t = x_1^4 + \cdots + x_q^4$ is fixed and seek to minimize s. We may suppose at the start that $x_1 \geq x_2 \geq \cdots \geq x_q \geq 0$.

Consider x_q as a function of x_1 and take partial derivatives with $x_2, \cdots, x_{q-1}$ fixed. This gives $\partial x_q/\partial x_1 = -x_1^3/x_q^3$ and $\partial s/\partial x_1 = 2x_1(x_q^2 - x_1^2)/x_q^2$. We suppose here $x_q > 0$. Since the derivative is negative, we decrease s by increasing x_1 and decreasing x_q, and this preserves the ordering $x_1 \geq x_2 \geq \cdots \geq x_q \geq 0$.

Using this it may be seen that if $t = i + \delta^4$, $0 \leq \delta < 1$, is the fixed value of t, then the minimum for s is obtained by taking $x_1 = x_2 = \cdots = x_i = 1$, $x_{i+1} = \delta$, $x_{i+2} = \cdots x_q = 0$. Thus $s = i + \delta^2$ and $t = i + (s - i)^2$, as asserted in (3.3).

This argument shows even more, that the minimum value of s can be obtained from $x_1, \cdots, x_q$ if and only if $x_1 = 1, \cdots, x_i = 1$ (when $t = i + \delta^4$).

We now show, using an argument like that of the first paragraph of this subsection, that the only possible location of a zero of $d^*(\cdot, \xi^*)$ (for ξ^* optimum) on the boundary segment $\{t = i + (s - i)^2, i \leq s \leq i + 1\}$, is at an end point. The polynomial $d^*((s, i + (s - i)^2), \xi^*)$ is a quadratic in s for $0 \leq s < \infty$ which, being equal to $f(x)M^{-1}(\xi^*)f(x)'$ with $x_1 = x_2 = \cdots = x_i = 1$, $x_{i+1} = s^{1/2}$, $x_{i+2} = \cdots = x_q = 0$, is nonnegative for $0 \leq s < \infty$ and goes to infinity with s, so that it cannot have a local maximum over the interval $i \leq s \leq i + 1$ at an interior point of the latter.

Finally, using the same type of argument, we show that no zero of $d^*(\cdot, \xi^*) - k$ can occur interior to the segment $\{qt = s^2, 0 \leq s \leq q\}$. This is so because $d^*((s, s^2/q), \xi^*)$ is a quadratic in s for $0 < s < \infty$ which, being equal to $f(x)M^{-1}(\xi^*)f(x)'$ with $x_1 = x_2 = \cdots = x_q = (s/q)^{1/2}$, goes to infinity with s, and can thus not attain its maximum over $0 \leq s \leq q$ at an interior point of the latter.

Our discussion has not yet excluded the possibility that the optimal design is rotatable, that is, that $d^*((s, t), \xi^*)$ has the form $as^2 + bs + c$ for an optimum ξ^*. Were this the case, then the design would be optimum for X replaced by the ball $K = \{x: \sum_1^q x_i^2 \leq q\}$, since the argument of the previous paragraph shows that if the optimum ξ^* is rotatable, then $d(x, \xi^*)$ takes on its maximum value $(q + 1)(q + 2)/2$ at the point $(1, 1, \cdots, 1)$ satisfying $\sum_1^q x_i^2 = q$. The moment matrix $M(\xi)$ for an optimal design ξ on K is uniquely determined and is known (Kiefer [6]) to put mass at $s = 0$ and on $s = q$, the moments of the conditional distribution on $s = q$ being those of the uniform measure on this surface.

But, in our problem, the design ξ must be concentrated in the cube (1.9), and the only points in common between the cube and the shell $s = q$ are the corners $(\pm 1, \cdots, \pm 1)$. It is easily seen that every symmetric ξ which is concentrated on the corners and at the origin makes $\int x_1^4 \xi(dx) = \int x_1^2 x_2^2 \xi(dx)$, which is not the case for the optimum rotatable design on the ball. Hence the optimal design for quadratic regression on the q-cube cannot be rotatable.

The discussion substantiates what was already known from simpler calculations in this special case, as indicated earlier. We now bring these ideas to bear on the problem of cubic regression on the cube. We shall return in sections 3.3–3.5 to quadratic regression and shall consider at length there the possible supporting sets for optimum designs.

3.2. *Cubic regression on the q-cube.* We first treat the case $q = 2$, for the sake of simplicity and explicitness of numerical results. The function $d(\cdot, \xi^*)$ for an optimum ξ^* is now a nonnegative polynomial of degree 6 in the variables x_1, x_2, having the following properties. From (1.4) and (1.6),

$$\bar{d}(\xi^*) \leq 10; \tag{3.6}$$

and d is symmetric in x_1, x_2 and invariant under sign changes.

It follows that if we write $t = x_1^4 + x_2^4$ and $s = x_1^2 + x_2^2$ as before, then we can define $d^*(\cdot, \xi)$ for symmetric ξ as in the paragraph below (3.1). This will now be a polynomial of degree 3 in s, t, of the form

$$P(s, t) = as^3 + bs^2 + cs + d + est + ft. \tag{3.7}$$

The domain X^*, from (3.3) and (3.4) with $q = 2$, is the closed bounded set whose boundary consists of the three curves

$$\begin{aligned} &\{s^2 = t, 0 \leq s \leq 1\}, \\ &\{(s-1)^2 + 1 = t, 1 \leq s \leq 2\}, \\ &\{s^2 = st, 0 \leq s \leq 2\}. \end{aligned} \tag{3.8}$$

We consider first the implication of assuming that for some (s_0, t_0) interior to this region $d((s_0, t_0), \xi^*) = 10$. This would be a local maximum interior to the domain X^*, so that $P(\sigma + s_0, \tau + t_0) = a\sigma^3 + b'\sigma^2 + c'\sigma + d' + e\sigma\tau + f\tau$ (say), defined in a neighborhood of $(\sigma, \tau) = (0, 0)$, would have a local maximum at $(0, 0)$. We would therefore have $e = f = 0$, and $P(s, t)$ would be a function only of s, which is the definition of ξ^* being rotatable. We also note that in this case we would have

$$P(s, t) = (as + b'')(s - s_0)^2 + 10 \tag{3.9}$$

with $as + b'' \leq 0$ for $0 \leq s \leq 2$ and with $a > 0$ (since $P \to \infty$ as $s \to \infty$).

Thus, if the design is not rotatable, then the polynomial $d^*(\cdot, \xi^*)$ can vanish only on the boundary curves of (3.8). Substitution of any one of the three relations of (3.8) for t into $P(s, t) - 10$ gives a cubic in s which does not change sign. Hence, any root s interior to the interval determined by the substitution would require the root to be a double root. Therefore, each of the three sections of boundary in the s, t plane can have at most two points at which the polynomial $P - 10$ vanishes. Since the boundary of the square X is mapped into the curve $\{(s - 1)^2 + 1 = t, 1 \leq s \leq 2\}$, it follows from the final paragraph of section 1.2 that this curve contains at least one point at which $P - 10$ vanishes.

We now eliminate the possibility of a rotatable design being optimum. The theory of optimum designs for polynomial regression on the ball has been developed by Kiefer [6] and in section 4 of the present paper. An optimum design ξ^* for the square, if rotatable, would have a $d^*(\cdot, \xi^*)$, given by (3.9), attaining its maximum on the square at the corners ($s = t = 2$) and satisfying $d^*((s, t), \xi^*) \leq 10$ on the image under h of the ball K of radius $2^{1/2}$. (This image is bounded by the curves $\{s^2 = 2t, 0 \leq s \leq 2\}$, $\{s^2 = t, 0 \leq s \leq 2\}$, and

$\{s = 2, 2 \leq t \leq 4\}$.) Hence ξ^* would be optimum for the problem of cubic regression on $K = \{x: x_1^2 + x_2^2 \leq 2\}$, with $d(x, \xi^*)$ attaining its maximum over K on the two circles $s = 2$ and $s = s_0$. Thus, $s_0 = 2\rho^2$, where ρ and 1 are the radii of circles where $d(x, \xi') = 10$, where ξ' is optimum for cubic regression when X is the unit ball of (1.8), to be discussed further in section 4. Although this ξ' is not unique, its moments up to those of order 4 (that is, the elements of $M(\xi')$), and the total masses β and $1 - \beta$ which it assigns to the circles of radii ρ and 1, respectively, are unique. (For the first of these facts, see just above (1.5); for the second, replace ξ' by the optimum design ξ'' defined by $\xi''(A) = \int \xi'(gA)\nu(dg)$, where ν is the invariant probability measure on the orthogonal group in two dimensions as in [6], and use the uniqueness of the masses in such a ξ'', proved in [6].) Since $\int_0^{2\pi} (2\pi)^{-1} \cos^4 \theta \, d\theta = \frac{3}{8}$, we would thus obtain

$$\int x_1^4\xi^*(dx) = 4 \int x_1^4\xi'(dx) = 4 \left(\tfrac{3}{8}\right) [(1 - \beta) + \rho^4\beta]. \tag{3.10}$$

Since ξ^* is supported within the square (1.9), we also have $\int x_1^4\xi^*(dx) \leq 1$. Hence, (3.10) yields

$$\frac{3(1 - \beta)}{2} \leq \left(\tfrac{3}{2}\right) [(1 - \beta) + \rho^4\beta] \leq 1, \tag{3.11}$$

or $\beta \geq \frac{1}{3}$. But (from table 4.1 of section 4) $\beta = .3077$. We conclude that the optimum design cannot be rotatable.

The following maximization problem was solved numerically. Put mass $p_1/4$ at each of $(1, 1)$, $(1, -1)$, $(-1, 1)$ and $(-1, -1)$. Put mass $p_2/8$ at each of the eight points $(\pm 1, \pm a)$ and $(\pm a, \pm 1)$, $0 < a < 1$. Put mass $(1 - p_1 - p_2)/4$ at each of the four points $(\pm b, \pm b)$, $0 < b < 1$. (In each of the above, the two $\pm$ signs act independently.) Our earlier discussion shows a design of this form to be a candidate for being optimum (although we did not yet eliminate certain other forms). The determinant of $M(\xi)$ was maximized on the Cornell CDC 1604 as a function of the four parameters involved, giving

$$\begin{aligned} a &= 0.35880, \\ b &= 0.48000, \\ p_1 &= 0.36770, \\ p_2 &= 0.46100. \end{aligned} \tag{3.12}$$

For this design ξ, the quantity $\bar{d}(\xi)$ was computed numerically and was found to be ≤ 10 to five decimal places.

We now leave the case $q = 2$ to discuss cubic regression on the cube (1.9) for general $q \geq 3$, where an analysis similar to the one just given for $q = 2$ may again be carried out. If ξ is symmetric, $d(x, \xi)$ is now a sixth degree symmetric polynomial in $x_1, \cdots, x_q$, in which any monomial term involving an odd exponent has zero coefficient. We now need three symmetric functions,

$$s = x_1^2 + \cdots + x_q^2,$$
(3.13) $$t = x_1^4 + \cdots + x_q^4,$$
$$u = x_1^6 + \cdots + x_q^6,$$

and define $h: X \to X^*$ by $h(x) = (s, t, u)$. (When $q = 2$, we do not need u since $2u = 3st - s^3$ in that case.) Again $d^*(\cdot, \xi)$ on X^* is well-defined for symmetric ξ by $d^*(h(x), \xi) = d(x, \xi)$, and now has the form

(3.14) $$P(s, t, u) = as^3 + bs^2 + cs + d + est + ft + gu.$$

If this polynomial has a local maximum in the interior of X^*, then one may show $e = f = g = 0$ as before, and therefore one may conclude that the design is rotatable. (In fact, this conclusion clearly holds if (1.9) is replaced by an arbitrary compact symmetric q-dimensional set.)

We now extend the argument we used when $q = 2$, to conclude again that an optimum design ξ^* cannot be rotatable. Such a ξ^* would, by the same argument as before, be optimum for cubic regression on $K = \{x: x_1^2 + \cdots + x_q^2 \leq q\}$. Let β and $1 - \beta$ again denote the total mass assigned to the spheres $\{\sum_1^q x_i^2 = \rho\}$ and $\{\sum_1^q x_i^2 = 1\}$, by each optimum design for cubic regression on the unit ball (1.8). The integral of x_1^4, with respect to the uniform probability measure on $\{\sum_1^q x_i^2 = 1\}$, is now $3/q(q + 2)$. Also, just as before, $\int x^4\xi^*(dx) \leq 1$. Thus, the analogue of (3.10) and the second inequality of (3.11) is that, if ξ^* is an optimum design for K, then

(3.15) $$q^2[3/q(q + 2)][(1 - \beta) + \beta\rho^4] = \int x_1^4\xi^*(dx) \leq 1.$$

If one writes down equations from which the parameters of an optimum rotatable design on (1.8) may be calculated, then one obtains (see section 4, equation (4.5))

(3.16) $$\frac{(q + 3)(q + 2)(q + 1)}{6} = \frac{q + 1}{1 - \beta} + \frac{(q - 1)(q + 2)}{2((1 - \beta) + \beta\rho^4)} + \frac{(q + 4)q(q - 1)}{6((1 - \beta) + \beta\rho^6)}.$$

Since $\rho < 1$, we may replace ρ^6 by ρ^4 and also drop the first term on the right in (3.16), and may then divide both sides by $(q + 1)/6$, obtaining

(3.17) $$q^2 + 5q + 6 > [q^2 + 5q - 6]/[(1 - \beta) + \beta\rho^4].$$

Substituting this inequality for $1 - \beta + \beta\rho^4$ into (3.15) yields

(3.18) $$0 > 3q[q^2 + 5q - 6] - (q + 2)[q^2 + 5q + 6] = 2q^3 + 8q^2 - 34q - 12.$$

The last polynomial is easily seen to be positive for $q \geq 3$. We have thus proved theorem 3.1.

THEOREM 3.1 *For cubic regression on the q-cube* (1.9) *with* $q \geq 2$, *an optimum design cannot be rotatable.*

Thus, the problem reduces to a study of the nature of the boundary of X^* in the (s, t, u)-space and of the solution of the appropriate maximization problem. We do not attempt to do this in the present paper.

3.3. *Optimum symmetric designs for quadratic regression on the q-cube.* We have already given a short proof in the first paragraph of section 3.1, that all optimum designs for $m = 2$ on the q-cube (1.9) are supported by a subset of the 3^q-array J. For $j = 0, 1, \cdots, q$, let J^j be the subset of J consisting of those $2^{q-j}\binom{q}{j}$ points with exactly j coordinates equal to zero. (Thus, J^j consists of the midpoints of all j-dimensional faces of X.) Kiefer [7] obtained optimum designs supported by $J^0 \cup J^1 \cup J^2$ when $q \leq 5$, showed that this set could not support an optimum design when $q \geq 6$, and described ([7], footnote 5) the method, obtained with Farrell, for obtaining optimum designs for each q on the union of three J^i's and certain subsets of such a union by solving (3.22) below. This joint work is the subject of the present subsection. Subsequently, Kono [9], citing this description in [7], also showed that optimum symmetric designs for each q can only be supported by a subset of J, and carried out detailed calculations for an optimum design on the union of three J^i's when $q \leq 9$, obtaining optimum symmetric designs on $J^0 \cup J^1 \cup J^q$, again by solving equations (3.22) below, for that choice of the J^i's.

We shall first characterize those sets of the form

$$J(j_1, j_2, j_3) = \bigcup_{i=1}^{3} J^{j_i} \tag{3.19}$$

which can support a symmetric optimum design. We shall take $j_1 < j_2 < j_3$. (It can be seen that two J^i's cannot suffice when $q \geq 2$, by noting that no α_{j_i} can be 0 in the demonstration which follows.) Such a design assigns probability $\alpha_{j_i}/2^{q-j_i}\binom{q}{j_i} > 0$ to each point of $J^{j_i} (i = 1, 2, 3)$, where $\sum_1^3 \alpha_{j_i} = 1$. The pertinent moments of such a design are computed, as in (4.1) of [7], to be

$$\begin{aligned} u(\xi^*) &= \int x_1^2 \xi^*(dx) = \int x_1^4 \xi^*(dx) = \sum_1^3 \alpha_{j_i}(q - j_i)/q, \\ v(\xi^*) &= \int x_1^2 x_2^2 \xi^*(dx) = \sum_1^3 \alpha_{j_i}(q - j_i)(q - j_i - 1)/q(q - 1). \end{aligned} \tag{3.20}$$

Write as in (4.5) of [7],

$$\begin{aligned} U_q &= \frac{(q + 3)}{4(q + 1)(q + 2)^2} \{(2q^2 + 3q + 7) \\ &\qquad + (q - 1)[4q^2 + 12q + 17]^{1/2}\}, \\ V_q &= \frac{(q + 3)}{8(q + 2)^3(q + 1)} \{(4q^3 + 8q^2 + 11q - 5) \\ &\qquad + (2q^2 + q + 3)[4q^2 + 12q + 17]^{1/2}\}. \end{aligned} \tag{3.21}$$

One computes $M^{-1}(\xi^*)$, as in (4.3) of [7], and then observes, exactly as in [7], that $\bar{d}(\xi^*) = k$ if and only if

$$u(\xi^*) = U_q, \qquad v(\xi^*) = V_q. \tag{3.22}$$

To solve these equations for the j_i and α_{j_i}, we may think of plotting in the (u, v)-plane, for fixed $q \geq 2$, the $q + 1$ points

$$(u_j, v_j) = ((q - j)/q, (q - j)(q - j - 1)/q(q - 1)), \qquad 0 \leq j \leq q, \tag{3.23}$$

and the point (U_q, V_q). Then clearly (3.22) is satisfied for nonnegative j_i if and only if (U_q, V_q) lies in the triangle with vertices (u_{j_i}, v_{j_i}). Even though we shall not use any such geometric considerations in the demonstration which follows, they help in understanding the results, and also in understanding what is involved in considering unions of more than three J^i's, which we shall forego. (We also remark to the reader that the idea of the demonstration which follows, without such tedious computational details, can be obtained by replacing U_q and V_q by their asymptotic values $2 - q^{-1} + 3q^{-2}/2$ and $1 - 2q^{-1} + 5q^{-2}$, and following through the argument for "large q".)

We first note, replacing $[4q^2 + 12q + 17]^{1/2}$ by the smaller value $(2q + 3)$, that it is easy to verify that $U_q > (q - 1)/q \geq u_j, j > 0$, from which we conclude that $j_1 = 0$. Substituting this fact into (3.22) (and using $\sum_1^3 \alpha_{j_i} = 1$), we obtain

$$\begin{aligned} \alpha_{j_2} &= [q/j_2(j_3 - j_2)]\{-(q - 1)V_q + (2q - j_3 - 1)U_q - (q - j_3)\}, \\ \alpha_{j_3} &= [q/j_3(j_3 - j_2)]\{(q - 1)V_q - (2q - j_2 - 1)U_q + (q - j_2)\}. \end{aligned} \tag{3.24}$$

For fixed q, the expression $\{(q - 1)V_q - (2q - j - 1)U_q + (q - j)\} = F_q(j)$ (say) is linear and (since $U_q < 1$) decreasing in j. We shall show in the next two paragraphs that

$$F_q(2) > 0 > F_q(3) \qquad \text{for} \quad q > 5. \tag{3.25}$$

It follows then from (3.24) that (3.22) can be satisfied for $q > 5$ (with positive α_{j_i}'s) if and only if

$$0 < j_2 < 3 \leq j_3. \tag{3.26}$$

The first inequality of (3.25) can be written as

$$q - 2 > (2q - 3)U_q - (q - 1)V_q, \tag{3.27}$$

or

$$\begin{aligned} 8(q - 2)(q + 2)^3(q + 1)/(q + 3) > 4q^4 + 12q^3 + 7q^2 - 6q \\ - 89 + (2q^3 - q^2 - 16q + 15)[4q^2 + 12q + 17]^{1/2}. \end{aligned} \tag{3.28}$$

A direct computation shows that the left side of (3.28) equals

$$\begin{aligned} 8\{q^4 + 2q^3 - 2q^2 - 10q - 4 + (2q - 4)/(q + 3)\} \\ > 8q^4 + 16q^3 - 16q^2 - 80q - 32. \end{aligned} \tag{3.29}$$

Using the fact that $[4q^2 + 12q + 17]^{1/2} < 2q + 3 + 4/(2q + 3)$ and that $4(2q^3 - q^2 - 16q + 15)/(2q + 3) < 2q(2q - 3)$, we obtain that the right side of (3.28) is less than $8q^4 + 16q^3 - 24q^2 - 30q - 44$. This last is less than the right side of (3.29) by $(q - 6)(8q + 2)$, proving (3.28) (and thus (3.27)) for $q \geq 6$.

The second inequality of (3.25) can be written as

$$q - 3 < (2q - 4)U_q - (q - 1)V_q, \tag{3.30}$$

or

$$8(q - 3)(q + 2)^3(q + 1)/(q + 3) < 4q^4 + 8q^3 - 7q^2 - 32q - 117 + (2q^3 - 3q^2 - 18q + 19)[4q^2 + 12q + 17]^{1/2}. \tag{3.31}$$

The left side of (3.31) is

$$8\{q^4 + q^3 - 6q^2 - 16q - 4 - 12/(q + 3)\} < 8q^4 + 8q^3 - 48q^2 - 128q - 32. \tag{3.32}$$

Using the fact that $[4q^2 + 12q + 17] > 2q + 3 + 4/(2q + 3) - 16/(2q + 3)^3$ and that $[4/(2q + 3) - 16/(2q + 3)^3](2q^3 - 3q^2 - 18q + 19) > 4q^2 - 12q - 22$, we obtain that the right side of (3.31) is greater than $8q^4 + 8q^3 - 48q^2 - 60q - 82$. The latter is clearly greater than the right side of (3.32), proving (3.31) and thus (3.30).

In the same manner that (3.26) was proved (or by direct calculation in the few cases $q < 6$), one can show that (3.26) is replaced by $0 < j_1 < 2 \leq j_2$ for $2 \leq q \leq 6$. To summarize, then, we have the following theorem.

THEOREM 3.2. *The set $J(j_1, j_2, j_3)$ of* (3.19) *supports a symmetric optimum design for quadratic regression on the q-cube, if and only if*

$$\begin{aligned} &j_1 = 0, \quad j_2 = 1, \quad 2 \leq j_3 \leq q, && \text{when } 2 \leq q \leq 5, \\ &j_1 = 0, \quad j_2 = 1 \text{ or } 2, \quad 3 \leq j_3 \leq q, && \text{when } q \geq 6. \end{aligned} \tag{3.33}$$

We remark that, among the sets $J(0, j_2, j_3)$ permitted by (3.33), the set $J(0, 1, q)$ consisting of vertices, midpoints of edges, and center, has the smallest number of points $(2^q + q2^{q-1} + 1)$ of any optimum symmetric design. In view of (1.12), such designs are quite unsatisfactory for large q, and in the remainder of section 3 we shall therefore seek asymmetric designs on fewer points.

The weights α_{j_i} for any optimum symmetric design on a set (3.19) permitted by (3.33) may be obtained from (3.24) and $\alpha_0 = 1 - \alpha_{j_2} - \alpha_{j_3}$. For $j_2 = 1$, $j_3 = 2$, and $q \leq 5$, $\alpha_{j_i}/2^{q-j_i}\binom{q}{j_i}$ is tabled in [7]; for $j_2 = 1$, $j_3 = q \leq 7$, the α_j are tabled in [9].

3.4. *Bounds on N for quadratic regression on the q-cube.* We now apply the considerations of section 1.3 regarding the minimum number N of points needed to support an optimum design on the q-cube when $m = 2$. We first note

THEOREM 3.3. *The optimum design for quadratic regression on the q-cube is unique if and only if $q \leq 2$, in which case the support is J.*

PROOF. The lack of uniqueness when $q \geq 3$ follows from (3.33). The uniqueness when $q = 1$ or 2 (the former of which is well known) can be proved by using (1.16) with $U = J$; the matrix $\{\phi_t(x), x \in J, 1 \leq t \leq p\}$ is easily seen to have rank 3 or 9 in these two cases. (We recall that the subscript value $t = 0$ need not be included in polynomial regression.)

We next improve the upper bound $\binom{q+4}{4}$ of (1.12) by use of (1.14). We shall take $U = W = J(0, 1, q)$ in the calculation which follows. For $x \in J(0, 1, q)$ (in fact, for $x \in J$), the relations

$$(3.34)\qquad \begin{aligned} x_i &= x_i^3, \\ x_i^2 &= x_i^4, \\ x_i x_j &= x_i x_j^3, \qquad i \neq j \end{aligned}$$

for $1 \leq i, j \leq q$ are satisfied. These are $q^2 + q$ linearly independent relations among the $\phi_t(1 \leq t \leq p)$. Among the set of ϕ_t which remain after deleting those on the right side of (3.34), the relations

$$(3.35)\qquad \begin{aligned} (1 - x_j^2)(x_r^2 - x_s^2) &= 0, \\ (x_i^2 - x_j^2)(x_r^2 - x_s^2) &= 0, \end{aligned}$$

are satisfied on $J(0, 1, q)$ with all subscripts unequal and between 1 and q, inclusive. (Equalities (3.35) are vacuous if $q < 3$.) This is so because either all $x_i^2 = 0$, or else at most one $x_i^2 = 0$. The relations (3.35) among the ϕ_t are not linearly independent when $q \geq 3$, so we must find the dimension of the vector space spanned by the ϕ_t. To this end, we write $L = q + 1$, $y_i = x_i^2$ for $1 \leq i \leq q$, and $y_{L+1} = 1$.

For $L \geq 4$, let Q be the vector space over the reals of all linear combinations of the polynomials $(y_i - y_j)(y_r - y_s)$ with i, j, r, s distinct integers between 1 and L, inclusive (a subspace of the quadratic polynomials in L variables). We shall show the following lemma.

LEMMA 3.4. *For $L \geq 4$, we have*

$$(3.36)\qquad \dim Q = L(L - 3)/2.$$

PROOF. All subscripts in the proof which follows are to be reduced mod L. We first show that the $L(L - 3)/2$ special polynomials

$$(3.37)\qquad (y_j - y_{j+1})(y_{j+i+1} - y_{j+i+2}),$$

with all subscripts distinct, span Q. (Note, for example, that $j = L$ is permitted.) We must show that any polynomial $(y_i - y_j)(y_r - y_s)$ is a linear combination of these special polynomials. There are two cases to which any other can be reduced by symmetry.

Case 1 $(i < j < r < s)$. Then

$$(3.38)\qquad (y_i - y_j)(y_r - y_s) = \sum_{0 \leq u < j-i, 0 \leq v < s-r} (y_{i+u} - y_{i+u+1})(y_{r+v} - y_{r+v+1}).$$

Case 2 $(i < r < j < s)$. Use the identity

$$(3.39)\qquad (y_i - y_j)(y_r - y_s) = (y_i - y_r)(y_j - y_s) - (y_r - y_j)(y_s - y_i)$$

to reduce to case 1.

To conclude the proof of (3.36), we need only show that the special polynomials (3.37) are linearly independent. To this end, we obtain an appropriate ordering of the special polynomials, say $\{g_\alpha, 1 \leq \alpha \leq L(L-3)/2\}$, and of the functions $y_i y_j (i \neq j)$, say $\{h_\beta, 1 \leq \beta \leq L(L-1)/2\}$; then writing $g_\alpha = \sum_\beta c_{\alpha\beta} h_\beta$, we show that $c_{\alpha\alpha} \neq 0$ and $c_{\alpha\beta} = 0$ for $\alpha > \beta$, which proves the desired result. The g_α are, in order, the special polynomials of (3.37) with $i = 1$ and $j = 0, 1, \cdots, L-1$; then, with $i = 2$ and $j = 0, 1, \cdots, L-1; \cdots, i = \{$greatest integer $\leq (L-3)/2\}$ and $j = 0, 1, \cdots, L-1$; if L is even, there are then $L/2$ additional special polynomials with $i = (L-2)/2$ and $j = 0, 1, \cdots, (L-2)/2$. Note that those functions in the ith collection of L polynomials (or of $L/2$ if L is even and $i = (L-2)/2$) have i as the minimum distance between subscripts γ, δ which appear in any term $y_\gamma y_\delta$ entering with nonzero coefficient in the special polynomial. Moreover, for fixed i, these $y_\gamma y_{\gamma+i}$ appear in the order $\gamma = 1, 2, \cdots, L$. Thus, when we order the h_β as $y_j y_{j+1}$ for $j = 1, 2, \cdots, L$, and then $y_j y_{j+2}$ for $j = 1, \cdots, L$, and so on, we see at once that the $c_{\alpha\beta}$ have the desired property. This completes the proof of the lemma.

Putting $L = q + 1$ in (3.36), we conclude that the ϕ_t on the left side of (3.35) span a real vector space of dimension $(q+1)(q-2)/2$ (which is also correct if $q = 2$). Adding this to the number $q^2 + q$ of restrictions (3.34), which are independent of (3.36), and subtracting the result from $p = \binom{q+4}{4}$ and using (1.14) with $\{0 \leq t \leq p\}$ replaced by $\{1 \leq t \leq p\}$, we obtain theorem 3.5.

THEOREM 3.5. *For quadratic regression on the q-cube,*

$$N \leq (q+1)(q^3 + 9q^2 - 10q + 48)/24. \tag{3.40}$$

3.5. *The use of orthogonal arrays to reduce the number of points of support.* We have already mentioned in section 1 how orthogonal arrays of strength 2 are used classically to reduce the number of points of support for an optimum design for linear regression on the q-cube. Similar techniques can be employed in other settings, as we now illustrate for quadratic regression on the q-cube. We shall consider the following particular scheme of application.

Suppose, for each positive integer r, that we can find a subset A_r of the 2^r corners of the r-cube ((1.9) with $q = r$), such that the uniform probability measure on A_r has the same moments of order ≤ 4 as the uniform probability measure on the 2^r corners. Suppose A_r has n_r points. Then, suppose we replace J^0 in $J(0, 1, q)$ by A_q (with α_0/n_q probability per point); replace the 2^{q-1} points of J^1 with $x_i = 0$ by the n_{q-1} points of the form

$$x_i = 0, \qquad (x_i, \cdots, x_{i-1}, x_{i+1}, \cdots, x_q) \in A_{q-1},$$

for $1 \leq i \leq q$ (with probability α_1/qn_{q-1} per point); retain J^q, with probability α_q. Since only moments of order ≤ 4 are present in $M(\xi)$, and because of the way in which zero coordinate values enter into the replacement of J^1, we obtain a design with the same M as the optimum symmetric design on $J(0, 1, q)$, and which is therefore also optimum. It is supported by

$$n_q + qn_{q-1} + 1 \tag{3.41}$$

points.

The classical construction of such an A_r is in terms of an *orthogonal array of strength 4 with 2 levels*, that is, an $n_r \times r$ matrix T_r with entries ± 1 such that every 4-row submatrix has the property that each of the 16 possible 4-vectors with entries ± 1 appears equally often in the submatrix. We then consider the n_r columns of the matrix T_r as the points of A_r and clearly obtain the required moment properties. The reader is referred to such references as [1] for detailed discussion of orthogonal arrays.

Orthogonal arrays of strength ≥ 3 have been considered extensively by Rao, Bose, Bush, Seiden, and others. The principal method of construction is geometric. A set C of r points in the finite projective space $PG(d-1, 2)$ of dimension $d-1$, no 4 of which lie on the same 2-dimensional flat, yields a T_r with $n_r = 2^d$; for, writing B for the $r \times d$ matrix whose rows are the points of C (each of the d coordinates of such a point being an element of the Galois field $GF(2)$), and writing D for the $d \times 2^d$ matrix whose columns are the different d-vectors with coordinate values in $GF(2)$, one sees easily that BD has the required properties of T_r, except that the elements ± 1 of T_r are replaced by 0, 1 (of $GF(2)$) in BD. (It is not always known when this geometric construction yields the maximum r for given d.)

For fixed r, Rao's lower bound on n_r, usually given in geometric terms, can be obtained for general orthogonal arrays $T_n = \{t_{ij}, 1 \leq i \leq r, 1 \leq j \leq n_r\}$ of strength 4 with elements ± 1, as follows. Let τ_0 be the row vector of n_r 1's; let τ_i be the i-th row of T_r, and for $1 \leq i < i' \leq r$ let $\tau_{i,i'} = (t_{i1}t_{i'1}, \cdots, t_{in_r}t_{i'n_r})$. The $1 + r + \binom{r}{2}$ vectors $\tau_0, \tau_1, \cdots, \tau_r, \tau_{12}, \cdots, \tau_{(r-1)r}$ are easily shown to be orthogonal, because of the properties of T_r. Hence,

$$n_r \geq (r^2 + r + 2)/2. \tag{3.42}$$

In the other direction, it is simple to give a geometric construction which yields an orthogonal array of strength 4 satisfying

$$n_r = \text{largest number } 2^d \text{ which is } \leq 1 + r + \binom{r}{2} + \binom{r}{3}. \tag{3.43}$$

For, in $PG(d-1, 2)$, if we have chosen j points, no 4 of which are coplanar, there are $\binom{j}{2}$ pairs of points each of which determines a line with one point outside the pair, and $\binom{j}{3}$ triples of points, each of which determines a plane with one point not on the lines just mentioned. Thus, as long as $j + \binom{j}{2} + \binom{j}{3} < 2^d - 1$ (equal to the number of points in $PG(d-1, 2)$), there remains a $(j+1)$*st* point which can be chosen without destroying noncoplanarity. Continuing in this way, we can obtain r points, where r is the smallest integer for which $r + \binom{r}{2} + \binom{r}{3} \geq 2^d - 1$. This yields (3.43).

The reader should have no trouble in writing down analogues of (3.42) and (3.43) for other Galois fields, and, in fact, for arrays of different strength.

When q is large, the use of (3.41) with (3.43) yields an optimum design on $\leq q^4(1 + o(1))/3$ points. This is $O(q^4)$ like p or the bound (3.40), but these last are both $q^4(1 + o(1))/24$. Thus, we do not know whether or not the orthogonal array approach can, for large q, yield a design with no more points than (3.40), let alone whether the order r^3 of (3.43) rather than the order r^2 of (3.42) (or neither) is the best possible as $r \to \infty$. We do know that there are some small values of q for which the method of using orthogonal arrays cannot yield a value of (3.41) which is less than (3.40) or even p. This is a consequence of the fact ([10], [12]) that the minimum possible values of n_r for $r = 4, 5, 6, 7, 8$ are known to be 16, 32, 32, 64, 64, so that the numbers listed in the last column of table I below, and which were obtained by using these values in (3.41),

TABLE I

q	k	p	(3.40)	Points in $J(0, 1, q)$	Achievable Using (3.41)
2	6	15	9	9	9
3	10	35	21	21	21
4	15	70	45	49	49
5	21	126	87	113	113
6	28	210	154	257	225
7	36	330	254	577	289
8	45	495	396	1281	577
9	55	715	590	2817	705
10	66	1001	847	6145	1409
11	78	1365	1179	13313	1537
12	91	1820	1599	28673	1793
13	105	2380	2121	61441	3585
16	153	4845	4454	589825	4353
17	171	5985	5544	1245185	4608
Asymptotic Value	$q^2/2$	$q^4/24$	$q^4/24$	$q2^{q-1}$	$\leq q^4/3$

cannot be improved upon by using orthogonal arrays for $q \leq 8$. We have also used the values $n_r = 128$ for $9 \leq r \leq 11$ and $n_r = 256$ for $12 \leq r \leq 17$ in this table. These are the best values obtainable geometrically [12], but it is not yet known whether a nongeometric construction can yield better orthogonal arrays in these cases. (For values like $q = 10$ or 13, where $n_{q-1} > n_{q-2}$, the number obtained from (3.41) is at its worst compared with p or (3.40); similarly, for $q = 8$, 9, 11, 12, and 17, the comparison is more favorable.)

We are indebted to Professor Esther Seiden for several communications concerning the construction of these orthogonal arrays of strength 4.

In view of the unattainability of p or (3.40) for some values q by using the method of this subsection, it is clear that further study is needed of designs which have less symmetry. For example, by considering nonuniform measures on smaller sets than A_r, and perhaps subsets of more than three J^i's, one should be able to do considerably better. Perhaps one can even reduce the number of points required from $0(q^4)$ to a smaller order such as $0(q^3)$. One other obvious attempt to obtain $0(q^3)$ is to seek an optimum design with equal mass on each point of J, plus additional masses on J^0 and J^q, and thus to replace the q arrays used with J^1 in (3.41) by an orthogonal array of strength 4 with three levels, which is used in place of J (the n_q and 1 being present in (3.41) as before). The analogue of (3.43) for $GF(3)$ shows that this three-level array again requires only $0(q^3)$ points, so that (3.41) would yield $0(q^3)$. Unfortunately, one cannot choose positive probabilities on J^0, J, and J^q so as to satisfy the analogue of (3.22) for large q.

4. The q-ball

We now suppose X to be the unit q-ball (1.8). For regression of degree m, a rough characterization of optimum designs was given by Kiefer [6]): every optimum ξ assigns measure one to $(m + 1)/2$ spherical shells centered at 0, where one of these shells is the boundary of X and where 0 counts as $\frac{1}{2}$ shell. Some weighted mixture of uniform measures on these shells is optimum (although other measures with the same first $2m$ moments are also optimum). The weights and radii of shells are hard to compute for $m > 2$; when $m = 2$, measure $2/(q + 1)(q + 2)$ is assigned to the origin and the remainder is assigned to the boundary of X.

Two problems of interest here are (1) to obtain at least approximate information on the radii and weights when $m > 2$, and (2) to obtain discrete measures on the shells supported by as few points as possible. In most of the remaining paragraphs of this section we shall indicate the type of treatment of problem (1) which is possible for $m > 2$, considering here the example $m = 3$. Problem (2) entails considerations related to those of section 3 and also to the extensive literature on the construction of rotatable designs. It differs from the latter in its specification of the radii and weights and in its allowing of unequal masses on points which may not be symmetrically spaced. The implementation of the resulting optimum designs of the approximate theory for specified sample sizes by discrete designs which approximate them, will yield nonrotatable designs which can be expected to involve fewer distinct points and to have better performance characteristics than the rotatable designs which are usually used. The payment for this in the form of a design matrix which is harder to invert may be worthwhile with modern computing equipment.

An optimum design ξ^* in the case $m = 3$ on the q-ball can be described in terms of two parameters: measure β is spread uniformly on a sphere of radius $\rho < 1$, and measure $1 - \beta$ is assigned to the unit sphere (equal to the boundary of X). For such a design, $d(x, \xi^*)$ depends only on $r^2 = \sum x_i^2$, say $d(x, \xi^*) = d^*(r, \xi^*)$. The optimum ρ and β can be found either by solving the two equations $d^*(\rho, \xi^*) = \binom{q+1}{3}$ $(=k)$ and $\partial d^*(r, \xi^*)/\partial r|_{r=\rho} = 0$ (see [6]), or else by maximizing $\det M(\xi)$ with respect to ρ and β. We shall exhibit the second method.

Grouping the functions into four sets as $\{1, x_1^2, \cdots, x_k^2\}$, $\{x_i, x_i^3, x_ix_j^2; i \neq j\}$,

$\{x_ix_j; i < j\}$, $\{x_hx_ix_j; h < i < j\}$, one sees that the product of two elements from different sets has zero integral. Thus, for ξ of the specified form, det $M(\xi)$ can be evaluated as the product of four determinants; one obtains, with C_q denoting a constant depending on q,

$$\log \det M(\xi) = C_q + 2q \log \rho + (q + 1) \log [\beta(1 - \beta)(1 - \rho^2)^2] + \frac{(q + 2)(q - 1)}{2} \log [(1 - \beta) + \beta\rho^4] + \frac{(q + 4)q(q - 1)}{6} \log [(1 - \beta) + \beta\rho^6]. \tag{4.1}$$

The two equations obtained by setting equal to zero the derivatives with respect to each of ρ and β, are not very manageable analytically. (This is also true of the equations obtained by the other approach mentioned in the previous paragraph.) These equations, however, can be solved easily by machine, and the results of this computation on the Cornell Computing Center CDC 1604 are given in table II below. We note here that the behavior of the maximizing values of ρ_q and β_q (say) as $q \to \infty$ are easily discernible from (4.1). A routine analysis shows that $\beta_q = hq^{-2} + o(q^{-2})$ and $\rho_q = \rho^* + o(1)$ where $0 < q < \infty$ and $0 < \rho^* < 1$ and where h and ρ^* maximize the coefficient of q in (4.1):

$$\begin{aligned} h^{-1} + (1 - \rho^{*6})/6 &= 0, \\ 2/\rho^* - 4\rho^*/(1 - \rho^{*2}) + h\rho^{*5} &= 0. \end{aligned} \tag{4.2}$$

Thus, as $q \to \infty$,

$$\begin{aligned} \rho_q^2 \sim \rho^{*2} &= (3^{1/2} - 1)/2 = .3660254, \\ \beta_q \sim hq^{-2} &= 4q^{-2}(1 + 3^{-1/2}) = 6.309401q^{-2}. \end{aligned} \tag{4.3}$$

A finer analysis can be used to produce further terms in an asymptotic expansion.

We digress in this paragraph to derive a result which was used in section 3.2. The matrix $M(\xi)$ may be inverted explicitly for ξ of the form we have been considering, the answer being expressed in terms of ρ_q and β_q. This allows one to write an expression for

$$\begin{aligned} d^*(r, \xi^*) = {} & \frac{(1 - \beta_q)(1 - r^2)^2 + \beta_q(r^2 - \rho_q^2)^2}{(1 - \beta_q)\beta_q(1 - \rho_q^2)^2} \\ & + q\,\frac{(1 - \beta_q)r^2(1 - r^2)^2 + \beta_q r^2\rho_q^2(r^2 - \rho_q^2)^2}{(1 - \beta_q)\beta_q\rho_q^2(1 - r_q^2)^2} \\ & + \frac{(q + 2)(q - 1)}{2}\,\frac{r^4}{(1 - \beta_q) + \beta_q\rho_q^4} \\ & + \frac{(q + 4)q(q - 1)}{6}\,\frac{r^6}{(1 - \beta_q) + \beta_q\rho_q^6}. \end{aligned} \tag{4.4}$$

Taking $r = 1$ gives (since $d^*(1, \xi^*) = k$)

$$\frac{(q + 1)(q + 2)(q + 3)}{6} = \frac{q + 1}{1 - \beta_q} + \frac{(q + 2)(q - 1)}{2(1 - \beta_q) + \beta_q\rho_q^4} + \frac{(q + 4)q(q - 1)}{6(1 - \beta_q) + \beta_q\rho_q^6}. \tag{4.5}$$

Equation (4.5) was used as (3.16) in section 3.2 to show that optimum designs for cubic regression on the q-cube could not be rotatable.

The following table of numbers for β_q and ρ_q were computed as described above. (Of course, for $q = 1$ we have the Guest-Hoel design.)

TABLE II

q	β_q	$q^2\beta_q$	ρ_q^2
1	0.5000	0.500	0.2000
2	0.3077	1.231	0.2657
3	0.2455	2.210	0.2970
4	0.1695	2.712	0.3142
5	0.1241	3.102	0.3249
6	0.09483	3.414	0.3321
7	0.07490	3.670	0.3373
8	0.06068	3.884	0.3412
9	0.05019	4.065	0.3442
10	0.04221	4.221	0.3465
10^2	0.6020×10^{-3}	6.020	0.364381
10^3	0.6279×10^{-5}	6.279	0.365866
10^4	0.6306×10^{-7}	6.306	0.366010
10^5	0.6309×10^{-9}	6.309	0.366024
∞		6.309401	0.3660254

From a practical point of view, what is important for other examples (for instance, larger m on the ball) is the indication that the use of the limiting values h and ρ^* for fairly small values of q leads to a value of $\max_x d(x, \xi)$ which is not too large; this aspect deserves further machine study in other contexts.

For $m \geq 4$, the same approach can be used, but of course the larger number of parameters makes the analysis messier, especially if $q > 2$. We remark that when $m = 4$, $q = 2$, the optimum weights are $\frac{1}{15}$, 0.343912, and 0.589422, at $r^2 = 0$, 0.460249, and 1, respectively.

In order to construct implementable optimum designs on the unit ball for any m, we replace the uniform distribution on each spherical shell by a distribution on a finite subset of the same shell, with the same moments. It will suffice to consider the shell of radius one. If the measure γ assigns mass α to each of the 2^q points having coordinates $\pm q^{-1/2}$, and mass β to each of the $2q$ points $(\pm 1, 0, \cdots, 0)$, $(0, \pm 1, \cdots, 0)$, $\cdots$, $(0, 0, \cdots, \pm 1)$, then the values $\alpha = q(q+2)^{-1}2^{-q}$, $\beta = 1/q(q+2)$ satisfy

$$\alpha + \beta = 1,$$

$$\int x_1^2 \gamma(dx) = 1/q, \tag{4.6}$$

$$\int x_1^2 x_2^2 \gamma(dx) = 1/q(q+2),$$

$$\int x_1^4 \gamma(dx) = 3/q(q+2).$$

Clearly all other moments of order >0 and <4 are zero.

The set of 2^q points with all coordinates $\pm q^{-1/2}$ may be replaced by a subset which is an orthogonal array of strength 4 consisting of $0(q^3)$ points, by (3.43). Thus we know how to construct optimum designs on $0(q^3)$ points. As in section 3.4, if the order of (3.42) were attainable, we could achieve $0(q^2)$ here; and perhaps less symmetric points and weights can also help to achieve a lower order than $0(q^3)$.

Using this method of construction with orthogonal arrays, and data provided by E. Seiden and described in section 3.4, we obtain the following table III giving the number of points of support for ξ in optimum designs for $q = 3, \cdots, 17$, when $m = 2$. There being one point at the origin, we obtain a design on

(4.7) $$2q + 1 + n_q$$

points in this case. The values, of course, compare favorably with those of table I, where the same values of k and p apply.

TABLE III

q	(4.7)
3	23
4	25
5	43
6	45
7	79
8	81
9	147
10	149
11	151
12	281
13	283
14	285
15	287
16	289
17	291

That these designs may not be the best possible, even among designs of quite symmetric construction, is illustrated by an example of Box and Behnken [2]. Using their construction for $q = 7$, one obtains a design with 56 points on the unit sphere plus one at the origin, for a total of 57 points.

REFERENCES

[1] R. C. Bose and K. A. Bush, "Orthogonal arrays of strength two and three," *Ann. Math. Statist.*, Vol. 23 (1952), pp. 508–524.

[2] G. E. P. Box and D. W. Behnken, "Simplex-sum designs: a class of second order rotatable designs derivable from those of first order," *Ann. Math. Statist.*, Vol. 31 (1960), pp. 838–864.

[3] P. G. Guest, "The spacing of observations in polynomial regression," *Ann. Math. Statist.*, Vol. 29 (1958), pp. 294–299.

[4] P. G. Hoel, "Efficiency problems in polynomial regression," *Ann. Math. Statist.*, Vol. 29 (1958), pp. 1134–1145.

[5] J. Kiefer, "Optimum experimental designs," *J. Roy. Soc. Ser. B*, Vol. 21 (1959), pp. 273–319.

[6] ———, "Optimum experimental designs V, with applications to systematic and rotatable designs," *Proceedings of the Fourth Berkeley Symposium on Mathematical Statistics and Probability*, 1961, Vol. 1, pp. 381–405.
[7] ———, "Optimum designs in regression problems II," *Ann. Math. Statist.*, Vol. 32 (1961), pp. 298–325.
[8] J. Kiefer and J. Wolfowitz, "The equivalence of two extremum problems," *Canad. J. Math*, Vol. 12 (1960), pp. 363–366.
[9] K. Kono, "Optimum design for quadratic regression on the k-cube," *Mem. Fac. Sci. Kyushu Univ. Ser. A.*, Vol. 16 (1962), pp. 114–122.
[10] E. Seiden and R. Zemach, "On orthogonal arrays," to be published in 1965.
[11] H. Uranisi, written communication concerning the special cubic on the *q*-simplex, 1962.
[12] R. Zemach, "On orthogonal arrays of strength four and their application," Michigan State University N. I. H. Report.

Reprinted from
Proc. Fifth Sympos. Math. Statist. and Probability
1, 113–138 (1967)

Actes, Congrès intern. Math., 1970. Tome 3, p. 249 à 254.

OPTIMUM EXPERIMENTAL DESIGNS

by J. KIEFER

1. Introduction.

From the mathematical viewpoint, optimum design theory studies the geometry of generalized moment spaces $\mathfrak{M}$ of certain collections Ξ of probability measures. From the practical viewpoint, the consequence is often to find experimental designs which are more efficient than those which were used classically in statistics ; such usage was motivated by considerations of computation which are much less relevant with modern machines, or by such intuitively attractive properties of certain combinatorial designs as symmetries which need not guarantee optimality.

We will try to mention some developments obtained since the subject was revived almost 20 years ago by Mood, Elfving, and Chernoff after a 30-year pause from the early results of K. Smith, and will select results which lead to interesting and open mathematical questions. For brevity, we will not list all references in detail, but only a few whose bibliographies list most of the others. Even so, there is not space to mention many important areas of development, in particular designs for nonlinear models and sequential designs.

Let $f = (f_1, f_2, \ldots, f_k)$ where the f_i are continuous real functions on a compact space $\mathcal{X}$. A point $\theta = (\theta_1, \theta_2, \ldots, \theta_k)$ in R^k is unknown and its value is the concern of the statistician, who must choose an element (= "exact design") $x^* = (x_1, x_2, \ldots, x_N)$ in a specified subset $\mathcal{X}^*$ of $\mathcal{X}^N$. He then observes an N-vector $Y(x^*)$ of uncorrelated *r.v.*'s $Y_1, Y_2, \ldots, Y_N$ with common (known or unknown) variance σ^2, and with Y_i having expectation $(f(x_i), \theta) = \sum_{j=1}^{k} \theta_j f_j(x_i)$. For general statistical decision problems, the exact specification of the possible probability laws of $Y(x^*)$, not just of the possible values of the first two moments, is necessary ; moreover, in even very simple settings, the choice of design x^* and the possibility that randomization among designs may be called for, will depend strongly on the decision space (see [5]). However, we restrict attention here to problems of *point estimation* of a collection of linear forms $\{(c, \theta), c \in \mathcal{C}\}$, where $\mathcal{C}$ is a specified datum of the problem. Moreover, we will be concerned with nonrandomized linear estimators $(Y(x^*), h_c) = \sum_{1}^{N} h_{ci} Y_i$ of the (c, θ), and of the first and second moments of these estimators ; this can be justified (see [5], [8]) in Gaussian and certain nonparametric settings by an appeal to invariance, minimax, or various other principles for optimality criteria based on risk functionals of expected quadratic forms in the estimation errors. These restrictions reduce considerations to the first two moments of $Y(x^*)$.

The matrix A_{x*} of elements

(1) $$a_{x*rs} = \sum_{i=1}^{N} f_r(x_i) f_s(x_i)$$

is called the *information matrix* of the design x^*. *We shall treat here only problems where the performance of a design x^* is measured by a functional of A_{x*}*. This is motivated by the well known fact that the variance $\sigma^2 v(c, x^*)$ of the best linear unbiased estimator (= least squares or Gauss-Markov estimtor) of (c, θ) is $\sigma^2(c, A_{x*}^{-1} c)$ if A_{x*} is nonsingular, with an obvious analogue if A_{x*} is singular but (c, θ) has a linear unbiased estimator. (If no such estimator exists, we define $v(c, x^*) = +\infty$). Thus, duplicating a standard notion of statistical decision theory, we define x^* to be *at least as good as* x^{**} for $\mathcal{C}$, abbreviated $x^* \succ x^{**} (\mathcal{C})$, if

(2) $$v(c, x^*) \leqslant v(c, x^{**}) \ \forall\ c \in \mathcal{C}.$$

In particular, for $\mathcal{C} = R^k$,

(3) $$x^* \succ x^{**}(R^k) \Leftrightarrow A_{x*} - A_{x**} \geqslant 0,$$

where we have written $A \geqslant 0$ if A is nonnegative definite. (In certain models considered by Box, Draper, and others, where restriction of the h_c to certain subspaces of R^N results in biased estimators, a functional of A_{x*} other than v is appropriate.)

Admissibility and complete classes of designs are defined as in decision theory, using maximality under the ordering $\succ$ of (2). If $\mathcal{C}$ is not a single element (or a set of multiples thereof), the partial ordering induced by (2) typically yields a "minimal complete class" consisting of more than one maximal A_{x*}, and in any particular application one must then choose the design x^* by reference to an optimality criterion induced by a real functional Φ on the space of variance functions $v(\cdot\,; x^*)$ on $\mathcal{C}$: We say x^* is *Φ-optimum for* $\mathcal{C}$ if it minimizes $\Phi(v(\cdot\,; x^*))$ over $\mathfrak{X}^*$.

The above developments, and those which follow, allow considerable generalization. Results like those below (for example, Theorem 2) can still be obtained when Y_i and $(f(x_i), \theta)$ are replaced by vectors, often when $\mathfrak{X}$ is not compact, and also when the cost per observation is not constant as implied above, and the covariance matrix of $Y(x^*_N)$ is not of the form $\sigma^2 I_N$. (Explicit characterization of Φ-optimum designs in terms of a given covariance function on $\mathfrak{X} \times \mathfrak{X}$, when $\mathfrak{X}$ is a real interval, is obtained in a series of penetrating papers by Sacks and Ylvisaker.) But the most interesting questions are in fact often best understood in terms less of generalities than of some very particular examples we shall discuss in the next two sections.

2. The approximate theory.

Often there is no restriction among the x_i's and hence $\mathfrak{X}^* = \mathfrak{X}^N$. Defining the discrete probability measure ξ_{x*} on $\mathfrak{X}$ by $N\xi_{x*}(x) =$ [number of i such that $x_i = x$], we see that $N^{-1} A_{x*}$ equals

$$M(\xi) = \int_{\mathfrak{X}} f(x)' f(x)\, \xi(dx) \tag{4}$$

with $\xi = \xi_{x*}$. Let Ξ denote the set of all probability measure on $\mathfrak{X}$ relative to a σ-field which includes all one-point sets, and let Ξ_N denote the probability measures with range contained in the multiples of $1/N$. In the discussion of (2), (3), and the paragraph following them, we can clearly replace $\{A_{x*}, x^* \in \mathfrak{X}^N\}$ by $\{M(\xi), \xi \in \Xi_N\}$. If we replace Ξ_N by Ξ in these considerations, we have what is called the ***approximate theory***, and an element of Ξ is called an ***approximate design***. Whereas the exact design theory usually presents difficult combinatorial questions with solutions which exhibit a fine-structure dependence on N, the approximate theory often admits simple computational algorithms as discussed below ; and a single Φ-optimum approximate design ξ^* can easily be translated into an exact design for each N which, for suitably regular Φ (such as Φ_1 and Φ_2 below), minimizes $\Phi((\nu(\cdot, x^*))$ to within a relative error of $O(N^{-2})$.

The set $\mathfrak{M} = \{M(\xi), \xi \in \Xi\}$ can be viewed as the convex moment space associated with the functions $f_i f_j$, $1 \leqslant i \leqslant j \leqslant k$ on $\mathfrak{X}$. The admissible designs are certain boundary points of $\mathfrak{M}$ which can generally be obtained (Elfving, also in [5] with misprints, and [3]) as measures supported by a subset of $\mathfrak{X}$ on which $(f(x), f(x)B)$ attains its maximum for some matrix $B \geqslant 0$; as in other such characterizations, if B is not *positive* definite additional criteria are required. Of great interest is the question of the minimum number L such that, given any ξ, there is a ξ' which is (i) at least as good or (ii) at least as Φ-good as ξ, supported by at most L points. Simple examples [1] show that L can be as bad as the obvious bound $k(k+1)/2$ for quite common Φ. In the case of certain well known moment spaces it is of course known that a smaller L suffices, and sometimes one can be even more precise about admissibility :

THEOREM 1 [5]. – *If* $\mathfrak{X} = \{x : -1 \leqslant x \leqslant 1\}$ *and* $f_i(x) = x^{i-1}$, *then* ξ *is admissible if and only if the open interval* $(-1, 1)$ *contains at most* $k-2$ *points of the support of* ξ.

This has been extended to weighted polynomials [3], splines, and other settings by Ehrenfeld, Karlin, Studden, Van Arman, Murty. But if f is a vector of polynomials on a Euclidean set $\mathfrak{X}$ of dimension > 1, the problem is still far from solved [2].

Turning to specific optimality criteria Φ, among those most often encountered are (expressed as functionals on $\mathfrak{M}$) :

$$\begin{aligned} &\Phi_1(M) = tr\, BM^{-1} && (B \geqslant 0), \\ &\Phi_2(M) = \det M^{-1} && (\text{“}D\text{-optimality”}), \\ &\Phi_3(M) = \max_{c \in \mathcal{C}}\ (c, cM^{-1}). && \end{aligned} \tag{5}$$

When B is of rank 1, a Φ_1-optimum approximate design yields optimum estimation of (c, θ) where $c'c = B$, and it was shown in [8] that this is equivalent to a Chebyshev approximation problem.

When $\mathcal{C} = \{f(x), x \in \mathfrak{X}\}$, a Φ_3-optimum approximate design minimizes the maximum over $\mathfrak{X}$ of the variance of the best estimator of $(f(x), \theta)$. In this case, we have [9].

THEOREM 2. – *If* $\mathcal{C} = \{f(x),\ x \in \mathcal{X}\}$, *then, in the approximate theory,*

$$\text{(6)} \quad \xi^* \text{ is } \Phi_3\text{-}optimum \Leftrightarrow \xi^* \text{ is } \Phi_2\text{-}optimum \Leftrightarrow \max_{c \in \mathcal{C}} (c, c M^{-1}(\xi^*)) = k.$$

This turns out to yield a useful computational technique, especially the last equivalence. In [7] an analogous technique was obtained when $\Phi_2(M)$ is replaced by the determinant of a principal $s \times s$ minor of M^{-1}, corresponding to concern with s out of the k θ_i's. An alternative technique was given in [3] (corrected in [1]), but it remains to obtain a more useful algorithm than either of these, which are often much more difficult to apply than Theorem 2 to which they reduce when $s = k$. (When $s = 1$ they reduce to the Chebyshev problem mentioned above.)

There are many interesting optimality questions in even the simple model where f consist of polynomials on a d-dimensional ball, cube, or simplex [2]. In particular, the simplex yields the most elegant characterizations, Φ_2-optimum designs seeming to possess a regularity as d increases which is absent for the ball or cube. Uranisi and Atwood obtained related results.

If we think of $\mathcal{X}$ as a subset of $\bar{\mathcal{X}}$ on which f is defined, and if $\tilde{\mathcal{X}}$ is a subset of $\bar{\mathcal{X}}$, then Φ_3-optimality with $\mathcal{C} = \{f(x),\ x \in \tilde{\mathcal{X}}\}$ refers to *extrapolatory* estimation of $(f(x), \theta)$ on $\tilde{\mathcal{X}}$. In [10] this was studied for the univariate polynomial model of theorem 1 when $\tilde{\mathcal{X}} = [-a, a]$, but the solution is difficult except when $a \to 0$ or $a \to 1$ or $a \to +\infty$. (This was recently extended to dimension $d > 1$). Hoel and Levine discovered the striking fact that the less symmetric problem $\tilde{\mathcal{X}} = \{a\}$ (or sometimes $\tilde{\mathcal{X}} = [-1, a]$) with $a > 1$ has a much more elegant solution, the Φ_3-optimum design being supported by the same "Chebyshev set" Λ_{k-1} (say) that supported the Φ_1-optimum design when $c = (0, \ldots, 0, 1)$. This result was extended to Chebyshev systems f in [11], where the set of all c for which the Φ_1-optimum design (with $B = c'c$) is supported by Λ_{k-1} is characterized. Among more recent results, we mention Studden's characterization for the univariate polynomial case (and certain other Chebyshev systems), that the optimum design for estimating θ_{k-i} is supported by Λ_{k-1} or Λ_{k-2} depending on whether i is odd or even ($i < k$).

We mention finally, in the approximate theory, that there is a simple invariance result [5], [6] : Often there is a compact group G which operates on $\{\mathcal{X}, f, \Phi\}$ in such a manner that, because of the convexity in α of

$$M^{-1}(\alpha\xi_1 + (1-\alpha)\xi_2)$$

and of $-\Phi$ (or an increasing function of it), there is a Φ-optimum ξ which is G-invariant ($\xi(A) = \xi(gA)$, $g \in G$). Thus, for f consisting of all polynomials of degree $\leq m$ on the d-ball, one concludes that there is a Φ_2-optimum design consisting of multiples of Lebesgue measure on $(m+1)/2$ spheres of dimension $d-1$, where the origin counts as half a sphere [6]. What remains is then the more difficult problem of implementing this with a design of small finite support on those spheres, and with the same relevant moments [2].

3. Exact theory.

We illustrate the ideas with a setting in which combinatorial block designs are often used, the model of *2-way heterogeneity without interactions.* Here $N = k_1 k_2$ where the k_i are positive integers, and $k = k_1 + k_2 + u$ with $u > 1$. We rewrite $\theta = (\alpha_1, \ldots, \alpha_{k_1}, \beta_1, \ldots, \beta_{k_2}, \gamma_1, \ldots, \gamma_u)$. The space $\mathfrak{X}$ consists of triples (r, s, t) of integers with $1 \leqslant r \leqslant k_1$, $1 \leqslant s \leqslant k_2$, $1 \leqslant t \leqslant u$. However, in $\mathfrak{X}^*$ there is only one t, say $t(r, s)$, for each (r, s). One thinks of x^* as being represented by a $k_1 \times k_2$ array with entries $t(r, s)$. The expectation of the Y_i corresponding to position (r, s) is $\alpha_r + \beta_s + \gamma_{t(r,s)}$, a sum of "row, column, variety effects" where γ_t represents the contribution to Y_i of this t^{th} of u "varieties" (perhaps of grains being planted in the N positions of the rectangular array). Thus, $f_i(r, s, t(r, s))$ is an appropriate vector of 0's and 1's. The object in this experiment is usually to estimate the "variety contrasts" $\Sigma\, a_i\, \gamma_i$ with $\Sigma\, a_i = 0$. Let $c^{(1)}, c^{(2)}, \ldots, c^{(u-1)}$ be orthonormal u-vectors orthogonal to $(1, 1 \ldots, 1)$. Let $\sigma^2 H_{x*}$ be the covariance matrix of best linear estimators of the $u - 1$ contrasts $(c^{(i)}, \gamma)$. An invariance analysis related to that at the end of the last section shows easily, with $\Phi_i'(x^*)$ denoting the functionals of (5) with M^{-1} replaced by H_{x*},

THEOREM 3. – *If $\bar{x}^*$ maximizes tr $H_{x^*}^{-1}$ and $H_{\bar{x}^*}^{-1} = const.\ I_{u-1}$, then $\bar{x}^*$ minimizes $\Phi_1'(x^*)$, $\Phi_2'(x^*)$, and $\Phi_3'(x^*)$ with $\mathcal{C} = \{c : (c, c) = 1\}$.*

The computational importance of this is that $tr\ H_{x*}^{-1}$ is easily computed from A_{x*} without inversion. In a less complex setting such as that where all α_i are assumed to be zero ("one way heterogeneity"), it is easily seen that a balanced block design with block size k_1 (appropriate generalization of balanced incomplete block design if $k_1 > u$) satisfies the hypothesis of Theorem 3. In our present context, though, the situation is much more complicated. A "generalized Youden square" (G Y S) is defined to be a balanced block design with respect to both rows and columns. In [4] we proved

THEOREM 4. – *If $u | k_1$ or $u | k_2$, and if a G Y S x^* exists, then it satisfies the hypothesis of Theorem 3.*

Recently we have shown that, although a G Y S is still Φ_3-optimum in the sense of Theorem 3 if the hypothesis of Theorem 4 is violated, *it need not be Φ_2-optimum.* The importance of this is that *exotic combinatorial designs with appealing symmetry properties need not be optimum for quite common criteria.*

In the above setting, even of one-way heterogeneity, there remains the important problem of design construction, on which we have made some progress over the work of Shrikhande and of Agrawal. The literature contains hundreds of papers on the case $k_1 < u$ for every one on $k_1 > u$.

Also of concern is the question of what to do when no balanced design exists. We have extended the idea of the first paragraph of Section 2 to give bounds on departure from optimality of certain unbalanced designs. This was also considered by Shah, and K. Takeuchi gave optimality proofs for certain partially balanced designs.

One of the most striking recent investigations concerns the possibility, in the settings of Section 2 (for example, of Theorem 1) that an exact design ξ' in Ξ_N with support identical to that of a Φ-optimum approximate design ξ^*, and with values appropriately "as close as possible" to those of ξ^*, is *exactly* Φ-optimum. The result is not always true, but Salaevskii showed for $\Phi = \Phi_2$ and the polynomial setting of Theorem 1 that it is always true for N sufficiently large ! Some special results for small N have been obtained by Granovskii.

REFERENCES

[1] Atwood C.L. — *Ann. Math. Statist.*, 40, 1969, 1570.
[2] Farrell R.H., Kiefer J. and Walbran A. — *Proc. 5th Berkeley Symp.*, 1, 1965, p. 1-13.
[3] Karlin S. and Studden W.J. — *Ann. Math. Statist.*, 37, 1966, p. 783.
[4] Kiefer J. — *Ann. Math. Statist.*, 29, 1958, p. 675.
[5] Kiefer J. — *J.R.S.S. Series B*, 21, 1959, p. 272.
[6] Kiefer J. — *Proc. 4th Berkeley Symp.*, 1, 1960, p. 381.
[7] Kiefer J. — *Can. J. Math.*, 14, 1962, p. 597.
[8] Kiefer J. and Wolfowitz J. — *Ann. Math. Statist.*, 30, 1959, p. 271.
[9] Kiefer J. and Wolfowitz J. — *Can. J. Math.*, 14, 1960, p. 363.
[10] Kiefer J. and Wolfowitz J. — *Ann. Inst. Stat. Math.*, 16, 1964, p. 79-295.
[11] Kiefer J. and Wolfowitz J. — *Ann. Math. Statist.*, 36, 1965, p. 1627.

Cornell University
Dept. of Mathematics,
White Hall
Ithaca, N.Y. 14850 (USA)

Reprinted from:
STATISTICAL DECISION THEORY AND RELATED TOPICS

Academic Press, Inc., New York and London

THE ROLE OF SYMMETRY AND APPROXIMATION IN EXACT DESIGN OPTIMALITY

By J. Kiefer*
Cornell University

1. *Introduction.* For brevity, we treat the simplest framework: Let $f = (f_1,\ldots,f_k)$ where the f_i are known real continuous functions on a compact space $\mathcal{X}$. The coefficient parameter space R^k is coordinatized by $\theta=(\theta_1,\ldots,\theta_k)$. In the *exact theory* the statistician must choose an element (= "exact design") $x^* = (x_1,\ldots,x_N)$ in a specified subset $\mathcal{X}^*$ of $\mathcal{X}^N$. He then observes an N-vector $Y(x^*)$ of uncorrelated rv's $Y_1,\ldots,Y_N$ with common (known or unknown) variance σ^2, and with Y_i having expectation

$$(f(x_i),\theta) = \sum_{j=1}^k \theta_j f_j(x_i)$$

We shall also restrict attention to problems of *point estimation* of a collection of linear forms $\{(c,\theta),\ c \in C\}$, where C is specified, and will consider only nonrandomized linear estimators

$$(Y(x^*),h_c) = \sum_1^N h_{ci} Y_i$$

of the (c,θ). Moreover, we shall be concerned only with functionals of the expected squared errors; and, although the results pertain also to certain settings where biased estimators are called for (Box and Draper, Hader et al, etc.), we treat here only unbiased estimators. Thus, $(Y(x^*),h_c)$

*Research performed under an NSF Grant.

may be assumed to be the Gauss-Markov estimator of (c,θ) if the latter is estimable under x^*, and it has variance

$$\sigma^2 v(c,x^*) = \sigma^2 (c, c\, A_{x^*}{}^{-1})$$

where

$$A_{x^*} = \sum_1^N f(x_i)'\, f(x_i)$$

(All expressions involving inverses have obvious meanings in singular cases.)

In block design settings there are usually restrictions which make $\mathcal{X}^* \neq \mathcal{X}^N$. When, as in common regression experiments with unlimited exactly repeatable observation expectations, there are no such restrictions, we can define the discrete probability measure ξ_{x^*} on $\mathcal{X}$ by $N\xi_{x^*}(x)$ = [number of i such that $x_i = x$], and we then see that $N^{-1}A_{x^*}$ equals

$$M(\xi) = \int_{\mathcal{X}} f(x)'\, f(x)\xi(dx)$$

with $\xi = \xi_{x^*}$. We are restricted here, in the exact theory, to the class Ξ_N of those ξ's whose range consists of integer multiples of $1/N$. If we omit this last restriction and consider the class Ξ of *all* probability measures on $\mathcal{X}$ relative to some σ-field which includes all one-point sets, we have the *approximate theory*, and any element of Ξ is called an *approximate design*.

We omit discussion of admissibility and complete classes to concentrate on specific optimality criteria. If Φ is a real functional on the space of variance functions $v(\cdot;x^*)$ on C, we say that $\bar{x}^*$ is *Φ-optimum for* C if it minimizes $\Phi(v(\cdot;x^*))$ over $\mathcal{X}^*$. The corresponding definition of an *approximate Φ-optimum design* $\bar{\xi}$ in the approximate theory is obvious. Among the most commonly employed optimality criteria are, expressed as functionals on the class M_k of k x k nonnegative definite matrices $M = A_{x*}$ above,

$$\Phi_1(M) = \text{tr } BM^{-1}, \quad \Phi_2(M) = \det BM^{-1},$$

$$\Phi_3(M) = \max_c (Hc', M^{-1}Hc'),$$

where B (in M_k) and H are specified. We hereafter assume Φ lower semi-continuous to avoid worrying about attainment of optimality.

2. *Symmetry*. A discussion of symmetry has been given in several papers. In the approximate theory, if a compact group G operates appropriately on $(\mathcal{X}, f, C, \Phi)$, which includes convexity in ξ of some increasing function of $\Phi(M(\xi))$, then there is an approximate Φ-optimum $\bar{\xi}$ which is G-invariant ($\bar{\xi}(gA) = \bar{\xi}(A)$ for all g in G and measurable A). Common applications of the Φ_i's listed above fall into this framework, and we have used this to treat regression problems on spheres, cubes, and simplices.

The corresponding result applies less often in the exact theory; we can still represent x^* as a probability measure ξ' on $\mathcal{X}^*$ (not on $\mathcal{X}$), but the "symmetrical" (G-invariant) design ξ'_G defined formally by $\xi'_G(x) = \int_G \xi'(gx)\mu(dg)$ (where μ is Haar probability measure), and which will improve on ξ' if Φ is invariant and convex, need not correspond to an element of $\mathcal{X}^*$. However, sometimes a particular $\bar{x}^*$ can be shown to be better than all competitors in $\mathcal{X}^*$, as follows: Suppose there are u linear parametric functions of interest, which we can take to be $\theta_1, \ldots, \theta_u$. Partition A_{x*} as $\binom{BC'}{CD}$ with B u x u, and write $\tilde{A}_{x*} = B - C'D^{-1}C$. The f_i need not be linearly independent,

111

especially in block design settings, and hence not all linear functions of $\theta_1,\ldots,\theta_u$ need be estimable. If $\theta_1,\ldots,\theta_u$ are all estimable for some design, we write $s = u$ and $A^* = \tilde{A}$. If only contrasts of $\theta_1,\ldots,\theta_u$ are estimable, all contrasts being estimable for some design, we write $s = u-1$ and $A^* = L\tilde{A}L'$ where L is $s \times u$ and has orthonormal rows, each summing to zero. (Other cases of s-u, such as arise when estimating all main-order contrasts in r-way analysis of variance, have a similar treatment.) Then $A^*_{x^*}$ is proportional to the inverse of the covariance matrix of the s parametric functions being estimated, and we suppose $\Phi(A_{x^*}) = \Phi^*(A^*_{x^*})$ for some Φ^*. We also suppose a compact group $G = \{g\}$ operates on the problem in such a way that $\xi' \in \mathcal{X}^*$ implies $\xi' g \in \mathcal{X}^*$ and that $\int_G A^*_{\xi' g}\mu(dg)$ is a multiple of the identity; this last is automatic if G is homomorphic to the orthogonal group $\tilde{G} = \{\tilde{g}\}$ operating on M_s as $A^*_{\xi' g} = \tilde{g}A^*_{\xi'}\tilde{g}'$, or in the case $s = u-1$ if G is the permutation group on the u coordinates in $\tilde{A}$ and if (often the case) all rows of $\tilde{A}_{x^*}$ sum to zero for each x^*. Our simple result is then:

In the above setting, suppose $\Phi^*(bA^*) \leq \Phi^*(A^*)$ *for all* $b > 1$, *that* Φ^* *is convex, and that* $\Phi^*(A^*_{\xi'}) = \Phi^*(A^*_{\xi' g})$ *for* $\xi' \in \mathcal{X}^*$ *and* $g \in G$. *If* $\bar{x}^*$ *maximizes* $\operatorname{tr} A^*_{x^*}$ *over* $\mathcal{X}^*$, *and if* $A^*_{\bar{x}^*}$ *is a multiple of the identity, then* $\bar{x}^*$ *is* Φ*-optimum.*

This tool yielded our 1958 results on A-, D-, and E-optimality (Φ_1-, Φ_2-, Φ_3-optimality for appropriate B,C,H) of balanced block designs (BBD), and also of generalized Youden squares (GYS) under certain conditions. For

112

estimating all treatment contrasts, all conditions for using this tool are satisfied if $\bar{x}^*$ is the usual "symmetric" design (a BBD or GYS if one exists), except for the condition on $\operatorname{tr} A^*_{\bar{x}^*}$ in the case of a GYS. In these settings $\tilde{A}$ has all row sums zero; and $\tilde{A}_{\bar{x}^*}$ has equal diagonal entries and equal off-diagonal entries, so it suffices to show that $\operatorname{tr}\tilde{A}_{x^*}$ is maximized by $\bar{x}^*$, which is a great simplification over looking directly at $\Phi(A_{x^*})$. In the usual setting of 2-way heterogeneity on a $k_1 \times k_2$ array of plots, with the common model of row-plus-column-plus-treatment-effects for expectations, it was shown in 1958 only that a GYS (if one exists) is optimum (where k_1 and k_2 might be $> u$, unlike the usual YS setting) if k_1 or k_2 is divisible by u.

3. *Counterexamples to "symmetry implies optimality".* Having proved E-optimality (minimization of the maximum eigenvalue of A^{*-1}) for the GYS without divisibility some dozen years ago, we were motivated to search for a D-optimality proof. We discovered recently that *the* GYS *is sometimes not* D-*optimum in the absence of the above divisibility property*. The exact determination of which cases yield an optimum GYS and which do not, seems difficult; there are some cases of optimality without divisibility; but there are infinitely many cases of nonoptimality in the absence of divisibility.

These are the first cases we know of where an "exotic design" with full symmetry has been proved nonoptimum for a symmetric estimation problem. (Nonoptimality results in my 1958 paper were concerned with hypothesis testing.)

The simplest case is that of a 6×6 array with 4 treatments, the setting that was used for the example on p. 690

113

of my 1958 paper, where the GYS d^* had rows (134324), (412233), (241342), (124123), (313412), (321441), $a_{ii} = 25/4$ $a_{ij} = -25/12 (i \neq j)$. This example illustrated that the proof of Section 2 above did not work in this setting, i.e., that the GYS d^* had $\text{tr}\,\tilde{A}_{d^*} < \text{tr}\,\tilde{A}_{d'}$ for some competitor d'. However, the d' considered there yielded $\det A^*_{d'} < \det A^*_{d^*}$, so no counterexample to D-optimality was obtained. If, instead, we consider the design d'' with rows (122334), (213344), (231442), (334122), (344213), (442231), we obtain for the entries of $\tilde{A}_{d''}$ the values $a_{11} = 5$, $a_{22} = a_{33} = a_{44} = 61/9$, $a_{12} = a_{13} = a_{14} = -5/3$, $a_{23} = a_{34} = a_{24} = -23/9$. The eigenvalues of $A^*_{d''}$ are seen to be 28/3, 28/3, 20/3, and $\det A^*_{d''} = 15680/27 > 15625/27 = \det A^*_{d^*}$; thus, d^* is not D-optimum.

The simplest other examples are obtained in a sequence of settings which use the same idea and exhibit the same phenomenon: with four treatments on a square array of side $6t$, t odd, one treatment is "short-changed" by 3 replications while each of the other 3 treatments receives one extra replication over the average number of replications $9t^2$. A suitable design (as symmetric as possible) with these parameters yields smaller generalized variance than the fully symmetric GYS.

The degree of departure of d^* from D-optimality in the above example is not great, but of course that is not the point. We do not yet know how bad fully symmetric designs can be in other settings.

4. *What to do when no exact theory symmetric design exists.*

A. *Block design problems.* For certain values of the

number of treatments u, of the block size, and of number of blocks, there do not exist BBD's; and, in the setting of two-way heterogeneity of the previous section there similarly do not always exist GYS's. These are two familiar examples of a common phenomenon. Very little has been done in such cases.

Let us fix attention on the setting where we know BBD's, if they exist, are optimum, since the discussion of the previous section shows in the setting of two-way heterogeneity that a phenomenon additional to nonexistence of fully symmetric designs may complicate matters and even lead to a choice of fairly asymmetric designs for some Φ. Since $\mathrm{tr}\,\tilde{A}_{x*}$ is the same value $(u-1)J$ (say) for all x^* in the BBD setting, it is tempting when no BBD exists to try to obtain a Φ-optimum design by finding a design which minimizes some simple measure Ψ of the departure of A^*_{x*} from JI_{u-1}.

It appears difficult to make this approach precise in general settings where a fully symmetric design does not exist. Among other difficulties, one can give examples where the Φ-optimum design depends on Φ. An obvious choice of Ψ, suggested by expanding a symmetric Φ about its ideal (unattainable) minimum, is $\Psi = \sum_1^{u-1}(\lambda_{x*i}-J)^2 = \mathrm{tr}(\tilde{A}_{x*})^2 - (u-1)J^2$ where the λ_{x*i} are the eigenvalues of A^*_{x*}. This was discussed by K.R. Shah (1960), but no applications were made. In a few examples we can show that Ψ is so small for some near-symmetric design x^*, that x^* is A-, D-, or E-optimum, but results here are very fragmentary.

The only proof of optimality through direct comparison of the values $\Phi(A_{x*})$ is due to Takeuchi (1961), who proved A-

and E-optimality for certain PBIBD's in some of the settings where no BIBD exists.

B. *Relation of approximate to exact theory when* $\mathcal{X}^* = \mathcal{X}^N$. The practical importance of the continuous theory is that, when N is large, an approximate Φ-optimum design ξ^* can be implemented in terms of an exact design ξ_N (in Ξ_N), close to ξ^*, and consequently close to being exactly Φ-optimum if Φ is smooth. To make this precise with brevity, consider ξ^* to be Φ-optimum with finite support S; such a ξ^* always exists, although a prescription like that which follows can also be given for other ξ^*. If $\Phi(M(\xi))$ for ξ with support S, considered as a function of the values $\xi(x)$ for x in S, is twice continuously differentiable in a neighborhood of ξ^*, then

$$\Phi(M(\xi)) = \Phi(M(\xi^*)) + O(\max_{x\in S} |\xi(x)-\xi^*(x)|^2) .$$

Consequently, if ξ_N is chosen in Ξ_N to minimize $\max_{x\in S} |\xi(x)-\xi^*(x)|$, we have $\Phi(M(\xi_N))/\Phi(M(\xi^*)) = 1 + O(N^{-2})$. This is the situation for A- or D-optimality; for E-optimality or G-optimality ($\Phi(M(\xi)) = \max_{x\in\mathcal{X}} (f(x), f(x)M^{-1}(\xi)))$, we obtain $1 + O(N^{-1})$ in the worst cases.

There remains a possibility that the ξ_N constructed as above, while typically *not an approximate* Φ*-optimum design* if $\xi_N \neq \xi^*$, *is nevertheless* Φ*-optimum in the exact theory*. This result is too much to hope for in general, and in fact one of the motivations for considering the approximate theory is the common occurrence of problems where there is a fine structure that makes the support of the exactly Φ-optimum ξ_N' from Ξ_N depend on N. A remarkable result of

Salaevskii is that, for the criterion of D-optimality in estimating all parameters in univariate polynomial regression of degree $k-1$ on an interval, where ξ^* is the well known Guest-Hoel design, *the* ξ_N *of the previous paragraph is exactly* D-*optimum for* N *sufficiently large!* Some special results for small N have been obtained by Granovskii.

Note that, in this last example, $\xi_N \neq \xi^*$ and $\det M(\xi_N) < \det M(\xi^*)$ unless k divides N. Thus, for cubic regression $(k = 4)$ on $[-1,1]$ with $N = 5$, the measure ξ^* assigns measure 1/4 to each of the points ± 1, $\pm 1/2$, while $5\xi_5$ puts two observations at any one of these points (it doesn't matter which one) and one observation at each of the other three. Thus, ξ_5 lacks symmetry in its weights. This brings us back to the role of symmetry: since the development of the first paragraph of Section 2 implied that there was a symmetric D-optimum *approximate* ξ^* in this setting, it is surely tempting to seek an exactly D-optimum design among the symmetric designs $5\xi_5''$ which assign one observation to each of 5 points $0, \pm b_1, \pm b_2$. (It is a common error to suppose designs on more than 4 points can be eliminated by invoking the moment-space result which applies to the approximate theory.) But there is nothing in the development of Section 5 to imply that this will succeed, and indeed the best symmetric design is inferior to ξ_5.

The determination of general conditions on χ, f, Φ which imply exact Φ-optimality for some symmetric design, or which which imply it for ξ_N, is an outstanding problem in exact design theory.

References

1. Box, G.E.P. and Draper, N.R., (1959). "A basis for the selection of a response surface design". *JASA*, 54, 622-654.

2. Hader, R.J., Karson, M.J., and Manson, A.R. (1969). "Minimum bias estimation and experimental design for response surfaces", *Technometrics*, 11, 461-475.

3. Kiefer, J. (1958). "On the nonrandomized optimality and randomized nonoptimality of symmetrical designs" *Ann. Math. Statist.* 29, 675-699.

4. Kiefer, J. (1959). "Optimum experimental designs, *JRSS* (B), 21, 272-319.

5. Kiefer, J., (1960). "Optimum experimental designs V, with applications to systematic and rotatable designs, *Proc. Fourth Berk. Symp.* , Vol. 1, 381-405.

6. Salaevskii, O.V., (1965). "The problem of the distribution of observations in polynomial regression", *Proc. Steklov Inst.* 79, *Amer. Math. Soc. Transl.*, (1966).

7. Shah, K.R., (1960). "Optimality criteria for incomplete block designs", *Ann. Math. Statist.* 31, 791-794.

8. Takeuchi, K., (1961). "On the optimality of certain type of PBIB designs", *Rep. Stat. Appl. Res. JUSE* 8, 140-145.

Reprinted from:
MULTIVARIATE ANALYSIS – III
© 1973
ACADEMIC PRESS, INC.

Optimum Designs for Fitting Biased Multiresponse Surfaces[1]

J. KIEFER
CORNELL UNIVERSITY

1. INTRODUCTION

Let $\tilde{F}$ be a $k \times m$ matrix of real-valued functions on a set $\mathscr{X} \cup \mathscr{Z}$ (not necessarily disjoint), and suppose $\tilde{F}$ is continuous on the compact set $\mathscr{X}$. Let $\tilde{G}$ be an $s \times m$ matrix of functions on $\mathscr{Z}$. If N multiresponse row m-vectors $\tilde{Y}_1(x_1), \ldots, \tilde{Y}_N(x_N)$ are observed corresponding to (not necessarily distinct) levels $x_1, \ldots, x_N$ of the "controlled variable" taking on values in $\mathscr{X}$, we suppose the $\tilde{Y}_i$ uncorrelated with

$$E\,\tilde{Y}_i(x) = \theta'\tilde{F}(x), \qquad \mathrm{Cov}\ \tilde{Y}_i(x) = \sigma^2 Q_x; \tag{1.1}$$

here θ is the unknown column k-vector of parameters and Q_x is a known positive definite $m \times m$ matrix, varying continuously on $\mathscr{X}$; σ^2 may be unknown. We suppose the law (1.1) to be meaningful for a "virtual" observation at x in $\mathscr{Z} - \mathscr{X}$, *but the experiment* $(x_1, x_2, \ldots, x_N)$ *is restricted to values in* $\mathscr{X}$. (The definition of Q_x for x in $\mathscr{Z} - \mathscr{X}$ will often be arbitrary, since its variation may be subsumed into that of v, defined below.)

The problem, to be made precise shortly, is to choose $X_N = (x_1, x_2, \ldots, x_N)$ in $\mathscr{X}$ in some "good" fashion for the purpose of estimating the response $\theta'\tilde{F}(x)$ for x in $\mathscr{Z}$. However, for economy of form we use *not* a linear combination of the rows of $\tilde{F}$, but, instead, a linear combination $t'\tilde{G}(x)$ of the rows of $\tilde{G}$.

If $\tilde{F} = \tilde{G}$ we have the classical problem of minimum variance curve-fitting. When $\mathscr{Z} = \mathscr{X}$ and $m = 1$, this was treated by various authors (see especially [15, 16, 12, 13, 3b, 4, 9]) under various optimality criteria. The multiresponse generalization $m > 1$ in this case was given in detail in [5]. For $\mathscr{Z} \neq \mathscr{X}$, we have extrapolation or interpolation problems, as treated in [17, 8, 18, 9, etc.] for $m = 1$.

[1] Prepared under NSF Grant GP24438.

When $m = 1$, $\mathscr{X} = \mathscr{Z}$, and $\tilde{G}$ is taken to be the first s rows $\tilde{F}_1$ of $\tilde{F}$ and $\theta' = (\theta_1', \theta_2')$ is partitioned correspondingly, we have the model of Box and Draper [1] for estimating $\theta'\tilde{F}$ by $t_1'\tilde{G} = t_1'\tilde{F}_1$. Further work on this model, with different recommendations (discussed in Section 2 below), was carried out by Karson, Manson, and Hader [10]. Recently Hader, Manson, and Cote considered this model when $\tilde{G}$ is not so restricted. Related work by Fedorov and Malyutov [6] treated particular $(\mathscr{X}, \tilde{F})$ in the case $m = 1$, $\mathscr{Z} = \mathscr{X}$, with the restricted form of $\tilde{G}$, but with the maximum of the mean squared error function replacing its integral as considered below.

A common formulation of a "biased estimation" problem, especially in fractional factorial settings, is to ask for estimates of $L\theta$ where $L = [L_1 \,\vdots\, L_2]$ and L_1 is $r \times s$, and where the estimates are of the form $L_1 t_1$ in a notation consistent with that of the previous paragraph; θ_2 may not be estimable. This is fitted into the previous formulation by letting $\mathscr{Z} = \{1, 2, \ldots, r\}$ and $[\tilde{G}(1), \ldots, \tilde{G}(r)] = L_1'$, $[\tilde{F}(1), \ldots, \tilde{F}(r)] = L'$, if $m = 1$.

If $m > 1$ this last formulation has an obvious extension, since $\mathscr{Z} \cap \mathscr{X} = \varnothing$. Thus, for example, for each i the m columns of $\tilde{F}(i)$ and $\tilde{G}(i)$ can be taken to be the same as the single column of the previous paragraph, or $m - 1$ of them can be taken as 0. Phrased another way, we can avoid this artificial representation by noting that if $\mathscr{X} \cap \mathscr{Z} = \varnothing$, there is no need for $\tilde{F}$ and $\tilde{G}$ on $\mathscr{Z}$ to have m columns.

Thus, our theory includes the simultaneous fitting of several response surfaces $\tilde{G}_i$ on what can even be different $\mathscr{Z}_i$'s.

If the observable components of $\tilde{Y}_i(x)$ (and the number of such components) vary with x, this is easily subsumed into our model by assigning the value zero to the corresponding elements of $\tilde{F}(x)$ (or of $F(x)$, below).

The present note is intended to unify and extend these considerations to arbitrary m, $\mathscr{Z}$, $\tilde{G}$ (Section 2). At the same time, a modification of the previous approaches is related to the author's general theory of approximate design optimization [14] (Section 3). We show by example that the recommendations of [1] and [10] can lead to procedures whose risk can be much improved upon (Section 4).

Before proceeding further, we simplify notation by a reduction. Write $[\tilde{Y}_i(x), \tilde{F}(x), \tilde{G}(x)] = [Y_i(x), F(x), G(x)]Q_x^{-1/2}$ where $Q_x^{1/2}$ is the symmetric positive definite square root of Q_x. Then

$$EY_i(x) = \theta' F(x), \qquad \text{Cov } Y_i(x) = \sigma^2 I. \tag{1.2}$$

We hereafter treat the problem in terms of Y_i, F, G; the results are then easily transformed back into the original formulation.

To minimize confusion in the sequel, we shall use the dummy variable x when referring to the choice of experimental points, and z when referring to the estimated response on $\mathscr{Z}$.

Suppose that the estimator $t'G$ is used to estimate $\theta'F$ on $\mathscr{X}$. Write Y for the Nm-vector $(Y_1, Y_2, \ldots, Y_N)$. *We restrict consideration to estimators which are linear in* Y. Thus, $T = CY'$, where C is $s \times Nm$. We denote by $\bar{F}(X_N)$ the $k \times Nm$ matrix $[F(x_1) \mid F(x_2) \mid \cdots \mid F(x_N)]$, and also write $D = D(X_N) = \bar{F}(X_N)C'$ and $CC' = H$. Hence, we have

$$Et = D'\theta, \qquad \mathrm{Cov}(t) = \sigma^2 H. \tag{1.3}$$

Now suppose we are given an $m \times m$ matrix-valued *loss measure* ν on $\mathscr{X}$, nonnegative definite in value, and such that F and G are square-integrable relative to ν. We define $\Gamma_{FF}(k \times k)$, $\Gamma_{GG}(s \times s)$, and $\Gamma_{FG}(k \times s)$ or Γ_{GF} by

$$\Gamma_{A_1 A_2} = \int_{\mathscr{X}} A_1(z)\, d\nu(z)\, A_2(z)'. \tag{1.4}$$

We assume the (quadratic) expected loss incurred if the design X_N and estimator t are used and θ is the true parameter value is then

$$\begin{aligned} R(\theta; X_N, t) &= E \int_{\mathscr{X}} [t'G(z) - \theta'F(z)]\nu(dz)[t'G(z) - \theta'F(z)]' \\ &= \sigma^2 \operatorname{tr} H\Gamma_{GG} + \theta'[D\Gamma_{GG} D' - D\Gamma_{GF} - \Gamma_{FG} D' + \Gamma_{FF}]\theta \\ &= \sigma^2(V + B) \qquad \text{(say)}, \end{aligned} \tag{1.5}$$

the last being a notation which specializes to that adopted by previous authors [1, 10] when certain restrictions on D are introduced, as described in the next section. As will be seen in the example of Section 4, the vector function (B, V) is often convenient to consider without combining as in (1.5).

We conclude this section by recording a simple quadratic minimization result, essentially the Gauss–Markov theorem. With the usual ordering of nonnegative definite matrices, we write $P_1 \prec P_2$ to mean $P_2 - P_1$ is nonnegative definite. Suppose the row space of $R(k \times h)$ is contained in that of $L(n \times h)$, so that there are matrices $U(k \times n)$ satisfying $UL = R$. Then, if

$$U_0 = R(L'L)^- L', \qquad U_0 U_0' = R(L'L)^- R', \tag{1.6}$$

we have $U_0 L = R$ and

$$U_0 U_0' \prec UU' \quad \text{if} \quad UL = R. \tag{1.7}$$

Here $(\ )^-$ denotes generalized inverse.

2. FORMULATION OF BOX AND DRAPER

Without further restrictive criteria, the unknown relative magnitudes of $\theta'\theta$ and σ^2 make the problem of "choosing X_N and t so as to make (1.5) small" too vague or unfruitful. In the approach of [1] and [10] this difficulty is overcome by first considering B alone for minimization. Before discussing this more precisely, we must recall for the reader the distinction between the *exact* and *approximate* design theory developments; failure to make clear which development is being used has created some confusion in the previous literature, in the author's opinion. The *exact theory* follows the previous lines, wherein a design X_N in the present setting is an N-tuple of points (not necessarily distinct) in $\mathscr{X}$; or, equivalently, it is a discrete probability measure ξ on $\mathscr{X}$ restricted to the family of probability measures $\{\xi$: values of ξ are integral multiples of $N^{-1}\}$, where the ξ corresponding to X_N is given by $N\xi(x) =$ [number of $x_i = x$]. In the *approximate theory* ξ is permitted to be any member of the family of *all* probability measures on $\mathscr{X}$ relative to a specified σ-field which contains at least all the finite subsets of $\mathscr{X}$. The use of the approximate theory makes tractable various unwieldy minimization problems of the exact theory, but then necessitates implementation in terms of exact designs which may only be approximately optimum. See [11, 15, 4] for discussion. We write $M(\xi) = \int_{\mathscr{X}} F(x)F(x)'\xi(dx)$ for the information matrix (per observation) under ξ.

We continue with our description of the work of [1] and [10] in the case $m = 1$, $G =$ first s elements of F. Motivated by illustrative examples which indicated that minimization of B alone yielded a risk close to the minimum of R (assuming knowledge of $\sigma^{-2}\theta_2{}^2$), Box and Draper (hereafter BD) assumed t to be the least squares estimator of θ_1 under the assumption $\theta_2 = 0$ and then found, *in what we must in general regard as the approximate theory for reasons described in the next paragraph*, the design which minimized B. It is not at all evident that "minimization of B" makes mathematical sense in view of the dependence of B on the vector θ, but in the present setting it turns out that a single achievable matrix of the quadratic form in θ of (1.5) (second line) yields, for each possible value of θ, the smallest value of B; that is, the matrix of the quadratic form B is smaller than all other possible ones in the sense of the ordering $\prec$ defined at the end of Section 1. In Section 4 the reader will find an illustration of what we regard as a fairly common occurrence, in the failures of the BD motivation (almost minimizing R) to be an accurate premise.

The original BD considerations always contemplated, I believe, an F consisting of monomials on a convex Euclidean $\mathscr{X}$ with $\mathscr{X} = \mathscr{Y}$ and $\nu =$ Lebesgue measure, and with a large enough N (for a given k and s), that the solution could be implemented as an *exact design*. But it is easily seen that lack of convexity or smallness of N can make this solution sensible in the approximate theory alone.

The work of Karson, Manson, and Hader (hereafter KMH) in the same setting assumes it is satisfactory to begin by "minimizing B" (made precise two paragraphs above). They do *not* restrict t to be the least squares estimator of θ_1, but instead find, for any X_N for which $\theta_1 + \Gamma_{FF}^{-1}\Gamma_{FG}\theta_2$ is estimable, a t

which minimizes B; *their striking observation is that the minimizing function B is the same for all such X_N*. In illustrative examples they then choose an exact X_N to minimize V for this t, subject to some simplifying symmetry restrictions on X_N; this must do as well as BD and can often do quite better, as we will see in Section 4. (In subsequent papers these and other authors have used other criteria than simple minimization of V; these will not be discussed here.) We observe that N is quite small in the illustrative examples, and this is simply a reflection of the difficulty of solving such extremum problems in the exact theory.

The fact that G consists of the first s rows of F produces a solution in the KMH formulation in which the minimizing function B does not depend on θ_1, but the corresponding result for general G does not have this feature and hence is often unrealistic in practical terms, as we will describe in the second paragraph below. For the moment, let us nevertheless give the extension of the KMH solution to our general setting, since it does seem sensible in some cases and will enable us to describe the resulting problem of minimizing V, using approximate theory algorithms, in this simplest approach. The matrix of the quadratic form (B) in θ in (1.5) can obviously be rewritten as

$$\Gamma_{FF} - \Gamma_{FG}\Gamma_{GG}^{-}\Gamma_{GF} + (D - \Gamma_{FG}\Gamma_{GG}^{-})\Gamma_{GG}(D - \Gamma_{FG}\Gamma_{GG}^{-})'; \qquad (2.1)$$

here $\Gamma_{GG}^{-}\Gamma_{GF}$ is well-defined and (2.1) holds even if Γ_{GG} is singular; note that

$$\begin{pmatrix} \Gamma_{FF} & \Gamma_{FG} \\ \Gamma_{GF} & \Gamma_{GG} \end{pmatrix}$$

is nonnegative definite. If X_N is such that $D(X_N)$ (that is, C) can be chosen to satisfy

$$\bar{F}(X_N)C' = D(X_N) = \Gamma_{FG}\Gamma_{GG}^{-}, \qquad (2.2)$$

then any such choice clearly minimizes the *matrix* (2.1) in the sense described earlier and defined at the end of Section 1, and yields

$$B = \sigma^{-2}\theta'[\Gamma_{FF} - \Gamma_{FG}\Gamma_{GG}^{-}\Gamma_{GF}]\theta \qquad (2.3)$$

for the minimum of the matrix. Of course, there is a C satisfying (2.2) if and only if $\theta'\Gamma_{FG}\Gamma_{GG}^{-}$ is estimable (has a linear unbiased estimator) when design X_N is used. For such an X_N, it follows from (1.6) that H achieves its (matrix) minimum among C satisfying (2.2) with the choice

$$C = \Gamma_{GG}^{-}\Gamma_{GF}(\bar{F}(X_N)\bar{F}(X_N)')^{-}\bar{F}(X_N), \qquad (2.4)$$

which yields

$$H = \Gamma_{GG}^{-}\Gamma_{GF}(\bar{F}(X_N)\bar{F}(X_N)')^{-}\Gamma_{FG}\Gamma_{GG}^{-}. \qquad (2.5)$$

As is usual in optimum design considerations, there will not generally be a design uniformly best in our matrix sense of making H smallest. However, if we continue along the KMH line, we must by (1.5) only minimize tr $\Gamma_{GG} H$. (The BD analogue will be clear; we shall omit it.)

Write $A = \Gamma_{FG} \Gamma_{GG}^{-} \Gamma_{GF}$. The minimization of tr $A[\bar{F}(X_N)\bar{F}(X_N)']^{-}$ is generally difficult in the exact theory. In the approximate theory, Fedorov [4] characterizes the solution, and obtains iterative methods for solving this problem; the optimum ξ^* (say) is characterized, if $M(\xi^*)$ is nonsingular, by

$$\max_{x \in \mathscr{X}} \operatorname{tr} F(x)' M^{-1}(\xi^*) A M^{-1}(\xi^*) F(x) = \operatorname{tr} A M^{-1}(\xi^*) \tag{2.6}$$

or by the fact that the left side of (2.6) cannot be decreased by altering ξ^*. (For $m = 1$, (2.6) is originally due to Elfving [3a] and Karlin and Studden [9]; Chernoff's work [2] is related.) There is a corresponding result for $M(\xi^*)$ singular. We omit further discussion, except to note that this approximate theory development often yields solutions without much difficulty, which can then be implemented to supply almost optimum solutions for even fairly small N.

We mentioned, two paragraphs above, that if $G = F_1$ = first s rows of F on $\mathscr{X}$, then the minimizing B does not depend on θ_1. In fact, abbreviating $\Gamma_{F_i F_j}$ by Γ_{ij} in this case, a trivial calculation shows the minimum to be

$$B = \sigma^{-2} \theta_2' (\Gamma_{22} - \Gamma_{21} \Gamma_{11}^{-} \Gamma_{12}) \theta_2. \tag{2.7}$$

In other cases, (2.3) need not simplify in this way. One can still use the generalized KMH line of development as described in the previous two paragraphs, but this may not even approximately reflect the experimenter's goals. The reason for this is that minimizing B may now so restrict C (and thus affect the choice of X_N) that it makes $V + B$ obtained by the KMH route much larger than the minimum of $V + B$ for a wide spectrum of parameter values. While (Section 4) the BD or KMH prescription may not be satisfactory even in the situation $G = F_1$ originally contemplated by BD, the motivation for considering B separately is now completely absent. For, if G is not of the form F_1, then F_1 and θ_1 no longer have the special meaning they had and which resulted in both BD and KMH using (possibly different) estimators which were unbiased when $\theta_2 = 0$. In the case of general G, only the (possibly zero) part of the row space of G contained in the row space of F has this meaning.

This motivates our return now to consideration of $V + B$, and of trying to make it small from the outset.

3. MINIMIZING V + B

We know we must make further assumptions before it can make sense to minimize (1.5). One possibility is to minimize the maximum of (1.5) subject to an assumption $\sigma^{-1}\theta \in S$ where S is a specified set in R^k. Somewhat simpler, but in the same direction, is the minimization of an integral of (1.5) with respect to a specified measure ψ on $\sigma^{-1}\theta$. This has obvious meaning to Bayesians, but it may appeal to others as a possible compromise. For, in order to inspect the risk functions of the *admissible* (X_N, t) (after the reduction to (1.5)), it suffices to consider the closure of such integral minimizers (more generally, without the invariance reduction to $\sigma^{-1}\theta$). Although our restriction to linear estimators could then well be questioned, nevertheless, as an illustration of the type of mathematics which arises, we now consider such a development.

Let us suppose, then, that $\int \sigma^{-2}\theta\theta'\psi(d\theta/\sigma) = \Phi$ is a specified nonnegative definite $k \times k$ matrix and that, in accordance with the discussion of the previous paragraph (and omitting the dependence of $\bar{F}$ on X_N), we seek to choose (X_N, t) to minimize

$$\bar{R} = \int \sigma^{-2} R\psi(d\theta/\sigma) = \mathrm{tr}\{\Gamma_{GG} CC' + \Phi[D\Gamma_{GG} D' - D\Gamma_{GF} - \Gamma_{FG} D' + \Gamma_{FF}]\}$$

$$= \mathrm{tr}\{\Gamma_{GG}^{1/2} C[I_{Nm} + \bar{F}'\Phi\bar{F}]C'\Gamma_{GG}^{1/2} - C\bar{F}'\Phi\Gamma_{FG} - \Gamma_{GF}\Phi\bar{F}C' + \Phi\Gamma_{FF}\}. \quad (3.1)$$

The minimum with respect to C of the *matrix* whose trace is taken in (3.1) is easily seen, by an argument like that used in conjunction with (2.1), to be attained when

$$C = \Gamma_{GG}^{-}\Gamma_{GF}\Phi\bar{F}[I_{Nm} + \bar{F}'\Phi\bar{F}]^{-1}. \quad (3.2)$$

The C of (3.2) yields

$$\bar{R} = \mathrm{tr}\{\Phi\Gamma_{FF} - \Phi\Gamma_{FG}\Gamma_{GG}^{-1}\Gamma_{GF}\Phi\bar{F}[I_{Nm} + \bar{F}'\Phi\bar{F}]^{-1}\bar{F}'\}. \quad (3.3)$$

(As before, matrix minimization with respect to $\bar{F}$ is unachievable, so we consider the trace.)

Since

$$\Phi^{1/2}\bar{F}[I_{Nm} + \bar{F}'\Phi\bar{F}]^{-1}\bar{F}'\Phi^{1/2} = I_k - [I_k + \Phi^{1/2}\bar{F}\bar{F}'\Phi^{1/2}]^{-1}, \quad (3.4)$$

we see that minimization of $\bar{R}$ with respect to X_N is equivalent to minimization of

$$\bar{\bar{R}} = \mathrm{tr}\, \Gamma_{FG}\Gamma_{GG}^{-1}\Gamma_{GF}\Phi^{1/2}[I_k + \Phi^{1/2}\bar{F}\bar{F}'\Phi^{1/2}]^{-1}\Phi^{1/2}$$

$$= \mathrm{tr}\, \Gamma_{FG}\Gamma_{GG}^{-1}\Gamma_{GF}[\Phi^{-1} + \bar{F}\bar{F}']^{-}, \quad (3.5)$$

where the last bracketed expression has the well-defined meaning obtained from the previous line in the obvious way.

So far we have not invoked the approximate theory. In a recent work on general approximate theory optimality criteria, the author [14] has shown that the analogue of (2.6) for a minimizing criterion of the form tr $A[\bar{A} + M(\xi)]^{-1}$, where A and $\bar{A}$ are symmetric nonnegative definite, is that the optimum ξ^* satisfy

$$\max_{x \in \mathscr{X}} \operatorname{tr} F(x)'[\bar{A} + M(\xi^*)]^{-1} A [\bar{A} + M(\xi^*)]^{-1} F(x)$$

$$= \operatorname{tr} M(\xi^*)[\bar{A} + M(\xi^*)]^{-1} A [\bar{A} + M(\xi^*)]^{-1}, \quad (3.6)$$

with an appropriate analogue if $M(\xi^*)$ is singular. This criterion differs from that of (2.6) or from the analogue for D-optimality, in that the left side of (3.6) need not be a minimum as was the case for (2.6); this is a consequence of the fact that $\operatorname{tr} A[\bar{A} + M]^{-1}$ is not homogeneous in M. However, (3.6) leads to iterative procedures for finding an optimum design.

We note that one is often led to recommend what in the limit *appears* as a singular matrix replacing Φ^{-1} in the last line of (3.5). (It is not justified simply to substitute such a Φ^{-1} in (3.5).) For example, in the special case we have discussed wherein $G = F_1$, the treatment of θ_1 and θ_2 in comparable terms in the ψ of (3.1) is of doubtful applicability except in the rarest Bayesian models. More likely is the original BD spirit, implemented here by assuming no restriction on θ_1 and letting $\int \sigma^{-2}\theta_2\theta_2'\psi_2(d\theta_2/\sigma) = \Phi_2$ (specified). This case can be developed by requiring the estimator $t'F_1$ to have bounded risk as a function of θ_1, which means being unbiased when $\theta_2 = 0$, and then going through a line of reasoning analogous to that of (3.1)–(3.6). Alternatively, one can let

$$\Phi = \begin{pmatrix} \lambda I & 0 \\ 0 & \Phi_2 \end{pmatrix}$$

and treat the result of letting $\lambda \to +\infty$ in our development above. Illustrations of this will be given elsewhere.

4. AN ILLUSTRATIVE EXAMPLE

This is not a practical example, but rather one chosen to be arithmetically trivial (not necessitating the use of (2.6) or (3.6)), at the same time that it exhibits various general features which occur in more meaningful settings. Suppose $m = 1$, $k = 2$, and that $\mathscr{X} = \mathscr{Y}$ consists of the two points $\{-1, a\}$ where $a > 0$. The problem is one of "linear regression" with the estimated regression *homogeneous*. This may be expressed as $F(x)' = (x, 1)$ with $s = 1$ and $G(x) = F_1(x) = x$. For arithmetical simplicity suppose $\nu(-1) = a/(1 + a)$ and $\nu(a) = 1/(1 + a)$, so that $\Gamma_{FF} = (\begin{smallmatrix} a & 0 \\ 0 & 1 \end{smallmatrix}) = \Gamma$ (say) is diagonal. (In practice this might arise as an approximation to an example where there is a uniform measure ν on a cluster of r_{-1} points near -1 and r_{-a} near a, in proportion $r_{-1}/r_a \sim a$.)

In the calculations which follow, σ^2 should be thought of as $N^{-1} \operatorname{Var}(Y_1)$.

We abbreviate $\xi(-1) = \alpha$, $\xi(a) = 1 - \alpha$. Thus,

$$M(\xi) = \begin{pmatrix} \alpha + (1-\alpha)a^2 & -\alpha + (1-\alpha)a \\ -\alpha + (1-\alpha)a & 1 \end{pmatrix}.$$

The BD prescription is $m_{11}^{-1}m_{12} = \gamma_{11}^{-1}\gamma_{12} = 0$, or $\alpha = a/(1 + a)$. (As we have remarked, this may not be even approximately implementable for small N.) This yields (from (2.2) or (2.7))

$$B_{\text{BD}} = \sigma^{-2}\theta_2^{\,2}, \qquad V_{\text{BD}} = \gamma_{11}(M^{-1})_{11} = 1. \tag{4.1}$$

On the other hand, the KMH prescription for t yields a bias term equal to that of (4.1) for every design ξ, and corresponding value $V = a/\alpha(1-\alpha)(1+a)^2$ if the value α is used for $\xi(-1)$. This is minimized by $\alpha = \frac{1}{2}$ and yields

$$B_{\text{KMH}} = \sigma^{-2}\theta_2^{\,2}, \qquad V_{\text{KMH}} = 4a/(1+a)^2. \tag{4.2}$$

(We note that this design is also not approximately implementable when N is small and odd.)

As $a \to 0$ or $+\infty$ we see that $V_{\text{KMH}}/V_{\text{BD}} \to 0$. The corresponding ratio $R_{\text{KMH}}/R_{\text{BD}}$ of course depends on $\sigma^{-2}\theta_2^{\,2}$, but if this is suspected to be small (which is reasonable where one adopts the biased curve-fitting rationale of this model, especially if N is small), the KMH procedure will often yield great improvement.

Suppose we consider instead the approach of Section 3. In accordance with the last paragraph of that section, we consider only linear estimators $tG(x)$ which are unbiased estimators of $(\theta_1 + c\theta_2)x$ for some c. From (1.3) and (1.5) we obtain $B = (1 + c^2a)\theta_2^{\,2}/\sigma^2$. For fixed α, the variance of the Gauss–Markov estimator of $\theta_1 + c\theta_2$ is proportional to $(1, c)M^{-1}(\xi)(1, c)'$. Thus, if we have a value of $\int \theta_2^{\,2}\sigma^{-2}\psi_2(d\theta_2/\sigma) = \Phi_2$ (say) to be considered, the last paragraph of Section 3 tells us to minimize

$$a(1, c)M^{-1}(\xi)(1, c)' + \Phi_2(1 + c^2a) \tag{4.3}$$

with respect to c and α. Rather than to do this for general Φ_2, let us see what happens when Φ_2 is very small; that is, let us minimize V alone. (Recall this does not mean making $V = 0$, since t is restricted to unbiased estimators of $\theta_1 + c\theta_2$.) The minimum with respect to c (for fixed α) is easily seen to occur at $c = [(1-\alpha)a - \alpha]/[\alpha + (1-\alpha)a^2]$, for which $V = a/[\alpha + (1-\alpha)a^2]$. This is minimized by $\alpha = 1$ or 0 if $a < 1$ or > 1 (or by any α if $a = 1$). The corresponding V and B are

$$\begin{aligned} B &= \sigma^{-2}\theta_2^{\,2}(a + 1), & V &= a, & &\text{if } a \le 1; \\ B &= \sigma^{-2}\theta_2^{\,2}(a^{-1} + 1), & V &= a^{-1}, & &\text{if } a \ge 1. \end{aligned} \tag{4.4}$$

Comparing (4.4) with (4.2), we see that for a close to 0 or very large, there is a kind of "subadmissibility" advantage of (4.4) over (4.2): the B's are almost the same, and the V of (4.4) is about $\frac{1}{4}$ that of (4.2). When $\sigma^{-2}\theta_2^{\,2} = N\theta_2^{\,2}/\mathrm{Var}(Y_1)$ is not suspected to be of larger order of magnitude than $\min(a, a^{-1})$, the resulting saving in $B + V$ can be large; and, even when $\sigma^{-2}\theta_2^{\,2}$ is large, the maximum possible relative increase in risk is small. The factor of decrease $\frac{1}{4}$ in the above example can be made arbitrarily close to 0 in suitable examples.

The moral of the above example is that a slight alteration in the design and estimator of KMH can often achieve a risk function R^* (say) such that R^*/R_{KMH} is much less than 1 over a large set of parameter values, while R^*/R_{KMH} is never more than very slightly above 1. Thus, the investigation of (and choice among) the risks R of many (X_M, t) seems preferable to reliance upon any simple single prescription.

The lack of reasonably close implementability of the approximate theory optimum for N small (and, in the above example, odd) reinforces these last comments; it may be preferable to depart from minimization of B in order to improve V over what it is for an inadequate discrete implementation of our KMH approximate theory optimum.

5. OTHER COMMENTS

Nonlinear models can be fitted into our scheme by the usual method of linearizing after a first stage of experimentation yields sufficiently accurate estimates.

Variable costs of observations (cost depending on x) can be included and then reduced to the present constant cost model in the manner used just above (1.2) to eliminate covariance variation. See, e.g., [2, 11, 15, 4] regarding the above.

The problem of curve-fitting is not really solved by these considerations. We have not treated such a difficult but more meaningful problem as that of deciding, for a given F, which rows of it to call G. A sensible decision-theoretic framework is then to let the loss function reflect both errors in estimation and also a penalty depending on the choice of G, for example, a function of the number of rows s (or coefficients θ_i estimated to be nonzero). There are not at present satisfactory prescriptions for this model even in the absence of the choice of design.

Analytical solutions of the problem described by (2.6) or (3.6) are often aided by invariance considerations or a conclusion about the support ξ^* must have in order that it assign all measure to the set of x where the maximum on the left side is attained. For example, if $\mathscr{X}$ is the q-cube, $m = 1$, and F consists of all monomials of degree ≤ 2, and if G is symmetric with respect to the cube (e.g., consists of monomials of degree ≤ 1, or these and the $x_i^{\,2}$, $1 \leq i \leq q$) along with $\mathscr{X}$ and v, then one concludes that there is a symmetric optimum design; the fact that the function maximized in (2.6) or (3.6) is a symmetric quartic then allows one to conclude, as in the D-optimality argument of [3b], that ξ^* is supported by a subset of the obvious 3^q array of points. This reduces the problem to a minimization in q variables, the weights on faces of various dimensions. Further reduction may then be possible, as in [3b].

Added in proof: Recent additional work on designs for biased curve fitting has been obtained in work of S. Stigler, C. Atwood, L. Jordan, and P. Huber, each using a different approach.

REFERENCES

1. Box, G. E. P. and Draper, N. R. (1959). A basis for the selection of a response surface design. *J. Amer. Statist. Assoc.* **54** 622.
2. Chernoff, H. (1953). Locally optimum designs for estimating parameters. *Ann. Math. Statist.* **24** 586.

3a. Elfving, G. (1959). Design of linear experiments. *Cramér Festschrift Volume*, 58. Wiley, New York.

3b. Farrell, R., Kiefer, J. and Walbran, A. (1965). Optimum multivariate designs. *Proc. Fifth Berkeley Symp. Math. Statist. Prob.* **1** 113.

4. Fedorov, V. V. (1972). *Theory of Optimal Experiments.* Academic Press, New York.
5. Fedorov, V. V. (1971). The design of experiments in the multiresponse case. *Theor. Probability Appl.* **16** 323.
6. Fedorov, V. V. and Malyutov, M. B. (1971). Preprint 18. LSM, Moscow State Univ., Moscow.
7. Hader, R. J., Manson, A. R. and Cote, R. (1971). Abstract. *Ann. Math. Statist.* **42** 2190.
8. Hoel, P. and Levine, A. (1964). Optimal spacing and weighting in polynomial prediction. *Ann. Math. Statist.* **35** 1553.
9. Karlin, S. and Studden, W. (1966). Optimal experimental designs. *Ann. Math. Statist.* **37** 783.
10. Karson, M. J., Manson, A. R. and Hader, R. J. (1969). Minimum bias estimation and experimental designs for response surfaces. *Technometrics* **11** 461.
11. Kiefer, J. (1959). Optimum experimental designs. *J. Roy. Statist. Soc. Ser. B.* **21** 272.
12. Kiefer, J. (1961). Optimal designs in regression problems, II. *Ann. Math. Statist.* **32** 298.
13. Kiefer, J. (1960). Optimum experimental designs with applications to systematic and rotatable designs. *Proc. Fourth Berkeley Symp. Math. Statist. Prob.* **1** 381.
14. Kiefer, J. (1972). Equivalence theory for general convex design criteria. To be published. (Abstract in *Bull. IMS* **1** 1972.)
15. Kiefer, J. and Wolfowitz, J. (1959). Optimum designs in regression problems. *Ann. Math. Statist.* **30** 271.
16. Kiefer, J. and Wolfowitz, J. (1960). The equivalence of two extremum problems. *Canad. J. Math.* **12** 363.
17. Kiefer, J. and Wolfowitz, J. (1964). Optimum extrapolation designs I and II. *Ann. Inst. Statist. Math.* **16** 79, 295.
18. Kiefer, J. and Wolfowitz, J. (1965). On a theorem of Hoel and Levine on extrapolation designs. *Ann. Math. Statist.* **36** 1627.

The Annals of Statistics
1974, Vol. 2, No. 5, 849-879

GENERAL EQUIVALENCE THEORY FOR OPTIMUM DESIGNS (APPROXIMATE THEORY)[1]

BY J. KIEFER

Cornell University

For general optimality criteria Φ, criteria equivalent to Φ-optimality are obtained under various conditions on Φ. Such equivalent criteria are useful for analytic or machine computation of Φ-optimum designs. The theory includes that previously developed in the case of D-optimality (Kiefer-Wolfowitz) and L-optimality (Karlin-Studden-Fedorov), as well as E-optimality and criteria arising in response surface fitting and minimax extrapolation. Multiresponse settings and models with variable covariance and cost structure are included. Methods for verifying the conditions required on Φ, and for computing the equivalent criteria, are illustrated.

1. Introduction. Let $f' = (f_1, f_2, \cdots, f_k)$ where the f_i are continuous real functions on a compact set $\mathscr{X}$. The expected value of an observation "at the level x in $\mathscr{X}$" is $\sum_1^k \theta_i f_i(x) = \theta' f(x)$. Observations are uncorrelated and have variance independent of x (an assumption relaxed in Section 5). We are concerned with the approximate design theory wherein the designs are a class Ξ of probability measures on $\mathscr{X}$ including all discrete measures, and the information matrix of a design ξ is $M(\xi) = \int_{\mathscr{X}} f(x)f(x)'\xi(dx)$. This has the usual meaning that $M^{-1}(\xi)$ is proportional to the covariance matrix of best linear estimators of θ (with the obvious analogue if M is singular). See Kiefer [17] or Fedorov [12] for further remarks on interpretation. We let $\mathscr{M} = \{M(\xi) : \xi \in \Xi\}$.

Let Φ be a function which is real or $+\infty$ on $\mathscr{M}$. One problem of optimum design theory is the characterization of designs ξ^* which are Φ-*optimum*; that is, for which

$$(1.1) \qquad \Phi(M(\xi^*)) = \min_{\xi \in \Xi} \Phi(M(\xi)) .$$

The most common examples of optimality criteria are

$$(1.2) \qquad \Phi_0(M) = \det M^{-1} \quad (D\text{-optimality}) ,$$

$$(1.3) \qquad \Phi_{1,C}(M) = \operatorname{tr} CM^{-1} \quad (L\text{-optimality; } A\text{-optimality if } C = I) ,$$

$$(1.4) \qquad \Phi_\infty(M) = \text{maximum eigenvalue of } \; M^{-1} \quad (E\text{-optimality}) ;$$

here C is a given nonnegative definite symmetric matrix, and (1.2) and (1.4) are to be regarded as infinite if M is singular (with the obvious analogue for (1.3)). The significance of the subscripts will be seen in Sections 4C—D. (We shall consider other criteria, later.)

Received January 1973; revised July 1973.

[1] Research supported by NSF Grant GP24438.

AMS 1970 *subject classification.* 62K05.

Key words and phrases. Optimum experimental designs, equivalence theory of designs, D-optimality, A-optimality, E-optimality, iterative design optimization, large eigenvalues.

The desired characterization just mentioned should aid in the computation of Φ-optimum designs. Thus, writing $\bar{d}_0(\xi) = \max_{x \in \mathscr{X}} f(x)'M^{-1}(\xi)f(x)$, Kiefer and Wolfowitz [23] showed in the case (1.2) that $M(\xi^*)$ is the same for all Φ_0-optimum ξ^* and that

$$(1.5) \qquad \xi^* \text{ is } \Phi_0\text{-optimum} \Leftrightarrow \bar{d}_0(\xi^*) = \min_\xi \bar{d}_0(\xi) \Leftrightarrow \bar{d}_0(\xi^*) = k .$$

This is also useful because, for a given ξ' which one guesses to be nearly optimum, one cannot usually assess the departure of det $M^{-1}(\xi')$ from the unknown minimum of det $M^{-1}(\xi)$; while the last statement of (1.5) gives both a verifiable condition for optimality and an indication (made precise in Section 6C) that $\bar{d}_0(\xi')$ near k implies det $M^{-1}(\xi')$ near the minimum. This last fact has also been implemented by a number of authors to obtain iterative schemes for computing ξ^*. This will be discussed in Section 6B. Another useful aspect of (1.5), implemented in [18], [19], [9], is that $f(x)'M^{-1}(\xi^*)f(x) = k$ on the support of ξ^*; this and the form of f often enable one to limit drastically the possible supports among which one must search, as indicated in Section 6A.

Subsequently Karlin and Studden [14], Theorems 8.1–8.2, and Fedorov [10], [12], studying what the latter called linear (in M^{-1}) optimality criteria, showed in the case (1.3) that, if $M(\xi^*)$ is nonsingular,

$$(1.6) \qquad \begin{aligned} \xi^* \text{ is } \Phi\text{-optimum} &\Leftrightarrow \bar{d}_1(\xi^*) = \min_\xi \bar{d}_1(\xi) \\ &\Leftrightarrow \bar{d}_1(\xi^*) = \operatorname{tr} CM^{-1}(\xi^*) , \end{aligned}$$

where $\bar{d}_1(\xi) = \max_x f(x)'M^{-1}(\xi)CM^{-1}(\xi)f(x)$. The analogy between (1.5) and (1.6) is obvious, and Fedorov's presentation makes it clear that the steps in his proof of (1.6) parallel those of the proof of (1.5).

Since neither Φ_0 nor Φ_1 is a special case of the other, this suggests that there is a larger class of Φ for which one can obtain equivalent characterizations of Φ-optimality analogous to the last two statements of (1.5) and of (1.6). This has undoubtedly occurred to a number of workers in the field, and I have mentioned it in talks and an abstract [21 a]; but perhaps the intuitive appeal and computational tractability of D- and A-optimality have continued to make them the main topics of concentration.

A number of people have asked me for my results on this subject, and the present paper is a selection of some of the material on general Φ-optimality characterizations which I have collected over the years. Since the completion of a comprehensive monograph (now in progress) seems several years off, it seemed appropriate to publish a collection of material which could be useful to other research workers *now*. While the present paper is long, many details, ramifications, and examples of a type previously published, have been omitted to yield, it is hoped, the best immediate guide the author can offer to aid others in solving optimum design problems. The basic Equivalence Theorem 1 is simple, but applying it can entail analytic labor.

Available literature on verifying convexity of such functions Φ is not too easy to find in applicable form, and therefore some space has been used to list and illustrate useful tools for this purpose. Development of equivalence criteria for Φ's which permit our treatment is the content of the longest section of the paper, Section 4. In addition to Φ_0 and Φ_1, they include the Φ_∞ of (1.4) and certain criteria which arise in response-surface problems in which one purposely fits biased surfaces of a simpler form than $\theta' f(x)$, discussed in [4], [16], [21], and in Section 4F.

For the sake of simplicity, Theorem 1 of the next section presents the basic theory in the case so far described, wherein (i) observations are univariate, (ii) observations have equal variance and cost, (iii) Φ is continuously differentiable at the $M(\xi^*)$ under consideration (which is automatic for some Φ, such as those of (1.2), (1.4) and (1.3) if C is nonsingular, for each of which any optimum $M(\xi^*)$ is nonsingular, implying differentiability). The modifications required to extend the results to other cases are described in Theorem 3 of Section 2, Section 3K, Section 4E, Section 5 and Section 7. A sequel to the present paper will treat illustrations of such modifications in detail.

We conclude this section by recording some additional notation and elementary results of matrix calculus. All matrices considered here have entries in the reals, R^1. We denote by $\mathscr{R}_{k_1,k_2}$ the $k_1 \times k_2$ matrices and by $\mathscr{S}_k$ the symmetric nonnegative definite $k \times k$ matrices. By $\mathscr{M}^+$, $\mathscr{R}^+_{k,k}$, and $\mathscr{S}^+$ we denote the nonsingular members of the corresponding classes without superscript; we write $(\mathscr{M}^+)^- = \{D: D^{-1} \in \mathscr{M}^+\}$.

In practice the criterion function Φ will often be defined not merely on the $\mathscr{M}$ of the application at hand, but on $\mathscr{S}_k$ or even an open subset of $\mathscr{R}_{k,k}$. The computations in this paper are typically carried out for a Φ defined on one of these larger domains. Where such a Φ is written in such fashion as to be defined on an open subset of $\mathscr{R}_{k,k}$, care must be taken in using the calculus. Specifically, in this case denote by $\Phi^{[s]} = \Phi \mid \mathscr{S}_k$ (s for "symmetric") the function Φ restricted to $\mathscr{S}_k$. It is usually easier to differentiate Φ with respect to the k^2 variables on $\mathscr{R}_{k,k}$ rather than to differentiate $\Phi^{[s]}$ with respect to the $k(k+1)/2$ variables on the submanifold $\mathscr{S}_k$ which is really of interest, but one must then note that, when the derivatives are meaningful,

$$(1.7) \qquad \frac{\partial}{\partial m_{ij}} \Big| \mathscr{S}_k = \left(\frac{\partial}{\partial m_{ij}} + \frac{\partial}{\partial m_{ji}}\right) \Big| \mathscr{R}_{k,k} \qquad \text{if } i \neq j ,$$

$$= \frac{\partial}{\partial m_{ii}} \Big| \mathscr{R}_{k,k} \qquad \text{if } i = j .$$

Consequently, a first derivative of $\Phi^{[s]}$ is a sum of one or two derivatives of Φ on $\mathscr{R}_{k,k}$, while a second derivative is a sum of one, two, or four.

This is particularly important for checking convexity of $\Phi^{[s]}$ (condition (2.11) below), since convexity of Φ on $\mathscr{R}_{k,k}$, which is easier to check in terms of the Hessian of second derivatives, may not hold although $\Phi^{[s]}$ is convex; this is the case for such a simple Φ as tr AM^{-1}. One must handle such cases by using (1.7);

or, less conveniently (and hence not hereafter used), by making sure that Φ has been chosen as that extension of $\Phi^{[s]}$ to $\mathscr{R}_{k,k}$ which satisfies $\Phi(M) = \Phi(M')$ (e.g., as 2 tr $A(M + M')^{-1}$ in place of tr AM^{-1} noted above) and by restricting consideration to nonnegative definite $M + M'$.

In Section 2 we will have to consider, for $\bar{M} \in \mathscr{M}$ and $M \in \mathscr{M}$, the function

$$(1.8) \qquad -\sum_{i\leq j} m_{ij}\, \partial\Phi^{[s]}(\bar{M})/\partial\bar{m}_{ij} = -\sum_{i,j} m_{ij}\, \partial\Phi(\bar{M})/\partial\bar{m}_{ij}\,,$$

the relation holding by (1.7) if Φ is an extension of $\Phi^{[s]}$ on an open subset of $\mathscr{R}_{k,k}$ corresponding to $\mathscr{M}$. The right side of (1.8) is more convenient computationally; when Φ is defined and differentiable on an open subset of $\mathscr{R}_{k,k}$, we define thereon the $k \times k$ matrix $\nabla\Phi$ (which is a shorter notation than the more proper grad Φ) by

$$(1.9) \qquad (\nabla\Phi)_{ij} = \partial\Phi(M)/\partial m_{ij}\,;$$

then (1.8) attains its most useful form, $-\text{tr}\, M\,\nabla\Phi(\bar{M})$. If $\Phi^{[s]}$ is only defined on $\mathscr{S}_k$, it will still be convenient to use this form, by letting $(\nabla\Phi^{[s]})_{ij} = \frac{1}{2}(1 + \delta_{ij})\,\partial\Phi^{[s]}/\partial m_{ij}$ for all i, j.

Let E_{ij} be the $k \times k$ matrix with 1 in the (i, j)th place and 0 elsewhere. If b is a positive integer, it is well known (chain rule, or see [6]) that, on $\mathscr{R}^{+}_{k,k}$,

$$(1.10) \qquad \frac{\partial A^b}{\partial a_{ij}} = \sum_{h=0}^{b-1} A^h E_{ij} A^{b-h-1}\,,$$
$$\frac{\partial A^{-b}}{\partial a_{ij}} = -\sum_{h=0}^{b-1} A^{-h-1} E_{ij} A^{-b+h}\,.$$

Thus,

$$(1.11) \qquad \frac{\partial^2 A^b}{\partial a_{ij}\,\partial a_{st}} = \sum_{p,q,r\geq 0;\,p+q+r=b-2} A^p[E_{ij} A^q E_{st} + E_{st} A^q E_{ij}]A^r\,,$$
$$\frac{\partial^2 A^{-b}}{\partial a_{ij}\,\partial a_{st}} = \sum_{p,q,r\geq 0;\,p+q+r=b-1} A^{-p-1}[E_{ij} A^{-q-1} E_{st} + E_{st} A^{-q-1} E_{ij}]A^{-r-1}\,.$$

Noted during revision: We indicated earlier the likelihood that general equivalence theory has also occurred to others, and this is born out by some current publications of which the author has recently been made aware. We now describe the relationship between these results and those of the present paper, using the notation of the latter. The survey paper of Fedorov and Malyutov [12a], Theorem 2.2, states our Theorem 1 omitting the conclusion (2.17)(c), for convex Φ; nonsingularity of $M(\xi^*)$ seems to be assumed, but no differentiability assumption is stated (see our Theorem 3 and comments on it). Peter Whittle [26b], treating Φ as a convex function of ξ rather than of M as we do, proves essentially our Theorem 3, with a saddle point interpretation of $\mathscr{D}(M(\xi^*), M(\xi^*))$, and also (2.18) in the differentiable case; thus (Section 3G, below), further assumptions are needed to derive (2.17)(c) from Whittle's results. He also presents material like that of our Section 6B. A geometric duality approach of Silvey and Titterington [26a], which originally treated D-optimality and yielded iterative methods, is, according to a letter from the former, extendable to other criteria. For D-optimality, other duality cosiderations have been given by R. Sibson (discussion of [28] and a forthcoming paper). These papers all contain additional material, but no overlap with our main results, the development of explicit equivalence theory criteria for such new particular cases as those of Section 4.

2. Basic equivalence results. We examine the proofs of [23] and [10], [12] and see which portions extend to other Φ. The natural analogues of the three statements in each of (1.5) and (1.6) will be stated in (2.1), (2.10) and (2.9). The first is of course

$$\xi^* \text{ is } \Phi\text{-optimum.} \tag{2.1}$$

The third statement is obtained by computing the obvious necessary condition for ξ^* to yield a local minimum,

$$\frac{\partial}{\partial\alpha}\Phi(M([1-\alpha]\xi^*+\alpha\xi))\bigg|_{\alpha=0^+} \geqq 0 \qquad \forall\xi \text{ in } \Xi\,; \tag{2.2}$$

we will have to assume that the differentiation in (2.2) makes sense as a right-hand derivative in α (automatic if Φ is convex), and in our derivation of Theorem 1 we in fact assume Φ continuously differentiable. Since optimality for many common Φ's, such as those of (1.2)—(1.4), entails $M(\xi^*)$ nonsingular, we will sometimes find it convenient to work in terms of the function ψ defined on $(\mathscr{M}^+)^-$ by

$$\psi(D) = \Phi(D^{-1})\,. \tag{2.3}$$

In order to state analogues of (1.5) and (1.6), we must define the function $\mathscr{D}$ of (2.4), which requires Φ to be defined and differentiable in a neighborhood in $\mathscr{P}_k$ of the $M(\xi^*)$ under consideration. This is no restriction in many examples, where $\dim(\mathscr{M}) = k(k+1)/2$; but it might be, in some isolated cases of Φ and of small and discrete $\mathscr{X}$. For the latter cases, the modifications needed in the proof of Theorem 1 are obvious, and for completeness the conclusions are stated as Theorem 2. But the most interesting, natural, and useful examples of Φ (and the most meaningful ones, from the viewpoint of "equivalence theory" parallel to (1.5) and (1.6)) are not so restricted. Hence,

THROUGHOUT THIS PAPER, UNLESS EXPLICITLY STATED TO THE CONTRARY, WE ASSUME Φ DEFINED AND DIFFERENTIABLE ON A NEIGHBORHOOD IN $\mathscr{P}_k$ OF THE $M(\xi^*)$ OR $\bar{M}$ UNDER CONSIDERATION.

For $\bar{M}$ as just described, and for $M \in \mathscr{M}$, define

$$\begin{aligned}\mathscr{D}(M,\bar{M}) = -\frac{\partial}{\partial\alpha}\Phi(\bar{M}+\alpha M)\bigg|_{\alpha=0^+} &= -\operatorname{tr}[M\,\nabla\Phi(\bar{M})]\\ &= \operatorname{tr}[\bar{M}^{-1}M\bar{M}^{-1}\,\nabla\psi(\bar{M}^{-1})]\,;\end{aligned} \tag{2.4}$$

the last two forms require Φ to be defined on an open subset of $\mathscr{R}_{k,k}$, as described in connection with (1.9), and the last form requires $\bar{M}$ to be nonsingular; if Φ is only defined on an open subset of $\mathscr{S}_k$, (2.4) becomes the first form of (1.8); the last equality of (2.4) depends on (1.7) and the chain rule $\nabla\Phi = -\bar{M}'^{-1}(\nabla\psi)\bar{M}'^{-1}$, which in turn depends on (1.10) for A^{-1}; of course, $\bar{M} + \alpha M \in \mathscr{S}_k$ for $\alpha > 0$. We also abbreviate, for ξ_x the measure assigning unit probability to the single point x, and for $M(\xi')$ of the form $\bar{M}$ described above (2.4) (the last two forms below holding when they did in (2.4)),

$$
\begin{aligned}
d(x, \xi') &= \mathscr{D}(M(\xi_x), M(\xi')) \\
&= -f(x)' \nabla\Phi(M(\xi'))f(x) \qquad (2.5)\\
&= f(x)'M^{-1}(\xi') \nabla\psi(M^{-1}(\xi'))M^{-1}(\xi')f(x) ,
\end{aligned}
$$

and

$$\bar{d}(\xi') = \sup_{x \in \mathscr{X}} d(x, \xi') . \qquad (2.6)$$

We also abbreviate

$$
\begin{aligned}
d^{*}(\xi') &= \mathscr{D}(M(\xi'), M(\xi')) \qquad (2.7)\\
&= -\operatorname{tr} M(\xi') \nabla\Phi(M(\xi')) \\
&= \operatorname{tr} M^{-1}(\xi') \nabla\psi(M^{-1}(\xi')) .
\end{aligned}
$$

We shall append a subscript Φ to $\mathscr{D}$, $\bar{d}$, or d^{*} whenever ambiguity might otherwise arise.

We can now rewrite (2.2) as

$$\mathscr{D}(M(\xi), M(\xi^*)) \leq d^{*}(\xi^*) \qquad \forall\xi \text{ in } \Xi . \qquad (2.8)$$

Since $\mathscr{D}(M, \bar{M})$ is linear in M, (2.8) is valid if and only if it is valid for all ξ of the form ξ_x. Also, there is equality in (2.8) when $\xi = \xi^*$. Thus, (2.2) is equivalent to

$$\bar{d}(\xi^*) = d^{*}(\xi^*) . \qquad (2.9)$$

(Of course, the linearity of $\mathscr{D}(M, \bar{M})$ in M implies $\bar{d} \geqq d^{*}$.)

The relation (2.9) generalizes the third statement of (1.5) and (1.6). The extension of the second statement is

$$\bar{d}(\xi^*) = \inf_{\xi} \bar{d}(\xi) . \qquad (2.10)$$

We now study the implications among (2.1), (2.9), and (2.10). Of course, we have already seen that (2.1) implies (2.9), which we hereafter call *local Φ-optimality* of ξ^*. The most useful general condition on Φ such that local Φ-optimality implies global Φ-optimality is that there exist a strictly increasing function G on $\Phi(\mathscr{M})$ which is continuously differentiable at $\Phi(M(\xi^*))$ and such that

$$G \circ \Phi \quad \text{is convex on} \quad \mathscr{M} , \qquad (2.11)$$

which means $G(\Phi([1 - \alpha]\bar{M} + \alpha M))$ convex in α, $0 < \alpha < 1$. (See also the first parts of Remark 3B.) If $\Phi(M(\xi')) < \Phi(M(\xi^*))$, then (2.11) implies

$$(2.12)\qquad 0 > \frac{\partial}{\partial\alpha} G(\Phi([1-\alpha]M(\xi^*) + \alpha M(\xi'))\Big|_{\alpha=0+}$$
$$= G'(\Phi(M(\xi^*)))\frac{\partial}{\partial\alpha}\Phi([1-\alpha]M(\xi^*) + \alpha M(\xi'))\Big|_{\alpha=0+},$$

violating (2.2). Thus, (2.11) and (2.2) imply (2.1).

We require a preliminary result before turning to (2.10). We shall find it convenient to invoke the condition on $\mathscr{M}' = \{M: M \in \mathscr{M}, \Phi(M) < \infty\}$,

$$(2.13)\qquad \Phi(M) = P(H(M)),$$

where H is positive and is *homogeneous of positive degree* h, and P is strictly decreasing and continuously differentiable on $H(\mathscr{M}')$, and such that $\log P^{-1}(\phi)$ is convex in ϕ. We will discuss these assumptions and the consequences of their violation (e.g., of $h < 0$) in Section 3. For the moment, we note that, abbreviating $M_0 = M(\xi_0)$ and $\phi_0 = \Phi(M(\xi_0))$, we have under (2.13),

$$(2.14)\qquad \begin{aligned} d^\sharp(\xi_0) = \mathscr{D}(M_0, M_0) &= -hH(M_0)P'(H(M_0)) \\ &= -hP'(P^{-1}(\phi_0))P^{-1}(\phi_0) \\ &= -h/[d\log P^{-1}(\phi)/d\phi|_{\phi=\phi_0}]. \end{aligned}$$

From the fact that $\log P^{-1}$ is decreasing and convex, we conclude that, on $\mathscr{M}'$,

$$(2.15)\qquad d^\sharp(\xi) \text{ is a non-decreasing function of } \Phi(M(\xi)).$$

The role of (2.13) is to insure (2.15); without knowing that Φ and $d^\sharp$ are functionally related, we cannot hope to relate (2.9) and (2.10).

Now assume ξ^* is Φ-optimum and hence optimum in the sense of (2.9), and that ξ^{**} satisfies (2.10) (with ξ^{**} for ξ^* there). Then, assuming (2.13) and using in order (2.10), (2.9), (2.15), and the trivial $d^\sharp \leqq \bar{d}$, we have

$$(2.16)\qquad \bar{d}(\xi^{**}) \leqq \bar{d}(\xi^*) = d^\sharp(\xi^*) \leqq d^\sharp(\xi^{**}) \leqq \bar{d}(\xi^{**}),$$

so that all members of (2.16) are equal. We conclude that ξ^* satisfies (2.10) and ξ^{**} satisfies (2.9) and (by (2.15), since $d^\sharp(\xi^*) = d^\sharp(\xi^{**})$) (2.1). Thus, under (2.13) we have shown that (2.1) and (2.10) are equivalent.

Two details remain to be treated, both just as in the D-optimality proof. Firstly, it is clear that any design ξ^* satisfying (2.9) assigns all measure to $\{x: d(x, \xi^*) = \bar{d}(\xi^*)\}$, since otherwise one obtains $d^\sharp(\xi^*) < \bar{d}(\xi^*)$. Secondly, if Φ satisfies (2.11), then any convex linear combination of designs satisfying (2.1) is also $G \circ \Phi$-optimum and hence Φ-optimum; and if (2.11) is strengthened by demanding that $G \circ \Phi$ be *strictly* convex, then all matrices $M(\xi)$ must be identical for Φ-optimum ξ.

We summarize:

THEOREM 1 ("Equivalence Theorem"). *For Φ continuously differentiable in a* *neighborhood of* $M(\xi^*)$,

$$(2.17)\qquad \begin{aligned}&\text{(a)}\quad (2.1) \Rightarrow (2.9)\,;\\ &\text{(b)}\quad \textit{Under}\quad (2.11): (2.9) \Rightarrow (2.1)\,;\\ &\text{(c)}\quad \textit{Under}\quad (2.13): (2.1) \Leftrightarrow (2.10)\,.\end{aligned}$$

Furthermore,

$$(2.18)\qquad \xi^* \quad \textit{satisfies} \quad (2.9) \Leftrightarrow \xi^*\{x: d(x, \xi^*) = \bar{d}(\xi^*)\} = 1\,.$$

Under (2.11), *the* Φ*-optimum* ξ^**'s (and corresponding* $M(\xi^*)$*'s) are convex; if* $G \circ \Phi$ *is strictly convex in a neighborhood in* $\mathscr{M}$ *of an optimum* $M(\xi^*)$, *then* $M(\xi^*)$ *is the unique optimum* M.

Recalling the remarks just below (2.4), we also state:

THEOREM 2. *If we do not assume* Φ *is defined in a neighborhood (in* $\mathscr{P}_k$*) of* $\mathscr{M}$, *then Theorem* 1 *is valid with the following alterations: In* (2.17) (a) *and* (b), *replace* (2.9) *by* (2.2) *or by*

$$(2.19)\qquad \inf_{x \in \mathscr{X}} \frac{\partial}{\partial \alpha} \Phi(M([1-\alpha]\xi^* + \alpha\xi_x))\Big|_{\alpha=0+} \geqq 0\,;$$

delete (2.17)(c); *replace* (2.18) *by*

$$(2.20)\qquad (2.19) \Rightarrow \xi^*\left\{x: \frac{\partial}{\partial \alpha} \Phi(M([1-\alpha]\xi^* + \alpha\xi_x))\Big|_{\alpha=0+} = 0\right\} = 1\,.$$

Finally, we turn to the modification of our theory in the event that Φ is not everywhere differentiable. A simple assumption with which to work is

(2.21) H is continuous on a neighborhood of $\mathscr{M}$, where Φ satisfies (2.11) and (2.13); Φ is no longer assumed differentiable.

Since Φ is convex and continuous, for fixed M and $\bar{M}$, the right-hand derivative of $\Phi((1-\alpha)\bar{M} + \alpha M)$ exists at $\alpha = 0$, the derivative $(\partial/\partial\alpha)\Phi((1-\alpha)\bar{M} + \alpha M)$ exists for almost all α, and the derivative at a convergent sequence of non-exceptional α's converges. This differentiability conclusion is seen also to hold for $H(\bar{M} + \alpha M) = (1+\alpha)^{-h}H((1+\alpha)^{-1}\bar{M} + \alpha(1+\alpha)^{-1}M)$, upon differentiating $P^{-1} \circ \Phi$. In conformity with the first relation of (2.4), we use the same definition with the derivative understood to be right-hand; or, alternatively, we can use

$$(2.22)\qquad \begin{aligned}\mathscr{D}(M, \bar{M}) &= -\lim_{\alpha \downarrow 0} \frac{\partial}{\partial \alpha} \Phi(\bar{M} + \alpha M)\\ &= -P'(H(\bar{M})) \lim_{\alpha \downarrow 0} \frac{\partial}{\partial \alpha} H(\bar{M} + \alpha M)\,,\end{aligned}$$

the limit being taken on an unexceptional sequence. Also, $d^*(\xi) = \mathscr{D}(M(\xi), M(\xi)) = -hP'(H(M(\xi)))H(M(\xi))$, as before. In (2.2) the evaluation at $\alpha = 0^+$ is replaced by taking $\lim_{\alpha \downarrow 0}$ on a non-exceptional sequence (which depends on

ξ, ξ^*), or by taking right-hand derivative at $\alpha = 0$. In place of (2.9) we obtain, using $H((1-\alpha)\bar{M} + \alpha M) = (1-\alpha)^{-h} H(\bar{M} + \alpha(1-\alpha)^{-1} M)$,

$$\sup_\xi \mathscr{D}(M(\xi), M(\xi^*)) = d^\sharp(\xi^*) . \tag{2.23}$$

With the left side of (2.23) replacing $\bar{d}(\xi^*)$, the demonstration of (2.16) still holds. Thus, we obtain:

THEOREM 3. *Under Assumption* (2.21), *Theorem* 1 *is valid with the following alterations*: *Delete* (2.18); *in* (2.17), *replace* (2.9) *by* (2.23) *and* (2.10) *by*

$$\xi' = \xi^* \quad \textit{minimizes} \quad \sup_\xi -\mathscr{D}(-M(\xi), M(\xi')) , \qquad \xi' \in \mathscr{M}' . \tag{2.24}$$

For comments on this theorem, see Section 3K.

3. Remarks on complements on equivalence theorems.

A. The useful natural partial ordering on $\mathscr{M}$ is well known [17] to be

$$M_1 \succ M_2 \Leftrightarrow M_1 - M_2 \quad \text{is nonnegative definite.} \tag{3.1}$$

Most sensible Φ are non-increasing in this ordering; that is,

$$M_1 \succ M_2 \Rightarrow \Phi(M_1) \leqq \Phi(M_2) ; \tag{3.2}$$

in fact, if Φ did not satisfy (3.2), one would have the anomaly of a less informative experiment being preferred to a more informative one (without any consideration of experimental costs). One could often implement this anomaly by "throwing away" some of the information in M_1 so as to obtain M_2, as discussed on page 286 of [17]; in terms of optimality, this amounts to replacing Φ by $\tilde{\Phi}(M) = \min \{\Phi(M') : M' \prec M\}$ and solving the $\tilde{\Phi}$-optimality problem where $\tilde{\Phi}$ now satisfies (3.2). Nevertheless, we must be careful to distinguish that a Φ, which is not sensible for *all* problems because it violates (3.2) when $\mathscr{M}$ is replaced by $\mathscr{S}_k$, can be useful in a particular problem where (3.2) is satisfied. An example is $\Phi(M) = \text{tr}\, M^2$ discussed in Section 4H.

B. The use of convexity of Φ is simply to make local optimality imply optimality, and more general conditions of *unimodality* can be used instead; such conditions are of course not generally as easy to verify as convexity.

Regarding the form of (2.11) used in the proof of Theorem 1, Φ-optimality obviously coincides with $G \circ \Phi$-optimality if G is strictly increasing. For $k = 1$, Φ strictly decreasing (a strict form of (3.2)) implies (2.11). For $k > 1$, (2.11) is not so automatic. Let $Q_\phi = \{M : M \in \mathscr{M}, \Phi(M) = \phi\}$. If, for any value ϕ, some convex mixture $\bar{M}$ of elements of Q_ϕ has $\Phi(\bar{M}) > \phi$, then clearly no rescaling $G \circ \Phi$ can be convex. On the other hand, if no such $\bar{M}$ exists for any ϕ and if (3.2) is strengthened by adding that $\Phi(\alpha M) < \Phi(M)$ for $\alpha > 1$ and $M \in \mathscr{M}'$ (so that the Q_ϕ are not k-dimensional), then it is easy to see that such a G exists. A simple example where no G exists is $\Phi(M(\xi)) = \sum_1^k [1 + \lambda_i^2(\xi)]^{-1}$ where the $\lambda_i(\xi)$ are the eigenvalues of $M(\xi)$; this satisfies the strengthened (3.2), but the condition on Q_ϕ fails for small diagonal M in Q_ϕ.

C. If we assume the strengthened form of (3.2) with $\mathscr{M}'$ replaced by $\mathscr{P}_k^+$ then the condition (2.13) is actually necessary as well as sufficient for (2.15) to hold on $\mathscr{P}_k^+$ (or on any $\mathscr{M}^+$ of full dimension). This is seen by defining g_1 by fixing M_0 and writing $g_1(\Phi(tM_0)) = \text{tr}\,(tM_0)\,\nabla\Phi(tM_0)$ and then solving (for Φ) the univariate differential equation $\text{tr}\,(tM)\,\nabla\Phi(tM) = g_1(\Phi(tM))$ in t for each fixed M. One obtains $\Phi(tM) = g_2^{-1}(\log[tc(M)])$ where $c(M)$ is an integration constant and $g_2(u)$ is an indefinite integral of $1/g_1$. This is easily translated into the form (2.13).

D. We now consider the relationship between (2.13) and (2.11). Convexity and monotonicity of Φ are not sufficient for (2.13), as is illustrated for $k = 1$, $\mathscr{M} = \mathscr{P}_1$, by $\Phi(x) = e^{-x}$, for which $d^* = -x\Phi'(x) = xe^{-x}$ is not a monotone function of Φ. In the other direction, convexity and monotonicity of Φ are also not necessary for (2.13), as is illustrated by $\Phi(M) = (m_{33}^2 + m_{11}^2)^{-1} + (m_{33}^2 + m_{22}^2)^{-1}$ on the diagonal 3×3 matrices; as in the example at the end of Remark B (hold m_{33} fixed), no $G \circ \Phi$ is convex; but, with $h = 1$ and $H = \phi^{-2} = P^{-1}(\phi)$, we have $P \downarrow$ and $\log P^{-1}(\phi)$ convex. We also note that, under (3.2) and (2.13), it is easily verified that, except in degenerate cases, H cannot take on both positive and negative values, since that would make P non-monotone; it would also violate (2.11).

E. A problem may sometimes be studied most conveniently in terms of the ψ of (2.3) rather than in terms of Φ. Since (see, e.g., [17] and Section 4B1 below)

$$(3.3) \qquad [(1-\alpha)M_1 + \alpha M_2]^{-1} \prec (1-\alpha)M_1^{-1} + \alpha M_2^{-1},$$

we see at once that (2.11) on $\mathscr{M}^+$ is a consequence of (3.2) and

$$(3.4) \qquad G \circ \psi(D) \quad \text{is convex in} \quad D \quad \text{on} \quad (\mathscr{M}^+)^- .$$

In the cases treated in [23] and [10], (3.4) is satisfied; in fact, in [10] ψ is linear. But, in general, (2.11) is weaker than (3.4). For example, if $\mathscr{M}^+$ consists only of diagonal matrices, then convexity of $\Phi(M) = \text{tr}\,(M^{-\frac{1}{2}})$ is obvious; but no rescaling $G \circ \psi$ of $\psi(D) = \text{tr}\, D^{\frac{1}{2}}$ can be convex on $\mathscr{T}^+$ when $k > 1$ because, in analogy with the example of Remark B above, ψ is not convex on mixtures of two diagonal matrices with the same (permuted) set of non-identical diagonal entries.

F. Of course, the use of (2.13) is unchanged if we make h *negative* and P *increasing*. On the other hand, if we reverse only *one* of the two conditions $h > 0$, $P \downarrow$, we obtain d^* *decreasing* in Φ, in place of (2.15). The argument of (2.16) fails, and in general we cannot hope for the equivalence (2.17)(c) to be valid. For example, the function $\Phi(M) = \text{tr}\, M^2$ treated in Section 4H will not generally be minimized by the same M that minimizes $\bar{d} = \min_x f(x)'Mf(x)$; these work in opposite directions. What is more interesting is the study of such criteria in settings which satisfy an additional restriction such as that of Section 4H ($\text{tr}\, M =$ constant). As illustrated in the example there, it is then often possible to achieve (2.17)(c). Since this is a rather special consideration, we shall return to it elsewhere.

G. If (2.11) is satisfied but (2.13) is violated (as in examples like that of Section 4F or, less important, in F above), the first two parts of (2.17) remain valid, and these are the most important parts of the equivalence theorem. For, with a rare exception such as the criterion of G-optimality [22], $\bar{d}_\Phi$ does not arise as an optimality criterion of interest in itself, but only as a tool for proving Φ-optimality; it serves the latter role in (2.9) and (2.17)(a)-(b) (and in the resulting computational techniques and bounds of Section 6), rather than in (2.10) and (2.17)(c).

H. Simple examples show that (2.13) cannot be completely dispensed with in proving (2.17)(c); one such example is given in Section 4F. There are many simple and natural settings in which minimizing Φ, $d^\sharp$, and $\bar{d}$ can lead to three different designs in the absence of (2.13). (The criterion function $\bar{d} - d^\sharp$, which a Φ-optimum design still trivially minimizes, is of even less general intrinsic significance than $\bar{d}$.) There are special cases where (2.17)(c) is satisfied in the absence of (2.13), but we do not know definitive results. Thus, it is obvious that (2.15) can be replaced in our proof that (2.1) $\Rightarrow$ (2.10) (respectively, the opposite), by the condition that the Φ-optimum design is $d^\sharp$-optimum (respectively, the opposite), but this last is not necessary. Further conditions will be discussed elsewhere. It is interesting from the game-theoretic point of view to note that, if a $\bar{d}$-optimum design $\bar{\xi}$ is not Φ-optimum, $d(x, \bar{\xi})$ cannot achieve its maximum on the support of $\bar{\xi}$; this does not contradict usual "minimax behavior," since $\bar{\xi}$ is not the maximin strategy of the other player for the game with payoff d.

I. The fact that constancy of $d^\sharp$ in the case (1.2) makes (2.9) easier to verify there than in the case (1.3) suggests we delimit those Φ for which there is a regular G_1 such that $d_{\tilde{\Phi}}^\sharp$ is constant, where $\tilde{\Phi} = G_1 \circ \Phi$. (If some $G \circ \Phi$ satisfies (2.11), so will $(G \circ G_1^{-1}) \circ \tilde{\Phi}$, so we need not worry about convexity.) Solving the differential equation tr $M \nabla\tilde{\Phi}(M) =$ constant (as in C above) yields $\tilde{\Phi}(M) = c_1 \log [c_2 H(M)]$ where H is homogeneous of degree h and $d_{\tilde{\Phi}}^\sharp = c_1 h$. This is exactly (2.13) with $P_\Phi(H) = G_1^{-1}(c_1 \log [c_2 H])$. (In terms of (2.14)—(2.15), $\log P_{\tilde{\Phi}}^\sharp$ is linear rather than *strictly* convex, and $d^\sharp$ is constant rather than strictly increasing.) Now, for any positive criterion function Φ^*, the chain rule always yields $\mathscr{D}_{\log \Phi^*}(M, \bar{M}) = \mathscr{D}_{\Phi^*}(M, \bar{M})/\Phi^*(\bar{M})$, and thus $\bar{d}_{\log \Phi^*}(\xi) = \bar{d}_{\Phi^*}(\xi)/\Phi^*(M(\xi))$. Hence, if the original Φ satisfies (2.13) and we put $\Phi^* = 1/P_\Phi^{-1}(\Phi)$ so that Φ^* is equivalent to Φ in the sense of (2.1) and Φ^* is homogeneous of degree $-h$, we have (by (2.14) and the first sentence of F, taking $P_{\Phi^*}(H) = H$) $d_{\Phi^*}^\sharp(\xi) = h\Phi^*(M(\xi))$, and thus $\bar{d}_{\log \Phi^*}(\xi) = h\bar{d}_{\Phi^*}(\xi)/d_{\Phi^*}^\sharp(\xi)$. This whole process, then, amounts to replacing Φ (satisfying (2.13)) by Φ^*, writing (2.9) for Φ^*, and dividing both sides by $d_{\Phi^*}^\sharp$, and it does not distinguish why (2.9) was genuinely simpler in the case (1.2). The reason for the latter is the multiplicative form of the determinant, which yields $\mathscr{D}_{\Phi_0}(M, \bar{M}) = \text{tr } \bar{M}^{-1}M/\det \bar{M}$.

J. In the manipulations of Section 3I it became evident that (2.17)(c) could be altered by replacing $\bar{d}_\Phi$ by $\bar{d}_{\tilde{\Phi}}$ in (2.10) for certain $\tilde{\Phi}$ equivalent to Φ in the sense of (2.1). This is quite a general result: if $\tilde{\Phi} = \tilde{G} \circ \Phi$ for some strictly increasing $\tilde{G}$, *and* $\tilde{\Phi}$ *(in place of* Φ*) satisfies* (2.13), then the derivation from (2.13) through the paragraph containing (2.16) proceeds as before, but with the criterion function Φ replaced by $\tilde{\Phi}$ everywhere. Since $\tilde{\Phi}$- and Φ-optimality in the sense of (2.1) are equivalent, we have proved

THEOREM 4. *If $\tilde{G}$ is strictly increasing and $\tilde{\Phi} = \tilde{G} \circ \Phi$ satisfies* (2.13) *then* (2.17) (c) *can be replaced by*

$$(3.5) \qquad \xi^* \ \textit{is } \Phi\textit{-optimum} \Leftrightarrow \xi^* \quad \textit{minimizes} \quad \bar{d}_{\tilde{\Phi}}(\xi) .$$

Of course, we cannot replace $\bar{d}_{\tilde{\Phi}}$ in (3.5) by the $\bar{d}_\Phi$ of Theorem 1 (full equivalence) without the original (2.13) for Φ itself.

Although simultaneous minimization by the same ξ^* of $\bar{d}_{\tilde{G}\circ\Phi}$ for various $\tilde{G} \circ \Phi$ satisfying (2.13) is not a completely obvious result, replacement of (2.9) by $\bar{d}_{\tilde{\Phi}} = d_{\tilde{\Phi}}^*$ in (2.17)(a)-(b) requires no proof, whether or not $\tilde{G} \circ \Phi$ satisfies (2.13). Thus, we have the interesting possibility of varying $\tilde{G}$ to achieve the most useful possible computational form of $\bar{d}_{\tilde{\Phi}} = d_{\tilde{\Phi}}^*$. (In the discussion of Section 3I we disposed of the possibility of making $d_{\tilde{\Phi}}^*$ constant.) This will be illustrated in the examples of Section 4C.

K. Theorem 3 will be used for such criteria as E-optimality ((1.4) and Section 4E). The critical difference from Theorem 1 is of course that the supremum in (2.23) need not be the same if we restrict ξ to the form ξ_x as we did in going from (2.8) to (2.9); that is, $\mathscr{D}(-M, \bar{M})$ is no longer linear in M. Since $\Phi((1 - \alpha)\bar{M} + \alpha M)$ is *convex* in M, this non-linearity will be in the direction of possibly making (2.2) valid if ξ is restricted to ξ_x but not valid for certain other ξ. This will be illustrated in Section 4E.

Of course, (2.23) is not as useful a criterion as (2.9), and may sometimes be as difficult to implement as (2.1) directly. The example of Section 4E illustrates how particular properties of ξ^* can simplify (2.23). In all cases, violation of (2.23) when ξ is restricted to the form ξ_x obviously implies non-Φ-optimality of ξ^*.

The right-hand derivative in the definition of $\mathscr{D}$ can easily differ from the left-hand one, and so the expression $(\partial/\partial\alpha)\Phi(\bar{M} - \alpha M)|_{\alpha=0+}$, which is equal to those of (2.4) under the differentiability assumption of Theorem 1, cannot be used in Theorem 3.

L. Much of the preceding material can be developed along game-theoretic lines in the manner of Karlin and Studden [14]. However, the present treatment seems more elementary and also seems to separate more clearly the conditions needed for the "minimax" criterion (2.10) to coincide with (2.1). Another aspect of the game-theoretic development will be mentioned in Section 7.

4. Computations and illustrations.

4A. *Transformations.* We now discuss the simple consequence of linear transformation on M and monotone transformation on Φ. Suppose $\mu = AMA' + B$ where μ is $k' \times k'$ and A is $k' \times k$, and where B is $k' \times k'$ nonnegative definite. This can be thought of as relating the given problem in terms of M to another problem with regression function k'-vector g on $\mathscr{X}$, where $\mu = B + \int gg'\xi(dx)$ and where $g = Af$, and where B is the information matrix available from previous experimentation (suitably normalized relative to $\xi(\mathscr{X}) = 1$). Suppose

$$\Phi(M) = G(\tilde{\Phi}(AMA' + B)) \tag{4.1}$$

relates the optimality function Φ on $\mathscr{M}$ to that, $\tilde{\Phi}$, on $\{\mu\}$. Then, since $\partial\Phi(\bar{M})/\partial\bar{m}_{ij} = G'(\tilde{\Phi}(\bar{\mu}))[A' \nabla\tilde{\Phi}(\bar{\mu})A]_{ij}$, we obtain

$$\begin{aligned} \mathscr{D}_\Phi(M, \bar{M}) &= -\operatorname{tr}\{M \nabla\Phi(\bar{M})\} = -G'(\tilde{\Phi}(\bar{\mu})) \operatorname{tr}\{MA' \nabla\tilde{\Phi}(\bar{\mu})A\} \\ &= G'(\tilde{\Phi}(A\bar{M}A' + B))\mathscr{D}_{\tilde{\Phi}}(AMA', A\bar{M}A' + B) . \end{aligned} \tag{4.2}$$

This allows $\mathscr{D}_\Phi$ to be computed in terms of $\mathscr{D}_{\tilde{\Phi}}$.

Similarly, if $\Phi(M)$ is rewritten $\psi(D)$ with $D = M^{-1}$, as in (2.3) and the last form of (2.4), and if

$$\psi(D) = G(\tilde{\psi}(ADA' + B)) , \tag{4.3}$$

we obtain for $\mathscr{D}$

$$\operatorname{tr}\{\bar{D}M\bar{D} \nabla\psi(\bar{D})\} = G'(\tilde{\psi}(A\bar{D}A' + B)) \operatorname{tr}\{A\bar{D}M\bar{D}A' \nabla\tilde{\psi}(A\bar{D}A' + B)\} . \tag{4.4}$$

4B. *Convexity tools.* We are mainly interested in verifying (2.11) and in computing $\mathscr{D}$ in the examples below. Usually (2.13) is of secondary interest, as we have mentioned earlier, and the status of (3.2) will usually be evident from the computation of $\mathscr{D}$, since (3.2) can be proved by showing that, if $M \in \mathscr{P}_k$ and $\Phi(M + \delta M)$ is defined for δ small positive and for $\delta = 0$,

$$0 \leqq \frac{\partial}{\partial\delta} \Phi(\bar{M} + \delta M)\bigg|_{\delta=0} = \operatorname{tr} M \nabla\Phi(\bar{M}) . \tag{4.5}$$

See [25b] for alternatives to (4.5).

Often the computation of the Hessian of $(k^2 + k)(k^2 + k + 2)/8$ second derivatives of Φ with respect to the $(k^2 + k)/2$ m_{ij}'s, in order to verify convexity, is tedious even in simple examples, as we shall illustrate briefly in C below, so that it is expeditious to invoke general convexity results instead. These are scattered in the literature (a recent list of some being in [2]) in such a way that the optimum design practitioner will often have difficulty finding what he needs. A forthcoming monograph by Marshall and Olkin [25a] should remedy this. Meanwhile, it seems useful to list three of the more useful tools in our setting.

1. If $\Gamma: \mathscr{M} \to \mathscr{P}_{k'}$ is convex in the ordering (3.2) on $\mathscr{P}_{k'}$ and Φ: convex span $(\Gamma(\mathscr{M})) \to R^1$ is convex and increasing, then $\Phi \circ \Gamma$ is convex. Also, R^1 may be replaced by $\mathscr{P}_{k''}$ in this statement. A familiar example on $\mathscr{P}_k^+$ is $\Gamma(M) = M^{-1}$, as described in Section 3E. The paper by Bendat and Sherman [3] discusses the classical work of Loewner (and his students Dobsch and Kraus)

on particular kinds of monotone matrix-valued functions $\tilde{\gamma}$ induced by functions $\gamma: R^1 \to R^1$ as

$$\tilde{\gamma}(M) = Q \operatorname{diag}[\gamma(\delta_1), \gamma(\delta_2), \cdots, \gamma(\delta_k)]Q' \tag{4.6}$$

where $\Delta = \operatorname{diag}[\delta_1, \cdots, \delta_k]$ is the diagonal matrix with diagonal entries $\{\delta_i\}$, Q is orthogonal, and $M = Q\,\Delta Q'$. These authors also extend work of Kraus on corresponding convex functions $\Gamma: \mathscr{P}_k \to \mathscr{P}_k$, which is our present interest.

2. There are known results on norms (convex by definition) on $\mathscr{P}_k$. For example, $(\operatorname{tr} A^p)^{1/p}$, $1 \leqq p < \infty$ (and maximum eigenvalue of A for $p = \infty$) is the L^p-norm.

3. A result of Ky Fan [8] (also related to results of von Neumann) states that, if $\lambda_1(A) \geqq \lambda_2(A) \geqq \cdots \geqq \lambda_k(A)$ are the eigenvalue of A, then $\sum_1^m \lambda_i(A)$ is convex on $\mathscr{P}_k$ $\forall m$. Hence, for $A, B \in \mathscr{P}_k$ and $0 < \alpha < 1$, if we define $x_i = \alpha\lambda_i(A) + (1-\alpha)\lambda_i(B)$ and $y_i = \lambda_i(\alpha A + (1-\alpha)B)$, we have the x_i and y_i nonnegative and increasing in i, and

$$\begin{aligned} \textstyle\sum_1^m x_i &\geqq \textstyle\sum_1^m y_i\,, \qquad 1 \leqq m < k\,, \\ \textstyle\sum_1^k x_i &= \textstyle\sum_1^k y_i\,, \end{aligned} \tag{4.7}$$

the last by linearity of the trace. But the "majorization" of $\{y_i\}$ by $\{x_i\}$ given by (4.7) is well known (e.g., [25c], [2]) to be equivalent to $\sum_1^k \gamma(x_i) \geqq \sum_1^k \gamma(y_i)$ for each real continuous convex function γ defined on some real interval. For each such γ, we conclude that $\sum_1^k \gamma \circ \lambda_i$ is convex on $\mathscr{P}_k$.

4C. *Simple trace criteria.* We recall the definition $M^m = Q\,\Delta^m Q'$ of (4.6) for $m > 0$ and $M \in \mathscr{P}_k$, or $m < 0$ and $M \in \mathscr{P}_k^+$. (To use (1.9), define M^m on $R_{k,k}^+$ through the exponential mapping.) We now consider the following parameters and family of functions on $\mathscr{P}_k$:

$$\begin{aligned} &r \text{ and } m \text{ are real and nonzero;} \\ &k' \text{ is a positive integer;} \quad k' \leqq k \text{ if } m < 0\,; \\ &C \in \mathscr{P}_{k'}\,; \quad A \in \mathscr{R}_{k',k}\,, \text{ of rank } k' \text{ if } m < 0\,; \\ &\Phi_{m,r,A,C}(M) = [\operatorname{tr} C(AMA')^m]^r \quad (\text{possibly } +\infty)\,. \end{aligned} \tag{4.8}$$

(A related, sometimes more useful family, will be introduced in (4.17).) As indicated in Remark 3B, changing the exponent r in (4.8) from 1 or -1 to a positive multiple thereof can only exhibit the G of (2.11), but does not change the optimum designs or validity of (2.17) (b). In what follows it will suffice to write $AMA' = \mu$ in (4.8) and to consider $[\operatorname{tr} \mu^m]^r$, since at worst this can mean *strict* convexity (not convexity itself) does not carry over from a function of μ to that of M; the question of strictness in M is then easy to answer.

We now illustrate the use of various techniques such as those mentioned in B above, to verify (2.11) in several cases. We assume $k' > 1$ since $k' = 1$ is trivial.

Case 1. Assume m integral and $r = 1$. First suppose $m > 0$, and write $\bar{E}_{ij} = E_{ij} + E_{ji}$ if $i \neq j$ and $\bar{E}_{ii} = E_{ii}$. From (1.11) and (1.7) we have, on $\mathscr{P}_{k'}^+$, for

$i \leqq j$ and $s \leqq t$,

$$\frac{\partial^2}{\partial \mu_{ij}\, \partial \mu_{st}} \operatorname{tr} C\mu^m = 2 \operatorname{tr} \textstyle\sum_{p,q,r\geqq 0;\, p+q+r=m-2} C\mu^p \bar{E}_{ij}\, \mu^q \bar{E}_{st}\, \mu^r \,. \tag{4.9}$$

Let x_{ij}, $i \leqq j$, be real, and write $\bar{X} = \sum_{i\leqq j} x_{ij} \bar{E}_{ij}$. Note that $\bar{X}$ is symmetric. We obtain, with the same right-side summation as in (4.9),

$$\textstyle\sum_{i\leqq j;\, s\leqq t} x_{ij}\, x_{st} \frac{\partial^2}{\partial \mu_{ij}\, \partial \mu_{st}} \operatorname{tr} C\mu^m = 2 \sum \operatorname{tr} C\mu^p \bar{X} \mu^q \bar{X} \mu^r \,. \tag{4.10}$$

Convexity as in (2.11) requires nonnegativity of (4.10) for all symmetric $\bar{X}$ in $\mathscr{R}_{k',k'}$. If $m = 1$, this is of course trivial. If $m = 2$, (4.10) becomes $2 \operatorname{tr} C\bar{X}^2$, which is nonnegative since C and $\bar{X}^2$ are in $\mathscr{P}_{k'}$. A similar derivation holds for $m = -1$. For other integral m and $C \neq$ const. $\times I$, convexity on $\mathscr{P}_{k'}$ fails, and changing r does not alter this. There is an open subset of $\mathscr{P}_{k'}$ (depending on C) where $\operatorname{tr}(C\mu^m)$ is convex, which may be relevant in particular examples; we shall not consider this further, here.

If $C = I$, the summand on the right side of (4.10) becomes $\operatorname{tr}[(\mu^{p+r})(\bar{X}\mu^q\bar{X})]$, which is nonnegative because each parenthesized matrix is in $\mathscr{P}_{k'}$. This and the analogue for $m < 0$ yields *convexity of* $\operatorname{tr}(\mu^m)$ *for all integers* m. (See also Case 3.)

Case 2. Assume m and C arbitrary, $|r| = 1$. A theorem in the work of Bendat and Sherman [3] alluded to in B1 above yields convexity of the function $\Gamma(\mu) = \mu^m$ on $\mathscr{P}_k^+$ for all k if and only if the function $(z^m - 1)/(z - 1)$ has nonnegative imaginary part in the upper half of the complex plane. This can be verified to be true if and only if $-1 \leqq m \leqq 0$ or $1 \leqq m \leqq 2$. Hence, $\Phi_{m,1,A,C}$ *satisfies* (2.11) *for all* k, k', A, *and* C, *provided* $-1 \leqq m \leqq 0$ *or* $1 \leqq m \leqq 2$. The example at the end of Section 3E indicates why, even when $C = I$, changing r *to another positive value* does not yield an increased range of m unless $k' = 1$. However, the Bendat–Sherman tool yields convexity of $-\mu^m$ for $0 \leqq m \leqq 1$; since $-1/\operatorname{tr}\mu$ is convex and increasing, the result at the start of B1 yields that $\Phi_{m,-1,A,C}$ *satisfies* (2.11) *for all* k, k', A, *and* C, *provided* $0 \leqq m \leqq 1$.

Case 3. Assume $C = I$, m arbitrary. (Integral m have also been treated in Case 1.) Using the tool of B3 above, we have that $\Phi_{m,1,A,I}$ *satisfies* (2.11) *for all* k, k', *and* m, *provided* $m \leqq 0$ *or* $m \geqq 1$. Alternatively, B2 can be used with the convex increasing nature of x^p, $p \geqq 1$, to yield (by B1) convexity of $\operatorname{tr}\mu^p$. Then convexity of μ^{-1} and B1 yields convexity of $\operatorname{tr}\mu^{-p}$; for the additional range $0 < p < 1$ this last was obtained in Case 2. Again, the example at the end of Section 3E shows why changing r cannot help in the case $0 < m < 1$ unless $k' = 1$.

We now abbreviate $k'^{-1/p}\Phi_{-p,1/p,A,I}$ on $\mathscr{P}_k^+$ (where we recall, from (4.8), that we are still treating the case $k' \leqq k$, rank $A = k'$) by

$$\Phi^*_{p,A}(M) = [k'^{-1} \operatorname{tr}(AMA')^{-p}]^{1/p} \,. \tag{4.11}$$

The techniques of the previous paragraph show that (4.11) satisfies (2.11) for all $p > 0$, even with G the identity. The normalization of (4.11) is convenient for comparing the effects of using various trace criteria, as we shall illustrate elsewhere. (Similarly, $(\mathrm{tr}\, \mu^p)^{-1/p}$ is convex for $0 < p \leqq 1$; for $p > 1$, this fails, since $(\mathrm{tr}\, \mu^m)^{-r}$ is easily seen not to be convex on diagonal matrices for $k = 2$, $m > 1$, $r > 0$.)

It is not hard to see that in all the previous cases of convexity except $m = 1$, the convexity is strict if C and A have rank k.

We now turn to considerations other than (2.11), for the functions Φ of (4.8).

As for (2.13), if $h > 0$ we have $H = \Phi^{h/mr} = P^{-1}(\Phi)$ homogeneous of degree h; P is decreasing and $\log P^{-1}$ is convex if $mr < 0$. Remark 3F covers the case $mr > 0$, where we cannot in general expect (2.17)(c) to hold.

We now compute $\nabla\Phi$, recalling the convention adopted just below (1.9) when such a computation is carried out using $\mathscr{P}_k^+$ alone. From (1.10) we have, for integral $m > 0$, using the symmetry of μ,

$$\begin{aligned}(4.12)\qquad \nabla\, \mathrm{tr}\, (C\mu^m) &= \textstyle\sum_{h=0}^{m-1} \mu^{m-h-1} C \mu^h\,, \\ \nabla\, \mathrm{tr}\, (C\mu^{-m}) &= -\textstyle\sum_{h=0}^{m-1} \mu^{-m+h} C \mu^{-h-1}\,.\end{aligned}$$

In particular, for all integral m,

$$(4.13)\qquad \nabla\, \mathrm{tr}\, \mu^m = m\mu^{m-1}\,.$$

The expression (4.13) *is in fact valid for all real m*, which is known from the theory of matrix functions (power series). The expressions corresponding to (1.10) for non-integral b, and thus to (4.10) for non-integral m, require more space to develop than is warranted here. Finally, from (4.2),

$$(4.14)\qquad \nabla\Phi^{(k)}_{m,r,A,C}(M) = r[\Phi^{(k')}_{m,1,I,C}(AMA')]^{r-1} A' \{\nabla\Phi^{(k')}_{m,1,I,C}(AMA')\} A\,,$$

where the superscript (j) denotes domain $\mathscr{P}_j^+$.

It remains to consider (3.2). From (4.5), (4.13) and (4.14), we see that, when $C = I$, the Φ of (4.8) satisfies (3.2) for all m and r for which $mr < 0$.

If $C \neq I$ and $m = -r = \pm 1$, the expression (4.12) is positive definite, so that (3.2) again holds. Otherwise, (4.12) is not positive definite for *all* μ if $C \neq$ const. $\times\, I$, and (as for (2.13) in Case 1 above) (3.2) holds only on a proper subset of $\mathscr{P}_k^+$, which depends on C.

We note that, in particular, the family (4.11) satisfies (2.11), (2.13) and (3.2) for all $p > 0$, and convexity is strict if A has rank k. For $p = 1$ and C non-singular, putting $A = C^{-\frac{1}{2}}$ in (4.11) yields the $\Phi_{1,C}$ of (1.3), which can be obtained for general C as $\Phi_{-1,+1,I,C}$, or from (4.18) below with $p = 1$.

To compute the $\mathscr{D}$ corresponding to (4.11), we use (4.13) and (4.14) to obtain

$$\begin{aligned}(4.15)\qquad d(x, \xi) &= k'^{-1/p}[\mathrm{tr}\, (AM(\xi)A')^{-p}]^{-1+1/p} f(x)'A'(AM(\xi)A')^{-p-1}Af(x)\,, \\ d^*(\xi) &= [k'^{-1}\, \mathrm{tr}\, (AM(\xi)A')^{-p}]^{1/p}\,,\end{aligned}$$

in terms of which (2.9) and (2.10) can be written. If, instead of using (4.11), we use the equivalent $p^{-1}\operatorname{tr}(AM(\xi)A')^{-p}$ (which also satisfies (2.13)), we obtain

$$(4.16)\qquad \begin{aligned} d(x,\xi) &= f(x)'A'(AM(\xi)A')^{-p-1}Af(x)\,,\\ d^{*}(\xi) &= \operatorname{tr}(AM(\xi)A')^{-p}\,. \end{aligned}$$

Of course, (2.9) from (4.15) is obviously the same as from (4.16), although our result on the equivalence of the two forms of (2.10) is not so obvious, as mentioned in Section 3J. Although one would presumably use (4.16) in practice, the more complex (4.15) has also been stated, for use in Section 4D below.

A variant of (4.8), sometimes more useful when $k' < k$ in view of the meaning of $AM^{-1}A'$ as proportional to a covariance matrix, is

$$(4.17)\qquad [\operatorname{tr} C(AM^{-1}A')^{m}]^{r}\,,$$

for which the conditions analogous to those of (4.8) should be obvious; the expressions (4.17), (4.18), and (4.21), can of course be meaningful if $M \notin \mathscr{P}_k^+$, but then we must use Theorem 3 (see also Section 7) rather than the formulas we shall develop here to implement Theorem 1. The general discussion of (2.11) is similar to that for (4.8); a main difference is that $AM^{-1}A'$ is not the matrix inverse of a linear function of M. We shall not take the space for a full discussion, except in the important case (analogous to (4.11))

$$(4.18)\qquad \Phi_{p,A}^{**}(M) = [k'^{-1}\operatorname{tr}(AM^{-1}A')^{p}]^{1/p}\,,$$

for which (as in the alternate proof of Case 2) we can use the convex increasing structure of $(\operatorname{tr} D^{p})^{1/p}$ and the convexity in M of $D = AM^{-1}A'$ to obtain (2.11) for (4.18); (2.13) is again obvious. We now use (4.4) in computing $\mathscr{D}$ for $M \in \mathscr{P}_k^+$ (recalling the comment below (4.17) for $M \notin \mathscr{P}_k^+$). The simpler equivalent $p^{-1}\operatorname{tr}(AM^{-1}A')^{p}$ to (4.18) yields, in analogy with (4.16),

$$(4.19)\qquad \begin{aligned} d(x,\xi) &= f(x)'M^{-1}(\xi)A'(AM^{-1}(\xi)A')^{p-1}AM^{-1}(\xi)f(x)\,,\\ d^{*}(\xi) &= \operatorname{tr}(AM^{-1}(\xi)A')^{p}\,; \end{aligned}$$

and the analogue of (4.15), obtained by using (4.18) itself, is achieved by multiplying the functions of (4.19) by $k'^{-1/p}[\operatorname{tr}(AM^{-1}(\xi)A')^{p}]^{-1+1/p}$.

In (4.35) we will discuss (4.19) further.

Of course, for $k' = 1$ all criteria satisfying (3.2) coincide. The Chebyshev equivalence criterion and related matters in this case have been treated extensively in [7], [22], [25], [13], [14], [15], and require no further discussion here.

4D. *D-optimality.* It is natural to define, for $AMA' \in \mathscr{P}_{k'}^+$,

$$(4.20)\qquad \Phi_{0,A}^{*}(M) = \lim_{p\to 0}\Phi_{p,A}^{*}(M) = [\det(AMA')]^{-1/k'}$$

and, for $M \in \mathscr{P}_k^+$,

$$(4.21)\qquad \Phi_{0,A}^{**}(M) = \lim_{p\to 0}\Phi_{p,A}^{**}(M) = [\det(AM^{-1}A')]^{1/k'}$$

(Recall the remark below (4.17) for $M \notin \mathscr{P}_k^+$.) These automatically satisfy (2.11), (2.13), (3.2), and $A \in \mathscr{R}_{k',k}^+$ implies strict convexity throughout $\mathscr{P}_k^+$. For

$A \in \mathscr{R}^{+}_{k',k}$ these criteria are of course equivalent to the D-optimality criterion (1.2). In terms discussed in Section 3I, a more useful $\bar{d}$, that of G-optimality when $A = I$, is obtained if we use $k' \log \Phi$ in place of each of (4.20) and (4.21). Clearly (2.13) and (3.2) are still satisfied, and it is well known that (2.11) can be obtained directly or by using $\log \det \mu = \lim_{p \to 0} p^{-1}(\operatorname{tr} \mu^p - k')$ appropriately. We obtain, from $\nabla \log \det \mu' = \mu^{-1}$ and (4.2) or (4.4),

$$d(x, \xi) = f(x)'A'(AM(\xi)A')^{-1}Af(x) \tag{4.22}$$

in the case of (4.20), and

$$d(x, \xi) = f(x)'M^{-1}(\xi)A'(AM^{-1}(\xi)A')^{-1}AM^{-1}(\xi)f(x) \tag{4.23}$$

in the case of (4.21), with $d^*(\xi') = k'$ in both cases. Note that these coincide with the results obtained formally by taking limits in (4.16) and (4.19).

If f is partitioned as $\binom{f_1}{f_2}$ with f_1 having k' components and M and $M^{-1} = D$ (say) are partitioned correspondingly, then "D-optimality for the first k' parameters" corresponds to putting $A = [I \mid 0]$ in (4.21), from which (4.23) becomes

$$\begin{aligned} d(x, \xi) &= (D_{11}f_1 + D_{12}f_2)'D_{11}^{-1}(D_{11}f_1 + D_{12}f_2) \\ &= f'M^{-1}f - f_2'(M_{22})^{-1}f_2 . \end{aligned} \tag{4.24}$$

The latter form was first given in [19].

4E. *E-optimality*. For the sake of explicitness, we relabel the eigenvalues of 4B3 as $\lambda_{\max}(A) = \lambda_1(A)$ and $\lambda_{\min}(A) = \lambda_k(A)$. In analogy with (4.21)—(4.22), we define, for A of rank k',

$$\Phi^*_{\infty,A}(M) = \lim_{p \to \infty} \Phi^*_{p,A}(M) = \lambda_{\max}([AMA']^{-1}) \tag{4.25}$$

and

$$\Phi^{**}_{\infty,A}(M) = \lim_{p \to \infty} \Phi^{**}_{p,A}(M) = \lambda_{\max}(AM^{-1}A') . \tag{4.26}$$

When $A = I$ these both reduce to the E-optimality criterion (1.4). Both of them satisfy (2.11), (2.13), and (3.2), but at almost all points of $\mathscr{P}_k$ strict convexity is not satisfied, even when $A = I$. Since (4.25) and (4.26) are not differentiable, we must now use Theorem 3 rather than Theorem 1.

We first consider (4.25). If $\bar{\mu} \in \mathscr{P}^+_{k'}$ with $\lambda_{\min}(\bar{\mu})$ of multiplicity q, let the rows of $Q_1(q \times k')$ be orthonormal eigenvectors of $\bar{\mu}$ corresponding to $\lambda_{\min}(\bar{\mu})$, so that $Q_1 \bar{\mu} Q_1' = \lambda_{\min}(\bar{\mu}) I_q$. A simple computation of $\det[\bar{\mu} + \varepsilon\mu - (\lambda_{\min}(\bar{\mu}) + \delta)I_{k'}]$ as $\varepsilon \downarrow 0$ verifies the well-known result that this determinant vanishes when $\delta = \varepsilon \times$ (any eigenvalue of $Q_1 \mu Q_1'$) $+ O(\varepsilon^2)$. Hence,

$$\lim_{\alpha \downarrow 0} \frac{\partial}{\partial \alpha} \lambda_{\min}(\bar{\mu} + \alpha\mu) = \lambda_{\min}(Q_1 \mu Q_1') . \tag{4.27}$$

If, in place of (4.25), we use the equivalent $-\lambda_{\min}(AMA')$, we thus obtain, for (2.23), with Q_1 corresponding to $\bar{\mu} = AM(\xi^*)A'$,

$$\sup_\xi \lambda_{\min}(Q_1 AM(\xi)A'Q_1') = \lambda_{\min}(AM(\xi^*)A') ; \tag{4.28}$$

and, for (2.24), that ξ^* minimizes the left side of (4.28). The corresponding results for the original (4.25) are obtained by dividing both members of (4.28) by $\lambda^2_{\min}(AM(\xi^*)A')$, a less useful form.

In an important special case, (4.28) simplifies considerably: if $\lambda_{\min}(\bar{\mu})$ is simple, $\lambda_{\min}(Q_1 \mu Q_1') = Q_1 \mu Q_1'$, linear in μ, yielding

THEOREM 5. *If* $\lambda_{\min}(AM(\xi^*)A')$ *is positive and simple* ($q = 1$), *with normalized row eigenvector* Q_1, *then* ξ^* *satisfies* (2.1), (2.23) *and* (2.24) *for* $\Phi(M) = -\lambda_{\min}(AMA')$, *iff*

$$\sup_x [Q_1 A f(x)]^2 = \lambda_{\min}(AM(\xi^*)A') . \tag{4.29}$$

Otherwise, (4.29) *is replaced by* (4.28).

It is easy to give examples which demonstrate the insufficiency of restricting consideration to M of rank 1 as in (2.9) (or, in fact, to any rank $< q$) when $q > 1$. Most obvious is the extreme case $q = k' \geqq 2$: for any $\bar{\mu}$ of the form $cI_{k'}$ with c a positive scalar, the minimum eigenvalue of $\bar{\mu} + \alpha\mu$ is again c if μ has rank $< k'$. For such a $\bar{\mu}$, the left side of (4.28) when ξ is restricted to the form ξ_x is even less than the right side! See also Section 3K.

We note that formally letting $p \to \infty$ in (4.15) yields an incorrect result in place of (2.23), both because of the incorrect restriction to ξ_x and also because of interchanging passage to the limit with differentiation.

We now turn to (4.26). Let $\lambda_{\max}(AM^{-1}(\xi^*)A')$ have multiplicity $\bar{q}$ and let $AM^{-1}(\xi^*)A'$ have corresponding normalized row eigenvectors $\bar{Q}_1(\bar{q} \times k')$. An analysis like that above (using also $(\bar{M} + \varepsilon M)^{-1} = \bar{M}^{-1} - \varepsilon \bar{M}^{-1} M \bar{M}^{-1} + O(\varepsilon^2)$) yields, from (4.26), for (2.23),

$$\sup_\xi \lambda_{\max}(\bar{Q}_1 AM^{-1}(\xi^*)M(\xi)M^{-1}(\xi^*)A'\bar{Q}_1') = \lambda_{\max}(AM^{-1}(\xi^*)A') , \tag{4.30}$$

with (2.24) being obtained from the left side. The seemingly more complex form obtained by dividing both sides of (4.30) by $\lambda^2_{\max}(AM^{-1}(\xi^*)A')$ also has a left side which can be used for (2.24), since this is the form obtained from the equivalent $\Phi = -\lambda_{\min}([AM^{-1}A']^{-1})$. We mention this alternate form because it, rather than (4.30) as it stands, is a more direct analogue of (4.28), since this alternate form reduces to the latter when $k = k'$, upon writing A'^{-1} for A here. Similarly, the alternate form mentioned below (4.28) is the analogue of (4.30). Which form is more convenient in each of the cases (4.25) and (4.26) may depend on the example at hand, but the nonanalogous forms (4.28) and (4.30) as stated seem the simplest choices unless $k' = k$, in which case (4.30) is less convenient.

As an illustration of (4.30) when $k' < k$, suppose we are interested in the accuracy of "standard" linear combinations of the first k' parameters without scale change; that is, we take $A = [I \mid 0]$ as in (4.24), and we partition M and $D = M^{-1}$ correspondingly. Write $N_{12}(\xi^*) = D_{11}^{-1}(\xi^*)D_{12}(\xi^*) = -M_{12}(\xi^*)M_{22}^{-1}(\xi^*)$. Then (4.30) reduces to

$$\sup_\xi \lambda_{\max}(\bar{Q}_1[I \mid N_{12}(\xi^*)]M(\xi)[I \mid N_{12}(\xi^*)]'\bar{Q}_1') = \lambda_{\max}(D_{11}(\xi^*)) . \tag{4.31}$$

Thus, we have

THEOREM 6. *If* $\lambda_{\max}(AM^{-1}(\xi^*)A')$ *is positive and simple* ($\bar{q} = 1$), *with normalized row eigenvector* $\bar{Q}_1$, *then* ξ^* *satisfies* (2.1), (2.23), *and* (2.24) *for the* Φ *of* (4.26), *iff*

$$\sup_x [\bar{Q}_1 AM^{-1}(\xi^*)f(x)]^2 = \lambda_{\max}(AM^{-1}(\xi^*)A') , \tag{4.32}$$

which reduces in the case of (4.31) *to*

$$\sup_x \{\bar{Q}_1[f_1(x) + N_{12}(\xi^*)f_2(x)]\}^2 = \lambda_{\max}(D_{11}(\xi^*)) . \tag{4.33}$$

Otherwise, (4.32) *and* (4.33) *are replaced by* (4.30) *and* (4.31).

It is interesting to compare the reductions of (4.19), (4.24), and (4.33) in the case $A = [I \mid 0]$. Since (4.24) is the most familiar of these, we express the other functions in terms of

$$\delta(x, \xi^*) = D_{11}^{\frac{1}{2}}(\xi^*)f_1(x) + D_{11}^{-\frac{1}{2}}(\xi^*)D_{12}(\xi^*)f_2(x) . \tag{4.34}$$

We then have

$$\begin{aligned} &D\text{-optimality } \bar{d} \text{ of } (4.19) = \sup_x \delta'\delta ; \\ &\bar{d} \text{ of } (4.24)(A\text{-optimality if } p = 1) = \sup_x \delta' D_{11}^{p}(\xi^*)\delta ; \\ &(4.33)(E\text{-optimality criterion if } \bar{q} = 1) = \lambda_{\max}^{-1}(D_{11}) \sup_x (\bar{Q}_1'\delta)^2 . \end{aligned} \tag{4.35}$$

(If we had used the logarithm of (4.26) for Φ, we would have obtained simply $\sup_x (\bar{Q}_1'\delta)^2$ in the last line of (4.35), but this does not seem simpler in applications than the other two forms discussed above.)

An example. Although one often encounters parametric families of matrices M over which $\lambda_{\min}(M)$ is maximized when $q > 1$, the simpler forms of Theorems 5 and 6 have frequent applicability. For example, in the case of quadratic regression $[-1, 1]$, with $f(x)' = (1, x, x^2)$ and $A = I$, either Theorem 5 or Theorem 6 can be used to show that the unique E-optimum design is given by $\xi^*(1) = \xi^*(-1) = \frac{1}{5}$, $\xi^*(0) = \frac{3}{5}$. For this design,

$$M(\xi^*) = \begin{pmatrix} 1 & 0 & .4 \\ 0 & .4 & 0 \\ .4 & 0 & .4 \end{pmatrix} \quad \text{and} \quad Q_1 = (5^{-\frac{1}{2}}, 0, -2(5^{-\frac{1}{2}}))$$

corresponding to the simple $\lambda_{\min} = \frac{1}{5}$; thus, (4.29) becomes the trivially verified $\sup_x (1 - 2x^2)^2/5 = \frac{1}{5}$.

More complex examples will be treated elsewhere, including detailed computations in higher-dimensional simplex experiments (generalizing the results obtained for dimensions 1 and 2 in the present example and that of Section 6A).

4F. *Trace criteria modified to include previous information.* For any of the trace criteria $\tilde{\Phi}$ of Sections 4C—4E, and for fixed B in $\mathscr{P}_k$, replacing the argument M by $M + B$ yields a new criterion $\Phi(M) = \tilde{\Phi}(M + B)$. Such Φ's arise in at least four different contexts of applications:

(i) The experimenter has available an information matrix B from a previous experiment (scaled relative to $\xi(\mathscr{X}) = 1$) and wants to combine it with the information $M(\xi)$ from a new experiment so as to minimize $\tilde{\Phi}(M(\xi) + B)$.

(ii) For certain normal Bayesian models which have appeared in the literature, minimizing the total expected loss for an appropriate loss function is often equivalent to minimization of such a corresponding $\tilde{\Phi}(B + M)$, where B is a parameter of the prior distribution.

(iii) In response surface theory (e.g., [4], [16], [21]) where one purposely fits surface of incorrect form for the sake of simplicity (fewer nonzero parameters in the fitted curve), the form $\tilde{\Phi}(B + M)$ arises with B a matrix in terms of which an assumption on the bias of the fit is expressed.

(iv) In some iterative methods for minimizing $\tilde{\Phi}(M)$, if B_n is the approximation to the minimizer after n stages, one tries, approximately, to minimize $\tilde{\Phi}((1 - \varepsilon_n)B_n + \varepsilon_n M_n)$ at the next stage, for a suitable sequence $\{\varepsilon_n\}$. (See Section 6B.)

We shall not pursue these uses here. The arithmetic of Sections 4C—4E can be modified for use here in accordance with the formula obtained from (4.2): $\mathscr{D}_\Phi(M, \bar{M}) = \mathscr{D}_{\tilde{\Phi}}(M, \bar{M} + B)$, which is valid also for the nondifferentiable E-optimality criteria. *We note that the condition for differentiability is now weakened*; for example, whenever we previously required $M(\xi^*) \in \mathscr{P}_k^+$, we now require only $B + M(\xi^*) \in \mathscr{P}_k^+$.

Here, then, we will only take the space to discuss the fact that (2.13) does not hold, so that one cannot expect (2.17)(c) to hold in general. (See also Section 3H.) We now give one such example, chosen for arithmetical simplicity. Suppose $\mathscr{X} = \{1, 2\}$, $k = 2$, $f(1)' = (1, 5)$, $f(2)' = (3, 16)$, and $B = \begin{pmatrix} 1 & 5 \\ 5 & 29 \end{pmatrix}$. The problem is to minimize $\Phi(M) = \Phi^*_{1,I}(B + M) = \operatorname{tr}(B + M)^{-1}$. As before, let ξ_i assign measure 1 to the point i, and abbreviate $f(i)f(i)' = M(\xi_i)$ by M_i. The criterion (2.9), by (4.15) or (4.16), is

$$\max_i \{\operatorname{tr}[B + M(\xi^*)]^{-2}M_i\} = \operatorname{tr}[B + M(\xi^*)]^{-2}M(\xi^*)\,, \tag{4.36}$$

and $\bar{d}(\xi^*)$ is the left side of (4.36). A direct computation yields

$$(B + M_1)^{-2} = \frac{1}{8}\begin{pmatrix} 377 & -70 \\ -70 & 13 \end{pmatrix}, \tag{4.37}$$

$$\operatorname{tr}(B + M_1)^{-2}M_2 = \tfrac{1}{8}\,, \qquad d^*(\xi_1) = \operatorname{tr}(B + M_1)^{-2}M_1 = \tfrac{1}{4} = \bar{d}(\xi_1)\,,$$

from which (4.36) yields Φ-*optimality of* ξ_1. On the other hand,

$$(B + M_2)^{-2} = \frac{1}{1681}\begin{pmatrix} 84034 & -15635 \\ -15635 & 2909 \end{pmatrix},$$

$$d^*(\xi_2) = \operatorname{tr}(B + M_2)^{-2}M_2 = \tfrac{50}{1681}\,, \tag{4.38}$$
$$\operatorname{tr}(B + M_2)^{-2}M_1 = \tfrac{409}{1681} = \bar{d}(\xi_2)\,,$$

so that $\bar{d}(\xi_1) > \bar{d}(\xi_2)$ and $d^*(\xi_1) > d^*(\xi_2)$. Thus, without taking the space to compute the $\bar{d}$- or the d^*-optimum design, we see that neither can be the Φ-optimum design ξ_1.

As remarked in Section 3H, it is also not hard to find examples where $\bar{d}$-, $d^{\#}$-, and Φ-optimality coincide; for example, let $B = I$ and $f(i)' = (2 - i, i - 1)$ above. More generally, under (2.11) if $\mathcal{X}$ has k points and a Φ-optimum design has $\xi^*(i) > 0$ for all i, then ξ^* is $\bar{d}$- and $d^{\#}$-optimum even without (2.13).

4G. *Compound criteria*. This heading will be used, loosely, to describe criteria "built up" from simpler criteria, usually for one of the following three reasons: (1) uncertainty about the loss or covariance structure; (2) incomparability of various parts of M^{-1}, in terms of simple loss considerations, (3) the desire to combine features of several Φ's (illustrated also in many examples which fall under both of the previous headings). These descriptions are of course imprecise, intended to give the rationale behind adoption of certain criteria, rather than to categorize them taxonomically. We shall concentrate on such rationale, rather than on detailed analysis of these criteria, here.

Ideally, the criterion Φ is known exactly, at least to either decision theorists or subjectivists. In practice, the customer is often vague or confused about his objectives and relative losses, and the statistician's discussion of the positive and negative features of various loss structures may aid in the choice of a criterion and resulting design which reflect the customer's aims. There are arguments (e.g., [5]) that only criteria of the $\Phi_{1,A}^{**}$ variety need be considered, and it is certainly arguable from a decision-theoretic point of view that expectation of a non-quadratic function of errors may in principle be more meaningful than are our possibly non-linear functions Φ of covariances. Nevertheless, we feel the general discussion of such Φ's can be a fruitful basis for constructing a variety of designs among which the practitioner can find one at least approximately achieving his goals.

Regarding (1), suppose $\{\Phi_\tau\}$ is a family of criteria, indexed by some set $T = \{\tau\}$. If various possible customers of a single experiment have different loss functions Φ_τ, the designer may want to consider, for example, $\sup_{\tau \in T} \Phi_\tau$ or some convex average $\int_T \Phi_\tau \, \eta(d\tau)$ as his optimality criterion. (We omit mention of other such combinations, and discussion of "rational behavior axioms" which justify using or not using any of these.) Such a combination could also arise because of the experimenter's uncertainty about which Φ_τ is appropriate for a single user; equivalently, the same mathematical framework arises even in the case of a single Φ, if there is uncertainty about the covariance structure (thus far assumed to be that specified in the first paragraph of Section 1), as we shall discuss in (5.1)—(5.2).

If each of the Φ_τ satisfies (2.11), or (3.2), or (2.13) with the same degree h of homogeneity, then so does a supremum or convex average. If the Φ_τ are equidifferentiable, then their convex average is differentiable and Theorem 1 applies to it. This is not the case for the supremum, for which the treatment of Section 4E is typical.

In order to introduce (2), we describe the rationale some designers have given for their choice of certain criteria. On the one hand, some of these designers find D-optimality does not reflect their aims because the choice of design may be governed to such a great extent by a few characteristic values of M, that other possible advantages are sacrificed. On the other hand, use of a criterion such as A-optimality often seems unsatisfactory because in a multifactor setting it may mean adding squared units of apples to those of dung, or even to those of water $\times$ sunshine. Ideally the matrix A in $\Phi_{1,A}^{**}$ reflects such differences in units and their relative importance; in practice, again, A is usually not even approximately known, and the practitioner may feel uneasy about treating different factors, or interactions of different orders, in additive terms as in A-optimality.

One possible response to the above considerations is to use a criterion which combines, in the determinental manner of D-optimality, the losses from factors measured in different units; but which combines the contributions from different levels of the same factor through a criterion such as A- or E-optimality among items measured in the same units. This means that if θ' is decomposed as $(\theta^{(1)'}, \theta^{(2)'}, \cdots, \theta^{(r)'})$ where $\theta^{(i)}$ is a k_i-vector, with a corresponding decomposition of M, and if A_i is a $k_i \times k$ matrix of 0's except for an I_{k_i} in the $(1 + \sum_1^{i-1} k_j)$th to $(\sum_1^i k_j)$th columns, we use a criterion such as

$$\Phi(M) = \prod_{i=1}^r \Phi_{p,A_i}^{**}(M) , \tag{4.39}$$

where the simplest choice of p is 1 or ∞. Modifications of (4.39) will be obvious.

Thus, for example, if $p = 1$ in (4.39) and we compute $\mathscr{D}_{\log \Phi}$ from (4.19) and (4.2), we obtain

$$\begin{aligned} d(x, \xi) &= \textstyle\sum_{i=1}^r (\operatorname{tr} A_i M^{-1}(\xi) A_i')^{-1} f(x)' M^{-1}(\xi) A_i' A_i M^{-1}(\xi) f(x) , \\ d^*(\xi) &= r ; \end{aligned} \tag{4.40}$$

of course, $A_i M^{-1} A_i'$ is proportional to the covariance matrix of best linear estimators of $\theta^{(i)}$.

We mention two other illustrations of compound criteria. Firstly, in problems of extrapolation, one can think of Φ_τ as the variance of estimated response at the point τ. The average of Φ_τ's has then been considered frequently in response surface design considerations, and $\max_\tau \Phi_\tau$ has arisen in extrapolation to more than one point [24] as well as in the formulation of G-optimality [22]. Secondly, the criterion

$$\Phi(M) = \max_i \Phi_{p,A_i}^{**}(M) \tag{4.41}$$

includes, in the form $\max_i (M^{-1})_{ii}$ when all $k_i = 1$, the criterion Elfving [7a] described as that of "minimaxing over single parameter variances," to distinguish it from E-optimality, which is "minimaxing over variances of standard parametric functions."

4H. *Shah's criterion*. As a final illustration of our theory, we consider the criterion $\Phi_{-2,I}^*(M) = \operatorname{tr} M^2$ mentioned in Sections 3A and F. Suppose $\operatorname{tr} M$ is a constant kc on $\mathscr{M}$. Then minimizing $\operatorname{tr} M^2$ may not be so foolish. To see this, suppose the optimality criterion $\tilde{\Phi}$ of primary interest, but which leads to more difficult computations than tr M^2, has its minimum over $\mathscr{N} = \{M: M \in \mathscr{P}_k$, tr $M = kc\}$ (which includes $\mathscr{M}$) at $M = cI_k$, which is close to but not in $\mathscr{M}$. If $\tilde{\Phi}$ on $\mathscr{N}$ is a twice-differentiable function of only the eigenvalues of M (i.e., is orthogonal-invariant), then the terms through second degree of its Taylor series development about $M = cI_k$ are $\Phi(cI_k) + c_1 \operatorname{tr}(M - cI_k)^2 = c_2 + c_3 \operatorname{tr} M^2$, where the c_i are constants. Thus, if cI_k is sufficiently close to $\mathscr{M}$, minimizing tr M^2 will come close to minimizing $\Phi(M)$. Explicit bounds can be given, but we shall not take the space to do so here. (There are obvious modifications of the above where cI_k is replaced by another matrix. This occurs in the example below.) The minimization of tr M^2 was first considered by Shah [26], and we hereafter call it *S-optimality*. The assumption that tr $M = ck$ on $\mathscr{M}$ is most applicable in incomplete block design settings where, also, the approximate theory is of negligible usefulness (and where J. Eccleston [6a] has recently made use of the tr M^2 criterion for exact theory optimality). It does occur in some reasonable regression settings, illustrated below.

We first note, from (3.13), that

$$d(x, \xi) = -2f(x)'M(\xi)f(x)\,, \qquad d^*(\xi) = -2 \operatorname{tr} M^2(\xi)\,. \tag{4.42}$$

Thus, (2.9) is

$$\inf_x f(x)'M(\xi)f(x) = \operatorname{tr} M^2(\xi)\,. \tag{4.43}$$

An example. A simple setting where $f'f$ is constant on $\mathscr{X}$, so that tr M is constant on $\mathscr{M}$, is that of linear regression on a subset of the unit $(k-2)$-sphere; there is an obvious trigonometric reformulation, which we consider at the same time. Here we treat in detail the case $k = 3$, with $\mathscr{X}$ the arc $\{(x_1, x_2): x_1 = \cos\theta,\ x_2 = \sin\theta,\ |\theta| \leqq \theta_0\}$ where θ_0 is specified, $0 < \theta_0 \leqq \pi$. Also, $f(x)' = (1, x_1, x_2)$. On grounds of symmetry, we try designs of the form $\xi_{(\alpha)}(\theta = 0) = 1 - 2\alpha$, $\xi_{(\alpha)}(\theta = \pm\theta_0) = \alpha$. Then

$$M(\xi_{(\alpha)}) = \begin{pmatrix} 1 & 1 - 2\alpha(1 - \cos\theta_0) & 0 \\ 1 - 2\alpha(1 - \cos\theta_0) & 1 - 2\alpha\sin^2\theta_0 & 0 \\ 0 & 0 & 2\alpha\sin^2\theta_0 \end{pmatrix},$$

$$\begin{aligned} \operatorname{tr} M^2(\xi_{(\alpha)}) = 4\{&2\alpha^2[\sin^4\theta_0 + 2(1 - \cos\theta_0)^2] \\ &- \alpha[2(1 - \cos\theta_0) + \sin^2\theta_0] + 1\}\,. \end{aligned} \tag{4.44}$$

Let the positive value q satisfy $2q^3 + 2q^2 + q - 1 = 0$, and let $\theta^* = \cos^{-1} q$. If $\theta_0 \geqq \theta^*$, so that $\cos\theta_0 \leqq q$, the value

$$\alpha = \frac{2(1 - \cos\theta_0) + \sin^2\theta_0}{4[(1 - \cos^2\theta_0)^2 + (1 - \cos\theta_0)^2]} = \frac{3 + \cos\theta_0}{4[2 - \cos^2\theta_0 - \cos^3\theta_0]} \tag{4.45}$$

is $\leqq \frac{1}{2}$ (so $1 - 2\alpha \geqq 0$), and this value minimizes tr $M(\xi_{(\alpha)})$; and (4.45) is always $\geqq 0$. Since $f(x)'M(\xi_{(\alpha)})f(x) = \operatorname{tr} M^2(\xi_{(\alpha)})$ for $\theta = 0, \pm\theta_0$ (this in fact being equivalent to (4.45) and thus another way of deriving it), and since $f(x)'M(\xi_{(\alpha)})f(x)$

can be written as a quadratic in x_1 for $1 \geqq x_1 \geqq \cos\theta_0$, we see that (4.43) is satisfied by $\xi_{(\alpha)}$ if and only if the coefficient of x_1^2 in this quadratic is $\leqq 0$. This coefficient, from (4.44), is $1 - 4\alpha\sin^2\theta_0$, which is $\leqq 0$ (from (4.45)) if and only if $\theta_0 \leqq 2\pi/3$. We conclude, as part of our solution, that $\xi_{(\alpha)}$ as given by (4.45) is S-optimum if $\theta^* \leqq \theta_0 \leqq 2\pi/3$.

The remaining parts of the solution are easy. If $2\pi/3 \leqq \theta_0 \leqq \pi$, assigning probability $\frac{1}{3}$ to each of three points $2\pi/3$ apart (and in $\mathscr{X}$) yields a constant $f(x)'Mf(x)$ and thus satisfies (4.43). If $\theta_0 \leqq \theta^*$, the design $\xi_{(\frac{1}{2})}$ which assigns probability $\frac{1}{2}$ to each of the values $\theta = \pm\theta_0$ is seen to be S-optimum upon checking that the quadratic $f(x)'Mf(x) = (1 - x_1^2)\sin^2\theta_0 + (1 + x_1\cos\theta_0)^2$ attains its minimum on $\mathscr{X}$ at $x_1 = \cos\theta_0$.

We turn to the considerations of the first paragraph of the present subsection. Although $cI_3 \notin \mathscr{M}$ for any θ_0, the fact that $m_{11} = 1$ and $\operatorname{tr} M = 2$ for all designs makes $\begin{pmatrix} 1 & 0 \\ 0 & \frac{1}{2}I_2 \end{pmatrix} = M_0$ (say) the analogue of I_k in our initial discussion; thus, when $\theta_0 \geqq 2\pi/3$, it is easily checked that the uniform 3-point design of the previous paragraph yields this M_0 and is also D-optimum (as well as A-optimum, etc.). If $\theta_0 < 2\pi/3$, assigning probability $\frac{1}{3}$ to each of the 3 points $\theta = 0, \pm\theta_0$ achieves $\bar{d} = 3$ and, thus, D-optimality. For $\theta^* < \theta_0 < 2\pi/3$, the efficiency for D-optimality of the S-optimum design (in terms of ratio of approximate numbers of observations needed to achieve the same generalized variance) is

$$(4.46) \qquad [\det M(\xi_{S\text{-OPT}})/\det M(\xi_{D\text{-OPT}})]^{\frac{1}{3}} = 3[\alpha^2(1 - 2\alpha)]^{\frac{1}{3}},$$

where α is given by (4.45). Thus, when M_0 is close to $\mathscr{M}$, θ_0 is close to $2\pi/3$, α is close to $\frac{1}{3}$, and the efficiency of (4.46) is $1 - O((\theta_0 - 2\pi/3)^2)$. Of course, $\theta_0 \leqq \theta^*$ yields efficiency 0, not surprising in view of the distance of M_0 from $\mathscr{M}$ in such cases.

5. Modification to vector observations and variable or unknown covariance or cost structure. The previous theory applies to multiresponse problems with changes that are essentially only notational, as described in [19] and [11], [12]. In terms of the first paragraph of Section 1, f is now a $k \times m$ matrix of continuous real component functions on compact $\mathscr{X}$. The expectation of a row m-vector (single multiresponse observation) corresponding to level x is $\theta' f(x)$. The m-vectors are uncorrelated, and, for the moment, each has the same covariance matrix $\sigma^2 I_m$ (altered below). Then the entire previous development of this paper is valid with only one obvious alteration: In (2.5) and all subsequent expressions $f'Bf$, a trace operation must be inserted at the beginning; note that the formula for $M(\xi)$ in terms of f is still valid.

Now suppose a multiresponse m-vector Y_x at level x costs an amount $c_x > 0$ and has positive definite covariance matrix Q_x; again, observational m-vectors are uncorrelated with each other. Assume c_x and Q_x vary continuously in x. If V_x^{-1} is the positive definite symmetric square-root of $c_x Q_x$, then replacing Y_x by $Y_x V_x$ and $f(x)$ by $\bar{f}(x) = f(x)V_x$ makes $\bar{M}(\eta) = \int \bar{f}(x)\bar{f}(x)'\eta(dx)$ the appropriate information matrix for the approximate theory problem of minimizing a functional Φ of the inverse of the covariance matrix of best linear estimators, subject to a given restriction on the total cost (rather than number) of observations. This $\bar{f}$ and $\bar{M}$ then replace f and M in the previous paragraph and in the expressions of Theorem 1. For an optimum design subject to an upper bound C on total cost, one then sets $\xi(dx) = c_x^{-1}\eta(dx)/\int_{\mathscr{X}} c_x^{-1}\eta(dx)$ in the original framework, and takes a number of observations costing approximately C.

This type of substitution is considered in [17] and [22]; the formulas in [12] for the case of variable Q_x (and constant c_x) are given in terms of f and Q, and can be obtained by making the substitution for $\bar{f}$ given above.

Next, suppose that V_x is unknown but that it is known that it is a member of some class $\{V_x^{(\tau)}, \tau \in T\}$ for some index set T. Thus, we have $\bar{M}^{(\tau)}$ in place of $\bar{M}$ in the previous paragraph. For simplicity, assume c_x constant; it will be clear how to alter this. In the language of Section 4G, $\Phi(\bar{M}^{(\tau)})$ cannot in general be rewritten as $\Phi^{(\tau)}(M)$. Rather, even when $m = 1$, we must consider $\Phi^{(\tau)}$ as a function of ξ or, what is slightly more convenient, of the vector measure $\mu_\xi(dx) = f(x)f(x)'\xi(dx)$. We then write

$$\Phi^{(\tau)}(\mu) = \Phi\left(\int_{\mathscr{X}} V_x^{(\tau)}\mu(dx)V_x^{(\tau)}\right). \tag{5.1}$$

For a criterion such as $\max_\tau \Phi^{(\tau)}(\mu)$, which is convex if Φ is, a development like that of Theorem 6 and (4.30) is now possible, but it is even harder to apply than these tools of Section 4E because of the dependence on μ (or ξ) rather than on a single matrix $M(\xi)$. However, simplification is possible for the criterion $\Phi^*(\mu) = \int_T \Phi^{(\tau)}(\mu)\eta(d\tau)$ where η is a probability measure and $\Phi^{(\tau)}([1-\alpha]\bar{\mu} + \alpha\mu)$ is such that integration over τ and differentiation with respect to α (near $\alpha = 0$) commute if $\bar{\mu}$ is Φ^*-optimum. (This condition is satisfied for many of the criteria Φ considered previously, e.g., the $\Phi_{p,A}^{**}$ for $p < \infty$, under natural assumptions on $\{V^{(\tau)}, \tau \in T\}$.) We observe that $\partial\Phi^*((1-\alpha)\bar{\mu} + \alpha\mu)/\partial\alpha|_{\alpha=0}$ is linear in μ; thus, writing $M^{(\tau)}(\xi) = \int_{\mathscr{X}} V_x^{(\tau)}\mu_\xi(dx)V_x^{(\tau)}$, we have for the equivalent (2.9) to Φ^*-optimality,

$$\begin{aligned}\max_x -\operatorname{tr} f(x)f(x)' \int_T V_x^{(\tau)}\, \nabla\Phi(M^{(\tau)}(\xi^*))V_x^{(\tau)}\eta(d\tau) \\ = -\operatorname{tr} \int_T M^{(\tau)}(\xi^*)\, \nabla\Phi(M^{(\tau)}(\xi^*))\eta(d\tau).\end{aligned} \tag{5.2}$$

6. Computational techniques.

6A. *Analytical demonstrations.* The tools used elsewhere (e.g., [17], [18], [19], [9], [1]) to prove designs optimum for particular criteria such as D-optimality can often be employed in the same manner for general Φ of the type we have considered. A common approach is to minimize Φ over a promising finite-dimensional subset of $\{\xi\}$ and then to use (2.9) to prove optimality of the design so obtained, relative to *all* competitors. But these computations can often be shortened greatly by the use of one or more of the following.

(i) *Invariance.* Let $G = \{g\}$ be a compact group of measurable transformations of $\mathscr{X}$ onto $\mathscr{X}$, and define ξ_g by $\xi_g(A) = \xi(gA)$. Suppose Φ has the invariance property

(6.1) $$\Phi(M(\xi)) = \Phi(M(\xi_g))$$

for all ξ and g. Then, if $\bar{\xi} = \int_G \xi_g \mu(dg)$ where μ is Haar probability measure, the design $\bar{\xi}$ is *invariant* under $G(\bar{\xi}(gA) = \bar{\xi}(A)$ for all g and $A)$ and, assuming some increasing function of Φ is convex on $\mathscr{M}$, we obtain $\Phi(M(\bar{\xi})) \leq \Phi(M(\xi))$. Thus, *there exists an invariant Φ-optimum design*. An alternate approach which is sometimes useful is to replace (6.1) and convexity by the single assumption that

(6.2) $$\bar{d}(\xi_g) = \bar{d}(\xi) .$$

We conclude that *there is a $\bar{d}$-optimum invariant design*, and under (2.13) it is also Φ-optimum. See [17], [18], [19], [9], [1] for discussion and examples.

(ii) *Nature of* $\{x: d(x, \xi) = \bar{d}\}$. Sometimes (2.18) and the nature of d can be used to describe limitations on the nature of Φ, especially if f consist of polynomials or functions with similar oscillatory properties. This has been used extensively (e.g., [18], [19], [9], [1]) in the case of polynomial regression on Euclidean sets $\mathscr{X}$, as illustrated below.

(iii) *Special properties of certain supports*. If the f_i are linearly independent on a set $B = \{x_1, \cdots, x_k\}$ of cardinality k, we can find A in $\mathscr{R}^+_{k,k}$ such that $(Af)_i(x_j) = \delta_{ij}$. This often simplifies greatly the computation of a Φ-optimum design *among those with support* B; simplest is Φ_0 (D-optimality), for which of course $\xi(x_i) = 1/k$. Similar computations can sometimes be made for B of larger cardinality.

(iv) *Uniqueness*. Suppose we are in a setting where the optimum $M(\xi)$ is unique. There is then the question of whether the optimum ξ is unique. Sometimes this has a trivial negative answer because one knows an optimum design whose support has cardinality $> 1+$ the dimension of $\mathscr{M}$. If one knows all optimum designs are supported by subsets of a set B, then the question is that of uniqueness of a nonnegative solution to the linear equations $\sum_{x \in B} \xi(x) f_i(x) f_j(x) = m_{ij}(\xi_{\text{OPT}})$ in the variables $\xi(x)$, and this has yielded uniqueness results in some cases [9].

An example. Suppose $\mathscr{X}$ is the 2-simplex $\{(x_1, x_2, x_3): \sum_1^3 x_i = 1, \text{all } x_i \geq 0\}$, that $k = 6$, and that the components of f are the functions x_i and $x_i x_j$, $i < j$: quadratic regression. If Φ is convex and invariant under permutations of the three variables x_i, we conclude by (i) that there is a symmetric Φ-optimum design. Now note (ii) that $d(x, \xi)$ is a quartic. It is often easy to see, as in the case of $\Phi^{**}_{p,I}$ for $p < \infty$, that this quartic, extended to the plane, approaches $+\infty$ at ∞. Consequently, on the line segment which is the intersection of any line with $\mathscr{X}$, we can have $d(x, \xi) = \bar{d}(\xi)$ at no more than one interior point of the segment. For an invariant design this means the support of an optimum design is a subset of the set B consisting of the three vertices, three midpoints of edges, and center $\bar{x}$ (say). In the case of D-optimality, we can try the six points other than the center; applying (iii), if this supports a D-optimum ξ^* we know $\xi^* = \frac{1}{6}$ at each of these other points. It is then automatic (D-optimality among designs on the 6-point set) that $d(x, \xi^*) = 6$ at each of these six points, and the single explicit computation required is to check that $d(\bar{x}, \xi^*) < 6$, which is true. Finally, (iv) uniqueness is obvious here. Thus, the computations needed to characterize the D-optimum designs in general have been reduced considerably by applying (i), (ii), (iii). For A- or E-optimality (the latter having support in B by a simple limiting argument), slightly more computation is needed. It is perphaps quickest to minimize $\Phi(M)$ in these cases with respect to the two variables ξ(vertex), $\xi(\bar{x})$. The interesting feature of the result is that the optimum ξ is now positive on all seven points of B, unlike the D-optimum ξ^*. This example and its higher-dimensional extensions will be treated elsewhere; I am indebted to R. J. Walker and Z. Galil for carrying out the computations.

6B. *Iterative methods.* There is a considerable literature on the determination of a sequence $\{\xi^{(n)}\}$ which converges to an optimum design, in the case of A- and D-optimality; e.g., [1a], [27], [28], [12]. As this aspect of the subject has developed, these authors have given increasing attention to the difficult problem of improving the obvious iterative techniques. The latter are all we will comment on here: if Φ is convex, we have at our disposal all the computational techniques for minimizing a convex function on a compact finite dimensional set whose extreme points $M(\xi_x)$ are readily available.

Simplest (and most used as a basis for D- and A-optimality in the past) are the descent methods for which $M(\xi^{(0)})$ is nonsingular and

$$\xi^{(n+1)} = (1 - \varepsilon_n)\xi^{(n)} + \varepsilon_n \xi_{x_n} \tag{6.3}$$

where x_n maximizes (or approximately maximizes) $d(x, \xi^{(n)}) - d^*(\xi^{(n)})$ (see (2.2)—(2.9)). There are many possible choices for the ε_n. Most general in applicability is any fixed sequence for which $1 > \varepsilon_n \downarrow 0$ and $\sum \varepsilon_n = +\infty$. For example, $\Phi(M(\xi^{(n)}))$ converges to the minimum for convex Φ with two bounded derivatives on a closed convex set to which (6.3) is limited by truncation and in whose interior all minima lie; under Theorem 1, if Φ is strictly convex, $M(\xi^{(n)})$ must also converge to the unique optimum value, but convergence of $\xi^{(n)}$ depends on the choice of x_n or considerations of 6A (iv).

By letting ε_n depend on $\xi^{(n)}$ and x_n, one can weaken regularity assumptions and speed convergence, but uses more computation. For D-optimality, det $M(\xi^{(n+1)})$ is easily minimized analytically with respect to ε_n, and this is the basis for procedures in the literature cited above. For A-optimality, Fedorov [12] obtains an upper bound b_n on the optimum choice of ε_n, and chooses $\varepsilon_n = cb_n$ where $0 < c < 1$. For general convex Φ a corresponding prescription is not always so easy, but the fact that

$$M^{-1}(\xi^{(n+1)}) = (1 - \varepsilon_n)^{-1}[I_k + \varepsilon_n(1 - \varepsilon_n)^{-1}M^{-1}(\xi^{(n)})f(x_n)f(x_n)']^{-1}M^{-1}(\xi^{(n)}) \tag{6.4}$$

yields useful bounds on the optimum ε_n, and hence reasonable analytic prescriptions for ε_n, for many common Φ. For example, a lower bound on $\partial^2\Phi(M(\xi^{(n+1)}))/(\partial\varepsilon_n)^2$, together with the evaluation (using (2.4)) of the first derivative at $\varepsilon_n = 0$, yields an upper bound on the optimum choice of ε_n.

With obvious modifications, the above comments can be applied to such non-differentiable cases as E-optimality, where the procedures are altered by replacing ξ_{x_n} by a design on more than one point; non-convex Φ of course require additional care. Further modifications are discussed in depth in [28] and [1a]. A principal need appears to be an efficient smoothing routine for $\xi^{(n_j)}$ to consolidate the information at certain stages n_j so that the support of $\xi^{(n)}$ is not of cardinality unbounded in n.

6C. *Bounds on departure from optimality.* The approach of [18], [1] in the case of D-optimality, for bounding the (relative) departure from optimality of a given ξ' (for example, in order to know when to stop the iterative scheme of B above), extends easily to general convex Φ. Thus, in the differentiable case one can use estimates of derivatives just as for Φ_0; for example, (2.4) and convexity yield at once the roughest (but useful) upper bound

$$\Phi(M(\xi')) - \min_\xi \Phi(M(\xi)) \leqq \bar{d}(\xi') - d^\sharp(\xi') . \tag{6.5}$$

Thus, if $\Phi > 0$ and $\bar{d}(\xi^{(n)}) - d^\sharp(\xi^{(n)}) < \varepsilon\Phi(M(\xi^{(n)}))$, one can stop an iterative process with the assurance that $\Phi(M(\xi^{(n)}))/\min_\xi \Phi(M(\xi)) < (1 - \varepsilon)^{-1}$.

7. The singular case. We have already seen, in Theorem 3 and its application to Theorems 5 and 6 (especially (4.30)—(4.31)), the complicated form by which the Φ-optimality equivalent $d^\sharp = \bar{d}$ must be replaced if Φ is not differentiable. Of special interest are cases where the $M(\xi^*)$ being tested for optimality is singular; for it then often occurs that a convex Φ which is differentiable on $\mathscr{M}^+$ is not differentiable (where finite) on $\mathscr{M}$, and $\mathscr{D}(M, \bar{M})$ is not linear in M. In particular, if $\Phi^{(k')}$ on $\mathscr{P}_{k'}^+$ denotes as smooth a criterion as $\Phi_{p,A}^{**}$, and (partitioning as in (4.24)) $M^* = M_{11} - M_{12}M_{22}^- M_{21}$ is the "information matrix for the first k' parameters" (well-defined even if M_{22} is singular), the criterion $\Phi(M) = \Phi^{(k')}(M^*)$ has this nature (if M is singular), due to the non-linearity in M of M^*.

In such cases Theorem 3 is still valid but is of course difficult to use. The problem, then, is to translate (2.24) into more useful terms. One would hope for an analogue of (4.32) of the nonsingular non-differentiable case, but we might sometimes expect to obtain an analogue of the less satisfactory (4.30).

In the case $\Phi^{(k')} = \Phi_0$, it was shown by Kiefer [20] and by Karlin and Studden [13] (as corrected in [1]) that, in rough terms for brevity, if $\bar{d}(\xi)$ is computed from (4.24), then the sufficient condition $\bar{d} = k'$ (for D-optimality) may not be realizable, but that it is always realizable for some transformed system Af, for some nonsingular A for which $(A'^{-1}\theta)^{(1)} = \theta^{(1)}$. In both treatments it is necessary to solve an auxiliary game to find the right A; or, equivalently, to take the infimum of $\bar{d}$ over all choices of A. In any event, it is too unwieldy an approach to be useful in many problems, although it has sometimes been applied with success [1]. The analogue of this approach for general Φ will be treated in a sequel to the present paper.

In the case of D-optimality for k' parameters, an extremely useful and simple sufficient condition was obtained by Atwood, and we can duplicate it for general convex Φ which is differentiable on $\mathscr{M}^+$: Suppose we want to demonstrate the Φ-optimality of a singular $\bar{M}$ at which Φ is continuous, and that M is any element of $\mathscr{M}$ such that $M + \bar{M}$ has rank k. Define $M(\xi_\varepsilon) = (1 - \varepsilon)\bar{M} + \varepsilon M$ for $0 < \varepsilon < 1$. Suppose

$$\lim_{\varepsilon \downarrow 0} [\bar{d}(\xi_\varepsilon) - d^*(\xi_\varepsilon)] = 0 . \tag{7.1}$$

Then, by (6.5) and continuity at $\bar{M}$, we conclude that $\bar{M}$ is Φ-optimum. The interchange of the operations $\lim_\varepsilon$ and $\sup_x$ in (7.1) can be treated as in [1]. Illustrations of the use of (7.1) will appear in the sequel.

8. Acknowledgments. Several people, over several years, contributed to the ideas in this paper. Leonard Gross helped, with several conversations about the analysis used. Bob Walker performed the computations mentioned in Section 6. Mark Brown suggested the problem of unknown V_x described above (5.1). A conversation with J. N. Srivastava about the rationale behind choosing a reasonable Φ, led me to consideration of criteria like (4.39). Ingram Olkin and Bill Studden contributed helpful suggestions. To these colleagues, the others whose help I may have forgotten, and the students who have had to suffer through my presentation of this material, go my sincere thanks.

REFERENCES

[1] Atwood, C. L. (1969). Optimal and efficient designs of experiments. *Ann. Math. Statist.* **40** 1570–1602.

[1a] Atwood, C. L. (1973). Sequences converging to D-optimal designs of experiments. *Ann. Statist.* **1** 342–352.

[2] Beckenbach, E. F. and Bellman, R. (1965). *Inequalities*. Springer-Verlag, Berlin.

[3] Bendat, J. and Sherman, S. (1955). Monotone and convex operator functions. *Trans. Amer. Math. Soc.* **79** 58–00.

[4] Box, G. E. P. and Draper, N. R. (1959). A basis for the selection of a response surface design. *J. Amer. Statist. Assoc.* **54** 622–654.

[5] Chernoff, H. (1953). Locally optimal designs for estimating parameters. *Ann. Math. Statist.* **24** 586–602.

[6] Dwyer, P. S. and MacPhail, M. S. (1948). Symbolic matrix derivatives. *Ann. Math. Statist.* **19** 517–534.

[6a] Eccleston, J. A. (1972). On the theory of connected designs. Thesis, Cornell Univ.

[7] Elfving, G. (1952). Optimum allocation in linear regression theory. *Ann. Math. Statist.* **23** 255–262.

[7a] Elfving, G. (1959). Design of linear experiments. *Cramér Festschrift Volume*. Wiley, New York. 58–00.

[8] Fan, K. (1959). On a theorem of Weyl concerning eigenvalues of linear transformations. *Proc. Nat. Acad. Sci. USA* **35** 652–000.

[9] Farrell, R., Kiefer, J. and Walbran, A. (1965). Optimum multivariate designs. *Proc. Fifth Berkeley Symp. Math. Statist. Prob.* **1** 113–138.

[10] FEDOROV, V. V. (1969). Design of experiments for linear optimality criteria. *Theor. Probability Appl.* **16** 189-000.

[11] FEDOROV, V. V. (1969). The design of experiments in the multiresponse case. *Theor. Probability Appl.* **16** 323-000.

[12] FEDOROV, V. V. (1972). *Theory of Optimal Experiments* (Transl. and ed. by E. M. Klienko and W. J. Studden). Academic press, New York.

[12a] FEDOROV, V. V. and MALYUTOV, M. B. (1972). Optimal designs in regression problems. *Math. Operationsforsch. und Statist.* **3** 281-308.

[13] HOEL, P. G. and LEVINE, A. (1964). Optimal spacing and weighting in polynomial prediction. *Ann. Math. Statist.* **35** 1553-1560.

[14] KARLIN, S. and STUDDEN, W. J. (1966a). Optimal experimental designs. *Ann. Math. Statist.* **37** 783-815.

[15] KARLIN, S. and STUDDEN, W. J. (1966b). *Tchebycheff Systems.* Interscience, New York.

[16] KARSON, M. J., MANSON, A. R. and HADER, R. J. (1969). Minimum bias estimation and experimental designs for response surfaces. *Technometrics* **11** 461-000.

[17] KIEFER, J. (1959). Optimum experimental designs. *J. Roy. Statist. Soc. Ser. B* **21** 272-319.

[18] KIEFER, J. (1960). Optimum experimental designs *V*, with applications to systematic and rotatable designs. *Proc. Fourth Berkeley Symp. Math. Statist. Prob.* **1** 381-405.

[19] KIEFER, J. (1961). Optimum designs in regression problems, II. *Ann. Math. Statist.* **32** 298-325.

[20] KIEFER, J. (1962). An extremum result. *Canad. J. Math.* **12** 597-601.

[21] KIEFER, J. (1972). Optimum designs for fitting biased multiresponse surfaces. *Proc. Multivar. Symp. III.* To appear.

[21a] KIEFER, J. (1972). General optimality equivalence theory for approximate designs (abstract). *Bull. Inst. Math. Statist.* **1** 258.

[22] KIEFER, J. and WOLFOWITZ, J. (1959). Optimum designs in regression problems. *Ann. Math. Statist.* **30** 271-294.

[23] KIEFER, J. and WOLFOWITZ, J. (1960). The equivalence of two extremum problems. *Canad. J. Matth.* **14** 363-366.

[24] KIEFER, J. and WOLFOWITZ, J. (1964). Optimum extrapolation designs I and II. *Ann. Inst. Statist. Math.* **16** 79-108, 295-303.

[25] KIEFER, J. and WOLFOWITZ, J. (1965). On a theorem of Hoel and Levine on extrapolation. *Ann. Math. Statist.* **36** 1627-1655.

[25a] MARSHALL, A. W. and OLKIN, I. (1974). Majorization in multivariate distributions. To appear in *Ann. Statist.*

[25b] MARSHALL, A. W., WALKUP, D. W. and WETS, R. J.-B. (1967). Order-preserving functions; applications to majorization and order statistics. *Pacific J. Math.* **23** 569-584.

[25c] MIRSKY, L. (1963). Results and problems in the theory of doubly stochastic matrices. *Z. Wahrscheinlichkeitstheorie und Verw. Gebiete* **1** 319-334.

[26] SHAH, K. R. (1960). Optimality criteria for incomplete block designs. *Ann. Math. Statist.* **31** 791-794.

[26a] SILVEY, S. D. and TITTERINGTON, D. M. (1973). A geometric approach to optimal design theory. *Biometrika* **60** 21-32.

[26b] WHITTLE, P. (1973). Some general points in the theory of optimal experimental design. *J. Roy. Statist. Soc. Ser. B* **35** 123-130.

27] WYNN, H. P. (1970). The sequential generation of *D*-optimum experimental designs. *Ann. Math. Statist.* **41** 1655-1664.

[28] WYNN, H. P. (1972). Results in the theory and construction of *D*-optimum experimental designs. *J. Roy. Statist. Soc. Ser. B* **34** 133-147.

DEPARTMENT OF MATHEMATICS
CORNELL UNIVERSITY
ITHACA, NEW YORK 14850

Discussion on the Paper "Planning Experiments for Discriminating Between Models" by A. C. Atkinson and D. R. Cox

J. KIEFER
Cornell University

Bad design procedure accounts for these comments being too late to be available when the paper was read; the planning of the mails requires no further comment.

The paper presents an interesting beginning on a practical problem of greatest importance: "What curve should one fit?" More often than not, the experimenter faces a choice among models as they appear in (3.3) rather than as in the examples of Section 4. That is, various possible models are obtained from others by specializing some parameter values to zero. Penalties are incurred for misestimation (variance plus bias) and, working in the opposite direction, for fitting a more complex function than needed (e.g. number of non-zero parameter estimates). This is an old problem, and recent attacks on it have been made by Lindley, Brooks and Halpern. Results are lacking for small sample sizes, and one hopes to get ideas of what to do in such cases from attacking simple models in the present manner. This may suggest plans whose properties can be compared numerically for the less tractable problems with many parameters and dimensions.

With reference to the last paragraph of Section 4.2, the "best movement" to the $(n+1)$th point is known in general not to be of the simple form described for D-optimality. However, this is remediable in a number of ways. If one is minimizing a strictly convex function $\Phi(\mathbf{M}_n)$ (or an increasing function thereof), where $\mathbf{M}_n = n^{-1}\mathbf{X}^{\mathrm{T}}\mathbf{X}$ in the authors' notation, then the iterative scheme motivated by the equivalence theory is to choose $\mathbf{M}_n = \varepsilon_n \mathbf{x}\mathbf{x}^{\mathrm{T}} + (1-\varepsilon_n)\mathbf{M}_{n-1}$ where $\mathbf{x}$ is a direction of steepest descent and ε_n might depend on $\mathbf{M}_n$ in some optimum way or, as in the present case, be chosen so that $\varepsilon_n \downarrow 0$, $\Sigma\varepsilon_n = +\infty$. (In the paper, $\varepsilon_n = n^{-1}$.) What might help in the present situation from a practical computational point of view is that $\mathbf{x}$ need not be chosen optimally; for example, if it is chosen to yield a direction in which the derivative of Φ is half the steepest value, that would suffice. As several of the authors who have worked in this area have pointed out, many iterative schemes for finding the minimum of a convex function are at our disposal.

The asymptotic data-dependent results are related to those of Chernoff and his students, and to those of Kiefer and Sacks. All of these have little known relevance to small sample size. If one is going to take N observations where N is large, then (for example) $N^{\frac{1}{2}}$ of these can be squandered in a rather inefficient first stage which will nevertheless distinguish among the models of (say) Table 6 with error probability of order $\exp(-cN^{\frac{1}{2}})$ *unless* some of the β_i's are very near 0 (in which case, from the point of view described two paragraphs above, several "models" would often be accurate enough, anyway). But then the remaining $N-N^{\frac{1}{2}}$ observations can be allocated to estimate the regression coefficient(s) as though the inferred model is indeed correct. As $N \to \infty$, 100 per cent efficiency results. But how well does this scheme work when $N = 25$ or 100?

The Annals of Statistics
1975, Vol. 3, No. 1, 109-118

BALANCED BLOCK DESIGNS AND GENERALIZED YOUDEN DESIGNS, I. CONSTRUCTION (PATCHWORK)[1]

BY J. KIEFER

Cornell University

The elementary constructive methods for BBD's and GYD's, which have been used in optimality considerations for a number of years, are listed and illustrated. These are not detailed algebraic or geometric prescriptions for listing each entry, but rather methods for combining LS's and known BIBD's to yield the desired products. Nevertheless, there results a large class of useful and previously unpublished designs.

1. Introduction. The original Youden square (YS) for v varieties was a $k \times v$ array obtained from a BIBD (v, b, k, r, λ) with $b = v > k$ by considering blocks as columns, arranged to make each variety appear once per row. Generalizations by Shrikhande [9] and Agrawal [1] allowed $b = mv$ for integral v. In the simpler setting of one-way heterogeneity, a number of authors have considered "binary" and "ternary" designs which were meant to generalize BIBD's to block size $k > v$.

It was noticed in design optimality proofs over the last fifteen years [3], [4], [5], [6], [7] that a restriction like $k > v$ seemed inessential and sometimes mathematically unnatural (despite the obvious practical motivation), and the BIBD and YS were generalized in [3] to the balanced block design (BBD) and generalized Youden design (GYD), which we now define. In the block design setting with v varieties and b blocks of size k, we again think of a design as a $k \times b$ array with blocks as columns, and let n_{ij} be the number of appearances of variety i in block j. Write $r_i = \sum_j n_{ij}$ and $\lambda_{ih} = \sum_j n_{ij} n_{hj}$, and let θ be the fractional part of k/v.

DEFINITION 1. A BBD is a design with all r_i equal, all λ_{ih} equal for $i < h$, and $|n_{ij} - k/v| < 1$ for all i, j.

The last condition is conveniently thought of and described as "n_{ij}'s as nearly equal as possible". Designs satisfying all but this last condition have by now been called "balanced" by some authors because they estimate every difference between two varieties, with the same variance. This nomenclature seems misleading combinatorially (the sets $\{n_{i1}\}$ and $\{n_{i2}\}$ may differ), and without the last condition of Definition 1 there is no relationship to optimality. We also use the notation (v, b, k, r, λ) or (v, b, k) for a BBD, with all $r_i = r = bk/v$ and all $\lambda_{ih} = \lambda$ in the above definition. The usual counting argument shows that

Received February 1973; revised February 1974.

[1] Research supported by NSF Grant GP24438.

AMS 1970 *subject classifications.* Primary 62K10; Secondary 05B15.

Key words and phrases. Experimental designs, balanced block designs, generalized Youden designs, two-way heterogeneity, combinatorial design construction.

$\lambda = b[k^2(v-1) - v^2\theta(1-\theta)]/v^2(v-1)$. Other formulas and resulting divisibility restrictions are detailed in [8].

DEFINITION 2. A $b_1 \times b_2$ array of the symbols $1, 2, \cdots, v$ is a GYD if it is a BBD when each of {rows} and {columns} is considered as blocks.

During the author's investigation of optimality properties of GYD's, designs of this type were constructed by piecing together other designs—what we shall call "patchwork methods". (Example 4.4 discusses a GYD of [3] not of this type.) The resulting designs were referred to when needed, but the methods were never published. Recently, Ruiz and Seiden [8] have given elegant geometric constructions of GYD's for certain parameter values involving prime powers. Since the patchwork techniques include additional parameter values which are of practical value and which are required in the optimality considerations [7], and since these techniques are quick to use, have not appeared in the literature, and use ideas which may be applicable in other design constructions, it seems worthwhile to list them in the present note; optimality considerations, which use some of the present results but contain mostly quite different ideas, will appear in the sequel.

It will be seen that the basic techniques are elementary and largely obvious. Nevertheless, if methodically applied they yield large families of previously unpublished designs of practical size.

We require some further definitions.

As usual, we use $m \mid n$ to mean that n/m is an integer.

DEFINITION 3. A GYD is *regular* if $v \mid b_1$ or $v \mid b_2$.

DEFINITION 4. A BIBD is *partly resolvable* (PR) if $k \mid v$ and there are v/k blocks whose union contains each variety once.

This last clearly generalizes resolvability. There is an extension to BIBD's or BBD's in which tv/k blocks contain t appearances of each variety, which for our application must be suitably balanced in a manner which will be indicated in connection with Proposition 6.

We denote by $r \times s$ $LS(v)$ an $rv \times sv$ array formed from $r \times s$ latin squares of order v. By $\operatorname{int}\{x\}$ we denote the greatest integer $\leq x$.

The author is indebted to Esther Seiden for many helpful discussions.

2. Regular GYD's. We shall see that no new methods are required in the regular case. The first proposition is evident:

PROPOSITION 1. *A* BBD, *after rearrangement in columns, is the union of b complete blocks in which each treatment appears* $\operatorname{int}\{k/v\}$ *times, and a* BIBD.

Thus, necessary for the existence of a BBD with given parameters is existence of an associated BIBD with corresponding parameters.

The methods used to make YS's from BIBD's, as devised by Hartley and Smith [2] or Shrikhande [9] and Agrawal [1] (using also systems of distinct representatives) can be extended to certain BBD's, as described in [4]. We state this along with a variant used later:

PROPOSITION 2. (i) *A* BBD *with* $v \mid b$ *can be made into a* $k \times b$ GYD *by rearrangement within columns (blocks).*

(ii) *If* $v \mid kb$, *any* $k \times b$ *array (not necessarily a* BBD) *with equal replications of varieties can be rearranged in columns so that each variety occurs as nearly equally as possible per row (and hence equally if* $v \mid b$).

The above simple rearrangement schemes can fail to yield a GYD if $v \nmid b$, even if one starts with a BBD [4]. In a sense, the nonregular GYD construction problem can be thought of as that of finding when a BBD with $v \nmid b$, k can be suitably rearranged within columns. But this point of view has not yet yielded any useful generalization of the methods of [2], [9], [1].

By applying Fisher's result to the BIBD of Proposition 1, we obtain

PROPOSITION 3. *A* BBD *with* $b < v$ *must have* $v \mid k$.

Hence, the only nonregular GYD's have $b_1, b_2 > v$.

Thus, for given b_1, b_2, v, the existence of a regular GYD is by Propositions 1 and 2 equivalent to that of a corresponding BIBD; when such a BIBD is known, these propositions yield a method of constructing the desired GYD. Proposition 3 delimits the additional cases we must study.

3. Patchwork methods. In what follows, t, a_i, b_i and c_i denote positive integers. When v is understood from the context, we shall often describe a GYD by its b_i alone. The simplest patchwork joins a regular GYD to a general GYD in obvious fashion:

PROPOSITION 4. *The union of rows of an* $a_1 v \times b_2$ GYD *and a* $c_1 \times b_2$ GYD *yield an* $(a_1 v + c_1) \times b_2$ GYD.

This is used to yield new designs when the $c_1 \times b_2$ GYD is nonregular. The conclusion of Proposition 4 is not generally true if the $a_1 v \times b_2$ GYD is replaced by a nonregular one.

In view of Proposition 1, a necessary condition for existence of a GYD with $b_i = a_i v + c_i$ is existence of the two BIBD's $B_i = (v, b_i, c_{3-i})$. The next two propositions are our main patchwork methods, which impose additional conditions on the B_i to obtain GYD's. (We also illustrate, in Example 4.4, that not all GYD's can be constructed by these methods.)

PROPOSITION 5. *Suppose there are* PRBIBD's *with parameters* $(v, a_1 v + c_1, c_2)$ *and* $(v, a_2 v + c_2, c_1)$ *where* $c_1 c_2 = v$. *Then there is an* $(a_1 v + c_1) \times (a_2 v + c_2)$ GYD.

Construction. We label the required array

$$G = \begin{pmatrix} G_{11} & G_{12} \\ G_{21} & G_{22} \end{pmatrix}.$$

Here G_{11} is any $a_1 v \times a_2 v$ GYD, for example (but not necessarily) $a_1 \times a_2$ LS(v). Next, (G_{21}, G_{22}) is the PRBIBD with block size c_1, the c_2 blocks which contain each treatment once being G_{22}. After a renumbering of varieties in the other PRBIBD, to make the c_1 blocks containing each treatment once be G'_{22}, that PRBIBD is taken to be (G'_{12}, G'_{22}). Finally, the columns of G_{21} and of G'_{12} are rearranged as described in the second part of Proposition 2 so that each row of G_{21} contains each variety a_2 times, and similarly for G'_{12}.

The dual role of G_{22} in the above construction makes clear a generalization of the notion of partial resolvability of the two designs used above: We require $c_1c_2 = tv$ and the existence of BBD's $(v, a_iv + c_i, c_{3-i})$ as before (no longer necessarily BIBD's), but with G_{22}, containing each variety t times, having columns which are c_2 of the blocks of the first BBD and rows which are c_1 of the blocks of the other. It is no longer automatic, as it was when $t = 1$, that existence in these BBD's of c_{3-i} blocks of size c_i containing each variety t times suffices to permit a renumbering of varieties which allows the two BIBD's to fit consistently in G_{22}. Illustrations of how this generalization works are contained in Section 4. We now state the principle formally, and then some simple sufficient conditions which will be useful for large b_i. We denote the columns of (G_{21}, G_{22}) by B_2, and those of (G'_{12}, G'_{22}) by B_1.

PROPOSITION 6. *Assume* $v \mid c_1c_2$. *Suppose* (i) *there are* BBD*'s* $B_i = (v, b_i, c_{3-i})$ *with* $v \mid b_i - c_i$ *and* (ii) *there are* c_2 *blocks of* B_2, *whose union contains each variety exactly* t *times, and such that* (iii) *the* $c_1 \times c_2$ *array formed with these blocks as columns has rows which are blocks of* B_1. *Then there is a* $b_1 \times b_2$ GYD. *Moreover,* (iii) *is satisfied if* (ii) *is satisfied and* (iv) B_1 *is composed of all the blocks of at least* c_1 BBD*'s; and* (ii) *and* (iii) *are satisfied if, in addition to* (iv), B_2 *is composed of all the blocks of at least* c_2 BBD*'s.*

Construction. The GYD is constructed from (i), (ii), (iii) as in Proposition 5. Next, assuming (ii), rearrange within the columns of G_{22} so that varieties are as nearly equally replicated as possible in each row (Proposition 2). Each of these c_1 rows can be taken as a block of a different (relabeled) component BBD of B_1, proving that (iv) implies (iii). If also B_2 has c_2 component BBD's this device also produces the columns of G_{22}.

The number of component BBD's needed for B_2 to satisfy (ii) can often be reduced from c_2 without much knowledge of B_2. For example, if c_2 is even, B_2 contains at least $c_2/2$ copies of the same sub-BIBD, and one knows two blocks of the latter with an even number of common varieties, that clearly suffices.

There are various BIBD operations which can be extended to BBD's, such as *derivation* and *residuation*. Because of Proposition 1, these yield nothing new for BBD's, and additional argument is needed to yield new GYD's. (It is well known that certain operations, such as identification of varieties in sets of q to form a design with v/q varieties, do not work at all except for very special parameter values.) As an illustration of the type of additional argument that is needed to obtain new GYD's, after an obvious consideration of the operation of *complementation*, we then give an application of it (Proposition 8).

PROPOSITION 7. *If a* $b_1 \times a_2 v$ GYD $\bar{G}$ *contains a* $b_1 \times b_2$ GYD G *as a rectangular subarray, then the complement* $\bar{G} - G$ *of the latter in the former is a* GYD.

If the $a_2 v$ is replaced by a number not divisible by v, the above conclusion is generally false.

The operation of complementation is in a sense the opposite of that of union, used in Proposition 4. Similarly, there is an opposite of the operation of Proposition 5, consisting of the removal of G_{21}, G_{22}, G_{12} from an $a_1 v \times a_2 v$ GYD G, under appropriate conditions.

It is not clear which GYD parameter values are obtainable from Proposition 7 but not from Propositions 4, 5, and 6, but the use of Proposition 7 sometimes avoids the more complex verifications needed in using Proposition 6.

PROPOSITION 8. *Suppose* $v = c_1 c_2$ *and that there exist*

(i) BIBD $(v, a_3 v - c_2, c_1)$,
(ii) PRBIBD $(v, a_2 v + c_2, c_1)$,
(iii) PRBIBD $(v, a_1 v + c_1, c_2)$.

Then there is an $(a_1 v + c_1) \times (a_3 v - c_2)$ GYD.

Construction. Let G be the $(a_1 v + c_1) \times (a_2 + c_2 v)$ GYD obtained by the method of Proposition 5; we shall use the same notation for the G_{ij}. Since each variety occurs once in G_{22}, so that rearrangement of its columns amounts merely to relabeling varieties, it follows from the second part of Proposition 2 that a reordering of the columns of the BIBD (i) can be used to produce an array G_{23} such that each of the rows of the $c_1 \times a_3 v$ array (G_{22}, G_{23}) has equal replication of each variety. Let H be an $a_1 v \times (v - c_2)$ array whose rows are complements in $\{1, 2, \cdots, v\}$ of the corresponding rows of G_{12}. By the second part of Proposition 2, we can reorder within the rows of H so that each of the columns of H has equal replication of each variety. Let G_{13} be the union of the columns of this reordered H with those of an $a_1 v \times (a_3 - 1)v$ GYD (or with nothing, if $a_3 = 1$). Then $\bar{G} = [G, \binom{G_{13}}{G_{23}}]$ is seen to be a regular GYD, and hence (Proposition 7) so is $\binom{G_{13}}{G_{23}}$.

Proposition 8 can be extended using Proposition 6.

We note that a useful condition implying (i) and (ii) is existence of a PRBIBD (v, b, c_1) with $b \mid a_3 v - c_2, a_2 v + c_2$.

Note that the construction cannot be simplified to mere application of Proposition 6 to $\begin{pmatrix} G_{12} & G_{13} \\ G_{22} & G_{23} \end{pmatrix}$ in place of $\bar{G}$; for, $\binom{G_{12}}{G_{22}}$ will not be a GYD unless $c_1 = v$ (Proposition 3).

4. Examples of nonregular GYD's. We do not attempt an exhaustive list of GYD's, but rather illustrate the patchwork methods in a representative selection of cases, including ones which have turned out to be important in optimality considerations [7]. Designs for many of the parameter values we consider can be obtained by using any of several methods (Propositions 4–8), but we give only one construction here in each case.

In optimality considerations, the values of θ_i = fractional part of b_i/v are important. In a sense, the cases where $\theta_i = \frac{1}{2}$ represent maximum departure of a GYD from the regularity of $a_1 \times a_2$ $LS(v)$, and the best chance for a GYD not to be optimum.

In these examples t, q, J_i, and J_i' will be positive integral parameters of design series. J_i' generally represents the number of replications of a sub-BIBD making up B_i, and is expressed in terms of the unrestricted J_i.

EXAMPLE 4.1. *The $\theta_1 = \theta_2 = \frac{1}{2}$ series.* It is easily seen that these designs can be represented in terms of three integer parameters $t > 0$, $J_1 > 0$, $J_2 > 0$, with

$$v = 4t\,, \qquad b_i = 2t(4t-1)(2J_i-1)\,. \tag{4.1}$$

The J_i' defined above will be seen to be $(2J_i - 1)t$.

EXAMPLE 4.1.1. *The $t = 1$ series.* A $6(2J_1 - 1) \times 6(2J_2 - 1)$ GYD with $v = 4$ is obtained by using Proposition 5 with (G_{21}, G_{22}) equal to $(2J_2 - 1)$ copies of the (4, 6, 2) BIBD and (G'_{12}, G'_{22}) equal to $(2J_1 - 1)$ copies.

EXAMPLE 4.1.2. *General considerations for $t > 1$.* We must now use Proposition 6, since each variety occurs t times in $G_{22}(2t \times 2t)$. The required BIBD's are most easily obtained as $(2J_1 - 1)t$ replicates of the BIBD

$$(4t,\ 2(4t-1),\ 2t,\ 4t-1,\ 2t-1)\,, \tag{4.2}$$

if it exists. Professor Seiden points out that this is always the case if an $8t \times 8t$ Hadamard matrix exists.

PROPOSITION 9. *For $t > 1$, if either $J_i \geqq 2$ and the* BIBD (4.2) *exists, then the* GYD *exists.*

Construction. Suppose (4.2) exists and $J_1 \geqq 2$, so that $(2J_1 - 1)t \geqq c_1$. Then (iv) of Proposition 6 is satisfied, and we need only verify (ii). In the design (4.2), the number $\binom{8t-2}{2}$ of pairs of different blocks is odd, and the sum of all block intersection numbers is $4t\binom{4t-1}{2}$, which is even. Hence, at least one intersection number is even, and the corresponding pair of blocks can be used as described in the paragraph following "Construction" of Proposition 6, since $(2J_2 - 1)t \geqq c_2/2$.

In the remainder of the discussion of the $\theta_1 = \theta_2 = \frac{1}{2}$ series, assuming (4.2) exists, it is thus only necessary to consider the values $J_1 = J_2 = 1$.

EXAMPLE 4.1.3. *The case $t = 2$.* Here there is a resolvable design (4.2) which easily yields

$$G_{22} = \left\| \begin{array}{cccc} \infty & 6 & 3 & 4 \\ 0 & 2 & 5 & 1 \\ 1 & 5 & 2 & 0 \\ 3 & 4 & \infty & 6 \end{array} \right\|$$

as a G_{22} whose rows and columns are blocks of (4.2); one does not even need the 2 copies of (4.2) available to us, in obtaining G_{22}.

EXAMPLE 4.1.4. *The case* $t = 3$. A resolvable design (4.2) can be developed mod 11 from the blocks $\beta_1 = (1, 2, 4, 5, 6, 10)$ and $\beta_2 = (3, 7, 8, 9, 0, \infty)$. There are many possible constructions of G_{22}. For example, to get equal replication numbers, add 0, 1 and 3 to β_1 and 0, 1, 3 to β_2. The resulting G_{22}, whose columns are $\beta_2 + 2, 4, 7$ and $\beta_1 + 2, 4, 7$, is

$$G_{22} = \left\| \begin{array}{rrrrrr} 2 & 4 & 10 & 1 & 5 & 6 \\ 0 & 7 & 5 & 3 & 6 & 2 \\ 5 & 2 & 7 & 4 & 8 & 9 \\ 9 & 0 & \infty & 7 & 3 & 8 \\ 10 & \infty & 4 & 8 & 9 & 1 \\ \infty & 1 & 3 & 6 & 10 & 0 \end{array} \right\|$$

The author is again grateful to Esther Seiden, this time for pointing out the existence of this resolvable design on Preece's list [7a]; the more common non-resolvable design on most BIBD lists had not led to a successful construction.

EXAMPLE 4.1.5. *The case* $t = 4$. There is a resolvable design (4.2), and G_{22} is easily constructed in a manner similar to that used in Example 4.1.3.

The above examples give constructions of the $\theta_1 = \theta_2 = \frac{1}{2}$ series for all $v \leqq 16$, which probably includes most "practical" cases.

EXAMPLE 4.2. *The series* $v = c_1 c_2$, $b_i = a_i v + c_i$. To avoid trivialities, we assume $c_1 \geqq 2$, $c_2 \geqq 3$; the case $c_1 = c_2 = 2$ falls under Example 4.1.1. We first consider a subseries for which the calculations are particularly simple.

EXAMPLE 4.2.1. *Assume each* $c_i - 1$ *relatively prime to* $c_1 c_2 - 1$ (always true if $c_1 = 2$). For a PRBIBD $(G_{21}, G_{22})(v, b', k', r', \lambda')$ to exist, we need $r' = \lambda'(v - 1)/(k' - 1) = \lambda'(c_1 c_2 - 1)/(c_1 - 1)$; hence, r' is of the form $J_2'(c_1 c_2 - 1)$ and $b' = vr'/k' = c_2 J_2'(c_1 c_2 - 1)$. Also, in order to use Proposition 5 we require $v \mid b' - c_2$, which yields $J_2' = J_2 c_1 - 1$. Thus, finally, the parameters of (G_{21}, G_{22}) are necessarily of the form

$$(4.3) \qquad (c_1 c_2, c_2(J_2 c_1 - 1)(c_1 c_2 - 1), c_1, (J_2 c_1 - 1)(c_1 c_2 - 1), (J_2 c_1 - 1)(c_1 - 1))$$

with $J_2 > 0$, and $a_2 = J_2 c_1 c_2 - c_2 - J_2$. Interchanging subscripts 1 and 2 in (4.3), we obtain the form (4.3)′ (say) of (G'_{21}, G'_{22}).

The design (4.3) can be obtained as $J_2 c_1 - 1$ copies of a $(c_1 c_2, c_2(c_1 c_2 - 1), c_1)$ BIBD if the latter exists, and similarly for (4.3)′. One must still check that PR designs result.

If

$$(4.4) \qquad J_2 c_1 - 1 \geqq c_2 \qquad \text{and} \qquad J_1 c_2 - 1 \geqq c_1 ,$$

we can use the tool of Proposition 6(iv) in this simpler $t = 1$ setting: There are enough copies of $(c_1 c_2, c_2(c_1 c_2 - 1), c_1)$ to take a relabeled block from each of c_2 different copies, as the columns of G_{22}; similarly for rows. Thus, a GYD always exists under (4.4), if the two BIBD's $(c_1 c_2, c_i(c_1 c_2 - 1), c_{3-i})$ exist.

If one (or both) of the inequalities (4.4) is violated but the corresponding design(s) (4.3) or (and) (4.3)′ is PR, of course we still obtain a GYD. This is the case when $c_1 = 2$, since the BIBD $(2c_2, c_2(2c_2 - 1), 2)$ is resolvable and $J_1 c_2 - 1 \geqq 2$ for $c_2 \geqq 3$. *Thus, we need only check the existence of the* BIBD $(2c_2, 2(2c_2 - 1), c_2)$ *to know that a* GYD *exists for all* $J_i > 0$, *if* $c_1 = 2$. As in Example 4.12, if a $4c_2 \times 4c_2$ Hadamard matrix exists, so does this BIBD.

If $c_1 = 3$ and c_2 is even, we are still in the framework of Example 4.2.1. On the other hand, if $c_1 = 3$ and c_2 is odd, we obtain an illustration of the fact that *the present example gives sufficient conditions for the existence of a design even if the* $c_i - 1$ *are not both relatively prime to* $c_1 c_2 - 1$, *but there may exist* GYD's *for other parameter values*. Rather than attempt an exhaustive study of the relevant divisibility considerations for general c_1, c_2, we treat only the cited illustration:

EXAMPLE 4.2.2. *The series* $c_1 = 3$, $c_2 = 2q + 1$. The possible values of r' in the development of Example 4.2.1 (and, hence, of b' and λ') are seen to be multiplied by $\frac{1}{2}$ in the present setting, for both (G_{21}, G_{22}) and (G'_{12}, G'_{22}). In the subsequent treatment of the requirement $v \mid b' - c_i$, we obtain that any even J_i' is of the form $2(J_i c_{3-i} - 1)$. The parameter values of (4.3) and (4.3)′ are unchanged, but one tries to achieve them as $2(3J_2 - 1)$ copies of $(6q + 3, (2q + 1)(3q + 1), 3)$ and $2[(2q + 1)J_1 - 1]$ copies of $(6q + 3, 3(3q + 1), 2q + 1)$. If these latter BIBD's exist, we see that there are indeed some GYD's which were obtainable as indicated in italics in the previous paragraph.

If J_i' is odd, we obtain that it is of the form $c_{3-i}(2J_i - 1) - 2$. We are led to seek $6J_2 - 5$ copies of $(6q + 3, (2q + 1)(3q + 1), 3)$ and $2(2q + 1)J_1 - (2q + 3)$ copies of $(6q + 3, 3(3q + 1), 2q + 1)$. Of course, the J_i' can have opposite parity.

The analogue of (4.4) is

$$(4.5) \qquad \begin{aligned} &J_i' \text{ even:} \quad && 2(c_{3-i}J_i - 1) \geqq c_i\,; \\ &J_i' \text{ odd:} \quad && c_{3-i}(2J_i - 1) \geqq 2 + c_i\,. \end{aligned}$$

For $i = 1$ and $q > 1$, these inequalities are valid for all J_1. For $q = 1$, we have $c_1 = c_2$ and the same problem in both directions. Thus, as in the paragraph following (4.4) (with obvious modifications), we are led to seek a PRBIBD $(6q + 3, (3q + 1)(2q + 1), 3)$.

This is the resolvable Steiner triple design, which is known to exist for all $q > 0$. We conclude:

If the BIBD $(6q + 3, 3(3q + 1), 2q + 1)$ *exists, then a* GYD *exists for all* J_1, J_2 *(with* J_i' *of both odd and even forms).*

For example, for $q = 1$ we obtain the $v = 9$, $\theta_1 = \theta_2 = \frac{1}{3}$ series for b_i of the form $12(6J_i - 2)$ or $12(6J_i - 5)$; that is, of the form $12(3J_i - 2)$. For $q = 2$, the design (15, 21, 5) does not exist, and one would have to try to work with a (15, 21h, 5) design for some $h > 1$; such a modification of our treatment can in general yield impractical parameter values. When $q = 3$ or 4, the required BIBD again exists.

EXAMPLE 4.3. *The series* $tv = c_1c_2$, $t > 1$. This extends Example 4.2 in the same way that Examples 4.1.2—4.1.5 extended 4.1.1. We must again use Proposition 6. We illustrate with an extension for the v of Example 4.2.2 with $t = 2$:

EXAMPLE 4.3.1. *The series* $v = 3(2q + 1)$, $c_1 = 3$, $c_2 = 2(2q + 1)$. We now obtain, for (G_{21}, G_{22}), that $r' = (3q + 1)J_2'$ and $J_2' = 3J_2 - 1$. Similarly, for (G'_{12}, G'_{22}), we obtain $r' = (6q + 2)J_1'$ and $J_1' = (2q + 1)J_1 - 2$. We thus try to apply the technique of Proposition 6 to J_2' copies of $(6q + 3, (2q + 1)(3q + 1), 3)$ and J_1' copies of $(6q + 3, 3(3q + 1), 2(2q + 1))$.

For $q > 1$ or $J_1 > 1$, we have $J_1' \geqq 3$ and thus the rows of G_{22} can be arbitrary (provided no variety occurs more than once per row and each variety occurs twice). Since $J_2' \geqq 2$, we have at least two Steiner triples making up (G_{21}, G_{22}), so we can choose the first $2q + 1$ columns of G_{22} from one Steiner triple so as to contain each variety once, and the last $2q + 1$ columns from cyclic permutation of the rows of the first $2q + 1$ columns.

There remains the case $q = 1$, $J_1 = 1$. The prescription of the previous paragraph then yields three rows of G_{22} which can be taken as blocks of the BIBD (9, 12, 6) (obtained as complements of the blocks of (9, 12, 3)). We conclude:

If the BIBD $(6q + 3, 3(3q + 1), 2(2q + 1))$ *exists, then there is a* $3[(2q + 1)J_1 - 2](3q + 1) \times (3J_2 - 1)(2q + 1)(3q + 1)$ GYD *with* $v = 6q + 3$, *for all positive* J_i.

EXAMPLE 4.4. *Nonisomorphic* GYD's. These can of course occur when nonisomorphic BIBD's with the same parameter values exist and are used in our construction. More interesting is the existence of GYD's which are nonisomorphic because one is obtained from one of our patchwork methods and the other cannot be so obtained. For example, when $v = 4$, the 6×6 GYD of [3], [6],

$$\begin{array}{cccccc} 1 & 4 & 2 & 4 & 3 & 2 \\ 2 & 1 & 4 & 3 & 3 & 4 \\ 2 & 3 & 1 & 3 & 4 & 2 \\ 1 & 3 & 3 & 1 & 2 & 4 \\ 4 & 1 & 4 & 2 & 1 & 3 \\ 3 & 2 & 1 & 4 & 2 & 1 \end{array} \tag{4.6}$$

has no subarray of 4 rows and columns which constitute a LS (4), since any such 4×4 array has at least one row or column with zero or two one's. Hence, this design is not isomorphic to (obtainable by row or column permutations or relabeling from) that of Example 4.1.1 with $J_1 = J_2 = 1$.

It appears that the methods of [8] are likely to yield designs of the form (4.6). Since all GYD's with the same v, k_1, k_2 have the same covariance matrix for variety contrasts, optimality properties of GYD's cannot vary between two nonisomorphic designs. However, our preliminary investigations indicate that counterexamples to optimality, where they exist, may be more easily obtainable from slight modification of one of two such designs [6], [7], [8].

REFERENCES

[1] AGRAWAL, H. (1966). Some generalizations of distinct representatives with applications to statistical designs. *Ann. Math. Statist.* **37** 525–528.

[2] HARTLEY, H. O. and SMITH, C. A. B. (1948). The construction of Youden squares. *J. Roy. Statist. Soc. Ser. B* **10** 262–263.

[3] KIEFER, J. (1958). On the nonrandomized optimality and randomized non-optimality of symmetrical designs. *Ann. Math. Statist.* **29** 675–699.

[4] KIEFER, J. (1959). Optimum experimental designs. *J. Roy. Statist. Soc. Ser. B* **21** 272–319.

[5] KIEFER, J. (1970). Optimum experimental designs. *Proc. Internat. Congress Math.* (Nice) **3** 249–254. Gauthiers-Villars, Paris.

[6] KIEFER, J. (1971). The role of symmetry and approximation in exact design optimality. *Statistical Decision Theory and Related Topics*, 109–118. Academic Press, New York.

[7] KIEFER, J. (1973). Generalized Youden Designs, II. Optimality. To appear.

[7a] PREECE, D. A. (1967). Incomplete block designs for $v = 2k$. *Sankhyā Ser. A* **29** 305–316.

[8] RUIZ, F. and SEIDEN, E. (0000). Generalized Youden designs. To appear.

[9] SHRIKHANDE, S. S. (1951). Designs with two-way elimination of heterogeneity. *Ann. Math. Statist.* **22** 235–247.

DEPARTMENT OF MATHEMATICS
CORNELL UNIVERSITY
ITHACA, NEW YORK 14850

J. N. Srivastava, ed., *A Survey of Statistical Design and Linear Models*

Construction and Optimality of Generalized Youden Designs

J. KIEFER*

Cornell University, Ithaca, N.Y. 14850, USA

1 Introduction and summary

In 1958 the present author [3] extended Wald's work [12] on the *D*-optimality of Latin square (LS) designs to the broader context where a generalized Youden design (GYD) is used. In several subsequent publications (e.g., [4], [5], [6]) we have indicated the unification of ideas used in proving such *exact theory* results in a framework which is more general both in design settings and also in optimality criteria. (See [4] or [5] for the distinction between *exact* and *approximate* theory.) In the present paper we present this theory as it applies to the GYD setting.

During these almost 15 years of work on the GYD problem, motivated by optimality considerations indicated herein, we have had to construct GYD's for various sets of parameter values not previously covered by LS or the classical YD considerations of [1], [2], [11]. We shall mention briefly, in Section 5, a principal method of construction, taking further space in [7] to describe other methods and illustrations. A paper by Ruiz and Seiden [10] gives some more elegant methods for construction in particular cases.

In the usual block design setting of one-way heterogeneity, positive integers b, v, and k are specified, we have v varieties and b blocks of size k, and a *design* can be thought of as a $k \times b$ array of variety labels, with blocks as columns. Let n_{dij} be the number of appearances design d assigns to variety i in block j. Write $r_{di} = \Sigma_j n_{dij}$ and $\lambda_{dih} = \Sigma_j n_{dij} n_{dhj}$. Also, let $\rho =$ fractional part of k/v.

In this setting we define a *balanced block design* (BBD) as a design d^* with all r_{d^*i} equal, all λ_{d^*ih} equal for $i < h$, and $|n_{d^*ij} - k/v| < 1 \; \forall \, i,j$. This last condition can be described as "all n_{d^*ij}'s as nearly equal as possible".

* Research supported by NSF Grant GP35816X.

In the setting of two-way heterogeneity, integers v, b_1, b_2, all $\geqq 2$, are given, and a design d is a $b_1 \times b_2$ array G of variety labels $1, 2, \cdots, v$. We write $\lambda_{dih}^{(Q)}$ and $\rho^{(Q)}$ with $Q = R$ or C for the quantities λ and ρ when rows (R) or columns (C) are considered as blocks. We say the setting is *regular* if $\rho^{(R)}$ or $\rho^{(C)} = 0$.

A design here is defined to be a GYD if it is a BBD when each of {rows} and {columns} is considered as the blocks.

In this setting, the $b_1 b_2$ observations will be assumed uncorrelated with common variance σ^2. The expectation of an observation on variety t in the unit at row r and column c is $\alpha_t + \beta_r + \gamma_c$. For a specified design d, the vector of these $b_1 b_2$ observation expectations can be written as

$$X_d \theta = [X_d^{(1)} \vdots X_d^{(2)}] \begin{pmatrix} \theta^{(1)} \\ \theta^{(2)} \end{pmatrix}$$

where $\theta^{(1)}$ is the v-vector of α_t's. Then $X'_d X_d$ is the coefficient matrix of the "normal equations" of usual least squares theory. If we are interested only in estimation of linear combinations $c'\theta^{(1)} = \Sigma_t c_t \alpha_t$, we diagonalize these equations in blocks and obtain $C_d = X_d^{(1)\prime} X_d^{(2)} (X_d^{(2)\prime} X_d^{(2)})^- X_d^{(2)\prime} X_d^{(1)}$ for the coefficient matrix of the reduced normal equations, the "information matrix" of d for $\theta^{(1)}$. Since $c'\theta^{(1)}$ is well-defined in this model only if $\Sigma_t c_t = 0$ (i.e., $c'\theta^{(1)}$ is a "contrast"), we are led to consider a $(v-1) \times v$ real matrix P whose rows are orthonormal and orthogonal to constant vectors; then $P\theta^{(1)}$ consists of $k - 1$ linearly independent functions each of which can be estimated if C_d has rank $v - 1$. (C_d always has zero row and column sums in the present example.) Moreover, $\sigma^2 (PC_d P')^{-1} = \sigma^2 V_d$ (say) is then the covariance matrix of the usual least squares (or "best linear unbiased") estimators of $P\theta^{(1)}$. Thus, it is natural to specify some *optimality functional* ψ on the $(v-1) \times (v-1)$ matrices and to pose the problem:

$$\text{Find } d \text{ to minimize } \psi((PC_d P')^{-1}). \tag{1.1}$$

A design solving this problem is said to be *ψ-optimal*. If ψ is orthogonal invariant, the solution has the desirable practical advantage of not depending on the choice of P. We shall not discuss here the implications, in practicality or "foundations", of assuming that loss can be described as a function of V_d.

(In the one-way setting, the β_r's are omitted. More complex problems in which the γ_c's and β_r's are also to be estimated can be treated by modifications of the methods herein described.)

Some commonly used optimality criteria are

$$D\text{-optimality: } \psi(V) = \det V;$$

$$A\text{-optimality: } \psi(V) = tr\ V; \tag{1.2}$$

$$E\text{-optimality: } \psi(V) = \text{maximum eigenvalue of } V.$$

The relationship among these is well known, and will be repeated briefly in the example following Proposition 2; in the two-way heterogeneity setting, D-optimality of a GYD implies A-optimality, and A-optimality implies E-optimality.

In the special setting $b_1 = b_2 = v$, Wald showed a GYD (= LS in this setting) was D-optimal. The author proved the stronger conclusion of universal optimality (defined in Section 2) for the GYD in the larger case of all 2-way *regular* settings, and proved E-optimality of the GYD in all 2-way settings. It was subsequently discovered [5], [6] that a GYD is not necessarily D-optimal, a somewhat surprising result (at least to the author) in view of the highly symmetric structure of the GYD and the usual optimality of such highly symmetric designs (for example, of the BBD). The conclusions known at this time in nonregular settings are:

(1.3)

(a) *A GYD is always A-optimal.*
(b) *A GYD is D-optimal unless* $v = 4$.
(c) *If* $v = 4$ *and* $b_1 = b_2$, *a GYD is never D-optimal.*
(d) *If* $v = 4$ *and* b_1/b_2 *is sufficiently near* 1, *a GYD is not D-optimal.*

A main purpose of the present paper and [9] is to prove these results; the proof of (1.3) (a) is contained in Sections 3 and 4 herein. The general tools used in exact design optimality proofs are described in Section 2. Counter-examples to D-optimality when $v = 4$ are considered in Section 6.

2. Optimality tools

It is sometimes convenient to write the optimality criterion ψ as a function Φ on the set of possible matrices C_d. We allow Φ to take on the value $+\infty$, as it will for the criteria of (1.2) if C_d has rank $< v-1$. If ψ is orthogonal-invariant, we can also write it as a function Φ^* on the nonnegative $(v-1)$-vectors

$$\lambda_d = (\lambda_{d1}, \cdots, \lambda_{d(v-1)}), \quad \text{where } \lambda_{d1} \geqq \lambda_{d2} \geqq \cdots \geqq \lambda_{d(v-1)} \geqq \lambda_{dv} = 0$$

are the eigenvalues of C_d.

The rationale for finding computationally simple sufficient criteria for optimality is obvious: we do not want to have to compute $(PC_dP')^{-1}$, or even λ_d, for every competing design d, and hope to find a more tractable computation that will suffice.

The simplest such computation also yields the widest optimality conclusions, and hence can be applied least frequently; nevertheless, it is useful. To describe it, suppose (in a more general context than that of 2-way heterogeneity) that $\mathscr{B}_v$ consists of the $v \times v$ nonnegative definite matrices, $\mathscr{B}_{v,0}$ consists of those elements of $\mathscr{B}_v$ with zero row and column sums, and $\Phi: \mathscr{B}_{v,0} \to (-\infty, +\infty]$ satisfies

$$\begin{aligned}&\text{(a)}\quad \Phi \text{ is convex,}\\ &\text{(b)}\quad \Phi(bC) \text{ is nonincreasing in the scalar } b \geqq 0,\\ &\text{(c)}\quad \Phi \text{ is invariant under each permutation of rows}\\ &\qquad\ \text{and (the same on) columns.}\end{aligned} \tag{2.1}$$

A design d^* will be termed *universally optimal* in the class $\mathscr{D}$ of designs under consideration if d^* minimizes $\Phi(C_d)$ for every Φ satisfying (2.1). We define C_{d^*} to be *completely symmetric* (c.s.) if it is of the form $\alpha I_v + \beta J_v$ where α, β are scalars and J_v consists of all 1's. (Some confusion exists in the literature in the use of "symmetry" or "balance" to refer sometimes to X_d, sometimes to C_d, sometimes to the diagonal elements of V_d; this confusion is compounded by the occasional tacit assumption that such notions are automatically synonymous with optimality.)

We have used the following simple tool repeatedly in earlier work:

Proposition 1. *Suppose a class* $\mathscr{C} = \{C_d, d \in \mathscr{D}\}$ *of matrices in* $\mathscr{B}_{v,0}$ *contains a* C_{d^*} *for which*

$$\begin{aligned}&\text{(a)}\quad C_{d^*} \text{ is c.s.,}\\ &\text{(b)}\quad \operatorname{tr} C_{d^*} = \max_{d\in\mathscr{D}} \operatorname{tr} C_d.\end{aligned} \tag{2.2}$$

Then d^* is universally optimal in $\mathscr{D}$. (Since $-\operatorname{tr} C$ satisfies (2.1), it follows that (2.2) (b) is necessary for universal optimality.)

Proof. Suppose $\Phi(C_{d'}) < \Phi(C_{d^*})$. If $\tau C_{d'}$ is obtained from $C_{d'}$ by permuting rows and columns according to τ, and if $\bar{C}_{d'} = \sum_\tau \tau C_{d'}/v!$, then by (2.1) (c) and (a) we have

$$\Phi(C_{d^*}) > \Phi(C_{d'}) = \Phi(\tau C_{d'}) \geqq \Phi(\bar{C}_{d'}). \tag{2.3}$$

Of course, $\tau C_{d'}$ and $\bar{C}_{d'}$ need not be in $\mathscr{C}$, but $\bar{C}_{d'}$ is completely symmetric and in $\mathscr{B}_{v,0}$, and is hence of the form bC_{d^*} for some $b \geqq 0$. Since $\operatorname{tr} \bar{C}_{d'} = \operatorname{tr} C_{d'}$, (2.2) (b) implies $b \leqq 1$. But then $\Phi(\bar{C}_{d'}) \geqq \Phi(C_{d^*})$ by (2.1) (b), which with (2.3) contradicts $\Phi(C_{d'}) < \Phi(C_{d^*})$.

Remark. If Φ is strictly convex (and hence, also, "nonincreasing" is "decreasing" in (2.1) (b)), it is seen that if C_{d*} satisfying (2.2) exists, then every Φ-optimal $\bar{d}$ has $C_{\bar{d}} = C_{d*}$. This and Proposition 1 obviously hold if Φ is defined only on $\{\tau C_{d'}$ and $\bar{C}_{d'}, d' \in \mathscr{D}\}$. Finally, if $\mathscr{B}_{v,0}$ is replaced by $\mathscr{B}_v$ (in a setting where all components of $\theta^{(1)}$ are estimable for some designs d, for which $\psi(C_d^{-1}) = \Phi(C_d)$) we have the even simpler

Proposition 1′. *If a class of matrices $\mathscr{C} = \{C_d, d \in \mathscr{D}\}$ contains a C_{d*} which is a multiple of I_v and which maximizes* tr C_d *for $d \in \mathscr{D}$, then d^* is universally optimal in $\mathscr{D}$.*

Proposition 1′ is the justification of the classical intuitive principle of looking for balanced "orthogonal" designs in multifactorial settings. Proposition 1 treats PC_dP' the way Proposition 1′ treats C_d; indeed, PC_dP' is a multiple of I_{v-1} if C_d is c.s. Proposition 1 was used by us in earlier papers in the BBD setting. The achievability of the conclusion of Proposition 1 in such a practical setting justifies introducing the additional nomenclature of "universal optimality".

In the next section we shall review briefly the proof [3] of universal optimality of the GYD in *regular settings*; we rewrite a sketch of this proof in the terminology introduced below, in order to help explain the need for additional tools in nonregular settings where universal optimality fails for the GYD although it may be A- or even D-optimal. (We do not know in which nonregular settings the GYD is still universally optimal.) In the next section we also review the brief proof of E-optimality of the GYD in all settings where a GYD exists. This last conforms with the general fact that E-, A-, and D-optimality of c.s. C_{d*} are, in that order, increasingly difficult to prove. This is consistent with the relationship indicated below (1.2), which is part of the following obvious relationship in any design setting $\mathscr{C} = \{C_d, d \in \mathscr{D}\}$:

Proposition 2. *If $\Phi_1 \leqq \Phi_2$ on $\mathscr{C}$, with equality at C_{d*}, and if C_{d*} is Φ_1-optimal, then C_{d*} is Φ_2-optimal.*

Example. A useful family of criteria in the $\mathscr{B}_{v,0}$ context is

$$\Phi_p^*(\lambda_d) = \left(\frac{1}{v-1} \sum_i \lambda_{di}^{-p}\right)^{1/p} \tag{2.4}$$

for $0 < p < \infty$, with the limiting values $\Phi_0^*(\lambda_d) = \Pi_i \lambda_{di}^{-1/(v-1)}$ and $\Phi_\infty^*(\lambda_d) = \max_i \lambda_{di}^{-1}$. Here $p < q \Rightarrow \Phi_p(\lambda_d) \leqq \Phi_q(\lambda_d)$ with equality iff all λ_{di} are equal. Hence, from Proposition 2,

$$C_{d*} \text{ c.s., } d^*\Phi_p^*\text{-optimal} \Rightarrow d^*\Phi_q^*\text{-optimal } \forall q > p. \tag{2.5}$$

Note that $p = 0, 1, \infty$ yield the criteria of (1.2). Also, the family of criteria (2.4) can be considered also for p negative, and (2.5) is seen to hold for all real p. The Φ_p^* for negative values of p are not important as criteria with much intuitive appeal for applications, and can fail to have the desired convexity property. (Such properties of Φ_p^* are well known in functional analysis, and will be summarized in a forthcoming paper on approximate theory optimality methods.) Nevertheless, these Φ_p^* yield useful sufficient conditions for more meaningful optimality criteria: Φ_{-1}^*-optimality is the same as (2.2) (b), and the Φ_{-2}^*-optimality criterion of Shah has proved useful in other contexts.

While the form (2.4) is convenient for comparison as p is varied, we will find it slightly more convenient analytically to work in the sequel with the equivalent

$$\begin{aligned} &\text{(a)} \quad \Phi_p^{**}(\lambda_d) = \sum_i \lambda_{di}^{-p}, \qquad 0 < p < \infty; \\ &\text{(b)} \quad \Phi_0^{**}(\lambda_d) = -\sum_i \log \lambda_{di}, \\ &\text{(c)} \quad \Phi_\infty^{**}(\lambda_d) = \Phi_\infty^*(\lambda_d), \end{aligned} \tag{2.6}$$

and of course with $-\sum_i \lambda_{di} = -\operatorname{tr} C_d$ in the case of Proposition 1.

The analogue of this example in the $\mathscr{B}_v$ context is obvious.

We now turn to the inequalities which can be used in the absence of universal optimality, to obtain weaker optimality results (again, avoiding computation of $(PC_dP')^{-1}$ or λ_d). We treat only the $\mathscr{B}_{v,0}$ context here; the $\mathscr{B}_v$ setting is again easier. Suppose Φ^* is of the form

$$\Phi^*(\lambda_d) = \sum_{i=1}^{v-1} f(\lambda_{di}) \tag{2.7}$$

where f is convex on $[0, +\infty)$. Fix C_d and choose P as before but also so that PC_dP' is the diagonal matrix with diagonal entries $\lambda_{d1}, \cdots, \lambda_{d(v-1)}$. Let $e_{ij} = p_{ij}^2$ for $1 \leqq i \leqq v-1$, $1 \leqq j \leqq v$. Then $\sum_{j=1}^{v} e_{ij} = 1$ and $\sum_{i=1}^{v-1} e_{ij} = (v-1)/v$. Since $\sum_{i=1}^{v-1} e_{ij}\lambda_{di} = c_{djj}$, Jensen's inequality yields

$$\frac{v-1}{v} f\left(\frac{v}{v-1} c_{djj}\right) = \frac{v-1}{v} f\left(\sum_i \left(\frac{v}{v-1}\right) e_{ij}\lambda_{di}\right) \leqq \sum_i e_{ij} f(\lambda_{di}). \tag{2.8}$$

Summing on j, we obtain

$$\frac{v-1}{v} \sum_{j=1}^{v} f\left(\frac{v}{v-1} c_{djj}\right) \leqq \sum_{i=1}^{v-1} f(\lambda_{di}), \tag{2.9}$$

with equality if all λ_{di} are equal, i.e., *if C_d is c.s.* Thus, we obtain

Proposition 3. *If Φ^* is given by (2.7) with f convex, and if C_{d*} is c.s. and d^* minimizes*

$$\sum_j f\left(\frac{v}{v-1} c_{djj}\right),$$

then d^* is Φ^*-optimal.

Example. In the case of (2.6) (i.e., (2.4)), we obtain, for $0 < p < \infty$,

$$C_{d*} \text{ is c.s. and minimizes } \Sigma_j c_{djj}^{-p} \Rightarrow d^* \text{ is } \Phi_p^*\text{-optimal.} \qquad (2.10)$$

Either by passing to the limit with p in (2.4) or else by working with (2.6) (b) for $p = 0$ or by using the slightly different argument of [3], Lemma 3.4 for $p = \infty$, we obtain

$$C_{d*} \text{ is c.s. and maximizes } \sum_j \log c_{djj} \Rightarrow d^* \text{ is } \Phi_0^*\text{-}(D\text{-}) \text{ optimal} \qquad (2.11)$$

and

$$C_{d*} \text{ is c.s. and maximizes } \min_j c_{djj} \Rightarrow d^* \text{ is } \Phi_\infty\text{-}(E\text{-}) \text{ optimal.} \qquad (2.12)$$

Also, from Proposition 1, maximization of $\operatorname{tr} C_d$ by a c.s. C_{d*} implies Φ_p^*-optimality of d^* for $0 \leqq p \leqq \infty$, and much more.

Remark. More general but perhaps less important orthogonal-invariant criteria, not of the form (2.7), can be treated by characterizing extreme points of the set of matrices $\| e_{ij} \|$ and using a majorization argument. One obtains [8], for $\tilde{\Phi}$ convex and symmetric on the nonnegative v-vectors, writing

$$\mu_{di} = v^{-1}[(v-i)\lambda_{v-i} + (i-1)\lambda_{v-i+1}] \text{ for } 1 \leqq i \leqq v \text{ (with } \lambda_{d0} = \lambda_{dv} = 0),$$

$$\tilde{\Phi}(c_{d11}, \cdots, c_{dvv}) \leqq \tilde{\Phi}(\mu_{d1}, \cdots, \mu_{dv}) \qquad (2.13)$$

in place of (2.9). For $\tilde{\Phi}(x_1, \cdots, x_v) = \Sigma_1^v \tilde{f}(x_i)$ with $\tilde{f}$ convex, (2.13) becomes

$$\sum_1^v \tilde{f}(c_{djj}) \leqq \sum_1^v \tilde{f}(v^{-1}[(v-j)\lambda_{d(v-j)} + (j-1)\lambda_{d(v-j+1)}]), \qquad (2.14)$$

which is stronger but more complicated than (2.9), and which yields it upon an obvious application of Jensen's inequality.

3. The GYD setting

3.1. Preliminaries

Throughout this section we consider for $\mathscr{D}$ the 2-way heterogeneity setting described in Section 1. For given v, b_1, b_2, our results are always of the nature, "if a GYD exists..."; some existence results are described in Section 5. We are concerned with the use of Propositions 1 and 3. Thus Φ^* is given by (2.7) or by Φ^*_∞. We assume f nonincreasing, in conformity with (2.1).

We use d^* to denote a GYD, throughout Sections 3–6.

We also write $\bar{r} = b_1 b_2/v$. Since we are only concerned with settings where a GYD can exist, $\bar{r}$ is an integer.

It is well known that the entries of C_d are

$$c_{dij} = \delta_{ij} r_{di} - \frac{\lambda^{(R)}_{dij}}{b_2} - \frac{\lambda^{(C)}_{dij}}{b_1} + \frac{r_{di} r_{dj}}{b_1 b_2}. \tag{3.1}$$

Because of combinatorial restrictions among the c_{djj} (both here and in other design problems), minimization of

$$\sum_j f\left(\frac{v}{v-1} c_{djj}\right)$$

may seem more difficult at first glance than is the verification of a more tractable sufficient condition which we now examine. Define

$$c(r) = \max_{\{d: r_{dj} = r\}} c_{djj}. \tag{3.2}$$

The invariance under variety relabeling of the present setting makes $c(r)$ independent of j. Since $\lambda^{(Q)}_{d11} = \Sigma_h (n^{(Q)}_{d1h})^2$ is minimized subject to $\Sigma_h n^{(Q)}_{d1h} = r$, for $Q = R$ or C, by taking all $n^{(Q)}_{d1h}$ as nearly equal as possible, such a choice clearly yields $c_{d11} = c(r)$ (and enables us to write out $c(r)$, below). In particular, $c(\bar{r}) = c_{d^*jj}$. Since f is nonincreasing, we conclude that

$$\sum_j f\left(\frac{v}{v-1} c_{djj}\right)$$

is minimized by d^* if

$$\min_H \sum_j f\left(\frac{v}{v-1} c(r_j)\right) = v f\left(\frac{v}{v-1} c(\bar{r})\right) \tag{3.3}$$

where

$$H = \{(r_1, r_2, \cdots, r_v): r_j \text{ nonnegative integers } \forall j;\ \Sigma_j r_j = v\bar{r}\}. \tag{3.4}$$

Thus, since C_{d^*} is c.s., we conclude from Proposition 3 that d^* *is* Φ^*-*optimal if*

$$\min_H \sum_j f\left(\frac{v}{v-1} c(r_j)\right) = v f\left(\frac{v}{v-1} c(\bar{r})\right) \tag{3.5}$$

(or, from Proposition 1, that d^* is universally optimal if (3.5) is satisfied with $f(x) = -x$).

Before reviewing previously known results and then proceeding with the Φ_1^*-(A-) optimality proof, we calculate $c(r)$. Define, with r, k, n_i restricted to nonnegative integers and int(x) denoting the integer part of x,

$$\begin{aligned} h(r,k) &= \min_{\{\Sigma_1^k n_i = r\}} \sum_1^k n_i^2 \\ &= [r - k\,\mathrm{int}(r/k)][1 + \mathrm{int}(r/k)]^2 + [k - r + k\,\mathrm{int}(r/k)][\mathrm{int}(r/k)]^2 \\ &= -k[\mathrm{int}(r/k)]^2 + (2r-k)[\mathrm{int}(r/k)] + r. \end{aligned} \tag{3.6}$$

Then, from (3.1)

$$g(r) \overset{\mathrm{def}}{=} b_1 b_2 c(r) = b_1 b_2 r - b_1 h(r, b_1) - b_2 h(r, b_2) + r^2. \tag{3.7}$$

From the last expression of (3.6) (or the fact that, in the second expression, one n_i is increased by 1 from the value int(r/k) when r is increased by 1), we have $h(r+1,k) - h(r,k) = 1 + 2\,\mathrm{int}\,(r/k)$. Hence

$$\Delta(r) \overset{\mathrm{def}}{=} g(r+1) - g(r) = b_1 b_2 + 1 + 2r - \sum_1^2 b_i[1 + 2\,\mathrm{int}(r/b_i)]. \tag{3.8}$$

Further properties of g are developed in Section 3.4.

3.2. *E*-optimality

In the case of Φ_∞^*-(E-) optimality, we use (2.12) directly. We now dispose of this case quickly (reviewing that part of [3] in the present terminology), verifying our earlier comment that this is the easiest Φ_p^*. (Although A-optimality implies E-optimality, it is the illustration of this earlier comment that motivates our giving both proofs, for comparison.)

The analogue of (3.5) is now

$$\max_H \min_j g(r_j) = g(\bar{r}). \tag{3.9}$$

This is clearly satisfied if $\Delta(r) \geqq 0$ for $r < \bar{r}$. From (3.8),

$$\Delta(r) \geqq b_1 b_2 + 1 + 2r - b_1 - b_2 - 4r = (b_1 - 1)(b_2 - 1) - 2r. \tag{3.10}$$

In nonregular GYD settings, it is easily seen that $b_i > v \geqq 4$ ([7], Proposition 3), from which the positivity of (3.10) follows at once for $r \leqq \bar{r} - 1 = v^{-1} b_1 b_2 - 1$. If $v = 2$ or 3, the stronger universal optimality result of the next subsection holds. We conclude ([3])

Theorem 1. *If a GYD exists, it is E-optimal.*

3.3. Universal optimality of the GYD in the regular case

The argument of [3] can be summarized as follows, in the present terminology: Suppose $v \mid b_1$. Then, from (3.1),

$$\sum_j c_{djj} = b_1 b_2 - b_2^{-1} \sum_{i,j} (n_{dij}^{(R)})^2 + b_1^{-1} \sum_i \left[\frac{r_{di}^2}{b_2} - \sum_j (n_{dij}^{(C)})^2 \right]. \quad (3.11)$$

The expression in square brackets is nonpositive, and is zero for $d = d^*$. The remaining sum is a minimum for d^*. Consequently, d^* maximizes (3.11). Thus, from Proposition 1,

Theorem 2. *In regular settings the GYD is universally optimal.*

The arithmetic corresponding to that used above, in nonregular settings where we seek to maximize

$$- \sum_j f(\lambda_{dj}),$$

will obviously not be so simple, which is why we were led to Proposition 3. We now continue with the next step of that development, illustrating it in the simple case $b_1 = b_2$ by the resulting proof of Theorem 2 which shows the role of regularity.

3.4. Concavity properties of g

We begin by discussing the nature of g

$$\left(\text{or of } - f\left(\frac{v}{v-1} c \right) \right),$$

as it affects our approach. From Proposition 1 (see just below (2.12)), universal optimality of a GYD is a consequence of

$$\max_H \Sigma g(r_j) = v g(\bar{r}), \quad (3.12)$$

and this would follow at once if g were concave. Unfortunately, g is not concave; for example, if $b_1 = b_2$ and $Hb_1 \leqq r \leqq (H+1)b_1 - 2$ for some integer H, we see from (3.8) that $\Delta(r+1) - \Delta(r) = 2$. (We use this special setting $b_1 = b_2$ to motivate our development in simplest terms.)

Let $\bar{g}$ be the concave envelope of g; i.e., $\bar{g}$ is defined on the set $\mathscr{G} = \{0, 1, \cdots, b_1 b_2\}$ to be the smallest function $\geqq g$ whose second differences $\bar{\Delta}(r+1) - \bar{\Delta}(r)$ are all $\leqq 0$, where $\bar{\Delta}(r) = \bar{g}(r+1) - \bar{g}(r)$. Then (3.12) will still hold if

$$\bar{g}(\bar{r}) = g(\bar{r}). \quad (3.13)$$

In the special case used as illustration two paragraphs above, one sees that g is *convex* on the set $Hb_1 \leqq r \leqq (H+1)b_1$. On the other hand, $g(Hb_1) = b_1^3 H - b_1^2 H^2$ is concave in H. We conclude that $\bar{g}(Hb_1) = g(Hb_1)$ and that $\bar{g}$ is linear on $Hb_1 \leqq r \leqq (H+1)b_1$.

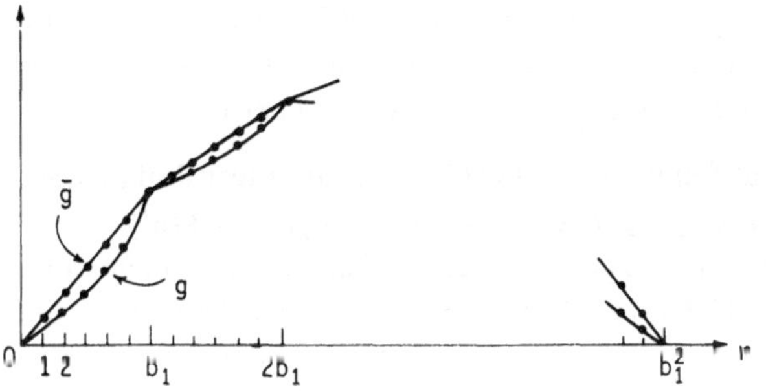

Fig. 1 (dots of graph joined for readability).

Thus, we see that (3.13) is satisfied in this example if $b_1 \mid \bar{r}$. In this simple example, that is a very special case, the symmetric multiple LS setting where $b_1 = b_2$ and $v \mid b_1$; nevertheless, the above discussion and the next two paragraphs illustrate the ideas of our approach to proving Φ^*-optimality in more complex settings. (When $b_1 \neq b_2$, the analysis of $\bar{g}$ is less simple, and not worth considering in place of the proof of Section 3.3; it is the treatment of more complex f that makes the present development worthwhile then.)

Even in the special case treated above, though, it is seen how the method can break down in nonregular cases, where $g(\bar{r}) < \bar{g}(\bar{r})$. This could conceivably reflect weakness of the method rather than existence of a design d' with larger tr C_d than that of d^*, but it was already seen in [3] that such d' can exist in nonregular cases, where d^* is thus not universally optimum. Subsequently [5], [6] a d' was found with smaller $\Phi_0^*(C_{d'})$ than that of d^*.

These facts motivate the next step, that of replacing $g(r)$ by

$$q(r) = -f\left(\frac{v}{v-1}\, c(r)\right)$$

in all of the above, and of showing that

$$\bar{q}(\bar{r}) = q(\bar{r}) \tag{3.14}$$

in order to prove Φ^*-optimality of d^*. The hope is that, although (3.13) fails to hold in some setting, the composition of the concave $-f$ and g may eliminate some of the convex pieces of the latter. Indeed, this works, and in Section 4 we use this idea with $f(x) = x^{-1}$ to prove A-optimality: D-optimality for $v \neq 4$ is proved in [9] by using $f(x) = -\log x$.

We end this section with an outline of the proof of A- and D-optimality in these cases.

(a) Since [7] any setting where a GYD exists is regular for prime v, by Theorem 2 we can hereafter limit consideration to $v \geqq 4$ and suppose we are in a nonregular setting in which a GYD exists.

(b) In what follow we write $[C, D]$ for an interval of (successive) integers. Let $\mathcal{N} = \{n: 0 \leqq n \leqq b_1 b_2; n = tb_1 \text{ or } tb_2, t \text{ integral}\}$. We note that $\mathcal{N}$ is symmetric about $b_1 b_2/2$, and let $\mathcal{M}$ denote the elements of $\mathcal{N}$ which are $\leqq b_1 b_2/2$. If $C, D \in \mathcal{N}$, $C < D$, and no integer between C and D is in $\mathcal{N}$, we call $[C, D]$ an *elementary interval*. Whenever we write $r \in [C, D]$, r is restricted to integer values. The *basic interval* $[C_0, D_0]$ is that elementary interval containing the integer $\bar{r}$. (The setting is nonregular, so $C_0 < \bar{r} < D$; we shall use the fact that $1 + C_0$, $\bar{r} \notin \mathcal{N}$.)

We also note that $D_0 \in \mathcal{M}$. (*Proof*: Suppose $b_1 \leqq b_2$. In a nonregular setting, $v < b_1$ ([7], Proposition 3). Since $v \geqq 4$, we obtain

$$D_0 < \bar{r} + b_2 = b_1 b_2 (v^{-1} + b_1^{-1}) < b_1 b_2/2.)$$

(c) The following properties of g are easily established from (3.7) and (3.8):

(i) For each elementary $[C, D]$, $\Delta(r)$ is linear in r and increasing for $C \leqq r < D$, i.e., g is a convex quadratic on each elementary interval.

(ii) g is increasing in each elementary interval $[C, D]$ with $D \leqq D_0$. (This was proved just below (3.10).)

(iii) g is symmetric about $b_1 b_2/2$.

(iv) If C_1, $C_2 \in \mathcal{N}$ with $C_1 < C_2$, then $\Delta(C_1) \geqq \Delta(C_2)$ and hence (by (iii)) $\Delta(C_1 - 1) \geqq \Delta(C_2 - 1)$. (*Proof*: It suffices to prove the result when C_1, C_2 are successive members C, D of $\mathcal{N}$. In the expression (3.8), when r is increased from C to D, the term $2r$ is increased by $2(D-C)$. But at least one term $2b_i \operatorname{int}(r/b_i)$ must be increased by $2 \min(b_1, b_2) \geqq 2(D-C)$.)

(v) g is nondecreasing on $\mathcal{M}$ and (by (iii)) nonincreasing on the remainder of $\mathcal{N}$. (*Proof*: From (i), we have for each elementary $[C, D]$ that $g(D) \geqq g(C)$ iff $\Delta(C) + \Delta(D-1) \geqq 0$. It follows from (iv) that if $g(C) \leqq g(D)$ for any two successive points of $\mathcal{N}$, then g is nondecreasing on the points of $\mathcal{N}$ which are $\leqq D$. A symmetric function of this form must have the asserted property.) *Remark*: g need not be *strictly* increasing on the first half of $\mathcal{N}$: if b_1 is even, b_2 is odd, and $b_2 \leqq b_1/2$, it is easily seen that $g(jb_2 + \frac{1}{2}b_1 b_2)$ is the same for $j = -1, 0, 1$.

(d) Let $\bar{D}$ be the first integer where g attains its maximum on $\mathscr{G}$. From (c), it follows that $\bar{D} \in \mathscr{M}$. Since $-f$ is monotone, it is enough to prove (3.14) with the domain $\mathscr{G}$ replaced by $\mathscr{G}' = \{r: r \in \mathscr{G}, r \leqq \bar{D}\}$. Note that in any elementary interval $[C, D]$ of $\mathscr{G}'$, we have $g(C) \leqq g(D)$ by (c) (v). For any such interval where g is not monotone (i.e., where $\Delta(C) < 0$), if r' satisfies $g(r') \leqq g(C)$, then $q(r') \leqq q(C) \leqq q(D)$. Hence, $\mathscr{G}'$ can be replaced by its subset $\mathscr{G}''$ which excludes such r', as domain of q in defining $\bar{q}$.

(e) A sufficient condition for (3.14) using $\mathscr{G}''$ is

$$q(r_1+1) - q(r_1) \geqq q(r_2) - q(r_2-1) \text{ for } r_1+1 \leqq \bar{r} \leqq r_2 - 1 < \bar{D}, \quad r_1 \text{ and } r_2 \in \mathscr{G}'' \tag{3.15}$$

(thought of intuitively as increasing $q(r_1) + q(r_2)$ by changing (r_1, r_2) to (r_1+1, r_2-1)). Note that $r_2 - 1$ need not be in $\mathscr{G}''$. To establish (3.15), we shall prove that

$$\begin{aligned} &\text{(a) } \min_{0 \leqq r_1 < \bar{r}}[q(r_1+1) - q(r_1)] = q(\bar{r}) - q(\bar{r}-1),\\ &\text{(b) } \max_{\bar{r} < r_2 \in \mathscr{G}''}[q(r_2) - q(r_2-1)] = q(\bar{r}+1) - q(\bar{r}),\\ &\text{(c) } q(\bar{r}+1) - q(\bar{r}) \leqq q(\bar{r}) - q(\bar{r}-1). \end{aligned} \tag{3.16}$$

We now show that (3.16) follows from

$$q(r+1) - q(r) \text{ is nonincreasing for } C_0 \leqq r < D_0 \tag{3.17}$$

(that is, from concavity of q on the basic interval).

Clearly, (3.17) implies (3.16) (c). Now suppose $r_1 \in [C, D]$ and $r_1 < D \leqq C_0$. From (c) (i), (ii), (iv) we have $\Delta(r_1) \geqq \Delta(C) \geqq \Delta(C_0) > 0$. By (c) (ii) and the concave nondecreasing nature of $-f$, this yields

$$q(r_1+1) - q(r_1) \geqq q(C_0+1) - q(C_0), \tag{3.18}$$

which with (3.17) yields (3.16) (a). Finally, if $r_2 \in [C, D]$ and $D_0 \leqq C < r_2 \leqq \bar{D}$, an analogous proof yields (3.16) (b) provided $g(r_2) \geqq g(C)$; it is irrelevant that $r_2 - 1$ need not be in $\mathscr{G}''$.

(f) To summarize, *in any nonregular setting* ($b_i > v \geqq 4$) *where a GYD* d^* *exists, if* f *is convex and nonincreasing, then* d^* *is* Φ^**-optimal provided* (3.17) *is satisfied for* $q(r) = -f(g(r))$. The detailed calculations in our A- and D-optimality proofs are needed to establish (3.17), which we do in the next section for $f(x) = 1/x$. We remark, finally, that it is tempting to try to establish (f) as follows: Define g on the *reals* $C_0 \leqq r \leqq D_0$ by (3.7). Sufficient for (3.17) is concavity of $-f \circ g$ on this real interval, established by differentiating

$$d^2f(g(r))/dr^2 \geqq 0, \qquad C_0 < r \text{ (real)} < D_0. \tag{3.19}$$

Although (3.19) is easier computationally than (3.17), and can be used for some parameter values (v, b_1, b_2), it sometimes fails in cases where (3.17) is valid, both in A- and D-optimality proofs. For example, when $v = 4$ and $b_1 = b_2 = 6$, one obtains $(C_0, D_0) = (6, 12)$ and $g(r) = r^2 + 144$, so $d^2[1/g(r)]/dr^2 = [6r^2 - 288]/(r^2 + 144)^3$, so (3.19) is false for r near 6; but (3.17) is true, as we shall now see. (In that proof, we shall in fact use the validity of (3.19) for $r \geqq C_0 + 1$.)

4. A-optimality of the GYD

In this section we complete the proof of

Theorem 3. *If a GYD exists, it is A-optimum.*

We may still restrict consideration to nonregular settings, so that $b_i > v \geqq 4$. (The proof of (3.17) is somewhat shorter if $b_1 = b_2$, but we prove the general result here. Also, A-optimality when $v \geqq 6$ follows from the D-optimality result (1.3) (b) [9], but we give the full proof here for the sake of unity and comparability.) In the present case of $f(x) = 1/x$, we can simplify (3.17) considerably: substituting $g(r-1) = g(r) + \Delta(r-1)$ and $g(r+1) = g(r) + \Delta(r)$, the condition (3.17) of convexity of $1/g(r)$ on $[C_0, D_0]$ can be written

$$\begin{aligned} C_0 < r < D_0 \Rightarrow 0 &\leqq \tfrac{1}{2}g(r-1)g(r)g(r+1)\left[\frac{1}{g(r+1)} - \frac{2}{g(r)} + \frac{1}{g(r-1)}\right] \\ &= \tfrac{1}{2}g(r)[\Delta(r-1) - \Delta(r)] + \Delta(r-1)\Delta(r) \\ &= \Gamma(r) \text{ (say)}. \end{aligned} \tag{4.1}$$

On the interval $[C_0, D_0]$, $g(r)$ is (by (b) (i), (ii) of the previous section) of the form $\alpha + \beta r + r^2$ with $\beta + 2r \geqq 0$ for $r \geqq C_0 + 1$. Thus,

$$\Gamma(r) = 3r^2 + 3\beta r + (\beta^2 - \alpha - 1). \tag{4.2}$$

Considering g to be this quadratic function on the *real* interval $C_0 \leqq r \leqq D_0$, we see that $d\Gamma(r)/dr = 3(\beta + 2r) \geqq 0$ for $r \geqq C_0 + 1$. Hence, (4.1) will follow from

$$\Gamma(C_0 + 1) \geqq 0. \tag{4.3}$$

We also record the values of α, β. Suppose for the remainder of this section that $C_0 = B_1 b_1 \geqq B_2 b_2 = b_2 \operatorname{int}(C_0/b_2)$. Then, from (3.7),

$$\alpha = b_1^2 B_1(B_1 + 1) + b_2^2 B_2(B_2 + 1), \qquad (4.4)$$
$$\beta = b_1 b_2 - (2B_1 + 1)b_1 - (2B_2 + 1)b_2.$$

It is convenient to divide the proof of (4.3) into two cases.

Case 1. $v \geqq 6$. To simplify (4.3), we first note that β and $-\alpha$ are both decreased if we replace B_2 by the larger $B_1 b_1 / b_2$. Also, since $C_0 < b_1 b_2 / 6$ and both $b_i > 6$, the resulting decreased value of β is $b_1 b_2 - 4C_0 - b_1 - b_2 > 0$, so that β^2 is also decreased. Thus, (4.2) is decreased if we make this substitution for B_2; abbreviating by $b_1 b_2 = \pi$ and $b_1 + b_2 = \sigma$, we obtain

$$\begin{aligned}\Gamma(C_0 + 1) &\geqq 3(C_0 + 1)^2 + 3[b_1 b_2 - 4C_0 - b_1 - b_2](C_0 + 1) \\ &\quad + \{[b_1 b_2 - 4C_0 - b_1 - b_2]^2 - 2C_0^2 - C_0(b_1 + b_2) - 1\} \\ &= 5C_0^2 + [-6 + 4\sigma - 5\pi]C_0 + \{2 + 3[\pi - \sigma] + [\pi - \sigma]^2\}. \qquad (4.4)\end{aligned}$$

As a function of C_0, the last expression of (4.4) has derivative < 0 on the domain $C_0 < b_1 b_2 / 6$ which includes any value of C_0 we can encounter. Hence, this expression is decreased if we substitute $b_1 b_2 / 6$ for C_0, and (4.3) is satisfied if the resulting expression is $\geqq 0$; that is (after multiplying by 36 and rearranging terms) if

$$11\pi^2 - (48\sigma - 72)\pi + 36(\sigma^2 - 3\sigma + 2) \geqq 0. \qquad (4.5)$$

In nonregular cases where $v \geqq 6$, we always have both $b_i \geqq 8$ and hence $\pi \geqq 4\sigma$. The derivative with respect to π of the expression on the left side of (4.5) is positive for $\pi \geqq 4\sigma$, and substituting the lower bound $\pi = 4\sigma$ yields $20\sigma^2 + 180\sigma + 72$, which is positive.

Case 2. $v = 4$. (This is all that remains, since $v = 2, 3, 5$ are regular.) Since the setting is nonregular, the b_i are divisible by 2 but not by 4; thus, $B_i = \mathrm{int}(b_1 b_2 / 4b_i) = (b_{3-i} - 2)/4$. Hence, in (4.4) we obtain $\beta = 0$ and $\alpha = [b_1^2 b_2^2 - 2b_1^2 - 2b_2^2]/8$. Also, $C_0 = b_1 B_1 = b_1(b_2 - 2)/4$; and, since $b_2 B_2 \leqq b_1 B_1$, we have $b_1 \leqq b_2$. From (3.7),

$$\begin{aligned}16g(C_0 + 1) &= 3[b_1(b_2 - 2) + 4]^2 - 2[b_1^2 b_2^2 - 2b_1^2 - 2b_2^2] - 16 \\ &= (b_1^2 + 4)b_2^2 + 12b_1(2 - b_1)b_2 + 16(b_1 - 1)(b_1 - 2). \qquad (4.6)\end{aligned}$$

Fix $b_1 \geqq 6$ (the least value for nonregularity). As a function of b_2, (4.6) has derivative which, at the minimum allowable value b_1 of b_2, is $2b_1(b_1^2 - 6b_1 + 16) > 0$. Hence, the convex (in b_2) expression (4.6) has its minimum on $\{b_2 \geqq b_1\}$ at $b_2 = b_1$, where (4.6) can be rewritten $(b_1 - 6)^2 b_1^2 + 8b_1(b_1 - 6) + 32 > 0$.

5. Some methods for constructing GYD

The most fruitful construction methods used by the author during the past 15 years are perhaps best described as "patchwork methods". since they are based largely on the piecing together of known combinatorial structures. By contrast, the original construction, an example of which appeared in [3] was not of this form; nor are the attractive methods of Ruiz and Seiden [10], which, although treating a smaller set of parameter values (v, b_1, b_2) than the patchwork methods, explicitly give the detailed algebraic description of new combinatorial schemes. In the present section we summarize briefly some of the methods of [7] and list some of the examples there. We shall not take the space to list the obvious number-theoretic (divisibility) restrictions on (v, b_1, b_2).

It is easily seen that any BBD can be rearranged in blocks (as columns) so that the resulting array has its first $v \operatorname{int}(k/v)$ rows "complete" (each variety appearing $\operatorname{int}(k/v)$ times per column), with the remaining rows constituting a BIBD. This suggests the following method for constructing a nonregular GYD (although it is not known whether every achievable parameter set (v, b_1, b_2) can be treated by this method): Suppose $0 < c_i < v$ and $b_i = a_i v + c_i$ with a_i positive integers. Represent the $b_1 \times b_2$ array G (with entries from $\{1, 2, \cdots, v\}$) of the design to be constructed as

$$G = \begin{Vmatrix} G_{11} & G_{12} \\ G_{21} & G_{22} \end{Vmatrix}$$

where G_{11} is an $a_1 v \times a_2 v$ array with equal replications of variety labels in rows and columns (e.g., obtained as an $a_1 \times a_2$ array of $v \times v$ LS's). The remainder of G is pieced together, when possible, as follows: (G_{21}, G_{22}) is a (v, b_2, c_1) BIBD each of whose rows has varieties replicated as nearly equally as possible, and similarly for the (v, b_1, c_2) BIBD (G_{12}', G_{22}'). The rearranging of columns of a BIBD to obtain nearly equal replications is easy; the critical feature is the fitting together of the two BIBD's in a consistent fashion in G_{22}. Conditions for this last are given in [7]; if $c_1 c_2 = v$, a slightly weaker condition than resolvability of the two BIBD's suffices; if $c_1 c_2 / v = t > 1$, there are slightly more involved combinatorial considerations. Often the BIBD's are conveniently obtained as unions of BIBD's with fewer blocks.

In [7], other related tools are given. Here we list some examples of the parameter values for which our methods achieve construction. Throughout, J_i, q, t denote positive integers.

(a) *The* $\rho^{(R)} = \rho^{(C)} = \frac{1}{2}$ *series.* All possibilities not eliminated by elementary divisibility considerations can be described in terms of three parameters t, J_1, J_2 with $v = 4t$ and $b_i = 2t(4t - 1)(2J_i - 1)$. For $t = 1$, all J_1, J_2 are achievable. For $t > 1$, construction is easy if either $J_i \geqq 2$ and the BIBD $(4t, 2(4t - 1), 2t)$ exists. When $J_1 = J_2$ and $t > 1$, the required additional considerations described earlier can be carried out for various values of t (e.g., $t = 2, 3, 4$, which with the above yield all possible parameter values in the practical range $v \leqq 16$).

(b) *Other designs with* $v = c_1 c_2$. Some examples where designs have been constructed are:

(i) $c_1 = 2$. If the BIBD $(2c_2, 2(2c_2 - 1), c_2)$ exists, so does the GYD with $a_i = J_i v - c_i - J_i$ with $J_i > 0$. (The form of a_i comes from divisibility restrictions.)

(ii) $c_1 = 3$, $c_2 = 2q + 1$. If the BIBD $(6q + 3, 3(3q + 1), 2q + 1)$ exists, all parameter values b_1, b_2 satisfying the divisibility requirements are achievable. For example, when $q = 1$ we obtain the series $v = 9$, $\rho^{(R)} = \rho^{(C)} = \frac{1}{3}$, with b_i of the form $12(3J_i - 2)$.

(c) *Other designs with* $c_1 c_2 = tv$, $t > 1$. An example is $c_1 = 3$, $c_2 = 2(2q + 1)$, $v = 3(2q + 1) = c_1 c_2 / 2$. It can be shown that, if the BIBD $(6q + 3, 3(3q + 1), 2(2q + 1))$ exists, then so does the GYD satisfying the divisibility form $b_1 = 3[(2q + 1)J_1 - 2](3q + 1)$ and $b_2 = (3J_2 - 1)(2q + 1)(3q + 1)$, for all $J_1, J_2 > 0$.

Nonisomorphic designs. There seems to be a great possibility of these, since they can arise not only from the use of nonisomorphic BIBD's in the patchwork construction, but also because other constructive methods yield different designs. Thus, our earliest nonregular example [3], the 6×6 GYD for $v = 4$, had an array G of the form

$$\begin{matrix} 1 & 4 & 2 & 4 & 3 & 2 \\ 2 & 1 & 4 & 3 & 3 & 4 \\ 2 & 3 & 1 & 3 & 4 & 2 \\ 1 & 3 & 3 & 1 & 2 & 4 \\ 4 & 1 & 4 & 2 & 1 & 3 \\ 3 & 2 & 1 & 4 & 2 & 1 \end{matrix} \tag{5.1}$$

This is easily seen to have no subarray of 4 rows and columns which constitute a LS, since any such 4×4 subarray has at least one row or column with zero or two 1's. Hence, this design is not isomorphic to (obtainable by permutation within rows, columns, and variety labels, from) any design of example (a) with $t = J_1 = J_2 = 1$:

$$\begin{matrix} 1 & 2 & 3 & 4 & 2 & 4 \\ 4 & 1 & 2 & 3 & 1 & 3 \\ 3 & 4 & 1 & 2 & 2 & 3 \\ 2 & 3 & 4 & 1 & 4 & 1 \\ 1 & 3 & 4 & 2 & 1 & 2 \\ 2 & 4 & 1 & 3 & 3 & 4 \end{matrix} \tag{5.2}$$

The methods of [10] appear to yield designs like (5.1) rather than (5.2). All GYD's with the same v, b_1, b_2 have the same c_{dij}, and thus optimality properties cannot vary between two nonisomorphic GYD's. However, investigations [5], [6], [10] of counterexamples to D-optimality when $v = 4$ indicate that examples of designs which compete favorably with GYD's may be easily constructed by slight modification of forms like (5.1). We now turn briefly to such examples.

6. Counterexamples to D-optimality when $v = 4$

The first counterexample d' to D-optimality of a GYD [5], [6] was obtained by modifying the d^* of (5.1) as follows: Change the three lower left-hand 1's, in positions (4,1), (5,2), (6,3), to 4, 2, and 3, respectively. Note that this choice has been made in such a way as to retain symmetry among varieties 2,3,4, while "short-changing" variety 1 to the equal benefit of the other three varieties; that is, for $i,j = 2,3$, or 4, all $r_{d'i}$ are equal, and so are all $\lambda^{(Q)}_{d'ii}$, all $\lambda^{(Q)}_{d'ij}$ for $i < j$, and all $\lambda^{(Q)}_{d'1i}$. One computes that $\lambda_{d^*} = (25/3, 25/3, 25/3)$, while $\lambda_{d'} = (28/3, 28/3, 20/3)$, and consequently that $\Pi\lambda_{d'i} = 15680/27 > 15625/27 = \Pi\lambda_{d^*i}$. In fact, the set of p for which d' is better than d^* in the sense of Φ_p^* must be open, and includes values up to approximately $p = .0922$.

The same idea can be used to construct designs d' better than d^* in other settings $v = 4$, where (Ex. 5(a)) $b_1 = b_2 = 6(2J-1) = 2P$ (say), with $J > 0$. But the possibility now arises of gypping variety 1 by $3w$ appearances for some value $w > 1$. (As will be seen below, this does not produce a counterexample when $J = 1$.) We shall indicate the results briefly, without consideration of the way in which the competing design can be constructed.

Thus, we denote by $d(w)$ a design with symmetry properties analogous to those of d' above. If $w \leqq P$, in $d(w)$ each of treatments 2, 3, 4 now appears $3J-2$ times in each of $P-w$ rows and columns, and $3J-1$ times in each of the remaining $P+w$ rows and columns; while if $w \leqq P/3$ (resp., $P/3 \leqq w \leqq P$), treatment 1 appears $3J-2$ times in each of $P+3w$ (resp., $3P-3w$) rows and columns and $3J-1$ (resp., $3J-3$) times in the remainder. The λ's are to have the obvious symmetries. For $w \leqq 2J+1$, one obtains, for $1<i<j$,

$$\begin{aligned} \lambda_{ii}^{(Q)} &= P[P^2+1+2w]/2, \\ \lambda_{ij}^{(Q)} &= \begin{cases} P[P^2+2w-\tfrac{1}{3}]/2 & \text{if } 0 \leqq w \leqq P/3, \\ w+P[P^2-1+2w]/2 & \text{if } P/3 \leqq w \leqq P. \end{cases} \end{aligned} \tag{6.1}$$

From this and (3.1), we obtain

$$\begin{aligned} c_{d(w)ii} &= [3P^2+2(w-1)+w^2P^{-2}]/4, \\ c_{d(w)ij} &= \begin{cases} [2-3P^2]/12+w[wP^{-2}-2]/4 & \text{if } w \leqq P/3, \\ [2(p-w)^2-(p^2+w)^2]/4P^2 & \text{if } P/3 \leqq w \leqq P. \end{cases} \end{aligned} \tag{6.2}$$

The $c_{d(w)1j}$ are then determined by the requirement of zero column sums, and this determines $c_{d(w)11}$.

The value of $\Pi\lambda_{di}$ is easily computed for a matrix C_d of the special form $C_{d(w)}$ to be $4(c_{d22}+2c_{d23})(c_{d22}-c_{d23})^2$. This yields, for $0 \leqq w \leqq P/3$,

$$27\Pi\lambda_{d(w)i} = [3P^2-2+9w^2P^{-2}-6w]\,[3P^2-2+3w]^2, \tag{6.3}$$

and consequently, since $d(0) = d^*$,

$$27[\Pi\lambda_{d(w)i}-\Pi\lambda_{d^*i}] = 9P^{-2}w^2[(4w-2)(3P^2-2)+9w^2-4w]. \tag{6.4}$$

The last is positive and strictly increasing for $0<w \leqq P/3$.

A straightforward but longer analysis of the range $P/3 \leqq w \leqq P$ shows that the form analogous to (6.4), and which coincides with it for $w = P/3$, drops sharply to a negative value at $w = P/3+1$, and is negative thereafter. Moreover, the analogue of (6.4) for $w \geqq P$ is even more negative. Thus, we conclude:

$\Phi_0^*(\lambda_{d(w)})$ is strictly decreasing for $0 \leqq w \leqq P/3$ but is $> \Phi_0(\lambda_{d(0)})$ for $w > P/3$; the Φ_0^*-best design in this family is $d(P/3)$.

The above construction can be carried out for certain other cases ($b_1 \neq b_2$) of $v = 4$. Writing $b_i = 6y_i$ with $y_1 \leqq y_2$, we find, in analogy with (6.4) (where $3w$ is again the number of appearances by which variety 1 is gypped, and the analogue of P is $3y_1$) that $\Pi\lambda_{d(y_1)i} - \Pi\lambda_{d(0)i}$ is proportional to

$$y_1^{-4}\{108y_1^4y_2^3-18y_1^4y_2^2-27y_1^2y_2^4-6y_1^2y_2^3-6y_1^4y_2+(y_1^2+y_2^2)^2\}. \tag{6.5}$$

Whenever this last expression is positive, $d(y_1)$ is Φ_0^*-(D-) better than d^*. This is the case for $y_1 = 1$, $y_2 = 1$ or 3, for $y_2 = 3$, $y_2 \leqq 35$, etc. For $y_i \to \infty$, we see that (6.5) is eventually positive if $\overline{\lim}\ y_2/y_1^2 < 4$.

On the other hand, we find that $\Pi\lambda_{d(1)i} - \Pi\lambda_{d(0)i}$ is proportional to

$$y_1^{-4}\{108y_1^3y_2^3 - 27y_1^2y_2^2(y_1^2 + y_2^2) + (y_1^2 + y_2^2 - 3y_1y_2)^2\}. \tag{6.6}$$

Thus, when $y_1 = 1$, the result coincides with that of the previous paragraph. However, when $y_1 = 3$, we find that (6.6) is positive only when $y_2 \leqq 11$: *Not only can one check that $d(y_1)$ is always D-better than $d(1)$ if $y_1 > 1$, as it was in the simpler case $y_1 = y_2$, but also the domain of values y_2 for which $d(y_1)$ is D-better than d^* exceeds the domain for which $d(1)$ is D-better than d^*, in contrast with the case $y_1 = y_2$ where the domains coincided.*

In fact, we see from (6.6) that $d(1)$ is better than d^* if the sum of terms other than the last (which is always positive) is nonnegative, i.e., if $y_2/y_1 \leqq 2 + 3^{\frac{1}{2}}$. As the $y_i \to \infty$, this bound gives the asymptotic domain of D-preferability of $d(1)$ over d^*, and is strikingly inferior to the result for $d(y_1)$.

We remark that the competing design constructed by Ruiz and Seiden [10] in the special case $y_1 = y_2$ considered by them is precisely $d(y_1) = d(P/3)$, which coincides when $y_1 = y_2 = 1$ with the $d' = d(1)$ of [3] described earlier in this section. We have seen the preferability of $d(y_1)$ in other cases.

This area of research is very incomplete, due to the computational difficulties for designs not of such a simple structure as $d(w)$. Thus, we do not know designs other than $d(w)$ which are D-better than d^*, and we do not know whether $d(y_1)$ has any optimal properties. Nor do we know in *exactly* which cases $(4, b_1, b_2)$ d^* is not D-optimal. As suggested by the construction above, if b_2/b_1 is sufficiently large (depending on b_1), we might guess that d^* is again D-optimal. Finally, it is not difficult to give an upper bound on the possible departure from D-optimality of d^* [9]; this shows that, from a practical point of view, little can be lost in using d^*. However, the practitioner should be warned that the conservative moral to be drawn from these counterexamples is not so much that exotic designs with complete symmetry may fail (slightly) to be optimum in the GYD setting, but rather that in other settings they may be *far* from optimum unless one proves the opposite.

References

1. Agrawal, H. (1966). Some generalizations of distinct representatives with applications to statistical designs. *Ann. Math. Statist.* **37**, 525–528.
2. Hartley, H. O. and Smith, C. A. B. (1948). The construction of Youden squares. *J. Roy. Statist. Soc., Ser. B* **10**, 262–263.
3. Kiefer, J. (1958). On the nonrandomized optimality and randomized non-optimality of symmetrical designs. *Ann. Math. Statist.* **29**, 675–699.
4. Kiefer, J. (1959). Optimum experimental designs. *J. Roy. Statist. Soc., Ser. B* **21**, 272–319.
5. Kiefer, J. (1970). Optimum experimental designs. *Proc. Intl. Congress Math.* **3**, Nice, Gauthiers-Villars, Paris, 249–254.
6. Kiefer, J. (1971). The role of symmetry and approximation in exact design optimality. *Statistical Decision Theory and Related Topics*, Academic Press, New York, 109–118.
7. Kiefer, J. (1973a). Balanced block designs and generalized Youden designs. I. Construction. (To appear).
8. Kiefer, J. (1973b). Optimality methods for exact designs. (To appear).
9. Kiefer, J. (1973c). *D*-optimality of certain generalized Youden designs. (To appear).
10. Ruiz, F. and Seiden, E. (1973). Generalized Youden Designs. (To appear).
11. Shrikhande, S. S. (1951). Designs with two-way elimination of heterogeneity. *Ann Math. Statist.* **22**, 235–247.
12. Wald, A. (1943). On the efficient design of statistical investigations. *Ann. Math. Statist.* **14**, 134–140.

Biometrika (1975), **62**, 2, *p.* 277
Printed in Great Britain

Optimal design: Variation in structure and performance under change of criterion

BY J. KIEFER

Department of Mathematics, Cornell University, Ithaca, New York

SUMMARY

The advisability of comparing designs on the basis of several different criteria of goodness is discussed. As an illustration, designs for quadratic regression on a simplex are compared in terms of a family of criteria that includes those of A-, D- and E-optimality, as well as one of Box & Draper. Efficiency robustness under change of region, as well as of criterion, is considered. The dependence of structure of good designs on dimension is determined.

Some key words: A-optimality; Biased regression; D-optimality; E-optimality; Mixture design; Optimal design; Response surface; Robust design; Simplex design.

1. INTRODUCTION

When statisticians use an optimality criterion in selecting a procedure, the precisely expressed criterion is usually only an approximation to some vague notion of 'goodness'. It thus seems appropriate, after one has computed an optimal design relative to some criterion, to see how it performs in other respects. Such an examination of the performance of competing designs can of course be found in various applied papers whose authors have usually avoided the specification of precise criteria. The present paper summarizes an attempt to study, perhaps more systematically, such a comparison of performance in a particularly simple family of settings.

Such a study is analogous to examination of risk functions in statistical decision theory. There, a procedure that is 'best' according to a Bayes, minimax, or other criterion, may be found inferior upon comparison with a second procedure which, at the sacrifice of a small relative loss for the specified criterion, performs much better in other respects. Since the prior law or minimax criterion are only approximate guides to optimality, the second procedure might then well be used in place of the first.

In theory, the examination of the totality of admissible risk functions could lead to the choice of a satisfactory procedure, but in practice this is too huge a prescription to carry out. Hence, one instead uses a particular criterion to select a procedure which, at least roughly, accomplishes the desired aim, and then looks at nearby risk functions to see whether some obvious improvement, such as that described in the previous paragraph, is possible.

In the design setting, suppose for simplicity that we are interested in best linear unbiased estimation and measure the goodness of a design in terms of some functional of the associated covariance matrix. If, in the usual notation, defined in full below, $M(\xi)$ is the information matrix of the design ξ, the admissible designs are among those that minimize

$$\Phi_A(\xi) = \operatorname{tr}\{AM^{-1}(\xi)\}$$

for some nonnegative definite A, analogous to the relationship between admissible and Bayes decision procedures. In the spirit of the previous paragraph, rather than to search among all these M we may choose a particular criterion Φ, find the ξ^* that minimizes $\Phi\{M(\xi)\}$, and then see whether the resulting design is indeed satisfactory, or whether there is another design that is only slightly less efficient in terms of Φ, but which is noticeably superior to ξ^* in terms of other criteria of interest.

The criteria considered in this study are mainly a family $\{\Phi_p, 0 \leqslant p \leqslant \infty\}$ that include the three most commonly used criteria, those of D-, A- and E-optimality, as those obtained when $p = 0, 1, \infty$. This family of criteria has been discussed in more detail by Kiefer (1974), and the intuitive meaning of the criteria, for example as various measures of the size of a confidence ellipsoid in the normal case, is well known. There are obviously examples in which performance in terms of $\{\Phi_p\}$ does not adequately reflect serious defects or advantages of procedures. Nevertheless, this family of criteria appears to be useful in many settings; this is the case for the examples studied herein, where the M of any design with reasonable symmetry has at most five distinct eigenvalues, which we list with multiplicities. Other criteria we have considered are the maximum $\bar{d}$ of the variance function and criteria associated with the model of Box & Draper (1959) in which account is taken of the bias due to incorrectness of the model used in fitting the regression function. This last approach and some of its possible defects are discussed in more detail in § 5. Section 6 contains discussion of a design's performance when the scale is changed from that under which the design was originally chosen to be used.

A notable conclusion of the study is that, in simple practical settings such as those considered here, many procedures remain much more efficient under variation in criterion than one could expect from general theoretical bounds. For example, if ξ_p^* is a design optimum under Φ_p, then for each $p > 0$ there are settings where $\Phi_0(\xi_p^*)/\Phi_0(\xi_0^*)$ is arbitrarily large; that is, the D-efficiency of ξ_p^* is arbitrarily close to zero. But in our examples of quadratic regression on a simplex of any dimension q, it turns out that ξ_p^* is fairly efficient. In particular, ξ_∞^* seems remarkably robust in its efficiency under variation of criterion.

A phenomenon occurs here that cannot be expected in general theoretically, but that is common in practical examples: although ξ_p^* varies noticeably with p, its performance characteristics often vary much less. Some of the literature of design comparisons, for example some of that on the Box–Draper approach and its variations, unfortunately emphasizes the dependence on criterion of an optimal design without adequate comparison of performance.

Also of interest is the qualitative structure of various 'good' designs; for example, how far into the simplex, as measured by dimensionality of the face in which a point lies, is it necessary to take observations? This is the type of information that can be useful in suggesting promising forms of design to be investigated in more complex settings where precise optimality results are more difficult to obtain. Thus, if observations must be restricted to a highly asymmetric subset of the simplex, a computer search for optimal designs might be prohibitive for large q; but we would still expect to find reasonably efficient designs among those with dimensionality characteristics suggested by the present study. The same is true for many approximately quadratic nonlinear regression models.

Another feature of this study is the simultaneous consideration of a family of related settings, here parameterized by the dimension q. The motivation is that one hopes for some regularity that will enable many of these settings to be served satisfactorily by a single simple design formula. This idea is implemented here by letting the results for small q suggest the form of a good design as $q \to \infty$. The limiting calculations are much easier to perform than those for large q, and yield a design whose efficiency is fairly high for moderate values of q.

2. Preliminaries

Our considerations are limited to the approximate theory, in order to elucidate the comparisons with a minimum of arithmetical complexity; the exact theory comparisons will exhibit similar phenomena, even for fairly small sample sizes. Thus a design is a probability measure ξ on the compact space $\mathscr{X}$ of possible values of the controllable variable. The measure ξ can be assumed discrete and $\xi(x)$ then represents the proportion of observations to be taken at the point x of $\mathscr{X}$. The reader may consult, e.g., Kiefer (1959, 1974) or Fedorov (1972, p. 58) for interpretation of these standard notions of optimal design theory. The observations are uncorrelated with variance σ^2. The expected value of an observation at x is $\theta' f(x)$, where θ is a column vector of k unknown real parameters and $f(x)$ is a column vector of k known continuous functions. The $k \times k$ information matrix per observation for unit variance is defined for ξ as

$$M(\xi) = \int_{\mathscr{X}} f(x) f(x)' \, \xi(dx),$$

and the corresponding normalized variance function for nonsingular $M(\xi)$ is

$$d(x, \xi) = f(x)' M^{-1}(\xi) f(x).$$

Thus, when ξ is used with N observations, $N^{-1}\sigma^2 M^{-1}(\xi)$ is the covariance matrix of best linear unbiased estimators of θ, and $N^{-1}\sigma^2 d(x, \xi)$ is the variance at x of the regression function fitted by least squares.

The optimality functionals we consider are infinite for singular $M(\xi)$, and we therefore limit consideration to nonsingular $M(\xi)$ in the sequel. Denote the eigenvalues of $M(\xi)$ by $\lambda_1(\xi), \ldots, \lambda_k(\xi)$. We define

$$\begin{aligned} \Phi_p(\xi) &= [k^{-1} \operatorname{tr}\{M^{-p}(\xi)\}]^{1/p} = \left\{k^{-1} \sum_{i=1}^{k} \lambda_i^{-p}(\xi)\right\}^{1/p} \quad (0 < p < \infty), \\ \Phi_0(\xi) &= \lim_{p \to 0+} \Phi_p(\xi) = \{\det M^{-1}(\xi)\}^{1/k}, \\ \Phi_\infty(\xi) &= \lim_{p \to +\infty} \Phi_p(\xi) = \max_i \{\lambda_i^{-1}(\xi)\}. \end{aligned} \tag{2·1}$$

As mentioned in the introduction, Φ_0, Φ_1 and Φ_∞ are the familiar D-, A- and E-optimality criteria, here normalized so as to make comparisons easy: all measure loss in the same scale per unit of variance, and take on the value c^{-1} if $M(\xi) = cI_k$. A Φ_p-optimal design ξ_p^* is one that minimizes $\Phi_p(\xi)$. Another common criterion, that of G-optimality, is

$$\bar{d}(\xi) = \max_{x \in \mathscr{X}} d(x, \xi). \tag{2·2}$$

The original equivalence theorem (Kiefer & Wolfowitz, 1960) asserts that G-optimality is equivalent to Φ_0-optimality and that $\bar{d}(\xi_0^*) = k$. The equivalence theory analogue of $\bar{d}$ for the case of Φ_p, discussed after (3·1), does not have intrinsic appeal as an optimality criterion the way (2·2) does, and will not be discussed here; its usefulness lies in helping to characterize or compute Φ_p-optimal designs (Kiefer, 1974).

A measure of the performance of a design ξ' with respect to various criteria is the absolute efficiency ratio based on multiplicative regret

$$\Phi_p(\xi')/\Phi_p(\xi_p^*) \quad (0 \leqslant p \leqslant \infty), \quad \bar{d}(\xi')/\bar{d}(\xi_0^*) = k^{-1}\bar{d}(\xi'). \tag{2·3}$$

A design ξ' for which these ratios are all near one may be judged adequate for many purposes, perhaps unstated or only vaguely understood. We shall concentrate herein on the efficiency ratio (2·3) for ξ' taken to be various ξ_{p_0}'s, especially for $p_0 = 0, 1$ or ∞. The finer comparison between two designs ξ' and ξ'' obtained by contrasting the entries of $M(\xi')$ and $M(\xi'')$, or the eigenvalues $\{\lambda_i(\xi')\}$ and $\{\lambda_i(\xi'')\}$, or by computing the eigenvalues of $M(\xi')\,M^{-1}(\xi'')$, does not, at least in our examples, seem worth the much greater effort required.

Since (Kiefer, 1960) $k^{-1}\bar{d}(\xi')$ is near one if $\Phi_0(\xi')/\Phi_0(\xi_0^*)$ is, we do not obtain much added information from the former unless the latter ratio is markedly different from one.

The setting we will consider is that of quadratic regression for experiments with mixtures, developed especially by Scheffé (1958). There are $q+1$ components, proportions, in a mixture x, so that $\mathscr{X}$ is the q-simplex $\{(x_1, \ldots, x_{q+1}) : \Sigma x_i = 1, \text{ all } x_i \geqslant 0\}$. A convenient basis for the polynomials on $\mathscr{X}$ of degree $\leqslant 2$ is the set of functions $\{x_i, x_i x_j : 1 \leqslant i < j \leqslant q+1\}$; there are $k = \frac{1}{2}(q+1)(q+2)$ functions in this set, which we take to be the components of f.

3. Reduction of the problem: Methods of solution

3·1. *Preliminary remarks*

Except when $q = 1$ or $p = 0$ the computation of a Φ_p-optimal design is algebraically intractable; we cannot hope for simple formulae. Hence, we must be satisfied with numerical results. However, the direct minimization of $\Phi_p(\xi)$ is too large a problem: the space of matrices $\mathscr{M} = \{M(\xi)\}$ has dimension $\frac{1}{12}q(q+1)(q^2+q+22)$, and minimization of Φ_p with respect to roughly $\frac{1}{12}q^4$ locations $x^{(i)}$ and weights $\xi(x^{(i)})$ is a huge problem. We now use various theoretical tools of optimal design theory to reduce the dimensionality of our problem. It will be useful to introduce some additional nomenclature. We define a barycentre of depth j $(0 \leqslant j \leqslant q)$ in $\mathscr{X}$ to be a point with $j+1$ coordinates equal to $1/(j+1)$ and the remaining coordinates equal to zero. By J_j we denote the set of $(q+1)!/\{(j+1)!\,(q-j)!\}$ barycentres of depth j. Thus J_0 consists of vertices of the q-simplex, J_1 consists of midpoints of edges, J_q is the centre, etc. Also, write $J = \cup_j J_j$.

3·2. *Invariance*

Suppose ξ' is Φ_p-optimum. Then, by the symmetry of $(\mathscr{X}, f, \Phi_p)$, so is ξ'_σ defined by $\xi'_\sigma(x) = \xi'(\sigma x)$ for any element σ of the symmetric group S_{q+1}. This is clear in the present example; see, e.g., Kiefer (1974) for further discussion. Hence, by convexity of Φ_p on $\mathscr{M}$ (Kiefer, 1974), if

$$\bar{\xi} = \frac{1}{(q+1)!} \sum_{\sigma \in S_{q+1}} \xi'_\sigma,$$

we have $\Phi_p(\bar{\xi}) \leqslant \Phi_p(\xi')$; this last must be an equality by the optimality of ξ', so $\bar{\xi}$ is also optimal. Since $\bar{\xi}(\sigma x) = \bar{\xi}(x)$ for all x and σ, we conclude that there is an optimal design invariant under S_{q+1}. Incidentally, for sufficiently large q, and for p correspondingly large, it is easy to conclude from our results in the sequel and dimensionality considerations that there are also Φ_p-optimal noninvariant designs. We will see that, for fixed p, all Φ_p-optimal designs have the same information matrix and are hence equivalent for the considerations of this paper, except in §5. We shall therefore not consider the computation of noninvariant designs. One interest in such designs is the possibility that they may require fewer different observation points than invariant ones.

3·3. *Equivalence theory and the nature of d*

For $0 \leqslant p < \infty$, the General Equivalence Theorem (Kiefer, 1974) states that ξ' is Φ_p-optimal if and only if, for all $x \in \mathscr{X}$,

$$f(x)' M^{-p-1}(\xi') f(x) \leqslant \operatorname{tr}\{M^{-p}(\xi')\}, \tag{3·1}$$

and that an optimal ξ' assigns measure one to the set of x for which equality holds in (3·1); the maximum over x of the left-hand side of (3·1) is the analogue of $\bar{d}(\xi')$ for the case $p = 0$. Note that the left-hand side of (3·1) is a quartic in x. It is easy to see that this function, extrapolated as a quartic on the q-dimensional Euclidean space $\Sigma x_i = 1$, goes to $+\infty$ at infinity. Hence, on the intersection L of any line with the simplex, this quartic can have at most one maximum interior to L. From the phrase following (3·1) we conclude that any optimal design that is invariant, known to exist by § 3·2, must be supported on a subset of J, since any other invariant design gives positive measure to a symmetric pair of points interior to some such L. We then conclude that nonsymmetric optimal designs must also be supported on a subset of J.

We therefore restrict attention to invariant designs on J. For such a design, we can replace ξ by the equivalent $\alpha = (\alpha_0, \alpha_1, \ldots, \alpha_q)$, where $\alpha_j = \xi(J_j)$. The 'design α' thus assigns measure $\alpha_j/[(q+1)!/\{(j+1)!\,(q-j)!\}]$ to each point of J_j $(0 \leqslant j \leqslant q)$.

For $p = \infty$ a somewhat more complex analysis is needed because (Kiefer, 1974) the analogue of (3·1) involves measures ξ instead of simply points x; this is due to the non-differentiability of Φ_∞. An alternative to this analysis is to note that any ξ_∞^* has as $p \to \infty$ $\lim \Phi_p(\xi_\infty^*)/\Phi_p(\xi_p) = 1$, this continuity being easy to verify in the present example; from this $\xi_\infty^*(J) < 1$ would yield $f(x)' M^{-p-1}(\xi_\infty^*) f(x)$ almost constant on some line segment for large p, a behaviour that can be shown impossible for the given f. Simpler is the fact that any convergent sequence $\{\xi_{p_i}^*, p_i \to \infty\}$ approaches a ξ_∞^* in the present setting; but in other examples there can be ξ_∞^*'s which do not arise in this way.

We also note that, although the strict convexity of Φ_p on the nonsingular M for $0 \leqslant p < \infty$ implies uniqueness of $M(\xi_p^*)$, not of ξ_p^*, the lack of strictness for Φ_∞ requires another argument for uniqueness, as mentioned later in this section.

3·4. *Further calculations*

Scheffé had suggested the design $\alpha_0 = 2/(q+2) = 1-\alpha_1$ that puts equal weight on the k points in $J_0 \cup J_1$. When $p = 0$, and only then, in general, it is easy to see for every $(\mathscr{X}, f)$ that, if an optimal design has support consisting of exactly k points, then it assigns mass k^{-1} to each of these points. Hence, it is natural to conjecture that Scheffé's design $\xi^{(s)}$ is Φ_0-optimal, and this was proved by Kiefer (1961) by showing that $\bar{d}(\xi^{(s)}) = k$ for that design. The discussion of § 3·3 regarding the quartic nature of d shows that a computationally easier proof can be obtained by simply computing $d(x^{(j)}, \xi^{(s)})$ for a single point $x^{(j)}$ in each J_j and showing that the resulting value is $\leqslant k$; it is automatically k for $j = 0$ or 1. Since we obtain $d(x^{(j)}, \xi^{(s)}) < k$ for $j > 1$, and since d is the same for all Φ_0-optimal designs, we conclude that $\xi^{(s)}$ is the unique, invariant or not, Φ_0-optimal design.

For $p > 0$, even after our reduction of §§ 3·2 and 3·3 we cannot hope to obtain an explicit algebraic form of ξ_p^* for which (3·1) can be verified, or, equivalently, which can be shown to minimize $\Phi_p(\alpha)$ on the basis of computation of derivatives of $\Phi_p(\alpha)$. For $0 < p < \infty$, our numerical method was based on formulae obtained for the $\lambda_i(\alpha)$. This was not too difficult in the present setting, in which every symmetric design on J has an M with at most seven different entries $a(\alpha), b(\alpha), \ldots, g(\alpha)$, all linear in α, namely the respective averages with respect to ξ of x_1^2, x_1x_2, $x_1^2x_2$, $x_1x_2x_3$, $x_1^2x_2^2$, $x_1^2x_2x_3$ and $x_1x_2x_3x_4$. In fact, only three of these averages are linearly independent. One can invert the matrix $M(\xi)$ or compute its determinant or eigenvalues in explicit algebraic terms. There are five distinct eigenvalues for $q \geqslant 3$; denoting by $L_j(\alpha)$ and $Q_j(\alpha)$ linear and quadratic functions of α, these eigenvalues are of the form

$$\begin{aligned} \mu_0(\alpha) &= L_0(\alpha), \quad \text{of multiplicity } \tfrac{1}{2}(q+1)(q-2); \\ \mu_1^{\pm}(\alpha) &= L_1(\alpha) \pm Q_1^{\frac{1}{2}}(\alpha), \quad \text{each of multiplicity } q; \\ \mu_2^{\pm}(\alpha) &= L_2(\alpha) \pm Q_2^{\frac{1}{2}}(\alpha), \quad \text{each of multiplicity } 1. \end{aligned} \tag{3·2}$$

Note that, for $q = 2$, μ_0 is absent; for the trivial case $q = 1$, a slightly different form holds. The Φ_p of (2·1) can now be computed easily in terms of these. Since Φ_p is convex in α, it is not difficult to compute all minima of (3·2) numerically; the computation, details of which will be given elsewhere, is aided by the fact that there are always 'basic solutions' in which at most four α_i's are positive. Note that we used the equivalence theory in our reduction of § 3·3, but not at the present stage, where the derivatives $\partial\Phi_p(\alpha)/\partial\alpha_i$ can be used directly to verify optimality at a point α^*. It is convenient to substitute $\alpha_0 = 1-(\alpha_1+\ldots+\alpha_q)$ in such computations.

For $p = \infty$, we use (3·2) to search for an α maximizing $\min\{\mu_0(\alpha), \mu_1^-(\alpha), \mu_2^-(\alpha)\}$. For $q \geqslant 3$, it turns out that these three roots are equal at the maximum, which was intuitively plausible, and was suggested by the closeness of the roots corresponding to ξ_p^* for p large. Thus, we found a solution by maximizing μ_0 subject to the two restraints $\mu_0 = \mu_1^- = \mu_2^-$, numerically our most delicate computation; then we checked the optimality and nonuniqueness in terms of the matrix of derivatives $\partial(\mu_0, \mu_1^-, \mu_2^-)/\partial\alpha_j$, $1 \leqslant j \leqslant q$; this last is really a feasibility consideration of linear programming. For $q = 2$ there is equality of the two roots of relevance. For $q = 1$, a single root is smaller than the other two at the optimum.

4. MAIN NUMERICAL RESULTS

4·1. *Dimensionality structure*

Motivated by the result of § 3·4 that the unique ξ_0^* is supported on $J_0 \cup J_1$, we would expect a similar result to hold for ξ_p^* when p is small, with the points of course no longer receiving equal weights. This turned out to be the case, and there is a critical largest value p'_{crit}, say, of p for which ξ_p^*, which is the unique optimal design as was ξ_0^*, is of this form. As p increases beyond p'_{crit}, we find that α_0, α_1 and α_2 are positive for the unique symmetric Φ_p-optimal design. This continues until p reaches a second critical value p''_{crit}, say, just beyond which it no longer suffices to have only three positive α_i's. The value p''_{crit} is absent when $q \leqslant 2$, as is also p'_{crit} when $q = 1$.

The pattern for larger p now changes in two respects. First, there is no additional critical point beyond which four positive α_i's do not suffice. Secondly, there is no longer a unique symmetric solution. For $p > p''_{\text{crit}}$, and the values of $q \leqslant 10$ that we studied in detail, there is for each $j \geqslant 3$ a Φ_p-optimal design with α_0, α_1, α_2 and α_j, and only these, positive, and every symmetric optimal design is a probabilistic average of these 'basic' solutions; α_0, α_1 and α_2 must be positive in every symmetric solution.

This last pattern persists when $p = \infty$. This is because no new solutions arise which are not limits of sequences $\{\xi_p^*, p \to \infty\}$. This also reflects the fact, verified in the feasibility computation described in § 3·4, that $M(\xi_\infty^*)$ is unique even though Φ_∞ is not strictly convex.

The critical values of p for $2 \leqslant q \leqslant 10$ are given in Table 1. Thus, for all $q \geqslant 3$ the unique Φ_1-(A-) optimal design is supported on $J_0 \cup J_1$, while every symmetric Φ_∞-(E-) optimal design requires observations not only on J_0, J_1 and J_2, but also on at least one additional J_j. This behaviour is not, however, more complex than for quadratic regression on the q-cube (Kiefer, 1960; Farrell, Kiefer & Walbran, 1965), where even the Φ_0-optimal designs have a dimensionality structure that changes with q.

Table 1. *Critical values of p*

q	p'_{crit}	p''_{crit}	q	p'_{crit}	p''_{crit}
2	0·864	—	7	1·610	8·317
3	1·056	5·669	8	1·713	8·781
4	1·223	6·500	9	1·807	9·198
5	1·368	7·201	10	1·893	9·578
6	1·496	7·796			

Table 2. *Selected values of optimal weights* α_i

q	p	α_0	α_1	α_2	α_3
1	0	$\frac{2}{3}$	$\frac{1}{3}$	—	—
	1	$5-2\sqrt{5}$	$2(\sqrt{5}-2)$	—	—
	∞	$\frac{17}{33}$	$\frac{16}{33}$	—	—
2	0	0·500	0·500	—	—
	1	0·425	0·562	0·013	—
	∞	0·435	0·379	0·186	—
3	0	0·400	0·600	—	—
	1	0·375	0·625	—	—
	∞	0·403	0·339	0·195	0·063

The behaviour of ξ_p^* as a function of p and q will be tabulated extensively elsewhere. As examples, the cases $q = 1, 2, 3$ for the three values $p = 0, 1, \infty$ of principal interest are listed in Table 2. When $q = 1$, the results differ from those for the common normalization for quadratic regression on $\mathscr{X} = [-1, 1]$, with $f(x)' = (1, x, x^2)$, for which ξ_1^* has $\alpha_1 = \frac{1}{2}$ and ξ_∞^* has $\alpha_1 = \frac{3}{5}$; of course, ξ_0^* is the same as here.

4·2. *Efficiency ratios*

As an example of the behaviour of efficiency ratios, Table 3 lists the ratio $\Phi_p(\xi')/\Phi_p(\xi_p^*)$ for $q = 3$, where ξ' is a D-, A- or E-optimal design. The $\bar{d}$ efficiency ratios are similar to those of the first row of Table 3. The behaviour for other values of q is quite similar to that for $q = 3$. As mentioned in the introduction, this robustness under change in criterion is strikingly better than the general upper bounds that hold for such ratios. For example, there are problems in which $\Phi_\infty(\xi_0^*)/\Phi_\infty(\xi_\infty^*)$, 1·44 here, is as large as k, 10 here; and the entries $\Phi_0(\xi_p^*)/\Phi_0(\xi_0^*)$ of the first row for $\xi' \neq \xi_0^*$, 1·00 and 1·17 here, can be arbitrarily large! It would be useful to find general conditions on $(\mathscr{X}, f)$ under which such theoretical bounds could be improved without resorting to explicit computations.

Table 3. *Selected absolute efficiency ratios when* $q = 3$

p	$\xi' = \xi_0^*$	$\xi' = \xi_1^*$	$\xi' = \xi_\infty^*$
0	1·00	1·00	1·17
0·1	1·00	1·00	1·18
0·5	1·00	1·00	1·19
1	1·00	1·00	1·17
2	1·01	1·02	1·12
10	1·24	1·27	1·04
∞	1·44	1·47	1·00

4·3. *Behaviour as* $q \to \infty$

The behaviour of ξ_p^* for fixed p as a function of q shows a fair amount of regularity, which we now illustrate. The pattern mentioned in §4·1 changes slightly for large q: the solution with only $\alpha_0, \alpha_1, \alpha_2, \alpha_3 > 0$ disappears, and some new ones, e.g. with $\alpha_0, \alpha_1, \alpha_3, \alpha_4 > 0$, arise and remain as $q \to \infty$. We consider now the simplest form that persists as $q \to \infty$, the ξ_∞^* on $J_0 \cup J_1 \cup J_2 \cup J_4$. For this design α_0 decreases slowly from 0·380 at $q = 4$ to 0·346 at $q = 10$, approaching 0·328 as $q \to \infty$; α_1 decreases very slowly from 0·308 at $q = 4$ to 0·277 at $q = 10$ and 0·276 at ∞; α_2 decreases from 0·250 at $q = 4$ to 0·216 at $q = 10$ and 0·132 at ∞; α_3 increases from 0·062 at $q = 4$ to 0·161 at $q = 10$ and 0·264 at ∞. As one might expect, α_3 and α_4 vary more for high q than do α_0 and α_1. The limiting values as $q \to \infty$ were much easier to compute than those for finite q because of the simplifying asymptotic expressions obtainable for the $\mu_i(\alpha)$. If one had simply computed this limiting design and used it for large q instead of computing ξ_∞^* exactly for such q, the results would have been quite satisfactory; for $q = 10$, for example, there is only an 8 % loss of efficiency compared with using the optimal ξ_∞^* for Φ_∞^*, while the loss is 12 % for $q = 7$ and 30 % for $q = 4$.

A finer calculation of this type, for example computing the asymptotic behaviour of the α_i to order q^{-1}, would presumably be even more satisfactory. Unfortunately, no useful theoretical bounds are known on the efficiency one can expect from even the simpler values obtained above, when used for such families of designs. We remark that such asymptotic behaviour was studied for cubic regression on the unit q-ball by Farrell *et al.* (1965), but no efficiencies were computed there.

5. COMPARISONS FOR THE BOX–DRAPER MODEL

Box & Draper's model, in the present setting, takes into account the unlikelihood that the quadratic model is exactly correct, by considering also the bias one suspects to be present from cubic or other terms. For simplicity of discussion, we hereafter limit consideration to cubic terms, as is also done in much of the associated literature. The reader may consult Box & Draper (1959), Karson, Manson & Hader (1969), and Kiefer (1973), for details of the arithmetic sketched briefly, just below, and for modifications of the original Box–Draper prescription. In that prescription, the least squares fit, assuming the quadratic model, when cubic terms are present, has mean squared error at x equal to

$$r(x, \beta, \sigma; \xi) = \sigma^2 N^{-1} d(x, \xi) + [\beta'\{M_c(\xi)\, M^{-1}(\xi) f(x) - f_c(x)\}]^2, \tag{5·1}$$

where f, M, d refer to the quadratic model as before, β is the vector of cubic parameters, f_c is the corresponding vector of basic cubic functions, and $M_c(\xi) = \int f_c f' d\xi$. Box & Draper propose integrating r over $\mathscr{X}$ with respect to Lebesgue measure; another measure ν could be used instead. This yields from (5·1) a sum of two terms generally denoted $\sigma^2(V+B)$, where

$$V(\xi)+B(\xi) = N^{-1}\operatorname{tr}\{\Gamma M^{-1}(\xi)\}+\sigma^{-2}\beta' H(\xi)\beta; \qquad (5{\cdot}2)$$

here Γ is the Lebesgue integral of ff' and H contains both ξ and Lebesgue integrals obtained from f and f_c. Since $\sigma^{-1}\beta$ is unknown, there is no way to choose ξ to minimize $V+B$ unless B is replaced by B', the integral or maximum over an assumed set of possible vectors $\sigma^{-1}\beta$. Box & Draper propose to minimize B alone. This turns out to make sense mathematically even though B is a function of the vector β, since there is a ξ_{BD} such that $H(\xi)-H(\xi_{BD})$ is nonnegative-definite for all ξ. The practical justification for doing this, according to Box & Draper, is that in typical settings the design ξ_{BD}, or more accurately a choice among the many possible ξ_{BD}'s satisfying the previous description, turns out to be close to the design that minimizes $V+B'$ for a wide range of assumptions on $\sigma^{-1}\beta$ used to obtain B'. This justification seems to me to be accepted by some practitioners of this approach, or of Karson *et al.*'s (1969) modification, in a sweeping generality which it does not deserve, and with which I doubt that Box & Draper would agree. Several points should be made. To reiterate, our discussion has concentrated on the approximate theory; but we believe the phenomena that arise in our criticism below are also present in exact design considerations that are more meaningful practically, and that are the subject of most papers in which Box–Draper designs are computed.

(i) In fitting a quadratic, even knowing that this is incorrect, the statistician presumably believes B is relatively small; otherwise he would fit a cubic or at least let the data determine what type of curve he should fit. Thus, there are certainly many practical examples in which it makes no sense to decide to fit a quadratic but then to choose ξ with regard to B, completely ignoring V.

(ii) There are many examples, a simple one being given by Kiefer (1973), in which minimization of B' penalizes V greatly, and in which great reduction in V from $V(\xi_{BD})$ can be achieved by a design that increases B' only slightly over its value for ξ_{BD}. Thus, in the spirit of the second paragraph of § 1, one should look at the performance characteristics of procedures more broadly than in terms of a single optimality criterion such as B.

(iii) Of course, the design ξ_{BD} depends on the choice of ν as Lebesgue measure, and a different ν would yield a different design. I submit that the choice of Lebesgue measure is usually purely out of convenience, and does not necessarily reflect, even approximately, the concerns of the experimenter regarding the seriousness with which he weighs various points of $\mathscr{X}$. It can well be replied that this criticism is inessential, since nobody can usually supply a ν guaranteed to be more meaningful, and Lebesgue measure with the Box–Draper approach gets the job done of choosing a useful design more satisfactorily than does consideration of V, or some other functional of M, alone. A point, though, is that this could also be said of the Box–Draper approach with many other possible choices of ν, and one such satisfactory choice would often be a measure that makes ξ_{BD} equal to some ξ_p^*. If ν is any measure whose moments up to degree six are the same as those of ξ_p^*, this is the case!

In the absence of firmer information upon which to base the choice of ν, I believe the above comment to provide an important justification in practice for the use, even within the Box–Draper rationale, of many designs obtained on the basis of optimality functionals of $M(\xi)$ alone. Other considerations should often also come into play, such as the possibility of higher degree bias or the inclusion of additional observation points beyond those of J

in order to discover certain model inadequacies; with such modifications, the basic idea remains that many designs are satisfactory from the Box–Draper point of view if one does not really know ν well, and it will often be satisfactory to check the nature of spread or moments of a ξ^* motivated by other considerations, rather than to compute a Lebesgue ξ_{BD}. We note, in particular, that choices of ξ^*_∞, among the ξ^*_p, can give enough weight to J_j's with j large, to give the same moments as highly diffuse ν's.

Of at least theoretical interest is the determination of that ν and corresponding $\Phi^{(\Gamma)}$-optimal design ξ^* for which the appropriate moments of ν coincide with those of ξ^*: this minimizes both B and V, for that ν. This will be discussed further elsewhere.

The above remarks apply also to Karson *et al.*'s modification and, at least qualitatively, to other formulations of fitting the wrong degree curve, such as that of Huber (1974).

(iv) Even in 'nice' examples, the choice of ξ_{BD} corresponding to ν being Lebesgue measure may lead to unsatisfactory consequences in terms of other criteria. For example, even assuming that ξ_{BD} has satisfactory efficiency in terms of $V(\xi)+B'(\xi)$, as discussed in (v), below, the behaviour of the maximum over $\mathscr{X}$ of (5·1) may be poor. In the simplex setting, where $\bar{d}(\xi^*_0) = k = \frac{1}{2}(q+1)(q+2)$ and $\bar{d}(\xi^*_p)$ is relatively close to that value for all p, it is easy to compute that $\bar{d}(\xi_{\text{BD}}) = k^2$; there is less disparity between the maxima of corresponding bias functions. Thus, for a wide practical range of values of $N^{-1}\sigma^2$ and β, the function r of (5·1) behaves poorly for ξ_{BD} when x is on or near J_0. The optimality computations that arise if integration of r with respect to ν is replaced by maximization with respect to x are intractable except in a few cases, in which ξ^*_0 has been shown to perform well in an unpublished Moscow report by V. V. Fedorov and M. B. Malyutov.

(v) If one accepts integration of r, and Lebesgue measure for ν, as appropriate, it is clear that the Box–Draper approach will lead to a reasonable design if N is sufficiently large. One may wonder how large that N must be. If h is an appropriate norm on $\sigma^{-1}\beta$, we may ask how large a value N_h, say, N must be in order that $V+B'$ for ξ_{BD} be better than for ξ^*_0, for example. In the simplex model, f_c can be taken to include all functions of the form $x^2_{i_1}x_{i_2}$ and $x_{i_1}x_{i_2}x_{i_3}$, with $i_1 < i_2 < i_3$; this is convenient despite some linear dependence. Of these, $x^2_{i_1}x_{i_2}$ has the larger maximum or integral of its square. Suppose $\sigma^{-1}\beta$ has coefficient h corresponding to $x^2_1x_2$ and has other coefficients zero, this representing the most troublesome form of bias for many choices of the norm. Then N_h will be of the form $c_q h^{-1}$. Our computations show that, although c_q is small for small q, it increases rapidly, and that N_h is in the thousands for moderate q and reasonable h; presumably h is less than one if one feels a quadratic should be fitted. If the maximum of r, discussed in (iv), were considered instead, N_h would of course be even larger.

(vi) As a final comment on the Box–Draper approach, let me make it clear that I would not have spent this much space criticizing some details of it, if I did not think it introduces ideas of a worthwhile point of view, in fact being the first meaningful quantification of such optimal design considerations that accounted for incorrectness of the model $(\mathscr{X}, f)$. But no single simple prescription can be expected to yield satisfactory designs in all, or even most, applications, and the Box–Draper formula is no exception.

6. Other changes of criterion

As mentioned in §1, every admissible design is $\Phi^{(A)}$-optimal for some nonnegative-definite A, and hence we cannot expect any single design ξ' to have efficiency ratio close to one relative to all such criteria, except in degenerate examples. Nevertheless, we investigated the efficiency ratio for certain restricted families of A's. One such family corresponds to a scale change for $\mathscr{X}$, perhaps a less natural occurrence in practice for the simplex than for the cube or ball. For the ball, we investigated the dependence on a combination of change in scale and in p by considering criteria $\Phi_{p,a}(\xi) = [k^{-1}\operatorname{tr}\{A_a M(\xi) A_a\}^{-p}]^{1/p}$, where A_a is diagonal with entry a^j if the corresponding element of f is of degree j. This is criterion Φ_p evaluated on a design $\xi^{(a)}$, say, for the polynomial problem $(a\mathscr{X}, f)$, in terms of a corresponding design ξ for $(\mathscr{X}, f)$, where $\xi^{(a)}(aC) = \xi(C)$ for each subset C of $\mathscr{X}$. Thus, a $\Phi_{0,a}$-optimal ξ is the same for all a, but $\Phi_{1,a}$-optimality can be looked upon either as depending on the size of the region for the average variance criterion or else, equivalently, as reflecting the dependence on a of the optimal design for the criterion $\Sigma_j a^{-j} \times$ (sum of variances of estimators of jth degree parameters) for the fixed region $\mathscr{X}$. As one would expect, the dependence on a of a $\Phi_{p,a}$-optimal design, and the variation of efficiency ratios with a, is least for p small; and the dependence on a for larger p in general yields larger ratios than we encountered in §4·2, from variation in p. These results will be published in detail in a paper concerned with regression on the ball. They indicate, roughly, that it is more important to be able to specify the relative importance of the sets of parameters of various degrees, than to decide on an appropriate p, unless one knows that p is near zero.

We remark that Draper & Lawrence (1966) considered the effect of using a procedure designed for a cubical region when the actual region is spherical, and vice versa, but their normalization is somewhat different and they do not consider the variation in scale mentioned above. Also, Kiefer & Wolfowitz (1964) considered dependence on a in extrapolation problems in which a design on $\mathscr{X} = [-1, 1]$ was used for estimating $\theta' f(x)$ on $[-a, a]$.

I am deeply indebted to R. J. Walker, who performed the early calculations, up to 1972. In recent years Zvi Galil has carried out the computations, as well as participating in theoretical developments; a joint paper containing more details of the simplex example will appear shortly. The work was supported by the National Science Foundation.

This, and three other papers in this issue, are based on talks given at a conference at Imperial College, June 1974.

References

Box, G. E. P. & Draper, N. R. (1959). A basis for the selection of a response surface design. *J. Am. Statist. Assoc.* **54**, 622–54.

Draper, N. R. & Lawrence, W. E. (1966). The use of second-order spherical and cuboidal designs in the wrong regions. *Biometrika* **53**, 596–9.

Farrell, R., Kiefer, J. & Walbran, A. (1965). Optimum multivariate designs. *Proc. 5th Berkeley Symp.* **1**, 113–38.

Fedorov, V. V. (1972). *Theory of Optimal Experiments*, Trans. and Ed. W. J. Studden and E. M. Klimko. New York: Academic Press.

Huber, P. J. (1974). Robustness and designs. In *A Survey of Statistical Designs in Linear Models*, Ed. J. N. Srivastava. Amsterdam: North Holland.

Karson, M. J., Manson, A. R. & Hader, R. J. (1969). Minimum bias estimation and experimental designs for response surfaces. *Technometrics* **11**, 461–76.

Kiefer, J. (1959). Optimum experimental designs. *J. R. Statist. Soc.* B **21**, 272–319.

Kiefer, J. (1960). Optimum experimental designs V, with applications to systematic and rotatable designs. *Proc. 4th Berkeley Symp.* **1**, 381–405.

Kiefer, J. (1961). Optimum designs in regression problems, II. *Ann. Math. Statist.* **32**, 298–325.

Kiefer, J. (1973). Optimum designs for fitting biased multiresponse surfaces. In *Multivariate Analysis*, Vol. 3, Ed. P. R. Krishnaiah, pp. 287–97. New York: Academic Press.

Kiefer, J. (1974). General equivalence theory for optimum designs (approximate theory). *Ann. Math. Statist.* **75**, 849–79.
Kiefer, J. & Wolfowitz, J. (1960). The equivalence of two extremum problems. *Can. J. Math.* **12**, 363–6.
Kiefer, J. & Wolfowitz, J. (1964). Optimum extrapolation designs I and II. *Ann. Inst. Statist. Math.* **16**, 79–108, 295–303.
Scheffé, H. (1958). Experiments with mixtures. *J. R. Statist. Soc.* B, **21**, 344–60.

[*Received August* 1974. *Revised December* 1974]

The Annals of Statistics
1976, Vol. 4, No. 6, 1113-1123

OPTIMAL DESIGNS FOR LARGE DEGREE POLYNOMIAL REGRESSION

BY J. KIEFER[1] AND W. J. STUDDEN[2]

Cornell University and Purdue University

Polynomial regression of degree n on an interval is considered. Optimal designs ξ_n are discussed for various optimality criteria. The behavior of ξ_n for large n is investigated and comparisons of ξ_n with the limiting design ξ_0 are made.

1. Introduction. Let $f' = (f_0, f_1, \cdots, f_n)$ be a vector of linearly independent functions on a space $\mathscr{X}$. For each x or "level" in $\mathscr{X}$ an experiment can be performed whose outcome is a random variable $Y(x)$ with mean value $\beta' f(x) = \sum \beta_i f_i(x)$ and variance σ^2, independent of x. The functions f_i, $i = 0, 1, \cdots, n$ are called the regression functions and are assumed known to the experimenter while the vector of parameters $\beta' = (\beta_0, \beta_1, \cdots, \beta_n)$ and σ^2 are unknown. An experimental design is a probability measure ξ on $\mathscr{X}$. If ξ concentrates mass ξ_i at the points x_i, $i = 1, 2, \cdots, r$ and $\xi_i N = n_i$ are integers, the experimenter takes N uncorrelated observations, n_i at each x_i, $i = 1, 2, \cdots, r$. The covariance matrix of the least squares estimates of the parameters β_i is then given by $(\sigma^2/N)M^{-1}(\xi)$ where $M(\xi) = (m_{ij}(\xi))$, $m_{ij}(\xi) = \int f_i(x) f_j(x)\, d\xi(x)$ is the information matrix of the experiment or design ξ.

In experimental situations it may be desirable to use a design ξ which minimizes a particular functional of the matrix $M(\xi)$. For instance we may consider

(i) $|M^{-1}(\xi)|$ = determinant of $M^{-1}(\xi)$;

(ii) $\sup_{x \in \mathscr{X}} f'(x) M^{-1}(\xi) f(x)$; here $f'(x)M^{-1}(\xi)f(x)$ is proportional to the variance of the estimate of the regression function at x;

(iii) $f'(x_0)M^{-1}(\xi)f(x_0)$, or $c'M^{-1}(\xi)c$ for any vector c, or more generally $\operatorname{tr} CM^{-1}(\xi)$ where C is symmetric and positive semidefinite.

All of these are used in regression theory and are explained more fully in Fedorov (1972). It is known, see Kiefer and Wolfowitz (1960), that the minimization problems in (i) and (ii) are equivalent. Other results of this type are given in Fedorov (1972) and Kiefer (1974).

In the present paper we wish to examine the case where $\mathscr{X}$ is the interval $[-1, 1]$ and $f'(x) = (1, x, \cdots, x^n)$, i.e., the situation where our regression function is polynomial $\sum_{i=0}^{n} \beta_i x^i$. A given minimization problem of the type described

Received June 1975; revised March 1976.

[1] This research was supported by the NSF Grant MPS72-04998 A02. Reproduction is permitted in whole or in part for any purposes of the United States Government.

[2] Research support from NSF Grant 33552x2 and MP575-08294.

AMS 1970 subject classification. Primary 62K05.

Key words and phrases. Optimal design, regression, limiting design, generalized variance, extrapolation.

above, e.g., minimization of $|M^{-1}(\xi)|$, then gives rise to a sequence of designs ξ_n, one for each degree n. That is, for a fixed degree we suppose ξ_n satisfies

$$\min_\xi |M^{-1}(\xi)| = |M^{-1}(\xi_n)| .$$

We wish to examine the designs ξ_n for large values of n, in particular the limiting design ξ_0, if it exists, and to compare, say,

$$|M^{-1}(\xi_n)| \quad \text{and} \quad |M^{-1}(\xi_0)| \qquad \text{or}$$
$$\text{perhaps} \quad f'(x)M^{-1}(\xi_n)f(x) \quad \text{and} \quad f'(x)M^{-1}(\xi_0)f(x) .$$

The paper is motivated by an attempt to find single designs, for various criteria, which would work reasonably well for all degrees. This search led naturally to investigating whether there was any regularity in ξ_n for large n. In practice the regression function will usually be of low degree. It turns out, as in various other investigations, that the limiting design ξ_0 performs reasonably well even when used in a linear regression. For example the D-efficiency, using the generalized variance and defined below after Lemma 2.2, is about 70% and increases to the value one with increasing degree. Other, seemingly natural, definitions of efficiency involving $\sup_x f'(x)M^{-1}(\xi_0)f(x)$ are decreasing with the degree. We have indicated in Theorem 2.2 how the limiting efficiency in this case can be improved. The material here should also provide information on what to look for in more complex situations (see for example, Kiefer (1975) where quadratic regression on an n-simplex is considered).

In some situations sequential experimentation is costly but a large number of observations can be taken nonsequentially. If the degree is not known, we can obtain some idea of the cost, from using a design which is optimum for a very large degree, over what one could achieve with an approximately optimum sequential design which would determine the right degree with high probability and act accordingly. The material should also be of some theoretical interest.

The paper is divided into three more sections. In Section 2 we consider the determinant of the information matrix $|M(\xi)|$ as mentioned above. The extrapolation problem minimizing $f'(x_0)M^{-1}(\xi)f(x_0)$ for a fixed x_0 outside of $\mathscr{X}$ is considered in Section 3. Results similar to those given in Section 2 and 3 for the optimal designs for estimating the separate coefficients β_i are discussed in Section 4.

2. The generalized variance. Consider the case where

$$f'(x) = (1, x, \cdots, x^n) \equiv \bar{f}_n'(x) \quad \text{(say)}, \qquad \mathscr{X} = [-1, 1],$$
$$M_n(\xi) = \int_{-1}^{1} f(x)f'(x)\, d\xi(x)$$

and maximize the determinant $|M_n(\xi)|$. The design ξ_n maximizing $|M_n(\xi)|$ is called D-optimal. The following theorem is given in Fedorov (1972), page 91. We include a short proof here for completeness.

THEOREM 2.1. *The sequence ξ_n, $n = 1, 2, \cdots$ of D-optimal designs converges weakly to ξ_0 where ξ_0 has density $1/\pi(1 - x^2)^{\frac{1}{2}}$.*

PROOF. The proof follows fairly readily from a number of known results. The D-optimal design ξ_n concentrates mass $(n+1)^{-1}$ on the $n+1$ zeros x_v, $v = 0, 1, \cdots, n$ of $(1-x^2)P_n'(x)$, where P_n is the nth Legendre polynomial. See Karlin and Studden (1966a). If θ_v are defined by $x_v = \cos\theta_v$, $0 \leqq \theta_v \leqq \pi$, it is known (see Erdös and Turán (1940)) that the θ_v become uniformly distributed on the half circle $0 \leqq \theta \leqq \pi$ in the sense that if $N(a, b)$ denotes the number of θ_v in $[c, d]$ then

$$n^{-1}N(c, d) \to \frac{|d-c|}{\pi}.$$

It then follows that $\xi_n \to \xi_0$ where ξ_0 is the distribution of $Y = \cos X$ and X is uniform on $(0, \pi)$. The statement of the theorem then follows.

Let $d_n(x, \xi) = \bar{f}_n'(x)M_n^{-1}(\xi)\bar{f}_n(x)$ and denote $\sup_x d_n(x, \xi)$ by $d_n(\xi)$. The quantity $d_n(x, \xi)$ is proportional to the variance of our estimate of the regression curve at the point x assuming our regression was of degree n and we used the design ξ. We will compare $d_n(x, \xi_0)$ and $d_n(x, \xi_n)$ for small values of n and consider whether ξ_0 is "asymptotically optimal" in some sense. We also compare $|M_n(\xi_n)|$ and $|M_n(\xi_0)|$.

A general calculation for $d_n(x, \xi_0)$ can be made. The quantity $d_n(x, \xi_0)$ is invariant under a change of basis for our functions $1, x, x^2, \cdots, x^n$. We use as a basis the polynomials which are orthonormal on $[-1, 1]$ with the weight or measure ξ_0. These are the polynomials $1, 2^{\frac{1}{2}}T(x), \cdots, 2^{\frac{1}{2}}T_n(x)$ where $T_k(\cos\theta) = \cos k\theta$ are the Chebyshev polynomials of the 1st kind. In this case we find that

$$\begin{aligned} d_n(x, \xi_0) &= 1 + 2\textstyle\sum_{k=1}^n T_k^2(x) \\ &= n + \tfrac{1}{2} + \frac{1}{2}\frac{\sin(2n+1)\theta}{\sin\theta} \qquad (2.1)\\ &= n + \tfrac{1}{2} + \tfrac{1}{2}U_{2n}(x) \end{aligned}$$

where $U_k(\cos\theta) = \sin(k+1)\theta/\sin\theta$ are the Chebyshev polynomials of the 2nd kind.

It is known that $d_n(\xi_n) = \sup_x d_n(x, \xi_n) = n + 1$ and that the sup is attained at the points where ξ_n concentrates its equal mass. These are the zeros of $(1-x^2)P_n'(x)$, where P_n denotes the Legendre polynomial. If we use ξ_0 instead of ξ_n we have that $d_n(\xi_0) = n + \frac{1}{2} + \frac{1}{2}\sup_x U_{2n}(x)$. Since $\sup_x U_k(x) \leqq k+1$ (see Davis (1963)) it follows that $d_n(\xi_0) = 2n+1$. This value is about double the value $n+1$. The sup here is reached only at the end points $x = \pm 1$.

We shall consider both functions $d_n(x, \xi_n)$ and $d_n(x, \xi_0)$ for $n = 1, 2$ and 3. For $n = 1$ the D-optimal design has equal mass at $x = \pm 1$. Simple calculations show that

$$d_1(x, \xi_1) = 1 + x^2 \qquad \text{and} \qquad d_1(x, \xi_0) = 1 + 2x^2$$

so that $d_1(x, \xi_1) \leqq d_1(x, \xi_0)$ for all x. The next case, $n = 2$, seems to be somewhat more indicative of the general situation. Here the D-optimal design has mass $\frac{1}{3}$

on points $x = 1, 0, 1$. Calculations then give

$$d_2(x, \xi_2) = 3 - \tfrac{9}{2}x^2 + \tfrac{9}{2}x^4$$
$$d_2(x, \xi_0) = 3 - 6x^2 + 8x^4$$

and we have $d_2(x, \xi_0) \leqq d_2(x, \xi_2)$ for $|x| \leqq (\frac{3}{7})^{\frac{1}{2}} = .655$. Thus the approximate D-optimal design is better for x in the middle of our interval $[-1, 1]$. For $n = 3$ the situation is more complicated, as expected. The D-optimal design has equal weight $\frac{1}{4}$ on ± 1 and $\pm 1/5^{\frac{1}{2}} = \pm .447$. More calculations give $d_3(x, \xi_3) = 3.248 + 8.261x^2 - 26.267x^4 + 18.756x^6$ while $d_3(x, \xi_0) = \frac{7}{2} + \frac{1}{2}(\sin 7\theta/\sin\theta)$, $x = \cos\theta$. The values are roughly comparable in the range $|x| \leqq 0.9$. A small table of values is given below. Both functions are symmetric about zero.

θ	90	80	70	60	50	40	30	20	10	0
$x = \cos$	0	.174	.342	.5	.643	.766	.866	.94	.98	1
$d(x, \xi_3)$	3.25	3.47	3.88	3.96	3.50	2.84	2.58	2.98	3.66	3
$d_3(x, \xi_0)$	3	3.33	3.91	4.00	3.39	2.73	3.00	4.44	6.20	7

If we define the G-efficiency (see Atwood 1969) of a design ξ as $(n + 1)/d_n(\xi)$ we then note that the limiting design ξ_0 has G-efficiency $(n + 1)/(2n + 1)$ which, unexpectedly, decreases to the value $\frac{1}{2}$. It is natural to inquire whether there is a design with limiting G-efficiency equal to one. In this regard we have

THEOREM 2.2. *For each $\varepsilon > 0$ there exists a design ξ_ε such that*

$$\liminf_{n\to\infty} \frac{n+1}{d_n(\xi_\varepsilon)} \geqq 1 - \varepsilon \,. \tag{2.2}$$

PROOF. The result is obtained using the following lemma from Szego (1959), page 31.

LEMMA 2.1. *Let $\rho(x)$ be a polynomial of degree l on $[-1, 1]$ and write $\rho(\cos\theta) = |h(e^{i\theta})|^2$ where $x = \cos\theta$, $h(z)$ is of degree l, $h(z) \neq 0$ for $|z| < 1$ and $h(0) > 0$. Let $h(e^{i\theta}) = c(\theta) + is(\theta)$ and $w(x) = (d/\pi)(1 - x^2)^{-\frac{1}{2}}/\rho(x)$ where d is such that $\int w(x)\,dx = 1$. Then the polynomials orthonormal with respect to $w(x)$ are given by $p_0(x) = 1$ and*

$$p_k(\cos\theta) = \left(\frac{2}{d}\right)^{\frac{1}{2}} \{c(\theta)\cos k\theta + s(\theta)\sin k\theta\} \qquad k \geqq 1 \,.$$

The idea is to change the measure ξ_0 by putting some mass near ± 1. This is accomplished by changing ξ_0 to a design ξ_ε with density $w(x)$ where $\rho(x)$ depends on $\delta(\varepsilon)$. For fixed $\delta = \delta(\varepsilon)$ we let $\rho(x)$ be a polynomial such that

$$\delta^{\frac{1}{2}} \leqq \rho(x) \leqq 1 + \delta \qquad \text{for} \quad -1 \leqq x \leqq 1 \,,$$
$$1 \leqq \rho(x) \leqq 1 + \delta \qquad \text{for} \quad -1 + \delta \leqq x \leqq 1 - \delta \,,$$
$$\delta^{\frac{1}{2}} \leqq \rho(x) \leqq 2\delta^{\frac{1}{2}} \qquad \text{for} \quad 1 - |x| < \delta/2 \,.$$

The proof of the theorem will be to show that $(n + 1)/d_n(\xi_\varepsilon) \geqq 1 - \varepsilon$ for large

n when $\delta = \delta(\varepsilon)$ is taken sufficiently small. Such a polynomial exists by the Weierstrass theorem and we denote its degree by l. Applying the above lemma we then have

$$d_n(x, \xi_\varepsilon) - 1 = \sum_1^n p_k^2(\cos\theta)$$
$$= \frac{2}{d}\{c^2(\theta)\sum_1^n \cos^2 k\theta + s^2(\theta)\sum_1^n \sin^2 k\theta + c(\theta)s(\theta)\sum_1^n \sin 2k\theta\}.$$

Inserting the values $\cos^2 k\theta = (1 + \cos 2k\theta)/2$ and $\sin^2 k\theta = (1 - \cos 2k\theta)/2$ and using $\rho(\theta) = c^2(\theta) + s^2(\theta)$ we obtain

$$\sum_1^n p_k^2(\cos\theta) = d^{-1}\{n\rho(\theta) + (c^2(\theta) - s^2(\theta))\sum_1^n \cos 2k\theta + 2c(\theta)s(\theta)\sum_1^n \sin 2k\theta\}.$$

From the choice of $\rho(\theta)$ it is easily seen that $d = d(\delta) \to 1$ as $\delta \to 0$ so that the first term is $d^{-1}n\rho(\theta) = n(1 + o(\delta))$. The other two terms can be readily handled. Since $\rho(\theta) = c^2(\theta) + s^2(\theta)$ it follows that $|c^2(\theta) - s^2(\theta)| \leqq 2\rho(\theta)$ and $|c(\theta)s(\theta)| \leqq \rho(\theta)$. The remaining two terms are then bounded by

$$2d^{-1}\rho(\theta)\{|\sum_1^n \cos 2k\theta| + |\sum_1^n \sin 2k\theta|\}.$$

Now for $1 - |x| < \delta/2$, $\rho(\theta)$ is small and the term in brackets is bounded by $2n$. For $1 - |x| > \delta/2$ the bracket can be bounded by a term depending on δ but not on n. Therefore the remaining two terms can be bounded by $f(\delta) + no(\delta)$. The result then follows.

We now turn to a comparison of the two determinants $|M_n(\xi_n)|$ and $|M_n(\xi_0)|$. The ratio can be calculated in a fairly explicit form. We let $D_n(\xi) = |M_n(\xi)|$.

THEOREM 2.3. *If ξ_n is the D-optimal design and ξ_0 has the* arcsin *density as described in Theorem* 2.1 *then*

$$\frac{D_{n+1}(\xi_{n+1})}{D_{n+1}(\xi_0)} = 2n^{\frac{1}{4}}e^{\delta_n} \tag{2.3}$$

where

$$\delta_n = \tfrac{1}{4}\left\{\sum_{k=1}^n \frac{1}{k} - \ln n\right\} - \sum_{k=2}^\infty (-1)^k \frac{\zeta_n(k)}{k(k+1)}\left(1 - \frac{1}{2^k}\right) \tag{2.4}$$

and

$$\zeta_n(k) = \sum_{l=1}^n \frac{1}{l^k}, \qquad \zeta(k) = \sum_{l=1}^\infty \frac{1}{l^k}.$$

Before proving Theorem 2.3 we shall state an additional relevant lemma and make a few additional remarks.

LEMMA 2.2 (Rubin). *If δ_n is defined above in* (2.4) *then*

$$\delta_n \to \delta = \tfrac{1}{2} - \tfrac{13}{12}\log 2 - 3\zeta'(-1)$$
$$\approx -.00464602.$$

The proof of Lemma 2.2 will be omitted. The quantity $\zeta'(-1)$ is the derivative of the zeta function at -1 and has value $\zeta'(-1) = -.165421142$. (See

Walther (1926)). If one wished to use the ratio (2.3), the only complicated quantity is δ_n. If we replace δ_n by δ and consider

$$\rho_{n+1} = \frac{D_{n+1}(\xi_0)}{2n^{\frac{1}{2}}e^{\delta}D_{n+1}(\xi_{n+1})}$$

then $\rho_2 \approx 1.00465$ and ρ_n seems to be decreasing to one. (The limit of course is equal to one.) Moreover the quantity $e^{\delta} \approx .99536$ so that the ratio in (2.3) is essentially $2n^{\frac{1}{2}}$.

In terms of efficiency, the appropriate quantity is the D-efficiency of ξ_0 (see Atwood (1969)) defined by

$$E_n = \left(\frac{D_n(\xi_0)}{D_n(\xi_n)}\right)^{1/(n+1)} .$$

The values of E_n for $n = 1, 2, 3, 4$ can be easily calculated to be 0.71, 0.75, 0.79 and 0.81 respectively. By Theorem 2.3 it is easy to see that ξ_0 has limiting D-efficiency equal to one. This immediately raises the question about the relationship between the G-efficiency and the D-efficiency of a design. We note that by the Kiefer–Wolfowitz equivalence theorem D-optimality or D-efficiency equal to one is equivalent to G-optimality or G-efficiency equal to one. Using inequalities (see Kiefer, 1960) of the type

$$\frac{D_n(\xi)}{D_n(\xi_n)} \geqq \exp\{n + 1 - d_n(\xi)\}$$

one can show that a limiting G-efficiency of one produces a limiting D-efficiency of one. The converse however is not true as our example shows. (A limiting G-efficiency of $1 - \varepsilon$ produces a limiting D-efficiency of $\exp -\varepsilon/(1 - \varepsilon)$.)

We note finally that the design ξ_ε (used in Theorem 2.2 to give a better G-efficiency than the limiting design ξ_0) should presumably have a better D-efficiency than ξ_0 calculated from equation (2.3). This however is not the case as will be seen from Theorem 2.4.

PROOF OF THEOREM 2.3. From Szego (1959), page 28, we find the value $D_n(\xi_0) = \prod_{r=1}^{n} k_r^{-2}$ where $k_n = 2^{\frac{1}{2}}2^{n-1}$ is the coefficient of x^n in $2^{\frac{1}{2}}T_n(x)$: these being the polynomials orthonormal to ξ_0. In this case $D_n(\xi_0) = 2^{-n^2}$. The value $D_n(\xi_n)$ is given by (see Karlin and Studden (1966b))

$$D_n(\xi_n) = 2^{n(n+1)}(\textstyle\prod_{v=1}^{n} v^v)^4 n^{-n} \prod_{v=1}^{2n} v^{-v} .$$

Letting

$$R_n = \frac{D_n(\xi_n)}{D_n(\xi_0)} ,$$

a straightforward calculation gives

$$\frac{R_{n+1}}{R_n} = \frac{(1 + 1/n)^{n+1}}{(1 + 1/2n)^{2n+1}} .$$

Then

$$\log \frac{R_{l+1}}{R_l} = (l+1)\log\left(1+\frac{1}{l}\right) - (2l+1)\log\left(1+\frac{1}{2l}\right)$$
$$= \frac{1}{4l} - \sum_{k=2}^{\infty} \frac{(-1)^k}{l^k k(k+1)}\left(1-\frac{1}{2^k}\right),$$

and

$$\log \frac{R_{n+1}}{R_1} = \sum_{l=1}^{n} \log \frac{R_{l+1}}{R_l}$$
$$= \frac{1}{4}\sum_{l=1}^{n} \frac{1}{l} - \sum_{k=2}^{\infty} \frac{\zeta_n(k)(-1)^k}{k(k+1)}\left(1-\frac{1}{2^k}\right).$$

Since $R_1 = 2$ it then follows that

$$\log \frac{R_{n+1}}{2n^{\frac{1}{4}}} = \frac{1}{4}\left\{\sum_{l=1}^{n} \frac{1}{l} - \ln n\right\} - \sum_{k=2}^{\infty} \frac{\zeta_n(k)(-1)^k}{k(k+1)}\left(1-\frac{1}{2^k}\right),$$

and Theorem 2.3 follows.

The asymptotic behavior of $D_n(\xi)$ for designs ξ with densities can be ascertained from results on Hankel determinants. The next result follows nearly immediately from Grenander and Szego (page 84) and Szego (page 142).

THEOREM 2.4. *If ξ has density $g(x)$ then*

$$D_n^{1/(n+1)}(\xi) \approx D_n^{1/(n+1)}(1)e^{G(g)},$$

where

$$G(g) = \int_{-1}^{1} \frac{\log g(x)}{\pi(1-x^2)^{\frac{1}{2}}}\, dx,$$

and

$$D_n(1) = 2^{-n(n-1)} \prod_{\nu=1}^{n} \nu^{3\nu-2n}(n+\nu)^{n-\nu}.$$

The approximation is taken in the sense that the ratio of both sides tends to one.

A variational argument shows that $G(g)$ is maximized by ξ_0 or $g_0(x) = 1/\pi(1-x^2)^{\frac{1}{2}}$. It then follows that g_0 is the only density which has asymptotic D-efficiency equal to one. Thus the designs ξ_ε used in Theorem 2.2 have asymptotic D-efficiency less than one.

3. Extrapolation. In this section we consider the minimization of

$$d_n(x_0, \xi) = \bar{f}_n'(x_0)M_n^{-1}(\xi)\bar{f}_n(x_0), \qquad |x_0| > 1. \tag{3.1}$$

Using the design ξ and assuming an nth degree regression, the quantity $d_n(x_0, \xi)$ is proportional to the variance of the least squares estimate of the regression at the point x_0. Since $|x_0| > 1$ and observations are confined to $[-1, 1]$ we have an extrapolation problem. It is known, see Hoel and Levine (1964) or Studden (1968), that the optimal design ξ_n minimizing $d_n(x_0, \xi)$ concentrates mass p_v on the points $s_v = \cos v\pi/n$, $v = 0, 1, \cdots, n$. The value p_v is given by

$$p_v = \frac{|l_v(x_0)|}{\sum_{v=0}^{n} |l_v(x_0)|}$$

where $l_v(x)$, $v = 0, 1, \cdots, n$ are the Lagrange polynomials of degree n specified by $l_v(s_\mu) = \delta_{\mu v}$, i.e., fixing v, $l_v(x)$ vanishes at all the values s_l except s_v where it has the value one.

THEOREM 3.1. *If ξ_n is the optimal extrapolation design then $\xi_n \to \xi_0$ where ξ_0 has density*

$$\frac{(x_0^2 - 1)^{\frac{1}{2}}}{\pi|x_0 - x|(1 - x^2)^{\frac{1}{2}}}.$$

PROOF. We consider only the case $x_0 > 1$. The measure ξ_n has weights proportional to $|l_v(x_0)|$ at $s_v = \cos v\pi/n$. The Lagrange polynomial l_v is given by

$$l_v(x_0) = \frac{\prod_{\mu=0}^{n} (x_0 - s_\mu)}{(x_0 - s_v) \prod_{\mu \neq v} (s_\mu - s_v)}.$$

Since the numerator is constant we see that ξ_n is proportional to a measure

$$\frac{1}{x_0 - x}\, d\rho_n(x)$$

where $d\rho_n$ has mass $[\prod_{\mu \neq v} (s_\mu - s_v)]^{-1} = \gamma_v$ at $s_v = \cos v\pi/n$. Substituting the values $s_v = \cos v\pi/n$ in γ_n and using the law

$$\cos A - \cos B = -2 \sin (A + B)/2 \sin (A - B)/2$$

a straightforward calculation reveals that the values $\gamma_0, \gamma_1, \cdots, \gamma_n$ are proportional to $1, 2, 2, \cdots, 2, 1$. It then follows as in Theorem 2.1 that the limiting measure has density proportional to

$$\frac{1}{(x_0 - x)(1 - x^2)^{\frac{1}{2}}}. \tag{3.2}$$

The CRC tables (*Handbook for Probability and Statistics*, 2nd edition, page 609, formula 215) give a value of $\pi/(x_0^2 - 1)^{\frac{1}{2}}$ for the integral of (3.2) over the range -1 to 1. The theorem then follows.

We will now make a comparison of $d_n(x_0, \xi_n)$ and $d_n(x_0, \xi_0)$. From Studden (1968) we know that $d_n(x_0, \xi_n) = T_n^2(x_0)$ where $T_n(x)$ is the Chebyshev polynomial of the first kind. In order to evaluate $d_n(x_0, \xi_0)$ we use the fact as noted above that the expression for it given in (3.1) is invariant under a basis change so that

$$d_n(x_0, \xi_0) = \textstyle\sum_{k=0}^{n} p_k^2(x_0) \tag{3.3}$$

if we let $p_k(x)$ denote the polynomials orthonormal with respect to the measure ξ_0.

In order to evaluate (3.3) we use Lemma 2.1. We apply this result with $\rho(x) = (x_0 - x)$. We find that $c(\theta) = (a + b \cos \theta)$, $s(\theta) = b \sin \theta$ where a and b satisfy the two conditions

$$a^2 + b^2 = x_0, \qquad 2ab = -1. \tag{3.4}$$

The restrictions on $h(z)$ in the lemma give

$$a^2 = \tfrac{1}{2}(x_0 + (x_0^2 - 1)^{\frac{1}{2}}), \qquad b^2 = \tfrac{1}{2}(x_0 - (x_0^2 - 1)^{\frac{1}{2}}). \tag{3.5}$$

This then gives

$$(3.6) \qquad p_k(\cos\theta) = \left(\frac{2}{c}\right)^{\frac{1}{2}} \{(a + b\cos\theta)\cos k\theta + b\sin\theta\sin k\theta\}$$
$$= \left(\frac{2}{c}\right)^{\frac{1}{2}} \{a\cos k\theta + b\cos(k-1)\theta\}$$

where $c = (x_0^2 - 1)^{\frac{1}{2}}$.

In order to evaluate $\sum_{k=0}^{n} p_k^2(x_0)$ we use (3.6), some half-angle trigonometric formulae, some series summation from Jolley (1961), and simplify the following expression:

$$\sum_{k=1}^{n} (a\cos k\theta + b\cos(k-1)\theta)^2$$
$$= \sum_1^n [a^2\cos^2 k\theta + b^2\cos^2(k-1)\theta + 2ab\cos k\theta\cos(k-1)\theta]$$
$$= b^2 + a^2\cos^2 n\theta + x_0\sum_{k=1}^{n-1}\cos^2 k\theta - \sum_{k=1}^{n}\cos k\theta\cos(k-1)\theta$$
$$= b^2 + a^2\cos^2 n\theta + x_0\left(\frac{n-1}{2} + \frac{\cos n\theta\sin(n-1)\theta}{2\sin\theta}\right)$$
$$- \tfrac{1}{2}\sum_{k=1}^{n}(\cos(2k-1)\theta + \cos\theta)$$
$$= b^2 + a^2\cos^2 n\theta + x_0\left(\frac{n-1}{2} + \frac{\cos n\theta\sin(n-1)\theta}{2\sin\theta}\right)$$
$$- \tfrac{1}{2}\left(\frac{1}{2}\frac{\sin 2n\theta}{\sin\theta} + n\cos\theta\right)$$
$$= b^2 + a^2\cos^2 n\theta - \frac{x_0}{2} - \tfrac{1}{2}\cos n\theta\cos(n-1)\theta\,.$$

Using equation (3.5) and $T_n(\cos\theta) = \cos n\theta$ we then have

$$\sum_{k=0}^{n} p_k^2(x_0) = \frac{1}{(x_0^2 - 1)^{\frac{1}{2}}}[(x_0 + (x_0^2 - 1)^{\frac{1}{2}})T_n(x_0) - T_n(x_0)T_{n-1}(x_0)]\,.$$

The following theorem can then be readily deduced.

THEOREM 3.2. *Let* $d_n(x_0, \xi)$ *be as in* (3.1) *and let* $r_n(x_0) = d_n(x_0, \xi_n)/d_n(x_0, \xi_0)$. *Then*

$$r_n^{-1}(x_0) = \frac{1}{(x_0^2 - 1)^{\frac{1}{2}}}\left\{x_0 + (x_0^2 - 1)^{\frac{1}{2}} - \frac{T_{n-1}(x_0)}{T_n(x_0)}\right\}.$$

For linear regression we have $r_1^{-1}(x_0) = 1 + (x_0^2 - 1)^{\frac{1}{2}}/x_0$; while for quadratic regression $r_2^{-1}(x_0) = 1 + 2x_0(x_0^2 - 1)^{\frac{1}{2}}/(2x_0^2 - 1)$. Note as expected that both of these values are near 1 for x_0 close to 1. Bounds and other limit relations can be obtained using the following two lemmas.

LEMMA 3.1. *The ratio* $a_n = T_n(x_0)/T_{n-1}(x_0)$ *is increasing to* $x_0 + (x_0^2 - 1)^{\frac{1}{2}}$.

PROOF. Since $T_{n+1}(x_0) = 2x_0T_n(x_0) - T_{n-1}(x_0)$ we have $a_{n+1} = 2x_0 - 1/a_n$ and then a_n is seen to be increasing using induction. The actual limit value follows from the fact that (see Szego (1959), page 189)

$$\lim_{n\to\infty}\frac{(x_0 + (x_0^2 - 1)^{\frac{1}{2}})^n}{2T_n(x_0)} = 1\,.$$

LEMMA 3.2. *For fixed x_0, the ratio $r_n(x_0)$ decreases to the value $\frac{1}{2}$.*

The proof follows immediately from Lemma 3.1. Note in particular that $\frac{1}{2} \leq r_n(x_0)$ and of course $r_n(x_0) \leq 1$. One can show further that $\lim_{x_0 \to 1} r_n(x_0) = 1$ and $\lim_{x_0 \to \infty} r_n(x_0) = \frac{1}{2}$.

The asymptotic behavior of $d_n(x_0, \xi)$ can also be evaluated for any design ξ with density using results from Szego (1959).

THEOREM 3.3. *If the design ξ has density g then*

$$d_n(x_0, \xi) \approx (2\pi)^{-1} D_g^{-2}(z_0^{-1}) \frac{z_0^{2n} - 1}{z_0^{2n}},$$

where $z_0 = x_0 + (x_0^2 - 1)^{\frac{1}{2}}$, and

$$D_g(z) = \exp\left\{\frac{1}{4\pi} \int_{-\pi}^{\pi} \log\left[g(\cos\theta)|\sin\theta|\right] \frac{1 + ze^{-i\theta}}{1 + ze^{-i\theta}}\, d\theta\right\}.$$

The approximation means the ratio converges to one.

PROOF. From Szego (1959), page 295 we find that if $p_n(x_0)$ denotes the nth polynomial orthonormal with respect to g then

$$p_n(x_0) \approx (2\pi)^{-\frac{1}{2}} z^n \{D(z^{-1})\}^{-1}.$$

The result then follows by a simple Abelian argument since

$$d_n(x_0, \xi) = \sum_{k=0}^{n} p_k^2(x_0).$$

COROLLARY. *Among designs which are absolutely continuous with respect to Lebesgue measure, the one which asymptotically minimizes $d_n(x_0, \xi)$ has density*

$$g_0(x) = \frac{(x_0^2 - 1)^{\frac{1}{2}}}{\pi |x_0 - x| (1 - x^2)^{\frac{1}{2}}}.$$

PROOF. This result follows by noting that $D_g(z_0)$ must be real. We therefore maximize $|D_g(z_0)|^2$. A variational argument shows this to be $g_0(x)$.

4. Individual coefficients. Analyses similar to Sections 2 and 3 can be given for the estimation of the individual coefficients. Since a more complete investigation of the asymptotic properties of the information matrix is presently being made, we shall be somewhat briefer and proofs will be omitted.

We consider the estimation of β_k in the model $\sum_{k=0}^{n} \beta_k x^k$. The variance of the LSE using a design ξ is denoted by $V(n, k, \xi)$ and the optimal design is devoted by $\xi(n, k)$.

It can be shown that for k fixed, $\xi(n, n-k)$ has limit density $g_0(x) = \pi^{-1}(1 - x^2)^{-\frac{1}{2}}$ while if $k/n = q$ and q is fixed with $0 < q < 1$ then $\xi(n, k)$ has limiting density proportional to $[q^2 + (1 - q^2)x^2]^{-1}(1 - x^2)^{-\frac{1}{2}}$. In the case $k = n$, let $\xi_n = \xi(n, n)$ and ξ_0 denote the design with density $g_0(x)$. Then

$$\frac{V(n, n, \xi_n)}{V(n, n, \xi_0)} \to \frac{\pi}{2} D^2(0) \tag{4.1}$$

where $D(z)$ is defined in Section 3. The value $D(0)$ is maximized by the design with density g_0. The efficiency given in (4.1) is constant and equal to $\frac{1}{2}$. This value "agrees" with the remark in Section 3 that $r_n(\xi_0) \to \frac{1}{2}$ (the case $x_0 \to \infty$ and the highest coefficient are equivalent problems in certain respects).

For k fixed the designs $\xi(n, k)$ degenerate to having all mass at the origin. Thus for estimating the slope at the origin or the coefficient β_1 the optimal design concentrates its mass closer and closer to zero as $n \to \infty$. The actual rate of this convergence has been ascertained.

Acknowledgments. We wish to thank Professor Herman Rubin for proving Lemma 2.2 and for other helpful conversations. We also thank Professor Szego for providing basic ideas and the referee and editors for their helpful comments.

REFERENCES

[1] DAVIS, PHILIPS J. (1963). *Interpolation and Approximation.* Blaisdell, Waltham, Mass.

[2] ERDÖS, P. and TURÁN, P. (1940). Interpolation III. *Ann. of Math.* **41** 510–553.

[3] FEDOROV, V. V. (1972). *Theory of Optimal Experiments.* Academic Press, New York.

[4] GRENANDER, ULF and SZEGO, GABOR (1958). *Toeplitz Forms and Their Applications.* Univ. of California Press.

[5] HOEL, P. G. and LEVINE, A. (1964). Optimal spacing and weighing in polynomial regression. *Ann. Math. Statist.* **35** 1553–1560.

[6] JOLLEY, L. B. W. (1961). *Summation of Series.* Dover, New York.

[7] KARLIN, S. and STUDDEN, W. J. (1966a). Optimal experimental designs. *Ann. Math. Statist.* **37** 783–815.

[8] KARLIN, S. and STUDDEN, W. J. (1966b). *Tchebycheff Systems: with Applications in Analysis and Statistics.* Interscience, New York.

[9] KIEFER, J. and WOLFOWITZ, J. (1960). The equivalence of two extremum problems. *Canad. J. Math.* **12** 363–366.

[10] KIEFER, J. (1966). Optimum experimental designs V. *Proc. Fourth Berkeley Symp. Math. Statist. Prob.* **1** 381–405, Univ. of California Press.

[11] KIEFER, J. (1974). General equivalence theory for optimum designs (approximate theory). *Ann. Math. Statist.* **2** 849–879.

[12] KIEFER, J. (1975). Optimal design: Variation in structure and performance under change of criterion. *Biometrika* **62** 277–288.

[13] STUDDEN, W. J. (1968). Optimal designs on Tchebycheff points. *Ann. Math. Statist.* **39** 1435–1447.

[14] SZEGO, S. (1959). *Orthogonal Polynomials.* Amer. Math. Soc. Colloq. Publ. **23** New York.

[15] WALTHER, ALWIN (1926). Anschauliches zur Riemannschen Zeta funktion. *Acta Math.* **48** 393–400.

DEPARTMENT OF MATHEMATICS
CORNELL UNIVERSITY
ITHACA, NEW YORK 14850

DEPARTMENT OF STATISTICS
PURDUE UNIVERSITY
LAFAYETTE, INDIANA 47907

Journal of Statistical Planning and Inference 1 (1977) 27–40.

COMPARISON OF ROTATABLE DESIGNS FOR REGRESSION ON BALLS, I (QUADRATIC)*

Z. GALIL and J. KIEFER
Cornell University, Ithaca, N.Y., U.S.A.

Received 2 June 1976; revised manuscript received 3 August 1976
Recommended by A.M. Herzberg

Designs for quadratic and cubic regression are considered when the possible choices of the controlable variable are points $x=(x_1,x_2,\ldots,x_q)$ in the q-dimensional ball of radius R, $B_q(R)=\{x:\sum_1^q x_i^2 \leqq R^2\}$. The designs that are optimum among rotatable designs with respect to the D-, A-, and E-optimality criteria are compared in their performance relative to these and other criteria, including extrapolation. Additionally, the performance of a design optimum for one value of R, when it is implemented for a different value of R, is investigated. Some of the results are developed algebraically; others, numerically. For example, in quadratic regression the A-optimum design appears to be fairly robust in its efficiency, under variation of criterion.

AMS Subject Classification: 62K05.
Key words and phrases:
Optimum designs, rotatable designs, experiments in balls, robust designs, quadratic regression

1. Introduction

The rationale for studying the performance under various optimality criteria of a design selected according to a particular criterion, was given in Kiefer (1975). Details of such an investigation for quadratic regression on the q-simplex and q-cube were given in Galil and Kiefer (1975a, b). Other results mentioned briefly in the earlier paper, for quadratic regression on the q-ball of radius R, denoted $B_q(R)$, are described in the present work; in Part II of this work, we treat cubic regression as well as results for extrapolation problems.

For brevity, we omit repetition of discussion of this approach, and do not take the space to record here all computations parallel to those of the simplex paper; thus, the Box–Draper criterion that was treated in detail earlier is not considered here, but we add consideration (Section 6) of the effect of implementing a design allocation optimum for one R on a ball of different radius that represents the actual experimental setting.

The optimality computations of this paper were made under the restriction to rotatable designs. As described in Section 2, the designs obtained are consequently not optimum among *all* designs when $q>1$, except in the case of D-optimality. Nevertheless, it has seemed worthwhile to record here our comparisons under the restriction to rotatable designs, for a number of reasons. Firstly, discussions with several practitioners indicated a continued insistence on the attractiveness of

*Research under NSF Grant MPS 72-04998 A02.

rotatability as a primary criterion over our various optimality criteria with respect to which rotatable designs are not optimum. (A strict believer in the practical usefulness of a single simple optimality criterion for design selection might therefore infer that such practitioners are at least subconsciously using something like D- or G-optimality or a special type of L-optimality described in Section 2; however, the other considerations of such writers as Box seem more relevant to these experimenters.) Secondly, a comparison of the performance of these rotatable designs with the optimum non-rotatable ones in a few cases (not recorded here) indicates that the loss in efficiency is often fairly small, and that some of the rotatable designs may even have preferable properties of robustness under variation of criterion. Finally, as in the other papers in this series, our aim has been to illustrate in simple examples the type of design comparison we believe to be fruitful in more complex settings where much more computation is needed; and, particularly in the case of cubic regression, the computation of optimum non-rotatable designs entails more computation than we believed warranted by this purpose, especially in view of the greater importance of exact over approximate design considerations for large parameter sets.

Under this restriction to rotatable designs, for most criteria the ball of course turns out to be much more tractable algebraically than the most common other design spaces (treated in other papers in this series), the simplex and cube, notable exceptions occurring in the simple rational solutions for D- and E-optimum designs, respectively, in these last two settings.

Consequently, many results can be obtained algebraically, and this development is carried out in Section 3 after the introduction of basic notation in Section 2. The structure of optimum designs for various criteria, obtained algebraically or numerically, is treated in Section 4. In Section 5 we discuss the efficiency of particular designs as the criterion is varied, an investigation of robustness under change in the loss structure. The results of Section 6 can also be interpreted in this vein.

The present class of design settings, under the restriction to rotatability, is admittedly a simple one, perhaps the simplest in which there occur nontrivial variations in design structure and efficiency. For this reason, it is feasible to present a more succinct summary of such variations than would be possible in more complex problems. The results are in some respects typical of what occurs in many more complex settings; in other respects, it has been found that noticeable differences appear when the region of interest is changed from a ball to a cube or simplex.

2. Preliminaries

Our considerations in this paper are limited to the approximate theory, so as to exhibit the comparisons without the obscuring arithmetical complexity and longer design lists of exact design theory as sample size varies. Comparisons of exact designs exhibit similar phenomena, even for small sample sizes.

A *design setting* is a pair $(\mathscr{X}, \boldsymbol{f})$ where $\boldsymbol{f}$ is a known (column) k-vector of continuous functions $f_1, \ldots, f_k$ on the compact space $\mathscr{X}$. A *design* ξ is a probability measure on $\mathscr{X}$. If the measure ξ is discrete, $\xi(\boldsymbol{x})$ represents the proportion of observations to be taken at the point $\boldsymbol{x}$ of $\mathscr{X}$. (The reader may consult, e.g., Kiefer (1959, 1974) or Fedorov (1969) for interpretation of these standard notions of optimum design theory.) The observations are uncorrelated with variance σ^2. The expected value of an observation at $\boldsymbol{x}$ is $\boldsymbol{\theta}'\boldsymbol{f}(\boldsymbol{x})$ where $\boldsymbol{\theta}$ is a column vector of k unknown real parameters. The $k \times k$ *information matrix* (per observation for unit variance) of ξ is defined as

$$\boldsymbol{M}(\xi) = \int_{\mathscr{X}} \boldsymbol{f}(\boldsymbol{x})\, \boldsymbol{f}(\boldsymbol{x})'\xi(\mathrm{d}\boldsymbol{x})$$

and the corresponding normalized variance function for nonsingular $\boldsymbol{M}(\xi)$ is

$$d(\boldsymbol{x}, \xi) = \boldsymbol{f}(\boldsymbol{x})'\boldsymbol{M}^{-1}(\xi)\boldsymbol{f}(\boldsymbol{x}).$$

Thus, when ξ is used with N observations, $N^{-1}\sigma^2\boldsymbol{M}^{-1}(\xi)$ is the covariance matrix of best linear unbiased estimators of $\boldsymbol{\theta}$, and $N^{-1}\sigma^2 d(\boldsymbol{x}, \xi)$ is the variance at $\boldsymbol{x}$ of the least-squares fitted regression function.

The optimality functionals considered herein are infinite for singular $\boldsymbol{M}(\xi)$, and we therefore limit consideration to nonsingular $\boldsymbol{M}(\xi)$ in the sequel. Denote the eigenvalues of $\boldsymbol{M}(\xi)$ by $\lambda_1(\xi), \ldots, \lambda_k(\xi)$. We define

$$\begin{aligned}
\Phi_p(\xi) &= (k^{-1}\operatorname{tr}\boldsymbol{M}^{-p}(\xi))^{1/p} = \left(k^{-1}\sum_1^k \lambda_i^{-p}(\xi)\right)^{1/p} \quad \text{for } 0 < p < \infty; \\
\Phi_0(\xi) &= \lim_{p \downarrow 0} \Phi_p(\xi) = (\det \boldsymbol{M}^{-1}(\xi))^{1/k}; \\
\Phi_\infty(\xi) &= \lim_{p \to +\infty} \Phi_p(\xi) = \max \lambda_i^{-1}(\xi).
\end{aligned} \tag{2.1}$$

It is easily seen that the criteria Φ_0, Φ_1, and Φ_∞ are the familiar D-, A-, and E-optimality criteria, here normalized so as to make comparisons easy: all Φ_p measure loss in the same scale per unit of variance, and take on the value c^{-1} if $\boldsymbol{M}(\xi) = cI_k$. A Φ_p-*optimum rotatable design* $\xi^{(p)}$ is one that minimizes $\Phi_p(\xi)$ among all rotatable designs ξ. (Throughout this paper we shall not exhibit the dependence of $\xi^{(p)}$ on q and R unless, as at the ends of Sections 2 and 6, it is required for lack of ambiguity.)

A measure of the performance of a rotatable design ξ' with respect to the family $\{\Phi_p\}$ of criteria is the (multiplicative regret, among rotatable designs) *absolute efficiency ratio*

$$e_p(\xi') = \Phi_p(\xi')/\Phi_p(\xi^{(p)}), \qquad 0 \leqq p \leqq \infty. \tag{2.2}$$

A design ξ' for which these ratios are all near 1 may be judged adequate for many purposes in applications. We shall concentrate herein on the efficiency ratio (2.2) for $\xi' =$ various $\xi^{(r)}$'s; especially for $r = 0, 1$ or ∞. The finer comparison between two designs ξ' and ξ'' obtained by contrasting the entries of $\boldsymbol{M}(\xi')$ and $\boldsymbol{M}(\xi'')$ in more detail, does not seem to warrant the much greater effort required, at least in the present setting.

Another common criterion, that of G-optimality, is

$$\bar{d}(\xi) = \max_{\boldsymbol{x} \in \mathscr{X}} d(\boldsymbol{x}, \xi).$$

The original "equivalence theorem" (Kiefer and Wolfowitz (1960)) asserts that $\bar{d}$- and Φ_0-optimality among all designs are equivalent, and that $\bar{d}(\xi^{(0)})=k$, a useful fact in computing $\xi^{(0)}$. (The parallel fact for $\xi^{(p)}$ is phrased in terms of a criterion that does not share the intrinsic appeal of d. See Kiefer (1974).) In the present paper we shall for brevity omit consideration of the analogue of (2.2) when Φ_p is replaced by $\bar{d}$, since this analogue, $k^{-1}\bar{d}(\xi')$, is known (Kiefer (1960)) to be near 1 if $\Phi_0(\xi')/\Phi_0(\xi^{(0)})$ is; thus, we do not obtain much added information from the former unless the latter ratio is markedly different from 1.

The settings we consider are those of quadratic or cubic regression for experiments in which the controlable variable is restricted to the q-dimensional ball of given radius R. We denote this set $\mathscr{X}$ by

$$B_q(R)=\left\{(x_1,x_2,\ldots,x_q):\sum_1^q x_i^2\leqq R^2\right\}.$$

A convenient basis for the polynomials on $\mathscr{X}$ of degree $\leqq 2$ is the set of functions

$$1;x_1^2,x_2^2,\ldots,x_q^2;x_1,x_2,\ldots,x_q;x_1x_2,\ldots,x_1x_q,x_2x_3,\ldots,x_{q-1}x_q. \tag{2.3}$$

In the case of quadratic regression we take $\boldsymbol{f}$ to consist of these $(q+1)(q+2)/2$ $(=k)$ functions, written in this order (for convenience in (2.5)), and separated into subvectors of orders $1, q, q$, and $q(q-1)/2$ by the semicolons.

We now discuss the restriction to rotatable designs, those for which $d(\boldsymbol{x},\xi)=d(\boldsymbol{Q}\boldsymbol{x},\xi)$ for all orthogonal transformations $\boldsymbol{Q}$. The invariance argument of Kiefer (1960) p. 399 shows that, for the G-optimality criterion (hence, for D-optimality) there is a design optimum among *all* designs that is rotatable, and which is supported by the center $\boldsymbol{O}$ of $B_q(R)$ and by its outer surface, the sphere

$$S_q(R)=\left\{\boldsymbol{x}:\sum_1^q x_i^2=R^2\right\};$$

one such design is that which, for an appropriate value α $(0<\alpha<1)$, assigns mass $\xi(\boldsymbol{O})=1-\alpha$ to the center and spreads the remaining mass *uniformly* on $S_q(R)$. This argument depends on the fact that, for an *arbitrary* design ξ, if ξ_Q is the design defined by $\xi_Q(A)=\xi(\boldsymbol{Q}^{-1}A)$ for all Borel sets A, we have $d(\boldsymbol{x},\xi_Q)=d(\boldsymbol{Q}^{-1}\boldsymbol{x},\xi)$; it follows easily that, for fixed ξ, the average ξ' of ξ_Q over the orthogonal group with respect to invariant probability measure on Q has $\bar{d}(\xi')\leqq\bar{d}(\xi)$, and ξ' is rotationally invariant. It is instructive for the present discussion to carry out an invariance argument directly in terms of Φ_0: if $\boldsymbol{C}_Q$ is the $k\times k$ matrix for which $\boldsymbol{f}(\boldsymbol{Q}\boldsymbol{x})=\boldsymbol{C}_Q\boldsymbol{f}(\boldsymbol{x})$, we obtain $\boldsymbol{M}(\xi_Q)=\boldsymbol{C}_Q\boldsymbol{M}(\xi)\boldsymbol{C}_Q'$; hence, if $\bar{\xi}$ minimizes $\Phi_0(\boldsymbol{M}(\xi))$, so does $(\bar{\xi})_Q$ (since $\det\boldsymbol{M}(\xi_Q)=\det\boldsymbol{M}(\xi)\det^2\boldsymbol{C}_Q$), but there is no reason why this should be true in general if Φ_0 is replaced by Φ_p. We remark that an argument similar to that above for G-optimality shows that if H is a rotationally invariant (probability) measure on $B_q(R)$, and if $\boldsymbol{L}=\int\boldsymbol{f}\boldsymbol{f}'\,\mathrm{d}H$, there is an L-optimum design ξ^* for minimizing $\operatorname{tr}\boldsymbol{L}\boldsymbol{M}^{-1}(\xi)=\int d(\boldsymbol{x},\xi)H(\mathrm{d}x)$, such that ξ^* is rotationally invariant. (The analysis in Kiefer (1974, 1975) shows that the Φ_p-optimum design among all designs is still supported by $\boldsymbol{O}$ and a subset of $S_q(R)$, and that one such design is invariant under reflections and permutations of coordinates.)

For the reasons given in Section 1, we compare herein designs that are Φ_p-optimum among rotatable designs, even though these are not optimum among all designs if $q>1$ and $p>0$. A result of Karlin and Studden (1966) implies that, for quadratic regression on $B_q(R)$, every design admissible among rotatable designs is supported by the center $\boldsymbol{O}$ of $B_q(R)$ and by its outer surface, the sphere

$$S_q(R)=\left\{\boldsymbol{x}:\sum_1^q x_i^2=R^2\right\}.$$

For fixed p, q, R, it is not hard to show that all of these admissible rotatable designs with a given value of $\xi(\boldsymbol{O})=1-\alpha$ (say) have the same information matrix. A convenient form of rotatable design to work with in the approximate theory is consequently that which spreads the remaining mass α *uniformly* on $S_q(R)$. Convexity considerations (and a slight additional argument when $p=\infty$) show that, for each p, q, R, a unique α minimizes $\Phi_p(\boldsymbol{M}(\xi))$ among rotatable ξ.

We denote by $\alpha^{(p)}$ the appropriate value of α for the spherically symmetric design of the form just described, that is Φ_p-optimum among such designs. Of course, $\alpha^{(p)}$ depends also on the unexhibited variables q and (except when $p=0$) R. In order to implement such a design, approximately, so as to obtain an exact design for given N that is near optimum among rotatable designs, one must obtain a discrete design that has approximately the same moment matrix $\boldsymbol{M}$, i.e., whose $S_q(R)$-portion is close to that of the uniform distribution, as represented by the appropriate moments of order $\leqq 4$. This implementation is not the subject of the present paper. However, our results apply to it, since all (approximate theory) designs that are Φ_p-optimum among rotatable designs have the same $\boldsymbol{M}$ and, in particular, the same value of $\xi(\boldsymbol{O})$. For computational simplicity we may therefore restrict further consideration to designs of the spherical structure described in the previous paragraph. We simplify notation by writing $\boldsymbol{M}(\alpha)$, $\lambda_i(\alpha)$, $\Phi(\alpha)$, etc.

Fix q and R. We now compute $\boldsymbol{M}(\alpha)$. The uniform probability measure μ on $S_q(R)$ is well known to have moments

$$\int x_1^2\,\mathrm{d}\mu=R^2/q, \qquad \int x_1^4\,\mathrm{d}\mu=3R^4/q(q+2), \qquad \int x_1^2x_2^2\,\mathrm{d}\mu=R^4/q(q+2), \tag{2.4}$$

and one obtains 0 for the corresponding integral of x_1, x_1x_2, $x_1x_2^2$, $x_1x_2x_3$, x_1^3, $x_1x_2x_3^2$, $x_1x_2x_3x_4$, or $x_1x_2^3$. The only nonzero moment of the probability measure δ concentrated at $\boldsymbol{O}$ is the zero-th. Hence, with the ordering and partitioning of (2.3), we obtain

$$\boldsymbol{M}(\alpha)=\boldsymbol{M}((1-\alpha)\delta+\alpha\mu)$$

$$=\begin{pmatrix} 1 & U & 0 & 0 \\ U' & T & 0 & 0 \\ 0 & 0 & [\alpha R^2/q]I_q & 0 \\ 0 & 0 & 0 & [\alpha R^4/q(q+2)]I_{q(q-1)/2} \end{pmatrix}, \tag{2.5}$$

where I_h is the $h \times h$ identity, $U(1 \times q)$ is a vector of elements $\alpha R^2/q$, and $T(q \times q)$ has diagonal elements $3\alpha R^4/q(q+2)$ and off-diagonal elements $\alpha R^4/q(q+2)$.

A simple computation yields, for the eigenvalue of $M(\alpha)$,

$$\begin{aligned}
&\alpha R^2/q \quad \text{of multiplicity } q, \\
&\alpha R^4/q(q+2) \quad \text{of multiplicity } q(q-1)/2, \\
&2\alpha R^4/q(q+2) \quad \text{of multiplicity } q-1, \\
&(R^4\alpha q^{-1}+1)/2 \pm [R^8\alpha^2 q^{-2} + 2R^4\alpha(2\alpha-1)q^{-1}+1]^{1/2}/2, \quad \text{each simple.}
\end{aligned} \tag{2.6}$$

Define, for $0<p<\infty$,

$$\begin{aligned}
c_p \equiv c_p(q,R) = q(R^2/q)^{-p} + [q(q-1)/2][R^4/q(q+2)]^{-p} \\
+ (q-1)[2R^4/q(q+2)]^{-p}.
\end{aligned} \tag{2.7}$$

Then (2.1), (2.6) and (2.7) yield

$$\begin{aligned}
k[\Phi_p(\alpha)]^p &= \sum_k \lambda_i^{-p}(\alpha) \\
&= c_p\alpha^{-p} + \sum_{j=0}^{1} 2^p\{R^4\alpha q^{-1} \\
&\quad + (-1)^j[R^8\alpha^2 q^{-2} + 2R^4\alpha(2\alpha-1)q^{-1}+1]^{1/2}\}^{-p}.
\end{aligned} \tag{2.8}$$

This is the expression that was minimized numerically with respect to α in our computations. The special algebraic solutions obtainable for $p=0, 1, \infty$, or asymptotically as $R \to 0$ or ∞, will be treated in the next section. Those results also served as a convenient check on the numerical computations.

Analogues of (2.3)–(2.8) for cubic regression will be given in Part II of this work.

We conclude this section by describing the meaning of variation in R. The most obvious interpretation is that the setting $\mathscr{X} = B_q(R)$ is being varied for a fixed criterion such as Φ_p. In Section 6 we treat the consequence of using on the ball $B_q(R)$ an allocation on $S_q(R)$ and $\boldsymbol{O}$ that was optimum when $R=1$. It should be understood that this does *not* mean using the optimum design on $S_q(1)$ and studying its performance on $S_q(R)$ for large or small R; this would be in the realm of extrapolation or interpolation criteria, to be treated in Part II.

A second interpretation of variation in R is that we keep $\mathscr{X}$ fixed at $B_q(1)$ but change the optimality criterion from Φ_p to

$$\Phi_{p,R}(\xi) = (k^{-1} \operatorname{tr}[A_R M(\xi) A_R]^p)^{1/p}, \qquad 0<p<\infty \tag{2.9}$$

where

$$A_R = \begin{pmatrix} 1 & 0 & 0 & 0 \\ 0 & R^2 I_q & 0 & 0 \\ 0 & 0 & R I_q & 0 \\ 0 & 0 & 0 & R^2 I_{q(q-1)/2} \end{pmatrix}, \tag{2.10}$$

and with an analogous form for $p=0, \infty$. Thus, variation in R for fixed $\mathscr{X}$ can be thought of as varying the criterion in terms of the seriousness with which the various elements of $M(\xi')$ are weighed. From (2.4) and (2.5) it will be seen that, if ξ is an allocation on $S_q(1)$ and O, and $\bar{\xi}$ is the same allocation on $S_q(R)$ and O, then $\Phi_{p,R}(\xi) = \Phi_p(\bar{\xi})$. Thus, our calculations of variation in R, discussed in Sections 4 and 6 primarily in terms of the interpretation of the previous paragraph, can be reinterpreted in the present context of variation in criterion $\Phi_{p,R}$ for fixed $\mathscr{X}$.

A third and perhaps less natural interpretation is to view $\mathscr{X} = B_q(1)$ and Φ_p as fixed, but to consider a variety of regression functions $A_R f$ where f is given by (2.3). Thus, instead of varying $\mathscr{X}$ in the design setting $(\mathscr{X}, f)$ (as in the first interpretation), we are now varying the f and observing the consequence of using a single design.

3. Special algebraic results

3.1. *The case* $p=0$

From (2.5) or (2.6), we have

$$\begin{aligned} [\Phi_0(\alpha)]^{-k} &= \det M(\alpha) \\ &= R^{2q(q+2)} 2^{q-1} q^{-q(q+3)/2} (q+2)^{-(q+2)(q-1)/2} \alpha^{q(q+3)/2} (1-\alpha). \end{aligned} \tag{3.1}$$

As is well known (Kiefer (1960)), the D-optimum solution, independent of R, is

$$\alpha^{(0)} = q(q+3)/(q+1)(q+2), \tag{3.2}$$

for which

$$\begin{aligned} &\Phi_0(\alpha^{(0)}) \\ &\quad = R^{-4q(q+1)} 2^{-2q/(q+1)(q+2)} (q+1)(q+2)^{2q/(q+1)} (q+3)^{-q(q+3)/(q+1)(q+2)}. \end{aligned} \tag{3.3}$$

3.2. *The case* $p=1$

The expression (2.8) now simplifies to

$$k\Phi_1(\alpha) = c_1/\alpha + 1/(1-\alpha) + qR^{-4}/\alpha(1-\alpha), \tag{3.4}$$

from which the A-optimum design is

$$\alpha^{(1)}=\begin{cases}(c_1-1)^{-1}\{qR^4+c_1-[(qR^{-4}+1)(qR^{-4}+c_1)]^{1/2}\} & \text{if} \quad c_1\neq 1,\\ \frac{1}{2} \quad \text{if} \quad c_1=1. & \end{cases} \tag{3.5}$$

For this design,

$$\Phi_1(\alpha^{(1)})=\{2(2qR^{-4}+c_1+1)+4[(qR^{-4}+1)(qR^{-4}+c_1)]^{1/2}\}/(q+1)(q+2). \tag{3.6}$$

3.3. *The case* $p=\infty$

Denote by $\lambda^*(\alpha)$ the eigenvalue of the last line of (2.6) obtained from choosing the minus sign. We must now find α to maximize the minimum of the $\lambda_i(\alpha)$ of (2.6). Since the second and third lines of (2.6) are absent when $q=1$, that dimension has a separate treatment.

(A) *The dimension* $q=1$. The E-optimum design is

$$\alpha^{(\infty)}=\begin{cases}2/(4+R^4) & \text{if} \quad R^2\leqq 2,\\ (R^2-1)/R^4 & \text{if} \quad R^2\geqq 2;\end{cases} \tag{3.7}$$

the first of these maximizes $\lambda^*(\alpha)$, where that root is $\leqq \alpha R^2/q$ (first line of (2.6)); the second occurs when $\lambda^*(\alpha)=\alpha R^2/q$. For this design,

$$\Phi_\infty(\alpha^{(\infty)})=\begin{cases}1+4R^{-4} & \text{if} \quad R^2\leqq 2,\\ 1+(R^2-1)^{-1} & \text{if} \quad R^2\geqq 2.\end{cases} \tag{3.8}$$

(B) *Dimensions* $q\geqq 2$. The E-optimum design is now

$$\alpha^{(\infty)}=\begin{cases}q(q+1)(q+2)/[(q+1)R^4+q(q+2)^2] & \text{if} \quad R^2\leqq q+2,\\ q(R^2-1)/R^2[R^2+q-1] & \text{if} \quad R^2\geqq q+2.\end{cases} \tag{3.9}$$

The first of these occurs where $\lambda^*(\alpha)=\alpha R^4/q(q+2)$ (second line of (2.6)), while the second occurs where $\lambda^*(\alpha)=\alpha R^2/q$ (first line of (2.6)); in each case, the other λ_i are larger. The corresponding Φ_∞ values are

$$\Phi_\infty(\alpha^{(\infty)})=\begin{cases}1+q(q+2)^2(q+1)^{-1}R^{-4} & \text{if} \quad R^2\leqq q+2\\ 1+q(R^2-1)^{-1} & \text{if} \quad R^2\geqq q+2.\end{cases} \tag{3.10}$$

3.4. *Behavior as* $R\to 0$

For fixed α with $0<\alpha<1$ and fixed p with $0<p<\infty$, as $R\to 0$ the root $\lambda^*(\alpha)$ is approximately $\alpha(1-\alpha)R^4/q$, and the other root of the last line of (2.6) is approximately 1. The expression (2.8) becomes

$$k[\Phi_p(\alpha)]^p=[R^4\alpha/q(q+2)]^{-p}(q-1)[2^{-p}+q/2]+[\alpha(1-\alpha)R^4/q]^{-p} +\mathrm{O}(R^{-2}). \tag{3.11}$$

From this we obtain that $\lim_{R\to 0}\alpha^{(p)}=\bar{\alpha}$ (say) for $0<p<\infty$ is the solution of

$$(1-\bar{\alpha})^{-p-1}(2\bar{\alpha}-1)=(q-1)(q+2)^p[2^{-p}+q/2]. \tag{3.12}$$

This is of course the result for fixed p, and the limit of the solution of (3.12) as $p\to 0$ (namely, $(q^2+q)/(q^2+q+2)$) or as $p\to\infty$ (namely, $q/(q+2)$) does not agree with the correct values of $\lim_{R\to 0}\alpha^{(0)}$ and $\lim_{R\to 0}\alpha^{(\infty)}$ obtained from (3.2) and (3.9) (but does agree with (3.7)):

$$\begin{aligned}&\lim_{R\to 0}\alpha^{(0)}=q(q+3)/(q+1)(q+2),\\ &\lim_{R\to 0}\alpha^{(\infty)}=\begin{cases}\frac{1}{2} & \text{if } q=1,\\ (q+1)/(q+2) & \text{if } q\geqq 2.\end{cases}\end{aligned} \tag{3.13}$$

The solution of (3.12) for $p=1$ agrees with the value obtained by letting $R\to 0$ in (3.5):

$$\lim_{R\to 0}\alpha^{(1)}=\begin{cases}\frac{1}{2} \text{ if } q=1,\\ h^{-1}\{h+1-[h+1]^{1/2}\} \text{ if } q\geqq 2, \text{ where}\\ \qquad h=(q+2)(q+1)(q-1)/2.\end{cases} \tag{3.14}$$

With slight additional effort, an improved approximation can be obtained.

3.5. *Behavior as* $R\to\infty$

The expression (2.8) is now asymptotically $[\lambda^*(\alpha)]^{-p}\approx(1-\alpha)^{-p}$, so we obtain $\lim_{R\to 0}\alpha^{(p)}=0$ for $0<p<\infty$, and for $p=\infty$ this is also in agreement with (3.7) and (3.9) as $R\to\infty$ there; once more, (3.2) gives a different result. It is not difficult to obtain the more useful next order approximation,

$$\alpha^{(p)}\approx\begin{cases}1/[1+q^{-1}R^{2p/(p+1)}] & \text{if } 0<p<\infty,\\ 1/[1+q^{-1}R^2] & \text{if } p=\infty.\end{cases} \tag{3.15}$$

This form is obtained by ignoring the terms of order R^{-4p} in (2.8); the asymptotically equivalent $qR^{-2p/(p+1)}$ is less accurate in the domain studied herein. Additional terms in the approximation can be obtained in obvious fashion.

4. Make up of optimum rotatable designs

Table 1 lists selected values of the optimum $\alpha^{(p)}$ for various q, p, and R. Of course, $\alpha^{(0)}$ does not vary with R.

The column labeled "$\to 0$" lists the limiting values of $\alpha^{(p)}$ as $R\to 0$, obtained from (3.12)–(3.14). These are remarkably close to the values for $R=0.1$, the greatest departures occurring for small p and q. The column labeled "10A" lists the values obtained from the asymptotic formula (3.15) with $R=10$. The agreement with the exact values for $R=10$ is not as close as was that for $R\to 0$ noted just above. However, it is adequate in terms of yielding reasonably efficient designs ξ' in the sense of the ratio

Table 1

Selected values of optimum $\alpha^{(p)}$

q	p	$\to 0$	0.1	0.5	1	2	10	10A
	0	0.667	0.667	0.667	0.667	0.667	0.667	0.667
	0.5	0.500	0.517	0.564	0.555	0.446	0.287	0.177
1	1	0.500	0.501	0.520	0.500	0.351	0.091	0.091
	2	0.500	0.500	0.496	0.449	0.281	0.044	0.044
	∞	0.500	0.499	0.492	0.400	0.188	0.010	0.010
	0	0.833	0.833	0.833	0.833	0.833	0.833	0.833
	0.5	0.734	0.740	0.757	0.752	0.665	0.493	0.301
2	1	0.726	0.726	0.729	0.710	0.564	0.169	0.167
	2	0.720	0.720	0.717	0.677	0.465	0.085	0.085
	∞	0.750	0.750	0.746	0.686	0.300	0.020	0.020
	0	0.900	0.900	0.900	0.900	0.900	0.900	0.900
	0.5	0.834	0.836	0.844	0.841	0.780	0.595	0.393
3	1	0.821	0.821	0.822	0.809	0.695	0.237	0.231
	2	0.808	0.808	0.807	0.782	0.600	0.122	0.122
	∞	0.800	0.800	0.797	0.759	0.432	0.029	0.029
	0	0.933	0.933	0.933	0.933	0.933	0.933	0.933
	0.5	0.885	0.887	0.891	0.889	0.846	0.651	0.463
4	1	0.872	0.871	0.872	0.863	0.779	0.297	0.286
	2	0.857	0.857	0.856	0.840	0.697	0.157	0.157
	∞	0.833	0.833	0.832	0.805	0.536	0.038	0.038
	0	0.952	0.952	0.952	0.952	0.952	0.952	0.952
	0.5	0.916	0.916	0.919	0.917	0.887	0.685	0.519
5	1	0.902	0.902	0.902	0.897	0.833	0.351	0.333
	2	0.887	0.887	0.886	0.876	0.765	0.189	0.188
	∞	0.857	0.857	0.856	0.837	0.616	0.048	0.048
	0	0.964	0.964	0.964	0.964	0.964	0.964	0.964
	0.5	0.935	0.935	0.937	0.936	0.914	0.708	0.564
6	1	0.922	0.922	0.922	0.918	0.870	0.400	0.375
	2	0.908	0.908	0.907	0.899	0.814	0.219	0.218
	∞	0.875	0.875	0.874	0.859	0.677	0.057	0.057
	0	0.972	0.972	0.972	0.972	0.972	0.972	0.972
	0.5	0.948	0.948	0.949	0.949	0.932	0.723	0.601
7	1	0.936	0.937	0.936	0.933	0.896	0.445	0.412
	2	0.923	0.923	0.922	0.917	0.849	0.247	0.245
	∞	0.889	0.889	0.888	0.877	0.725	0.065	0.065
	0	0.978	0.978	0.978	0.978	0.978	0.978	0.978
	0.5	0.957	0.958	0.958	0.958	0.944	0.751	0.633
8	1	0.947	0.947	0.947	0.944	0.915	0.486	0.444
	2	0.934	0.934	0.933	0.929	0.875	0.273	0.271
	∞	0.900	0.900	0.899	0.890	0.763	0.074	0.074

Table 1 (continued)

q	p	$\to 0$	0.1	0.5	1	2	10	10A
	1	0.982	0.982	0.982	0.982	0.982	0.982	0.982
	0.5	0.964	0.965	0.965	0.964	0.954	0.783	0.660
9	1	0.955	0.954	0.954	0.953	0.929	0.523	0.474
	2	0.942	0.942	0.942	0.939	0.894	0.303	0.295
	∞	0.909	0.909	0.909	0.901	0.793	0.082	0.083
	0	0.985	0.985	0.985	0.985	0.985	0.985	0.985
	0.5	0.970	0.970	0.970	0.970	0.961	0.810	0.683
10	1	0.961	0.961	0.961	0.959	0.940	0.558	0.500
	2	0.949	0.949	0.949	0.946	0.909	0.338	0.317
	∞	0.917	0.917	0.916	0.910	0.817	0.091	0.099

(2.2), and is easily improved by adding another term in the asymptotic approximation. (This type of efficiency consideration is related to that of Section 6.)

These asymptotic results thus confirm, in this simple setting, a prescription given earlier in Kiefer (1975) and Galil and Kiefer (1975a) and which, it is hoped, may simplify optimality calculations in more complex settings: when a family of similar design problems can be parametrized in terms of a parameter for which optimality calculations become simpler as some limiting parameter value is approached, one can often invoke these simpler limiting computations to obtain adequately efficient designs for a wide range of parameter values. In those earlier papers the parameter value treated in this fashion was *dimension* ($q \to \infty$), while here we have considered variation in R in order to obtain an illustration of a different character; a variation of *degree* of regression ($\to \infty$) is studied in Kiefer and Studden (1976).

5. Efficiency ratios under variation of criterion

The absolute efficiency ratio $e_p(\xi')$ defined in (2.2) was computed for ξ' equal to the Φ_r-optimum rotatable design, for various values of r (and of p, q, and R). The maximum over p of $e_p(\xi')$ was computed as an indication of the overall attractiveness of ξ' when its operation is considered under various criteria; we emphasize that a more complete operating characteristic of competing designs ξ' (such as (2.2)) than that given by such a single number should be examined when choosing a design. Also, we remind the reader that these computations compare only rotatable designs. Table 2 lists selected results for $\xi' = \xi^{(0)}$, $\xi^{(1)}$, $\xi^{(\infty)}$, the designs obtained from the D-, A-, and E-optimality criteria most often employed in practice.

The maximum over p of $e_p(r)$ was always attained at (or very near) 0 for $\xi^{(\infty)}$ and $+\infty$ for $\xi^{(0)}$ and $\xi^{(1)}$, except in the cases of $\xi^{(1)}$ marked by an asterisk, where it was attained at 0.

The results of main interest are that, among these three designs, $\max_p e_p(\xi')$ is smallest for $\xi' = \xi^{(\infty)}$ if R is small, in agreement with the pattern found for the simplex in Kiefer (1975) and Galil and Kiefer (1975a); but that, for large R (how large depending upon q), the design $\xi^{(1)}$ is much superior to the other two in terms of this criterion. Thus, the

Table 2

Selected values of $\max_{0 \le p \le \infty} e_p(\alpha^{(r)})$

q	r	R = 0.1	0.5	1	2	10
1	0	1.13	1.14	1.37	2.39	2.97
	1	1.06*	1.04*	1.06*	1.24	2.70*
	∞	1.06	1.06	1.16	1.73	11.33
2	0	1.35	1.37	1.69	4.06	5.88
	1	1.03	1.03*	1.06	1.56	2.89*
	∞	1.02	1.02	1.06	1.84	16.77
3	0	1.77	1.80	2.15	5.47	9.69
	1	1.09	1.10	1.21	1.82	2.72*
	∞	1.04	1.04	1.07	1.63	17.40
4	0	2.23	2.26	2.63	6.71	14.42
	1	1.24	1.26	1.36	1.73	2.49*
	∞	1.05	1.05	1.07	1.47	16.36
5	0	2.71	2.74	3.13	7.81	20.07
	1	1.39	1.40	1.50	2.24	2.28*
	∞	1.05	1.05	1.07	1.37	14.96
6	0	3.17	3.20	3.60	8.71	26.38
	1	1.53	1.54	1.63	2.42	2.11*
	∞	1.05	1.05	1.06	1.30	13.67
7	0	3.66	3.69	4.10	9.57	33.69
	1	1.66	1.67	1.76	2.58	1.97*
	∞	1.05	1.05	1.06	1.25	12.47
8	0	4.14	4.17	4.59	10.32	41.66
	1	1.79	1.80	1.88	2.72	1.85*
	∞	1.05	1.05	1.06	1.21	11.48
9	0	4.59	4.62	5.03	10.93	49.97
	1	1.90	1.91	2.00	2.85	1.92
	∞	1.05	1.05	1.06	1.18	10.55
10	0	5.14	5.17	5.60	11.76	60.30
	1	2.02	2.03	2.12	2.97	2.06
	∞	1.05	1.05	1.05	1.16	9.82

variation in efficiency of a design as setting and criterion change depends on details of the dependence on these variables of the λ_i, and it may require calculations as well as insight to guess the results. In the present instance, it is perhaps not obvious why the results for $R = 0.1$ and $R = 10$ are so different, and one must examine the way in which R enters into Φ_p to understand the results.

Slight variations from $\xi' = \xi^{(1)}$ or $\xi^{(\infty)}$ will sometimes reduce $\max_p e_p(\xi')$ slightly (but minimizing this quantity is not recommended as particularly natural). For example, for $R = 10$ the choice of r that minimizes $\max_p e_p(\xi^{(r)})$ is somewhat < 1 for $q \leqq 3$, slightly < 1 for $4 \leqq q \leqq 8$, and slightly > 1 for $q = 9$, 10; presumably some ξ' of a different form than $\xi^{(r)}$ might do better in this respect.

Theoretical bounds on efficiency ratios are derived in Galil and Kiefer (1975a). Although these are "sharp" when one considers all design problems, they are not very close in the present setting.

6. Efficiency ratio under variation of R

We have mentioned in Section 4 the possibility of using the asymptotic prescription as $R \to 0$ or ∞ for constructing designs when R is small or large. In the present section

Table 3

Selected efficiency ratios if the "$R = 1$" allocation is used on $S_q(R')$ and **0** when $\mathscr{X} = B_q(R')$

q	p	0.1	0.5	2	10
1	1	1.00	1.00	1.09	1.67
	∞	1.04	1.04	1.34	1.65
2	1	1.00	1.00	1.10	2.42
	∞	1.09	1.09	2.16	3.12
3	1	1.00	1.00	1.08	3.12
	∞	1.05	1.05	2.29	4.04
4	1	1.00	1.00	1.06	3.72
	∞	1.03	1.03	2.31	4.94
5	1	1.00	1.00	1.04	4.21
	∞	1.02	1.02	2.28	5.82
6	1	1.00	1.00	1.03	4.57
	∞	1.02	1.02	2.23	6.69
7	1	1.00	1.00	1.02	4.84
	∞	1.01	1.01	2.17	7.56
8	1	1.00	1.00	1.02	5.01
	∞	1.01	1.01	2.11	8.40
9	1	1.00	1.00	1.01	5.08
	∞	1.01	1.01	2.05	9.24
10	1	1.00	1.00	1.01	5.12
	∞	1.01	1.01	1.99	10.07

we treat the consequence of using the value of α optimum when $R=1$, on the ball of radius R' when the latter is the actual $\mathscr{X}$. The reader may refer to the end of Section 2 for the meaning of such a use; in the first interpretation given there, we suppose that a design allocation optimum for one $\mathscr{X}$, suitably transformed to conform to the geometry of another $\mathscr{X}$, is employed in the latter setting. Thus, if $\xi^{(p)}$ is the Φ_p-optimum design for $\mathscr{X}=B_q(R')$ and if ξ' is the design that spreads the weight $\alpha^{(p)}$ *that was optimum for* $R=1$ on $S_q(R')$ (and puts its complement at 0), we compute the ratio (2.2) for the quadratic model with $\mathscr{X}=B_q(R')$.

For $p=0$, $\alpha^{(p)}$ is independent of R, and the resulting efficiency ratio is thus 1. For $p=1$ and ∞, selected results are tabled in Table 3. Excepting one instance ($R=10, q=1$), the inefficiency is worse for the E-optimality criterion.

If one adopts the first or third interpretation of variation in R mentioned at the end of Section 2, the above consideration of avoiding calculation or tabling of designs by using one prescription for many settings differs in the nature from that of Section 5, where for a fixed $\mathscr{X}$ we studied an aspect of the operating characteristic of different designs. However, the latter was in the same spirit as the second interpretation of variation in R given at the end of Section 2, in that it considered variation in p of the criterion Φ_p $=\Phi_{p,1}$ of (2.9) and we are now considering variation in R' of $\Phi_{p,R'}$, again for fixed $\mathscr{X}$ $=B_q(1)$. The recording of the more complicated variation of both p and R in $\Phi_{p,R}$ requires more space than is warranted here; again, the design $\xi^{(1)}$ appears to perform fairly well under such wide variation of criterion.

Added in proof

We recently became aware of extensive optimum design calculations carried out independently by Dr. Leon Pesotchinsky and which, it is hoped, will be published in the future. These have overlap with the present paper, and include optimum designs without the rotability restriction, described briefly in Section 2.

References

Fedorov, V.V. (1969). *Theory of Optimal Experiments*. Academic Press, New York. [Transl. 1972.]

Galil, Z. and J. Kiefer (1975a). Comparison of quadratic simplex designs. *Technometrics*, to appear.

Galil, Z. and J. Kiefer (1975b). Comparison of designs for quadratic regression on cubes. *J. Statist. Planning Inf.*, to appear.

Karlin, S. and W.J. Studden (1966). *Tchebycheff Systems*. Interscience, New York.

Kiefer, J. (1959). Optimum experimental designs. *J. Roy. Statist. Soc. Ser. B* 21, 272–319.

Kiefer, J. (1960). Optimum experimental designs V, with applications to systematic and rotatable designs. *Proceedings of the 4th. Berkely Symposium, I*, pp. 381–405.

Kiefer, J. (1974). General equivalence theory for optimum designs (approximate theory). *Ann. Math. Statist.* 75, 849–879.

Kiefer, J. (1975). Optimal design: Variation in structure and performance under change of criterion. *Biometrika* 62, 277–288.

Kiefer, J. and W. Studden (1976). Optimum designs for large degree polynomial regression. *Ann. Math. Statist.*

Kiefer, J. and J. Wolfowitz (1960). The equivalence of two extremum problems, *Canad. J. Math.* 12, 363–366.

Journal of Statistical Planning and Inference 1 (1977) 121–132.

COMPARISON OF DESIGN FOR QUADRATIC REGRESSION ON CUBES*

Z. GALIL and J. KIEFER
Cornell University, Ithaca, NY, U.S.A.

Designs for quadratic regression are considered when the possible choices of the controllable variable are points $x=(x_1, x_2, \ldots, x_q)$ in the q-dimensional cube of side 2. The designs that are optimum with respect to such criteria as those of D-, A-, and E-optimality are compared in their performance relative to these and other criteria. Some of the results are developed algebraically; others, numerically. The possible supports of E-optimum designs are much more numerous than the D-optimum supports characterized earlier. The A-optimum design appears to be fairly robust in its efficiency, under variation of criterion.

AMS Subject Classification: 62K05.
Key words and phrases:
Optimum designs, regression on cubes, D-optimality, A-optimality, E-optimality, robust designs, quadratic regression

1. Introduction

The rationale for studying the performance under various optimality criteria of a design selected according to a particular criterion, and a summary of some of the results in the present sequence of papers, were given in Kiefer (1975). Further details of that investigation, for quadratic regression on the q-simplex and ball, are given in Galil and Kiefer (1977). Corresponding results for quadratic regression on the q-cube

$$C_q = \{(x_1, \ldots, x_q) : |x_i| \leqq 1, \forall i\}$$

are described in the present work. For brevity, we omit repetition of discussion of this approach, or even of the by-now-familiar interpretation of design theory notions, and do not take the space to record here all computations parallel to those of the earlier work. For example, the variation of optimum design with the width of the experimental region (considered for the ball) and the Box–Draper criterion (considered for the simplex) are not treated here; but we list all eigenvalues of the competing designs to give an illustration of other operating characteristic reductions than the family $\{\Phi_p\}$ of functionals considered in the earlier work (and also recorded herein). Also, considerable space is spent on the wide choice possible for optimum design supports, especially for the E-optimality criterion.

The cube turns out to be fairly tractable computationally because the information matrices of symmetric designs depend only on two simple linear functions of the weights, as described by Kiefer (1960). What is perhaps surprising is that the E-

*Research under NSF Grant MCS 75-22481.

optimality developments are so much simpler algebraically than those for the simplex or even the ball. These special algebraic results are discussed in Section 2, after basic notation and computations are introduced in the remainder of the present section. Optimum designs for various criteria are given in Section 3, where we also discuss the efficiency of particular designs as the criterion is varied, an investigation of robustness under change in the loss structure.

Although the present class of design settings is a simple one, just as were the simplex and ball also considered in this sequence of papers, it is consequently more feasible to present a succinct summary of such performance variations than would be possible in more complex problems. The results are in some respects typical of what occurs in many more complex settings; in other respects, it has been found that noticeable differences appear when the region of interest is changed from a cube to a ball or simplex.

The considerations of this paper are limited to the approximate theory, in order to exhibit the comparisons without the obscuring arithmetical complexity and longer design lists of exact design theory as sample size varies. The comparison of exact designs exhibits similar phenomena, even for small sample sizes.

A *design setting* is a pair $(\mathscr{X}, f)$ where f is a known (column) k-vector of continuous functions $f_1, \ldots, f_k$ on the compact space $\mathscr{X}$. A *design* ξ is a probability measure on $\mathscr{X}$. If the measure ξ is discrete, $\xi(x)$ represents the proportion of observations to be taken at the point x of $\mathscr{X}$. The observations are uncorrelated with variance σ^2. The expected value of an observation at x is $\theta' \boldsymbol{f}(x)$ where θ is a column vector of k unknown real parameters. The $k \times k$ *information matrix* (per observation for unit variance) of ξ is defined as $M(\xi) = \int_{\mathscr{X}} \boldsymbol{f}(x) \boldsymbol{f}(x)' \xi(dx)$, and the corresponding normalized variance function for nonsingular $M(\xi)$ is $d(x, \xi) = \boldsymbol{f}(x)' M^{-1}(\xi) \boldsymbol{f}(x)$. Thus, when ξ is used with N observations, $N^{-1}\sigma^2 M^{-1}(\xi)$ is the covariance matrix of best linear unbiased estimators of θ, and $N^{-1}\sigma^2 d(x, \xi)$ is the variance at x of the least-squares fitted regression function.

The optimality functionals considered herein are infinite for singular $M(\xi)$, and we therefore limit consideration to nonsingular $M(\xi)$ in the sequel. Denote the eigenvalues of $M(\xi)$ by $\lambda_1(\xi), \ldots, \lambda_k(\xi)$. We define

$$
\begin{aligned}
\Phi_p(\xi) &= (k^{-1} \operatorname{tr} M^{-p}(\xi))^{1/p} \\
&= \left(k^{-1} \sum_1^k \lambda_i^{-p}(\xi) \right)^{1/p} \quad \text{for } 0 < p < \infty; \qquad (1.1) \\
\Phi_0(\xi) &= \lim_{p \downarrow 0} \Phi_p(\xi) = (\det M^{-1}(\xi))^{1/k}; \\
\Phi_\infty(\xi) &= \lim_{p \to +\infty} \Phi_p(\xi) = \max_i \lambda_i^{-1}(\xi).
\end{aligned}
$$

Thus, Φ_0, Φ_1, and Φ_∞ are the familiar D-, A-, and E-optimality criteria, here normalized so as to make comparisons easy: all Φ_p measure loss in the same scale per unit of variance, and take on the value c^{-1} if $M(\xi) = cI_k$. A Φ_p*-optimum design* $\xi^{(p)}$ is one that minimizes $\Phi_p(\xi)$. (Throughout this paper we shall not exhibit the dependence of $\xi^{(p)}$ or related quantities on q unless it is required for lack of ambiguity.)

A measure of the performance of a design ξ' with respect to the family $\{\Phi_p\}$ of criteria is the (multiplicative regret) *absolute efficiency ratio*

$$e_p(\xi')=\Phi_p(\xi')/\Phi_p(\xi^{(p)}), \quad 0\leqq p\leqq\infty. \tag{1.2}$$

A design ξ' for which these ratios are all near 1 may be judged adequate for many purposes in applications. We shall concentrate herein on the efficiency ratio (1.2) for ξ' = various $\xi^{(r)}$'s, especially for $r=0, 1$, or ∞. The finer comparison between two designs ξ' and ξ'' obtained by contrasting the entries of $M(\xi')$ and $M(\xi'')$ in more detail, does not seem to warrant the much greater effort required, at least in the present setting. However, we shall list the eigenvalues of Φ_p-optimum designs in Section 3 as a possible additional reduced operating characteristic that can be consulted by the design selector who does not believe in any particular Φ_p as the single criterion for design choice.

Another common criterion, that of G-optimality, is $\bar{d}(\xi)=\max_{x\in\mathscr{X}} d(x,\xi)$. The original "equivalence theorem" (Kiefer and Wolfowitz (1960)) asserts that $\bar{d}$- and Φ_0-optimality are equivalent, and that $\bar{d}(\xi^{(0)})=k$, a useful fact in computing $\xi^{(0)}$. (The parallel fact for $\xi^{(p)}$ is phrased in terms of a criterion that does not share the intrinsic appeal of $\bar{d}$. See Kiefer (1974).) For brevity, in the present paper we omit consideration of the analogue of (1.2) when Φ_p is replaced by $\bar{d}$, since this analogue, $k^{-1}\bar{d}(\xi')$, is known (Kiefer (1960)) to be near 1 if $\Phi_0(\xi')/\Phi_0(\xi^{(0)})$ is; thus, we do not obtain much added information from the former unless the latter ratio is markedly different from 1.

The settings we consider are those of quadratic regression for experiments in which the space $\mathscr{X}$ of the controllable variable is the q-dimensional cube C_q defined earlier. A convenient basis for the polynomials on $\mathscr{X}$ of degree $\leqq 2$ is the set of functions

$$1;\quad x_1^2, x_2^2,\ldots,x_q^2;\quad x_1,x_2,\ldots,x_q;\quad x_1x_2,\ldots,x_1x_q,x_2x_3,\ldots,x_{q-1}x_q. \tag{1.3}$$

We take f to consist of these $(q+1)(q+2)/2$ $(=k)$ functions, written in this order (for convenience in writing (1.7)), and separated into subvectors of orders $1, q, q$, and $q(q-1)/2$ by the semicolons.

Much of our discussion is phrased in terms of certain simple subsets of C_q, which we now define. A *barycenter of depth* j $(0\leqq j\leqq q)$ is a point with j coordinates equal to 0 and the remaining coordinates equal to ± 1. We denote the set of $\binom{q}{j}2^{q-j}$ barycenters of depth j by J_j, and write $J=\bigcup_j J_j$. Thus, J_0 consists of the vertices of C_q, while J_1 consists of midpoints of edges, J_2 consists of the centers of 2-dimensional cubical faces, J_q is the single center point, etc.

It was shown by Kiefer (1960, 1974) and by Farrell, Kiefer and Walbran (1965) that *every* Φ_p-optimum design is supported by a subset of J, and that for every p there is an optimum $\xi^{(p)}$ that is symmetric under all permutations of coordinates and multiplication of any x_i by -1. Any such symmetric design, whether or not it is optimum, can be described in terms of a probability $(q+1)$-vector $\boldsymbol{\alpha}=(\alpha_0,\alpha_1,\ldots,\alpha_q)$ that assigns measure $\alpha_j/\binom{q}{j}2^{q-j}$ to each point of J_j. For such designs we write $M(\boldsymbol{\alpha})$, $\lambda_i(\boldsymbol{\alpha})$, etc. We denote by $\xi_{\boldsymbol{\alpha}}$ the probability measure on C_q induced by $\boldsymbol{\alpha}$.

(This conclusion concerning the 3^q-array J means that the results herein can also be interpreted as conclusions for the design problem when $\mathscr{X}=J$, the factorial setting with 3 levels for each of q factors.)

While (for some p and q) there can be asymmetric $\xi^{(p)}$, it can be shown that, for each p and q, all $M(\xi^{(p)})$ are identical. Thus, *we hereafter limit consideration to symmetric designs $\xi^{(p)}$ on J, of the form described just above.*

For designs of this structure, the moments of order $\leqq 4$ that appear in $M(\xi)$ for general ξ are linearly dependent and may be expressed in terms of a constant and two linearly independent moments (labeled u, v, below). There may be several optimum $\alpha^{(p)}$ for some p, q; thus, it is convenient first to find the optimum $(U^{(p)}, V^{(p)})$, which is unique, and then to see what $\alpha^{(p)}$'s yield those entries. We note that the design $\boldsymbol{\alpha}$ with $\alpha_j = 1$ has values $(u(\boldsymbol{\alpha}), v(\boldsymbol{\alpha}))$ given by

$$(u_j, v_j) = ((q-j)/q, (q-j)(q-j-1)/q(q-1)), \quad 0 \leqq j \leqq q. \tag{1.4}$$

Hence, if $(U^{(p)}, V^{(p)})$ minimizes $\Phi_p(M(\alpha))$, we see that a symmetric $\boldsymbol{\alpha}$ is Φ_p-optimum if and only if

$$\sum_j \alpha_j (u_j, v_j) = (U^{(p)}, V^{(p)}). \tag{1.5}$$

This leads to the characterization of the simplest $\boldsymbol{\alpha}^{(p)}$'s as ones for which only three α_j's are positive, any such design corresponding to a triangle containing $(U^{(p)}, V^{(p)})$ which has the relevant three points (u_j, v_j) as vertices. (For rare p there may be an $\boldsymbol{\alpha}^{(p)}$ with only two α_j positive; such a p is ∞, for many q, as discussed in Section 2.) A main feature of our work is the study of these sets $L^{(p)} = L_q^{(p)}$ of minimal support of optimum symmetric $\boldsymbol{\alpha}^{(p)}$'s. In this respect, the situation for the cube is considerably more complicated than for the simplex, since the former but not the latter already exhibits a variety of possible supports even when $p=0$. Moreover, for larger p (in particular, $p=\infty$) the variety in structure of the different $L_q^{(p)}$ is far greater than for the simplex and also for the cube when $p=0$, obtained by Farrell et al. (1965). Where there is such a wide choice of optimum designs, of course a secondary criterion of interest can sensibly be used to choose among them; we shall not pursue this possibility here.

Fixing q, we now compute $M(\boldsymbol{\alpha})$. We define

$$\begin{aligned} u(\boldsymbol{\alpha}) &= \int_{C_q} x_1^2 \xi_{\boldsymbol{\alpha}}(dx) = \int_{C_q} x_1^4 \xi_{\boldsymbol{\alpha}}(dx) = 1 - \sum_{j=0}^{q} \alpha_j j/q, \\ v(\boldsymbol{\alpha}) &= \int_{C_q} x_1^2 x_2^2 \xi_{\boldsymbol{\alpha}}(dx) = \sum_{j=0}^{q} \alpha_j (q-j)(q-j-1)/q(q-1). \end{aligned} \tag{1.6}$$

A simple computation yields

$$M(\boldsymbol{\alpha}) = \begin{pmatrix} 1 & h & 0 & 0 \\ h' & H & 0 & 0 \\ 0 & 0 & uI_q & 0 \\ 0 & 0 & 0 & vI_{q(q-1)/2} \end{pmatrix}, \tag{1.7}$$

where I_r is the identity matrix of order r, h is the row q-vector with all entries u, and H has diagonal entries u and off-diagonal entries v. The eigenvalues of M are then found to be

$$\begin{aligned}
&u, \quad \text{of multiplicity } q;\\
&v, \quad \text{of multiplicity } q(q-1)/2;\\
&u-v, \quad \text{of multiplicity } q-1;\\
&\tfrac{1}{2}[u+(q-1)v+1]\pm\tfrac{1}{2}\{[u+(q-1)v+1]^2-4[u+(q-1)v-qu^2]\}^{1/2}, \quad \text{each of multiplicity } 1.
\end{aligned} \tag{1.8}$$

(These formulas are also valid when $q=1$.) The functionals of (1.1) are easily computed from (1.8); of course, the result for $p=0$ was known from earlier work, cited below.

2. Special values

For $p=0, 1, \infty$, further algebra can be used to simplify the computation of optimum $\alpha^{(p)}$'s. For $p=0$, this was carried out by Kiefer (1960), Kono (1962), and Farrell et al. (1967), by differentiating det M with respect to u and v to obtain the solution

$$\begin{aligned}
U^{(0)} &= \frac{(q+3)}{4(q+1)(q+2)^2}\{(2q^2+3q+7)+(q-1)[4q^2+12q+17]^{1/2}\},\\
V^{(0)} &= \frac{(q+3)}{8(q+2)^3(q+1)}\{(4q^3+8q^2+11q-5)+(2q^2+q+3)\\
&\qquad [4q^2+12q+17]^{1/2}\}.
\end{aligned} \tag{2.1}$$

This yielded, in the manner outlined in Section 1, the existence of an optimum $\alpha^{(0)}$ with only three positive α_j, say for $j\in L_q^{(0)}$, if and only if

$$\begin{aligned}
&L_q^{(0)}=\{0,1,j\} \quad \text{for any } j \text{ with } 2\leqq j\leqq 5, \quad \text{if } 2\leqq q\leqq 5;\\
&L_q^{(0)}=\{0,1 \text{ or } 2,j\} \quad \text{for any } j \text{ with } 3\leqq j\leqq q, \quad \text{if } q\geqq 6.
\end{aligned} \tag{2.2}$$

For p a positive integer, U_q and V_q cannot be expressed in such a simple algebraic form as (2.1). Thus, in the simplest case $p=1$, differentiation of Φ_1 yields a pair of 6th degree equations which must be solved numerically. For any p and q, analogues of (2.2) can be obtained once the minimizing $U^{(p)}$, $V^{(p)}$ (or, equivalently, any optimum $\alpha^{(p)}$) are found, as described just below (1.5).

The case $p=\infty$ is perhaps the most interesting, since it yields the simplest solution of all. (This is in contrast to what occurs when $\mathscr{X}$ is the simplex, for which setting the E-optimum designs required the greatest amount of computing time!) Write $t=u-v$, and define $g(t,v)$ to be the smaller root of the last line of (1.8), expressed in terms of t and v.

B

Equating the eigenvalues t and $g(t,v)$, we obtain

$$t=\tfrac{1}{2}\{-3v+[4v+5v^2]^{1/2}\},$$

and this last has its maximum at $v=\frac{1}{5}$, where $u=\frac{2}{5}$. Thus, all eigenvalues of (1.8) are seen to be $\geqq\frac{1}{5}$ at

$$(U^{(\infty)},V^{(\infty)})=(\tfrac{2}{5},\tfrac{1}{5}), \tag{2.3}$$

and that this is indeed the required solution follows from the fact that the region $g(t,v)\geqq\frac{1}{5}$ contains $\{(t,v):t\geqq\frac{1}{5},v\geqq\frac{1}{5}\}$. (The larger root of the last line of (1.8) is $1+(q/5)$, at (2.3).) The simple solution (2.3) was illustrated in Kiefer (1974).

The striking result is that the E-optimum structure, in terms of $(U^{(\infty)},V^{(\infty)})$, is independent of q. Of course, the actual symmetric E-optimum designs are obtained by solving the equations (1.5) in $\boldsymbol{\alpha}$ for the moments appearing in M to be those given by (2.3), in the manner described earlier.

Because of the rationality of (2.3), the interesting possibility now emerges that $L_q^{(\infty)}$ can consist of only two elements rather than three. This is resolved as follows: if the equations

$$u_q(\boldsymbol{\alpha})=\tfrac{2}{5},\quad v_q(\boldsymbol{\alpha})=\tfrac{1}{5} \tag{2.4}$$

are satisfied for an $\boldsymbol{\alpha}$ with only two positive coordinates α_i and α_j (it being trivial that one coordinate never suffices), we find upon eliminating α_i and $\alpha_j=1-\alpha_i$ from (2.4) that

$$5ij-3qi-3qj+q(2q+1)=0. \tag{2.5}$$

Thus, we are reduced to the problem of finding all integer solutions (i,j,q) to (2.5) with $0\leqq i<j\leqq q$, and it can be shown that the α_i,α_j corresponding to any such solution are positive. The complete solution to (2.5) can be found by standard methods of Diophantine analysis (e.g., Mordell (1969)). For $q\leqq 25$, that perhaps includes the "practical" range, the solutions are given in Table 1. We note that solutions exist for 18 out of the 25 values of q; the values of q with no solutions become sparser as q increases. (We will be happy to provide any interested readers with all solutions for $q\leqq 100$.)

The value of α_i for any such solution is

$$\alpha_i=[q(q-1)/5-(q-j)(q-j-1)]/[2q(j-i)+i(i+1)-j(j+1)], \tag{2.6}$$

with α_j being found by interchanging i and j in (2.6).

All solutions of (2.5) are members of arithmetic "series" of the form $(i,j)=(aq+b,cq+d)$, for q in an appropriate arithmetic progression, and each of those of Table 1 is a member of one of the seven series given by

$$(a,b,c,d)=(0,0;\tfrac{2}{3},\tfrac{1}{3}),(\tfrac{1}{2},-\tfrac{1}{2};1,0),(\tfrac{2}{5},0;\tfrac{4}{5},1),(\tfrac{2}{5},-1;\tfrac{4}{5},0),$$

$$(\tfrac{1}{5},0;\tfrac{7}{10},\tfrac{1}{2}),(\tfrac{1}{5},-2;\tfrac{7}{10},0),(\tfrac{1}{3},-\tfrac{4}{3};\tfrac{3}{4},0).$$

Table 1

Solutions (i,j) of (2.5) for $q \leqq 25$

q	(i,j)	q	(i,j)
1	(0, 1)	14	—
2	—	15	(3, 11); (5, 12); (6, 13); (7, 15)
3	(1, 3)	16	(0, 11); (4, 12)
4	(0, 3)	17	(8, 17)
5	(1, 4); (2, 5)	18	—
6	—	19	(0, 13); (9, 19)
7	(0, 5); (3, 7)	20	(2, 14); (7, 16); (8, 17)
8	—	21	(10, 21)
9	(4, 9)	22	(0, 15)
10	(0, 7); (3, 8); (4, 9)	23	(11, 23)
11	(5, 11)	24	—
12	—	25	(0, 17); (5, 18); (9, 20); (10, 21); (12, 25)
13	(0, 9); (6, 13)		

The first of these, for which $(q,i,j)=(3m+1,0,2m+1)$, $m=0,1,2,\ldots$, is of special interest because part of the support consists of the set of vertices, J_0. The second, for which $(q,i,j)=(2m+1,m,2m+1)$, $m=0,1,2,\ldots$, is of interest because $\alpha_j=\alpha_q$ is concentrated on the single point J_q, the center of the q-cube. See further Kiefer (1977).

Solutions to (2.5), such as those illustrated in the previous paragraph, are attractive to implement because only two J_r's are involved. However, for a given q it need not be that the number of points in some symmetric two-J_r solution (if one exists) is smaller than the number in another symmetric solution. For example, for $q=9$ the only symmetric solution of Table 1, with $(i,j)=(4,9)$, is supported by $\binom{9}{4}2^5+1=4033$ points, while the solution $(\alpha_0,\alpha_6,\alpha_9)=(\frac{2}{15},\frac{12}{15},\frac{1}{15})$ to (1.5) is on $2^9+\binom{9}{6}2^3+1=1185$ points. (There is always a solution $(\alpha_0,\alpha_i,\alpha_q)$ if $\frac{1}{2}(q-1)\leqq i\leqq\frac{1}{3}(2q+1)$; the equalities here are attained when (2.5) is satisfied, for the two-J_r series $(a,b,c,d)=(\frac{1}{2},-\frac{1}{2};1,0)$ and $(0,0;\frac{2}{3},\frac{1}{3})$ of the previous paragraph, respectively.)

Nevertheless, a convenient design of the form (2.5) can often be found. For example, when $q=3$ the irreducible symmetric design other than that given in Table 1 is on $J_0\cup J_2\cup J_3$, and it requires 15 points; while the solution of Table 1, on $J_1\cup J_3$, requires 13. This last design cannot be reduced further in the set of *all* (symmetric or not) designs, in the sense that no proper subset of $J_1\cup J_3$ can support an E-optimum design; this follows from a result of Pesotchinsky (1975, p. 337). The 13 points are to be contrasted with the 21 required by the well-known symmetric D-optimum design of $J_0\cup J_1\cup J_3$, which Pesotchinsky (1975) showed to have minimum support among *all* designs.

We remark that the methods employed by Farrell et al. (1965) for the case $p=0$ can be used to reduce the support of $\alpha^{(p)}$ in some cases. For example, for $q=11$ the portion J_5 of the design of Table 1 can be halved in the number of points it contains, through the use of orthogonal arrays of strength four. The clever and more striking but more delicate method employed by Pesotchinsky (1975) to construct subsets of $J_0\cup J_1$ in the case $p=0, q \leqq 6$, can be extended to certain examples for $p>0$. We shall return elsewhere to these constructions.

3. Optimum designs

Optimum $\boldsymbol{\alpha}^{(p)}$ for $2 \leqq q \leqq 10$ and for selected values of p and the supporting J_r's, are listed in Table 2. Some explanations:

(A) We have omitted the case $q=1$, since it is identical with the $q=1$ case for the ball and the $q=1^*$ case for the simplex, both recorded previously.

(B) Our selection of the supporting J_r's (that is, of which α_r's are positive) is guided by the desire to have as few different α_r's as possible in the table for each q, so as to make the variation in p of $\boldsymbol{\alpha}^{(p)}$ as easy as possible to perceive; in particular, $\boldsymbol{\alpha}^{(\infty)}$ is chosen to be a limit of $\boldsymbol{\alpha}^{(p)}$)s. We always list $\boldsymbol{\alpha}^{(p)}$'s with only two or three positive components. For $q=2,3,4,6$ it is possible to find a set of three J_r's that suffices for all p. For the other tabled values of q, four are needed, and one should be aware that the resulting large "jump" in $\boldsymbol{\alpha}$ when the set of three supporting J_r's changes is not due to any strange behavior of Φ_p, and could be made to disappear if we introduced an appropriate transitional phase where four components α_r were positive.

(C) As a consequence of the selection described in (B), the listed solutions are not always the symmetric optimum designs with smallest support. In listing the five eigenvalues of $M^{-1}(\boldsymbol{\alpha}^{(p)})$ (w^+ and w^- being the reciprocals of the two eigenvalues of M listed on the last line of (1.8)) we have included an extra decimal place in the first two columns so that these values of $1/U^{(p)}$ and $1/V^{(p)}$ can be used to solve (1.5) for designs $\boldsymbol{\alpha}^{(p)}$ with other than the tabled support. Thus, for example, when $q=6$ the tabled $J_0 \cup J_2 \cup J_5$ has 314 points, but for $p=0$ the best symmetric choice is $J_0 \cup J_1 \cup J_6$ with 257 points, while for $p=1$ it is $J_0 \bigcup J_3 \bigcup J_6$ with 225 points. These last two design supports do not occur in Table 2 because neither is a set that supports an $\boldsymbol{\alpha}^{(p)}$ for all p as $J_0 \cup J_2 \cup J_5$ does.

(D) The $\alpha_i^{(p)}$ have been rounded off from more accurate calculations but have not been altered further to make $\sum \alpha_i = 1$; the sum may be off by 0.001. For practical purposes of implementation, any reasonable adjustment of the α_i's suffices.

(E) The largest set of p's is tabled for $q=2$ and 3, to indicate the somewhat surprisingly slow approach of $\boldsymbol{\alpha}^{(p)}$ to $\boldsymbol{\alpha}^{(\infty)}$. Since larger values of q will arise less often in practice, we have further reduced the tabled p-values for $q>5$.

(F) The tabling of eigenvalues in addition to Φ_p-values is meant to provide an additional basis for choosing among competing designs; one may regard these five numbers and the corresponding multiplicities as given in (1.8) as a more detailed operating characteristic (although still a reduction from M itself) than is given by the Φ_p. In fact, we do not here table $\Phi_p(\alpha^{(p_0)})$ as a function of p for fixed p_0, but this may be computed from the eigenvalues, and was used to obtain Table 3 (discussed in (H), below).

(G) Regarding the desirability of achieving a small support for an optimum design, we note the possibility that, even if a symmetric optimum design must have three positive $\alpha_i^{(p)}$'s, it may be that one of these components is so small that reassigning it to the other two J_r's achieves a design that is very close to optimality. Similarly, reductions such as those mentioned at the end of Section 2, when not strictly available to the desired extent, may sometimes be approximated by small sets supporting almost-optimum measures. Corresponding reductions are of course even more important in small sample "exact theory" implementation.

(H) As in many other settings, the variation in p of $\Phi_{p_0}(\alpha^{(p)})$, for fixed p_0, is often less than the variation in $\alpha^{(p)}$ itself, evident in Table 2. This affects favorably the possibility that some $\boldsymbol{\alpha}^{(p)}$'s may be fairly robust under change of criterion. Table 3 gives the maximum over p of the efficiency ratio $e_p(\xi')$ of (1.2), for ξ' equal to various $\boldsymbol{\alpha}^{(p_0)}$'s. The behavior of $\boldsymbol{\alpha}^{(0)}$ is rather poor, and for $q \leqq 7$ the design $\boldsymbol{\alpha}^{(1)}$ is superior to $\boldsymbol{\alpha}^{(\infty)}$ in this respect, while $\boldsymbol{\alpha}^{(\infty)}$ is better for larger q. This contrasts with the corresponding results for the ball, where $\boldsymbol{\alpha}^{(1)}$ was always better, and for the simplex, where $\boldsymbol{\alpha}^{(\infty)}$ was. An $\boldsymbol{\alpha}^{(p_0)}$, with a suitable value of p_0 between 1 and ∞, is somewhat better than either $\boldsymbol{\alpha}^{(1)}$ or $\boldsymbol{\alpha}^{(\infty)}$ in the present setting, and $\boldsymbol{\alpha}^{(5)}$ was quite good for all tabled q, in terms of e_p.

Table 2

Selected Φ_p-optimum designs and corresponding eigenvalues

weights					eigenvalues				
q	p	α_0	α_1	α_2	u^{-1}	v^{-1}	$(u-v)^{-1}$	w^+	w^-
	0	0.583	0.321	0.096	1.345	1.715	6.24	0.45	10.07
	0.25	0.500	0.356	0.144	1.476	2.002	5.61	0.49	7.93
	0.5	0.443	0.375	0.182	1.585	2.255	5.34	0.52	6.93
	1	0.376	0.391	0.233	1.750	2.661	5.11	0.56	6.05
	2	0.314	0.400	0.286	1.946	3.185	5.00	0.61	5.49
	5	0.261	0.403	0.336	2.163	3.837	4.96	0.65	5.18
2	10	0.238	0.404	0.358	2.272	4.195	4.96	0.67	5.09
	15	0.229	0.403	0.368	2.322	4.364	4.96	0.68	5.06
	20	0.222	0.403	0.374	2.352	4.468	4.97	0.69	5.05
	50	0.212	0.401	0.386	2.422	4.713	4.98	0.70	5.02
	100	0.201	0.401	0.392	2.454	4.827	4.99	0.71	5.01
	∞	0.200	0.400	0.400	2.500	5.000	5.00	0.71	5.00
		α_0	α_1	α_3					
	0	0.510	0.424	0.066	1.261	1.534	7.07	0.33	14.44
	0.25	0.400	0.492	0.108	1.373	1.772	6.10	0.36	10.39
	0.5	0.326	0.531	0.143	1.471	1.989	5.65	0.39	8.59
	1	0.235	0.569	0.196	1.627	2.353	5.27	0.43	7.02
	2	0.152	0.594	0.254	1.825	2.858	5.05	0.48	5.99
	5	0.078	0.605	0.316	2.076	3.571	4.96	0.54	5.37
3	10	0.047	0.606	0.347	2.218	4.016	4.95	0.57	5.18
	15	0.035	0.605	0.360	2.282	4.229	4.96	0.58	5.12
	20	0.028	0.604	0.368	2.321	4.359	4.96	0.59	5.09
	50	0.014	0.602	0.384	2.408	4.663	4.98	0.61	5.03
	100	0.008	0.601	0.391	2.447	4.801	4.99	0.62	5.02
	∞	—	0.600	0.400	2.500	5.000	5.00	0.62	5.00

Table 2 (continued)

q	p	α_0	α_1	α_3		u^{-1}	v^{-1}	$(u-v)^{-1}$	w^+	w^-
4	0	0.498	0.408	0.095		1.209	1.425	7.92	0.26	19.86
	0.25	0.394	0.437	0.169		1.309	1.632	6.61	0.28	13.22
	0.5	0.331	0.433	0.236		1.399	1.827	5.98	0.31	10.38
	1	0.264	0.395	0.341		1.549	2.166	5.44	0,34	7.99
	2	0.218	0.316	0.466		1.750	2.662	5.11	0.39	6.47
	5	0.194	0.197	0.608		2.023	3.413	4.96	0.46	5.54
	10	0.192	0.127	0.680		2.184	3.908	4.95	0.49	5.25
	15	0.193	0.096	0.711		2.258	4.148	4.96	0.50	5.17
	20	0.194	0.078	0.728		2.302	4.294	4.96	0.52	5.12
	∞	0.200		0.800		2.500	5.000	5.00	0.56	5.00

q	p	α_0	α_1	α_3	α_5	u^{-1}	v^{-1}	$(u-v)^{-1}$	w^+	w^-
5	0	0.497	0.383	0.119		1.174	1.352	8.89	0.21	26.30
	0.25	0.412	0.359	0.229		1.264	1.537	7.12	0.23	16.35
	0.5	0.374	0.292	0.334		1.349	1.716	6.31	0.25	12.28
	1	0.361	0.132	0.507		1.494	2.037	5.60	0.28	8.97
	2	0.332		0.644	0.024	1.697	2.525	5.18	0.33	6.92
	5	0.236		0.670	0.094	1.984	3.301	4.97	0.39	5.70
	10	0.194		0.673	0.133	2.160	3.831	4.95	0.43	5.33
	15	0.177		0.673	0.150	2.241	4.090	4.96	0.45	5.21
	20	0.168		0.672	0.160	2.288	4.247	4.96	0.46	5.15
	∞	0.133		0.667	0.200	2.500	5.000	5.00	0.50	5.00

q	p	α_0	α_2	α_5		u^{-1}	v^{-1}	$(u-v)^{-1}$	w^+	w^-
6	0	0.617	0.379	0.004		1.149	1.301	9.84	0.18	33.76
	0.5	0.405	0.517	0.079		1.312	1.636	6.63	0.21	14.26
	1	0.292	0.556	0.151		1.452	1.942	5.76	0.24	9.93
	5	0.105	0.515	0.380		1.955	3.216	4.98	0.35	5.84
	15	0.064	0.458	0.478		2.227	4.044	4.95	0.40	5.25
	∞	0.040	0.400	0.560		2.500	5.000	5.00	0.45	5.00

q	p	α_0	α_1	α_3	α_7	u^{-1}	v^{-1}	$(u-v)^{-1}$	w^+	w^-
7	0	0.329	0.648		0.022	1.130	1.262	10.80	0.15	42.23
	0.5	0.491		0.504	0.004	1.283	1.574	6.94	0.18	16.30
	1	0.366		0.592	0.042	1.420	1.868	5.91	0.21	10.88
	5	0.118		0.701	0.182	1.930	3.147	4.99	0.31	5.96
	15	0.048		0.706	0.246	2.215	4.006	4.95	0.36	5.28
	∞			0.700	0.300	2.500	5.000	5.00	0.42	5.00

q	p	α_0	α_1	α_4	α_8	u^{-1}	v^{-1}	$(u-v)^{-1}$	w^+	w^-
8	0	0.430	0.484	0.086		1.115	1.232	11.77	0.13	51.71
	0.5	0.305	0.376	0.319		1.260	1.525	7.24	0.16	18.41
	1	0.383	0.070	0.547		1.393	1.809	6.06	0.18	11.83
	5	0.174		0.700	0.127	1.910	3.089	5.00	0.28	6.08
	15	0.100		0.707	0.193	2.205	3.975	4.95	0.33	5.31
	∞	0.050		0.700	0.250	2.500	5.000	5.00	0.38	5.00

Table 2 (continued)

q	p	α_0	α_1	α_5	α_9	u^{-1}	v^{-1}	$(u-v)^{-1}$	w^+	w^-
9	0	0.376	0.569	0.055		1.103	1.208	12.74	0.12	62.19
	0.5	0.139	0.640	0.221		1.241	1.485	7.54	0.14	20.57
	1	0.139	0.468	0.393		1.371	1.760	6.20	0.16	12.76
	5	0.209		0.718	0.072	1.892	3.040	5.01	0.25	6.19
	15	0.132		0.727	0.141	2.196	3.947	4.95	0.30	5.34
	∞	0.080		0.720	0.200	2.500	5.000	5.00	0.36	5.00
		α_0	α_1	α_5	α_{10}					
10	0	0.375	0.567	0.058		1.094	1.189	13.72	0.11	73.67
	0.5	0.170	0.578	0.252		1.225	1.452	7.83	0.13	22.78
	1	0.241	0.298	0.461		1.352	1.718	6.31	0.15	13.60
	5	0.174		0.717	0.108	1.876	2.997	5.02	0.23	6.29
	15	0.093		0.727	0.180	2.189	3.923	4.95	0.28	5.37
	∞	0.040		0.720	0.240	2.500	5.000	5.00	0.33	5.00

Table 3

Values of $\max_p e_p(\alpha^{(p_0)})$ for selected values of p_0

q	0	0.5	1	5	15	∞
2	2.01	1.39	1.21	1.23	1.29	1.36
3	2.89	1.72	1.40	1.33	1.44	1.57
4	3.98	2.08	1.60	1.41	1.58	1.75
5	5.26	2.45	1.79	1.49	1.70	1.93
6	6.75	2.85	1.99	1.55	1.80	2.08
7	8.45	3.26	2.18	1.61	1.90	2.22
8	10.34	3.68	2.37	1.65	1.98	2.35
9	12.44	4.11	2.55	1.70	2.06	2.46
10	14.73	4.56	2.74	1.73	2.13	2.57

References

Farrell, R.H., J. Kiefer and A. Walbran (1965). Optimum multivariate designs. *Proc. 5th Berk. Symp.* 1, 113–138.

Galil, Z. and J. Kiefer (1976). Comparison of quadratic simplex designs. To appear.

Galil, Z. and J. Kiefer (1976). Comparison of designs for quadratic regression on balls. To appear.

Kiefer, J. (1960). Optimum experimental designs V, with applications to systematic and rotatable designs. *Proc. 4th Berk. Symp.* 1, 381–405.

Kiefer, J. (1974). General equivalence theory for optimum designs (approximate theory). *Ann. Math. Statist.* 75, 849–879.

Kiefer, J. (1975). Optimal design: Variation in structure and performance under change of criterion. *Biometrika* 62, 277–288.

Kiefer, J. (1977). Some combinatorial problems in optimum design. *Ann. Math. Statist.* To appear.

Kiefer, J. and J. Wolfowitz (1960). The equivalence of two extremum problems. *Canad. J. Math.* 12, 363–366.

Kono, K. (1962). Optimum designs for quadratic regression on the k-cube. *Mem. Fac. Sci., Kyusan Univ., Ser. A*, 16, 114–122.
Mordell, J. (1969). *Diophantine Equations*. Academic Press, New York.
Pesotchinsky, L.L. (1975). D-optimum and quasi-D-optimum second-order designs on a cube. *Biometrika* 62, 335–340.

Comparison of Simplex Designs for Quadratic Mixture Models

Z. Galil and J. Kiefer

Cornell University
Ithaca, New York 14853

Designs for quadratic regression are considered when the possible values of the controlable variable are *mixtures* $x = (x_1, x_2, \cdots, x_{q+1})$ of nonnegative components x_i with $\sum_1^{q+1} x_i = 1$. The designs that are optimum with respect to the D-, A-, and E-optimality criteria are compared in their performance relative to these and other criteria. Computational routines for obtaining these designs are developed, and the geometry of optimum structures is discussed. Except when $q = 2$, the A-optimum design is supported by the vertices and midpoints of edges of the simplex, as is the case for the previously known D-optimum design. Although the E-optimum design requires more observation points, it is more robust in its efficiency, under variation of criterion; but all three designs perform reasonably well in this sense.

KEY WORDS

Optimum designs
Simplex designs
Experiments with mixtures
Robust designs

1. INTRODUCTION

A first step away from the traditional choice of a design to satisfy some principle of intuition or symmetry, is to base the choice on a specific criterion. (For example, one may choose a design to minimize the generalized variance, the average variance, or the largest eigenvalue of the covariance matrix, of best linear estimators; these criteria are called D-, A-, and E-optimality, respectively.) Realistically, though, such a criterion is usually at best only an approximate reflection of some vague notion of "goodness". Hence, it seems prudent to check that a design, selected in this fashion, performs reasonably well in other respects, relative to other possible designs. While such an examination of the performance of competing designs occurs in various applied papers, it is almost always imprecise or unsystematic. The present study attempts what is perhaps a more systematic study in a particular family of simple settings.

Such a study of characteristics of performance of designs is analogous to the examination of risk functions of various procedures in a statistical decision framework (with fixed experiment). There, a procedure that is "best" according to a Bayes, minimax, or other criterion, may be found inferior upon comparison with a second procedure which, at the sacrifice of a small relative loss for the specified criterion, performs much better in other respects. (Such phenomena are termed subadmissibility, subminimaxity, etc., in the literature.) Since the prior law or minimax criterion are only approximate guides to optimality, the second procedure might then well be used in place of the first.

Although the examination of the totality of (admissible) risk functions in a decision-theoretic setting could lead to the choice of a satisfactory procedure, in practice this is too huge a prescription to carry out. One might therefore use a particular criterion to select a procedure which, at least roughly, accomplishes the desired aim, and might then look at nearby risk functions to see whether some obvious improvement, such as that described in the previous paragraph, is possible. We now seek a corresponding program in the design setting.

Suppose for simplicity that we restrict consideration to the "approximate" design theory and to best linear unbiased estimation, and measure the goodness of a design in terms of some functional of the associated covariance matrix. If (in the usual notation, defined in full below) $M(\xi)$ is the information matrix per observation of the design ξ, in a setting where the total number of observations is fixed, the admissible designs are well known (e.g. [6], [10]) to be among those that minimize $\Phi_A(\xi) = \text{tr}\, AM^-(\xi)$ for some nonnegative definite A (analogous to the relationship between admissible and Bayes decision procedures). In the spirit of the previous para-

Research under NSF Grant MPS 72-04998 A02.
AMS Subject Classification 62K05.

Received March 1976; revised May 1977

graph, rather than to search among all these M we may choose a particular criterion Φ, find the ξ^* that minimizes $\Phi(M(\xi))$, and then see whether the resulting "Φ-optimum" design is indeed satisfactory, or whether there is another design that is only slightly less efficient in terms of Φ, but which is noticeably superior to ξ^* in other terms of interest.

Ideally, and in parallel with the comparison of risk functions in simple decision-theoretic models, one should compare the entire matrix $M(\xi^*)$ with competitors $M(\xi)$. It is difficult to visualize and interpret the various differences among matrices, especially when the number of parameters is large. We might therefore simplify the comparison by looking, instead, only at a "reduced operating characteristic" such as the list of eigenvalues of each $M(\xi)$. (More satisfactory, but much more complex computationally, would be the examination of the eigenvalues of $M^{-1}(\xi^*)M(\xi)$ for each ξ^* and ξ.) The reduced operating characteristic we have chosen in the present study, as a convenient basis for comparison, is a one-parameter family $\{\Phi_p, 0 \leq p \leq \infty\}$ of real functionals of M, essentially the *L_p-norms of the set of eigenvalues of M^{-1}*. At the same time, the basic Φ used to select ξ^* is most often taken to be a member of this family. This family includes the three most commonly used criteria, those of D-, A-, and E-optimality, as those obtained when $p = 0, 1, \infty$; the family has been discussed in more detail in Kiefer [10], and the intuitive meanings of the criteria are listed in Kiefer [6]; for example, the D-optimum design minimizes the volume of the usual confidence ellipsoid in the normal case, while the E-optimum design minimizes its maximum diameter.

There are obviously examples in which performance in terms of $\{\Phi_p\}$ does not adequately reflect serious defects or advantages of designs. Nevertheless, this family of criteria appears to be useful in many settings; this is the case for the examples of simplex (mixture) designs studied herein, where the M of any design with reasonable symmetry has at most five distinct eigenvalues (which we list with multiplicities). Other criteria we have considered are the maximum $\bar{d}$ of the variance function and criteria associated with the model of Box and Draper [1], in which systematic account is taken for the first time of the bias due to incorrectness of the model used in fitting the regression function; at the editor's suggestion, results on the latter appear in a separate paper [5].

One of the conclusions of our study (Section 5) is that, in simple practical settings such as those considered here, many procedures remain much more efficient under variation in criterion than one could expect from general theoretical bounds. For example, it is shown in Section 5 that if $\xi^{(p)}$ is a design optimum under Φ_p, then for each $p > 0$ there are settings where $\Phi_0(\xi^{(p)})/\Phi_0(\xi^{(0)})$ is arbitrarily large; that is, the D-efficiency of $\xi^{(p)}$ is arbitrarily close to zero. But in our examples of quadratic regression on a simplex of any dimension q, it turns out that $\xi^{(p)}$ is fairly efficient. (In particular, $\xi^{(\infty)}$ seems remarkably robust in its efficiency under variation of criterion.) This is related to the fact that, in these examples, although $\xi^{(p)}$ varies noticeably with p, its performance characteristics often vary much less. Some of the literature of design comparisons unfortunately emphasizes the dependence on criterion of the *makeup* of an optimum design without adequate comparison of actual *performance*.

Although the symmetric E-optimum designs for large q are supported on more points than will be feasible in many applications, it will often be possible to reduce the support greatly by using an asymmetric design with the same (or approximately the same) matrix M. Some techniques for doing this have been presented in [14] for regression on cubes. We shall treat this topic in more detail elsewhere.

In our examples we point out certain qualitative geometric features of the structure of various "good" designs; for example, in Section 4 we describe how far into the simplex (as measured by dimensionality of the face in which a point lies) it is necessary to take observations. We feel that such information can be useful in suggesting promising forms of designs to be investigated in more complex settings where precise optimality results are harder to obtain. If observations must be restricted to some asymmetric subset of the simplex, a computer search for optimum designs might be prohibitive for large q; but we would still hope to find reasonably efficient designs among those with dimensionality characteristics suggested by the present study. The same is true for many nonlinear regression models.

A preliminary report of a few results of this study, with additional discussion of such matters as the theory behind the reduction of the optimization problem (as mentioned in Section 3, below) and the Box-Draper philosophy [1], was given at the 1974 Imperial College Conference; see Kiefer [11]. Many of the longer formulas used in the developments (especially of Section 3) are omitted here but are available from the authors. Without presenting all details of our computations, we have tried to outline enough of the theoretical and numerical developments, and the difficulties encountered therein, that future workers on such developments in other settings may benefit from our experience.

2. PRELIMINARIES

The considerations of this paper are limited to the approximate theory, so as to exhibit the comparisons without the obscuring arithmetical complexity and longer design lists of exact design theory as sample size varies. Comparisons of exact designs exhibit similar phenomena, even for small sample sizes.

A *design setting* is a pair $(\mathfrak{X}, \mathbf{f})$ where f is a known (column) k-vector of continuous functions $f_1, \cdots, f_k$ on the compact space $\mathfrak{X}$. A *design* ξ is a probability measure on $\mathfrak{X}$. The measure ξ can be assumed discrete, and $\xi(x)$ then represents the proportion of observations to be taken at the point x of $\mathfrak{X}$. (The reader may consult, e.g., Kiefer [6], [10] or Fedorov [4] for interpretation of these standard notions of optimum design theory.) The observations are uncorrelated with variance σ^2. The expected value of an observation at x is $\theta'\mathbf{f}(x)$ where θ is a column vector of k unknown real parameters. The $k \times k$ *information matrix* (per observation for unit variance) of ξ is defined as

$$M(\xi) = \int_{\mathfrak{X}} \mathbf{f}(x)\mathbf{f}(x)'\xi(dx),$$

and the corresponding normalized variance function for nonsingular $M(\xi)$ is $d(x, \xi) = \mathbf{f}(x)'M^{-1}(\xi)\mathbf{f}(x)$. Thus, when ξ is used with N observations, $N^{-1}\sigma^2 M^{-1}(\xi)$ is the covariance matrix of best linear unbiased estimators of θ, and $N^{-1}\sigma^2\, d(x, \xi)$ is the variance at x of the least-squares fitted regression function.

The optimality functionals considered herein are infinite for singular $M(\xi)$, and we therefore limit consideration to nonsingular $M(\xi)$ in the sequel. Denote the eigenvalues of $M(\xi)$ by $\lambda_1(\xi), \cdots, \lambda_k(\xi)$. We define

$$\Phi_p(\xi) \equiv \bar{\Phi}_p(M(\xi)) = (k^{-1} \operatorname{tr} M^{-p}(\xi))^{1/p} = \left(k^{-1}\sum_1^k \lambda_i^{-p}(\xi)\right)^{1/p} \quad \text{for } 0 < p < \infty;$$

$$\Phi_0(\xi) \equiv \bar{\Phi}_0(M(\xi)) = \lim_{p\downarrow 0} \Phi_p(\xi) = (\det M^{-1}(\xi))^{1/k}; \tag{2.1}$$

$$\Phi_\infty(\xi) \equiv \bar{\Phi}_\infty(M(\xi)) = \lim_{p\to+\infty} \Phi_p(\xi) = \max_i \lambda_i^{-1}(\xi).$$

(We use the more succinct notation Φ_p rather than $\bar{\Phi}_p$ except where the nature of dependence on the elements of $M(\xi)$ must be made explicit.) As mentioned in the introduction, Φ_0, Φ_1, and Φ_∞ are the familiar D-, A-, and E-optimality criteria, here normalized so as to make comparisons easy: all Φ_p measure loss in the same scale per unit of variance, and take on the value c^{-1} if $M(\xi) = cI_k$. A Φ_p*-optimum design* $\xi^{(p)}$ is one that minimizes $\Phi_p(\xi)$. (Throughout this paper we shall not exhibit the dependence of $\xi^{(p)}$ on q unless, as in Section 4B, it is required for lack of ambiguity.)

A measure of the performance of a design ξ' with respect to the family $\{\Phi_p\}$ of criteria is the (multiplicative regret) *absolute inefficiency ratio*

$$\Phi_p(\xi')/\Phi_p(\xi^{(p)}), \quad 0 \le p \le \infty. \tag{2.2}$$

A design ξ' may be judged adequate for many purposes in applications if these ratios are all near 1. (All ratios ≤ 1.3 would certainly seem satisfactory; but in complex settings where no ξ' exists with all ratios so small, one will have to settle for less.) We shall concentrate herein on the efficiency ratio (2.2) for $\xi' =$ various $\xi^{(r)}$'s, especially for $r = 0$, 1, or ∞. The finer comparison between two designs ξ' and ξ'' obtained by contrasting the entries of $M(\xi')$ and $M(\xi'')$, or the eigenvalues $\{\lambda_i(\xi')\}$ and $\{\lambda_i(\xi'')\}$, or by computing the eigenvalues of $M(\xi')M^{-1}(\xi'')$, does not (at least in our examples) seem to warrant the much greater effort required.

Another common criterion, that of G-optimality, is that of selecting ξ to minimize $\bar{d}(\xi) = \max_{x\in\mathfrak{X}} d(x, \xi)$. The original "equivalence theorem" (Kiefer and Wolfowitz [13]) asserts that $\bar{d}$- and Φ_0-optimality are equivalent, and that $\bar{d}(\xi^{(0)}) = k$, a useful fact in computing $\xi^{(0)}$. (The parallel fact for $\xi^{(p)}$ is phrased in terms of a criterion that does not share the intrinsic appeal of $\bar{d}$. See Kiefer [10].) In the present paper we shall mention briefly the analogue of (2.2) when Φ_p is replaced by $\bar{d}$, at the end of Section 5; this analogue, $k^{-1}\bar{d}(\xi')$, is known [7] to be near 1 if $\Phi_0(\xi')/\Phi_0(\xi^{(0)})$ is, so we do not obtain much added information from the former unless the latter ratio is markedly different from 1.

The settings we consider are those of quadratic regression for *experiments with mixtures,* developed especially by Scheffé [15]. An excellent summary of past work, and development of new models for this setting, is contained in Draper and St. John [3]. There are $q + 1$ components (proportions) in a mixture x, so that $\mathfrak{X}$ is the q-simplex $S_q = \{(x_1, x_2, \cdots, x_{q+1}) : \sum_1^{q+1} x_i = 1,\ \text{all } x_i \ge 0\}$. A convenient basis for the polynomials on $\mathfrak{X}$ of degree ≤ 2 is the set of functions $\{x_i, x_ix_j : 1 \le i, j \le q + 1;\ i < j\}$; there are $k = (q + 1)(q + 2)/2$ functions in this set, which we take to be the components of f.

It will turn out that much of our discussion is phrased in terms of certain simple subsets of S_q, which we now define. A *barycenter of depth* j ($0 \le j \le q$) is a point with $j + 1$ coordinates equal to $1/(j + 1)$ and the remaining coordinates equal to 0. We denote the set of $\binom{q+1}{j+1}$ barycenters of depth j by J_j, and write $J = \cup_j J_j$. Thus, J_0 consists of the vertices of S_q, while J_1 consists of midpoints of edges, J_2 consists of the centers of triangular 2-dimensional faces, J_q is the single center point, etc. These are components of Scheffé's "simplex centroid design" [15], in which an equal number of observations on all points of $J_0 \cup J_1 \cup \cdots \cup J_m$ is used for regression of degree $m + 1$; that design is D-optimum for $m = 0$ or 1.

3. BASIC COMPUTATIONAL METHODS

Except when $q = 1$ or $p = 0$, the computation of a Φ_p-optimum design is algebraically intractable, and we must be satisfied with numerical results. However, the direct computer minimization of $\Phi_p(\xi)$ is too large a problem, since the space of matrices $\mathfrak{M} = \{M(\xi)\}$ has dimension $q(q+1)(q^2+q+22)/12$, and minimization of Φ_p with respect to roughly $q^4/12$ locations $x^{(i)}$ and corresponding weights $\xi(x^{(i)})$ is a gigantic problem. However, various theoretical tools of optimum design theory can be used to reduce the dimensionality of our problem. These tools make use of the symmetry of the problem, the convexity of Φ_p, the general equivalence theory, the quartic nature (in x) of quadratic forms in f(x) if f itself is quadratic, and some additional considerations when $p = \infty$. They are treated in Kiefer [10], and their implementation in our simplex setting is sketched in Kiefer [11]. The conclusion is that, for each p, *there is a* $\xi^{(p)}$ *which is supported by* J *and which is symmetric under permutation of coordinates of* S_q. Any such design can be described in terms of a probability $(q+1)$-vector $\alpha = (\alpha_0, \alpha_1, \cdots, \alpha_q)$, where $\xi^{(p)}$ assigns measure $\alpha_j \Big/ \binom{q+1}{j+1}$ to each point of J_j. For such designs we write $M(\alpha)$, $\lambda_i(\alpha)$, etc.

While the general equivalence theory is useful in helping to characterize the form of solution in the manner just indicated, thus reducing tremendously the problem of searching for a solution, we do not use the most obvious computational technique suggested by it. That technique, which involves computing a maximum of a function on S_q (as in the computation of $\bar{d}$), has obvious shortcomings for large q.

While (for some p and q) there can be asymmetric $\xi^{(p)}$, it can be shown that, for each p and q, all $M(\xi^{(p)})$ are identical. Thus, *we hereafter limit consideration to symmetric designs* $\xi^{(p)}$ *on* J, *of the form described just above.*

For designs of this structure, the seven moments that appear in $M(\xi)$ for general ξ are linearly dependent and may be expressed in terms of a constant and three linearly independent moments (e.g., b, c, e, below) or in terms of α_j's. The former may seem preferable, since the optimum b, c, e are unique, while there may be several optimum $\alpha^{(p)}$ for some p, q; thus, it might appear convenient first to find the optimum entries of M and then to see what $\alpha^{(p)}$'s yield those entries. However, the entries of M satisfy restrictions due to their dependence on the nonnegative α_i's, so that it was in fact just as convenient to invoke these restrictions by working in terms of nonnegative α_i's in the manner described later in this section. We have always substituted $\alpha_0 = 1 - \sum_1^q \alpha_i$ to eliminate that linear restriction, and thus work with $\{(\alpha_1, \cdots, \alpha_q) : \alpha_i \geq 0 \ \forall i; \sum_1^q \alpha_i \leq 1\}$. It is to be understood that this substitution has been made before differentiation as in (3.6) and (3.8). The moments of ξ that appear in $M(\xi)$, designated $a(\alpha)$, $b(\alpha)$, $\cdots$, $g(\alpha)$, are, respectively, the moments with respect to ξ of x_1^2, x_1x_2, $x_1^2x_2$, $x_1x_2x_3$, $x_1^2x_2^2$, $x_1^2x_2x_3$, $x_1x_2x_3x_4$. They satisfy

$$(q+1)a - 1 = -\sum_1^q j(j+1)^{-1}\alpha_j = -q(q+1)b,$$

$$q(q+1)b = \sum_1^q j(j+1)^{-1}\alpha_j,$$

$$q(q+1)c = \sum_1^q j(j+1)^{-2}\alpha_j,$$

$$(3.1)\quad (q-1)q(q+1)d = \sum_1^q j(j-1)(j+1)^{-2}\alpha_j = q(q+1)(b-2c),$$

$$q(q+1)e = \sum_1^q j(j+1)^{-3}\alpha_j,$$

$$(q-1)q(q+1)f = \sum_1^q j(j-1)(j+1)^{-3}\alpha_j = q(q+1)(c-2e),$$

$$(q-2)(q-1)q(q+1)g = \sum_1^q j(j-1)(j-2)(j+1)^{-3}\alpha_j = q(q+1)(b-5c+6e).$$

The coefficient vectors of α_j in these sums, multiplied by $(j+1)^3 j^{-1}$, are constant, linear, or quadratic in j, from which the described dependence, exhibited in the last form in each line, follows. Thus, if H_1 is the polyhedral region $\{(b, c, e) : \sum_1^q \alpha_i = 1\}$, the actual region $H = \{(b, c, e)\}$ of interest is $\{\beta H_1, 0 \leq \beta \leq 1\}$. ($H_1$ is of dimension 3 if $q > 3$; H is of dimension 3 if $q \geq 3$.)

The entries of f are conveniently written $\{x_i, 1 \leq i \leq q+1\}$, followed by $\{x_{i_1}x_{i_2}, i_1 < i_2\}$, with corresponding labeling i, (i_1, i_2) for the rows and columns of M. We then partition M as

$$(3.2)\qquad M = \begin{pmatrix} A & B \\ B' & C \end{pmatrix}$$

with A $(q+1) \times (q+1)$ and with entries

$$(A)_{ij} = \begin{cases} a & \text{if } i = j, \\ b & \text{if } i \neq j; \end{cases}$$

$$B_{i,(j_1,j_2)} = \begin{cases} c & \text{if } i = j_1 \text{ or } j_2, \\ d & \text{if } i \neq j_1 \text{ or } j_2; \end{cases}$$

$$(3.3)\qquad C_{(i_1,i_2),(j_1,j_2)} = \begin{cases} e & \text{if } i_1 = j_1 \text{ and } i_2 = j_2, \\ f & \text{if exactly one of } i_1, i_2 = \text{exactly one of } j_1, j_2, \\ g & \text{if both } i_1, i_2 \neq \text{both } j_1, j_2. \end{cases}$$

Using the well known formula for the determinant of a partitioned matrix of the form (3.2), and a simple computation of the determinant of a matrix of the type C, one obtains, for $q \geq 2$ (a simpler form holding when $q = 1$),

$$(3.4)\quad \det M = [e + g - 2f]^{(q+1)(q-2)/2} \times \{(a - b)[e + (q - 3)f - (q - 2)g] - (q - 1)(c - d)^2\}^q \times \{(a + qb)[e + (q - 1)(q - 2)g/2 + 2(q - 1)f] - q[2c + (q - 1)d]^2/2\}.$$

From the same expression with a, e replaced by $a - \lambda$, $e - \lambda$, we obtain $\det(M - \lambda I)$ and thus the eigenvalues of M, of which at most five are distinct:

(3.5)

$\lambda = e + g - 2f$ of multiplicity $(q + 1)(q - 2)/2$;

the zeros of

$$\lambda^2 - [a - b + e + (q - 3)f - (q - 2)g]\lambda + \{(a - b)[e + (q - 3)f - (q - 2)g] - (q - 1)(c - d)^2\},$$

each of multiplicity q;

the zeros of

$$\lambda^2 - [a + qb + e + (q - 1)(q - 2)g/2 + 2(q - 1)f]\lambda + \{(a + qb)[e + (q - 1) \cdot (q - 2)g/2 + 2(q - 1)f] - q[2c + (q - 1)d]^2/2\},$$

each of multiplicity 1.

(For $q = 2$, the first of these is absent; a slightly different form holds in the trivial case $q = 1$.)

The references cited earlier (also [8]) show that the unique symmetric Φ_0-optimum design has $\alpha_0 = 2/(q + 2) = 1 - \alpha_1$ and $\alpha_i = 0$ for $i > 1$. (In fact, this symmetric design is uniquely D-optimum among *all* designs, symmetric or not.) This algebraic result is a convenient check on the computer routine for $p > 0$, and also motivates searching for a solution with $\alpha_i = 0$ for $i > 1$ if p is sufficiently small.

Our routine for $0 < p < \infty$ was based on a Broyden minimization technique (e.g., [2]), using the formulas obtained from (3.5) for the $\lambda_i(\alpha)$ and their derivatives to find an $\alpha^{(p)}$ that minimizes $\Sigma_1^k \lambda_i^{-p}(\alpha)$. Since $\Phi_p(M(\alpha))$ is convex in α, a local minimum is also a global minimum. It is easy to see that α^* is a solution ($\alpha^{(p)}$) if

$$(3.6)\quad \partial\Phi_p(\alpha)/\partial\alpha_i|_{\alpha=\alpha^*} \begin{cases} = 0 & \text{for } \alpha_i > 0, \\ \leq 0 & \text{for } \alpha_i = 0. \end{cases}$$

For each q, we began with small p, seeking a solution for which only $\alpha_0, \alpha_1 > 0$. There was then, for $2 \leq q \leq 10$, a largest value $p_{crit}'(q)$ of p for which such a solution was obtained, as checked by (3.6). For p slightly larger, a solution was obtained with only $\alpha_0, \alpha_1, \alpha_2 > 0$. For $3 \leq q \leq 10$, there was then a largest value $p_{crit}''(q)$ of p for which the solution was of this form. For *all larger p*, a solution was found in which only $\alpha_0, \alpha_1, \alpha_2, \alpha_3 > 0$. (The possibility of other symmetric solutions will be discussed in Section 4B, and a different structure that holds for sufficiently large q will be described in the forthcoming paper [12].) These critical values of p at which the form of solution changes are given in Table 1.

TABLE 1—*Critical values of p.*

q	p'_{crit}	p''_{crit}
2	.864	——
3	1.056	5.669
4	1.223	6.506
5	1.368	7.201
6	1.496	7.796
7	1.610	8.317
8	1.713	8.781
9	1.807	9.198
10	1.893	9.577

For $p = \infty$, a different computational method was needed because $\Phi^{(\infty)}$ is not differentiable at its minimum. As discussed in [10], this changes the equivalence theory developments from what they are for $p < \infty$. However, in our simplex settings it can be shown that all $\xi^{(\infty)}$ are obtainable as limits of convergent sequences of $\xi^{(p)}$'s as $p \to \infty$, so we can still work with design vectors α on J. A helpful aspect of the structure of $\xi^{(\infty)}$ was guessed from looking at $\xi^{(p)}$ for large p, for which it was noted that the first listed eigenvalue λ_1 (say) of (3.5), and each of the eigenvalues λ_2 and λ_3 (say) obtained as a smaller root of one of the quadratics of (3.5), were almost equal. We therefore sought a solution to the problem of finding $\alpha_1, \alpha_2, \alpha_3 > 0$ (with $\alpha_i = 0$ for $i > 3$) to

$$(3.7)\quad \text{maximize } \lambda_1(\alpha) \text{ subject to } \lambda_1(\alpha) = \lambda_2(\alpha) = \lambda_3(\alpha).$$

(For $q = 2$, λ_1 is absent and we maximize λ_2 subject to $\lambda_2 = \lambda_3$. For $q = 1$, a single eigenvalue of M is smaller than the other two, at the optimum.) Once such a solution α^* of (3.7) was obtained, it was necessary to check that it was also a (local, hence global) solution $\alpha^{(\infty)}$ of the original design problem. We could no longer use (3.6), but instead proceeded as follows: Since $1/\Phi_\infty(\alpha) = \min_{1 \leq i \leq 3} \lambda_i(\alpha)$ is concave in α, this function can be increased over its value at α^* only if there is a local first order (tangential) improvement. Thus, if we compute the $3 \times q$ matrix $L(\alpha^*)$ with entries

$$L_{ij}(\alpha^*) = \partial\lambda_i(\alpha)/\partial\alpha_j|_{\alpha=\alpha^*}, \quad 1 \le i \le 3, \quad 1 \le j \le q, \tag{3.8}$$

we see that in order that the value $\alpha = \alpha^* + \epsilon c$ (for some q-vector c) yield exactly an ϵ-order increase in the value of $1/\Phi_\infty(\alpha)$ from its value at α^* (for ϵ small and positive) it is necessary that

$$(Lc')_i > 0 \quad \text{for} \quad i = 1, 2, 3. \tag{3.9}$$

The α^* we obtained as a numerical solution to (3.7) always had $\alpha_0^* > 0$. In order for $\alpha^* + c\epsilon$ to be a probability vector for ϵ small and positive, we must have

$$c_i \ge 0, \; 4 \le i \le q; \tag{3.10}$$

The c_i have arbitrary sign for $i = 1, 2, 3$ because $\alpha_0^* > 0$. Thus, our problem of showing α^* is an α^∞ is reduced to the nonfeasibility problem of linear programming, of showing that there is no solution c to (3.9) subject to (3.10). We remark that the first three rows of L are linearly dependent, because the restriction $\lambda_1 = \lambda_2 = \lambda_3$ of (3.7) is a curve in $(\alpha_1, \alpha_2, \alpha_3)$-space, and thus λ_1 has directional derivative 0 at α^* along the curve.

It can also be verified that, along this curve, λ_1 is *strictly* concave in a neighborhood of α^*, at which point the directional first derivative vanishes but the second derivative is negative. Since (b, c, e) is obtained by nonsingular linear transformation from $(\alpha_1, \alpha_2, \alpha_3)$, it follows that the optimum $M(\alpha^{(\infty)})$ is unique in our present setting, even though Φ_∞ is not *strictly* convex.

The solution to (3.7) can be sought by various iterative methods. We found it difficult to implement common procedures in the literature successfully. Some progress was made with an iterative scheme in which the search was reduced from 3 to 2 dimensions by eliminating one variable algebraically, but this also encountered convergence difficulties when many-place accuracy was desired. Thus, after obtaining an initial guess from the solution $\alpha^{(p)}$ for large p, and a refinement from one of the differential schemes just mentioned, we used a (more expensive) grid search, with increasing refinement, to obtain our most accurate approximation to $\alpha^{(\infty)}$. The matrix L can be computed either analytically or from differences on the grid.

We believe the computation of Φ_∞-(E-) optimum designs invites interesting research in numerical analysis. Our computations of $\alpha^{(\infty)}$ consumed perhaps 90% of the computer time in this study.

4. MAIN NUMERICAL RESULTS

A. *Basic solutions.* The $\alpha^{(p)}$ of the form described in connection with the discussion of critical p-values are listed in Table 2 for $1 \le q \le 10$ and selected values of p. For comparison we have also listed under "$q = 1^*$" the results for the commonly employed 1-dimensional model in which $\mathcal{X} = [-1, 1]$ and $f(x) = (1, x, x^2)$; the $\alpha^{(0)}$ result, which is invariant under linear transformations, is the same as for our $q = 1$ model, but the other $\alpha^{(p)}$ are different for the two models.

For $q \le 4$, the tables are believed to be accurate as given. For $q \ge 5$, there may be error of 1 or 2 units in the last decimal place, particularly for large p.

The pattern of these solutions is interesting. As one might expect, $\alpha_0^{(p)}$ and $\alpha_1^{(p)}$ are always substantial, although for large q they vary oppositely with p. However, $\alpha_2^{(p)}$, which becomes sizable between $p_{crit}'(q)$ and $p_{crit}''(q)$, drops strikingly shortly thereafter, when $\alpha_3^{(p)}$ becomes large.

We have not taken the space to list the five different $\lambda_i(\alpha^{(p)})$'s, which can be computed from (3.5). The joint effect of their variation in p is perhaps most usefully summarized in the efficiency ratios to be discussed in Section 5C.

B. *Other symmetric solutions.* We have remarked that, for each q and p, the quantities $b(\alpha^{(p)})$, $c(\alpha^{(p)})$, $e(\alpha^{(p)})$ that determine M take on the same values $\bar{b}^{(p)}$, $\bar{c}^{(p)}$, $\bar{e}^{(p)}$ for all $\alpha^{(p)}$. Once we have determined a basic solution (as given in part A, above) and the corresponding $\bar{b}^{(p)}$, $\bar{c}^{(p)}$, $\bar{e}^{(p)}$, we can then by (3.1) determine all other symmetric p-optimum designs by finding the solutions α to the linear programming problem

$$\begin{aligned} &\alpha_1, \cdots, \alpha_q \ge 0, \quad \sum_1^q \alpha_i \le 1, \\ &\sum_1^q j(j+1)^{-1}\alpha_j = q(q+1)\bar{b}, \\ &\sum_1^q j(j+1)^{-2}\alpha_j = q(q+1)\bar{c}, \\ &\sum_1^q j(j+1)^{-3}\alpha_j = q(q+1)\bar{e}. \end{aligned} \tag{4.1}$$

This yields the following conclusions for $1 \le q \le 10$:

(i) For $p \le p_{crit}'$, the unique symmetric design $\alpha^{(p)}$ is supported by $J_0 \cup J_1$. As in the case $p = 0$ mentioned earlier, this symmetric design is uniquely Φ_p-optimum among *all* designs in this case, which includes Φ_1-(A-) optimality for $a \ge 3$.

(ii) For $p_{crit}' < p \le p_{crit}''$, the unique symmetric $\alpha^{(p)}$ is supported by $J_0 \cup J_1 \cup J_2$.

(iii) For $p > p_{crit}''(q)$, every symmetric $\alpha^{(p)}$ has α_0, α_1, α_2, > 0, and at least one other $\alpha_j > 0$. Consequently, every symmetric $\alpha^{(p)}$ can be represented uniquely as a convex mixture $\Sigma\, \beta_j\alpha^{(p,j)}$ (where $\beta_j \ge 0$, $\Sigma\, \beta_j = 1$) of "extreme" solutions $\alpha^{(p,j)}$. Here each

$\alpha^{(p,j)}$ is a Φ_p-optimum probability vector $(\alpha_0, \alpha_1, \cdots, \alpha_q)$ for which *only* the four components $\alpha_0, \alpha_1, \alpha_2, \alpha_j$ are positive, the other components α_i being zero. Such solutions $\alpha^{(p,j)}$ exist for each q and $p > p_{crit}''(q)$, for all j, $3 \le j \le q$. (Note that the "basic solution" of Part A is $\alpha^{(p,3)}$.) Thus, for $p > p_{crit}''(q)$, the $\alpha^{(p,j)}$'s are the *simplest* symmetric $\alpha^{(p)}$'s, in the sense of having only four positive components.

The above results can be understood in terms of the geometry of points $h = (b, c, e)$ of H. The extreme points of H are the points $h^{(i)}$, $0 \le i \le q$, obtained from the α for which $\alpha_i = 1$. For p small, the Φ_p-optimum h lies on the edge $\overline{h^{(0)}h^{(1)}}$ of H, and since $h^{(i)}$ is not on that edge for $i > 1$, no such α_i can be positive in an optimum $\alpha^{(p)}$. Similarly, for $p_{crit}'(q) < p \le p_{crit}''(q)$, the optimum point h^* is in the interior of the triangle T (2-face) of H determined by $h^{(0)}, h^{(1)}, h^{(2)}$, and no α_i for $i > 2$ can be positive in an $\alpha^{(p)}$. For p slightly larger than $p_{crit}''(q)$, the optimum h is near h^* but not on T, and is contained in each of the 3-simplices determined by T and $h^{(j)}$, for $j > 2$. That this last pattern persists even for much larger p is a special phenomenon of the cases $q \le 10$ studied, and in Section 6 we shall see that the pattern does not hold for sufficiently large q, for which some $\alpha^{(p,j)}$ can fail to exist for large p.

As an illustration of these results, to see how $\alpha^{(p,j)}$ varies in j, we list these vectors in Table 3 for $q = 5$,

TABLE 2—*Basic Φ_p-optimum designs for selected values of p and q.*

q	p	α_0	α_1	α_2	α_3
1*	0	.667	.333		
	.1	.637	.363		
	.5	.555	.445		
	1	.500	.500		
	2	.449	.551		
	10	.400	.600		
	∞	.400	.600		
1	0	.667	.333		
	.1	.636	.364		
	.5	.562	.438		
	1	.528	.472		
	2	.516	.434		
	10	.515	.485		
	∞	.515	.485		
2	0	.500	.500		
	.1	.475	.525		
	.5	.433	.567		
	1	.425	.562	.013	
	2	.423	.509	.068	
	10	.430	.417	.152	
	∞	.435	379	.186	
3	0	.400	.600		
	.1	.383	.617		
	.5	.364	.636		
	1	.375	.625		
	2	.381	.532	.087	
	4	.387	.448	.165	
	5	.390	.426	.184	
	10	.396	385	.196	.023
	∞	.403	.339	.195	.063
4	0	.333	667		
	1	.340	.660		
	2	.356	.556	.088	
	4	365	444	.191	
	5	.368	.416	.216	
	10	376	.366	.222	.036
	∞	.385	324	.168	.124
5	0	.286	.714		
	1	.314	.686		
	2	.339	.581	.080	
	4	.350	.446	.204	
	7	.360	.370	.270	
	10	.364	.353	.243	.040
	∞	.375	.318	134	.173
6	0	.250	.750		
	1	.294	.706		
	2	.327	.604	.069	
	4	.339	.451	.210	
	7	.351	364	.285	
	10	.356	.343	.264	.038
	∞	.369	.316	.104	.211
7	0	.222	.778		
	1	.278	.722		
	2	.318	.626	.056	
	4	.331	.457	.212	
	7	.343	.362	.295	
	10	.349	.334	.283	.033
	∞	.364	315	078	.243
8	0	.200	.800		
	1	.264	.736		
	2	.310	.647	.043	
	4	.325	.464	.212	
	7	.337	361	.302	
	10	.345	327	.302	.027
	∞	.351	.316	.047	.286
9	0	.182	.818		
	1	.253	.747		
	2	.304	666	.030	
	4	.319	.471	210	
	7	332	.361	307	
	10	.341	.321	.320	.019
	∞	.351	.317	029	.303
10	0	.167	.833		
	1	.243	757		
	2	.299	.684	.017	
	4	.315	.473	.207	
	5	.320	.426	.254	
	7	.328	.362	.310	
	∞	.345	.318	.007	.330

TABLE 3—*Selected extreme designs* $\alpha^{(p,j)}$

$q = 5$:

j	$p = 10$: α_0	α_1	α_2	α_j	$p = \infty$: α_0	α_1	α_2	α_j
3	.364	.353	.243	.040	.375	.318	.134	.173
4	.364	.348	.268	.019	.376	.296	.244	.084
5	.364	.346	.277	.013	.376	.285	.280	.058

$q = 10$:

j	$p = 10$: α_0	α_1	α_2	α_j	$p = \infty$: α_0	α_1	α_2	α_j
3	.337	.316	.343	.004	.345	.318	.007	.330
4	.337	.315	.345	.004	.346	.277	.216	.161
5	.337	.315	.346	.002	.347	.256	.285	.111
6	.337	.315	.347	.001	.348	.244	.320	.088
7	.337	.315	.347	.001	.348	.236	.341	.075
8	.337	.315	.347	.001	.348	.230	.355	.067
9	.337	.315	.347	.001	.349	.225	.365	.061
10	.337	.315	.347	.001	.349	.222	.372	.057

10 and $p = 10, \infty$. (For other q and p, they may be obtained easily from (4.1).) It is interesting to see how little $\alpha_0^{(p,j)}$ varies in j: for $p = 10$, it is the same to three decimal places, for all j. In fact, for $p = 10$ the whole vector $\alpha^{(p,j)}$ varies little, especially for $q = 10$. There is a 7-dimensional set of optimum $\alpha^{(p)}$'s in this case, but it is a fairly small set, with α_0, α_1, α_2 almost constant and $\Sigma_3^q\ \alpha_j$ small; this is because p is near enough to p_{crit}" that $\alpha_j^{(10,j)}$ is small. The variation for $p = \infty$ is much greater.

An important practical feature of the nonuniqueness of $\xi^{(p)}$ is that, if Φ_p is the criterion of primary interest, one can choose an $\alpha^{(p,j)}$ or a combination of such designs that perform as satisfactorily as possible according to some secondary criterion involving features other than $M(\xi)$, such as a measure of performance if the model is cubic rather than quadratic (treated in another paper [5]).

5. INEFFICIENCY RATIOS

As mentioned in the introduction, a convenient assessment of the operating characteristic of a design ξ' may be obtained by calculating, for a large family $\{\Phi\}$ of criteria, the ratio $\Phi(\xi')/\min \Phi(\xi)$. In this section we make such calculations when ξ' is $\xi^{(r)}$ for a specified r in $[0, +\infty]$ and the family of criteria $\Phi^{(p)}$ is considered. Thus, we study the ratios

$$e_p(r) = \Phi^{(p)}(\xi^{(r)})/\Phi^{(p)}(\xi^{(p)}). \tag{5.1}$$

We first derive theoretical bounds on such ratios for arbitrary design settings, and then examine the calculated ratios for our simplex model.

A. *Theoretical bounds.* We restrict consideration here to the extreme but perhaps practically most important cases $p = 0, \infty$ in (5.1). In both cases one can obtain sharp bounds (unimprovable in the "worst" cases) when consideration is taken of the class $\mathcal{L}_k$ of all k-parameter linear models $(\mathfrak{X}, f)$. Write $h = k - 1$, and for $0 \le \beta \le 1$ define

$$g(\beta, k) = \max_{0 \le x \le 1} \frac{1 + hx^\beta}{1 + hx}. \tag{5.2}$$

We shall state the results without proof.

Theorem 5.1. For $0 \le r < \infty$, the inefficiency ratio $e_\infty(r)$ satisfies

$$\sup_{\mathcal{L}_k} e_\infty(r) = g(r/(r + 1), k). \tag{5.3}$$

Intuitively, (5.2) arises here because the poorest performance of $\xi^{(r)}$ in terms of Φ_∞ turns out to occur in examples where $\mathfrak{X} = \{1, 2, \cdots, k\}$ and $\|f_i(j), 1 \le i, j \le k\|$ has diagonal elements $1, A, A, \cdots, A$ and zero off-diagonal elements, where $A > 1$. For this setting, straightforward computations yield $\xi^{(r)}$, $\xi^{(\infty)}$, and the value $e_\infty(r) = (1 + hA^{-2r/(r+1)})/(1 + hA^{-2})$, the maximum of which (over A) is (5.2).

The quantity $g(\beta, k)$ does not have a simple expression in terms of elementary functions. The maximizing value of x in (5.2) is the root in $[0, 1]$ of $h(1 - \beta)x^\beta - \beta x^{\beta - 1} + 1 = 0$. When $\beta = 0$,

$$e_\infty(0) = k; \tag{5.4}$$

and as $\beta \to 0$, we obtain by standard asymptotics,

$$e_\infty(r) \sim k - h\beta \log \beta^{-1}. \tag{5.5}$$

Also, $\lim_{r \to \infty} e_\infty(r) = 1$ and, for the A-optimum design,

$$e_\infty(1) = (k^{1/2} + 1)/2. \tag{5.6}$$

The upper bound for e_0 is of even simpler form:
Theorem 5.2. For $0 < r \le \infty$,

$$\sup_{\mathcal{L}_k} e_0(r) = +\infty. \tag{5.7}$$

This means that an A- or E-optimum deisgn can have arbitrarily poor efficiency in terms of the D-optimality criterion, in *some models* (such as that described just below (5.3), with A large); what happens in our quadratic model on the simplex, will be discussed next.

B. *Results for the quadratic simplex model.* As is often the case in applied mathematics, results in important examples are much better than those predictable a priori from universal ($\mathcal{L}_k$, here) bounds. We have computed $e_p(r)$ for a large set of values p, r and for $q \le 10$, and the values obtained are much smaller than those implied by (5.3) (let alone (5.7)!). A possible measure of the performance of $\xi^{(r)}$ for considerations other than those given by Φ_r (really a measure of robustness under change of criterion in the family $\{\Phi_p\}$) is $\max_p e_p(r)$, which expresses the performance of $\xi^{(r)}$ under the criterion Φ_p for which $\xi^{(r)}$ is least efficient. This is recorded in Table 4 for D-, A-, and E-optimum designs ($r = 0, 1, \infty$), one column being devoted to each of these designs. For example, the first column states that, for dimension $q = 5$, the D-optimum design $\xi^{(0)}$ has ratio $e_0(r)$ of (5.1) at most equal to 1.83, for all r. Except when $q = 1$, we see that the E-optimum design performs better than the D- and A-optimum designs in these terms. Incidentally,

TABLE 4—*Selected Values of* $\max_{0 \le p \le \infty} e_p(r)$.

q \ r	0	1	∞
1	1.11	1.04	1.05
2	1.25	1.26	1.10
3	1.44	1.47	1.19
4	1.63	1.61	1.24
5	1.83	1.73	1.29
6	2.03	1.83	1.35
7	2.24	1.92	1.39
8	2.44	2.01	1.44
9	2.65	2.08	1.48
10	2.86	2.16	1.52

the supremum in $\sup_p e_p(0)$ occurs at $p = \infty$ for all q, and $\sup_p e_p(1)$ is attained at $p = \infty$ if $q > 1$ (and at $p = 0$ if $q = 1$); while $e_p(\infty)$ is maximized at or near $p = 0$.

We have not bothered to compute the ξ^r that minimizes $\max_p e(r)$ (or the possibly more general ξ that minimizes $\max_p [\Phi_p(\xi)/\Phi_p(\xi^{(p)})]$), since this criterion is at best an imprecise guide to satisfactory performance. However, we record here some values of r that improve upon the minimum of the three values recorded in Table 4 for each q: For $2 \le q \le 10$, various values of r between 1 and ∞ (5 is always an example) yield a smaller value of $\sup_p e_p(r)$ than $\sup_p e_p(\infty)$. (When $q = 1$, $\sup_p e_p(.5)$ improves upon $\sup_p e_p(1)$.)

We remark that the design $\xi^{(\infty)}$ had the largest ratio $\bar{d}(\xi^{(r)})/\bar{d}(\xi^{(0)}) = 2\bar{d}(\xi^{(r)})/(q+1)(q+2)$. This ratio, discussed in the paragraph below (2.2), goes from 1.375 when $q = 1$ to 1.54 when $q = 4$, to 1.83 when $q = 7$, to 1.96 when $q = 10$. These values are to be compared with the corresponding, somewhat smaller, values of $\Phi_0(\xi^{(\infty)})/\Phi_0(\xi^{(0)})$ in the last column of Table 4.

In summary, all of the $\xi^{(r)}$'s perform reasonably well in the sense considered here; $\xi^{(\infty)}$ (which, from its concentration on one or a few eigenvalues, we had thought might prove least satisfactory) is remarkably efficient under Φ_p.

REFERENCES

[1] BOX, G. E. P. and DRAPER, N. R. (1959). A basis for the selection of a response surface design. *J. Amer. Statist. Assoc. 54*, 622–654.
[2] BROYDEN, C. G. (1967). Quasi-Newton methods and their application to function minimisation. *Math. Comp. 21*, 368–381.
[3] DRAPER, N. R. and ST. JOHN, R. C. (1974). Models and designs for experiments with mixtures. Univ. of Wisc. Technical Reports, MRC 1435, 1436, 1437.
[4] FEDOROV, V. V. (1969, Transl. 1972), *Theory of Optimal Experiments*. Academic Press (New York).
[5] GALIL, Z. and KIEFER, J. (1977). Comparison of Box-Draper and D-optimum designs for experiments with mixtures. *Technometrics 19*, 441–444.
[6] KIEFER, J. (1959). Optimum experimental designs. *J. Roy. Statist. Soc. B 21*, 272–319.
[7] KIEFER, J. (1960). Optimum experimental designs V, with applications to systematic and rotatable designs. *Proc. 4th. Berk. Symp. 1*, 381–405.
[8] KIEFER, J. (1961). Optimum designs in regression problems, II. *Ann. Math. Statist. 32*, 298–325.
[9] KIEFER, J. (1973). Optimum designs for fitting biased multiresponse surfaces. *Proc. Multivar. Symp. III*, 287–297. Academic Press, N.Y.
[10] KIEFER, J. (1974). General equivalence theory for optimum designs (approximate theory). *Ann. Math. Statist. 75*, 849–879.
[11] KIEFER, J. (1975). Optimal design: Variation in structure and performance under change of criterion. *Biometrika, 62*, 277–288.
[12] KIEFER, J. (1977). Asymptotic approach to families of design problems, to appear in *Commun. in Statist.*
[13] KIEFER, J., and WOLFOWITZ, J. (1960). The equivalence of two extremum problems, *Canad. J. Math. 12*, 363–6.
[14] PESOCHINSKY, L. (1975). D-optimum and quasi-D-optimum designs on a cube, *Biometrika* 62, 335–340.
[15] SCHEFFÉ, H. (1958). Experiments with mixtures. *J. Roy. Statist. Soc., B, 21*, 344–360.

Comparison of Box-Draper and D-Optimum Designs for Experiments with Mixtures

Z. Galil and J. Kiefer
Cornell University
Ithaca, New York 14853

In continuation of the study of [10] and [5], designs for quadratic regression are considered when the possible values of the controlable variable are *mixtures* $x = (x_1, x_2, \cdots, x_{q+1})$ of nonnegative components x_i with $\sum_1^{q+1} x_i = 1$. The "all-bias" design of Box and Draper [1] for guarding against cubic bias, and the design that is optimum with respect to the D-optimality criterion ignoring the possibility of such bias, are compared in terms of the average and the maximum of the variance and bias functions of the fitted regression. The D-optimum design performs well in terms of the Box-Draper criterion unless the sample size is fairly large, and is superior in terms of maximum variance and bias.

KEY WORDS

Optimum designs
Simplex designs
Experiments with mixtures
Robust designs
Biased regression
Box-Draper model

1. INTRODUCTION

This paper continues the study of [5], of detailed numerical comparisons of the performances of various designs for the setting of experiments with mixtures considered by Scheffé [11] and others. The reader is referred to [5] for discussion of this model in our context of comparison of several designs with respect to a number of different optimality criteria.

The present paper treats departures from the exact quadratic model assumed in the computations of [5], in the form of a model introduced by Box and Draper (BD). The BD model takes into account the unlikelihood that the quadratic model is exactly correct, by considering also the bias one suspects to be present from cubic or other terms. For simplicity of discussion, we hereafter limit consideration to cubic terms, as is also done in much of the BD literature. The reader may consult the pioneering paper of Box and Draper [1], and also Kiefer [8], for details of the arithmetic, which will be sketched briefly, just below. The modifications of the original Box-Draper prescription introduced by Karson, Manson, and Hader [6] are omitted here for the sake of brevity, as are exact design considerations; these exhibit results similar to those we shall list herein.

Research under NSF Grant MPS 72-04998 A02.
AMS Subject Classification 62K05.

Received May 1977

As in [5], we suppose the space $\mathfrak{X}$ of values of the controlable variable $x = (x_1, \cdots, x_{q+1})$ consists of mixtures of $q + 1$ components, so that $\mathfrak{X}$ is the q-simplex $S_q = \{(x_1, \cdots, x_{q+1}): \sum_{1}^{q+1} x_i = 1, \text{ all } x_i \geq 0\}$. We again denote by $\mathbf{f}$ the column vector of $(q + 1)(q + 2)/2$ functions x_i and $x_i x_j$ with $1 \leq i, j \leq q + 1$ and $i < j$; all second degree polynomials on S_{q+1} can be expressed as linear combinations of these monomials. We let $\mathbf{g}$ denote the vector of $(q + 5)(q + 1)q/6$ functions $x_i^2 x_j (i \neq j)$ and $x_h x_i x_j$ $(h < i < j)$. We now suppose the actual expectation of an observation at the point x of S_{q+1} is $\theta'\mathbf{f}(x) + \beta'\mathbf{g}(x)$, where the vectors θ and β are coefficient vectors of unknown parameters corresponding to the elements of $\mathbf{f}$ and $\mathbf{g}$. Together, $\mathbf{f}$ and $\mathbf{g}$ span the space of polynomials of degree ≤ 3 on S_q. There is some redundancy in $\mathbf{g}$, in that the dimension of the space of polynomials of degree ≤ 3 exceeds that of the space of polynomials of degree ≤ 2 by only $(q + 2)(q + 1)q/6$. However, the computations required for certain comparisons are more easily performed in terms of $\mathbf{g}$ rather than in terms of a minimal set of cubic functions. Write

$$(1.1) \qquad \begin{pmatrix} M(\xi) & L(\xi) \\ L'(\xi) & K(\xi) \end{pmatrix} = \int_{\mathfrak{X}} \begin{pmatrix} \mathbf{f}(x) \\ \mathbf{g}(x) \end{pmatrix} \begin{pmatrix} \mathbf{f}(x) \\ \mathbf{g}(x) \end{pmatrix}' \xi(dx),$$

where, as in [5], ξ is a probability measure which, in the "approximate design theory", specifies the proportion of observations to be taken at each point of $\mathfrak{X}$. Thus, M and K are square matrices of order $(q + 1)(q + 2)/2$ and $(q + 5)(q + 1)q/6$, respectively, and M is the information matrix of the design (per observation and unit of variance) when the quadratic model is true ($\beta = 0$); the study of [5] compared various designs ξ in terms of certain optimality functions of $M(\xi)$.

Following the BD development, suppose that we use the method of least squares, *assuming the quadratic model* $\beta = 0$, to fit a quadratic surface on S_q to the data. If, in fact, $\beta \neq 0$, the mean-squared error of the fitted surface at x from the true regression $\theta'\mathbf{f}(x) + \beta'\mathbf{g}(x)$, if N observations are taken according to the prescription ξ and the observations are uncorrelated with variance σ^2, is then

$$(1.2) \qquad \sigma^2 N^{-1} d(x, \xi) + \{\beta'[L'(\xi)M^{-1}(\xi)\mathbf{f}(x) - \mathbf{g}(x)]\}^2;$$

the first term of (1.2) is the variance function of the design as in [5].

Box and Draper suggest measuring performance of ξ in terms of the integral of (7.2) with respect to the uniform probability measure $\xi^{(U)}$ on $\mathfrak{X}$. This integration yields

$$(1.3) \quad \sigma^2 r_N(\xi, \sigma^{-1}\beta) = \sigma^2\{V_N(\xi) + \sigma^{-2}\beta' B(\xi)\beta\},$$

where

$$(1.4) \qquad NV_N(\xi) = V_1(\xi) = \operatorname{tr} M(\xi^{(U)})M^{-1}(\xi)$$

and

$$(1.5) \quad B(\xi) = \{K(\xi^{(U)}) - L'(\xi^{(U)})M^{-1}(\xi^{(U)})L(\xi^{(U)})\} + [L'(\xi)M^{-1}(\xi) - L'(\xi^{(U)})M^{-1}(\xi^{(U)})] M(\xi^{(U)})[L'(\xi)M^{-1}(\xi) - L'(\xi^{(U)})M^{-1}(\xi^{(U)})]'.$$

The minimum of r_N with respect to ξ depends on $\sigma^{-1}\beta$, which is unknown. Box and Draper suggest minimizing the bias term alone, and it follows from (1.5) that any ξ for which

$$(1.6) \qquad L'(\xi)M^{-1}(\xi) = L'(\xi^{(U)})M^{-1}(\xi^{(U)})$$

minimizes $\beta' B(\xi)\beta$ for all β. The practical justification for minimizing $B(\xi)$ alone is said to be that, for typical values of $\sigma^{-1}\beta$ that one could often expect in applications, a design that minimizes $B(\xi)$ alone comes close to minimizing r_N. This will be the subject of part of our numerical study.

Since (1.6) is satisfied if $L(\xi) = L(\xi^{(U)})$ and $M(\xi) = M(\xi^{(U)})$, a convenient prescription is to let ξ be any design for which all moments through those of sixth degree (more common in practice than fifth degree, which is sufficient) agree with those of the uniform measure $\xi^{(U)}$. We will denote any such design by $\xi^{(BD)}$; all properties of $\xi^{(BD)}$ considered here do not depend on any other properties of the design.

2. COMPARISON OF DESIGNS

The comparison of $\xi^{(BD)}$ with other designs in terms of r_N would be too exhaustive if we were to try to table the performance for a large collection of values of $\sigma^{-1}\beta$. One possibility is to compare integrals of r_N with respect to some measure on $\sigma^{-1}\beta$. The different nature of the components $x_i^2x_j$ and $x_hx_ix_j$ of $\mathbf{g}$ makes the meaning of such an integration more difficult to grasp intuitively than does the comparison that follows, which is perhaps often more favorable to $\xi^{(BD)}$ since it weighs B more heavily compared with V_1 than do many such integrations. A second possibility is to compute the maximum eigenvalue of $B(\xi)$, but we replace this formidable computation by an easier one of similar spirit: We consider the performance of r_N when $\sigma^{-1}\beta$ has *only one nonzero component*, which we can take to be either $\sigma^{-1} \times$ (coefficient in β of $x_1^2x_2$) $= \sigma^{-1}\beta_{112} = \rho_{112}$ (say), or else $\sigma^{-1} \times$ (coefficient in β of $x_1x_2x_3$) $= \rho_{123}$ (say). The corresponding diagonal elements of $B(\xi)$ we denote by $\gamma_{112}(\xi)$ and $\gamma_{123}(\xi)$. Our computations showed $\gamma_{112}(\xi) > \gamma_{123}(\xi)$ for $2 \leq q \leq 10$ and $\xi = \xi^{(0)}$ or $\xi^{(BD)}$. (For $q = 1$ there is no γ_{123}.) Thus, for both of these ξ's, assuming ρ_{112} = a given value h always yields a larger value of B than does assuming $\rho_{123} = h$. It does not give as large a value ($h^2\gamma_{112}$) as the maximum of $\sigma^{-2}\beta' B\beta$ on the sphere $\sigma^{-2}\beta'\beta = h^2$ (which is the value associated with the maximum eigenvalue of B), but seems satisfactory for yielding a meaningful comparison. It may, of course, give a larger value of r_N than do various other B satisfying $\sigma^{-2}\beta'\beta = h^2$.

We have chosen a particular design, the D-optimum design *assuming* the quadratic model $\beta = 0$, to compare with $\xi^{(BD)}$ in terms of r_N. This design, denoted by $\xi^{(0)}$ here as in [5], was proved in [7] to assign equal numbers $2N/(q + 1)(q + 2)$ of observations to each of the vertices and midpoints of edges of S_q. This choice of $\xi^{(0)}$ as a design with which to compare $\xi^{(BD)}$ was not made because of any intuitive feeling that $\xi^{(0)}$ should do well in terms of r_N; indeed, minimization of $V_1(\xi)$, an "L-optimality" functional ([4], [9]), is closer in spirit to the computation of $\xi^{(1)}$ of [5] than of $\xi^{(0)}$. The design $\xi^{(0)}$ was chosen because of the frequent use of the D-optimality criterion when β is assumed 0, and because of its equivalent meaning of G-optimality under the quadratic model, discussed in [5] (minimization of the maximun of the variance function,

$$N^{-1}\sigma^2\bar{d}(\xi) = N^{-1}\sigma^2 \max_{x \in S_q} \mathbf{f}(x)M^{-1}(\xi)\mathbf{f}(x)).$$

Suppose N observations are taken according to $\xi^{(0)}$ or $\xi^{(BD)}$ and we assume $\rho_{112} = h$ and all other components of β are 0. If we ask for the smallest value of N for which $r_N(\xi^{(BD)}, \sigma^{-1}\beta) \leq r_N(\xi^{(0)}, \sigma^{-1}\beta)$ for such a β, we obtain that N is the smallest integer greater than or equal to N_q/h, where

$$(2.1) \qquad N_q = \{[V_1(\xi^{(BD)}) - V_1(\xi^{(0)})]/[\gamma_{112}(\xi^{(0)}) - \gamma_{112}(\xi^{(BD)})]\}^{1/2}.$$

TABLE 1—*Values of* N_q.

q	N_q
1	26.8
2	85.8
3	172.6
4	287.2
5	433.6
6	616.6
7	840.7
8	1110.5
9	1430.1
10	1804.1

From Table 1 we see that, even if $h = 1$, which means that *the cubic bias coefficient is as large as the standard error* σ, the value N_q of N that makes $\xi^{(BD)}$ worth while using over $\xi^{(0)}$ in terms of the *BD* criterion r_N, is in the hundreds for moderate q. If a realistic concern is that the cubic bias coefficient is $\leq \sigma/2$, the value of N is at least doubled; and so on.

The above compares $\xi^{(BD)}$ with $\xi^{(0)}$ only in terms of a single optimality functional, but it is the functional that motivated use of $\xi^{(BD)}$, and in terms of other criteria such as $\bar{d}$ or Φ_0 the performance of $\xi^{(BD)}$ is worse (a matter to be discussed further, in the second paragraph below). Even in terms of r_N, we see that *a fairly large sample size must be present to justify using* $\xi^{(BD)}$, if β_{112} is of the same order as σ.

Some other comments about the rationale of the BD approach are indicated in [10]. We repeat here only one remark, which is of some practical relevance: There is often nothing particularly compelling about the use of the uniform probability measure $\xi^{(U)}$ in integrating (1.2) to get (1.3). Suppose, instead, one used some other measure ν, for example, one whose moments through those of fifth degree were the same as for the design $\xi^{(P)}$ considered in [5], where we recall that $\xi^{(0)}$, $\xi^{(1)}$, and $\xi^{(\infty)}$ are, respectively, designs that are D-, A-, and E-optimum. (Such a measure ν can be quite different from $\xi^{(P)}$.) Then the resulting B would be minimized by $\xi^{(P)}$, which in addition has optimality features discussed in Section 4 of [5]. This is perhaps some reassurance to the person who uses $\xi^{(P)}$ and is upset by the point quite properly raised by Box and Draper, that departures from the model with $\beta = 0$ should be considered. He may find some comfort in the fact that his design also minimizes B, provided integration with respect to $\xi^{(U)}$ is altered in the manner described.

It would also seem relevant to compare designs in terms of the *maximum* over S_q of (1.2), rather than in terms of the average (1.3). Of course, the result depends on β, and the computation is much more difficult than those of the preceding paragraphs. We have therefore replaced that computation by a simpler one that will still give an indication of how large (1.2) can be for various designs: We look *separately* at $\bar{d}(\xi) = \max_{x \in S_q} d(x,\xi)$ and at

$$\bar{\gamma}_{112}(\xi) = \max_{x \in S_q} |[L'(\xi)M^{-1}(\xi)\mathbf{f}(x) - \mathbf{g}(x)]_{112}|, \tag{2.2}$$

where $[\ ]_{112}$ denotes the component corresponding to β_{112}. Thus, if $\beta_{112} = h\sigma$ and the other elements of β are zero, the second term of (1.2) has maximum $[h\sigma\bar{\gamma}_{112}(\xi)]^2$. The choice of the term corresponding to the single coefficient β_{112} was made for reasons similar to those in the earlier analysis.

The computation of (2.2) for $\xi^{(0)}$ is quite simple and yields the same value for all q:

$$\bar{\gamma}_{112}(\xi^{(0)}) = 3^{1/2}/36 = .0481. \tag{2.3}$$

The computation of $\bar{\gamma}_{112}(\xi^{(BD)})$ is more formidable because $L'M^{-1}$ has so many nonzero elements for $\xi^{(BD)}$. We therefore computed a *lower bound* $\bar{\bar{\gamma}}_{112}(\xi^{(BD)})$, obtained by replacing the maximum over S_q in (2.2) by the maximum over the *line* $\{x: x_1 + x_2 = 1\}$. (It is conceivable that $\bar{\gamma}_{112}(\xi^{(BD)}) = \bar{\bar{\gamma}}_{112}(\xi^{(BD)})$; the corresponding result is true for $\xi^{(0)}$.) The result, to be compared with (2.4) below, is given in Table 2.

The comparison of $\bar{d}$ values is simpler, and one obtains

$$\bar{d}(\xi^{(BD)}) = [\bar{d}(\xi^{(0)})]^2 = [(q+1)(q+2)/2]^2. \tag{2.4}$$

Thus, $\xi^{(BD)}$ is inferior to $\xi^{(0)}$ for all q in terms of both $\bar{\gamma}_{112}$ and $\bar{d}$, and increasingly so as q grows. It may well be pointed out that $\xi^{(BD)}$ was designed to minimize an average, not a maximum, of (1.2). Nevertheless, for

TABLE 2—*Values of* $\bar{\bar{\gamma}}_{112}(\xi^{(BD)})$.

q	$\bar{\bar{\gamma}}_{112}(\xi^{(BD)})$
1	.05
2	.25
3	.40
4	.52
5	.62
6	.70
7	.76
8	.82
9	.87
10	.91

all the sense in considering model departures in the manner begun by Box and Draper, the above comparison offers additional weight against the mechanical adoption of the particular design $\xi^{(BD)}$ in all simplex experiments where there is concern about cubic bias in a quadratic fit on S_q. We note that while $\xi^{(0)}$ minimizes $\bar{d}$, presumably one can find other designs that perform even better in terms of $\bar{\gamma}_{112}$ than (2.3).

In conclusion, we remark that the papers [2], [3] of Draper and Lawrence contain some comparisons of $\xi^{(BD)}$ with other designs, especially for the *first degree* model with *quadratic bias*. However, for the quadratic model with cubic bias that we have considered, the algebraic complexities we have mentioned led Draper and Lawrence to compare $\xi^{(BD)}$ only with the special family of designs obtained by scaling $\xi^{(BD)}$, with results differing from those for the first degree model, and with no comparison at all with $\xi^{(0)}$, the subject of our computations.

REFERENCES

[1] BOX, G. E. P. and DRAPER, N. R. (1959). A basis for the selection of a response surface design. *J. Amer. Statist. Assoc. 54*, 622–654.

[2] DRAPER, N. R., and LAWRENCE, W. L. (1965). Mixture designs for three factors. *J. Roy. Statist. Soc., B27*, 450–465.

[3] DRAPER, N. R., and LAWRENCE, W. L. (1965). Mixture designs for four factors. *J. Roy. Statist. Soc., B27*, 473–478.

[4] FEDOROV, V. V. (1969, Transl. 1972), *Theory of Optimal Experiments*. Academic Press (New York).

[5] GALIL, Z. and KIEFER, J. (1977). Comparison of simplex designs (approximate theory). *Ann. Math. Statist. 75*, 849–879.

[6] KARSON, M. J., MANSON, A. R., and HADER, R. J. (1969). Minimum bias estimation and experimental designs for response surfaces. *Technometrics, 11*, 461–476.

[7] KIEFER, J. (1961). Optimum designs in regression problems, II. *Ann. Math. Statist. 32*, 298–325.

[8] KIEFER, J. (1973). Optimum designs for fitting biased multi-response surfaces. *Proc. Multivar. Symp. III*, 287–297. Academic Press, N. Y.

[9] KIEFER, J. (1974). General equivalence theory for optimum designs (approximate theory). *Ann. Math. Statist. 75*, 849–879.

[10] KIEFER, J. (1975). Optimal design: Variation in structure and performance under change of criterion. *Biometrika, 62*, 277–288.

[11] SCHEFFÉ, H. (1958). Experiments with mixtures. *J. Roy. Statist. Soc.* (Ser. B), *21*, 344–360.

COMMUN. STATIST.-THEOR. METH., A7(14), 1347-1362 (1978)

ASYMPTOTIC APPROACH TO FAMILIES OF DESIGN PROBLEMS

J. Kiefer

Cornell University

Key Words and Phrases: near-optimum designs; computational techniques for finding optimum designs.

ABSTRACT

Examples are given that demonstrate the usefulness, for obtaining near-optimum designs in analytically intractable settings, of embedding such settings in a regular family of designs whose limit is easier to treat.

1. INTRODUCTION

By now it is well known that the computation of optimum designs can lead to formidable and analytically intractable computations, often the minimization of messy functions of many variables. This is true even in the approximate theory in which a design ξ is considered to be a probability measure on the space $\underline{X}$ of the controlable variable, to be implemented when N observations are taken by replacing $N\xi$ by a nearby measure taking on only integer values (which describes the number of observations to be taken at each point). See Kiefer (1959, 1974) or Fedorov (1972) for further explanation of terms used herein.

An old practice in mathematics is that of replacing a complicated function g_t which is to be minimized, by a simpler function g_0 which can be handled more easily. Often g_t can be thought of as belonging to a natural family $\{g_t\}$ for which $\lim_{t\to 0} g_t = g_0$, and if the functions and the paramétrization of this family are sufficiently smooth the minimizing argument u_0 for g_0 can be expected to come close to minimizing g_t for t small. It is possible to improve the technique by modifying u_0 by a simple correction in t, and to extend it to vectorial t.

It has been suggested (e.g., in Section 4.3 of Kiefer, 1975) that this technique can be employed profitably in many design settings. As in the case with other such approximation techniques in mathematics, easily obtainable a priori bounds on the goodness of the results are much more pessimistic than what one observes in practice; thus, it would seem valuable to try to improve upon rudimentary bounds of the type illustrated in Section 2, below. For the present, numerical results on the goodness of such approximations are given as the proof of the pudding in a few simple examples. In more complex settings in which the technique would be employed as a substitute for numerical minimization of g_t (so that we would of course not know the result of the latter computation with which to compare the minimization of g_0 as in the illustrations below), and in which no satisfactory a priori bounds are available, a useful estimate of the goodness of the method can often be obtained from a quadratic approximation through numerical evaluation of the first two derivatives of $g_t(u)$ with respect to t and u at $(t,u) = (0,u_0)$, when this makes sense.

Suppose, then, that we have a family $\underline{S} = \{S_t, t \in T\}$ of design problems, where for simplicity we consider herein only the case where T is a subset of the positive reals, with 0 as accumulation point. (The choice of 0 is an arbitrary one.) The design problem S_t consists of a specification of the space $\underline{X}_t$ of the controlable variable, the collection P_t of possible probability and decision models, and the optimality function Φ_t that is to be minimized (and which will be assumed positive or $+\infty$). In the present paper we suppose P_t, Φ_t are concerned with a linear regression model for which the observations are uncorrelated with equal variance, and for which best linear unbiased estimators will be used; thus, P_t is replaced by this covariance assumption and the k_t-vector f_t of functions on $\underline{X}_t$, the expected value of an observation at the point x of $\underline{X}_t$ being $\theta_t' f_t(x)$ where θ_t is the k_t-vector of unknown parameters, and the decision side of the problem is replaced by the assumption that least squares estimators are used. For a design (measure) ξ_t, the information matrix per observation and unit of variance is then $M_t(\xi_t) = \int_{\underline{X}_t} f_t(x)f_t(x)'\xi_t(dx)$, and Φ_t is a real-valued functional on the space of M_t's.

Our treatment depends on there being some regularity in the family of problems of minimizing $\Phi_t(M_t(\xi_t))$ for t in T. Often one can prove that $\Phi_t(M_t(\xi_t))$ is minimized (not necessarily uniquely) by some ξ_t^* in a family $\{\xi_{t,u}\}$ of measures on $\underline{X}_t$ that can be parametrized in terms of a parameter u taking on values in some finite-dimensional set U for all t. Suppose, furthermore, that Φ_t can be renormalized by means of a strictly increasing function h_t on $(0,+\infty]$ so that the function

$$g_t(u) = h_t(\Phi_t(M_t(\xi_{t,u}))) \tag{1}$$

approaches a limit $g_0(u)$ for each u in U, as $t \to 0$ in T. The method sketched earlier then amounts to minimizing g_0 and, if the minimum is attained at u_0, using ξ_{t,u_0} in place of ξ_t^* (presumed hard to compute) for the problem S_t. The ratio

$$e_t = h_t(\Phi_t(M_t(\xi_{t,u_0})))/h_t(\Phi_t(M_t(\xi_t^*))) \tag{2}$$

(or the corresponding expression with h_t omitted), which is always ≥ 1, gives a measure of inefficiency of ξ_{t,u_0} for the problem S_t. Finally, replacement of g_0 by a better approximation to g_t that is still tractable, for example by $g_0+t(\partial g_t/\partial t)|_{t=0+}$ if that makes sense, with minimum u_t, leads to using ξ_{t,u_t}, usually an improvement on ξ_{t,u_0} as a substitute for ξ_t^*.

The parametrization in terms of U, and the normalization in terms of g_t, deserve comment. The importance of choosing these in such a way as to yield useful results, and the method of implementing the approach, can be illustrated in the problem of finding a D-optimum design for quadratic regression on the unit ball $\overline{X}_t = \{(x_1,\ldots,x_q):\Sigma_1^q x_i^2 \leq 1\}$ of dimension $q = t^{-1}$, with f_t the vector of functions 1, x_i, $x_i x_j$ $(1 \leq i \leq j \leq q)$, for which $k_t = (q+1)(q+2)/2$. An invariance argument (Kiefer, 1960) shows that the optimum ξ_t^* is a member of the family $\{\tilde{\xi}_{t,v}, v>0\}$ that assigns measure v to the center of the ball and the remaining measure $1-v$ uniformly to the surface (the unit $(q-1)$-sphere). It is not hard to compute (Galil and Kiefer, 1977a) that

$$\det M^{-1}(\tilde{\xi}_{t,v}) = 2^{1-q}q^{q(q+3)/2}(q+2)^{(q+2)(q-1)/2}v^{-1}(1-v)^{-q(q+3)/2} \tag{3}$$

which, at the minimum $v = v_t = 1/k_t$, attains the value

$$\det M^{-1}(\xi_t^*) = 2^{-q}(q+1)^{(q+1)(q+2)/2}(q+2)^{q(q+2)}(q+3)^{-q(q+3)/2}. \tag{4}$$

Thus, $\min_v[\det M^{-1}(\tilde{\xi}_{t,v})]^{1/k_t} \sim t^{-2}$ as $t \to 0$, so that, even after the customary reduction of taking the k_tth root of the generalized variance, we have the criterion becoming unbounded for every $\tilde{\xi}_{t,v}$ as $t \to 0$. Thus, $h_t(\Phi_t(M_t)) = [\det M_t^{-1}]^{1/k_t}$ would not work in this example; we might try $h_t(\Phi_t(M_t)) = t^2[\det M_t^{-1}]^{1/k_t}$ instead. With this last choice we would obtain

$$\tilde{g}_t(v) = t^2 \, [\det M_t^{-1}(\tilde{\xi}_{t,v})]^{1/k_t}$$
$$= c_t v^{-1/k_t}(1-v)^{1-1/k_t} \tag{5}$$

where $\lim_{t \to 0} c_t = 1$. This would yield

$$\tilde{g}_0(v) = \begin{cases} 1-v & \text{if } 0 < v < 1, \\ +\infty & \text{if } v = 0 \text{ or } 1, \end{cases} \tag{6}$$

and thus the minimum is not attained but is approached as $v \to 0$, with $v = 0$ yielding the useless $\tilde{g}_t(0) = +\infty$ for all t. The parametrization in terms of v was thus unproductive; if we instead write $u = v/2t^2$ and $\xi_{t,u} = \tilde{\xi}_{t,2t^2u}$, and set

$$g_t(u) = k_t \log \tilde{g}_t(u/k_t) - \log k_t, \tag{7}$$

we obtain $g_0(u) = u - \log u$, with a minimum at $u = u_0 = 1$. The design ξ_{t,u_0} puts measure $2t^2$ at 0 (which only makes sense for $q \geq 2$). The resulting value of $[\det M_t^{-1}(\xi_{t,u_0})/\det M_t^{-1}(\xi_t^*)]^{1/k_t}$ is close to 1, being already 1.3 when $q = 2$ and 1.1 when $q = 3$, and even $\det M_t^{-1}(\xi_{t,u_0})/\det M_t^{-1}(\xi_t^*)$ approaches 1 in this case. If we had instead defined $u = k_t v$, we would still have obtained $u_0 = 1$, and thus ξ_{t,u_0}, putting mass $1/k_t$ at 0, would have been an optimum ξ_t^*; but such a fortuitous accident can of course not be expected in more complex examples.

The straightforward use of this approach will not always yield useful results. For example, if $T = \{n^{-1}, n \text{ positive integer}\}$ and $\underline{X}_t = [-1,1]$ for all t, with the problem being D-optimum estimation for quadratic regression if t^{-1} is odd and for cubic regression if t^{-1} is even, the family of problems does not behave regularly enough as $t \to 0$ for the method to be useful. More interesting is the consideration in Kiefer and Studden (1976) of polynomial regression of degree n on [-1,1] for $t = n^{-1}$, in which the G- (or D-) optimum design measures for successive degrees approach a limit ξ_0 whose efficiency is only about 1/2 (inefficiency measure 2) for large n, and which requires a rather subtle modification of ξ_0 to yield a sequence that is asymptotically efficient. Nevertheless, even in such cases of using the asymptotic approach as the degree of the regression increases, we obtain, perhaps with further analysis, interesting and useful results. Work of this nature has been pursued further by Studden.

The examples we shall discuss further herein are ones in which the degree of the regression is fixed but the problems change with t in one of three ways: (a) the dimension of $\underline{X}_t$ is t^{-1}; (b) the regression is on $\underline{X}_t = t\underline{X}_1$ or $t^{-1}\underline{X}_1$, a homothetic transformation of a fixed region $\underline{X}_1$; or (c) the observations are taken in $\underline{X}_1$ for all t, but the problem is one of extrapolation to $t^{-1}\underline{X}_1$ or interpolation to $t\underline{X}_1$. Of these, (c) is expressed and (b) can be rewritten in our examples as problems in which, for a fixed $\underline{X}_t$ and f_t, we are faced with a set of different functionals Φ_t of $M(\xi)$. Problems of this last nature, in which the optimality criterion is varied, are the main concern of Kiefer (1975), Galil and Kiefer (1977a,b,c), etc., where various designs are compared in terms of variation in p of the criterion $\Phi^{(p)}$, defined on k×k information matrices M with eigenvalues $\lambda_1(M),\ldots,\lambda_k(M)$ by

$$\Phi^{(p)}(M) = (k^{-1}\mathrm{tr}\, M^{-p})^{1/p} = [k^{-1}\Sigma_1^k\lambda_i^{-p}(M)]^{1/p} \text{ for } 0<p<\infty,$$

$$\Phi^{(0)}(M) = \lim_{p\to 0}\Phi^{(p)}(M) = [\det M]^{-1/k} \quad \text{(D-optimality)},$$

$$\Phi^{(\infty)}(M) = \lim_{p\to+\infty}\Phi^{(p)}(M) = 1/\min_i\lambda_i(M) \quad \text{(E-optimality)}; \tag{8}$$

although the emphasis there is different, the approximation of

$\Phi^{(p)}$-optimum designs for p near a value p_0 for which the $\Phi^{(p_0)}$-optimality computation is analytically tractable can make use of the present viewpoint. It is not hard to list other settings $\underline{S}$ where the approach can be implemented; also, a combination of considerations such as those of (a) and (c) leads to consideration of vectorial t. The example of the next section illustrates other products of the approach in suggesting optimality structure for $t > 0$ and in helping to determine the change in that structure as t varies.

2. EXAMPLE OF DIMENSIONAL VARIATION: QUADRATIC REGRESSION ON THE q-SIMPLEX

This is an example which, in outline, is again of type (a), like the illustration of Section 1, but for which the arithmetic is much less simple.

We again write $t = q^{-1}$ for q a positive integer, and $\underline{X}_t$ is the q-dimensional simplex $\{(x_1, x_2, \ldots, x_{q+1}): x_j \geq 0\ \forall j$ and $\Sigma_1^{q+1} x_i = 1\}$. The problem is that of quadratic regression for "mixtures" of $q+1$ ingredients as described in Scheffé (1958), Kiefer (1961), Galil and Kiefer (1977b). The components of f_t can be taken as x_i and $x_i x_j$, with $1 \leq i, j \leq q+1$ and $i < j$; these are linearly independent and, once more, $k_t = (q+1)(q+2)/2$. It has been shown in Kiefer (1974,1975) that all $\Phi^{(p)}$-optimum designs are supported by the set J_q of <u>barycenters</u> of $\underline{X}_t$; here $J_q = \cup_0^q J_{q,j}$ where $J_{q,j}$ is the set of <u>barycenters of depth</u> j, each point of which has $j+1$ coordinates equal to $1/(j+1)$ and the remaining coordinates equal to 0. Moreover, the cited references show that there is a <u>symmetric</u> Φ_p-optimum design (not necessarily the only optimum design) on J_q, one for which the total measure $\alpha_{q,j}$ on $J_{q,j}$ is spread uniformly over the $\binom{q+1}{j+1}$ points of $J_{q,j}$. We write $\underline{\alpha}_q$ for the probability $(q+1)$-vector $(\alpha_{q,0}, \alpha_{q,1}, \ldots, \alpha_{q,q})$, and hereafter only consider symmetric designs on J_q; the $\Phi^{(p)}$-optimum $\underline{\alpha}_q$ will be denoted $\underline{\alpha}_q^{*(p)}$.

We define certain functions a,b,c,d,e,f,g of $\underline{\alpha}_q$ by

$$(q+1)a-1 = -\Sigma_1^q\, j(j+1)^{-1}\alpha_{q,j} = -q(q+1)b,$$

$$q(q+1)b = \Sigma_1^q\, j(j+1)^{-1}\alpha_{q,j}$$

$$q(q+1)c = \Sigma_1^q\, j(j+1)^{-2}\alpha_{q,j},$$

$$(q-1)q(q+1)d = \Sigma_1^q\, j(j-1)(j+1)^{-2}\alpha_{q,j} = q(q+1)(b-2c),$$

$$q(q+1)e = \Sigma_1^q\, j(j+1)^{-3}\alpha_{q,j},$$

$$(q-1)q(q+1)f = \Sigma_1^q\, j(j-1)(j+1)^{-3}\alpha_{q,j} = q(q+1)(c-2e),$$

$$(q-2)(q-1)q(q+1)g = \Sigma_1^q\, j(j-1)(j-2)(j+1)^{-3}\alpha_{q,j}$$

$$= q(q+1)(b-5c+6e). \qquad (9)$$

These functions are thus all linear functions of b,c,e, but we have exhibited them all in order to write out, as succinctly as possible, the eigenvalues of $M(\underline{\alpha}_q)$:

$$\lambda = e+g-2f, \text{ of multiplicity } (q+1)(q-2)/2;$$

the zeros of

$$\lambda^2-[a-b+e+(q-3)f-(q-2)g]\lambda$$
$$+\{(a-b)[e+(q-3)f-(q-2)g]-(q-1)(c-d)^2\},$$

each of multiplicity q;

the zeros of

$$\lambda^2-[a+qb+e+(q-1)(q-2)g/2+2(q-1)f]\lambda$$
$$+\{(a+qb)[e+(q-1)(q-2)g/2+2(q-1)f]-q[2c+(q-1)d]^2/2\},$$

each of multiplicity 1. (1

(For $q = 2$, the first of these is absent; a slightly different form holds in the trivial case $q = 1$.)

Except when $p = 0$, the case treated in Kiefer (1961), the exact minimization of $\Phi^{(p)}(M(\underline{\alpha}_q))$ is analytically intractable, and numerical methods are used to obtain $\underline{\alpha}_q^{(p)}$ in Galil and Kiefer (1977b). Especially hard is the accurate computation of the E-optimum design $\underline{\alpha}_q^{(\infty)}$, which stems from the nondifferentiability of $\Phi^{(\infty)}$ and the consequent failure of descent methods in the numerical analysis and design literature that are based on differentiability. Consequently, although computations of the type carried out in this section can be used for other criteria, we give the development here only for $\Phi^{(\infty)}$; thus, the Φ_t of Section 1 is $\Phi^{(\infty)}$ for the t^{-1}-simplex. How useful the approach is will be measured not only by how far the obtained approximation $\bar{\underline{\alpha}}_q^{(\infty)}$ (the ξ_{t,u_0} of Section 2) is from the optimum $\underline{\alpha}_q^{*(\infty)}$, but on the inefficiency ratio of eq. (2), which we now write as

$$\bar{e}_q = e_{1/t} = \min_i \lambda_i(M(\underline{\alpha}_q^{*(\infty)}))/\min_i \lambda_i(M(\bar{\underline{\alpha}}_q^{(\infty)})). \tag{11}$$

Let b,c,e be fixed values compatible with (9) for all sufficiently large q, and for which the right-hand expressions of (12) below are positive. We define $\lambda_1,\lambda_2,\lambda_3$, to be the first eigenvalue listed in (10) and the smaller root of each of the two quadratics in λ there; thus, $\min_{1\leq i\leq k} \lambda_i(M(\underline{\alpha}_q)) = \min_{1\leq i\leq 3}\lambda_i(M(\underline{\alpha}_q))$. A straightforward computation from (10) and (9) then shows that the three eigenvalues of interest satisfy, as $q \to \infty$,

$$\begin{aligned} q^2\lambda_1 &\sim e, \\ q^2\lambda_2 &\sim c-e-c^2/(1-b), \\ q^2\lambda_3 &\sim [b-c-b^2]/2. \end{aligned} \tag{12}$$

The problem of maximizing the minimum of the three right-hand expressions of (12) where b,c,e, are obtained from probability vectors $\underline{\alpha}_{\infty}$ by setting $q = \infty$ in (9), is easily verified to be solved by minimizing e subject to equality of the three right-hand expressions of (11). That minimization is a simple algebraic calculation, in contrast with the complicated problem of maximizing $\min_i \lambda_i(M(\underline{\alpha}_q))$ for fixed q. We obtain

$$b = 7/16,\ c = 9/64,\ e = 27/512 \tag{13}$$

for the solution; it is easy to verify that this indeed uniquely maximizes $\lim_{q\to\infty} q^2 \min(\lambda_1,\lambda_2,\lambda_3)$, in a manner described in Section 3 of Galil and Kiefer (1977b) for the corresponding problem for fixed q. It can be verified that the optimum (b,c,e) for dimension q must approach those of (13) as $q\to\infty$, and that a corresponding convergent sequence of $\underline{\alpha}_q^{*(\infty)}$'s will approach an $\bar{\underline{\alpha}}^{(\infty)}$ (say) determined by (2.5), whose possible values we now discuss.

From (13) and (9) with $q = \infty$, the components of $\bar{\underline{\alpha}}^{(\infty)} = (\bar{\alpha}_0, \bar{\alpha}_1, \bar{\alpha}_2, \ldots)$ must satisfy

$$\Sigma_{j\geq 1}\ j(j+1)^{-1}\bar{\alpha}_j = 7/16,$$
$$\Sigma_{j\geq 1}\ j(j+1)^{-2}\bar{\alpha}_j = 9/64,$$
$$\Sigma_{j\geq 1}\ j(j+1)^{-3}\bar{\alpha}_j = 27/512. \tag{14}$$

Since $j(j+1)^{-1} \geq 1/2$, it follows from the first equation of (14) that $\bar{\alpha}_0 \geq 1/8$ in every solution. Since for $j \geq 2$ we have $j(j+1)^{-2}/j(j+1)^{-3} \geq 3 > (9/64)/(27/512)$, it follows that we cannot have $\bar{\alpha}_1 = 0$ in any solution. Thus, we need only seek solutions $\bar{\underline{\alpha}}^{(\infty)}$ with $\bar{\alpha}_0, \bar{\alpha}_1 > 0$.

We look for basic solutions in which only four $\bar{\alpha}_i$'s are ≥ 0, say $\bar{\alpha}_0$, $\bar{\alpha}_1$, $\bar{\alpha}_j$, $\bar{\alpha}_h$, with $j < h$. It is clear that every $\bar{\underline{\alpha}}^{(\infty)}$ is a convex combination of these basic solutions. Solving the linear equations (13) for $\bar{\alpha}_j$, $\bar{\alpha}_h$, one finds that these are nonnegative if and only if $h \geq 4$ and $j \leq 3$ (that is, $j = 2$ or 3). The corresponding $\bar{\alpha}_1$ and $\bar{\alpha}_0$ are then verified to be positive in all these cases. The two families of solutions are, for $j = 2$, $h \geq 4$,

$$\bar{\alpha}_0 = (171h^3-526h^2+381h-26)/512h(h-1)(h-2),$$

$$\bar{\alpha}_1 = (9h^2-h-34)/64(h-1)(h-2),$$

$$\bar{\alpha}_2 = 27(9h-31)/512(h-2),$$

$$\bar{\alpha}_h = 13(h+1)^3/256h(h-1)(h-2); \qquad (15)$$

and, for $j = 3$, $h \geq 4$,

$$\bar{\alpha}_0 = (63h^3-253h^2+193h-3)/192h(h-1)(h-3),$$

$$\bar{\alpha}_1 = (9h^2-34h+21)/32(h-1)(h-3),$$

$$\bar{\alpha}_3 = (9h-31)/24(h-3),$$

$$\bar{\alpha}_h = (h+1)^3/64h(h-1)(h-3). \qquad (16)$$

The above results imply the interesting fact that the $\alpha_q^{*(\infty)}$ obtained in Section 3 of Galil and Kiefer (1977b), and for which only $\alpha_{q,0}, \alpha_{q,1}, \alpha_{q,2}, \alpha_{q,3} > 0$, which existed for $q \leq 10$, cannot exist for sufficiently large q. On the other hand, for sufficiently large q there must be solutions with only $\alpha_{q,0}, \alpha_{q,1}, \alpha_{q,3}, \alpha_{q,h} > 0$ (for each $h \geq 4$), which were found not to exist for $q \leq 10$. We note also that the computations of $\underline{\bar{\alpha}}^{(\infty)}$ and of Galil and Kiefer (1977b) lead one to suspect that designs $\alpha_q^{*(\infty)}$ with only $\alpha_{q,0}, \alpha_{q,1}, \alpha_{q,2}, \alpha_{q,h} > 0$, with $h \geq 4$, exist for all $q \geq h$.

We consider here only the simplest implementation of our asymptotic approach, defining the design $\underline{\bar{\alpha}}_q^{(\infty)}$ to be $\underline{\bar{\alpha}}^{(\infty)}$ for all q, without any correction that depends on q. We assume $q \geq 4$, in order that a design of the series (15) or (16) exist. The ratio (11), computed by comparing $\bar{\underline{\alpha}}_q^{(\infty)}$ with the $\underline{\alpha}_q^{*(\infty)}$ obtained numerically, is 1.30 for $q = 4$, 1.12 for $q = 7$, and 1.08 for $q = 10$. The design $\bar{\underline{\alpha}}_q^{(\infty)}$ also performs reasonably well in terms of other criteria $\Phi^{(p)}$, behaving approximately the way the E-optimum design $\underline{\alpha}_q^{*(\infty)}$ does in the inefficiency ratios studied in Kiefer (1975) and Galil and Kiefer (1977b). A finer calculation, for example including terms of order q^{-1} on the right side of (11) with a resulting correction of order q^{-1} to the $\bar{\underline{\alpha}}_q^{(\infty)}$ obtained above, would presumably yield even better results.

We note that, in addition to providing useful approximations to E-optimum designs, which was the original aim of the approach, it also yielded two important theoretical aids: the nature of the optimum as occurring when the right sides of (12) are equal, which correctly suggested that the more difficult optimization problem for fixed q would be solved when $\lambda_1 = \lambda_2 = \lambda_3$; and the conclusions and conjectures about the possible dimensional structures of $\underline{\alpha}_q^{*(\infty)}$ for large q, described just below (16).

3. QUADRATIC REGRESSION ON THE q-BALL

We now fix the dimension q (not exhibiting it in this section) and consider an extrapolation problem of the type (c) mentioned near the end of Section 1. We suppose f_t is the vector of functions $1, x_i, x_i x_j, 1 \leq i \leq j \leq q$, where $\underline{X}_t$ is the unit q-ball $\{(x_1,\ldots,x_q): \Sigma_1^q x_i^2 \leq 1\}$ for all t. Let $B_{t^{-1/2}}$ be the larger q-ball of radius $t^{-1/2}$ for $t < 1$. The problem is that of choosing a design ξ_t to minimize the maximum over x in B_t of the "variance function" $f_t(x)'M^{-1}(\xi_t)f_t(x) = d(x,\xi_t)$ (say) of the estimator of the extrapolated regression at x. It is not hard to show (Kiefer, 1974, Galil and Kiefer, 1977a) that there is a design that assigns mass v at 0 and $1-v$ spread uniformly on the surface of $B_{t^{-1/2}}$, that is now optimum among <u>all</u> designs. It is shown in Galil and Kiefer (1977a) that the optimum design <u>for</u> t <u>sufficiently near</u> 0 ($t < .6$ suffices for all q, and the required upper bound on t is near 1 for large q) is obtained by choosing v to minimize the variance function on the surface of $B_{t^{-1/2}}$, where it is the constant

$$D_t(v) = \frac{1}{v} + [\frac{q}{1-v} - \frac{2}{v}]t^{-1} + [\frac{q(q-1)}{2(1-v)} + \frac{q}{v(1-v)} - \frac{(q-1)}{1-v}]t^{-2}. \qquad (17)$$

Of course, this can be solved explicitly, yielding for the optimum design

$$v_t^* = \{1+[qt^{-1}+q(q+1)t^{-2}/2]^{1/2}|t^{-1}-1|\}^{-1},$$

$$D_t(v_t^*) = \{|t^{-1}-1|+[qt^{-1}+q(q+1)t^{-2}/2]^{1/2}\} \qquad (18)$$

The simplicity of this arithmetic makes it easy to test the effectiveness of our asymptotic approach in this problem. The simplest parametrization and normalization, with $g_t(v) = t^2D_t(v)$, leads to choosing v to minimize the coefficient of t^{-2} in (17), yielding the value $v_t' = \{1+[q(q+1)/2]^{1/2}\}^{-1}$, independent of t. The better approximation described just below (2), which here amounts to minimizing the sum of the terms in t^{-2} <u>and</u> t^{-1} in (17), is $v_t'' = \{1+[(q(q+1)+2qt)/(2-4t)]^{1/2}\}^{-1}$. (The second of these is of course only valid for $t < .5$.) For extrapolation to the ball $B_{1.5}$ ($t = 1/2.25$, only slightly $< .5$), the inefficiency ratio $D_t(v)/D_t(v_t^*)$ at $q = 1,5,10$ is 1.14, 1.05, 1.03 for $v = v_t'$ and 1.06, 1.03, 1.02 for $v = v_t''$. For extrapolation to B_2 ($t = 1/4$), the corresponding values are 1.04, 1.01, 1.01 and 1.001, 1.0004, 1.0003. To illustrate the type of a priori bound computation mentioned in Section 1, we consider v_t'. We obtain $\partial g_t(v)/\partial t|_{v=v_t'} = \{1+[q(q+1)/2]^{1/2}\} \times [2t^{-1}(q+2)/q+1)-1] = h(t)$ (say). Since this is positive, the optimum v_t^* is $< v_t'$. From the convexity in v of $g_t(v)$, we thus obtain $[g_t(v_t')-g_t(v_t^*)] \leq v_t'h(t)$. From these computations, for example when $q = 10$, we obtain $g_t(v_t')/g_t(v_t^*) \leq 1.10$ when $t = 1/2.25$ and ≤ 1.06 when $t = 1/4$, compared with the actual values of 1.03 and 1.02 of the inefficiency ratio. Use of more derivatives will sharpen such bounds.

For "interpolation" to a ball $B_{t^{1/2}}$ of radius smaller than $\underline{X}_t$ (again B_1) with $t < 1$, we replace t^{-1} by t throughout (17). The roughest limiting approach considers only the term v^{-1}, which leads to the useless result $v = 1$, analogous to that obtained in the first parametrization in the example of Section 1. Consideration of a better parametrization, for example by minimizing v^{-1} plus the term in t of the altered (17), leads to developments similar to those for v_t' in the previous paragraph, which we shall consequently not give here.

In the problem of type (c) of Section 1 we let $\underline{X}_t$ be $B_{t^{-1/2}}$ and consider the maximum $\bar{d}$ of d on that set, or some $\Phi^{(p)}$ for observations taken in $\underline{X}_t$. Within the class of rotationally invariant designs, it is not hard to see that for each p the $\Phi^{(p)}$-optimum design or minimizer of $\bar{d}$ is again of the structure considered in the ball example just above (or in Section 1), with mass $v_t^{(p)}$ at 0 and $1-v_t^{(p)}$ spread uniformly on the surface of $\underline{X}_t$. (For $p > 0$, the $\Phi^{(p)}$-optimum design in this class, although "admissible" among all designs, is not $\Phi^{(p)}$-optimum among all designs; nevertheless, for reasons given in Galil and Kiefer (1977a), it seems worthwhile to study designs subject to this restriction.) The D-optimum choice of $v_t^{(0)}$ (which also minimizes $\bar{d}$) does not depend on t, but for $p > 0$ the optimum choice of $v_t^{(p)}$ does depend on t. One finds in the simplest calculation that $v_t^{(p)} \to 1$ as $t \to 0$ for fixed $p > 0$, so that, as in the illustration of Section 1, it is more productive to reparametrize, e.g., in terms of $t^{-p/(p+1)}(1-v_t^{(p)})$. (For the problem on the sphere of radius $t^{1/2}$, $\lim_{t\to 0} v_t^{(p)}$ is not 0 or 1, so this reparametrization is unnecessary.) Since the computations and an indication of their usefulness in approximating the $\Phi^{(p)}$-optimum design on $\underline{X}_t$ is given in Galil and Kiefer (1977a), we do not repeat them here.

In another paper (Galil and Kiefer, 1977c) this approach is used for extrapolation in cubic regression on the unit ball, where the optimum design is more complicated than for quadratic regression, so that the usefulness of the asymptotic approach is more important. In addition to the analogue of v_t' for the simplest asymptotic approach, one must now determine the radius (< 1) of a sphere on which the mass v_t' is spread uniformly, the remaining mass $1-v_t'$ again being spread on B_1. The inefficiency ratio of the asymptotic procedure for extrapolation to $B_{1.5}$ for $q = 1,5,10$ is now 1.10, 1.09, 1.07, and for B_2 it is now 1.02, 1.02, 1.02, comparable to what these values were in the simpler quadratic case recorded in the paragraph below (18).

ACKNOWLEDGEMENT

The author is grateful to Zvi Galil for performing computations recorded herein. This research was supported by NSF Grant MCS 75-22481.

BIBLIOGRAPHY

Fedorov, V. V. (1972). Theory of Optimal Experiments. New York: Academic Press.

Galil, Z. and Kiefer, J. (1977a). Comparison of rotatable designs for regression on balls, I (Quadratic). J. Statist. Planning and Inf. 1, 27-40.

Galil, Z. and Kiefer, J. (1977b). Comparison of designs for experiments with mixtures. To appear in Technometrics 19.

Galil, Z. and Kiefer, J. (1977c). Cubic regression on balls (to appear).

Kiefer, J. (1959). Optimum experimental designs. J. R. Statist. Soc. B 21, 272-319.

Kiefer, J. (1960). Optimum experimental designs V, with applications to systematic and rotable designs. Proc. 4th Berkeley Symp. Math. Statist. Prob. Berkeley: Univ. of Calif. Press.

Kiefer, J. (1961). Optimum designs in regression problems II. Ann. Math. Statist. 32, 298-25.

Kiefer, J. (1974). General equivalence theory for optimum designs (approximate theory). Ann. Math. Statist. 45, 849-79.

Kiefer, J. (1975). Optimal design: Variation in structure and performance under change of criterion. Biometrika 62, 277-88.

Kiefer, J. and Studden, W. (1976). Optimal designs for large degree polynomial regression. Ann. Statist. 4, 1113-23.

Scheffé, H. (1958). Experiments with mixtures. J. R. Statist. Soc. B 21, 344-60.

Received June, 1977.

Refereed anonymously.

A DIOPHANTINE PROBLEM IN OPTIMUM DESIGN THEORY

J. Kiefer*

ABSTRACT. **An aspect of optimum experimental design theory for the model of quadratic regression on the q-cube is reduced to finding all integer solutions of $5xy-3qz-3qy+q(2q+1) = 0$ subject to $0 \le x < y \le q$. The solutions are characterized and are shown to fall into certain arithmetic progressions, termed linear series.**

1. *Introduction.*

In the design of statistical investigations, a theory that has proved fruitful in the choice of an efficient experiment is based on the following model: N uncorrelated observations $Y_1, Y_2, \ldots, Y_N$, with common variance σ^2, are to be taken with Y_j corresponding to value z_j of a controllable variable taking value in a compact space Z. The expectation of Y_j is

$$\theta' f(z_j) = \sum_{i=1}^{k} \theta_i f_i(z_j) \ ,$$

where f is the column vector of known continuous real-valued functions on Z and θ is the unknown column vector of coefficients of this "linear model". A design $(z_1, z_2, \ldots, z_N)$ is to be chosen by the experimenter, and it is convenient to represent such a design as the probability measure ξ given by $\xi(z) = N^{-1}$ (numbers of $z_j = z$). The k×k "information matrix" of this design is $M(\xi) = \int f(z)f'(z)\xi(dz)$; its importance is that, when the design measure ξ is employed, the covariance matrix of best linear unbiased (least squares) estimators of θ is $\sigma^2 N^{-1} M^{-1}(\xi)$ if M is nonsingular (with a corresponding interpretation otherwise).

An optimality functional Φ on the set of possible M is given, and ξ is to be chosen to minimize $\Phi(M(\xi))$. Typical Φ's are the trace, determinant, and maximum eigenvalue of $M^{-1}(\xi)$. The discreteness of possible values of ξ (multiples of N^{-1}) usually make this intractable algebraically, but a useful "approximate theory", in which

* Research supported under NSF Grant MCS 75-22481.

UTILITAS MATHEMATICA Vol. 14 (1978), pp. 81-98.

theoretical characterizations and computational algorithms for optimum designs exist, is obtained by enlarging the set of allowable ξ to include at least all the discrete probability measures. The optimum approximate theory ξ that minimizes $\Phi(M(\xi))$ can be implemented in an obvious way for given N to yield an actual design that is close to optimum when N is sufficiently larger than k. Further details and interpretation of this subject and of other developments in the remainder of this section can be found in Kiefer [3], [5]. In the remainder of this section we consider only the approximate theory.

The setting we treat here is that of quadratic regressions on the q-cube

$$Z = \{z = (z^{(1)},\ldots,z^{(q)}) : -1 \le z^{(h)} \le 1 \ \forall\ h\}$$

and the $k(=(q+1)(q+2)/2)$ components f_i of f are the monomials 1, $z^{(h)}$, $z^{(g)}z^{(h)}$, $1 \le g \le h \le q$. For many Φ, including those mentioned in the previous paragraph, it can be shown that *every* optimum design is supported by a subset of the 3^q-array $J^{(q)}$ of points with all coordinates in $\{-1,0,1\}$, and that there is an optimum ξ invariant under the group G generated by reflections and permutation of coordinates. We write

$$J^{(q)} = \bigcup_{i=0}^{q} J_i^{(q)} ,$$

where $J_i^{(q)}$, the set of "barycenters of depth i", consists of points with j zero coordinates and the remaining coordinates ± 1.

Unless q is small, the optimum design will not be unique, and there is obvious practical interest in finding a design with support of minimal size. That difficult problem (the class of such problems for general Z and f being NP-complete) is not the concern of this paper. Rather, we discuss the similar and related question, which also has a practical justification of its own, of finding an optimum design with support in the smallest possible number of $J_i^{(q)}$; that is, with support in as few barycentric depths as possible. Since, for the convex symmetric criteria Φ of interest, any optimum ξ may be transformed under any element of G and then averaged to yield an invariant design that is still optimum, it will suffice to study this question for designs ξ invariant under G.

Details of results stated in the present and next paragraph, concerning G-invariant ξ, can be found in [4], [1], [2]. For such ξ, $M(\xi)$ depends only on

$$u(\xi) = \int_J (z^{(1)})^2 \xi(dz)$$

and

$$v(\xi) = \int_J (z^{(1)} z^{(2)})^2 \xi(dz) .$$

For the invariant ξ_i (say) supported by $J_i^{(q)}$, we clearly have

$$u(\xi_i) = (q-1)/q$$

$$v(\xi_i) = (q-i)(q-i-1)/q(q-1) .$$

Thus, if a general invariant ξ is represented as $\sum_0^q \alpha_i \xi_i$ where $(\alpha_0, \alpha_1, \ldots, \alpha_q)$ is a probability vector, we have

$$(u(\xi), v(\xi)) = \sum_0^q \alpha_i (u(\xi_i), v(\xi_i)) .$$

For most commonly used criteria, such as $\det M^{-1}(\xi)$ and $\operatorname{tr} M^{-1}(\xi)$, it turns out that the values u^* and v^* that minimize $\Phi(M(\xi))$ are such that (u^*, v^*) cannot lie on one of the finitely many line segments between two different points $(u(\xi_i), v(\xi_i))$ in the u-v plane if $q \geq 2$. (We hereafter usually disregard the trivial degenerate case $q = 1$.) By convex representation of (u^*, v^*), there will always exist triples (i_1, i_2, i_3) that suffice to give

$$(u^*, v^*) = \sum_{j=1}^{3} \alpha_{i_j} (u(\xi_{i_j}), v(\xi_{i_j})), \sum_1^3 \alpha_{i_j} = 1 .$$

For example, for the criterion $\det M^{-1}(\xi)$, it is necessary and sufficient that the i_j be 0,1, and any value greater than or equal to 2 if $2 \leq q \leq 5$; and that they be 0, 1 or 2, and any value greater than or equal to 3 if $q > 5$.

For the criterion $\Phi(M(\xi))$ = [maximum eigenvalue of $M^{-1}(\xi)$], which is the basis for the remainder of this paper, it turns out that $(u^*, v^*) = (2/5, 1/5)$ for every q, quite different from the irrational solutions for the trace and determinant criteria (which also depend on q).

- 83 -

There is thus more hope that there may be q for which an optimum design exists that is supported by barycenters of only *two* depths x and y (say), $J_x^{(q)}$ and $J_y^{(q)}$. (One depth alone is never possible.) For this to be the case, it is necessary and sufficient that

$$(u^*,v^*) = \sum_{i=x,y} \alpha_i(u(\xi_i),v(\xi_i)) ,$$

and the fact that $\alpha_x = 1-\alpha_y$, together with the values of $u^*,v^*,u(\xi_i),v(\xi_i)$ given above, show this to be equivalent to

$$(1.1) \qquad 5xy - 3qx - 3qy + 2q^2 + q = 0$$

for integers x,y with $0 \le x \le y \le q$. (This is also meaningful when $q = 1$, when $x = 0$, $y = 1$ is indeed the solution.) If (1.1) is satisfied, the corresponding α_x is

$$(1.2) \qquad \alpha_x = [q(q-1)/5-(q-y)(q-y-1)]/[2q(y-x)+x(x+1)-y(y+1)] ,$$

with α_y being found by interchanging x and y in (1.2) (when $q = 1$, the solution is $\alpha_0 = 2/3$, $\alpha_1 = 1/3$.)

A main purpose of this paper is to show that all solutions of (1.1) subject to the stated restriction

$$(1.3) \qquad 0 \le x < y \le q, \quad x,y,q \text{ integers},$$

are members of infinite linear "series" with y in an arithmetic progression and x,q linear in y. For fixed q, it is simple to obtain all solutions to (1.1) by writing

$$(1.4) \qquad x = (3q-X)/5, \quad y = (3q+Y)/5 .$$

This reduces (1.1) to

$$(1.5) \qquad XY = q(q+5)$$

with X,Y positive integers. From this and (1.3) we obtain

THEOREM 0. *For fixed* q, *the solutions to* (1.1) *are obtained from* (1.4), *and* (1.5) *with* $Y \equiv 2q (\text{mod } 5)$ *and* $(q+5)/3 \le Y \le 2q$, *so that the number of solutions* (x,y) *is the number of factors of* $q(q+5)$ *in the interval* $[(q+5)/3, 2q]$ *that are* $\equiv 2q \pmod 5$.

In particular, if $5|q$, say $q = 5Q$, write $X' = X/5$, $Y' = Y/5$ (integers). Then (1.5) becomes $X'Y' = Q(Q+1)$ with $(Q+1)/3 \le Y' \le 2Q$, and the number of solutions is the number of factors of $Q(Q+1)$ in the interval $[(Q+1)/3, 2Q]$.

For q a prime greater than 5, without the reduction (1.4), consideration of (1.1) modulo q shows that either $y = q$ (and hence $x = (q-1)/2$) or else $x = 0$ (and hence $y = (2q+1)/3$, a solution only if $q \equiv 1 \pmod 3$).

It is expeditious to develop, in Theorem 1, the solution in another form than that of Theorem 0, for use in developing the linear series of Section 4 and for discussion of the special solutions of Section 3.

The problem considered here typifies a class of problems of growing concern in optimum design theory, in which one seeks designs whose supports are small in an appropriate sense, and in which the problem can be reduced to one of diophantine analysis under inequality restrictions.

The author is grateful to the referee, R. G. Stanton, for many substantial improvements.

2. *Reduction and Characterization.*

We first transform (1.1) into a homogeneous diagonal form in four restricted variables. The solutions to the corresponding unrestricted problem can then be censored to yield the solutions we seek.

The nonhomogeneous equation (1.1) can be diagonalized by the transformation

$$(2x + 3,\ 2y + 3,\ 2q + 5) = (u,v,w) \begin{pmatrix} 1 & 1 & 1 \\ 1 & 1 & 2 \\ 1 & -1 & 0 \end{pmatrix} \tag{2.1}$$

into

$$u^2 + v^2 = 5(w^2 + 1) , \tag{2.2}$$

where u, v, w are integers (even though the matrix of the transformation has determinant 2) because the additive constants on the left side of (2.1) transform properly to yield

$$(u,v,w) = (2x + 2y - 2q + 1, -x-y + 2q + 2, x - y) .$$

Thus, any solution of (1.1) yields a solution of (2.2); the restrictions needed to reverse the direction will be considered below.

From (2.2) we have $u^2 + v^2 \equiv 0 \pmod 5$ and hence, modulo 5, the integers (u,v) in some order must be congruent to one of the pairs (0,5), (1,2), (1,3), (2,4), (3,4). Consequently, either $2u + v$ and $u - 2v$ are congruent to zero, or else $2u - v$ and $u + 2v$ are. We conclude that u and v can be expressed in terms of *integers* s and t, through the substitution of *either*

$$s = (2u + v)/5, \quad t = (u - 2v)/5,$$

hereafter referred to as

$$\text{(2.3)} \qquad \text{Case I: } u = 2s + t, \qquad v = s - 2t ,$$

or

$$s = (2u - v)/5, \quad t = (u + 2v)/5 ,$$

hereafter referred to as

$$\text{(2.4)} \qquad \text{Case II: } u = 2s + t, \qquad v = -s + 2t .$$

Both of these substitutions transform (2.2) into

$$\text{(2.5)} \qquad s^2 + t^2 = w^2 + 1 .$$

We recall (e.g., [6, p.15]) that the full integer solution of the quaternary $s^2 + t^2 = w^2 + z^2$ is given by $2s = d_1X + d_2Y$, $2t = d_1Y - d_2X$, $2w = d_1Y + d_2X$, $2z = d_1X - d_2Y$, where X,Y,d_1,d_2 are integers satisfying the obvious parity conditions. We set $z = 1$ to obtain the solutions of (2.3), and it is convenient to define $b = d_2Y$, $a = b + 2 = d_1X$, $c = d_2X$, $d = -d_1Y$, which entails the restriction

$$\text{(2.6)} \qquad b(b + 2) + cd = 0 .$$

The formal solutions obtained from (2.1) and either (2.3) or (2.4) are then

$$(2.7)\quad \text{Case I}\quad \begin{cases} x = (3b + c)/2\ , \\ y = (3b + d)/2\ , \\ q = 2b + \dfrac{3(c+d) - 2}{4}\ ; \end{cases}$$

and

$$(2.8)\quad \text{Case II:}\quad \begin{cases} x = (b-2-c-2d)/2\ , \\ y = (b-2-2c-d)/2\ , \\ q = -5(c+d+2)/4\ . \end{cases}$$

Because any solution of (1.1) in integers yields an integer solution of (2.2) and then of (2.5), we need now only determine which choices of b,c,d satisfying (2.6) yield integers x,y,q in (2.7) or (2.8), and then determine which of these solutions also satisfy (1.3). For the first of these, it is evident in both (2.7) and (2.8) that is is necessary and sufficient that

$$(2.9)\qquad \begin{aligned} c + d &\equiv 2 \pmod 4\ , \\ b \equiv c &\equiv d \pmod 2). \end{aligned}$$

If b,c,d are odd, the first line of (2.9) is automatic because $cd \equiv -b(b + 2) \equiv 1 \pmod 4$ so that c and d are both $\equiv 1$ or both $\equiv 3$ (mod 4). We hereafter refer to the case where b,c,d are odd as "Parity O" and that where b,c,d are even with $c + d \equiv 2 \pmod 4$ as "Parity E"; this dichotomy in combination with that into Cases I and II thus yields four possibilities.

As for the restrictions (1.3), we see in both (2.7) and (2.8) that $x < y \iff c < d$. From (2.7) and (2.8), the remaining inequalities, $x \geq 0$ and $q-y \geq 0$, become

$$(2.10)\qquad \text{Case I:}\ \begin{cases} 3b + c \geq 0\ , \\ 2b + 3c + d \geq 2\ , \end{cases}$$

and

$$(2.11)\qquad \text{Case II:}\begin{cases} b - c - 2d \geq 2\ , \\ -2b - c - 3d \geq 6\ . \end{cases}$$

We summarize in

THEOREM 1. *The set of all triples of integers* (x,y,q) *satisfying*

(1.1) *and* (1.3) *is the set of all* (x,y,q) *obtainable through the formulas* (2.7) *or* (2.8), *respectively, from integers* b,c,d *satisfying* $c < d$, (2.6), (2.9), *and also either* (2.10) *or* (2.11), *respectively.*

We also remark that, given an (x,y,q) satisfying (1.1) and (1.3), we can determine which case(s) and b,c,d correspond to it by solving (2.7) or (2.8). In fact, since x,y and (1.1) determine at most two values $q = [(3x + 3y-1) \pm R]/4$, where

$$(2.12) \qquad R \equiv R(x,y) = [(3x + 3y - 1)^2 - 40xy]^{1/2} ,$$

while (2.6) determines at most four possible sets (b,c,d,q), we may at the same time determine which values of these parameters are possible for the given (x,y) known to correspond to at least one q. The desired conclusion is obtained by first computing $R(x,y)$, necessarily an integer, and then verifying for Case I whether either choice of sign makes $b = [3x + 3y - 1 \pm R]/10$ an integer. If so, set $c = 2x - 3b$, $d = 2y - 3b$ (and (2.6) is then satisfied); if (2.9) is satisfied, the last line of (2.7) completes specification of a Case I (x,y,q), provided $q \geq y$. Similarly, if $b = [x + y - 7 \pm 3R]/10$ is an integer, we set $c = [2x - 4y + b - 2]/3$, $d = [2y - 4x + b - 2]/3$, and if (2.9) is satisfied we use the last line of (2.8) to provide a Case II (x,y,q), if $q \geq Y$. We sometimes label a case with sign (for example, Case I+) depending on the sign appropriate in the formula for b.

3. *Special Values.*

In this section we list a number of facts of theoretical or practical interest concerning the set of achievable values (x,y,q) or (b,c,d).

A. All eight possible combinations of case number, sign and parity (O or E) occur. There are values of (x,y,q) in which each combination occurs as the unique description, and other (x,y,q) for which pairs (I+,II+) or (I−,II−) occur with both elements of the pair having the same parity (either O or E).

B. There are infinitely many triples (x,y,q) for which $R = 0$, which follows from Gauss's theorem on the infinitude of solutions of certain quadratic equations (e.g., [6, p.57]). Temporarily ignoring the restriction $x < y$ and solving for y in terms of x in the equation $R = 0$ then yields that $10x^2 + 30x$ is a perfect square, so that, for some integer L, either $x = 5L$ and $2L(5L+3)$ is a square, or else $x = 5L-3$ and $2L(5L-3)$ is a square. The first of these reduces, after some elimination of cases by divisibility considerations, to solving $w_1^2 - 10s_1^2 = 6$ with w_1 even and s_2 odd, and then setting $x = 5s_1^2$ and $y = [11x+3\pm 10s_1w_1]/9$ (only one sign yields an integer), with $q = (3x+3y-1)/4$. The second reduces to solving $w_2^2 - 10s_2^2 = -6$ and setting $x = 5s_2^2 - 3$ and $y = [11x+3\pm 10s_2w_2]/9$. It is not hard to see that, if w_1, s_1 satisfy the first equation, then $w_2 + (10s_1 \pm w_1)/3$ and $s_2 = (w_1 \pm s_1)/3$ are integers for some choice of sign, and satisfy the second equation, and that the resulting (x,y) is simply the (y,x) obtained from (s_1, w_1); all this is true with the equations (and subscripts 1,2) interchanged. Consequently, it suffices to find the solutions of $w_1^2 - 10s_1^2 = 6$ and to study the resulting solutions. In these cases (x,y,q) for which $R = 0$, the I+ and I- cases degenerate into a single case, as do the II+ and II- cases, always with parity E. The first three cases are $s_1 = 1, 5, 43$ (the last with $q = 10{,}535$), so one soon leaves the realm of practicality.

C. There are many pairs (x,y) for which (1.1) is satisfied for two values q. For example, a simple linear series of the type considered in Section 4, containing infinitely many such pairs, is $(x,y) = (2j, 4j+1)$ for $j = 2,3,\ldots$, for which $q = y$ (Case I+E) or $q = 5j$ (Case II-E). (When $j = 1$, the two q's coincide.) The subseries of this for which $j \equiv 1 \pmod 5$ also falls into the other two cases of parity E, so that these (x,y) yield two q's and correspond to four triples (b,c,d). For another example, the family with $(x,y) = h^2+2h-3$, $y = 2h^2+3h-2$ for $h = 3,4,\ldots$, yields $q=5h(h+1)/2-5$ or $q=2h^2+5h-3$; the former is Case II+ if $h \le 5$ and Case II- if $h \ge 6$, while the latter is Case I- if $h \le 5$ and I+ if $h \ge 6$. (This change of sign of case with h also indicates, as we shall see in Section 4, that it is sometimes more natural to deal with a family in terms of $\pm R$ rather than R.) The subfamily with $h \equiv 2$ or $3 \pmod 5$ yields four cases.

D. Because $5|q$ in Case II, such values of q typically yield more (x,y) than do other q's. The value of $q \le 300$ with

- 89 -

most (x,y) (namely, 15) is 175. Similarly, some values of y are "rich", such as 225, which occurs in C above and also with $x = 116$, $q = 250$ or 261. The value $x = 0$ occurs infinitely often (just below, in E). A list of all solutins for $q \le 50$ is appended in Section 5.

E. The two series of greatest practical importance are perhaps $(x,y,q) = (0,2i+1,3i+1)$ for $i=0,1,2,\ldots$, and $(x,y,q)=(2j,4j+1,4j+1)$ mentioned earlier. For the first of these, $J_0^{(q)}$ consists of the corners of the q-cube, while for the second $J_q^{(q)}$ consists of a single center-point.

F. The following facts regarding (a,b,c,d) are easily established:

(i) $b \neq -1$; and the possible cases (one of I, then two of II) where $ab = 0 = cd$ are

(a) $b=0=c$, $d=4i+2$, for which $(x,y,q) = (0,2i+1,3i+1)$, $i \ge 0$,

(b) $b=0=d$, $c=-4i-2$, for which $(x,y,q) = (2i,4i+1,5i)$, $i \ge 1$,

(c) $a=0=d$, $c=-4i-2$, for which $(x,y,q) = (2i-1,4i,5i)$, $i \ge 1$.

Thus, it cannot occur that $a = 0 = c$.

(ii) In the remaining instances $b(b+2) > 0$. If $b+2 < 0$, (2.7) yields $x < 0$, so Case I cannot occur but Case II can. For both cases, it can occur that $b > 0$.

G. We conclude this section by rewriting the inequalities (2.10) and (2.11) when $b(b+2) > 0$, in a form that will be used in Section 4. First consider Case I, for which $b > 0$ from the previous paragraph. The second inequality of (2.10) becomes, upon writing $d = -b(b+2)/c$,

$$b(1+bc^{-1})(3b^{-1}c-1) \ge 2(1+bc^{-1}) , \tag{3.1}$$

which is easily seen to hold (for $b > 0 > c$) if and only if $-1 \le c/b < 0$. This last entails the first inequality of (2.10). We conclude that, when $b > 0$, (2.10) is equivalent to

$$-1 \le c/b < 0 . \tag{3.2}$$

Similarly, in Case II we obtain that (2.11) is equivalent to

$$\begin{aligned} -c - 3b &\ge 6 \quad \text{if} \quad b > 0 , \\ c/b &\le -2 \quad \text{if} \quad b + 2 < 0. \end{aligned} \tag{3.3}$$

The only instances of Case I not covered by (3.2) are those of F(i)(a), when $b = c = 0$. In Case II, the first line of (3.3) still holds when $b = 0$ (F(i)(b)), and the second still holds when $a = 0$ (F(i)(c)).

4. *Linear Series.*

The multiplicative domain of the homogeneous quaternary $s^2 + t^2 = w^2 + z^2$ is of course destroyed by setting $z = 1$, and a restriction such as (1.3) can further interrupt a regular pattern of solutions for other quadratic equations similar to (1.1). In the present case, though, some regularity persists, in the form of certain simple infinite families of solutios which we now define.

A *linear series of pairs* (LSP) is a family F of pairs (x,y) each of which satisfies (1.1) and (1.3) for some q, and such that the values of y for pairs in F form an arithmetic progression, with the corresponding values x satisfying

$$x = Ky + L \tag{4.1}$$

for some rational K,L and all (x,y) in F.

A *linear series of triples* (LST) is a family of G of triples (x,y,q) each of which satisfies (1.1) and (1.3), with $F_G \overset{\text{def}}{=\!=} \{(x,y):(x,y,q)\epsilon G\}$ an LSP, and with the corresponding q satisfying

$$q = My + N \tag{4.2}$$

for some rational M,N and all (x,y,q) in G.

It will be shown as part of the developments of this section that every LSP is the F_G of some LST G; we have introduced the two definitions despite this, because a succinct description can be given of all LSP's containing a given (x,y) (Theorem 2, below), while examples like those of Section 3 can be used to illustrate that some (K,L) correspond to a single (M,N) and others correspond to two. For an example of an (x,y) with the greatest possible number, four, of LST's containing it, we note that $(x,y) = (12,25)$ is in the LSP with $x = (18y-30)/35$ (with LST's $q = (15y-25)/14$ and $q = 6y/5$) and also in the LSP with $x = (y-1)/2$ (with LST's $q = y$ and $q = 5(y-1)/4$). The relationship (4.3) of the next paragraph shows that when $K \geq 1/2$ it is possible to have two values M corresponding to K; put another way, when $1 \leq M \leq 5/4$ it is possible that another M in that interval will correspond to the same K.

Necessarily, the values q in an LST form an arithmetic progression, and so do the values x in an LSP unless $K = 0$, which we permit. It is evident from substitution of (4.1) and (4.2) into (1.1) with $y \to \infty$ that

$$K = (2M^2-3M)/(3M-5) , \tag{4.3}$$

which is meaningful for $1 \leq M \leq 3/2$ (since $0 \leq K \leq 1$). This achieves its maximum value $K = [11-(40)^{1/2}]/9 = .51949$ at $M = [10-(10)^{1/2}]/6 = 1.1396$; thus

$$0 \leq K < [11-(40)^{1/2}]/9 \overset{\text{def}}{=\!=} \bar{K} ,$$

this upper bound being unattainable since K is rational, but being approached by the LS's containing (x,y) with $R = 0$. This follows from setting $w \sim s(10)^{1/2}$ in Section 3B and computing $x/y \sim \bar{K}$ there (the other choice of sign interchanges x and y), from which (4.8) yields that the K of a LS containing such an (x,y) approaches this same value $\bar{K}$ as $y \to \infty$. The first two instances of $R = 0$ mentioned in Section 3B give $K = .5$ and $K = .51948$, the latter very close to the supremum. As for the other extreme values, $K = 0$, $M = 3/2$, and $M = 1$ all occur in the series of Section 3E.

We have allowed a subset of an LSP or LST, obtained by considering values y in a sub-progression of those in the original series, also to be defined as an LS. Thus, only maximal LS or the corresponding K,L,M,N are of concern for describing the LSP's or LST's containing a given pair or triple.

A pair (x_0,y_0) with $R = 0$ (see Section 3B) can be verified (from (4.8), below) to be contained in only one maximal LSP, corresponding to two LST's with the same (K,L) but different (M,N)'s. These instances where $R = 0$ turn out to be the only exceptional (x,y), in that we shall show that every pair with $R \neq 0$ is contained in LSP's corresponding to two (and no more) different pairs (K,L); the two pairs degenerate into one when $R = 0$. Because of the possibility mentioned two paragraphs above that a (K,L) may correspond to either one or two (M,N), we shall limit our discussion in the remaining paragraphs to the (K,L)'s associated with linear series containing a given (x_0,y_0,q_0). When an (x_0,y_0,q_0) falls into both cases, it can be verified that the two K's (and associated L's) of Case I LST's given by (4.11) below are identical with those of the Case II LST's given by (4.12), in the opposite order.

THEOREM 2. *Each* (x_0,y_0) *satisfying* (1.1) *and* (1.3) *for some* q_0, *and for which* $R \neq 0$, *is contained in LSP's with exactly two different* K,L, *given by* (4.8) *below and by* $L = x_0 - Ky_0$ *(or by* (4.7)*).*

Proof. We first show that at most two (K,L)'s satisfying (4.1) are possible for a given (x_0,y_0) that satisfies (1.1) and (1.3) and is in an LSP with that (K,L). Define, for any rational K,

$$(4.4) \qquad r_K = [9K^2 - 22K + 9]^{1/2} ,$$

obviously not zero. Substituting (4.1) into (2.12) yields

$$(4.5) \qquad \begin{aligned} R &= [r_K^2 y^2 + 2(9KL - 3K - 11L - 3)y + (3L-1)^2]^{1/2} \\ &= \pm r_K[y + r_K^{-2}(9KL - 3K - 11L - 3)] + O(y^{-1}) \end{aligned}$$

as $y \to \infty$. (The + sign holds for large y.) It follows that R can be an integer for all y in an arithmetic progression only if r_K is rational and the $O(y^{-1})$ term in (4.5) is zero. Thus

$$(4.6) \qquad \pm R = r_K y + [9KL - 3K - 11L - 3]/r_K$$

(the - sign sometimes being needed for small y), which with the first line of (4.5) yields (with ± sign not necessarily that of (4.6))

$$(4.7) \qquad L = [3K - 3 \pm r_K]/2 .$$

From $x_0 = Ky_0 + L$ we then obtain $(2x_0 + 3) - K(2y_0 + 3) = \pm r_K$ and hence

- 93 -

(4.8) $$K = [2x_0y_0 + 3x_0 + 3y_0 - 1 \pm R_0]/2y_0(y_0+3) ,$$

where $R_0 = R(x_0,y_0)$. Thus, at most two values K are possible, and for each of them only one choice of sign in (4.7) can make $Ky_0 + L = x_0$.

We now show that both choices of sign in (4.8) do yield LSP's. This includes showing the pairs in the series correspond to integers q satisfying (1.1), and that the inequalities (1.3) are satisfied. Because of the various "cases" of Section 2 that arise, it is simpler arithmetically to do this in terms of the (a_0,b_0,c_0,d_0) corresponding to x_0,y_0, rather than to work directly with (4.7) and (4.8).

It is convenient to separate from the main discussion the possibility that $a_0b_0 = 0$ and to exhibit two LST's with different (K,L) for any such (x_0,y_0,q_0) with $R_0 \neq 0$. Referring to Section 3F(i), one verifies easily the following results for the three possibilities mentioned there.

(a) If $(x_0,y_0,q_0) = (0,2i_0 + 1, 3i_0 + 1)$ belongs to the series of F(i)(a) (for which $K = 0$), then

$$(x,y,q) = ([3i_0+1]j,[2i_0+1][j(i_0+2)+1],[3i_0+1][j(i_0+2) + 1])$$

for $j \geq 0$ is an LST containing (x_0,y_0,q_0) for $j = 0$, and with $K = (3i_0+1)/(2i_0+1)(i_0+2) \neq 0$.

(b) If $(x_0,y_0,q_0) = (2i_0,4i_0+1,5i_0)$ belongs to the series of F(i)(b) (for which $K = 1/2$), then

$$(x,y,q) = (i_0[2i_0+3]j+2i_0,[4i_0+1][1+(i_0+1)j],5i_0[1+(i_0+1)j])$$

for $j \geq 0$ is an LST containing (x_0,y_0,q_0) for $j = 0$, and with $K = i_0(2i_0+3)/(4i_0+1)(i_0+1)$; thus, $K \neq 1/2$ unless $i_0 = 1$, when $R_0 = 0$.

(c) If $(x_0,y_0,q_0) = (2i_0 - 1,4i_0,5i_0)$ belongs to the series of F(i)(c) (for which $K = 1/2$), then

$$(x,y,q) = ([2i_0-1][1+(i_0+1)j],i_0[4+j(4i_0+3)],5i_0[1+(i_0+1)j])$$

for $j \geq 0$ is an LST containing (x_0,y_0,q_0) for $j = 0$, and with $K = (2i_0-1)(i_0+1)/i_0(4i_0+3) \neq 1/2$.

According to F(ii), we may hereafter assume (x_0,y_0,q_0) to be such that $a_0b_0 > 0$. For any such (a_0,b_0,c_0,d_0), we next exhibit two formally different arithmetic progressions of (x,y,q) containing (x_0,y_0,q_0), and clearly satisfying (1.1). We then verify (1.3). Lastly, we show the two resulting LST's have different K's and are hence indeed distinct.

The two families of (a,b,c,d) are constructed as follows.

(A) For non-negative integral i (to be limited further, below), define

$$(4.9) \qquad b = b_0-b_0c_0i, \quad c = c_0-c_0^2i, \quad d = d_0+b_0^2i, \quad a = b+2 = a_0-b_0c_0i .$$

For $i = 0$ this is the given configuration (a_0,b_0,c_0,d_0), and (2.6) is evident for all i. Whatever the case of (x_0,y_0,q_0) (select and fix one if it is not unique), we then use the appropriate one of (2.7) and (2.8) corresponding to that case to obtain (x,y,q) corresponding to (4.9). Since the expressions of (4.9) are linear in i and those of (2.7), (2.8) are linear in (b,c,d), these (x,y,q) form arithmetic progressions, and (since x_0,y_0,q_0 are integers) they will be integers if the change in x,y,q is integral when i is increased from one value to the next, say by an integer Δ. All these changes are seen to be integral multiples of $\Delta/4$ in both Cases I and II, so that (4.9) with $i \equiv 0 \pmod 4$ yields an arithmetic progression of triples of integers (x,y,q) satisfying (1.1). (Of course, a finer progression often suffices, but yields the same K,L,M,N.)

(B) Essentially interchanging the roles of b and a, we now define

$$(4.10) \qquad a = a_0-a_0c_0i, \quad c=c_0-c_0^2i, \quad d = d_0+a_0^2i, \quad b = a-2 = b_0-a_0c_0i .$$

It follows as in (A) that for non-negative integral $i \equiv 0 \pmod 4$ we obtain an arithmetic progression of triples (x,y,z) satisfying (1.1), equal to (x_0,y_0,z_0) for $i = 0$.

For the progression of (4.9), we have $c/b = c_0/b_0$, so the inequalities of (2.14) and the second line of (2.15) are satisfied in these respective cases, which are the same cases as for b_0,c_0. Since $c+3b = (c_0+3b_0)(1-c_0i)$, the first line of (2.15) also holds, in Case II with $b,b_0 > 0$. The corresponding demonstration for (4.10) proceeds similarly, upon substituting $b = a-2$ in (2.14) and (2.15).

Thus, we have shown that each of (4.9) and (4.10) yields an LST, and it remains only to show that they are distinct. The respective values of K, obtained by dividing the change in x resulting from a change of Δ in i by the corresponding change in y, are, in Case I,

$$(4.11)\qquad \begin{aligned} K &= (3+b_0^{-1}c_0)/(3-b_0c_0^{-1}) \quad \text{for } (4.9)\ , \\ K &= (3+a_0^{-1}c_0)/(3-a_0c_0^{-1}) \quad \text{for } (4.10), \end{aligned}$$

and, in Case II,

$$(4.12)\qquad \begin{aligned} K &= (b_0+c_0)(2b_0-c_0)/(b_0+2c_0)(b_0-c_0) \quad \text{for } (4.9)\ , \\ K &= (a_0+c_0)(2a_0-c_0)/(a_0+2c_0)(a_0-c_0) \quad \text{for } (4.10). \end{aligned}$$

One can show directly that the two lines of (4.11) are unequal, as are the two lines of (4.12). Somewhat neater computationally is the argument that follows. In Case I, each of the two lines of (4.11) must equal one of the two distinct values of (4.8). We may assume $x_0 > 0$, since the only occurrences of $x_0 = 0$ can be seen to be in the family of F(i)(a) treated earlier. For $x_0 > 0$, the two lines of (4.11) are distinct if and only if their product equals the product $x_0(x_0+3)/y_0(y_0+3)$ of the two values in (4.8); that is, if and only if

$$(4.13)\qquad (3b_0+c_0)(3a_0+c_0)/(3a_0-a_0b_0/c_0)(3b_0-a_0b_0/c_0) = x_0(x_0+3)/y_0(y_0+3)\ ,$$

which is evident upon putting $-a_0b_0/c_0 = d_0$ and using (2.7). Similarly, in Case II we must show that the product of the two lines of (4.12) is $x_0(x_0+3)/y_0(y_0+3)$. Upon multiplying out the numerator and denominator of the left side of (4.14) below and substituting $a_0b_0 = -c_0d_0$, we obtain from (2.8)

$$(4.14)\qquad (b_0+c_0)(2a_0-c_0)/(b_0-c_0)(a_0+2c_0) = (x_0+3)/y_0\ ,$$

and the remaining factors from (4.12) yield $x_0/(y_0+3)$. □

5. *Solutions for* $q \le 50$.

At the suggestion of the referee, a list of all solutions for

$q \leq 50$ is given in Table 1. A simple computer program can be used with Theorem 1 to print out solutions (x,y,q) together with (a,b,c,d) and "case". Alternatively, the (x,y,q) can be obtained from Theorem 0, or else by simply checking whether, for each q and y, the solution $x = q(1+2q-3y)/(3q-5y)$ of (1.1) yields an integer satisfying (1.3). Whichever method is used, the LS containing (x,y,q) can then be obtained from (4.8), (4.1), (4.3), (4.2). The fact that (x;y) satisfying (1.1) and (1.3) lies on a hyperbolic segment of positive slope, from $(0,(2q+1)/3)$ to $((q-1)/2,q)$, shows that we may restrict consideration to $x \leq (q-1)/2$ and $y \geq (2q+1)/3$ in checking solutions. The printout for $q \leq 300$ was used to obtain some of the examples noted in Section 3, where solutions of particular interest have been mentioned. For each $q \leq 50$ we list all solutions (x,y).

TABLE 1

q	Solutions (x,y)
1	(0,1)
2	NONE
3	(1,3)
4	(0,3)
5	(1,4)(2,5)
6	NONE
7	(0,5) (3,7)
8	NONE
9	(4,9)
10	(0,7) (3,8) (4,9)
11	(5,11)
12	NONE
13	(0,9) (6,13)
14	NONE
15	(3,11) (5,12) (6,13) (7,15)
16	(0,11) (4,12)
17	(8,17)
18	NONE
19	(0,13) (9,19)
20	(2,14) (7,16) (8,17)
21	(10,21)
22	(0,15)
23	(11,23)
24	NONE
25	(0,17) (5,18) (9,20) (10,21) (12,25)
26	NONE
27	(9,21) (13,27)
28	(0,19) (8,21)
29	(14,29)
30	(4,21) (11,24) (12,25)
31	(0,21) (15,31)
32	NONE
33	(16,33)
34	(0,23)
35	(7,25) (13,28) (14,29) (17,35)
36	NONE
37	(0,25) (18,37)
38	NONE
39	(13,30) (19,39)
40	(0,27) (6,28) (12,30) (15,32) (16,33) (18,36)
41	(20,41)
42	NONE
43	(0,29) (21,43)
44	(11,32)
45	(9,32) (12,33) (17,36) (18,37)(21,42) (22,45)
46	(0,31)
47	(23,47)
48	NONE
49	(0,33) (21,42) (24,49)
50	(8,35) (19,40) (20,41)

REFERENCES

[1] R. H. Farrell, J. Kiefer, and A. Walbran, *Optimum multivariate designs*, Proc. 5th Berk. Symp. *1* (1965), 113-138.

[2] Z. Galil and J. Kiefer, *Comparison of designs for quadratic regression on cubes*, Statist. Planning and Inference *1* (1977), 121-132.

[3] J. Kiefer, *Optimum experimental designs*, Roy Statist. Soc., Series B, *21* (1959), 272-319.

[4] J. Kiefer, *Optimum designs in regression problems - II*, Ann. Math. Statist. *32* (1961), 298-325.

[5] J. Kiefer, *General equivalence theory for optimum designs (approximate theory)*, Ann. Math. Statist *75* (1974), 849-879.

[6] L. J. Mordell, *Diophantine Equations*, Academic Press, New York, 1969.

Cornell University
Ithaca
New York 14853

Received January 31, 1978; revised March 10, 1978.

Comment

J. KIEFER*

I have little familiarity with the rationale for choosing among methods of simulation, but the results of Schruben and Margolin appear interesting and useful. My comments will outline a mathematical development that has been used in optimum experimental design and linear estimation theory (Kiefer 1959, 1974) and which can also be implemented in the present work to show more clearly what is going on by simplifying the proofs while strengthening the conclusions.

The central notion is one of a domination that can be introduced between some pairs of positive semidefinite symmetric (PSDS) matrices. For such matrices P and Q, we write

$$P \gg Q \quad \text{if} \quad P - Q \text{ is PSDS} . \tag{1}$$

(This is the partial ordering generated by defining the nonnegative cone $P \gg 0$ to mean P is PSDS, in the space of symmetric matrices.)

If P and Q are information matrices (inverses of the covariance matrices $\mathrm{Cov}(\hat{\beta})$ of Section 2.3) for two different experimental designs or linear estimators, this $P \gg Q$ notion means that any linear parametric function $c'\beta$ estimable under Q is also estimable under P, with variance at least as small under the latter. Also, P and Q invertible means $P \gg Q \Leftrightarrow P^{-1} \ll Q^{-1}$ (from which we get $c'P^{-1}c \leq c'Q^{-1}c$ for all c, demonstrating the variance ordering of the previous sentence).

In comparing designs or linear estimators, the notion $\gg$ is like that of admissibility in decision theory; if $P \gg Q$, then P is better for *any* reasonable criterion and there is no point in using a particular criterion unless one is faced with a collection of matrices that cannot be compared merely by using $\gg$. Where possible, it is more instructive and stronger to write $P \gg Q$ rather than $\det P \geq \det Q$ or $\mathrm{tr} P^{-1} \leq \mathrm{tr} Q^{-1}$.

In these terms, the relationship between the members of (2.3), which can be regarded as a consequence of the Gauss–Markov theorem for a fixed design, is the easily proved

$$X'V^{-1}X \gg (X'X)(X'VX)^{-1}(X'X) , \tag{2}$$

or, equivalently,

$$(X'V^{-1}X) \ll (X'X)^{-1}X'VX(X'X)^{-1} , \tag{3}$$

and the conclusion $D_W \leq D_O$ at the end of Section 2.3 is a special case of the following: Call Φ, a real-valued function on the space of covariance matrices, *nondecreasing* if $M \gg N \Rightarrow \Phi(M) \geq \Phi(N)$ (or, alternatively, define the related Ψ on information matrices by the condition $A \gg B \Rightarrow \Psi(A) \leq \Psi(B)$). Then the result of Section 2.3 just mentioned follows from (3) upon noting that the determinant is nondecreasing on PSDS matrices. But other functions Φ such as the trace are also nondecreasing, so we can conclude that WLS is at least as good as OLS in these other senses as well.

The point is that we can separate the strong notion of domination given by $\gg$ (when that domination is present) from the weaker and incomplete statement of domination given by a particular criterion. It is reasonable to state $\det\begin{pmatrix}1&0\\0&3\end{pmatrix} < \det\begin{pmatrix}2&0\\0&2\end{pmatrix}$ because the two matrices can't be compared by $\ll$. But it is misleading merely to say $\det\begin{pmatrix}1&0\\0&1\end{pmatrix} < \det\begin{pmatrix}2&0\\0&2\end{pmatrix}$, instead of the stronger $\begin{pmatrix}1&0\\0&1\end{pmatrix} \ll \begin{pmatrix}2&0\\0&2\end{pmatrix}$. At the same time, whenever one can show $M \ll N$, one automatically obtains $\Phi(M) \leq \Phi(N)$ for all nondecreasing criteria *without separate calculations.* Such Φ include not only the determinant (D-optimality) and trace (A-optimality, referring to average variance), but also the maximum eigenvalue (E-optimality), the maximum diagonal element, and many others of interest.

In Section 2.4, part of the result is that if X is saturated, then $\mathrm{Cov}(\hat{\beta}_{WLS}) = \mathrm{Cov}(\hat{\beta}_{OLS})$. This equality of matrices is more basic than that of determinants, and the two equalities in Lemma 1 follow from the equality of matrices. The inequality in Lemma 1 does not follow from the ordering $\ll$, since (for example) $C_\rho \overset{\text{def}}{=} \begin{pmatrix}1&\rho\\\rho&1\end{pmatrix}$ is neither $\ll$ nor $\gg C_0$. There are (less interesting) criteria Φ for which $\Phi(C_\rho) > \Phi(C_0)$ for $\rho \neq 0$. One can give the inequality of Lemma 1 for a restricted class of Φ's more general than the determinant, but not as general as the nondecreasing Φ's mentioned earlier. In any event, it seems worthwhile to state the equalities of Lemma 1 in the stronger form.

In terms of equality of matrices, (3.6) is a sufficient condition for

$$X'V^{-1}X = X'X(X'VX)^{-1}X'X , \tag{4}$$

as may be seen by multiplying the two sides out.

The matrices of (3.8)–(3.10) can be compared as

$$\begin{aligned} &(3.9) \ll (3.8) \quad \text{always} , \\ &(3.9) \ll (3.10) \quad \text{if} \quad \rho_+ \leq 2\rho_- , \end{aligned} \tag{5}$$

with (3.8) and (3.10) not comparable. This means (3.9) is the best of the three for all nondecreasing criteria Φ (D-, A-, E-optimality, etc.) if $\rho_+ \leq 2\rho_-$. The article gets at this in the paragraph following (3.10), but the $\ll$ notion makes it more precise: better for "estimating main effects" refers not only to diagonal elements of (3.9), etc.,

* J. Kiefer is Horace White Professor of Mathematics at Cornell University, Ithaca, NY 14853.

but also to quantities like $c' \operatorname{Cov}(\hat{\beta})c$, referring to linear combinations of main effects.

Theorem 1, its proof and corollaries, are similarly better described (where (C.9′) means the covariance matrix for (C.9)) as

$$(C.4) \ll (C.7) \ ,$$

$$(C.4) \ll (C.9')$$

$$\text{if} \quad (N-1)\rho_+ - 2N^{-1}N_1N_2(\rho_+ + \rho_-) < 0$$

$$\text{(hence, always if } N_1 = N_2 \text{ and } \rho_+ = \rho_-) \ . \quad (6)$$

This then gives the result for other criteria as well.

There is also a notion of $\ll$ for submatrices corresponding to subsets of parameters. Thus, (C.7) and (C.9′) are not directly comparable in terms of $\ll$ for estimating *all* parameters, but in terms of the main effects (ignoring the first parameter, the grand mean), one sees $(C.7)^* \ll (C.9')^*$, where C^* is the lower right $p \times p$ submatrix of C.

Theorem 2 is a very striking result, but even more striking in terms of $\gg$. (And Corollary 3, E-optimality, etc., are then automatic.) The proof in Appendix D can be shortened considerably by simply noting, in terms of the paragraph containing (D.3), that since $[c_1(X_1'u_1)(u_1'X_1) + \ldots] \gg 0$ (the matrix is PSDS), automatically one gets $A \ll (1-\rho)^{-1}X'X = X'V^{(O)-1}X = A^{(O)}$ (say). Hence, $\Phi(A^{(O)-1}) \leq \Phi(A^{-1})$ for all A (all V) and all nondecreasing Φ. Replacing the last line of Appendix D, one now has for the covariance matrices,

$$\operatorname{Cov}_O(X, V^{(O)}) = \operatorname{Cov}_W(X, V^{(O)}) \ll \operatorname{Cov}_W(X, V) \ll \operatorname{Cov}_O(X, V) \ ; \quad (7)$$

the middle of these relations follows from $A \ll A^{(O)}$, just shown; the other two were shown earlier.

Perhaps these remarks will help support the authors' conclusions regarding optimality of certain schemes of simulation, by demonstrating the optimality in a wider, more natural sense than that of the single D-optimality criterion. Schruben and Margolin have opened up an interesting new area of optimality studies. Their systematic approach and the method of these comments should be applicable to many other simulation comparisons.

REFERENCES

Kiefer, J., (1959), "Optimum Experimental Designs," *Journal of the Royal Statistical Society*, Ser. B, 21, 272–319 (with discussion).

——— (1974), "General Equivalence Theory for Optimum Designs (Approximate Theory)," *Annals of Statistics*, 2, 849–879.

Reprinted from
J. Amer. Statist. Assoc. **73** (1978), 523-524

Journal of Statistical Planning and Inference 3 (1979) 27–38.

EXTRAPOLATION DESIGNS AND Φ_p-OPTIMUM DESIGNS FOR CUBIC REGRESSION ON THE q-BALL

By Z. GALIL and J. KIEFER

Department of Mathematics, Cornell University, Ithaca, New York and Department of Statistics, University of California, Berkeley

Received 18 October 1977

Recommended by J.N. Srivastava

Abstract: This paper continues earlier work of the authors in carrying out the program discussed in Kiefer (1975), of comparing the performance of designs under various optimality criteria. Designs for extrapolation problems are also obtained. The setting is that in which the controllable variable takes on values in the q-dimensional unit ball, and the regression is cubic. Thus, the ideas of comparison are tested for a model more complex than the quadratic models discussed previously. The E-optimum design performs well in terms of other criteria, as well as for extrapolation to larger balls. A method of simplifying the calculations to obtain approximately optimum designs, is illustrated.

AMS classification number: 62K

Key words: A-optimality; Cubic Regression; D-optimality; Designs on Balls; E-optimality; Extrapolation Design; Interpolation Design; Optimum Design; Response Surface; Rotable Design.

1. Introduction

A rationale for studying the performance under various optimality criteria of a design selected according to a particular criterion, was given in Kiefer (1975). The details of such an investigation for quadratic regression on the q-simplex, q-cube, and q-ball have been given in Galil and Kiefer (1977a, b, c, d). The present paper treats the more complicated model of cubic regression on the q-ball, perhaps less important per se than as an illustration of the difficulties encountered in complex multiparameter models. It also includes consideration of designs for extrapolation problems of the type considered for regression of any degree on the line in Kiefer and Wolfowitz (1964), and more generally in Kiefer (1977a).

For brevity, we shall omit repetition of discussion of our approach. Also, we shall not take the space to record here all computations parallel to those of the simplex paper. For example, the criterion of Box and Draper (1959) that was treated in detail in Galil and Kiefer (1977b) is mentioned only briefly in Section 4 without computations; and we also omit discussion of the effect of implementing a design allocation optimum for a ball of one radius on a ball of a different radius that represents the actual experimental setting, as treated in Galil and Kiefer (1977d).

All the optimality computations of this paper were made under the restriction to rotatable designs. As described in Galil and Kiefer (1977d), the designs obtained are consequently not optimum among *all* designs when $q>1$, except in the case of D-optimality or in the extrapolation problems of Section 4; but the practical and theoretical (computational) reasons given in that earlier paper on the quadratic model, for recording comparisons under the restriction to rotatable designs, are all the more cogent in the less tractable cubic model.

Even under the restriction to rotatable designs, there are no longer such algebraically simple results as were obtained for particular criteria in the quadratic models. In Section 2 we introduce notation and compute the expressions for measures of design performance, that are to be minimized and compared. Section 3 gives the rotatable designs that are best in terms of the D-, A-, and E-optimality criteria, and compares the performance of these designs under these and other criteria, including those appropriate to extrapolation problems. Such extrapolation and interpolation problems are treated in detail in Section 4. In Section 5 we illustrate an approach that has proved useful in algebraically intractable problems, computing approximately optimum designs by carrying out the simple calculations when the dimension of the ball, or its radius, grows large.

2. Preliminaries

The considerations of this paper are limited to the approximate theory. Thus, a *design setting* is a pair $(\mathscr{X}, \boldsymbol{f})$ where $\boldsymbol{f}$ is a known (column) k-vector of continuous functions $f_1, \ldots, f_k$ on the compact space $\mathscr{X}$. A *design* ξ is a probability measure on $\mathscr{X}$. If the measure ξ is discrete, $\xi(\boldsymbol{x})$ represents the proportion of observations to be taken at the point $\boldsymbol{x}$ of $\mathscr{X}$. (See, e.g., Kiefer (1959, 1974) or Fedorov (1969) for further discussion of these standard notions of optimum design theory and their implementation.) The observations are uncorrelated with variance σ^2. The expected value of an observation at $\boldsymbol{x}$ is $\boldsymbol{\theta}'\boldsymbol{f}(\boldsymbol{x})$ where $\boldsymbol{\theta}$ is a column vector of k unknown real parameters. The $k\times k$ *information matrix* (per observation for unit variance) of ξ is defined as $\boldsymbol{M}(\xi)=\int_{\mathscr{X}}\boldsymbol{f}(\boldsymbol{x})\boldsymbol{f}(\boldsymbol{x})'\xi(\mathrm{d}\boldsymbol{x})$, and the corresponding normalized variance function for nonsingular $\boldsymbol{M}(\xi)$ is $d(\boldsymbol{x}, \xi)=\boldsymbol{f}(\boldsymbol{x})'\boldsymbol{M}^{-1}(\xi)\boldsymbol{f}(\boldsymbol{x})$. Thus, when ξ is used with N observations, $N^{-1}\sigma^2\boldsymbol{M}^{-1}(\xi)$ is the covariance matrix of best linear unbiased estimators of $\boldsymbol{\theta}$, and $N^{-1}\sigma^2\,\mathrm{d}(\boldsymbol{x}, \xi)$ is the variance at $\boldsymbol{x}$ of the least-squares fitted regression function.

The optimality functionals considered herein are infinite for singular $\boldsymbol{M}(\xi)$, and we therefore limit consideration to nonsingular $\boldsymbol{M}(\xi)$ in the sequel. Denote the eigenvalues of $\boldsymbol{M}(\xi)$ by $\lambda_1(\xi), \ldots, \lambda_k(\xi)$. We define

$$\Phi_p(\xi)=k^{-1}\mathrm{tr}(\boldsymbol{M}^{-p}(\xi))^{1/p}=\left(k^{-1}\sum_{1}^{k}\lambda_i^{-p}(\xi)\right)^{1/p} \quad \text{for} \quad 0<p<\infty;$$

$$\Phi_0(\xi)=\lim_{p\downarrow 0}\Phi_p(\xi)=(\det \boldsymbol{M}^{-1}(\xi))^{1/k}; \tag{2.1}$$

$$\Phi_\infty(\xi)=\lim_{p\to+\infty}\Phi_p(\xi)=\max_i \lambda_i^{-1}(\xi).$$

As discussed in the earlier papers, Φ_0, Φ_1, and Φ_∞ are the familiar D-, A-, and E-optimality criteria, here normalized so as to make comparisons easy: all Φ_p measure loss in the same scale per unit of variance, and take on the value c^{-1} if $M(\xi)=cI_k$. A *Φ_p-optimum rotatable design* $\xi^{(p)}$ is one that minimizes $\Phi_p(\xi)$ among all rotatable designs ξ. Throughout this paper we shall not exhibit the dependence of $\xi^{(p)}$, or the $\xi^{[R]}$ of Section 4, on q.

A measure of the performance of a rotatable design ξ' with respect to the family $\{\Phi_p\}$ of criteria is the (multiplicative regret, among rotatable designs) *inefficiency ratio*

$$e_p(\varepsilon')=\Phi_p(\xi')/\Phi_p(\xi^{(p)}),\quad 0\leqq p\leqq\infty. \tag{2.2}$$

A design ξ' for which these ratios are all near 1 may be judged adequate for many purposes in applications. As in the earlier papers, we shall concentrate in Section 3 on the inefficiency ratio (2.2) for $\xi'=$ various $\xi^{(r)}$'s, especially for $r=0$, 1, or ∞. We shall then write $\bar{e}(r)=\max_{0\leq p\leq\infty} e_p(\xi^{(r)})$ for the maximum inefficiency of $\xi^{(r)}$ overall Φ_p criteria.

Another common criterion, that of G-optimality, uses the optimality functional $\bar{\mathrm{d}}(\xi)=\max_{x\in\mathscr{X}}\mathrm{d}(\boldsymbol{x},\xi)$. This will be discussed below in connection with extrapolation problems.

We write $B_q(R)=\{(x_1,\ldots,x_q)\colon \sum_1^q x_i^2\leqq R^2\}$ for the ball of radius R, and $S_q(R)=\{(x_1,\ldots,x_q)\colon \sum_1^q x_i^2=R^2\}$ for the spherical shell of radius R. The settings we consider are those in which $\mathscr{X}$ is the unit ball $B_q(1)$ and $\boldsymbol{\theta}'\boldsymbol{f}$ is an arbitrary cubic. This cubic regression is conveniently expressed in terms of the following partition $\boldsymbol{f}'=(\boldsymbol{f}_1', \boldsymbol{f}_2', \boldsymbol{f}_3', \boldsymbol{f}_4')$, with respective numbers of components $q+1$, $q(q+1)$, $\frac{1}{2}q(q-1)$, $\frac{1}{6}q(q-1)(q-2)$:

$$\boldsymbol{f}_1(\boldsymbol{x})'=(1, x_1^2,\ldots,x_q^2);$$

$$\boldsymbol{f}_2(\boldsymbol{x})'=(x_1, x_1^3, x_1x_2^2,\ldots,x_1x_q^2;\ldots; x_q, x_q^3, x_qx_1^2,\ldots,x_qx_{q-1}^2);$$

$$\boldsymbol{f}_3(\boldsymbol{x})'=\{x_ix_j,\ i<j\};$$

$$\boldsymbol{f}_4(\boldsymbol{x})'=\{x_hx_ix_j,\ h<i<j\}.$$

In what follows we assume the $k(=\frac{1}{6}(q+1)(q+2)(q+3))$ components of $\boldsymbol{f}$ to be written in this order.

Under our restriction to rotatable designs, a result of Karlin and Studden (1966) implies that every design ξ admissible among rotatable designs and with nonsingular $\boldsymbol{M}(\xi)$, is supported by $S_q(1)$ and an $S_q(\rho)$ for some ρ satisfying $0<\rho<1$. All rotatable designs with the same values of ρ and of $\xi(S_q(\rho))=\alpha$ (say) have the same information matrix, so we may conveniently consider only designs that spread mass α (resp., 1-α) *uniformly* on $S_q(\rho)$ (resp., $S_q(1)$). It is not hard to show that, for each q and p, a unique pair $\alpha^{(p)}$, $\rho^{(p)}$ minimizes $\Phi_p(\boldsymbol{M}(\xi))$ among all such designs. The implementation of such designs in terms of approximately optimum exact designs, is discussed in the earlier references.

For a rotatable design of the above structure, a straightforward computation shows that $\int_{B_q(1)} \boldsymbol{f}_i\boldsymbol{f}_j'\,\mathrm{d}\xi=0$ for $i\neq j$, so that the nonzero elements of $\boldsymbol{M}(\xi)$ are contained in four square diagonal blocks $\mathbf{A}_i(\xi)$, $1 \leq i \leq 4$, of the same orders as the f_i. Writing I_m for the $m \times m$ identity and

$$\mu_i(\xi) = \int_{B_q(1)} x_1^i \, d\xi, \qquad \mu_{ij}(\xi) = \int_{B_q(1)} x_1^i x_2^j \, d\xi,$$
$$\mu_{hij}(\xi) = \int_{B_q(1)} x_1^h x_2^i x_3^j \, d\xi, \tag{2.4}$$

we obtain

$$\begin{aligned} \mathbf{A}_3(\xi) &= \mu_{22} I_{q(q-1)/2}, \\ \mathbf{A}_4(\xi) &= \mu_{222} I_{q(q-1)(q-2)/6}, \\ \mathbf{A}_1(\xi) &= \begin{pmatrix} 1 & U \\ U' & T \end{pmatrix}, \end{aligned} \tag{2.5}$$

where $U(1 \times q)$ has all elements μ_2 and $T(q \times q)$ has diagonal elements μ_4 and off-diagonal elements μ_{22}. Finally, the nonzero elements of $\mathbf{A}_2$ consists of q identical diagonal blocks $\bar{\mathbf{A}}_2((q+1) \times (q+1))$, where

$$\bar{\mathbf{A}}_2 = \begin{pmatrix} \mu_2 & \mu_4 & v \\ \mu_4 & \mu_6 & w \\ v' & w' & z \end{pmatrix} \tag{2.6}$$

has $v(1 \times (q-1))$ with all elements μ_{22}, w has all elements μ_{42}, and z has diagonal elements μ_{42} and off-diagonal elements μ_{222}.

An easy computation yields, for the uniform probability measure on $S_q(1)$, the values $\mu_2 = 1/q$, $\mu_{22} = 1/q(q+2)$, $\mu_{222} = 1/q(q+2)(q+4)$. Consequently, for ξ of the type we are considering, we have the following moments, with notation we adopt for them:

$$\begin{aligned} \mu_2 &= [(1-\alpha) + \alpha\rho^2]/q = B/q = b, \\ \mu_{22} &= [(1-\alpha) + \alpha\rho^4]/q(q+2) = C/q(q+2) = c, \\ \mu_{222} &= [(1-\alpha) + \alpha\rho^6]/q(q+2)(q+4) = D/q(q+2)(q+4) = d, \end{aligned} \tag{2.7}$$

and from these the other moments can be computed from the fact that, for an arbitrary rotatable design,

$$\mu_4 = 3\mu_{22}, \qquad \mu_{42} = 3\mu_{222}, \qquad \mu_6 = 15\mu_{222}. \tag{2.8}$$

From these facts we compute that the eigenvalues of $\boldsymbol{M}(\xi)$ are

c, of multiplicity $\frac{1}{2}q(q-1)$;

d, of multiplicity $\frac{1}{6}q(q-1)(q-2)$;

$2c$, of multiplicity $q-1$;

$2d$, of multiplicity $q(q-2)$, only if $q>2$;

λ_0^+ and λ_0^-, given by $\frac{1}{2}\{1+(q+2)c \pm [[1-(q+2)c]^2 + 4qb^2]^{1/2}\}$,

each of multiplicity 1; (2.9)

$\bar{\lambda}_1, \bar{\lambda}_2, \bar{\lambda}_3,$ the zeros of $\lambda^3-[b+(q+16)d]\lambda^2$
$+[(6q+24)d^2+(q+16)db-(q+8)c^2]\lambda-6d[(q+4)bd-(q+2)c^2],$
each of multiplicity q, only if $q>1$;
$\lambda_1^+, \lambda_1^- = \frac{1}{2}\{15d+b\pm[(15d-b)^2+36c^2]^{1/2}\}$, each of multiplicity 1,
only present if $q=1$.

From this we computed and minimized $k[\Phi_p(\boldsymbol{M}(\xi))]^p=\sum_1^k \lambda_i^{-p}(\xi)$ numerically for $0<p<\infty$ and $q\leqq 10$. For $p=0$, We maximized

$$\det \boldsymbol{M}(\xi)=2^{q^2-q-1}c^{(q+2)(q-1)/2}d^{q(q+5)(q-2)/6}[(q+2)c-qb^2]\times[(6q+24)bd^2-(6q+12)c^2d]^q \tag{2.10}$$

for $q>1$, the last factor is replaced by $15bd-9c^2$ if $q=1$. (This formula is equivalent to a different form given in equation (4.1) of Farrell et al. (1965).) For $p=\infty$, since $d<c$ we must maximize

$$1/\Phi_\infty(\boldsymbol{M}(\xi))=\min(d, \lambda_0^-, \bar{\lambda}_1, \bar{\lambda}_2, \bar{\lambda}_3); \tag{2.11}$$

when $q=2$, d is omitted in (2.11), and (2.11) is replaced by $\min(\lambda_0^-, \lambda_1^-)$ when $q=1$.

The matrix $\boldsymbol{M}$ can be inverted explicitly for designs of the type considered above. Writing $z=\sum_1^q x_i^2=\|\boldsymbol{x}\|^2$ and $\gamma=\rho^2$, and recalling that the variance function of a rotatable design depends only on z, we obtain

$$\begin{aligned} d(\boldsymbol{x}, \xi) &= [\alpha(1-\alpha)\gamma(1-\gamma)^2]^{-1}\{\gamma C+[qD-2B\gamma]z \\ &\quad +[\gamma-2qC+(q-1)(q+2)\alpha(1-\alpha)\gamma(1-\gamma)^2/2C]z^2 \\ &\quad +[qB+(q+4)(q-1)q\alpha(1-\alpha)\gamma(1-\gamma)^2/6D]z^3\} \\ &= D(z, \alpha, \gamma) \quad \text{(say)}. \end{aligned} \tag{2.12}$$

The maximum of the variance function of any design ξ, over the ball $B(R)$ of radius $R=T^{1/2}$, and of designs of the special form considered above, are denoted

$$\begin{aligned} \bar{d}_R(\xi) &= \max_{x\in B_q(R)} d(\boldsymbol{x}, \xi), \\ \bar{D}_R(\alpha, \gamma) &\ \max_{0\leqq z\leqq R^2} D(z, \alpha, \gamma). \end{aligned} \tag{2.13}$$

Of course, the meaningfulness of (2.12) when $z>1$ and of (2.13) when $R>1$ depends on the validity of the cubic model for the ball $B_q(R)$ and not just the original $B_q(1)$. Consideration of the fitted regression on $B_q(R)$ is termed *extrapolation* if $R>1$ and *interpolation* if $R<1$; see Kiefer and Wolfowitz (1964).

For the sake of brevity, we omit in this paper any detailed comparison of designs with respect to several optimality criteria for extrapolation, analogous to our earlier consideration of the various Φ_p's. The most obvious criteria other than $\bar{d}_R$ are averages of $d(\boldsymbol{x}, \xi)$ with respect to measures ν over B_R, which can be expressed as $\mathrm{tr}AM^{-1}(\xi)$ for an appropriate A depending on ν. When $R=1$, the relative merits of these criteria ($\bar{d}_1$ being the "G-optimality" criterion) have been discussed extensively in the literature, including the references cited earlier. It is not hard to give examples in which a design that minimizes tr$AM^{-1}(\xi)$ performs very badly in terms of $\bar{d}_R$, and our own view is that this is a more serious defect than the opposite one for a design minimizing $\bar{d}_R$. In the present context, the two designs that minimize $\bar{d}_R$ and tr AM^{-1} for ν = Lebesgue measure on $B_q(R)$, are almost the same for large R. In any event, we hereafter restrict consideration to $\bar{d}_R$ as the optimality criterion in extrapolation (and interpolation) problems, and denote by $\alpha^{[R]}$, $\rho^{[R]}$ the values minimizing $\bar{d}_R(\alpha, \rho^2)$. The inefficiency ratio of a design ξ' for extrapolation to $B_q(R)$ is then, in analogy with (2.2),

$$E_R(\xi') = \bar{d}_R(\xi')/\bar{D}_R(\alpha^{[R]}, (\rho^{[R]})^2). \tag{2.14}$$

We abbreviate $E_R(\xi^{(r)})$ by $E_R(r)$ (in Table 2), and also write

$$E_\infty(r) = \lim_{R\to\infty} E_R(r). \tag{2.15}$$

3. The Φ_p-optimum designs

The values of the optimum $\alpha^{(p)}$, $\rho^{(p)}$ are listed in Table 1 for $1 \leqq q \leqq 10$ and $p = 0, 0.5, 1, 3, \infty$. We remark that the less accurate numerical results for $p = 0$ given in Farrell, et al. (1965) are seen to be slightly off from the present values, especially in terms of relative error for the small $\alpha^{(0)}$ when $q \geqq 3$. However, the function Φ_0 is smooth enough that there is little loss in Φ_0-efficiency from using the less accurate computations.

The D-optimum design $\xi^{(0)}$ has its inefficiency ratio $e_p(\xi^{(0)})$ increasing in p. The A-optimum $\xi^{(1)}$ has $e_p(\xi^{(1)})$ decreasing for $p<1$ and increasing for $p>1$, with maximum at $p=0$ for $q \leqq 2$ and at $p=\infty$ for $q>2$; for $q>2$, $e_0(\xi^{(1)})<1.10$, decreasing with q to 1.06 when $q=10$. The E-optimum design $\xi^{(\infty)}$ has $e_p(\xi^{(\infty)})$ decreasing in p. The maximum $\bar{e}(r)$ of $e_p(\xi^{(r)})$ is given in the third column of Table 2 for $r = 0, 1, \infty$. In terms of $\bar{e}(r)$, the A-optimum design is the best of the three for $q \leqq 4$; but the E-optimum design is best for $q>4$, and also performs well for $q \leqq 4$. The value of r that minimizes $\bar{e}(r)$ is slightly <1 for $q \leqq 2$, and is >1 for $q>2$, but is not of much practical interest, as discussed in earlier papers.

The remainder of Table 2 compares the D-, A-, and E-optimum designs with the optimum design for extrapolation to $B_q(R)$ (discussed in the next section), in terms of the criterion $E_R(r)$ of (2.14). Of course, $E_1(0) = 1$ because of the equivalence of D- and G- optimality, Kiefer and Wolfowitz (1960). We see that the E-optimum design performs extremely well except when R is quite small, with inefficiency ratio very near 1 for $R \geqq 2$. The A-optimum design has a smaller value of $\max_{R \geqq 1} E(r)$ than does $\xi^{(\infty)}$, due to its better performance for R near 1; but it is less good for very small or large R. The D-optimum design is the least good of the three, by a wide margin for large q.

Table 1

Selected Φ_p-optimum rotatable designs

q	p	$\alpha^{(p)}$	$\rho^{(p)}$	q	p	$\alpha^{(p)}$	$\rho^{(p)}$
1	0	0.500	0.447	6	0	0.088	0.576
	0.5	0.645	0.455		0.5	0.182	0.583
	1	0.699	0.464		1	0.244	0.579
	3	0.738	0.490		3	0.344	0.567
	∞	0.747	0.500		∞	0.441	0.551
2	0	0.308	0.515	7	0	0.070	0.580
	0.5	0.466	0.526		0.5	0.153	0.587
	1	0.536	0.528		1	0.210	0.583
	3	0.605	0.527		3	0.309	0.572
	∞	0.614	0.525		∞	0.417	0.556
3	0	0.208	0.544	8	0	0.058	0.584
	0.5	0.351	0.553		0.5	0.130	0.590
	1	0.425	0.553		1	0.183	0.586
	3	0.508	0.544		3	0.280	0.575
	∞	0.522	0.541		∞	0.385	0.560
4	0	0.150	0.560	9	0	0.048	0.586
	0.5	0.275	0.568		0.5	0.112	0.592
	1	0.346	0.565		1	0.162	0.589
	3	0.440	0.554		3	0.255	0.578
	∞	0.495	0.544		∞	0.375	0.565
5	0	0.113	0.569	10	0	0.041	0.588
	0.5	0.221	0.577		0.5	0.098	0.594
	1	0.288	0.574		1	0.144	0.591
	3	0.387	0.562		3	0.234	0.581
	∞	0.468	0.549		∞	0.356	0.565

The results described in the previous two paragraphs indicate that the E-optimum design is a most acceptable choice in terms of its performance for extrapolation to larger balls as well as its performance under the various Φ_p. When interpolation to much smaller balls $B_q(R)$ is known to be of primary concern, other designs might be considered; no design can perform efficiently both for R very large and also for R very small. The prescription to use $\xi^{(\infty)}$ of course depends on the setting and criteria considered, as our earlier papers illustrate; for example, when $\mathscr{X}$ is taken to be $B_q(R)$ rather than $B_q(1)$, one obtains $\bar{e}(1)$ much smaller than $\bar{e}(\infty)$ for large q and R, as shown in the quadratic case in Galil and Kiefer (1977d). This last also illustrates that the scaling of $\mathscr{X}$, which is irrelevant for D-optimality, of course matters for A- and E-optimality. The considerations of the next section do not depend on scaling: the problem of extrapolation from $\mathscr{X} = B_q(R_0)$ to $B_q(RR_0)$ has the same optimum choice of α and ρ/R_0, regardless of the value of R_0.

Table 2

Inefficiency ratios of Φ_r-optimum designs

q	r	$\bar{e}(r)$	$E_{0.1}(r)$	$E_{0.5}(r)$	$E_1(r)$	$E_2(r)$	$E_5(r)$	$E_\infty(r)$
	0	1.34	2.38	1.77	1.00	1.07	1.15	1.17
1	1	1.09	1.87	1.27	1.66	1.04	1.01	1.01
	∞	1.16	2.03	1.27	1.97	1.12	1.04	1.03
	0	1.52	4.04	2.31	1.00	1.25	1.40	1.43
2	1	1.11	2.52	1.38	1.46	1.00	1.02	1.02
	∞	1.20	2.24	1.22	1.74	1.03	1.00	1.00
	0	1.98	5.95	2.96	1.00	1.46	1.71	1.75
3	1	1.11	3.10	1.52	1.35	1.02	1.07	1.08
	∞	1.22	2.50	1.27	1.61	1.01	1.01	1.01
	0	2.58	8.11	3.67	1.00	1.67	2.00	2.06
4	1	1.25	3.68	1.68	1.28	1.05	1.12	1.14
	∞	1.28	2.49	1.23	1.63	1.01	1.00	1.00
	0	3.25	10.52	4.41	1.00	1.86	2.28	2.35
5	1	1.40	4.28	1.83	1.23	1.07	1.16	1.19
	∞	1.32	2.53	1.22	1.62	1.01	1.00	1.00
	0	3.97	13.17	5.17	1.00	2.03	2.53	2.63
6	1	1.55	4.89	1.98	1.19	1.10	1.21	1.23
	∞	1.34	2.58	1.21	1.59	1.01	1.00	1.00
	0	4.74	16.05	5.94	1.00	2.18	2.76	2.87
7	1	1.71	5.53	2.13	1.16	1.12	1.25	1.27
	∞	1.34	2.67	1.21	1.55	1.02	1.00	1.00
	0	5.56	19.17	6.71	1.00	2.31	2.97	3.10
8	1	1.86	6.18	2.26	1.14	1.14	1.28	1.31
	∞	1.34	2.73	1.21	1.52	1.02	1.00	1.00
	0	6.39	22.52	7.48	1.00	2.44	3.17	3.31
9	1	2.01	6.85	2.39	1.13	1.16	1.31	1.34
	∞	1.33	2.87	1.21	1.49	1.02	1.00	1.00
	0	7.25	26.09	8.25	1.00	2.55	3.34	3.50
10	1	2.16	7.53	2.52	1.11	1.17	1.33	1.36
	∞	1.32	2.92	1.21	1.46	1.02	1.00	1.00

4. Extrapolation and interpolation

We begin by describing the routine used to compute the values $\alpha^{[R]}$, $\rho^{[R]}$ that minimize (2.13). For fixed q and R we first computed the values α^*, ρ^* that minimized $D_{\mathbf{R}}(R, \alpha, \rho^2)$, the variance function on the boundary $S_q(R)$ of $B_q(R)$. Since $D_{\mathbf{R}}(z, \alpha, \rho^2)$ is a cubic in z, it is then easy to check whether $D_{\mathbf{R}}(z, \alpha^*, (\rho^*)^2)$ attains its maximum over $0 \leqq z \leqq R$ at the value $z = R$. If so, obviously $\alpha^{[R]}$, $\rho^{[R]} = \alpha^*$, p^*. If not, another structure for the optimum design must be sought, and it turns out in the present setting always to be obtained by minimizing $D_{\mathbf{R}}(R, \alpha, \rho^2)$ subject to $D_{\mathbf{R}}(R, \alpha, \rho^2) = D_{\mathbf{R}}(z_{\alpha,\rho}, \alpha, \rho^2)$, where $z_{\alpha,\rho}$ is the local maximum of the cubic $D_{\mathbf{R}}(z, \alpha, \rho^2)$ in the interior $(0, R)$ of the domain of z; this interior maximum is always present for the optimum $\alpha^{[R]}$, $\rho^{[R]}$. This form of solution is intuitively what one would expect from graphing D_R and from our knowledge that the solution for $R=1$ (the G-optimum design) is of that form.

The first form of solution holds in the exterior of an interval $[R_q', R_q'']$ containing the value $R=1$ in its interior; the second form of solution holds inside the interval. When $q=1$ this interval is [0.3, 1.1], so that R_1'' is already near 1; R_q' increases to 0.6 for $q=2$ and 0.8 for $q=3$. The interval narrows rapidly in q, being [0.90, 1.05] for $q=5$ and [0.95, 1.03] for $q=9$. Most important, this indicates that in other more complex extrapolation settings in which f grows appropriately, one can often minimize $\max_{x\in\mathscr{X}'} d(x, \xi)$ for $\mathscr{X}'$ at least as large as a set only slightly larger than $\mathscr{X}$, by minimizing the maximum of $d(x, \xi)$ on the boundary of $\mathscr{X}'$. This was known for polynomial regression in one dimension, from the results of Kiefer and Wolfowitz (1964) for two-sided extrapolation from an interval $\mathscr{X}$, and from those of Hoel and Levine (1964) and Kiefer and Wolfowitz (1965) for one-sided extrapolation.

The optimum values $\alpha^{[R]}$, $\rho^{[R]}$ are listed in Table 3 for $R=0.1$, 0.5, 1, 2, 5, and $1\leqq q\leqq 10$. From our previous remarks, only the entries for $R=1$ and for $q=1$, $R=0.5$ represent the second form of solution. The last column gives the limiting design as $R\to\infty$ for fixed q, discussed further in the next section. The evident closeness of $\alpha^{[R]}$, $\rho^{[R]}$ to the $\alpha^{(\infty)}$, $\rho^{(\infty)}$ of the E-optimum design explains the good performance of the latter, noted in the previous section.

We have not considered here the possibility of using the approach of Box and Draper (1959) or of Karson et al. (1969) for responding to the possibility that the cubic model is incorrect and there is, for example, small quartic bias from fitting a cubic. In the case of quadratic regression on the q-simplex with $\mathscr{H}'=\mathscr{H}$ and hence no extrapolation, it was shown in Galil and Kiefer (1977a) that the "all bias" Box–Draper design may perform rather poorly in terms of its variance function or maximum bias. For first degree regression on the ball $B_q(1)$ Draper and Herzberg (1974) showed that the uniform measure on $S_q(1)$ minimizes the Lebesgue

Table 3

Optimum designs for extrapolation from $B_q(1)$ to $B_q(R)$

q \ R	0.1 $\alpha^{[0.1]}$	$\rho^{[0.1]}$	0.5 $\alpha^{[0.5]}$	$\rho^{[0.5]}$	1 $\alpha^{[1]}$	$\rho^{[1]}$	2 $\alpha^{[2]}$	$\rho^{[2]}$	5 $\alpha^{[5]}$	$\rho^{[5]}$	∞ $\alpha^{[\infty]}$	$\rho^{[\infty]}$
1	0.969	0.211	0.911	0.434	0.500	0.447	0.602	0.486	0.657	0.498	0.667	0.500
2	0.958	0.250	0.858	0.516	0.308	0.515	0.541	0.510	0.603	0.513	0.614	0.514
3	0.947	0.270	0.797	0.533	0.208	0.544	0.490	0.524	0.554	0.525	0.565	0.525
4	0.939	0.288	0.752	0.544	0.150	0.560	0.448	0.535	0.513	0.534	0.524	0.534
5	0.933	0.302	0.717	0.553	0.113	0.569	0.411	0.543	0.476	0.541	0.487	0.541
6	0.928	0.313	0.685	0.560	0.088	0.576	0.380	0.549	0.444	0.547	0.454	0.546
7	0.923	0.324	0.658	0.565	0.070	0.580	0.352	0.555	0.415	0.552	0.426	0.551
8	0.919	0.332	0.634	0.569	0.058	0.584	0.329	0.559	0.390	0.556	0.401	0.556
9	0.915	0.339	0.612	0.573	0.048	0.586	0.309	0.562	0.368	0.559	0.378	0.559
10	0.912	0.346	0.591	0.575	0.041	0.588	0.291	0.565	0.348	0.562	0.358	0.562

integrals of both the variance and squared bias terms over the region $1 \leqq z \leqq R^2$ for $R > 1$; this is in fact a special case of the fact that this design is optimum for *all* rotationally invariant convex criteria. Further discussion of extrapolation designs for higher degree polynomial regression is contained in Kiefer (1977a).

5. Designs asymptotically optimum in q or R

In a number of works on design theory, use has been made of an approach discussed in Kiefer (1977b): where an optimum design computation is algebraically intractable, the design settings may sometimes fit naturally into a family of settings depending on a parameter, and for which a suitably normalized asymptotic optimality computation is much simpler in the limit for the parameter, here $q \to \infty$ or $R \to \infty$ or 0. The result of this simpler computation can then be implemented for the settings at hand and will often work well. A computation that retains higher order terms can be used where the first order approximation is inadequate. This technique will have greatest importance in settings more complex than that of this paper; here, we can compare the asymptotic results with those of Sections 3 and 4 to obtain an idea of the adequacy of the approach for other settings in which such a comparison is impossible and we are in fact using the approach to obtain, we hope, adequately efficient designs.

We begin with the considerations of Section 3. First fixing $p = 0$ and letting $q \to \infty$ in (2.10), we find easily as in Farrell et al. (1965) that the values $\alpha = 4(1 + 3^{-1/2})q^{-2}$, $\rho = [(3^{1/2} - 1)/2]^{1/2}$ give a first order approximation to D-optimality. Only when $q \geqq 3$ is $\alpha \leqq 1$ so that this is meaningful as a design $\bar{\xi}(0)$. One obtains $e_0(\bar{\xi}^{(0)}) = 1.57$, 1.14, 1.06 for $q = 3$, 4, 5, with rapid decrease to 1.004 at $q = 10$. Thus, the design $\bar{\xi}^{(0)}$ is quite acceptable for moderate q in terms of D-optimality. A bonus that could not be expected in other settings is that $e_\infty(\bar{\xi}^{(0)})$ and thus $\bar{e}(\bar{\xi}^{(0)})$ are better than $e_\infty(\xi^{(0)})$ and $\bar{e}(\xi^{(0)})$; for example, the former values are 4.76 at $q = 10$, while the latter, from Table 2, are 7.25.

The case $p = 1$ can be treated similarly.

The corresponding calculation for $p = \infty$, letting $q \to \infty$ in (2.11), is slightly more complicated. The optimum is seen to be obtained if one maximizes d asymptotically, subject to $d = \min(\lambda_1, \lambda_2, \lambda_3)$; λ_0^- and the other two roots of the cubic of (2.9) are $> d$ at this solution, which is also the pattern for $\xi^{(\infty)}$ for $q > 2$. The resulting solution $\alpha = 2q^{-1}[\frac{5}{3} + 3^{1/2}]$, $\rho = [\frac{1}{2}(3^{1/2} - 1)]^{1/2}$ can only be implemented as a design $\bar{\xi}^{(\infty)}$ for $q > 6$, which yields $e_\infty(\bar{\xi}^{(\infty)}) = 7.77$, 3.21, 2.26, 1.85 for $q = 7, 8, 9, 10$; for $q = 10$, $\bar{e}(\bar{\xi}^{(\infty)}) = 2.29$, compared with $\bar{e}(\xi^{(\infty)}) = 1.32$. This slower attainment in q of satisfactory performance of $\bar{\xi}^{(\infty)}$ compared with $\bar{\xi}^{(0)}$ is an accident of the form used for the asymptotic expansion: $(q + h)^{-1}$ is asymptotically the same for every constant h, but for small q the choice of h matters. Thus, we may expect in more complex settings that a finer asymptotic approximation, that also results in a proper choice of the analogue of h, will be worthwhile.

Alternately, a useful h could be chosen by minimizing $\Phi(\tilde{\xi}^{(\infty)})$ with respect to h for a particular q.

We turn now to the extrapolation problems of Section 4. First letting $R\to\infty$, we see that the values $\alpha^{[\infty]}$, $\rho^{[\infty]}$ that minimize $\lim_{R\to\infty} R^{-6}\bar{D}_R(\alpha,\rho^2)$ are those that minimize the coefficient of z^3 in (2.12). These are tabled in the last column of Table 3. The efficiency of these designs, when used for the problem of interpolation to $B_q(R)$ for moderate R, is very good: as q goes from 1 to 10, the inefficiency ratio $E_R(\xi^{[\infty]})$ of (2.14) ranges from 1.10 to 1.07 for $R=1.5$, is 1.02 for $R=2$, and for $R=5$ is 1.001 for $q=1$ and 1.000 for $q>1$.

As $R\to 0$, minimization of the term independent of z in (2.12) yields the useless $\alpha=1$, $\sigma=0$, a design which is useful only for estimating $\boldsymbol{\theta}'\boldsymbol{f}(0)$. A higher order approximation, minimizing the sum of the above term and that in z at $z=R$, can be obtained as in Kiefer and Wolfowitz (1964). We omit details.

Acknowledgment

This research was sponsored by the National Science Foundation.

References

Box, G.E.P., and Draper, N.R. (1959). A basis for the selection of a response surface design. *J. Am. Statist. Assoc.* 54, 622–654.

Draper, N.R. and Herzberg, A.M. (1974). An investigation of first order designs for extrapolation outside a sphere. University of Wisconsin Stat. Dept. Tech. Report 356.

Farrell, R., Kiefer, J., and Walbran, A. (1965). Optimum multivariate designs. *Proc. Fifth Symp. Math. Statist. Prob.* 1, 113–138.

Fedorov, V.V. (1972). *Theory of Optimal Experiments.* Academic Press, New York.

Galil, Z. and Kiefer, J. (1977a). Comparison of simplex designs for quadratic mixture models. *Technometrics* 19.

Galil, Z. and Kiefer, J. (1977b). Comparison of Box-Draper and D-optimum designs for experiments with mixtures. *Technometrics* 19.

Galil, Z. and Kiefer, J. (1977c). Comparison of designs for quadratic regression on cubes. *J. Statist. Planning and Inference* 1.

(Galil, Z. and Kiefer, J. (1977d). Comparison of rotatable designs for regression on balls, I (quadratic). *J. Statist. Planning and Inference* **1**.

Hoel, P.G. and Levine, A. (1964). Optimal spacing and weighting in polynomial prediction. *Ann. Math. Statist.* 33, 1553–1560.

Karlin, S. and Studden, W. (1966). Optimal experimental designs. *Ann. Math. Statist.* 37.

Karson, M.J., Manson, A.R. and Hader, R.J. (1969). Minimum bias estimation and experimental designs for response surfaces. *Technometrics* **11,** 461–476.

Kiefer, J. (1959). Optimum experimental designs. *J. Roy Statist. Soc. Ser. B* 21, 272–319.

Kiefer, J. (1960). Optimum experimental designs V, with applications to systematic and rotatable designs. *Proc. Fourth Berkeley Symp. Math. Statist. Prob.* 1, 381–405.

Kiefer, J. (1974). General equivalence theory for optimum designs (approximate theory). *Ann. Statist.* 2, 849–879.

Kiefer, J. (1975). Optimal design: Variation in structure and performance under change of criterion. *Biometrika* 62, 277–288.

Kiefer, J. (1977a). Designs for extrapolation. Paper invited for December 1977 ISI meeting in India.
Kiefer, J. (1977b). Asymptotic approach to families of design problems. *Comm. Statist.* (to appear).
Kiefer, J. and Wolfowitz, J. (1960). The equivalence of two extremum problems. *Can. J. Math.* 14, 363–366.
Kiefer, J. and Wolfowitz, J. (1964). Optimum extrapolation designs I and II. *Ann. Inst. Statist. Math.* 16, 79–108, 295–303.
Kiefer, J. and Wolfowitz, J. (1965). On a theorem of Hoel and Levine on extrapolation. *Ann. Math. Statist.* 36, 1627–1655.

Annals of Discrete Mathematics 6 (1980) 225–241

OPTIMAL DESIGN THEORY IN RELATION TO COMBINATORIAL DESIGN

J. KIEFER*
Cornell University and University of California, Berkeley, CA, U.S.A.

1. Introduction

The usefulness of combinatorial arrays for designing statistical experiments was evident to R.A. Fisher, and under his energetic leadership the construction and application of such designs enjoyed tremendous expansion. Fisher and his students and followers were responsible for much of the development in the 1930's and 1940's. Mathematicians such as R.C. Bose gave general structure to classes of these designs, and invented general methods for constructing them. It would be presumptuous of me to recite more of this history to the audience of this conference.

The spirit of much of this construction, of designs that were in some sense as symmetric as possible in their treatment of the statistical parameters of interest (for example, balanced incomplete block designs (BIBD) or latin squares (LS)), perhaps stemmed from three factors: such designs yielded "information matrices" (coefficient matrices of normal equations for least squares estimation) that, especially in that pre-computer age, made statistical calculations easy; the designs had aesthetic appeal to mathematicians and often had algebraic or geometric representations that helped one to understand and construct them; and the symmetric treatment of parameters of interest seemed a reasonable property that made such designs yield statistical estimators that looked intuitively as accurate as possible for the given number of observations.

The work of Neyman and of Wald made statisticians increasingly formulate statistical problems and criteria for "goodness" of procedures in a precise manner. A pioneering example of this in design theory is the paper by Wald [45], who considered the standard 2-way heterogeneity model with v treatments and the possibility of allowing any $v \times v$ matrix of symbols from $\{1, 2, \ldots, v\}$ as a design, and proved that a LS design was optimum in a precise sense. The result is not too surprising in view of the intuition indicated in the previous paragraph, but it turns out (Section 4) that intuition is not always so reliable; what is important, though, is that Wald gave a structure for choice of a design on well-formulated grounds that in no way rested on the motivation employed earlier for using such designs.

* Research under NSF Grant MCS75-22481 A02.

Authors such as Mood [33] extended such considerations to the setting of weighing designs; and subsequently other settings, such as that where BIBD's are often employed, were considered, as in Kiefer [23]. In a companion development, problems were also treated in which these combinatorial structures are not relevant because there is a continuum of design choices, as in curve-fitting settings in which the independent variable may be choosen to lie in an interval. This is discussed, for example, in Kiefer [24, 26] and Kiefer and Wolfowitz [30].

The present paper is devoted to indicating how, in recent years, the relationship between combinatorial and optimal design developments has begun to exhibit a motivational reversal. Instead of constructing exotic designs on intuitive grounds and then proving that they are indeed optimum, we encounter settings in which optimality considerations motivate the construction of new combinatorial structures, or the selection of a particular subset of the known ones for special emphasis and study. An interesting example (to be discussed further in Section 2) is the setting of "one-way heterogeneity", that is, of block designs. If the parameter values (block size k, number of blocks b, number of varieties v) were such that a BIBD did not exist, the early design developments sought designs that were still quite symmetric, as in the construction of PBIBD's by Bose and Nair [3]. After optimality properties of BIBD's were proved by Kiefer [23], work by Takeuchi [43, 44] showed that certain PBIBD's were also E-optimum (see (1.1)). This result (pioneering in giving the first optimality conclusion for designs lacking maximum combinatorial symmetry) was strengthened and extended by Cheng [5], who was able to delimit a particular class of PBIBD's that possess strong optimum properties. Finally, the consideration of that class led to the graph-theoretic characterization in the paper by Cheng and Gray [10] presented at this conference. Several other illustrations of such developments will be considered in this paper, but they are by no means exhaustive; for example, the work of Seiden, Hedayat, and others on various "repeated measurement" models, will not be touched upon; nor will Srivastava's search design considerations, which in part entail finding, as in Section 5, the "smallest" design that achieves an appropriate aim.

Throughout this paper we will restrict attention to a setting (slightly modified in Section 5) in which there is a class D of designs available to the statistician and in which, for each d in D, there is an $n \times m$ matrix X_d such that the expectation of the observed n-vector Y, if design d is used, is $X_d\theta$; here θ is the m-vector of unknown parameters in this "linear model". The components of Y are assumed uncorrelated, with equal variances σ^2. The "information matrix" of the design d is $X_d'X_d$ (primes denote transposes), and the normal equations (NE's) for obtaining least squares (best linear unbiased) estimators are $(X_d'X_d)t = X_d'Y$. This means that, if $c'\theta$ is a linear parametric function for which *some* linear unbiased estimator of $c'\theta$ exists when design d is used, then for any solution t of the NE's $c't$ is the best linear unbiased estimator of $c'\theta$. If X_d has rank m, then $\sigma^2(X_d'X_d)^{-1}$ is the covariance matrix of t, with a similar interpretation otherwise. Thus, we want to choose a design whose information matrix $X_d'X_d$ is "large" in some sense, and since the symmetric nonnegative definite (SND) $m\times m$ matrices are only partially ordered by the natural (and only useful definition) that $A\geq B$ if $A-B$ is SND, an "optimality functional" Φ on the SND matrices must be considered. We then choose d to minimize $\Phi(X_d'X_d)$ over D.

Often only the first v out of the m components of θ are of interest. Partitioning $X_d=(X_d^{(1)} \vdots X_d^{(2)})$ and $\theta'=(\theta^{(1)'}, \theta^{(2)'})$ accordingly, the analogue of $X_d'X_d$ is Bose's "C-matrix"

$$C_d=X_d^{(1)'}X_d^{(1)}-X_d^{(1)'}X_d^{(2)}(X_d^{(2)'}X_d^{(2)})^{-}X_d^{(2)'}X_d^{(1)};$$

then C_d^{-} is the upper $v\times v$ submatrix of $(X_d'X_d)^{-}$. This is the "information matrix for $\theta^{(1)}$". The optimality criterion will then typically reduce to

$$\Phi(X_d'X_d)=\Phi^{(1)}(C_d).$$

In some settings, such as that where a chemical balance is used in sufficiently many "weighings" (Section 3), C_d can be nonsingular. Typical $\Phi^{(1)}$'s in such settings (interpretable as size measures of C_d^{-1} of a "confidence ellipsoid" asserted to contain θ if Y is normal) are

$$\begin{aligned} &\log\det C_d^{-1} && (D\text{-optimality}),\\ &\operatorname{tr} C_d^{-1} && (A\text{-optimality}),\\ &\text{max eigenvalue of } C_d^{-1} && (E\text{-optimality}). \end{aligned} \tag{1.1}$$

Sometimes it is possible to avoid specification of a particular functional. Thus, if C_{d^*} is a multiple of the identity and $\operatorname{tr} C_{d^*}=\max_{d\in D}\operatorname{tr} C_d$, it follows that d^* minimizes $\Phi^{(1)}(C_d)$ for every orthogonally invariant convex $\Phi^{(1)}$ such that $\Phi^{(1)}(bC)$ is nonincreasing in the scalar b. (The $\Phi^{(1)}$'s of (1.1) are obvious examples.) A d^* optimum for all such $\Phi^{(1)}$ is termed *universally optimum.*

In other settings C_d is singular for all d in D. For example, in the setting of one-way heterogeneity (Section 2) in which we must assign v treatments to b blocks of size k, we have $n=bk$ for all allowable designs, and $m=v+b$, an observation on treatment i in block j having expectation $\alpha_i+\beta_j$, with $\theta^{(1)}=\boldsymbol{\alpha}$ and $\theta^{(2)}=\boldsymbol{\beta}$. Then C_d is the information matrix for treatment effects. It can have rank at most $v-1$, since only "contrasts" $\sum_1^v c_i\alpha_i$ with $\sum_i^v c_i=0$ can be estimated. In fact, C_d has zero row and column sums. In that case if P is any $(v-1)\times v$ real matrix with rows orthonormal and orthogonal to $(1,1,\ldots,1)$, we may replace C_d by $\bar{C}_d=PC_dP'$ in (1.1); the statistical meanings of the resulting criteria are parallel to those of (1.1), and are of course independent of P. For universal optimality, orthogonal invariance of

$$\Phi^{(1)}(C_d)=\bar{\Phi}^{(1)}(PC_dP')$$

can be replaced by invariance under (common) relabeling of components of $\theta^{(1)}$ (that is, of rows and columns of C_d), and the condition that $\bar{C}_{d^*}$ is a multiple of the identity can be rewritten conveniently as "all diagonal elements of C_{d^*} equal, all off-diagonal elements equal", which we term *complete symmetry* of C_{d^*}. When all C_d have zero row and column sums, we thus obtain

$$C_{d^*} \text{ completely symmetric, } \operatorname{tr} C_{d^*} = \max_{d \in D} \operatorname{tr} C_d \Rightarrow d^* \text{ universally optimum} \tag{1.2}$$

The use of this tool to prove optimality of BIBD's, or of LS's in their setting, is described further in Kiefer [23, 28]. We shall see that there are familiar settings where (1.2) cannot be used because the obvious candidate d^* does not satisfy one of the two conditions. The usefulness of (1.2) of course lies in the simplicity of evaluating $\operatorname{tr} C_d$ rather than $\Phi^{(1)}(C_d)$ for each d. When (1.2) cannot be applied, more difficult computations and comparisons must be made, as we shall describe.

We hereafter denote the eigenvalues of C_d by $\mu_{d1} \geqslant \mu_{d2} \geqslant \cdots \geqslant \mu_{dv}$, and $\boldsymbol{\mu}_d = (\mu_{d1}, \ldots, \mu_{d(v-1)})$, in the setting of the last paragraph where $\mu_{dv} = 0$ for all d, but where such criteria as the $\Phi^{(1)}$ corresponding to (1.1) can be expressed in terms of $\boldsymbol{\mu}_d$.

We note that, of the three optimality criteria considered explicitly in (1.1), it is generally easiest to verify that a design is E-optimum and most difficult to verify that it is D-optimum. This is not surprising, since for C_{d^*} or $\bar{C}_{d^{**}}$ a multiple of the identity ($C_{d^{**}}$ completely symmetric) it is obvious that D-optimality implies A-optimality implies E-optimality for d^* or d^{**}. The opposite implications are not valid.

Finally, we remark that many optimality proofs in the literature have diminished value because of restrictions that are put on d to simplify the proofs. For example, in the one-way heterogeneity setting with $k < v$, D is sometimes restricted to designs for which each variety appears at most once per block; of course, a BIBD d^* has this property, but in using (1.2) one need not restrict the competitors of d^*. Again, other optimality proofs in the literature make such restrictions on D as equal replication of varieties or even complete symmetry of C_d. Such restrictions assume away many of the mathematical difficulties and, in the example of (3.1) or that of the generalized Youden design setting of Section 4, also eliminate consideration of less symmetric designs that are better!

The author thanks C.S. Cheng for helpful comments.

2. Block design settings for which no BBD exists

We consider the one-way heterogeneity setting described in Section 1, with b, v, k given. We do not require $k < v$. Let n_{dij} be the number of appearances of variety i in block j, $r_{di} = \sum_j n_{dij}$, $\lambda_{dih} = \sum_j n_{dij} n_{dhj}$. A *balanced block design* has all r_{di} equal, all λ_{dih} equal, and all n_{dij} as nearly equal as possible (i.e., $|n_{dij} - k/v| < 1$); if $k < v$, this reduces to a BIBD. Construction of BBD's can easily be reduced to construction of BIBD's with block size $k - v \operatorname{int}(k/v)$, where $\operatorname{int}(x)$ denotes the integral part of x.

If b, v, k are such that no BBD exists, we cannot use (1.2) to find an optimum design; there are designs that maximize tr C_d, and others for which C_d is completely symmetric, but none that achieve both. Suppose $\Phi^{(1)}$ is of the form

$$\Phi_f^{(1)}(C_d) \overset{\text{def}}{=} \sum_{1}^{v-1} f(\mu_{di}) \tag{2.1}$$

where f is convex and nonincreasing. (This includes the D- and A-optimality criteria, and the E-optimality criterion can be obtained as a limit of such criteria.) Then, (1.2) can be looked at geometrically in terms of finding the value

$$\mu^* = \max_d \sum_{l}^{v-1} \mu_{di}$$

and then noting that, on the hyperplane $H = \{\boldsymbol{\mu}: \sum \mu_{di} = \mu^*\}$ in R^{v-1}, the "minimum" of (2.1) occurs at the point $\boldsymbol{h}$ where all μ_i are equal. But if no BBD exists, there is no d^* to which $\boldsymbol{h}$ corresponds.

One possible intuitively appealing approach in this case is to find, on H, the actual d for which $\boldsymbol{\mu}_d$ is closest to $\boldsymbol{h}$ in some sense. If the sense is Euclidean distance, this was the approach suggested by Shah [40] under the restriction (see Section 1) to a class of designs for which $\sum \mu_{di} = \text{tr}\, C_d$ is a constant. Without this last restriction, it is the approach taken in the more general setting of unequal block sizes by Eccleston and Hedayat [12], and which led to construction considerations. Such designs were termed (M, S)-optimum designs by them. Although the method may produce reasonably efficient designs in some case, it is clear from the fact that, in Shah's case, it amounts to minimizing tr C_d^2, that it does not guarantee optimality in any of the senses of (1.1); in fact, settings can be given in which it leads to very inefficient designs, even to a d^* with $\bar{C}_{d^*}$ singular. In geometric terms, there may be an actual d' on $\{\boldsymbol{\mu}_d: \sum \mu_{di} = \mu^* - \varepsilon\}$ for some small $\varepsilon > 0$, that is much closer to $\boldsymbol{h}$ than is any point on H; additionally, even if the resulting distance $\|\boldsymbol{\mu}_{d'} - \boldsymbol{h}\|$ is small, we would have to justify that geometric "closeness" of $\boldsymbol{\mu}_{d'}$ to $\boldsymbol{h}$ implies smallness of $\Phi^{(1)}$ at d', as is suggested by the quadratic approximation of sufficiently smooth $\Phi^{(1)}$ near $\boldsymbol{h}$. Cheng [6] has given conditions under which (M, S)-optimum designs are actually optimum in senses such as those of (1.1). The results follow from the more basic conclusions in Cheng [5], which will be discussed below.

The definition of PBIBD is extended to PBBD in an obvious way when $k \geq v$. For a group divisible (GD) PBBD with two groups, write $\lambda_{dij} = \lambda_1$ if i, j are in the same associate class, and $\lambda_{dij} = \lambda_2$ otherwise. Cheng defines a GD PBBD with 2 groups to be *most balanced* (*MB*) *of type 1* if $\lambda_2 = \lambda_1 + 1$, and *MB of type 2* if $\lambda_1 = \lambda_2 + 1 > 1$. We write MBD for MBGDPBBD.

Example. For $v = 4$, $k = 2$, let $A = \{(1, 2), (3, 4)\}$ and $B = \{(1, 3)(1, 4), (2, 3), (2, 4)\}$. Then B or $(2B) \cup A$ is type 1, and $B \cup (2A)$ is type 2.

Takeuchi's result (Section 1) is that, when $k<v$, if no BIBD exists but a MBD of type 1 exists (even with more than 2 groups), it is E-optimum. Conniffe and Stone [11] proved, when $k<v$, that a MBD of type 1 is A-optimum among designs for which all r_{di} are equal and all $n_{dij}=0$ or 1. Despite these restrictions, their method of proof used tools that Cheng [5] extended and sharpened to obtain much stronger results, some of which we now describe. (Another result obtained under restriction to equal replications is that of D-, A-, E-optimality of linked block designs [41].)

Suppose $\Phi_f^{(1)}$ is of the form (2.1) where f is defined on $[0, \max_{d\in D} \operatorname{tr} C_d]$, is strictly decreasing and strictly convex thereon (possibly with $f(0)=+\infty$), and is continuously differentiable on the interior of that interval. If also f' is strictly concave, $\Phi_f^{(1)}$ is said to be a criterion of *type 1*; if f' is strictly convex, it is of *type 2*. *Generalized type i criteria* are pointwise limits of these. The criteria of (1.1) are all generalized type 1; although there exist type 2 criteria, they are of less statistical interest (for example, $f(x)=\varepsilon x^3-x$ on $[0,c)$ with ε sufficiently small). A delicate calculation by Cheng replaces (1.2) by the assertion that, for $v>2$, if C_{d*} has two distinct nonzero eigenvalues, the larger of multiplicity 1, and if $\operatorname{tr} C_{d*}^2<(\operatorname{tr} C_{d*})^2/(v-2)$ and

$$C_{d*} \text{ maximizes} \begin{cases} \operatorname{tr} C_d \\ \operatorname{tr} C_d - \left(\dfrac{v-1}{v-2}\right)^{1/2}\left[\operatorname{tr} C_d^2 - \dfrac{(\operatorname{tr} C_d)^2}{v-1}\right]^{1/2}, \end{cases} \tag{2.2}$$

then d^* is optimum wrt all generalized type 1 criteria. An analogous result is obtained for generalized type 2 criteria if the *smaller* eigenvalue has multiplicity 1 and the last line of (2.2) is replaced by

$$\operatorname{tr} C_d-[(v-1)(v-2)(\operatorname{tr} C_d^2-(\operatorname{tr} C_d)^2/(v-1))]^{1/2}.$$

From these results and a further computation it follows that MBD's of type i are optimum (sometimes uniquely so) wrt all generalized type i criteria. In fact, these conclusions hold for the (two corresponding types of) regular graph designs (RGD's) studied by John and Mitchell [22], and this fact (together with a study of the list of designs obtained by John and Mitchell) led to the conjecture that RGD's (or the corresponding regular graphs) with appropriate eigenvalue structure must be MBD's. This was verified by Cheng and Gray [10], as mentioned in Section 1.

A remarkable feature of this result is the separation of the two types of combinatorial structures on the basis of the optimality criterion. Moreover, in earlier work that used (1.2), only convexity of f, not the sign of f''', arose. It is interesting that when one leaves cases where a BBD exists, the set of convex f of universal optimality becomes split in this way into two (nonexhaustive) sets of criteria that depend on the third derivative. This is all in the case where, although a BBD does not exist, there is a design which is perhaps as "close" combinatorially as one can imagine to a BBD. When also a MBD does not exist, one would then expect any optimality results to be much more difficult, involving a further split among criteria relative to which *different* designs are optimum, even for the criteria of (1.1). Indeed, it is easy to give settings where this occurs, and almost no general optimality results are known in such cases.

3. Weighing designs and fractional factorial designs

In the setting of weighing with a chemical balance, $n \geq m = v$, we are interested in all parameters, and the (i, j)th elements x_{dij} of X_d is 1, -1, or 0 depending on whether, in the ith of n weighings, the jth object (with weight θ_j) is in the left pan, in the right pan, or absent. Sometimes x_{dij} is restricted to be ± 1. If $2^v \mid n$, one universally optimum design obviously takes $n/2^v$ observations corresponding to each v-vector of ± 1's. More interesting, if $v \mid n$ and there is a $v \times v$ Hadamard matrix, then n/v copies of it yield a universally optimum design. There is a considerable literature of optimality results obtained under the restriction that $X_d'X_d$ is completely symmetric; much of this is reviewed in Raghavarao [37]. Mood [33], besides discussing the Hadamard cases, had enumerated a few D-optimum designs for small n and v, but the first general results in the non-Hadamard case were obtained by Ehlich [13], who proved, under only the restriction $x_{dij} = \pm 1$, that if $n = v$ and $X_{d^*}'X_{d^*} = (n-1)I_n + J_n$ (where J_n consists entirely of 1's), then d^* is D-optimum. Using methods like those employed in obtaining the results of Section 2, Cheng [7] showed, more generally (and among other results), that, for general n and v, a d^* with $X_{d^*}'X_{d^*} = (n-1)I_v + J_v$ is optimum among all designs wrt all generalized type 1 criteria $\sum_1^v f(\mu_{di})$. Similarly, if one restricts x_{dij} to be ± 1, a d^* with $X_{d^*}'X_{d^*} = (n+1)I_v - J_v$ is optimum wrt all generalized type 2 criteria. Existence of designs of these last two types entails $n \equiv 1 \pmod 4$ and $n \equiv 3 \pmod 4$, respectively the former are sparse, requiring $2n-1$ to be a square if $v = n$ (see [37]; the latter exist for $n = v$ whenever there is an $(n+1) \times (n+1)$ Hadamard matrix.

Ehlich also showed that, when $n = v$ is even and all x_{dij} are restricted to be ± 1, d^* is D-optimum if $X_{d^*}'X_{d^*}$ consists of two $(\frac{1}{2}n) \times (\frac{1}{2}n)$ blocks $(n-2)I_{n/2} + 2J_{n/2}$ and all-zero off-diagonal blocks. This optimality result led to the construction problem for such designs, considered, e.g., by Yang [46]. This result also gives a good illustration, cited by Cheng, of the effect of assuming complete symmetry: If

$$X_d = \begin{bmatrix} 1 & 1 & -1 & 1 & 1 & 1 \\ 1 & 1 & 1 & 1 & 1 & -1 \\ 1 & 1 & 1 & -1 & 1 & 1 \\ 1 & 1 & 1 & 1 & -1 & 1 \\ 1 & -1 & 1 & 1 & 1 & 1 \\ 1 & -1 & -1 & -1 & -1 & -1 \end{bmatrix}, \quad X_{d'} = \begin{bmatrix} 1 & 1 & -1 & 1 & 1 & 1 \\ 1 & 1 & 1 & -1 & -1 & 1 \\ 1 & 1 & 1 & -1 & -1 & -1 \\ 1 & 1 & 1 & -1 & 1 & -1 \\ 1 & -1 & 1 & 1 & 1 & 1 \\ 1 & -1 & -1 & -1 & -1 & -1 \end{bmatrix}, \quad (3.1)$$

then $X_d'X_d = 4I_6 + 2J_6$, which is D-optimum among designs with completely symmetric information matrix; but $X_{d'}'X_{d'}$ is of the Ehlich block form, and has a 56% larger determinant.

Recently Galil and Kiefer [20, 21], using a combination of theoretical results based on work of Ehlich [13, 14, 15] and of Ehlich and Zeller [16], and machine computations using an improvement of the method of Mitchell [32], greatly extended the list of known D-optimum weighing designs. This work shows $X_d'X_d = (n+1)I_v - J_v$ is D-optimum if $2v-5 \leq n \equiv 3 \pmod 4$, and characterizes optimum designs for all $n \geq v$ if $v \leq 12$; the previous best complete list was for $v \leq 5$, due to Payne [34].

This is clearly an area that invites much more combinatorial and optimality work. Another such area, closely related, is that of fractional factorial designs of the 2^s complete s-factor 2-level factorial design. Here the x_{dij} are restricted to be 0 or 1, and for the model in which interactions of more than t out of the s factors are assumed zero, there are $m = \sum_{i=0}^{t} \binom{s}{i}$ parameters. A "fractional factorial design d of resolution $2t+1$" allows all these parameters to be estimated' i.e., $X_d'X_d$ is nonsingular. The analogue of an Hadamard matrix is now an orthogonal array (OA) of size n, s constraints, 2 levels, strength $2t$; when it exists, it is universally optimum. From a statistical viewpoint, it is not even clear that the invariant criteria of universal optimality reflect the experimenter's aim; perhaps higher order interaction parameters are not as important as main effect parameters. In any event, a natural design symmetry sometimes assumed in this setting is that $X_d'X_d$ be invariant under interchange of two rows and columns representing interactions of the same order. Such designs are the balanced arrays (BA's) of Chakravarti [4], and in a number of papers Srivastava and his students (e.g., [42]) have constructed the A-optimum design among BA's of resolution V in the model with $t=2$, for a wide range of values of s and n. These designs are of considerable practical value, since prior to their construction experimenters were in the position of almost always having to restrict n to be a power of 2. Although, as in the settings treated earlier, one cannot suppose the A-optimum BA's are necessarily optimum among *all* designs, the work of Cheng [7] shows that many of them are. A BA of size n with q constraints, 2 levels, and strength L and index set $(\alpha_0, \alpha_1, \ldots, \alpha_L)$ can be considered as an $n \times q$ matrix X_d of entries 0 and 1 such that each $n \times L$ submatrix contains α_i rows equal to each L-vector with i 1's and $L-i$ 0's. (If all α_i are equal, this is an OA of strength L.) A typical result of Cheng is that a BA of strength $2t$ and index set of the form $(\alpha, \alpha, \ldots, \alpha, \alpha+1)$ is optimum wrt all generalized type 1 criteria, among all designs of resolution $2t+1$. This and the other results on optimality of BA's (e.g., for designs with $\alpha_{2t} = \alpha - 1$, $\alpha \pm 2$, or $\alpha + 3$ instead of $\alpha + 1$ as above) again motivate construction of combinatorial arrays with particular parameter values.

4. Latin squares, Youden designs, and generalizations

In the setting of "two-way heterogeneity", for given integers $v, b_1, b_2 \geq 2$, a design is a $b_1 \times b_2$ array of entries from the set $\{1, 2, \ldots, v\}$. Here $n = b_1 b_2$ and

$m = v + b_1 + b_2$. The row of X_d corresponding to cell (i, j), if that cell has entry h, consists of all zeros with the exception of a 1 in the hth, $(v+i)$th, and $(v+b_1+j)$th cell. This means variety h is planted or treated in that cell, and the expected value of the resulting observation is $\alpha_h + \beta_i + \gamma_j$, the three terms being respectively variety, row, and column effects. The vector $\theta^{(1)}$ consists of the α_h, and as in the case of one-way heterogeneity C_d can have rank at most $v-1$. The (i, j)th element of C_d is in fact

$$C_{dij} = \delta_{ij} r_{di} - \frac{\lambda_{dij}^{(R)}}{b_2} - \frac{\lambda_{dij}^{(C)}}{b_1} + \frac{r_{di} r_{dj}}{b_1 b_2} \tag{4.1}$$

where the $\lambda_{dij}^{(R)}$ are the quantities λ of Section 2 (first paragraph) computed by considering the rows of the design as blocks, and similarly for $\lambda^{(C)}$ from columns.

If $b_1 - b_2 = v$, we have the setting in which Wald [45] proved any LS to be D-optimum, and in which the tool described in Section 1 can be used to show a LS is universally optimum. It was noticed in the proof of the latter that it applied equally well (with only slightly more work to verify maximization of tr C_d in using (1.2)) to proving the optimality of a Youden square when $v \mid b_1$ and there exists a BIBD with $k = b_2$ and $b = b_1$ (and, hence, a Youden square). This suggested consideration of general $b_1 \times b_2$ arrays, and the definition of a generalized Youden design (GYD) as a $b_1 \times b_2$ array that is a BBD (not necessarily incomplete) when the rows are considered as blocks and also when the columns are considered as blocks. Thus, as described in Section 1, optimality considerations led to new combinatorial constructions. GYD's are constructed by patching together BIBD's in Kiefer [27], and by more elegant, largely geometric methods in work by Seiden and her students [38, 39]. Ash [1] constructed GYD's for essentially all practical parameter values.

When $v \mid b_1$, the setting is termed *regular* and if a GYD exists it can again be proved universally optimum by using (1.2). The situation in nonregular cases is much more difficult. In the simplest case, $v = 4$, $b_1 = b_2 = 6$ (where at least two nonisomorphic GYD's exist), it was noticed in Kiefer [23] that a GYD no longer maximizes tr C_d, so (1.2) cannot be used; later, another design was discovered that was better than any GYD, in the sense of D-optimality, in all nonregular cases $v = 4$, $b_1 = b_2$ (where necessarily $b_1 \equiv 6 \pmod{12}$).

This was perhaps surprising, an example of a most symmetric, intuitively appealing design (in the Fisherian tradition described in Section 1) which was not D-optimum. The simple universal optimality tool would not apply, and a new technique must be used to prove whatever optimum properties a GYD has. Such a device was described in Kiefer [28], and as we might expect from the last paragraph of Section 1, it is usually fairly simple to use to prove E-optimality and fairly difficult to use to prove D-optimality. The device was used in that reference to prove any GYD is always A-optimum (hence, E-optimum), and the proof announced there of the more difficult fact that a GYD is D-optimum unless $v = 4$ and the setting is nonregular, appears at the end of this section.

A natural generalization is to consider the q-way setting in which a design is a $b_1 \times b_2 \times \cdots \times b_q$ array and there are t (not necessarily distinct) variety symbols in each cell. Kishen [31] in fact constructed latin hypercubes in the case $b_1 = b_2 = \cdots = b_q$ with $t=1$ and $v = b_1^{q'}$ for some integer $q' < q$. For general q, b_i, and t, Cheng [8] extended the methods of the three previously cited papers on construction of GYD's, to construct certain Youden hyperrectangles (YHR); a YHR is an array which, for each i, is a BBD when the union of all cells with the same ith coordinate is considered to be a block. If all b_i are equal it is called a Youden hypercube (YHC). There are notions of regularity as in the GYD case, the only one of which we use in succeeding paragraphs is that $v \mid (t\prod_{j \neq i} b_j)$ for $2 \leq i \leq q$ (which specializes to the earlier definition of regularity in the GYD setting). If the above divisibility also holds for $i=1$, Cheng shows a YHR always exists; as in the GYD case, construction is easiest in regular settings. It is not yet known exactly which parameter values $v, \{b_i\}, t$ permit the construction of a YHR, even in the original GYD setting $q=2$, $t=1$.

It is then natural to ask, in the q-way setting with the parameters such that a YHR d^* exists, in what sense d^* is optimum. Cheng [9] extended the universal optimality result for regular GYD's to regular YHR's. Moreover, using the device of Kiefer [28] mentioned above, he proved that YHR's (regular or not) are always E-optimum. Among other cases considered by Cheng, perhaps most interesting is the case of Youden (not necessarily latin) hypercubes d^* (with general t), in which he used the D-optimality device of the end of this section to show that d^* is always D-optimum, if $q>2$. Thus, the nonregular cases $q=2$, $v=4$ mentioned earlier, in which a GYD is not D-optimum, are quite special. A further comment on this is contained in the last paragraph of this section.

Another direction of generalization suggested by the above results and those of Section 2 is consideration, when v, $\{b_i\}$, t are such that no YHR exists, of arrays that are only *partially balanced* designs in each of the r directions. There has been little investigation of construction of such designs, particularly in nonregular cases. If the setting is regular and d^* is a BBD when considered as a block design in each of the last $q-1$ directions, but is a MBD of type i when considered in the first direction, Cheng used the results mentioned in Section 2 to prove optimality of d^* for all type i criteria.

We now give the proof that any GYD d^* is D-optimum when $v \neq 4$. Since the result is known in regular cases and any GYD is regular if v is prime, it suffices to consider $v \geq 6$. The device of Kiefer [28], replacing maximization of $\operatorname{tr} C_d$ as in (1.2), is now to show that d^* maximes $\sum_{j=1}^{v} \log c_{djj}$, from which D-optimality follows since C_d is completely symmetric. This is a more tedious computation than are those where (1.2) can be used, but again it avoids computing

$$\Phi^{(1)}(C_d) = -\sum_{1}^{v-1} \log \mu_{dj}.$$

Direct maximization of $\sum_1^v \log c_{djj}$ is also complicated because of combinatorial restrictions on the c_{dij}, but it is easily seen that it suffices to show, with

$$c(r) \stackrel{\text{def}}{=} \max_{\{d:\, r_{dj}=r\}} c_{dij},$$

that the maximum of $\sum_1^v \log c(r_j)$ subject to $\{r_j$ nonnegative integers, $\sum_1^v r_j = b_1 b_2\}$ is

$$\sum_1^v \log c_{d^*jj} = vc(\bar{r}),$$

where $\bar{r} = b_1 b_2 / v$. Noting (4.1), we have

$$g(r) \stackrel{\text{def}}{=} b_1 b_2 c(r) = b_1 b_2 r - b_1 h(r, b_1) - b_2 h(r, b_2) + r^2 \tag{4.2}$$

where

$$h(r, k) = -k[\text{int}(r/k)]^2 + (2r - k)[\text{int}(r/k)] + r. \tag{4.3}$$

We also write $\Delta(r) = g(r+1) - g(r)$. It then follows from the above development and simple properties of g (as outlined in Kiefer [28]) that it suffices to show that $\log g(r+1) - \log g(r)$ is nonincreasing ($\log g$ concave) in the "basic interval" of integers $C_0 \leqslant r < D_0$, where $[C_0, D_0]$ is the largest closed interval containing $\bar{r}$ but whose interior does not contain any integral multiples of b_1 or b_2. (The reduction to this interval is a consequence, in part, of the fairly simple fact that

$$\log[g(r+1)/g(r)] \geqslant \log[g(C_0+1)/g(C_0)] \quad \text{for} \quad r < C_0.)$$

Thus, we must show

$$g^2(r) - g(r-1)g(r+1) \geqslant 0 \quad \text{for} \quad C_0 < r < D_0. \tag{4.4}$$

Substituting $g(r-1) = g(r) - \Delta(r-1)$ and $g(r+1) = g(r) + \Delta(r)$, (4.4) thus becomes

$$C_0 < r < D_0 \Rightarrow 0 \leqslant \Delta(r)\Delta(r-1) + g(r)[\Delta(r-1) - \Delta(r)] = \Gamma_0(r) \quad \text{(say)}. \tag{4.5}$$

(Incidentally, the greater ease of proving A-optimality appears here explicitly, in that the expression $\Gamma(r)$ of (4.1) of Kiefer [28], which for A-optimality corresponds to (4.5) here, has the negative term in square brackets in (4.5) multiplied by $\frac{1}{2}$.) Define B_1, B_2, α, β by

$$\begin{aligned} C_0 &= B_1 b_1 \geqslant B_2 b_2 = b_2 \,\text{int}(C_0/b_2), \\ \alpha &= b_1^2 B_1(B_1+1) + b_2^2 B_2(B_2+1), \\ \beta &= b_1 b_2 - (2B_1+1)b_1 - (2B_2+1)b_2. \end{aligned} \tag{4.6}$$

From (4.5), we have on $[C_o+1, D_0-1]$,

$$\Gamma_0(r) = 2r^2 + 2\beta r + \beta^2 - 2\alpha - 1. \tag{4.7}$$

Considered as a quadratic function in *real* r, this has nonnegative derivative for

$r \geq C_0+1$ since $\beta+2r \geq 0$ for $r \geq C_0+1$. It is easy to see (as in the derivation of (4.3) of the earlier paper) that (4.5) will follow from

$$\Gamma_0(C_0+1) \geq 0. \tag{4.8}$$

We now proceed in a development similar to that used for $v \geq 6$ ("Case 1") in proving A-optimality. Simple inequalities show that (4.7) is decreased if B_2 is replaced by $B_1 b_1/b_2$. This yields

$$\Gamma_0(C_0+1) \geq 6C_0^2+[-4+4\sigma-6\pi]C_0+\{1+2[\pi-\sigma]+[\pi-\sigma]^2\} \tag{4.9}$$

where $\pi = b_1 b_2$ and $\sigma = b_1+b_2$ with both $b_i \geq 8$ (by nonregularity when $v \geq 6$).

The expression on the right-side of (4.9) has negative derivative with respect to C_0 for $C_0 < \frac{1}{6}\pi$, since $\sigma-\pi-1<0$. However, substitution of $\frac{1}{6}\pi$ for C_0 in that expression no longer yields a nonnegative expression (as it did in the case of the A-optimality proof, in (4.4) of the previous reference), again reflecting the greater difficulty of proving D-optimality. Instead, we must substitute the sharper upper bound for C_0,

$$C_0 \leq \tfrac{1}{6}(\pi-2b_1), \tag{4.10}$$

which we now justify.

Since $B_1 = C_0/b_1 < \bar{r}/b_1 = b_2/v$, (4.10) will follow from

$$\operatorname{int}(b_2/v) \leq \tfrac{1}{6}(b_2-2). \tag{4.11}$$

If $b_2 \equiv 0$ or 1 (mod 6), it follows from the nonregularity condition $v \nmid b_2$ that $v \geq 8$, and then (4.11) is satisfied for $b_2 \geq 8$. For other congruences (mod 6) of b_2, (4.11) is obvious. Thus, (4.10) is proved, and consequently (4.9) will follow from the nonnegativity of the expression obtained by substituting $C_0 = \frac{1}{6}(\pi-2b_1)$ into the right-side of (4.9). This expression, after mutiplication by 6 and rearrangement of terms, is

$$[b_1^2-8b_1+6]b_2^2+12(b_1-1)b_2+\{2b_1^2-4b_1+6\}, \tag{4.12}$$

which is clearly positive since both $b_i \geq 8$. This completes the proof.

One can see in (4.11) why this proof does not work when $v=4$. On the other hand, in Cheng's proof of the YHC D-optimality for *all* v when $q>2$, the bound analogous to (4.11) is

$$\operatorname{int}(b_1^{q-1}t/v) \leq (b_1^{q-1}t-q)/(2q+2),$$

which follows from $v \geq 2^q \geq 2q+2$ for $q \geq 3$. This is why $q=2$ yields the only cases where a GYD is not D-optimum.

5. Minimal support designs

Some design settings are of the special form of allowing repetition of rows of X_d as much as is wanted. In such problems, there is a space $\mathscr{X}$ of allowable row

m-vectors, and for a given n a design d is a choice $(\boldsymbol{x}_{d1}, \boldsymbol{x}_{d2}, \ldots, \boldsymbol{x}_{dn})$ of n not necessarily distinct elements of $\mathscr{X}$ or, equivalently, of a discrete probability *design measure* ξ on $\mathscr{X}$ defined by

$$\xi_d(\boldsymbol{x}) = n^{-1} \text{ (number of } \boldsymbol{x}_{di}\text{'s equal to } \boldsymbol{x}). \tag{5.1}$$

The settings of Sections 2 and 4 are not of this nature, but in the weighing design setting of Section 3 (unless we prohibit repetition of a weighing or factor combination, which there is no *physical* reason to do) we have the above setup with $\mathscr{X} = \{-1, 0, 1\}^m$ or $\{-1, 1\}^m$. In the fractional factorial setup, $\mathscr{X}$ consists of the range of a certain m-vector of polynomials on $\{0, 1\}^s$ of degree $\leq t$. In such a setting, the information matrix is

$$X_d'X_d = n \sum_{\boldsymbol{x} \in \mathscr{X}} \boldsymbol{x}'\boldsymbol{x}\xi_d(\boldsymbol{x}) \overset{\text{def}}{=} nM(\xi_d), \tag{5.2}$$

and $M(\xi_d)$ is called the *information matrix* (per observation and unit of variance) of ξ_d. For a given n, ξ_d must be a member of $\Xi_n = \{\xi: \xi$ is a probability measure on $\mathscr{X}$ taking only integral multiples of n^{-1} as values$\}$. If the optimality criterion Φ can be rewritten

$$\Phi(nM(\xi)) = a_n + b_n\Phi^*(M(\xi))$$

with $b_n > 0$ (as is the case for the Φ of (1.1)), we may instead consider the problem of choosing ξ in Ξ_n to minimize $\Phi^*(M(\xi))$. This, in general, is a tedious discrete programming problem, whose solution ξ^* (for a fixed $\mathscr{X}$) typically exhibits a fine structure in n, even when n is large, in which the support of ξ^* (as well as, of course, its value at each $\boldsymbol{x}$) varies. Because of this, it is useful to consider the *approximate theory* in which we choose ξ to minimize $\Phi^*(M(\xi))$ over $\Xi = \{$all discrete probability measures on $\mathscr{X}\}$. We call any solution a *Φ^*-optimum approximate design measure.* There are then useful theoretical characterizations and computer algorithms for determining an optimum ξ^*, which can be recorded for a given $\mathscr{X}$ without any dependence on n (see, e.g., [23, 26] for discussion and references); and the resulting ξ^* can often be implemented, at least for n sufficiently large compared with m, by taking an approximation to ξ^* in Ξ_n, so as to yield a design which is reasonably close to optimality for the n-observation problem.

If φ^* is strictly convex, as is the case for the D- and A-optimality criteria when $v = n$ (and often, also, when Φ^* is the E-optimality criterion), $M(\xi)$ must be the same for all Φ^*-optimum ξ in Ξ. It is obviously useful to find such a ξ that has minimal support size, since such a ξ can typically be more satisfactorily approximated in Ξ_n than one with large support, and also since (from a practical viewpoint) it may be economically advantageous to construct an experiment with as few distinct $\boldsymbol{x}_{di}$ as possible. If $\bar{\xi}$ is *some* Φ^*-optimum approximate design measure, this means finding $\tilde{\xi}$ in Ξ such that $M(\tilde{\xi}) = M(\bar{\xi})$ and $\tilde{\xi}$ has minimal support size. Assuming here that $\mathscr{X}$ is finite (as in our examples), say with N

points, list the points $\boldsymbol{x}^{(1)}, \ldots, \boldsymbol{x}^{(N)}$ (say) of $\mathscr{X}$ in any order and write ξ as an N-vector $(\xi_1, \xi_2, \ldots, \xi_N)'$ with $\xi_i = \xi(\boldsymbol{x}^{(i)})$. Write $\boldsymbol{x} = (x(1), \ldots, x(m))$ in coordinate form on $\mathscr{X}$, and list the $\frac{1}{2}m(m+1)$ ($=p$, say) functions $x(i)x(j)$ on $\mathscr{X}$, $1 \leqslant i \leqslant j \leqslant m$, in any order; call them $y_1, y_2, \ldots, y_p$. Let G be the $p \times N$ matrix of elements $y_i(\boldsymbol{x}^{(j)})$ $(1 \leqslant i \leqslant p,\ 1 \leqslant j \leqslant N)$, and let $\bar{g} = G\bar{\xi}$. The problem is then to find an N-probability vector $\tilde{\xi}$ satisfying

$$\begin{aligned} &G\tilde{\xi} = \bar{g}, \\ &\text{card}(\text{supp}(\tilde{\xi})) = \text{minimum}. \end{aligned} \tag{5.3}$$

Sometimes the problem can be rewritten with much-reduced p and N (see [17]).

For an example, consider the weighing problem of Section 3. Even if $\mathscr{X} = \{-1, 0, 1\}^m$, it is not hard to see that a universally optimum approximate design $\bar{\xi}$ assigns measure 2^{-m} to each point of $\{-1, 1\}^m$, and *every* design with the same information matrix $M(\bar{\xi}) = I_m$ has support a subset of $\{-1, 1\}^m$. Thus, $N = 2^m$, and G is a matrix of ± 1's, while $\bar{g}$ contains m 1's and $p - m$ 0's. If there exists an $m \times m$ Hadamard matrix, it corresponds to a $\tilde{\xi}$ satisfying (5.3) and with $\text{card}(\text{supp}(\tilde{\xi})) = m$, clearly a minimum. Otherwise, the solution will be supported by more than m points of $\mathscr{X}$. The existence of an $m \times m$ Hadamard matrix in this example is in fact equivalent to there being a $\tilde{\xi}$ on m points, satisfying (5.3); any such $\tilde{\xi}$ must assign measure m^{-1} to each point of support. More generally, one could ask, for each n, whether there are solutions to $G\xi = \bar{g}$ with ξ in Ξ_n. For example, when $m = 3$, any $\tilde{\xi}$ satisfying (5.3) has 4 support points, e.g., the points $(1, \pm 1, \pm 1)$ (with independent signs) with uniform measure on each point; if $n = v = m = 3$, this approximate solution of course differs from any exact Ξ_3 solution.

Another example, that has received considerable theoretical and practical effort, is that of the 3^s factorial setup with $t = 2$. In this case $m = \frac{1}{2}(s+1)(s+2)$ and $\mathscr{X}$ is the range of the m-vector F (say) of monomials of degree $\leqslant 2$ on $\{-1, 0, 1\}^s$ (The problem that originates as "quadratic regression on the s-cube" can be reduced to this form for the Φ^*'s usually considered (see [26]).) There are no longer universally optimum designs; Φ^* matters. If $J_i^{(s)}$ consists of the points of $\{-1, 0, 1\}^s$ with i coordinates 0 and $s - i$ coordinates ± 1, it can be shown that for $s \geqslant 2$ there is a D-optimum approximate design supported on the union of three sets $F(J_i^{(s)})$, which in experimental terms means taking observations corresponding to a fraction of the 3^s factorial points, namely, at points in these three $J_i^{(s)}$'s. The set of three values of i for such solutions, obtained in [25] and [17], is

$$\begin{aligned} &0, 1 \text{ any } i \text{ with } 2 \leqslant i \leqslant s, \text{ if } 2 \leqslant s \leqslant 5, \\ &0, 1 \text{ or } 2, \text{ any } i \text{ with } 3 \leqslant i \leqslant s, \text{ if } 6 \leqslant s. \end{aligned} \tag{5.4}$$

The corresponding design assigns equal measure to each point of $J_i^{(s)}$ for a fixed i, but that measure varies with i. It was also shown in the last reference, by rank considerations of G in (5.3), that the resulting design $\bar{\xi}$ on 9 points for $s = 2$ and on 21 points for $s=3$ (taking $i=3$ in (5.4)), satisfies (5.3). (The case $s=1$ is trivial.) Recently Pesotchinsky [35] showed how to reduce the corresponding designs for $s=4, 5, 6$, from support on 49, 113, or 257 points, to support on 45, 61, or 89 points respectively. These reductions, which use a clever combination of OA's of strength 2, improved considerably on the reductions obtained in the earlier reference by an obvious use of OA's of strength 4. Moreover, Pesotchinshy was able to show that his designs do have minimal support, and for moderate n they are implementable by nearby designs in Ξ_n that appear to be fairly efficient compared with other actual n-observation designs. There remains the problem of extending this work to larger s; in particular, of constructing the analogues of Pesotchinsky's combinatorial arrangements.

For the E-optimality criterion, it was shown by Galil and Kiefer [19] that this problem on the 3^s-array has an arithmetically simpler solution that, for certain s, allows a solution to be obtained on only *two* $J_i^{(s)}$'s, say with $i=i_1$ and i_2. It can then be computed, from knowledge of one optimum $\bar{\xi}$ and the resulting $\bar{g}$, that i_1, i_2 must satisfy

$$5i_1i_2-3s(i_1+i_2)+s(2s+1)=0, \tag{5.5}$$

which on writing $i_1=\frac{1}{5}(3s-u)$, $i_2=\frac{1}{5}(3s+w)$ becomes $uw=s(s+5)$, from which the possible values of s and of i_1, i_2 can easily be obtained. It was also shown in Kiefer [29] that all solutions (i_1, i_2, s) lie in "infinite series" $(i, ai+b, ci+d)$ of such solutions with i in arithmetic progression, and that, with a few distinguishable exceptions, each (i_1, i_2, s) lies in such series for exactly two different pairs (a, b) and two or four (c, d)). It can occur for some s that no design with support contained in only two $J_i^{(s)}$ has minimal support, but the full picture, including an analogue of Pesotchinsky's work for D-optimality, is incomplete. In any event, this example illustrates how optimality considerations can lead to a diophantine problem.

We mention that equations of the form (5.3) have also arisen in the work of Foody and Hedayat [18]. There one is interested, for given b, k, v, in finding a BIBD with a minimum number of *distinct* blocks. $\mathscr{X}$ is replaced by the set of possible blocks, with $N=\binom{v}{k}$. The $p=\binom{v}{2}$ subsets of size 2 of $\{1, 2, \ldots, v\}$ correspond to rows of G, and the (i, j)th element of G is 1 if pair i occurs in block j and 0 otherwise. Then $\bar{g}$ is the vector $\bar{c}$ with all components equal to λ/b. Here of course one is only interested in solutions to (5.3) for which, also, $b\bar{\xi}$ has integral components. In the case $b=56$, $k=3$, $v=8$ considered in detail by Foody and Hedayat, a minimal support design has recently been determined by Pesotchinsky [36]. Aside from the combinatorial interest in nonisomorphic BIBD's produced by solutions to $G\xi=\bar{c}$ with more than minimal support, there is an "optimality" consideration: among the (universally optimum) BIBD's one may want to select the design that minimizes some secondary loss functional, which from cost considerations could well be the number of distinct blocks. The primary loss consideration here of course amounted to showing the existence of some BIBD (equivalent to showing $G\xi = \bar{c}$ has a solution in Ξ_b), or, in the absence of such a solution, to undertaking a development like that of Section 2.

If b, k, v in the block design setting are such that no BIBD, or MBD, or any similar structure exists, finding a design that minimizes $\Phi^{(1)}(C_d)$ is a very difficult computational problem. The best "search" scheme currently available is that of Mitchell [32], but Mitchell showed that, even in simple settings such as the weighing problem on $\{-1, 1\}^m$ with $m = 8$ and $n = 16$, where there is a Hadamard solution, his scheme (with a random initial design) rarely found it. There are local minima, "holes" that it is hard to climb out of. Galil and the present author have made some slight progress in trying to improve Mitchell's method, but better ideas are needed. Similarly, even when we know *some* optimum design $\bar{\xi}$, there are no satisfactory computer routines for solving (5.3). In fact, as John Hopcroft showed the author, the family of such problems is NP-complete. Both for the design optimization problem and for the minimal support problem, at least in a large class of settings, one would hope to find polynomial-time algorithms that yield solutions close to the optimum, of the type that exist in other optimization problems.

References

[1] A.S. Ash, Construction of generalized Youden designs, Thesis, Univ. of Illinois at Chicago Circle (1977).

[2] A.S. Ash, Optimality of GYD's, unpublished paper (1978).

[3] R.C. Bose and K.R. Nair, Partially balanced incomplete block designs, Sankhya 4 (1939) 337–372.

[4] I.M. Chakravarti, On some methods of construction of partially balanced arrays, Ann. Math. Statist. 32 (1961) 1181–1185.

[5] C.-S. Cheng, Optimality of certain asymmetrical experimental designs, Ann. Statist. 6 (1978) 1239–1261.

[6] C.-S. Cheng, A note on (*M. S*)-optimality, Commun. Statist., to appear.

[7] C.-S. Cheng, Optimality of some weighing and 2^n fractional factorial designs, Ann. Statist. 8 (1980) 436–446.

[8] C.-S. Cheng, Construction of Youden hypercubes, to appear.

[9] C.-S. Cheng, Optimal designs for the elimination of multi-way heterogeneity, Ann. Statist. 6 (1978) 1262–1272.

[10] C.-S. Cheng and L.J. Gray, A characterization of group-devisible designs and some related results, Ann. Discrete Math. 6 (1980).

[11] D. Conniffe and J. Stone, Some incomplete block designs of maximum efficiency, Biometrika 62 (1975) 685–686.

[12] J.A. Eccleston and A. Hedayat, On the theory of connected design characterization and optimality, Ann. Statist. 2 (1974) 1238–1255.

[13] H. Ehlich, Determinantenabschätzungen für binäre Matrizen, Math. Z. 83 (1964) 123–132.

[14] H. Ehlich, Determinantenabschätzungen für binäre Matrizen mit $m \equiv 3 \bmod 4$, Math. Z. 84 (1964) 438–447.

[15] H. Ehlich, Personal communication (1978).

[16] H. Ehlich and K. Zeller, Binäre Matrizen, Z. Angew. Math. Mech. 42 (1962) T20–T21.

[17] R.H. Farrell, J. Kiefer and A. Walbran, Optimum multivariate designs, Proc. 5th Berkeley Symp. 1 (1965) 113–138.

[18] W. Foody and A. Hedayat, On theory and applications of BIB designs with repeated blocks, Ann. Statist. 5 (1977) 932–945.

[19] Z. Galil and J. Kiefer, Comparison of designs for quadratic regression on cubes, J. Statist. Planning and Inference 1 (1977) 121–132.
[20] Z. Galil and J. Kiefer, D-optimum weighing designs, in Ann. Statist. (1980).
[21] Z. Galil and J. Kiefer, Time- and space-saving computer methods, related to Mitchell's Detmax, for finding D-optimum designs, to appear in Technometrics (1980).
[22] J.A. John and T.J. Mitchell, Optimal incomplete block designs, J. R. Statist. Soc., Ser. B 39 (1977) 39–43.
[23] J. Kiefer, On the nonrandomized optimality and randomized non-optimality of symmetrical designs, Ann. Math. Statist. 29 (1958) 675–699.
[24] J. Kiefer, Optimum experimental designs, J. R. Statist. Soc., Ser. B 21 (1959) 272–319.
[25] J. Kiefer, Optimum designs in regression problems—II, Ann. Statist. 32 (1961) 298–325.
[26] J. Kiefer, General equivalence theory for optimum designs (approximate theory), Ann. Statist. 2 (1974) 849–879.
[27] J. Kiefer, Balanced block designs and generalized Youden designs, I. Construction (patchwork), Ann. Statist. 3 (1975) 109–118.
[28] J. Kiefer, Construction and optimality of generalized Youden designs, in: J.N. Srivastava (ed.), A Survey of Statistical Designs and Linear Models (North-Holland, Amsterdam, 1975) 333–353.
[29] J. Kiefer, A diophantine problem in optimum design theory, Utilitas Math., to appear.
[30] J. Kiefer and J. Wolfowitz, Optinum designs in regression problems, Ann. Math. Statist. 30 (1959) 271–294.
[31] K. Kishen, On the construction of Latin and Hypergraeco-Latin cubes and hypercubes, J. Indian Soc. Agric. Statist. 2 (1949) 20–48.
[32] T.J. Mitchell, An algorithm for the construction of D-optimal experimental designs, Technometrics 16 (1974) 203–210.
[33] A.M. Mood, On Hotelling's weighing problem, Ann. Math. Statist. 17 (1946) 432–446.
[34] S.E. Payne, On maximizing $\det(A^{\mathrm{T}}A)$, Discrete Math. 10 (1974) 145–158.
[35] L.L. Pesotchinsky, D-optimum and quasi-D-optimum second-order designs on a cube, Biometrika 62 (1975) 335–340.
[36] L.L. Pesotchinsky, Remarks on the lower bound of number of different blocks in BIB designs, to appear.
[37] D. Raghavarao, Constructions and Combinatorial Problems in Design of Experiments (John Wiley and Sons, New York, 1971).
[38] F. Ruiz, and E. Seiden, On construction of some families of generalized Youden designs, Ann. Statist. 2 (1974) 503–519.
[39] E. Seiden and C.-J. Wu, A geometric construction of generalized Youden designs for v a prime power, Ann. Statist. 6 (1978) 451–460.
[40] K.R. Shah, Optimality criteria for incomplete block designs, Ann. Math. Statist. 31 (1960) 791–794.
[41] K.R. Shah, D. Raghavarao and C.G. Khatri, Optimality of two and three factor designs, Ann. Statist. 4 (1976) 419–422.
[42] J.N. Srivastava and D.V. Chopra, Balanced optimal 2^m fractional factorial designs of resolution V, $m<6$, Technometrics 13 (1971) 257.
[43] K. Takeuchi, On the optimality of certain type of PBIB designs, Rep. Statist. Appl. Res., JUSE, 8 (1961) 140–145.
[44] K. Takeuchi, A remark added to "On the optimality of certain type of PBIB designs", Rep. Statist. Appl. Res., JUSE, 10 (1963) 47.
[45] A. Wald, On the efficient design of statistical investigations, Ann. Math. Statist. 14 (1943) 134–140.
[46] C.H. Yang, On designs of maximal $(+1,-1)$ matrices of order $n \equiv 2 \pmod 4$, Math. Comp. 22 (1968) 174–180.

P. R. Krishnaiah, ed., *Multivariate Analysis-V*
© North-Holland Publishing Company (1980) 79–93.

DESIGNS FOR EXTRAPOLATION WHEN BIAS IS PRESENT

J. KIEFER*
Cornell University and University of California (Berkeley)

1. Introduction

We specialize and adapt the notation of Kiefer (1973) to the present setting. Let f be a k vector of real valued functions on $\mathcal{X}^* = \mathcal{X} \cup \mathcal{Z}$, with $f = \binom{g}{h}$, this and the subsequent decompositions being into s and k-s components. The expectation of an observation corresponding to value x (in $\mathcal{X}^*$) of a controllable variable is $\theta' g(x) + \beta' h(x)$. We are permitted to take an N-vector $Y = (Y_1, \ldots, Y_N)'$ of uncorrelated observations Y_i with common variance σ^2, the ith at x_i in $\mathcal{X}$. We must estimate $\theta' g + \beta' h$ on $\mathcal{Z}$ by a linear function $t'g$ of the components of g, where the s-vector t is a linear homogeneous estimator, $t = CY$. The problem is to choose the design $X_N = (x_1, x_2, \ldots, x_N)$ and estimation matrix C to achieve some goal of accuracy, described further, below.

When $s = k$, this is the well known problem of extrapolation of the regression (assuming the model $\theta' g$ is correct) to $\mathcal{Z}$ based on observations on $\mathcal{X}$, often called interpolation if $\mathcal{Z} \subset \mathcal{X}$; see Kiefer and Wolfowitz (1964). The estimator t is usually taken to be the best linear unbiased, or least squares (LS), estimator. In the case of polynomial regression in one variable with $\mathcal{X} = [-1, 1]$ and $\mathcal{Z}$ a point (or possibly a half-infinite interval), an elegant solution was given by Hoel and Levine (1964) and generalized by Kiefer and Wolfowitz (1965). Similar problems for higher dimensional $\mathcal{X}$ with $\mathcal{Z}$ a point were considered by Studden (1971). Galil and Kiefer (1978) considered extrapolation when $\mathcal{X}$ is a q-dimensional ball and $\mathcal{Z}$ is a larger or smaller ball, in the case of cubic regression; for quadratic regression this is also discussed in Kiefer (1978). We write $B_q(R)$ for the q-ball of radius R centered at O and $S_q(R)$ for the surface of $B_q(R)$, i.e., the $(q-1)$-sphere of radius R.

In the case $\mathcal{Z} = \mathcal{X}$, the path-breaking paper of Box and Draper (1959) initiated consideration of the possibility that the "assumed model" $\theta' g$

*Research sponsored by the National Science Foundation.

might be incorrect, so that the estimated regression $t'g$ might contain bias as well as variance, due to the presence of the contaminating term $\beta' h$ in the actual, true model. Write

$$F = F(X_N) = \left[f(x_1) \vdots f(x_2) \vdots \cdots \vdots f(x_N) \right] = \begin{pmatrix} G \\ H \end{pmatrix} (k \times N)$$

and $D = D(X_N) = FC'(k \times s)$. Also write

$$Q = N^{-1} FF' = \begin{pmatrix} M & L \\ L' & K \end{pmatrix},$$

$$\Gamma = \begin{pmatrix} \Gamma_{gg} & \Gamma_{gh} \\ \Gamma_{hg} & \Gamma_{hh} \end{pmatrix} = \int_{\mathfrak{X}} f(z) f(z)' \nu(dz) \tag{1.1}$$

for a specified measure ν that reflects the "importance" at various points of $\mathfrak{X}$ of the mean-square error (MSE) at z when X_N and CY are used. This MSE, integrated over $\mathfrak{X}$ wrt ν, is $J = \sigma^2(V + B)$, where, for any $s \times (k\text{-}s)$matrix p satisfying $\Gamma_{gg} p = \Gamma_{gh}$,

$$V = \operatorname{tr} CC' \Gamma_{gg},$$

$$\sigma^2 B = \int_{\mathfrak{X}} \{ (\theta', \beta') [Dg(z) - f(z)] \}^2 \nu(dz)$$

$$= (\theta', \beta') \left[D - \begin{pmatrix} I_s \\ p' \end{pmatrix} \right] \Gamma_{gg} \left[D - \begin{pmatrix} I_s \\ p' \end{pmatrix} \right]' \begin{pmatrix} \theta \\ \beta \end{pmatrix}$$

$$+ \beta' \left[\Gamma_{hh} - \Gamma_{hg} \Gamma_{gg}^- \Gamma_{gh} \right] \beta \overset{\text{def}}{=} \sigma^2 B_1 + \sigma^2 B_2. \tag{1.2}$$

Note that B_2 is unaffected by our choice of design or estimator. In the BD approach, one chooses t to be the LS estimator of θ under the assumed model $\beta = 0$; that is, $C = (GG')^- G$. Then $\sigma^2 B_1$ and V reduce to

$$\sigma^2 B_{1,\text{BD}} = \beta' (L'M^- - p') \Gamma_{gg} (L'M^- - p')' \beta,$$

$$V_{\text{BD}} = N^{-1} \operatorname{tr} \Gamma_{gg} M^-. \tag{1.3}$$

The BD prescription is then that, since one cannot choose X_N to minimize J (because $\sigma^{-1}\beta$ is unknown), one chooses X_N (if possible) to minimize B, by making $L'M^- = p' = \Gamma_{hg} \Gamma_{gg}^-$, often achieved by making $L = \Gamma_{gh}$ and $M = \Gamma_{gg}$ (the design moments of certain orders thus matching those of the loss matrix Γ). The rationale for this choice, in that it is supposed to yield a value of J close to the minimum one could achieve if one knew $\sigma^{-1}\beta$, has been discussed with examples in Kiefer (1973, 1975) and Galil and Kiefer (1977a); justification for the BD approach additional to that given by Box and Draper is contained in Draper and Herzberg (1973) (hereafter DH), as described below.

The departure from the BD approach of Karson, Manson, and Hader (1969) (hereafter KMH) begins by allowing arbitrary C. Then B is minimized for each design for which $(\theta'\Gamma_{gg}+\beta'\Gamma_{hg})\Gamma_{gg}^{-}$ is estimable by choosing $C=[I_s \vdots p](FF')^{-}F$ so that $D=\binom{I_s}{p'}$ and thus $B_1=0$. Then X_N is chosen to satisfy some other criterion, usually (as hereafter) to minimize the resulting

$$V_{\mathrm{KMH}}(X_N)=N^{-1}\operatorname{tr}\begin{bmatrix}\Gamma_{gg} & \Gamma_{gh}\\ \Gamma_{hg} & \Gamma_{hg}\Gamma_{gg}^{-}\Gamma_{gh}\end{bmatrix}\begin{bmatrix}M & L\\ L' & K\end{bmatrix}^{-1}. \tag{1.4}$$

Although $J_{\mathrm{KMH}} \leqslant J_{\mathrm{BD}}$ if $LM^{-}=p'$ is satisfied, examples in the present paper illustrate that, if no design can satisfy $LM^{-}=p'$, the BD design may sometimes be better. Possible shortcomings of the KMH approach are also illustrated in Kiefer (1973). Some remarks in DH, addressed to the BD vs. KMH approach controversy, criticize the KMH approach (p. 268) on the grounds that its estimators are biased for the model fitted and that it does not permit use of certain standard least squares procedures. This does not seem completely convincing, in view of the motivation (for BD, too) of consideration of such procedures because of concern about bias: if one modifies the design to this end, why not also change the estimator? We also note the description on p. 272 of DH to the effect that the "typical" situation contemplated by BD was that in which $V/B \approx 1$; in terms of the discussion of Galil and Kiefer (1977a), this means the variable h there (a component of $\sigma^{-1}\beta$) should be of order $N^{-1/2}$ rather than of order 1 in the comparisons, strongly reinforcing the conclusion that, at least in the example considered there, other designs such as the D-optimum design may have preferred behavior in terms of J and of the maximum of the MSE function. In any event, the remainder of the present paper is devoted primarily to consideration of the BD and KMH approaches, as an early exploratory attempt to investigate procedures for extrapolation with bias. Comparisons of the MSE function, for example in terms also of its maximum, should be carried out for various designs as they were in Galil and Kiefer (1977a) for $\mathfrak{Z}=\mathfrak{X}$.

Draper and Herzberg's paper was the first to treat extrapolation with bias (i.e., considerations of the type developed in the previous two paragraphs, but no longer with $\mathfrak{Z}=\mathfrak{X}$). The previous formulas have been written in such a form that they apply also when $\mathfrak{Z}\neq\mathfrak{X}$. In DH, first degree regression is considered when $\mathfrak{X}$ is a q-ball and $\mathfrak{Z}$ is the line segment from a point z outside the ball to the closest point of $\mathfrak{X}$. The designs considered are limited to a certain subclass of designs over which B is minimized, assuming knowledge of which of two regions β lies within. In a subsequent unpublished paper (1976, hereafter DH2), which the authors have kindly sent the present author, $\mathfrak{X}$ is a q-ball, which we take to be $B_q(1)$, $\mathfrak{Z}$ is the region outside $\mathfrak{X}$ and inside a larger ball $B_q(R)$, ν is uniform measure on $B_q(R)-B_q(1)$, and either (a) $s=q+1$ and g is first order while h is quadratic, or else (b) g is quadratic and h is cubic. In case (a), designs are restricted to those with "equal scaling" in all coordinates, and if the BD estimator is used the best design, which minimizes both B and J independently of R, takes all observations as "far out" as possible; in the approach of the present paper, where the "scaling"restriction is not adopted, this means taking all observations on $S_q(1)$ with resulting M the same as for the uniform probability measure on $S_q(1)$. In case (b), detailed computations are carried out only for central composite rotatable designs obtained from hypercube corners, cross-polytope points, and a center-point, and it is found that a particular choice of the design parameters again minimizes both V and B independently of R; since we do not restrict the form of design we consider, our results (Section 4) will be somewhat different. Additionally, we consider also the KMH approach in both cases (a) and (b).

All our developments will be obtained in the approximate theory, in which a design measure ξ is an arbitrary probability measure on $\mathfrak{X}$ and

$$Q=\begin{pmatrix} M & L \\ L' & K \end{pmatrix}=\int_{\mathfrak{X}} f(x)f(x)'\xi(dx). \tag{1.5}$$

Throughout the remainder of the paper, $\mathfrak{X}=B_q(1)$, and in Sections 3 and 4 we let $\mathfrak{Z}=B_q(R)$ with ν Lebesgue measure. This slight difference from DH2 (in which $\mathfrak{Z}=B_q(R)-B_q(1)$), described in the example of Section 3.1, makes the arithmetic slightly simpler without usually affecting the character of the results materially; from a practical viewpoint, it seems likelier that we are interested in the regression on $B_q(R)$ rather than only on $B_q(R)-B_q(1)$.

The examples also illustrate such phenomena as nonuniqueness of the

BD design (Section 4.1) and possible reduction of the biased regression problem to a related problem for unbiased regression.

The restriction when $q=1$ to symmetric designs in Sections 3 and 4 is justified as in Kiefer (1974). For $q>1$ (Section 3.2), p. 30 of Galil and Kiefer (1977b) discusses the possible slight loss in efficiency due to restriction to rotationally invariant ξ: although, for uniform ν, the unrestricted minimum of V alone can be attained by a rotationally invariant ξ, there are noninvariant ξ satisfying $L'M^{-}=\Gamma_{hg}\Gamma_{gg}^{-1}$, over which V must now be minimized. Nevertheless, for the example of Section 3.2 we restrict consideration to rotationally invariant ξ for the reasons given in the cited paper.

One may well wonder about the usefulness of the model considered here for extrapolation problems. As the originators of this approach have pointed out (e.g., DH, p. 272), even when $\mathfrak{X}=\mathfrak{Z}$ caution is required in using this approach unless B/V is small. Obviously, if the data indicates B/V to be large, one would reconsider only fitting $t'g$, and what is really wanted is a procedure that decides whether or not the fit $t'g$ is satisfactory. When we are extrapolating to a larger region, the breakdown of the approach is then obviously all-the-more possible. Thus, the present study is only a fragmentary beginning.

2. Extrapolation or interpolation to a point

The case where $\mathfrak{Z}$ is a point z seems a less important problem than that in which $\mathfrak{Z}$ is larger, so our illustrations are brief.

For extrapolation to a point, the KMH development reduces the problem to a known one without bias, as we now indicate. We can suppose $\mathfrak{X}$, z, f arbitrary, subject only to $g(z)$ not being the 0-vector. Since ν is supported by the single point z, the integral form of $\sigma^2 B$ in (1.2) attains its zero minimum for any D satisfying $Dg(z)=f(z)$ (for example, $D=[g(z)'g(z)]^{-1}f(z)g(z)'$). For any such D, one computes easily from (1.3) that $\sigma^{-2}J$ reduces to $V=N^{-1}f(z)'Q^{-}(\xi)f(z)$; the problem is thus that of unbiased estimation of $\binom{\theta}{\beta}'f(z)$, using the usual LS estimator of that linear parametric function. In simple polynomial settings in one variable it is the Hoel–Levine problem mentioned in Section 1.

The BD development is more complex. For simplicity, let us limit consideration here to the example of first degree regression in one variable with quadratic bias: $\mathfrak{X}=[-1,1]$, $g(x)=(1,x)'$, $h(x)=x^2$. The first line of

(1.3) now becomes (with $\nu(z)=1$)

$$\beta^2\{(\mu_2-\mu_1^2)^{-1}[\mu_2^2-\mu_1\mu_3+z(\mu_3-\mu_1\mu_2)]-z^2\}^2 \tag{2.1}$$

where $\mu_i=\int_{-1}^{1}x^i\xi(dx)$; necessarily $\mu_2-\mu_1^2\neq 0$ for $\theta' g(z)$ to be estimable, if $z>1$. One can show in the case $z>1$ that there are no designs ξ for which $\theta' g(z)$ is estimable and for which (2.1) is zero. Thus, this is an example (the first of several herein) in which the BD sufficient condition $L'M^-=\Gamma_{hg}\Gamma_{gg}^-$ cannot be satisfied. The calculation of a ξ that minimizes (2.1) by maximizing the expression inside braces ignoring z^2 (it suffices to let the support of ξ contain at most three points) is intractable algebraically, but one can obtain a computer solution. We omit details. The treatment of DH assuming a restricted form of design was mentioned in Section 2. In the more general problem $\mathfrak{X}=B_q(1)$ and $z=(z_1,0,0,\ldots,0)$ with $z_1>1$, these authors restrict consideration to (exact) designs that put some number of observations at $(1,0,0,\ldots,0)$ and the remainder, equally divided, among the q vertices of a regular (q-1)-simplex perpendicular to (and centered on) the first coordinate axis. That amounts, in our present one-dimensional setting, to consideration only of designs on $\{c,1\}$ with $-1\leqslant c<1$, so that $\xi(c)+\xi(1)=1$. As one sees in the expression for B' on p. 270 of DH, the one-dimensional case under this restriction leads to the degenerate solution $\xi(1)=1$, for which $\theta' g(z)$ is not even estimable. (This is *not* the minimizer of (2.1) without such a design restriction. Also, the problem actually treated in DH, as described in Section 1, is not this one.) If $|z|<1$, a large class of designs makes (2.1) vanish, and V can be minimized among them.

3. Extrapolation to $B_q(R)$ for linear g, with quadratic bias

3.1. *The case $q=1$*

We suppose $\mathfrak{X}=[-1,1]$ and $\mathfrak{Z}=[-R,R]$, with $g(x)'=(1,x)$ and $h(x)=x^2$. At the outset, we let ν be any *symmetric* probability measure on $\mathfrak{Z}$. We may restrict consideration to symmetric ξ. (See Section 1.) We write $\gamma_i=\int_{\mathfrak{Z}}z^{2i}\nu(dz)$ and $\mu_j=\lambda_{j/2}=\int_{\mathfrak{X}}x^j\xi(dx)$. The regression is $\theta_0+\theta_1x+\beta x^2$, with t_0+t_1x being the fitted regression. Thus,

$$\Gamma=\begin{pmatrix}1&0&\gamma_1\\0&\gamma_1&0\\\gamma_1&0&\gamma_2\end{pmatrix},\qquad Q=\begin{pmatrix}1&0&\lambda_1\\0&\lambda_1&0\\\lambda_1&0&\lambda_2\end{pmatrix}. \tag{3.1}$$

For the KMH development, we obtain from (1.4)

$$NV_{\text{KMH}}=\frac{\lambda_2-2\lambda_1\gamma_1+\gamma_1^2}{\lambda_2-\lambda_1^2}+\frac{\gamma_1}{\lambda_1}, \tag{3.2}$$

with an obvious interpretation if Q is singular (which it cannot be when γ_1 is sufficiently large, if (3.2) is to be finite). Since (1.4) is of the form tr AQ^{-1} for nonnegative definite A where Q is the design information matrix for the problem of *quadratic* regression on $\mathcal{X}$, it follows from symmetry of ξ and the characterization of admissible designs in Kiefer (1959) that V is minimized by a design of the form $\xi(1)=\xi(-1)=\alpha/2$, $\xi(0)=1-\alpha$. We obtain $\lambda_1=\lambda_2=\alpha$ and thus, from (3.2), $\alpha=\{1+|\gamma_1-1|[\gamma_1(\gamma_1+1)]^{-1/2}\}^{-1}$ for the minimizing value. Thus, the KMH procedure yields

$$\sigma^2B_{\text{KMH}}=\sigma^2B_2=\beta^2(\gamma_2-\gamma_1^2), \qquad NV_{\text{KMH}}=\{|\gamma_1-1|+[\gamma_1(\gamma_1+1)]^{1/2}\}^2. \tag{3.3}$$

As for the BD development, consider first an arbitrary (not necessarily symmetric) ξ. Then the BD prescription $L'M^-=p'$ (see (1.3)) becomes (for nonsingular M, with an analogue otherwise)

$$(\mu_2-\mu_1^2)^{-1}(\mu_2^2-\mu_1\mu_3,\mu_3-\mu_1\mu_2)=(\gamma_1,0) \tag{3.4}$$

which, upon substitution of the second component equation of (3.4) into the first, becomes $\mu_2=\gamma_1$. Thus, if $\gamma_1\leqslant 1$, we can make $B_{1,\text{BD}}=0$ by taking $\alpha=\gamma_1$ in the design family above; if $\gamma_1>1$, we have another illustration of the unsatisfiability of $L'M^-=\Gamma_{hg}\Gamma_{gg}^-$, and it is easily shown that $\alpha=1$ is best (no asymmetric design being better in either case). Thus, we obtain from (1.3)

$$\left.\begin{aligned}\sigma^2B_{\text{BD}}&=\beta^2(\gamma_2-\gamma_1^2)\\ NV_{\text{BD}}&=2\end{aligned}\right\}\quad \text{if } \gamma_1\leqslant 1$$

$$\left.\begin{aligned}\sigma^2B_{\text{BD}}&=\beta^2[\gamma_2-\gamma_1^2+(\gamma_1-1)^2]\\ NV_{\text{BD}}&=1+\gamma_1.\end{aligned}\right\}\quad \text{if } \gamma_1>1 \tag{3.5}$$

We see from (3.3) and (3.5), in the case $\gamma_1\leqslant 1$ where the prescription $L'M^-=\Gamma_{hg}\Gamma_{gg}^-$ can be satisfied, confirmation of the main KMH conclusion that $B_{BD}=B_{KMH}$ and $V_{BD}\geqslant V_{KMH}$, with strict inequality unless $\gamma_1=1$. However, for $\gamma_1>1$, the fact that the original BD sufficient condition cannot be satisfied works in favor of BD! In fact, for $\gamma_1>1$ we *always* have $V_{BD}<V_{KMH}$ and $B_{BD}<B_{KMH}$, so the interesting question is, when is $J_{BD}<J_{KMH}$? We find, for $\gamma_1>1$,

$$J_{BD}<J_{KMH}\Leftrightarrow\frac{N\beta^2}{\sigma^2}<\frac{2\gamma_1+2[\gamma_1(\gamma_1+1)]^{1/2}}{\gamma_1-1}, \tag{3.6}$$

and the ratio J_{BD}/J_{KMH} may be written down from (3.3) and (3.5).

One may specialize these considerations for ν's of usual interest. For $R<1$ (interpolation), we are always in the case $\gamma_1<1$. For $R>1$, the ν of DH (uniform measure on $[-R,R]-[-1,1]$) yields $\gamma_1=(R^3-1)/3(R-1)>1$, while the uniform measure on $[-R,R]$ yields $\gamma_1=R^2/3$. Consequently, for fixed $N\beta^2/\sigma^2$, the end points of the interval of R-values for which $J_{BD}<J_{KMH}$, for the DH ν, are to the left of those for the uniform ν on $[-R,R]$. For R large (where, however, usefulness of the model is most in doubt) and either ν, we see from (3.6) that $N\beta^2/\sigma^2\approx 4$ divides the regions where each of the two methods is better, and a first approximation to J_{KMH}/J_{BD} (as $R\to\infty$ and $R^2N\beta^2/\sigma^2\to\infty$) is $(4/9)+(20\sigma^2/9N\beta^2)$. Such conclusions can also be obtained for ν's other than the uniform measure.

3.2. *The case $q>1$*

We abbreviate the previous development and, as described in Section 1, restrict consideration to rotationally invariant ξ. We also take ν to be uniform probability measure on $B_q(R)$, the ν of DH2 (uniform on $B_q(R)-B_q(1)$) requiring only minor modification. If $x^{(i)}$, $1\leqslant i\leqslant q$, are the coordinate functions, we order the components of $g(x)'$ as $(1,x^{(1)},\ldots,x^{(q)})$ and those of $h(x)'$ as $(x^{(1)2},\ldots,x^{(q)2},x^{(1)}x^{(2)},\ldots,x^{(q-1)}x^{(q)})$. Writing $\int_{\mathfrak{X}}(x^{(1)})^i\xi(dx)=\mu_i$, an effect of rotational symmetry is that $\int_{\mathfrak{X}}(x^{(1)}x^{(2)})^2\xi(dx)=\mu_4/3$. We obtain, with 1_q denoting a q-vector of 1's and J_q denoting a $q\times q$ matrix of 1's,

$$Q=\begin{bmatrix} 1 & 0 & \mu_2 1_q' & 0 \\ 0 & \mu_2 I_q & 0 & 0 \\ \mu_2 1_q & 0 & \mu_4\left(\frac{2}{3}I_q+J_q\right) & 0 \\ 0 & 0 & 0 & \frac{1}{3}\mu_4 I_{q(q-1)/2} \end{bmatrix}, \tag{3.7}$$

from which pattern Γ can also be obtained. The exact KMH computations are now somewhat messy, and we give only the asymptotic behavior for large R. We obtain designs ξ consisting of mass $1-\alpha$ at 0 and mass α spread uniformly on $S_q(1)$. The optimum BD design for R large (in fact, for $R>1$ for the ν of DH2) has $\alpha=1$; more general than the DH2 conclusion mentioned in Section 1 is the optimality of this design for a general class of convex invariant optimality criteria. For the KMH design, putting $\gamma_i=\int_{\mathfrak{X}}(x^{(1)})^{2i}\nu(dx)=R^2/(q+2)$, we find the optimum $\alpha\to 1/2$ as $\gamma_1\to\infty$ (as when $q=1$, just above (3.3)). In order to compare the two designs, we consider now the *special case in which all components of β are zero except for the coefficient of* $(x^{(1)})^2$, which is $\bar{\beta}$ (say); other comparisons can be made similarly, for other β. Since $\gamma_2/\gamma_1^2=3(q+2)/(q+4)$, we obtain, as $R\to\infty$,

$$\sigma^2 B_{\text{KMH}}\approx\frac{2(q+1)}{(q+4)}\gamma_1^2\bar{\beta}^2,\qquad N V_{\text{KMH}}\approx 4q^2\gamma_1^2,$$

$$\sigma^2 B_{\text{BD}}\approx\frac{3(q+2)}{(q+4)}\gamma_1^2\bar{\beta}^2,\qquad N V_{\text{BD}}=1+q^2\gamma_1. \tag{3.8}$$

The analogue of the last expression of Subsection 3.1 (again, with $R\to\infty$ and $R^2N\bar{\beta}^2/\sigma^2\to\infty$) is now

$$J_{\text{KMH}}/J_{\text{BD}}\approx\left[2(q+1)+4q^2(q+4)\sigma^2/N\bar{\beta}^2\right]/3(q+2), \tag{3.9}$$

and the KMH design is better if (approximately, for large R) $N\bar{\beta}^2/\sigma^2>4q^2$.

4. Quadratic regression, cubic bias

To conserve space, we consider only the case $q=1$ here; the arithmetical complexity for general q is similar to that of Galil and Kiefer (1978). However, we consider an additional ν in Subsection 4.1, because of some additional features that are well illustrated by it, including that mentioned just below (4.4).

4.1. Extrapolation to a shell

For general q, ν is now uniform measure on $\mathfrak{X}=S_q(R)$; for $q=1$, we have $\nu(R)=\nu(-R)=1/2$. Now $\theta' q(x)=\theta_0+\theta_1 x+\theta_2 x^2$ and $h(x)=x^3$. We have

$$Q=\begin{bmatrix} 1 & 0 & \mu_2 & 0 \\ 0 & \mu_2 & 0 & \mu_4 \\ \mu_2 & 0 & \mu_4 & 0 \\ 0 & \mu_4 & 0 & \mu_6 \end{bmatrix} \tag{4.1}$$

and Γ of the same form with μ_{2i} replaced by R^{2i}. Thus, $L'M^{-}(\xi)=(0,\mu_4/\mu_2,0)$ and $p'=(0,R^2,0)$; also, $B_2=0$. Thus, from (1.3),

$$NV_{\mathrm{BD}}=(R^2/\mu_2)+[\mu_4-2R^2\mu_2+R^4]/[\mu_4-\mu_2^2],$$

$$\sigma^2 B_{1,\mathrm{BD}}=R^2[R^2-\mu_4/\mu_2]^2. \tag{4.2}$$

Since $m_4(\xi)/m_2(\xi)$ has the unit interval as range, we can (as in Section 3) make $B_{1,\mathrm{BD}}=0$ iff $R\leqslant 1$. In that case, substituting $\mu_4=R^2\mu_2$ into (4.2), we obtain $NV_{\mathrm{BD}}=2R^2/\mu_2=2\mu_4/\mu_2^2\geqslant 2$, with equality iff ξ is symmetric on two points $\pm c$; hence, $c=R$ to satisfy the previous relations, and this yields $\xi(R)=\xi(-R)=1/2$ as the unique BD design minimizing V. (Although the formula for V_{BD} in (4.2) is meaningless in this case because $\mu_4-\mu_2^2=0$, the limiting value of V_{BD} obtained here for the singular M with $\xi(\pm R)=1/2$ is correct; this is of course the only symmetric design with singular M for which $\theta' g(\pm R)$ is estimable.)

If $R>1$, $B_{1,\mathrm{BD}}$ is minimized by taking a design of the form $\xi(-1)=\xi(1)=\alpha/2>0$, $\xi(0)=1-\alpha$, for which $m_4/m_2=1$. For such a design,

$$NV_{\mathrm{BD}}=(R^2/\alpha)+[\alpha-2R^2\alpha+R^4]/\alpha(1-\alpha),$$

minimized by

$$\alpha=[1+(R^2-1)/R(1+R^2)^{1/2}]^{-1}.$$

Thus, we obtain

$$NJ_{\mathrm{BD}}=\begin{cases} 2 & \text{if } R\leqslant 1,\\ [R(1+R^2)^{1/2}+R^2-1]^2+(N\beta^2/\sigma^2)R^2[R^2-1]^2 & \text{if } R>1. \end{cases} \tag{4.3}$$

The KMH approach in this case, from (1.4), since $\Gamma_{hg}\Gamma_{gg}^{-}\Gamma_{gh}=R^6=\Gamma_{hh}$ and because of the special form of Γ and of Q for ξ symmetric, yields

$$NJ_{\text{KMH}}=NV_{\text{KMH}}=\operatorname{tr}\Gamma Q^{-1}=(1,R,R^2,R^3)Q^{-1}(1,R,R^2,R^3)'. \tag{4.4}$$

Thus, the KMH approach in this case of quadratic regression with cubic contamination reduces to that of finding the *optimum design for unbiased extrapolation to* $\pm R$ *when the cubic model is assumed and a cubic is fitted*, a problem considered in Kiefer and Wolfowitz (1964). (Although such a reduction also occurs in the analogous linear-quadratic model of Section 3 for the present ν, it does not occur for higher degree problems such as the cubic with quartic bias.) For $R\leqslant 1$, the solution is again the two-point design of the BD approach. For $R>1$, the solution may be found numerically; for $R>1.1$ the solution here coincides with that for extrapolation to $[-R,R]$ in the cubic without bias, as tabled by Galil and Kiefer (1978). As $R\to\infty$, the design approaches the "Tchebycheff design" optimum for estimating the cubic coefficient alone, for which $\xi(\pm 1/2)=1/3$ and $\xi(\pm 1)=1/6$; moreover, $J_{\text{KMH}}=V_{\text{KMH}}\approx 8R^6/N$ as $R\to\infty$. We conclude from (4.3) that, as $R\to\infty$, if also $R^2N\beta^2/\sigma^2\to\infty$, the BD design is preferable to the KMH design iff (approximately) $N\beta^2/\sigma^2<8$.

4.1. *Extrapolation to $B_q(R)$*

We now suppose ν to be uniform probability measure on $B_q(R)$, once more treating in detail only the case $q=1$. We now obtain $\sigma^2\beta^{-2}B_{1,\text{BD}}=(3R^2/5-\mu_4/\mu_2)^2R^2/3$, so we can have $B_{1,\text{BD}}=0$ iff $R^2\leqslant 5/3$ (replacing the value 1 of Section 4.1). Substituting $\mu_4=3\mu_2R^2/5$ in this case, we obtain from (1.3)

$$NV_{\text{BD}}=2R^2(R^2-\mu_2)/5\mu_2\left[3R^2/5-\mu_2\right]. \tag{4.5}$$

The treatment of (4.5) is somewhat different from that of the corresponding expression $2R^2/\mu_2$ of Section 4.1, the optimum no longer being unique. We begin by formal minimization of (4.5) wrt μ_2 (paying no attention to the restriction $\mu_4/\mu_2=3R^2/5$). The minimum, attained at $\mu_2=R^2[1-(2/5)^{1/2}]$, is $NV_{\text{BD}}=[14+(160)^{1/2}]/9$. We must check that there is indeed a design ξ on $[-1,1]$ with this μ_2 and with $\mu_4/\mu_2=3R^2/5$. In fact, there are many such designs. Two of them are ξ_1 and ξ_2 defined in terms of

$\delta=3R^2/5$ by

$$\xi_1(\pm\delta^{1/2})=1/[2+(8/5)^{1/2}]=[1-\xi_1(0)]/2,$$

$$\xi_2\left(\pm\left[\delta(1-\delta)/(1+(2/5)^{1/2}-\delta)\right]^{1/2}\right)$$
$$=[(2/5)^{1/2}+1-\delta]^2/[2+(8/5)^{1/2}][(2/5)^{1/2}+(1-\delta)^2]$$
$$=\tfrac{1}{2}-\xi_2(\pm 1). \tag{4.6}$$

(The design ξ_2 degenerates into ξ_1 when $\delta=1$.) It is straightforward to verify that both of these designs are indeed symmetric probability measures on $[-1,1]$ and that both satisfy $\mu_2=R^2[1-(2/5)^{1/2}]=\mu_4/\delta$ as desired. Thus, these designs yield the same performance in terms of J_{BD}. However, if one looks at higher moments one finds $\mu_6(\xi_2)>\mu_6(\xi_1)$. Although μ_6 does not appear in J_{BD}, if (for example) one contemplated the *possibility* of fitting a cubic with the resulting data, which one could do with the design $\beta\xi_1+(1-\beta)\xi_2$ for $0\leqslant\beta<1$, consideration of μ_6 becomes important. It follows from results in Kiefer (1959) that, within this family with $\delta<1$, only the design for $\beta=0$ (that is, ξ_2) is admissible for unbiased estimation problems in the cubic model, ξ_2 being "better than" other members of the family. If also quartic regression might be considered, both ξ_1 and ξ_2 are admissible (although neither can yield an unbiased estimator of the entire quartic), while $\beta\xi_1+(1-\beta)\xi_2$, with seven support points if $0<\beta<1$, is admissible only if sextic regression is considered.

For $R^2>5/3$, the optimum design is again seen to be of the form $\xi(\pm 1)=\alpha/2$, $\xi(0)=1-\alpha$, with

$$\alpha=1/\left\{1+\left[(R^4/5-2R^2/3+1)/(R^4/5+R^2/3)\right]^{1/2}\right\}$$

yielding the minimum of V_{BD}. We thus obtain (since $B_2=4R^6/175$)

$$NJ_{\mathrm{BD}}=\begin{cases}[14+(160)^{1/2}]/9+(N\beta^2/\sigma^2)4R^6/175 & \text{if } R^2\leqslant 5/3,\\ \left\{[R^4/5+R^2/3]^{1/2}+[R^4/5-2R^2/3+1]^{1/2}\right\}^2 & \\ \quad +(N\beta^2/\sigma^2)[4R^2/175+(1-3R^2/5)^2R^2/3] & \text{if } R^2>5/3.\end{cases} \tag{4.7}$$

The KMH approach is more complicated, and does not lead to a reduction such as (4.4) because Γ now has rank 4. As $R \to \infty$, though, the variance of the estimator of the cubic coefficient in the cubic model without bias becomes dominant as it did in the case of (4.4), and we obtain $NJ_{KMH} \approx 24R^6/25 + (N\beta^2/\sigma^2)4R^6/175$. If $R^2(N\beta^2/\sigma^2) \to \infty$, once more we obtain preference for the BD procedure for large R iff (approximately) $N\beta^2/\sigma^2 < 8$.

We have not here compared other designs and estimators with those of BD and KMH as was done for the $\mathfrak{X} = \mathfrak{F}$ (no extrapolation) on the q-simplex in Galil and Kiefer (1977a). We mention that, for sufficiently large R, the design $\xi^{(R)}$ that minimizes the maximum of the variance function on $[-R, R]$ assigns mass $[1 + (R^2 + R^4)^{1/2}/(R^2 - 1)]^{-1}$ to 0 and the remainder equally to ± 1. This design behaves very much like the BD design as $R \to \infty$. However, as we have indicated earlier, extrapolation for large R, when the model $\theta' g$ is even slightly suspect, is a problem for which any treatment may be difficult to take seriously. Even for moderate R, the KMH or BD design and estimator appear to behave reasonably well in terms of the maximum (rather than average) of the MSE over $\mathfrak{Z}$, compared with $\xi^{(R)}$. This contrasts with the result for the simplex with $\mathfrak{X} = \mathfrak{Z}$ cited above, in which the D-optimum design (with LS estimator) was superior to that of BD in terms of J (and, especially, in terms of the maximum MSE) unless $N\beta^2/\sigma^2$ was fairly large. The difference stems both from a difference between the simplex and ball moment space geometries, and also from the increased role of bias in extrapolation, as exhibited in the last line of (4.7) in the powers R^4 and R^6 in which R entered into V_{BD} and B_{BD}.

5. Adaptive methods

Once the design ξ is chosen, one can look at J as a function of C and the vector $\rho = \sigma^{-1}\beta$, and formally minimize it with respect to C. If $C_{\rho,\xi}$ is the coefficient vector of the resulting linear estimator $t = C_{\rho,\xi} Y$, one could then conceivably estimate ρ from the data Y, say by an estimator $\hat{\rho}$, and then use the adaptive (nonlinear) estimator $\hat{t} = C_{\hat{\rho},\xi} Y$. Unless the observations are taken in at least two stages, the estimator $\hat{\rho}$ cannot be used to help choose the design, but one can consider the problem of finding which design ξ performs "best" in some sense (such as integrated MSE) in terms of its estimator $Y' C'_{\hat{\rho},\xi} g$. The arithmetic of computing $C_{\rho,\xi}$ is essentially that used in Section 3 of Kiefer (1973) for the problem of minimizing the average of $V + B$ in a Bayesian-like setup; but the ensuing computation of the performance of $C_{\hat{\rho},\xi}Y$, and even consideration of the choice of the estimator $\hat{\rho}$, are more difficult. This is being investigated in more meaningful contexts by J. Sacks and the present author; here we shall indicate only the first step, of calculation of $C_{\rho,\xi}$.

A natural restriction we impose is that the risk should be bounded in θ if $\beta=0$; this entails $D'=[\,I_s \,\vdots\, d\,]$ for some $s\times(k-s)$ matrix d. Thus, $GC'=I_s$ and $HC'=d'$. From (1.2), then,

$$V+B_1=\operatorname{tr}\Gamma_{gg}\{CC'+(CH'-p)\rho\rho'(HC-p')\}. \tag{5.1}$$

Upon writing

$$U=\Gamma_{gg}^{\frac{1}{2}}\{C[I_N+H'\rho\rho'H]^{\frac{1}{2}}-p\rho\rho'H[I_N+H'\rho\rho'H]^{-\frac{1}{2}}\},$$

$$R=\Gamma_{gg}^{\frac{1}{2}}\{I_s-p\rho\rho'H[I_N+H'\rho\rho'H]^{-1}G'\},$$

$$L=[I_N+H'\rho\rho'H]^{-\frac{1}{2}}G', \tag{5.2}$$

we find that the problem of minimizing (5.1) subject to $GC'=I_s$ is equivalent to the problem of minimizing tr UU' subject to $UL=R$ (assuming G of rank s so that this last can be satisfied). In fact, the *matrix minimum* of UU' subject to $UL=R$ is $U=R(L'L)^{-}L'$ (see Kiefer (1973), (1.6)), from which the minimizing $C_{\rho,\xi}$ (independent of Γ_{gg}, reflecting its yielding a matrix minimum) can be written in terms of $P_\rho=[I_N+H'\rho\rho'H]^{-1}$ as

$$C_{\rho,\xi}=p\rho\rho'P_\rho\{I_N-G'[GP_\rho G']^{-1}GP_\rho\}+[GP_\rho G']^{-1}GP_\rho. \tag{5.3}$$

This can be simplified slightly using such relations as $P_\rho=I_N-(1+\rho'HH'\rho)^{-1}H'\rho\rho'H$, but the resulting estimator is not simple: the coefficient of p in (5.3) does not generally vanish (even if $GH'=0$), and even the simpler last term in (5.3) depends in rational rather than polynomial fashion on ρ.

Another "adaptive" estimator of a different form is obtained by simply writing $\tilde{Y}=Y-H'\beta$ and finding the usual LS estimator for θ in the model $E\tilde{Y}=G'\theta$, yielding $(GG')^{-1}GY-(GG')^{-1}GH'\beta$ as estimator t in the fitted $t'g$; an adaptive estimator $\hat{\beta}$ of β can then be used. The resulting $\hat{t}$ generally differs from that of the previous paragraph, being of simpler form here. (Making GH' vanish, by changing the h of the original model through orthogonalization, of course yields the BD estimator t_{BD}, but this results in a different fitted $t'g$.) One might use for $\hat{\beta}$ the LS estimator of β in the "full" model $\theta' g + \beta' h$ (which does not generally yield $\hat{t}$ equal to the LS estimator t^* of θ in that full model), but one could also use other estimators, such as ridge regression or Stein estimators for $\hat{\beta}$.

The comparison of performance of the adaptive estimators of the types mentioned in the previous two paragraphs, of t_{BD}, and of t^*, for various actual configurations (β, σ), remains to be carried out. Ultimately this is only a beginning on the old problem (where β is not known to be small) of choosing what curve to fit.

References

[1] Box, G. E. P., and Draper, N. R. (1959). A basis for the selection of a response surface design. *J. Am. Statist. Assoc.* **54**, 622–654.

[2] Draper, N. and Herzberg, A. (1973). Some designs for extrapolation outside a sphere. *JRSS (B)* **35**, 268–276.

[3] Draper, N. and Herzberg, A. (1976). An investigation of first-order and second-order designs for extrapolation outside a hypersphere. (Unpublished).

[4] Galil, Z. and Kiefer, J. (1977a). Comparison of Box-Draper and D-optimum designs for experiments with mixtures. *Technometrics* **19**.

[5] Galil, Z. and Kiefer, J. (1977b). Comparison of rotatable designs for regression on balls, I (quadratic). *J. Statist. Planning and Inference* **1**.

[6] Galil, Z. and Kiefer, J. (1978). Extrapolation designs and Φ_p-optimum designs for cubic regression the q-ball. *J. Statist. Planning and Inference*. To appear.

[7] Hoel, P. G. and Levine, A. (1964). Optimal spacing and weighting in polynomial prediction. *Ann. Math. Statist.* **33**, 1553–1560.

[8] Karson, M. J., Manson, A. R., and Hader, R. J. (1969). Minimum bias estimation and experimental designs for response surfaces. *Technometrics* **11**, 461–476.

[9] Kiefer, J. (1959). Optimum experimental designs. *J. Roy. Statist. Soc., Ser.* **B 21**, 272–319.

[10] Kiefer, J. (1973). Optimum designs for fitting biased multiresponse surfaces. In: *Multivariate Analysis III* (P. R. Krishnaiah, Ed.). Academic Press, New York, 287–297.

[11] Kiefer, J. (1974). General equivalence theory for optimum designs (approximate theory). *Ann. Statist.* **2**, 849–879.

[12] Kiefer, J. (1975). Optimal design: Variation in structure and performance under change of criterion. *Biometrika* **62**, 277–288.

[13] Kiefer, J. (1978). Asymptotic approach to families of design problems. *Commun. in Statist.* (to appear).

[14] Kiefer, J. and Wolfowitz, J. (1964). Optimum extrapolation designs I and II. *Ann. Inst. Statist. Math.* **16**, 79–108, 295–303.

[15] Kiefer, J. and Wolfowitz, J. (1965). On a theorem of Hoel and Levine on extrapolation. *Ann. Math. Statist.* **36**, 1627–1655.

[16] Studden, W. J. (1971). Optimal designs for multivariate polynomial extrapolation. *Ann. Math. Statist.* **42**, 828–832.

The Annals of Statistics
1980, Vol. 8, No. 6, 1293–1306

D-OPTIMUM WEIGHING DESIGNS

BY Z. GALIL[1,2] AND J. KIEFER[1]

Universities of Tel Aviv and Cornell, and University of California (Berkeley)

For the problem of weighing k objects in n weighings ($n \geq k$) on a chemical balance, and certain related problems, we obtain new results and list the designs which have been proved D-optimum up to this time. While some of these optimality results have been known for some time, others are fairly recent. In particular, in the most difficult case $n \equiv 3 (\text{mod } 4)$ we prove a result characterizing optimum designs when $n \geq 2k - 5$. In addition, by a combination of theoretical bounds and computer search we find previously unknown optimum designs in the cases $(k, n) = (9, 11)$, $(11, 15)$, and $(12, 15)$, and establish the optimality of Mitchell's $(10, 11)$ design. In some cases the optimum $X'X$ is not unique. Thus, we find two optimum $X'X$'s for the $(6, 7)$, $(8, 11)$, $(10, 11)$, and $(10, 15)$ cases. As a consequence of these results and other constructions, D-optimum designs are now known in all cases $k \leq 12$ (for all $n \geq k$), and in many other cases. Essentially complete listings for all $n \geq k$ had been given previously only for $k \leq 5$.

1. Introduction. Let k and n be positive integers with $k \leq n$, and let $\mathscr{X} \equiv \mathscr{X}(k, n)$ denote the set of all $n \times k$ matrices $X = \{x_{ij}\}$ consisting entirely of entries ± 1. If $\bar{X}$ maximizes $\det(X'X)$ over $\mathscr{X}$, then $\bar{X}$ or $\bar{X}'\bar{X}$ is said to be D-optimum.

The problem of characterizing such $\bar{X}$ arises in two statistical settings, both with uncorrelated homoscedastic observations. In both cases $1/\det(X'X)$ is proportional to the generalized variance of the least squares estimators of the parameters $\theta_1, \theta_2, \cdots, \theta_k$ of interest.

Firstly, there is the setting of finding the weights θ_j ($1 \leq j \leq k$) of k objects with n weighings. In one model, in which a chemical balance is used with each object present on each weighing, we let $x_{ij} = 1$ or -1 depending on whether the jth object is on the left or right pan in the ith weighing. That weighing model may be altered to allow the x_{ij} to be 1, -1, or 0; i.e., all k objects need not be present in each weighing. The development of the next paragraph shows that every $\bar{X}$ optimum for the previous model is optimum for this one. Also, when $k = n = r$ the optimality results for $x_{ij} = \pm 1$ are well-known to correspond to optimality results for $k = n = r - 1$ with $x_{ij} = 0$ or 1, the "spring-balance" model; see Mood (1946). The equivalences of the various D-optimality problems for the two settings is also treated by Hedayat and Wallis (1978), pages 1206 and 1220, when $k = n$.

Secondly, there is the setting of estimating the parameters of the first order regression model on the p-dimensional cube $[-1, 1]^p$ with $p = k - 1$, the ith observation being at $(z_{i1}, z_{i2}, \cdots, z_{ip})$ with expectation $\theta_k + \sum_1^p z_{ij}\theta_j$, which we can write $\sum_1^k z_{ij}\theta_j$ by defining $z_{ik} = 1$. Expanding $\det(Z'Z)$ into a sum of $\binom{n}{k}$ squares of $k \times k$ determinants (Cauchy-Binet expansion) each linear in its entries, we conclude that, as a function of a single $z_{i_0 j_0}$ ($j_0 < k$) for all other z_{ij} fixed, $\det(Z'Z)$ is quadratic in $z_{i_0 j_0}$ with nonnegative coefficient of $(z_{i_0 j_0})^2$. Hence, $z_{i_0 j_0}$ can be changed to one of the values 1 or -1 without decreasing $\det(Z'Z)$. Making such changes, one by one, for each $\bar{z}_{i_0 j_0}$ of a D-optimum $\bar{Z}$, we conclude that there is a D-optimum $\bar{Z}$ in $\mathscr{X}$. Conversely, each X in $\mathscr{X}$ can be transformed, by multiplication of each row by 1 or -1, into an element of $\mathscr{X}$ with the same determinant and all $x_{ik} = 1$. (The reduction to $\mathscr{X}$ need not yield all $\bar{Z}$; thus, for $p = 2$, $n = 3$, the three points $(1, 1)$, $(1, -1)$, $(-1, z_{3,2})$ constitute a D-optimum design for every $z_{3,2}$ in $[-1, 1]$, with $\det(X'X) = 16$.)

Received November 1978; revised August 1979.
[1] Research sponsored by the National Science Foundation.
[2] Research supported in part by the Israel Commission of Basic Research.
AMS 1970 *subject classification.* Primary 62K5, 62K15, 05B20.
Key words and phrases. Optimum designs, weighing designs, first-order designs, D-optimality, fractional factorials.

If $[-1, 1]^p$ is replaced by $\{-1, 1\}^p$ in the above, we obtain the even simpler correspondence of the weighing problem to the first order (resolution III) fractional 2^p-factorial problem.

The cases $k = n$ are called *saturated.*

The problem of finding an X is the subject of many papers, two early ones being those of Hotelling (1944) and Mood (1946). For reference to the many contributions of Kishen, Banerjee, Raghavarao, and others, see Raghavarao (1971), who also gives typical results. Many of the known results characterize a D-optimum $\bar{X}$ *subject to the restriction* to X's in $\mathscr{X}$ for which $X'X$ is permutation invariant (has all diagonal elements equal and all off-diagonal elements equal). The imposition of this restriction simplifies the optimization problem considerably, but for many k and n it yields designs that, although often fairly efficient, are not optimum in $\mathscr{X}$. This is known, for example, from the saturated cases $n = 6$ or 7 in Mood (1946) and $n \equiv 2 (\mathrm{mod}\ 4)$ in Ehlich (1964a). The matter is discussed in Cheng (1980) and Kiefer (1978). In the present paper we are concerned with finding a D-optimum $\bar{X}$ in $\mathscr{X}$ without any such restriction.

We note that recent combinatorial literature often refers to "weighing matrices" as square orthogonal matrices with entries from $\{0, 1, -1\}$. This should not be confused with our weighing designs X.

Other optimality criteria, such as $\mathrm{tr}(X'X)^{-1}$, have also been considered, and we shall refer in Subsections 2.0 and 2.1 to some results on them; but our main concern here is with $\det(X'X)$.

There are some commonly employed rules, discussed by various authors such as Mood (1946) and Mitchell (1974b), for augmenting or reducing Hadamard matrices to yield designs in cases other than the saturated case of Case 0 of (2.0). However, the optimality or nonoptimality of the designs produced by such recipes is not always clear in the literature. The paper of Payne (1974) clarified and vastly expanded the state of knowledge of optimality of such designs. In Section 2 (in terms of n) and in Section 5 (in terms of k) we summarize the current state of knowledge, incorporating some new constructive devices and optimality results of the present paper. We shall occasionally refer to the searches, more extensive than those of Mitchell (1974b), which have been carried out using our modification of his pioneering technique. Details were reported in Galil and Kiefer (1980).

Results of this and previous papers imply that certain forms of X and $X'X$, both of which forms we term "regular", are optimum if they exist. Previous work, including Payne's for $n \geq 5$, assumes the existence of Hadamard matrices for constructions. Although we are able to avoid assuming such matrices always exist, and to extend constructions to such cases as $k = n - 1$ of Case 2 of (2.0) not previously considered, other existence questions are generally more difficult, and are not much treated here. We show that easily constructed optimum designs are regular except in certain saturated instances of Case 1 and in Case 3 for $n < 2k - 5$, for all "practical" values (k, n). The (9, 9) and (11, 11) optimum designs gave difficulty because they are not of this regular nature, and their optimality was proved, respectively, by Ehlich and Zeller (1962) and by Ehlich in unpublished work mentioned in Subsections 2.1 and 2.3, using a combination of theoretical and computer developments. Regular optimum designs for (9, 11), (11, 15), (12, 15) were found by us by extensive computer search and (with Mitchell's (10, 11) design) are proved optimum herein by means of our development that stems from the work of Ehlich (1964b).

Ehlich is the pioneer and chief contributor of ideas to this subject of finding D-optimum designs in the non-Hadamard cases. We are grateful for the inspiration of his work and for the communication of Ehlich (1978).

2. Listing of *D*-optimum designs. For further discussion, we divide the values of n into four cases:

$$
\begin{aligned}
&\text{Case 0: } n \equiv 0 (\mathrm{mod}\ 4);\\
&\text{Case 1: } n \equiv 1 (\mathrm{mod}\ 4);\\
&\text{Case 2: } n \equiv 2 (\mathrm{mod}\ 4);\\
&\text{Case 3: } n \equiv 3 (\mathrm{mod}\ 4).
\end{aligned}
\tag{2.0}
$$

We denote by $\mathcal{N}$ the nonnegative integers and by $\mathcal{N}_j$ the set of all integers in $\mathcal{N}$ that fall in Case j. We have included $n = 0$ in Case 0 to shorten the discussion, below. Although $X'X$ is $k \times k$, the value of n appears to be more critical than that of k in determining optimality structure, since n has more important influence on the values of the entries of $X'X$. This accounts for the classification (2.0). (The case $k = 2$ is separated by its triviality: every D-optimum X has orthogonal columns if n is even and columns with inner product ± 1 if n is odd.)

The term "normalization" will be used herein to refer to the following operations on X: multiplying on the left by a diagonal matrix of ± 1's and/or a permutation matrix (which permutes or reflects the points of $\{-1, 1\}^k$ that are rows of X, but does not affect $X'X$); the same operations on the right, which permutes rows and corresponding columns of $X'X$ and multiplies some entries of $X'X$ by -1, but leaves $\det(X'X)$ unchanged. The optimum structure of $X'X$ is generally listed in a convenient form "after normalization". For example, in the regular case in Subsection 2.1 all off-diagonal elements of an optimum $X'X$ can be taken as -1 after normalization. Without normalization, some could be of each sign.

2.0 Case 0. An $n \times n$ Hadamard matrix H_n is a member of $\mathcal{X}(n, n)$ with $H_n' H_n = nI_n$. A necessary condition for H_n to exist is that n be 1, 2, or $\equiv 4$ in Case 0, and we also include the empty matrix H_0 for use in further discussion. There is much more literature on the existence of H_n than on all other aspects of the subject of weighing designs; see, e.g., Hedayat and Wallis (1978). By now H_n are known to exist in Case 0 for all $n \leq 200$, and for infinitely many other n. There is an $\bar{X}$ in $\mathcal{X}(k, n)$ with $\bar{X}'\bar{X} = nI_k$ if H_n exists (namely, k columns of H_n), and such $\bar{X}$ can in fact often be found much more easily, as we describe in the next paragraph. Such an $\bar{X}$ is not only well known to be D-optimum, but also minimizes $\Phi(X'X)$ over $\mathcal{X}$ for every nonincreasing convex orthogonally invariant extended real-valued Φ defined on the nonnegative definite symmetric $k \times k$ matrices; see Kiefer (1975). It also minimizes the individual variances of best unbiased estimators of the θ_i (diagonal element of $(X'X)^{-1}$), as was shown by Hotelling (1944).

In Case 0 the "regular" $\bar{X}$'s in $\mathcal{X}(k, n)$ are those with $\bar{X}'\bar{X} = nI_k$, for which $\det(\bar{X}'\bar{X}) = n^k$. We now make a simple observation about the existence of such designs. For fixed k it is unnecessary to assume the Hadamard conjecture of existence of an H_n for all n in $\mathcal{N}_0$ as Payne and others do, in order to give optimality results for all such n. (Payne mentions that the assumption can be avoided when $k \leq 4$.) For, we only need k orthogonal n-vectors for the columns of $\bar{X}$, not n of them. One way of constructing such an $\bar{X}$ is to adjoin vertically sufficiently many $n_i \times k$ submatrices of H_{n_i}'s. If $n_1 + n_2 + \cdots + n_L = n$ and $n_i \geq k$ and H_{n_i} exists for $1 \leq i \leq L$, such an $\bar{X}$ can obviously be constructed. A convenient sufficient condition for this to be possible for given k and *all* $n \geq k$ is the

Proposition. *For $k \geq 4$, let $n_k = \min\{j: j \geq k, j \in \mathcal{N}_0\}$. Suppose H_j exists for all j in $\mathcal{N}_0$ satisfying $n_k \leq j < 2n_k$. Then, for all $n \geq k$ with $n \in \mathcal{N}_0$, there is an $\bar{X}$ in $\mathcal{X}(k, n)$ with $\bar{X}'\bar{X} = nI_k$. (A construction can similarly be given in terms of orthogonal arrays of strength* 2.)

Thus, for example, when $k = 6$ our knowledge of H_8 and H_{12} suffices, and for $k \leq 100$ (which presumably includes all "practical" values) the proposition may be used.

The other three cases of (2.0) are not so simple, and their investigation in the saturated case was pioneered by Ehlich (1964a, b). (See also Wojtas (1964).)

2.1 Case 1. Ehlich showed that an X in $\mathcal{X}(n, n)$ with $\bar{X}'\bar{X} = (n-1)I_n + J_n$ (where J_n consists entirely of 1's) is D-optimum. Unfortunately, such an $\bar{X}$ can exist only if $2n - 1$ is the square of an integer. Such designs are known for the "practical" values $n = 1, 5, 13, 25$.

It is perhaps somewhat surprising at first glance that the unsaturated case of Case 1 is easier to handle than the saturated case. It was shown by Cheng (1980) that any $\bar{X}$ in $\mathcal{X}(k, n)$ with $\bar{X}'\bar{X} = (n-1)I_k + J_k$ and $\det(\bar{X}'\bar{X}) = (n-1)^{k-1}(n-1+k)$ (the "regular" designs of Case 1) is not only D-optimum, but also optimum with respect to a large subclass of the Φ's of the previous paragraph, including all those of common interest. (The D-optimality in the unsaturated case, obtained by Payne (1974), can also be obtained by a simple modification of Ehlich's saturated case proof; but the more general results require Cheng's analysis.) Moreover, for $k < n$ such an $\bar{X}$ can always be obtained when the regular design of Case 0 in $\mathscr{X}(k, n-1)$ exists, by adjoining a row of 1's to that design, whose construction can often be obtained from the Proposition of the previous paragraph even if an H_{n-1} is not known. Although such an adjoining is a common practice in the literature of weighing designs, the D-optimality over $\mathscr{X}$ (without the additional symmetry restriction) of the resulting $\bar{X}$ was evidently unknown before Payne's paper. Thus, Mitchell (1974b) made computer searches in several of these cases, always obtaining such an $\bar{X}$, and remarking that Mood had suggested such designs would be "very efficient." For values of $n \leq 20$ in Case 1, we are left without knowledge of an optimum design only in the saturated cases $k = n = 9, 17$. Ehlich and Zeller (1962) state that for $k = n = 9$ the nonregular design obtained by them, for which the above-diagonal elements of $X'X$ are all 1 except for a single 5, can be proved optimum. A normalization of the design given in Table 4b of Mitchell (1974b) is of this form, and such a design can also be constructed using a method of Williamson (1946, page 433). Ehlich (1978) has indicated to us that the method of proof of optimality is similar to, but simpler than, that mentioned in Subsection 2.3 below for the $k = n = 11$ case. The method also shows no other form of $X'X$ can be optimum for $k = n = 9$. While the $k = n = 11$ case required machine help, the calculations by Ehlich and Zeller (1962) in the $k = n = 9$ case were done by hand.

2.2 Case 2. Here Ehlich (1964a) and Wojtas (1964) showed in the saturated case that any X for which $X'X = \begin{pmatrix} M & 0 \\ 0 & M \end{pmatrix}$, where $M = (k-2)I_{k/2} + 2J_{k/2}$, is D-optimum. Ehlich constructed such X of the form $\begin{pmatrix} A & B \\ -B' & A \end{pmatrix}$ with A and B circulants, in all cases $k \leq 38$ except $k = 22$ and 34. Other optimum designs in these cases were obtained by Yang (1968), who in references cited by him there also obtained optimum X for $k = 42, 46, 48, 52$.

For general $k \leq n \in \mathscr{N}_2$, we define the regular $\bar{X}$ in $\mathscr{X}(k, n)$ to be those for which $\bar{X}'\bar{X} = \begin{pmatrix} L & 0 \\ 0 & M \end{pmatrix}$, where for k even $L = M = (n-2)I_{k/2} + 2J_{k/2}$, and for k odd L and M are $(n-2)I_{(k\pm1)/2} + 2J_{(k\pm1)/2}$; thus, $\det(\bar{X}'\bar{X}) = (n-2)^{k-2}(n-2+k)^2$ or $(n-2)^{k-2}(n-1+k)(n-3+k)$, for k even or odd. These designs were proved D-optimum by Payne for $k \leq n-2$ using the work of Wojtas, and one can also see that Ehlich's proof requires only simple modifications to apply to Case 2 for $n \geq k$. (In fact, Payne's proof also applies for $n \geq k$, but he does not say so because he gives constructive methods only when $k \leq n-2$ and H_{n-2} exists.)

When $k \leq n-2$ and one knows an H_{n-2} or, more generally, a regular $\bar{X}$ in $\mathscr{X}(k, n-2)$ constructed from the Proposition of Case 0, a regular optimum $\bar{X}$ for Case 2 is achieved by using one of Mood's devices, discussed and employed by Mitchell and by Payne. This $\bar{X}$ is obtained by adjoining to $\bar{X}$ two rows, one consisting entirely of ones and the other consisting of $k/2$ (respectively, $(k-1)/2$) 1's following by $k/2$ (respectively, $(k+1)/2$) -1's, depending on whether k is even or odd. It seems not to have been observed by the cited authors that, when $k = n-1$ with $n \in \mathscr{N}_2$, removing a column from an optimum saturated regular X of Ehlich in $\mathscr{X}(n, n)$ (mentioned two paragraphs above) yields an optimum regular design in $\mathscr{X}(n-1, n)$. Thus, just as the construction problem in Case 1 was much simpler for $k < n$ than in the saturated case, so in Case 2 it is simpler for $k < n-1$ than in the saturated or near-saturated ($k = n-1$) case.

That these results seem unknown is indicated by the fact that Mitchell's computer search for optimal designs included the cases $(k, n) = (5, 6), (9, 10)$ (also excluded in Payne's work), in which cases the previous paragraph implies that such search could be dispensed with. Unlike the D-optimum designs of Case 1, those of Case 2 are not yet known to have other optimum properties, except that Cheng (1980) has shown they are among the E-optimum designs (that is, they maximize the minimum eigenvalue of $X'X$), not all of which need have this $X'X$ structure.

2.3 CASE 3. This is well known to be the most difficult case, and we devote the next two sections to new results for it. We first summarize, here, the previously known results. If one knows an H_{n+1} or, more generally, a regular $\bar{X}$ in $\mathcal{X}(k, n+1)$ from the Proposition of Case 0, deletion of one row of $\bar{X}$ yields an $\bar{X}$ in $\mathcal{X}(k, n)$ with $\bar{X}'\bar{X} = (n+1)I_k - J_k$. However, such an $\bar{X}$ was until recently known to be optimum for $k > 2$ only when $n = 3$ (e.g., Mood (1946)). For $k = n = 7$, the optimum design X is not of this form. It was found by Williamson (1946) and discussed by Mood (1946), and the exceptional above-diagonal elements (3's) of that $X'X$ other than -1 can be put into positions (1, 2), (3, 4), (5, 6) by normalization. Designs for the cases $k < n = 7$ have been obtained through computer search by Mitchell (1974b) but their optimality was not previously verified theoretically. His computer search yielded, after normalization, the $\bar{X}$ described just above, and Payne (1964) proved these optimum for $k \leq 5$. The optimality for $k = 6$ is proved herein.

For $n = k = 11$ (not treated by Mitchell), an $X'X$ was obtained through computer search combined with some algebra by Ehlich and Zeller (1962), in which paper the optimality of the design was indicated to be questionable. This design was subsequently verified by Ehlich to be optimum, as described to us in Ehlich (1978). This $X'X$ has exceptional above-diagonal elements 3 rather than -1 in positions that (by normalization) can be taken to be (1, 2), (2, 3), (3, 4), (4, 5), (6, 7), (8, 9), (10, 11). An X of this character, obtained by the computer search method of Galil and Kiefer (1978), is listed in Table 9a herein.

Not knowing this $X'X$ had been proved optimum in earlier unpublished work of Ehlich, and viewing this evidently "nonregular" case as a good example on which to investigate properties of variations of our computing method, we ran several thousand trials in this case and found other designs that gave the same value of det($X'X$) but two other forms of $X'X$. Two such designs are listed in Tables 9b and 9c. The first of these has an $X'X$ that differs from that described above only in having an additional above-diagonal 3 in position (1, 3). The second has an $X'X$ that is a "block matrix" whose description is deferred until that concept is defined. (All three forms of $X'X$ were obtained several times, in a total of about 1% of the computer trials. Thus, the designs are not "easy" to find in this case by use of our general search method, which does not make use of the special theoretical devices used by Ehlich in this case.)

Subsequently, Professor Ehlich (1978) very kindly sent us a description of the ingenious combination of theoretical developments and computer search by means of which he obtained designs with all three of these structures of $X'X$, proved them optimum, and proved no other structures of $X'X$ could be optimum. Thus, the design of Table 9a should be viewed as that of Ehlich and Zeller, and those of Tables 9b and 9c as Ehlich's, but we list them here because there is no indication that they are to appear elsewhere in print. These $X'X$ all have determinant $B = (5 \times 2^{16})^2$.

Ehlich's indication of the proof of optimality is that a simple inequality eliminates all $X'X$ with off-diagonal elements other than -1, -5, or 3. The computer program searches among all normalized 11×11 matrices C with such elements (and diagonal elements 11) and shows none with a -5 can have det $C \geq B$. Using the upper bound of Ehlich (1964b) listed in Table 1 herein, and further computer search, only 7 possible C's with det $C \geq B$ are found to within normalization. Finally, a systematic search for X's that realize one of these seven C's as $X'X$ yields the three structures of Table 9.

Designs for $k < n = 11$ have not previously been proved optimum theoretically except for Payne's treatment when $k \leq 5$. Of the designs found by Mitchell (1974b, Tables 4d and 4e) for $k = 9$ and 10, the former is improved upon by the design of Table 5 herein, found in our more extensive search. Both the design of Table 5 and Mitchell's design for $k = 10$, as well as $\bar{X}$ with $\bar{X}'\bar{X} = 12I_k - J_k$ for $k \leq 8$, are proved optimum herein. The size of these designs is such that computer search using our modifications of the Detmax procedure invented by Mitchell (1974a), which of course requires more computing for these larger values of k and n, is also decreasingly successful. For example, as stated above, only about 1% of the starts (involving some random element) reached an optimum solution when $k = n = 11$.

Payne (1974) showed that an $\bar{X}$ with $\bar{X}'\bar{X} = (n+1)I_k - J_k$ (and determinant $(n+1)^{k-1}(n$

$+ 1 - k)$) is D-optimum provided n is sufficiently large compared with k. He gives $n > (5/2)3^k k^2 \binom{k}{[k/2]}$ as a crude sufficient bound for which his proof works, and remarks that numerical evidence suggests that $n > 7k/2$ might suffice, and that the proof is likely to fail in general for $k \leq n < 3k$. Our own early numerical investigations indicated that $n \geq 2k$ might suffice, so that the evidence cited by Payne is a commentary on his method of proof rather than on the definitive results. In Section 3 we show $n \geq 2k - 5$ suffices.

Because of this complex situation, we have chosen not simply to define $\bar{X}$ of the above form as "regular" in Case 3. Rather, we are guided by a development of Ehlich (1974b), which we now describe. Thus material through (2.5) is also used in proving the Theorem of the next section.

Let $\mathscr{C} = \mathscr{C}_{k,n}$ be the class of all symmetric $k \times k$ matrices with diagonal entries n and off-diagonal entries -1 or 3, where $n \in \mathscr{N}_3$. Let

$$\Psi(k, n) = \max_{A \in \mathscr{C}_{k,n}} \det A. \tag{2.1}$$

Ehlich shows $\max_{X \in \mathscr{X}} \det(X'X) \leq \Psi(k, n)$ in Case 3.

A *block* of size r is an $r \times r$ matrix with diagonal elements n and off-diagonal elements 3. A *block matrix* in $\mathscr{C}_{k,n}$ with block sizes $r_1, r_2, \cdots r_s$ satisfying $\sum_1^s r_i = k$ is a $k \times k$ matrix with diagonal blocks of those sizes and with all other elements equal to -1. As Ehlich shows, any such block matrix C has

$$\begin{aligned} \det C &= (n-3)^{k-s}\{1 - G\} \textstyle\prod_1^s (n - 3 + 4r_i), \\ G &= \textstyle\sum_1^s r_i/(n - 3 + 4r_i). \end{aligned} \tag{2.2}$$

Ehlich also shows that there is a block matrix in $\mathscr{C}_{k,n}$ which has maximum determinant in $\mathscr{C}_{k,n}$ and which is a member of the subset $\mathscr{B}_{k,n}$ of $\mathscr{C}_{k,n}$ which consists of block matrices with *blocks of only one size or blocks of only two contiguous sizes*, u of size r and v of size $r + 1$, where consequently

$$u + v = s, \qquad ur + v(r + 1) = sr + v = k. \tag{2.3}$$

For any block matrix C_s in $\mathscr{B}_{k,n}$ with s blocks, (2.2) and (2.3) yield

$$\begin{aligned} \det C_s \equiv D_{k,n}(s) &= (n-3)^{k-s}(n - 3 + 4r)^u(n + 1 + 4r)^v\{1 - G\} \\ &= (n-3)^{k-s}(n - 3 + 4r)^{(r+1)s-k}(n + 1 + 4r)^{k-sr}\{1 - G\}, \\ G &= [k(n-3) + 4sr(r+1)]/(n + 4r + 1)(n + 4r - 3). \end{aligned} \tag{2.4}$$

Ehlich's last-cited result is thus $\Psi(k, n) = \max_s D_{k,n}(s)$. Of course, s uniquely determines r except when $s | k$. In that case, the block matrix with $r = r_0$, $u = u_0$, $v = 0$ is identical to that with $r = r_0 - 1$, $u = 0$, $v = u_0$, and either yields the same result in (2.4). The $\bar{X}'\bar{X} = (n + 1)I_k - J_k$ discussed earlier has $s = k$.

In Case 3 we call X in $\mathscr{X}(k, n)$ "regular" if $X'X = C_s^*$ is of the form C_s in $\mathscr{B}_{k,n}$ described in (2.3) and (2.4) and

$$\det C_s^* = D_{k,n}(s) = \max_t D_{k,n}(t). \tag{2.5}$$

If $s = k$ maximizes $D_{k,n}(s)$, we call the resulting X and $X'X = (n + 1)I_k - J_k$ "very regular." As we shall see, the s maximizing $D_{k,n}(s)$ need not be unique.

In Section 3 we characterize cases where very regular designs maximize $D_{k,n}(s)$, and the construction problem of D-optimum designs in $\mathscr{X}(k, n)$ is then handled in all practical cases by the simple construction described at the outset of the discussion of Case 3, above. For other k with $n \in \mathscr{N}_3$, we encounter difficult construction problems of whether C_s^* is realizable as an $X'X$. If not, we have no optimality characterization. Many saturated or near-saturated cases give evidence that such C_s^* are not realizable as $X'X$. As we mention in Section 4, that is always so when $k = n < 91$. In Section 4 we discuss $D_{k,n}(s)$ further and give a few positive results in regular cases that are not very regular. As more becomes known about the form of D-optimum designs in Case 3, it may become convenient to alter the definition of regularity.

2.4 UNIQUENESS. In Case 0, if an optimum $X'X$ exists that is regular, then it is well known that every D-optimum $X'X$ must have that same form. The same conclusion holds (with possible normalization) in Case 1 and Case 2, as can be seen by examining the modification of the uniqueness part of the proofs of Ehlich (1964a) needed to make them apply when $k < n$. Finally, when $n > 2k - 5$ in Case 3, the same conclusion applies if a D-optimum $X'X$ exists that is very regular, as one can see by tracing through the inequalities in the reductions in Ehlich (1964b) that are described in the development of (2.1)–(2.4) relating to $\mathscr{C}_{k,n}$ and $\mathscr{B}_{k,n}$, together with the proof of Section 3 below, as these apply when $n > 2k - 5$.

When $n \leq 2k - 5$, no general Case 3 uniqueness results for $X'X$ are known, and we now describe examples of nonuniqueness. (Uniqueness for the optimum $X'X$ for $k = n = 7$ may be obtainable from Williamson's development.) Table 1 and Section 4 describes the lack of uniqueness that is possible among regular block designs that are optimum in $\mathscr{X}(k, n)$. In the borderline case $n = 2k - 5$ of case 3, the theorem of Section 3 shows that an optimum design can have the very regular $X'X$ consisting entirely of blocks of size 1 ($s = k$) or can have one block of size 2 and the rest of size 1 ($s = k - 1$). The smallest possible example is $(k, n) = (6, 7)$, and in Table 2 we give an optimum X for which $X'X$ has $s = 5$, the above-diagonal 3 being in position (1, 2). We remark herein that our theorem obviates the need for Payne's longer proof of optimality of the very regular $X'X$ in this case. In addition, we have settled his question about uniqueness in the negative. In the next case $(k, n) = (8, 11)$, we also obtained, and list in Table 3, an optimum X ($X'X$ with $s = 7$, above-diagonal 3 in position (1, 2)) other than the very regular one. For the case (10, 15), we list in Table 4 an optimum X whose $X'X$ has $s = 9$, the above-diagonal 3 again being in position (1, 2).

For $n < 2k - 5$, there are four cases of (k, n) in which we know a regular optimum $X'X$ at the current writing. For the case (9, 11), Table 1 shows that $s = 6$ or 7 is possible, and the first of these (Table 5) was found frequently in our search. However, the second of these was not obtained in 1500 trials. (Mitchell's Table 4d gives a block design with $s = 5$, not optimum.) For (10, 11), where Mitchell's design (proved optimum in our Section 4) has $s = 5$, we have also found an alternate with $s = 6$, given in Table 6. (Mitchell states that he did not attempt to list more than one maximizer of $\det(X'X)$.) For (11, 15), Table 1 shows that $s = 8$ (Table 7) gives the unique block design optimum in $\mathscr{X}(11, 15)$. For (12, 15) we have only found an optimum $X'X$ with $s = 6$ (Table 8), not one with $s = 7$. Ehlich (1978) indicates that he has found the latter.

Perhaps most interesting is the presence of three different forms $X'X$ mentioned in Subsection 2.3 as being optimum in the case $k = n = 11$. Two of these, not block matrices, were described earlier (Tables 9a and 9b). The third, yielded by the X of Table 9c, is a block matrix with one block of size 5 and three of size 2.

Thus, lack of uniqueness of the optimum $X'X$ is quite possible, and we do not yet know the general situation. It is well known from simple examples that the uniqueness of the D-optimum "information matrix" (analogue of $X'X$) in "approximate" design theory does not persist in the exact theory, as has been illustrated in the example of $k = n = 3$ on $[-1, 1]^2$ in Section 1. Still, these first examples of nonunique D-optimum $X'X$ in the simple standard weighing design setting are somewhat surprising. (For nonuniqueness of the E-optimum design, even possible in the approximate setting, see Cheng (1980) and also his references to earlier work of Takeuchi.) We emphasize that we are not treating here the more detailed consideration of nonisomorphic X with the same $X'X$—for example, of nonisomorphic H_n.

3. Optimum very regular designs in Case 3.

We now prove

THEOREM. *If* $k \leq n \in \mathscr{N}_3$ *and* $n \geq 2k - 5$, *then* $\max_t D_{k,n}(t) = D_{k,n}(k) = (n + 1)^{k-1}(n - k + 1)$, *and hence any very regular* X (*with* $X'X = (n + 1)I_k - J_k$) *is D-optimum in* $\mathscr{X}(k, n)$. *If* $n > 2k - 5$, *the value* $t = k$ *uniquely maximizes* $D_{k,n}(t)$, *while if* $n = 2k - 5$ *the one other maximizing value is* $t = k - 1$, *for which* $X'X$ (*if it exists*) *differs from* $(n + 1)I_k - J_k$ *by having* 3*'s in positions* (1, 2) *and* (2, 1) *after normalization. If* $n < 2k - 5$, $D_{k,n}(t)$ *is not maximized by* $t = k$.

PROOF. Suppose $k \leq n \in \mathscr{N}_3$ and $n \geq 2k - 5$, and that $t = s < k$ maximizes $D_{k,n}(t)$.

Continuing with the nomenclature and developments of (2.1)–(2.4), we adopt the second representation just after (2.4) in the cases $s \mid k$. This means we may assume $v > 0$ for all s, and hence

$$sr \le k - 1. \tag{3.1}$$

With this choice, the parameters u, v, r, s, G hereafter refer to C_s^* satisfying (2.5) with $s < k$ and hence $r > 0$.

Let C^{**} be obtained from C^* by replacing one block of length $r + 1$ (recall $v > 0$) by a block of length r and a block of length 1 (perhaps now yielding blocks of three lengths in C^{**}). We shall show the resulting contradiction $\det C^{**} > \det C^*$ except in the single case $r = 1$, $v = 1$, $n = 2k - 5$, when $\det C^{**} = \det C^*$. Writing $L = n + 4r + 1$ to simplify calculations, we have from (2.2) after some simple arithmetic,

$$\begin{aligned} &L^{1-v}(L-4)^{-u}(L-4r-4)^{s+1-k}[\det C^{**} - \det C^*] \\ &= (L-4)(L-4r)\left\{1 - G + \frac{r+1}{L} - \frac{1}{L-4r} - \frac{r}{L-4}\right\} - (L-4r-4)L\{1-G\} \\ &= 16r\{1-G\} + 8r\{-1 + 2(r+1)L^{-1}\} = 8r\{1 - 2G + 2(r+1)L^{-1}\}. \end{aligned} \tag{3.2}$$

From (2.4) and (3.1),

$$\begin{aligned} &L(L-4)\{1 - 2G + 2(r+1)L^{-1}\} \\ &\ge L(L-4) - 2[k(L-4r-4) + 4(k-1)(r+1)] + 2(r+1)(L-4) \\ &= L[L - 2 - 2k + 2r] = L[n + 6r - 1 - 2k] \ge L[n + 5 - 2k]. \end{aligned} \tag{3.3}$$

This last, and hence (3.2), is nonnegative provided $n \ge 2k - 5$.

We also see that the left side of (3.3) can be zero only if $r = 1$, $v = 1$, and $n = 2k - 5$. In all other cases we have obtained the contradiction $\det C^{**} > \det C^*$. In the single case $n = 2k - 5$, we have also proved that $s = k$ is optimum in $\mathscr{C}_{k,n}$, and have also shown that $s = k - 1$ ($r = 1$, $v = 1$) is optimum (and no other s is) in that case.

Finally, if C^* has $s = k - 1$ and $r = 1$, so that C^{**} has $s = k$, the left side of (3.3) is $L(L - 2k)$, which is <0 if $n < 2k - 5$. □

4. Case 3 with $n < 2k - 5$. When $k \le n \le 3$ the optimum design is always very regular. When $n = 7$, the only value $k \le 7$ not covered by the theorem above is $k = 7$. In particular, Payne's intricate proof in the case $k = 5$ is unnecessary. The saturated case $k = 7$ was treated by Williamson: As mentioned earlier, an optimum X is not regular since $\Psi(7, 7)$ is not a square. It is of interest that Ehlich's inequality $\det X'X \le \max_s D_{k,n}(s)$ (see just below (2.1) and (2.4)) can be used to give a much shorter proof of Williamson's result: One computes easily that $\Psi(7, 7) = D_{7,7}(5) = 84 \times 2^{12}$, but since $\det(X'X)$ is trivially seen to be a square divisible by 2^{12} in this case, we must have $\det(X'X) \le 81 \times 2^{12}$. Williamson's design attains this bound, and is thus D-optimum. The upper bound of Ehlich's inequality is actually achieved for $k = n = 3$, but the method just used obviously fails for the X's proved optimum by Ehlich when $k = n = 11$.

Before going on to the case $n = 11$, we list in Table 1 the values $s_{\max}$ of s that maximize $D_{k,n}(s)$, together with $\Psi(k, n)$ for $k \le 15$ and $n < 2k - 5$. (A printout for larger k, is available to interested readers.)

The values $D_{n,n}(s)$ were studied extensively by Ehlich. We note that, as was the case for $\Psi(7, 7)$ in the previous paragraph, $\Psi(n, n)$ is not a square for $n = 11$ or 15 and thus Ehlich's upper bound is not attainable in $\mathscr{X}(n, n)$ (there is no regular optimum X) in these cases. (When $k < n$, $\det(X'X)$ need not be a square, and this simple unattainability argument fails.) In fact, an analysis of Ehlich's results for $\Psi(n, n)$ show that it is a square (for $n \in \mathscr{N}_3$) for infinitely many (but sparse) n, the only values <200 being 91 and 147; it seems quite unlikely that the bound is attainable in those large cases. Of course, whenever $\Psi(k, n)$ is not attainable by an $X'X$, there is no reason for the D-optimum $X'X$ to be a block matrix (with smaller

TABLE I

Optimum block matrices for $n < 2k - 5$, $n \leq 15$ (Case 3)

k	n	s_{max}	$\Psi(k, n)$
7	7	5	$.3441 \times 10^{6}$
9	11	6, 7	$.1359 \times 10^{10}$
10	11	5, 6	$.1288 \times 10^{11}$
11	11	5, 6	$.1203 \times 10^{12}$
11	15	8	$.5617 \times 10^{13}$
12	15	6, 7	$.7644 \times 10^{14}$
13	15	6, 7	$.1032 \times 10^{16}$
13	19	9, 10	$.2899 \times 10^{17}$
14	15	6	$.1387 \times 10^{17}$
14	19	7, 8	$.5130 \times 10^{18}$
15	15	6	$.1855 \times 10^{18}$
15	19	7, 8	$.9029 \times 10^{19}$
15	23	11	$.1906 \times 10^{21}$

$D_{k,n}(s)$ than $\Psi(k, n)$). Indeed, when $k = n = 11$ a D-optimum $X'X$ need not even be a block matrix (Tables 9a and 9b) as it was in the nonregular case $k = n = 7$.

The approach of finding new D-optimum designs in Case 3 by using a large number of trials with our computational search scheme succeeded in four cases where $n < 2k - 5$ (in addition to a number of very regular cases before we knew the Theorem of Section 3), as well as in the three cases $n = 2k - 5$ where it found nonunique optimum $X'X$. These are the cases $(k, n, s) = (9, 11, 6), (10, 11, 6), (11, 15, 8), (12, 15, 6), (6, 7, 5), (8, 11, 7), (10, 15, 9)$, listed in Table 1, mentioned in Section 2.4, and given in Tables 5, 6, 7, 8, 2, 3, 4. In all of these, our search succeeded in finding a D-optimum X in $\mathcal{X}(k, n)$, achieved for a block matrix $X'X$ satisfying det $X'X = \Psi(k, n)$. In these seven tables the designs X have been written so that the blocks of length 2 occur at the beginning, the above-diagonal 3's appearing in positions (1, 2), (3, 4), $\cdots$, $(2v - 1, 2v)$, where $v = 3, 4, 3, 6, 1, 1, 1$ in the respective cases. As mentioned in Subsection 2.3, we have also listed, in Table 9, three X's for $k = n = 11$ obtained by our computer search, which yield matrices $X'X$ for designs X found earlier by Ehlich and Zeller (1962) or Ehlich (1978), and which were proved optimum by the latter.

5. Summary of results for small k. The case $k = 2$, mentioned at the start of Section 2, is trivial and has been known for many years. For $k = 3$ and 4, Payne gave detailed calculations that characterize the D-optimum designs for all $n \geq k$. For these values of k, the developments of Sections 2 and 3 also guarantee existence and optimality of a regular X, always very regular in Case 3. The same is true for $k = 5$, which was treated almost completely by Payne (except for $n = 6$). Payne did not treat cases $k > 5$ so completely. For $k \geq 5$ his development assumes all H_n exist in Case 0, an assumption we have seen how to weaken, and to eliminate in all practical cases.

For $k = 6$, our results again imply existence and optimality of a regular X for all $n \geq k$, always very regular in Case 3. For $k = 8$ there is a regular X that is optimum for $n \geq k$, always very regular in Case 3. Tables 2 and 3 illustrate the lack of uniqueness of $X'X$ for $(k, n) = (6, 7)$ and $(8, 11)$.

For $k = 7$ we again have a regular D-optimum X for all $n \geq k$, except that in the saturated case the Williamson block design with $s = 4$, not regular (Table 1), is optimum. For all other values n in Case 3 we have very regular designs.

For $k = 9$ the saturated case was settled by Ehlich and Zeller (1962), as discussed earlier. Our developments yield regular optimum X for all $n > 9$, the only Case 3 design that is not very regular being that of Table 4 for $n = 11$.

For $k = 10$ we always have regular optimum X, always very regular in Case 3 except for $n = 11$. The design of Table 4e of Mitchell (1974b), after adjunction of a column of +1's and

TABLE 2

D-optimum X for $(k, n) = (6, 7)$

($X'X$ *has* $s = 5$, *first block of size* 2; *very regular X is also optimum*; $\det(X'X) = 2^{16} = .6554 \times 10^5$.)

+	+	+	−	−	+
−	−	+	−	−	+
+	−	−	−	+	+
−	−	+	+	+	+
−	−	−	+	−	+
−	+	−	−	+	+
+	+	−	+	−	+

TABLE 3

D-optimum X for $(k, n) = (8, 11)$

($X'X$ *has* $s = 7$, *first block of size* 2; *very regular X is also optimum*; $\det(X'X) = 3^7 2^{16} = .1433 \times 10^9$.)

−	−	+	−	−	+	+	+
−	−	−	+	+	+	−	+
+	+	−	+	−	−	+	+
+	−	+	+	−	+	−	+
−	−	−	+	−	−	+	+
−	+	+	+	+	−	−	+
−	+	−	−	−	+	−	+
+	−	−	−	+	−	−	+
+	+	−	−	+	+	+	+
−	−	+	−	+	−	+	+
+	+	+	−	−	−	−	+

TABLE 4

D-optimum X for $(k, n) = (10, 15)$

($X'X$ *has* $s = 9$, *first block of size* 2; *very regular X is also optimum*; $\det(X'X) = 2^{37} 3 = .4123 \times 10^{12}$.)

+	−	−	−	−	−	+	+	+	+
−	+	−	−	−	−	+	+	+	+
+	+	−	+	+	+	−	+	+	+
−	−	−	−	+	+	−	+	−	+
+	+	+	+	−	−	−	+	−	+
−	+	−	+	+	−	+	−	−	+
+	+	+	−	−	+	+	−	−	+
−	−	−	+	−	+	−	−	+	+
+	−	+	−	+	−	−	−	+	+
−	−	+	−	+	+	+	+	−	+
+	+	−	−	+	−	−	−	−	+
−	−	+	+	+	−	+	−	+	+
−	−	+	+	−	−	−	+	−	+
+	−	−	+	−	+	+	−	−	+
−	+	+	−	−	+	−	−	+	+

TABLE 5

D-optimum X for $(k, n) = (9, 11)$

($X'X$ *has* $s = 6$; *first three blocks have size* 2; $\det(X'X) = 3^4 2^{24} = .1359 \times 10^{10}$.)

+	−	+	+	−	−	−	+	+
+	−	−	−	+	+	+	−	+
−	−	−	+	+	−	+	+	+
−	+	−	−	+	+	−	+	+
−	−	+	−	+	−	−	−	+
+	+	+	+	+	+	−	−	+
−	−	+	−	−	+	+	+	+
+	+	−	−	−	−	+	−	+
−	−	−	+	−	+	−	−	+
−	+	+	+	−	−	+	−	+
+	+	−	−	−	−	−	+	+

TABLE 6

D-optimum X for $(k, n) = (10, 11)$

($X'X$ *has* $s = 6$; *first four blocks have size* 2; *Mitchell's design with* $s = 5$ *is also optimum*; $\det(X'X) = 3 \times 2^{32} = .1288 \times 10^{11}$.)

+	−	−	−	+	+	+	−	+	+
−	−	+	−	−	+	+	+	−	+
−	+	−	−	+	+	−	+	+	+
+	+	−	+	−	+	−	−	−	+
−	−	−	−	−	−	−	−	+	+
−	−	−	+	+	−	+	+	−	+
+	+	−	−	−	−	+	+	−	+
+	−	+	+	−	−	−	+	+	+
−	−	+	+	+	+	−	−	−	+
−	+	+	+	−	−	+	−	+	+
+	+	+	−	+	−	−	−	−	+

TABLE 7

D-optimum X for $(k, n) = (11, 15)$

($X'X$ *has* $s = 8$; *first three blocks have size* 2; $\det(X'X) = 2^{28}5^2 3^3 31 = .5617 \times 10^{13}$.)

+	−	−	−	−	+	+	−	+	+	+
+	+	−	+	+	−	−	−	−	+	+
+	+	−	−	−	−	+	−	+	−	+
−	+	+	−	−	−	+	+	−	+	+
+	−	+	+	−	+	−	+	−	−	+
−	+	−	+	−	+	−	+	+	+	+
+	−	+	−	+	−	−	+	+	+	+
−	−	+	+	−	−	+	−	−	+	+
−	+	+	−	+	+	−	−	+	−	+
−	−	−	−	+	+	+	+	−	−	+
−	−	+	+	−	−	−	−	+	−	+
−	−	−	+	+	−	+	+	+	−	+
+	+	−	−	−	−	−	+	−	−	+
−	−	−	−	+	+	−	−	−	+	+
+	+	+	+	+	+	+	−	−	−	+

TABLE 8

D-optimum X for $(k, n) = (12, 15)$

($X'X$ *has* $s = 6$ *blocks of size* 2; $\det(X'X) = 3^6 5^5 2^{25} = .7644 \times 10^{14}$.)

−	−	+	+	−	−	−	−	+	+	+	+
−	+	−	−	+	−	−	+	−	+	−	+
−	−	−	−	−	−	+	+	+	+	+	+
+	−	−	+	−	+	+	+	−	+	−	+
−	−	−	+	+	+	−	+	+	−	+	+
+	+	−	+	+	−	+	−	−	−	+	+
−	−	+	−	+	+	+	−	−	+	+	+
+	+	+	−	−	+	−	+	−	−	+	+
−	+	−	−	−	+	+	−	+	−	−	+
+	−	+	−	+	−	+	+	+	−	−	+
+	+	−	−	+	+	−	−	+	+	+	+
+	−	−	−	−	−	−	−	−	−	+	+
−	−	+	+	+	+	−	−	−	−	−	+
+	+	+	+	−	−	−	−	+	+	−	+
−	+	+	+	−	−	+	+	−	−	+	+

TABLE 9

D-optimum X's *for* $(k, n) = (11, 11)$; $\det(X'X) = 5^2 2^{32} = .1074 \times 10^{12}$

TABLE 9a. $X'X$ *has above-diagonal* 3's *in positions* (1, 2), (2, 3), (3, 4), (4, 5), (6, 7), (8, 9), (10, 11).

+	+	+	+	+	+	+	−	−	−	+
−	+	+	+	−	−	−	+	+	−	+
−	−	−	−	−	+	+	−	−	−	+
−	−	+	−	−	+	+	+	+	+	+
−	−	−	+	+	+	−	+	−	+	+
+	−	−	−	+	−	−	+	+	−	+
−	+	+	−	+	−	−	−	−	+	+
+	+	−	−	−	−	+	+	−	+	+
+	+	−	−	−	+	−	−	+	+	+
+	−	+	+	−	−	−	−	−	+	+
−	−	−	+	+	−	+	−	+	+	+

TABLE 9b. $X'X$ *has above-diagonal* 3's *in positions* (1, 2), (1, 3), (2, 3), (3, 4), (4, 5), (6, 7), (8, 9), (10, 11)

−	−	−	+	−	+	+	+	+	+	+
+	+	+	+	−	−	−	+	−	−	+
−	−	−	−	+	+	+	+	−	−	+
+	+	−	−	+	−	−	+	+	+	+
−	+	+	+	+	−	+	−	−	+	+
+	−	−	−	−	−	+	−	−	+	+
+	+	+	−	−	+	+	−	+	−	+
−	−	−	+	+	−	−	−	+	−	+
+	−	+	+	+	+	−	−	−	+	+
−	+	−	−	−	+	−	−	−	+	+
−	−	+	−	−	−	−	+	+	+	+

TABLE 9c. *$X'X$ is a block matrix with $s = 4$ and blocks of size* 5, 2, 2, 2 *in that order*

−	+	+	+	+	−	−	−	−	+	+
+	−	+	−	−	+	+	−	−	+	+
+	+	−	−	+	−	−	+	+	+	+
−	−	−	+	−	+	+	+	+	+	+
−	−	−	−	−	+	−	−	+	−	+
+	−	−	+	−	−	−	−	−	+	+
−	+	−	−	+	+	+	−	−	+	+
−	−	+	−	−	−	−	+	+	+	+
+	+	+	+	+	−	+	−	+	−	+
+	+	+	+	+	+	−	+	−	−	+
−	−	−	−	−	−	+	+	−	−	+

normalization, is proved by our developments to be optimum, since it is a block design with $(k, n, s) = (10, 11, 5)$ (see Table 1). Our search found Mitchell's design frequently, but also succeeded in finding a design with the alternate value $s = 6$, given in Table 6. Table 4 illustrates the lack of uniqueness of $X'X$ for $(k, n) = (10, 15)$.

For $k = 11$ there are the three $X'X$'s proved optimum by Ehlich, as described in Subsection 2.3 (see Table 9). For all $n > k$ we have a regular optimum X, the only Case 3 design that is not very regular being that for $n = 15$ given in Table 7.

For $k = 12$ our development yields regular optimum X for all $n \geq 12$, the only Case 3 value of n for which there is no very regular design being 15. An optimum design for $n = 15$ with $s = 6$ is given in Table 8. Ehlich (1978) has found the optimum design with $s = 7$ in this case.

For larger values of k of practical interest, the Proposition of Section 2 and our other developments yield regular D-optimum designs in all cases except $n \in \mathcal{N}_3$ with $k \leq n < 2k - 5$. The last 7 lines of Table 1 list these unknown cases for $k = 13, 14, 15$.

Added in proof. Additional Case 3 results obtained by us, to appear in the *Proceedings 1979 Tokyo Conference*, include D-optimum designs for the cases (12, 15, 7) and $(2k - 5, k, k - 1)$ for all k; and unattainability of $\Psi(k, n)$ for the cases (13, 15), (14, 15), (9, 11, 7).

REFERENCES

CHENG, C.-S. (1980). Optimality of some weighing and 2^n fractional-factorial designs. *Ann. Statist.* **8** 436–446.

EHLICH, H. (1964a). Determinantenabschätzungen für binäre Matrizen. *Math. Z.* **83** 123–132.

EHLICH, H. (1964b). Determinantenabschätzungen für binäre Matrizen mit $n \equiv 3 \bmod 4$. *Math. Z.* **84** 438–447.

EHLICH, H. (1978). Personal communication.

EHLICH, H. and ZELLER, K. (1962). Binäre Matrizen. *Z. Angew. Math. Mech.* **42** T20–T21.

GALIL, Z. and KIEFER, J. (1980). Time- and space-saving computer methods, related to Mitchell's Detmax, for finding D-optimum designs. *Technometrics.* To appear.

HEDAYAT, A. and WALLIS, W. D. (1978). Hadamard matrices and their applications. *Ann. Statist.* **6** 1184–1238.

HOTELLING, H. (1944). Some improvements in weighing and other experimental techniques. *Ann. Math. Statist.* **15** 297–306.

KIEFER, J. (1975). Construction and optimality of generalized Youden designs. In *A Survey of Statistical Designs and Linear Models.* 333–353. (J. N. Srivastava, ed.) North Holland, Amsterdam.

KIEFER, J. (1978). Optimal design theory in relation to combinatorial design. *Proc. Ft. Collins Conference on Design.* To appear.

MITCHELL, T. J. (1974a). An algorithm for the construction of "D-optimal" experimental designs. *Technometrics* **16** 203–210.

MITCHELL, T. J. (1974b). Computer construction of "D-optimal" first-order designs. *Technometrics* **16** 211–220.

MOOD, A. M. (1946). On Hotelling's weighing problem. *Ann. Math. Statist.* **17** 432–446.

PAYNE, S. E. (1974). On maximizing $\det(A^TA)$. *Discrete Math.* **10** 145–158.

RAGHAVARAO, D. (1971). *Construction and Combinatorial Problems in Design of Experiments.* Wiley, New York.

WILLIAMSON, J. (1946). Determinants whose elements are 0 and 1. *Amer. Math. Monthly* **53** 427–434.
WOJTAS, W. (1964). On Hadamard's inequality for the determinants of order non-divisible by 4. *Colloq. Math.* **12** 73–83.
YANG, C. H. (1968). On designs of maximal (+1, −1) matrices of order $n \equiv 2 \pmod 4$. *Math. Comp.* **22** 174–180.

DEPARTMENT OF MATHEMATICS
TEL AVIV UNIVERSITY
RAMAT AVIV, TEL AVIV, ISRAEL

STATISTICS DEPARTMENT
UNIVERSITY OF CALIFORNIA
BERKELEY, CALIFORNIA

Time- and Space-Saving Computer Methods, Related to Mitchell's DETMAX, for Finding D-Optimum Designs

Z. Galil
Department of Mathematical Sciences
University of Tel Aviv
Ramat Aviv, Tel Aviv, Israel

J. Kiefer
Department of Statistics
University of California, Berkeley
Berkeley, CA 94720

We develop a family of computer search methods for finding optimum designs, that generalize and improve upon Mitchell's exchange and DETMAX algorithms. Our emphasis is on time- and space-saving considerations that permit us to handle larger problems, or to run each search "try" for the optimum more rapidly, than was previously possible. This last means that more tries can be attempted for a given total cost, with consequent greater chance of finding an optimum or near-optimum design. Indeed, we have found a number of new optimum or improved designs using these methods. For $k = 6$ to 12 parameters and with n observations and $k \leq n \leq 2k$, our methods are typically 15 to 50 times faster than DETMAX (more as n and k increase), with comparable success rates. Numerical studies in linear and quadratic regression examples treat also the effect of amount of initial randomization on the success of a try.

KEY WORDS

Optimum designs
Computer search
Numerical optimization
D-optimality
DETMAX
Linear or quadratic regression
Fractional factorial designs
Weighing designs

1. INTRODUCTION

In a pathbreaking sequence of papers, Mitchell (1974a,b) developed and implemented a general technique, termed DETMAX, for obtaining D-optimum experimental designs in a wide variety of settings. It is a search technique. Exhaustive search in typical applications involves too many possible design matrices X. Moreover, in attempting to maximize $\det(X'X)$, all known usable techniques that move from an X to a nearby "better" X can get trapped in a neighborhood of a local maximum that is not the desired global maximum, and perhaps not even moderately efficient. One therefore introduces some randomization into the search technique, both in the initial guess, and also in later "tie-breaking", so that different "tries" can lead to different X's; thus, with enough tries, one can hope to find an X that is optimum or close to it. Mitchell's technique seems to us by far the most successful general method that has appeared for solving such problems.

We had hoped, over a period of several years, to improve upon DETMAX, e.g., by finding a way of "jumping" far enough out of a local maximum to escape from it in a usually favorable direction; or by adding, subtracting, or exchanging more than the one point per step that Mitchell does. These hopes proved fruitless, the first from lack of the right idea, the second because of astronomically increased computer time. Thus, with renewed respect for Mitchell's method, we were led to try to modify it in more modest ways, in terms of the actual computational steps it performs.

This involved a careful analysis of the individual operations it performs, especially in the updating of such entities as $(X'X)^{-1}$, that are used in improving the design successively. Section 2 describes a collection of methods capable of both time and space saving. We tested, principally, the method described in Sections 2.2, 2.5, and 2.7, and found in our examples that it was typically 15 to 50 times faster than the original DETMAX in problems with 5 to 10 parameters and 10 to 20 observations, on each "try" in

Received December 1978; revised February 1980

which a design X was found. One could therefore perform that many more tries than did DETMAX, for a given expenditure of computer time or money, and thereby increase greatly the chance of finding an improved solution in many problems. We did often achieve such solutions; this is discussed in Section 4. In Section 3 we list some of the characteristics of our methods, as exhibited in experimental run.

This paper is written entirely in terms of the D-optimality criterion of maximizing $\det(X'X)$. We shall not discuss the meaning and usefulness of this criterion, which has by now been the subject of a large number of papers. (See St. John and Draper, 1975, for a review.) From the viewpoint of computation emphasized herein, $\det(X'X)$ has advantages stemming from the simple "updating" formula treated in (2.1) and thereafter. The next simplest criteria are the "L-optimality" criteria of the form $trL(X'X)^{-1}$, where L is a given positive definite matrix; if L is the identity, this is "A-optimality". Some of the considerations of Section 2 can be carried over to such criteria.

Other competitors of DETMAX or the simple "exchange" method have been devised (e.g., Johnson and Last, 1975, Jones, 1976, and Rechtshaffner, 1967) for use in various settings. An excellent summary of such techniques appears in St. John and Draper (1975). We have not carried out detailed comparisons with all of these procedures, but believe our methods would compare favorably.

2. DESCRIPTION OF METHODS

2.0. *Introductory Comments and Definitions.*

We are given two integers q and n and a finite set $S = \{x^1, \cdots, x^N\}$ of N points in R^q. We assume that $N \gg n$. The ith coordinate of x is written x_i, and that of x^j is x_i^j. We are also given a vector $f = (f_1, \cdots, f_k)$ of k functions $f_1, \cdots, f_k$: $R^q \to R$ where $k \leq n$. Any choice of n (not necessarily distinct) points in S yields an nxk matrix X whose rows are the vectors $f(x) = (f_1(x), \cdots, f_k(x))$ for the n chosen x's. Our goal is to find n points in S that maximize, or come close to maximizing, the determinant of $M = X'X$ where X is the resulting matrix. (Of course, it is usually out of the question to look at all possible N^n choices.) We assume that the functions $\{f_i\}$ are easy to compute; more specifically we assume that it takes c ($c = 1$ or 2 in our examples) operations to compute and store $f_i(x)$. (Thus f_i depends only on a few components of x; it will be obvious what changes are needed if c varies greatly with i.)

Our modifications of DETMAX work for general problems, but we tested them on two particular examples:

1. *The linear example* on the corners of the cube, where $N = 2^q$ and the x^i are all q-vectors with coordinates ± 1. Here $k = q + 1$ and $f_1(x) \equiv 1$, $f_i(x) = x_{i-1}$ for $2 \leq i \leq q + 1$, $S = \{(\alpha_1, \cdots, \alpha_q) | \alpha_i \epsilon \{-1, 1\}\}$.

2. *The quadratic example* on the 3^q array, where $N = 3^q$ and the x^i are the points of $S = \{(\alpha_1, \cdots, \alpha_q) | \alpha_i \epsilon \{-1, 0, 1\}\}$, with coordinates -1, 0, or 1. Here $k = (q + 1)(q + 2)/2$ and the first $q + 1$ functions f_i are those of the linear case, while the remaining $q(q + 1)/2$ functions f_i are the quadratic monomials $x_s x_t$ for $1 \leq s \leq t \leq q$.

We will deal with a *family* of computational procedures that use the following approach for searching for the optimal n points: First an initial choice of n points is made (as described in Section 2.5) which yields a certain M. Then a sequence of "excursions" is performed. Each such excursion yields a possible new choice of n points which in turn yields a possibly new matrix M. The excursion is successful if the new M has a larger determinant. If the excursion is successful we start with the resulting new M and subsequently try to improve on it in the same way. Otherwise we may try a different excursion from the original M.

An excursion starts with a choice of n points and a certain matrix M. It either starts by adding a point or by subtracting a point. Then it either adds or subtracts another point, and so on, and it ends this excursion whenever there are exactly n points. So except at the beginning and the end of an excursion the number of chosen points n' at any state of the excursion is either always larger or always smaller than n.

Assume we are in the middle of an excursion and we have n' points. We will allow excursions that do not let n' differ from n by much (e.g., $|n' - n| \leq 6$). The reason for this restriction is to eliminate the possibility of very long excursions; also, Mitchell's experiments with DETMAX indicated longer excursions were usually not worthwhile. By subtracting a point we mean finding the point among the n' current points whose deletion will yield the largest value of $\det(M)$ among the n' possible choices. (In our examples we will subtract a point only if $n' > k$, but this may be altered, as discussed in Section 2.7). By adding a point we mean finding the point among the N which when added to the current n' points will yield the largest $\det(M)$ value.

Note that we did not specify the following: (1) how to choose the initial set of n points, (2) how to decide what to do next in an excursion (i.e., whether to add a point or subtract a point); (3) how to decide whether to terminate in case of an unsuccessful excursion, or to try another excursion; and (4) what to do in case of ties (nonuniqueness of best x^i to add or drop). The reason for not specifying these is that we are dealing with a family of procedures that may differ in the unspecified parts mentioned above. Examples of procedures that belong to the family are Mitchell's DETMAX (1974a) and its specialization to excursions of length 1, his "exchange method" (evidently developed independently by several other authors, motivated by the work of Wynn, 1970). Also, we shall describe a variety of methods for obtaining the initial matrix M, choice of which seems to affect the success rate and speed considerably. The notion of an excursion is due to Mitchell. Obviously, procedures of the above type are not guaranteed to find the optimum. They just find an approximation to the optimum, a "local optimum" whose efficiency might be good or bad. To improve the performance of such procedures one runs them several times, usually each time with a different starting choice. Such attempts are called *tries*. This motivates our work of trying to modify Mitchell's very successful method (we know of no better type of approach) to achieve faster tries, so that more attempts can be made for the same total cost.

We will call the search in an excursion, for the best point to add or subtract, a *step*. From the viewpoint of computational efficiency, we will be mainly interested in the case of *adding* points, because then we have to choose the best point among N candidates, whereas in subtracting a point we have to choose one out of only n' points where $n' \ll N$. If the number of "adding" steps is m, then the time required in an excursion is essentially proportional to mN. Since N is typically huge in such simple examples as the linear and quadratic ones if q is very large, it follows that we can solve problems with this approach only with q quite small. For example, Mitchell's DETMAX required 50 computer seconds per try on an IBM 360/91, for $q = 9$, $n = 12$ in the linear case. In the quadratic case the largest q considered by Mitchell and Bayne (1976) using DETMAX is 5.

What is even more frustrating is that when we increase q we increase N (in the examples N depends on q exponentially), and even if the machine has enough time per try we now usually need to make more tries in order to have a chance of getting a final M that is reasonably efficient, since there are so many more local maxima in which one can get "trapped" for large q. It is now clear why each "adding" step should be implemented with extreme caution so as to save every possible operation.

We describe below several ways of implementing a step. In order to save time we sometimes use large storage space to eliminate the need to recompute the same results several times. Consequently, some of the ways described, although faster, may break down because they need too much space. As a result we will not be able to point at one of the ways and claim that it is "best" for all problems. The best way will depend on the machine used and on the time or space available, as well as on details of the problem that determine the relative effect of employing additional storage space or of performing some additional computational task. Even in our simple linear example, one would vary the program as q becomes larger.

Throughout the descriptions that follow we use $O(L)$ to mean any function whose ratio to L is bounded as $L \to \infty$; typically L will depend on n, k, N in our usage.

2.1 The Simple Method.

We store a huge matrix of size $N \times q$ that includes all points in S. In each step we consider each point x of S and the corresponding row $y' = f(x)$ adjoined to X. Then we compute the determinant of the new M. We compare the N different determinants and choose the largest.

A very naive way to compute the determinant is to compute $M = X'X$ and to calculate its determinant. This approach takes $O(k^3 + nk^2)$ operations to compute one determinant and thus $O(nk^2N)$ per step. Obviously, one can use the known formula for computing the determinant of $M_y \equiv \begin{pmatrix} X \\ y' \end{pmatrix}' \begin{pmatrix} X \\ y' \end{pmatrix}$ from that of M, when a row $y' = f(x)$, corresponding to the point x, is adjoined to X:

$$\det(M_y) = \det(M)\,(1[+]y'M^{-1}y), \tag{2.1}$$

where $[+]$ is $+$ and should be replaced by minus in the case of deleting a row y from X. (We use $[-]$ for $-[+]$ similarly, below.) In order to be able to use (2.1) we need to maintain M^{-1} and update it as we add (or delete) a row y' to (or from) X:

$$M_y^{-1} = M^{-1}[-](M^{-1}y)(M^{-1}y)'/(1[+]y'M^{-1}y). \tag{2.2}$$

For each x in S we compute $\det(M_y)$ using (2.1) with $y' = f(x)$. When we find y that maximizes $\det(M_y)$ we use (2.2) to update M^{-1}. So this approach takes bk^2N operations per step, where b is a constant. (Actually, only about $bk^2N/2$ operations are needed because we can take advantage of the symmetry of the matrices involved.)

2.2 The Improved Method.

Obviously, by (2.1), finding x that maximizes $\det(M_y)$ is equivalent to finding y that maximizes (minimizes in case of subtracting a point) $y'M^{-1}y$ where $y' = f(x)$. In order to achieve this faster we maintain an array z of length N where $z_i = (y^i)'M^{-1}y^i$, and $y^i = (f_1(x^i) \ldots f_k(x^i))'$. (Recall, x^i is the ith point in S.) As a result the problem of finding the x that maximizes $\det(M_y)$ becomes trivial because it involves one scan of the array z (with N operations per step). But the main cost is now to update z. Following (2.2), we have

$$(y^i)'M_y^{-1}y^i = (y^i)'M^{-1}y^i[-]\frac{((y^i)'M^{-1}y)^2}{1[+]y'M^{-1}y}. \tag{2.3}$$

So once we find the y that we want to add to X so as to change M to M_y, we can compute $u \equiv M^{-1}y$ and

$\alpha \equiv 1 [+] y'M^{-1}y$ which are fixed with respect to y. To update z we need to subtract $(u'y^i)^2/\alpha$ from the entry corresponding to x^i, which takes about k operations, and thus we require a total of approximately $b'kN$ operations per step where b' is a constant.

The above method, when used with Mitchell's excursion rules (together with subsequent variations mentioned in Sections 2.5 and 2.7) will be termed modified DETMAX (*MD*). It was the procedure we used mostly, and numerical results herein are based on it, except for large q in the linear case.

Note that the methods of Sections 2.1 and 2.2 can be slightly modified by using a larger ($N \times k$) matrix to store $(f_1(x) \ldots f_k(x))$ for every x in S by computing these Nk elements once (at the beginning). This modification saves a factor c in the time bound. (Recall, the cost of computing $f_i(x)$ was c.) However for the linear case $c = 1$ and thus the modification does not save time. In the quadratic case since $c = 2$ (for most of the f_i) the modification saves time; but the space requirement that originally was Nq becomes $N(q + 1)(q + 2)/2$, which makes this version useless for moderate values of q. Nevertheless, this modification will be useful for small values of q in the quadratic case, and for certain applications other than the two examples that we consider. In our examples even the space required by the first two methods (namely, Nq) becomes prohibitive quickly (e.g., around $q = 8$ in the quadratic example). The next methods we will discuss need much less space but are not always applicable.

2.3 Space-efficient Versions of the First Two Methods

These versions are applicable when x^i is easily computable for $1 \le i \le N$. (By "easily" we mean using ($O(k)$ operations.) This is indeed the case for the linear and quadratic examples. In the linear example we produce x^i by writing $(i - 1)$ in binary form and replacing each 0 by -1. For example, if $q = 3$, the point $x^5 = (+1, -1, -1)$ because the binary representation of 4 is 100. Similarly, in the quadratic example to get x^i we write $(i - 1)$ in ternary form (base 3) and subtract one from each digit. Thus, if $q = 4$, the point x^{22} is $(-1, 1, 0, -1)$ because 21 written in ternary form is 0210. This assumption of easily computable x^i's holds for other applications when the points of S form a regular grid; what is essential is the presence of a simple way of "listing" the x^i's in terms of an easily computable function $i \to x^i$ from $\{1, 2, \cdots, N\}$ onto S.

Since the ith point of S can be computed at will, we do not need to store the $N \times q$ matrix, but compute each of its rows when needed. Consequently, this version needs only $O(k^2 + qn)$ units of space if the method of Section 2.1 is used. (The k^2 is for the $k \times k$ matrices and the qn is for the n chosen points.) In fact the space needed is only $O(k^2 + n)$ units because we need not store the points x^i but only their indices i. If the method of Section 2.2 is used, the space required is $O(N + k^2 + n)$ (the N being for the vector z).

The space-efficient versions of the first two methods are slightly lower than the unmodified methods (e.g., about 15% slower in the linear example for q near 10). So, if space permits, the latter are preferable.

The space-efficient version of the faster method may prove to be insufficient because we still have to store the huge vector z. In the next section we explain how in some cases we need not store z, and how we can thus reduce the required space considerably.

2.4 Another Space-efficient Method.

This method is applicable when (a) x^i is easily computable and (b) $(y^{i+1})'$ (the row added to X which corresponds to x^{i+1}) does not differ by much from $(y^i)'$ "on the average." (We intentionally use a vague phrase here, but its precise meaning will be indicated in an example.) The conditions (a) and (b) hold for the linear and quadratic examples. We explain below the linear example, and the extension to the quadratic example is clear.

To implement the method of Section 2.2 without storing the $N \times q$ matrix and even without storing the vector $z = \{(y^i)'M^{-1}y^i, 1 \le i \le N\}$, we will compute $(y^{i+1})' M^{-1}y^{i+1}$ from $(y^i)' M^{-1}y^i$ as follows. Instead of M^{-1}, we maintain the matrix M^y, whose (s,t)th coordinate is defined by

$$M_{st}^y = y_s y_t (M^{-1})_{st} \tag{2.4}$$

where y_s is the sth component of the candidate y for adjoining to X. Note that

$$z_h = \sum_{s,t} M_{st}^{y_h}. \tag{2.5}$$

Initially $y_1 = (1, -1, \cdots, -1)'$ and we compute z_1 and M^y directly. We now show how to compute $M^{y_{h+1}}$ and z_{h+1} from y_h and z_h.

Consider the binary representation of h. It must end with $011 \cdots 1$, a zero followed by r 1's for some $r \ge 0$; that is, r is defined to be the number of 1's after the last 0 in this representation. Thus the binary representation of $h + 1$ must end with $100 \cdots 0$, a 1 followed by r zeros; and the other digits must be the same as those of h. Consequently, the only changes from y^h to y^{h+1} in this linear example are that the last $r + 1$ components are multiplied by minus one.

For fixed i, consider two k-vectors u and v such that $u_j = v_j$ for $j \ne i$ and $u_i = -v_i$. We then have, from (2.4),

$$\begin{aligned} M_{ji}^u &= M_{ij}^u = -M_{ij}^v \quad \text{for} \quad j \ne i, \\ M_{st}^u &= M_{st}^v \quad \text{for} \quad i \notin \{s, t\}, \end{aligned} \tag{2.6}$$

and consequently, from (2.5), writing $z_{(y)} = y'M^{-1}y$, we have

$$z_{(v)} = z_{(u)} - 2 \sum_{j \neq i} M_{ij}^{u} . \tag{2.7}$$

So, to compute $M^{y_{h+1}}$ and z_{h+1} ($= z_{(y_{h+1})}$) from M^{y_h} *and* z_h ($= z_{(y_h)}$) we just repeat (2.7) for $i = k, k - 1, \cdots k - r$, because the last $r + 1$ components of y^{h+1} are (-1) times the corresponding components of y^h, while the other components are the same. (In these $r + 1$ uses of (2.7), at each stage v is a vector one digit less different from y^{h+1} than is u, and the v of one stage becomes the u of the next.)

To help the update, we will maintain the binary representation of h. Obviously (2.7) involves $O(k)$ operations. This bound has to be multiplied by $r + 1$ (the number of bit changes). Although r can be as large as $q - 1$, the sum of the $2^q - 1$ values of $(r + 1)$ as h goes from 0 to $2^q - 2$ (for which the largest value of $h + 1$ is $h + 1 = 2^q - 1$), is $\sum_{r=0}^{q-1}(r + 1)2^{q-1-r} < 2^{q+1} = 2N$. (Thus, the "average" change referred to earlier is approximately 2 for moderate or large q.) Consequently we obtain running time of $O(kN)$ per step and space $O(k^2 + n)$ as before.

In the quadratic case a similar but more involved scheme may be devised. In our experiments with the quadratic example, we found it unnecessary to use because in this case we ran into time trouble first. Similar methods may be devised for other cases when the points of S are distributed "regularly".

This method is slightly (approximately twice) slower than the space-efficient version of the method of Section 2.2, as given in Section 2.3. Hence if one has enough space to use the latter he should use it, but the present method is useful for problems where space is the limiting factor.

2.5 *The Choice of Initial Design.*

Mitchell (1974a, Section 2.5) mentions having studied various nonrandom designs but having abandoned them in favor of a completely random start on each try. (Results such as those for $n = 12$, $q = 7$, described in Section 3.4, may typify what motivated Mitchell's conclusion.) We studied a family of starting procedures which vary from completely random to completely determined (except for ties). The determined procedure starts with a row k-vector X and adjoins rows to X, one by one, in a manner that attempts to maximize the determinant at the stage where X is square. Formally, this procedure operates as follows: after j rows of X have been chosen for $j < k$, yielding a $j \times k$ design matrix $X^{(j)}$ (say), choose the next point x^{j+1} so that the corresponding $y' = f(x^{j+1})$ maximizes (for $j = 0, 1, \cdots, k-1$)

$$\det\begin{pmatrix} X^{(j)} \\ y' \end{pmatrix}\begin{pmatrix} X^{(j)} \\ y' \end{pmatrix}' \tag{2.8}$$

Thus, roughly, we consider $X^{(j)}X^{(j)\prime}$ rather than $X^{(j)\prime}X^{(j)}$ because the latter is singular for $j < k$. The formulas of the earlier part of this section, such as (2.1), can be implemented with appropriate changes. (The updating formulas are altered because the matrix size is no longer constant.) Since the k steps needed to reach a nonsingular $X^{(k)}$ (together with the $n - k$ additional steps needed to reach an "initial" $X^{(n)}$ by considering $X^{(j)\prime}X^{(j)}$ for $j \geq k$ and adding X^{j+1} as described earlier) are usually of smaller magnitude than the number of steps in the excursions of a typical try, this nonrandom start does not noticeably increase the time per try over what it would be for a randomly chosen initial $X = X^{(n)}$. In fact, the statistics of Section 3.4 indicate that random starts generally yield more steps per try, and thus tries that take longer.

Except for the randomness introduced if several (x^{j+1})'s yield y's maximizing (2.8) (or, later, (2.1)), so that one must choose among them, there is no random element in the above start and subsequent excursions. Without randomness, a method is deterministic, and all tries lead to the same final X. Thus, it seems advisable to consider choosing some random number n_r of points at the outset (we generally choose $n_r \leq k$) and then using (2.8) and (2.1) for $j = n_r$ and thereafter. (The opposite procedure, of using (2.8) up to some stage j and *then* adding additional random points, did not appear as useful.) In Section 3.4 we shall give some numerical evidence that, at least in the linear examples, a start that is deterministic or close to it is usually fairly successful. We found that n_r should be chosen considerably smaller than k for moderate k in most examples, to achieve much more success than that obtained from a completely random start, and to achieve it more quickly.

We remark that, in the linear example, when $k < n$ one could work entirely with the n-vectors that are *columns* of X and, letting $X^{[j]}$ be an $n \times j$ matrix, with $j < k$, build up from $X^{[j]\prime} X^{[j]}$ to $X^{[j+1]\prime} X^{[j+1]}$ using an analogue of (2.8). This has the disadvantage of replacing $N = 2^k$ by $N = 2^n$, thinking of columns of X as points of $\{-1, 1\}^n$. Excursions up or down, as mentioned in Section 2.7, could be employed. For models other than that of the linear example, this device of interchanging rows and columns could not be employed without extensive further modifications, since the columns of X do not represent *points*, and going from $X^{[j]}$ to $X^{[j+1]}$ involves adjoining *functions*. Nevertheless, the method might at least be useful for starting in the linear example if n is not too large. It could be used to construct the initial $n \times k$ matrix X on which the methods of Sections 2.1–2.4 are used. This is in the spirit of trying an occasional alternate attack rather than of using more tries with a more economical method which may be proving unproductive of efficient solutions.

2.6 Families of Designs for Varying n.

It will often be the case in practice that for a given problem (S, $\{f_i\}$) and thus a fixed k and set of possible y's, one would like to list good designs for an interval of values $n \geq k$. Clearly, considerable time is saved if one uses the result for the problem (k, n) in the determination of a design for (k, n') where $n' > n$. This can be done in several ways. For example, if n' is near n (say $n' - n = 1$ or 2), the best value of $\det(X^{(n')\prime}X^{(n')})$ over all excursions for the n-observation problem can serve as a solution or the start of a try for the n'-observation problem.

Other variations are obvious. Although it will usually only be feasible to treat n' near n in this way (due to limits on excursion length), more distant n'' can then be based on the results obtained for the n'-observation problem with n' near n, etc.

The excursion rules of Mitchell (1974a) that we used are geared to a given value of n (e.g., in deciding when an excursion is "successful" and $X^{(n)}$ is to be replaced by a new value), so routines of the type just described need not yield results for $n' > n$ equivalent to those obtained if one uses our *MD* methods for n' observations. This is so even if, in the latter, one uses the initial $X^{(n)}$, which could conveniently be employed in building up the initial $X^{(n')}$ (which is often what occurs in the first excursion for the n-observation problem). The numerical results in this paper do not use this device; the conclusions thus have the extra precision of not being based on correlated tries.

2.7 Other Differences from Mitchell's Study.

Since our main interest was in achieving more tries per unit of time or money than did previous methods, we concentrated on the computational features described in Sections 2.1–2.4, and did not modify our methods to include such features as those found in Mitchell (1974a, Sections 2.6–2.8), of "protecting a subset of runs" (augmenting a given, fixed design) or "the sequential option" of obtaining a sequence of designs, one for each n, each $X^{(n)}$ being obtained by augmenting the $X^{(n-1)}$ of the sequence by addition of a row. Since the discussion of Sections 2.1–2.4 herein is concerned with time or cost reduction for a single step, it is clear that our methods can be employed also in these and other variations discussed by Mitchell and other authors. In our illustrations we have not included printouts of the final $(X'X)^{-1}$ or maximum of variance function, although in actual use one would obviously include such features. We have emphasized in previous papers, as have many other authors, that choice of a design for actual use should not be based on a single criterion.

We have also, with Mitchell, omitted long discussions of how to approach problems when S is infinite. Some sort of additional search routine, perhaps on a grid as suggested by Mitchell (1974a, Section 2.3), perhaps with additional introduction of some "descent" or other continuum search method, would have to be used with the methods we have described.

Finally, we have not used the technique of Mitchell (1974a, Section 2.4) of avoiding singularities of $X'X$, on excursions that lead to an intermediate X with fewer than k rows, by adding a small nonsingular matrix A to $X'X$ and choosing the next point (to add or subtract) with reference to $\det[A + \binom{x}{y}'\binom{x}{y}]$ in place of (2.1). Once the process was started (as described in Section 2.5), we generally never let the number of rows of X be reduced to less than k; in particular, in the saturated case our excursions were only "one-sided," increasing the number of rows from the k rows of $X^{(k)}$ and eventually returning to a new $X^{(k)}$, but never taking an excursion among $X^{(j)}$ with $j < k$. This did not seem to have a negative effect on the success rate of the searches with our "modified DETMAX," as was shown by comparisons with Mitchell's results (Section 3). Computationally, we thus used a simpler routine in cases where $n - k$ was small (close to the saturated case $n = k$) than in those where $n - k$ was large; interestingly, the former cases are those which are much harder to treat algebraically, in terms of *proving* optimality of particular designs. If desired, excursions among $X^{(j)}$ with $j < k$ are easily injected into the program without using Mitchell's device, by simply using (2.8) (or an obvious modification when a point is to be dropped) in place of (2.1).

3. COMMENTS ON THE METHODS OF SECTION 2, AND COMPARISONS WITH DETMAX

3.1 General Remarks.

Mitchell used the simple method of Section 2.1 to attack the linear and the quadratic examples. As a result, in the time he allowed he could solve problems with at most $q = 9$ in the linear case and $q = 5$, $n = k$ (saturated) in the quadratic case. As we noted, as q grows we generally need more tries to obtain reasonably good results. It turned out that because he could not afford to run many tries Mitchell did not find the optimum for the linear case $(q, n) = (8, 11)$ and the $q = 5$ saturated quadratic.

We used the improved method described in Sections 2.2, 2.5, 2.7, with Mitchell's rules for limiting excursions, to attack the linear and quadratic examples. As a result we could treat problems with $q \leq 12$ in the linear example and $q \leq 7$ in the quadratic example, with computing facilities and time comparable to Mitchell's. In the linear (8, 11) case we achieved a design better than Mitchell's (that in Galil and Kiefer, 1980, we show must be optimal). In the $q = 5$ quadratic saturated case we matched a design obtained algebraically by Notz (1978) that improves upon that of Mitchell and Bayne (1975). In the quadratic $q = 6, 7$ saturated cases we obtained designs better than those of Notz (which used to be the best known). The reason we could not increase q over 12 in the linear case was that the space requirement became prohibitive. So in order to treat larger q we used a space-efficient version of the method of Section 2.2, as described in Section 2.3. We did not need the full savings of the method of Section 2.4. Consequently we were able to treat the linear case with $q = 13, 14, 15$. For example in the $(q, n) = (13, 14)$ case (the saturated case) we obtained the design proved optimum by Ehlich (1964a). (For $q \le n \le 14$ we knew the theoretical optimum [Galil and Kiefer, 1980]; but this range still provided useful test cases.)

We shall describe some more detailed numerical results concerning the methods in the remainder of Section 3.

3.2 Speed of the Methods.

Our computations were performed primarily on Cornell's IBM 370/168 computer. Because of possible differences in machine setup, our results would not have been easy to compare with Mitchell's were it not for the fact that he kindly loaned us his program so that we could try the original DETMAX and our MD method (Sections 2.2, 2.5, last paragraph of 2.7) on the same machine. A typical result is that the average length of time per trial in the linear example with $q = 5$, $k = 6$, $n = 12$ was about 1.855 seconds for DETMAX and .10 seconds for our modification. Roughly speaking, in this example the factor by which we improve on DETMAX is proportional to k, so it is about 3.1 k for moderate to large k. Thus, when $k = 12$ we could afford about 37 times as many tries for the same expenditure of money or time, which is why we could attack such cases with MD when they would have been too expensive for DETMAX.

The time trial just described was based on 100 tries of our procedure (of which 51 found the absolute optimum $\det(X'X)$, comparable to the 44.5% of the larger study of Section 3.3), and 10 tries by us of DETMAX (of which 1 found the optimum, compared with 2 out of 10 in Mitchell's study). Our tries chose $n_r = 1, 2$ or 3 at random (see Section 2.5), which saved slightly over Mitchell's completely random choice (discussed at the end of Section 3.4). The comparison of success rates here and in Section 3.3 makes us conclude that the indicated time comparison contrasts DETMAX with a faster MD method that is at least as successful as DETMAX in terms of the designs obtained.

Incidentally, Mitchell's machine gave 2.3 seconds per trial in place of the above-mentioned 1.855, so that time comparisons between results on the two machines should turn out to be at least somewhat meaningful in other examples.

We remark that, perhaps surprisingly, the simple exchange method often took as much as 60% as much time as DETMAX. Evidently DETMAX sometimes spends much of its time on many very short excursions.

Further time comparisons, relating to the choice of n_r, are contained in section 3.4.

3.3 Success Rates for Modified DETMAX and Modified Exchange Methods.

Mitchell (1974b) intelligently chose the linear examples with $n = 8, 12, 16, 20$ for assessment of the success rate of his DETMAX, these being cases in which an optimum design X is known for all $k \le n$, being any k (orthogonal) columns of an $n \times n$ Hadamard matrix H_n. Table 1 of Mitchell, repeated here for convenience as Table 1a, lists his results for 10 tries in each case (20 in 2 cases), and Table 1b herein gives our comparable results for 200 tries of our modified DETMAX (MD) method, extending somewhat the range of cases studied. For $q = 2$ or 3 we were always successful, just as Mitchell was. For larger q, the rates are generally comparable, with each being better in some cases. The effect of n_r, discussed in Section 3.4, accounts for part of this difference. Perhaps our larger number of tries makes our results slightly more meaningful, for the choices of n_r we made. We remind the reader that $k = q + 1$.

One feature to be noted, especially for $q = 6$ and 7, is the tremendously oscillatory behavior in n. This indicates how problem-dependent the success of any search method will be. Even within this regular family of "nice" Hadamard cases, the existence of simple forms of H_8 and H_{16} compared with H_{12} and H_{20} evidently has a number-theoretic relationship with the fact that the "geometry" of the graph of $\det(X'X)$ on S is much more complicated in the latter two cases, exhibiting local maxima from which it is more difficult for search excursions to extract the process than is the case when $n = 8$ or 16. Further aspects of this are discussed in connection with the choice of n_r in Section 3.4. When we investigate some of the less regular cases (k, n), let alone other examples, we

TABLE 1a—*Number of successes in 10 tries of DETMAX (Table 1 of Mitchell, 1974b); * signifies number of successes in 20 tries.*

n \ q	4	5	6	7	8	9
8	10	7	10	10	-	-
12	7	2	1	2	2	1
16	7	4	1	1*	0*	-
20	9	2	1	1*	-	-

TABLE 1b—*Number of successes in 200 tries of MD and 10 of ME in linear example. Main entry is number of successes in 200 tries with MD; * signifies only 100 tries were made; (e, e^k) refers to efficiency (3.1) of best design obtained using MD when optimum design was not obtained in any trial; [j] gives number of successes in 10 tries with ME.*

n \ q	4	5	6	7	8	9	10	11
8	195 [8]	139 [7]	119 [7]	141 [5]	-	-	-	-
12	166 [8]	89 [2]	14 [0]	9 [0]	7 [0]	2 [0]	33 [2]	65
16	194 [10]	116 [5]	180 [10]	153 [8]	21 [0]	14 [1]	(.988,.875) [0]	(.978,.772)*
20	194 [7]	109 [4]	10 [0]	10 [1]	1 [0]	(.996,.960) [0]	(.993,.922) [0]	(.983,.811)*

must be aware that the numerical results tabled in this paper are at best rough guides.

The 5 entries in the lower-right corner of Table 1b refer to cases in which our method was never successful. In those cases, the efficiency of the best design $\tilde{X}$ obtained is defined as

$$e(\tilde{X}) = [\det(\tilde{X}'\tilde{X})/\max_X \det(X'X)]^{1/k}. \quad (3.1)$$

The reason for taking the kth root in (3.1) is to reduce the comparison to units of variance, as one automatically does when using criteria such as average or maximum variance. This seems more meaningful than the ratio e^k of the determinants, themselves, but in Table 1b we list (e,e^k) in these 5 cases. (Throughout this paper we have avoided the common practice, when the denominator of (3.1) is unknown, of replacing it by its upper bound obtained from the "approximate theory." This last gives a lower bound on e which is often so unattainable as to be misleading.)

Also listed, in square brackets in Table 1b, are the success rates for 10 tries in each of these cases except $q = 11$, using our modification (Section 2.2) of the exchange method (ME) described in Section 2.0, in which, once the initial $X^{(n)}$ is obtained, each subsequent change (excursion of length only 1) amounts to making an exchange of one point of S already in use, for another; i.e., it reduces to exchanging the row of X which is least "beneficial" to det $(X'X)$ for a new row y. This method compared surprisingly well with DETMAX in terms of our Table 1b, or Mitchell's (as it also did for n outside the "Hadamard" cases), and required slightly less time per try and a simpler program (eliminating various excursion rules). Thus, we would certainly recommend that those approaching a new family of design problems investigate the efficacy of the exchange method relative to DETMAX in a few cases, before deciding between the methods. It may be, as in our linear examples, that the local maxima are such that the longer excursions of DETMAX have little more success in "escaping" from them than does the simpler exchange method; or the problem may be such that longer excursions do help.

In the tries used for Table 1b, n_r was chosen at random between 1 and min(q,6). The upper bound on n_r was used because of earlier studies that indicated use of larger n_r to be less useful (Section 3.4). Two groups of 100 tries each were performed separately before combining them, as a partial check on effects of the random number generation; the two groups yielded similar results.

3.4 Effect of Amount of Initial Randomization.

We described our routine for starting the program, to obtain an initial $X^{(n)}$, in Section 2.5. To obtain a better idea of the effect of the value of n_r, the number of points in S chosen at random before invoking the prescription described by using (2.8), we made an extensive study encompassing approximately 3300 tries in the saturated linear example $k = n = 11$. We chose that case because it is a more difficult one theoretically than the Hadamard cases (n divisible by 4) of Table 1b. For each possible value of n_r, $1 \le n_r \le 11$, we ran 300 tries; three sets of 100 tries performed at different times were checked for homogeneity before combining them. (A total of 8 out of the 3300 tries failed because of machine problems.) In this simple linear example, we note that choosing $n_r = 1$ has essentially the same effect as using no randomization ($n_r = 0$), because the symmetry of the problem as given by the functions f_i and space $S = \{-1,1\}^{11}$ is such that all choices of the first point x^1 lead to equivalent final X's (with the same value of det$(X'X)$) if there is no subsequent randomization. The reason that there is any spread in values when $n_r = 0$ (or 1) is because of the randomization in case of ties in the start and subsequent excursions. If such randomization were avoided by always choosing (say) the *first* y

that maximized (2.8) for a fixed ordered listing of S, the choice $n_r = 0$ (first point always the same) would yield no dispersion of results, while $n_r = 1$ might still yield a dispersion.

Table 2 gives, in each column, the frequency distribution of values obtained for $\det(X'X)$ on the 300 tries for a given value of n_r. The theoretical maximum, verified by Ehlich (1978), is $25 \times 2^{32} = 1.074 \times 10^{11}$. We therefore grouped the outcomes into eleven intervals $[j \times 10^{10}, (j+1) \times 10^{10}]$, $0 \leq j \leq 10$. In addition, we give, in the bottom row, the number of times the actual maximum 1.074×10^{11} was achieved for each value of n_r.

Several interesting features are exhibited in Table 2. At the start, we note that this was a very difficult problem to *solve completely* (attaining the maximum) by our MD search (or, we believe, by any similar method): about 1% of the tries achieved the maximum, and for no n_r was this rate better than 2%. However, the usefulness elsewhere of DETMAX or MD will be in examples where we do not know the maximum as we did in the present example (chosen to permit exact comparisons). Hence, the frequency of values of $\det(X'X)$ that come reasonably close to the maximum is likely to be much more important than that of achievement of the maximum.

What suffices as "close" matters here, but it turned out in the example that, qualitatively, there is little difference between comparing values of n_r in terms of the proportion of tries that achieved at least 95% or 90% or even 75% of the maximum. The three rows 8-9, 9-10, and 10-11 of the table indicate this, and the last of these is most revealing. Values of $n_r < 3$ show up very well, and there appears to be little reason to use a value $n_r > 6$. In particular, the fully randomized start ($n_r = 11$), although it achieved the actual maximum once, performed much worse than the MD method with $n_r \leq 3$. Most of the results in the 10-11 row for small n_r were at the value 1.040×10^{11}, almost 97% of the maximum (efficiency $e = .997$).

As was indicated earlier, the spread in the case $n_r = 1$ (linear example) is due to randomization in the case of ties. The extreme bimodality of the distribution, which is almost as extreme for $n_r = 2$, may indicate the need for caution in other examples. For instance, one could easily encounter examples in which very few ties arise, and in which the choice $n_r = 0$ or 1 (or even $n_r = 2$) almost always yields the same result, perhaps much less than the theoretical maximum. Also, as mentioned earlier, the details of the S and f_j in the problem being studied yield details of the geometry of the graph of $\det(X'X)$ on S that will affect the success of MD; for example, there are local maxima in the 6-7 row of Table 2 that are easily achieved (and easy for a search excursion to become "stuck" in) for all n_r in the present example, while those in the 10-11 row are difficult to attain with much initial randomization; other local maxima are only achieved frequently as n_r increases.

For a further study of the effect of initial randomization, which illustrates the problem dependence just mentioned, we looked at 100 tries for each value $1 \leq n_r \leq \min(10,k)$ in the Hadamard cases $n = 12,16$ of Table 1b, for $7 \leq q \leq 9$. We have not combined the

TABLE 2—*Frequency of determinant values for various values of n_r, linear example, $k = n = 11$. (*296 tries; **298 tries; ***299 tries; all other n_r had 300 tries.)*

$10^{-10}\det(X'X)$ \ n_r	0,1*	2	3	4	5	6	7	8***	9***	10	11**
0-1					1		3		11	24	76
1-2											
2-3					2	2	1	4	2	2	3
3-4			1	5	9	14	20	21	28	33	22
4-5			10	18	36	53	56	66	70	48	52
5-6		3	13	50	56	63	62	69	65	69	48
6-7	133	147	109	103	90	98	103	79	78	78	54
7-8	4	21	52	50	46	42	29	44	25	24	17
8-9	1	3	18	28	31	18	19	6	17	14	14
9-10	1	4	4	12	13	2	5	7	2	7	9
10-11	157	122	93	34	16	8	2	3	1	1	3
OPTIMUM ATTAINED	3	1	6	6	4	5	1	1			1

success rates with those of Table 1b, but they were consistent. In particular, the rate of achieving the actual maximum when $n = 16$ was much higher for $n_r \leq 4$ than for larger n_r.

For the case $n = 12$, whose greater difficulty we described in Section 3.3, the results were somewhat different. We shall describe in detail only the case $q = 7$ ($k = 8$), $n = 12$, although similar results were noted when $q = 9$ (with less such effect when $q = 8$). The number of times the theoretical maximum was achieved for $n_r = 1,2,\ldots,8$ was 0,0,8,4,7,16,10,7 (the last out of 64 tries); the "success-rate" for attainment of $e = 1$ in the range $1 \leq n_r \leq 4$ was thus only 27% of what it was in the range $5 \leq n_r \leq 8$.

An analysis of the output of each individual try in this example revealed that the final $\det(X'X)$ obtained was 2.684×10^8 ($e = .943$, $e^k = .624$), compared with the maximum of 4.300×10^8, *in 100% of the cases* for $n_r = 1$, 88% of them for $n_r = 2$, 50% for $n_r = 3$. Thus, although a sequence of eight q-vectors *can* be chosen in order with use of (2.8) so that subsequent use of (2.1) yields an optimum 12×8 design X, the choice among ties for small n_r *usually* (and *always* when $n_r = 1$) resulted in a sequence that at some stage "blocked" against eventual production of an X consisting of eight orthogonal 12-vectors, instead leading to formation of an easily attained X with $\det(X'X) = 2.684 \times 10^8$. The difference between this case, or that with $q = 6$, and the nearby case $q = 5$, is striking (see Table 1b and the discussion of Section 3.2). The geometry for $n = 12$ makes search much less successful for medium q than for large or small q, and the geometry for $n = 16$ allows the fully orthogonal X to be attained more easily in that case than for $n = 12$; evidently ties do not yield as much actual randomness in the final X when $n = 12$ as they do when $n = 16$ (or in the earlier $k = n = 11$ study). The other side of the randomization picture (illustrated also in the last column of Table 2) is that, when $q = 7$, the choice $n_r = 8$ yielded half the tries with $e \leq .4$ ($e^k \leq 1/4000$), many of them with zero determinant; however, if enough tries are undertaken, as was the case here, the presence of many such inefficient outcomes is less objectionable than are the results described for $n_r = 1$.

Repeating our look at the oscillatory behavior in n for $q = 6$ or 7 in Table 1b, it is striking to realize that (2.8) *starts* all these cases in the same way for a given n_r value. Thus, it is the rarity of paths leading to an optimum X that is accentuated in the effect of choice of n_r in these cases, as n varies.

A good choice of n_r is thus *very* problem-dependent. The considerations discussed above make it advisable, in other problems, to perform a few preliminary tries with varying n_r, if costs permit this, to affirm or alter the conclusions of Table 2 or those just described for the linear $q = 7$ cases or the $n = 12$ cases. What if one cannot afford such preliminary experimentation? We feel that cases like the $n = 12$ or $q = 6$ or $q = 7$ cases are *not* what one usually encounters, and other linear and quadratic examples indicate that choosing n_r approximately equally frequently among values $0 \leq n_r \leq 4$ or 5 will usually be satisfactory, at least when k is not too large (say $k \leq 15$), while larger n_r often lead to results like those shown in Table 2; tries with larger n_r are often costlier, too, as we indicate at the end of this subsection. For k as large as 30, we would tend to let n_r be slightly larger, perhaps $0 \leq n_r \leq 6$ or 7.

This last is based on our results in the quadratic examples, particularly the saturated cases with $q = 5,6,7$ ($n = k = 21,28,36$). We generally chose n_r at random between 1 and $k/2$ in our exploratory work. We do not know the theoretical maximum in any of these three quadratic cases, but complete or nearly complete initial randomization (n_r near k) always yielded poor results. For $k = 21$, a relatively "nice" case theoretically (as discussed in Section 4.2), 200 tries with n_r chosen at random between 1 and 11 yielded 25 results within 1% of the largest value of $\det(X'X)$ achieved, and 15 of those 25 tries were for $n_r \leq 3$ (three of them being for $n_r = 1$). Comparable results held for frequencies of slightly less efficient outcomes. For $k = 28$, with n_r chosen at random between 1 and 14, in 230 tries the maximum $\bar{m}$ attained was achieved once, when $n_r = 6$, and a determinant at least 70% as large (e at least .987 as large as for $\bar{m}$) was achieved 3 other times, once for $n_r = 3$ and twice for $n_r = 1$; of the 19 tries with resulting determinant at least 40% as large as $\bar{m}$ (e at least .967 as large), 13 had $n_r \leq 5$, of which 5 had $n_r = 1$. For $k = 36$, with $1 \leq n_r \leq 17$, values $n_r \leq 3$ were most successful, the best determinant being achieved for $n_r = 1$. We note that, in the quadratic example, not all choices of a first point are "equivalent" as they were in the linear case. The use of $n_r = 1$ (quite successful) with probability $(2/3)^q$ chooses a first point with all coordinates $= \pm 1$ (corner of q-cube), of which we know there will be many in a good design.

Unfortunately, too little is known about the geometry of $\det(X'X)$ on S, even in the linear example, for us to be more definite about an "efficient" prescription for choosing n_r. Choosing all possible n_r's with equal frequency avoids disaster at considerable cost. Choosing n_r near k will almost always be poor if k is large, but for $k \leq 15$ our examples indicate that choosing n_r small (1,2,3) half the time and large (k,k-1,k-2) half the time would have protected against the somewhat opposite bad values of n_r of Table 2 and of the $n = 12$ case. But there is no reason to think that prescription is always reasonably productive, either.

The cost of different values of n_r is also important. The total length of time required for 600 tries, 100 for each of the cases $7 \leq q \leq 9$, $n = 12,16$, increased, slightly more rapidly than linearly, from 9:48 minutes for $n_r = 1$ to 14:14 minutes when $n_r = 7$. Of this, over 60% of the time was spent on the $q = 9$ cases. In some cases of larger k, the time seems to increase sharply for larger n_r, indicating that very random starts may sometimes lead to long excursions.

4. EXAMPLES

We now discuss briefly the results of our use of MD in the two examples listed in Section 2.0.

4.1 Linear Example.

As a consequence of some theoretical developments together with computer search, we were able in Galil and Kiefer (1980) to give a list of theoretically-proven optimum designs for $k \le 12$ and all $n \ge k$. This considerably extends all previous lists of optimum designs. In particular, we were able to prove that the design of Mitchell (1974*b*) for $(k,n) = (10,11)$ was indeed optimum, to find a design for (9,11) by MD that is better than Mitchell's in that case and in fact is optimum, and to find new designs for the (11,15) and (12,15) cases by MD search, which can be proved optimum. The optimality proofs rest on computing an upper bound $\psi(k,n)$ based on work of Ehlich (1964*b*), and on finding a design X with $\det(X'X) = \psi(k,n)$. Unfortunately, the bound is not always attainable (for example, the optimum design for $k = n = 9$, mentioned below, has $\det(X'X) = (49/68)\psi(9,9)$); but it was attainable in these cases. In these "non-Hadamard" cases, frequency of success expectedly decreases as k and n become large, although, as we saw in Table 1b in Hadamard cases, a fine structure prevents the success rate from being monotone. Thus, Mitchell's (9,9) design (which yields a $\det(X'X)$ value obtained first by Ehlich and Zeller, 1962, and proved optimum by them in unpublished work) was attained by MD 215 times in 1000 tries (with n_r chosen at random between 1 and 4, and an average of .45 seconds per try being required), while, as we saw in Table 2, the success rate was only about 1% in the (11,11) case (which required about 2.7 seconds per try). The optimum was achieved, among 1000 tries for each case, in 64 tries for (9,11), in 94 tries for (10,11) (Mitchell's design), and in only 4 tries for (11,15) and (12,15). In the cases (11,11), (11,15), (12,15), there were always local maxima very close to the global maximum ($e^k \ge .97, .99, .995$) and which were attained very frequently; we have already discussed this in the (11,11) case. Those of the above-mentioned optimum designs not previously tabled are displayed (or prescriptions for their construction from known designs are given) in Galil and Kiefer (1980).

For $n = 15$ and $k = 13, 14, 15$, we have not yet obtained a design we can prove optimum, but in all cases have obtained designs for which (from the bounds of Galil and Kiefer, 1980) $e > .9$. These are all complicated settings combinatorially, in which the optimum design is considerably better than that yielded by dropping a row from an orthogonal design for $n = 16$. These problems are large (taking approximately one minute per try for (15, 15)). The space requirements are such that, as mentioned in Section 3.1, it begins to become worthwhile to use the method of Section 2.3.

4.2. Quadratic Example.

A number of workers have published results on this model for $2 \le q \le 5$ (the case $q = 1$ being trivial). The designs of Rechtshaffner (1967), which coincide with some of those of Hoke (1974) when $k = n$, were easy to construct (for all q) and as efficient as any obtained up to that time. Mitchell and Bayne (1976) used DETMAX to construct designs that tied Rechtshaffner's for $q = 2$ and improved upon them for $3 \le q \le 5$. Notz (1978) devised a general method of design construction which, in the present quadratic example, made a clever use of balanced arrays to prescribe a construction of a saturated design for any q, which tied Rechtshaffner's for $q \le 4$ but which improved upon even Mitchell and Bayne's (by a factor of 1.16 in $\det(X'X)$) for $q = 5$, and on Rechtshaffner's for larger q. Lucas (1976), using an unpublished method of R. E. Wheeler, treats the quadratic model on the cube $[-1, 1]^q$, not necessarily the 3^q array.

Our search using MD tied the best of these (Bayne and Mitchell for $q \le 4$, Notz for $q = 5$) for $q \le 5$. For $q = 6$ (using the method of Section 2.3), we achieved a determinant with 1.801 times that obtainable by Notz's method, and for $q = 7$ we achieved 3.912 times his determinant. Although for $q = 5$ the design we achieved contains the same 16-point balanced array on $\{-1, 1\}$ as does Notz's (the other 5 points containing some zero coordinates), in the other two cases our designs included more points containing some zero coordinates than did his (11 vs. 6, 10 vs. 7). We alluded earlier to the fact that $q = 5$ seems relatively tractable; that is because the balanced array is even an orthogonal array in that case.

In the unsaturated case ($n > k$), Mitchell and Bayne have the best published designs. We ran MD on all the cases they did. In *all* cases we obtained at least as good a design, and we obtained a better design in 4 out of 11 cases for $q = 3$, in 7 out of 14 cases for $q = 4$, and in 8 out of 10 cases for $q = 5$. Table 3 summarizes the improved values. We do not take the space to give the designs here, but will make them available to interested readers.

We have just received the book by Vuchkov et al. (1978), in which Tables 14–16 list designs for the settings of our Table 3. These designs are inferior to our MD designs in all cases, and to the MB designs in

TABLE 3—*Improvement of MD results over those of Mitchell and Bayne (MB) for quadratic regression on the q-cube. The value of det(X'X) is listed.*

q	n	MB	MD
3	16	0.4194×10^{9}	0.4499×10^{9}
3	17	0.81125×10^{9}	0.8320×10^{9}
3	18	0.1488×10^{10}	0.1527×10^{10}
3	20	0.4529×10^{10}	0.4736×10^{10}
4	17	0.1397×10^{14}	0.1521×10^{14}
4	18	0.4315×10^{14}	0.4985×10^{14}
4	24	0.6379×10^{16}	0.6566×10^{16}
4	25	0.1328×10^{17}	0.1427×10^{17}
4	26	0.1970×10^{17}	0.2665×10^{17}
4	27	0.3383×10^{17}	0.4819×10^{17}
4	28	0.8296×10^{17}	0.8651×10^{17}
5	21	0.4079×10^{21}	0.4612×10^{21}
5	22	0.2137×10^{22}	0.2158×10^{22}
5	23	0.5677×10^{22}	0.6585×10^{22}
5	25	0.4052×10^{23}	0.4869×10^{23}
5	26	0.8409×10^{23}	0.1132×10^{24}
5	27	0.2566×10^{24}	0.2698×10^{24}
5	28	0.5237×10^{24}	0.5930×10^{24}
5	29	0.1056×10^{25}	0.1283×10^{25}

most cases. While such "sequential option" designs (Section 2.7 herein) cannot be expected to be efficient for $n > k$, the designs of Vuchkov et al. are not too efficient even in the saturated case $k = n$ where there is no sequential option; for example, for $q = 5$, $n = 21$, the ratio of their determinant to ours is .191 (efficiency ratio .92).

We will also make available a listing of the MD program, upon request to the second author.

5. SUMMARY

We have presented a family of computer search methods for finding optimum designs. Several ideas are introduced for saving time and/or space over the methods that had been used previously. Which of our techniques will prove most useful in a givin setting depends on the detailed structure of the problem; thus, a few exploratory runs, using various of the methods, may be advisable in large problems. Applied to Mitchell's DETMAX and exchange algorithms, our techniques yield modifications that are typically 15 to 50 times faster than the original methods. This improvement has enabled us to search enough times to find an optimum design in several settings in which it was previously not feasible to search enough to find the optimum.

6. ACKNOWLEDGEMENTS

The research of both authors was sponsored by the National Science Foundation. Galil was supported in part by the Israel Commission for Basic Research.

We are most grateful to Toby Mitchell for inspiration and continued suggestions over a period of many years, including the making available of his program. We also thank Nick Gimbrone for his aid at the Cornell Computing Center, Henry Wynn for helpful conversations, and C. F. Wu for useful comments.

REFERENCES

EHLICH, H. (1964a). Determinantenabschätzungen für binäre Matrizen. *Math. Zeitschr., 83,* 123–132.

EHLICH, H. (1964b). Determinantenabschätzungen für binäre Matrizen mit $m \equiv 3 \bmod 4$. *Math. Zeitschr., 84,* 438–447.

EHLICH, H. (1978). Personal communication.

EHLICH, H. and ZELLER, K. (1962). Binäre Matrizen. *Zeit. Angewandte Math. Mech., 42,* T20–T21.

GALIL, Z. and KIEFER, J. (1980). D-Optimum weighing designs. *Ann. Math. Statist.* (to appear).

HOKE, A. T. (1974). Economical second-order designs based on irregular fractions of the 3^n factorial. *Technometrics, 17,* 375–384.

JOHNSON, J. D. and LAST, K. W. (1975). Constrained experimental design. Unpublished; abstract in *A Survey of Statistical Design and Linear Models,* ed. by J. N. Srivastava, North Holland (Amsterdam), p. 674.

JONES, B. (1976). An algorithm for deriving optimal block designs. *Technometrics, 18,* 451–458.

LUCAS, J. M. (1976). Which response surface design is best: a performance comparison of several types of quadratic response surface designs in symmetric regions. *Technometrics, 18,* 411–417.

MITCHELL, T. J. (1974a). An algorithm for the construction of "D-optimal" experimental designs. *Technometrics, 16,* 203–210.

MITCHELL, T. J. (1974b). Computer construction of "D-optimal" first-order designs. *Technometrics, 16,* 211–220.

MITCHELL, T. J. and BAYNE, C. K. (1976). D-optimal fractions of three-level factorial designs. Oak Ridge National Laboratory report.

NOTZ, W. I. (1978). Two problems in optimal design theory. Ph.D. thesis, Cornell University.

RECHTSHAFFNER, R. L. (1967). Saturated fractions of 2^n and 3^n factorial designs. *Technometrics, 9,* 569–575.

ST. JOHN, R. C. and DRAPER, N. R. (1975). D-optimality for regression designs: a review. *Technometrics, 17,* 15–24.

VUCHKOV, I. N., YONCHEV, H. A., DAMGALIEV, D. L., TSOCHEV, V. K. and DIKOVA, T. D. (1978). *Catalogue of Sequentially Generated Designs* (in Russian with short English summaries). Higher Institute of Chemical Technology, Sofia, Bulgaria.

WYNN, H. P. (1970). The sequential generation of D-optimum experimental designs. *Ann. Math. Statist., 41,* 1655–1661.

Recent Developments in Statistical Inference and Data Analysis; K. Matusita, editor
© North-Holland Publishing Company, 1980

OPTIMUM WEIGHING DESIGNS

J. Kiefer[1] and Z. Galil[1]

University of California, Berkeley
and
Tel Aviv University

Suppose $k \leq n$, and consider $n \times k$ matrices X with entries ± 1. Such X arise in weighing and factorial design problems etc. New D-optimality results have been obtained by us by using a combination of theoretical work, an improvement on the Detmax search method of T.J. Mitchell, and a computer search for a solution to $X'X = \bar{C}$ where $\bar{C}$ is a candidate for maximizing det C. Our results list all D-optimum X'X for $k \leq 12$. For some (k,n) there are several nonisomorphic D-optimum (X'X)'s, which we compare according to other criteria.

1. Introduction. Let k and n be positive integers satisfying $k \leq n$. Denote by $\chi = \chi(k,n)$ the set of all $n \times k$ matrices $X = \{x_{ij}\}$ whose entries are all ± 1. In the "chemical balance" problem in which each of the k objects under consideration appears in each of the n weighings, use of the design X means that the jth object appears on the left or right pan of the balance in the ith weighing according to whether $x_{ij} = 1$ or -1. If the observations are uncorrelated with equal variances σ^2 and the jth object weighs θ_j, then the measured weight of left minus right pan on the ith weighing has expectation $\Sigma_j x_{ij}\theta_j$. If X has rank k, all θ_j are estimable, and their best unbiased estimators have covariance matrix $\sigma^2(X'X)^{-1}$.

The problem of choosing X is an old one. The most used criterion is that of choosing X to maximize det(X'X). We shall refer to both X and X'X as D-optimum if X yields such a maximum. The history of the problem, and the relationship of the above model to ones in which the x_{ij} can assume other values, is discussed by Galil and Kiefer (1980a). In particular, we recall that when $n \equiv 0 \pmod 4$ a design that is optimum for essentially all commonly used optimality criteria is obtained by letting X consist of k columns of a Hadamard matrix H_n, yielding $X'X = nI_k$; for all practical k, designs with such an X'X can be constructed for all $n \equiv 0 \pmod 4$ even though we do not know all H_n exist. In other cases of n, much of the literature finds optimum designs only under the restriction to X'X for which all off-diagonal elements are equal (the diagonal elements automatically being equal). This simplifies the problem greatly. See Raghavarao (1971) for results and references. Our previous work, the present paper, and the references of the next paragraph do not suffer from this restriction.

The first systematic work on this problem when $n \not\equiv 0 \pmod 4$ is contained in the pioneering work of Ehlich (1964a,b), who studied D-optimality in the saturated case $k = n$. In our previous work we noted that many of Ehlich's developments can be extended to the case $n > k$. We shall repeat here neither the consequent D-optimality results, nor the results of Cheng (1978) which yield optimality in many senses when $n \equiv 1 \pmod 4$. Selected results are mentioned in the sections that follow, when appropriate.

In the present paper we continue our study, in two directions, in certain cases where $n \equiv 3 \pmod 4$, which are generally the most difficult cases to treat theoretically. Firstly, in Section 2, we give a construction method additional to those we described earlier, and results of certain machine computations, which complete the list of D-optimum designs for certain cases (k,n). Secondly, in Section 3, we obtain results in certain cases in which there are two non-isomorphic information matrices $X'X$ that are D-optimum, as to which of them is optimum in the sense of other criteria.

A few remarks on the latter considerations will conclude the present section. Examples of such considerations are not common in the optimum design literature, partly because in the approximate theory all D-optimum information matrices are identical. (A few examples have been considered when D-optimality is considered for only a subset of s of the parameters, in which case the approximate D_s-optimum $X'X$ need not be unique.) The comparisons in the present paper will be limited to criterion functions (to be minimized) that are increasing functions of

$$\Sigma_1^k f(\lambda_i(X)) \tag{1.1}$$

where f is nonincreasing and convex and the $\lambda_i(X)$ are the eigenvalues of $X'X$, or to limits of sequences of such criteria. In particular, we consider the criteria

$$\Phi_p(X'X) = [\tfrac{1}{k}\Sigma_1^k(\lambda_i(X))^{-p}]^{1/p},\quad 0 < p < \infty;$$

$$\Phi_0(X'X) = \lim_{p\to 0}\Phi_p(X'X) = [\det(X'X)]^{-1/k}; \tag{1.2}$$

$$\Phi_\infty(X'X) = \lim_{p\to\infty}\Phi_p(X'X) = \max_i[1/\lambda_i(X)].$$

The last two are, respectively, the D- and E-optimality criteria, and Φ_1 is the A-optimality criterion. In Section 3 we find the Φ_p-optimum $X'X$ within the class of Φ_0-optimum $X'X$ for certain cases where $n \equiv 3 \pmod 4$. We emphasize that this is quite different from finding the Φ_p-optimum $X'X$ over all X in $X(k,n)$, which has been done only in the cases $n \equiv 0$ or $1 \pmod 4$ mentioned earlier.

2. New D-optimality results. In the cases $n \equiv 3 \bmod 4$, to which we hereafter restrict attention, Ehlich developed a theory that yielded an upper bound on $\det(X'X)$, which was extended in our earlier paper. The main result is that, if $\mathscr{C} = \mathscr{C}(k,n)$ is the class of positive definite symmetric $k \times k$ matrices with diagonal elements n and off-diagonal elements $\equiv 3 \bmod 4$, then $\det C$ is maximized over $\mathscr{C}$ by, and only by, a "block matrix" C^* which, for some s, is of the following form: C_s has u diagonal $r \times r$ blocks with off-diagonal elements 3, and $v = s - u$ diagonal $(r+1) \times (r+1)$ blocks of this form; all other off-diagonal elements of C_s are -1. Here $r > 0$, $u > 0$, and $v \geq 0$. Thus, $k = rs + v$. The determinant of such a C_s is

$$D_{k,n}(s) = (n-3)^{k-s}(n-3+4r)^u(n+1+4r)^v\{1-\frac{k(n-3)+4sr(r+1)}{(n+4r+1)(n+4r-3)}\}. \tag{2.1}$$

Thus, C^* is a C_s for which s maximizes $D_{k,n}(s)$; we denote such an s by s_{opt}. The only lack of uniqueness of the determinant-maximizing C^*, except for row-and-column permutations, occurs when more than one value of s maximizes (2.1). The importance of the above results is that every X in $\underline{X}$ can be transformed, by multiplying some of its columns by -1, into an $\overline{X}$ (say) for which $\overline{X}'\overline{X} \in \mathscr{C}$. Hence, Ehlich's bound $\max_s D_{k,n}(s)$ is an upper bound on $\max_{X\in\underline{X}} \det(X'X)$; and if we can find an X such that $X'X = C^*$, that X is D-optimum, uniquely so except for the possible lack of uniqueness of C^* and X (including isomorphisms of X obtained by permuting rows or columns of X or multiplying some of them by -1).

In the saturated case, Ehlich's bound seems unfortuantely never to be attainable except in the simplest case $k = n = 3$ where $s_{opt} = 3$, $r = 1$, $u = 3$, $v = 0$.

For $k = n = 7$ the optimum design was first obtained by Williamson (1946), while Ehlich (1978) obtained the three nonisomorphic $X'X$ that are D-optimum by a computer routine when $k = n = 11$. For $k = n = 15$, Ehlich has constructed a promising $X'X$ whose optimality has not yet been verified.

However, Ehlich's bound is more fruitful when $n > k$, and it was shown in Galil and Kiefer (1980a) that, for $n \equiv 3 \pmod 4$ and $k = k_n \overset{d}{=} (n+5)/2$, we have $s_{opt} = k_n$ or $k_n - 1$ (two nonisomorphic C^*'s), and that an X corresponding to $s_{opt} = k_n$ can be obtained as follows: let H_{n+1} be a Hadamard matrix, assumed here to exist. We can assume the last row of H_{n+1} consists entirely of 1's. Delete that row and select as X any k_n columns of the resulting matrix. This is then a D-optimum design. We hereafter label any such X as X_0, so that $X_0'X_0 = (n+1)I_{k_n} - J_{k_n}$ where J_k is the $k \times k$ matrix of 1's.

We now give a simple construction of an $n \times k_n$ matrix X_1 such that $X_1'X_1$ differs from $X_0'X_0$ only in having a 3 in place of a -1 in positions (1,2) and (2,1); thus, X_1 is D-optimum, corresponding to $s_{opt} = k_n - 1$. (We compare X_0 and X_1 further, in Section 3.) Since H_{n+1} is orthogonal, we have $x_{1j}x_{2j} = 1$ for $(n+1)/2$ columns j, which we relabel as columns j of X_0 above for $3 \leq j \leq k_n$. Since $n + 1 > 2$, Hadamard H_{n+1} is known to have two columns, which we now take to be the first two of X_0, such that $x_{11} = x_{22} = 1$ and $x_{12} = x_{21} = -1$. Let X_1 differ from X_0 only in that the elements x_{11} and x_{21} of X_0 are multiplied by -1 to obtain those elements in X_1. The inner product of columns 1 and 2 of X_1 is thus increased by 4 over what it was for X_0, from the value -1 to the value 3. All other inner products are left unchanged, as we want. We note that deletion of any number of the last $(n+1)/2$ columns of the above X_1 yields, for $2 \leq k \leq k_n$, a design whose information matrix contains all -1's off the diagonal except for a single symmetric pair of 3's. However, when $k < k_n$ this is of no consequence for D-optimality: k columns of X_0 give the unique (except for trivial isomorphisms) D-optimum information matrix $(n+1)I_k - J_k$, for such k.

We now turn to the results we have obtained recently with the aid of a computer. These differ in method of construction from our earlier results, which were obtained as applications of the search methods of Galil and Kiefer (1980b). Those methods, themselves extensions of the "Detmax" algorithm of Mitchell (1974) and of the even simpler "exchange" algorithm, are suitable for use in general experimental design settings. Hence, in the particular setting of weighing designs we can hope to use computer time more efficiently by making use of the particular algebraic structure of the problem. In fact, this is what Ehlich (1978) has done in the work mentioned earlier. Moreover, search methods of the type we used earlier can never demonstrate nonexistence of a design with given structure, while the present approach can.

To give an example, consider the problem of finding whether there is an X with $X'X = C^*$ in the case $(k,n) = (13,15)$. In that case $s_{opt} = 6$ or 7. Let U_i be the ith row of X'. Without loss of generality (that is, using the symmetries of the problem), we can take U_1 to consist entirely of 1's and U_2, in the same block of size 2 with U_1, to have 1's in the first nine positions and -1's in the next 6. The inner product $\langle U_1, U_2 \rangle$ is then 3. We can take U_3, which has inner product -1 with both of these, to have its signs in order consist of 4 +'s, 5 −'s, 3 +'s, 3 −'s. For U_4, which has inner product -1 with U_1 and U_2 and 3 with U_3, it then turns out that there are 4 choices, even after reduction taking account of symmetries; they correspond to the number of + signs in the last group of 3 (equal to the number of − signs in the next to last group of 3). If U_5 and U_6 yield the next block of size 2, we find a different number of possible U_5's for each U_4. These and the vectors obtained from them by permuting certain elements to undo the symmetry reductions used thus far, give a list L of all possible candidates for the remaining U_i's, constrained to have inner product -1 with U_1, U_2, U_3, U_4. The U_6 possibilities are chosen for each U_4, U_5 to have inner product 3 with U_5.

One now observes a possibility for cutting down subsequent computations: for a given U_4, there are sometimes two (or more) U_5's, say U_5^* and U_5^{**}, and a $k \times k$ matrix T which is a permutation matrix with some 1's replaced by -1's, such that $U_5^* = U_5^{**}T$ and such that the set of U_6's, U_7's, etc. that are candidates for use with U_5^{**} are transformed by the same T to the corresponding set for U_5^*. It is then only necessary, in further steps in the tree of possible solutions, to pursue the U_5^* branch rather than both the U_5^* and U_5^{**} branches. An analogous phenomenon sometimes occurs at subsequent stages, e.g., for several U_6's for the given U_4, U_5. The computer can search rapidly through the inner product list at each new branch, and each new inner product condition that is adjoined as one proceeds from U_i to U_{i+1} rapidly cuts L in size and makes the computation much shorter than one would expect from a priori bounds on the number of paths. In the cited example, after U_{12} is chosen to yield 6 blocks of size 2, we branch into trying for a U_{13} that either makes the last block of size 2 into a block of size 3 ($s_{opt} = 6$), or else that constitutes a block of size 1 ($s_{opt} = 7$).

This computation showed that for $(k,n) = (13,15)$ Ehlich's bound is not attainable: for neither value of s_{opt} is there an X with $X'X = C^*$. When $(k,n) = (14,15)$, $s_{opt} = 6$. The corresponding design X does not exist in this case since, if it did, deletion of an appropriate column would yield the design for $(k,n) = (13,15)$ corresponding to $s_{opt} = 6$. These are the first <u>nonsaturated</u> cases in which unattainability of Ehlich's bound has been shown; for many saturated cases, unattainability follows much more easily, from the fact that $D_{k,k}(s_{opt})$ is not a square. While we have fairly efficient candidates for D-optimality in the $(13,15)$ and $(14,15)$ cases, we have not proved them optimum; the technique followed by Ehlich (1978) to verify the optimality of a design in the 11×11 case, which is summarized in Galil and Kiefer (1980a), would require much more computation when $n = 15$.

In the smaller case $(k,n) = (9,11)$, a design attaining Ehlich's bound is given in Galil and Kiefer (1980a) for $s_{opt} = 6$. The value $s_{opt} = 7$ is also theoretically possible here, but the computer routine described two paragraphs above showed quickly that this is unattainable.

In the case $(k,n) = (12,15)$, our previous paper gave a design attaining Ehlich's bound with $s_{opt} = 6$. The value $s_{opt} = 7$ is also possible in this case, and the computer routine described above found a design achieving this configuration. We give it in Table 1, below.

<u>Table 1</u>. D-optimum X for $(k,n) = (12,15)$.

−	−	−	+	+	−	−	+	+	+	+	+
−	−	+	−	−	+	+	−	+	+	+	+
+	+	−	+	−	+	−	−	−	+	+	+
+	+	+	−	+	−	−	−	+	−	+	+
+	−	−	−	+	+	+	+	−	−	+	+
−	+	+	+	−	−	+	+	−	−	+	+
−	−	−	−	−	−	−	−	−	−	+	+
−	+	−	−	−	−	+	+	+	+	−	+
+	−	+	+	−	−	−	−	+	+	−	+
−	−	+	−	+	+	−	+	−	+	−	+
+	+	−	−	+	−	+	−	−	+	−	+
+	+	−	−	−	+	−	+	+	−	−	+
−	−	−	+	+	+	+	−	+	−	−	+
+	−	+	+	−	−	+	+	−	−	−	+
−	+	+	+	+	+	−	−	−	−	−	+

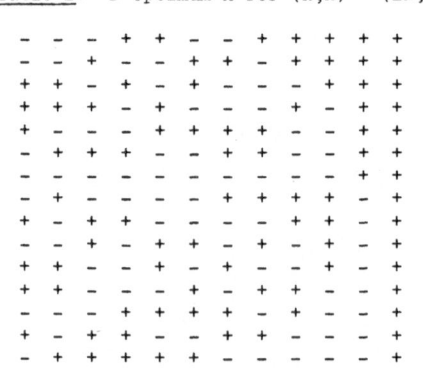

The above $X'X$ has $s_{opt} = 7$, the first 5 blocks being of size 2. The design with $k = 6$ (Galil and Kiefer (1980a)) is also D-optimum. Both designs have $\det X'X = 3^6 5^5 2^{25} = .7644 \times 10^{14}$.

The results of the previous two paragraphs, together with those of our earlier paper and the construction of X_1 given earlier in this section, yield a complete listing of all nonisomorphic D-optimum information matrices for all $k \leq 12$ and all $n \geq k$.

It is to be hoped that future research will lead to a better understanding of these discrete optimization problems, and, in particular, of the cases (k,n) for which Ehlich's bound is attainable.

3. Comparison of two D-optimum designs. For even $k \geq 6$ and $n = 2k - 5$ (so that $n \equiv 3 \pmod 4$ and $k = k_n$), we consider the two D-optimum designs X_0 and X_1 of Section 2. The formula (2.1) can be used with n replaced by $n - \lambda$, to determine the eigenvalues $\lambda_i(X_k)$ for $k = 0,1$. We obtain

$$\{\lambda_i(X_0)\} = \begin{cases} 2k - 4, & \text{of multiplicity } k - 1, \\ k - 4, & \text{of multiplicity } 1, \end{cases} \tag{3.1}$$

and

$$\{\lambda_i(X_1)\} = \begin{cases} 2k - 4, & \text{of multiplicity } k - 3, \\ 2k - 8, & \text{of multiplicity } 1, \\ \frac{1}{2}[3k - 4 \pm (k^2+8k-16)^{1/2}], & \text{each of multiplicity } 1. \end{cases} \tag{3.2}$$

For criteria of the form (1.1) we may ignore the $k - 3$ appearances of the eigenvalue $2k - 4$ in the two sets (3.1) and (3.2) and simply compare $\Sigma_1^3 f(\lambda_i(X_k))$ for $k = 1,2$, where, written in ascending order for each k,

$$\begin{aligned} (\lambda_1(X_0),\lambda_2(X_0),\lambda_3(X_0)) &= (k-4,2k-4,2k-4); \\ (\lambda_1(X_1),\lambda_2(X_1),\lambda_3(X_1)) &= (\tfrac{1}{2}[3k-4-(k^2+8k-16)^{1/2}],2k-8, \\ &\qquad \tfrac{1}{2}[3k-4+(k^2+8k-16)^{1/2}]). \end{aligned} \tag{3.3}$$

It is easily seen that

$$\lambda_1(X_0)<\lambda_1(X_1)<\lambda_2(X_1)<\lambda_2(X_0)=\lambda_3(X_0)<\lambda_3(X_1). \tag{3.4}$$

Note that neither of the two sets of eigenvalues majorizes the other. (Nor do the sets of logarithms.) Thus, we cannot expect either X_0 or X_1 to dominate the other for all convex criteria. However, we have the

Theorem. *For* $0 < p \leq \infty$,

$$\Phi_p(X_1'X_1) < \Phi_p(X_0'X_0). \tag{3.5}$$

The strict inequality for $p = \infty$ makes use of (3.4) and the fact that $\lambda_2(X_0) = 2k - 4$ is the value of $k - 3$ eigenvalues omitted from both lists in (3.3). Indeed, this E-optimality of X_1 over X_0 implies the rest of (3.5); this follows from the

Lemma. *Suppose* $\{\lambda_i^{(k)}, i = 1,2,3\}$, $k = 0,1$ *are two sets of positive values, nondecreasing in* i, *and such that*

$$\Sigma_1^3\lambda_i^{(0)} = \Sigma_1^3\lambda_i^{(1)} \quad \text{and} \quad \Pi_1^3\lambda_i^{(0)} = \Pi_1^3\lambda_i^{(1)}. \tag{3.6}$$

If also

$$\lambda_1^{(0)} < \lambda_1^{(1)}, \tag{3.7}$$

then $\Sigma_1^3(\lambda_i^{(1)})^{-p} < \Sigma_1^3(\lambda_i^{(0)})^{-p}$ *for* $0 < p < \infty$.

Proof. The demonstration consists of showing that, for each positive p, $\Sigma_1^3\lambda_i^{-p}$ is strictly decreasing along the curve from $\{\lambda_i^{(0)}\}$ to $\{\lambda_i^{(1)}\}$ determined by $\Sigma_1^3\lambda_i = \Sigma_1^3\lambda_i^{(0)}$ and $\Pi_1^3\lambda_i = \Pi_1^3\lambda_i^{(0)}$. To this end we compute the derivative on the interior of this curve. We may first divide by λ_1 to obtain

$1 < A < B$ as the coordinates of a general interior point of the curve. (It is easily seen that we can only have $A = B$ at an endpoint, such as the point given in the first line of (3.3).) From that point, increasing 1 by $\eta > 0$ (to move in the direction indicated by (3.7)) and decreasing A by δ and B by ε is easily seen, from the conditions of keeping the sum and product of the three values constant, to require

$$\text{(3.8)}\qquad \begin{aligned} \delta/\eta &= (1 - B^{-1})/(B^{-1} - A^{-1}) + o(1) , \\ \varepsilon/\eta &= (1 - A^{-1})/(A^{-1} - B^{-1}) + o(1). \end{aligned}$$

We then obtain, as $\eta \downarrow 0$,

$$\text{(3.9)}\qquad \begin{aligned} &(B-A)(p\eta)^{-1}\{[(1+\eta)^{-p}+(A+\delta)^{-p}+(B+\varepsilon)^{-p}]-[1+A^{-p}+B^{-p}]\} \\ &\quad \sim\{(B-1)A^{-p}-(A-1)B^{-p}+A-B\}= Q(\text{say}). \end{aligned}$$

To show that $Q < 0$ for $1 < A < B$, we note that $\lim_{B\downarrow A} Q = 0$ and that, for $B > A$,

$$\text{(3.10)}\qquad \begin{aligned} \frac{\delta Q}{\delta B} &= -1 + A^{-p} + pB^{-p-1}(A - 1) \\ &< -1 + A^{-p} + pA^{-p-1}(A - 1) \\ &= A^{-p}[(p + 1)A^{0} - pA^{-1} - A^{p}] < 0, \end{aligned}$$

the last by convexity of A^x. This completes the proof.

Remark 1. Note that $\Phi_p(X_0) - \Phi_p(X_1)$, which we proved positive for $0 < p \le \infty$, is zero at $p = -1$ and 0 (see (3.6)). The proof and conclusion of the Lemma can be modified for $p < 0$, as follows: the first inequality of (3.10) is unchanged for $p < -1$ and is reversed for $-1 < p < 0$; also, $p+1-pA^{-1}-A^{p}$ is strictly concave in p and vanishes at $p = -1$ and 0. We conclude that $Q < 0$ if $p < -1$ and $Q > 0$ if $-1 < p < 0$, so that X_1 is again better than X_0 in the former case, while X_0 is better in the latter. The Φ_p for negative p are not common criteria, and in fact a heuristic of Shah suggests, when two designs have the same value of Φ_{-1}, that one might choose the design that maximizes Φ_{-2} so as to minimize $\text{tr}[(X'X)^2]$. However, the fact that X_0 is better for $-1 < p < 0$ suggests that there will be f "near" the function -log of D-optimality, that are convex and decreasing and such that X_0 is better than X_1 in the sense of (1.1) for those f. This is indeed the case. For example, if $b > 0$ and $f(x) = -\log(x + b)$, minimizing $\Sigma_1^k f(\lambda_i)$ means maximizing $\Pi_1^k(\lambda_k + b)$. A simple computation yields, with $C_k = 4(k - 2)^2(k - 4)$,

$$\text{(3.9)}\qquad \begin{aligned} \Pi_1^3[\lambda_i(X_0)+b]&=C_k+(8k^2-40k+48)b+(5k-12)b^2+b^3, \\ \Pi_1^3[\lambda_i(X_1)+b]&=C_k+(8k^2-40k+40)b+(5k-12)b^2+b^3, \end{aligned}$$

the first of which is larger for all k and all $b > 0$.

Remark 2. In Section 2 and our previous paper we mentioned three (k,n) with $n \equiv 3 \pmod 4$, other than the ones with $n = 2k - 5$ treated above, in which designs exist for which D-optimality ties occur. For $(k,n) = (10,11)$ there can be 5 or 6 blocks and, for $(12,15)$, 6 or 7. In both cases a proof essentially like that of the theorem above shows that the design with fewer blocks s is again better for $0 < p \le \infty$. These results and consideration of other theoretically possible ties, some involving comparison of more than three eigenvalues, will appear elsewhere. We also consider there the third known case of D-optimality ties, the $(11,11)$ case, where Ehlich's bound is not attained but where he found three nonisomorphic $X'X$ that are D-optimum. The eigenvalue comparison there is more complex. (The conclusion is that the design of Table 9c of Galil and Kiefer (1980a) is Φ_p-better than that of Table 9b, which in turn is Φ_p-better than that of Table 9a, for $0 < p \le \infty$.)

FOOTNOTE

[1]Research supported by National Science Foundation Grant MCS78-25301.

REFERENCES

[1] Cheng, C.-S. (1978). Optimality of some weighing and 2^n fractional-factorial designs. Ann. of Statist. (to appear).

[2] Ehlich, H. (1964a). Determinantenabschätzungen für binare Matrizen. Math. Zeitschr. **83**, 123-143.

[3] Ehlich, H. (1964b). Determinantenabschätzungen für binare Matrizen mit $n \equiv 3 \pmod 4$. Math. Zeitschr. **84**, 438-447.

[4] Ehlich, H. (1978). Personal communication.

[5] Galil, Z. and Kiefer, J. (1980a). D-optimum weighing designs. To appear in Ann. of Statist.

[6] Galil, Z. and Kiefer, J. (1980b). Time- and space-saving computer methods, related to Mitchell's Detmax, for finding D-optimum designs. To appear in Technometrics.

[7] Mitchell, T.J. (1974). An algorithm for the construction of 'D-optimal' experimental designs. Technometrics **16**, 203-210.

[8] Raghavarao, P. (1971). Construction and Combinatorial Problems in Design of Experiments. John Wiley and Sons, New York.

[9] Williamson, J. (1946). Determinants whose elements are 0 and 1. Amer. Math. Monthly **53**, 427-434.

The Annals of Statistics
1981, Vol. 9, No. 4, 737–757

OPTIMUM BALANCED BLOCK AND LATIN SQUARE DESIGNS FOR CORRELATED OBSERVATIONS

BY J. KIEFER[1] AND H. P. WYNN[1, 2]

Cornell University, University of California, Berkeley, and Imperial College, London

In this paper designs are found which are optimum for various models that include some autocorrelation in the covariance structure V. First it is noted that the ordinary least squares estimator is quite robust against small perturbations in V from the uncorrelated case $V_0 = \sigma^2 I$. This "local" argument justifies our use of such estimators and restriction to the class of designs $\mathscr{X}^*$ (balanced incomplete block or Latin squares) optimum under V_0. Within $\mathscr{X}^*$ we search for designs for which the least squares estimator minimizes appropriate functionals of the dispersion matrix under various correlation models V. In particular, we consider "nearest neighbor" correlation models in detail. The solutions lead to interesting combinatorial conditions somewhat similar to those encountered in "repeated measurement" designs. Typically, however, the latter need not be BIBD's and require twice as many blocks. For Latin squares, and hypercubes, the conditions are less restrictive than those giving "completeness."

1. Introduction; literature. The relevant literature concentrates on the estimation of the fixed effects of independent variables, or factors, rather than the estimation of the underlying error process. In other words, the emphasis is on correcting the usual estimators and finding designs for which estimators do well under given error assumptions.

The two papers by Papadakis (1937) and Bartlett (1938) are the first in the field. Papadakis produced estimators which are close to the weighted least squares estimators for certain designs. A thorough analysis of the method was carried out by Atkinson (1969). R. M. Williams (1952) studied designs with treatments laid out in a one-dimensional array, under similar models, with the first-order autocorrelation in that dimension. Recent work on the analysis of spatial patterns by Bartlett (1975), Besag (1974), Ripley (1977), and others has a bearing on the design and analysis problems. The recent paper by Bartlett (1978) stimulated a lively and useful discussion of the issues.

There have been other approaches to guarding against effects from neighboring plots in experimental design. The classical methods of randomization, due primarily to Fisher, should certainly be mentioned. For example, selecting a Latin square at random from all, or a subclass, of Latin squares has been advocated.

There is a considerable literature on designs for so-called "change-over" or "residual" effects where fixed treatment effects are carried over to neighboring plots. A good presentation and source of references for this subject is Hedayat and Afsarinejad (1975). Important early papers in the area are by E. J. Williams (1949, 1950) and Patterson (1950, 1951, 1952).

In the field of optimum design Kiefer (1960) extends the work of R. M. Williams (1952). The papers by Sacks and Ylvisaker (1966, 1968, 1969) and the extensions by Wahba (1971,

Received April 1979; revised June 1980.
[1] Research sponsored by the National Science Foundation.
[2] Research also sponsored by the Grants Council.
AMS 1970 *subject classifications.* Primary 62K05; secondary 05B20.
Key words and phrases. Optimum designs, correlated observations, difference sets, balanced incomplete block designs, Latin squares and hypercubes.

1974) study the estimation of fixed effects under very general continuous time error processes. Bickel and Herzberg (1979) develop some asymptotic theory for similar processes with simple linear regression models for the fixed effects. O'Hagan (1978) gives a Bayesian approach to the same kind of model.

Closest to the approach presented in the present paper are the papers by Berenblut and Webb (1974) and by Duby et al. (1977), and the recent more comprehensive thesis by Martin (1977). These authors investigate the behavior of different classical designs under various correlation assumptions. The stationary correlation model considered by them turns out to be less tractable than the "nearest neighbor" structure we treat in detail; thus, Duby et al. give numerical illustrations of design comparisons rather than general combinatorial conclusions of the type we reach. The present paper may be seen as giving some theoretical backing to (1) the choice of classical designs as a class for investigation, (2) the use of the ordinary least squares estimators, and (3) the use of various "equineighbored" conditions on combinatorial designs.

Further references to optimum design theory and combinatorial analysis will be given as they are required.

2. Least squares versus BLU estimators. We assume the linear regression model

$$E(Y) = X\theta,$$

where $Y = (Y_1, \ldots, Y_n)^T$ is a vector of observations and X is an $n \times p$ design matrix belonging to a class $\mathscr{X}$ of such matrices. The parameter vector $\theta = (\theta_1, \ldots, \theta_p)^T$ is unknown. The covariance matrix of the observations $\operatorname{Cov}(Y) = V$ belongs to a class $\mathscr{V}$ of covariance matrices which contains that of the standard case, $V_0 = \sigma^2 I$, where I is the $n \times n$ identity matrix. We assume every V in $\mathscr{V}$ is positive definite.

Let $t = B\theta$, with B $v \times p$, be a v-vector of estimable functions of θ. A well-known necessary and sufficient condition for this estimability is that $B^T = X^T XC$ for some $p \times v$ matrix C. Let $\hat{t}_0$ be the unique minimum variance linear unbiased estimator (BLUE) of t under $V = V_0$, which we shall refer to as the least squares (LS) estimator. Thus $\hat{t}_0 = B\hat{\theta}_0$, where $\hat{\theta}_0$ is any solution of the normal equations. Let M^- be the Moore-Penrose g-inverse of M. The covariance matrix of $\hat{t}_0$ under $V = V_0$ is

$$\operatorname{Cov}(\hat{t}_0 \mid V_0) = \sigma^2 B(X^T X)^- B^T.$$

The estimability of t is unaffected by the choice of V in $\mathscr{V}$. The BLUE for t under a general V in $\mathscr{V}$ is the weighted least squares (WLS) estimator

$$\hat{t}_V = B(X^T V^{-1} X)^- X^T V^{-1} Y,$$

whose covariance matrix under V is

$$\operatorname{Cov}(\hat{t}_V \mid V) = B(X^T V^{-1} X)^- B^T.$$

(It should cause no confusion that $\hat{t}_{V_0}$ has been abbreviated $\hat{t}_0$.) We may also calculate the covariance matrix of the LS estimator $\hat{t}_0$ under V, which is

$$\operatorname{Cov}(\hat{t}_0 \mid V) = B(X^T X)^- X^T V X (X^T X)^- B^T.$$

(Note that $\hat{t}_0$ is an unbiased estimator of t under V.) Writing $A_0 = X^T X$ and $A_V = X^T V^{-1} X$ and $V = \sigma^2 (I + \Gamma)$, we obtain (as proved by Strand (1973) in the case $B = I$)

$$\begin{aligned} &\sigma^{-2}[\operatorname{Cov}(\hat{t}_0 \mid V) - \operatorname{Cov}(\hat{t}_V \mid V)] \\ &\qquad = \sigma^{-2}(A_0^- X^T V - A_V^- X^T) V^{-1} (A_0^- X^T V - A_V^- X^T)^T B^T \qquad (2.1) \\ &\qquad = B A_0^- X^T \Gamma [I - X A_0^- X^T] \Gamma X A_0^- B^T + O(\Gamma^3) = O(\Gamma^2), \end{aligned}$$

the last as $\Gamma \to 0$.

If we take, for example, the case of $\Gamma = \rho L$ where L is a fixed symmetric matrix and ρ is sufficiently small, then $O(\Gamma^2)$ becomes $O(\rho^2)$.

A well-known necessary and sufficient condition that $\mathrm{Cov}(\hat{t}_0 \mid V) = \mathrm{Cov}(\hat{t}_V \mid V)$, or, equivalently, that $\hat{t}_0 = \hat{t}_V$ (for all Y), is that

$$VXA_0^- B^T \subset R(X) \tag{2.2}$$

where $R(U)$ is the column space (range) of a matrix U. This condition, and similar ones, can be found in the work of a number of authors, such as Zyskind (1967), Rao (1967), Watson (1967, Theorem 1), McElroy (1967), Kruskal (1968), and others. It is interesting to note that condition (2.2) in turn is a necessary and sufficient condition for the first term in the third form of (2.1) to be zero, which is equivalent to

$$[I - XA_0^- X^T]\Gamma XA_0^- B^T = 0.$$

Thus we cannot choose X, Γ and B to get a better approximation than $O(\Gamma^2)$ in (2.1) without forcing $\mathrm{Cov}(\hat{t}_0 \mid V) = \mathrm{Cov}(\hat{t}_V \mid V)$ and hence $\hat{t}_0 = \hat{t}_V$.

The expansion (2.1) says that in an approximate sense we are justified in using the ordinary least squares estimate if we feel that any autocorrelation present is small. Furthermore it supplies our basic motivation for the approach adopted in this paper. Thus assuming V is unknown, but that we are primarily concerned with V's whose perturbation from $V = V_0$ is small, we suggest the following robustness argument. It is somewhat similar in style to the approach taken by Box, Draper, and others in the search for designs robust against the possible presence of higher order polynomial terms in regression models, that is, the so-called variance-bias methods. Our motivation is also two-stage:

(1) Find the class of designs in $\mathscr{X}$, say $\mathscr{X}^*$, which are optimum (in some specified sense) under $V = V_0$, using $\hat{t}_0$ as our estimator of t.

(2) Among all the designs $\mathscr{X}^*$ from stage (1), find the class of designs $\mathscr{X}^{**}$ which are optimum, again using $\hat{t}_0$ but under a specific structure V or class of V that may be present. (These V can be quite far from V_0.)

Thus at stage (1) we seek to minimize specified functional(s) Φ of $\mathrm{Cov}(\hat{t}_0 \mid V_0)$ (or of its g-inverse, and at stage (2) we minimize functionals Ψ of $\mathrm{Cov}(\hat{t}_0 \mid V)$. We now consider this process in more detail.

3. Optimum designs. In the settings considered in this paper we take the components $t_1, \ldots, t_v$ of t to be certain estimable contrasts $t_i = \sum_j b_{ij}\theta_j$; that is, $\sum_j b_{ij} = 0$ $\forall i$. (It may help the reader to think of these in terms of (4.2) in the BIBD setting.) To stress the dependence of the covariances on the design, we define, for X in $\mathscr{X}$,

$$D(X, V) = \mathrm{Cov}(\hat{t}_0 \mid V).$$

We will always be dealing with choices of the t_i for which

$$D(X, V) \text{ has zero row and column sums } \forall X, \tag{3.1}$$

and hereafter *assume* that (3.1) holds. Define

$$C(X, V) = D^-(X, V). \tag{3.2}$$

This, too, has zero row and column sums. When $V = V_0$ and the $t_1, \ldots, t_v$ are appropriately defined (as in settings considered later herein), $C(X, V_0)$ is the so-called "C-matrix" of experimental design theory. We emphasize that, when $V \neq V_0$, $C(X, V)$ is the inverse of the covariance matrix of the LS estimator $\hat{t}_0$, *not* of some $\hat{t}_{V_1}$.

Kiefer (1975) pointed out that the class of convex decreasing functionals Φ on the set of C is more general than the class of convex increasing functionals Ψ on the matrices $D = C^-$, since $\Phi(C) = \Psi(C^-)$ is convex if Ψ is. The strict inclusion of one class in the other is illustrated by the D-optimality criterion (as discussed below). Thus it is more satisfactory to know that a design is optimum relative to all suitably symmetric convex criteria on the class of C-matrices than to know such a result on the class of D-matrices. This is the spirit of the definition employed in Kiefer (1975) in the settings of

(i) two factors, blocks and treatments with a fixed number b of blocks and fixed block size k (with v treatments and the t_i appropriate treatment contrasts),

(ii) three factors, rows, columns, and treatments in a $v \times v$ array with one observation per cell (with the t_i contrasts of effects of any one of the factors).

A design is called *universally optimum relative to* $\mathscr{X}$ if it minimizes $\Phi(C(X, V_0))$ over $\mathscr{X}$ for every Φ which is convex and invariant under permutations of coordinates (rows and columns of C) and has the property that

$$\Phi(bC) \leq \Phi(C) \qquad \forall b > 1$$

(or for every Φ which is an increasing function of such a function).

In the settings (i) and (ii) above the BIBD's (if they exist) and the Latin squares are the only universally optimum designs (and in fact are the only optimum designs for any single strictly convex Φ). They will therefore exactly comprise the class $\mathscr{X}^*$ referred to at the end of Section 2. Note that when $k = v$ in (i), $\mathscr{X}^*$ is exactly a set of b complete blocks.

We now turn to the second stage of optimization referred to in Section 2. The computation of $C(X, V)$ is often complicated, even for $X = X^* \subset \mathscr{X}^*$. In essence the procedure involves first computing $D(X^*, V)$ and then obtaining $C(X^*, V)$ as its Moore-Penrose inverse, which may be a formidable task if it is to be exhibited as a specific function of X, Γ and B. We try to avoid this inversion by use of an analogue of the method used to prove universal optimality, but cannot achieve quite such a strong optimality result. Call X^{**} *weakly universally optimum relative to* $\mathscr{X}^*$ for a covariance matrix V if it minimizes $\Psi(D(X^*, V))$ over $\mathscr{X}^*$ for every convex Ψ invariant under permutation of coordinates and such that

$$\Psi(bD) \geq \Psi(D) \qquad \forall b > 1$$

(or for every Ψ which is an increasing function of such a function). An even simpler argument than that used to prove universal optimality then shows that X^{**} is weakly universally optimum relative to $\mathscr{X}^*$ for covariance matrix V if

(3.3) (a′) $D(X^{**}, V)$ is completely symmetric (CS),

(b′) trace $D(X^{**}, V) = \min X^* \in \mathscr{X}^*$ trace $D(X^*, V)$

(If Ψ is strictly convex and X^{**} satisfies (3.3), no other D can be Ψ-optimum.)

The criteria Ψ covered by this definition include (with λ_i = eigenvalues)

$$\Psi_p(D) = \textstyle\sum_1^{v-1} \lambda_i^p(D) \qquad \text{for } p \geq 1$$

(which for $p = 1$ is the A-optimality criterion, equivalent to $\sum_{i,j} \operatorname{Var}(\hat{t}_{0i} - \hat{t}_{0j} \mid V)$),

$$\Psi_\infty(D) = \max_i \lambda_i(D) \qquad (E\text{-optimality}),$$

but not, unfortunately,

$$\Psi_0(D) = \textstyle\sum_1^{v-1} \log \lambda_i(D) \qquad (D\text{-optimality}).$$

(The criteria Ψ_p, which depend only on the λ_i, are not merely permutation invariant, but are also orthogonally invariant. Two other criteria, which are of the former but not the latter form, are mentioned at the end of this section.)

Conditions (3.3) (a′) and (b′) also imply, by a similar averaging argument to that used to prove (weak) universal optimality, the optimality of d^* for every Ψ which is a nondecreasing Schur-convex function of the ordered eigenvalues of D. See Giovagnoli and Wynn (1980) or Constantine (1980) for a full discussion in the more usual context of $\Phi(C)$. The use of Schur convexity in design optimality was suggested to one of the present authors by I. Olkin about 1970, and design majorization arguments have appeared in Cheng (1979) and Kiefer (1975, page 339). D-optimality is also excluded in these terms: a competitor of d^* satisfying (3.3) could in general have larger trace(D) but smaller det(D).

In some settings, however, D-optimality and more can be concluded for V near V_0 (and V in $\mathscr{V}$); the following two conditions will be shown to suffice: (1) Suppose (as is the case for $\mathscr{X}^*$ consisting of all BIBD's or all Latin squares) that $D(X, V_0) = D_0$ (say) is the same for all X in $\mathscr{X}^*$ and that $\mathscr{X}^*$ is finite. (2) Suppose furthermore, that for fixed σ there is a constant $c_1 > 0$ such that as $D \to D_0$

$$\Psi(D) = \Psi(D_0) + c_1 \operatorname{trace}(D - D_0) + o(D - D_0)$$

for D satisfying (3.1) (as is the case for Ψ_0 and other suitably regular and symmetric Ψ when D_0 is CS, which is true for D_0 in the examples we have mentioned). Under these conditions since (Section 2) $\sigma^{-2}\operatorname{Cov}(\hat{t}_0 \mid V) = B(A_0^- + A_0^- X^T \Gamma X A_0^-)B^T$, we have

$$\sigma^{-2} \operatorname{trace} D(X, V) = \operatorname{trace} BA_0^- B^T + \operatorname{trace}(XA_0^- B^T BA_0^- X^T)\Gamma,$$

where, as above, $V = \sigma^2(I + \Gamma)$. It then follows from the finiteness of $\mathscr{X}^*$ and the fact that $D(X, V) - D_0 = O(\Gamma)$ as $\Gamma \to 0$ that only an A-optimum X in $\mathscr{X}^*$ can possibly minimize $\Psi(D(X, V))$ for V sufficiently near V_0. Moreover, if all A-optimum X^* yield the same $D(X^*, V)$ for V sufficiently near V_0, we conclude that all these X^*, and only these, are also Ψ-optimum for V in a neighborhood of V_0 (and in $\mathscr{V}$). This last is true in some of the examples of interest.

The results obtained in Sections 4 and 5 are of the following kind. At stage (1) the Latin squares, BIBD's, etc., are universally optimum under V_0 for the $\mathscr{X}$ under consideration, and thus constitute the class $\mathscr{X}^*$. At stage (2) within $\mathscr{X}^*$ the weakly universally optimum designs for given V (or Γ) are found (and the optimality can be extended locally as indicated in the previous paragraph). The form of $\hat{t}_0$, whose components are the LS estimators of some standard treatment contrasts, are simple because of the balanced or orthogonal nature of the designs in $\mathscr{X}^*$. This makes calculation of $D(X^*, V)$ a relatively easy exercise. This means, in turn, that minimization of a $\Psi(D(X^*, V))$ is *easier* than minimization of a $\Psi(\operatorname{Cov}(\hat{t}_V \mid V))$ notwithstanding the simple form of $\operatorname{Cov}(\hat{t}_V \mid V)$ given in Section 2.

It will be a feature of the examples discussed in Sections 4 and 5 that trace$D(X^*, V)$ is a constant for all X^* in $\mathscr{X}^*$. Thus with respect to minimizing the A-optimality criterion trace(D), all such X^* are equally good. Furthermore if we can find an X^* for which all the diagonal elements of $D(X^*, V)$ are equal and hence all $\operatorname{Var}(\hat{t}_{0i} \mid V)$ are equal, then obviously this X^* minimizes $\max_i \operatorname{Var}(\hat{t}_{0i} \mid V)$. (This certainly holds for any X^* with the CS property (3.3)(a′), but this last condition is not necessary, just as it was not for A-optimality.) Although this last optimality criterion is not of as much interest as the Ψ_p because of the meaning of the $\hat{t}_{0i}$ (see (4.2)), it has been considered in the literature, e.g., in Duby et al. (1977). (A-optimality is more meaningful because of its equivalent meaning, given shortly below (3.3).) Of greater interest is the criterion $\max_{i,j} \operatorname{Var}(\hat{t}_{0i} - \hat{t}_{0j} \mid V)$, the maximum of variances of estimators of "principal contrasts" $\alpha_i - \alpha_j$; when all X^* in $\mathscr{X}^*$ have the same value of tr$D(X^*, V)$ and (3.1) holds, any X^* satisfying (3.3)(a′) is optimum in this sense. The last two criteria mentioned are additional to the orthogonally invariant Ψ_p, but both also fall within weakly universal optimality.

4. M-way layouts. In this paper the examples will be confined to special M-way layouts, for models with M factors. In the present section we consider settings where balanced orthogonal designs (complete blocks, Latin squares or hypercubes, etc.) exist.

Consider a model with M factors labelled $1, \ldots, M$. Let

$$N(i_1, \ldots, i_M)$$

be the number of observations at the i_jth level of factor j ($j = 1, \ldots, M$). We shall assume for simplicity when $M \geq 3$ that $N(i_1, \ldots, i_M) = 0$ or 1 for all $i_1, \ldots, i_M$. Factor j is assumed to have n_j levels ($j = 1, \ldots, M$).

Factor 1 will be called the treatment factor and its levels called *treatments*. We put $n_1 = v$ in accordance with standard notation. The usual linear additive model is

$$E(Y_{(i_1,\ldots,i_M)}) = \alpha_{i_1}^{(1)} + \cdots \alpha_{i_M}^{(M)},$$

where $Y_{(i_1,\ldots,i_M)}$ is the observation (if there is one) at level i_j of factor j ($j = 1, \ldots, M$). If Y is the vector of $Y_{(i_1,\ldots,i_M)}$ in some order (e.g., lexicographic), then let $V = \mathrm{Cov}(Y)$. In this section for $M \geq 3$ we shall take V to be such that

$$\begin{aligned} \mathrm{Cov}(Y_{(i_1,\ldots,i_M)}, Y_{(i_1',\ldots,i_M')}) &= \sigma^2 \qquad \text{if } \textstyle\sum_{j=2}^M |i_j - i_j'| = 0 \\ &= \rho\sigma^2 \qquad \text{if } \textstyle\sum_{j=2}^M |i_j - i_j'| = 1 \qquad (4.1) \\ &= 0 \qquad \text{otherwise.} \end{aligned}$$

This is a "nearest neighbor" (NN) correlation structure in that observations are considered to be positioned by the levels of their last $M-1$ factors, and an observation has correlation ρ with its nearest neighbors but not with observations "further away". The NN structure is described slightly differently, below and in Section 5, for $M = 2$; this is because we are considering (for example) only one Latin square for $M = 3$, but several blocks for $M = 2$.

The dot notation will be used for summation (rather than averaging) here, for convenience in considering N as well as Y:

$$N(i_1, \cdot, \ldots, \cdot) = \textstyle\sum_{i_2=1}^{n_2} \cdots \sum_{i_M=1}^{n_M} N(i_1, \ldots, i_M),$$

etc. Similarly,

$$Y_{(i_1,\cdot,\ldots,\cdot)} = \textstyle\sum_{i_2=1}^{n_2} \cdots \sum_{i_M=1}^{n_M} Y_{(i_1,\ldots,i_M)},$$

etc.

Define the standard treatment contrasts (chosen to make $t_i - t_j = \alpha_i^{(1)} - \alpha_j^{(1)}$),

$$t_i = \alpha_i^{(1)} - v^{-1} \textstyle\sum_1^v \alpha_j^{(1)} \qquad i = 1, \ldots, v. \qquad (4.2)$$

We shall suppress the superscript so that $\alpha^{(1)} \equiv \alpha$ in all that follows. Thus $t = (t_1, \ldots, t_v)^T$ is the vector t of the previous sections.

When $M = 2$ we consider $\mathscr{X}$ to be the class of designs for which

(i) $n_1 = v$, $n_2 = b$,

(ii) $N(\cdot, j) \leq v$.

(Often (ii) is strengthened to assume equality.) That is, we have b "blocks", each permitted to be of size $\leq v$. A treatment may appear more than once in a block. It is well known that the complete block designs, with b blocks of size v and each treatment repeated once per block, are uniquely universally optimum (Kiefer, 1958, 1975), and we take the class of such designs to be $\mathscr{X}^*$. Of course $\hat{t}_{0i} = b^{-1}Y_{(i,\cdot)} - (bv)^{-1}Y_{(\cdot,\cdot)}$. To save space, we postpone detailed calculations of $D(X^*, V)$ until the block design considerations of Section 5 for general block size k. The NN covariance structure (as in Section 5) is that observations in different blocks are uncorrelated, and observations in the same block are correlated only if they are neighbors. (The position in a block can refer to time or to position in a linear array of plots, etc.) Theorem 5.1, specialized to the present case, then shows that a design X^* in $\mathscr{X}^*$ is weakly universally optimum (for every possible value $\rho \neq 0$) if and only if, for all pairs i, i' with $1 \leq i < i' \leq v$, the quantities

$$v\#\{j \mid i, i' \text{ are NN's in block } j\} + \#\{j \mid i \text{ is at an end of block } j\} + \#\{j \mid i' \text{ is at an end of block } j\}$$

are the same. (If such an X^* exists, no X not satisfying this condition can be optimum for strictly convex Ψ.) A simple counting argument (special case of (5.5)) shows that a necessary condition for this is that $v \mid b$ when v is odd and $v \mid 4b$ when v is even; however, when $4 \mid v$ the value $b = v/4$ is too small for the last displayed expression to be independent of i, i', since at most half the treatment pairs can occur as NN's. We thus find that $b = v$ is the smallest number of blocks for v odd and that $b = v/2$ is the smallest number when v is even. Designs satisfying these conditions, and which are thus complete block designs of minimal size to be weakly universally optimum for the NN model, can be obtained by taking the blocks to be the columns of the Latin square design of Theorem 4.2 below when v is odd, and to be the first $v/2$ columns of that Latin square design when v is even.

When $M = 3$ define the class of designs $\mathscr{X}$ with

(i) $n_1 = n_2 = n_3 = v$,

(ii) $N(i, j, k) = 0$ or 1 and $N(\cdot, j, k) = 0$ or 1 $(1 \le i, j, k \le v)$.

(Often one assumes all $N(\cdot, j, k) = 1$, but this is not necessary.) Kiefer (1958, 1975) proves that Latin squares, namely those designs for which

$$N(i, j, \cdot) = N(i, \cdot, k) = 1$$

$(1 \le i, j, k \le v)$ are the class of universally optimum designs, $\mathscr{X}^*$, within this $\mathscr{X}$. (Of course, such designs do have all $N(\cdot, j, k) = 1$.)

Because of the orthogonal and balanced nature of Latin squares the LS estimators of the t_i have a simple form, namely,

$$\hat{t}_{0i} = v^{-1} Y_{(i,\cdot,\cdot)} - v^{-2} Y_{(\cdot,\cdot,\cdot)}.$$

From this we can calculate for a Latin square X^*

$$D(X^*, V_0) = v^{-1}\sigma^2(I_v - v^{-1}J_{v,v}).$$

Let $N_{ii'}$ denote the number of times that treatments i and i' are adjacent in the square X^* in any direction (vertically, horizontally but *not* diagonally, and counting each adjacency just once). For any Latin square X^* we have

$$\sum_{i'(\neq i)} N_{ii'} = 4(v - 1), \tag{4.3}$$

$$\sum_{i<i'} N_{ii'} = 2v(v - 1).$$

Moreover if all the $N_{ii'}$ are equal for $i \neq i'$, their common value is 4.

Making use of (4.3), after some manipulation we find that with V of (NN) form (4.1),

$$D(X^*, V) = v^{-1}\sigma^2(I - v^{-1}J) - (v - 1)v^{-3}4\rho\sigma^2 J + v^{-2}\rho\sigma^2 N, \tag{4.4}$$

where N is the matrix whose i, i'th entry is $N_{ii'}$ and whose diagonal entries are 0.

Notice that since the diagonal entries of $D(X^*, V)$ are the same for all Latin squares X^*, all such X^* are equally good with respect to the A-optimality and $\min_{X^*}\max_i \operatorname{Var}(\hat{t}_{0i} \mid V)$ criteria, as described in Section 3. Also there are no "edge effects" which typically arise with some other types of V matrices and $\hat{t}$. (See Martin, 1977.)

We can immediately see that (b′) of (3.3) is automatically satisfied and that (a′) is satisfied if and only if all $N_{ii'}$ are equal. We shall soon see that such designs always exist. Thus we have

THEOREM 4.1. *Among the class $\mathscr{X}^*$ of Latin square designs and under the nearest neighbor covariance V (for each $\rho \neq 0$), the class of weakly universally optimum designs is the class of designs with all $N_{ii'}$ equal* $(1 \le i, i' \le v;\ i \neq i')$.

The condition "all $N_{ii'}$ equal" (= 4) is much weaker than the condition that a square be "complete", that is, row- and column-complete. See Dénes and Keedwell (1974) for a discussion of the latter. However the method, first proposed by E. J. Williams (1949) and rediscovered by a number of authors, which is used to construct complete Latin squares when v is even, can be used to construct squares with all $N_{ii'}$ equal for all values of v. The construction is given by

THEOREM 4.2. *Form a* $v \times v$ *square whose* (j, k) *cell* $(1 \leq j, k \leq v)$ *contains the treatment number*

$$\sum_{r=1}^{j} (-1)^r (r-1) + \sum_{r=1}^{k} (-1)^r (r-1),$$

reduced (mod v). *This has all* $N_{ii'}$ *equal.*

EXAMPLE 4.1. For $v = 5$, identifying $0 \equiv 5$, we obtain

5	1	4	2	3
1	2	5	3	4
4	5	3	1	2
2	3	1	4	5
3	4	2	5	1

A common alternate description of this construction places successive integers in alternate rows of the first column on the way down and continues this process on the way back, completes the design cyclically in rows, and then reorders columns so that the first row and first column are the same. Notice that these squares have the property that in any row (column) the differences between *successive* treatment numbers produce every nonzero residue (mod v) exactly twice. This property is discussed extensively in Section 5 for BIBD's.

The method of Theorem 4.2 when v is even can be used to give designs with columns of size v but rows of size only $v/2$ which still have the property that each i, i' pair of adjacent treatments occurs once in every column. The reason that we may use a number of columns equal to only half of the number of rows, compared to the full number used in repeated measurement designs, is that we do not distinguish the order of adjacent pairs. As an example consider the design on page 232 of Hedayat and Afsarinejad (1975) for $v = 6$. The first three columns are $(1, 6, 2, 5, 3, 4)^T$, $(2, 1, 3, 6, 4, 5)^T$ and $(3, 2, 4, 1, 5, 6)^T$. Each column gives successive differences equivalent to (5, 2, 3, 4, 1) (mod 6), that is, all nonzero residues (mod 6). Each adjacent i, i' pair occurs just once. Notice the somewhat different order to that in Theorem 4.2. Also, the order of columns is immaterial in the Hedayat-Afsarinejad design, so that the latter setting is closer to that of block designs discussed in Section 5.

The results for Latin squares can be extended to higher way layouts and in particular to Latin cubes and hypercubes.

Suppose, then, for $M \geq 3$ and for a given positive integer $p < M - 1$, that $\mathscr{X}$ is the class of designs with

(i) $n_1 = v = k^p$, $n_j = k (j = 2, \ldots, M)$,

(ii) $N(i_1, \ldots, i_M) = 0$ or 1 and $N(\cdot, i_2, \ldots, i_M) = 0$ or 1 $(1 \leq i_1 \leq v, 1 \leq i_2, \ldots, i_M \leq k)$.

The previous optimality considerations can be extended to M-way layouts (Kiefer, 1958, page 690), and in fact a similar argument to that in Kiefer (1975) shows that the fully balanced and orthogonal designs are universally optimum; these are the designs X^* for which

$$N(i_1, i_2, \cdot, \ldots, \cdot) = N(i_1, \cdot, i_3, \ldots, \cdot) = \cdots = N(i_1, \cdot, \cdot, \ldots, i_M) = k^{M-p-2}.$$

(This implies that all $N(\cdot, i_2, \ldots, i_M) = 1$.) That is, every $k^p \times k$ two-way layout formed by taking treatments and one other factor (and ignoring the others) is a complete two-way layout with k^{M-p-2} observations in each cell.

We hereafter let $\mathscr{X}^*$ be the class of such designs X^*.

It is convenient to think of the k^{M-1} cells of the $(M-1)$-cube, with k levels in each direction, as specifying the last $M-1$ factors. Then a design in $\mathscr{X}$ is an assignment of treatment labels, one to each cell, or with none to a cell for which $N(\cdot, i_2, \ldots, i_M) = 0$. The designs X^* in $\mathscr{X}^*$ were called Latin hypercubes by Kishen (1949) when $M > 3$. (For discussion of other definitions see Kerr et al., 1973.) When $M = 3$, $p = 1$, the formulation will be seen to reduce to the earlier Latin square setup.

By a calculation similar to that for Latin squares, again using the NN structure V, one can show that, for $i \neq i'$, a Latin hypercube with $n = k^{M-1}$, $v = k^p$, and $r = n/v$ replications of each treatment, has

$$\mathrm{Cov}(\hat{t}_{0i}, \hat{t}_{0i'}) = -n^{-1}\sigma^2 + \rho\sigma^2 r^{-2}[N_{ii'} - v^{-1}(N_{i\cdot} + N_{i'\cdot}) + v^{-2}N_{\cdot}],$$

where $N_{ii'}$ counts the number of times treatments i and i' are neighbors in any of the $M-1$ coordinate directions (not diagonally) and where $N_{i\cdot} = \sum_{i'} N_{ii'}$ and $N_{\cdot} = \sum_i N_{i\cdot}$; we note that the sum over i' includes N_{ii}, which need not be 0 for a Latin hypercube (as it was automatically for a Latin square) unless $p = M - 2$. A simple counting argument yields

$$N_{\cdot} = \textstyle\sum_{i,i'} N_{ii'} = 2(M-1)k^{M-2}(k-1).$$

Thus, equality of all $N_{i\cdot}$ and of all $N_{ii'}$ $(i \neq i')$ implies (3.3)(a′), and since (3.1) entails $\mathrm{tr}\, D = -\sum_{i \neq i'} \mathrm{Cov}(\hat{t}_{0i}, \hat{t}_{0i'})$, we see that (3.3)(b′) will be satisfied if the design minimizes

$$\sigma^{-2}\,\mathrm{tr}\, D = n^{-1}v(v-1) - \rho r^{-2}(v^{-1}N_{\cdot} - \textstyle\sum_i N_{ii}), \qquad (4.5)$$

which reduces to what one obtains from (4.4) in the Latin square case.

Our $\mathscr{X}^*$ consists of the Kishen hypercubes, for which we have mentioned that it is not automatic that $N_{ii} = 0$ unless $p = M - 2$. (Greater "strength" of the array, restricting designs to a small subset of the designs universally optimum under V_0, would be required.) Hence we can no longer minimize (4.5) without knowledge of sgn ρ. There are three options, which parallel those considered in Kiefer (1960) in a simpler setting:

(α) Assume it known that $\rho \geq 0$, so that we do want all $N_{ii} = 0$;

(β) assume it known that $\rho \leq 0$, so we want $\sum_i N_{ii}$ maximized;

(γ) not knowing sgn ρ, take a minimax approach, by minimizing $|N_{\cdot} - v\sum_i N_{ii}|$.

The most interesting and perhaps the most natural of these is (α), which is the only formulation we consider in the remainder of this section.

Under assumption (α), then, the weakly universally optimum designs in $\mathscr{X}^*$ are the Latin hypercubes with all $N_{ii} = 0$ and with all $N_{ii'}$ equal for $i \neq i'$. This requirement places rather rigid constraints on the parameter values, since we must equate the total number of adjacencies with the common value of $N_{ii'}$ times the total number of i, i' pairs. That is, $N_{\cdot}/2$ must be a multiple of the total number of treatment pairs, $k^p(k^p - 1)/2$. When $p = 1$ this is always achievable. When $p > 1$ we obtain that $2(M-1)$ is a multiple of $k^{p+2-M}(k^p - 1)/(k-1)$, and since k^{M-p-2} is relatively prime to $(k^p - 1)/(k-1)$ this means that

$$2(M-1) = h(k^p - 1)/(k-1) \qquad (4.6)$$

where h is a positive integer; and each treatment pair then occurs hk^{M-p-2} times as a neighboring pair, if the desired design is achieved. When $p = 2$, we obtain that $M-1$ is a multiple of $(k+1)/2$ for k odd ($k = 1, 3, 5, \ldots$) and of $k+1$ for k even ($k = 2, 4, 6, \ldots$). (For $k = 3$, $M-1$ must not only be a multiple of 2 but must also be ≥ 3 to satisfy $p < M - 1$; that is, $h \geq 2$.) For larger p, except for the case $p = 3$, $k = 2$, when $M-1$ must be at least 7, we find that the number of observations $n = k^{M-1}$ is always at least 2^{15} (and $M - 1 \geq 13$); see Table 1. This removes such designs, if they exist, from the realm of practicality for almost all applications, and suggests the usefulness of future investigation of smaller designs which are not balanced but are close to it.

When $p = 1$ we can obtain a design of the required form by an extension of the method of Theorem 4.2 to more than 2 dimensions.

TABLE 1
Minimum value of $M - 1$ satisfying (4.6)

		p			
		2	3	4	5
k	2	3	7	15	31
	3	2	13	20	121
	4	5	21	85	341
	5	3	31	78	781

THEOREM 4.3. *The following design with v treatments, each replicated v^{M-2} times, with one observation in each cell of an $(M-1)$-dimensional hypercube of side v, has all $N_{ii'}$ $(i \neq i')$ equal and all $N_{ii} = 0$. In the $(i_2, \ldots, i_M)$ cell of the hypercube place the treatment number*

$$\sum_{j=2}^{M} \sum_{r=1}^{i_j} (-1)^r (r-1),$$

reduced (mod v).

When $p > 1$ it is difficult to find designs with the right structure. Here are two of the simplest examples. Both constructions can be described in the geometric terms of Kishen (1949), and we represent them in the resulting notation without giving detailed description of the geometric configurations.

EXAMPLE 4.2. For $k = 2$, $p = 2$, $M - 1 = 3$, the levels of the last 3 factors are represented as $x_1, x_2, x_3 = 0$ or 1. The treatment in cell (x_1, x_2, x_3) is $(x_1 + x_3, x_2 + x_3)$ mod (2, 2). Rewriting the resulting treatments mod (2, 2) as 0, 1, 2, 3 (by considering them as binary numbers), we obtain

0	1		3	2
2	3		1	0

for the two "layers" for $x_3 = 0, 1$.

EXAMPLE 4.3. For $k = 3$, $p = 2$, $M - 1 = 4$, letting the levels of the last 4 factors be represented as $x_1, x_2, x_3, x_4 = 0$, 1 or 2, and letting the treatment in cell (x_1, x_2, x_3, x_4) be $(x_1 + 2x_3 + 2x_4, x_2 + x_3 + 2x_4)$ mod (3, 3), we obtain an equineighbored design in which each treatment pair is adjacent six times. Rewriting the treatments 0, 1, $\cdots$, 8 by considering them as ternary numbers, and letting x_1, x_2 be the rows and columns of "little" squares below and x_3, x_4 be the "large" rows and columns, we obtain

012	867	453
345	201	786
678	534	120
786	345	201
120	678	534
453	012	867
534	120	678
867	453	012
201	786	345.

Not all choices of parallel pencils of 2-flats for treatments in Kishen's construction will yield an equineighbored design. For example, the assignment $(x_1 + x_2 + x_4, x_2 + x_3 + x_4)$ of treatments will yield a hypercube, but it is not equineighbored.

Despite considerable trial and error, the authors have been unable to find a solution to the case $k = 5, p = 2, M - 1 = 3$. Simple trial and error methods break down after a few "layers" have been obtained.

A direct analogue for $M > 2$ of the consideration of several blocks when $M = 2$ is the consideration of several uncorrelated Latin squares or hypercubes. The use of several hypercubes extends the parameter values k, M, p for which designs with the right structure for weak universal optimality can be obtained. We shall not pursue this here. Similarly, the considerations can be extended to the larger but less practical design settings in which block size (for $M = 2$) or possibly unequal Latin rectangle or hyperrectangle sides are multiples of v, so that $\mathscr{X}^*$ still consists of designs with classical orthogonality and balance properties under V_0. The considerations of R.M. Williams (1952) and of Kiefer (1960) can be viewed as treating, for $M = 2$ and a different covariance structure, a single block ($b = 1$) of length k much greater than v. For the approach of the present paper, k would be a multiple of v and $\mathscr{X}^*$ would consist of designs for which each treatment appears k/v times. No design in $\mathscr{X}^*$ satisfies (3.3)(a′), but (for example) designs with all $N_{ii} = 0$ and all $N_{ii'}$, as nearly equal as possible will be close to weak universal optimality when k is large, if $\rho \geq 0$ is assumed.

5. Non-orthogonal settings.

5.1. *Optimality*. Consider the designs with $M = 2$ factors. Take $\mathscr{X}$ to be the class of block designs with b blocks, v treatments and block size k. Thus, as in the case $M = 2$ of Section 4, we assume

(i) $n_1 = v, \qquad n_2 = b$
(ii) $N(\cdot, j) \leq k \qquad (1 \leq i \leq v, 1 \leq j \leq b)$,

where (ii) is strengthened to equality if we impose complete use of all plots in each block, and where $N(i, j)$ is often restricted to be 0 or 1 if $k \leq v$. With or without such restrictions, the universally optimum designs, $\mathscr{X}^*$, are the balanced block designs with parameters b, v, k, when such designs exist; see Kiefer (1958, 1975). We shall only consider the case of $k < v$ here, the case of BBD's when $k > v$ requiring only slight modification in our development and being of less practical interest; the case $k = v$ was treated in Section 4. As usual, we use r and λ to denote the replication and treatment intersection numbers.

For BIBD let T_i be the sum of the observations on treatment i, i.e., $T_i = Y_{(i,\cdot)}$; and let B_j be the sum of observations on block j, $B_j = Y_{(\cdot,j)}$. LET A_i be the set of blocks in which treatment i occurs: $A_i = \{j \mid N(i, j) = 1\}$. Defining t_i as in (4.2), we have by the standard analysis of the BIBD

$$\hat{t}_{0i} = Q_i/\lambda v,$$

where

$$Q_i = kT_i - \textstyle\sum_{j \in A_i} B_j.$$

As in Section 4, $\hat{t}_{0i}$ is then the LS estimator of $\alpha_i^{(1)} - v^{-1} \sum_1^v \alpha_j^{(1)}$.

It is slightly more convenient to work with $Q = (Q_1, \cdots, Q_V)^T$ rather than $\hat{t}_0$, so we define

$$\bar{D}(X^*, V) = \mathrm{Cov}(Q \mid V).$$

We distinguish the position of an observation in a block. So let $g(j, r)$ be the treatment number of the rth observation in the jth block. We now list the observations $Y_{(i,j)}$ by lexicographic order of (j, r), not (i, j). Assuming a fixed covariance structure determined only by position in a block, but assuming zero covariance between blocks, we may write

$$\begin{aligned}\operatorname{Cov}(Y_{(i,j)}, Y_{(i',j')}) &= \sigma^2 && \text{if } |i-i'|+|j-j'|=0,\\ &= \rho_{rs}\sigma^2 && \text{if } j=j' \text{ and } i=g(j,r),\ i'=g(j,s),\\ &= 0 && \text{otherwise.}\end{aligned}$$

Define the $k \times k$ matrix $V^* = \{V^*_{rs}\}$ by $V^*_{rs} = \sigma^2\rho_{rs}$, where $\rho_{rr} = 1$. We assume this V^* is positive definite. Then the overall covariance structure is given by

$$V = I_b \otimes V^*.$$

Define $k \times v$ matrices P_j ($j = 1, \cdots, b$) by

$$\begin{aligned}\{P_j\}_{ri} &= 1 && \text{if } g(j,r)=i,\\ &= 0 && \text{otherwise.}\end{aligned}$$

Then we obtain

$$\bar{D}(X^*, V) = k^2 \textstyle\sum_{j=1}^{b} P_j^T W P_j, \tag{5.1}$$

where

$$W = (I_k - k^{-1}J_{k,k})V^*(I_k - k^{-1}J_{k,k}).$$

For $j \in A_i$ define the position of treatment i by $h(i, j) = r$ if $g(j, r) = i$. Then the (i, i')th entry of (5.1) is

$$\operatorname{Cov}(Q_i, Q_{i'}) = k^2 \textstyle\sum_{j\in A_i\cap A_{i'}} W_{h(i,j),h(i',j)}. \tag{5.2}$$

Thus

$$\operatorname{trace} \bar{D}(X^*, V) = bk^2 \operatorname{trace} W.$$

This means trace $\bar{D}(X^*, V)$ is independent of the choice of BIBD X^* so that every BIBD is equally good in terms of A-optimality for every V, and condition (b′) of (3.3) is automatically satisfied.

Now specialize to the NN covariance structure within a block. Put

$$\rho_{rs} = \begin{cases} 1 & \text{if } r = s,\\ r & \text{if } |r-s| = 1,\\ 0 & \text{otherwise.}\end{cases}$$

From (5.2) we then have

$$\begin{aligned}\sigma^{-2}\operatorname{Var}(Q_i) &= r[k(k-1) - 2\rho(k+1)] + 2\rho k e_i, \\ \sigma^{-2}\operatorname{Cov}(Q_i, Q_{i'}) &= -\lambda[k - 2\rho(k+1)] + k\rho[e_{ii'} + kN_{ii'}], \qquad i \neq i',\end{aligned} \tag{5.3}$$

where

(i) e_i is the number of blocks with treatment i at an end, that is,

$$e_i = \#\{j \mid g(j,1) = i \quad \text{or} \quad g(j,k) = i\};$$

(ii) $e_{ii'}$ is the number of blocks in $A_i \cap A_{i'}$ in which i occurs at an end plus the number where i' occurs at an end, that is,

$$\begin{aligned}e_{ii'} = &\#\{j \mid j \in A_i \cap A_{i'}, g(j,1) = i \quad \text{or} \quad g(j,k) = i\}\\ &+ \#\{j \mid j \in A_i \cap A_{i'}, g(j,1) = i' \quad \text{or} \quad g(j,k) = i'\},\end{aligned}$$

a block counting twice if i and i' are at the two ends of it; and

(iii) $N_{ii'}$ is the number of times i and i' are adjacent in a block:

$$N_{ii'} = \#\{j \mid g(j,r) = i, g(j,s) = i', |r-s| = 1\}.$$

Simple enumeration gives

$$\textstyle\sum_{i'(\neq i)} e_{ii'} = 2r + (k-2)e_i \tag{5.4}$$

and

(5.5) $$\sum_{i'(\neq i)} N_{ii'} = 2r - e_i.$$

From (5.2), (5.4), (5.5), and the fact that $\sum_i e_i = 2b$ for every design in $\mathscr{X}^*$, we thus have

THEOREM 5.1. *A* BIBD *is weakly universally optimum for the* NN *model if all the quantities*

$$e_{ii'} + kN_{ii'} \qquad (i \neq i')$$

are equal.

(See just below (3.3), regarding uniqueness.)

A sufficient condition is that all the $N_{ii'}$ are equal and all the $e_{ii'}$ are equal. To make all $\mathrm{Var}(\hat{t}_{0i})$ equal we only need all the e_i equal; this is relevant for handling criteria of the form $\sum_i f(\mathrm{Var}(\hat{t}_{0i}))$ with f convex.

The condition of Theorem 5.1 imposes an extra condition on the parameters of a BIBD for the theorem to be usable to show it is weakly universally optimum. Further enumeration gives $\sum_{i<i'} e_{ii'} = 2b(k-1)$ and $\sum_{i<i'} N_{ii'} = rv - b$. Thus equality of all $e_{ii'} + kN_{ii'}$ implies

(5.6) $$v(v-1) \mid 2b(k-1)(k+2).$$

Since $2b(k-1)(k+2) = v(v-1)[2\lambda(k+2)/k]$, (5.6) is equivalent to $k \mid 4\lambda$. If all $N_{ii'}$ are equal, condition (5.6) certainly holds since equality of the $N_{ii'}$ implies

(5.7) $$k \mid 2\lambda$$

(in which case all $N_{ii'} = 2\lambda/k$); in fact, (5.7) is equivalent to (5.6) if k is odd, and, more simply then, to $k \mid \lambda$.

Our main interest in the remainder of this section is in the construction, for given k and v, of BIBD's which satisfy the condition of Theorem 5.1 and for which b *is as small as possible*; the obvious design with $b = v(v-1) \cdots (v-k+1)/2$ (half the permutations of length k) is of little practical interest. We hereafter assume $k \geq 3$ to avoid trivialities.

It is clear that not all (v, b, k) for which a BIBD exists will satisfy (5.6). The familiar (7, 7, 3) is an example. Even when a BIBD satisfies (5.6) there need not exist a BIBD satisfying the condition of Theorem 5.1, as the following example shows.

EXAMPLE 5.1. For $v = b = 7$ and $k = 4$ the condition of Theorem 5.1 requires that all $e_{ii'} + 4N_{ii'}$ equal 6. Thus $N_{ii'} = 1$ and $e_{ii'} = 2$. Since all BIBD's with $v = b = 7$ and $k = 4$ are isomorphic if order within blocks (and of blocks) is ignored, we may assume the first ordered block is (1, 2, 3, 4). Since $\lambda = 2$, treatments 2 and 3 occur together in one other block. To make $e_{23} = 2$ that block must be $(2, x, y, 3)$ or its reverse. But by the known structure of this BIBD, neither x nor y can be 1 or 4. So we may take the block to be (2, 5, 6, 3). Repeating the argument, to make $e_{56} = 2$ we require a block $(5, x', y', 6)$, or its reverse, and the known structure of the BIBD forces $\{x', y'\} = \{1, 4\}$. Picking any fourth block of the BIBD and repeating the argument, one finds that one must be led either to the repetition of one of the first three blocks, which is not allowed, or else to another cycle of three blocks, with one block hence left over. (The latter actually never occurs.) In either case, the design cannot be completed with the desired structure.

An argument that sometimes works to construct designs with minimum b for which all $e_{ii'}$ and $N_{ii'}$ are equal, when $k = 4$, will now be illustrated for the above example. Take (1, 2, 5, 3) as the initial block. This is a difference set with the additional property that the difference ± 1, ± 3, ± 2 (mod 7) between pairs of *successive* treatments give all the nonzero residues (mod 7) exactly once. It is easy to see that all the $N_{ii'}$ are equal in the BIBD with $b = 7$ obtained by developing this initial block in the usual fashion, the jth block being obtained by adding $j - 1$ (mod 7) to each element of (1, 2, 5, 3). This method is considered more generally later. Moreover for any pair $\{i, i'\}$ with $i - i' \equiv \pm 3$ (mod 7) the pair occurs once as the middle pair of a block, contributing 0 to $e_{ii'}$, and once where one of i, i' is at an end. Thus for the 7 such $\{i, i'\}$ we have $e_{ii'} = 1$. Similarly for the 7 pairs with $i - i' \equiv \pm 2 \pmod 7$ we have $e_{ii'} = 3$ and for the remaining 7 pairs with $i - i' \equiv \pm 1 \pmod 7$ we have $e_{ii'} = 2$. Now consider the 7 blocks obtained from developing (5, 1, 3, 2). All the $N_{ii'}$ are still equal but the roles of $i - i' \equiv \pm 2$ and $i - i' \equiv \pm 3 \pmod 7$ have been reversed. Hence in the BIBD with all 14 blocks we have all $N_{ii'} = 2$ and all $e_{ii'} = 4$. Moreover the previous paragraph implies that $b = 14$ is the smallest value of b for which the condition of Theorem 5.1 can be satisfied when $v = 7$, $k = 4$.

The case $k = 3$ always yields a solution. The result of Hanani (1961) asserts that the usual necessary conditions for existence of a BIBD with given (v, b, k, r, λ) are sufficient when $k = 3$. Thus a design exists if and only if

$$\lambda(v-1) \equiv 0 \pmod 2, \qquad \lambda v(v-1) \equiv 0 \pmod 6.$$

We also recall that, as noted just below (5.7), the latter is equivalent to (5.6) for k odd as in the present case.

Thus Hanani's conditions and (5.6) reduce to the consideration of a family of designs depending on two nonnegative integers m and v with

$$k = 3, \quad \lambda = 3m, \quad b = mv(v-1)/2, \quad r = 3m(v-1)/2, \tag{5.8}$$

with m even or m and v odd.

THEOREM 5.2. *For $k = 3$ condition* (5.8) *is necessary and sufficient for the existence of a design satisfying the condition of Theorem* 5.1.

PROOF. When $k = 3$, each adjacent pair of varieties i, i' in the same block contributes 1 to each of $N_{ii'}$ and $e_{ii'}$; thus it suffices to order the blocks of a BIBD satisfying (5.8) in such a way that all $N_{ii'}$ are equal. Since $\pi_{ii'} = \lambda - N_{ii'}$ is the number of blocks in which both i and i' occur at the ends, it is sufficient to make all $\pi_{ii'}$ equal. Given a BIBD satisfying (5.8), write each block as the triple $\{\tau_1, \tau_2, \tau_3\}$ of subsets of size two contained in it. In what follows, two blocks of the design that contain identical elements are considered different. Let S_τ be the set of all blocks containing a fixed subset τ of size 2. For any p different subsets $\tau_1, \cdots, \tau_p$ of size 2 $\left(1 \le p \le \binom{v}{2}\right)$ the number of distinct blocks in $\cup_{i=1}^{p} S_{\tau_i}$ is at least $p\lambda/3 = mp$, since each τ occurs in λ blocks of the design and there are three τ's per block. Using a theorem of Agrawal (1966) on m-ple systems of distinct representatives, we conclude that we can select a collection H_i of m blocks in each S_{τ_i}, $1 \le i \le \binom{v}{2}$, with the H_i disjoint. Each of the m blocks in H_i is then ordered so that the pairs of treatments in τ_i occur at the ends. This makes all the $\pi_{ii'} = m$. □

When $k = 4$, although Hanani's work again shows that the usual necessary conditions for existence of a BIBD are sufficient, those conditions together with (5.5) no longer guarantee the existence of a design satisfying the condition of Theorem 5.1, as Example 5.1 shows. Thus although we have a comprehensive picture of the combinatorial considerations associated with our optimality criteria for the NN model when $k = 2$, 3 or v (Section 4), the combinatorics seem much more difficult for other values of k.

5.2. *Equineighbored* BIBD *construction.* For large k the contribution of $N_{ii'}$ in (5.3) is much larger than that of $e_{ii'}$, since $kN_{ii'}$ is, on average, $k/2$ times as large as $e_{ii'}$. Also, the equality of the $N_{ii'}$ implies equality of the e_i by (5.5). We shall therefore devote most of the rest of this section to the construction of BIBD's with all the $N_{ii'}$ equal, which we call *equineighbored* BIBD's or EBIBD's. Such designs, besides being A-optimum, make all the $\mathrm{Var}(\hat{t}_i)$ equal and can be expected to perform fairly efficiently for criteria other than A-optimality.

The NN model leads us to seek the equal appearance of unordered pairs of neighbors, not of ordered ones. This distinguishes the present work from the problems treated by such authors as E.J. Williams (1949), Hedayat and Afsarinejad (1975, 1978), and Hedayat and Magda (1979) in the construction of repeated measurement designs, who distinguish the order of pairs. In seeking to minimize b for given v and k we may typically construct designs with half the value of b required by those authors. However, it must be emphasized that we restrict our designs to be BIBD's, which Hedayat and Afsarinejad (1975) do not. For example, their design with $v = 5$, $k = 3$, $b = 10$ is not a BIBD, and no BIBD with those parameters can have equal appearance of all ordered pairs, although by our Theorem 5.2 a BIBD of our type exists. Similarly for $v > k$ their design has $b \geq 2v$, which is not necessarily the case for our EBIBD's, which often exist with $b = v$. On the other hand, they construct designs with $k = (v + 1)/2$ and $b = 2v$ for all odd v, but these are not always BIBD's except in special cases, for example, if v is a prime power $\equiv 3 \pmod 4$. In this last case their design will of course automatically be an EBIBD of our type. The main aim of Hedayat and Afsarinejad (1975) is the construction, for given v, of designs with the minimum conceivable value $2v$ of b, which forces $k = (v + 1)/2$. On the other hand, for fixed v and k, we want to construct equineighbored designs with minimum b. Obviously any equineighbored design can be replicated with blocks in reversed order to yield a design of the type Hedayat and Afsarinejad consider. If $N_{ii'} = 1$ for the equineighbored design, the resulting Hedayat and Afsarinejad design has minimum b for given v, k and minimum k for given v and b.

The designs of Hwang (e.g., 1973) and others have NN structure for *circular* (cyclic) block designs with $k = (v - 1)/2$ because of the extra pair of neighbors per block. Those designs are easily seen to be optimum in their setting, by our argument. Powers of a primitive element can be used for prime power v construction there, but not in our problem. C.S. Cheng has recently shown how Hwang's designs can sometimes be modified to yield designs for our problem, but they do not always attain the minimum b we seek.

Assuming (v, b, k) are such that a BIBD exists and that (5.7) is satisfied, we want to construct an EBIBD, and, for given v and k, to do this for smallest possible b. One may try to make all the $N_{ii'}$ equal by reordering treatments within each block of a given BIBD, but we have no general algorithm for doing this even for the case $k = 3$ (which is treated in alternate fashion in Theorem 5.2). We shall give methods of generating EBIBD's based on the difference set technique of developing cyclic designs.

Let $\{a_1, \cdots, a_k\}$ be a difference set (mod v), i.e., a set of integers such that amongst all the difference $\pm(a_i - a_j)$ $(1 \leq i, j \leq k, i \neq j)$ each nonzero residue (mod v) $1, \cdots, v - 1$ appears the same number of times. When v is odd and $k = (v + 1)/2$ define an *equineighbored difference set* (mod v) as a difference set $\{a_1, \cdots, a_k\}$ (mod v) with the additional property that amongst the $2(k - 1)$ *successive differences*

$$\pm(a_i - a_{i+1}), \qquad 1 \leq i \leq k - 1,$$

each nonzero residue appears exactly once. We recall that for the analogous case $v = k$ ($M = 2$ in Section 4), the property is similar to that held by the first row and column of the squares generated by Theorem 4.1. We give the following without proof:

Theorem 5.3. *If* $\boldsymbol{a} = \{a_1, \cdots, a_k\}$ *is an equineighbored difference set* (mod v) *where* $k = (v + 1)/2$, *then the symmetric* BIBD *formed by developing* $\boldsymbol{a}$ *as the initial block so that the* $(1 + j)$th *block is*

$$(a_1 + j, a_2 + j, \cdots, a_k + j) \pmod v$$

$(0 \leq j \leq v - 1)$ *is an* EBIBD *with all* $N_{ii'} = 1$.

Difference sets are known for many cases where $k = (v + 1)/2$. (See Hall, 1967; Baumert, 1971.) Since the complement of a difference set is a (smaller) difference set of $k - 1$ elements (mod v), it is usually listed in these references. For all $v \leq 35$ of the form $4\lambda - 1$, such sets my be constructed by one of the following methods:

(1) If $v = p^m$, p prime, $v \equiv 3 \pmod 4$, then all the nonzero quadratic residues (mod v) yield a difference set with $k = (v - 1)/2$ elements (cyclic if $m = 1$).

(2) If $v = pq$ where p and $q = p + 2$ are twin primes, a slightly more complex recipe (Hall, page 141) yields a difference set of $k = (v - 1)/2$ elements.

These are the only two methods we discuss here.

In case (1) with $m = 1$, the element 0 may be adjoined to the nonzero quadratic residues to yield a difference set (mod v) of k elements which, in their natural order, yield an equineighbored difference set.

THEOREM 5.4. *If v is prime and $v \equiv 3 \pmod 4$, then*

$$\left\{0^2, 1^2, 2^2, \cdots, \left(\frac{v-1}{2}\right)^2\right\} \pmod v$$

is an equineighbored difference set.

PROOF. It is well known that this is a difference set. Moreover,

$$\pm(a_{i+1} - a_i) \equiv \pm(2i - 1),$$

$1 \leq i \leq (v - 1)/2$, and these are exactly the $(v - 1)$ nonzero residues (mod v). □

We remark that it is in general necessary to establish that $\boldsymbol{a}$ is a difference set, as well as that $\pm(a_{i+1} - a_i)$ are distinct. For example (0, 1, 6, 2) (mod 7) has distinct successive differences (±1, ±5, ±4), but is not a difference set because of the twelve differences $a_i - a_j$ $(i \neq j)$ there are three appearances of ±1, two of ±2, and one of ±3. When developed, this block thus does not yield a BIBD.

In case (1) with $m > 1$ the quadratic residues of $GF(p^m)$ with 0 included again form a difference set with $k = (v + 1)/2$ elements. However, there is no obvious ordering as there was in the case of Theorem 5.4. It is tempting to let x be a primitive element and consider the successive differences of the set $x^2, x^4, \cdots, x^{(v-1)}$; however, the difficulty is that adjoining the remaining element 0 anywhere in the sequence need not work, as is shown by the case $v = 11$, $x = 2$, $(x^2, x^4, x^6, x^8, x^{10}) = (4, 5, 9, 3, 1) \pmod{11}$. The case (2), above, also yields no simple mechanism. We construct a difference set of $(v - 1)/2$ elements (mod v) by the cited method, and its complement is then a difference set of the required size, $(v + 1)/2$. But again there is no obvious ordering scheme that produces an equineighbored difference set. This inability to find a scheme analogous to that of Theorem 5.4 in the cases of the previous two paragraphs has led us to the following simple routine.

Start with a given unordered difference set $c_1, \cdots, c_{(v+1)/2}$. We shall try to arrange the elements of this set in a special sequence $a_1, \cdots, a_{(v+1)/2}$ by a simple iterative method which adjoins one element at a time to the ends of the sequence already constructed. First arrange all the pairs $\{c_i, c_j\}$ $(i \neq j)$ in a list consisting of $(v - 1)/2$ groups in such a way that group h $(1 \leq h \leq (v - 1)/2)$ consists of all those pairs $\{c_i, c_j\}$ with $c_i - c_j = \pm h$. Begin with some pair $\{c_i, c_j\} = \{a_{2,1}, a_{2,2}\}$. In general when there are s out of the c_i in the sequence label them

$$a_{s,1}, \cdots, a_{s,s}. \tag{5.9}$$

At this stage we must be sure that the following are deleted from the list of pairs (since none is available to yield a difference on a subsequent step):

(1) any group h for which

$$a_{s,r} - a_{s,r+1} = \pm h \pmod v$$

for some r with $1 \le r \le s - 1$;

(2) any pair $\{c_i, c_j\}$ for which c_i or $c_j = a_{s,r}$ for some r with $2 \le r \le s - 2$;

(3) the pair $\{a_{s,1}, a_{s,s}\}$ for $s > 2$.

To add the next, $(s + 1)$st, element to the sequence (5.9), we proceed as follows. Choose an undeleted group that contains the *smallest* number of undeleted pairs $\{c_i, c_j\}$ among undeleted groups. If the chosen group contains an undeleted $\{c_i, c_j\}$ one member of which, say c_i, is either $a_{s,1}$ or $a_{s,s}$, then adjoin the other member c_j to the appropriate end to form $a_{s+1,1}, \cdots, a_{s+1,s+1}$. If no such $\{c_i, c_j\}$ exists then we are "blocked" and we choose instead a group with the next smallest number of undeleted $\{c_i, c_j\}$. Special rules may be adopted if a previously "blocked" group is still undeleted at some stage, if there are several "tied' groups, and so on.

It is easy to construct examples where this simple routine can fail. However, in every case of the form $v = b = 4\lambda - 1$ and $k = r = 2\lambda$, that we have been discussing for $\lambda \le 9$, the method has worked on the first try after less systematic attempts had failed. For larger values of λ, when there are only four sparse groups left, the various combinations of remaining additions at both ends were considered, to avoid failure. The routine is also simple to program on a computer and can be extended to the case of more than one initial block and to generalized difference sets, discussed later. The equineighbored difference sets we obtained in the cases $v = 15$, 27 and 35 for $\lambda \le 9$, not covered by Theorem 5.4, are as follows.

(5.10)
v = 15: (11, 14, 4, 12, 8, 2, 3, 1) (mod 15),
v = 27: (020, 102, 111, 202, 221, 121, 022, 021, 001, 120, 000, 110, 211, 100) (mod (3, 3, 3)),
v = 35: (19, 6, 30, 5, 8, 31, 23, 24, 26, 20, 2, 32, 18, 22, 15, 34, 25, 10) (mod 35).

In the case $v = 27$ the differencing is carried out component-wise and the difference set is not cyclic. In all cases we started with an unordered difference set. The only case in this family of parameter values with $v \le 50$ not covered by Theorem 5.4 or (5.10) is $v = 39$ for which it is a consequence of a theorem of Hall and Ryser (see Baumert, 1971, page 25) that no cyclic difference set exists. There is a symmetric BIBD in that case but we have not attempted to reorder it (by necessarily different methods), and thus do not know whether there is a symmetric EBIBD for $\lambda = 10$.

Since the classification of all difference sets is itself incomplete we cannot hope for a simple listing of all equineighbored sets. We now make some brief remarks on the classification of equineighbored difference sets, for the simple case when v is prime for the family considered above.

We consider equineighbored difference sets to be equivalent if they can be transformed into each other by one or more of the following transformations: (1) addition to each element $a_1, \cdots, a_{(v+1)/2}$ of an arbitrary integer $\boldsymbol{a}$ (and reduction (mod v)); (2) multiplication of each element by an arbitrary integer $c \not\equiv 0$ (mod v), and reduction (mod v) (with c not restricted to be a "multiplier" in the technical sense of the difference set literature, as defined by Hall or Baumert); (3) reversal of the order $a_1, \cdots, a_{(v+1)/2}$ to $a_{(v+1)/2}, \cdots, a_1$. Note that under (1) the successive differences $d_1, \cdots d_{(v-1)/2}$ of an equineighbored difference set are preserved in the same order, where $d_i = a_{i+1} - a_i$ $(i = 1, \cdots, (v - 1)/2)$. Since we distinguish between d_i and $-d_i$, (2) and (3) do not preserve the order of the d_i's; nevertheless, it seems natural to regard difference sets which can be obtained from each other through these transformations, as equivalent.

EXAMPLE 5.2. In the case $\lambda = 2$, $v = 7$, the two sets of successive differences (1, 3, 5) and (1, 5, 4) identify two equivalence classes of equineighbored difference sets. Typical members of each are respectively (0, 1, 4, 2) and (1, 2, 0, 4). Thus neither of these can be transformed into the other by tranformations (1), (2), or (3), or by combinations of them. Under (2) and (3), (1, 3, 5) and (1, 5, 4) together generate half of the $2^3 3!$ possible difference triples $(\pm\alpha, \pm\beta, \pm\gamma)$ where (α, β, γ) is a permutation of (1, 2, 3); the other half cannot occur. For example, (1, 2, 3) occurs but (1, 2, 4) does not, nor indeed does any permutation of $\pm(1, 2, 4)$. For large v the picture is more complex.

We now give a few additional examples to illustrate the extension to settings where more than one initial block must be developed. There is often more than one possible set of initial blocks, but we give only one in each example. For brevity we include among these examples only one of the common infinite "families" of BIBD's (Example 5.9), but it is easy to extend the method to others. All the examples yield the smallest b for which (5.7) is satisfied, except Example 5.5.

All of these examples will be treated by the use of an elementary development we next set forth.

Suppose, for a given v, that we have a set of $\bar{b}$ initial blocks of size k, that together constitute a difference set. We now describe a simple device for obtaining a set $\mathscr{B}$ (say) of $\bar{b}k$ or $\bar{b}k/2$ initial blocks when k is odd or even, respectively, so that $\mathscr{B}$ is an equineighbored difference set which can be used as an initial block set to generate an EBIBD with $b = v\bar{b}k$ or $v\bar{b}k/2$. To this end, we recall from the discussion of the case $M = 2$ of complete blocks in Section 4, but with block size now k, that the k rows of a $k \times k$ Latin square given by Theorem 4.2 for k odd, or the first $k/2$ rows for k even, yield an equineighbored complete block design $\mathscr{D}$ (say). For each of the $\bar{b}$ initial blocks of size k, substitute its k symbols into $\mathscr{D}$ to yield k or $k/2$ blocks of size k; we call this the $\mathscr{D}$-development of the initial block. Doing this for all $\bar{b}$ initial blocks yields the set $\mathscr{B}$. From the nature of $\mathscr{D}$ it follows that, although any of the $\bar{b}$ initial blocks need not be a difference set, every difference occurring in the initial block occurs proportionally often as a NN difference in the $\mathscr{D}$-development of that block. Since the $\bar{b}$ initial blocks are a difference set, we have

THEOREM 5.5. *The set $\mathscr{B}$ of blocks just described is an equineighbored difference set.*

EXAMPLE 5.3. Even when $k = 3$, it is convenient to have a simple explicit formula for the construction of EBIBD's. Consider the case $v = 7$, $k = 3$. It follows from (5.7) that the minimum b is 21. The initial blocks (1, 2, 4), (4, 1, 2), (2, 4, 1) may be constructed by using Theorem 5.5, and contain, among them, each pair of successive differences (± 1, ± 2, ± 3) twice. Thus these blocks developed by adding all residues (mod 7) yield an EBIBD. Note that for this small value of k it was possible to choose the initial blocks to be cyclic permutations of (1, 2, 4).

EXAMPLE 5.4. Consider the case $k = 4$, $v = 13$. The familiar design with $\lambda = 1$ is obtained by developing (0, 1, 3, 9) (mod 13). A second initial block is needed to achieve (5.7). From Theorem 5.5 we take it to be (1, 9, 0, 3), and then each successive difference pair $\pm i$ $(1 \le i \le 6)$ appears once in the set of two initial blocks. Clearly since $k = 4$ the second initial block could not be obtained by a cyclic permutation of the first.

EXAMPLE 5.5. Here is an example based on an acyclic difference set. For $v = 16$, $k = 6$, the usual symmetric design with $\lambda = 2$ is developed from an initial block consisting of a difference set of six suitably ordered elements mod (2, 2, 2, 2), listed as the first line of (5.11), below. By Theorem 5.5 we obtain 3 initial blocks which can be developed into an EBIBD with $b = 48$. (We do not know whether a BIBD with $b = 24$, which exists, can be ordered to be equineighbored.)

$$\begin{array}{ll} & ((0010), (0001), (0011), (1100), (1000), (0100)), \\ (5.11) & ((0001), (1100), (0010), (0100), (0011), (1000)), \\ & ((0011), (0010), (1000), (0001), (0100), (1100)). \end{array}$$

EXAMPLE 5.6. Here are two illustrations in which the basic design (non-EBIBD with minimum b) which we expand, is based on more than one initial block. For $v = 13$, $k = 3$, $b = 26$, a BIBD with $\lambda = 1$ is often constructed by developing (mod 13) the two initial blocks (1, 3, 9) and (2, 5, 6). It is easily seen that the six initial blocks obtained from the cyclic permutation of these two blocks yields an equineighbored difference set (as in Example 5.3). This amounts to using Theorem 5.5 on the original two initial blocks.

For $v = 9$, $k = 4$, $b = 18$, a BIBD with $\lambda = 3$ is often obtained by developing the two initial blocks (1, 4, 0, 2) and (1, 0, 4, 6) (mod 9). As in Example 5.4, a set consisting only of cyclic permutations of these will not work, but we may use Theorem 5.5 to adjoin the two blocks (4, 2, 1, 0) and (0, 6, 1, 4), to obtain an equineighbored difference set and thus an EBIBD with the minimum b value of 72. Interestingly, each of the last two initial blocks contains a repeated neighbored difference.

EXAMPLE 5.7. For an example involving generalized differences including an "∞" treatment, we consider the case $v = 12$, $k = 4$, $b = 33$ where a design with $\lambda = 3$ is often obtained by developing (mod 11) the three initial blocks (0, 3, 7, 1), (0, 1, 3, 9), (∞, 0, 1, 5). If we use Theorem 5.5 to adjoin (3, 1, 0, 7), (1, 9, 0, 3) and (0, 5, ∞, 1), we obtain a set of six initial blocks with equineighbored successive differences, which can be developed to yield an EBIBD with the minimum b of 66. Once more, each of the last two initial blocks contains a repeated neighbored difference.

EXAMPLE 5.8. We conclude our examples with an illustration involving mixed difference sets. The T_1 system of BIBD's consists of designs with (for t a positive integer) $v = 6t + 3$, $k = 3$, $b = (2t + 1)(3t + 1)$, and $\lambda = 1$. These are often obtained by developing, mod $(2t + 1)$, the initial set of $3t + 1$ blocks

$$(5.12)\quad \begin{array}{ll} (i_1, (2t+1-i)_1, 0_2) & 1 \le i \le t, \\ (i_2, (2t+1-i)_2, 0_3) & 1 \le i \le t, \\ (i_3, (2t+1-i)_3, 0_1) & 1 \le i \le t, \\ (0_1, 0_2, 0_3). & \end{array}$$

It is easily verified that the blocks of (5.12) and their cyclic permutations (which arise from using Theorem 5.5) yield a set of $3(3t + 1)$ equineighbored initial blocks which can be developed into an EBIBD with the minimum $b = (6t + 3)(3t + 1)$.

We close with brief mention of nonorthogonal layouts for $M \geq 3$. For $M = 3$, optimality of Youden designs (YD's) or generalized Youden designs under V_0 is considered by Kiefer (1958, 1975), and for $M > 3$ results for Youden hyperrectangles are obtained by Cheng (1978). We limit discussion here to the case $M = 3$. If treatments are assigned to a $k \times mv$ array with $k < v$ and m an integer, there exists a "regular" YD if there is a BIBD with $b = mv$. The LS estimators for such a YD are in fact those of the BIBD consisting of the columns of the YD. The YD's are now our $\mathscr{X}^*$. If $m > 1$, it is possible that $N_{ii} > 0$ in a YD. The expression for $D(X^* \mid V)$ is slightly more complicated than it is for a Latin square, and will not be given here. As was the case for hypercubes, if $\rho > 0$ one is led to make all $N_{ii} = 0$ and to look for designs for which all $N_{ii'}$ are equal for $i \neq i'$. Unfortunately, simple number-theoretic considerations show that *no regular* YD *can be equineighbored*. In the nonregular setting in which there are $k_1 \times k_2$ generalized YD's with both $k_i > v$ and neither k_i divisible by v, this simple nonexistence argument no longer applies; we do not know the extent to which equineighbored generalized YD's may exist.

REFERENCES

AGRAWAL, H. (1966). Some generalisations of distinct representatives with applications to statistical designs. *Ann. Math. Statist.* **37** 525–526.

ATKINSON, A. C. (1969). The use of residuals as a concomitant variable. *Biometrika* **56** 33–41.

BARTLETT, M. S. (1938). The approximate recovery of information from replicated field experiments with large blocks. *J. Agric. Sci.* **28** 418–427.

BARTLETT, M. S. (1975). *The Statistical Analysis of Spatial Pattern*. Chapman and Hall, London.

BARTLETT, M. S. (1978). Nearest neighbour models in the analysis of field experiments (with discussion). *J. R. Statist. Soc. B* **40** 147–174.

BAUMERT, L. (1971). *Cyclic difference sets. Lecture Notes in Mathematics* No. 182. Springer, Berlin.
BERENBLUT, I. I. AND WEBB, G. I. (1974). Experimental design in the presence of autocorrelated errors. *Biometrika* **61** 427–437.
BESAG, J. (1974). Spatial interaction and the statistical analysis of lattice systems. *J. R. Statist. Soc. B* **36** 192–236.
BICKEL, P. J. AND HERZBERG, A. M. (1979). Robustness of designs against autocorrelation in time. I. Asymptotic theory for location and linear regression. *Ann. Statist.* **7** 77–95.
CHENG, C.-S. (1978). Optimal designs for the elimination of multi-way heterogeneity. *Ann. Statist.* **6** 1262–1272.
CHENG, C.-S. (1979). Optimal incomplete block designs with four varieties. *Sankhya, Ser. B* **41** 1–14.
CONSTANTINE, G. M. (1980). On Schur-optimality. Preprint, Univ. of Indiana.
DÉNES, J. AND KEEDWELL, A. D. (1974). *Latin Squares and Their Applications.* English Universities, London.
DUBY, C., GUYON, X. AND PRUM, B. (1977). The precision of different experimental designs for a random field. *Biometrika* **64** 59–66.
GIOVAGNOLI, A. AND WYNN, H. (1980). A majorization theorem for the C-matrices of binary designs. *J. Statist. Planning and Inf.* **4** 145–154.
HALL, M. JR. (1967). *Combinatorial Theory.* Blaisdell, Waltham, Mass.
HANANI, H. (1961). The existence and construction of balanced incomplete block designs. *Ann. Math. Statist.* **32** 361–386.
HEDAYAT, A. AND AFSARINEJAD, K. (1975). Repeated measurement designs. I. In *A Survey of Statistical Design and Linear Models* (J. N. Srivastara, ed.), 229–242. North-Holland, Amsterdam.
HEDAYAT, A., AND AFSARINEJAD, K. (1978). Repeated measurement designs. II. Characterizations, construction and optimality. *Ann. Statist.* **6** 61–62.
KERR, J. R., PEARCE, S. C. AND PREECE, D. A. (1973). Orthogonal designs for three-dimensional experiments. *Biometrika* **60** 349–358.
KIEFER, J. (1958). On the nonrandomized optimality and randomized nonoptimality of symmetric designs. *Ann. Math. Statist.* **29** 675–699.
KIEFER, J. (1960). Optimum experimental designs. V, with applications to systematic and rotatable designs. *Proc. Fourth Berk. Symp.* **2** 381–405.
KIEFER, J. (1975). Construction and optimality of generalised Youden designs. In *A Survey of Statistical Design and Linear Models* (J. N. Srivastara, ed.), 333–353. North-Holland, Amsterdam.
KISHEN, K. (1949). On the construction of Latin and hyper-graeco-latin cubes and hypercubes. *J. Ind. Soc. Ag. Statist.* **2** 2–48.
KRUSKAL, W. (1968). When are Gauss-Markov and least squares estimators identical? A coordinate free approach. *Ann. Math. Statist.* **39** 70–75.
MARTIN, R. J. (1977). Spatial models with applications in sampling and experimental design. Ph.D. thesis, London School of Economics, University of London.
O'HAGAN, A. (1978). Curve fitting and optimal design for prediction (with discussion). *J. R. Statist. Soc. B* **40** 1–42.
PAPADAKIS, J. S. (1937). Méthode statistique pour des expériences sur champ. *Bull. Inst. Amél. Plantes à Salonique*, No. 23.
PATTERSON, H. D. (1950). The analysis of change-over trials. *J. Agric. Sci.* **40** 375–380.
PATTERSON, H. D. (1951). Change-over trials. *J. R. Statist. Soc. B* **13** 256–271.
PATTERSON, H. D. (1952). The construction of balanced designs for experiments involving sequences of treatments. *Biometrika* **39** 32–48.
RAO, C. R. (1967). Least squares theory using an estimated dispersion matrix and its application to measurement of signals. *Proc. V Berk. Symp.* **1** 355–372.
RIPLEY, B. D. (1977). Modelling spatial patterns (with discussion). *J. R. Statist. Soc. B* **39** 172, 212.
SACKS, J. AND YLVISAKER, D. (1966). Designs for regression problems with correlated errors. *Ann. Math. Statist.* **37** 66–89.
SACKS, J. AND YLVISAKER, D. (1968). Designs for regression problems with correlated errors; many parameters. *Ann. Math. Statist.* **39** 49–69.
SACKS, J. AND YLVISAKER, D. (1969). Designs for regression problems with correlated errors. III. *Ann. Math. Statist.* **41** 2057–2074.
WAHBA, G. (1971). On the regression design problem of Sacks and Ylvisaker. *Ann. Math. Statist.* **42** 1035–1053.
WAHBA, G. (1974). Regression design for some equivalence classes of kernels. *Ann. Statist.* **5** 925–934.
WILLIAMS, E. J. (1949). Experimental designs balanced for the estimation of pairs of residual effects of treatments. *Aust. J. Sci. Res.* **2** 149–164.
WILLIAMS, E. J. (1950). Experimental designs balanced for pairs of residual effects. *Aust. J. Sci. Res.* **3** 351–363.

WILLIAMS, R. M. (1952). Experimental designs for serially correlated observations. *Biometrika* **39** 151–167.
WATSON, G. S. (1967). Linear least squares regression. *Ann. Math. Statist.* **38** 1679–1699.
ZYSKIND, G. (1967). On canonical forms, non-negative covariance matrices and best and simple least squares linear estimators in linear models. *Ann. Math. Statist.* **38** 1092–1109.

DEPARTMENT OF MATHEMATICS
IMPERIAL COLLEGE OF LONDON
LONDON SW7 2B7, ENGLAND

DEPARTMENT OF STATISTICS
UNIVERSITY OF CALIFORNIA
BERKELEY, CALIFORNIA 94720

The Canadian Journal of Statistics
Vol. 9, No. 1, 1981, Pages 1–10
La Revue Canadienne de Statistique

The interplay of optimality and combinatorics in experimental design*

Jack C. KIEFER†
University of California, Berkeley

Key words and phrases: Optimal designs, combinatorial designs, weighing designs, incomplete block designs.
AMS 1980 *subject classification:* Primary 62K05.

ABSTRACT

The design of statistical experiments, as developed by R. A. Fisher and his followers, often used combinatorial structures that yielded simple calculation of estimates and/or symmetric variances and covariances. Examples are block designs with balance, regression experiments with equally spaced observations, etc. More recently, considerations of optimality (choosing a design that achieves most accurate inference in some sense) have sometimes justified the traditional designs, but have sometimes led to new combinatorial investigations. Illustrations are given.

1. INTRODUCTION

For many years now, there has been a considerable amount of research activity devoted to the invention of experimental designs of a certain kind, which one could loosely call exotic combinatorial designs, such as Latin squares, balanced incomplete block designs (BIBDs), Hadamard matrices and things of that sort. The designs of this type created and used in the earlier years were fairly simple and were the direct result of R. A. Fisher's pioneering contributions to statistical inference. He emphasized intuitively appealing properties of designs that led to simple computation of estimators and gave equal precision to all estimates of interest; optimality in the current sense was not an issue but Fisher's rationale led to the adoption of aesthetically attractive combinatorial designs, and the mathematical tools needed for their construction came to be of intrinsic interest. By now, consequently, we have situations in which notions purely of combinatorial aesthetics lead to the creation of designs which people sometimes use for these intuitive reasons, but where notions of optimality different from the original Fisherian notions lead to the development and use of new combinatorial structures that do not belong to the class of simple "exotic" designs mentioned earlier. There has been a "back-and-forth" between design criteria and combinatorial constructions, as new ideas came to each. It is this interplay of optimality and combinatorics in experimental design that is the main theme of this paper.

---

* Specially invited paper based on an invited lecture given at the Pacific Northwest Statistics Meeting held at the University of Victoria, 7 November 1980. [Assistance provided by the University of Victoria in tape-recording the lecture is gratefully acknowledged. *Ed.*]

† Deceased 10 August 1981.

2. TRADITIONAL OPTIMALITY FOR STATISTICAL APPLICATIONS

When we look at the dominant motivations in the construction of experimental designs in the early days, we see that, in addition to theoretical precision of statistical inference, a substantial role was played by computational ease (since the problem of inverting a 10×10 matrix by hand is an entirely different level of magnitude than that of plugging the matrix into a computer for it to invert). There were also aesthetic considerations like symmetry. We shall illustrate the point by considering two specific examples.

The first example is of a weighing design in which there are k objects to be weighed, the weight of the jth object being θ_j; there are m weighings ($m \geq k$); and the weighing is to be done on a chemical balance. In each weighing several weights can be in each pan. We assume that the weighings have equal precision and are independent of one another. The mathematical model appropriate to this problem really has broad applicability, because its solution also solves the problem of linear regression on the $(k - 1)$-dimensional cube. Every design d is characterized by its design matrix $\mathbf{X}_d$; for the weighing problem $\mathbf{X}_d = \{x_{ij}, i = 1, \ldots, m; j = 1, \ldots, k\}$ is defined by $x_{ij} = 1, -1$ or 0 according as, in the ith weighing, the jth object is in the left pan, right pan, or absent. If $\mathscr{X}$ is the class of all available design matrices, the problem is to choose the optimal elements of this class. In the weighing problem, $\mathscr{X}$ is the class of all $m \times k$ matrices whose elements are 0 or ± 1. If Y_i is the weight added to the right pan for perfect balance, and $\mathbf{Y}^{\mathrm{T}} = \{Y_i, \ldots, Y_m\}$, then $\mathscr{E}\mathbf{Y} = \mathbf{X}_d\boldsymbol{\theta}$, where $\boldsymbol{\theta}^{\mathrm{T}} = (\theta_1, \ldots, \theta_k)$.

The other example we shall consider is that of an incomplete block design for testing v drugs on Kb experimental animals (K from each of b litters), in which the litters are the blocks. Assuming the additive model without interaction (response = drug effect + litter effect), the parameter vector is $(v + b)$-dimensional and the observation vector Kb-dimensional. The design matrix $\mathbf{X}_d$, in this case, consists of 0's and 1's only, having exactly two 1's in each row, and $\mathbf{X}_d^{\mathrm{T}}\mathbf{X}_d$ is singular. In this respect, this example differs from the previous one; it also differs in that we are now usually interested in only some of the parameters, namely, differences among drug effects. But in the main aspect the two examples agree, in that the expected values of the observations are linear functions of the parameters.

Let us now restrict ourselves to such linear models, and temporarily let us assume that we are interested in all the parameters, so that the class $\{\mathbf{X}_d\}$ of design matrices is a certain class of $m \times k$ matrices for which $\mathbf{M}_d = \mathbf{X}_d^{\mathrm{T}}\mathbf{X}_d$ is a nonsingular matrix. This $\mathbf{M}_d$ is the "information matrix" of the design in the sense that if the observations are independent and have a common variance σ^2, then the covariance matrix of the least-squares estimators of the parameters is $\sigma^2\mathbf{M}_d^{-1}$; this matrix has added significance in describing confidence ellipsoids if the observations are normally distributed. These facts and the intuitive appeal of having equal variances for the estimators of the different parameters, together with the simplicity of inverting a diagonal matrix (necessary for computing the least-squares estimators), led to the early emphasis on balanced orthogonal designs (that is to say, designs for which $\mathbf{M}_d$ is proportional to the identity; an analogue in the block design setting motivates using BIBDs). Going back to the weighing design problem, an information matrix which is a multiple of the identity has these appealing properties. This insistence on "balance" in experimental design is a prominent feature of Fisher's rationale mentioned earlier, and of many constructions of Fisher, Yates, Bose and others. Now we shall look at experimental designs without these traditional restrictions. (Indeed, there are simple examples which show that the restriction to balanced designs can be costly, increasing variances considerably; nevertheless, Fisher's excellent intuition showed foresight in yielding designs that are often optimum from a modern viewpoint.)

3. SOME OPTIMALITY CRITERIA

We agreed to limit ourselves to linear models, and let us also agree to use least-squares estimates. Now some people in design theory worry these days about robustness and what terrible consequences can arise from using simple least-squares theory if some of the experimental plots have been contaminated by the wanderings of a stray cow that has, minimally, produced some outliers. Also, there is justifiable concern that the assumed regression model might be at least slightly incorrect. We shall not worry about such problems here, important as these considerations are, since we want to emphasize the optimality-combinatorial aspect. We shall be concerned here, then, with questions of optimality under the simplifying assumption that the observations are uncorrelated with equal variance and that the regression model we have postulated is correct. If we now grant that the information matrix $\mathbf{M}_d$ need not be a multiple of the identity, then we must agree on a notion of optimality. There are two possible approaches for this.

One way to compare experimental designs is in terms of the order relation $\geq$ for positive definite symmetric matrices: $\mathbf{A} \geq \mathbf{B} \Leftrightarrow \mathbf{A} - \mathbf{B}$ is nonnegative definite. Design d is at least as good as design d' if and only if $\mathbf{M}_d \geq \mathbf{M}_{d'}$. This induces a notion of admissibility, and of complete classes of design, analogous to what one has in decision theory; and usually, one does not have a design which is better than all others, so one must still introduce a further rationale for selecting the design to be used (as in the other approach, just below). We shall not pursue this first approach further, here.

The other approach is to say, from the outset: "Since it does not generally happen that one design is, in some natural sense, better than all the others, we will choose some real-valued optimality functional of $\mathbf{M}_d$ with which to compare designs, choosing a design that minimizes this functional". Three of the most commonly used functionals are:

(1) the D-optimality criterion, in which the functional used is the generalized variance or the determinant of $\mathbf{M}_d^{-1}$;
(2) the A-optimality or "average variance" criterion, in which the functional used is the trace of $\mathbf{M}_d^{-1}$;
(3) the E-optimality criterion, in which the functional used is the maximum eigenvalue of $\mathbf{M}_d^{-1}$.

These all have very simple geometric meanings in the normal case in which one might give an ellipsoidal confidence set instead of a point estimate: the determinant is proportional to the square of the volume; the maximum eigenvalue, to the square of the maximum diameter; and the trace, to the sum of squares of lengths of the principal axes.

A fourth criterion, which made its appearance quite early, is the G-optimality criterion, which is particularly sutiable when there is a response surface involved in the problem and the emphasis is on accuracy of the estimated surface rather than of the estimates of θ_i's. For example, in the problem of linear regression on a cube (equivalent to a weighing design problem), the approach would be: supposing we are interested in fitting the least-squares surface (the response surface), find the maximum value (over all points of the cube) of the mean squared error between the true response surface and the least-squares estimated surface. The design that minimizes this quantity is called *G*-optimal; it is a global optimality criterion.

The *G*-optimality criterion was introduced in 1918 by a lady called Smith who came from Scandinavia to London and wrote a massive paper, which is perhaps the first instance of attempting to find an optimal design with respect to a functional. She was interested in polynomial regression up to degree 6 on an interval, and computed by brute force what the best way was to allocate observations to fit such a polynomial. (This is slighty inaccurate. In the curve-fitting settings we are discussing, as distinct from the settings where the exotic combinatorial structures arise, a useful device is to compute the optimum allocation, ignoring the fact that the number of observations at each point must be a whole number. For moderately large sample sizes, this is implemented in an obvious way to yield an actual design that is very close to optimum. This device yields optimality algorithms that would otherwise be unavailable.)

Then nothing happened for a long time. In 1944 Hotelling published a paper on weighing designs which is related to some of what we will discuss in this context, and in 1943 Wald wrote a paper which proved *D*-optimality of Latin square designs. In 1952, Elfving started developing what one can think of as the theory of algorithms for finding optimum regression designs in certain senses. A pioneering coincidence occurred in 1958 when Hoel and Guest both looked at the problem of polynomial regression on an interval of the line from the point of view of how one should space the values of the independent variate. Hoel used the *D*-optimality criterion, the generalized variance of the parameter estimators, reducing the solution to one obtainable from work of Stieltjes or of Schur on the equilibrium spacing of point charges on an interval. Guest used the Smith criterion, the maximum of the variance function for the fitted response. Their independent solutions to the problem turned out to be the same. This was noticed, and created a new interest in this subject—perhaps there were other coincidences which would lead to the development of algorithms for finding optimum designs in the sense of different criteria. Since that time the first results are those of Kiefer and Wolfowitz (1960), and a number of us have been working on this sort of thing.

4. CONSTRUCTION OF OPTIMAL WEIGHING DESIGNS

Having considered some criteria of optimality and noted that, in some cases, different criteria lead to the same solultion, we now consider the problem of constructing optimum designs or of showing that certain designs are optimum. In theory, all you need is to look at every possible matrix $\mathbf{X}$, compute $(\mathbf{X}^\tau\mathbf{X})^{-1}$ and compare the values of the particular function one is using as the optimality criterion; but in practice, that is just too much work. So, in choosing an optimality tool, the first requirement is that it should not require inverting matrices; secondly, it is desirable that one does not need a different calculus for proving optimality under different criteria. In some simple situations, there is such a tool. For example, the trace of the information matrix is easy to compute. In the weighing design problem, there is sometimes a design matrix $\mathbf{X}_{d^*}$ which maximizes the trace of the information matrix and, in addition, has the nice property that the information matrix is a multiple of the identity. A design having these properties is usually optimum in any reasonable sense, including the four senses we considered earlier. In such circumstances, if you are unsure which optimality criterion is appropriate, you do not face the customary corresponding choice among designs: one design is optimum in all these senses. Such a design, when it exists, is referred to as a universally optimum design. The optimality of such a design holds for every optimality functional having three properties: (1) monotonicity, which means that as the information matrix increases, there is a decrease in the amount of penalty which the optimality functional represents; (2) rotational invariance, so that the value of the functional is not affected by orthogonal transformations of the parameters; (3) convexity with respect to mixing of designs. [Sometimes one uses Schur convexity in place of (3), and permutation invariance in place of (2) in the block design example.]

In less simple cases, there will not be one design that is optimum in every possible "reasonable" sense; some work has been done on developing algorithms for more complicated situations (Kiefer 1958, 1975), and also some work of Cheng (1978) which will be mentioned later is relevant. Cheng developed a more complex optimality criterion suitable for less symmetric situations in which it is not possible to find a design matrix such that $\mathbf{M}_d$ (or $\tilde{\mathbf{M}}_d$ in the block setting treated below) has maximal trace and is a multiple of the identity.

Now, going back to the weighing design problem, let us omit the possibility of 0's in the design matrix, that is to say, the possibility of an object not appearing in a weighing. It was Hotelling who had the original idea that one could do well by including all objects on each weighing. It is really striking that there is a way to weigh 12 objects in 12 weighings (all objects being included in each weighing) which is so very efficient: the variance of the estimate of each weight is 1/12 of what it is for estimates based also on 12 weighings, but in which each weighing is of the more obvious form that weighs each of the objects separately. The elements of the design matrix of any weighing scheme in which every object is included in every weighing has its elements ± 1, and the trace of $\mathbf{X}^T\mathbf{X}$ is mk for every design, which automatically makes it maximal. You can see that inclusion of 0's would yield a smaller trace. Thus the problem reduces to finding a matrix whose elements are ± 1 and such that $\mathbf{X}^T\mathbf{X}$ is a multiple of the identity. In the mathematically simplest possible situation, $m = 2^k$; here the 2^k different k-vectors made up of ± 1's give all possible weighings and hence a matrix whose rows are these vectors has the desirable property, namely that $\mathbf{X}^T\mathbf{X}$ is a multiple of the identity. However, this solution is impractical, because 2^k is a large number; usually there will be few weighings per object, and for $k = 10$ it would be unusual to take 1024 weighings. Most often m is close to k ($m = k$ is called the *saturated* case). Hotelling realized that a satisfactory solution $\mathbf{X}$ that is sometimes available in the saturated case consists of a Hadamard matrix $\mathbf{H}_m$, that is to say a square matrix of ± 1's whose rows are mutually orthogonal. (These are named after Hadamard although Sylvester considered them much earlier.) It is not hard to see what to do if $m > k$: start with an $\mathbf{H}_m$-matrix and delete some of the columns. So, from a Hadamard matrix, we can get design matrices $\mathbf{X}$ which have the right structure. Unfortunately, an $\mathbf{H}_m$ exists only if $m = 1, 2$ or is divisible by 4. This is a necessary condition, but it is still an open question whether "m divisible by 4" is sufficient for the existence of Hadamard matrices. Literally thousands of papers have been written about Hadamard matrices.

If m is not divisible by 4, then a universally optimal design is not easy to find and may not even exist. Hence we will single out a particular optimality criterion—the D-optimality criterion, which requires the determinant of the information matrix to be a maximum. This has a simple geometric meaning: $\det(\mathbf{X}^T\mathbf{X})$ is the square of the volume of the k-dimensional parallelopiped in $\mathbb{R}^m$ spanned by the k column vectors of $\mathbf{X}$, each of length $\sqrt{m}$; we maximize this volume by choosing the vectors to be, in a sense, as close as possible to orthogonal (as they could be only when $4 \mid m$, if $m > 3$). In 1946, Williamson constructed optimal designs for samples of size 7 and less. The path-breaking papers are those of Ehlich in 1964. In the whole history of Hadamard matrices, Ehlich was almost the only person up to that time who considered systematically matrices of sizes which are not divisible by 4. Wojtas (1964) also worked on this problem, both he and Ehlich considering only the saturated case $m = k$. More recently Galil and Kiefer (1980) have worked on the case $m \geq k$. There are three cases other than that of Hadamard matrices, and these three cases correspond to the three residue classes modulo 4. In a sense, in all cases we are trying to achieve an information matrix "close" to the unachievable $m\mathbf{I}_k$.

In case 2, where $m \equiv 2 \pmod 4$, Ehlich showed (and this is also valid when $m > k$) that for the best design, a sufficient condition is that $\mathbf{X}^T\mathbf{X}$ has a block-diagonal form, consisting of a 2×2 partition, with 0's everywhere in the two off-diagonal blocks, and, in the 2 diagonal blocks, m's along the diagonal and 2's everywhere else; making the diagonal blocks of equal order (of orders differing by one if k is odd) then maximizes the determinant. Yang (1968) also worked on the construction of such D-optimal matrices in case 2. It is interesting that restricting attention to a symmetric solution in the intuitive Fisherian spirit described earlier (and in which all of the off-diagonal elements are equal) is quite poor, especially for small m; thus, in the case $m = k = 6$, the determinant of Ehlich's information matrix is 56% larger than for the best matrix with all off-diagonal elements equal.

The most interesting cases are case 1 and case 3 [$m = k \equiv 1$ and $m = k \equiv 3 \pmod 4$, respectively]. Case 1 at first appears relatively easy, because the design is easily proved D-optimal if $\mathbf{X}^T\mathbf{X}$ has 1's in all the off-diagonal positions. Unfortunately, although such an $\mathbf{X}$ can be obtained from an $\mathbf{H}_{m-1}$ if $m > k$, such an $\mathbf{X}$ exists in only a few saturated cases: it is necessary that $2n - 1$ be a perfect square. For $m = k = 9$, Ehlich and Zeller were able to find the optimal matrix, using a considerable amount of computation, and also algebraic and number-theoretic ingenuity; the case $m = k = 17$ was recently studied by Kounias and Moyssiadis.

Case 3 is mathematically the most interesting. Here $m = 4l - 1$; if we start with an $\mathbf{H}_{m+1}$ in which the first row and column consist only of +1's, and delete the 1st row and column and choose any k columns of the result, we end up with an $m \times k$ matrix which has the property that $\mathbf{X}^T\mathbf{X}$ has m's along the main diagonal and −1's off the diagonal. Let's call this $\mathbf{X}_e$ (e for "easy", since it's so easy to construct). Unfortunately, $\mathbf{X}_e$ is almost never optimal if m is close to k. Thus, in an interesting aesthetic sense, cases 1 and 3 are dual to each other. In case 1, the matrix is optimal if it exists, but it almost never exists when $m = k$; in case 3, one always exists but is almost never optimal. Ehlich developed a very ingenious theory for this case, showing what types of $k \times k$ matrices $\mathbf{C}$ with the same congruence properties as $\mathbf{X}^T\mathbf{X}$ could have largest determinant; but usually there do not exist design matrices $\mathbf{X}$ for which $\mathbf{X}^T\mathbf{X}$ has the form $\mathbf{C}$ required by the Ehlich theory, if $m = k$. So, we started here with the result that for $m = k$, Hadamard matrices $m\mathbf{I}_k$ are universally optimal; for the cases in which they don't exist, we have developed a nice theory for D-optimality, and in the end we find that these optimal matrices $\mathbf{C}$ are almost never achieved as $\mathbf{X}^T\mathbf{X}$, either.

But all is not lost. In practice, we are not restricted to $m = k$, and cases of practical interest are really those in which $m > k$, but m is usually not substantially larger than k. For $m > k$, Galil and Kiefer were able to extend Ehlich's theory and develop methods of construction to show that optimum designs often do exist. In 1974 Payne showed that if m is larger than something like 5^k, then $\mathbf{X}_e$ is optimum. Galil and Kiefer were able to reduce the size of m, and show that these easily obtained $\mathbf{X}_e$ are optimum if $m \geq 2k - 5$. Between this bound and saturation, the problem seems to be quite difficult. We have been able to solve it in some cases. An interesting aspect of the problem is that there are situations in which the optimal solution is not unique. One then uses the other optimality criteria to break ties between D-optimal designs.

Before leaving the subject of weighing designs, we emphasize how criteria of optimality led to solutions which would not be acceptable if we were limited to considerations of symmetry, that those solutions sometimes required new combinatorial devices, and that even nonexistence of solutions of a certain form one was seeking produced some interesting combinatorics when one sought solutions of an alternate structure.

5. OPTIMAL INCOMPLETE BLOCK DESIGNS

Finally, coming now to the second example mentioned at the beginning, we are looking at block designs to compare v "varieties" or treatments (say drugs). To minimize extraneous bias, we would like to test all drugs on animals from the same litter (block). We assume, for simplicity, that the animals cooperate and all have litters of the same size. If we can get litters big enough for this, we have complete block designs, but usually we have to make do with incomplete block designs. In the early days of this subject, a constraint was imposed on these designs on account of the requirements we discussed earlier, such as simplicity of the computational technique to be used for analysis of the data. This resulted in the constraint of balance—that every variety should appear in the same number of blocks (r) and every pair of varieties should appear together in the same number of blocks (λ). A common way of constructing a balanced incomplete block design is to start with one or more initial blocks and develop the others from these by systematic operations. R. C. Bose used this technique extensively and developed the method of symmetrically repeated differences. It gives a very simple method of construction where it works, as we shall illustrate shortly. Often such designs do not exist, because of the restrictions on the parameters (numbers of varieties, blocks, etc.) that need to be satisfied for the existence of a balanced incomplete block design.

We are now interested only in differences of the effects α_i ($1 \leq i \leq v$) of different drugs. The analogue of $\mathbf{M}_d$ is now a matrix $\tilde{\mathbf{M}}_d$ such that $\sigma^2\tilde{\mathbf{M}}_d^{-1}$ is the covariance matrix of any $v - 1$ orthogonal "contrasts" [e.g., $(\alpha_1 - \alpha_2)/\sqrt{2}$, $(2\alpha_3 - \alpha_1 - \alpha_2)/\sqrt{6}$, etc.]. Let $\{\mu_{di}\}$ be the eigenvalues of $\tilde{\mathbf{M}}_d$ (which do not depend on the choice of the $v - 1$ contrasts, this consequently also being the case for the optimal solution below). For simplicity we consider only optimality criteria of the form $\sum_1^{v-1} f(\mu_{di})$; for example, if $f(\mu) = 1/\mu$, this is proportional to the average variance of least-squares estimators of all principal contrasts $\alpha_i - \alpha_j$.

A block design is described by its incidence matrix $\mathcal{N}$ whose elements are integers, usually 1's and 0's—the element in the ith row and jth column is the number of times the ith variety appears in the jth block. For a balanced incomplete design, the matrix $\mathcal{N}\mathcal{N}^{\mathsf{T}}$ has r's along the main diagonal and λ's everywhere else. If such a design exists, it is optimum for *every* criterion function of the above form with f convex and decreasing. Where a balanced incomplete block design is impossible, one looks for the next best thing—a partially balanced incomplete design with two associate classes, which is a special type of group-divisible design. Here, the varieties are divided into two associate classes and the design is constructed so that each variety is paired in the same block λ_1 times with every member of its class and λ_2 times with every member of the other class. The matrix $\mathcal{N}\mathcal{N}^T$ has a 2×2 partitioned form in which the two off-diagonal blocks have all elements $= \lambda_2$, and the two diagonal blocks have r's along the main diagonal and λ_1's everywhere else.

Let us consider a specific example: Suppose we have 4 varieties to compare and the block size is 2. If we can use only 6 blocks, we can construct a balanced incomplete design, but if we want to use ten or eight blocks, we have to be satisfied with partially balanced designs; two such designs are:

1	3	1	1	1	1	2	2	2	2
2	4	3	3	4	4	3	3	4	4

1	1	3	3	1	1	2	2
2	2	4	4	3	4	3	4

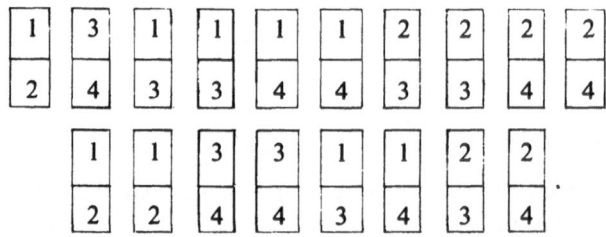

Treatments 1 and 2 are in the first class; treatments 3 and 4, in the second. In the first case $\lambda_1 = 1$ and $\lambda_2 = 2$, whereas in the second, $\lambda_1 = 2$ and $\lambda_2 = 1$. In general, which of these can be achieved is determined by the values of the parameters v, b and K. Cheng (1978) was able to prove an interesting fact, namely, that what is optimum in these two cases depends on the optimality criterion. For example, designs with $\lambda_2 = \lambda_1 + 1$ are optimal among all designs using the same number of blocks if the f of the optimality functional is decreasing convex and has a negative third derivative (A-, D-, and, in a limiting sense, E-optimality are of this form); whereas designs with $\lambda_1 = \lambda_2 + 1$ are optimal if the f of the optimality functional is decreasing convex and has a positive third derivative. This is a remarkable dichotomy, since usually in statistics convexity of the cost function is all that is needed; but there is such a fine structure here that the sign of the third derivative of f determines whether a given design, which is close to a balanced incomplete block design, is optimum or not.

It turns out that the optimality of these designs is equivalent to certain graph-theoretic considerations—the incidence matrices of these designs are related to those of regular graphs. So it turns out that questions of optimality of designs lead to graph-theoretic questions, some of which Cheng was able to answer.

Finally, we mention one more example of connections between experimental designs and combinatorics. If we depart from the classical model of independent or uncorrelated observations, we run into new problems. Here we shall discuss only one of the many correlation structures that can be treated. There are many situations in practice, for example in psychological experiments, where "neighbours" are not independent but interactive. Here "neighbourliness" may mean physical closeness (as for rats in cages or plants in plots) or closeness in time, etc. In this situation, observations from neighbouring plots in a block are correlated, whereas if two plots are separated by one or more plots, then the corresponding observations are independent. We will take a kind of robustness approach: correlations are small or absent, so we will use the usual least-squares estimator and a BIBD optimum in that case; but we will choose among BIBDs one for which performance under correlation is best. It turns out that, for the usual case of positive correlation, designs that are efficient among balanced incomplete block designs (BIBDs) are those that have the nearest-neighbour structure; that is to say, not only have varieties to be paired together in the same block the same number of times, but all pairs have to appear as nearest neighbours equally often. Such structures have been investigated by Kiefer and Wynn (1981).

Thus, for example, if we had $v = 7$ drugs and $b = 7$ litters of rats, of size 4 each, a classical BIBD developed from a symmetrically repeated difference set is

1	2	3	4	5	6	7
2	3	4	5	6	7	1
3	4	5	6	7	1	2
5	6	7	1	2	3	4

,

obtained as mentioned earlier by developing the first block (litter) by adding to all of its entries (treatment labels) the same value $j - 1$ (mod 7) to produce block j. This works because the 12 differences of pairs in the first block [$2 - 1 = 1$, $1 - 2 = 6$ (mod 7), etc.] repeat every zero element (mod 7) twice. But the differences of *neighbouring* pairs in the first block repeat ± 1 twice, ± 2 once, and ± 3 not at all. In the resulting design, for example, treatments 1 and 2 appear twice as nearest neighbours, but 1 and 5 never appear as neighbours. This design is not efficient under the nearest-neighbour correlation structure. It is easily checked that the BIBD developed from the initial block

1	2	5	3

has the right form, because the *neighbouring* pairs in this initial block yield each nonzero difference once. Here, then, we have another example of a question about design efficiency that has created a new problem in combinatorial construction.

ACKNOWLEDGEMENTS

The author is indebted to Sudhish Ghurye and Sarah Kiefer for their assistance.

RÉSUMÉ

Les modèles d'expériences statistiques, qui ont été développés par R.A. Fisher et ses disciples, ont souvent employé des structures combinatoires qui conduisaient à un calcul simple des estimations, et/ou des matrices symétriques de covariance. Par exemple des plans de blocs équilibrés, des expériences de régression avec espacement uniforme des observations, etc. Récemment, des études d'optimalité (consistant à choisir un «design» qui atteindra d'une certaine façon l'inférence la plus précise) ont justifié les modèles traditionnels, mais ils ont aussi quelquefois mené à de nouvelles études combinatoires. Des exemples sont inclus.

REFERENCES

Cheng, C.S. (1978). Optimality of certain asymmetrical experimental designs. *Ann. Statist.*, 6, 1239–1261.

Ehlich, H. (1964a). Determinantenabschätzungen für binäre Matrizen. *Math. Z.*, 83, 123–132.

Ehlich, H. (1964b). Determinantenabschätzung für binäre Matrizen mit $n \equiv 3$ mod 4. *Math. Z.*, 84, 438–447.

Elfving, G. (1952). Optimum allocation in linear regression theory. *Ann. Math. Statist.*, 23, 255–262.

Galil, Z., and Kiefer, J. (1980). *D*-optimum weighing designs. *Ann. Statist.*, 8, 1293–1306.

Guest, P.G. (1958). The spacing of observations in polynomial regression. *Ann. Math. Statist.*, 29, 294–299.

Hoel, P.G. (1958). Efficiency problems in polynomial estimation. *Ann. Math. Statist.* 29, 1134–1146.

Hotelling, H. (1944). Some improvements in weighing and other experimental techniques. *Ann. Math. Statist.*, 15, 297–306.

Kiefer, J., and Wolfowitz, J. (1960). The equivalence of two extremum problems. *Canad. J. Math.*, 14, 363–366.

Kiefer, J., and Wynn, H. P. (1981). Optimum balanced block and Latin square designs for correlated observations. *Ann. Statist.*, 9, 737–757.

Smith, Kirstine (1918). On the standard deviations of adjusted and interpolated values of an observed polynomial function and its constants and the guidance they give towards a proper choice of the distribution of observations. *Biometrika*, 12, 1–85.

Wald, A. (1943). On the efficient design of statistical investigations. *Ann. Math. Statist.*, 14, 134–140.

Received 2 January 1981
Revised 29 March 1981

Department of Statistics
University of California
Berkeley, California 94720, U.S.A.

Journal of Statistical Planning and Inference 5 (1981) 213–219
North-Holland Publishing Company

RELATIONSHIPS OF OPTIMALITY FOR INDIVIDUAL FACTORS OF A DESIGN

J. ECCLESTON

Department of Statistics, School of Mathematics, University of New South Wales, Kensington, NSW 2033, Australia

J. KIEFER*

Department of Statistics, University of California, Berkeley, CA 94720, USA

Received 12 December 1980
Recommended by A.S. Hedayat

Abstract: The optimality of two-factor experimental designs is studied in the dual senses of estimating contrasts in the parameters for each of the factors. The outline of comparison employed allows one to judge the performance of different designs for estimating contrasts of one set of parameters directly with the performance of the complementary set without going through a common intermediary step of considering all the parameters. The results hold for a wide class of optimality criteria (not merely D-, A- and E-optimality), which must satisfy a functional equation obtained in connection with our method. Also we investigate the optimality of row–column designs which satisfy an 'adjusted orthogonality' condition. Our point of departure is the paper by Shah, Raghavarao and Khatri (1976) and that of Mitchell and John (1977).

AMS Subject Classification Number: 62K05.

Key words: p-optimality; 2-way layouts; Block Designs; Adjusted Orthogonality.

1. Introduction and summary

An interesting paper by Shah, Raghavarao and Khatri (1976) studied the relationship between optimality in two different senses of one design d_1 over another d_2 (i) for estimating contrasts in the parameters for one of several factors, and (ii) for estimating contrasts for all the parameters of all factors. The designs that are compared must satisfy certain conditions described below. A corollary in the case of two factors, obtained by considering factor 1 (treatments) vs. both and then factor 2 (blocks) vs. both, is that, under the appropriate conditions, d_1 is better than d_2 for factor 1 iff it is better for factor 2. Here 'optimality' and 'better' are of course relative to certain criteria. Shah, Raghavarao and Khatri (1976) study D-, A- and E-optimality and it is easily seen that their presentation requires an explicit calculation of the eigenvalues and optimality functionals of the relevant information matrices. Their approach makes an extension of the results to other criteria appear difficult.

* Research partially supported by NSF Grant MCS78-25301.

Mitchell and John (1977), in the block design setting, noted that the eigenvalues or the information matrix for estimating each of the two factors, are related in such a way that A-, D-, or E-optimality of a design for one factor implies such optimality for the other factor.

The object of this paper is to extend the results of Shah, Raghavarao and Khatri and Mitchell and John to other optimality criteria and at the same time make the arithmetic of comparison more transparent. We do not consider in detail the original Shah–Raghavarao–Khatri comparison of (i) and (ii) above, but only that between two factors. As in Mitchell and John (1977), we consider the relationship between the two sets of eigenvalues of d_1 and d_2 and note which optimality criteria are consistent with the domination of one set over the other, thus employing a method of comparison other than that of Shah, Raghavarao and Khatri. Specifically, in section 2 we compare the performance for estimating contrasts of one set of parameters *directly* with that of the complementary set without going through the intermediate step of considering all the parameters as in Shah, Raghavarao and Khatri (1976). This method yields optimality results for a wide variety of functionals satisfying (2.4) and (2.5) (or limits of such functionals), not merely D-, A- and E-optimality, and it can be implemented in a main application of Shah, Raghavarao and Khatri, that of incomplete block designs with treatment and block parameters taken as the two sets. In section 3 we treat row–column designs satisfying an 'adjusted orthogonality' condition.

We omit here details of an analogous development of a functional equation for the Shah–Raghavarao–Khatri comparison (i)(ii), parallel to that of (2.4) below. The equation is

$$f(cv(d-v)) = a_{c,d}[f(v) + f(d-v)] + b_{c,d} \tag{1.1}$$

for constants $a_{c,d} > 0$ and $b_{c,d}$. This includes the A- and D-criteria and, in a modified form, the E-criterion; but not such other common criteria as Φ_p. The greater complexity of (1.1) than of (2.4) explains why the original Shah–Raghavarao–Khatri comparison (i)(ii) is more limited in criteria than is that between two factors.

It is to be emphasized that the optimality of any design d^* inferred from our undertaking is only relative to the specified class of competitors. Thus, like Shah, Raghavarao and Khatri, we prove a linked-block design is D-, A- and E-optimum for treatment contrasts among equireplicated designs for the given number of blocks b, number of treatments v, and (constant) block size k, but further obtain in section 2 a more general optimality result under the same restrictions. The more difficult and satisfactory optimality results without this restriction have been considered for certain parameter values only, by Cheng (1978). Clearly a design which is optimal for some restricted class of designs may not be optimal overall.

2. Complementary parameter sets in block designs

We consider a block design setting with a specified number of treatments v and b blocks of sizes k_j, $j=1,2,\dots,b$. The ith treatment is replicated r_i times, so $\sum_j k_j = \sum_i r_i$. Denote by R the diagonal matrix $\mathrm{diag}(r_1,\dots,r_v)$ and by K the diagonal matrix $\mathrm{diag}(k_1,\dots,k_b)$. Let $N=\{n_{ij}, 1\le i\le v, 1\le j\le b\}$ be the incidence matrix of a design, where n_{ij} is the number of times treatment i appears in block j. We impose no restriction on n_{ij}, nor that $k_j \le v$. We assume that usual linear model for a block design, namely, one in which an observation of treatment i in block j has expectation $\alpha_i+\beta_j$, where $\{\alpha_i, 1\le i\le v\}$ represent the treatment parameters and $\{\beta_j, 1\le j\le b\}$ represent the block parameters. The observations are assumed to be uncorrelated with equal variance. The coefficient matrices for the treatments adjusted for blocks and for the blocks adjusted for treatments are, respectively,

$$C_{\mathrm{T|B}} = R - NK^{-1}N' \quad \text{and} \quad C_{\mathrm{B|T}} = K - N'R^{-1}N. \tag{2.1}$$

Since the treatments and blocks enter symmetrically in the discussion that follows, we may assume $v\le b$. Let γ_i $(1\le i\le v)$ be the eigenvalues of NN'. Therefore v of the eigenvalues of $N'N$ are γ_i $1\le i\le v$, and the remainder are 0.

Suppose we now restrict consideration to designs for which

$$k_j = k, \quad \text{for all } j, \tag{2.2a}$$

$$r_i = r, \quad \text{for all } i. \tag{2.2b}$$

Then, under (2.2) since $C_{\mathrm{T|B}} = rI - k^{-1}NN'$ and $C_{\mathrm{B|T}} = kI - r^{-1}N'N$, the eigenvalues $\mu_i(C_{\mathrm{B|T}})$ and $\mu_i(C_{\mathrm{T|B}})$ of these matrices are

$$\mu_i(C_{\mathrm{T|B}}) = k^{-1}(rk-\gamma_i), \quad 1\le i\le v;$$

$$\mu_i(C_{\mathrm{B|T}}) = \begin{cases} r^{-1}(rk-\gamma_i), & 1\le i\le v, \\ k, & v<i\le b. \end{cases} \tag{2.3}$$

We note that, since $C_{\mathrm{T|B}}$ is singular, at least one $\gamma_i = rk$; if all the elementary treatment contrasts are estimable, the design is connected and exactly one $\gamma_i = rk$.

Let $\boldsymbol{F}$ be the class of all functions f from $[0,+\infty)$ into $(-\infty,+\infty]$ (that is, f may take on the value $+\infty$). In this section we shall impose on f the condition that for each $c>0$ there are $a_c>0$ and b_c such that

$$f(cx) = a_c f(x) + b_c. \tag{2.4}$$

On the class of $m\times m$ nonnegative definite matrices C of rank $\le m-1$, suppose $\mu_1(C)$ is a zero eigenvalue, and for f in $\boldsymbol{F}$ write

$$\Phi_f^{(m)}(C) = \sum_2^m f(\mu_i(C)). \tag{2.5}$$

For fixed m, let $\boldsymbol{F}^{(m)}$ denote the class of all increasing functions of $\Phi_f^{(m)}$'s or limits of sequences of such functions. Let $\boldsymbol{F}^{(m)*}$ be the class obtained by restricting f to satisfy

(2.4). We associate $\Phi^{(m')}$ and $\Phi^{(m'')}$ if they come from the same sequence of increasing functions of criteria of type (2.5) for the same f. As usual, we say a design is $\Phi^{(v)}$-optimum among designs in a class $\mathscr{C}$, for treatment contrasts, if it minimises $\Phi^{(v)}(C_{T|B})$ over all designs in $\mathscr{C}$; similarly for block contrasts.

From (2.3) and (2.5):

Proposition 1. *Under* (2.2a), *for every $\Phi^{(v)}$ in $F^{(v)*}$ and associated $\Phi^{(b)}$ in $F^{(b)*}$, a design is $\Phi^{(v)}$-optimum for treatment contrasts, among designs in any class $\mathscr{C}$ satisfying* (2.2b), *if and only if it is $\Phi^{(b)}$-optimum for block contrasts among designs in $\mathscr{C}$.*

The criteria covered here include the common *D*-, *A*- and *E*-criteria (the last as a limit), and also the other Φ_p-criterion given by $f(x) = x^{-p}$. The *E*-criterion can also be treated separately by simply looking at (2.3) without taking a limit. Also it suffices that f be defined on the domain of possible eigenvalues, never larger than $[0, \max(r, k)]$. Obviously two designs may be compared by letting $\mathscr{C}$ consist of the two designs in question only.

The optimality for treatment contrasts of balanced block designs without restriction (2.2b) was proved by Kiefer (1958, 1975) for a class of $\Phi_f^{(m)}$'s which we will denote by $F^{(m)**}$. Cheng (1978) extended these results to two classes of partially balanced designs. These results can be used to obtain the optimality for block contrasts of the above designs, under (2.2b). (Even though it is not used in the results just cited, we require (2.2b) in our conclusion because there is no simple relationship between the $\mu_i(C_{T|B})$ and $\mu_i(C_{B|T})$ otherwise.) Alternatively if a design is the dual of one of the above mentioned balanced or partially balanced block designs, then from its optimality for block contrasts we obtain its optimality for treatment contrasts, subject to (2.2b). In particular, we have

Corollary 1. *Under* (2.2a), *if b, v, r and k are such that a linked block design exists, then it is $\Phi^{(v)}$-optimum for treatment contrasts among designs satisfying* (2.2b), *for all $\Phi^{(v)}$ in $F^{(m)*} \cap F^{(m)**}$.*

This result was proved for the *A*-, *D*- and *E*-criteria in Shah, Raghavarao and Khatri (1976) (Corollary 1), by the methods described in section 1.

3. Three factor designs; adjusted orthogonality

We consider 3-factor designs of v treatments applied to a b_1 (rows) by b_2 (columns) array and with one treatment per cell. Such designs are usually referred to as Row–Column Designs, RCD.

In this setting, optimality results for Generalised Youden Squares, GYDs (in particular, Youden Squares and Latin Squares), were obtained by Kiefer (1958,

1975, 1980), for particular v, b_1, b_2 for which such designs exist. The developments of this section are of specific interest when no GYD exists. It should be noted that Youden Squares and Latin Squares exhibit the adjusted orthogonality condition defined here; for a detailed discussion of adjusted orthogonality see Eccleston and Russell (1977).

We will consider designs that satisfy the analogue of (2.2b):

$$\text{each treatment is replicated } r=(b_1b_2)/v \text{ times.} \tag{3.1}$$

Let N_{TR}, N_{TC}, N_{RC} be the indicated incidence matrices among treatments, rows and columns; of course $N_{RC}=J_{b_1b_2}$. In parallel with (2.1) we write

$$\begin{aligned} C_{T|C}&=rI_v-b_1^{-1}N_{TC}N'_{TC},\\ C_{T|R}&=rI_v-b_2^{-1}N_{TR}N'_{TR}, \end{aligned} \tag{3.2}$$

as the information matrices for the treatments adjusted for columns ignoring rows and for the treatments adjusted for rows ignoring columns, respectively. The information matrix for treatments adjusted for rows and columns is easily seen to be (in a form that will be useful in the sequel)

$$\begin{aligned} C_{T|RC}&=rI_v-b_1^{-1}N_{TC}N'_{TC}-b_2^{-1}N_{TR}N'_{TR}+rv^{-1}J_{v,v}\\ &=rI_v-rv^{-1}J_{v,v}-[rI_v-C_{T|R}-rv^{-1}J_{v,v}]-[rI_v-C_{T|C}-rv^{-1}J_{v,v}]\\ &=rI_v-rv^{-1}J_{v,v}-A-B \quad \text{(say).} \end{aligned} \tag{3.3}$$

A design is said to have 'rows-adjusted-for-treatments orthogonal to columns-adjusted-for-treatments' (hereafter 'adjusted orthogonality') if $N_{RT}N_{TC}=rJ_{b_1b_2}$; under this condition Raghavarao and Federer (1975) showed $C_{T|RC}$ has rank $v-1 \Leftrightarrow$ $C_{T|R}$ and $C_{T|C}$ have rank $v-1$. For further discussion of adjusted orthogonality see Eccleston and Russell (1977). In a recent paper Raghavarao and Shah (1980) give a method of construction of adjusted orthogonal designs and note their high efficiency. Let $\mathscr{D}$ be the class of adjusted orthogonal designs satisfying (3.1) and with $C_{T|RC}$ of rank $v-1$. For such a design, it is seen from (3.2) that A and B of (3.3) satisfy $AB=0$. Also, $JA=JB=0$. Consequently, there is an orthogonal matrix Q such that QAQ', QBQ' and QJQ' are diagonal with nonzero entries in distinct locations. Hence, if A has p nonzero eigenvalues and B has q nonzero eigenvalues, we have $p+q\le v-1$. Thus, $C_{T|R}$ has its trivial zero eigenvalue ($JC_{T|R}=0$), $v-1-p$ eigenvalues equal to r, and p eigenvalues $\mu_i(C_{T|R})=r-\mu_i(A)$, $1\le i\le p$, and similarly for the eigenvalues of $C_{T|C}$. Moreover, on applying Q we see from (3.3) that $QC_{T|RC}Q'$ is also diagonal. Taking QJQ' to have its nonzero element in the (1, 1) position, we have for the remaining $v-1$ diagonal elements, from the positions of QAQ' and QBQ',

$$\begin{pmatrix} \mu_2(C_{T|RC}) \\ \cdot \\ \cdot \\ \cdot \\ \cdot \\ \cdot \\ \cdot \\ \cdot \\ \cdot \\ \cdot \\ \cdot \\ \cdot \\ \cdot \\ \cdot \\ \mu_v(C_{T|RC}) \end{pmatrix} = \begin{pmatrix} -r \\ \cdot \\ \cdot \\ \cdot \\ \cdot \\ \cdot \\ \cdot \\ \cdot \\ \cdot \\ \cdot \\ \cdot \\ \cdot \\ \cdot \\ \cdot \\ -r \end{pmatrix} + \begin{pmatrix} \mu_1(C_{T|R}) \\ \cdot \\ \cdot \\ \cdot \\ \mu_p(C_{T|R}) \\ r \\ \cdot \\ \cdot \\ \cdot \\ r \\ r \\ \cdot \\ \cdot \\ \cdot \\ r \end{pmatrix} + \begin{pmatrix} r \\ \cdot \\ \cdot \\ \cdot \\ r \\ r \\ \cdot \\ \cdot \\ \cdot \\ r \\ \mu_1(C_{T|C}) \\ \cdot \\ \cdot \\ \cdot \\ \mu_q(C_{T|C}) \end{pmatrix}; \tag{3.4}$$

the last two vectors give the eigenvalues of $C_{T|R}$ and $C_{T|C}$, with $\mu_i(C_{T|R})$ and $\mu_i(C_{T|C})$ in distinct positions. This can also be obtained from Shah (1977), but for our purpose one must take care, as we have, to list the eigenvalues of various diagonalized matrices in corresponding order.

Consequently, for Φ of the form (2.5), we have from (2.5) and (3.4),

$$\Phi(C_{T|RC}) = \Phi(C_{T|R}) + \Phi(C_{T|C}) - (v-1)f(r). \tag{3.5}$$

Thus, we have

Proposition 3. *An* RCD *d^* in $\mathscr{D}$ is Φ-optimum for treatment effects among all designs in $\mathscr{D}$ for any Φ satisfying* (2.5) *for which d^* minimizes $\Phi(C_{T|R}) + \Phi(C_{T|C})$ over $\mathscr{D}$. In particular, d^* is Φ-optimum in $\mathscr{D}$ for treatment effects if it is Φ-optimum in $\mathscr{D}$ for treatment effects in each of the treatment-row and treatment-column two-factor models. This last result also holds for E-optimality.*

The E-optimality result can be obtained by a limiting argument, or can be derived directly from (3.4) and (3.2).

These results illustrate still another relationship among optimality results for a design in several factors, other than those of section 2, which is one reason we include them despite the fact that the restriction to consideration of the class $\mathscr{D}$ may eliminate designs that are optimum among all designs. This does not mean $v \mid b_1$ or $v \mid b_2$ is necessary for adjusted orthogonality, as Shah (1977) illustrates with a design due to Khirasagar with $v=9$, $b_1 = b_2 = 6$. Additionally, there is recent work of Cheng on pseudo-Youden designs, and earlier work on 'nonregular GYD's' for which $b_1 > v$, $b_2 > v$ and neither is divisible by v, and which have optimum properties.

Another example relevant to this section is that of Shah, Raghavarao and Khatri

1976, p. 422. For $v=2$, $b_1=4$, $b_2=8$, these authors give two designs which have the same value of the nonzero eigenvalue $\mu_2(C_{T|RC})$, although, as they remark, the information matrices $\bar{C}$ for the whole designs are not of such structure (3.1) that the Shah-Raghavarao-Khatri theorem applies. But both designs are in $\mathscr{D}$, so (3.5) can be used to compare the two designs, from the values of $\Phi(C_{T|R})$ and $\Phi(C_{T|C})$ for those two designs. Of course, Shah, Raghavarao and Khatri (1976) gave this example to indicate that designs with the same performance for treatments need not have the same performance for all parameters; an obvious design, different from both of these, is optimum for treatments.

Acknowledgement

We thank K.R. Shah for helpful comments.

References

Cheng, C.S. (1978). Optimality of certain asymmetrical experimental designs. *Ann. Statist.* 6, 1239–1261.

Eccleston, J.A. and K.G. Russell (1977). Adjusted orthogonality in non-orthogonal designs. *Biometrika* 64 (2), 339–345.

Kiefer, J. (1958). On the nonrandomised optimality and randomised non-optimality of symmetrical designs. *Ann. Math. Statist.* 29, 675–699.

Kiefer, J. (1975). Construction and optimality of generalised Youden designs. In: J.N. Srivastava, ed., *A Survey of Statistical Designs and Linear Models,* North Holland, Amsterdam, 333–353.

Kiefer, J. (1980). Combinatorics and optimal design. In: J.N. Srivastava, ed., *Proc. 1978 Fort Collins Meeting.* To appear.

Mitchell, T.J. and J.A. John (1977). Optimal incomplete block designs. *JRSS,* Ser. B 39, 39–43.

Raghavarao, D. and W. Federer (1975). On connectedness in elimination of two-way heterogeneity designs. *Ann. Statist.* 3, 730–735.

Raghavarao, D. and K.R. Shah (1980). A class of D_0 designs for two-way elimination of heterogeneity. *Commun. Statist. Theor. Meth.* A9(1), 75–80.

Shah, K.R. (1977). Analysis of designs with two-way elimination of heterogeneity. *J. Statist. Planning and Inference* 1, 207–216.

Shah, K.R., D. Raghavarao and C.G. Khatri (1976). Optimality of two and three factor designs. *Ann. Statist.* 4, 419–422.

ON THE CHARACTERIZATION OF D-OPTIMUM WEIGHING DESIGNS FOR $n \equiv 3 \pmod 4$

Z. Galil[1]

Department of Mathematics
Tel Aviv University
Tel Aviv, Israel

J. Kiefer[1]

Department of Statistics
University of California at Berkeley
Berkeley, California, U.S.A.

I. INTRODUCTION

Let k and n be positive integers with $k \leq n$, and let $\mathcal{X} = \mathcal{X}(k,n)$ be the set of all $n \times k$ matrices $X = \{x_{ij}\}$ consisting entirely of entries ±1. We seek a D-optimum X*, i.e., one that maximizes $\det(X'X)$ over $\mathcal{X}$. The weighing, fractional factorial, and first order regression settings where this problem arises, a partial history, and a list of known results, are given in Galil and Kiefer [2] (hereafter denoted GK). As described there, the most interesting unsolved cases are those for which $n \equiv 3 \pmod 4$, and the present article is devoted entirely to improving knowledge on the structure of a D-optimum X* in those cases.

WE HEREAFTER ASSUME $n \equiv 3 \pmod 4$.

Ehlich [1] developed the pioneering approach to this subject for the saturated case $k = n$, and its main aspects are easily

[1]This research was supported by the National Science Foundation under grant no. MCS 78-25301. The first author was also supported by the Israel Commission for Basic Research.

Statistical Decision Theory and Related Topics III, Vol. 1

ISBN 0-12-307501-7

extended to the general case $k \leq n$. In the remaining paragraphs of this section, except for the last one in which we summarize this paper's contents, we list, from GK, the main results of that development which we shall use herein. Let $\mathcal{C} = \mathcal{C}(k,n)$ be the class of all symmetric k x k matrices with diagonal elements n and off-diagonal entries -1 or 3. (The conclusions that follow are valid if in addition C is restricted to positive-definite matrices.) Let

$$(1.1) \qquad \Psi(k,n) = \max_{C \in \mathcal{C}(k,n)} \det C.$$

It can be shown that letting symmetric C have off-diagonal entries $\equiv 3 \pmod 4$ other than -1 or 3 can only reduce det C. Every X in $\mathcal{X}$ can be transformed, by multiplying certain columns by -1, into an $\tilde{X}$ for which the entries of $\tilde{X}'\tilde{X}$ are all $\equiv 3 \pmod 4$, and hence

$$\max_{X \in \mathcal{X}} \det(X'X) \leq \Psi(k,n).$$

Even if we find a C with $\det C = \Psi(k,n)$, we must still either find an X that satisfies $X'X = C$ or show that no such X exists and pursue a further development - see Section 7. Nevertheless, a first step is to determine $\Psi(k,n)$ and the C's achieving it.

A *block* of size r is an r x r matrix with diagonal entries n and off-diagonal entries 3. A *block matrix* in $\mathcal{C}(k,n)$ with s blocks, of sizes $r_1, r_2, \ldots, r_s$ satisfying $\Sigma_1^s r_i = k$, has diagonal blocks of those sizes and all other elements -1. Such a block matrix C has

$$(1.2) \qquad \det C = (n-3)^{k-s}\{1 - G\}\Pi_1^s(n - 3 + 4r_i),$$

$$G = \Sigma_1^s r_i/(n - 3 + 4r_i).$$

The Ehlich development shows that $\Psi(k,n)$ is achieved only by a block matrix. Let $\mathcal{B} = \mathcal{B}(k,n)$ be the subset of $\mathcal{C}(k,n)$ consisting of block matrices with *blocks of only one size or blocks of only two contiguous sizes*. The Ehlich theory shows that if $C \in \mathcal{C}$ and $\det C = \Psi(k,n)$, then $C \in \mathcal{B}$. (The original method of proof in Ehlich ([1], p. 443) only shows that *some* maximizing C is in $\mathcal{B}$, but this is easily strengthened by showing (i) if there are 3 or more different r_i's we can alter some pair of them so as to improve det C *strictly* and (ii) with $s \geq 3$ and only two block sizes with $r_2 - r_1 \geq 2$, we can either *strictly* improve det C, or can obtain the same value of det C with blocks of at least 3 sizes, except for some possibilities with $r_1 = 1$, $r_2 = 3$ that are handled separately.)

For any C_s in $\mathcal{B}$ having s blocks, u of size r and v of size r + 1, we have

$$u + v = s, \quad ur + v(r + 1) = sr + v = k. \tag{1.3}$$

From (1.2) and (1.3),

$$\begin{aligned} D_{k,n}(s) &\overset{\text{def}}{=} \det C_s \\ &= (n-3)^{k-s}(n - 3 + 4r)^{u}(n + 1 + 4r)^{v}\{1 - G\} \\ &= (n-3)^{k-s}(n - 3 + 4r)^{(r+1)s-k}(n + 1 + 4r)^{k-sr}\{1 - G\}, \\ 1-G &= 1 - ur/(n - 3 + 4r) - v(r + 1)/(n + 1 + 4r) \\ &= [k(n - 3) + 4sr(r + 1)]/(n + 4r + 1)(n + 4r - 3). \end{aligned} \tag{1.4}$$

Thus,

$$\Psi(k,n) = \max_s D_{k,n}(s). \tag{1.5}$$

We use the notation of ceiling and floor functions, $\lfloor x \rfloor$ and $\lceil x \rceil$, for the greatest integer $\leq x$ and least integer $\geq x$. For C_s in $\mathcal{B}$ having blocks of sizes r and r + 1 as in (1.3), we define the maximum block size to be R. Thus, we have R = r + 1 unless v = 0, when R = r. Furthermore

$$r = \begin{cases} \lfloor k/s \rfloor & \text{if } u > 0, \\ (k/s) - 1 & \text{if } u = 0, \end{cases}$$

(1.6)

$$R = \lceil k/s \rceil .$$

An ambiguity occurs in the definition of r, u, and v only if $s|k$. We may remove the ambiguity either by assuming always that

(a) $u > 0$ (hence $r = k/s$ and $v = 0$ if k/s = integer)

(1.7)

(b) $v > 0$ (hence $r + 1 = k/s$ and $u = 0$ if k/s = integer).

For computational simplicity we use (1.7)(a) in Section 2 and (1.7)(b) in Section 4, as we shall repeat there. Thus, s determines R and, once the convention (1.7)(a) or (b) is chosen, s determines r; but not conversely.

We denote by s_{OPT}, r_{OPT}, R_{OPT}, any set of values that maximizes $D_{k,n}(s)$. Sometimes there is more than one maximizing set. In GK we showed that a solution was

(1.8) $\{s_{OPT} = k,\ r_{OPT} = R_{OPT} = 1\} \Leftrightarrow n \geq 2k - 5$

and that k is the unique value of s_{OPT} if $n > 2k - 5$. When $n = 2k - 5$ the single other solution is

(1.9) $s_{OPT} = k - 1,\ r_{OPT} = 1,\ R_{OPT} = 2.$

IN THE REMAINDER OF THIS PAPER WE MAY HENCE ASSUME $n \geq 2k - 5$.

In Section 2 we follow Ehlich to derive simple inequalities on s_{OPT}, r_{OPT}, R_{OPT}. (The value of R_{OPT} is crucial for our construction methods.) Exact determination of these quantitites is not simple, as one sees from the extensive computer study of $D_{n,n}(s)$ needed by Ehlich (described in Comment 6 of Section 2) to show $s_{OPT} = 7$ for $n \geq 63$ when $k = n$. In the comments of Section 2 we show how better inequalities can be obtained by studying in detail the various cases of congruence, and how s_{OPT} can sometimes be determined exactly in this way. (As an illustration, we show by this method that $s_{OPT} = 7$ if $n - k = 3$ and $n \equiv 7$ or $11 (\mathrm{mod}\ 12)$, or if $n - k = 4$ or 5 and $n \equiv 3 (\mathrm{mod}\ 8)$, there being four cases (k,n) of "ties" where also $s_{OPT} = 8$.) However, Ehlich's and our approach, of first obtaining simple inequalities and then supplementing these by a finer calculation comparing $D_{k,n}(s)$ among s-values not eliminated by the inequalities, is useful because the number of such s-values is often small. The number of such s-values to be compared is roughly $n/(2k - n)$, and the calculations following (2.15) imply that, as $n \to \infty$ with $k/n \to 1 - \lambda > 1/2$, the numbers of such s-values is within 1 of $(1 - 2\lambda)^{-1}$ as $n \to \infty$. Thus, in all "practical" cases except when k/n is near 1/2, the number of such s-values is reasonably small. In Section 3 we treat the simpler asymptotic determination of s_{OPT} as $k,n \to \infty$. In Section 4 we determine exactly those pairs of (k,n) for which there is a "tie" for neighboring s-values in the sense that $D_{k,n}(s) = D_{k,n}(s + 1)$ for some s. This must be the case when two neighboring values of s_{OPT} exist, as they did in the special case (1.8) - (1.9), and it turns out that neighboring s_{OPT}-values are the only cases of neighboring ties, i.e., $D_{k,n}$ can have no other "flat" places on the domain of integral s. The pairs (k,n) of this type turn out to be certain ones satisfying $2k - n = d$ for a fixed small d. For these and other small values of d, we determine s_{OPT} for the "series" $\{(k,n)\colon\ 2k - n = d\}$, in Section 5. (Although, as we just mentioned, k/n near 1/2 means more s-values to be compared, small values of d allow this comparison to be carried out easily analytically.) In Section 6 we develop an approximate formula for s_{OPT}, for general k,n. In Section 7 we discuss briefly the problem of construction of an optimum X, described earlier.

II. SIMPLE INEQUALITIES

Throughout this section we adopt the convention (1.7)(a). The basic inequalities are on values of r or R that *cannot* be optimum, and we use whichever of these gives the better inequality on s_{OPT}, and also obtain inequalities on r_{OPT} and R_{OPT} from them.

We shall only detail an extension of the argument in Ehlich's Lemmas 3.1 and 3.2, and then comment briefly on finer arguments that yield slight improvements.

An upper bound on s_{OPT}, *lower bound on* r_{OPT} *or* R_{OPT}. Suppose $\bar{s}$ is a proposed value of s_{OPT} and $C_{\bar{s}}$ is the corresponding block matrix with blocks of two adjacent sizes (or one if $\bar{s}|k$). Of course, $\bar{s} \le k$. Let $\rho = 2r$ if $v = 0$ and $\rho = 2r + 1$ if $v > 0$, and consider the family $\{C(x), 0 \le x \le r\}$ of block matrices of the following form: Of the $\bar{s}$ blocks in $C_{\bar{s}}$ keep $\bar{s} - 2$ intact, say blocks number $3, 4, \ldots, \bar{s}$. The first two blocks, of size r, r (respectively, r, r + 1) if $\rho = 2r$ (respectively, 2r + 1) are eliminated and blocks of size x, $\rho - x$ are introduced. As x varies, it is easily seen from (1.2) that det C(x) is proportional to

$$f(x) = (n-3+4x)(n-3+4\rho-4x)\left(1-\beta-\frac{x}{n-3+4x}-\frac{\rho-x}{n-3+4\rho-4x}\right),$$

(2.1)

$$\beta = \Sigma_3^{\bar{s}} r_i/(n-3+4r_i).$$

Since f is quadratic and $f(0) = f(\rho)$, if $f'' \equiv -16(1-2\beta) > 0$, decreasing x to 0 from the value r it has in $C_{\bar{s}}$ will increase det C(x) so $s_{OPT} \ne \bar{s}$. Changing s need not mean changing r. However, if $(1-2\beta) < 0$ whenever $r = \bar{r}$ (regardless of the value of $\bar{s}$), we conclude that $r_{OPT} \ne \bar{r}$.

If $v = 0$ $(R = r)$, we have

$$(n-3+4r)(1-2\beta) = n-3+4r - 2(\bar{s}-2)r$$
$$= n-3+8r - 2k.$$

If $v > 0$ $(R = r + 1)$, we have

$$(n+1+4r)\beta = (u-1)r(n+1+4r)/(n-3+4r)+(v-1)(r+1)$$
$$\geq (u-1)r+(v-1)(r+1)$$

and thus

$$(n+1+4r)(1-2\beta) \leq (n+1+4r) - 2(u-1)r - 2(v-1)(r+1)$$
$$= n+3+8r-2k.$$

From the previous paragraph, we conclude that s can be reduced in both cases while increasing the determinant, if $\bar{s}$ corresponds to r for which

(2.2) $\quad r < (2k-n-3)/8$

or, alternatively (the first case being more restrictive in terms of R than the second), if $\bar{s}$ corresponds to R for which

(2.3) $\quad R < (2k-n+3)/8.$

Thus, since (2.2) and (2.3) are conditions that hold for all $\bar{s}$ with such r or R, we have

(2.4)
$$r_{OPT} \geq (2k-n-3)/8,$$
$$R_{OPT} \geq (2k-n+3)/8.$$

Since $r \leq k/\bar{s}$, (2.2) is satisfied if $\bar{s} > 8k/(2k-n-3)$, so that

(2.5) $\quad s_{OPT} \leq 8k/(2k-n-3).$

(The inequality on s obtained from (2.3) is less good.) Of course, we can actually write $r_{OPT} \geq \lceil (2k-n-3)/8 \rceil$ and similarly for R_{OPT}, and thus $s_{OPT} \leq \lfloor k/\lceil (2k-n-3)/8 \rceil \rfloor$.

Comments and refinements. 1. We know (Section 4) exactly when two *successive* values of s can both be optimum, and we know s_{OPT}, r_{OPT}, and R_{OPT} in all those cases. In all other cases, (2.5) holds with strict inequality; and the strict inequality holds for the smallest s_{OPT} in any event. The question of strict inequality in (2.4) requires further consideration: if the unimodality conjecture described in (6.1) is false (which we doubt), it is conceivable that two separated values of s_{OPT} could exist, and these might yield the same or different r_{OPT} and R_{OPT} values (since r is not strictly decreasing in s); but the strict inequality in (2.4) holds for the largest r_{OPT} or R_{OPT} even if there are such ties.

2. The inequalities can be sharpened by considering dependence on v in the second case, $v > 0$. If $1 \le v \le \bar{s} - 2$, (2.2) is improved slightly by taking $\rho = 2r$; but for $v = \bar{s} - 1$, $u = 1$, the treatment above gives the best one can do with this argument, so (2.2) is binding if the inequality is not to depend on v. We shall not use this minor improvement in the sequel, but considerations of this kind are important if one wants to determine better bounds. We now give an example of this treatment.

When $k = n$ the proof of Ehlich (Lemma 3.1) that $s_{OPT} \le 7$ requires detailed calculation of $1 - 2\beta$ when $\bar{s} = 8$ and $n \equiv 3 \pmod 8$ $(u = 5)$ or $n \equiv 7 \pmod 8$ $(u = 1)$. The analogue for showing, *sometimes*, that $s_{OPT} = 7$ for the nonsaturated case $k < n$, is more complicated because k can have any congruence $q \pmod 8$. (We can at best hope to obtain $s_{OPT} \le 8$ from (2.5).) Suppose, then, that $\bar{s} = 8$ (hence $k \ge 8$) and $k = n-B = 8r + v$, r and v integers, $r \ge 1$, $0 \le v \le 7$. Then the largest possible β, attained when min(v,6) of the 6 blocks have size $r + 1$, is

$$(2.6) \qquad \beta = \frac{(6-v)r}{12r+B-3+v} + \frac{v(r+1)}{12r+B+1+v}, \quad 0 \le v \le 6,$$

$$\beta = \frac{6(r+1)}{12r+B+8}, \qquad v = 7.$$

This is $\ge 1/2$ iff

$$(2.7) \qquad 4r(v+9-3B) \ge (B-3+V)(B+1-v), \quad 0 \le v \le 6,$$

$$B \le 4, \qquad v = 7.$$

An examination of the various possible cases shows that this is satisfied $\forall r \ge 1$ if $0 \le B \le 3$ (for both $n \equiv 3$ and $7 \pmod 8$); and, for $B = 4$ or 5, if $v = 11 - B$ $(n \equiv 3 \pmod 8)$. Among these cases, we find $\beta = 1/2$ in (2.6) only when $B = 4$ or 5 or when $B = 3$ and $v = 0$, so we might have a "tie" ($s_{OPT} = 8$ as well as 7) in those cases. However, a check of the cases of ties (Section 4) shows that in the above domain s_{OPT} can be 8 as well as 7 only when $(n,k) = (19,15)$, $(19,14)$, $(27,22)$, or $(11,8)$. In all the cases where (2.7) is satisfied, improvement if $\bar{s} \geq 9$ is easily checked. Thus, we have obtained $s_{OPT} \leq 7$ for $0 \leq n - k \leq 3$ and for "half-the cases" $n-k = 4$ or 5 (namely, those with $n \equiv 3 (\mathrm{mod}\ 8)$), with s_{OPT} also 8 for the four values (n,k) noted above. In comment 5 on the lower bound, below, the two bounds obtained by this finer argument will be combined. This extends Ehlich's inequality for the saturated case to "near-saturated" cases, and it is seen how finer calculations were required than those that yielded (2.5). Of course, even this refinement does not yield *necessary* conditions for $s_{OPT} \leq 7$, since it was based on the single mode of improvement of det C through altering only two block sizes. (Illustrations of how more block sizes must be altered in finer arguments will be encountered in the verification of the cases $d = 13, 15, 17$ of (5.1), and in Section 6.) When k/n is near 9/10 and (2.5) yields only $s_{OPT} \leq 9$, an analogous argument can be used to delimit some cases in which $s_{OPT} \leq 8$, etc.; the argument is now longer because there are many more combinations (B,v) that are possible for $\bar{s} = 9$. We omit details.

Lower bound on s_{OPT}, *upper bound on* r_{OPT} *or* R_{OPT}. We now consider $\tilde{C}(x)$, obtained from $C_{\bar{s}}$ by replacing one block of size R in $\tilde{C}_{\bar{s}}$ by two blocks, of size x and R-x. From (1.2), the determinant of C(x) is proportional to

$$g(x) = (n-3+4x)(n-3+4R-4x)\ \{1-\gamma- \frac{x}{n-3+4x} - \frac{R-x}{n-3+4R-4x}\},$$

$$\gamma = \frac{ur}{n-3+4r} + \frac{(v-1)R}{n-3+4R}.$$

(Note that $\gamma = (\bar{s}-1)r/(n-3+4r)$ if $v = 0$, $R = r$.)

We suppose $R \geq 2$, which means $\tilde{C}(1) \neq C_{\bar{s}}$. (The case $R = 1$ is covered by (1.8).) Since g is quadratic and $g(0) = g(R)$, $\det \tilde{C}(1) > \det C_{\bar{s}}$ if $0 < g'(0) = 8R(1-2\gamma)$. If $v = 0$ ($R=r, r\bar{s}=k$), we have

$$\begin{aligned}(n-3+4r)(1-2\gamma) &= n-3+4r-2(\bar{s}-1)r \\ &= n-3-2k+6r.\end{aligned}$$

If $v > 0$, we have

$$\begin{aligned}(n-3+4r)(1-2\gamma) &\geq n-3+4r-2ur-2(v-1)(r+1) \\ &= n-1-2k+6r.\end{aligned}$$

We conclude, as with the earlier bounds, that we can improve on $C_{\bar{s}}$ if

(2.9) $\quad r > (2k-n+3)/6$

or, alternatively, if

(2.10) $\quad R > (2k-n+7)/6.$

In particular,

$$(2.11) \qquad \begin{aligned} r_{OPT} &\leq (2k-n+3)/6, \\ R_{OPT} &\leq (2k-n+7)/6. \end{aligned}$$

As before, we obtain from $\bar{R} \geq k/\bar{s}$ and (2.10) (which yields a better inequality than (2.9))

(2.12) $\quad s_{OPT} \geq 6k/(2k-n+7).$

Again, (2.9), (2.10), and (2.11) can be written in a stronger form, corresponding to that described below (2.5).

Comments and refinements. 3. We cannot replace the inequality in (2.12) by strict inequality even in the case that s_{OPT} is known to be unique, because the existence of the new matrix $\tilde{C}(1)$ with $\det \tilde{C}(1) = \det C'_{\bar{s}}$ does not guarantee existence of a *block matrix in* with $s > \bar{s}$ unless $R = 2$. (The upper bound argument for $\bar{s}$ gave a smaller number of blocks, and a further argument gives a block matrix with at most $\bar{s}$ - 1 blocks.)

4. In place of differentiating g, one can obtain the results from evaluation of $C(1) - C_{\bar{s}}$ by modifying appropriately the calculations (3.2) - (3.3) of GK.

5. Within the case $v > 0$, the value $v = 1$ shows that this kind of argument can't yield a better bound that does not depend on v. The analogue of (2.6) is now, from (2.8) with $\bar{s} = 6$, $k = 6r + v = n-B$ $(r \geq 1,\ 0 \leq v \leq 5)$,

$$\gamma = \frac{(6-v)r}{10r+B-3+v} + \frac{(v-1)(r+1)}{10r+B+1+v}, \quad 1 \leq v \leq 5, \tag{2.13}$$
$$\gamma = \frac{5r}{10r+B-3}, \qquad v = 0.$$

This is < 1/2 iff

$$r(10B-18-2v) > (v-3)^2-B^2, \quad 1 \leq v \leq 5, \tag{2.14}$$
$$B > 3, \qquad v = 0.$$

Thus, for $r \geq 1$ we obtain $\gamma < 1/2$ for all $B > 3$ and all possible cases of congruence $n \equiv 3 \pmod 4$; and for $B = 3$ we have $\gamma < 1/2$ if $v > 0$, which means $n \equiv 7$ or $11 \pmod{12}$. (In addition, we have $\gamma = 1/2$ if $B = 3$ and $v = 0$, or if $B = 2$ and $v = 1$; but note Comment 3 above on the inapplicability of the argument in this case.)

Thus, in the saturated case $B = 0$ (or cases of small B) we cannot obtain $s_{OPT} \geq 7$ by this argument, and indeed Ehlich resorted to another argument to obtain this conclusion (Comment 6, below). However, for $B \geq 4$ we always have $s_{OPT} \geq 7$ for at least some of the possible congruences. Combining this with the results of Comment 2 on the upper bound, we have shown $s_{OPT} = 7$ for all (k,n) satisfying some of the possible congruences, when $3 \leq n - k \leq 5$; namely, for $n-k = 3$ if $n \equiv 7$ or $11 \pmod{12}$ and for $n-k = 4,5$ if $n \equiv 3 \pmod 8$; and the value $s_{OPT} = 7$ is unique in this domain except for a tie with $s_{OPT} = 8$ when $(n,k) = (19,15)$, $(19, 14)$, $(27,22)$, and $(11,8)$.

On reading Ehlich's proof that $s_{OPT} = 7$ for $k = n \geq 63$ (see Comment 6, below), it is perhaps surprising that one can obtain s_{OPT} exactly in the cases noted above (each an infinite family of n, k) by these simple refined versions of Ehlich's method. (As we have mentioned, he also used division into cases by congruence, but when $k = n$ this did not determine s_{OPT} from the β- and γ- inequalities.) However, since s_{OPT} can not always be determined exactly in this way, we now discuss other tools. (For larger s_{OPT}-values, the method can again be used, as for the upper bound. But obtaining agreement of the two bounds becomes less possible as s_{OPT} increases. We omit details.)

6. Having used inequalities of the previous type and a few small n calculations of $1 - 2\beta$ to show $s_{OPT} = 6$ or 7 for $k = n > 39$, Ehlich completes his proof that $s_{OPT} = 7$ for $n \geq 63$ by the following two steps: (A) He shows analytically that $D_{n,n}(7) > D_{n,n}(6)$ if $n \geq 10^4$; (B) he verifies $D_{n,n}(7) > D_{n,n}(6)$ by computer for $63 \leq n < 10^4$.

This suggests trying to improve the argument of (A) to reduce the values 10^4 and, thus, the computing needed in (B); and then to carry out analogous computations in nonsaturated cases, sometimes for other values of s_{OPT}. An improvement of (A) is in fact possible if one takes account of congruences of n (mod 7 and mod 6) in computing a lower bound on $D_{n,n}(7)$ and upper bound on $D_{n,n}(6)$: it turns out that the effect of the congruence on the value of $D_{k,n}(s)$ is very slight. Specifically, if one writes $\log D_{k,n}(s)$ from (1.4) in terms of n, k, v, s with the substitution $r = (k-v)/s$, and expands in usual fashion as $n \to \infty$, one finds that the coefficient of terms in v is $O(n^{-2})$, the $O(n^{-1})$ coefficient vanishing. Of course, k and s determine v, but for the sake of obtaining the bounds now under discussion this calculation shows that the actual fine structure dependence on congruence is slight. However the dependence enters when one tries to bound $D_{k,n}(s)$ by considering (1-G) and the remaining factor separately (as Ehlich did), because the former is decreasing in v and the latter is increasing, whereas these effects largely cancel in the product. With care in estimating $D_{n,n}(s)$ by not bounding the two factors separately, one can thus obtain a sequence c_n that approaches 0 more rapidly than in Ehlich's estimate and such that

$$(1+c_n)D_{n,n}(7)/D_{n,n}(6) > 2\cdot 3^6\cdot 11^6\cdot 5^{-5}\cdot 7^{-7} = 1.0036$$
$$= \lim_n D_{n,n}(7)/D_{n,n}(6).$$

A corresponding computation can be carried out for $k < n$ and other values of s_{OPT}, but it seems an overwhelming task to try to determine s_{OPT} for every n, k in this way. We shall determine s_{OPT} exactly for certain cases near the "very regular" border $n = 2k - 5$ (see (1.8)-(1.9)), in Section 5, and asymptotically as $n \to \infty$ and $k/n \to 1-\lambda > 1/2$, in Section 3. For other "practical" values, the bounds can be combined with a small computer search, to obtain s_{OPT} for a given n,k: compute $\log D_{k,n}(s)$ from (1.4) for s between the values given by (2.5) and (2.12). A list of the values of s_{OPT} for $k \leq n \leq 100$, obtained by this procedure, is found in Table 1. An indication of the number of values s that must be compared can be seen upon writing $k/n = 1 - \lambda$ in (2.5) and (2.12), yielding

$$(2.15) \qquad \frac{6(1-\lambda)}{1-2\lambda+7/n} \leq s_{OPT} \leq \frac{8(1-\lambda)}{1-2\lambda-3/n}.$$

For n large, as λ increases from 0 toward 1/2, the lower and upper bound in (2.15) increase by unity each time $w = (1-\lambda)/(1-2\lambda)$ passes through a value near an integer multiple of 1/6 or 1/8, respectively. Thus, the set of s values to be searched oscillates slightly as λ increases, but has size roughly $2w = 1 + (1-2\lambda)^{-1}$, which increases with λ. The first 10 of the limiting intervals $\lceil 6w \rceil$ to $\lfloor 8w \rfloor$ of s-values that must be

Table 1

Optimum Number of Blocks for $k \leq n \equiv 3 \pmod 4$.

Ties occur when $2k-n = 5; 7$ (every other case); 9, 11; 17 (every third case).

Key: $\overline{j}$ means a tie, $s_{OPT} = j$ or $j + 1$.

k \ n	7	11	15	19	23	27	31	35	39	43	47	51	55	59	63
6	$\overline{5}$														
7	7														
8		$\overline{7}$													
9		$\overline{6}$													
10		$\overline{5}$	$\overline{9}$												
11		$\overline{5}$	8												
12			$\overline{6}$	$\overline{11}$											
13			$\overline{6}$	$\overline{9}$											
14			6	$\overline{7}$	$\overline{13}$										
15			6	$\overline{7}$	11										
16				7	$\overline{8}$	$\overline{15}$									
17				6	$\overline{8}$	$\overline{12}$									
18				6	8	$\overline{9}$	$\overline{17}$								
19				6	7	$\overline{9}$	14								
20					7	9	$\overline{10}$	$\overline{19}$							
21					7	8	$\overline{10}$	$\overline{15}$							
22					6	$\overline{7}$	10	$\overline{11}$	$\overline{21}$						
23					6	7	9	$\overline{11}$	17						
24						7	8	11	$\overline{12}$	$\overline{23}$					
25						7	8	9	$\overline{12}$	$\overline{18}$					
26						6	7	9	11	$\overline{13}$	$\overline{25}$				
27						6	7	9	10	$\overline{13}$	20				
28							7	8	$\overline{9}$	12	$\overline{14}$	$\overline{27}$			
29							7	8	9	11	$\overline{14}$	$\overline{21}$			
30							6	7	8	10	13	$\overline{15}$	$\overline{29}$		
31							6	7	8	10	11	$\overline{15}$	23		
32								7	9	9	11	14	$\overline{16}$	$\overline{31}$	
33								7	8	9	10	12	$\overline{16}$	$\overline{24}$	
34								6	7	8	10	$\overline{11}$	15	$\overline{17}$	$\overline{33}$
35								6	7	8	9	11	13	$\overline{17}$	26

Table 1 (continued).

Optimum Number of Blocks for $k \leq n \equiv 3 \pmod 4$.

Ties occur when 2k-n = 5;7 (every other case); 9, 11; 17 (every third case).

Key: $\overline{j}$ means a tie, $s_{OPT} = j$ or $j + 1$.

k \ n	39	43	47	51	55	59	63	67	71	75	79	83	87	91	95	99
36	7	8	9	10	12	16	$\overline{18}$	$\overline{35}$								
37	7	7	8	10	12	14	$\overline{18}$	$\overline{27}$								
38	6	7	8	9	11	13	17	$\overline{19}$	$\overline{37}$							
39	6	7	8	9	10	12	14	$\overline{19}$	29							
40		7	8	8	10	11	$\overline{13}$	18	$\overline{20}$	$\overline{39}$						
41		7	7	8	9	11	13	15	$\overline{20}$	$\overline{30}$						
42		6	7	8	9	10	12	14	19	$\overline{21}$	$\overline{41}$					
43		6	7	8	9	10	11	14	16	$\overline{21}$	32					
44			7	7	8	9	11	12	15	19	$\overline{22}$	$\overline{43}$				
45			7	7	8	9	10	12	14	16	$\overline{22}$	$\overline{33}$				
46			7	7	8	9	10	11	13	$\overline{15}$	20	$\overline{23}$	$\overline{45}$			
47			6	7	8	8	9	11	12	15	17	$\overline{23}$	35			
48				7	7	8	9	10	12	13	16	21	$\overline{24}$	$\overline{47}$		
49				7	7	8	9	10	11	13	16	18	$\overline{24}$	$\overline{36}$		
50				7	7	8	8	10	10	12	14	17	22	$\overline{25}$	$\overline{49}$	
51				6	7	8	8	9	10	12	13	16	$\overline{19}$	$\overline{25}$	38	
52					7	7	8	9	10	11	13	15	$\overline{17}$	23	$\overline{26}$	$\overline{51}$
53					7	7	8	9	9	11	12	14	17	19	$\overline{26}$	$\overline{39}$
54					7	7	8	8	9	10	11	13	15	18	24	$\overline{27}$
55					6	7	7	8	9	10	11	12	14	18	20	$\overline{27}$
56						7	7	8	9	9	11	12	14	16	19	25
57						7	7	8	8	9	10	11	13	15	18	$\overline{21}$
58						7	7	8	8	9	10	11	12	14	16	$\overline{19}$
59						6	7	7	8	9	10	10	12	13	15	19
60							7	7	8	8	9	10	11	12	15	17
61							7	7	8	8	9	10	11	12	14	16
62							7	7	8	8	9	9	10	12	13	15
63							7	7	7	8	9	9	10	11	13	14
64								7	7	8	8	9	10	11	12	13
65								7	7	8	8	9	10	11	11	13
66								7	7	7	8	9	9	10	11	13
67								7	7	7	8	8	9	10	11	12

Table 1 (continued).

Optimum Number of Blocks for $k \leq n \equiv 3 \pmod 4$.

Ties occur when 2k-n = 5;7 (every other case); 9, 11; 17 (every third case).

Key: $\bar{j}$ means a tie, s_{OPT} = j or j + 1.

k \ n	71	75	79	83	87	91	95	99
68	7	7	8	8	9	10	10	11
69	7	7	8	8	9	9	10	11
70	7	7	7	8	8	9	10	11
71	7	7	7	8	8	9	10	10
72		7	7	8	8	9	9	10
73		7	7	8	8	9	9	10
74		7	7	7	8	8	9	10
75		7	7	7	8	8	9	9
76			7	7	8	8	9	9
77			7	7	7	8	8	9
78			7	7	7	8	8	9
79			7	7	7	8	8	9
80				7	7	8	8	8
81				7	7	7	8	8
82				7	7	7	8	8
83				7	7	7	8	8
84					7	7	7	8
85					7	7	7	8
86					7	7	7	8
87					7	7	7	8
88						7	7	7
89						7	7	7
90						7	7	7
91						7	7	7
92							7	7
93							7	7
94							7	7
95							7	7
96								7
97								7
98								7
99								7

searched for large n are given in Table 2.

Table 2.
Interval of s-values to which (2.15) limits
search for s_{OPT}, for n large.

Interval of λ-values	$\lceil 6w \rceil$ to $\lfloor 8w \rfloor$
(0, 1/10)	[7, 8]
(1/10, 1/8)	[7, 9]
(1/8, 1/6)	[8, 9]
(1/6, 1/5)	[8, 10]
(1/5, 3/14)	[9, 10]
(3/14, 1/4)	[9, 11]
(1/4, 5/18)	[10, 12]
(5/18, 2/7)	[10, 13]
(2/7, 3/10)	[11, 13]
(3/10, 5/16)	[11, 14]

Thus, for example, if $\lambda = 1/9$, s_{OPT} is 7, 8, or 9 for n sufficiently large; examining (2.15) more closely, we see that [7, 9] is in fact the interval if $n > 63$. We next turn to the exact determination of s_{OPT} for fixed λ as $n \to \infty$.

III. ASYMPTOTICS

We think of $k,n \to \infty$ while s is fixed or bounded; when this last is not the case, as in the verification of (3.3), further care is needed. Write $k/n = b > \frac{1}{2}$, and substitute into (1.4). We obtain

$$D_{k,n}(x) = (n-3)^k s^{-s}(s+4b)^{s-1}(s+4b-bs)[1+O(n^{-1})]. \tag{3.1}$$

Thus, asymptotically we need only maximize $(1+4b/s)^{s-1}(1-b+4b/s)$ with respect to s for each b, and Table 2 gives the values of s to which we can limit our search. One obtains $s_{OPT} = 7$ for λ near 0, then $s_{OPT} = 8$, etc.; the first few intervals of λ-values (the "near-saturated" cases) that yield a given s_{OPT}-value are, as $n \to \infty$, with $k/n \to 1-\lambda$,

$$(3.2) \qquad \lim s_{OPT} = \begin{cases} 7 & \text{for} \quad 0 \leq \lambda < .08837, \\ 8 & \text{for} \quad .08837 < \lambda < .17027, \\ 9 & \text{for} \quad .17027 < \lambda < .22494, \end{cases}$$

and one can continue in this manner. In particular, for $k/n \to 1$ in any way as $n \to \infty$, it is easy to see that s_{OPT} is eventually 7.

For comparison, when n is 99 we have $s_{OPT} = 7$ for $0 \leq \lambda \leq .111$, $s_{OPT} = 8$ for $.121 \leq \lambda \leq .192$, $s_{OPT} = 9$ for $.202 \leq \lambda \leq .242$; for fixed λ, (3.2) may overestimate s_{OPT} slightly.

When λ is near 1/2, we approach the domain (1.8) where $s_{OPT} = k$, $R_{OPT} = 1$. Writing $b = 1/2 + \varepsilon$ with $\varepsilon > 0$, and noting that $s \to \infty$ as $\varepsilon \to 0$ by (2.12), we find that the logarithm of $(1 + 4b/s)^{s-1}(1-b+4b/s)$ (see just under (3.1)) is

$$2-\log 2 + 2\varepsilon - 2\varepsilon^2-10/3s^2 + 4\varepsilon/s + O((\varepsilon+s^{-1})^3)$$

as $\varepsilon \to 0$ and $s \to \infty$. From this one can show that the maximum is given at

$$(3.3) \qquad s_{OPT} \approx 5/3\varepsilon$$

as $\varepsilon \to 0$. This is meaningful if, having first let $n \to \infty$, we then let $\varepsilon \to 0$. A closer examination of the $O(n^{-1})$ term in (3.1) shows that its behavior as $n \to \infty$ and $s \to \infty$ is $3/n + O(1/ns+s^2/n^2)$, and one can then verify that (3.3) is valid for $\varepsilon = o(n^{-1/2})$.

The approximation (3.3) is reasonably accurate for moderate n and even fairly large ε. For example, when $\varepsilon = 1/5$, (3.3) gives 11.7; from Table 1, for $n = 31$ and $k = 22$, we have $s_{OPT} = 10$, and for $n = 59$, $k = 41$, we have $s_{OPT} = 11$. Even for $2k-n = d$, an integer ≥ 5, so that $\varepsilon \approx d/2n$, the approximation gives approximately the right ratio of k/s_{OPT} and thus of R_{OPT}, although the absolute error from s_{OPT} is unbounded: for $d = 5, 7, 9, 11, 17,$

the approximation gives for $k/s_{OPT} \approx 3k/5 \approx 3d/20$ the values .75, 1.05, 1.35, 1.65, 1.95, 2.25, 2.55, compared with the actual values 1, 1.33, 2.00, 2.00, 2.25, 2.77, 3.00, recorded in Section 5.

Improvement of the estimate (3.1) means inclusion of terms of order n^{-2} in the expansion of $\log D_{k,n}(s)$, which includes dependence on v and, thus, greater complexity.

IV. NEIGHBORING TIES

In this section we determine the pairs (k,n) for which there is an s with

$$D_{k,n}(s) = D_{k,n}(s+1). \tag{4.1}$$

In Section 5 it is then shown that $s_{OPT} = s$ or $s + 1$ (only) for the values satisfying (4.1). The five "tied series", as we will call them, turn out to be pairs (k, n) and corresponding s in (4.1) for which

$$\begin{gathered}(2k-n,s) = (5,k-1);\ (7,3(k-1)/4) \text{ when } n \equiv 3 (\text{mod } 8); \\ (9,k/2);\ (11,(k-1)/2);\ (17,(k-1)/3) \text{ when } n \equiv 3 (\text{mod } 12).\end{gathered} \tag{4.2}$$

The proof that the pairs (k,n) satisfying (4.2), and only those, satisfy (4.1), is divided into two cases, of which (A) below is simplest. Of the two possible conventions of u, v we choose (1.7)(b), with v > 0, throughout this section.

(A) *Cases where* (4.1) *holds with the same value of* R *for* s *as for* s + 1. We let u refer to the design with s blocks, which thus has u blocks of size r and s-u of size r+1; the design with s+1 blocks has u+r+1 blocks of size r and s-u-r of size r+1. We write n-3 = 4m. Then, from (1.4), (4.1) says

$$4^{k-2}m^{k-s}(m+r)^{u-1}(m+r+1)^{s-u-1}A' = 4^{k-2}m^{k-s-1}(m+r)^{u+r} \times (m+r+1)^{s-u-r-1}B', \tag{4.3}$$

where

$$A' = 4A = 4\{4(m+r)(m+r+1)-mk-(k+u)r\},$$
(4.4)
$$B' = 4B = 4\{4(m+r)(m+r+1)-mk-(k+u+r+1)r\}.$$

Simplifying, we obtain from (4.3),

$$m(m+r+1)^r A = (m+r)^{r+1} B. \tag{4.5}$$

Since m+r and m+r+1 are relatively prime, (4.5) implies $(m+r)^{r+1} | mA$ and hence $mA \geq (m+r)^{r+1}$, which from (4.4) implies

$$4m(m+r+1) > (m+r)^r. \tag{4.6}$$

It is easily seen that (4.6) is false for $r = 3$ $\forall m$; and, since $(m+r)^r/(m+r+1)$ is increasing (has positive logarithmic derivative) for $r \geq 1$, it follows that (4.6) and hence (4.3) is false for all $r \geq 3$, $\forall m$.

The remainder of the solution in case (A) is found by checking that u is an integer such that $u \geq 0$, $s-u > 0$, $u+r+1 \geq 0$, and $s-u-r > 0$. For $r = 1$ this implies $0 \leq u < k-2$ (the last from $s-u-1 > 0$ and $s \leq k-1$), and the formula for u, from (4.5), is $u = (m+1)(2m+6-k)$. Thus, the condition $0 \leq u < k-2$ becomes $2m+4 < k \leq 2m+6$, or $(n+5)/2 < k \leq (n+9)/2$. The only solutions are $k = (n+9)/2$ and $k = (n+7)/2$, and in the latter case the integrality of $s = (u+k)/2$ shows we must have $n \equiv 3 \pmod 8$. These are of course the series $2k-n = 9$ and 7 of (4.2).

There remains the possibility $r = 2$ in (A). (This can be disposed of in a number of ways.) We now have, from (4.5),

$$u = \frac{(4m+12-k)(m+2)}{2} - \frac{3(m+2)^3}{3m+8}, \tag{4.7}$$

so that $3(m+2)^3/(3m+8)$ must be an integral multiple of 1/2. This is possible only if $m = 0$, $n = 3$, for which (4.7) gives $u = 9-k$, impossible for $k \leq n$.

(B) *Cases in which, if* R *is the maximum block size for* s *blocks, then* R-1 *is the maximum for* s+1 *blocks.* We now suppose the s blocks consist of v of size r+1 and u = s-v of size r; and that the s+1 blocks consist of s+v-r+1 of size r and r-v = u+r-s of size r-1. Thus, $v \leq r$. (It is easily checked that these are the right values.) Again, u and v always refer to the s blocks. In order for (4.1) to hold we have, in place of (4.5),

$$(4.8) \qquad m(m+r)^{r-2v-1}(m+r+1)^{v-1}A = (m+r-1)^{r-v-1}B^*$$

where A (rewritten from (4.4) in terms of v) and B* (after some simplification) are given by

$$(4.9) \qquad \begin{aligned} A &= 4(m+r)(m+r+1)-k(m+r+1)+v(r+1), \\ B^* &= 4(m+r)(m+r-1)-k(m+r-1) - (r-v)(r-1). \end{aligned}$$

If $r-2v-1 < 0$, the power of m+r in (4.8) can be transferred to the other side to make it a positive power, and similarly for $(m+r-1)^{-1}$ if r = v. In any event, m+r is relatively prime to m+r+1 and m+r-1, and the last two have at most a factor 2 in common, so that, writing $[x]^+ = \max(x,0)$, we see that $(m+r+1)^{v-1}$ and $(m+r-1)^{r-v-1}$ (when $r-v-1 \geq 0$) have at most a factor $2^{[r-3]^+}$ in common. (This maximum factor occurs when v = 2 and $2^{[r-3]^+} | (m+r+1)$, or when v = r-2 and $2^{[r-3]^+} | (m+r-1)$; in fact, the common factor is 1 if m+r is even or if v = 1 or r or r-1, and is at most $2^{\max([v-1]^+,[r-v-1]^+)}$ otherwise.) Thus, we have, from (4.8) upon transferring terms if necessary to have positive powers on both sides,

$$(4.10) \qquad (m+r)^{|r-2v-1|}(m+r+1)^{v-1}(m+r-1)^{|r-v-1|} \,\Big|\, 2^{[r-3]^+}mAB^*,$$

and hence it is necessary that the left side of (4.10) is

$$(4.11) \qquad \leq 2^{[r-3]^+}16(m+r)^2(m+r-1)(m+r+1)m$$

in order that (4.8) be satisfied.

The total of exponents on the left side of (4.10) is at least r-2 (attained when v = (r-1)/2, r odd). Also, for $r \geq 10$, we have $(m+r)^2(m+r-1)(m+r+1)m < (m+r-1)^5$. Consequently, from (4.10) and (4.11), (4.8) can be satisfied for $r \geq 10$ only if $(m+r-1)^{r-7} < 2^{r+1}$. This is impossible for $r \geq 10$ if $m \geq 2$, and in the trivial cases m = 0, 1 we have $7 \geq n \geq r$. Thus, (4.10) is impossible for $r \geq 10$. For $r \leq 9$, certain other values can be eliminated by the same method with the use of the improved factor noted above (4.10); for example, for r = 8, 9 values $v \leq r-5$ can be eliminated, as can the values v = 1, r-1, r for $r \geq 6$, 5, 4, respectively.

For $r \leq 9$, the various possibilities can be listed; for each r, there are some values of v that can be eliminated by using (4.10)-(4.11), perhaps with the possible improvement on the factor $2^{[r-3]^+}$ noted just above (4.10). The remaining values of v can then be checked in each case by solving (4.8) for k seeing whether the resulting expression can be an integer. The calculations are tedious but straightforward, and, rather than to take the space to give them all, we treat in some detail here, for illustration, only the value r = 6. When r = 6, the values v = 1, 5, 6 are eliminated by the remark at the end of the previous paragraph, and the value v = 4 can be handled similarly. There remain the values v = 2 and 3, which can not be eliminated by using (4.10). Instead, we solve (4.8) for k. The general form of this (valid for all r and m), from (4.9) with (4.8) being abbreviated $\alpha A = \beta B^*$, is

$$(4.12) \qquad k = 4(m+r) - \frac{\alpha v(r+1) + \beta(r-v)(r-1)}{\beta(m+r-1) - \alpha(m+r+1)}$$

$$= 4(m+r) - F \text{ (say)}.$$

Thus, a necessary condition for (4.8) to hold is that F be an integer. For our case r = 6, v = 2, we obtain

$$F = \frac{14m(m+6)(m+7)+20(m+5)^3}{(m+5)^4-m(m+6)(m+7)^2} = \frac{34m^3+482m^2+2088m+2500}{17m^2+206m+625}$$

$$= 2m + 4 + \frac{2m^2+14m}{17m^2+206m+625},$$

which is obviously never an integer for $m \geq 1$. A similar development holds for $v = 3$, and also for the other cases that must be studied in this manner for $4 \leq r \leq 9$.

For each of the values $r = 1$, 2, and 3, respectively, the analogous development yields a single series of solutions and no others, namely, the series of (4.2) for $2k-n = 5$, 11, and 17 (the last with $n \equiv 3 \pmod{12}$).

V. CHARACTERIZATION OF s_{OPT} FOR THE "SERIES" $k = (n+d)/2$, $d \leq 17$

We fix d at any odd integral value ≥ 5 and consider the series of settings $(k,n) = ((n+d)/2, n)$ with $n \geq d$. (For $d < 5$, see (1.8).)

The basic idea is that, in order to show $s_{OPT} = g(n,d)$ uniquely, we show that altering C_s (the block matrix with blocks of at most two sizes) to some C^* increases the determinant from the value $D_{k,n}(s)$, if either (a) $s > g(n,d)$ or (b) $s < g(n,d)$; in the case of the five series of (4.2) in which two values $s = g(n,d)$ and $g(n,d)-1$ yield the same value $D_{k,n}(s)$, we show both values (and no others) are s_{OPT} by the same demonstration with (b) replaced by (b') $s < g(n,d)-1$.

The demonstration becomes more complicated as d increases. In GK, part (b') for $d = 5$ was handled by letting C^* be obtained from C_s by replacing a block of size R in the latter by one each of sizes $R-1$ and 1. This no longer always suffices, as one can check when $d = 13$ and $s < g(n,13)$ (given in (5.1)) with $r = 2$. What does work is to replace 2 blocks of size 3 by 3 blocks of size 2. Unfortunately, the corresponding calculation when s_{OPT}/n is smaller and r_{OPT} is larger (which occurs as d grows) is more tedious to carry out exactly, and having characterized s_{OPT} exactly for $d \leq 17$, we shall then outline an argument for general d but will only take the space here to give $\lim_{n\to\infty} g(n,d)/n$ for general d. Our exact result is

THEOREM. *For* $5 \leq d \leq 17$ (*d odd*), $(k,n) = ((n+d)/2,n)$, $d \leq n \equiv 3 \pmod 4$, *the only values of* s_{OPT} (*with accompanying* r_{OPT} *and* R_{OPT}) *are:*

(5.1)

$$\underline{d=5}: \quad s_{OPT} = \begin{cases} k=(n+5)/2, & r_{OPT} = 1^*, \quad R_{OPT} = 1, \\ k-1=(n+3)/2, & r_{OPT} = 1, \quad R_{OPT} = 2. \end{cases}$$

$\underline{d=7}$: *If* $n \equiv 3 \pmod 8$,

$$s_{OPT} = \begin{cases} (3k+1)/4 = (3n+23)/8, & r_{OPT} = 1, \; R_{OPT} = 2, \\ (3k-3)/4 = (3n+15)/8, & r_{OPT} = 1, \; R_{OPT} = 2, \end{cases}$$

if $n \equiv 7 \pmod 8$,

$$s_{OPT} = (3k-1)/4 = (3n+19)/8, \quad r_{OPT} = 1, \; R_{OPT} = 2.$$

$$\underline{d=9}: \quad s_{OPT} = \begin{cases} (k+2)/2 = (n+13)/4, & r_{OPT} = 1, \; R_{OPT} = 2, \\ k/2 = (n+9)/4, & r_{OPT} = 2^*, \; R_{OPT} = 2. \end{cases}$$

$$\underline{d=11}: \quad s_{OPT} = \begin{cases} (k+1)/2 = (n+13)/4, & r_{OPT} = 1, \; R_{OPT} = 2, \\ (k-1)/2 = (n+9)/4, & r_{OPT} = 2, \; R_{OPT} = 3. \end{cases}$$

$$\underline{d=13}: \quad s_{OPT} = \lfloor (4k+3)/9 \rfloor = \lfloor (2n+29)/9 \rfloor, \; r_{OPT}=2, \; R_{OPT}=3.$$

$$\underline{d=15}: \quad s_{OPT} = \lfloor (13k+25)/36 \rfloor = \lfloor (13n+245)/72 \rfloor, \; r_{OPT}=2, R_{OPT}=3.$$

(5.1) (continued)

d=17: *if* $n \equiv 3 (\mathrm{mod}\ 12)$,

$$s_{OPT} = \begin{cases} (k+2)/3 = (n+21)/6, & r_{OPT} = 2,\ R_{OPT} = 3, \\ (k-1)/3 = (n+15)/6, & r_{OPT} = 3,\ R_{OPT} = 4; \end{cases}$$

if $n \equiv 7 (\mathrm{mod}\ 12)$,

$$s_{OPT} = k/3 = (n+17)/6, \qquad r_{OPT} = 3^*,\ R_{OPT} = 3;$$

if $n \equiv 11 (\mathrm{mod}\ 12)$,

$$s_{OPT} = (k+1)/3 = (n+19)/6,\ r_{OPT} = 2, \quad R_{OPT} = 3.$$

(*In the three values marked* *, *the value is given for* r_{OPT} *under the convention* $u > 0$ *of* (1.7)(a), *and must be reduced by* 1 *for the value under the convention* $v > 0$ *of* (1.7)(b).)

The value $d = 5$ was covered in (1.8)-(1.9), but we include it here for unity of exposition.

Proof. We divide the demonstration into several parts.

1. *Series with ties*. We treat the values $d = 5, 7, 9, 11, 17$ first because these are simplest; we remarked, just above the statement of the theorem, on the difficulty of proof in other cases, but in these five cases the proof mentioned there can be used to improve C_s to C^* if s is too small.

(a) *Lower bound on* s_{OPT}. We first show s_{OPT} is $\geq$ the value given in (5.1). It is convenient in this proof to adopt the convention of (1.7)(b), so that always $R = r+1$ and $v > 0$ and thus

$$(5.2) \qquad k-sr-1 \geq 0.$$

Using (1.7)(b) means reducing by 1 the values r with * as described at the end of the theorem. We compare $D_{k,n}(s)$ with $D^* = \det C^*$ where C^* is obtained from C_s by replacing a block of size $R = r+1$ by blocks of size 1 and r. From (1.2) we obtain, after some simple arithmetic, and with G of (1.4) referring to

$D_{k,n}(s)$ and $L = n + 4r + 1$,

$$L^{2-v}(L-4)^{1-u}(L-4r-4)^{s+1-k}[D^*-D_{k,n}(s)]$$

$$= L(L-4)^2(L-4r)\{1-G+\frac{r+1}{L} - \frac{1}{L-4r} - \frac{r}{L-4}\}$$

$$(5.3) \quad - L^2(L-4)(L-4r-4)\{1-G\}$$

$$= 8rL(L-4)\{1-2G+2(r+1)L^{-1}\}$$

$$= 8r[L(L-2-2k+2r) + 8(k-1-sr)(r+1)]$$

$$= 8r[(2k-d+4r+1)(6r-d-1)+8(k-1-sr)(r+1)].$$

If $d = 5, 7, 9, 11$ and $r \geq 2$, we have $6r - d - 1 \geq 0$ and thus (noting (5.2)) that (5.3) is > 0 when $d \leq 9$ or when $d = 11$, $r = 2$, and $s < (k-1)/2$. The latter shows $s_{OPT} \geq (k-1)/2$ when $d = 11$. For $d = 5, 7, 9$, it remains still to consider only $r = 1$, for which (5.3) is > 0 iff

$$(5.4) \qquad s < k(1- \frac{d-5}{8}) + \frac{(d-5)^2}{16} - 1.$$

This becomes $s < k-1$ for $d = 5$; $s < \frac{3}{4}(k-1)$ for $d = 7$, $s < k/2$ for $d = 9$. Thus, $s_{OPT} \geq k-1, \frac{3}{4}(k-1), k/2$ in the respective cases; for $d = 7$ and $n \equiv 7 \pmod 8$ this implies $s_{OPT} \geq (3k-1)/4$.

When $d = 17$ and $r \geq 3$, we have $6r-d-1 \geq 0$ and thus (5.3) is > 0 if $r > 3$ or $r = 3$ and $s < (k-1)/3$. Hence, $s_{OPT} \geq (k-1)/3$. This implies $s_{OPT} \geq k/3$ if $n \equiv 7 \pmod{12}$ and $s_{OPT} \geq (k+1)/3$ if $n \equiv 11 \pmod{12}$.

This completes the proof that s_{OPT} is at least the smaller of the two values of s in each of the cases $d = 5, 7, 9, 11, 17$, or is at least the single value given in the subcases of $d = 7, 17$ when there is a unique s_{OPT}.

(b) *Upper bound on* s_{OPT}. When $d = 5$, $s_{OPT} \leq k$ is trivial. We next show that if $d > 5$ and s is too large we can increase the determinant from the value $D_{k,n}(s)$ as follows: Firstly, by (1.8) we know $s_{OPT} \leq k-1$ if $d > 5$, so (with the $v > 0$ convention) $r_{OPT} \geq 1$ then. In the cases $d = 7, 9, 11$, for $k > s >$ the value(s) of s_{OPT} in the theorem's statement, we have $r = 1$ and $s \geq \frac{1}{2}(k+1)+1$, from which $u = 2s-k \geq 3$. We shall now shown that if $d = 7, 9, 11$ and if s is such that $r = 1$ and $u \geq 2$, so that there are at least 2 blocks of size 1, then combining 2 blocks of size 1 into a block of size 2 yields a C^{**} with determinant $D^{**} > D_{k,n}(s)$, if $s > s_{OPT}$ of the theorem's statement. In fact, if we put $r = 1$ and $s-1$ for s in (5.3), $[D^*-D_{k,n}(s)]$ there is $[D_{k,n}(s)-D^*]$ here, which is thus proportional to

$$(2k-d+5)(5-d) + 16(k-s), \tag{5.5}$$

which is negative (D^* larger) if

$$s > k(1-\frac{d-5}{8}) + \frac{(d-5)^2}{16}. \tag{5.6}$$

Hence, for $d = 7, 9, 11$, we have $s_{OPT} \leq (3k+1)/4$, $(k+2)/2$, $(k+1)/2$, and $s_{OPT} \leq (3k-1)/4$ if $d = 7$ and $n \equiv 7 (\mathrm{mod}\ 8)$. This yields the desired result except for $d = 17$.

For $d = 17$, (5.6) always holds, so $r_{OPT} > 1$ and $s_{OPT} \leq k/2$. For $k/2 \geq s > (k+2)/3$ there are at least 3 blocks of size 2 in C_s, and C_{s-1} is obtained by replacing the 3 blocks of size 2 by 2 blocks of size 3. This yields (with $r = 2$ and G referring to C_s, and with the substitution $k = (n+d)/2$ for use in the other cases of d below)

$$\begin{aligned}
&(n-3)^{s-k}(n+5)^{4-u}(n+9)^{1-v}[D_{k,n}(s)-D_{k,n}(s-1)] \\
&= (n+5)^4(n+9)\{1-G\} - (n-3)(n+5)(n+9)^3 \times \{1-G+\frac{6}{n+5}-\frac{6}{n+9}\} \\
&= (48n+368)(n+5)(n+9)\{1-G\} - 24(n-3)(n+9)^2 \\
&= [24n+184][2(n+5)(n+9)-(n-3)(n+d)-\frac{3(n-3)(n+9)^2}{3n+23} - 48s].
\end{aligned} \tag{5.7}$$

This is negative (C_{s-1} better than C_s) iff

$$(5.8) \qquad s > [\frac{71n}{3} + \frac{1073}{9} + \frac{512}{9(3n + 23)} - d(n-3)]/48.$$

For d = 17 the right side of (5.8) is easily seen to be $< (n+21)/6$ if $n \geq 17$ (the smallest possible value); thus, C_{s-1} is better than C_s if $s > (n+21)/6$, so $s_{OPT} \leq (n+21)/6$. This (and the implications $s_{OPT} \leq (n+17)/6$ or $\leq (n+19)/6$ when $n \equiv 7$ or 11 (mod 12)) complete the proof for d = 17.

2. *The cases* d = 13, 15. We consider the expression (5.7) when s is such that this formula applies: $k/2 \geq s > s-1 \geq k/3$. From $u = 3s-k$ we see that there are always at least 3 blocks of size 2 among the s blocks in this range, and for the case of s-1 blocks $v = k-2(s-1)$ shows that there are at least 2 blocks of size 3 among the s-1 blocks. Thus, the exchange of (5.7) is always possible.

For d = 13, (5.7) is negative (C_{s-1} better than C_s) iff

$$(5.9) \qquad s > [2n + 89/3 + 32/3(3n + 23)]/9,$$

and is positive if the opposite inequality holds. Since $32/3(3n + 23) < 1/3$ for $n \geq 13$, we conclude that (5.7) is $\{\lessgtr\}$ 0 iff $s\{\gtrless\} \lfloor (2n + 29)/9 \rfloor$. For d = 15, the analogue of (5.9) is

$$(5.10) \qquad s > [13n + 739/3 + 256/3(3n + 23)]/72.$$

For $n \geq 15$, we have $256/3(3n + 23) < 5/3$ and thus the right side of (5.10) is $y = [13n + 245 + x]/72$ where $4/3 < x < 3$. Hence, if $n = 4m+3$, we have $y = [13m + 62 + x/4]/18$ and thus $\lfloor y \rfloor = \lfloor (13m + 62)/18 \rfloor$. We conclude that (5.7) is $\{\lessgtr\}$ 0 iff $s\{\gtrless\} \lfloor (13n + 245)/72 \rfloor$.

The above calculation shows that the s_{OPT} stated in (5.1) gives the unique maximum of $D_{k,n}(s)$ over the crucial interval $k/3 + 1 \leq s \leq k/2$. Values of $s > k/2$ or $\leq k/3$ are excluded by the fact that $r_{OPT} \geq 2$ implies $s_{OPT} \leq k/2$ and that $R_{OPT} \leq 3$, $r_{OPT} \leq 2$ implies $s_{OPT} > k/3$.

It is interesting that successive values of $g(4m+3,d) - g(4m-1,d)$ (where again $g(n,d) = s_{OPT}$) go through a cycle of length 9 when $d = 13$ and of length 18 when $d = 15$.

VI. GENERAL APPROXIMATION TO s_{OPT}

Some ideas for the determination of s_{OPT} in general arise from the calculations of the previous section on the series $k = (n+d)/2$, $d \leq 17$. The simple inequalities of Section 2 did not suffice, and we had to study $D_{k,n}(s)$ in detail, either in a crucial interval of s values with constant R and with s_{OPT} in its interior, or else in two neighboring intervals with different R values and s_{OPT} on their boundary. At the same time, we had to rule out s-values outside the crucial interval. For general (k,n) the analogue to this last is intuitively plausible but we do not have a general proof. A sufficient condition for our program of characterizing s_{OPT} to work is the

(6.1) *Unimodality conjecture:* $D_{k,n}(s)$ *is unimodal in* s.

Indeed, this was seen to be valid in all our numerical investigations. But a complete proof, even of unimodality in the interval of s-values not excluded by the simple inequalities of Section 2, would entail an analysis of $D_{k,n}(s) - D_{k,n}(s+1)$ involving calculations like those of Sections 4 and 5. In particular, the case where R differs for s and s+1 (part (B) of Section 4) seems the messiest to analyze.

Assuming this unimodality, we want a formula, or at least an accurate estimate, for s_{OPT}. The expression (5.7) suggests an approach for finding a general expression for s_{OPT} that holds for larger values of d than those of the previous section, but the details are more difficult to carry out because, as r increases, the analogue of (5.7) in the crucial interval $k/(r+1)+1 \leq s \leq k/r$ becomes more complicated. We cannot guarantee, as we did in each of (5.8), (5.9), (5.10), that the influence of a vanishing remainder term (of the form $c/(3n + 23)$ in the above) is also absent for all small n. Thus, although one can carry out a detailed analysis for each pair (k,n), we still do not have a simple presentation that covers all cases. We now indicate the idea of the development, an approximation to s_{OPT}, and an asymptotic formula.

Suppose, for fixed k, n, r, that s is an integer satisfying

$$k/(r+1) \le s-1 < s \le k/r. \tag{6.2}$$

Write

$$f(n,r) = (n+4r-3)^{r+1} - (n-3)(n+4r+1)^r. \tag{6.3}$$

Then from (1.4), with G, u, and v again referring to the value s, and by a calculation like that used earlier, where now we replace r+1 of the blocks of size r by r of size r+1 or vice versa, we have

$$\begin{aligned}[(n-3)^{s-k}(n+4r-3)^{r-u+2}(n+4r+1)^{1-v}/f(n,r)] \\ [D_{k,n}(s)-D_{k,n}(s-1)] \\ = (n+4r+1)(n+4r-3)\{1-G\}-4r(r+1)(n-3) \\ \times (n+4r+1)^r/f(n,r).\end{aligned} \tag{6.4}$$

This expression is $\{\substack{<\\=\\>}\}$ 0, respectively, iff

$$\begin{aligned} s\{\substack{>\\=\\<}\} & \frac{(n+4r+1)(n+4r-3)-k(n-3)}{4r(r+1)} - \frac{(n-3)(n+4r+1)^r}{f(n,r)} \\ &= \frac{2(n+4r+1)(n+4r-3)-(n+d)(n-3)}{8r(r+1)} - \frac{(n-3)(n+4r+1)^r}{f(n,r)} \\ &= p(k,n,r) \quad \text{(say)}, \end{aligned} \tag{6.5}$$

where we have again written $d = 2k-n$.

Write

$$h(k,n,r) = \lfloor p(k,n,r) \rfloor. \tag{6.6}$$

In the next two paragraphs (A) and (B), we make some brief

remarks about (6.5). We mostly ignore "ties" in what follows, having covered these in Section 4.

(A) If $s_0 = h(k,n,r)$ satisfies $k/(r+1)+1 \leq s_0 \leq k/r-1$, it follows from (6.5) that, on the interval $k/(r+1) \leq s \leq k/r$, $D_{k,n}(s)$ is nondecreasing up to s_0 and (from (6.5) with s replaced by s+1) nonincreasing thereafter, so that it has an *internal* maximum on this interval at s_0 (unique except for a possible "tie").

(B) On the other hand, if $p(k,n,r) \geq k/r$, $D_{k,n}(s)$ is nondecreasing for $k/(r+1) \leq s \leq k/r$; and if $p(k,n,r-1) - 1 \leq k/r$, $D_{k,n}(s)$ is nonincreasing for $k/r \leq s \leq k/(r-1)$. Hence, if $p(k,n,r-1) - 1 \leq k/r \leq p(k,n,r)$, and if $k/(r+1) + 1 \leq \lfloor k/r \rfloor \leq k/(r-1)-1$, then $D_{k,n}(s)$ has an internal maximum on the interval $k/(r+1) \leq s \leq k/(r-1)$ at $s = \lfloor k/r \rfloor$ (unique except for a possible "tie").

In both of the preceding paragraphs, the maximizing value of s is indeed s_{OPT} if the unimodality conjecture (6.1) is valid.

Expanding f(n,r) in powers of n for fixed r, we have

$$f(n,r)/8r(r+1) = \Sigma_{j\geq 0} n^{r+1-j} \binom{r+1}{j} [(4r-3)^j - (4r-4j+1) \times (4r+1)^{j-1}]/8r(r+1)$$

$$= n^{r-1} + \frac{(r-1)(12r-1)}{3} n^{r-2}$$

$$+ \frac{(r-1)(r-2)(48r^2-8r+3)}{6} n^{r-3}$$

$$+ \frac{(r-1)(r-2)(r-3)(320r^3-80r^2+60r-7)}{30} n^{r-4}$$

$$+ O(n^{r-5}),$$

and hence, by a straightforward but tedious computation, that

$$(6.7)\quad \frac{8r(r+1)(n-3)(n+4r+1)^r}{f(n,r)} = n^2 + \frac{16r-10}{3}n + \frac{4r^2-140r+1}{9} - \frac{128(r-1)(2r^2+5r+2)}{135}n^{-1} + O(n^{-2}).$$

Thus, for fixed r,

$$(6.8)\quad p(k,n,r) = \frac{1}{8r(r+1)}\left\{\frac{(32r+7-3d)n}{3} + 3d + \frac{284r^2-4r-55}{9} + \frac{128(r-1)(2r^2+5r+2)}{135n} + O(n^{-2})\right\}.$$

Suppose $n \to \infty$, $k \to \infty$, with $d = 2k-n$ fixed. Then, under the unimodality conjecture (6.1), we obtain from (6.8) and from (A), (B) respectively for the two lines of (6.9) below,

Asymptotic formula for s_{OPT}: For fixed d, let r be the unique integer satisfying one of the two inequalities of (6.9). Then

$$(6.9)\quad \lim_n \frac{s_{OPT}}{n} = \begin{cases} \dfrac{32r+7-3d}{24r(r+1)} & \text{if } \dfrac{20r-5}{3} \le d < \dfrac{20r+7}{3}, \\[2ex] 1/2r & \text{if } \dfrac{20r-13}{3} \le d \le \dfrac{20r-5}{3}. \end{cases}$$

For successive values of r the two lines of (6.9) mesh to include all odd values of $d \ge 5$.

One can check the agreement of this formula with the results of the theorem of Section 5, the asymptotic ratios obtained from (5.1) for d = 5,...,17 being precisely those for (r, line of (6.9)) = (1,2), (1,1), (2,2), (2,2), (2,1), (2,1), (3,2). In fact, for these small values of d, in the three "line 1 of (6.9)" cases the integer part of the approximation of (6.8), with or without the n^{-1} term, is always the correct value of s_{OPT}. (The asymptotic condition $20r-5 < 3d < 20r+7$ of (6.9) of course does not generally correspond to (A) when we look at nonasymptotic behavior.) An approximation to s_{OPT} that is fairly accurate for both the situations (A) and (B) is suggested by those paragraphs:

Let $\hat{p}$ be the approximation to p through terms of order n^0 (or n^{-1}) in (6.8). Put $\bar{r} = \min\{r: k^{-1}(1+r)\hat{p}(k,n,r) \geq 1\}$. Then $s^* = \lfloor \hat{p}(k,n,\bar{r}) \rfloor$ approximates s_{OPT}. (Obvious slight modifications are possible.) A numerical investigation for values $n \leq 199$, with $(n+5)/2 \leq k \leq n$, shows that this approximation is quite good, usually yielding s_{OPT} exactly and rarely being off by more than 1. Replacing $\hat{p}$ by $\hat{p}-1$ in the definition of $\bar{r}$, and increasing s^* where necessary to make it monotone in k, improve somewhat; inclusion of $O(n^{-1})$ terms matters less. Although the above calculations leading to an approximation for s_{OPT} are based on an expansion in powers of n for fixed r, the approximation turns out to be quite good even for values of (k,n) near saturation, for which cases r (corresponding to s_{OPT}) is of order n and the successive terms in the expansion (6.8) are all of the same order in n. This is evidently because the terms after the first three decrease rapidly; for example, for $k = n = d = 7r$ (about what occurs near saturation), the term in n^{-1} is only about .001 times the sum of the first three terms.

VII. CONSTRUCTION OR NONCONSTRUCTIBILITY

When $k = n$, the Ehlich theory leading to (1.5) is rarely implementable in the sense of there existing an X with $\det(X'X) = \Psi(n,n)$. Specifically, $\Psi(n,n)$ is infrequently a square, which is necessary for such an X to exist; the only two values of $n < 200$ for which $\Psi(n,n)$ is square, other than the trivial value $n = 3$, are 91 and 147 (misprinted 47 in GK). It is not known whether an X with $\det(X'X) = \Psi(n,n)$ is constructible for any $n > 3$.

When $k < n$ we do not of course have squareness of $D_{k,n}(s_{OPT})$ as a condition for constructibility of an X with $\det(X'X) = \Psi(k,n)$. Nonconstructibility in the cases $(k,n) = (13,15)$ (both $s_{OPT} = 6$ and 7 in (5.1), $d = 11$), (14,15), and (9,11) with

$s_{OPT} = 7$, was demonstrated by a computer search of the tree of possibilities, somewhat reduced by observing certain symmetries, as described in Galil and Kiefer [3].

The pairs (k,n) for which X attaining $\Psi(k,n)$ had been obtained by 1979 are listed in Galil and Kiefer [2], [3]. These were sometimes obtained by computer search (see also Mitchell [6]) and sometimes by combinatorics; the construction of an X in the case (1.8) as a submatrix of a Hadamarad matrix (or of a union of such matrices - see GK) is well known, and that of an X satisfying (1.9) is described in Galil and Kiefer [3].

Recently we obtained new combinatorial methods that yield X's attaining $\Psi(k,n)$ for infinitely many (k,n), including many of those listed in (5.1). This is described in Galil and Kiefer [4].

When $\Psi(k,n)$ is not attainable by an X, one must both find a likely optimality candidate X* and must also methodically show no X exists which is better. Williamson [7] carried this out for $k = n = 7$, although a simpler proof of optimality is possible then - see GK, p. 1300. For $k = n = 11$ Ehlich, in unpublished work described in outline by GK, found and proved optimality of X's yielding three nonisomorphic X'X forms.

In cases of "ties" among the determinants of nonisomorphic X'X's, such as that just described or those of (5.1) where X's yielding both values of s_{OPT} exist, these X's are compared further in terms of other optimality criteria in Galil and Kiefer [3], [5]. For example, if $\det C_s = \det C_{s+1}$ with s and s+1 both optimum, it can be shown that C_s is better than C_{s+1} in terms of the Φ_p-optimality criterion $\Phi_p(C) = (k^{-1}\mathrm{tr}\, C^{-p})^{1/p}$, $0 < p < \infty$, and $\Phi_\infty(C)$ = maximum eigenvalue of C^{-1}.

REFERENCES

[1] Ehlich, H. (1964). Determinantenabschätzungen für binäre Matrizen mit $n \equiv 3 \bmod 4$. *Math. Z.* *84*, 438-447.

[2] Galil, Z. and Kiefer, J. (1980a). D-optimum weighing designs. *Ann. Statist.* *8*, 1293-1306.

[3] Galil, Z. and Kiefer, J. (1980b). Optimum weighing designs. In *Recent Developments in Statistical Inference and Data Analysis* (K. Matusita, ed.). North Holland, Amsterdam.

[4] Galil, Z. and Kiefer, J. (1981a). Construction of optimum block matrices for weighing designs with $n \equiv 3 (\bmod 4)$. To appear in *Ann. Statist.*

[5] Galil, Z. and Kiefer, J. (1981b). Comparison among D-optimum designs. To appear.

[6] Mitchell, T. J. (1974). Computer construction of "D-optimal" first-order designs. *Technometrics* *16*, 211-220.

[7] Williamson, J. (1946). Determinants whose elements are 0 and 1. *Amer. Math. Monthly* 53, 527-534.

The Annals of Statistics
1982, Vol. 10, No. 2, 502–510

CONSTRUCTION METHODS FOR *D*-OPTIMUM WEIGHING DESIGNS WHEN $n \equiv 3 \pmod 4$

BY Z. GALIL[1] AND J. KIEFER[1,2]

Tel Aviv University and University of California, Berkeley

In the setting where the weights of k objects are to be determined in n weighings on a chemical balance (or equivalent problems), for $n \equiv 3 \pmod 4$, Ehlich and others have characterized certain "block matrices" C such that, if $X'X = C$ where $X(n \times k)$ has entries ± 1, then X is an optimum design for the weighing problem. We give methods here for constructing X's for which $X'X$ is a block matrix, and show that it is the optimum C for infinitely many (n, k). A table of known constructibility results for $n < 100$ is given.

1. Introduction. Let k and n be positive integers with $k \leq n$, and let $\mathscr{X} = \mathscr{X}(n, k)$ be the set consisting of every $n \times k$ matrix X whose entries are all ± 1. Our goal is to find a D-optimum X, i.e., one that maximizes $\det(X'X)$ over $\mathscr{X}$. A discussion of the settings (weighing, fractional factorial, first order regression) where this problem arises, some history, and statements of results obtained up to that time are contained in Galil and Kiefer (1980a, b). The most interesting unsolved cases noted there are ones for which $n \equiv 3 \pmod 4$.

Throughout this paper $n \equiv 3 \pmod 4$. In the present paper we give methods for construction of D-optimum X's for infinitely many pairs (n, k) for which an optimum X was not previously known.

Section 2 gives methods for constructing X's for which $X'X$ is a "block matrix" of a type we now describe. A *block* of size r is an $r \times r$ matrix with diagonal elements n and off-diagonal elements 3. A $k \times k$ *block matrix* with s blocks of sizes $r_1, r_2, \cdots, r_s$, satisfying $\sum_1^s r_i = k$, was diagonal blocks of those sizes and elements -1 everywhere else. If the blocks are of two neighboring sizes (one size R if $s \mid k$), the sizes are $R = \lceil k/s \rceil$ and $R - 1$, where $\lceil x \rceil$ is the least integer $\geq x$. (We shall also later use $\lfloor x \rfloor$ for the greatest integer $\leq x$.) The X's constructed in Section 2, denoted $X(n, k, s, R)$ or $X(n, k, s)$, have $X'X$ of the last form for certain n, k, s, R, although (Remark 3) the method also yields block matrices $X'X$ with blocks of more than two sizes.

The reason for constructing X's of this form is that for each (n, k) there are particular values (usually one, sometimes two) s_{OPT} such that, if an $X(n, k, s_{\text{OPT}})$ exists, it is D-optimum. This is a consequence of a theory originated by Ehlich (1964) and further developed in Galil and Kiefer (1980a, 1981a), which shows that, for each pair (n, k),

$$\max_{X \in \mathscr{X}} \det(X'X) \leq \max_s \det C_s \tag{1.1}$$

where C_s is a block matrix with s blocks of at most two neighboring sizes. Thus, if s_{OPT} is a value maximizing $\det C_s$, a sufficient condition for X to be D-optimum is that $X = X(n, k, s_{\text{OPT}})$. The condition is not necessary because X's of this form do not always exist.

In Section 3 we use our knowledge of s_{OPT} to give examples of pairs (n, k) for which the methods of Section 2 yield D-optimum X's. Although s_{OPT} is not known exactly for *every*

Received May 1981; revised October 1981.

AMS 1970 *subject classification.* Primary 62K5, 62K15, 05B20.

Key words and phrases. Optimum designs, weighing designs, construction methods, D-optimality, first order designs, fractional factorials.

[1] Research supported by NSF Grant MCS 78-25301. The contents are part of an invited hour address given by one of the authors at the August 1980 AMS-IMS Annual Meeting in Ann Arbor. The first author was also supported by the Israel Commission for Basic Research.

[2] Professor Kiefer died on August 10, 1981.

(n, k), it is known for many (n, k). The following information is taken from the above references; other values of s_{OPT} are listed here, or are obtainable by the methods of those papers.

(A) If $2k - n < 5$, $s_{OPT} = k$ uniquely.

(B) Two neighboring values of s_{OPT} exist iff

$$2k - n = 5,\ 7\ (\text{and } n \equiv 3 (\text{mod } 8)),\ 9,\ 11,\ 17\ (\text{and } n \equiv 3 (\text{mod } 12)).$$

(C) For $d = 2k - n = 5, 7, 9, 11, 13, 15$ and 17, s_{OPT} and $R_{OPT} = \lceil k/s_{OPT} \rceil$ are given by

$$d = 5: \quad (s_{OPT}, R_{OPT}) = (k, 1) \quad \text{or} \quad (k-1, 2);$$

$$d = 7: \quad (s_{OPT}, R_{OPT}) = \begin{cases} ((3k+1)/4, 2) \quad \text{or} \quad ((3k-3)/4, 2) & \text{if} \quad n \equiv 3 (\text{mod } 8), \\ ((3k-1)/4, 2) & \text{if} \quad n \equiv 7 (\text{mod } 8); \end{cases}$$

$$d = 9: \quad (s_{OPT}, R_{OPT}) = ((k+2)/2, 2) \quad \text{or} \quad (k/2, 2);$$

$$d = 11: \quad (s_{OPT}, R_{OPT}) = ((k+1)/2, 2) \quad \text{or} \quad ((k-1)/2, 3);$$

$$d = 13: \quad (s_{OPT}, R_{OPT}) = (\lfloor (4k+3)/9 \rfloor, 3);$$

$$d = 15: \quad (s_{OPT}, R_{OPT}) = (\lfloor (13k+25)/36 \rfloor, 3);$$

$$d = 17: \quad (s_{OPT}, R_{OPT}) = \begin{cases} ((k+2)/3, 3) \quad \text{or} \quad ((k-1)/3, 4) & \text{if} \quad n \equiv 3 (\text{mod } 12), \\ (k/3, 3) & \text{if} \quad n \equiv 7 (\text{mod } 12), \\ ((k+1)/3, 3) & \text{if} \quad n \equiv 11 (\text{mod } 12). \end{cases}$$

We refer to $\{(n, k) : 2k - n = d\}$ as "the series $2k - n = d$".

(D) For all (n, k),

$$6k/(2k - n + 7) \leq s_{OPT} \leq 8k/(2k - n - 3),$$

$$(2k - n + 3)/8 \leq R_{OPT} \leq (2k - n + 7)/6.$$

(E) For $n \to \infty$ with $k/n \to 1 - \lambda$, we have

$$\lim s_{OPT} \begin{cases} = 7 & \text{if} \quad 0 \leq \lambda \leq .08837, \\ = 8 & \text{if} \quad .08838 \leq \lambda \leq .17027, \\ = 9 & \text{if} \quad .17028 \leq \lambda \leq .22494, \\ > 9 & \text{if} \quad .22495 \leq \lambda. \end{cases}$$

In Table 1 we list the values of s_{OPT} for $n < 100$, omitting (A), and also summarize the known results, including those of this paper, on D-optimum designs X. These constructibility results, in the form of symbols x, e, # and * as described there, are based on the results of Section 3 herein and also the following facts from the earlier and other references.

(F) Whenever H_{n+1} (a Hadamard matrix of order $n + 1$) exists, normalize it by letting the first row consist of 1's, and delete that row. Any k columns of the resulting matrix yield an X with $X'X = C_k$, optimum for (A) above and for the first value, k, of s_{OPT} in (C) with $d = 5$. This exists for all "practical" n including all $n < 100$, and is omitted from Table 1 in the domain of (A). It is denoted in Table 1 as constructible by the methods of this paper for $d = 5$ because this simple and old construction scheme is a degenerate case of our methods. Even when existence of H_{n+1} is unknown, an X with $X'X = C_k$ can be constructed for sufficiently small k by adjoining several H_j's; see Galil and Kiefer (1980a, top of page 1297).

TABLE 1

Values of s_{OPT} *for* $(n + 5)/2 \le k \le n < 100$, $n \equiv 3 (\mathrm{mod}\ 4)$, *with some constructibility results*

KEY: $\bar{j}$ indicates a tie, $s_{\mathrm{OPT}} = j$ or $j + 1$, which occurs when $2k - n = 5$, 7 (every other n), 9, 11, 17 (every third n); * a design attaining the bound (1.1) is constructible by the methods of the present paper; # a design is constructed in an earlier paper; see (G) and (H) of Section 1; *x* design is not constructible, as described in (I) and (J) of Section 1; *e* a design attaining the bound (1.1) is

k \ n	7	11	15	19	23	27	31	35	39	43	47	51
6	$\overline{5}$#*											
7	5*e*											
8		$\overline{7}$#*										
9		$\overline{6}$#*x*										
10		$\overline{5}$##	$\overline{9}$#*									
11		$\overline{5}$*ee*	8#									
12			$\overline{6}$##	$\overline{10}$#*								
13			$\overline{6}$*xx*	$\overline{9}$								
14			6*x*	$\overline{7}$**	$\overline{13}$#*							
15			6*x*	$\overline{7}$*	11							
16				7	$\overline{8}$	$\overline{15}$#*						
17				6	$\overline{8}$	$\overline{12}$*						
18				6	8	$\overline{9}$**	$\overline{17}$#*					
19				6*x*	7	$\overline{9}$*	14					
20					7	9	$\overline{10}$	$\overline{19}$#*				
21					7	8	$\overline{10}$	$\overline{15}$**				
22					6	$\overline{7}$	10	$\overline{11}$**	$\overline{21}$#*			
23					6*x*	7	9	$\overline{11}$*	17			
24						7	8	11	$\overline{12}$	$\overline{23}$#*		
25						7	8	9	$\overline{12}$	$\overline{18}$**		
26						6	7	9	11	$\overline{13}$**	$\overline{25}$#*	
27						6*x*	7	9	10	$\overline{13}$*	20	
28							7	8*	$\overline{9}$	12	$\overline{14}$	$\overline{27}$#*
29							7	8*	9	11	$\overline{14}$	$\overline{21}$**
30							6	7	8	10	13	$\overline{15}$**
31							6*x*	7	8	10	11	$\overline{15}$*
32								7	8	9	11	14
33								7	8	9	10	12*
34								6	7	8	10	$\overline{11}$**
35								6*x*	7	8	9	11*
36									7	8	9	10*
37									7	7	8	10*
38									6	7	8	9
39									6*x*	7	8	9
40										7	8	8
41										7	7	8
42										6	7	8
43										6*x*	7	8
44											7	7
45											7	7
46											7	7
47a											6*x*	7
48												7
49												7
50												7
51												6*x*
52												
53												
54												
55												
56												

TABLE 1 continued

not constructible but the optimum design is known; see (I) of Section 1. The pair of constructibility symbols after a tie $\overline{j}$ refer in order to j and $j + 1$. If only the symbol * is given, it refers to j, while $\overline{\ }$* alone refers to $j + 1$. Absence of a constructibility symbol denotes ignorance.

55	59	63	67	71	75	79	83	87	91	95	99	n / k
	7	7	8*	8	9	10	11	13	15	18	21*	57
	7	7	8*	8	9	10	11	12	14	16	$\overline{19}$**	58
	$6x$	7	7	8	9	10	10	12	13	15	19*	59
		7	7	8	8	9	10	11	12	15	17*	60
		7	7	8	8	9	10	11	12	14	16*	61
		7	7	8	8	9	9	10	12	13	15	62
		$7x$	7	7	8	9	9	10	11	13	14	63
			7	7	8	8	9	10	11	12	13	64
			7	7	8	8	9	10	11	11	13	65
			7	7	7	8	9	9	10	11	13	66
			$7x$	7	7	8	8	9	10	11	12*	67
				7	7	8	8	9	10	10	11*	68
				7	7	8	8	9	9	10	11*	69
				7	7	7	8	8	9	10	11*	70
				$7x$	7	7	8	8	9	10	10*	71
					7	7	8	8	9	9	10*	72
					7	7	8	8	9	9	10*	73
					7	7	7	8	8	9	10*	74
					$7x$	7	7	8	8	9	9	75
						7	7	8	8	9	9	76
						7	7	7	8	8	9	77
						7	7	7	8	8	9	78
						$7x$	7	7	8	8	9	79
							7	7	8	8	8*	80
$\overline{29}$#*							7	7	7	8	8*	81
23							7	7	7	8	8*	82
$\overline{16}$	$\overline{31}$#*						$7x$	7	7	8	8*	83
$\overline{16}$	$\overline{24}$**							7	7	7	8*	84
15	$\overline{17}$**	$\overline{33}$#*						7	7	7	8*	85
13	$\overline{17}$*	26						7	7	7	8*	86
12	16	$\overline{18}$	$\overline{35}$#*					$7x$	7	7	8*	87
12	14	$\overline{18}$	$\overline{27}$**						7	7	7	88
11	13	17	$\overline{19}$**	$\overline{37}$#*					7	7	7	89
10	12	14	$\overline{19}$*	29					7	7	7	90
10	11	$\overline{13}$	18	$\overline{20}$	$\overline{39}$#*				7	7	7	91
9	11	13	15*	$\overline{20}$	$\overline{30}$**					7	7	92
9	10	12	14*	19	$\overline{21}$**	$\overline{41}$#*				7	7	93
9	10	11	14*	16	$\overline{21}$*	32				7	7	94
8	9	11	12*	15	19	$\overline{22}$	$\overline{43}$#*			$7x$	7	95
8	9	10	12*	14	16	$\overline{22}$	$\overline{33}$**				7	96
8	9	10	11	13	$\overline{15}$	20	$\overline{23}$**	$\overline{45}$#*			7	97
8	8	9	11	12	15	17	$\overline{23}$*	35			7	98
7	8	9	10	12	13	16	21	$\overline{24}$	$\overline{47}$#*		$7x$	99
7	8	9	10	11	13	16	18*	$\overline{24}$	$\overline{36}$**			
7	8	8	10	10	12	14	17*	22	$\overline{25}$**	$\overline{49}$#*		
7	8	8	9	10	12	13	16*	19	$\overline{25}$*	38		
7	7	8	9	10	11	13	15*	$\overline{17}$	23	$\overline{26}$	$\overline{51}$#*	
7	7	8	9	9	11	12	14*	17	19	$\overline{26}$	$\overline{39}$**	
7	7	8	8*	9	10	11	13	15	18	24	$\overline{27}$**	
$6x$	7	7	8*	9	10	11	12	14	18	20	$\overline{27}$*	
	7	7	8*	9	9	11	12	14	16	19	25	

(G) Assuming H_{n+1} exists, a construction of D-optimum X with $X'X = C_{k-1}$ when $d = 5$, the other optimum structure for that case of (C) is given in Galil and Kiefer (1980b, page 185). (For $n = 7, 11, 15$, these X had been found earlier by computer search.) That case $s_{\mathrm{OPT}} = k - 1$ is not covered by the methods of the present paper.

(H) D-optimum X's with $X'X = C_{s_{\mathrm{OPT}}}$, obtained by computer search, are given in Mitchell (1974) for $(n, k, s_{\mathrm{OPT}}) = (11, 10, 5)$; in Galil and Kiefer (1980a) for (11, 9, 6), (11, 10, 6), (15, 11, 8), (15, 12, 6); and in Galil and Kiefer (1980b) for (15, 12, 7). These are not obtainable by the methods of the present paper.

(I) For the saturated cases, $k = n$, $\det(X'X)$ is a square, and hence equality in (1.1) is rarely achieved. Thus, one can easily check that with the possible exception of $k = n = 91$, an X with $X'X = C_{s_{\mathrm{OPT}}}$ does not exist in all saturated cases in Table 1. When $k = n = 7$, a D-optimum X was first found and proved optimum by Williamson (1946), and when $k = n = 11$, three X's yielding the three possible nonisomorphic matrices $X'X$ that are D-optimum were first found and proved optimum in unpublished work of Ehlich, and are given in Mitchell (1974) (who found one by computer search) and in Galil and Kiefer (1980a).

(J) Nonattainability of (1.1) for $(n, k, s_{\mathrm{OPT}}) = (11, 9, 7), (15, 13, 6), (15, 13, 7), (15, 14, 6)$ was shown by Galil and Kiefer (1980b) using a computer search of the tree of all possibilities, reduced somewhat by taking account of certain symmetries. For $(n, k) = (11, 9)$, we know a D-optimum design from (H); for (15, 13) and (15, 14), optimum designs are still unknown at this writing.

In the cases mentioned in (B) and (I) in which the D-optimum $X'X$ may not be unique to within obvious isomorphisms, these designs may be compared according to other criteria. For example, in the cases listed in (B) in which $\det C_s = \det C_{s+1}$ for some s, it is always true that C_s is better than C_{s+1} in terms of giving a smaller value of the Φ_p-criterion for $0 < p \le \infty$, where $\Phi_p(C) = (k^{-1}\mathrm{tr}(C^{-p}))^{1/p}$ for $0 < p < \infty$, and $\Phi_\infty(C)$ = maximum eigenvalue of C^{-1}. See Galil and Kiefer (1980b, 1981b).

2. Construction methods. We again write H_q for a Hadamard matrix of order q. Let H_j have first row $e_j = (1, 1, \cdots, 1)$, and write G_j for the $(J - 1) \times J$ submatrix of H_j consisting of the last $J - 1$ rows. Let $\bar{H}_{M+4}$ be an $(M + 4) \times M$ matrix of ± 1's with orthogonal columns and first row e_M. Our basic construction is the $(JM + 4) \times JM$ matrix

$$Z \equiv Z(J, M) = \begin{bmatrix} e_J \otimes \bar{H}_{M+4} \\ G_J \otimes H_M \end{bmatrix}. \tag{2.1}$$

Here $\otimes$ denotes the Kronecker (tensor) product.

If we denote the ith column of H_J by $\begin{bmatrix} 1 \\ g_i \end{bmatrix}$, we have

$$\begin{bmatrix} \bar{H}_{M+4} \\ g_i \otimes H_M \end{bmatrix}' \begin{bmatrix} \bar{H}_{M+4} \\ g_{i'} \otimes H_M \end{bmatrix} = \begin{cases} (JM + 4)I_M & \text{if } i = i', \\ 4I_M & \text{if } i \ne i'. \end{cases} \tag{2.2}$$

We hereafter write $L = JM + 4$. Hence, if P is the $J \times J$ matrix with diagonal entries L and off-diagonal entries 4, we have

$$Z'Z = P \otimes I_M. \tag{2.3}$$

Let Y be obtained from Z by deleting the first row e_{JM} of the latter. Then $Y'Y = Z'Z - E_{JM}$, where $E_m = e_m' e_m$ is a matrix of 1's; $Y'Y$ has diagonal entires $L - 1$ and off-diagonal entries 3 or -1. We permute the columns of Y to form an $(L - 1) \times JM$ matrix $\bar{X} \equiv \bar{X}(J, M)$, as follows: for i and h integers, $0 \le i \le J - 1$, $1 \le h \le M$, the $((h - 1)J + (i + 1))$th column of $\bar{X}$ is the $(iM + h)$th column of Y. Then, denoting by $B(\rho, n)$ the $\rho \times \rho$ "block matrix" with diagonal entries $n = L - 1$ and off-diagonal entries 3, we obtain

$$
\bar{X}'\bar{X} = \begin{bmatrix} B(J,n) - E_J & \cdots - E_J \\ -E_J & B(J,n) & \\ \vdots & \ddots & \vdots \\ -E_J & \cdots & B(J,n) \end{bmatrix}, \tag{2.4}
$$

a block matrix with M blocks of size J.

Finally, for $s \leq M$, $R \leq J$, and $1 \leq v \leq s$, write $k = sR - s + v$, and let $X \equiv X(n, k, s, R)$ be the $n \times k$ matrix obtained by selecting R columns from the ith set of J contiguous columns of $\bar{X}$, $1 \leq i \leq v$, and $R - 1$ columns from the ith set for $v < i \leq s$ (there are none of the latter, if $v = s$). We have obtained

(2.5) $X(n, k, s, R)$, $n \times k$, with $X'X$ having s blocks, maximum block size R, all blocks of size R or $R - 1$.

Of course, n, k, s determine R, but n, k, R do not determine s. Given n, k, s, R, and $n \equiv 3 \pmod 4$, write $n' = n + 1$. The above method of construction yields an $X(n, k, s, R)$ if and only if, for some J and M for which H_J, H_M, $\bar{H}_{M+4}$ exist,

$$
n' = JM + 4, \quad k \leq JM, \quad s \leq M, \quad \lceil k/s \rceil = R \leq J. \tag{2.6}
$$

Suppose (2.6) is satisfied for $(J, M) = (J_1, M_1)$ and that $J_2 M_2 = J_1 M_1$ and $J_2 < J_1$ and the required H's and $\bar{H}$ exist. If $R \leq J_2$, then since $s \leq M_1 < M_2$ we see that (2.6) is satisfied for (J_2, M_2). Similarly, if $J_3 M_3 = J_1 M_1$ and $s \leq M_3 < M_1$, we have (2.6) satisfied for (J_3, M_3). In summary,

(2.7) If the above construction works to yield an $X(n, k, s, R)$, then it works for the smallest J for which the required H's and $\bar{H}$ exist and for which $R \leq J$, and it also works for the smallest M for which the required H's and $\bar{H}$ exist and for which $s \leq M$.

We now turn to the role of existence of H's and $\bar{H}$. The trivial case $J = 1$ produces blocks of size 1, the design of (F) of Section 1. We hereafter assume $R \geq 2$ so $J \geq 2$. If $\bar{H}_{M+4}$ is obtained as M columns of an H_{M+4}, then $4 \mid M$, say $M = 4m$.

In the present and next paragraph we treat the construction with $R = 2$. When $R = 2$, we obtain $X(n, k, s, 2)$ by the above construction with $J = 2$, provided that, for some positive integer m, H_{4m} and H_{4m+4} exist and

$$
n = 8m + 3, \quad k \leq 8m, \quad s \leq 4m, \quad R = 2. \tag{2.8}
$$

Since $s \leq k - 1$ for $R = 2$, a sufficient condition for (2.8) when $R = 2$ is $n = 8m + 3$, $k \leq 4m + 1$.

Because of (2.7), when $R = 2$, one cannot do better by our method of construction than with $J = 2$. When $J = 2$ there remains the single additional case $M = 2$: the possibility that H_{4m} and $\bar{H}_{4m+4}$ but not H_{4m+4} exist is eliminated by a result of Vijayan (1976); but for $M = 2$ we know that H_2 exists and H_6 does not, but $\bar{H}_6$ does (a column of 1's, a column of three 1's and three -1's). This yields a construction of $X(7, k, s, 2)$ for $k \leq 4$; $s \leq 2$.

For $R > 2$, we need $J > 2$ and hence $4 \mid J$, say $J = 4j$. We obtain an $X(n, k, s, R)$ by our method if, for some positive integers m and j, H_{4j}, H_{4m}, and H_{4m+4} exist and

$$
n = 16mj + 3, \quad k \leq 16mj, \quad s \leq 4m, \quad R \leq 4j. \tag{2.9}
$$

For given $n \equiv 3 \pmod{16}$, write $n^* = n - 3$. If j_R is the smallest divisor of $n^*/16$ which is $\geq R/4$, we see by (2.7) that X can be constructed (for some j and m) by this method if and only if (assuming all necessary H's exist)

$$
k \leq n^*, \quad s \leq n^*/4j_R. \tag{2.10}
$$

Since $s \leq (k - 1)/(R - 1)$, a sufficient condition for (2.10) is $k \leq (R - 1)n^*/4j_R + 1$. Additionally, corresponding to the special case described in the previous paragraph when $R = 2$ there, we now obtain, for $M = 2$, the designs $X(8j + 3, k, s, R)$ for $k \leq 8j$, $s \leq 2$, $R \leq 4j$.

REMARK 1. We have phrased the results, using (2.6), in terms of given R. Similarly, one can work in terms of s, an m_s, and the second inequality of (2.10) replaced by $R \leq n^*/4m_s$.

REMARK 2. The H_M's (resp., $\bar{H}_{M+4}$'s) are the same across each set of M contiguous rows of Z, but can vary from one set of M rows to the next.

REMARK 3. It is clear that submatrices X of $\bar{X}$ can also be selected to yield block matrices $X'X$ with blocks of more than two sizes. This may be of interest, e.g., for optimality criteria other than D-optimality.

3. Construction of optimum designs. We now apply the methods of the previous section to construct optimum designs by implementing (2.6), usually in the form (2.8) or (2.9), for value(s) of s_{OPT} listed in Section 1 (A) through (E) for various (n, k).

I. *Series* $2k - n = d$, $5 \leq d \leq 17$; see (C).

(a) $d = 5$: The "very regular" case $s_{OPT} = k$ is a trivial construction, but as noted earlier, $R = J = 1$ in (2.6) formally includes it. The construction for $s_{OPT} = k - 1$ with $R = 2$ requires $J \geq 2$ and thus $s_{OPT} \leq k/2$, which is false. However, a design for $s_{OPT} = k - 1$ is listed in (G) of Section 1.

(b) $d = 7$: Here $R = 2$, and it turns out here and in the next two series that only (2.8), with $J = 2$, can be used; the inequality $s_{OPT} \leq (n - 3)/4$ required for (2.9) is always false. We find $(3k - 3)/4 = 3(n + 5)/8 \leq (n - 3)/2 \Leftrightarrow n \geq 27$ and $3(n + 5)/8 + 1 \leq (n - 3)/2 \Leftrightarrow n \geq 35$. Thus, for $n \equiv 3 \pmod 8$ we obtain designs for both s_{OPT} values if $n \geq 35$, but only for the smaller s_{OPT} value when $n = 27$. We note that in this and the next two series, the method gives a construction once it starts, only for every other value of $n \equiv 3 \pmod 4$, since it requires $n \equiv 3 \pmod 8$.

(c) $d = 9$: Again $R = 2$ and we find both inequalities $s_{OPT} \leq (n - 3)/2$ are satisfied for $n \equiv 3 \pmod 8$ when $n \geq 19$. Thus, we get both designs.

(d) $d = 11$: For $s_{OPT} = (k + 1/2$, $R = 2$, the method of (2.8) works for $n \geq 19$, $n \equiv 3 \pmod 8$. However, for $s_{OPT} = (k - 1)/2$, we have $R = 3$, and the method does not work since we never have $s_{OPT} \leq (n - 3)/4$.

(e) $d = 13$: We use (2.9) with $j = 1$, with $n \equiv 3 \pmod{16}$. We have $s_{OPT} = \lfloor (2n + 29)/9 \rfloor$, which is $\leq (n - 3)/4$ if $n \geq 115$ and $n \equiv 3 \pmod{16}$, and thus we obtain an optimum design for these values, all of which fall outside Table 1.

(f) $d = 15$: We again use (2.9) with $j = 1$. Now $s_{OPT} = \lfloor (13n + 245)/72 \rfloor$, which is $\leq (n - 3)/4$ if $n \geq 51$ and $n \equiv 3 \pmod{16}$, so we obtain optimum designs in all these cases.

(g) $d = 17$: Once more we use (2.9) with $j = 1$ and $n \equiv 3 \pmod{16}$. From (C) we obtain the four expressions for s_{OPT} in the three cases (mod 12), three with $R = 3$ and one with $R = 4$. We use (2.9) with $j = 1$ and check $s_{OPT} \leq (n - 3)/4$, and find that both possible values of s_{OPT} satisfy this condition when $n \geq 51$ for $n \equiv 3 \pmod{48}$, i.e., $n \equiv 3 \pmod{16}$ and $n \equiv 3 \pmod{12}$; and find that the condition is satisfied when $n \geq 67$ for $n \equiv 19 \pmod{48}$ and when $n \geq 83$ for $n \equiv 35 \pmod{48}$. In summary, then, when $n \equiv 3 \pmod{16}$, designs for all values of s_{OPT} are constructible if $n \geq 51$.

II. Other constructions when $n < 100$. In addition to the parameter values that fall into the series of part I, we obtain constructions in the following cases, using (2.9):

$$n = 35, \quad k = 28, 29 \quad (m = 2, j = 1).$$

$$n = 51, \quad k = 35, 36, 37 \quad (m = 3, j = 1).$$

$$n = 67, \quad k = 43, 44, 45 \quad (m = 4, j = 1);$$

$$k = 54(1)58 \quad (m = 2, j = 2).$$

$$n = 83, \quad k = 51, 52, 53 \quad (m = 5, j = 1).$$

$$n = 99, \quad k = 59, 60, 61 \quad (m = 6, j = 1);$$
$$k = 67(1)74 \quad (m = 3, j = 2);$$
$$k = 80(1)87 \quad (m = 2, j = 3).$$

The above designs and those for the series of part I above for $n < 100$ are designated by an asterisk (*) in Table 1.

In the above listing, the large number of cases covered when $n = 67$ and 99 is a reflection of divisibility properties of $n - 3$. In the actual construction, one really needs to use only a subset of the columns of H_M and $\bar{H}_{M+4}$ in some cases; for example, we need use only 16 of the 24 columns for $n = 99$, $k = 61$, $s = 16$, $R = 4$. We can only use (2.9), never (2.8), for cases outside the series of Part I, since we never have $R = 2$ for $d > 11$.

III. Constructibility without knowing s_{OPT}. Suppose we use only the simple inequalities (D) which do not determine R_{OPT} or s_{OPT}. Sometimes these inequalities are nevertheless sufficient, in that the conditions (2.6) are sufficiently weak that we can conclude constructibility of an optimum design by our methods *whatever* s_{OPT} turns out to be. We give only one example, since the development is straightforward, and since one would be unable to use it without knowing s_{OPT} exactly. Nevertheless, the technique could be useful for showing, for a given (k, n), that it is worthwhile working out s_{OPT} exactly because it will definitely be possible to construct an optimum design.

Suppose $d = 41$. From (D), $R_{\mathrm{OPT}} \le 8$, so we can use (2.9) with $j = 2$. Using our method, we see we must have $s_{\mathrm{OPT}} \le (n - 3)/8$. But (D) gives $s_{\mathrm{OPT}} \le 8k/(d - 3)$, so we can use (2.9) if $k/(n - 3) \le 19/32$. We conclude that, for $n \equiv 3 (\mathrm{mod}\ 32)$ and $n \ge 259$, with $k = (n + 41)/2$, the method works with $j = 2$ and $m = (n - 3)/32$ in (2.9).

IV. Asymptotics. As $n \to \infty$ with $k/n \to 1 - \lambda$, considerations like those of III above can be obtained from (D), but we now consider the more precise results obtainable from (E).

If $s_{\mathrm{OPT}} = 7$, using (2.9) we need $R \le 4j$ and $m \ge 2$, so that $k/n \le 7R/(16mj + 3) \le 7/8$. From (E), $s_{\mathrm{OPT}} = 7$ for n large only if $k/n > .91$, so we conclude that our construction method can never work in that domain for n large. In fact, from looking at the results for $n < 100$, we doubt that the method applies to any cases for which $s_{\mathrm{OPT}} = 7$, except for the single case $(n, k) = (19, 14)$.

If $s_{\mathrm{OPT}} = 8$ and $n \equiv 3 (\mathrm{mod}\ 32)$, for $m = 2$ we obtain, if $.83 < k/n < .91$ (see (E)) with $j = (n - 3)/32$, that $R/4j \approx (k/8)/4j \approx k/(n - 3) < 1$ and $n \to \infty$, so the method works as $n \to \infty$. In fact, Table 1 (for $n = 35, 67, 99$) indicates that the method might *always* work for $s_{\mathrm{OPT}} = 8$ and $n \equiv 3 (\mathrm{mod}\ 32)$. For $m \ge 3$, one finds that the construction only works if $k/n < .8$, so this cannot work asymptotically. Again, examination of part II above indicates that this may be the case for all n of the right congruence, when $s_{\mathrm{OPT}} = 8$.

When $s_{\mathrm{OPT}} = 9$, we need $m \ge 3$ in (2.9) and hence $k/n \approx 9R/n \le 36j/16mj \le .75$, whereas asymptotically, from (E), $k/n > .78$ for s_{OPT} to be 9. So the method does not work for large n. Again, from Table 1 (for $n = 51$ or 99) it seems likely that the method never works when $s_{\mathrm{OPT}} = 9$.

When $s_{\mathrm{OPT}} = 10$, 11, or 12, for $n \equiv 3 (\mathrm{mod}\ 32)$, $m = 3$, and $j = (n - 3)/48$, we obtain $R/4j \approx k/4js_{\mathrm{OPT}} \approx (k/n)(12/s_{\mathrm{OPT}}) < 1$ asymptotically, since from (E) $k/n < .78$ asymptotically in order that $s_{\mathrm{OPT}} > 9$. Thus the method works for large $n \equiv 3 (\mathrm{mod}\ 48)$ when $s_{\mathrm{OPT}} = 10$, 11, or 12, and Table 1 for $n = 51, 99$ indicates that it might work for all n. We omit discussion of larger s_{OPT} values, which are handled similarly.

REFERENCES

EHLICH, H. (1964). Determinantenabschätzungen für binäre Matrizen mit $n \equiv 3$ mod 4. *Math. Z.* **84** 438–447.

GALIL, Z. and KIEFER, J. (1980a). *D*-optimum weighing designs. *Ann. Statist.* **8** 1293–1306.

GALIL, Z. and KIEFER, J. (1980b). Optimum weighing designs. In *Recent Developments in Statistical Inference and Data Analysis* (K. Matusita, ed.). North Holland, Amsterdam.

GALIL, Z. and KIEFER, J. (1981a). On the characterization of D-optimum weighing designs for $n \equiv 3 \pmod 4$. To appear in *Proceedings of* 1981 *Purdue Conference*.

GALIL, Z. and KIEFER, J. (1981b). Comparison among D-optimum designs. Unpublished manuscript.

MITCHEL, T. J. (1974). Computer construction of "D-optimal" first-order designs. *Technometrics* **16** 211–220.

VIJAYAN, K. (1976). Hadamard matrices and submatrices. *J. Austral. Math. Soc.* **22** (Series A) 469–475.

WILLIAMSON, J. (1946). Determinants whose elements are 0 and 1. *Amer. Math. Monthly* **53** 427–434.

Z. GALIL
SCHOOL OF MATHEMATICAL SCIENCES
RAMAT-AVIV, TEL-AVIV 69978
ISRAEL

Autocorrelation-robust Design of Experiments

J. Kiefer and H. P. Wynn

1. BACKGROUND

There has been considerable recent research interest in designing experiments under error processes which have spatial autocorrelation. The classical paper on the subject is Williams (1952) which was extended by Kiefer (1960).

The basic model is the linear model

$$E(Y) = X\alpha, \qquad \mathrm{Cov}(Y) = V \tag{1}$$

where the parameter vector α or some special set of contrasts θ, such as all treatment contrasts, is to be estimated. The matrix V derives from some autocorrelated process whose geometry is related to the design region and hence to the design matrix X. The published work in the field divides roughly into the following approaches.

(i) V is known and we look for optimum designs for the ordinary least squares estimate $\hat{\theta}_0$ of θ but under model (1), above.

(ii) V is known and we look for optimum designs but use the best linear unbiased estimate $\hat{\theta}_V$ under (1).

(iii) V is known and investigate designs for which $\hat{\theta}_0 \equiv \hat{\theta}_V$.

(iv) V is unknown and consider design and analysis.

(v) V is unknown but look for designs robust against a range of possible V matrices, e.g. minimax.

Scientific Inference,
Data Analysis, and Robustness

ISBN 0-12-121160-6

Thus (i) above was treated by Kiefer and Wynn (1980) under a simple neighbor model for V. That paper also contains an extensive list of references. Approaches (ii) and (v) were taken by Kiefer and Wynn (1981). Some study of (iii) is carried out by Martin (1981).

2. ONE DIMENSION: ANALYSIS

The paper by Kiefer and Wynn (1981) gives a thorough foundation for optimum treatment design in one dimension in the presence of autocorrelated errors. A brief resumé is necessary to provide a statistical background for the new combinatorial constructions of this paper.

Let $\{Y_t\}$ be a p-th order stationary autoregressive process on the line in discrete time. Thus

$$Y_t + \rho_1 Y_{r-1} + \cdots + \rho_p Y_{r-p} = \varepsilon_t$$

$(t = 0,1,2,\ldots)$, where $\{\varepsilon_t\}$ is an uncorrelated zero mean innovation process with variance σ^2. The actual observation at time t is

$$W_t = Y_t + \alpha_{[t]}$$

where $\alpha_{[t]} \in \{\alpha_1,\ldots,\alpha_k\}$ is the unknown parameter for one of k treatments allocated at time t. We are interested in estimating the symmetric contrast vector

$$\theta = (I - \frac{1}{k} J)\alpha \quad ,$$

where I is the k×k identity and J the k×k matrix of ones. We assume that $\rho = (\rho_1,\ldots,\rho_p)$ is known and work with the best linear unbiased estimate $\hat{\theta}$ of θ (under finite sample size N). The design optimality criterion is to minimize, through the choice of treatment allocation to time points the quantity

$$\Phi \{\mathrm{cov}(\hat{\theta})^-\} \qquad (2)$$

where A^- is the Moore-Penrose g-inverse of a matrix A, $\mathrm{cov}(\hat{\theta})$ is the covariance of $\hat{\theta}$ under the model and Φ is a member of a special class of functionals on k×k non-negative definite matrices. This class comprises all Φ which are (i) convex, (ii) invariant under permutations of coordinates and (iii) non-increasing: G-H non-negative definite => $\Phi(G) \leq \Phi(H)$. If a design minimizes (2) for all such Φ (fixed ρ) then it is said to be universally optimum.

Let $C = \sigma^2(\text{cov}(\hat{\theta}))^-$. Then sufficient conditions for universal optimality can be given in terms of the form of C. These derive from Proposition 1 of Kiefer (1975). They are

(a) C is completely symmetric: i.e. C is a multiply of $I - \frac{1}{k} J$.

(b) trace (C) is a maximum.

(Note, this is a simpler version than in Kiefer and Wynn, 1981, which is allowed if we assume equal treatment replication.)

To realize (a) and (b) we need to convert them into conditions on the design.

It turns out that all that is needed is the adjacencies counts (or proportions), namely the number of times treatments are next to themselves up to lag p (where p, we recall, is the order of the process).

To make the description easier to understand label the treatments A,B,C... Define

$$\pi_{r,N} = \frac{1}{N}\{\text{number of times A is r steps from A, or B is r-steps from B}, \ldots\} \qquad (r = 1,\ldots,p)$$

For example consider the sequence

ABAABACAABBCABCA ,

with N = 16. If p = 3 we calculate

$$\pi_{1,N} = \frac{3}{16}, \quad \pi_{2,N} = \frac{3}{16}, \quad \pi_{3,N} = \frac{3}{8}.$$

It is very convenient to take an asymptotic approach. Thus we call a sequence allowable if

$$\pi_{r,N} \to \text{a limit}, \ \pi_r$$

as $n \to \infty$. With this notation we can evaluate an asymptotic version of trace (C) which is

$$\frac{1}{N}\text{trace (C)} = \sum_{r=0}^{p} \rho_r^2 - \frac{1}{k}\left(\sum_{r=0}^{p} \rho_r\right)^2 + 2\sum_{r=1}^{p} \pi_r \sum_{s=0}^{p-r} \rho_s \rho_{s+r} .$$

Condition (b) says maximize

$$\sum_{r=1}^{p} \pi_r \sum_{s=0}^{p-r} \rho_s \rho_{s+r} \qquad (3).$$

Condition (a) is achieved by "complete symmetrization" namely taking sequences in which the adjacency counts between different treatments (up to lag p) are unaltered by a permutation of letters (treatment labels).

Define for k treatments

$$\Pi_p(k) = \{(\pi_1,\ldots,\pi_p)\}$$

that is, the set of $(\pi_1,\ldots,\pi_p)$ vectors attainable in the limit. It can be shown, and is fully discussed in the next section, that $\Pi_p(k)$ is a polyhedron whose vertices are obtained from purely periodic sequences with finite period. Since (3) is linear in the π_r it is maximised at a vertex and thus if the extreme periodic sequences can be found one of them will be optimum for a given ρ. In principle, then, the optimisation problem is solved. For example, with k = 2 and p = 3 the extreme π-vectors with their sequences are:

Sequence	π-vector
A . . .	(1, 1, 1)
AB . . .	(0, 1, 0)
AAB . . .	$(\frac{1}{3}, \frac{1}{3}, 1)$
AABB . . .	$(\frac{1}{2}, 0, \frac{1}{2})$
AAABBB . . .	$(\frac{2}{3}, \frac{1}{3}, 0)$

Just a full cycle is given above, it being assumed that the cycle is repeated indefinitely.

The extreme sequences for k = 2, p = 2 are A (1,1), AB (0,1) and AABB $(\frac{1}{2}, 0)$. For p = 3 and k = 3 and ≥ 4 are given in Kiefer and Wynn (1981). A theorem, there, says that all $\Pi_p(k)$ are identical for $k \geq p+1$.

The next two sections develop the theory of Π-regions for one and more dimensions.

3. STATIONARY BINARY PROCESSES IN ONE DIMENSION

We shall restrict ourselves to designs with 2 treatments k = 2. For the moment label $A \equiv 1$ and $B \equiv -1$. Then the design can be written as a real sequence $\ldots X_{-1}, X_0, X_1, \ldots$ with $X_t = \pm 1$. Then

$$2\pi_r - 1 = \frac{1}{N} \sum_{r=p}^{N} X_r X_{r-p} \qquad (4)$$

If we subtract $(\frac{1}{N}\sum_{r=p}^{N} X_r)^2$ we have the sample autocovariance function c(r) of the sequence $\{X_t\}$. Assume that $\lim \frac{1}{N}\Sigma X_r = 0$. Then the condition that $\pi_{r,N} \to$ a limit π_r $(r = 1,\ldots,p)$ is then a statement that the sample autocovariance function c(r) converges $(r = 1,\ldots,p)$. Now suppose that the binary sequence $\{X_t\}$ is a single realization of a stationary ergodic zero mean binary process then (with probability one) the sample autocovariance function will converge to the actual autocovariance function, by the Ergodic Theorem. The key fact is that the converse holds: for any $c(r) = 2\pi_r - 1$ with $\pi_r \in \Pi_p(2)$ there is a stationary ergodic binary process with c(r) as its auto covariance function. Thus there is an exact one-one correspondence between the c(r) and π_r.

There is a limited literature in this area. Hobby and Ylvisaker (1964) look at stationary discrete time finite state processes. There are some unpublished notes of the late Walter Weissblum (Bell Telephone Laboratories). The early study of covariances for binary processes is due to Shepp (1967), see also Masry (1972). We have also used the work of Martins de Carvalho and Clark (1977, 1981).

It is essentially easier to find the extreme distributions among stationary binary distributions on the line rather than the extreme covariances. It is not true that every extreme distribution gives an extreme covariance. The problem of finding a characterisation of extreme distributions was essentially solved by Hobby and Ylvisaker (1964) and given a precise formulation in the unpublished notes of Weissblum (Zaman, 1981, has also exploited this work independently of the authors in the study of Markov exchangeable sequences). We now explain this theory.

Let $X_0, X_1, X_2, \ldots$ be a stationary binary process. Consider a window of length p+1 covering p+1 adjacent time points, say $t = 0,1,\ldots,p+1$. The process induces distributions at any finite set of points. So define

$$P_{p+1}(x_0,\ldots,x_p) = \text{Prob}(X_0 = x_1,\ldots,X_p = x_p)$$

Then

$$P_p(x_1,\ldots,x_p) = \text{Prob}(X_1 = x_1,\ldots,X_p = x_p)$$

can be expressed in two ways

$$P_p(x_1,\ldots,x_p) = \sum_{x=0,1} P(x,x_1,\ldots,x_p) = \sum_{x=0,1} P(x_1,\ldots,x_p,x) \tag{5}$$

Thus we get a bank of 2^p equations (5) one for each p-tple $(x_1,\ldots,x_p)$. It is notationally convenient in writing out examples to drop the P notation and just write $P(x_1,\ldots,x_p) \equiv x_0\ldots\ldots x_p$ identifying the probability with the binary string itself. Then, for example, for p = 2 we obtain the 4 equations

$$\begin{array}{ccccccc} 0\,0\,0 & + & 1\,0\,0 & = & 0\,0\,0 & + & 0\,0\,1 \\ 0\,0\,1 & + & 1\,0\,1 & = & 0\,1\,0 & + & 0\,1\,1 \\ 0\,1\,0 & + & 1\,1\,0 & = & 1\,0\,0 & + & 1\,0\,1 \\ 0\,1\,1 & + & 1\,1\,1 & = & 1\,1\,0 & + & 1\,1\,1 \end{array} \tag{6}$$

By a "solution" to (5) we mean any collection of $P(x_0,\ldots,x_p) \geq 0$ with $\Sigma P(x_0,\ldots,x_p) = 1$ satisfying the equations. Any such solution is a distribution P on the vertices of the 2^{p+1} unit cube $\bigotimes_{p+1}[0,1]$. The Hobby-Ylvisaker-Weissblum theorem (H-Y-W) then characterises the extreme distributions among all those satisfying (5).

Theorem 1 (H-Y-W). The extreme distributions in the (p+1)-tples amongst all those obtained from stationary binary sequences are precisely all those for periodic sequences whose cycles do not repeat a p-tple (together with a uniform random phase shift).

Proof. We sketch the main ideas of the proof using the p = 2 example. Let $\underset{\sim}{P}$ be a solution to (5). Select one equation for which on the left hand side $P(x_0,\ldots,x_p) > 0$. Label $x^{(1)} = (x_0,\ldots,x_p)$ and $P(x_0,\ldots,x_p) = \alpha_1 > 0$. Then on the right hand side of this equation there must be an $x^{(2)} = (x_1,\ldots,x_p,x')$ with $P(x_1,\ldots,x_p,x') = \alpha_2 > 0$. Then find $x^{(2)}$ again in one (possibly new) equation on the LHS. This means in turn that there is an $x^{(3)} = (x_2,\ldots,x_p,x',x'')$ on the RHS of this (new) equation with $P(x_2,\ldots,x_p,x',x'') = \alpha_3 > 0$. Continue in this way until we have returned to the same equation for the first time. Then we have a cycle

$$x^{(r)},\ x^{(r+1)},\ \ldots\ ,\ x^{(r+\ell_1)}$$

with $x^{(r)}$ and $x^{(r+\ell_1)}$ in the same equation (on LHS). Let

$$\alpha^{(1)} = \min_{1 \leq s \leq \ell_1} \alpha_{r+s}$$

Then "extract" this cycle. That is, express the solution $\underset{\sim}{P}$ as

$$\underset{\sim}{P} = \alpha^{(1)}\ell_1 \ \underset{\sim}{P}_1 + (1 - \alpha^{(1)}\ell_1)\underset{\sim}{P}_1'$$

where $\underset{\sim}{P}_1$ is the distribution which attaches mass $\frac{1}{\ell_1}$ to each of the (p+1)-tples $x^{r+1},\ldots,x^{r+\ell_1}$. Now clearly the "residual" $\underset{\sim}{P}_1'$ also satisfies the equation (5). Thus we may extract another cycle $\underset{\sim}{P}_2$ and write

$$\underset{\sim}{P}_1' = \alpha^{(2)}\ell_2 \ \underset{\sim}{P}_2 + (1 - \alpha^{(2)}\ell_2)\underset{\sim}{P}_2'$$

and so on. Thus P is decomposed into cycles

$$\underset{\sim}{P} = \sum_{i=1}^{M} \beta_i \underset{\sim}{P}_i$$

where $\beta_i \geq 0$ $\Sigma\beta_i = 1$. It is straightforward to show that every such cycle $\underset{\sim}{P}_i$ is extreme. Then, finally, we see that every $\underset{\sim}{P}_i$ is realizable as the (p+1)-window distribution from a stationary ergodic binary sequence. Stringing together the (p+1)-tples we get a cycle of x_i (= 0 or 1) of length ℓ. Take the process which is obtained by taking with probability $\frac{1}{\ell}$ each of the ℓ possible phase positions of the strictly period process formed by repeating the x_i cycle indefinitely. This process has the required $\underset{\sim}{P}_i$. The defining property of the $\underset{\sim}{P}_i$ cycles is that they do not repeat p-tples. Thus we have Theorem 1.

The theorem is close to theorems in transport theory where the (p+1)-tples are nodes and the $P(x_0,\ldots,x_p)$ are flows.

To make Theorem 1 clear consider the p = 2 case. The possible cycles in Theorem 1 are, referring to (6):

$$
\begin{array}{rl}
\text{(i)} & 0\ 0\ 0 \to (0\ 0\ 0) \\
\text{(ii)} & 1\ 1\ 1 \to (1\ 1\ 1) \\
\text{(iii)} & 0\ 1\ 0 \to 1\ 0\ 1 \to (0\ 1\ 0) \\
\text{(iv)} & 0\ 0\ 1 \to 0\ 1\ 0 \to 1\ 0\ 0 \to (0\ 0\ 1) \\
\text{(v)} & 1\ 1\ 0 \to 1\ 0\ 1 \to 0\ 1\ 1 \to (1\ 1\ 0) \\
\text{(vi)} & 0\ 0\ 1 \to 0\ 1\ 1 \to 1\ 1\ 0 \to 1\ 0\ 0 \to (0\ 0\ 1)
\end{array} \qquad (7)
$$

The extreme periodic sequences are (i) 0... (ii) 1... (iii) 01... (iv) 001... (v) 110... (vi) 0011... . These have π-vectors ($p = 2$) (i) and (ii) (1,1),... (iii) (0,1), (iv) and (v) $(\frac{1}{3}, \frac{1}{3})$ and (vi) $(\frac{1}{2}, 0)$.

Notice that the π-vector $(\frac{1}{3}, \frac{1}{3})$ (from (iv) and (v)) is not extreme. A further reduction is possible in an attempt to find the extreme covariances. Consider moving a p+1 window <u>along</u> a sequence and evaluating from each position of the window a contribution to the π-vector. For example for $p = 3$ we might have

...001[1100]011 ...

the contribution from the window [1100] evaluated from the left hand entry is $\frac{1}{N}(1,0,0)$. However the same contribution would be obtained from [0011] (reversing the 0's and 1's). If $\underset{\sim}{x}$ is a (p+1)-tple we call $\bar{\underset{\sim}{x}}$ the (p+1)-tple with the 1's and 0's reversed ($\bar{x}_i \equiv x_i + 1 \bmod 2$). We call $\bar{\underset{\sim}{x}}$ the dual of $\underset{\sim}{x}$. We say, then, that $(1,x_1,\ldots,x_p)$ and $(0,\bar{x}_1,\ldots,\bar{x}_p)$ are in the same equivalence class because they both "contribute" to the same π-vector. That is, given a distribution $\underset{\sim}{P}$ on the (p+1)-tples the π-vector is

$$\pi = \sum_{x_1,\ldots,x_p=0,1} (x_1,\ldots,x_p)\{P(1,x_1,\ldots,x_p) + P(0,\bar{x}_1,\ldots,\bar{x}_p)\}$$

Thus changing all $(0,\bar{x}_1,\ldots,\bar{x}_p)$ in the equations (5) to $(1,x_1,\ldots,x_p)$ we get a $\frac{1}{2}$ reduction of the original equations. We may also delete the initial 1 from each (p+1)-tple to get equations in p-tples. For example the equation (6) becomes

$$
\begin{array}{l}
1\,1 + 0\,0 = 1\,1 + 1\,0 \\
1\,0 + 0\,1 = 0\,1 + 0\,0
\end{array} \qquad (8)
$$

The p-tple $x_1,\dots,x_p$ now means $P(0,x_1,\dots,x_p) + P(0,\bar{x}_1,\dots,\bar{x}_p)$. We then apply Theorem 1 to the reduced system to identify the extreme distributions as cycles in the reduced system. For example the equations (8) have cycles

$$\begin{aligned} &1\ 1 \to (1\ 1) \\ &0\ 1 \to (0\ 1) \\ &0\ 0 \to 1\ 0 \to (0\ 0) \end{aligned} \tag{9}$$

The reduced system cycles can be traces back to a cycle in the original system and hence to a periodic sequence. The resulting sequences have the property that they do not (within a cycle) repeat a p-tple or go through a p-tple and its dual. The reduced cycles (9) correspond to (i) and (ii) (iii), and (vi) of equations (7), respectively. The non-extreme covariance cycles (iv) and (v) have been eliminated by the reduction; (iv), say, gives the reduced cycle

$$1\ 0 \to 0\ 1 \to 0\ 0 \to (1\ 0)$$

which is decomposed into

$$1\ 0 \to 0\ 0 \to (1\ 0) \quad \text{and} \quad 0\ 1 \to (0\ 1).$$

The original sequence (iv) has 01 and 10 in a cycle.

Theorem 2. If the covariance function c(1)...c(p) (or $(\pi_1,\dots,\pi_p)$-vector) of a stationary ergodic binary process is extreme then it must be obtainable from a strictly periodic process whose cycle (i) does not repeat a p-tple and (ii) does not contain a p-tple and its dual.

As a final example in this section consider the case p = 3. The 16 original equations reduce to

$$\begin{aligned} 0\,1\,1 + 1\,0\,0 &= 0\,0\,0 + 0\,0\,1 \\ 0\,1\,0 + 1\,0\,1 &= 0\,1\,0 + 0\,1\,1 \\ 0\,0\,1 + 1\,1\,0 &= 1\,0\,0 + 1\,0\,1 \\ 0\,0\,0 + 1\,1\,1 &= 1\,1\,0 + 1\,1\,1 \end{aligned}$$

The reduced cycles are

$$\begin{array}{l} 1\ 1\ 1 \to (1\ 1\ 1) \\ 0\ 1\ 0 \to (0\ 1\ 0) \\ 0\ 1\ 1 \to 0\ 0\ 1 \to 1\ 0\ 1 \to (0\ 1\ 1) \\ 1\ 0\ 0 \to 0\ 0\ 1 \to (1\ 0\ 0) \\ 1\ 0\ 0 \to 0\ 0\ 0 \to 1\ 1\ 0 \to (1\ 0\ 0) \\ 0\ 1\ 1 \to 0\ 0\ 0 \to 1\ 1\ 0 \to 1\ 0\ 1 \to (0\ 1\ 1) \end{array} \tag{10}$$

The first five are traced back to the sequences in section 1. However the last gives $\pi = (\frac{1}{2}, \frac{1}{2}, \frac{1}{2})$ and is not extreme. Thus Theorem 2 only gives a necessary condition.

The main technique to find the Π-region is to first use the reduction described here to eliminate as many as possible of the points. The rest are tested by a straight forward method to discover whether a "new" point is in the convex hull of the old points. This is essentially the method adopted by Martins de Carvalho and Clark (1981) but without explicitly describing the reduction of Theorems 1 and 2. They find the extreme covariances up to lag $p = 6$. (For $p = 6$ the longest cycle has length 14!)

An important result of Shepp (1967) allows us to explicitly mix together two sequences with π vectors π_1 and π_2 to obtain a new binary sequence with $\pi = (1-\alpha)\pi_1 + \alpha\pi_2$ for any $0 \le \alpha \le 1$. This allows us finally to claim that the Π-region is indeed the convex null of the extreme π. The method consists of putting end to end increasingly large segments of either sequence. Note that such a mixed sequence need not be periodic.

An interesting point is the point $\pi = (\frac{1}{k}, \ldots, \frac{1}{k}) = (\frac{1}{2}, \frac{1}{2}, \ldots, \frac{1}{2})$ when $k = 2$. This is clearly obtainable (with probability one) as the realisation of a Bernoulli sequence. However it can also be derived from a full length cyclic error correcting code (also called pseudo-random sequences or de Bruijn sequences). This π-vector is shown in Kiefer and Wynn (1981) to be minimax. That is, it achieves the minimum over the choice of design sequence of

$$\max_{\rho} \Phi(c)$$

for all Φ in the class of section 1. It guards against extreme processes and seems natural when ρ is completely unknown. We return to pseudo-random schemes in section 4.

4. HIGHER DIMENSIONS

The theory of first and second and higher order adjacencies in the last section can be extended to higher dimensions. We confine ourselves to a straightforward exposition of the combinatorial theory. The statistical motivation here is less well developed. One of the problems with the latter is the general difficulty with defining and elucidating the covariance structure of auto-regressive processes on multi-dimensional lattices. We hope to return to a full generalisation of the statistical theory of Kiefer and Wynn (1981) in a later paper. Even with the combinatorial theory there seems to be an explosion of complexity and computer solutions need to be devised for many of the problems. However the "by hand" constructions given here point the way and we have been able to "solve" the first order problem.

Proceeding as far as possible as in one-dimension we ask first for a generalisation of the equations (5). Let X_{st} be a stationary binary process on the infinite 2-dimensional integer lattice: $s,t = 0, \pm1, \pm2, \ldots$, $X_{st} = 0$ or 1. Stationarity requires that for any subsets $\{i_1,\ldots,i_n\}$ and $\{i_1,\ldots,j_n\}$) $(X_{i_r,j_r}, \ldots, X_{i_n,j_n})$ has the same distribution as $(X_{i_r+i,j_r+j},\ldots,X_{i_n+i,j_n+j})$ for all integer pairs (i,j).

Now consider a window of side $p+1$ containing $(p+1)^2$ entries. Placing this in any position say $i,j = 0,\ldots,p$ we see that the stationary process induces a distribution $\underset{\sim}{P}$ in the window. Thus for $p = 1$ we have a distribution on all 16 2×2 squares:

$$\begin{matrix}0&0\\0&0\end{matrix}\ ,\quad \begin{matrix}1&0\\0&0\end{matrix}\ ,\ \ldots,\quad \begin{matrix}1&1\\1&1\end{matrix}\ . \qquad (11)$$

For example, if the process is completely random: $\text{Prob}(X_{ij} = 1) = \frac{1}{2}$ independently for all i,j then the probability of any square (11) is $\frac{1}{16}$.

Proceeding as in the one-dimensional case let a typical $(p+1) \times (p+1)$ square be written X. Partition X into its first column x' and the rest X' thus

$$\underset{\sim}{X} = [x':X'].$$

Let P(X) be the probability that the induced distribution P assigns to $\underset{\sim}{X}$. Stationarity under horizontal shifts gives us the analogue of equations (5). Namely

$$\text{(H)} \quad P(\underset{\sim}{X}') = \sum_{x'} P[x':X'] = \sum_{x'} P[X':x']$$

We obtain also under vertical shifts and a horizontal partition

$$\underset{\sim}{X} = \begin{bmatrix} y' \\ \cdots \\ Y' \end{bmatrix},$$

$$\text{(V)} \quad P(Y') = \sum_{y'} P\begin{bmatrix} y' \\ Y' \end{bmatrix} = \sum_{y'} P\begin{bmatrix} Y' \\ y' \end{bmatrix},$$

In (H) (and (V)) (which stand for horizontal and vertical) the summation is over all binary p+1-tples x' (and y').

We can use an analogous notation to that used in (6) etc. dropping the P. Thus for $p = 1$ we obtain for (H)

$$\begin{matrix}00\\00\end{matrix}+\begin{matrix}00\\10\end{matrix}+\begin{matrix}10\\00\end{matrix}+\begin{matrix}10\\10\end{matrix}=\begin{matrix}00\\00\end{matrix}+\begin{matrix}00\\01\end{matrix}+\begin{matrix}01\\00\end{matrix}+\begin{matrix}01\\01\end{matrix}$$

$$\begin{matrix}00\\01\end{matrix}+\begin{matrix}00\\11\end{matrix}+\begin{matrix}10\\01\end{matrix}+\begin{matrix}10\\11\end{matrix}=\begin{matrix}00\\10\end{matrix}+\begin{matrix}00\\11\end{matrix}+\begin{matrix}01\\10\end{matrix}+\begin{matrix}01\\11\end{matrix}$$

$$\begin{matrix}01\\00\end{matrix}+\begin{matrix}01\\10\end{matrix}+\begin{matrix}11\\00\end{matrix}+\begin{matrix}11\\10\end{matrix}=\begin{matrix}10\\00\end{matrix}+\begin{matrix}10\\01\end{matrix}+\begin{matrix}11\\00\end{matrix}+\begin{matrix}11\\01\end{matrix}$$

$$\begin{matrix}01\\01\end{matrix}+\begin{matrix}01\\11\end{matrix}+\begin{matrix}11\\01\end{matrix}+\begin{matrix}11\\11\end{matrix}=\begin{matrix}10\\10\end{matrix}+\begin{matrix}10\\11\end{matrix}+\begin{matrix}11\\10\end{matrix}+\begin{matrix}11\\11\end{matrix} \qquad (12)$$

and for (V) the equations with every square replaced by its transpose (we call these (12) (H) and (12) (V)).

Now these two sets of equations (H) and (V) represent <u>necessary</u> conditions for stationarity. We shall investigate distribution $\underset{\sim}{P}$ which satisfy (H) and (V) and are therefore <u>candidates</u> for distributions actually realised under an original process X_{st}. We must emphasize that we have <u>not</u>

proved in general that (H) and (V) are sufficient for a process on the infinite lattice giving such a $\underset{\sim}{P}$ to exist. That is (H) and (V) are only local equations. It is quite conceivable that a particular solution can only be realised by a distribution on some other 2-dimensional manifold. This seems to us a profoundly interesting topological problem but one we shall defer to further research. Fortunately for all the examples considered here we can exhibit processes on the planar (or higher dimensional) lattice with the required solution P. Thus existence will not be a problem.

We shall proceed now largely to examples, and begin with (12) (H) and (V) for p = 1. The extreme distributions are found by the direct analogue of Theorem 1 and consist of (horizontal) cycles which do not visit the LHS of any equation more than once. That is they must not repeat a vertical 2-tple and are of length ≤ 4.

The one-dimensional theory connecting c(r) with π_r goes over to the two-dimensional case in the following sense. Define the values for the four points of the compass: $\pi = (\pi^N, \pi^E, \pi^{NW}, \pi^{NE})$ then the contribution to π from $P(\underset{\sim}{X})$ where

$$X = \begin{bmatrix} x_{11} & x_{12} \\ x_{21} & x_{22} \end{bmatrix}$$

is

$$P(x)(1+x_{11}+x_{21},\ 1+x_{11}+x_{12},\ 1+x_{12}+x_{21},\ 1+x_{11}+x_{22})$$

where the vector is reduced (mod 2). The contribution of $\begin{bmatrix} 1 & x_{12} \\ x_{21} & x_{22} \end{bmatrix}$ is the same as that of $\begin{bmatrix} 0 & \bar{x}_{12} \\ \bar{x}_{21} & \bar{x}_{22} \end{bmatrix}$, its dual. Thus we can form reduced equations by replacing every square in (12) H which has $x_{11} = 0$ by its dual. We get 4 equations:

$$\text{(H')}\quad \begin{array}{l} \begin{matrix}10\\00\end{matrix} + \begin{matrix}10\\10\end{matrix} + \begin{matrix}11\\01\end{matrix} + \begin{matrix}11\\11\end{matrix} = \begin{matrix}10\\10\end{matrix} + \begin{matrix}10\\11\end{matrix} + \begin{matrix}11\\10\end{matrix} + \begin{matrix}11\\11\end{matrix} \\[2ex] \begin{matrix}11\\00\end{matrix} + \begin{matrix}11\\10\end{matrix} + \begin{matrix}10\\01\end{matrix} + \begin{matrix}10\\11\end{matrix} = \begin{matrix}10\\00\end{matrix} + \begin{matrix}10\\01\end{matrix} + \begin{matrix}11\\00\end{matrix} + \begin{matrix}11\\01\end{matrix} \end{array}$$

We now need only consider cycles of length 2. They can easily be listed together with their π-vectors

Cycle	π-vector
$\begin{matrix}11\\11\end{matrix} \to \begin{bmatrix}11\\11\end{bmatrix}$	$(1, 1, 1, 1)$
$\begin{matrix}10\\10\end{matrix} \to \begin{bmatrix}10\\10\end{bmatrix}$	$(1, 0, 0, 0)$
$\begin{matrix}11\\00\end{matrix} \to \begin{bmatrix}11\\00\end{bmatrix}$	$(0, 1, 0, 0)$
$\begin{matrix}10\\01\end{matrix} \to \begin{bmatrix}10\\01\end{bmatrix}$	$(0, 0, 1, 1)$
$\begin{matrix}10\\00\end{matrix} \to \begin{matrix}11\\10\end{matrix} \to \begin{bmatrix}10\\00\end{bmatrix}$	$(\frac{1}{2}, \frac{1}{2}, 0, 1)$
$\begin{matrix}11\\01\end{matrix} \to \begin{matrix}10\\11\end{matrix} \to \begin{bmatrix}11\\01\end{bmatrix}$	$(\frac{1}{2}, \frac{1}{2}, 1, 0)$
$\begin{matrix}10\\00\end{matrix} \to \begin{matrix}10\\11\end{matrix} \to \begin{bmatrix}10\\00\end{bmatrix}$	$(\frac{1}{2}, \frac{1}{2}, \frac{1}{2}, \frac{1}{2})$
$\begin{matrix}11\\01\end{matrix} \to \begin{matrix}11\\10\end{matrix} \to \begin{bmatrix}11\\01\end{bmatrix}$	$(\frac{1}{2}, \frac{1}{2}, \frac{1}{2}, \frac{1}{2})$

Now the cycles for the corresponding vertical reduced equations (V') will give exactly the same set of π-vectors. Thus the extreme π-vectors for solution (12) (H) and (V) are precisely the first 6 in the above list.

To show existence we need to trace back to a process on the lattice which gives each π. But we can easily exhibit such processes they are period patterns in the plane (with a random position if we want the stationary process). We list below the standard pattern for each, it being understood that the patterns are to be repeated indefinitely vertically and horizontally (we exclude dual patterns)

pattern	π-vector
1 1 1 1	(1, 1, 1, 1)
1 1 0 0	(0, 1, 0, 0)
1 0 1 0	(1, 0, 0, 0)
1 0 0 1	(0, 0, 1, 1)
1 1 0 0 1 0 0 1 0 0 1 1 0 1 1 0	($\frac{1}{2}$, $\frac{1}{2}$, 0, 1)
1 1 0 0 0 1 1 0 0 0 1 1 1 0 0 1	($\frac{1}{2}$, $\frac{1}{2}$, 1, 0)

A generalisation of the Shepp result on mixing allows us to claim that the full Π-set is the convex hull of the 6 extreme π-vectors. These extreme patterns (processes) do not seem to have been exhibited before.

This whole analysis can be extended to higher dimensions. Consider the case of three dimensions and p = 1. Now the method works with little cubes of side 2 and 8 elements $\{x_{ijk}\}$. We will have vertical, horizontal and "sideways" (S) equations obtained by "moving" the cubes along the three axes. Each block of equations will have 16 equations one for each two dimension planar square perpendicular to the axis in question. Even with the "reduction" this still leaves 8 equations with 16 terms on either side. We shall not present the full solution here, rather we shall show how symmetry conditions can be used to simplify the problem; another kind of reduction.

Let us first generalise the "points of the compass" directions to 3 dimensions. Label a π-coordinate by its direction measured from (0, 0, 0). That is $\pi(i,j,k)$ is the coordinate in the direction $(0, 0, 0) \leftrightarrow (i,j,k)$ where $i,j,k = 0,\pm1$, $(i,j,k) \neq (0, 0, 0)$, and $\pi(i,j,k) = \pi(-i,-j,-k)$. Thus there are 13 possible π-coordinates: 3 axial directions, 3 pairs of short (planar) diagonals and four long diagonals. For m-dimensions there are $\frac{1}{3}(3^m-1)$ coordinates (= 4 when m = 2). The finite Euclidean symmetry condition is that

$$\pi(i,j,k) = \pi(|i|,|j|,|k|) \tag{13}$$

Another way of considering this is that if $i,j,k = 0,1$ and $i',j',k' = 0,1$ then the π in the direction $(i,j,k) \leftrightarrow (i',j',k')$ is the same as that $(i+r, j+s, k+t) \leftrightarrow (i'+r, j'+s, k'+t)$ for any $r,s,t = 0,1$ where all coordinates are reduced (mod 2). Thus the π's are invariant over the finite Euclidean group over GF(2). This reduces the distinct number of coordinates to 7: 3 axial direction, 3 planar diagonals, and one large diagonal (in general 2^m-1 coordinates).

This condition leads to a spectacular reduction in the equations (H)'. We now show that any solution to (H)' ((V') or (S')) with finite Euclidean symmetry can be decomposed into cycles of maximum length 2. The proof runs as follows (and applies to any number of dimensions). Consider a cycle of 2×2×2 cubes in the (H) direction

$$X_1 \to X_2 \to X_3 \; . \; . \; . \to X_L \to (X_1) \tag{14}$$

Then consider the 2-cycle

$$X_i \to \tilde{X}_i \to (X_i)$$

in the horizontal direction, where $\tilde{X}_i$ is the reflection of X_i in its leading 2×2 plane. Then we can realise the same π-vector as (14) by mixing together the 2-cycles with equal proportions $(i = 1,\dots,L)$. Furthermore each such 2-cycle distribution can actually be realised in 3 dimensions by a pattern which repeats X_i and its reflection $\tilde{X}_i$ alternatively in every axial directions. Now the contribution of X_i and $\tilde{X}_i$

to a π-vector with finite Euclidean symmetry is the same. Thus the set of all distributions is just those obtained from all cubes X_i. The extreme subset is easily found. Below, then, is listed the extreme π-vectors with Euclidean symmetry with the corresponding cube (of zeros and ones). The vertices of the cube are listed in order

(0,0,0), (1,0,0), (0,1,0), (0,0,1), (1,1,0), (1,0,1), (0,1,1), (1,1,1)

and the π-vector coordinates also in the corresponding standard order $\pi((0,0,0) \to (i,j,k))$ in the position of (i,j,k) in the above list (excluding $i = j = k = 0$).

Cube	π-vector	
(1,1,1,1,1,1,1,1)	(1,1,1,1,1,1,1)	
(1,1,1,0,1,0,0,0)	(1,1,0,1,0,0,0)	
(1,1,0,1,0,1,0,0)	(1,0,1,0,1,0,0)	
(1,0,1,1,0,0,1,0)	(0,1,1,0,0,1,0)	
(1,1,0,0,0,0,1,1)	(1,0,0,0,0,1,1)	
(1,0,1,0,0,1,0,1)	(0,1,0,0,1,0,1)	
(1,0,0,1,1,0,0,1)	(0,0,1,1,0,0,1)	
(1,0,0,0,1,1,1,0)	(0,0,0,1,1,1,1)	(15)

(Notice that the cubes are the complete 2^3 factorial plus all $\frac{1}{2}$ orthogonal fractions hinting at a kind of 2^m spatial factorial theory.)

We may impose a further symmetry, namely rotation invariance

$$\pi(i,j,k) = h(i^2 + j^2 + k^2)$$

for some function h. This is called isotropy for spatial processes. The extreme π under this restriction are quickly obtained from the Euclidean case. There are just three distinct coordinates (m in general): one axial, one short-diagonal and one long-diagonal. We merely average the coordinates in (15) appropriately. The extreme isotropic π-vectors are

$$(1,\ 1,\ 1),\quad (\tfrac{2}{3},\ \tfrac{1}{3},\ 0),\quad (\tfrac{1}{3},\ \tfrac{1}{3},\ 1),\quad (0,\ 1,\ 0).$$

Notice that these are not the same set as for p = 3 in one-dimension, $((\frac{1}{2}, 0, \frac{1}{2})$ is missing). These extreme π-vectors (covariance functions) for an isotropic process in 3-dimensions seem to be new. To obtain the actual processes we must mix together the patterns for the Euclidean case (and set $c(r) = 2\pi_r - 1$).

5. PSEUDO-RANDOM PATTERNS

In Kiefer and Wynn (1981) we mention how full length error correcting codes can be used to generate sequences with $\pi_r = \frac{1}{k}$ $(r = 1,\ldots,p)$ that is the same π-values as would be obtained (with probability one) by an infinite Bernoulli sequence. Briefly we take a primitive polynomial h(x) of degree p+1 over GF (q^m) where q is prime. That is all the elements of the cyclic group of the polynomial field GF (k^p) (where $k = q^m$) can be generated as powers of a solution x (in $GF(k^p)$) of $h(x) = 0$. Then take h(x) and use it as a shift register to generate a full length code. Add an additional zero and repeat the code indefinitely.

For example take q = 2, m = 2, k = 4. Label the elements of $GF(2^2)$: {0, 1, W, W+1}, where we assume that $w^2+w+1 = 0$. Then we seek a primitive polynomial with coefficients in $GF(2^2)$ of degree 2 (we take p = 1). We may take $z^2 + wz + w = 0$. The shift register then says $V_t = V_{t-1} + V_{t-2}$ where $V_t = 0,1,W$ or W+1, and arithmetic is over the addition group of $GF(2^2)$. This leads to the code:

$$0\ 1\ w\ 1\ 1\ 0\ w\ (w+1)\ w\ w\ 0\ (w+1)1(w+1)(w+1)\ . \quad (16)$$

Add an addition 0 and repeat. We get a first order pseudo-random sequence, every member of $GF(2^2)$ appears next to each other an equal number of times. This can be seen by moving a window of length 2 along the sequence. The process obtained by giving this sequence a random phase shift is locally uncorrelated.

A trivial method of extending the $\pi_r = \frac{1}{k}$ property to higher dimensions is to take the Kronecker sum of the codes with respect to the group $GF(q^m)$ (or any group on the set of q^m elements). That is for two dimensions if the 2-dimension pattern is V_{ij}

$$V_{ij} = V_i + V_j \quad (\text{over } GF(q^m))$$

Using the cyclic group on four elements (16) leads to

A	A	B	C	B	B	A	C	D	C	C	A	D	B	D	D
A	A	B	C	B	B	A	C	D	C	C	A	D	B	D	D
B	B	C	D	C	C	B	D	A	D	D	B	A	C	A	A
C	C	D	A	D	D	C	A	B	A	A	C	B	D	B	B
B	B	C	D	C	C	B	D	A	D	D	B	A	C	A	A
B	B	C	D	C	C	B	D	A	D	D	B	A	C	A	A
A	A	B	C	B	B	A	C	D	C	C	A	D	B	D	D
C	C	D	A	D	D	C	A	B	A	A	C	B	D	B	B
D	D	A	B	A	A	D	B	C	B	B	D	C	A	C	C
C	C	D	A	D	D	C	A	B	A	A	C	B	D	B	B
C	C	D	A	D	D	C	A	B	A	A	C	B	D	B	B
A	A	B	C	B	B	A	C	D	C	C	A	D	B	D	D
D	D	A	B	A	A	D	B	C	B	B	D	C	A	C	C
B	B	C	D	C	C	B	D	A	D	D	B	A	C	A	A
D	D	A	B	A	A	D	B	C	B	B	D	C	A	C	C
D	D	A	B	A	A	D	B	C	B	B	D	C	A	C	C

This technique does not in general lead to windows which have the 2×2 window property. Not every possible $(2^2)^4$ window "view" appears equally often. The harder problem of extending the window property to higher dimensions has been studied by MacWilliams and Sloane (1976), Van Lint, MacWilliams and Sloane (1979) and Gilbert (1980). Note however that this theory being based on codes usually excludes the zero view (all entries zero) and therefore produces a process with small constant negative autocovariance. The theory does however produce many interesting examples of two-dimensional patterns which simultaneously have the window property for windows of different size, e.g. 4×1, 2×2, 1×4.

Finally we return to the $k = 2$, $p = 1$ two-dimensional problem solved in the last section, (13). A two-dimensional pattern with $\pi = (\frac{1}{2}, \frac{1}{2}, \frac{1}{2}, \frac{1}{2})$ is easily seen by taking the Kronecker sum of the code 011(0). Rearranging this gives (repeated indefinitely)

1 1 0 0
1 1 0 0
0 0 1 1
0 0 1 1

However this does not have the 2×2 window property (there is a predominance of the square $\begin{smallmatrix}1 & 0\\ 0 & 1\end{smallmatrix}$). The following remarkable 4×4 square has the 2×2 window property (including the zero view).

1 1 1 0
1 1 0 1
0 1 0 0
1 0 0 0

Van Lint et al (1979) report that it is not possible to find a 4×16 array with the 4×1, 2×2 and 1×4 window properties. Note that our square does not have the 2×1 and 1×2 window property.

In Memoriam

The second author would like to express his great sadness at the death of Professor Jack Kiefer. This paper was prepared directly from rough notes written in July 1981.

Acknowledgements

Thanks are due to David Brillinger, Perci Diaconis and Larry Shepp for help with the literature in coding and communication theory. The research was partially supported by National Science Foundation Grant MCS78-25301.

REFERENCES

Gilbert, E.N. (1980). Random colorings of a lattice of squares in the plane. Siam., J., Alg. Disc. Math. 2, 152-159.

Hobby, C. and Ylvisaker, N.D. (1964). Some structure theorems for stationary probability measures on finite state spaces. Ann. Math. Statist., 35, 550-556

Kiefer, J. (1960). Optimum experimental designs, V, with applications to systematic and rotatable designs. Proc. Fourth Berk. Symp., 2, 381-405.

Kiefer, J. and Wynn, H.P. (1980). Optimum balanced block and Latin square designs for correlated observations. Ann. Statist., 9, 737-757.

Kiefer, J. and Wynn, H.P. (1981). Optimum and minimax exact treatment designs for one-dimensional autoregressive error processes. Ann. Statist., (submitted).

MacWilliams, F.J. and Sloane, N.J.A. (1976). Pseudo-random sequences and arrays. Proceedings I.E.E.E. 64, 1715-1729.

Martin, R.J. (1981). Some aspects of experimental design and analysis when errors are correlated. Biometrika, (to appear).

Martins de Carvalho, J.L. and Clark, J.M.C. (1981). Characterising the autocorrelations of binary sequences. Mimeo, Imperial College, London.

Masry, E. (1972). On covariance functions of unit processes. Siam. J. Appl. Math. 23, 28-33.

Shepp, L.A. (1967). Covariances of unit processes. Proc. Working Conference Stochastic Processes, Santa Barbara, California.

Williams, R.M. (1952). Experimental designs for serially correlated observations. Biometrika, 39, 151-167.

Van Lint, J.H., MacWilliams, F.J. and Sloane, N.J.A. On pseudo-random arrays. Siam. J. Appl. Math. 36, 62-72.

Zaman, A. (1981). An approximation theorem for finite Markov exchangeability. Technical Report 176, Department of Statistics, Stanford University.

Department of Statistics
University of California at Berkeley
Berkeley, CA 94720

Department of Mathematics
Imperial College
South Kensington
London S.W. 7, England

Journal of Statistical Planning and Inference 8 (1983) 103–116
North-Holland

COMPARISON OF DESIGNS EQUIVALENT UNDER ONE OR TWO CRITERIA*

Z. GALIL*

Tel-Aviv University, Tel-Aviv 69978, Israel

J. KIEFER**

University of California, Berkeley, CA 94720, USA

Received 27 October 1981; revised manuscript received 23 July 1982
Recommended by J. Sacks

Abstract: Two designs equivalent under one or two criteria may be compared under other criteria. For certain configurations of eigenvalues of the information matrices, we decide which design is the better of the two for many other such criteria. The relationship to universal optimality (in the case of equivalence under one criterion) is indicated. For two criteria, applications are given to weighing and treatment-with-covariate settings.

AMS Subject Classification: Primary 62K05, 62K15.

Key words and phrases: Optimum designs; Design comparison; Universal optimality; Φ_p-optimality; Weighing designs.

1. Introduction

In a linear model design of experiments setting, suppose $\mathcal{M}=\{M_d\}$ is the class of $m\times m$ information matrices for available designs d. Thus, if all parameters are estimable, the covariance matrix of the best linear unbiased estimators of parameters is assumed to be cM_d^{-1} if d is used, where c is the same scalar for all d. This includes the exact and approximate theories, but our examples will be of the former kind. Estimation of a subset of the parameters can be considered similarly.

A first task of strict optimality theory is to characterize, for a given functional Φ on $\mathcal{M}$, the Φ-optimum designs, defined to be those that minimize $\Phi(M_d)$ over $\mathcal{M}$. A secondary task might then be, given several Φ-optimum designs, to choose

* Research supported by National Science Foundation Grant MCS 78-25301. Research also supported in part by the Israel Commission of Basic Research.

** Research supported by National Science Foundation Grant MCS 78-25301 and in part by the Miller Research Institute.

Professor Kiefer died on August 10, 1981.

among them on the basis of a secondary criterion $\tilde{\Phi}$. We now indicate why one can view the methods often used for solving these two tasks as being variants of the same technique.

Let $\lambda_d = (\lambda_{d1}, \ldots, \lambda_{dm})$ be the vector of eigenvalues of M_d in nondecreasing order. For simplicity, we consider here criteria of the form (written as a function of λ_d)

$$\Phi_h(\lambda_d) = \sum_1^m h(\lambda_{di}). \tag{1.1}$$

Other Φ such as limits of sequences of such criteria, and Schur convex functions, can also be treated. Suppose $\bar{f}$ is strictly monotone on an interval U containing all λ_{di} $\forall d, i$, and that $\bar{g}$ on U is such that $g = \bar{g} \circ \bar{f}^{-1}$ is convex. Write $\mu_{di} = \bar{f}(\lambda_{di})$. As usual, we say $\mu_{d'}$ majorizes $\mu_{d''}$ if $\sum_1^k (\mu_{d'i} - \mu_{d''i}) \le 0$ for $1 \le k < m$ and $= 0$ for $k = m$. Since $\Phi_g(\mu_d) = \Phi_{\bar{g}}(\lambda_d)$, we have

$$\mu_{d'} \text{ majorizes } \mu_{d''} \quad \Rightarrow \quad \Phi_{\bar{g}}(\lambda_{d''}) \le \Phi_{\bar{g}}(\lambda_{d'}). \tag{1.2}$$

(See, e.g., Marshall and Olkin (1979).) If, additionally, $\bar{f}$ is nondecreasing and g is nonincreasing, we have

$$\mu_{d'} \text{ majorizes } c\mu_{d''} \text{ for some } c \le 1 \quad \Rightarrow \quad \Phi_{\bar{g}}(\lambda_{d''}) \le \Phi_{\bar{g}}(\lambda_{d'}). \tag{1.3}$$

In particular, if $\bar{f}$ is identical to e and we consider $\lambda_{d''}$ with all components equal and maximum trace, we have

$$\text{all } \lambda_{d''i} \text{ equal and } \sum_i \lambda_{d''i} = \max_d \sum_i \lambda_{di}$$

$$\Rightarrow \quad \Phi_g(\lambda_{d''}) \le \Phi_g(\lambda_d) \ \forall d \text{ and convex nonincreasing } g. \tag{1.4}$$

The last is a special case (in the restriction of the form of Φ and the simplifying feature of estimating all parameters) of the simple universal optimality criterion of Kiefer (1958, 1971, 1975); this criterion was extended in Cheng's more intricate development (1978a).

Referring to (1.2), or to (1.3) in the simplest case $c = 1$, we see that we are comparing, according to Φ_g, two designs with equal $\Phi_{\bar{f}}$-values. Although one usually does not think of Φ_e as an optimality criterion, but only as a tool for use in the universal optimality condition (1.4), the arithmetic in (1.2) and (1.3) is seen to be the same whether or not $\bar{f} = e$.

We now illustrate the use of (1.2) for a comparison of designs when $\bar{f} \ne e$, in a weighing design example. The allowable M_d are of the form $X_d' X_d$ where $m \times m$ X_d consists entirely of entries 1, -1 or 0. Suppose $m \equiv 2 \pmod 4$. It is then known from the work of Ehlich (1964a) that, if $X_{d'}$ exists for which $M_{d'}$ consists of two diagonal $\frac{1}{2}m \times \frac{1}{2}m$ blocks, each with diagonal elements m and off-diagonal elements 2, where $M_{d'}$ is 0 outside the blocks, then $M_{d'}$ is D-optimum, i.e. it is $\Phi_{\bar{g}}$-optimum with $\bar{g}(x) = -\log x$. (It is uniquely so modulo isomorphisms.) On the other hand, C.S. Cheng has pointed to us that $M_{d''} = (m-1)I_m$ is E-optimum (i.e., minimizes the maximum eigenvalue of M_d^{-1}) and that d' and d'' yield the same value of $\Phi_{\bar{g}_1}$ where $\bar{g}_1(x)=1/x$. How does this fit into (1.2)? To see this, we first note that the criteria being considered here are members of the family of 'Φ_p-criteria',

$$\begin{aligned} \Phi_p(\lambda_d) &= \left(m^{-1}\sum^{i} \lambda_{di}^{-p}\right)^{1/p}, \quad p\neq 0, \\ \Phi_0(\lambda_d) &= \left(\prod^{i} \lambda_{di}\right)^{-1/m}, \\ \Phi_\infty(\lambda_d) &= \lim_{p\to\infty} \Phi_p(\lambda_d) = 1/\lambda_{d1}, \\ \Phi_{-\infty}(\lambda_d) &= \lim_{p\to-\infty} \Phi_p(\lambda_d) = 1/\lambda_{dm}. \end{aligned} \tag{1.5}$$

(Here the criteria are normalized so that they all equal c^{-1} if all $\lambda_{di}=c$, but this normalization is not relevant to our development.) Note that, for $-\infty<p<\infty$, Φ_p is strictly increasing in $\Phi_{\bar{g}_p}$ where $\bar{g}_p(x)=(\operatorname{sgn} p)x^{-p}$ for $p\neq 0$ and $-\log x$ for $p=0$. The use of the notation Φ_h of (1.1) and Φ_p of (1.5) should cause no confusion.

In Cheng's example, neither of $\lambda_{d'}$, $\lambda_{d''}$ majorizes the other. But $\Phi_{\bar{g}_1}(\lambda_{d'})=\Phi_{\bar{g}_1}(\lambda_{d''})$, so that with $\bar{f}(x)=1/x$ one can see that $\mu_{d'}$ majorizes $\mu_{d''}$. Also, $\bar{g}_p(\bar{f}(x))$ is convex for $p>1$ and concave for $p<1$. Thus, (1.2) tells us that d'' is Φ_p-better than d' for $p>1$ and Φ_p-worse for $p<1$ (the values $p=\pm\infty$ being obvious). This conclusion does not tell us about Φ_p-optimality of either design for any p, but shows how their Φ_1-equivalence and the simple structure of $\lambda_{d''}$ yields the comparison of the two designs.

The developments in the remainder of the present paper were motivated by weighing design settings in which two designs d' and d'' were both found to be optimum for a single criterion (Φ_0, with which we were mainly concerned). At first glance this is an example of the type considered above, but examination shows difference in two respects: on the one hand, there is no majorization in the examples we encountered, even after transformation by $\bar{f}$ as above; on the other hand, there is in fact a second criterion, Φ_{-1}, with respect to which the two designs are equivalent. The latter is easy to overlook (it is Φ_e, discussed earlier) because it merely reflects the trivial fact that all design information matrices M_d in any one of the problems of Section 5.1 below have the same trace. But it is the feature of equivalence under *two* criteria that allows us to compare d' and d'' relative to many *other* criteria for certain eigenvalue configurations, without majorization. In fact, having seen this, we carry out the comparison when the two criteria (under which d' and d'' are equivalent) are fairly general, not merely in the most common setting of Φ_{-1} and one other Φ_p. (The examples of Sections 4 and 5 show that it is not common to encounter designs equivalent under three of the criteria (1.5), in our settings.)

2. Basic inequalities

In what follows, differentiablity of a function at an end point of an interval refers to a one-sided derivative.

Lemma 1. *Suppose f and g are real C^3 (thrice continuously differentiable) functions on an interval U containing $x<y<z$, and that, on U,*

$$f''>0, \qquad g''/f'' \text{ is decreasing.} \tag{2.1}$$

Then

$$\begin{aligned}\psi(x, y, z; f, g) := [f'(z)-f'(y)]g'(x) + [f'(y)-f'(x)]g'(z) \\ + [f'(x)-f'(z)]g'(y) < 0.\end{aligned} \tag{2.2}$$

Proof. Fix x and y. Then, for $z>y$,

$$\begin{aligned}\frac{\partial\psi}{\partial z} &= f''(z)g'(x) + [f'(y)-f'(x)]g''(z) - f''(z)g'(y) \\ &< f''(z)\{g'(x) + [f'(y)-f'(x)]g''(y)/f''(y) - g'(y)\} \\ &= f''(z)h(x, y) \quad \text{(say)},\end{aligned} \tag{2.3}$$

since $f'(y)-f'(x)>0$ and $g''(y)/f''(y)>g''(z)/f''(z)$. Clearly $h(x, y)\to 0$ as $y\downarrow x$, and for $y>x$

$$\frac{\partial h}{\partial y} = [f'(y)-f'(x)]\frac{\partial}{\partial y}[g''(y)/f''(y)] < 0. \tag{2.4}$$

Hence $h(x, y)<0$ for $x<y$ and thus, from (2.3), $\partial\psi/\partial z<0$ for $z>y$. Since $\psi\to 0$ as $z\downarrow y$, (2.2) is proved. □

Lemma 2. *With f and g as in Lemma* 1 *except that* (2.1) *is replaced by*

$$f''>0, \qquad g''/f'' \text{ is increasing,} \tag{2.5}$$

we have

$$\psi(x, y, z; f, g) > 0. \tag{2.6}$$

Proof. The inequalities in (2.3) and (2.4) in the proof of Lemma 1 are reversed.

Finally, the following is obvious, upon replacing f by $f \operatorname{sgn} f''$.

Corollary 1. *If f and g are real C^3 functions on an interval U containing $x<y<z$ and f'' is of one sign on U while g''/f'' is strictly monotone, then*

$$\psi(x, y, z; f, g) \begin{Bmatrix} > \\ < \end{Bmatrix} 0 \quad \text{if } (g''/f'')\operatorname{sgn} f'' \text{ is } \begin{Bmatrix} \text{increasing} \\ \text{decreasing} \end{Bmatrix}. \tag{2.7}$$

3. Comparison of designs equivalent under two criteria

We suppose three real-valued functions $\bar{f}$, $\bar{\bar{f}}$, $\bar{g}$ to be defined on an interval J in which lie all eigenvalues of the two information matrices of the two designs d', d'' to be compared according to the criteria Φ_h of (1.1), where $h=\bar{f}$, $\bar{\bar{f}}$ or $\bar{g}$, and $\lambda_d=(\lambda_{d1},\dots,\lambda_{dm})$ again consists of the nondecreasingly ordered eigenvalues of design d $(=d'$ or $d'')$. For the sake of comparing d' and d'' according to any Φ_h, we may suppose $\lambda_{d'i}\neq\lambda_{d''j}$ $\forall i,j$, or delete a common value from each set. (Such deletions are not made when we consider, as we shall, limits of sequences of increasing functions of criteria of the form (1.1), such as Φ_∞ of (1.5).) A design d' is said to be Φ-better than d'' if $\Phi(\lambda_{d'})<\Phi(\lambda_{d''})$.

We are interested in comparing $\Phi_{\bar{g}}(\lambda_{d'})$ with $\Phi_{\bar{g}}(\lambda_{d''})$ when

$$\Phi_{\bar{f}}(\lambda_{d'})=\Phi_{\bar{f}}(\lambda_{d''}) \quad\text{and}\quad \Phi_{\bar{\bar{f}}}(\lambda_{d'})=\Phi_{\bar{\bar{f}}}(\lambda_{d''}). \tag{3.1}$$

It is simplifying to assume, and usually the case in practice, that $\bar{f}$ is continuous and strictly monotone. We assume that and write $\mu_{di}=\bar{f}(\lambda_{di})$ and $g=\bar{g}\circ\bar{f}^{-1}$, $f=\bar{\bar{f}}\circ\bar{f}^{-1}$. Then (3.1) becomes

$$\sum \mu_{d'i}=\sum \mu_{d''i}, \tag{3.2a}$$

$$\sum f(\mu_{d'i})=\sum f(\mu_{d''i}), \tag{3.2b}$$

and we want to compare $\sum g(\mu_{di})$ for $d=d'$ and d''. We will work with this transformed problem, assuming f and g defined on an interval I containing all μ_{di} under consideration, and with $\mu_{d'i}\neq\mu_{d''j}$ $\forall i,j$. Although f and g are typically monotone in practice, we need not assume that.

To make the development easier to follow, we relabel

$$\mu_{d'}=(A'_1,\dots,A'_t,B'_1,B'_1,\dots,C'_s)$$

and similarly

$$\mu_{d''}=(A''_1,\dots,C''_s),$$

both in nondecreasing order. Thus $m=t+s+2$. We shall compare d' and d'' only when

$$\begin{aligned} &\mu_{d'1}<\mu_{d''1}, && \mu_{d'm}<\mu_{d''m},\\ &A'_i<A''_i \ \ \forall i, && C'_j<C''_j \ \ \forall j. \end{aligned} \tag{3.3}$$

Without this form of ordering, the comparison between d' and d'' is inconclusive even in the simple examples that illustrate and motivate this work; this will be discussed just before Corollary 3. The first line of (3.3) can be rephrased

$$t>0 \text{ if } m\geq 3, \quad s>0 \text{ if } m>3, \quad B''_2>B'_2 \text{ if } m=3.$$

The reason for assuming this in addition to the second line of (3.3) is that, if the second line were satisfied but the first were not (for any labeling), we would necessarily have one of $\mu_{d'}, \mu_{d''}$ majorizing the other, which is impossible by (3.2) under our assumption below that f'' is of constant sign. Of course, that case of majorization is the one in which designs were easily compared for usual criteria in Section 1. With $t>0$ and $s>0$, the two values $\operatorname{sgn}(B_i'-B_i'')$ are restricted only by the fact that at least one is $+1$, by (3.2a). The simple case $m=3$ is considered further in Lemma 3.

Theorem 1. *Suppose f and g are real C^3 functions on an interval U containing all $\mu_{d'i}$ and $\mu_{d''i}$. Furthermore, suppose f'' is of one sign on U and that* (3.2) *and* (3.3) *are satisfied. Then*

$$\Phi_g(\mu_{d''})-\Phi_g(\mu_{d'})\begin{Bmatrix}<\\>\end{Bmatrix}0$$

$$\text{if } (g''/f'')\operatorname{sgn} f'' \text{ is strictly } \begin{Bmatrix}\text{decreasing}\\ \text{increasing}\end{Bmatrix} \text{ on } U. \tag{3.4}$$

Proof. We shall only prove the part assuming $f''>0$. We begin by defining a smooth curve C from $\mu_{d'}$ to $\mu_{d''}$. In $\mathbb{R}^m$, let

$$Q=\left\{\mu:\ \sum_1^m f(\mu_{di})\le \sum_1^m f(\mu_{d'i})\right\}.$$

This is a convex set that contains the points $\mu_{d'}$ and $\mu_{d''}$ on its boundary q; and the interior of the line segment L between these points is contained in the interior of Q, by strict convexity of f. The point p with all coordinates equal to $m^{-1}\sum_1^m \mu_{d'i}$ is in the interior of Q for the same reason. Every halfline from p through a point of L intersects q in exactly one point. This collection of points, the projection of L onto q from p, constitutes the curve C, which has endpoints $\mu_{d'}$ and $\mu_{d''}$. All points μ of C have ordered coordinates ($\mu_i<\mu_{i+1}$ $\forall i$) and $\sum_1^m f(\mu_{di})=\sum_1^m f(\mu_{d'i})$. Moreover, since C is the projection of L, as one moves along C from $\mu_{d'}$ to $\mu_{d''}$ the first t and last s coordinates of μ increase strictly, because of (3.3).

We now prove the first half of (3.4), still assuming $f''>0$, by showing that, from any point μ of $C\setminus\mu_{d''}$, moving (infinitesimally) on C toward $\mu_{d''}$ decreases $\sum_1^m g(\mu_i)$; the remainder of the theorem is proved similarly, using Corollary 1 in place of Lemma 1 in the proof. Write

$$\mu=(A_1,\ldots,A_t,B_1,B_2,C_1,\ldots,C_s)$$

and, for the nearby point on C to which we move,

$$\mu^*=(A_1+\alpha_1,\ldots,A_t+\alpha_t,B_1+\beta_1,B_2+\beta_2,C_1+\gamma_1,\ldots,C_s+\gamma_s).$$

Then it follows from the previous paragraph that

$$\alpha_i>0\quad \forall i,\qquad \gamma_j>0\quad \forall j. \tag{3.5}$$

From $\sum_1^m \mu_i = \sum_1^m \mu_i^*$ and $\sum_1^m f(\mu_i) = \sum_1^m f(\mu_i^*)$ we obtain, as all α_i, β_h and $\gamma_j \downarrow 0$,

$$\begin{aligned} &\sum_i \alpha_i + \sum_h \beta_h + \sum_j \gamma_j = 0, \\ &\sum_i \alpha_i f'(A_i) + \sum_h \beta_h f'(B_h) + \sum_j \gamma_j f'(C_j) \sim 0, \end{aligned} \tag{3.6}$$

and hence

$$\begin{aligned} \beta_1 &\sim \frac{\sum_i \alpha_i [f'(A_i) - f'(B_2)] + \sum_j \gamma_j [f'(C_j) - f'(B_2)]}{f'(B_2) - f'(B_1)}, \\ \beta_2 &\sim \frac{\sum_i \alpha_i [f'(A_i) - f'(B_1)] + \sum_j \gamma_j [f'(C_j) - f'(B_1)]}{f'(B_1) - f'(B_2)}. \end{aligned} \tag{3.7}$$

Using (3.7) to substitute for β_1 and β_2 in the second form just below, and recalling the definition (2.2) of ψ, we have

$$\begin{aligned} &\sum_1^m g(\mu_i^*) - \sum_1^m g(\mu_i) \sim \sum_i \alpha_i g'(A_i) + \sum_h \beta_h g'(B_h) + \sum_j \gamma_j g'(C_j) \\ &\quad = [f'(B_2) - f'(B_1)]^{-1} \{ \sum_i \alpha_i \psi(A_i, B_1, B_2; f, g) + \sum_j \gamma_j \psi(B_1, B_2, C_j; f, g) \}. \end{aligned} \tag{3.8}$$

This is negative by Lemma 1 and (3.5). □

An obvious corollary is the following.

Corollary 2. *If $\Phi^* = \lim_j F_j \circ \Phi_{g_j}$ on an interval U containing all $\mu_{d'i}$ and $\mu_{d''i}$, where each F_j is nondecreasing and the assumptions of Theorem 1 are satisfied with g_i replacing g, then*

$$\Phi^*(\mu_{d''}) \begin{Bmatrix} \le \\ \ge \end{Bmatrix} \Phi^*(\mu_{d'}) \quad \textit{if } (g_j''/f'') \operatorname{sgn} f'' \textit{ is } \begin{Bmatrix} \textit{decreasing} \\ \textit{increasing} \end{Bmatrix} \textit{ on } U \ \forall j.$$

The most common such Φ^*, the E-optimality criterion μ_1, is of course treated in hypothesis (3.3). But other criteria Φ^*, in particular Φ_g's with g nondifferentiable, also arise. An extension to certain Schur-convex Φ is also possible.

It is also evident that, in imitation of (1.3) when $c<1$, we can obtain conclusions when (3.1) is weakened to assume inequalities, under appropriate monotonicity conditions. This gives less interesting conclusions in typical examples; for example, one will obtain only one of the inequalities in (3.5).

A further excursion could replace Φ_f by a limit Φ^{**} of criteria, in the hypothesis (3.2b). This is somewhat messier and more delicate to carry out since we could have $\Phi^{**}(\mu_{d'}) = \Phi^{**}(\mu_{d''})$ without the corresponding equality of Φ_{f_i} values. Thus, on the one hand, one needs further details such as replacing $\mu_{d''}$ and C by sequences. On the other hand, we shall see at the end of the next section that the conclusion one obtains through such a process is not always so simple to express. We shall not take the space here to carry this development further.

We have mentioned the need for some condition such as (3.3) to obtain general conclusions. The simplest case where this condition fails is $m=4$, e.g., with $\mu_{d'}=(A', B_1', B_2', C')$ and $\mu_{d''}=(A'', B_1'', B_2'', C'')$, but with $A'<A''$, $B_1'>B_1''$, $B_2'<B_2''$, $C'>C''$ instead of (3.3). The last expression of (3.8) reduces to one term in each sum, but the first ψ is negative while the second is positive; how they balance depends on fine details of the spacing and of f and g.

Nevertheless, the spacing of (3.3) is not necessary to the conclusions. An obvious but useful fact (employed in the examples of Section 5.2) is

Corollary 3. *If* $\{\mu_{di}\}$ *for* $d=d', d''$ *can be partitioned into* $\mu_d^{(r)}=\{\mu_{di}, i\in A^{(r)}\}$ *where the* $A^{(r)}$ *are a partition of* $\{1,\ldots,m\}$, *and if the assumptions of Theorem* 1 *or Corollary* 2 *for the given* f, g (*or* $\{F_j\}$, $\{g_j\}$) *hold for each pair* $\mu_{d'}^{(r)}$, $\mu_{d''}^{(r)}$ *replacing* $\mu_{d'}$, $\mu_{d''}$ *in* (3.2), (3.3), *then the conclusion of Theorem* 1 *or Corollary* 2 *holds for* $\mu_{d'}$, $\mu_{d''}$.

This can be used, for example, if the $\mu_d^{(r)}$ are the same for each r and $\mu_{d'}^{(1)}$, $\mu_{d''}^{(1)}$ satisfy the hypotheses; in such cases, the original $\mu_{d'}$, $\mu_{d,}$ will generally violate (3.3).

We conclude this section with a simple but useful tool.

Lemma 3. *If* $m=3$, *if* (3.2) *is satisfied with* f'' *of one sign, and if* $\mu_{d'1}<\mu_{d''1}$, *then* (3.3) *is satisfied.*

Proof. If $\mu_{d'3}\geq\mu_{d''3}$ it is easy to see that $\mu_{d'}$ majorizes $\mu_{d''}$, contradicting (3.2b) for f, of one sign. Hence $\mu_{d'3}<\mu_{d''3}$ and $\mu_{d'2}>\mu_{d''2}$ by (3.2a). □

4. Examples of criteria

Although it is easy to list, endlessly, applications of Theorem 1 and Corollary 2, we shall for the sake of brevity limit consideration here to the most common family of criteria that occur in practice. On $(0,\infty)$, for $-\infty<p<\infty$ let

$$h_p(x)=\begin{cases} x^{-p}\operatorname{sgn}(p) & \text{if } p\neq 0,\\ -\log x & \text{if } p=0; \end{cases} \tag{4.1}$$

this was termed $\bar{g}_p$ below (1.5) in the context of Section 1, where we remarked that the common Φ_p-criterion of (1.5) is a strictly increasing function of what we have termed Φ_{h_p}, namely, $[m^{-1}(\operatorname{sgn} p)\Phi_{h_p}]^{1/p}$ for $p\neq 0$ and $\exp(m^{-1}\Phi_{h_0})$ for $p=0$. The Φ_0, Φ_1- and Φ_∞-criteria are known as the D-, A- and E-optimality criteria. The role of Φ_{-1} was mentioned earlier, while $\Phi_{-\infty}$ arises in work of Cheng (1978a). The Φ_p-criteria for $p\geq 0$ are widely used. Those with $p<0$ are less attractive, although minimization of Φ_{-1} (maximization of $\sum_1^m \lambda_i$) together with other conditions was seen in Section 1 to be useful, e.g. for proving Φ_p-optimality for $p>0$. Among designs with the same (minimum) value of Φ_{-1}, that with *maximum* Φ_{-2}-value is often suggested, beginning with Shah (1960), to be good; see also Eccleston and Hedayat (1974) and Cheng (1978b).

Suppose (3.1) is satisfied for $\tilde{f}=h_{\bar{p}}$ and $\tilde{f}=-h_{\tilde{p}}$, with $\bar{p}\neq\tilde{p}$; the choice of sign in the latter is made to make f and g decreasing, where $\bar{g}=h_p$. We have (where we can assume p, $\tilde{p}$, $\bar{p}$ distinct)

$$g(y)=\bar{g}(\tilde{f}^{-1}(y))=\begin{cases}(-y\operatorname{sgn}\tilde{p})^{p/\tilde{p}}\operatorname{sgn}p & \text{if } \tilde{p}\neq 0,\ p\neq 0;\\ \tilde{p}^{-1}\log(-y\operatorname{sgn}\tilde{p}) & \text{if } \tilde{p}\neq 0,\ p=0;\\ e^{-py}\operatorname{sgn}p & \text{if } \tilde{p}=0,\ p\neq 0.\end{cases} \tag{4.2}$$

(Note that for y in the range of $\tilde{f}$, $-y\operatorname{sgn}\tilde{p}>0$.)

The corresponding values of $f(y)=\bar{f}(\tilde{f}^{-1}(y))$ are obtained by substituting $\bar{p}$ for p in (4.2). The values of $\operatorname{sgn}f''$ is easily seen to be $\operatorname{sgn}(\bar{p}-\tilde{p})$ in all cases. Checking the bevavior of $(g''/f'')\operatorname{sgn}f''$ in the various cases, we find that

$$(g''/f'')\operatorname{sgn}f'' \text{ is } \begin{Bmatrix}\text{increasing}\\ \text{decreasing}\end{Bmatrix} \quad \text{if } \operatorname{sgn}[(p-\tilde{p})(\bar{p}-p)] \begin{Bmatrix}>\\ <\end{Bmatrix} 0. \tag{4.3}$$

Recall that we chose $\tilde{f}$ to be increasing. This means the λ_{di} and corresponding μ_{di} are in the same order for each d. In the statement of Theorem 2 we can choose $\bar{p}<\tilde{p}$ without loss of generality. From Theorem 1 and (4.3) we then obtain the following (the cases $p=\infty$ and $-\infty$ of (4.5) following from (3.3) directly).

Theorem 2. *Suppose* $-\infty<\bar{p}<\tilde{p}<\infty$ *and that*

$$\Phi_{\bar{p}}(\lambda_{d'})=\Phi_{\bar{p}}(\lambda_{d''}), \qquad \Phi_{\tilde{p}}(\lambda_{d'})=\Phi_{\tilde{p}}(\lambda_{d''}), \tag{4.4}$$

and that $\lambda_{d'}=(A_1',\ldots,C_s')$ *and* $\lambda_{d''}=(A_1'',\ldots,C_s'')$ *satisfy* (3.3) *with* λ *replacing* μ. *Then, for* $-\infty\leq p\leq\infty$,

$$\Phi_p(\lambda_{d''}) \begin{Bmatrix}<\\ >\end{Bmatrix} \Phi_p(\lambda_{d'}) \quad \text{if} \begin{Bmatrix}p<\bar{p} \text{ or } p>\tilde{p}\\ \bar{p}<p<\tilde{p}\end{Bmatrix}. \tag{4.5}$$

The application that has arisen repeatedly in our study of weighing designs (Section 5.1) is $\bar{p}=-1$, $\tilde{p}=0$:

Corollary 4. *Suppose* (3.3) *is satisfied with* λ *replacing* μ, *and*

$$\sum_i \lambda_{d'i}=\sum_i \lambda_{d''i}, \qquad \prod_i \lambda_{d'i}=\prod_i \lambda_{d''i}. \tag{4.6}$$

Then d'' *is* Φ_p*-better than* d' *for* $p>0$ *or* $p<-1$, *and is* Φ_p*-worse for* $-1<p<0$.

This was proved for $m=3$ in the Theorem and Remark 1 of Galil and Kiefer (1980b). The first condition of (4.6) is automatically satisfied in many settings, while the second arises because D-optimality is a convenient and common subject of study. Application of Corollary 4, additional to those of the above reference, will be given in the next section.

We also note, from Theorem 1, using an argument similar to that used to obtain Corollary 4, that d' is Φ_g-better than d'' for $g(x) = -\log(x+b)$ with $b>0$, an example given in the above reference to illustrate a decreasing convex g for which, under these conditions, d' is better, in contrast to the conclusion that d'' is Φ_p-better for $p>0$.

We remarked in the previous section on the greater difficulty attending replacement of Φ_f by a limit of functions in (3.2b). This is well illustrated in the present context, where $\Phi_\infty = \lim_{p\to\infty} \Phi_p$. Suppose d' and d'' satisfy the hypotheses of Theorem 2, that $\lambda_0 < \lambda_{d'1}$, and that d^* and d^{**} satisfy $\boldsymbol{\lambda}_{d^*} = (\lambda_0, \boldsymbol{\lambda}_{d'})$ and $\boldsymbol{\lambda}_{d^{**}} = (\lambda_0, \boldsymbol{\lambda}_{d''})$. We have now altered (3.3) by adjoining λ_0 at the left end. Of course, we know that $\Phi_p(\boldsymbol{\lambda}_{d^*}) = \Phi_p(\boldsymbol{\lambda}_{d^{**}})$ for $p = \bar{p}, \tilde{p}, \infty$. But suppose that, in parallel with (4.4), we once more state the equality of Φ_p-values for only two values $p = p', \infty$, where we might choose p' to be either $\bar{p}$ or $\tilde{p}$. The character of our conclusion depends on this choice, in the sense that we can no longer simply state, in parallel with (4.5), that one design is better for p in $(-\infty, p')$ and the other is better in (p', ∞).

5. Design applications

5.1. D-optimum weighing designs

Let n and k be positive integers with $k \le n$. For the problem of weighing k objects in n weighings on a chemical balance, and for certain other problems such as (after reduction) that of first-degree regression on a $(k-1)$-cube, a design d can be thought of as any $n \times k$ matrix X_d with entries ± 1. Then $\boldsymbol{\lambda}_d$ consists of the eigenvalues of $X_d' X_d$. When $n \equiv 0 \pmod 4$, k columns of an $n \times n$ Hadamard matrix yield a design optimum in many of the usual senses (in particular, Φ_p-optimum $\forall p \ge 0$). For $n \equiv 1$ or $2 \pmod 4$, D-optimum designs are known in many cases. Most interesting are the cases where $n \equiv 3 \pmod 4$, both because of the varied structure of D-optimum designs as a function of n and k, and also because there are sometimes two nonisomorphic D-optimum designs d' and d'', satisfying (4.5).

The extensive study of D-optimality for $k = n \equiv 3 \pmod 4$ was initiated by Ehlich (1964b), and it was extended to $k < n \equiv 3 \pmod 4$ in Galil and Kiefer (1980a, 1982a). Briefly, every $X_d' X_d$, after multiplying certain rows and corresponding columns by -1, is a member of $\mathscr{C}_{n,k} = \{C : C$ is symmetric positive definite $k \times k$ with $c_{ii} = n\ \forall i$ and $c_{ij} \equiv 3 \pmod 4)\ \forall i, j\}$. The theory characterizes C^* with maximum determinant in $\mathscr{C}_{n,k}$. This C^* is not always obtainable as $X_d' X_d$; if it is not, we must look further to find a D-optimum design d^*; but for many cases of interest, methods have been obtained in Galil and Kiefer (1982b) for constructing such an X_{d^*}. For the sake of brevity, we limit discussion hereafter to C^*.

Every optimum C^* is shown by the theory to be of the following form (except for row-and-column permutations): for some positive integer s, there are s diagonal square 'blocks', u of order $r=\lfloor k/s \rfloor$ and v of order $r+1$ where $u+v=s$, $ur+v(r+1)=k$; each of these blocks has all off-diagonal entries 3 (and, of course, diagonal entries n); all c_{ij} outside the blocks are -1. Thus, a value s_{OPT} of s characterizes any C^*.

There are five infinite families or 'series' of (k, n) each with $2k-n$ constant, and for which two contiguous integer values s and $s+1$ both yield a D-optimum C^*, and there are no other (k, n) for which this is so. The five series (all with $n \equiv 3 \pmod 4$) are

(a) $2k-n=5$ $(7 \le n)$: $s_{\mathrm{OPT}}=k$ or $k-1$;
(b) $2k-n=7$ $(11 \le n \equiv 3 \pmod 8)$: $s_{\mathrm{OPT}}=\frac{1}{4}(3k+1)$ or $\frac{1}{4}(3k-3)$;
(c) $2k-n=9$ $(11 \le n)$: $s_{\mathrm{OPT}}=\frac{1}{2}(k+2)$ or $\frac{1}{2}k$;
(d) $2k-n=11$ $(11 \le n)$: $s_{\mathrm{OPT}}=\frac{1}{2}(k+1)$ or $\frac{1}{2}(k-1)$;
(e) $2k-n=17$ $(27 \le n \equiv 3 \pmod{12})$: $s_{\mathrm{OPT}}=\frac{1}{3}(k+2)$ or $\frac{1}{3}(k-1)$.

The determinant of a matrix C^* with s blocks, as described above, is

$$\begin{aligned} D_{k,n}(s) &= (n-3)^{k-s}(n-3+4r)^{(r+1)s-k}(n+1+4r)^{k-sr}\{1-G\}, \\ G &= [k(n-3)+4sr(r+1)]/(n+4r+1)(n+4r-3). \end{aligned} \tag{5.1}$$

Hence, the equation $0=D_{k,n-\lambda}(s)$ yields the eigenvalues λ_i of C^*. From this we obtain the following eigenvalues, expressed here as functions of k, in the five series listed above, *with d' in each case referring to the C^* with the larger s_{OPT} and d'' to that with the smaller s_{OPT}*:

(a)
$$\{\lambda_{d'i}\} = \begin{cases} 2k-4, & \text{of multiplicity } k-1, \\ k-4, & \text{of multiplicity } 1; \end{cases}$$

$$\{\lambda_{d''i}\} = \begin{cases} 2k-4, & \text{of multiplicity } k-3, \\ 2k-8, & \text{of multiplicity } 1, \\ \frac{1}{2}[3k-4 \pm (k^2+8k-16)^{1/2}], & \text{each of multiplicity } 1. \end{cases}$$

(b)
$$\{\lambda_{d'i}\} = \begin{cases} 2k-10, & \text{of multiplicity } \frac{1}{4}(k-1), \\ 2k-2, & \text{of multiplicity } \frac{1}{4}(k-5), \\ 2k-6, & \text{of multiplicity } \frac{1}{2}(k-1), \\ \frac{1}{2}[3k-8 \pm (k^2+24)^{1/2}], & \text{each of multiplicity } 1; \end{cases}$$

$$\{\lambda_{d''i}\} = \begin{cases} 2k-10, & \text{of multiplicity } \frac{1}{4}(k+3), \\ 2k-2, & \text{of multiplicity } \frac{1}{4}(k-1), \\ 2k-6, & \text{of multiplicity } \frac{1}{2}(k-5), \\ \frac{1}{2}[3k-8 \pm (k^2-8)^{1/2}], & \text{each of multiplicity } 1. \end{cases}$$

(c)
$$\{\lambda_{d'i}\} = \begin{cases} 2k-12, & \text{of multiplicity } \tfrac{1}{2}(k-2), \\ 2k-\ \ 4, & \text{of multiplicity } \tfrac{1}{2}(k-4), \\ 2k-\ \ 8, & \text{of multiplicity } 1, \\ \tfrac{1}{2}[3k-12 \pm (k^2-8k+48)^{1/2}], & \text{each of multiplicity } 1; \end{cases}$$

$$\{\lambda_{d''i}\} = \begin{cases} 2k-12, & \text{of multiplicity } \tfrac{1}{2}k, \\ 2k-\ \ 4, & \text{of multiplicity } \tfrac{1}{2}(k-2), \\ k-\ \ 4, & \text{of multiplicity } 1. \end{cases}$$

(d)
$$\{\lambda_{d'i}\} = \begin{cases} 2k-14, & \text{of multiplicity } \tfrac{1}{2}(k-1), \\ 2k-\ \ 6, & \text{of multiplicity } \tfrac{1}{2}(k-3), \\ \tfrac{1}{2}[3k-16 \pm (k^2-8k+32)^{1/2}], & \text{each of multiplicity } 1; \end{cases}$$

$$\{\lambda_{d''i}\} = \begin{cases} 2k-14, & \text{of multiplicity } \tfrac{1}{2}(k+1), \\ 2k-\ \ 6, & \text{of multiplicity } \tfrac{1}{2}(k-5), \\ \tfrac{1}{2}[3k-8 \pm (k^2+8k-32)^{1/2}], & \text{each of multiplicity } 1. \end{cases}$$

(e)
$$\{\lambda_{d'i}\} = \begin{cases} 2k-20, & \text{of multiplicity } \tfrac{1}{3}(2k-2), \\ 2k-\ \ 8, & \text{or multiplicity } \tfrac{1}{3}(k-7), \\ 2k-12, & \text{of multiplicity } 1, \\ \tfrac{1}{2}[3k-20 \pm (k^2-8k+80)^{1/2}], & \text{each of multiplicity } 1; \end{cases}$$

$$\{\lambda_{d''i}\} = \begin{cases} 2k-20, & \text{of multiplicity } \tfrac{1}{3}(2k+1), \\ 2k-\ \ 8, & \text{of multiplicity } \tfrac{1}{3}(k-7), \\ \tfrac{1}{2}[3k-12 \pm (k^2+8k-48)^{1/2}], & \text{each of multiplicity } 1. \end{cases}$$

In each of the four series (a), (c), (d), (e) (the first of which was the main consideration of Galil and Kiefer (1980b)), $\{\lambda_{d'i}\}$ and $\{\lambda_{d''i}\}$ have $k-3$ elements in common (counting multiplicities). The remaining three elements of the two sets are easily verified, most easily by using Lemma 3, to satisfy the hypothesis of Corollary 4 (that is, (3.3) with λ replacing μ). That hypothesis is also satisfied for the set of *four* elements remaining after deleting the k-4 common elements of $\{\lambda_{d'i}\}$ and $\{\lambda_{d''i}\}$ in the series (b). (This series and the considerations of the next example require the development of the present paper, that of our previous paper (Galil and Kiefer (1980b)) only covering, for the Φ_p-criteria alone, the comparison of two sets of eigenvalues of size three.) Thus, we have obtained:

Theorem 3. *In each of the 5 series (a)–(e) in which there are two C^*'s with neighboring numbers of blocks that minimize Φ_0, the one with fewer blocks is Φ_p-better if $p>0$ or $p<-1$, and is Φ_p-worse if $-1<p<0$.*

From a practical viewpoint, this result suggests strong preference for the D-optimum C^* with fewer blocks. As indicated earlier, Galil and Kiefer (1980b, 1981b) consider implementation of such a C^* in terms of an X_d.

A well known case of (n, k), in which C^* cannot be obtained as $X_d'X_d$, is $n=k=11$, which has been investigated thoroughly by Ehlich, who determined the D-optimum X_d through a combination of algebra and computer search. As described in Galil and Kiefer (1980a), there are three nonisomorphic $X_d'X_d$ forms. All have diagonal elements 11 and all off-diagonal elements -1 except for a few symmetric pairs of entries 3, the above-diagonal positions of which are (for the X_d listed in Tables 9a, 9b, 9c of Galil and Kiefer (1980a))

(a) (1,2), (2,3), (3,4), (4,5), (6,7), (8,9), (10,11);

(b) (1,2), (1,3), (2,3), (3,4), (4,5), (6,7), (8,9), (10,11);

(c) in blocks of size 5, 2, 2, 2.

Note that the last is a 'block matrix', but the blocks are not of contiguous sizes. The eigenvalues of $X_d'X_d$ for these three designs are

$$\text{(a)}\quad \begin{cases} 16, & \text{of multiplicity } 3,\\ 8, & \text{of multiplicity } 4,\\ 4.80326,\ 6.09016,\ 12.25725,\ 17.84932, & \text{each of multiplicity } 1.\end{cases}$$

$$\text{(b)}\quad \begin{cases} 16, & \text{of multiplicity } 3,\\ 8, & \text{of multiplicity } 4,\\ 5.22504,\ 6.18250,\ 10.23449,\ 19.35796, & \text{each of multiplicity } 1.\end{cases}$$

$$\text{(c)}\quad \begin{cases} 16, & \text{of multiplicity } 2,\\ 8, & \text{of multiplicity } 8,\\ 25, & \text{of multiplicity } 1.\end{cases}$$

(The noninteger eigenvalues in (a) and (b) are of the form $11-x$ where x is a zero of $x^4-3x^3-51x^2+151x+262$ or $x^4-3x^3-59x^2+279x-178$, respectively.) After deleting common eigenvalues, we must now compare sets of four values in comparing (a) with (b) and sets of five values in comparing (b) with (c). Using Corollary 4, we obtain:

Theorem 4. *Among Ehlich's D-optimum designs for $n=k=11$, (c) is Φ_p-better than (b) which is Φ_p-better than (a) if $p>0$ or $p<-1$, and the domination is in reverse order if $-1<p<0$.*

Thus, from a practical point of view, design (c) seems preferable.

5.2. Covariate models

Lopes-Troya (1982) has considered various linear models in which one of V varieties is tested on each observation, there being K covariates which, after a reduction, are limited to the values ± 1 for each. There are thus $m=V+K$ parameters. An example that occurs when $V\equiv 0 \pmod 4$, $K=\frac{1}{2}V$, and the total number of observations N is such that $N/V=R$, an odd integer, is the following: there are two D-optimum designs, comparison of which for other criteria was first considered by Lopes-Troya, and which we shall not describe here in detail (in terms of the covariate levels), mentioning only the interesting difference that each variety is replicated R times in d'', while in d' half the varieties are replicated $R+1$ times and half are replicated $R-1$ times. (Lopes-Troya gives other examples in which unequal replication is in fact D-better.) The eigenvalues of d' are $R-1$, $R+1$ and N, each of multiplicity $\frac{1}{2}V$. Those of d'' are

$$R \quad \text{and} \quad \tfrac{1}{2}R(V+1) \pm [\tfrac{1}{4}R^2(V-1)^2 + V]^{1/2},$$

also each of multiplicity $\frac{1}{2}V$. Thus, (4.6) is satisfied. Although (3.3) is not satisfied, we can employ Corollary 3 and Lemma 3. Since

$$R-1 < \tfrac{1}{2}R(V+1) - [\tfrac{1}{4}R^2(V-1)^2 + V]^{1/2},$$

we conclude from Corollary 4 that d'' is Φ_p-better than d' for $p>0$ and $p<-1$.

References

Cheng, C.-S. (1978a). Optimality of certain asymmetrical experimental designs. *Ann. Statist.* 6, 1239–1261.

Cheng, C.-S. (1978b). A note on M-S optimality. *Comm. Statist.* A7 (14), 1327–1338.

Cheng, C.-S. (1980). Optimality of some weighing and 2^n fractional factorial designs. *Ann. Statist.* 8, 436–446.

Eccleston, J.A. and A. Hedayat (1974). On the theory of connected design characterization and optimality. *Ann. Statist.* 2, 1238–1255.

Ehlich, H. (1964a). Determinantenabschätzungen für binäre Matrizen. *Math. Z.* 83, 123–132.

Ehlich, H. (1964b). Determinantenabschätzungen für binäre Matrizen mit $n\equiv 3$ mod 4. *Math. Z.* 84, 438–447.

Galil, Z. and J. Kiefer (1980b). Optimum weighing designs. In: K. Matusita, ed., *Recent Developments in Stat. Inf. and Data Anal.*, 183–189. North-Holland, Amsterdam.

Galil, Z. and J. Kiefer (1982a). On the characterization of D-optimum weighing designs for $n\equiv 3$ (mod 4). In: S. Gupta and J. Berger, eds., *Statist. Dec. Theory and Related Topics*, 1-35. Academic Press, New York.

Galil, Z. and J. Kiefer (1982b). Construction of optimum block matrices for weighing designs with $n\equiv 3$ (mod 4). *Ann. Statist.* 10, 502–510.

Kiefer, J. (1958). Optimum experimental designs. *J. Roy. Statist. Soc. Ser. B* 21, 272–319.

Kiefer, J. (1958). Optimum experimental designs. *J. Roy. Statist. Soc. Ser. B* 21, 272–319.

Kiefer, J. (1971). The role of symmetry and approximation in exact design optimality. In: S. Gupta and J. Yackel, eds., *Statist. Dec. Theory and Related Topics,* 109–118. Academic Press, New York.

Kiefer, J. (1975). Construction and optimality of generalized Youden designs. In: J.N. Srivastava, ed., *A Survey of Statistical Designs and Linear Models,* 333–353. North-Holland, Amsterdam.

Lopes-Troya, J. (1982). Optimal designs for covariate models. *J. Statist. Plann. Inference* 6, 373-419.

Marshall, A. and I. Olkin (1979). *Inequalities: Theory of Majorization and its Applications.* Academic Press, New York.

Shah, K.R. (1960). Optimality criteria for incomplete block designs. *Ann. Math. Statist.* 31, 791–794.

The Annals of Statistics
1984, Vol. 12, No. 2, 414–450

OPTIMUM AND MINIMAX EXACT TREATMENT DESIGNS FOR ONE-DIMENSIONAL AUTOREGRESSIVE ERROR PROCESSES[1]

By J. Kiefer and H. P. Wynn

University of California, Berkeley and Imperial College, London

A theory is developed following work by Williams (1952) and Kiefer (1960) for exact treatment designs in one dimension in which the errors are a stationary process. It is shown that the designs which achieve the minimax value of any of a wide class of functionals on the information matrix for estimation of treatment differences have a special property. If the process is autoregressive of order p then a random piece of the design of length $p + 1$ exhibits uncorrelated treatment values. Such designs can be formed using full length cyclic error-correcting codes of a suitable order. A new technique is developed for classifying the ergodic combinatorial structure of exact designs of arbitrary or infinite length. It is shown that all designs are, to pth order, generated by a finite number of sequences with finite length. The classification is given explicitly up to order 3. The method is used to find asymptotically optimum designs for different processes. It is also shown that the designs can be achieved to within an arbitrarily good approximation as the realization of an ergodic Markov chain of sufficiently high order.

1. Introduction. The problem of laying treatments out in a line, or in time, so as to obtain good estimation of between-treatment contrasts when the error process is autocorrelated was first systematically studied by Cox (1951) and Williams (1952). One of the present authors, Kiefer (1960), extended the results of Williams and proved a number of optimality results, some of which were conjectures from these earlier papers. In particular, first and second order autoregressive processes were studied in some detail. The present paper extends these results using a flexible formulation in which a general minimax property can be established for certain special designs. Briefly, it is sufficient that in every "window" of certain fixed length, every vector of treatments occurs once (or the same number of times). We shall show that such designs can be formed using full-length cyclic error correcting codes. These are an example of a pseudo-random sequence used in communication theory.

Most of the related work in experimental design theory has been on two- and higher-dimensional designs, often with added row, column, or block effects. We mention particularly a recent paper by the authors, Kiefer and Wynn (1981). The approach taken in that paper was somewhat different. The analysis was based on the ordinary least squares estimator and restricted to a nearest neighbor model equivalent to a first order moving average model. The models were more complex, involving row, column and treatment effects. Here we carry out the full

Received May 1981; revised December 1983.

[1] Research partially supported by National Science Foundation grant MCS78-25301.

AMS 1980 *subject classifications.* 62K05, 62K15, 62M10, 05B15, 62J05.

Kew words and phrases. Optimum experimental designs, exact design, dependent observations, stationary processes, binary sequences, error correcting codes, linear machines, linear programming, pseudo-random sequences, Markov chains.

best linear unbiased (BLU) estimation but confine ourselves to an in-depth study of the one-dimensional case. We should particularly like to acknowledge the work of Martin (1977, 1979), and papers by Berenblutt and Webb (1974) and Duby et al. (1977). A representative paper on sampling for continuous processes in which there is moderate literature is Bickel and Herzberg (1979).

To start with, we formulate the problem on the circle rather than the line. The extension to the line can be carried out straightforwardly but is best treated in an asymptotic way and is left until Sections 6 and 7.

Consider N units labeled by the letter t $(t = 1, \cdots, N)$ situated in order around a circle. To unit t attach a random variable Y_t $(t = 1, \cdots, N)$. Let the Y_t have a joint distribution having the autocorrelation structure

$$Y_t + \rho_1 Y_{t-1} + \cdots + \rho_{N-1} Y_{t-N+1} = \varepsilon_t \quad (t = 1, \cdots, N) \tag{1}$$

where because of the circulant nature of the set-up we identify unit $-i$ with unit $N-i$ $(i = 1, \cdots, N)$. The "innovation" vector $\varepsilon = (\varepsilon_1, \cdots, \varepsilon_N)^T$ has mean and covariance matrix $E(\varepsilon) = 0$, $\mathrm{Cov}(\varepsilon) = \sigma^2 I_{N\times N}$. Throughout this paper we shall assume that the ρ_r $(r = 1, \cdots, N-1)$ are known. (Lemma 1 and the end of Section 7 consider the ρ_r as variable. From a practical point of view, this might mean estimating the ρ_r from the data and using those estimates to "adapt" the BLU estimate for fixed ρ_r). At this stage we make no restriction on the ρ_r, but later, for a corresponding process on the line instead of on the circle, a stationarity condition is required. Whether σ^2 is known is immaterial. The model is said to be pth *order* (or less) if $\rho_{p+1} = \rho_{p+2} = \cdots = \rho_{N-1} = 0$. Usually, we shall treat the order $p(0 \le \rho \le N-1)$ as fixed and note here that it will play a crucial role in the construction of designs. Write $Y = (Y_1, \cdots, Y_N)^T$ and take X_p to denote the $N \times N$ circulant matrix

$$X_p = \begin{bmatrix} 1 & \rho_{N-1} & \rho_{N-2} & \cdot & \cdot & \cdot & \rho_1 \\ \rho_1 & 1 & \rho_{N-1} & \cdot & \cdot & \cdot & \cdot \\ \rho_2 & \rho_1 & 1 & \cdot & \cdot & \cdot & \cdot \\ \cdot & \cdot & \cdot & \cdot & \cdot & \cdot & \cdot \\ \cdot & \cdot & \cdot & \cdot & \cdot & \cdot & \cdot \\ \cdot & \cdot & \cdot & \cdot & \cdot & 1 & \rho_{N-1} \\ \rho_{N-1} & \cdot & \cdot & \cdot & \cdot & \rho_1 & 1 \end{bmatrix}$$

with $\rho_i = 0$ for $i > p$. Then assuming the model is pth order (1) can be written

$$X_p Y = \varepsilon. \tag{2}$$

Now assume that there are k treatments labeled $1, \cdots, k$, one applied to each unit. With treatment i, associate an unknown constant α_i $(i = 1, \cdots, k)$. Let $[t]$ denote the label of the treatment applied to unit t. The actual observation on unit t is then assumed to be $W_t = Y_t + \alpha_{[t]}$. Denote by U the $N \times k$ matrix describing the allocation of treatments so that its entries are

$$U_{ti} = 1 \quad \text{if } [t] = i$$
$$= 0 \quad \text{otherwise.}$$

Thus, U has exactly one unity in each row, the other entries being zero. If

$W = (W_1, \cdots, W_N)^T$, the full treatment model becomes

$$W = Y + U\alpha, \tag{3}$$

where $\alpha = (\alpha_1, \cdots, \alpha_k)^T$.

We shall be interested in considering the BLU estimates of contrasts among the α_i based on the observations W. Following Kiefer and Wynn (1981), we consider a vector θ of standardized treatment contrasts

$$\theta = (I_k - k^{-1}J_k)\alpha$$

where I_k is the $k \times k$ identity matrix and J_k the $k \times k$ matrix of ones. Write $P = I_k - k^{-1}J_k$ and notice that this is the usual projection operator associated with obtaining the treatment effects in one-way analysis of variance. Multiplying (3) through by X_p and using (2) we obtain

$$X_p W = X_p U\alpha + \varepsilon.$$

This is now in standard regression form and the covariance matrix for $\hat{\alpha}$, the BLU estimate of α, is

$$\text{Cov}(\hat{\alpha}) = \sigma^2(U^T X_p^T X_p U)^{-1},$$

where we have assumed that $\text{Rank}(X_p U) = k$. Define $A = U^T X_p^T X_p U$. Thus the BLU estimate of θ, $\hat{\theta}$, has covariance matrix

$$\text{Cov}(\hat{\theta}) = \sigma^2 P A^{-1} P.$$

It is more tractable to work with

$$C = PAP$$

(ignoring σ^2). However, C is not in general an inverse of $PA^{-1}P$. Writing B^- to denote the Moore-Penrose g-inverse of a matrix B, we have (Zyskind, 1967, Theorem 2)

$$\bar{C} \stackrel{\text{def}}{=} (PA^{-1}P)^- = C - D \tag{4}$$

where D is a nonnegative definite matrix which is zero when P and A (or P and A^{-1}) commute. Notice that both C and $\bar{C}$ have row and column sums equal to zero.

2. Optimum design. The design is defined to be the allocation of treatments to units and enters through $[t]$ or the matrix U. It is labeled by d. Where we are concerned to denote the dependence of the information matrix on d we write A_d or C_d. However, the vector of ρ_i values $\rho = (\rho_1, \cdots, \rho_p)^T$ (assuming a pth order model) also affects C through X_p so occasionally we may write $C_d(\rho)$ or $A_d(\rho)$.

We shall restrict slightly the definition of universally optimum designs of Kiefer (1958, 1975) and give a slighty different version of the key result (Kiefer, 1975, Proposition 1). This version is required to cope with the matrix D in (4).

Let $\mathscr{L}$ be the class of all nonnegative definite matrices with row and column sums zero. A design (here) is called universally optimum among the class of all designs if it minimizes $\Phi(\bar{C})$ for every real valued function Φ on $\mathscr{L}$ which is (i) convex, (ii) invariant under permutations of coordinates (rows and columns simultaneously), (iii) nonincreasing in the usual (Loewner) sense: $G - H$ nonnegative definte $\Rightarrow \Phi(G) \leq \Phi(H)$. Notice that we assume a fixed value of ρ.

We now show that a sufficient condition for a design d^* to be universally optimum is that

(a) A_{d^*} is a multiple of $aI_k + bJ_k$ for constants a, b.

(b) $\text{trace}(C_{d^*}) = \text{trace}(PA_{d^*}P) = \max_d \text{trace}(C_d)$.

In proving this, we use Proposition 1 of Kiefer (1975) on the matrices C, which we may do since our condition (iii) is stronger than Kiefer's. From (a) and (b) we thus conclude that $\Phi(C_d)$ is minimized by d^*. By (4) and (iii) we have $\Phi(\bar{C}_d) \geq \Phi(C_d)$ with equality when $D = 0$. But (a) implies that A_{d^*} commutes with P and hence that $D = 0$ for d^*. This completes the proof.

Condition (a) is summarized by saying A_{d^*} is completely symmetric (CS). For individual values of ρ, the problem of finding designs d^* is fairly difficult, particularly because of the complex nature of the class of designs. We give a "large N" solution of the problem in Section 7. We also discuss, at the end of that section, "complete classes" of designs in terms of the function (18) of the ρ_r, which may be thought of as analogous to the risk function of a procedure (here, of a design, in terms of its π_r's).

An alternative to finding optimum designs for given ρ, or a complete class, is to take a minimax approach. A design d^* is said to be universally minimax if

$$\max_\rho \Phi(C_{d^*}) = \min_d \max_\rho \Phi(C_d),$$

for all Φ satisfying (i), (ii) and (iii) above. The maximum is taken over all vectors ρ in R^p.

It will now be shown that if d^* is completely symmetric and satisfies the corresponding maximin condition on its *trace* then it is minimax. We note from the form of X_ρ that $\text{trace}(C_d(\rho))$ is a nonnegative quadratic in ρ, so that $\min_\rho C_d(\rho)$ is always attained.

LEMMA 1. *Suppose d^* is a design for which $A_{d^*}(\rho)$ is completely symmetric for all ρ and*

$$\min_\rho \text{trace}(C_{d^*}(\rho)) = \max_d \min_\rho \text{trace}(C_d(\rho)). \tag{5}$$

Then d^* is universally minimax.

PROOF. Let ρ^* be the value ot ρ achieving $\min_\rho \text{trace}(C_{d^*}(\rho))$. Since $C_{d^*}(\rho)$ is completely symmetric for all ρ, all these $C_{d^*}(\rho)$'s (as ρ varies) are multiples of $C_{d^*}(\rho^*)$ and hence, by (iii),

$$\Phi(C_{d^*}(\rho^*)) = \max_\rho \Phi(C_{d^*}(\rho)), \quad \text{for all } \Phi. \tag{6}$$

Now take any other d. By (5) there is a value of ρ, say ρ', such that

$$\text{trace } C_d(\rho') \leq \text{trace}(C_{d^*}(\rho^*)).$$

Using the argument following (a) and (b) in Section 1 applied to the *pair* of matrices $C_d(\rho')$ and $C_{d^*}(\rho^*)$ we have

$$\Phi(C_d(\rho')) \geq \Phi(C_{d^*}(\rho^*))$$

for all Φ. Thus

$$\max_\rho \Phi(C_d(\rho)) \geq \Phi(C_{d^*}(\rho^*)). \tag{7}$$

Combining (7) with (6) gives the result. □

The lemma helps considerably in finding minimax designs. The method is first to get the trace maximin and then hope that we can find a design which in addition makes C_d completely symmetric.

3. The structure of designs on the circle. Consider the treatments arranged on their units around the circle. Consider a "window" of length $p + 1$ placed on the circle so that when it is in position t the locations $1, \cdots, p + 1$ in the window show the adjacent units $t - p, t - p + 1, \cdots, t - 1, t$, respectively. The treatments on these units will be $[t - p], [t - p + 1], \cdots, [t - 1], [t]$, respectively. Now suppose that the window, instead of being in a fixed position, is placed at random with probability $1/N$ attached to each of the N possible positions. The vector of treatments in the random window is now a random vector $T_1, \cdots, T_{p+1}$ of treatment labels. Note that it is stationary under shifts; that is, $T_{i_1}, \cdots, T_{i_r}$ has the same joint distribution as $T_{i_1+\tau}, \cdots, T_{i_r+\tau}$ for $1 \leq i_1 < \cdots < i_r \leq p + 1$ and $0 < \tau \leq p + 1 - i_r$ (all integers).

Define $T_1^{(i)}, \cdots, T_{p+1}^{(i)}$ by

$$T_r^{(i)} = 1 \quad \text{if} \quad T_r = i \ (i = 1, \cdots, k; r = 1, \cdots, p + 1),$$
$$= 0 \text{ otherwise}$$

so that $T_r^{(i)}$ is the indicator for the ith treatment in the rth location in the window.

From this *randomized* version of the design we can recapture certain useful design quantities. Thus define

$$n_i = \#\{t \mid [t] = i\} \quad (i = 1, \cdots, k),$$

that is, the number of times treatment i occurs. We put $m_i = n_i/N$ and call the design treatment-balanced if $n_1 = n_2 = \cdots = n_k = N/k$. Define, for $r = 1, \cdots, p$,

$$\pi_r^{(ij)} = N^{-1}\#\{t \mid [t] = i, [t - r] = j, 1 \leq t \leq N\} \quad (i, j = 1, \cdots, k).$$

This is the proportion of times treatment j is r steps "behind" treatment i. (Here, $t = 1, \cdots, N$ but $t - r$ is identified, when ≤ 0, with $N + t - r$.) When $i = j$ define $\pi_r^{(i)} = \pi_r^{(ii)}$. Of considerable importance will be

$$\pi_r = \sum_{i=1}^k \pi_r^{(i)} \ (r = 1, \cdots, p),$$

which is the proportion of times the same treatment occurs r units apart, regardless of treatment label. We may connect the $T_r^{(i)}$ variables with these design quantities: since $T_r^{(i)}T_r^{(i')} = \delta_{ii'}T_r^{(i)}$ where $\delta_{ii'}$ = Kronecker symbol,

$$E(T_r^{(i)}) = m_i, \quad E(T_r^{(i)}T_r^{(i')}) = \delta_{ii'}m_i,$$

and, whether or not $i = i'$,

$$E(T_j^{(i)}T_{j+r}^{(i')}) = \pi_r^{(ii')} \quad \text{for } r > 0.$$

We may also express the matrix C in terms of these quantities. If we consider the individual contributions to the entries A_{ij} of A_d from different positions of the window, we can obtain these entries as expectations with respect to the window process $\{T_j^{(i)}\}$. Thus, writing $\rho_0 = 1$, whether or not $i = i'$ we have

$$\text{(8)} \quad \begin{aligned} N^{-1}A_{ii'} &= E\{(\textstyle\sum_{r=0}^{p} \rho_r T_{p+1-r}^{(i)})(\sum_{r=0}^{p} \rho_r T_{p+1-r}^{(i')})\} \\ &= \delta_{ii'}m_i(\textstyle\sum_{r=0}^{p} \rho_r^2) + 2\sum_{r=1}^{p} \pi_r^{(ii')} \sum_{s=0}^{p-r} \rho_s\rho_{s+r}. \end{aligned}$$

Since $\sum_i T_j^{(i)} = 1$, we have $N^{-1}\sum_{i=1}^{k}\sum_{i'=1}^{k} A_{ii'} = (\sum_{r=0}^{p} \rho_r)^2$. Consequently,

$$\text{(9)} \quad \begin{aligned} N^{-1}\text{trace}(C) &= N^{-1}\text{trace}(PAP) = N^{-1}\text{trace } A - (Nk)^{-1}\textstyle\sum_{i=1}^{k}\sum_{i'=1}^{k} A_{ii'} \\ &= \textstyle\sum_{r=0}^{p} \rho_r^2 - k^{-1}(\sum_{r=0}^{p} \rho_r)^2 + 2\sum_{r=1}^{p} \pi_r \sum_{s=0}^{p-r} \rho_s\rho_{s+r}. \end{aligned}$$

4. Minimax designs on the circle. We are now in a position to translate the conditions of minimaxity in Lemma 1 into conditions on the design. First consider the condition on the trace, (5). Write the covariance matrix of $T^{(i)} = (T_{p+1}^{(i)}, \cdots, T_1^{(i)})^T$ in the form

$$\Gamma^{(i)} = \begin{bmatrix} \tilde{\Gamma}^{(i)} & \gamma^{(i)} \\ \gamma^{(i)T} & m_i(1 - m_i) \end{bmatrix}.$$

We also write Γ, $\tilde{\Gamma}$, γ for the sum over i of the corresponding quantities. Also write $\bar{\rho}^T = (1, \rho^T) = (1, \rho_1, \cdots, \rho_p)$ and $\bar{T}^{(i)} = \bar{\rho}^T T^{(i)}$. In place of (8) we can also write

$$\begin{aligned} N^{-1}A_{ii} &= E(\bar{T}^{(i)})^2 = E(\bar{\rho}^T T^{(i)})^2 \\ &= \bar{\rho}^T[\Gamma^{(i)} + (ET^{(i)})(ET^{(i)})^T]\bar{\rho} \\ &= \bar{\rho}^T[\Gamma^{(i)} + m_i^2 J_{p+1}]\bar{\rho}. \end{aligned}$$

From this and the first line of (9), with e the p-vector $(1, 1, \cdots, 1)^T$ and $S = \sum_i(m_i - k^{-1})^2$, we obtain

$$\begin{aligned} N^{-1}\text{trace } C &= \bar{\rho}^T[\Gamma + SJ_{p+1}]\bar{\rho} \\ &= 1 - k^{-1} + 2\rho^T[\gamma + Se] + \rho^T[\tilde{\Gamma} + SJ_p]\rho. \end{aligned}$$

When $h = [\gamma + Se]$ is in the column space of $H = [\tilde{\Gamma} + SJ_p]$, the minimum of $2\rho^T h + \rho^T H\rho$ is $-h^T H^- h$, attained at $\rho = -H^- h$ (not uniquely if H is singular). Thus,

$$\text{(10)} \quad N^{-1}\min_\rho \text{trace}(C(\rho)) = 1 - k^{-1} - [\gamma + Se]^T[\tilde{\Gamma} + SJ_p]^-[\gamma + Se].$$

This is clearly maximized when $S = 0$ and $\gamma = 0$. The other case gives the same values.

Since $\Gamma^{(i)}$ and hence Γ are circulant, $\gamma = 0$ implies that all of the off-diagonal elements of Γ are zero. Thus max min trace(C) is achieved when

$$m_1 = m_2 = \cdots = m_k = k^{-1}\text{(treatment balance)},$$
(11)
$$\Gamma \equiv \textstyle\sum_i \Gamma^{(i)} = (1 - k^{-1})I_{p+1}.$$

Then it is easily seen that (11) is equivalent to $m_1 = \cdots = m_k = k^{-1}$ and $\pi_r = k^{-1}$ $(r = 1, \cdots, p)$. Now (11) says that the average covariance matrix of the separate indicators for the separate treatments in the window process is a nonzero multiple of the identity matrix. It is, of course, sufficient for (11) that

(12) $$\Gamma^{(i)} = k^{-1}(1 - k^{-1})I_{p+1}$$

$(i = 1, \cdots, k)$ so that each of the indicator processes (considered by itself) is uncorrelated. We shall return to this in a moment.

The condition in Lemma 1 that $A_{d^*}(\rho)$ be completely symmetric for all ρ is more restrictive. Assume that $m_1 = \cdots = m_k = k^{-1}$ already holds. Then it is easy to see, looking at (8), that all $A_{ii'}$ equal and A_{ii} all equal forces

(13) $$\pi_r^{(i_1,i_1')} = \pi_r^{(i_2,i_2')} (r = 1, \cdots, p)$$

for all $i_1 \neq i_1'$ and $i_2 \neq i_2'$, as well as

(14) $$\pi_r^{(i)} = \pi_r^{(i')}$$

for all $i \neq i'$ $(r = 1, \cdots, p)$. This says that any pair of different treatments (i, i') $(i \neq i')$ is r units apart the same number of times as any other pair of different treatments and any pair of identical treatments is r units apart the same number of times as any other pair of identical treatments. We call a design with properties (13) and (14) *completely balanced*. Actually, since $m_i = \pi_r^{(i)} + \sum_{j \neq i} \pi_r^{(ij)}$, complete balance implies treatment balance, although it does not by itself imply $\pi_r = k^{-1}$ $(r = 1, \cdots, p)$. Furthermore (11) together with complete balance implies (12). Thus also $\pi_r^{(i)} = 1/k^2$ $(r = 1, \cdots, p)$. We summarize these results as follows.

THEOREM 1. *A sufficient condition for a design d^* to be universally minimax is that*
(1) *it is completely balanced,*
(2) $\pi_r = 1/k$ $(r = 1, p)$.

This theorem generalizes to arbitrary order Theorems 3.1.2 and 3.1.4 of Kiefer (1960).

In the next section we give a method of constructing designs having the properties in Theorem 1.

5. Cyclic codes as minimax designs. It is simple to show that complete balance and the "window uncorrelated property" ((1) and (2) of Theorem 1) can be obtained by allowing each of the k^{p+1} possible $(p + 1)$-tuples of treatment numbers to occupy our window of length $p + 1$ exactly once, as it moves around the circle. For example, let $k = 3$, $p = 2$ and label the treatment 0, 1, 2 rather than 1, 2, 3. Take the design with 27 units and treatments laid out as follows:

(15) 001101021222100220201211120.

The circle is completed by making the last "0" a neighbor of the first. As we move a window of length 3 along the sequence every triple i, j, k $(i, j, k = 0, 1, 2)$ occurs just once (200 and 000 cover the join of the two ends). Properties (1) and (2) of Theorem 1 are easily verified. Such a sequence in fact has a stronger property, namely, the random window exhibits not just uncorrelated entries but an *independent* 3-state process. A sequence with this property is called pseudo-random in the communications theory literature. The sequence was generated in the following way. Let $V_t = [t]$, the treatment label (now 0, 1 or 2) of the tth unit. Then the sequence was generated by the recurrence relation

$$V_t = V_{t-1} + 2V_{t-2} + 2V_{t-3} \tag{16}$$

where all integers are identified with their residues mod 3. To start with, take 001. Continuing up to the end of the first cycle excludes 000 so one "dummy" zero must be added at the end before joining up to form the circle. This is a special case of a full length cyclic error-correcting code. There is a huge literature on the subject and several excellent books (see, for example, Berlekamp, 1968, MacWilliams and Sloane, 1977); and, we confine ourselves to a brief summary.

Let $k = q^m$ where q is prime. Consider a general recurrence relationship

$$V_t = \textstyle\sum_{i=1}^{p} \alpha_i V_{t-1} \tag{17}$$

in which all treatment values V_t and coefficients α_i are identified with members of the Galois field $\mathrm{GF}(q^m)$ of order q^m. In order for the recurrence (17) to generate a full length cyclic code with the required property that every $(p + 1)$-tuple of elements appears just once (except for $00 \cdots 0$), the α_i must be of a special form.

Consider the extended Galois field consisting of all polynomials of degree p whose coefficients lie in $\mathrm{GF}(q^m)$. This is itself a field if every polynomial of higher order than p is identified with its residue mod some fixed polynomial of degree $p + 1$ which is irreducible over $\mathrm{GF}(q^m)$. The polynomial field (which can be identified with $\mathrm{GF}(k^p)$) has nonzero elements which form a cyclic group under multiplication. An irreducible polynomial $h(x)$ of degree $p + 1$ over $\mathrm{GF}(q^m)$ is called *primitive* if all the elements of the cyclic group are generated as powers of a solution x in $\mathrm{GF}(k^p)$ of $h(x) = 0$. The element x is said to be a primitive element of $\mathrm{GF}(k^p)$. A primitive polynomial always exists.

A necessary and sufficient condition that the recurrence (17) generate a full length code with the required property is that

$$h(x) = x^k - \alpha_1 x^{k-1} - \alpha_2 x^{k-2} - \cdots - \alpha_p$$

is a primitive polynomial over $\mathrm{GF}(q^m)$. Thus in our example (16), the polynomial is $x^3 - x^2 - 2x - 2$ $(\equiv x^3 + 2x^2 + x + 1)$, which is primitive over GF(3). The polynomial is not unique; $x^3 + 2x + 1$ would do equally well, producing a different sequence.

Lists of primitive polynomials are available in many books on algebra, coding theory and combinatorial theory. In the communications literature, the designs are called m-sequences or pseudo-random sequences (see MacWilliams and Sloane (1977), Chapter 14). Techniques are available for generating pseudo-random sequences for values of k other than prime powers (see Shedd and Sarware, 1979).

We have shown, then, that it is possible to find minimax designs exactly for certain values of N and k.

6. Asymptotic designs for the line. There is no obvious way of extending the results on the circle to results on the line in an exact fashion for *finite* N. For example, we may form a design with $N = 29$ by adding 00 to the end of (15) so that triples 200 and 000 are included. However, since the first two observations are included, the conditions for minimaxity are upset. Moreover, for fixed N there can be at most a finite number of different designs and so the set of possible design quantities m_i, π_r, $\pi_r^{(i)}$, etc. is finite. This means that some values of these quantities which are useful from an optimization point of view may not be achievable. We have seen in Section 5, for example, that we have to choose the value of N carefully to get an exactly minimax design. Following Kiefer (1960) we therefore allow N to tend to infinity and discuss allowable asymptotic values of the design quantities.

First we shall redefine the quantities to emphasize the dependence on N. The designs will be on the line although the cyclic nature of some designs will play an important role.

Let d_N be a design on the line; that is, $d_N = (V_1, \cdots, V_N)$ where, as before, $V_t = [t]$ is one of the integers $1, \cdots, k$ $(t = 1, \cdots, N)$. Let $\mathscr{D}_N$ be the set of all d_N so that $\#(\mathscr{D}_N) = k^N$. For any $d_N \in \mathscr{D}_N$ let

$$\pi_{r,N}^{(i)} = (N-p)^{-1}\#\{t \mid V_t = V_{t-r} = i, \quad p+1 \le t \le N\}$$

and

$$\pi_{r,N} = \textstyle\sum_{i=1}^{k} \pi_{r,N}^{(i)}.$$

If there is a sequence of designs $\{d_N \in \mathscr{D}_N, N \in \mathscr{N}\}$ where $\mathscr{N}$ is an unbounded set of natural numbers, for which, for $1 \le r \le k$,

$$\pi_{r,N} \to \text{a limit } \pi_r \text{ as } N \to \infty, \quad N \in \mathscr{N},$$

we say that $\pi_1, \cdots, \pi_p$ or simply the vector $\pi = (\pi_1, \cdots, \pi_p)$ is *asymptotically achievable* (aa). Throughout this section the π_r will be (aa) values.

The rest of this section is devoted to the difficult problem of finding the set of asymptotically achievable π_r values $(1 \le r \le p)$. We do not need to use the individual $\pi_{r,N}^{(i)}$ because a symmeterization procedure to be described shortly will yield sequences for which $\pi_{r,N}^{(i)} \to (\pi_r/k)$ $(N \to \infty)$ $(i = 1, \cdots, k; r = 1, \cdots, p)$. Let $\Pi_p \equiv \Pi_p(k)$ be the set of all (aa) vectors π for given k. The development is based on a connection between the (aa) π-vectors and the covariance of stationary discrete time processes. The technique is discussed in some detail in Kiefer and Wynn (1983) for the binary $(k = 2)$ case.

Consider the sequence for a given i, $\{Z_t\}_0^\infty$:

$$Z_t = 1 \quad \text{if} \quad V_t = i$$
$$= -1, \quad \text{otherwise.}$$

Then if

$$C_{r,N} = \frac{1}{N-p} \sum_{t=p+1}^{N} Z_t Z_{t-r},$$

we have $\pi_{r,N} = ½ (1 + C_{r,N})$. Now $C_{r,N}$ is essentially the sample covariance function of the $\{Z_t\}$ values. If $\pi_{r,N} \to \pi_r$ $(r = 1, \cdots p)$ then $C_{r,N} \to 2\pi_r - 1$. If the $\{Z_t\}$ values are generated as a realization of a stationary ergodic zero mean binary process then $C_{r,N} \to C_r$ with probability one where the C_r is the covariance function of the process. If $(1/N) \sum Z_t \to 0$ the set of π_r values $(r = 1, \cdots, p)$ is contained in the set $½ (C_r + 1)$ values allowable as the covariance of a zero mean stationary binary process. Work by Shepp (1967) and Masry (1972) show that the two sets are the same and that the zero mean condition is irrelevant. We follow the results of Hobby and Ylvisaker (1964) to characterize stationary k state processes in terms of the distribution induced in a window of $p + 1$ neighbouring time points. This theory can be considered as the correct way to "unwrap" the structure of designs on a circle discussed in Section 3 onto the real line (discrete time).

Let $\{Z_t\}_0^\infty$ be a stationary k state process. Consider a window of $p + 2$ neighbouring time points, say $t = 0, 1, \cdots, p + 1$. In this window $\{Z_t\}$ induces a distribution $Z_0, \cdots, Z_{p+1}$. The stationarity property says that we can express the marginal joint distribution of $Z_1, \cdots, Z_p$ in two ways, summing over Z_0 or Z_{p+1}.

$$\text{Prob}(Z_1 = i_1, \cdots, Z_p = i_p)$$
$$= \sum_{i_0=1}^{k} \text{Prob}(Z_0 = i_0, Z_1 = i, \cdots, Z_p = i_p)$$
$$= \sum_{i_{p+1}=1}^{k} \text{Prob}(Z_1 = i_1, \cdots, Z_{p+1} = i_{p+1}).$$

The equations are necessary conditions on the quantities $\text{Prob}(Z_0 = i_0, \cdots, Z_p = i_p)$ which characterise the distribution induced on the $p + 1$ window. In fact the conditions are also sufficient. Moreover they lead to a very simple characterisation of the *extreme* distributions, analogous to the characterisation of exchangeable distributions. The results are also found in unpublished notes of the late Walter Weissblum (Bell Telephone Laboratories.)

THEOREM (Hobby-Ylvisaker-Weissblum). *The set of all extreme distributions induced in a $p + 1$ window by a stationary k state discrete process are all those for periodic processes which do not repeat a p-tuple within a cycle (together with a uniform random phase shift).*

To explain the theorem, consider the case $p = 2$, $k = 2$. We may list all the strings which do repeat a neighbouring pair of integers (0, 1). They are (up to permutation of the labels 0, 1):

sequences	**π-vector**
$\overline{0}$	(1, 1)
$\overline{01}$	(0, 1)
$\overline{001}$	(1/3, 1/3)
$\overline{0011}$	(1/2, 0)

Here the "bar" notation means we repeat the sequence infinitely often. The periodic processes obtained by giving each periodic sequence a uniform random phase shift is a stationary process. It is clear that no extreme cycle can have period greater than k^p.

The (aa) π_r values are merely the convex hull of the π-vectors from each of the periodic sequences. Every extreme π-vector derives from an extreme distribution but not vice-versa (see the example above). We summarize the result as

COROLLARY. *The set $\Pi_p(k)$ is the convex hull of π-vectors generated by a finite number of cyclic designs.*

The convexity of $\Pi_p(k)$ can also be established by working directly with infinite design sequences. Let π and π' be two vectors in Π_p. Let $\{d_N, N \in \mathcal{N}\}$ and $\{d'_N, N \in \mathcal{N}'\}$ be their defining sequences. Let $\alpha = r/s$ be rational, $0 \le \alpha \le 1$, with r and s integers. Take d_N with $N \in \mathcal{N}$ and $d'_{N'}$ with $N' \in \mathcal{N}'$. Form a design $d''_{N''}$ with $N'' = Nr + N'(s - r)$ formed by writing $N'(s - r)$ copies of d_N followed by Nr copies of $d'_{N'}$. Let $N, N' \to \infty$, $N \in \mathcal{N}$ and $N' \in \mathcal{N}'$. Then it can be shown with a little analysis that the sequence $\{d''_N, N \in \mathcal{N}''\}$, where $\mathcal{N}'' = \{N''\}$, gives the (aa) value $(1 - \alpha)\pi + \alpha\pi'$. The extension to irrational α is again simple. This is an alternative to the ingenious method of Shepp (1967) for "mixing" together two infinite sequences which preserves the π_r (C_r) values for all $1 \le r < \infty$.

Given any sequence (design) with a given π_r vector, it is always possible to find a symmetric version for which $\lim \pi_{r,N}^{(i)} = \pi_r/k$ ($i = 1, \cdots, k$). This is done by mixing, using our method or the method of Shepp, all the sequences obtained by permuting the treatment labels.

We prove one more general result on Π_p before analyzing the structure of Π_1, Π_2 and Π_3. The form of Π_p depends not just on p but also on k, the number of treatments. However, for large k this dependence ceases.

LEMMA 2. *For all $k \ge p + 1$ the $\Pi_p(k)$ are identical.*

PROOF. Let π be a point in $\Pi_p(k)$ for $k > p + 1$. Let it be generated by $\{d_N, N \in \mathcal{N}\}$. Consider a typical $d_N = \{V_1, V_2, \cdots, V_N\}$. We shall convert d_N to d'_N which has all $V_t \le p + 1$ without changing its $\pi_{r,N}$ values. If no $V_t > p + 1$, we are done. Consider the first unit t for which $V_t > p + 1$. Select a treatment number j with $1 \le j \le p + 1$ but $j \ne V_{t-s}$, $s = 1, \cdots, p$. (Ignore any values $t - s < 1$.) This is obviously possible. Now interchange the treatment label j with the treatment label V_t both at unit t and also wherever they occur after t. This does not effect $\pi_{r,N}$. Performing this operation repeatedly on this and on all other d_N, $N \in \mathcal{N}$, we create a sequence $\{d'_N, N \in \mathcal{N}\}$ generating π but which by definition is now a point of $\Pi_p(k)$ with $k = p + 1$. The reverse inclusion is trivial. □

We state Π_p for $p = 1, 2$ and different values of k.

$$\Pi_1 = \{\pi_1 \mid 0 \le \pi_1 \le 1\} \qquad \text{for } k \ge 2,$$

$$= \{1\} \qquad \text{for } k = 1.$$

$$\Pi_2 = \{(\pi_1, \pi_2) \mid 0 \le \pi_2 \le 1, 0 \le 2\pi_1 \le 1 + \pi_2\} \qquad \text{for } k \ge 3,$$

$$= \{(\pi_1, \pi_2) \mid \pi_2 \le 1, 1 - \pi_2 \le 2\pi_1 \le 1 + \pi_2\} \qquad \text{for } k = 2,$$

$$= \{(1, 1)\} \qquad \text{for } k = 1.$$

Finding Π_1 is trivial. Kiefer (1960) essentially discovered the Π_2 region.

Extensive computations have yielded $\Pi_3(k)$ for $k = 2$, 3 and ≥ 4. The case $k = 2$ (binary) is covered by Martins de Carvalho and Clark (1983) up to $p = 5$. The case $p = 3$, $k = 2$ was known to Shepp (private communication). The essential method is to search among the extreme distributions for the extreme π-vectors by testing "new" vectors to see if they lie in the convex hull of the "old" ones.

Consider now for $p = 3$ the following designs $d^1, \cdots, d^{10}$ written in cyclic notation and using capital letters. The π-vectors follow each design.

d^1:	$\overline{A}$	(1, 1, 1)
d^2:	$\overline{AB}$	(0, 1, 0)
d^3:	$\overline{AAB}$	(1/3, 1/3, 1)
d^4:	$\overline{AABB}$	(1/2, 0, 1/2)
d^5:	$\overline{AAABBB}$	(2/3, 1/3, 0)
d^6:	$\overline{ABC}$	(0, 0, 1)
d^7:	$\overline{AABBCC}$	(1/2, 0, 0)
d^8:	$\overline{ACBC}$	(0, 1/2, 0)
d^9:	$\overline{ABBCAABCC}$	(1/3, 0, 0)
d^{10}:	$\overline{ABCD}$	(0, 0, 0).

These generate the following ten half-spaces $H_1, \cdots, H_{10}$.The designs whose

π-vectors lie on the planes (sometimes more than 3) are given in brackets.

H_1,	$\pi_1 \geq 0$	(d^2, d^6, d^8, d^{10})
H_2,	$\pi_2 \geq 0$	$(d^4, d^6, d^7, d^9, d^{10})$
H_3,	$\pi_3 \geq 0$	$(d^2, d^5, d^7, d^8, d^9, d^{10})$
H_4,	$\pi_1 - \pi_2 - \pi_3 + 1 \geq 0$	(d^1, d^2, d^3, d^6)
H_5,	$-\pi_1 - \pi_2 + \pi_3 + 1 \geq 0$	(d^1, d^2, d^5)
H_6,	$-2\pi_1 + \pi_2 + 1 \geq 0$	(d^1, d^4, d^5, d^7)
H_7,	$-\pi_1 + \pi_2 - \pi_3 + 1 \geq 0$	(d^1, d^3, d^4, d^6)
H_8,	$3\pi_1 + 2\pi_2 + \pi_3 - 1 \geq 0$	(d^6, d^8, d^9)
H_9,	$\pi_1 + \pi_2 + \pi_3 - 1 \geq 0$	(d^2, d^4, d^5)
H_{10},	$2\pi_1 + \pi_2 - 1 \geq 0$	(d^2, d^3, d^4).

$k \geq 4$. The extreme points are given by d^1, d^2, d^4, d^5, d^6, d^7 and d^{10}, and the determining half-spaces are H_1, H_2, H_3, H_4, H_5, H_6 and H_7. See Figure 1.

$k = 3$. The point d^{10}, $ABCD$, (0, 0, 0) is no longer available, and the extreme points are those for $k > 3$ with d^{10} deleted and d^8 and d^9 adjoined. The half-spaces are those of the previous case with the addition of H_8. Each of the eight determining planes is seen to be generated by 3 or 4 of the extreme points. See Figure 2.

$k = 2$. The points d^6, d^7, d^8, d^9 of the case $k = 3$ (and, of course, d^{10} of $k \geq 4$) are no longer available. The remaining four points and d^3 are now extreme: d^1, d^2, d^3, d^4, d^5. Since the π-vector for $AAB = d^3$ is on the segment from the π-vectors for $A = d^1$ to $ABC = d^6$, we still obtain the planes determining H_4 and H_7, and similarly H_5 and H_6. But the planes associated with H_1, H_2, H_3 and H_8 are no longer determined by these extreme points and are lost. Two new half-spaces are adjoined and we have Π_3 determined by H_4, H_5, H_6, H_7, H_9 and H_{10}. See Figure 3.

We complete this section with a brief discussion of random allocation. It is obvious that complete random allocation independently at each time point (Bernoulli) gives with probability one $\pi_r = 1/k$ $(r = 1, \cdots, p)$. Alternatively we may use a length of a pseudo random code as explained in Section 5 ignoring edge effects.

Random allocation of treatments to units has been widely advocated and is probably the most frequently used procedure, expecially in clinical trials (see for example Zelen, 1974). Under classical assumptions completely random allocation

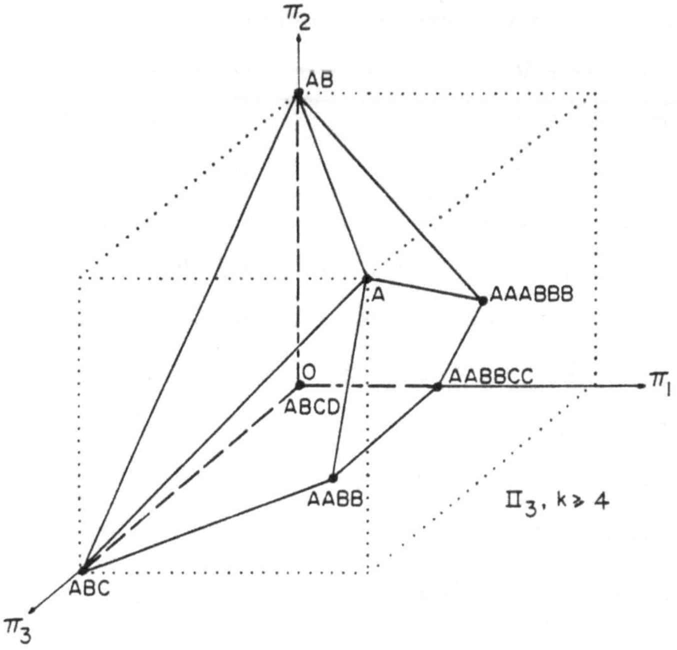
π2
AB
A
AAABBB
O
AABBCC
π1
ABCD
AABB
Π3, k ≥ 4
ABC
π3

FIG. 1

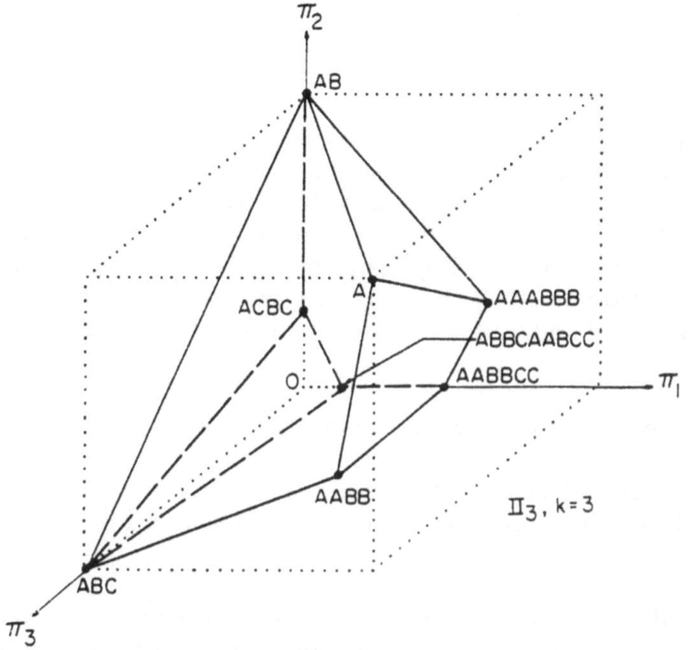
π2
AB
A
AAABBB
ACBC
ABBCAABCC
O
AABBCC
π1
AABB
Π3, k = 3
ABC
π3

FIG. 2

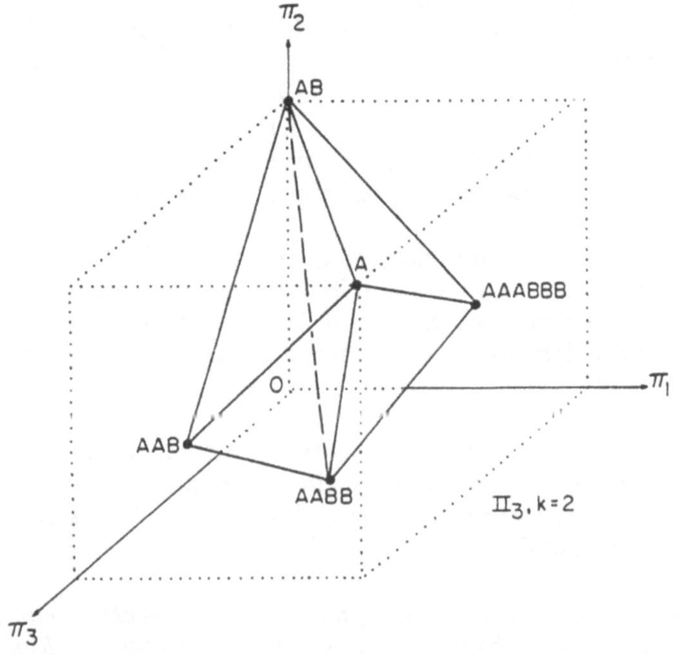

FIG. 3.

has been shown to be minimax by Blackwell and Hodges (1957) in the case of two treatments. They refer to the designs as "truncated binomal." Work by Stigler (1969), Efron (1971), and more recently, Wei (1977, 1978) advocates "adaptive" or Markovian methods of allocation. Complete randomization or a more restricted version certainly seems attractive if only to guard against deliberate or unwitting treatment/unit partiality. See also recent work by Atkinson (1982), Smith (1983).

Suppose that, indeed, the treatments are allocated according to a Markov chain of particular order, say L. We may call such a procedure purely random if the chain considered as a first order Markov chain on the states consisting of all L^k of the L-tuples is irreducible and aperiodic. Applying the Ergodic Theorem, we can say that the chain will, with probability one, give a realization which, considered as an infinite design $\{V_1, V_2, \cdots\}$, generates an $(a\ a)$ π-vector. Moreover, for the same chain every such π-vector is the same. The Markov chain thus uniquely determines the π-vector and, of course, the vector is easily obtained from the limiting distribution of the chain, over the L-tuples. We shall be interested in the extent to which the converse is the case: Can we with probability one obtain a *given* π-vector using the sample path of a Markov chain of sufficiently high (but finite) order?

Consider such a sample path. Then with probability zero it will be an infinite cyclic design. However, more strongly, since it gives positive limiting probability to every L-tuple (being nontransient), it must give it likewise to every π-vector corresponding to the $\pi_{r,L}$ of an L-tuple. This argument shows that *no* extreme point of a Π_p can be obtained in this way. In fact, no purely random Markovian sampling scheme can be optimum in the sense of Section 7. As a simple illustration let $L = 2, p = 1$ and suppose the second-order chain attaches limiting probabilities α_1, α_2, α_3 and α_4 to the pairs 11, 12, 21 and 22 respectively. Then with probability one any sample path gives a π-vector (now a single value) $0 < \pi_1 = \alpha_1 + \alpha_4 < 1$, which is never extreme. The details are straightforward in the general case.

A cyclic design is made stationary by giving it a random shift. However, such a random process is not an irreducible Markov aperiodic chain of any order, but a purely deterministic process. (See Doob (1953) for details of these distinctions.) In sampling terminology we have a systematic sample following a random start. We can obtain the same process by a periodic chain with a random initial distribution but this rather destroys our idea of a purely random sampling scheme.

Despite these careful distinctions between deterministic and nondeterministic schemes it is possible to obtain an irreducible aperiodic Markovian scheme whose π-vector is within ε of any given π-vector. We write $\| \ \|$ for the L_∞-norm in what follows. The proof is omitted.

THEOREM 3. *Let π in Π_p be an asymptotically achievable vector generated by a sequence $\{d_N, N \in \mathcal{N}\}$. Then given $\varepsilon > 0$, there is an irreducible Markov aperiodic chain of finite order such that with probability one its sample path generates a π' in Π_p with $\| \pi - \pi' \| < \varepsilon$.*

7. Optimum designs on the line. Now that the class of π-vectors has been defined for infinite designs (or technically *sequences* of designs) we are in a better position to carry out the optimization and minimax procedures of Section 2. To get asymptotic results as $N \to \infty$, $N \in \mathcal{N}$ we need to do two things: (1) use the limiting interpretation of the π_r from Section 6; and (2) choose $\rho = (\rho_1, \cdots, \rho_p)$ which makes the process $\{Y_t\}$ satisfying

$$Y_t + \rho_1 Y_{t-1} + \cdots + \rho_p Y_{t-p} = \varepsilon_t \quad (t = 1, 2, \cdots)$$

stationary in the usual wide sense. Here $\{\varepsilon_t\}$ is an uncorrelated sequence with finite variance σ^2. The details of the asymptotic development, which we omit, allow us to write "$+ O(N^{-1})$, $N \in \mathcal{N}$" after the key expression (8) and (9). Identifying d with the design sequence $\{d_N, N \in \mathcal{N}\}$ we define C_d as $\lim(N^{-1}C_{d_N})$ as $N \to \infty$, N in $\mathcal{N}$, when the limit exists. All the results go through and it will be sufficient for C_d to have the form satisfying (a) and (b) of Section 2 for optimality and (1) and (2) of Theorem 1 for minimaxity, the π_r now being interpreted as limiting π_r values. Both cases require complete balance and we turn to the idea of asymptotic symmetrization.

If we have a design sequence $\{d_N, N \in \mathcal{N}\}$ with given π_r values (say from a minimax or optimization argument) it is an easy matter to construct a sequence of $\{\tilde{d}_N, N \in \tilde{\mathcal{N}}\}$ with the same π_r values but which is asymptotically completely symmetric. Let $d_N^s (s = 1, \cdots, k!)$ be all designs of length N constructed from d_N by permutation of treatment labels. Let $\tilde{N} = k!N$ and write all the d_N^s in sequence so that $\tilde{d}_{\tilde{N}} = d_N^1 d_N^2 \cdots d_N^{k!}$. Then clearly letting $\tilde{\mathcal{N}}$ be $\{\tilde{N}\}$ we obtain a sequence with the required property. Note that it is sometimes possible to symmetrize using less than $k!$ permutations. Indeed the design may already be completely balanced.

The advantage of using codes for $k = q^m$ in Section 5 to construct minimax designs is that N is fairly small. Repetition of the code (with the adjoined dummy "O" treatment) will give an asymptotically minimax design. This seems an excellent procedure where it is possible. However, since the point $(1/k, \cdots, 1/k)$ always lies in Π_p, as can be easily seen, we can *always* construct asymptotically minimax designs by first finding a $\pi = (1/k, \cdots, 1/k)$ by suitable mixing of cyclic extreme point designs and then symmetrizing. Stopping at any $N \in \tilde{\mathcal{N}}$ then gives an almost (to within $O(N^{-1})$) minimax design. The details are left to the reader.

We want to concentrate now on optimization. Let $\rho = (\rho_1, \cdots \rho_p)$ be fixed. We have to maximize trace(C_d) (in the limit). But this maximum is given by a design with (1) $m_1 = \cdots = m_k$ and (2) maximum value of

$$\sum_{r=1}^{p} \pi_r \sum_{s=0}^{p-r} \rho_s \rho_{s+r} \quad (\rho_0 = 1). \tag{18}$$

Point (1) we can ignore because symmetrization will yield it automatically. Thus we are left with a *linear programming* problem to maximize (18) over Π_p. This can be carried out either by using a standard algorithm on the relevant half-space constraints, or, as the extreme points (vertices) are listed (not usually the case in LP), by just running through them.

For designs on the circle, we may select any value for the vector ρ and if we are lucky enough to have N just a multiple of the length of the correct extreme symmetric cyclic design, then that design or a repetition of it will be optimum. For designs on the line we need ρ to satisfy the stationarity condition that the roots of

$$x^p + \rho_1 x^{p-1} + \cdots + \rho_p = 0$$

lie inside the unit circle. Then with that value of ρ we (1) find the optimum π-vector and the corresponding design, (2) repeat it a reasonable number of times (to make the edge effect small), (3) symmetrize it if it is not already symmetric.

As a simple example take $p = 3$ and $\rho = (-0.4, -0.3, -0.1)$. Negative ρ-values correspond to positive autocorrelation. We suspect that a reasonable amount of separation between occurences of the same treatment will be optimum. Expression (18) becomes

$$-0.25\pi_1 - 0.26\pi_2 - 0.1\pi_3.$$

When $k \geq 4$ this is maximized uniquely in Π_3 at d_{10}, $\overline{ABCD}$ when $\pi = (0, 0, 0)$. Thus with symmetrization taking all permutations of 1234, we obtain for large N an approximately optimum design 12341234 $\cdots$ 21342134 $\cdots$, etc. When $k =$ 3 this design is not available. The unique optimum design is d^9, $\overline{ABBCAABCC}$. Symmetrization yields 122311233 $\cdots$ 211322133 $\cdots$ etc. When $k = 2$ the unique optimum design is d_4, *AABB*, given by 11221122 $\cdots$ without further symmetrization. The full computer implementation of the ideas of this paper is still being carried out. The steps are (1) choice of autocorrelation, possibly after preliminary estimation, (2) choice of the number of treatments and sample size N, (3) elucidation of the extreme points and associated region for given p and k, (4) choice of optimum (cyclic) extreme design, and (5) packing of the sample length N with a suitably symmetrized version of the extreme designs.

As a final exercise we show that sometimes an optimum design can be found for large values of p, despite the fact that Π_p has not yet been fully worked out for large p. Consider, then, the first order moving average model on the circle given by

$$Y_t = \varepsilon_t - \lambda\varepsilon_{t-1} \quad (t = 1, \cdots, N)$$

where $|\lambda| < 1$ and the ε_t are uncorrelated with variance σ^2, as before. Rewriting this in an auto-regressive formulation,

$$\rho = \frac{1}{1 - (-\lambda)^N} (1, -\lambda, \lambda^2, \cdots, (-\lambda)^{N-1}).$$

Then (18) becomes

$$[1 - (-\lambda)^N]^{-1} \textstyle\sum_{r=1}^{N-1} \pi_r \sum_{s=0}^{N-r} (-\lambda)^{2s+r}. \tag{19}$$

An informal analysis runs as follows. Letting $N \to \infty$ expression (19) becomes

$$-\alpha_1\pi_1 + \alpha_2\pi_2 - \alpha_3\pi_3 + \cdots, \lambda > 0,$$

$$\alpha_1\pi_1 + \alpha_2\pi_2 + \alpha_3\pi_3 + \cdots, \lambda < 0,$$

where $\alpha_i \geq 0$, $i = 1, 2, \cdots$. Thus since this must be maximized and $0 \leq \pi_r \leq 1$ for all r, when $\lambda > 0$, a π-vector $(0, 1, 0, 1 \cdots)$ would be (asymptotically) optimum. Fortunately such a π-vector always lies in Π_p, namely, that arising from AB (repeated). Thus an approximately optimum design for the first order MA model is to take $\overline{AB}$ symmetrized, that is, 121212 $\cdots$ 131313 $\cdots$ 232323 $\cdots$, etc. This design is intuitively appealing for a model only having a 1-step local autocorrelation.

Similarly, when $\lambda < 0$, the point A, with π-vector $(1, 1, 1, \cdots)$ yields an approximately optimum design 111 $\cdots$ 11222 $\cdots$ 22 $\cdots$ $kkk \cdots kk$.

We have discussed thus far in this section maximization of (18) for a particular vector ρ. One can also compare approximately balanced designs in terms of the integral of (18) wrt a positive measure on ρ's (approximately Bayes designs). Related is the notion of asymptotic admissibility of design sequences in terms of the behavior of (18) on the space of ρ's, of concern when ρ is unknown. Since (18) is linear in π, every (approximately balanced) design is seen to be approximately admissible, for large N. In practice when ρ is unknown and N is large, one would presumably replace the BLUE (MVUE) by an adaptive estimator that estimates ρ.

Acknowledgements. This paper benefitted from the advice of David Brillinger, Larry Shepp and Persi Diaconis on the communications theory literature.

REFERENCES

ATKINSON, A. C. (1982). Optimum biased coin designs for sequential clinical trials with prognostic factors. *Biometrika* **69** 61–67.

BERENBLUTT, I. I. and WEBB, G. I. (1974). Experimental designs in the presence of autocorrelated errors. *Biometrika* **61** 427–437.

BERLEKAMP, E. R. (1968). *Algebraic Coding Theory*. McGraw-Hill, New York.

BICKEL, P. J. and HERZBERG, A. M. (1979). Robustness of designs against autocorrelation in time. I. Asymptotic theory for location and linear regression. *Ann. Statist.* **7** 77–95.

BLACKWELL, D. H. and HODGES, J. L. (1957). Design for the control of selection bias. *Ann. Math. Statist.* **28** 449–460.

COX, D. R. (1951). Some systematic experimental designs. *Biometrika* **38** 312–323.

DOOB, J. L. (1953). *Stochastic Processes*. Wiley, New York.

DUBY, C., GUYON, X. and PRUM, B. (1977). The precision of different experimental designs for a random field. *Biometrika* **64** 59–66.

EFRON, B. (1971). Forcing a sequential experiment to be balanced. *Biometrika* **58** 403–417.

HOBBY, C. and YLVISAKER, N. D. (1964). Some structure theorems for stationary probability measures on finite state spaces. *Ann. Math. Statist.* **35** 550–556.

KIEFER, J. (1958). On the randomized optimality and randomized non-optimality of symmetric designs. *Ann. Math. Statist.* **29** 675–699.

KIEFER, J. (1960). Optimum experimental designs. *V*, with applications to systematic and rotatable designs. *Proc. Fourth Berk. Symp. Math. Statist. Probab.* **2** 381–405.

KIEFER, J. (1975). Construction and optimality of generalized Youden designs. In *A Survey of Statistical Design and Linear Models* 333–353. (J. N. Srivastava, ed.) North-Holland, Amsterdam.

KIEFER, J. and WYNN H. P. (1981). Optimum balanced block and latin square designs for correlated observations. *Ann. Statist.* **9** 737–757.

KIEFER, J. and WYNN, H. P. (1983). Autocorrelation-robust design of experiments. *Scientific Inference and Data Analysis and Robustness*. Academic, New York.

MACWILLIAMS, F. J. and SLOANE, N. J. A. (1977). *The Theory of Error-Correcting Codes*. North-Holland, Amsterdam.

MARTIN, R. J. (1977). Spatial models with applications in sampling and experimental design. Ph.D. thesis, London School of Economics, University of London.

MARTIN, R. J. (1979). A subclass of lattice processes applied to a problem in planar sampling. *Biometrika* **66** 209–217.

MARTINS DE CARVALHO, J. L. and CLARK, J. M. C. (1983). Characterizing the autocorrelations of binary sequences. *IEEE Trans. Inform. Theory* **IT-29** 502–508.

MASRY, E. (1972). On covariance functions of unit processes. *Siam J. Appl. Math.* **23** 28–33.

SHEDD, D. A. and SARWATE, D. V. (1979). Construction of sequences with good correlation properties. *IEEE Trans. Inform. Theory.* **IT-25** 94–97.

SHEPP, L. A. (1967). Covariances of unit processes. *Proc. Working Conf. on Stochastic Process.* 205–218. Santa Barbara, Calif.

SMITH, R. L. (1983). Sequential treatment allocations using biased coin designs. *J. Roy. Statist. Soc. B.*, to appear.

STIGLER, S. M. (1969). The use of random allocation for control of selection bias. *Biometrika* **56** 553–560.

WEI, L. J. (1977). A class of designs for sequential clinical trials. *J. Amer. Statist. Assoc.* **72** 382–386.

WEI, L. J. (1978). The adaptive biased coin design for sequential experiments. *Ann. Statist.* **6** 92–100.

WILLIAMS, R. M. (1952). Experimental designs for serially correlated observations. *Biometrika* **39** 151–167.

ZELEN, M. (1974). The randomization and stratification of patients to clinical trials. *J. Chronic Diseases* **27** 365–375.

ZYSKIND, G. (1967). On canonical forms, non-negative covariance matrices and best and simple least squares linear estimators in linear models. *Ann. Math. Statist.* **38** 1092–1109.

DEPARTMENT OF STATISTICS
EVANS HALL
UNIVERSITY OF CALIFORNIA
BERKELEY, CALIFORNIA 94720

DEPARTMENT OF MATHEMATICS
IMPERIAL COLLEGE
180 QUEENS GATE
LONDON SW7 2BZ
ENGLAND

D-Optimality of the GYD for $v \geq 6$

J. KIEFER[1]
Cornell University

In "Construction and Optimality of Generalized Youden Designs," *Proc. 1973 Ft. Collins Conference*, there appeared part of the derivation of optimality properties of nonregular GYD's, announced earlier by the author. (Nomenclature and notation will be taken from that paper.) A-optimality was proved in all nonregular cases, complementing the stronger universal optimality that holds in regular cases. In the course of this, a method was developed for treating other optimality criteria.

The present note, to be included in a longer paper on exact optimality, is a response to several inquiries about the proof of D-optimality for $v \geq 6$ and completes this proof based on the development just mentioned.

That development, summarized in Section 3(f) of the earlier paper, reduces the proof of D-optimality (by substitution of $q(r) = \log g(r)$ into (3.17)) to showing that the function g of (3.17) is such that $\log g(r+1) - \log g(r)$ is nonincreasing in the "basic interval" of integers; that is, that

$$g^2(r) - g(r-1)g(r+1) \geq 0 \quad \text{for} \quad C_0 < r < D_0. \tag{1}$$

Substituting $g(r-1) = g(r) - \Delta(r-1)$ and $g(r+1) = g(r) + \Delta(r)$ (see (3.8)), this becomes

$$\begin{aligned} C_0 < r < D_0 \Rightarrow 0 &\leq \Delta(r)\Delta(r-1) + g(r)[\Delta(r-1) - \Delta(r)] \\ &= \Gamma_0(r) \text{ (say)}. \end{aligned} \tag{2}$$

(Incidentally, the greater ease of proving A-optimality appears here explicitly, in that the expression $\Gamma(r)$ of (4.2), which for A-optimality corresponds to (2) here, has the negative term in square brackets in (2) multiplied by $\frac{1}{2}$.) From (2) we have on

[1]Research under NSF Grant GP35816X.

$[C_0+1, D_0-1]$

$$\Gamma_0(r) = 2r^2 + 2\beta r + \beta^2 - 2\alpha - 1, \tag{3}$$

where α, β are as defined in the earlier paper. Considered as a quadratic function in *real* r, this has nonnegative derivative for $r \geq C_0 + 1$ since again $\beta + 2r \geq 0$ for $r \geq C_0 + 1$. Thus, as in the derivation of (4.3), (2) will follow from

$$\Gamma_0(C_0+1) \geq 0. \tag{4}$$

We now proceed in a development similar to that used for $v \geq 6$ ("Case 1") in proving A-optimality. The argument of replacing B_2 by $B_1 b_1 / b_2$ to obtain (4.4) again applies and now yields

$$\Gamma_0(C_0+1) \geq 6C_0^2 + [-4 + 4\sigma - 6\pi]C_0 + \{1 + 2[\pi - \sigma] + [\pi - \sigma]^2\} \tag{5}$$

where again $\pi = b_1 b_2$ and $\sigma = b_1 + b_2$ with both $b_i \geq 8$ (by nonregularity when $v \geq 6$).

The expression on the right side of (5), like that of (4.4), has negative derivative with respect to C_0 for $C_0 < \pi/6$, since $\sigma - \pi - 1 < 0$. However, substitution of $\pi/6$ for C_0 in that expression no longer yields a nonnegative expression (as it did in the case of (4.4)), again reflecting the greater difficulty of proving D-optimality. Instead, we must substitute the sharper upper bound for C_0,

$$C_0 \leq (\pi - 2b_1)/6, \tag{6}$$

which we now justify.

Since $B_1 = C_0/b_1 < \bar{r}/b_1 = b_2/v$, (6) will follow from

$$\text{int}(b_2/v) \leq (b_2 - 2)/6. \tag{7}$$

If $b_2 \equiv 0$ or 1 (mod 6), it follows from the nonregularity condition $v | b_2$ that $v \geq 8$, and then (7) is satisfied for $b_2 \geq 8$. For other congruences (mod 6) of b_2, (7) is obvious. Thus, (6) is proved, and consequently (5) will follow from the nonnegativity of the expression obtained by substituting $C_0 = (\pi - 2b_1)/6$ into the right side of (5). This expression, after multiplying by 6 and rearranging terms, is

$$[b_1^2 - 8b_1 + 6]\, b_2^2 + 12(b_1 - 1)b_2 + \{2b_1^2 - 4b_1 + 6\}, \tag{8}$$

which is clearly positive since both $b_i \geq 8$.

Optimality Criteria for Designs

J. KIEFER[1]

This summarizes a few of my remarks at the Ft. Collins Conference's panel discussion on the subject of the title. The other participants were Chernoff, Srivastava, and Kempthorne (Chairman). A number of questions and comments were made from the floor; unfortunately, I recall only a few of my responses to these or to remarks of the other panelists.

In the limited time, it seemed appropriate to pinpoint three or four interesting facets of this broad topic, rather than to attempt a general discussion starting with "foundations." Many such features are best understood in the simplest, most idealized (if, often, unrealistic-in-practice) problems. Thus, no complex models involving such important ideas as extrapolation or robustness will be considered in my elementary comments.

Think of a usual linear regression model where $\mathscr{D}$ denotes the allowable designs and M_d is the $k \times k$ "information matrix" of the design d (= coefficient matrix of the normal equations) for the k unknown parameters of the model. Thus, M_d^{-1} is proportional to the covariance matrix of least squares estimators, and we suppose the optimality criterion depends only on M_d and (reflecting interest in estimating all parameters, a simplifying assumption) is finite only if M_d is nonsingular. If the criterion is orthogonal-invariant, an assumption which receives further comment later, then it can be expressed as a function of the eigenvalues $\lambda_{d1} \geq \lambda_{d2} \geq \cdots \geq \lambda_{dk}$ of M_d. The simplest such criteria are

$$\Phi_p(M_d) = \left(k^{-1}\Sigma_i \lambda_{di}^{-p}\right)^{1/p}, \qquad 0 < p < \infty;$$

$$\Phi_\infty(M_d) = \lim_{p\to\infty}\Phi_p(M_d) = \lambda_{dk}^{-1};$$

$$\Phi_0(M_d) = \lim_{p\to 0}\Phi_p(M_d) = \left(\det M_d^{-1}\right)^{1/k}.$$

[1]Prepared under NSF Grant GP 35816X.

1. The role of an optimality criterion. In problems with no choice of design, there are people of many persuasions, e.g., Bayesians, invariance protagonists, minimaxers, unbiased level 0.05 test users, etc. If the practitioner is sensible (a possibility those of each persuasion may deny for those of another), he does not blindly grind out a procedure by using his criterion; rather, having computed the procedure his criterion dictates, he compares its performance characteristic or risk function with those of other procedures to see whether some sort of subadmissibility or subminimax phenomenon opts for a different procedure. In fact, an atheistic decision theorist would possibly look at all risk functions to start with, if computations allowed this, and would choose among them without using a single, definite optimality criterion (notwithstanding the Bayesian's assertion that he is doing so subconsciously). In the absence of easy computations, one can at least compute risk functions of a number of such admissible procedures, and choose among them; for, even fairly orthodox Bayesians seem often to understand that their choice of a prior law is less to be taken literally than to be used as an instrument for computing a procedure with—hopefully—desirable risk function characteristics.

The analogous procedure in the choice of a design is rarely followed, although it should be. Thus, when $k=2$, if the common "E-optimality" criterion of minimizing $\Phi_\infty = \lambda_{d2}^{-1}$ resulted in choosing d' as in the diagram below in a setting where d'' was

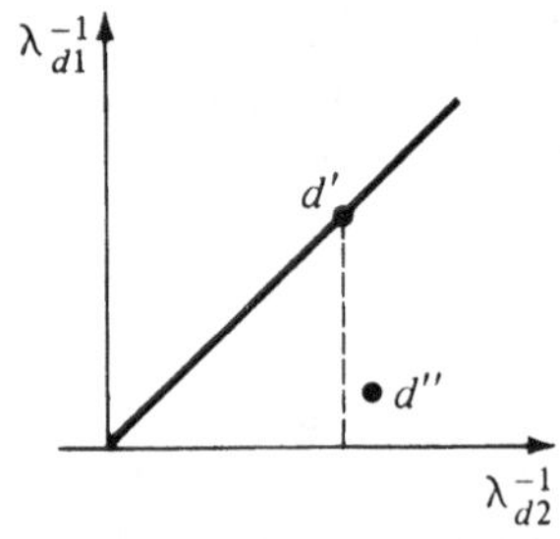

also a possible design, most people would choose the latter, happily sacrificing 10% in the value of λ_{d2}^{-1} to reduce λ_{d1}^{-1} by 70%. Similarly, if the "A-optimality" criterion of minimizing $\Phi_1 = \lambda_{d1}^{-1} + \lambda_{d2}^{-1}$ resulted in choosing d''', one could well prefer d^{iv}, thus increasing Φ_1 only slightly while almost halving Φ_∞.

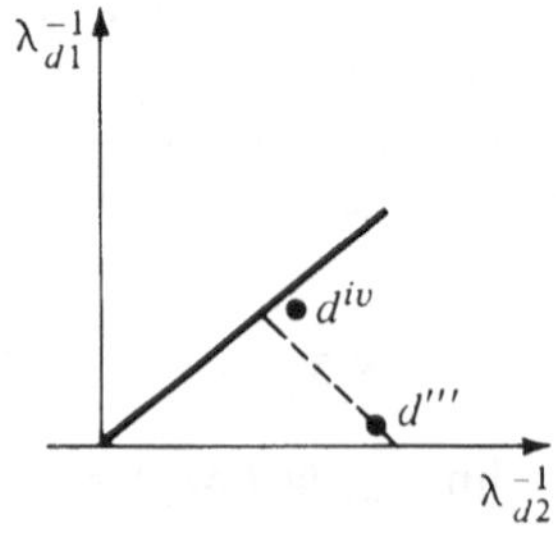

My point is not to rule out any criterion; rather, it is to emphasize that, for each criterion, one can construct *some* example for which it seems foolish. One can no more codify this part of statistics in terms of a single recipe designed to answer all questions of choosing a procedure, than one can for statistical inference in general. The moral is, *don't be dogmatic about the criterion, but look at the comparative performance of your suggested procedure in terms of other criteria as well.* This may not give a unique answer to "what design should I use," but it will often point out designs that you *shouldn't* use. Of course, one is often led to more striking vetos where such considerations as model robustness are involved.

2. Some specific criteria. Although one should not feel wedded to any single criterion, it is helpful to know the intuitive meaning, advantages, and disadvantages of several commonly used ones. In listening to practical people to get a feeling for the most important features in their choice of a procedure, the theoretician can help by disspelling myths and explaining what different criteria actually accomplish. Thus, it is useful for the experimenter to know that, in the normal case, the volume, average of principal axis lengths, and diameter of the usual confidence ellipsoid will be minimized, respectively, by the D-, A-, and E-optimality criteria. This observation could awaken the experimenter's realization of the sense in which he wants to achieve "maximum accuracy" to the point that he can make an intelligent formalization of criterion and, thus, choice of design.

There are commonly heard pros and cons for the usual criteria. For example, D-optimality proponents have pointed out that D-optimality is invariant under linear transformations of the model: The same design is D-optimum for estimating $A\theta$ as for estimating θ, if A is nonsingular. Chernoff has observed that, from a foundational (or large sample) viewpoint, one is led to a criterion $\operatorname{tr} CM_d^{-1}$, and that this should *not* lead to the orthogonal-invariant choice $C = I$ nor to the same choice of design if one is instead interested in $A\theta$. A D-optimality advocate would counter that in practice it is rare that one would know the matrix C of this criterion, and that for the computational viewpoint it is easier to table the single D-optimum design for a variety of settings than to have to list the dependence on C as well; moreover, although the elements of C are in units that remove the difficulty of adding (in a trace criterion) variances which are in different units (feet^2, feet^4, seconds^2, etc.), practitioners sometimes have an intuitive distaste for such arithmetic. On the other hand, there are examples where D-optimality can sacrifice much in the values of all but one or two diagonal elements of M_d^{-1}, in order to make $\Pi\lambda_{di}^{-1}$ small.

Faced with such a variety of criticisms, one can try to develop a compromise criterion which eliminates the worst objections. For example, in a factorial setting where the main effect factors are equally important and in the same units, suppose V_{dq} is the average variance of the estimates of interactions of combinations of q treatments. Then the criterion $\Pi_q V_{dq}$ eliminates some of the objections of both D- and A-optimality. Such "compound criteria" are considered further in my forthcoming paper on approximate theory and equivalence theory.

One should mention that optimality problems are often computationally intractable for even the common criteria, and that it is important to develop simple schemes for obtaining *nearly-optimum* designs in such problems. For example, we shall discuss elsewhere the criterion $\operatorname{tr} M_d^2$ of K. R. Shah, in cases where $\operatorname{tr} M_d$ is (almost) independent of d and the usual criteria lead to impossible computations.

3. Variation of criterion. An aspect of learning what the practitioner's (possibly subconscious) preferences are is to vary the criterion in a theoretical problem and see how the optimum design changes. Then, observing what design he used in practice, one can (if he judges the practitioner to be sensible) infer something about his optimality criterion. (I believe this often to be a more realistic inference about the customer than that asserted possible by Savage's existence theorem in Bayesian foundations.)

Here is a simple example: Quadratic regression on the 2-dimensional simplex $\{(x_1, x_2, x_3): \Sigma_1^3 x_i = 1, \text{ all } x_i \geq 0\}$. The D-optimum design takes all observations on the 6 points $\{(1,0,0),(1/2,1/2,0)$, and permutations of these$\}$. The A-optimum design takes a small proportion of observations at the center $(1/3,1/3,1/3)$ in addition, and the E-optimum design takes lots there. There are other reasons for taking observations in the center, e.g., distrust of the model, and that's ruled out here for the sake of brevity. It is then possible to infer, from the extent to which an experimenter takes observations in the center, whether he is closer to being an E-, A-, or D-optimizer.

4. Answers to some of the questions. Dr. Yates responsed today in the same way he did 14 years ago at my painful RSS venture. He simply does not seem to understand or tolerate the role of theory. He must really feel that, if Chernoff or Srivastava or I knew the solution to some simple problem like that of quadratic regression cited above, we would impose that pretty mathematics on the practitioner without listening to what the poor victim had to say. Let me repeat once more, then: The practical problems are often complex and we do not come close to knowing how to solve them. But by solving simpler analogues, we can hope to get a feeling for what good (if not optimum) procedures will look like in more complex settings. We hope thereby to do better than your know-nothing rules of thumb, and maybe, someday, we will even learn how to attack the hard problems.

Dr. Manson's comments on bias and robustness reflect the hobby-horse syndrome from which we all occasionally suffer. I stated why I wasn't including models where bias might be present in the regression, so he asserted that I hadn't included such models. It reminds one of the elephant and the Polish question.

Commentary on Papers [22], [55], [60], [61]

CHING-SUI CHENG[1]

University of California, Berkeley

In [22], Kiefer laid the foundation for the theory of optimal design. Although there had been some uses of optimality criteria before this paper appeared, for instance Smith (1918), Wald (1943), Ehrenfeld (1955), and others, it was Kiefer who consolidated the earlier works, put them in perspective, and opened up the whole field.

In this 1958 paper, Kiefer discussed several important optimality criteria and their statistical meanings. He also gave a unified method to show the D-, A-, and E-optimality of symmetric designs such as balanced incomplete block designs, Latin squares, Youden squares, etc. He further showed that in the problem of hypothesis testing, these symmetric designs are nonoptimal over randomized designs.

The most important among these results is perhaps the optimality proof of symmetric designs. Later it was extended to the so-called "universal optimality," a simple and powerful optimality tool which has had tremendous impact on the exact theory of optimal design. The proof, however, failed on generalized Youden designs, an invention of Kiefer's, and a generalization of Latin squares and Youden squares. The problem of proving or disproving the optimality of generalized Youden designs is an extremely difficult combinatorial problem and was finally solved in the early 1970's. The optimal properties of generalized Youden designs and the result on universal optimality were reported and fully discussed in [55], [61]. Paper [60] was devoted to the construction of generalized Youden designs.

History: The following optimality results were published before [22]:

1. D-optimality of Latin squares (Wald (1943)).
2. E-optimality of Latin squares and orthogonal weighing designs (Ehrenfeld (1955)).

[1] Work supported by National Science Foundation Grant No. MCS-8200909.

3. E-optimality of balanced incomplete block designs (Masuyama (1957)).
4. E-optimality of Youden and Shrikhande square designs (Masuyama and Okuno (1957))

In the same issue of the *Annals of Mathematical Statistics* where [22] appeared, two other papers also dealt with the optimality of some of the symmetric designs listed above: Mote (1958) showed the E-optimality of balanced incomplete block designs and Kshirsagar (1958) proved the A- and D-optimality of balanced incomplete block designs and Youden squares. The works of Masuyama and Okuno were published in a less-known Japanese journal and apparently were unknown to Kiefer, Mote, and Kshirsagar.

Kiefer's work was far more extensive than the others. He also proved his results under minimal restrictions. For example, he did not restrict himself to binary designs as most of the other authors did.

Universal Optimality and Schur-Optimality: Proposition 1 in [61] contains the universal optimality result for settings such as the elimination of one or multi-way heterogeneity in which all the information matrices have zero row and column sums. Here the information matrices are singular and a symmetric design can be defined as one such that all the eigenvalues of its information matrix except the trivial one (zero) are equal. Such a matrix is of the form $a\mathbf{I}+b\mathbf{J}$, where $\mathbf{I}$ is the identity matrix and $\mathbf{J}$ is the matrix of ones, and is called *completely symmetric*.

Let $\mathscr{B}_{v,0}$ be the set of all $v\times v$ symmetric matrices with zero row and column sums and $\mathscr{C}$ be a subset of $\mathscr{B}_{v,0}$. The proposition states that, if there exists a matrix $\mathbf{C}^*$ in $\mathscr{C}$ such that

(i) $\mathbf{C}^*$ completely symmetric,
(ii) $\mathbf{C}^*$ maximizes $\operatorname{tr}\mathbf{C}$ over $\mathscr{C}$,

then $\mathbf{C}^*$ is *universally optimal* over $\mathscr{C}$, i.e., it minimizes $\Phi(\mathbf{C})$ over $\mathscr{C}$ for any $\Phi:\mathscr{B}_{v,0}\to(-\infty,+\infty]$ such that

(iii) Φ is convex,
(iv) $\Phi(b\mathbf{C})$ is nonincreasing in the scalar $b\geq 0$,
(v) Φ is invariant under each simultaneous permutation of rows and columns.

Thus, for a completely symmetric $\mathbf{C}^*$, the universal optimality (which covers the commonly used A-, D-, and E-optimality) follows from the maximization of $\operatorname{tr}\mathbf{C}$, an extremely simple condition to verify. A key step in the proof involves taking the average of an arbitrary matrix $\mathbf{C}$ in $\mathscr{C}$ with respect to all simultaneous permutations of rows and columns. Two crucial things can be said about the averaged version of $\mathbf{C}$, denoted here by $\mathbf{C}_A$. First, $\mathbf{C}_A$ is completely symmetric and hence can easily be compared with $\mathbf{C}^*$; indeed it follows from (ii) and (iv) that $\Phi(\mathbf{C}^*)\leq\Phi(\mathbf{C}_A)$. Second, the convexity of Φ implies that $\Phi(\mathbf{C}_A)\leq\Phi(\mathbf{C})$. The proof is completed by piecing together the above two inequalities.

This argument can be modified for the case where the information matrices are nonsingular, which is the concern of Proposition 1′ in [61]. However, in this case, the information matrix of a symmetric design is of the form $a\mathbf{I}_v$ and one has to average an arbitrary matrix with respect to the *orthogonal transformations* to bring it to the desired form; the group of permutations is too small. Thus in Proposition 1′ one has to replace the "permutation-invariance" of Φ with "orthogonal invariance." This was not mentioned in the paper and was apparently an oversight since a correct statement was given in [55] which appeared four years before [61]. Other ways of rectifying Proposition 1′ can be found in Sinha and Mukerjee (1982) and Giovagnoli and Wynn (1983).

The simple idea of averaging used in the proof of Proposition 1 suggests an interesting link to the concept of Schur-optimality. Let $\lambda(\mathbf{C}_A)$, $\lambda(\mathbf{C})$, and $\lambda(\mathbf{C}^*)$ be the vectors of the eigenvalues of $\mathbf{C}_A$, $\mathbf{C}$, and $\mathbf{C}^*$, respectively. Then it is clear that $\lambda(\mathbf{C}_A) \prec \lambda(\mathbf{C})$ and $\lambda(\mathbf{C}^*) \overset{W}{\prec} \lambda(\mathbf{C})$, where $\prec$ is the ordinary majorization, and $\overset{W}{\prec}$ is upper weak majorization (see Marshall and Olkin (1979)). By Theorem A.8 in Chapter 3 of Marshall and Olkin (1979), one has

Proposition 1″. *Under* (i) *and* (ii), $\mathbf{C}^*$ *minimizes* $\Phi(\lambda(\mathbf{C}))$ *for any nonincreasing Schur-convex function* Φ *defined on* R^{v-1}.

A design minimizing $\Phi(\lambda(\mathbf{C}))$ for all nonincreasing Schur-convex functions is called *Schur-optimal.* The use of majorization in the context of design optimality appeared, e.g., in p. 339 of [61], Cheng (1979), Giovagnoli and Wynn (1980, 1981, 1983), and Constantine (1980). The connection between Schur-optimality and Proposition 1″ was obtained as a by-product of Cheng (1979). The results in Cheng (1979) actually hold for all nonincreasing Schur-convex criteria, although, for simplicity, only the optimality with respect to criteria of the form $\Sigma_{i=1}^{v-1} f(\lambda_i)$, where f is convex and nonincreasing, was stated in the paper. This appears more restrictive, but it indeed implies Schur-optimality (see Proposition B.2 in Chapter 4 of Marshall and Olkin (1979)). As opposed to Proposition 1″, the results in Cheng (1979) provide examples of Schur-optimal designs which are *not* symmetric. Schur-optimality was also formulated independently in Giovagnoli and Wynn (1981, 1983), and Constantine (1980); the latter also proved Proposition 1″.

Schur-optimality is weaker than universal optimality. First, the criteria covered by Schur-optimality are functions of eigenvalues and hence are orthogonally invariant. Second, if $\lambda_1(\mathbf{C}) \geq \lambda_2(\mathbf{C}) \geq \cdots \geq \lambda_{v-1}(\mathbf{C})$ are the $v-1$ nontrivial eigenvalues of $\mathbf{C}$, then $-\Sigma_{i=k}^{v-1}\lambda_i(\mathbf{C})$ is a convex function of $\mathbf{C}$ (see Fan (1949)) and satisfies (iv), (v) for $k=1,2,\ldots,v-1$. Thus if $\mathbf{C}_0$ is universally optimal over $\mathscr{C}$, then it minimizes $-\Sigma_{i=k}^{v-1}\lambda_i(\mathbf{C})$ for all k, which implies that $\lambda(\mathbf{C}_0) \overset{W}{\prec} \lambda(\mathbf{C})$ for any $\mathbf{C} \in \mathscr{C}$ and the Schur-optimality of $\mathbf{C}_0$ over $\mathscr{C}$ follows. We state this as

Theorem 1. *If a design is universally optimal, then it is Schur-optimal.*

Proposition 1″ is in fact a simple consequence of Proposition 1 in [61] and the above theorem. We note that Theorem 1 was also obtained independently by Bondar (1982).

The method of averaging has been used by many authors to solve various design problems. Some examples are Constantine (1981), Majumdar and Notz (1983), Jacroux and Notz (1983). A weaker form of universal optimality appeared in [88] and Cheng and Li (1983).

Application to Graph Theory: The D-optimality of a balanced incomplete block design proved in [22], on translation into the language of graph theory, yields the result that a graph G in which there are an equal number of lines joining each pair of vertices has the maximum number of spanning trees among all the graphs with the same numbers of vertices and lines as G. This result was rediscovered later by graph theorists, see the discussion in Cheng (1981).

Optimal Properties of Generalized Youden Designs: In the setting of two-way heterogeneity, a generalized Youden design does not necessarily maximize the trace of the information matrix, and hence Proposition 1 is not applicable. Proposition 3 of [61] is a useful tool for showing the optimality of symmetric designs which do not maximize the trace of the information matrix. In [61] we find remarkably ingenious arguments for attacking the optimality of generalized Youden designs. A generalized Youden design was shown to be A- and E-optimal, but a counterexample was given to show that it is not always D-optimal when v (the number of varieties) $= 4$. This is the first case where an "exotic" design with full symmetry is proved nonoptimal for a symmetric estimation problem.

This nonoptimality result for $v = 4$ is rather unique since a generalized Youden design with $v \neq 4$ is always D-optimal; a proof can be found in [83]. Cheng (1978b) extended Kiefer's result to the setting of multi-way heterogeneity. There is no such counterexample for higher-dimensional designs.

Optimality of Asymmetrical Designs: One line of active research that followed naturally from Kiefer's work on the optimality of symmetric designs is the search for optimum designs when symmetric optimum designs do not exist. Takeuchi (1961) and Ehlich (1964a, b) are two important pioneers. Some more recent papers include, e.g., Conniffe and Stone (1975), John and Mitchell (1977), Cheng (1978a, 1980), and [85, 95]. Universal optimality usually fails here (except for the Schur-optimality results in Cheng (1979) mentioned earlier) and optimal designs often depend on the criteria used.

Construction of Generalized Youden Designs: Paper [60] is perhaps the only paper Kiefer wrote on the combinatorial construction of designs. He devised ingenious "patchwork" methods to construct generalized Youden designs. These are not detailed listings of designs, but rather very general recipes for piecing together known Latin squares and balanced incomplete block designs to yield the desired designs. Ruiz and Seiden (1974), and Seiden and Wu (1978) gave geometric constructions of generalized Youden designs for certain parameter values, which can also be obtained by Kiefer's patchwork methods.

REFERENCES

Bondar, J. V. (1982). On universal optimality of designs—definitions and a simple necessary and sufficient condition. Unpublished manuscript.

Cheng, C. S. (1978a). Optimality of certain asymmetrical experimental designs. *Ann. Statist.* **6**, 1239–1261.

Cheng, C. S. (1978b). Optimal designs for the elimination of multi-way heterogeneity. *Ann. Statist.* **6**, 1262–1272.

Cheng, C. S. (1979). Optimal incomplete block designs with four varieties. *Sankhyā Ser. B* **41**, 1–14.

Cheng, C. S. (1980). Optimality of some weighing and 2^n fractional factorial designs. *Ann. Statist.* **8**, 436–446.

Cheng, C. S. (1981). Maximizing the total number of spanning trees in a graph: two related problems in graph theory and optimum design theory. *J. Comb. Theory Ser. B* **31**, 240 248.

Cheng, C. S. and Li, K. C. (1983). Optimality criteria in survey sampling. Mimeograph Series #83-40, Department of Statistics, Purdue University.

Conniffee, D. and Stone J. (1975). Some incomplete block designs of maximum efficiency. *Biometrika* **61**, 685–686.

Constantine, G. M. (1980). On Schur-optimality. Unpublished manuscript.

Constantine, G. M. (1981). Some *E*-optimal block designs. *Ann. Statist.* **9**, 886–892.

Ehlich, H. (1964a). Determinantenabschätzungen für binäre Matrizen. *Math. Z.* **83**, 123–132.

Ehlich, H. (1964b). Determinantenabschätzungen für binäre Matrizen mit $n \equiv 3 \bmod 4$. *Math. Z.* **84**, 438–447.

Ehrenfeld, S. (1955). On the efficiency of experimental designs. *Ann. Math. Statist.* **26**, 247–255.

Fan, K. (1949). On a theorem of Weyl concerning eigenvalues of linear transformations. I. *Proc. Nat. Acad. Sci. USA* **35**, 652–655.

Giovagnoli, A. and Wynn, H. P. (1980). A majorization theorem for the *C*-matrices of binary designs. *J. Statist. Plann. Inference* **4**, 145–154.

Giovagnoli, A. and Wynn, H. P. (1981). Optimum continuous block designs. *Proc. R. Soc. Lond. A* **377**, 405–416.

Giovagnoli, A. and Wynn, H. P. (1983). Schur-optimal continuous block designs. To appear in *Proc. Neyman–Kiefer Memorial Conference.*

Jacroux, M. and Notz, W. (1983). On the optimality of spring balance weighing designs. *Ann. Statist.* **11**, 970–978.

John, J. A. and Mitchell, T. J. (1977). Optimal incomplete block designs. *J. R. Statist. Soc. B* **39**, 39–43.

Kshirsagar, A. M. (1958). A note on incomplete block designs. *Ann. Math. Statist.* **29**, 907–910.

Majumdar, D. and Notz, W. (1983). Optimal incomplete block designs for comparing treatments with a control. *Ann. Statist.* **11**, 258–266.

Marshall, A. W. and Olkin, I. (1979). *Inequalities: Theory of Majorization and Its Applications.* Academic Press, New York.

Masuyama, M. (1957). On the optimality of balanced incomplete block designs. *Rep. Statist. Appl. Res. Un. Japan Sci. Engrs.* **5**, 4–8.

Masuyama, M. and Okuno, T. (1957). On the optimality of Latin-, Youden-, and Shrikhande square designs. *Rep. Statist. Appl. Res. Un. Japan Sci. Engrs.* **5**, 17–19.

Mote, V. L. (1958). On a minimax property of a balanced incomplete block design. *Ann. Math. Statist.* **29**, 910–914.

Ruiz, F. and Seiden, E. (1974). On construction of some families of generalized Youden designs. *Ann. Statist.* **2**, 503–519.

Seiden, E. and Wu, C. J. (1978). A geometric construction of generalized Youden designs for v a power of a prime. *Ann. Statist.* **6**, 452–460.

Sinha, B. K. and Mukerjee, R. (1982). A note on the universal optimality criterion for full rank models. *J. Statist. Plann. Inference* **7**, 97–100.

Smith, K. (1918). On the standard deviations of adjusted and interpolated values of an observed polynomial function and its constants and the guidance they give towards a proper choice of the distribution of observations. *Biometrika* **12**, 1–85.

Takeuchi, K. (1961). On the optimality of certain types of PBIB designs. *Rep. Statist. Appl. Res. Un. Japan Sci. Engrs.* **8**, 140–145.

Wald, A. (1943). On the efficient design of statistical investigations. *Ann. Math. Statist.* **14**, 134–140.

Commentary on Papers [72], [74], [75], [76], [80]

ZVI GALIL[1]

Tel-Aviv University and
Columbia University

These five papers study the performance under various optimality criteria of a design selected according to a particular criterion in three different settings. The rationale for this study is given in [63] and also in [74].

A first step away from the traditional choice of a design to satisfy some principle of intuition or symmetry is to base the choice on a specific criterion. (For example, one may choose a design to minimize the generalized variance, the average variance, or the largest eigenvalue of the covariance matrix of the best linear estimators; these criteria are called *D*-, *A*-, and *E*-optimality, respectively.) Realistically, though, such a criterion is usually at best only as an approximate reflection of some vague notion of "goodness." Hence, it seems prudent to check that a design, selected in this fashion, performs reasonably well in other respects, relative to other possible designs. While such an examination of the performance of competing designs occurs in various applied papers, it is almost always imprecise or unsystematic. The present study attempts what is perhaps a more systematic study in particular families of simple settings.

The considerations in these papers are limited to approximate theory, so as to exhibit the comparisons without the obscuring arithmetical complexity and longer design lists of exact design theory as sample size varies. Comparisons of exact designs exhibit similar phenomena, even for small sample sizes.

A *design setting* is a pair $(\mathscr{X}, f)$, where f is a known (column) k-vector of continuous functions $f_1, \ldots, f_k$ on the compact space $\mathscr{X}$. A design ξ is a probability measure on $\mathscr{X}$. If the measure ξ is discrete, $\xi(x)$ represents the proportion of observations to be taken at the point x of $\mathscr{X}$. The observations are uncorrelated with variance σ^2. The expected value of an observation at x is $\theta' f(x)$ where θ is a column vector of k unknown real parameters. The $k \times k$ *information matrix* (per

[1]Research supported by NSF Grant MCS 83-03139.

observation for unit variance) of ξ is defined as

$$M(\xi) - \int_{\mathscr{X}} f(x)f(x)'\xi(dx),$$

and the corresponding normalized variance function for nonsingular $M(\xi)$ is

$$d(x,\xi) = f(x)'M^{-1}(\xi)f(x).$$

Thus, when ξ is used with N observations, $N^{-1}\sigma^2M^{-1}(\xi)$ is the covariance matrix of best linear unbiased estimators of θ, and $N^{-1}\sigma^2d(x,\xi)$ is the variance at x of the least-squares fitted regression function.

The optimality functionals considered are infinite for singular $M(\xi)$, and therefore consideration is limited to non-singular $M(\xi)$. Denote the eigenvalues of $M(\xi)$ by $\lambda_1(\xi),\ldots,\lambda_k(\xi)$. Define

$$\Phi_p(\xi) = \left(k^{-1}\mathrm{tr}\,M^{-p}(\xi)\right)^{1/p} = \left(k^{-1}\sum_1^k \lambda_i^{-p}(\xi)\right)^{1/p}$$

for $0 < p < \infty$;

$$\Phi_0(\xi) = \lim_{p\to 0}\Phi_p(\xi) = \left(\det M^{-1}(\xi)\right)^{1/k};$$

$$\Phi_\infty(\xi) = \lim_{p\to +\infty}\Phi_p(\xi) = \max_i \lambda_i^{-1}(\xi).$$

It is easily seen that the criteria Φ_0, Φ_1, and Φ_∞ are the familiar D-, A-, and E-optimality criteria, here normalized so as to make comparisons easy: all Φ_p measure loss in the same scale per unit of variance, and take on the value c^{-1} if $M(\xi) = cI_k$. A *Φ_p-optimum design* $\xi^{(p)}$ is one that minimizes $\Phi_p(\xi)$ among all designs ξ.

A measure of the performance of a design ξ' with respect to the family $\{\Phi_p\}$ of criteria is the (multiplicative regret) *absolute efficiency ratio*

$$e_p(\xi') = \Phi_p(\xi')/\Phi_p(\xi^{(p)}) \qquad 0 \le p \le \infty.$$

A design ξ' for which these ratios are all near 1 may be judged adequate for many purposes in applications. The ξ' chosen are $\xi^{(0)}$, $\xi^{(1)}$, and $\xi^{(\infty)}$ simply because they are most often employed in practice. Other ξ' can undergo a similar treatment.

The settings considered are:

— Rotatable designs for quadratic and cubic regressions on the q-dimensional ball of radius R;
— Designs for quadratic regressions on the q-dimensional cube of width 2;
— Designs for quadratic regressions on the unit simplex.

Methods are used for computing the various optimal designs. Combinatorial and algebraic manipulations are used first to reduce the number of parameters involved. They are followed by numerical methods such as simple search or modified methods from numerical analysis (e.g., Broyden (1967)). As a by-product, the five papers list many optimal designs for the setting considered.

As an example we now describe the treatment of rotatable designs for quadratic regressions on balls ([72]). The set χ is the q-dimensional ball of radius R; f consists of the functions $1, x_i, x_i^2, x_i x_j$ $(k = (q+1)(q+2)/2)$. Only *rotatable* designs are considered, i.e., those for which d is invariant under rotation. The choice of rotatable designs is justified by the fact that it enables a more complete analysis. A simple argument implies that the best design is one that depends on one parameter $0 \le \alpha \le 1$ in which $1-\alpha$ of the observations are located at the center of the ball and α are distributed uniformly on its surface. The various moments are computed. They are used to write the information matrix $M(\alpha)$. The latter has a simple structure which enables the computation of the eigenvalues and of $(\Phi_p(\alpha))^p$, which is minimized with respect to α. The special cases $p = 0, 1, \infty$ are treated algebraically, or asymptotically as $R \to 0$ or ∞. The other cases are treated numerically: the minimization is achieved by a simple search strategy. (The interval was divided to subintervals. One subinterval was chosen and was subdivided, etc.) The values of the best α (which yields the best design) are listed for $1 \le q \le 10$, $p \in \{0, 0.5, 1, 2, \infty\}$ and $R \in \{0.1, 0.5, 1, 2, 10\}$. In almost all cases, the values for $R = 0.1$ (10) are very close to those obtained asymptotically for $R \downarrow 0$ ($R \uparrow \infty$). Next, the efficiency ratio $e_p(\xi')$ is computed for $\xi' = \xi^{(r)}$ $r \in \{0, 1, \infty\}$. The maximum over p is used as an indication of the overall attractiveness of ξ'. For small radius $\xi^{(\infty)}$ is found to be very attractive while for large R (how large depending on q) $\xi^{(1)}$ is found to be superior.

The results in the four settings are not uniform and are not always predictable: For quadratic regression on the ball and the cube, the A-optimum design appears to be fairly robust in its efficiency, while for quadratic regression on the simplex and cubic regression on the ball the E-optimum design is more robust. Even in the same setting there is not always one criterion which dominates the others. The techniques used are quite ad hoc and quite different in the various settings. The symmetries of the specific settings were taken into account in different ways. Nevertheless, such a comparative study is worthwhile when one has to choose among several optimality criteria. Perhaps, the techniques used will be found useful for other cases (e.g., higher order regression, or other domains).

Paper [76] treats the Box and Draper (1959) model where the bias due to cubic terms is considered. The D-optimum design ignoring the possibility of such a bias and Box and Draper's suggested design are compared. The former design performs well in terms of the Box–Draper criterion unless the sample size is fairly large, and is superior in terms of maximum variance and bias.

References

Broyden, C. G. (1967). Quasi-Newton methods and their application to function minimization. *Math. Comp.* **21**, 368–381.

Box, G. E. P., and Draper, N. R. (1959). A basis for the selection of a response surface design, *J. Amer. Statist. Assoc.* **54**, 622–654.

Commentary on Paper [86]

TOBY MITCHELL
Union Carbide Corporation

In the early 1970's several computer algorithms appeared for deriving D-optimal experimental designs (Wynn (1972), Fedorov (1972), Mitchell (1974)). I had written DETMAX primarily as an aid to industrial statisticians seeking designs for nonstandard models under nonstandard constraints. It could be useful even when it converged to a suboptimal design, or so I rationalized. However, I think Kiefer was more interested in it as a tool for deriving optimal exact designs in standard settings, in conjunction with theoretical approaches; suboptimal designs were not acceptable. He therefore became fascinated with the fundamental problem that still plagues iterative methods for combinatorial optimization: convergence to a suboptimum. I had attempted to overcome this problem in DETMAX by (1) introducing the idea of "excursions", which systematically make multiple point exchanges that cannot necessarily be expressed as a sequence of optimal single point exchanges, and (2) by making several tries, each with a new random starting design. I was only partially successful. Kiefer and Galil searched for ways to modify the iterative procedure en route to better avoid entrapment at local optima. They eventually abandoned this and focussed their attention on improving the random start approach. They did this first by modifying the approach so that only n_r of the initial design points are chosen randomly from the set of candidate points, the remainder systematically. They found that one could often do better by choosing relatively low n_r rather than choosing a completely random starting design. They also discovered ways to speed up various computations in DETMAX so that more tries could be made. Their main improvement is based on holding the variances of the estimated response at the candidate design points in an array that is updated after each addition or removal of a point from the design during the iterative optimization.

The immediate impact of [86] was to make it possible to obtain better designs in cases that had already been considered (e.g., Table 3) and to obtain new designs in cases that had previously been avoided because they required too much computation. By applying theoretical bounds to designs obtained by computer search, Galil and Kiefer were then able to derive new D-optimal designs and prove their optimality in [85].

Little progress with exchange methods has been made since Galil and Kiefer's work. The general question of how to keep an iterative combinatorial optimization procedure from getting stuck at a local but not global optimum is still being explored. An interesting approach, called "simulated annealing" was recently introduced by Kirkpatrick, Gelatt, and Vecchi (1983). If applied to exchange algorithms for deriving experimental designs, their method would choose a potential exchange randomly at each iteration, accepting it with a probability that depends on the change in the determinant. The authors draw an interesting analogy with annealing in solids to suggest ways of controlling the acceptance rule to increase the probability of arriving at the global optimum.

In the only major departure from the exchange approach, Welch (1982) developed a branch-and-bound algorithm for deriving optimal exact designs. Its great advantage is that it does find the global optimum. In fact, Welch used his algorithm to prove the optimality of Galil and Kiefer's designs for quadratic regression on the 3-cube (Table 3 of [86]). One would expect to pay a price for the branch-and-bound method in terms of computing time, but to my knowledge, no extensive comparisons have been made.

References

Fedorov, V. V. (1972). *Theory of Optimal Experiments.* (translated and edited by W. J. Studden and E. M. Klimko), Academic Press, New York.

Kirkpatrick, S., Gelatt, C. D., Jr., and Vecchi, M. P. (1983). Optimization by Simulated Annealing. *Science* **220**, 671–680.

Mitchell, T. J. (1974). An algorithm for the construction of "D-optimal" experimental designs. *Technometrics* **16**, 203–210.

Welch, W. J. (1982). Branch-and-bound search for experimental designs based on D-optimality and other criteria. *Technometrics* **24**, 41–48.

Wynn, H. P. (1972). Results in the theory and construction of D-optimum experimental designs. *J. Royal Statist. Soc. Ser. B.* **34**, 133–147.

Commentary on Papers [23], [29], [31], [33], [34], [43], [44], [58], [61]

FRIEDRICH PUKELSHEIM
Universität Augsburg, Federal Republic of Germany

1. The Essence

In the exact design theory, a design, ξ_n, for n observations determines, in a set of experimental conditions $\mathscr{X}$, a finite number of points $x_1,\ldots,x_l$, and assigns to these points weights of the form $0, 1/n,\ldots,(n-1)/n$, 1 such that $\sum_{i=1}^{l}\xi_n(x_i)=1$. Then $\xi_n(x_i)=n_i/n$, say, directs the experimenter to realize, in a sample of size n, n_i observations under experimental conditions $\{x_i\}$.

In the approximate design theory, a design ξ simply is taken to be a probability measure with finite support on the set $\mathscr{X}$. While the support points $x_1,\ldots,x_l$, say, of ξ still determine a finite number of experimental conditions for experimentation, the weights $\xi(x_i)$ may now attain any value between 0 and 1. Thus, a design ξ, in general, only provides an approximation to a design ξ_n which is realizable.

Paper [58] is a comprehensive collection of Kiefer's views and insights on optimum design in the context of the approximate theory. Though meant to be preliminary, Kiefer's sudden death makes the paper the definitive presentation of his approach. In Section 2 he singles out which real-valued functions Φ, called *optimality criteria*, may reasonably serve to evaluate and compare the "goodness" of a design, and this section also contains the basic equivalence results for optimality. The remaining Sections 3–7 offer an invaluable brain-storm attack in order to (3) identify the essentials of the theory, (4) list the tools which may aid the analysis, (5) outline possible modifications of the underlying model, (6) illustrate techniques of computation, and (7) comment on the complications arising from nondifferentiable criteria Φ and from designs ξ whose information matrix is singular. On the whole this is a sometimes eclectic collection of comments, but it would seem that for any problem in the field which can be cast into final form, Kiefer's remarks show the way.

A brief and concise derivation of the type of equivalence theorems which play a role here is given by Whittle (1973). Pukelsheim (1980) treats a class of optimality criteria, called *information functionals*, which also include nondifferentiable criteria and where the problem of characterizing optimality of designs whose information matrix is singular is amenable to a complete solution. The approach taken there is based on the duality theory of convex analysis which corresponds to the game theoretic approach as first considered in [31]; alternative formulations which are closer to classical differential calculus are presented in Pukelsheim and Titterington (1983), and Gaffke (1983).

The precise relation of the class of information functionals to the class of criteria considered on page 855 of [58] is not entirely clear at this time.

In the case of block designs, in the linear model with additive effects, it turns out that the particular optimality criterion is of ten of little relevance, since the classical designs turn out to be optimal with respect to a wide class of optimality criteria. This is illustrated, in the context of the exact theory, in [61], where Kiefer also coins the notion of *universal optimality*. Other examples of this type have been discussed by Cheng (1978).

The general theory thus treats optimality in a much wider setting than originally was the case when "optimality" typically meant D-optimality, i.e., maximality of the determinant of the information matrix of the design.

2. The Sources

The first Equivalence Theorem was announced in a footnote on page 292 of the joint work of Kiefer and Wolfowitz ([23]) and then presented as a short, separate paper [29]. It establishes, in the approximate theory and for estimation of all k parameters, the "equivalence" of D-optimality and of G-optimality, the latter being a kind of global optimality with an appealing statistical interpretation; "equivalence" here means that the two criteria lead to the same class of optimal designs. It is natural, then, to search for further equivalences to D-optimality, and in [33] Kiefer finds two more criteria with this property.

The general result was preceded by the special case of polynomial regression where the G-optimal designs of Guest (1958) and the D-optimal designs of Hoel (1958) were observed to coincide. This case as a surprise to people working in the field, or, as Kiefer paraphrases it in his leisurely compiled Purdue lectures [104]: "In fact the startling coincidence is that these two people have the same first two initials (P. G.) and you can compute the odds of that!!"

As it stands now, in the much more elaborate theory, an Equivalence Theorem for a general optimality criterion Φ seeks to exhibit a necessary and sufficient condition for Φ-optimality which is easy to verify; the pleasant feature that such a condition arises as an optimality criterion in its own right no longer applies.

Another line of generalization which was taken up almost immediately was the extension to s out of k parameters. Simple differential calculus fails here, and, in [31] and [34], Kiefer uses game theoretic methods to derive his results. Incidentally, each of the "big" papers ([23], [31], [44]) on polynomial regression and Chebyshev vectors is followed by a short, separate extract ([29], [34], [43]) of a problem *perhaps of interest per se*, as indeed proved to be true over the years, at least in the first two cases.

The transition from the exact theory to the approximate theory is based on the simple technical trick to make a discrete variable continuous, as is useful in many other mathematical problems. In design theory this had been employed by Elfving (1952), but it would seem that Kiefer was first to visualize the extent to which experimental design theory could reach a statistically satisfying level using this tool.

3. A Grain of Salt

A question, which in the work of Kiefer (and of other authors) is not always treated adequately, is whether there exists an optimal design. Whereas for D-optimality existence poses no problem, there are other criteria for which an optimal design fails to exist; see the examples in Pukelsheim (1980). To date it is an open problem whether the Equivalence Theorem of the general theory continues to hold when no optimal design exists; this is seen best by working through the proof of Whittle's (1973) Equivalence Theorem and is commented upon by Pukelsheim and Titterington (1983). This does not affect any major statistically relevant model, but to fill this gap seems to be a mathematically challenging and intricate open problem.

REFERENCES

Cheng, C. -S. (1978). Optimality of certain asymmetrical experimental designs. *Ann. Statist.* **6**, 1239–1261.

Elfving, G. (1952). Optimum allocation in linear regression theory. *Ann. Math. Statist.* **23**, 255–262.

Gaffke, N. (1983). Directional derivatives of optimality criteria at singular matrices in convex design theory. *Submitted for publication*.

Guest, P. G. (1958). The spacing of observations in polynomial regression. *Ann. Math. Statist.* **29**, 294–299.

Hoel, P. G. (1958). Efficiency problems in polynomial estimation. *Ann. Math. Statist.* **29**, 1134–1145.

Pukelsheim, F. (1980). On linear regression designs which maximize information. *J. Statist. Plann. Inference* **4**, 339–364.

Pukelsheim, F. and D. M. Titterington (1983). General differential and Lagrangian theory for optimal experimental design. *Ann. Statist.* **11**.

Whittle, P. (1973). Some general points in the theory of optimal experimental design. *J. Roy. Statist. Soc. Ser. B* **35**, 123–130.

Commentary on Papers [57], [63], [76], [84]

JEROME SACKS

Northwestern University

In 1959 Box and Draper (BD) had broken ground on the issue of how one should design for fitting a curve or surface (say, a quadratic) while at the same time guarding against the possibility that a different model holds (say, a cubic). If least-squares estimation is used for the quadratic model, then the effect of the presence of a cubic is to introduce bias. BD noted a number of examples in which there is a design which minimizes the integrated bias term, $B = 1/\sigma^2 \int (E\hat{f}(x) - f(x))^2\,dx$. ($\hat{f}$ is the fitted quadratic, f is the cubic), and which came close to minimizing the total integrated mean-square error, $\sigma^2(V+B)$ ($V = \int \operatorname{Var}\hat{f}(x)\,dx$), all the while restricting to least-squares estimates. Karson, Manson, and Hader (1969) (KMH) observed that there is a linear estimate (not necessarily least-squares) which minimizes B simultaneously for all designs in a class. Minimizing V for this estimate over the class of designs does as well as the BD approach and is often better in terms of $V+B$.

In [57] Kiefer gave an incisive general discussion of the BD and KMH approaches. He pointed out examples where modifications to either approach can produce procedures with more appealing properties and noted why neither approach can be generally satisfactory for minimizing $V+B$. Kiefer went further and gave a formulation to directly address the problems of BD and KMH. Implementation of Kiefer's analysis leads to computational questions similar to those found in fitting unbiased surfaces (no bias due to model departures) and apparently has not been pursued.

In [76] Galil and Kiefer noted that the BD approach may have little utility in the important practical setting of mixture experiments. In particular, they show that, for mixture experiments involving three or more factors, the BD approach for minimizing $V+B$ does better than the D-optimum design (which, of course, arises from minimizing V alone) only if the sample size is enormous or if the proportion of possible bias to variance is very large. Moreover, the maximum mean-square error of BD is much greater than that of the D-optimum design (thus, the BD design is not "criterion-robust"). Allusion to the results in [76] and discussion of them was given in [63]. Whether D-optimum designs themselves are model robust for designs on a simplex (i.e., guard adequately against departures from an assumed model) is uncertain at this time.

In [84] Kiefer discusses similar issues concerning model misspecification for extrapolation. Draper and Herzberg (1973) had taken initial steps in this direction largely within the BD context. Kiefer expanded on the issue by describing and comparing the BD and KMH formulations, and gave evidence that the KMH formulation is often preferable for extrapolating linear (quadratic) regression from one ball to a larger one when quadratic (cubic) departures are present. He notes that BD and KMH do well with respect to D-optimum designs for the criterion of maximum mean-square error, perhaps reflecting the differing geometries of the simplex and ball and almost certainly reflecting the extrapolation, which brings bias questions to greater prominence.

The final section of [84] points to possible further investigations (in and out of extrapolation contexts) of an adaptive sort. Since good designs guarding against bias depend on an (unknown) parameter, ρ, the ratio of a bias parameter to σ^2, one could conceivably compare the *performance* of a design using estimators which incorporate an estimate of ρ. Of course, unless the design proceeds by stages, estimation of ρ cannot be used in selecting the design. Kiefer did not pursue this very far; it remains an interesting open matter.

The issues of model robust design raised in BD, KMH, and [57] have stimulated other approaches in the last several years which are tied less to formulations requiring assumption of finite dimensional model departures from an assumed model. A variety of papers, including Huber (1975), Marcus and Sacks (1978), Agarwal and Studden (1980), Li (1981), Li and Notz (1982), Pesotchinsky (1982), Sacks and Ylvisaker (1984a, b), have appeared, many of them stimulated by Kiefer, some in theses under his guidance. Conclusive recommendations and directions have yet to emerge.

The major part of [63] takes up related issues of how one should regard results about optimality in design. In an introduction, well worth reading, Kiefer expresses his view that one should compare the performance of a procedure, found to be good under one criterion, with respect to other criteria and to the performance of "nearby" competitors. Thus, he observes that E-optimum design for quadratic regression on a simplex behaves robustly with respect to the class of Φ_p-criteria in particular, with respect to the D-optimum criterion. On the other hand, the D-optimum design behaves inefficiently with respect to the Φ_∞-criterion which produces the E-optimum design. (A more extensive commentary of Kiefer's subsequent work in comparing performance is given by Zvi Galil.)

Kiefer's attitude (not only in the design context) towards optimality, that it is not to be taken as a prescription, harks back to [1], his first published paper and, while not always explicitly stated, appeared prominently in his overall philosophy about statistical theory and practice.

References

Agarwal, G. and Studden, W. (1980). Asymptotic integrated mean square error using least squares and bias minimizing splines. *Ann. Statist.* **8**, 1307–1325.

Box, G. E. P. and Draper, N. R. (1959). A basis for the selection of a response surface design. *J. Amer. Statist. Assoc.* **54**, 622–654.

Draper, N. and Herzberg, A. (1973). Some designs for extrapolation outside a sphere. *J. Roy. Statist. Soc. Ser. B* **35**, 268–276.

Huber, P. (1975). Robustness and designs. *A Survey of Statistical Design and Linear Models.* (ed. by J. Srivastava) North Holland, Amsterdam, pp. 287–303.

Karson, M. J., Manson, A. R., and Hader, R. J. (1969). Minimum bias estimation and experimental designs for response surfaces. *Technometrics II*, 461–476.

Li, Ker-Chau (1981). Robust regression designs when the design space consists of finitely many points. Purdue Mimeo. Series #81-45.

Li, K. -C. and Notz, W. (1982). Robust designs for nearly linear regression. *J. Statist. Plann. Inference* **6**, 135–152.

Marcus, M. B. and Sacks, J. (1977). Robust designs for regression problems, in *Statistical Decision Theory and Related Topics*, II. (ed. by S. S. Gupta and D. S. Moore) Academic Press, New York, pp. 245–268.

Pesotchinsky, L. (1982). Optimal robust designs: linear regression in R^k. *Ann. Statist.* **10**, 511–525.

Sacks, J. and Ylvisaker, D. (1984a). On model robust design in regression. To appear in *Ann. Statist.*

Sacks, J. and Ylvisaker, D. (1984b). Model robust design in regression: Bayes theory. To appear in *Proceedings of the Neyman–Kiefer Memorial Conference* (ed. by L. LeCam and R. Olshen).

Commentary on Papers [40], [41], [44], [80], [84]

W. J. Studden

Purdue University

Kiefer wrote or co-authored five papers ([40], [41], [44], [80], [84]) which dealt mainly with designing experiments for extrapolation. The notation used will be the same as that used in the first few paragraphs of [44]. Given a regression vector $f = (f_0, f_1, \ldots, f_m)$ where the f_i's are real functions and defined on some space X, one observes a response with mean $\Sigma\theta_i f_i(x)$. The extrapolation problem, in the design context, is to allocate observations in X in order to provide good estimates of $\Sigma\theta_i f_i(y)$ for values y not in X. The usual least squares estimators are used, and the variance of the least squares estimate of the response at y is denoted by $\sigma^2 n^{-1} V(f(y), \xi)$ for a given allocation ξ.

Paper [44] was motivated by the work of Hoel and Levine (1964). Kiefer expressed considerable interest in the "elegant main result" of Hoel and Levine. This result stated that, for polynomial regression of degree m on $X = [-1,1]$, if interest was in a fixed $y \notin [-1,1]$, then observations could be restricted to the same $m+1$ points chosen independent of y. The proportion of observations, of course, depended on the point y.

Paper [44] was written in an algebraic setting and related the problem at hand to corresponding problems in approximation theory. It later became apparent that the algebraic results had a geometric setting, see Studden (1968). Define $\mathcal{R}$ = convex hull of the points $\pm f(x)$ for $x \in X$. Since $\mathcal{R}$ contains the origin, every $m+1$ dimensional vector a protrudes through the boundary of $\mathcal{R}$ at some point βa and hence βa can be written as $\beta a = \Sigma\varepsilon_i p_i f(x_i)$ where $\varepsilon_i = \pm 1$ and $p_i \geq 0, \Sigma p_i = 1$. A geometric result of Elfving (1952) (see Karlin and Studden (1966)) says that the optimal design for estimating the linear combination $a'\theta$ concentrates mass p_i at each x_i. The set $\mathcal{R}$ is symmetric about the origin. Kiefer and Wolfowitz discovered that, for the polynomial case, the set $\mathcal{R}$ has four, or two pairs, of opposing m dimensional faces (and no others). One pair of these faces corresponds to the first function f_0 being $f_0 \equiv 1$. The other, more interesting, pair is spanned by the points which Kiefer and

Wolfowitz termed "Chebyshev points." The geometric interpretation of the result of Hoel and Levine was that all the vectors $f(y)$, $|y|>1$ protrude through the set $\mathscr{R}$ on this same face.

Paper [44], among other things, extends this result to systems of functions other than the polynomials. At about the same time, Hoel (1965) was extending the polynomial case for an interval to the case where $X=[-1,\alpha]\cup[\beta,1]$, $-1<\alpha<\beta<1$ and $y\in[\alpha,\beta]$ or as before $|y|>1$. In these situations more than two types of m dimensional faces are present, and some of the elegance of the result is lost.

The results of papers [40], [41] were motivated by another result in the Hoel and Levine paper. This result says that if one considers

$$\varphi(\xi)=\sup_{y\in[-1,y_0]} V(f(y),\xi)$$

then the design minimizing $\varphi(\xi)$ and $V(f(y_0),\xi)$ are the same. Papers [40], [41] address the problem of minimizing

$$\varphi_a(\xi)=\sup_{|y|\le a} V(f(y),\xi)$$

when $a\to 0$ and $a\to\infty$. The general problem of characterizing the minimum of $\varphi_a(\xi)$ for fixed a is still unsolved.

References

Elfving, G. (1952). Optimum allocation in linear regression theory. *Ann. Math. Statist.* **23**, 255–262.

Hoel, P. G. and Levine, A. (1964). Optimal spacing and weighting in polynomial prediction. *Ann. Math. Statist.* **33**, 1553–1560.

Hoel, P. G. (1965). Optimum designs for polynomial extrapolation. *Ann. Math. Statist.* **36**, 1483–1493.

Karlin, S. and Studden, W. J. (1966). Optimum experimental designs. *Ann. Math. Statist.* **37**, 783–815.

Studden, W. J. (1968). Optimal designs on Tchebycheff points. *Ann. Math. Statist.* **39**, 1435–1447.

Commentary on Papers [23], [26], [28], [31], [88], [96], [98]

H. P. WYNN

Imperial College, London

Paper [23] is the first "Kiefer–Wolfowitz" design paper and introduces the basic generalization of a design to a measure ξ on the design space $\mathcal{X}$. This simple but startling idea in experimental design arises naturally out of the idea of "mixed strategies" in decision theory and immediately makes the optimum design problem *convex*: if ξ_1 and ξ_2 are designs so is $(1-\alpha)\xi_1+\alpha\xi_2$ for $0\le\alpha\le 1$. Otherwise, in terms of hard and useful results the paper does not get very far. It concentrates on some game theoretic developments and what is now called c-optimality: minimizing the variance of the least squares estimator $c\cdot\hat{\vartheta}$ of a *single* linear function $c\cdot\vartheta$ of the parameters. As such it follows closely the work of Elfving (1952). It is interesting that only comparatively little further use of game theory was made by Kiefer who preferred more direct analysis. Much of the subsequent theory can now be set in a general convex analysis framework (see the comments by Pukelsheim). Paper [23] contains an announcement of the General Equivalence Theorem (GET) and a proof of one part of it, (5.5).

The first full-length paper laying the theoretical foundations for optimum design is [31]. The GET had already been proved in [29], and Kiefer gave full rein to the results, extending the theory to the s out of k parameter case, a highly technical situation only recently receiving a definitive solution (Pukelsheim and Titterington, 1983). The examples in [23] are a spectacular application of the theory and show Kiefer at his computational best. The basic technique is (1) to expand the variance function in orthonormal polynomials $g_i(x)$ with respect to the design: thus $d(x,\xi)=\sum_{i=1}^{k}g_i^2(x)$; (2) guess at the form of the g_i using symmetry arguments for a likely class of designs; and (3) show that $d(x,\xi)$ reaches a maximum at the design points (an equivalent condition to D-optimality). The technique remains very powerful, and paper [46] shows one of the hardest cases—general quadratic regression. It is interesting that, although classical orthogonal designs can form the *support* of the D-optimum designs, the actual weights on the design points can render the design very unclassical. The section on simplex experiments (part II in [31]) is a paper in itself and laid the foundation for a pleasant subculture in design theory (see later work by Kiefer and Galil).

In this early sequence of papers the paper to the Royal Statistical Society [26] stands out partly for the almost orchestral attack on it by the British School of experimental design stemming from the work of R. A. Fisher and F. Yates. This is one of the few published clashes between the "Fisherian" British school and the "Waldian" decision school for which J. Kiefer was ambassador on that occasion. Needless to say, the British had met their match; his reply is masterly.

The paper itself is a discursive thesis on optimum design with particularly useful sections on invariance and the support size of a design.

Kiefer continued his elegant exploration of examples in [28] with a study of rotatable designs, showing that designs which placed uniform measures on certain concentric spheres are optimum.

Of lasting importance in that paper was the first analytic attack on the difficult problem of designs for treatment allocation on a line, or in time, when the error process is autocorrelated. This followed the earlier work of Williams (1952) and required the study of strings of treatments $ABAB\ldots$, $ABAB\ldots$. In the last published paper [98], Kiefer and Wynn tackled the problem for general pth order autoregressive processes. There had been no real advance in the intervening twenty years. There turns out to be a close connection with coding theory and stationary k-state processes. In higher dimensions the authors studied the neighbor properties of patterns of points and actually derived, as a by-product, the covariance function of an isotopic stationary binary process on the plane (up to lag one in any direction). As an indication of the flavour of this work it is worth mentioning the "magic" square

$$\begin{matrix} 1 & 1 & 1 & 0 \\ 1 & 1 & 0 & 1 \\ 0 & 1 & 0 & 0 \\ 1 & 0 & 0 & 0 \end{matrix}$$

which, if repeated over the whole plane, represents a pseudorandom pattern (every one of the 16 possible 2×2 windows appears equally often).

Paper [88] is closer in spirit to paper [22] (see the comments by Cheng) and discusses the optimality of Latin square and BIB designs for autocorrelated errors. The paper contains two approximations (1) the use of the least squares estimator rather than the best linear unbiased estimator, and (2) a simple nearest neighbor autocorrelation model between plots. Both [88] and [98] use the ideas of "universal optimality" which have proved so powerful in the more combinatorial papers. (See the comments by Cheng). Cheng (1983) has taken up the idea of nearest neighbor balanced incomplete block designs and proved additional existence theorems in some important cases. There will be an expanding literature on neighbor designs.

REFERENCES

Cheng, C-S (1983). Construction of optimal balanced incomplete block designs for correlated observations. *Ann. Math. Statist.* **11**, 240–246.

Elfving, G. (1952). Optimum allocation in linear regression. *Ann. Math. Stat.* **23**, 255–262.

Williams, R. M. (1952). Experimental designs for serially correlated observations. *Biometrika* **39**, 151–167.

Permissions

Springer-Verlag would like to thank the original publishers of the papers of Jack Kiefer for granting permission to reprint specific papers in this collection:

[22] Reprinted from *Ann. Math. Statist.* **29,** © 1958 by The Institute of Mathematical Statistics.
[23] Reprinted from *Ann. Math. Statist.* **30,** © 1959 by The Institute of Mathematical Statistics.
[26] Reprinted from *J. Roy. Statist. Soc.,* **Ser. B, 21,** © 1959 by Royal Statistical Society.
[28] Reprinted from *Proc. Fourth Berkeley Symp. on Mathematical Statistics and Probability* **1,** © 1960 by The University of California Press.
[29] Reprinted from *Canad. J. Math.* **12,** © 1960 by Canadian Journal of Mathematics.
[31] Reprinted from *Ann. Math. Statist.* **32,** © 1961 by The Institute of Mathematical Statistics.
[33] Reprinted from *Ann. Math. Statist.* **33,** © 1962 by The Institute of Mathematical Statistics.
[34] Reprinted from *Canad. J. Math.* **14,** © 1962 by Canadian Journal of Mathematics.
[40] Reprinted from *Ann. Inst. Statist. Math.* **16,** © 1964 by Annals of the Institute of Statistical Mathematics.
[41] Reprinted from *Ann. Inst. Statist. Math.* **16,** © 1964 by Annals of the Institute of Statistical Mathematics.
[43] Reprinted from *Proc. Amer. Math. Soc.* **16,** © 1965 by American Mathematical Society.
[44] Reprinted from *Ann. Math. Statist.* **36,** © 1965 by The Institute of Mathematical Statistics.
[46] Reprinted from *Proc. Fifth Berkeley Symp. on Mathematical Statistics and Probability* **1,** © 1967 by The University of California Press.
[53] Reprinted from *Actes du Congrès International des Mathématiciens, Nice,* © 1970 by Editions Bordas, Dunod.
[55] Reprinted from *Statistical Decision Theory and Related Topics (Proc. Symp.),* edited by S.S. Gupta and J. Yackel, © 1971 by Academic Press.
[57] Reprinted from *Multivariate Analysis III. Proc. of the Third International Symp. on Multivariate Analysis,* edited by P.R. Krishnaiah, © 1973 by Academic Press.
[58] Reprinted from *Ann. Statist.* **2,** © 1974 by The Institute of Mathematical Statistics.
[59] Reprinted from *J. Roy. Statist. Soc.,* **Ser. B, 36,** © 1974 by Royal Statistical Society.
[60] Reprinted from *Ann. Statist.* **3.** © 1975 by The Institute of Mathematical Statistics.